# Procedural Medicine for Spaceflight

# Procedural Medicine for Spaceflight


Edited by

**Dana R. Levin**
Space Medicine Operations, Mission Operations Vast Inc., Long Beach, CA, United States
Department of Emergency Medicine, University of California, Los Angeles, CA, United States
Division of Space, Ecological, Arctic, and Resource-Limited Medicine, Massachusets General Hospital, Boston, MA, United States

**Tovy Haber Kamine**
Trauma, Acute Care Surgery, and Surgical Critical Care, University of Massachusetts Chan Medical School—Baystate, Springfield, MA, United States

**Michael Pohlen**
Department of Radiology, University of California, San Diego, CA, United States



ELSEVIER

ACADEMIC PRESS
An imprint of Elsevier

Academic Press is an imprint of Elsevier
125 London Wall, London EC2Y 5AS, United Kingdom
50 Hampshire Street, 5th Floor, Cambridge, MA 02139, United States



ISBN: 978-0-443-27552-4

For Information on all Academic Press publications
visit our website at https://www.elsevier.com/books-and-journals

*Publisher:* Megan Ball
*Acquisitions Editor:* Elizabeth Brown
*Editorial Project Manager:* Kshitij Jha
*Production Project Manager:* Gomathi Sugumar
*Cover Designer:* Mark Rogers
*Cover Image Creator:* Segan Friend

Typeset by MPS Limited, Chennai, India

# Contents

## 11. Thoracic procedures

*Matthew Marc Cooper*



## 12. Abdominal procedures

*Margaret Siu*



## 13. Gynecologic evaluation and procedures

*Ruth M. Farrell and Jon G. Steller*

# Section IV
# Post procedure care

# Section V
# Other considerations

## 24. Diagnostic tools

*Christopher Zahner and Michael Pohlen*



## 25. Preflight training and curriculum development for procedural skills

*Chanel Fischetti, Emily Frisch, Laurel Kaye and Alaina Brinley Rajagopal*



## 26. Procedural Guidance

*Andrew W. Kirkpatrick, Jessica L. McKee, Smith L. Johnston, III and Douglas R. Hamilton*

## 30. The future of human habitation in space

*Ryan Zahalka and Dana R. Levin*

# List of contributors

**Yousef Ahmed** Department of Anesthesiology, Mass General Brigham, Harvard Medical School, Boston, MA, United States

**Peter Alexandrov** University of Florida, Gainesville, FL, United States

**Austin Almand** Department of Surgery, MedStar Georgetown University Hospital – Washington Hospital Center, Washington, DC, United States; 123d Special Tactics Squadron, Air Force Special Operations Command, Louisville, KY, United States

**Arian Anderson** Department of Emergency Medicine, University of Colorado Anschutz, Aurora, CO, United States

**Erik Antonsen** Department of Emergency Medicine, SPEAR Medicine Division, Massachusetts General Hospital, Boston, MA, United States

**Luke Apisa** Department of Emergency Medicine, SPEAR Division, Massachusetts General Hospital, Boston, MA, United States

**Amran Asadi** Space Exploration Technologies (SpaceX), Hawthorne, CA, United States; Department of Anesthesiology and Perioperative Medicine, University of California Los Angels, Los Angeles, CA, United States

**Tina Bayuse** Pharmacy Operations, KBR Inc., NASA Johnson Space Center, League City, TX, United States

**Daniel Berenson** Department of Anesthesiology, Mass General Brigham, Harvard Medical School, Boston, MA, United States

**Khushi Bhatt** University of California Irvine, Orange, CA, United States

**Linde Bischak** Department of Emergency Medicine, University of Colorado Anschutz, Aurora, CO, United States

**Carlo Andre Canepa** Department of Emergency Medicine, Alaska Native Tribal Health Consortium, Anchorage, AK, United States

**Kenny Chao** Space Medicine Operations, Mission Operations Vast Inc., Long Beach, CA, United States; Department of Emergency Medicine, University of California, Los Angeles, CA, United States

**Rowena Christiansen** Department of Medical Education, The University of Melbourne Medical School, Melbourne, VIC, Australia; School of Science Computing and Emerging Technologies, Swinburne University of Technology, Melbourne, VIC, Australia

**Matthew Marc Cooper** Peak Aerospace Medicine & Cardiovascular Surgery, Lake Elmo, MN, United States

**Benjamin Easter** Department of Emergency Medicine, University of Colorado School of Medicine, Aurora, CO, United States

**Kareem El-Tayeb** Department of Anesthesiology, Mass General Brigham, Harvard Medical School, Boston, MA, United States

**Ruth M. Farrell** Obstetrics and Gynecology Institute, Center for Bioethics, Cleveland Clinic, Cleveland, OH, United States

**Chanel Fischetti** Department of Emergency Medicine, Massachusetts General Hospital, Boston, MA, United States

**Paul J. Ford** Neuroethics Program, Cleveland Clinic, Cleveland, OH, United States

**Arthur Formanek** Department of Anesthesiology, Mass General Brigham, Harvard Medical School, Boston, MA, United States

**Emily Frisch** Department of Obstetrics and Gynecology, Stanford Hospital, Palo Alto, CA, United States

**Taania Girgla** Department of Anesthesiology, Mass General Brigham, Harvard Medical School, Boston, MA, United States

**Nicolai Goettel** Department of Anesthesiology, Mass General Brigham, Harvard Medical School, Boston, MA, United States

**Douglas R. Hamilton** Clinical Associate Professor Medicine Adjunct Professor Electrical Engineering, University of Calgary, Calgary, AB, Canada; General Internal Medicine, University of Calgary Foothills Medical Center - Health Sciences Center Calgary, AB, Canada

**Katie Harris** University of Toronto, Toronto, ON, Canada

**Nicolas Heft** Department of Emergency Medicine, Space Medicine Section, The University of Texas Health Science Center at Houston, Houston, TX, United States

**Michael Hodapp** General Dentist, Private Practice, Houston, TX, United States

**Michelle S. Hong** McGovern Medical School, The University of Texas Health Science Center at Houston, Houston, TX, United States

**Laurence Tang Hou** Hackensack University Medical Center, Hackensack, NJ, United States

**Megan Hunt** Department of Anesthesiology, Mass General Brigham, Harvard Medical School, Boston, MA, United States

**Sanghee Im** Department of Physical Medicine and Rehabilitation, Baylor College of Medicine, Houston, TX, United States; Department and Research Institute of Rehabilitation Medicine, Severance Hospital, Yonsei University College of Medicine, Seoul, South Korea

**Smith L. Johnston, III** Retired NASA JSC Medical Officer FMR CMO AXIOM Space, Inc, Houston, TX, United States; Professor of Aerospace Medicine at the School of Medicine Central Florida University Orlando, FL Director of Health Innovation Solamed Solutions, LLC Woodstock, GA, United States

**Tovy Haber Kamine** Trauma, Acute Care Surgery, and Surgical Critical Care, University of Massachusetts Chan Medical School—Baystate, Springfield, MA, United States

**Ravjot Kaur** Ohio State University College of Dentistry, Columbus, OH, United States

**Laurel Kaye** Department of Emergency Medicine, Massachusetts General Hospital, Boston, MA, United States

**Shahnawaz Khijmatgar** School of Dentistry, Royal College of Surgeons in Ireland, Dublin, Ireland

**Andrew W. Kirkpatrick** TeleMentored Ultrasound Supported Medical Interventions (TMUSMI) Research Group, Calgary, AB, Canada; Professor of Critical Care Medicine, Departments of Critical Care Medicine and Surgery Foothills Medical Centre, Calgary, AB, Canada; Surgery Departments of Critical Care Medicine and Surgery Foothills Medical Centre, Calgary, AB, Canada

**Saranjeev Lalh** Faculty of Medicine & Dentistry, University of Alberta, Edmonton, AB, Canada

**Stefanie P. Lazow** Department of Surgery, Division of Acute Care Surgery, University of Vermont Medical Center, Burlington, VT, United States

**Dana R. Levin** Space Medicine Operations, Mission Operations Vast Inc., Long Beach, CA, United States; Department of Emergency Medicine, University of California, Los Angeles, CA, United States; Division of Space, Ecological, Arctic, and Resource-Limited Medicine, Massachusets General Hospital, Boston, MA, United States

**Matthew A. Levin** Department of Anesthesiology, Perioperative and Pain Medicine, Icahn School of Medicine at Mount Sinai, New York, NY, United States; Department of AI and Human Health, Icahn School of Medicine at Mount Sinai, New York, NY, United States; Department of Genetics and Genomic Sciences, Icahn School of Medicine at Mount Sinai, New York, NY, United States

**Curtis Lowery III** Emergency Department, Minerva Medical Group, University of Texas Health Houston, Houston, TX, United States

**David Mampre** Department of Neurosurgery, University of Florida, Gainesville, FL, United States

**Jessica L. McKee** Clinical Director and Project Manager Consultant Zoll Medical Corporation TeleMentored Ultrasound Supported Medical Interventions (TMUSMI) Research Group Edmonton, AB, Canada

**Richard Meehan** Rheumatology, National Jewish Health, Denver, CO, United States

**Elana Meer** Department of Ophthalmology, University of California San Francisco, San Francisco, CA, United States

**Ariana M. Nelson** Department of Anesthesiology & Perioperative Medicine, University of California Irvine, Irvine, CA, United States

**Gwen E. Owens** Department of Anesthesiology, Critical Care and Pain Medicine, Harvard University, Cambridge, MA, United States

**Shawna Pandya** Space Medicine Group, International Institute for Astronautical Sciences, Boulder, CO, United States; Faculty of Medicine & Dentistry, University of Alberta, Edmonton, AB, Canada

**Prashant Parmar** Rheumatology, Veros Health, Denver, CO, United States

**Lonnie G. Petersen** Department of Aeronautics and Astronautics, Massachusetts Institute of Technology, Cambridge, MA, United States

**Amanda N. Pine** Department of Neurosurgery, University of Florida, Gainesville, FL, United States

**Michael Pohlen** Department of Radiology, University of California, San Diego, CA, United States

**Amanda N. Prine** Department of Emergency Medicine, University of Florida, Gainesville, FL, United States

**Alaina Brinley Rajagopal** California Institute of Technology, Pasadena, CA, United States

**Mark Rosenberg** Department of Neurology, Medical University of South Carolina, Charleston, SC, United States

**Meagan Rosenberg** General Surgery, University of Massachusetts Chan Medical School—Baystate, Springfield, MA, United States

**Connor Rupp** Department of Neurosurgery, University of Florida, Gainesville, FL, United States

**Terje Saehle** Civil Aviation Authority, Bodø, Norway

**Margaret Siu** Fellow of Surgical Critical Care, R Adams Cowley Shock Trauma Center, University of Maryland School of Medicine, Baltimore, MD, United States

**Kaleigh Stabenau** Department of Otolaryngology-Head and Neck Surgery, University of California Irvine, Orange, CA, United States; Arizona Program for Exploration (APEX) Medicine and Surgery, University of Arizona, Phoenix, AZ, United States

**Jon G. Steller** University of California Irvine, Irvine, CA, United States

**Faye Chiou Tan** Department of Physical Medicine and Rehabilitation, Baylor College of Medicine, Houston, TX, United States

**James Yacyshyn** Faculty of Medicine & Dentistry, University of Alberta, Edmonton, AB, Canada

**Ryan Zahalka** Division of Space, Ecological, Arctic, and Resource-Limited Medicine, Massachusets General Hospital, Boston, MA, United States

**Christopher Zahner** Department of Pathology, University of Texas Medical Branch, Galveston, TX, United States

**Cassandra Zhi** Department of Anesthesiology, Maimonides Medical Center, Brooklyn, NY, United States

# Foreword

Human spaceflight has always advanced through a willingness to attempt what has not yet been fully defined. Every operational capability we now consider routine in space—from rendezvous and docking to long-duration habitation—began as an imperfect first effort accompanied by uncertainty, debate, and the courage to begin anyway.

Medical care during spaceflight remains in a similar stage of evolution.

As missions extend farther from Earth and crews become more autonomous, the need to understand how medical procedures can be adapted to the space environment becomes increasingly important. The operational realities of microgravity, communication delays, constrained equipment, limited personnel, and altered human physiology challenge many assumptions that underpin terrestrial medicine. There are still more questions than answers.

Having spent 197 days aboard the International Space Station, I witnessed firsthand how spaceflight changes even routine tasks in unexpected ways. In that environment, operational practicality matters as much as theory, and preparation becomes inseparable from adaptability. Advancing medical capability in space will require contributions from clinicians, engineers, researchers, operators, and astronauts alike. It will also require the willingness to openly examine ideas before the field itself is fully mature.

This book is an early and necessary step toward building a framework for discussion, investigation, and operational experience in an area that will become essential to the future of exploration. The authors of this text deserve credit for assembling perspectives across an emerging discipline. They are helping move the conversation forward and encouraging others to refine, challenge, expand, and improve upon what is presented here. That is how every meaningful operational field develops.

As humanity prepares for longer missions to the Moon, Mars, and beyond, procedural medicine in space will evolve from an academic interest into an operational necessity. My hope is that this book serves not as an endpoint but as the beginning of a much larger effort—one that future crews, flight surgeons, and scientists will continue to build upon.

**Serena Auñón-Chancellor**
Flight Engineer and Crew Medical Officer, International Space Station Expedition
56/57 Vast Medical, P.C. Division of Space Medicine, Long Beach, CA, United States
Naresh K. Vashisht College of Medicine, Texas A&M University Health Science Center,
Department of Family Medicine, Bryan, TX, United States

Section I

# Introduction

Chapter 1

# General approaches to surgical procedures in spaceflight

**Linde Bischak and Arian Anderson**
*Department of Emergency Medicine, University of Colorado Anschutz, Aurora, CO, United States*

## Abbreviations

| | |
|---|---|
| **AR** | Augmented reality |
| **CMO** | Crew medical officer |
| **CPR** | Cardiopulmonary resuscitation |
| **ECLSS** | Environmental control and life support systems |
| **EVA** | Extravehicular activity |
| **ICL** | IMPACT conditions list |
| **IMPACT** | Information mission planning via analysis of complex tradespaces |
| **IR** | Interventional radiology |
| **ISS** | International Space Station |
| **LEO** | Low Earth orbit |
| **NBP** | Neutral body posture |
| **VR** | Virtual reality |
| **WMS** | Waste management system |

As human spaceflight becomes more accessible and targets missions farther from Earth and its physical and intellectual resources, the medical capabilities available to crewmembers in the space environment must expand. In-mission medical care currently follows a telehealth model in which clinical expertise is provided by flight surgeons at mission control, who guide crewmembers through medical evaluation, management, and evacuation from low Earth orbit, if necessary (Johnston, Smart and Pattarini, 2020). As the number of space travelers increases and missions venture deeper into space, evacuation will no longer be an option.

Communication with mission control will be hindered by bandwidth and channel capacity limitations, anticipated and unanticipated network demands, satellite interruptions, and signal degradation over long distances. The resulting communication delays and quality deterioration make the telehealth model impractical for many future spaceflight missions, particularly in emergent situations where telehealth providers can only analyze and convey information to the crew but cannot perform procedures. As the person-years on orbit grow and evacuation becomes less pragmatic, advancing in-flight medical capabilities is more important than ever.

Unanticipated medical events requiring procedural intervention may occur anywhere including in deep space. Crew medical officers (CMOs) must be prepared to provide treatment without Earth-based support (Levin et al., 2023). A medical *procedure* is defined here as performing a physical act on a patient beyond an exam but does not involve prolonged invasive techniques that require complex teams, tools, or interventions. For the purposes of this chapter, we will focus primarily on surgical and other interventional procedures that may be necessary to address medical events, such as central line placement, chest tube insertion, tracheostomy, or peritoneal lavage.

Key challenges of performing surgical procedures during spaceflight include the microgravity environment, the limited availability of medical supplies and equipment, the constrained medical training and skill of crewmembers who cannot promptly consult experts on Earth, and the delays or absence of resupply and/or evacuation. Preparing for potential medical emergencies is a critical and difficult aspect of mission planning. Consequently, developing capabilities that can address a diverse set of medical events in human spaceflight missions remains a priority.

**Procedural Medicine for Spaceflight. DOI:** https://doi.org/10.1016/B978-0-443-27552-4.00005-1

Clinicians have many roles in space medicine beyond the flight surgeon, who is responsible for overseeing all aspects of astronaut health including their initial screening, preflight fitness optimization and maintenance, in-flight medical concerns via operation of the console in-mission control, immediate postflight care, and long-term health monitoring. The internal spacecraft environment significantly impacts astronaut health, and medical providers must be crucial advocates for human-centered design, such as limiting noise levels to mitigate hearing loss (Avcı, 2024), using circadian lighting to improve sleep (Zong, Fei and Liu, 2025), and incorporating exercise equipment to counteract muscle mass and bone density loss and maintain cardiovascular fitness (Loehr et al., 2015). Beyond clinical roles, clinicians can support human systems integration by performing usability testing, error analysis, and system testing, helping to optimize interface design and operational performance.

## 1.1 Modifying procedures for the space environment

Any spaceflight mission requires extensive planning and risk analysis to balance mission goals with resource constraints. Every subsystem, from propulsion to environmental control and life support, is constrained by mass, volume, power availability, and human or computational resources. Since launch costs scale directly with mass, and spacecraft have limited physical space, engineers must make trade-offs between functionality, redundancy, and efficiency and risk management becomes a vital and formidable aspect of mission planning. Decisions are made by weighing the likelihood and consequence of failures against mission priorities such as crew safety, scientific research, and schedule. Some systems, such as environmental control and life support, are crucial enough to warrant redundancy despite the increased mass and volume requirements. Often, what might be considered a robust or ideal design on Earth is impractical in space, where maintenance is difficult, resources are finite, and environmental conditions are extreme. For example, most cooling systems on Earth rely on the dissipation of heat to the surrounding air through conduction and convection. To overcome the lack of atmosphere in space, innovative designs such as heat pipes (Gerasimov et al., 1975) or fluid loops (Iwata et al., 2022) are used to transfer heat generated inside the spacecraft to an external radiator, where it can be emitted into space via thermal radiation.

The medical subsystem also necessitates substantial adaptations to tackle the distinct challenges of providing medical care in an extreme environment. Here, we review the challenges of performing procedures in space, which is only one element of supporting astronaut health. Diminished gravitational forces, enclosed habitats, atmosphere management, finite equipment, and limited human resources like skillset or cognitive load require the modification of conventional Earth-based techniques and the development of novel methods to ensure the provision of safe and effective medical care in space.

### 1.1.1 Microgravity

In the effective absence of gravitational forces, healthcare providers must confront unique challenges that are not encountered on Earth. Primarily, any unsecured object will float in a microgravity environment, including the provider and the patient. The neutral body posture (NBP) is the posture the human body naturally assumes in microgravity and represents the most rested condition for a musculoskeletal system developed in Earth gravity. For most, the NBP entails flexion of the knees, hips, elbows, and neck and some arm extension (Fig. 1.1). Volitionally adopting any posture other than NBP requires muscle activation and energy expenditure. While NBP is likely ideal for the medical provider during a procedure, the optimal positioning of the patient is more variable. Traditionally, patients are situated to enhance a provider's access to the location of the procedure, such as having a patient extend their arm to repair a laceration, laying them prone on an operating table for a spinal surgery, or using Trendelenburg or reverse Trendelenburg for abdominal surgeries. Positioning also depends on the nature of the procedure and its expected duration, complexity, and need for sedation and anesthesia. Short procedures, such as a laceration repair or incision and drainage, with an alert and cooperative patient, would likely need minimal restraints and could forego prioritization of NBP. In comparison, NBP would be crucial for extended procedures such as major surgery. Ideally, the design of a restraint system will balance function, efficiency, and microgravity ergonomics for both the patient and the medical providers. Additionally, NBP is variable between individuals, and any restraint system would need to be able to accommodate different crew body sizes and positions.

The International Space Station (ISS) Crew Health Care System incorporates a medical restraint system, which consists of a rigid backboard equipped with straps to secure the patient and the medical care provider. This system is designed for the stabilization and transportation of critical injuries within a 24-hour period and was not intended to facilitate major surgical procedures during flight (Eichstadt, 1994). Advanced interventions may require a more robust medical restraint system capable of accommodating the NBP assumed in a weightless environment while maintaining access to essential

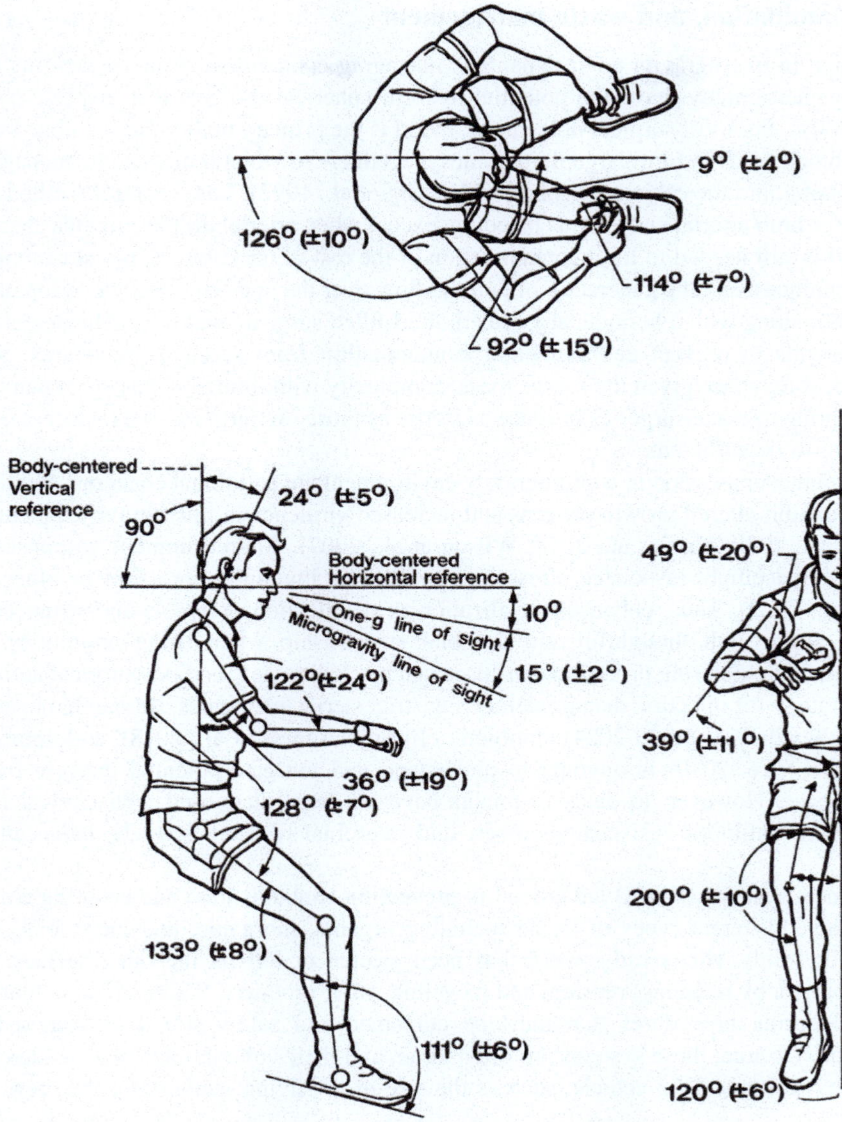


FIGURE 1.1 **Neutral body posture.** The average neutral body posture shown here was created from measurements of 12 people in the microgravity environment onboard Skylab.
Recreated from Mount, F., Whitmore, M. & Stealey, S. (2003). Evaluation of neutral body posture on shuttle mission STS-57 (SPACEHAB-1). 20040200967, NASA Technical Reports Server.

medical equipment. Prototypes of such restraint systems, which are readily deployable and ensure proximity among the CMO, patient, and supplies, have been tested in both parabolic and space flights (Taddeo and Armstrong, 2008). For instance, the Health Maintenance Facility, developed for Space Station Freedom (the design predecessor of the ISS), featured a microgravity surgical workstation that included an adjustable procedure table, along with patient and operator restraints, digitized x-ray capabilities, a ventilator, a defibrillator, monitors, an intravenous pump, a medical computer, storage for medical and surgical supplies, and a microgravity suction unit (Campbell and Billica, 2020).

In addition to the patient and medical provider, all medical equipment and supplies must be secured while remaining accessible. Three distinct types of surgical instrument restraints have been assessed during parabolic flight surgical procedures (Campbell et al., 2001). A minor surgical kit, composed of a soft pack with Velcro attachments that can be deployed on the cabin wall or floor, along with restrained tools featuring supply pockets, magnetic pads, elastic straps, and Velcro, was found to enhance the efficiency of performing surgical procedures in weightlessness when compared to other methods. Although the need for restraint of the provider, patient, and equipment can increase operating times (Pantalone, 2023), maintaining control over surgical instruments is vital to ensuring a sterile field during procedures and preventing injuries from sharp instruments.

### 1.1.2 Sterility, decontamination, and waste management

Another procedural challenge in microgravity environments is managing contamination and preserving sterility. Bodily fluids, like other liquids in microgravity, have the potential to form spherical droplets that disperse by floating in the environment (McQuillen, 2004). For most surgical procedures, blood is the primary bodily fluid of concern. However, due to its high surface tension, blood tends to form large fluid domes that adhere to wounds and can be managed with sponges or suction rather than dispersing into the cabin atmosphere (McCuaig et al., 1992). Local control methods for bleeding in microgravity are sufficient for both arterial and venous bleeding, except when arterial droplet streams are allowed to form. Thus, a closed surgical field is still needed to limit contamination of the rest of the cabin. Surgical canopies, containment chambers, laser surgical techniques, and the generation of laminar flow over the operative field to sweep up surgical debris (Campbell and Billica, 2020), along with a hermetically sealed, fluid-filled surgical enclosure (Hayden et al., 2013), have all been considered as methods to prevent environmental contamination from surgical procedures. Several of these proposed systems are not ideal for space given their large mass, complexity with limited ability for repairs, and significant power requirements. Attempting to make surgeries in space as sterile as those on Earth may be unnecessary in this context if disinfection can be shown to be sufficient.

Microbial growth and immunosuppression in a microgravity environment are additional concerns. Multiple studies have shown that some microbes exhibit altered growth patterns and increased virulence in microgravity, potentially raising the risk of infection (Huang et al., 2018; Kim et al., 2013; Wilson et al., 2007). Altered immune responses resulting from microgravity, radiation, and spaceflight-associated stressors may lead to immunosuppression or other changes to the immune system (Jacob et al., 2023). Thus, enhanced sterilization or disinfection protocols and stringent environmental monitoring are necessary, along with thoughtful antimicrobial stewardship when using antibiotics. As mentioned previously, maintaining a sterile field during procedures in space can be challenging, and postsurgical antibiotics may be a more practical mitigation strategy for infection than implementing strict sterile techniques. Research on pathogen genomes (Irby and Broddrick, 2024; Szydlowski et al., 2024), antibiotic efficacy (Aunins et al., 2018), and antimicrobial surface coatings (Armstrong and Monteiro, 2019) is ongoing to understand and mitigate potential dangers from microbes in confined environments in space. However, to date, astronauts have not been diagnosed with clinical findings of viral reactivation or infections from antibiotic resistant microbes, and microbial research must be balanced with clinically significant endpoints.

Effective waste management systems (WMS) are crucial in preventing contamination and ensuring crew safety. WMS are complex and must address different types of waste including organic, inorganic, and hazardous, as well as their different forms as liquids or solids. Historically, waste has been ejected or left on the lunar surface, but research is underway to convert trash burden by reducing, reusing, and recycling waste products. WMS are particularly important for biological waste generated during procedures that must be contained and safely stored or disposed of to prevent environmental contamination. Of note, these systems are often large, and their volume must be considered in the vehicle design and may impact other aspects of the vehicle, such as the size of the living quarters for the crew.

### 1.1.3 Medical resource limitations

In the confined environment of a spacecraft or habitat, the type and quantity of medical equipment that can be carried are limited due to mass, volume, and power constraints. When preparing for a mission, planners must strategically decide which medical equipment to bring based on the anticipated risks of the operation. Specific risks and their probabilities depend on various factors, including crew size, mission objectives, and duration. For example, astronauts on a mission to Mars will face an extended period of exposure to microgravity, so health maintenance countermeasures, such as large resistive exercise equipment, may be prioritized to mitigate musculoskeletal effects like muscle and bone loss, whereas astronauts on a lunar mission may engage in numerous extravehicular activities (EVA) and require medical capabilities that focus on trauma care.

Even the most well-trained medical providers are limited in the care they can deliver based on the medical equipment available. Ideally, a few essential tools would suffice to address all medical events. Although this is impossible, mission planners can focus on selecting necessary yet versatile tools for a wide range of procedures while minimizing waste. Consider the adaptability of a scalpel, wire, and a tube or drain, which can be used for chest tube placement, visceral abscess drainage, paracentesis, cricothyrotomy, or hematoma evacuation. The Seldinger technique (Seldinger, 1953) is one of the most widely used interventions in medicine owing to its versatility in accessing blood vessels and hollow organs with minimal equipment (needle, wire, catheter, etc.), making it ideal for human spaceflight.

Unfortunately, some medical events require specialized tools that occupy valuable space. One option for increasing the range and capability of available tools within mass and volume constraints is to provide a 3D printer that can create the necessary medical instruments for a given procedure (Wong and Pfahnl, 2016). However, 3D printing requires power, raw materials, and a database of designs, and the manufacturing process can be slow, which may not be feasible in an emergency. An additional consideration is the use of single-use disposable tools versus providing methods to sterilize tools for reuse (Kovalski et al., 2023). The trade-off between increased waste and additional mass, power, and volume requirements for decontamination methods, as well as the opportunity for resupply, is unique to each mission and its constraints. An orbital mission might prioritize single-use disposable tools given the relative ease of resupply compared to a mission to Mars, where the emphasis might be on decontamination and reuse, although decontamination supplies are also finite and affect mass and volume constraints. The mission planner's task becomes even more challenging in deciding not only what equipment to bring but also the quantity based on the anticipated frequency of different medical events. Single-use surgical tools may be more logistically efficient if only two major surgical events are expected, versus five incidents, which might favor decontamination of tools for reuse. Both options also produce waste, which must be carefully managed and considered during mission planning.

### 1.1.4 Atmosphere management in an enclosed environment

Another environmental challenge in space is atmosphere management, which can be influenced by laparoscopic procedures that require insufflation gases, such as carbon dioxide, for inflating the abdominal cavity or supplemental oxygen for sedation. In an enclosed environment, any gas leakage poses a risk of interfering with environmental control and life support systems and can present hazards to the crew, such as carbon dioxide toxicity or oxygen toxicity. NASA cautions against using enriched (>21%) oxygen atmospheres due to the increased risk of fire (Stroud et al., 2008). Supportive oxygen therapy, anesthetic gases, and the use of electrocautery tools are standard terrestrial approaches to many surgical procedures but, in spaceflight, present a significant threat to the crew.

Several potential options are available to mitigate these risks during surgical procedures. First, procedures such as laparoscopy may be bypassed altogether in favor of open procedures, which do not require insufflation gases, among other advantages (see Section 3.3.3). Alternatively, air revitalization systems would need to accommodate the gases used during medical procedures (Komorowski et al., 2013), but they also come with their own challenges. For example, single-use options for carbon dioxide removal, such as lithium hydroxide canisters, require allocating mass and volume for carrying additional canisters, whereas a regenerable system, like the Carbon Dioxide Removal Assembly used on the ISS, may need adjustments to compensate. Another option is to use a different insufflation gas, such as a nitrogen mixture or other inert gas, if oxygen is not displaced in the cabin. If sedation is required for a procedure, intravenous sedatives such as benzodiazepines or propofol may be considered in combination with local anesthetic instead of inhaled gases.

### 1.1.5 Limited crew size and training

The level of medical expertise among crewmembers is a critical factor in determining the feasibility of performing procedures in space. Historically, most astronaut crewmembers are not medical professionals, making simplified and highly standardized procedures a necessity. Current CMO training paradigms for missions to the ISS are limited to approximately 14 hours of preflight medical skills practice, which includes intravenous and urinary catheter placement, point-of-care ultrasound, and basic medical evaluation skills (Blue et al., 2014). Simulation-based training on Earth can prepare crewmembers for space-specific challenges, incorporating virtual reality or augmented reality systems to enhance realism. These training methods may include simulated microgravity environments and task-specific drills for anticipated medical scenarios. However, existing techniques may be less efficient or effective in environments with reduced gravitational forces. For example, a comparison of four airway management techniques found that the time to initial ventilation with both the i-gel® supraglottic airway device and endotracheal intubation by direct laryngoscopy was significantly longer in simulated microgravity compared to 1 g; there was no significant difference in time for a laryngeal tube or a different second-generation laryngeal mask (Warnecke et al., 2021). Furthermore, a systematic review of cardiopulmonary resuscitation (CPR) in hypogravity conditions simulating a lunar or Martian environment found that adequate CPR quality can only be achieved with increased physical effort to account for functional weight loss (Overbeek et al., 2022). Several alternative methods of providing effective CPR in microgravity and hypogravity environments have been explored (Overbeek et al., 2024), but they would require additional crew training.

Given the operational limitations of space missions, balancing the need for advanced medical skills with the constraints of crew training time and cognitive workload is essential. Just-in-time training, where the medical provider learns or reviews the steps of the procedure immediately before performing it, can help bridge the gap between crew training and the medical skills needed in-flight (Lerner et al., 2022). In the near future, decision-support systems, such as onboard medical AI or telemedicine platforms, may provide step-by-step guidance during procedures. These systems must be user-friendly and capable of operating autonomously when communication delays with Earth-based medical experts occur, ensuring crewmembers can still effectively respond to medical emergencies. As we plan for longer-duration missions, such as to Mars, a larger proportion of medical decision-making must remain self-contained to the crew and resources in the habitat. Less than 10% of astronauts who have traveled to space have also been physicians. The need for in situ medical expertise and development of Earth independent medical operations will continue to grow as mission complexity, mission duration, and crew size increase (Anderson et al., 2025). Most astronauts are cross-trained as medical providers, but in specific mission profiles it may be more beneficial to prioritize physicians cross-trained as astronauts (Asrar et al., 2025).

Conducting medical procedures in space requires addressing the unique challenges posed by microgravity and the confined environment of a spacecraft. Innovations in tools, protocols, and training are crucial for ensuring crew safety and mission success. By anticipating these challenges and implementing targeted modifications, space missions can be more effectively prepared to manage medical emergencies in deep space.

## 1.2 Anticipated conditions and procedures

Human spaceflight exposes astronauts to novel medical challenges arising from microgravity, radiation, and confined living conditions. For a Mars mission, researchers estimate that there will be an average of one surgical emergency every 2.4 years for a crew of seven (Summers et al., 2005). NASA has identified the medical conditions most likely to occur during various human spaceflight missions to ensure that appropriate medical capabilities are available. NASA employs empirical data (Tables 1.1–1.3) and probabilistic risk analysis to assess the likelihood of medical events occurring for a given mission based on parameters such as crew size, mission destination, and mission duration. The Information Mission Planning via Analysis of Complex Tradespaces (IMPACT) tool was designed to quantify differences in total mission risk for 119 medical conditions (Table 1.4) across different crew and mission parameters and to help optimize medical resource allocation (Kerstman et al., 2024). This information guides the selection of medical equipment and supplies to be carried, influencing the procedural capabilities for a given mission. In-flight medical events may range from simple cases, such as a small laceration needing repair, to complex situations, such as perforated appendicitis requiring surgery. In the latter scenario, some surgical procedures are better suited for microgravity or partial gravity than others.

**TABLE 1.1 In-flight medical events for US astronauts during the space shuttle program (STS-1 through STS-89, April 1981 to January 1998).**

| | Astronauts experiencing medical event | | |
|---|---|---|---|
| **Medical event or system by ICD-9 category** | **Number** | **Percent of total events** | **Incidence/ 14 days** |
| Space adaptation syndrome | 788 | 42.2 | 2.48 |
| Nervous system and sense organs | 318 | 17.0 | 1.00 |
| Digestive system | 163 | 8.7 | 0.52 |
| Skin and subcutaneous tissue | 151 | 8.1 | 0.48 |
| Injuries or trauma | 141 | 7.6 | 0.44 |
| Musculoskeletal system and connective tissue | 132 | 7.1 | 0.42 |
| Respiratory system | 83 | 4.4 | 0.26 |
| Behavioral signs and symptoms | 34 | 1.8 | 0.11 |
| Infectious diseases | 26 | 1.4 | 0.08 |
| Genitourinary system | 23 | 1.2 | 0.07 |
| Circulatory system | 6 | 0.3 | 0.02 |
| Endocrine, nutritional, metabolic, and immunological disorders | 2 | 0.1 | 0.01 |

**Source:** Data from Billica, R. (2000). In-flight medical events for US astronauts during space shuttle program STS-1 through STS-89. Institute of Medicine Committee on Creating a Vision for Space Medicine During Travel Beyond Earth Orbit. February 22, Johnson Space Center, Houston.

**TABLE 1.2** Medical events and recurrences among astronauts of all nationalities on Mir, March 14, 1995, through June 12, 1998.

| Event | Number of events | Recurrences |
|---|---|---|
| Superficial injury | 43 | 2 |
| Arrhythmia | 32 | 98 |
| Musculoskeletal | 29 | Not reported (NR) |
| Headache | 17 | 8 |
| Sleeplessness | 13 | 9 |
| Fatigue | 17 | 4 |
| Contact dermatitis | 5 | 3 |
| Surface burn | 5 | NR |
| Conjunctivitis | 4 | 2 |
| Acute respiratory infection | 3 | NR |
| Asthenia | 3 | 2 |
| Ocular foreign body | 3 | NR |
| Globe contusion | 2 | NR |
| Dental | 2 | NR |
| Constipation | 1 | NR |

**Source:** Data from Marshburn, T. (2000). Pathophysiology of long-duration spaceflight. Presentation to the Institute of Medicine Committee on Creating a Vision for Space Medicine During Travel Beyond Earth Orbit.

**TABLE 1.3** Medical events among seven NASA astronauts on Mir, March 14, 1995, through June 12, 1998.

| Event | Number of events | Incidence/100 days |
|---|---|---|
| Musculoskeletal | 7 | 0.74 |
| Skin | 6 | 0.63 |
| Nasal congestion, irritation | 4 | 0.42 |
| Bruise | 2 | 0.21 |
| Eyes | 2 | 0.21 |
| Gastrointestinal | 2 | 0.21 |
| Psychiatric | 2 | 0.21 |
| Hemorrhoids | 1 | 0.11 |
| Headaches | 1 | 0.11 |
| Sleep disorders | 1 | 0.11 |

**Source:** Data from Marshburn, T. (2000). Phase 1/Mir clinical experience. Presentation to the Institute of Medicine Committee on Creating a Vision for Space Medicine During Travel Beyond Earth Orbit.

**TABLE 1.4** Information Mission Planning via Analysis of Complex Tradespaces (IMPACT 1.0) medical condition list.

| **Abdominal wall hernia** | **Dental fracture/ exposed pulp** | **Gravity well—entry motion sickness** | **Space adaptation—back pain** |
|---|---|---|---|
| Abnormal uterine bleeding | Dental luxation/avulsion (tooth loss) | Gravity well—neurovestibular disturbance | Space adaptation—constipation |
| Acute coronary syndrome | Dislocation—finger | Gravity well—orthostatic intolerance | Space adaptation—epistaxis |
| Acute radiation syndrome | Dislocation—shoulder | Headache | Space adaptation—headache |
| Allergic reaction (mild to moderate) | Diverticulitis, acute | Headache—$CO_2$ Induced | Space adaptation—insomnia |
| Altitude sickness | Dust exposure—lunar | Hearing loss | Space adaptation—nasal congestion |
| Anaphylaxis | Ebullism | Hearing loss—noise-related | Space adaptation—space motion sickness |
| Appendicitis | Epistaxis | Hemorrhoids | Space adaptation—urinary retention |
| Arthritis, acute | EVA-related decompression sickness | Herpes zoster reactivation (shingles) | Space adaptation—urinary incontinence |
| Atrial fibrillation/atrial flutter | EVA-related dehydration | Mouth ulcer | Spaceflight-associated neuro-ocular syndrome (SANS) |
| Barotrauma (ear/sinus block) | EVA-related fingernail delamination | Nephrolithiasis | Sprain/strain—back |
| Benzodiazepine or opioid overdose | EVA-related hand injury | Neuropathy—central, impingement related | Sprain/strain—lower extremity |
| BHP—adjustment disorder | EVA-related heat illness | Otitis externa | Sprain/strain—neck |
| BHP—anxiety | EVA-related paresthesia | Otitis media | Sprain/strain—upper extremity |
| BHP—depression | EVA-related shoulder injury | Pancreatitis, acute | Streptococcal pharyngitis |
| BHP—grief reaction | EVA-related soft contact injury | Pregnancy, first trimester | Sudden cardiac arrest |
| BHP—psychosis secondary to depression | Eye—retinal injury | Pregnancy, risk for | Tendinopathy/enthesopathy/bursitis/ over-use injuries—lower extremity |
| BHP—sleep disturbance | Eye foreign body | Prostatitis, acute | Tendinopathy/enthesopathy/bursitis/ over-use injuries—upper extremity |
| BHP—Spaceflight-related relationship problems | Eye irritation/corneal abrasion/ulceration | Rash, spaceflight-associated | Toxic dermal exposure |
| Burn—chemical eye | Eyelid and anterior eye infection | Reactive airway | Toxic inhalation exposure |
| Burn—chemical skin | Fracture—arm | Respiratory failure | Toxic inhalation exposure—combustion products |
| Burn—mild, thermal | Fracture—cervical spine | Respiratory tract infection—lower | Trauma—abdominal injury (blunt) |
| Burn—moderate to severe, thermal | Fracture—distal leg | Respiratory tract infection—upper | Trauma—chest injury (blunt) |
| Cerebrovascular accident | Fracture—femur | Sepsis | Trauma—minor head |
| Cerumen impaction | Fracture—hand | Shock—cardiogenic | Trauma—severe head |

*(Continued)*

**TABLE 1.4** (Continued)

| Abdominal wall hernia | Dental fracture/ exposed pulp | Gravity well—entry motion sickness | Space adaptation—back pain |
|---|---|---|---|
| Choking/obstructed airway | Fracture—wrist | Skin abrasion | Traumatic hypovolemic shock |
| Cholelithiasis/biliary colic, acute | Fracture—thoracic/lumbar spine | Skin infection—bacterial | Urinary tract infection |
| Dental abscess | Gastritis/reflux/esophagitis | Skin infection—viral/fungal | Vaginal yeast infection |
| Dental crown loss | Gastroenteritis/acute diarrhea | Skin laceration | Venous thromboembolism |
| Dental filling loss | Glaucoma, acute angle-closure | Small bowel obstruction | |

*BHP*, Behavioral health and performance; *EVA*, extravehicular activity; $CO_2$, carbon dioxide; *SANS*, spaceflight-associated neuro-ocular syndrome.
**Source:** Data from Kreykes, A. J., Suresh, R., Levin, D., & Hilmers, D. C. (2023). Selecting medical conditions relevant to exploration spaceflight to create the IMPACT 1.0 medical condition List. Aerospace Medicine and Human Performance, 94(7), 550–557. https://doi.org/10.3357/AMHP.6199.2023.

## 1.3 Advantages/limitations of available modalities for major invasive procedures

Performing major invasive procedures in space poses unique challenges due to limitations in mass, volume, and power, the microgravity environment, and crew training. Each surgical modality—open surgery, laparoscopic surgery, robotic-assisted surgery, and interventional radiology (IR)/percutaneous techniques—offers distinct advantages and disadvantages in the space environment. Of note, on Earth, these procedures utilize several specially trained personnel (e.g., primary surgeon, assistant, scrub nurse technologist, circulating nurse, anesthesiologist). In space, a single provider may need to play multiple roles, and simultaneously performing the procedure and monitoring the patient is extremely difficult. Consider the standard of care for conscious sedation in the emergency department, such as for a musculoskeletal reduction, which requires a dedicated provider to monitor the patient during the procedure. In this context, crew training is essential to help disperse the burden on the primary provider during complex procedures.

### 1.3.1 Open surgery

Open surgery provides direct visualization and access to internal organs, enabling comprehensive intervention during life-threatening emergencies. It serves as a definitive treatment option when less invasive techniques are inadequate or unavailable. Open surgery utilizes conventional surgical tools, whereas laparoscopic and robotic surgeries require specialized equipment and training (Patil et al., 2024). However, microgravity complicates the ergonomics for both the patient and the medical provider, necessitating restraint systems that account for NBP. Internal organs, such as small bowel, also float and can obscure the view of the surgical field. Establishing and maintaining a sterile field in microgravity is complex, which increases the risk of infection and waste. Postoperative care and recovery take longer than with other surgical methods (Spanjersberg et al., 2015), reducing the time crew members can dedicate to mission objectives. Additionally, open surgery typically requires a team of personnel, including a surgeon, assistants, and anesthesia providers, which may exceed the capabilities of the available crew.

### 1.3.2 Laparoscopic surgery

Laparoscopic surgery is minimally invasive, resulting in reduced recovery times and fewer complications (Guend et al., 2015). It employs existing technologies that could potentially be adapted for space, such as specialized insufflation methods; however, as discussed, handling insufflation gases in microgravity presents challenges, including managing carbon dioxide diffusion and organ movement. Maintaining a stable visual field and precise instrument control may be difficult in microgravity, which could compromise surgical outcomes. Logistically, this approach is less viable in the near term since the equipment requires significant mass, volume, and power, as well as substantial skill, all of which are constrained in exploration paradigms. Nevertheless, some early research on adapted laparoscopy, such as hand-assisted laparoscopic surgery, shows promise for facilitating minimally invasive surgery by less surgically proficient/experienced CMOs (Broderick et al., 2005).

### 1.3.3 Robotic-assisted surgery

Robotic systems enhance surgical precision and reduce the physical demands on surgeons. They enable ground-based control, allowing experts on Earth to assist or perform procedures in real-time or with a time delay, which is particularly advantageous given the limited medical expertise onboard spacecraft. Robotics can also lessen the cognitive load on humans during high-stress situations (Herron et al., 2008). However, robotic surgery relies on advanced, often costly systems with substantial mass, volume, and power requirements, which may be impractical for space missions, particularly if resupply is not feasible. Robotic systems can malfunction and may be difficult to repair without real-time ground-based support (Pantalone et al., 2021). Communication latency and degradation pose significant challenges for long-distance operations, necessitating the development of semiautonomous systems capable of functioning with minimal real-time input or performing complex operations with minimally trained crewmembers. Development of fully autonomous capabilities of surgical robots may one day enhance surgical capabilities on human spaceflight missions, but this remains likely decades away from being a reality (Haidegger, 2019).

### 1.3.4 Interventional radiology/percutaneous techniques

IR and percutaneous techniques are minimally invasive procedures that often have a lower skill threshold compared to other surgical methods. IR also employs less equipment that is often smaller in volume and with multiple potential applications (e.g., ultrasound device, catheter, needle, etc.), making it ideal for the logistical limitations of deep-space missions. Efforts are ongoing to develop an IR "toolbox" sufficient to treat eleven different medical scenarios in deep space (Vidal et al., 2023). General anesthesia is often unnecessary for IR, reducing associated risks and typically resulting in shorter recovery times. However, the limited availability of advanced imaging tools for real-time guidance in space can hinder the effectiveness of these procedures. Success relies heavily on operator skill, and without specialized training, outcomes may be suboptimal (Lerner and Parmet, 2013). Additionally, percutaneous techniques can only treat a subset of the conditions that can be fully addressed with open surgery.

Ultimately, each surgical modality offers distinct advantages and limitations when considered for use in space. Factors such as equipment requirements, the challenges posed by the microgravity environment, the need for specialized training, and the constraints of space missions must be evaluated thoroughly to identify the most appropriate surgical approach for a specific medical event during spaceflight missions.

## 1.4 Ethics of intervention in space

The ethics of medical intervention in space are shaped by the unique challenges of the environment, including resource limitations, isolation, and the need to balance individual and collective well-being. Space medicine must adhere to principles of austere medicine and requires making difficult decisions in contexts where traditional standards of care are not feasible (Rajput et al., 2023). Additionally, many Earth standards of care may be less effective in partial gravity or microgravity environments, such as the previously mentioned CPR, which further complicates treatment decisions. These ethical considerations extend across triage, resource allocation, temporizing management, palliative care, and collaborative decision-making.

### 1.4.1 Austere medicine standards of care

In space, medical interventions often require a shift from the traditional goal of providing optimal care to offering limited, targeted interventions aimed at stabilizing the patient and ensuring mission success. Triage becomes a critical ethical concern, as the medical officer must assess the severity of injuries or illnesses and prioritize treatment based on the likelihood of survival and the impact on mission objectives (Kahn et al., 2014). For example, a minor injury might receive immediate attention to prevent escalation, while a condition with a low probability of survival might receive only palliative measures to conserve resources for the survival of the remaining crew. Several triage models, such as SALT (Sort, Assess, Life-saving intervention, Treatment/Transport) and START (Simple Triage Algorithm and Rapid Treatment), exist for mass casualty incidents or situations where current resources are not adequate to treat all patients to the recognized standard of care. These algorithms focus on identifying patients who have the greatest need for immediate medical intervention using the available resources, but often with the expectation that more equipment and personnel will be mobilized. No such algorithm yet exists for astronauts to apply to medical emergencies in space where there will be no access to additional resources or higher levels of care, and priorities must include the survival of the rest of the crew.

The concept of targeted care prioritizes stabilization over comprehensive treatment, acknowledging the constraints of space missions. Optimal care, such as definitive surgical interventions or extensive posttreatment rehabilitation, may be impractical due to limitations in medical expertise, equipment, and time. This compromise raises ethical questions about what level of care is acceptable in life-threatening situations and how to ensure fairness in decision-making.

### 1.4.2 Resource allocation

Space missions operate under strict resource constraints, compelling medical personnel to make ethical decisions about how to allocate supplies. Resources such as medications, oxygen, and diagnostic tools must be managed to ensure the crew's survival as a whole rather than concentrating solely on the needs of an individual. For example, treating a crew member's severe injury might deplete supplies that are crucial for the group's overall health, raising ethical dilemmas about whose needs should take precedence. Balancing individual care with the collective well-being of the crew necessitates a utilitarian approach, where decisions seek to maximize overall survival and mission success. This perspective is important during the spacecraft design phase, as essential equipment meeting medical operational needs must be incorporated into the vehicle with appropriate mass, volume, and power allocations. For example, should a ventilator be fixed and incorporated into the atmosphere control system, or portable with gas cannisters, or something in between? The mission profile and risk analysis will drive design decisions for medical equipment, recognizing that unexpected events will occur and that the ideal equipment may not be available, potentially leading to lethal outcomes.

### 1.4.3 Temporizing management: immediate versus potential benefit

Immediate medical interventions in space must weigh the potential benefits against the risks posed to the entire crew. Treating an infectious condition, for example, might involve isolating the affected individual to prevent contagion, even if this negatively impacts their mental and emotional health. Similarly, using limited medical supplies for one patient could hinder future interventions. The ethical principle of maintaining mission-critical redundancy—ensuring there are enough skilled crewmembers to complete the mission—is another important consideration. The loss of a key crew member might jeopardize the mission or the crew's safe return, creating a moral conflict between individual and collective interests.

Medical interventions can also have lasting psychological and emotional impacts on the crew. Witnessing or participating in significant medical procedures, or coping with the loss of a colleague, can strain interpersonal relationships and mental well-being. Ethical medical practice in space must address these concerns by prioritizing mental health alongside physical care and recognizing the unique stressors of the environment.

### 1.4.4 Palliative care

When recovery is unlikely or impossible, palliative care becomes an ethical necessity, ensuring that the patient's final moments are as comfortable and dignified as possible. In the limited environment of space, palliative care must balance compassion with practicality, minimizing the impact on mission resources while addressing the patient's pain and emotional needs. This approach requires ethical sensitivity, especially when end-of-life decisions must be made without the possibility of returning the patient to Earth.

### 1.4.5 Collaborative decision-making process

The ethical framework for medical intervention in space relies heavily on collaborative decision-making. In emergencies, input from the patient, the medical officer, and ground control plays a critical role. Effective communication and real-time collaboration between the crew and Earth-based medical experts are essential; however, delays or interruptions in communication can complicate the process. Shared decision-making models provide a structured approach to ethical dilemmas, incorporating medical expertise, mission priorities, patient considerations, and autonomy.

In certain scenarios, nonconsensual interventions may be ethically justified, such as when a patient is unconscious or when their condition poses a risk to others. These decisions must balance individual rights with crew safety and mission success. Ultimately, the final decision-making authority may rest with the mission commander or the medical officer, guided by preestablished protocols and ethical guidelines developed in coordination with ground-based experts.

### 1.4.6 Confidentiality

Another ethical consideration is maintaining the confidentiality and privacy of astronaut health data. Astronauts undergo extensive medical evaluations and monitoring, both on Earth and during missions, to ensure their health and safety.

However, this creates a potential conflict between respecting individual privacy and ensuring mission success and crew safety. Sharing health data with mission control, medical experts, and even other crewmembers may be necessary to address medical emergencies or to make critical decisions about mission operations. Ethically, maintaining the confidentiality of such data requires clear policies that balance the astronaut's right to privacy with the operational need for transparency. Astronaut health data also provides a wealth of information valuable for advancing scientific research and space medicine. However, the relatively small population of astronauts makes anonymizing health data difficult. Thus, consent regarding astronaut health data is crucial; astronauts should be informed about how their data will be used and who will have access to it. Moreover, the secure handling of health data is imperative to prevent unauthorized access or misuse, particularly as space missions involve international collaboration and potentially divergent privacy laws. Ultimately, establishing trust through ethical data governance is essential to safeguard astronauts' privacy while ensuring mission readiness.

Medical intervention in space presents profound ethical challenges, requiring a balance between individual care and the collective well-being of the crew. Austere medical practices, careful resource allocation, and temporizing management strategies must all be informed by ethical principles adapted to the constraints of the space environment. Collaborative decision-making ensures that medical care aligns with both patient needs and mission objectives, preserving the integrity and safety of the crew under extraordinary circumstances.

## References


Anderson, A., Stratton, E., Nelson, A., Lemery, J., Berens, K., Hilmers, D., et al., 2025. Development of progressively earth-independent medical operations to enable NASA exploration missions. Wilderness & Environmental Medicine 36 (1_suppl). Available at https://doi.org/10.1177/10806032241310386.

Armstrong, J., Monteiro, M., 2019. Anti-microbial polymer development for spacecraft cabin disease & system contamination. In: Proceedings of the International Astronautical Congress, IAC 00741795 International Astronautical Federation, IAF Australia. Available at http://conference.researchbib.com/?action=viewEventDetails&eventid=17342&uid=re90b8 2019.

Asrar, F.M., Sieber, M.A., Possnig, C., Mukai, C., Morgan, A.R., 2025. Physician astronauts advancing deep-space exploration. The Lancet 406 (10509), 1217–1218. Available at https://doi.org/10.1016/S0140-6736(25)01805-7, https://www.sciencedirect.com/science/journal/01406736.

Aunins, T.R., Erickson, K.E., Prasad, N., Levy, S.E., Jones, A., Shrestha, S., et al., 2018. Spaceflight modifies Escherichia coli gene expression in response to antibiotic exposure and reveals role of oxidative stress response. Frontiers in Microbiology 9. Available at https://doi.org/10.3389/fmicb.2018.00310, https://www.frontiersin.org/articles/10.3389/fmicb.2018.00310/full.

Avcı, A.U., 2024. Hearing loss in space flights: a review of noise regulations and previous outcomes. Journal of International Advanced Otology 20 (2), 171–174. Available at https://doi.org/10.5152/iao.2024.231434, https://advancedotology.org/Content/files/sayilar/127/171-174.pdf.

Blue, R., Bridge, L., Chough, N., Cushman, J., Khpal, M., 2014. NASA Technical Reports Server Identification of Medical Training Methods for Exploration Missions.

Broderick, T.J., Privitera, M.B., Parazynski, S.E., Cuttino, M., 2005. Simulated hand-assisted laparoscopic surgery (HALS) in microgravity. Journal of Laparoendoscopic and Advanced Surgical Techniques - Part A 15 (2), 145–148. Available at https://doi.org/10.1089/lap.2005.15.145.

Campbell, M.R., Billica, R.D., 2020. Surgical capabilities. Principles of Clinical Medicine for Space Flight. pp. 233–252. Available at https://doi.org/10.1007/978-1-4939-9889-0_7.

Campbell, M.R., Dawson, D.L., Melton, S., Hooker, D., Cantu, H., 2001. Surgical instrument restraint in weightlessness. Aviation, Space, and Environmental Medicine 72 (10), 871–876.

Eichstadt, F., 1994. SAE technical papers. SAE International Microgravity medical restraint: an adaptive approach. https://doi.org/10.4271/941321 26883627, http://papers.sae.org/.

Gerasimov, Y.F., Maidanik, Y.F., Shchegolev, G.T., Filippov, G.A., Starikov, L.G., Kiseev, V.M., et al., 1975. Low-temperature heat pipes with separate channels for vapor and liquid. Journal of Engineering Physics 28 (6), 683–685. Available at https://doi.org/10.1007/BF00867371.

Guend, H., Lee, D.Y., Myers, E.A., Gandhi, N.D., Cekic, V., Whelan, R.L., 2015. Technique of last resort: characteristics of patients undergoing open surgery in the laparoscopic era. Surgical Endoscopy 29 (9), 2763–2769. Available at https://doi.org/10.1007/s00464-014-4007-7, https://rd.springer.com/journal/464.

Haidegger, T., 2019. Autonomy for surgical robots: concepts and paradigms. IEEE Transactions on Medical Robotics and Bionics 1 (2), 65–76. Available at https://doi.org/10.1109/tmrb.2019.2913282.

Hayden, J.A., Pantalos, G.M., Burgess, J.E., Antaki, J.F., 2013. A hermetically sealed, fluid-filled surgical enclosure for microgravity. Aviation, Space, and Environmental Medicine 84 (12), 1298–1303. Available at https://doi.org/10.3357/ASEM.3751.2013, http://docserver.ingentaconnect.com/deliver/connect/asma/00956562/v84n12/s14.pdf.

Herron, Maron, A., Advincula, Aggarwal, Palese, Broderick, et al., 2008. A consensus document on robotic surgery. Surgical Endoscopy and Other Interventional Techniques 22 (2), 313–325. Available at https://doi.org/10.1007/s00464-007-9727-5.

Huang, B., Li, D.G., Huang, Y., Liu, C.T., 2018. Effects of spaceflight and simulated microgravity on microbial growth and secondary metabolism. Military Medical Research 5 (1), 20549369. Available at https://doi.org/10.1186/s40779-018-0162-9, http://mmrjournal.biomedcentral.com/about.

Irby, I., Broddrick, J.T., 2024. Microbial adaptation to spaceflight is correlated with bacteriophage-encoded functions. Nature Communications 15 (1). Available at https://doi.org/10.1038/s41467-023-42104-w, https://www.nature.com/ncomms/.

Iwata, N., Nakanoya, S., Nakamura, N., Takeda, N., Tsutsui, F., 2022. Thermal performance evaluation of space radiator for single-phase mechanically pumped fluid loop. Journal of Spacecraft and Rockets 59 (1), 225–235. Available at https://doi.org/10.2514/1.a35030.

Jacob, P., Oertlin, C., Baselet, B., Westerberg, L.S., Frippiat, J.P., Baatout, S., 2023. Next generation of astronauts or ESA astronaut 2.0 concept and spotlight on immunity. npj Microgravity 9 (1). Available at https://doi.org/10.1038/s41526-023-00294-z, https://www.nature.com/npjmgrav/.

Johnston, S.L., Smart, K.T., Pattarini, J.M., 2020. Medical evacuation risk and crew transport. Principles of Clinical Medicine for Space Flight 327–353. Available at https://doi.org/10.1007/978-1-4939-9889-0_10.

Kahn, J., Liverman, C.T., McCoy, M.A., 2014. Health Standards for Long Duration and Exploration Spaceflight: Ethics Principles, Responsibilities, and Decision Framework 1–172. Available at https://doi.org/10.17226/18576.

Kerstman, E., Parmar, P., Hilmers, D., Boley, L., Arellano, J., Vega Gina, O., et al., 2024. Millenia 2024 Impact Real World System Validation.

Kim, W., Tengra, F.K., Young, Z., Shong, J., Marchand, N., Chan, H.K., et al., 2013. Spaceflight promotes biofilm formation by *Pseudomonas aeruginosa*. PLoS ONE 8 (4). Available at https://doi.org/10.1371/journal.pone.0062437, http://www.plosone.org/article/fetchObjectAttachment.action?uri=info%3Adoi%2F10.1371%2Fjournal.pone.0062437&representation=PDF.

Komorowski, M., Watkins, S.D., Lebuffe, G., Clark, J.B., 2013. Potential anesthesia protocols for space exploration missions. Aviation, Space, and Environmental Medicine 84 (3), 226–233. Available at https://doi.org/10.3357/ASEM.3427.2013Germany, http://docserver.ingentaconnect.com/deliver/connect/asma/00956562/v84n3/s8.pdf.

Kovalski, E., Salazar, L., Levin, D., Kamine, T.H., 2023. A 3D-printed portable sterilizer to be used during surgical procedures in spaceflight. Aerospace Medicine and Human Performance 94 (11), 857–860. Available at https://doi.org/10.3357/AMHP.6294.2023, http://ingentaconnect.com/content/asma/amhp;jsessionid=13tlejf2f0ojk.alice.

Lerner, D.J., Parmet, A.J., 2013. Interventional radiology: the future of surgery in microgravity. Aviation, Space, and Environmental Medicine 84 (12), 1304–1306. Available at https://doi.org/10.3357/ASEM.3790.2013, http://docserver.ingentaconnect.com/deliver/connect/asma/00956562/v84n12/s15.pdf.

Lerner, D.J., Pohlen, M.S., Apland, R.C., Parivash, S.N., 2022. Just-in-time training with remote guidance for ultrasound-guided percutaneous intervention. Aerospace Medicine and Human Performance 93 (12), 882–886. Available at https://doi.org/10.3357/AMHP.6152.2022, http://ingentaconnect.com/content/asma/amhp;jsessionid=13tlejf2f0ojk.alice.

Levin, D.R., Steller, J., Anderson, A., Lemery, J., Easter, B., Hilmers, D.C., et al., 2023. Enabling human space exploration missions through progressively Earth Independent Medical Operations (EIMO). IEEE Open Journal of Engineering in Medicine and Biology 4, 162–167. Available at https://doi.org/10.1109/OJEMB.2023.3255513, https://ieeexplore.ieee.org/xpl/RecentIssue.jsp?punumber=8782705&_ga=2.23991482.1496039865.1593798416-2073569860.1593798416.

Loehr, J.A., Guilliams, M.E., Petersen, N., Hirsch, N., Kawashima, S., Ohshima, H., et al., 2015. Physical training for long-duration spaceflight. Aerospace Medicine and Human Performance 86. Available at https://doi.org/10.3357/AMHP.EC03.2015.

McCuaig, K., Lloyd, C.W., Gosbee, J., Snyder, W.W., 1992. Simulation of blood flow in microgravity. The American Journal of Surgery 164 (2), 119–123. Available at https://doi.org/10.1016/S0002-9610(05)80368-X.

McQuillen, J., 2004. NASA Technical Reports Server Gas-Liquid Flows and Phase Separation.

Overbeek, R., Liebold, F., Johnson Kolaparambil Varghese, L., Adams, N.B., Schmitz, J., Neumann, M., et al., 2024. Alternative techniques for cardiopulmonary resuscitation in extreme environments – a scoping review. Resuscitation Plus 20. Available at https://doi.org/10.1016/j.resplu.2024.100762, https://www.sciencedirect.com/science/journal/26665204.

Overbeek, R., Jan Schmitz, Rehnberg, L., Benyoucef, Y., Dusse, F., Russomano, T., et al., 2022. Effectiveness of CPR in hypogravity conditions—a systematic review. Life 12 (12), 1958. Available at https://doi.org/10.3390/life12121958.

Pantalone, D., Faini, G.S., Cialdai, F., Sereni, E., Bacci, S., Bani, D., et al., 2021. Robot-assisted surgery in space: pros and cons. A review from the surgeon's point of view. Nature Research, Italy npj Microgravity 7 (1). Available at https://doi.org/10.1038/s41526-021-00183-3, https://www.nature.com/npjmgrav/.

Pantalone, D., 2023. Surgery in the next space missions. Life 13 (7), 1477. Available at https://doi.org/10.3390/life13071477.

Patil, M., Gharde, P., Reddy, K., Nayak, K., 2024. Comparative analysis of laparoscopic versus open procedures in specific general surgical interventions. Cureus 16 (2). Available at https://doi.org/10.7759/cureus.54433.

Rajput, S., Mayor, I., Diamond, M., Rosenberg, M., Cole, V., Bhakare, N., et al., 2023. Medical ethics of long-duration spaceflight. NPJ Microgravity 9 (1). Available at https://doi.org/10.1038/s41526-023-00333-9, https://www.nature.com/npjmgrav/.

Seldinger, S.I., 1953. Catheter replacement of the needle in percutaneous arteriography: a new technique. Acta Radiologica 39 (5), 368–376. Available at https://doi.org/10.3109/00016925309136722.

Spanjersberg, W.R., van Sambeeck, J.D.P., Bremers, A., Rosman, C., van Laarhoven, C.J.H.M., 2015. Systematic review and meta-analysis for laparoscopic versus open colon surgery with or without an ERAS programme. Surgical Endoscopy 29 (12), 3443–3453. Available at https://doi.org/10.1007/s00464-015-4148-3.

Stroud, K., Pickett, L., Tillman, B., 2008. NASA Human Integration Design Handbook (HIDH): revitalization of space-related human factors, environmental, and habitability data and design guidance. NASA Technical Reports Server 20080010620.

Summers, R.L., Johnston, S.L., Marshburn, T.H., Williams, D.R., 2005. Emergencies in space. Annals of Emergency Medicine 46 (2), 177–184. Available at https://doi.org/10.1016/j.annemergmed.2005.02.010, http://www.elsevier.com/inca/publications/store/6/2/3/2/7/8/index.htt.

Szydlowski, L.M., Bulbul, A.A., Simpson, A.C., Kaya, D.E., Singh, N.K., Sezerman, U.O., et al., 2024. Adaptation to space conditions of novel bacterial species isolated from the International Space Station revealed by functional gene annotations and comparative genome analysis. Microbiome 12 (1). Available at https://doi.org/10.1186/s40168-024-01916-8, https://microbiomejournal.biomedcentral.com/.

Taddeo, T.A., Armstrong, C.W., 2008. Spaceflight medical systems. Principles of Clinical Medicine for Space Flight 69–100. Available at https://doi.org/10.1007/978-0-387-68164-1_4.

Vidal, V., Boyer, L., Luciani, A., Tacher, V., Kobeiter, H., Longere, B., et al., 2023. Bringing interventional radiology to Mars!. CardioVascular and Interventional Radiology 46 (4), 425–427. Available at https://doi.org/10.1007/s00270-023-03392-3, https://www.springer.com/journal/270.

Warnecke, T., Dauth, L., Ahlbäck, A., DuCanto, J., Fleischhammer, E., Glatz, C., et al., 2021. Time to ventilation and success rate of airway devices in microgravity: a randomized crossover manikin-trial using an underwater setting. Acta Anaesthesiologica Scandinavica 65 (5), 681–687. Available at https://doi.org/10.1111/aas.13780.

Wilson, J.W., Ott, C.M., Bentrup, K.H.Z., Ramamurthy, R., Quick, L., Porwollik, S., et al., 2007. Space flight alters bacterial gene expression and virulence and reveals a role for global regulator Hfq. Proceedings of the National Academy of Sciences of the United States of America 104 (41), 16299–16304. Available at https://doi.org/10.1073/pnas.0707155104, http://www.pnas.org/cgi/reprint/104/41/16299. United States.

Wong, J.Y., Pfahnl, A.C., 2016. 3D printed surgical instruments evaluated by a simulated crew of a Mars mission. Aerospace Medical Association, Canada Aerospace Medicine and Human Performance 87 (9), 806–810. Available at https://doi.org/10.3357/AMHP.4281.2016, http://docserver.ingentaconnect.com/deliver/connect/asma/23756314/v87n9/s9.pdf.

Zong, H., Fei, Y., Liu, N., 2025. Circadian disruption and sleep disorders in astronauts: a review of multi-disciplinary interventions for long-duration space missions. International Journal of Molecular Sciences 26 (11), 5179. Available at https://doi.org/10.3390/ijms26115179.

Chapter 2

# General considerations of the space environment

Nicolas Heft[1] and Rowena Christiansen[2,3]

[1]*Department of Emergency Medicine, Space Medicine Section, The University of Texas Health Science Center at Houston, Houston, TX, United States,* [2]*Department of Medical Education, The University of Melbourne Medical School, Melbourne, VIC, Australia,* [3]*School of Science Computing and Emerging Technologies, Swinburne University of Technology, Melbourne, VIC, Australia*

## 2.1 Introduction

The practice of medicine has always been shaped by the environment. On Earth, sophisticated healthcare systems have been developed in places where resources are plentiful. Robust infrastructure and large communities support a variety of specialist providers working in facilities where patients can be stabilized and/or quickly transferred to higher levels of care when needed. In remote areas of the Earth, medical care is limited to what resources and personnel are available. Beyond the systems we build, humans are intrinsically linked to and shaped by the physical environment. Reduced oxygen at high altitudes, increased pressure under the sea, cold, heat, availability and duration of sunlight, and many other variables drive responses at every level of human physiology from the expression of genes to altered regulation of cellular protein activity to macro physiological systems such as reflexes and blood circulation control. Spaceflight, with changes in gravity, spacecraft environments with altered pressure, acoustics, and lighting, and radiation is a similar biological challenge with it's own unique physiological responses.

The spacecraft protects the human, but the incredible amount of energy and engineering required to launch and operate it forces us to carefully consider everything we place on board. This often requires uncomfortable choices, compromises, and trades based on predicted risks to ensure everything astronauts need fits within the mass, volume, power, and operational constraints of the vehicle and launch system. In this way, space is both an extreme and an austere (resource-constrained) environment, and its relative absence of gravity, lack of a breathable atmosphere, isolation, confinement, and (for some missions) the distance from Earth combine to create operational and physical constraints significant enough to prompt mission planners to reconsider even the most basic assumptions of medical care.

Today, we maintain a continuous human presence in space on board orbital platforms such as the International Space Station and routinely fly missions lasting 6–12 months at a time. In the near term, the spaceborne population is likely to grow as commercial companies and a variety of government agencies build larger, more capable spacecraft including stations in Earth and Lunar orbit and vehicles capable of reaching other planets (Vast Space, 2025; Axiom Space, 2025; SpaceX, 2025). Looking to the future, exploration-class missions will take us back to the Moon, and eventually to Mars (NASA Moon to Mars Architecture, 2025; SpaceX, 2025). These longer and more distant journeys will inevitably carry a greater risk of more severe medical conditions developing in-flight. Although major invasive medical procedures in space have not yet been necessary and will likely remain beyond our near and mid-term capabilities, minor procedures are performed with frequency. Over the past seven decades of spaceflight, our experience with, and ability to manage, medical conditions requiring in-flight procedures has increased significantly. As space agencies and commercial companies eye a sustainable economy in low Earth orbit (LEO), a return to the Moon, and ambitious missions to Mars over the next 10–15 years, our need for procedural competence in space is likely to grow substantially.

Despite this, there is still no dedicated resource covering procedural medicine in space. This book is meant to close that gap. This text collates clinical experience and publications on procedural medicine in spaceflight from the past 65 years and offers a structured reference for what is practical and possible in the resource-constrained environment of space. These lessons will not only be valuable for astronauts and flight surgeons, but also for the engineers, mission planners, and educators who will shape the medical systems that support future crews.

**Procedural Medicine for Spaceflight. DOI:** https://doi.org/10.1016/B978-0-443-27552-4.00002-6

## 2.2 The space environment

### 2.2.1 Space environment

To date, NASA has identified five major hazards of human spaceflight, which are collectively known as "RIDGE" hazards: radiation, isolation and confinement, distance from Earth, gravity (or the lack thereof), and hostile/closed environments (National Aeronautics and Space Administration, 2021). Together, these hazards define the major biomedical challenges that any space-based healthcare system must overcome.

The physiological adaptations our bodies make to the space environment change the safety and feasibility of medical and surgical procedures that we take for granted here on Earth. The major stressors that are most associated with spaceflight include microgravity, space radiation, and prolonged isolation. Less often discussed, but just as important, are the physiological alterations (called "adaptations") that occur over various periods of time spent in space. Immune system changes (Hicks et al., 2023), shifts in the human and environmental microbiome (Gonzalez et al., 2024), and nutritional limitations (Tang et al., 2022) are examples of some of the physiological manifestations of spaceflight that may affect a patient's state prior to any procedure being undertaken. Collectively, these factors present unique challenges that make even the most extreme polar expeditions or early circumnavigations of the globe seem simple by comparison.

Microgravity is the most recognizable, and arguably the most influential, of these stressors. Its effects begin almost immediately after postlaunch main engine cutoff and continue for the duration of the mission. The absence of gravitational loading rapidly causes fluid to shift towards the head and chest, producing facial swelling, jugular venous distention, and an equilibration of blood pressure along the length of the body, as opposed to the normal blood pressure gradient from head to toe. For clinicians, these changes may likely complicate tasks as routine as placing an arterial line or securing an airway. At the same time, the cardiovascular system adapts in ways that reduce blood volume and lower stroke volume (Shibata et al., 2023). Consequently, patients in microgravity may have different physiological reserve, and clinicians may need to change the threshold for initiating fluid resuscitation or inotropic medications. Musculoskeletal deconditioning is also rapid and profound, and astronauts typically lose approximately 1-3% of their bone density per month in weight-bearing bones and experience marked muscle atrophy in postural muscle groups (Sibonga et al, 2017). Fortunately, these effects correlate with mission duration and can be partially mitigated through regular exercise and other countermeasures such as diet or pharmacologic adjuncts (Roberts et al., 2025). However, these technologies may pose unique risks of their own if misused or if equipment was to break during use and the architecture of bone may still be altered even if the density is preserved. This atrophy leaves astronauts with an increased susceptibility to fractures upon return to gravitational loading (Coulombe, Senwar and Ferguson, 2020). Furthermore, the calcium leached from an imbalance in bone remodeling, with bone resorption dominating over formation, can precipitate the development of kidney stones (Thamer and Buckey, 2022).

Microgravity also affects how fluids and gases behave. On Earth, blood will pool in a surgical field, vomit can be suctioned away, and irrigation fluid drains predictably. In space, those same substances may adhere to surfaces due to surface tension, or form into spheres and drift through the atmosphere, either contaminating the cabin interior or, at worst, being inhaled by other crew members. A small amount of bleeding that would be easily contained on Earth could easily contaminate the cabin environment in space, and for this reason, specialized surgical enclosures, containment systems, and modified suction devices will likely prove essential for protecting both patients and crew members.

Radiation is another hazard, particularly beyond LEO where Earth's magnetic field no longer provides moderate shielding. It is also a different form of radiation, including more high energy particles, neutrons, and other more exotic forms than are typically encountered on Earth. (European Space Agency, 2025). The effect of this radiation on organisms, cells, and stored complex biomolecules such as food and medications is still under evaluation.

Even with careful psychological screening, long-duration exploration class missions will test the limits of human resilience in Isolated, Contained, Extreme environments. Although severe psychological conditions are likely to be detectedduring astronaut selection, certain traits such as anxiety, irritability, or depressive symptoms are not only possible, but even expected to a certain degree and may still impact mission success. Disrupted sleep cycles and circadian misalignment degrade cognitive performance, slow reaction times, and reduce vigilance, which could further affect individuals already negatively affected by the stressors of space. Any physician who has worked overnight shifts can attest to the detrimental effects these factors can have on clinical judgment and coordination. In space, these deleterious effects could rapidly erode team cohesion and trust and magnify the impact of interpersonal conflicts within a small, tightly knit crew.

At the cellular level, potential immune dysregulation has been documented during spaceflight. Although the etiology is still under investigation and the clinical significance is not clear, numerous studies have shown that the distribution of peripheral leukocytes is altered, there is diminished function of specific leukocyte subpopulations, as well as changes in cytokines (Crucian et al., 2018). Clinically, there has also been reactivation of latent viruses such as herpes virus,

Epstein–Barr virus, and varicella zoster virus during flight, which suggests a causal relationship between immune dysregulation and viral reactivation (Mehta et al., 2013). Generally, the research to date seems to suggest that these changes are mostly subclinical and crews generally do not suffer significant illness during flight apart from a few well documented cases such as the urinary tract infection during Apollo 13 or prostatitis during Salyut-7 (Jones et al., 2005). Alas, the confidential nature of astronaut medical data results in most medical events being known only to the astronaut and their flight surgeon, and therefore the literature evidence detailing in-flight clinical data is sparse (Crucian et al., 2016).

Nutrition adds another layer of complexity. Even in LEO, astronauts rely heavily on shelf-stable, prepackaged foods. On longer missions, the resupply of the fresh produce that astronauts in LEO may occasionally receive will be even more limited, or even nonexistent. Lack of sunlight contributes to a vitamin D deficiency, and the astronaut diet often results in discrepancies such as too much sodium and not enough folate and omega-3 fatty acids. Nutrients are essential for immune function, collagen synthesis, wound healing, and coagulation, and their reduced intake may also lead to a prolonged postoperative recovery period. In addition, astronauts tend to eat less overall during missions, especially during early phases of the mission when space adaptation may induce symptoms similar to terrestrial motion sickness. As a result, the nutritional baseline of a patient undergoing a procedure in space may be compromised before an incision is ever made. Taken together, these environmental stressors create a different clinical environment from anything encountered on Earth. As with any clinical practice, it is important to account for these differences when planning and training to practice space medicine.

### 2.2.2 Engineered environment

While the natural environment of space changes the physiology of the human body, and even the basic physical laws upon which we rely, the engineered environment inside a spacecraft creates its own unique set of obstacles for medical care. A spacecraft is never just a vehicle; it is simultaneously a life-support system, a home, a kitchen, a toilet, a laboratory, and, if necessary, an operating room or intensive care unit. Every aspect of its design, from the composition of its atmosphere to the level of background noise and lighting, influences how medicine can be practiced in that setting.

Spacecraft engineers have always been driven by the need to minimize mass, volume, and power, while maximizing safety and redundancy. The result is a habitat that is compact, efficient, and multipurpose, but one that leaves little to no room for a dedicated medical facility. To date, in space there are no operating rooms, no recovery suites, and no intensive care units, although the conceptual *Space Station Freedom*, which was never built, originally had plans for a dedicated medical suite, in addition to a hyperbaric chamber. Any procedure must take place in a multiuse space, and a workstation used for science in the morning may need to be cleared and repurposed for a medical intervention in the afternoon. This reality complicates sterility, ergonomics, and workflow. Lighting may be uneven. Storage space is limited. The orientation of the patient and provider may depend on straps and restraints rather than gravity, and every tool and instrument must be secured, since floating equipment may pose a danger to the patient, crew, or even the spacecraft itself.

The same systems that sustain the crew also shape the boundaries of what is medically possible. Life-support systems maintain oxygen, carbon dioxide, humidity, pressure, and temperature within narrow ranges, but even small deviations can have major effects on both physiology and procedural safety. Elevated carbon dioxide, for example, has been a recurring issue aboard the ISS and is associated with headaches, impaired concentration, and altered respiratory drive. Humidity is another concern. Several space stations, such as Salyut-6, Salyut-7, Mir, and even the ISS, have been negatively impacted by the formation of biofilms on hardware, which has degraded equipment performance and even required components of the water recovery system to be returned to Earth for unclogging and reprocessing before being returned to ISS (Flores et al., 2023). From a medical perspective, these environmental conditions could potentially contaminate food supplies, increase the risk of infections, and/or affect the sterility of medical equipment. Fluids also pose hazards of their own, and blood, vomit, or irrigation solutions that escape containment can contaminate filters, sensors, and circulation systems critical for life-support, which could turn a simple medical complication into a vehicle-wide emergency.

Atmospheric composition is another variable that cannot be overlooked. While the ISS maintains sea-level conditions (pressure of 14.7 psi and 21 percent oxygen), future exploration-class missions may operate within different pressures or oxygen concentrations. One exploration-class paradigm involves crews living at a reduced pressure with enriched oxygen concentrations to reduce the risk for decompression sickness from extravehicular activities (Sanders et al., 2023). These changes alter partial pressures of gases, impose new physiological stressors, and may affect the administration of anesthesia (Puri et al., 2008).

Noise and vibration, while less obvious, are no less impactful. The ISS generally maintains an ambient daytime background noise level of approximately 60–70 decibels, with additional peaks near certain hardware and equipment (Limardo et al., 2021). This is comparable to a busy office or domestic washing machine, and for astronauts who are exposed to it perpetually, this constant sound becomes both unavoidable and fatiguing. A 1974 report by the U.S.

Environmental Protection Agency (EPA) concluded that maintaining a 24-hour equivalent continuous sound level below 70 dB ($L_{eq(24)}$) is protective against noise-induced hearing loss as well as chronic health problems such as high blood pressure, heart disease, attention and focus problems, stress, and fatigue (U.S. Environmental Protection Agency, Office of Noise Abatement and Control, 1974). However, a recent study on space station noise levels found spacecraft background noise levels above 65 dBA had substantial effects on astronaut cognitive load and operational efficiency (Li et al., 2026). Chronic noise disrupts sleep and contributes to cognitive strain and sympathetic activation, while acute noise during procedures interferes with communication and concentration. Both can degrade mission performance, and exposure to high levels of background noise for months to years may contribute to high blood pressure, heart disease, and other chronic illnesses.

Looking ahead, the designed medical environment of future spacecraft and planetary surface habitats must move toward a more patient-centered philosophy. In the near term, habitats are expected to be resource-limited and compact; however, crew medical needs cannot simply be addressed as an afterthought and must instead be integrated early into the overall system development (Patel et al., 2017). While space architecture defines the habitable environment and procedural ergonomics, Human Systems Integration (HSI) is one design level tool that treats humans as a system within the spacecraft and seeks to account for human capabilities and limitations throughout the design, implementation and operation of hardware and software (National Aeronautics and Space Administration, Johnson Space Center, 2021). Key domains of HSI include human factors engineering, resource maintainability, and supportability, as well as crew training and overall safety. Since humans cannot be redesigned in the way that engineered systems can, HSI helps ensure subsystems that interact with humans such as lighting, noise generating elements, controls, interfaces, environmental control, life support, waste management, and others are built to work with the fixed elements of human biology.

### 2.2.3 Operational environment

In addition to the natural hazards of space itself, as well as the engineered constraints of the spacecraft, the operational environment in which astronauts operate adds another layer of complexity to medical care in space. This includes factors such as how crews are composed, how communication occurs across vast distances, the management of scarce resources, and how cultural and ethical considerations may potentially influence medical decisions during multinational missions.

Crew composition is the first major factor. On Earth, medical providers rarely work independently, and instead function within a diverse team, each with clearly defined roles and professional experiences. In space, that team may be located hundreds to thousands to millions of miles (or kilometers) away. Medical care on the ISS has typically depended on the Crew Medical Officer (CMO), who is not always a trained physician, and typically receives a few dozen hours of medical instruction before launch. Even when a physician-astronaut is aboard, procedural expertise is not guaranteed, as medical backgrounds vary considerably. This reality forces a shift toward standardized checklists, just-in-time training, and the redistribution of tasks among crew members with little or no medical background. Complex procedures must be distilled into discrete steps that can be performed by an astronaut with little to no medical training under remote guidance by a flight surgeon.

Communication further complicates this picture. On the ISS, real-time telemedicine allows flight surgeons on the ground to guide astronauts step-by-step, even interpreting ultrasound images in real time. Under certain circumstances, a one-to-two second delay is tolerable, and near real-time care remains possible; however, significant challenges begin to emerge with longer delays. In fact, increasing communication delays within the Deep Space Network (NASA's global system of antennas and communication facilities supporting two-way links with spacecraft beyond LEO) are anticipated due to the higher operational tempo associated with Artemis missions as compared to Apollo. Whereas the round-trip communication time between Earth and the Moon was approximately three seconds during Apollo, delays during Artemis are expected to range from five to fourteen seconds unless system upgrades are implemented (Parisi et al., 2023). For the journey to Mars, a latency of between four and twenty-four minutes each way, together with some periods of "blackout," makes real time consultation impossible. Store-and-forward telemedicine will not be an option to treat medical emergencies in which seconds matter. In those scenarios, the crew will need to rely on their collective training and individual skills, together with available on-board resources. Advanced technologies for Earth-Independent Medical Operations (EIMO) are likely to include integrated artificial intelligence (AI) and clinical decision support systems, point-of-care diagnostics, passive monitoring, advanced AI for data analysis and prediction, and innovative training tools. These technologies will support crews in handling medical situations autonomously by assisting with diagnosis, treatment, and prevention, reducing reliance on mission control and mitigating risks during deep space missions.

Logistics and resources are another defining feature of the operational environment. On Earth, hospitals are stocked with equipment, sterile processing departments clean and repackage instruments, and blood banks can provide blood transfusions to patients in need. None of these replenishable critical resources is currently practical in space, and instruments and consumables must be severely rationed. A single set of supplies may need to last the duration of a mission, and a procedure as simple as starting an intravenous drip may require weighing the pros and cons of consuming these components far more than it would terrestrially. In space, a single patient could require care equivalent to a mass casualty incident due to the resources necessary to treat them, and innovations such as 3D printing and multipurpose tools will likely become essential to prevent the rapid consumption of medical equipment and support missions into deep space. With distance and isolation, ethical questions also play a larger role. With limited consumables, should there be a limit to how much any one patient can consume? Would palliation become the more ethical choice if a procedural complication arises that demands prolonged or more intensive care? The answers to these questions may vary depending on each mission, the medical capabilities available, and the distance from Earth at which the crew is operating.

Two-plus decades of ISS operations have shown that multinational crews bring diverse cultural, linguistic, and professional backgrounds into close quarters, and that these differences should be given due attention. Differences in medical culture, namely how illness is understood, who makes decisions, or how authority is distributed, can complicate care in high-stakes scenarios. Although less likely within a cohesive multinational crew, language barriers may also slow communication in an emergency. Attitudes toward risk, medical autonomy, or end-of-life care may differ among individual crew members. In such a small team, conflict, mistrust, or poor cohesion can quickly erode performance at the worst possible moments, and therefore crew selection, training, and cohesion become just as important to the practice of medicine as the medical kits themselves.

Fatigue and circadian disruption may also magnify these issues, and astronauts often experience fragmented sleep, frequent awakenings, and shortened total sleep time. Data from *Mir* suggests that the duration of sleep is reduced, not as deep, and perhaps even physiologically different from sleep on Earth (Institute of Medicine (US) Committee on Creating a Vision for Space Medicine During Travel Beyond Earth Orbit; Ball JR & Evans CH Jr. (Eds.), 2001). Poor sleep results in abnormal stress responses, diminished performance, and behavioral changes, all of which can increase the risk of an inadvertent medical error. In fact, a study performed at Michigan State University's Sleep and Learning Lab demonstrated that sleep deprivation doubles the odds of making place-keeping errors and triples the number of lapses in attention (Stepan et al., 2020). These consequences may be negligible for simple tasks such as taking a patient's vitals; however, they could be much more serious in an activity that requires following multiple complicated steps in a specific order. Any professional who has worked while fatigued will understand firsthand the toll this takes on decision-making, and in space these negative effects could have disproportionate consequences.

As missions expand to include planetary surface operations, such as on the Moon or Mars, new challenges will emerge. Lunar and Martian regolith are abrasive, electrostatically charged, and may vary in their biological effects. In these partial gravity environments, such as the Moon's 0.16G and Mars' 0.38G, dust particles do not settle at the same rate as they would on Earth and will be more easily agitated and aerosolized. This creates the possibility of dust-contaminated wounds, visual impairment, and respiratory irritation, which may impact patient and provider alike (Pohlen et al., 2022).

As the preceding discussion demonstrates, space medicine is not simply Earth medicine practiced in space, but rather its own discipline, requiring the careful navigation of altered physiology, new pathologies that terrestrial medicine physicians do not encounter, the use of prophylactic strategies and interventions, and the engineering constraints within which crews must operate.

## 2.3 The uniqueness of space medicine

### 2.3.1 Terrestrial comparisons

Terrestrial medicine practiced in hospitals and clinics generally assumes a state of abundance, specifically an abundance of personnel, resources, and diagnostic equipment. Even in prehospital and austere settings, where resources are limited, the fundamental objective remains stabilization and transport to a higher level of care. Spaceflight, however, occupies a unique and precarious position, whereby crews are resource-limited and isolated yet may need to provide prolonged medical care without the safety net of evacuation or definitive surgical capability. There is no trauma center "down the road," no capability for rapid blood transfusions, and no option to request meaningful additional resources, beyond that of telemedicine.

Terrestrial medicine also benefits from gravity. On Earth, clinicians rely on gravity without conscious thought: fluid settles and flows "down", blood pools, air rises and circulates, and instruments stay where they are placed. A chest tube can be open either to gravity or to suction, and its drainage behaves predictably. A "Foley" fluid collection bag is placed below a patient to allow the bladder to drain. A patient may be placed in a head-down/"Trendelenburg" position when placing a central line to prevent the inadvertent aspiration of air and subsequent air embolism during initial canalization. In microgravity, since there is no "up" or "down," movement of fluids and gases changes, driven primarily by pressure gradients other than gravity, and new processes will need to be developed to ensure fluids and gasses and tools behave in the manner required to effect care. Equipment that functions reliably on Earth may in fact be inoperable in orbit, and thus require complete redesign.

The constraints of spacecraft design, particularly mass, volume, and power, impose hard limits on what can be achieved. In a terrestrial hospital setting, physicians may employ ultrasound, X-ray, computerized tomography (CT), magnetic resonance imaging (MRI), or angiography. However, in space, imaging availability is heavily constrained. As of the writing of this chapter, ultrasound remains one of few imaging modalities available to help guide medical therapy, although the recent *Fram-2* mission did achieve the first space-based X-ray (Dinner, 2025), it remains to be seen how this new technology will help space medicine practitioners in the future. Unfortunately, the development of space-based CT or MRI capabilities remains distant due to mass, magnetic interference, maintenance requirements and other operational barriers.

The full range of diagnostic tests may not be available due to the size and support required to run laboratory-based analyzers; however, point-of-care testing will likely play a growing role in space-based medicine. To date, two technology demonstrations have been completed on ISS, specifically HemoCue, which counts white blood cells from a single fingerstick sample (Welch, 2021) and rHealth One, a flow cytometer (NASA Glenn Research Center, 2022). Other testing, such as peripheral blood smears or tissue samples that require assessment and interpretation by a pathologist may not be possible either, or at best, may still require telemedical support from an Earth-based pathologist. Simple tests such as urine dipsticks remain one of the few studies that can be performed in real time to help diagnose a medical condition with concrete conclusions.

### 2.3.2 Human factors challenges

Building on the operational constraints described previously, the "humans in the machine" remain the priority for any mission, and equipment used in space may need to be completely redesigned for microgravity compared to Earth. On the ISS, for instance, ultrasound devices are color-coded to guide less-experienced operators and reduce cognitive load. Lighting that can illuminate an entire surgical field from multiple angles on Earth may be challenging to reproduce in-flight in the near term and would require astronauts to adapt to design limitations to overcome specific challenges. Although the operational environment can be tailored to optimize certain procedures, every design decision comes with trade-offs, and a design decision that may benefit one task may at times directly hinder another. These trade-offs must often be negotiated between all major mission design stakeholders to determine which functions take priority.

Automation, while also often introduced to mitigate risk and reduce workload, can also introduce new forms of vulnerability if human-system integration is poorly understood. The Boeing 737 MAX 8 illustrates this risk. A redesign of the airframe altered the center of gravity. to compensate engineers developed the Maneuvering Characteristics Augmentation System (MCAS), designed to operate in the background and automatically push the aircraft's nose down to make it behave as the airframe did before the redesign and thus require less training for pilots. However, incomplete or absent training on the new system meant many pilots were not aware of the technology, or its function. This contributed to two fatal crashes in 2018 and 2019 (Federal Aviation Administration, 2020), with the combined loss of 349 lives and grounding the entire fleet world wide while an investigation was conducted. Certain medical systems, particularly those employing automation, such as tele-robotic surgery, may carry very similar risks. Without proper training or the ability for a human operator to directly intervene when anomalies arise, these technologies can fail in ways that are both unexpected and potentially catastrophic for patients.

A similar lesson emerged from the collision of the USS *John S. McCain* in 2017 near the Strait of Malacca (National Transportation Safety Board, 2019). There, a touchscreen navigation interface intended to simplify operations instead introduced confusion about control authority. During a high-tempo transit, an inadvertent transfer of steering control between stations led to the vessel turning unintentionally and a subsequent fatal collision that claimed the lives of 10 sailors on board. The system's design lacked intuitive feedback to show who was "in control," and the crew's mental model diverged from the machine's actual state. In spaceflight, the same risk exists when multiple crew members, remote flight surgeons, or automated systems share operational control of life-support, navigation, or medical systems. Ambiguity in authority, unclear hand-offs, or nonintuitive interface design could be equally disastrous in-flight, where time and resources are extremely limited and the ability to recover from off-nominal situations is severely diminished.

Together, these examples emphasize that technology alone does not ensure safety. Whether piloting an aircraft, steering a maritime vessel, or performing a medical procedure in microgravity, the boundary between human and automation must remain explicit, intuitive, and trainable. System design should always prioritize clarity of control, real-time feedback, and the ability for human operators to reassert authority when needed.

## 2.4 The perioperative journey in space

During space missions, patients may present under widely varying circumstances. For example, an unwell crew member may first present while within the spacecraft, or a medical condition may present while the crew member is performing an extravehicular activity, either in microgravity or on a planetary surface. Each environment imposes its own challenges for accessing the patient, stabilizing any initial insults, and performing initial interventions. A simple in-cabin emergency may allow immediate evaluation, whereas a medical event during EVA or surface operations could require complex extrication, recompression due to the pressure differential between spacesuits and the cabin, and possible decontamination, before an assessment or intervention is even possible.

While many presentations, just as on Earth, will be able to be managed conservatively, at times various types of procedural, or even surgical, interventions may be required. However, the scope of such interventions will be determined by limiting factors such as the medical equipment and supplies on board and the training and capabilities of the CMO. With the current level of technology development for surgical capabilities during spaceflight, it is unlikely that major invasive procedures will form part of available medical care, but as is currently the case, some minor procedures can be planned for and supported.

The concept of perioperative care refers to a patient's procedural or surgical journey. This can be divided into three specific phases, namely, the preoperative, intraoperative, and postoperative phases. On Earth, these phases are supported by multidisciplinary teams across a variety of specialized settings such as the operating room, Post-Anesthesia Care Unit (PACU), Intensive Care Units (ICU), and rehabilitation centers. In space, each of these various locations must therefore be condensed into a single multipurpose setting. The entire scope becomes not just a sequence of medical events separated by distinct settings with coordinated hand-offs and overlapping care from multiple teams, but rather an amalgamation of all these locations condensed into one.

While the full scope of perioperative care can be anticipated for many terrestrial inpatient procedures, for instance a routine joint replacement, certain segments of the process are less relevant or less intensive for simple procedures, such as inserting a urinary catheter. The perioperative model is inherently flexible and adaptable, depending on circumstances, while providing a useful guide and checklist of sorts for planning purposes.

### 2.4.1 Preoperative phase

The preoperative phase in space begins long before the moment a crew member presents with a condition that might require a procedure. On Earth, because there are dedicated medical facilities that are continually operational, a patient needs only to present to a hospital or clinic to receive care. In space, when a patient is unwell and needs a medical evaluation, there is generally an interruption to the operational flow, and the crew must pivot from their current tasks to triage the ill crew member, initiate a private medical conference, perform a diagnostic workup, and then treat the patient. The objectives include preparing the patient for a medical procedure in addition to ensuring that the spacecraft, crew, and environment can safely sustain the procedure from start to finish. Prior to starting a procedure, the crew will need to convert the spacecraft itself, or an area of it, into a dedicated medical facility. While the technical aspects of performing a procedure are crucial, accurately predicting immediate and long-term outcomes is just as important, and failure to do so dramatically increases the risk of postoperative mortality (Shaydakov & Tuma, 2025).

### 2.4.2 Intraoperative phase

The intraoperative phase represents one of the most technically complex and operationally fragile portions of a patient's care. On Earth, the operating room (OR) is the most "controlled" environment a patient can be in, and generally involves the coordinated efforts of anesthesiologists, surgeons, and nurses working within a predictable environment with little to no outside interference. An emergency department (ED) presents a more dynamic and less sterile environment than an OR; however, it is still a magnitude of order above what is achievable in a prehospital setting. In space, the same objectives must be achieved in a setting that is extremely dynamic, and where a procedure may be interrupted by unexpected emergencies such as cabin depressurization, fire, or other atmospheric contamination.

Analgesia can be provided terrestrially in many ways, including intravenously, by inhalation, topically, or through regional (localized) anesthesia. Given the challenges of providing intravenous medications in the setting of microgravity, and the risks that volatile (vaporized) anesthetics would play in an enclosed environment, regional anesthesia will likely play a large role in providing anesthesia to patients in space for the foreseeable future. Intravenous medications pose a risk of air embolisms due to the heterogeneous air-fluid mixing in the setting of microgravity, resulting in greater challenges in eliminating air bubbles. Volatile anesthetics are incompatible with closed spacecraft systems, as even minimal leaks could contaminate the cabin atmosphere and life-support systems, and potentially affect other crew members. Regional and local anesthesia, including nerve blocks, may be preferred for certain procedures to conserve resources and minimize systemic effects. These approaches, however, require skill in ultrasound-guided techniques, and an understanding of how pharmacokinetics and fluid distribution in microgravity may affect dosing and anesthetic diffusion. Careful consideration must be given to analgesic and anesthetic drug choices to maximize safety, utility, and flexibility. For instance, ketamine, a popular choice for prehospital medicine, has also often been suggested for spaceflight.

The subsequent medical procedure itself occurs within an environment that will bear little resemblance to a terrestrial OR or ED. There will be no fixed orientation, no stable surface, no natural gravitational drainage, and both the operator and patient alike must be restrained. A surgical barrier will need to surround the operative field, preventing blood, irrigation fluid, or aerosols from escaping into the cabin. Every instrument, sponge, and device will need to remain tethered or constrained in some way to prevent unintentional drift, as a single lost item could compromise both the patient and the spacecraft systems.

Physiological monitoring during space-based procedures or surgery may also be altered. Standard patient monitoring devices, including capnography, invasive arterial lines, and end-tidal gas analysis, are currently unavailable in space, and research is ongoing to identify equipment that could operate in these new environments. Currently, serial assessments would rely on repeating vital signs, physical exams, and ultrasound-based imaging. Given communication delays during exploration-class missions, autonomous decision-making becomes essential, and although ground-based flight surgeons may provide input based on what the crew communicates, the real-time interventions depend entirely on the crew themselves.

Throughout the intraoperative period, the spacecraft itself will also become an active participant in the procedure. Life-support systems will need to accommodate transient changes in oxygen consumption, carbon dioxide production, temperature, and humidity.

### 2.4.3 Postoperative phase

On Earth, a patient recovers in a PACU surrounded by specialists and continuous monitoring. In orbit or on a planetary surface, the compact, multipurpose module must simultaneously serve as an OR, PACU, and *in extremis*, as an ICU. On Earth, the average mortality associated with inpatient surgery fluctuates from 0.5% to 7%, and approximately 2–3 out of every 10 patients develop a postoperative complication. In space, the patient will need to recover in proximity to the rest of the crew, and perhaps near the kitchen, lavatory, and sleeping stations, which presents an infection-control challenge. Infections must be managed empirically using oral or intravenous antibiotics, but both drug selection and dosing are complicated by uncertain pharmacokinetics and the limited stability of stored medications. Space radiation, humidity, and repeated temperature cycling have been shown to degrade pharmaceutical potency, creating uncertainty in therapeutic effect. Drainage systems, wound dressings, enteric feeding (using a feeding tube), and even basic hygiene routines will need to be adapted for microgravity to prevent contamination of the cabin's air and surfaces.

Prolonged rehabilitation presents its own set of challenges. As the human body atrophies so quickly in space, the process by which an unwell or postoperative astronaut may recover their strength and health may be significantly delayed and potentially arduous. The same exercise countermeasures that protect healthy astronauts against deconditioning may potentially become the instruments of recovery. However, intensive exercise can disrupt sutures, reopen incisions, or increase intra-abdominal pressure, so exercise prescriptions must be individualized and administered judiciously and mindfully.

Return to duty would also occur sooner and under greater constraints than on Earth. As every crew member plays an integral role within the overall mission, a prolonged convalescence would burden other crew members with additional tasks and could even threaten overall mission objectives. Therefore, patients may need to resume certain critical mission tasks while still recovering, thus exposing them to the risk of re-injury or delayed healing. The space medicine physician's role, therefore, becomes one of "trade analysis," balancing the need of a crew member to rest and rehabilitate versus the overall operational needs. A shortened rehabilitation period may directly translate to increased operational capacity if all goes well; however, if this process occurs too quickly, this could backfire and lead to an even longer period at reduced operational capacity.

## 2.5 Common surgical and procedural emergencies

On Earth, the most common surgical pathologies include hernias, appendicitis, fractures, and cholecystitis (Fehlberg et al., 2019), whereas less invasive procedures performed in acute care setting such as in the ER or ICU include intubations, central venous catheters, arterial lines, incision and drainage, and suturing. Unlike medical facilities on Earth, in space the anticipated burden of disease shifts away from chronic pathology and leans more heavily toward acute, trauma-related, and environmentally-driven conditions.

### 2.5.1 Trauma

Both extravehicular and in-cabin activities carry a non-negligible risk of mechanical injury, especially on planetary surfaces. During EVA, high-energy impacts from micro-meteoroids, falls, or suit depressurization could cause various injuries (Pantalone et al., 2022). Within the cabin, unrestrained tools or equipment can become projectiles, and even minor collisions can produce lacerations or fractures in a microgravity environment where weight may be negligible but inertial mass is maintained.

### 2.5.2 Abdominal

Abdominal emergencies such as appendicitis or cholecystitis remain possible despite rigorous preflight screening, and even with careful selection, the extended duration of exploration-class missions increases the likelihood that such conditions will occur spontaneously. Although the risk–benefit profile of prophylactic appendectomy has been studied, and on occasions adopted as a policy by some countries in preparation for trips to Antarctica (Davis et al., 2016), there are ethical dilemmas associated with this, together with a risk of small bowel obstruction after a laparoscopic appendectomy. The decision to intervene procedurally must weigh the incidence of the potential illness, the severity of it's consequences against the incidence and severity of complications related to each potential treatment, as well as the logistical and environmental risks of surgery, and similar to the approach adopted on-board submarines, conservative therapy with antibiotics may serve as a temporizing measure until a return to Earth is possible (Talan, 2017). For example, while not optimal on Earth, in space, less invasive techniques, such as percutaneous cholecystotomies, may offer a better risk /benefit profile when employed as a temporizing measure while a patient is evacuated back to Earth.

### 2.5.3 Urological

Renal and urological issues present another recurrent threat. Microgravity induces bone demineralization and raises blood calcium levels (Smith et al., 2015), both of which increase the risk of renal stone formation. Cases of kidney stones have already been documented among astronauts, and even a small obstructing stone can cause severe pain, urinary retention, or infection. Management options are limited, lithotripsy devices to break up stones are impractical, evacuation will not be immediately possible from the Moon, and will be highly constrained for Mars missions. Here again, percutaneous nephrostomies (inserting a catheter into the kidney to drain urine) may help temporize a patient's pathology to prevent hydronephrosis (kidney swelling) from developing and preserve an astronaut's renal function until a return to Earth is possible.

### 2.5.4 Dental

Dental caries (tooth decay), cracked fillings, or abscesses can cause debilitating pain and infection in a setting where definitive dental care is unavailable. Dental emergencies have been identified by NASA as the most likely cause of an early mission termination from the ISS (Menon, 2012). Preventive measures together with simplified dental kits capable of extractions, temporary fillings, and abscess drainage are important management strategies, and the crews will need to be trained to be comfortable with their proper use.

### 2.5.5 Thoracic

Pneumothoraces (collapsed lung) may also pose a considerable challenge. In microgravity, thoracostomy (insertion of a chest drainage tube) would be complicated by the lack of passive drainage and would require active suctioning to ensure unidirectional flow away from a patient's thoracic cavity.

## 2.6 Looking forward

Performing a medical procedure in space is nearing the point where this will no longer remain a purely theoretical concept and is rapidly becoming an inevitability. As human missions extend beyond the Moon, communication delays, isolation, and logistical constraints will force exploration-class crews to function as self-contained procedural and critical care teams. The future of space medicine will depend not only on clinical training, but also on how effectively emerging technologies can extend human capability in environments where telemedical guidance from Earth may arrive too late, as can occur when Mars is eclipsed by the Sun in a process called conjunction, and a radio blackout can last weeks (Chappell et al., 2024).

Robotic systems may be able to fill a gap, and tele-robotic surgery has already been demonstrated under short-latency conditions on Earth. Future iterations may even function semi-autonomously or under intermittent human supervision. However, guidance from Earth will still become impractical once certain distances are reached due to our present understanding of the physical limits for transferring information.

Manufacturing capabilities will also redefine medical logistics, and processes such as additive manufacturing, or 3D printing, promise on-demand production of surgical instruments, consumables, and perhaps even biocompatible implants. Experiments aboard the ISS have shown that polymers, metals, and biological scaffolds can be printed in microgravity, and these, combined with sterilization and recycling processes, could close the logistical loop for long-duration missions.

The concept of artificial gravity simultaneously offers both a potentially effective countermeasure and a potentially significant stressor (Clément, Bukley and Paloski, 2015). Even partial gravity may restore directional flow of fluids, simplify patient restraint, and likely reduce both procedural risk and crew workload. Artificial gravity also holds potential as a rehabilitative tool, counteracting musculoskeletal atrophy and facilitating postoperative recovery. Ultimately, the future of procedural medicine in space will depend on our ability to integrate these technologies into a coherent, human-centered system.

## References


Axiom Space, 2025. Building era-defining space infrastructure. Available at https://www.axiomspace.com/.

Chappell, M.B., Chai, P.R., Rucker, M.A., 2024. Mars communications disruption and delay (Moon-to-Mars Architecture Concept Review White Paper). National Aeronautics and Space Administration.

Clément, G.R., Bukley, A.P., Paloski, W.H., 2015. Artificial gravity as a countermeasure for mitigating physiological deconditioning during long-duration space missions. Frontiers in Systems Neuroscience 9 (June). Available at https://doi.org/10.3389/fnsys.2015.00092, http://journal.frontiersin.org/article/10.3389/fnsys.2015.00092/pdf.

Coulombe, J.C., Senwar, B., Ferguson, V.L., 2020. Spaceflight-induced bone tissue changes that affect bone quality and increase fracture risk. Current Osteoporosis Reports 18 (1). Available at https://doi.org/10.1007/s11914-019-00540-y, http://www.springerlink.com/content/1544-1873/.

Crucian, B., Babiak-Vazquez, A., Johnston, S., Pierson, D.L., Ott, C.M., Sams, C., 2016. Incidence of clinical symptoms during long-duration orbital spaceflight. International Journal of General Medicine 9, 383–391. Available at https://doi.org/10.2147/IJGM.S114188, https://www.dovepress.com/getfile.php?fileID=33386.

Crucian, B.E., Choukèr, A., Simpson, R.J., Mehta, S., Marshall, G., Smith, S.M., et al., 2018. Immune system dysregulation during spaceflight: potential countermeasures for deep space exploration missions. Frontiers in Immunology 9. Available at https://doi.org/10.3389/fimmu.2018.01437, https://www.frontiersin.org/articles/10.3389/fimmu.2018.01437/full.

Davis, C.R., Trevatt, A.E.J., Dixit, A., Datta, V., 2016. Systematic review of clinical outcomes after prophylactic surgery. The Annals of The Royal College of Surgeons of England 98 (6), 353–357. Available at https://doi.org/10.1308/rcsann.2016.0089.

Dinner, J., 2025. SpaceX's Fram2 mission captures 1st human X-ray in Earth orbit. Available at Space.com.

European Space Agency, 2025. Swarm reveals growing weak spot in Earth's magnetic field. ESA.

Federal Aviation Administration, 2020. Summary of the FAA's review of the Boeing 737 MAX – return to service of the Boeing 737 MAX aircraft [PDF].

Fehlberg, T., Rose, J., Guest, G.D., Watters, D., 2019. The surgical burden of disease and perioperative mortality in patients admitted to hospitals in Victoria, Australia: a population-level observational study. BMJ Open 9 (5), e028671. Available at https://doi.org/10.1136/bmjopen-2018-028671.

Flores, P., McBride, S.A., Galazka, J.M., Varanasi, K.K., Zea, L., 2023. Biofilm formation of Pseudomonas aeruginosa in spaceflight is minimized on lubricant impregnated surfaces. NPJ Microgravity 9 (1). Available at https://doi.org/10.1038/s41526-023-00316-w, https://www.nature.com/npjmgrav/.

Gonzalez, E., Lee, M.D., Tierney, B.T., Lipieta, N., Flores, P., Mishra, M., et al., 2024. Spaceflight alters host-gut microbiota interactions. NPJ Biofilms and Microbiomes 10 (1). Available at https://doi.org/10.1038/s41522-024-00545-1.

Hicks, J., Olson, M., Mitchell, C., Juran, C.M., Paul, A.M., 2023. The impact of microgravity on immunological states. ImmunoHorizons 7 (10), 670–682. Available at https://doi.org/10.4049/immunohorizons.2200063, https://journals.aai.org/immunohorizons/article/7/10/670/266090/The-Impact-of-Microgravity-on-Immunological-States.

Institute of Medicine (US) Committee on Creating a Vision for Space Medicine During Travel Beyond Earth Orbit; Ball JR & Evans CH Jr. (Eds.), 2001. Risks to astronaut health during space travel. In Safe Passage: Astronaut Care for Exploration Missions (Chapter 2). National Academies Press.

Jones, J.A., Jennings, R., Pietryzk, R., Ciftcioglu, N., Stepaniak, P., 2005. Genitourinary issues during spaceflight: a review. International Journal of Impotence Research 17 (1), S64. Available at https://doi.org/10.1038/sj.ijir.3901431.

Li, Jiangyong, Gan, Yu, Sima, Yangyu, Hu, Jiawei, 2026. Spaceflight Noise Effects on Human Performance in a Microgravity Environment. Aerospace Medicine and Human Performance 97, 36–41. Available at https://doi.org/10.3357/amhp.6785.2026.

Limardo, J.G., Allen, C.S., Danielson, R.W., Boone, A.J., 2021. Status – International Space Station (ISS) crewmembers' noise exposures. In: Proceedings of INTER-NOISE 2021, 1–5 August 2021, Washington, DC.

Mehta, S.K., Crucian, B.E., Stowe, R.P., Simpson, R.J., Ott, C.M., Sams, C.F., et al., 2013. Reactivation of latent viruses is associated with increased plasma cytokines in astronauts. Cytokine 61 (1), 205–209. Available at https://doi.org/10.1016/j.cyto.2012.09.019.

Menon, A., 2012. Review of spaceflight dental emergencies (NASA/TM-2012-217368) [Technical Memorandum]. National Aeronautics and Space Administration.

NASA Glenn Research Center, 2022. NASA manages astronaut health with effective diagnostics research. NASA.

NASA Moon to Mars Architecture, 2025. Available at https://www.nasa.gov/moontomarsarchitecture/.

National Aeronautics and Space Administration, 2021. The human body in space. Available at https://www.nasa.gov/humans-in-space/the-human-body-in-space/.

National Aeronautics and Space Administration, Johnson Space Center, 2021. Human systems integration (NASA Human Health and Performance).

National Transportation Safety Board, 2019. Collision between USS John S. McCain and tanker Alnic MC, Singapore Strait, 5 Miles Northeast of Horsburgh Lighthouse, August 21 2017: Marine accident report NTSB/MAR-19/01. NTSB, Washington, DC.

Pantalone, D., Chiara, O., Henry, S., Cimbanassi, S., Gupta, S., Scalea, T., 2022. Facing trauma and surgical emergency in space: hemorrhagic shock. Frontiers in Bioengineering and Biotechnology 10. Available at https://doi.org/10.3389/fbioe.2022.780553, http://journal.frontiersin.org/journal/bioengineering-and-biotechnology#archive.

Parisi, M., Panontin, T., Wu, S.-C., McTigue, K., Vera, A., 2023. Effects of communication delay on human spaceflight missions. In: Paper presented at the 14th International Conference on Applied Human Factors and Ergonomics, NASA Technical Reports Server, San Francisco, CA.

Patel, Z.S., Polk, J.D., Young, L.R., 2017. Systems engineering for space exploration medical capabilities(NASA/TM-2017-21828). National Aeronautics and Space Administration.

Pohlen, M., Carroll, D., Prisk, G.K., Sawyer, A.J., 2022. Overview of lunar dust toxicity risk. NPJ Microgravity 8 (1). Available at https://doi.org/10.1038/s41526-022-00244-1, https://www.nature.com/npjmgrav/.

Puri, G.D., Jayant, A., Dorje, M., Tashi, M., 2008. Propofol-fentanyl anaesthesia at high altitude: anaesthetic requirements and haemodynamic variations when compared with anaesthesia at low altitude. Acta Anaesthesiologica Scandinavica 52 (3), 427–431. Available at https://doi.org/10.1111/j.1399-6576.2007.01561.x.

Roberts, B.M., Deane, C.S., Szewczyk, N.J., Fajardo, V.A., Maden-Wilkinson, T., Bagley, J.R., 2025. Musculoskeletal responses to spaceflight: mechanisms, countermeasures, and key gaps. American Journal of Physiology. Cell Physiology 329 (6), C1985. Available at https://doi.org/10.1152/ajpcell.00649.2025.

Sanders, R., Ray, K., Law, J., Reyes, D., 2023. Medical lessons learned from the Exploration Atmospheres Study. In: Presentation at the 2023 Aerospace Medical Association Conference [NASA Technical Reports Server Document 20220015741], NASA.

Shaydakov, M.E., Tuma, F., 2025. Operative risk. StatPearls [Internet]. StatPearls Publishing, Treasure Island, FL. Available from: https://www.ncbi.nlm.nih.gov/books/NBK532240/ (accessed 23.05.02).

Shibata, S., Wakeham, D.J., Thomas, J.D., Abdullah, S.M., Platts, S., Bungo, M.W., et al., 2023. Cardiac effects of long-duration space flight. Journal of the American College of Cardiology 82 (8), 674–684. Available at https://doi.org/10.1016/j.jacc.2023.05.058, www.elsevier.com/locate/jacc.

Sibonga, J.D., Evans, H.J., Smith, S.A., 2017. Risk of early onset osteoporosis due to spaceflight (NASA evidence report). National Aeronautics and Space Administration.

Smith, S.M., Heer, M., Shackelford, L.C., Sibonga, J.D., Spatz, J., Pietrzyk, R.A., et al., 2015. Bone metabolism and renal stone risk during International Space Station missions. Elsevier Inc., United States Bone 81, 712–720. Available at https://doi.org/10.1016/j.bone.2015.10.002, www.elsevier.com/locate/bone.

SpaceX, 2025. Making life multiplanetary. Available at https://www.spacex.com/.

Stepan, M.E., Altmann, E.M., Fenn, K.M., 2020. Effects of total sleep deprivation on procedural placekeeping: more than just lapses of attention. Journal of Experimental Psychology: General 149 (4), 800–806. Available at https://doi.org/10.1037/xge0000717, http://www.apa.org/pubs/journals/xge/.

Talan, D.A., 2017. Antibiotics may be effective for managing appendicitis. ACEP Now.

Tang, H., Rising, H.H., Majji, M., Brown, R.D., 2022. Long-term space nutrition: a scoping review. Nutrients 14 (1). Available at https://doi.org/10.3390/nu14010194, https://www.mdpi.com/2072-6643/14/1/194/pdf.

Thamer, S., Buckey, J.C., 2022. First void urinary calcium for tracking bone loss and kidney stone risk in space. Aerospace Medicine and Human Performance 93 (7), 546–550. Available at https://doi.org/10.3357/AMHP.5979.2022, http://ingentaconnect.com/content/asma/amhp;jsessionid=13tlejf2f0ojk.alice.

U.S. Environmental Protection Agency, Office of Noise Abatement and Control, 1974. Information on levels of environmental noise requisite to protect public health and welfare with an adequate margin of safety (EPA/ONAC 550/9-74-004).

Vast Space, 2025. Building next-generationspace stations. Available at https://www.vastspace.com/.

Welch, N., 2021. Protecting astronaut health with the prick of a finger. NASA.

Chapter 3

# Systems medicine and engineering for procedural care

Kenny Chao[1,2] and Erik Antonsen[3]
[1]*Space Medicine Operations, Mission Operations Vast Inc., Long Beach, CA, United States,* [2]*Department of Emergency Medicine, University of California, Los Angeles, CA, United States,* [3]*Department of Emergency Medicine, SPEAR Medicine Division, Massachusetts General Hospital, Boston, MA, United States*

## 3.1 Introduction

Procedural medicine describes a set of capabilities required to physically intervene on the human body in a safe and ethical manner. This set of capabilities ranges from minor procedures, such as wound care and suturing, to major procedures including open surgery. Determining whether it is worth including certain procedural medicine capabilities when designing a spacecraft or planning for a human spaceflight mission is a unique challenge that requires broader systems medical and engineering understanding beyond typical training for physicians. This includes a basic understanding of the spaceflight environment and how it is different from Earth, how our bodies change in the spaceflight environment, how spacecraft systems are designed, how humans are integrated into those designs, and how risk is understood not only at the patient level, but at the entire systems levels (Hanson et al., 2019; Mcginnis and Harris, 1998; Mindock et al., 2017; Shelhamer and Antonsen, 2023).

Any procedural capability considered for inclusion in a spaceflight system has a common set of base elements—hardware and resources required for the procedure, and knowledge, skills, and abilities (KSA) of the person (or robot) performing the procedure. In some cases, the provision or maintenance of KSA's requires software to provide training or guidance, or audio/visual hardware to provide remote guidance. The capability to perform a medical procedure encompasses all these features and acknowledges the reality that ongoing practice or training is required for those who perform medical procedures (Hanson et al., 2019; Blue, 2014; Levin et al., 2025).

A system is a collection of interdependent elements that aim to achieve shared goals. Systems science studies this interdependency, the complex interplay of nested systems, and the resulting holistic impact as the elements within change (Shelhamer and Antonsen, 2023; Meadows, 2008). In the context of human spaceflight, systems engineering and medicine play a central role in analyzing and dealing with the complexity inherent in spaceflight vehicles, habitats, and spacesuits (Antonsen, 2025b, a; Hirshorn, 2017; Klaus, 2024). In the context of procedural medicine in spaceflight, systems questions move beyond whether the hardware and KSAs are present to perform a procedure and ask whether the vehicle or habitat system can support all aspects of the procedure, including preparation, system resource requirements, procedural performance, environmental effects of the procedure, after care of the patient, and removal and disposal of waste products from the procedure. Each of these systems effects can produce additional risk to the astronauts or the system if not considered up front.

Determining the total risk human astronauts will face in a given mission with a given set of systems is a challenge (Antonsen et al., 2022). Asking whether that risk will be increased or decreased by the inclusion of specific medical capabilities is an ongoing subject of discussion and research (Anderson and Fenbert, 2024; Anderson et al., 2024). It is not a given that medical procedural capabilities must be included in all space vehicles or habitats. To justify the inclusion of procedural medical capabilities in any spaceflight system, several questions must be answered:

1. For a given mission or mission set, which medical conditions are likely to occur that would warrant a need for procedural capabilities?
2. How likely are those medical conditions to occur?
3. How likely are the proposed procedures to be successful in the space mission environment?

**Procedural Medicine for Spaceflight. DOI:** https://doi.org/10.1016/B978-0-443-27552-4.00006-3

4. What other important items or capabilities will medical procedural capabilities displace if included?
5. How does that trade affect the total mission risk—that is, is it worth including?
6. If included, how should procedural medical capabilities be integrated into the system as a whole?

These six questions correspond to six steps in a process that enables assessment and inclusion of procedural medical capabilities within the medical system of a spacecraft. The first three questions are medical in nature but require insight into the realities of the spaceflight environment and the specifics of the mission or vehicle design under consideration. The last four questions sit at the intersection of medicine, engineering, and design and require partnership with those sets of expertise complementary to medical personnel. These questions must be asked and answered from the perspective of systems medicine and as part of the systems engineering process (Shelhamer and Antonsen, 2023; Hirshorn, 2017; Holbrook et al., 2021).

*Systems medicine* acknowledges that the procedural care envisioned is part of a complex system and looks at the practice and delivery of medicine as the sum of many dynamically interwoven parts. The field is not fully defined and is evolving, but general concepts are described in this chapter and elsewhere (Shelhamer and Antonsen, 2023). Procedural medicine inclusion affects multiple systems and subsystems in any space vehicle or habitat. These include the Medical System, which is part of the larger Crew Health and Performance (CHP) system, but also the Communications Systems to support real-time guidance and consultation with Earth, the Environmental Control and Life Support System (ECLSS) that defines the vehicle conditions including airflow and clean water access, and other systems (Hanson et al., 2019; Klaus, 2024; Mindock et al., 2022). How procedural capabilities become included in the system is the domain of *systems engineering* (Hanson et al., 2019; Hirshorn, 2017). Finally, the extent to which those procedural capabilities are ultimately effective in use by human astronauts is determined by *human systems integration* processes (Antonsen, 2025b; Holbrook et al., 2021). In essence, the decision to include procedural capabilities for a specific mission impacts every vehicle system, vehicle, and habitat we design. This chapter provides an overview of the systems approaches that enable procedural medicine to be integrated into human spaceflight systems.

## 3.2 Systems considerations for procedural medicine

It is reasonable to ask why systems engineering and medicine are worth discussing in a book about procedural medicine. The simple answer is that medical experts do not always decide what goes into spacecraft systems. They work as part of a team in a much larger group of experts. In large and complex projects like human spaceflight, performing systems medicine analyses demonstrates that you've done your homework and have evidence-backed reasons for your recommendations (Ward et al., 2024; Antonsen et al., 2023). Systems engineering is the process by which you bring forward that evidence and have intelligent discussions with your engineering counterparts about the pros and cons of your recommendations in the larger systems framework (Hirshorn, 2017). Any hardware and software that are critical to performing procedural medicine must be considered in these processes. If a rational and well-supported argument is not provided for why procedural hardware should be included in a highly constrained mass and volume allocation, then that hardware will not make it onto the vehicle, and the procedural capability will not be available when needed (Canga et al., 2016).

As clinicians, we often function in systems (hospitals, clinics, etc.) where the hardware and software needed to do our jobs are provided by others. Think about who is responsible for moving and positioning a patient, for attaching electrocardiogram electrodes, for placing intravenous catheters and drawing blood, for cleaning and sterilizing supplies needed to perform safe procedures, for ensuring that vacuum for suction is functioning and that the wall-mounted oxygen supply is full and operating. In operational space medicine, these responsibilities by default are assigned to the crew medical officer and remaining crew. Long before an astronaut crew is put in this position somewhere in space, engineers and designers make decisions about what will be available and how it is designed to function in the space environment, with input from clinicians who also determine what KSAs and training allow the crew to perform their medical duties. Without advocacy throughout this process, all medical and procedural capabilities are relegated to a pre-set mass/volume allocation for the vehicle determined by engineers. The medical capability is often reductively viewed as a medical "kit" rather than a critical part of the integrated system (Antonsen et al., 2016).

A medical provider for spaceflight has a specialized role. However, the systems approach focuses not only on understanding your immediate role but also on how it fits into the overall mission alongside the priorities and responsibilities of different groups. Comprehending the interdependency among subsystems helps to avoid the silo trap, with the hope that broader insight leads to careful forethought, design, and successful integration of the medical system. When designing and implementing medical procedures and processes, considerations must extend beyond just the medical team.

## 3.3 Medical system design for spaceflight

Designing a medical system for spaceflight requires addressing the unique challenges of the space environment while ensuring the health and safety of crew members. This necessitates careful consideration of physiological, operational, and technological factors. Additionally, resource allocation presents a fundamental challenge for spaceflight, particularly during long-duration and exploratory missions, where the needs and limitations of resources can significantly compound.

This is often described as a "knapsack problem" (Fréville, 2004; Knapsack Problem - an overview | ScienceDirect Topics, n.d.). With limited carrying capacity, the problem explores what items of value you would want to bring on an excursion, or in this case, a mission. A variety of factors are considered, including the goals of your trip, the type of trip, the length of the trip, the number of individuals going, the size of your backpack, the resupply options, and many others. Medical systems for terrestrial exploration generally receive greater resources (such as medications, equipment, expertise, and personnel) in comparison to space medicine because of the significant cost of sending mass and volume to space (Mindock et al., 2022; McGuire, 2021; Kim, 2025). The medical system aboard a spacecraft constitutes a more isolated system than that which most clinicians are accustomed to encountering. As lunar and exploratory missions progress, the demand for medical autonomy correspondingly increases, necessitating an increased reliance on the resources and knowledge available onboard (Antonsen et al., 2022). In light of potential delays or challenges in obtaining definitive care, crew members may be required to perform medical procedures on one another. These procedures must be specifically designed to accommodate the altered gravity of the environment. Furthermore, the selection of equipment and supplies to be transported, along with their respective quantities and methods of utilization, necessitates thorough preplanning.

These decisions follow extensive analysis and deliberation to identify the medical conditions that may pose the highest risks and to assess which interventions can effectively minimize those risks using the available resources and technologies (Shelhamer and Antonsen, 2023). This is essential in determining what to include in your medical knapsack.

When considering system design in a spacecraft, there are several standard steps that clinicians must understand, and these correspond to the questions posed at the beginning of the chapter:

1. Developing a concept of operations
2. Mapping known planned medical needs
3. Developing expected and accepted medical conditions lists
4. Assessing risk and prioritizing medical conditions and procedures
5. Participating in the systems engineering process
6. Participating in the human systems integration process.

### 3.3.1 Concept of operations

The medical events likely to occur during a 3-day LEO mission for space tourism are very different from those on a 3-year mission to Mars. A Concept of Operations (ConOps) for a given mission helps as a starting point to understand the needs of the mission (Holbrook et al., 2021; Mindock et al., 2022). Once mission attributes are known (mission objectives, mission duration, expected astronaut tasks, etc.), then a medical team can identify what operational paradigm constraints they must consider during design (Ward et al., 2024).

For example, missions to Low Earth Orbit (LEO) may include the possibility for resupply or evacuation to Earth. Missions that include Extravehicular Activity (EVA) on the surface of the Moon are likely to have different injury expectations than a mission in LEO that has no EVAs. Some missions, vehicles, and habitats may allow the possibility of additive manufacturing to create supplies or tools "on demand." But this also requires significant planning, including having a 3D printer, adequate raw materials, proven design specifications for the printer, and so on. The ConOps provides insight into the mission attributes and the expectation of what the mission will look like (Parker and Krihak, 2024). This enables anticipation of both planned and unplanned medical needs and an initial vision of the medical capabilities. This often provides example medical scenarios that provide a shared mental model of how the system is expected to function, allowing functional requirements to be identified. This is necessary for the medical system as a whole and in particular for procedural medical capabilities.

### 3.3.2 Planned medical needs

Once a ConOps is in place, the medical team considers planned medical needs. Planned medical events are scheduled to occur throughout a mission and are health maintenance-oriented (Hanson et al., 2019; McGuire, 2021; Blue, Nusbaum and Antonsen, 2022). These include periodic medical and psychological evaluations, private medical conferences, periodic dental evaluations, and preventive medical plans. Health maintenance also overlaps with other parts of the CHP system (Mindock et al., 2022). This includes monitoring of exercise regimens for effectiveness and potential injuries, monitoring of available pharmaceuticals for insight into use rates and availability throughout a mission, dietary monitoring, and environmental monitoring to correlate any medical symptoms or complaints with any environmental changes.

If physiologic or other human research studies are to be integrated into the schedule, they should be included as planned medical events to ensure physician awareness of any issues that arise or develop. Inclusion of a standardized process to inform the clinical team of incidental medical findings is critically important (Marshall-Goebel et al., 2019). Historically, preplanned medical events do not include invasive procedures unless they are part of a research study.

Because these are preplanned mission events, clinicians must be prepared to detail the full scope of what is needed in terms of physical resources (hardware, consumables, monitoring sensors, data storage drives, cameras, medical devices, research devices, computers, etc.), software resources (video/audio, data bandwidth expectations, etc.), and a data governance and data sharing proposal that determines who should have access to what data during and after the mission (Mindock et al., 2022).

### 3.3.3 Development of expected and accepted medical condition lists

Planned medical events become a predictable part of the mission timeline and cadence. Procedural medical capabilities are more likely to be associated with unplanned medical events. These are medical events we hope do not occur but that have a reasonable probability of occurring during the mission. It is impossible to consider the full range of medical possibilities that may emerge during a human spaceflight mission. Therefore, we defer to medical condition lists and evidence from prior spaceflight missions to narrow consideration, and probabilistic risk assessment (PRA) to help quantify and prioritize those conditions that may require preventive, diagnostic, procedural, and prognostic capabilities during a mission (Blue, Nusbaum and Antonsen, 2022; Kreykes et al., 2023; Keenan, 2015; Walton and Antonsen, 2017).

The Expected Medical Condition List is often taken from NASA's medical condition lists and prior spaceflight experience and modified based on subject matter expert consideration of the ConOps at hand (Kreykes et al., 2023). The Accepted Medical Condition List is a list of those medical conditions that are prioritized by consideration of three variables—probability of occurrence, complexity of care, and futility of care (Blue, Nusbaum and Antonsen, 2022). Developing this product provides a traceable, repeatable, and consensus-based list of which unplanned medical conditions will have diagnostic and treatment options (including procedural medicine capabilities) supplied within the medical system of the spacecraft. To accomplish this, the risk-prioritization methods described below must be used.

### 3.3.4 Risk assessment and prioritization

Medical risk can be defined as the product of likelihood (the probability of a condition occurring) and consequence (the severity of the condition's impact on health, performance, and mission success; Antonsen et al., 2023; Romero and Francisco, 2020). This equation provides a metric to gauge the potential significance of a medical condition occurring.

NASA provides a model for approaching risk management at the organizational level. The agency defines human system risk as an undesirable flight crew health or performance outcome resulting in a clear consequence that may negatively affect an astronaut's ability to perform critical tasks related to mission objectives (Antonsen, 2020). An internal group called the Human Systems Risk Board, composed of experts in engineering, medicine, life sciences, human factors, and spaceflight operations, is tasked with assessing human system risks and communicating them to agency decision-makers (Antonsen et al., 2023; Antonsen, 2020). The Likelihood times Consequence (LxC) risk matrix is not a precise tool for evaluating risk; rather, it is used to compile the best available evidence and communicate a general risk status in an intuitive manner.

PRA provides a structured and quantitative approach to evaluating risk using the same LxC concept. PRA has been applied in many areas at NASA to assess safety from both human and hardware perspectives (Keenan, 2015; Boyer and Hamlin, 2011; Conkin, 2014; Hamlin et al., 2011). Medical PRA models, such as the Integrated Medical Model (IMM) and MEDPRAT, developed at NASA, provide a quantitative, evidence-based approach to mission planning and are used by clinicians and engineers during the systems engineering process to evaluate medical risk in a like-to-like manner alongside other mission sources of risk (McIntyre and Myers, 2024; Keenan, 2015; Walton and Antonsen, 2017; Prelich, 2025). These models start with the exploration or accepted medical condition lists described above and incorporate crew data and mission parameters, an evidence archive of data from past spaceflights, and terrestrial and epidemiological study results. Monte Carlo analysis is performed to simulate the mission thousands of times, resulting in probability distributions of mission outcomes such as the probability of loss of crew life, the probability of removal to definitive care, crew health index, quality time loss, and loss of crew mission (Keenan, 2015; Lumpkins, 2019). This allows a quantitative assessment of the likelihood that medical conditions requiring procedures will occur, and a description of the worst-case scenario for the likely medical outcome if those conditions were to go untreated. This is critical when deciding whether a procedural capability is worth the mass/volume allocation during the vehicle planning design phase.

Quantifying risk is foundational in guiding policies and countermeasures to ensure astronaut health and ultimately mission success. It allows for several benefits:

- **Comparison of medical condition risks**. Establishing a consistent scale enables the comparison of risks across different medical conditions. For example, a condition with a high incidence but low consequence (e.g., mild skin rash) may have an overall lower risk score than a condition with low incidence but high consequence (e.g., severe sepsis). This helps decision-makers prioritize resources toward managing conditions with higher risks.
- **Risk aggregation**. Risks from individual medical conditions can be pooled to evaluate total medical risk. This holistic perspective informs mission planning and risk mitigation strategies, ensuring a balance between astronaut safety, operational goals, costs, and resource expenditure.
- **Tracking changes in risk**. Risk is dynamic. The incidence of medical events can vary based on a crew's medical history and preexisting conditions. It can also change with the mission type, duration, preventive measures, and many other factors. Consequences may increase or decrease depending on the countermeasures and treatment options available in flight to mitigate the severity of medical conditions. Risk calculations can change by simply incorporating updated data that affects incidence and consequences.
- **Communication and stakeholder engagement are vital**. The language of risk offers a unified quantitative approach and language for diverse stakeholders. It aids in signaling and aligning priorities while clarifying trade-offs, facilitating informed decision-making on resource allocation.
- **Application of evidence-based medicine.** It involves quantitative risk assessments that rely on data, promoting an evidence-based approach to minimizing medical risks. Terrestrial and analog data can supplement the relatively limited historical spaceflight data, helping to reduce dependence on solely subject matter expert input.

### 3.3.5 Risk tolerance

Risk tolerance is the propensity for an organization or an astronaut to accept risk (Antonsen, 2020; Ocampo and Klaus, 2016). How much risk is too much? Does high risk imply a no-go for the mission? Completely mitigating all risks would likely be cost-prohibitive and logistically unfeasible. Some degree of risk is unavoidable. PRA analysis helps quantify the degree of risk present when inducing or excluding medical procedural capabilities.

For a short 15-minute suborbital flight, the incidence of medical events and the short time to return to definitive care greatly simplify the medical risk analysis, and it is likely not worthwhile to perform an in-depth analysis of medical system needs. In these cases, a clinician's subject matter expertise is likely sufficient for planning. However, for a 3-year Mars mission, the time for medical events to occur, the complexity of the mission profile, the duration of the mission, and the mass and volume constraints of the vehicle make the medical risk analysis exceedingly complex. In this case, it is critically important to perform medical PRA as part of an extensive planning and integration process (Antonsen et al., 2022). Most spaceflight missions fall somewhere between these two extremes, and a clinician must select the approach that is most appropriate in discussion with the organization that is legally responsible for the outcomes of the spaceflights (Zahalka and Iwry, 2024). It is important to note that at the time of writing, legal precedents for medical accidents or malpractice in space have not been set, and there is no clear governing body for adjudicating these issues. As commercial spaceflight evolves, it is likely that more clarity will come over time and as medical issues arise in spaceflight.

### 3.3.6 Participation in the systems engineering process

Once medical conditions are considered, PRA is performed, and a plan is made to recommend certain procedural medical capabilities for inclusion in the spacecraft medical system; the clinicians must work closely with engineers and designers to ensure that appropriate requirements are in place. Representatives from the CHP system must participate in the systems engineering design process as early as possible (Mindock et al., 2017; McGuire, 2021; NASA SP-2017-633, 2018). This includes interpretation and recommendation of standards, development of requirements, and consideration of tradespace realities.

The CHP system has the task of maintaining astronaut health, and it is imperative to ensure that this system is called out in high-level requirements (Hanson et al., 2019; Mindock et al., 2022). Once these are written, sub-parts of the CHP system can be written as child requirements underneath the CHP system umbrella.

The CHP system does not function alone in caring for astronaut health. Overall health maintenance is affected by other subsystems, for example, ECLSS, nutrition, acoustics, waste management, and many more (Klaus, 2024; Mindock et al., 2022). The medical system likewise impacts other systems, utilizing mass, volume, cost, electrical power, bandwidth, and time that may otherwise be dedicated to other subsystems, as well as influencing their design to support and enhance the human experience and crew health.

Clinicians and those involved in designing and achieving a procedural capability must be active participants in the development of functional and technical requirements (Hanson et al., 2019). These serve as the cornerstone for designing a spaceflight medical system, as well as other systems and projects. They clearly define what the system or a component of a system must achieve, creating a baseline blueprint and facilitating cross-communication among all involved parties to build towards incorporating necessary functions, performance parameters, and allowable constraints.

In developing requirements, a balance needs to be struck between addressing the needs of today's missions versus those of the future (NASA SP-2017-633, 2018). Engineering and design decisions made years in advance and incorporated into a spacecraft and its systems may affect medical decisions on mission years into the future. However, requirements that offer flexibility in addressing future needs may introduce additional complexity, resources, and costs.

Minimizing design complexity ensures that systems meet the basic requirements of a project. Minimizing potential points of failure and reducing unnecessary complexity are important for spacecraft design. It helps to prevent the diversion of resources from other relevant projects, cost increases, over-design, and maintenance challenges (Valinia, 2023). If medical treatments and procedures are required for mission risk mitigation, it is critical to explain the functions that need to be performed, the baseline benefits they should provide, and the maximum resource costs involved.

Tradespace analysis is the process by which the program managers and systems engineers work to resolve the differences in allocations, costs, and interfaces between different parts of the spacecraft throughout the systems engineering lifecycle. While clinicians struggle with what to include in the medical system within the constraints of mass, power, and volume for their subsystem, program managers struggle to make the mass, power, and volume allocations work across all systems (Hirshorn, 2017). They consider the total system costs, resources, and schedule.

Tradespace analysis is a critical part of medical system design, and it is helpful in a procedural sense to identify where specific hardware and consumables can be multipurposed. NASA developed a tradespace tool called IMPACT (Informing Mission Planning via Analysis of Complex Tradespaces) based on IMM that utilizes optimization algorithms and simulations to iteratively include and exclude medical items within the constraints of the medical system footprint, assessing their downstream effects on crew health (McGuire, 2024; Kreykes et al., 2023; Lumpkins, 2019). The resulting optimal medical hardware recommendations provide important input to design and engineering efforts (McGuire, 2024). KSAs are also essential but intangible components of the medical system and may benefit from tradespace analysis to inform training and/or recruitment priorities (Levin, 2023).

### 3.3.7 Participation in the human systems integration process

Throughout the systems engineering lifecycle, clinicians must consider system costs, resources, and schedule. However, there are additional considerations at the system and subsystem level when considering procedural medicine. These include time required for hardware or software development, training of users to actualize the medical capabilities, operational realities of working in the spaceflight environment, usability, maintainability, cleaning, waste, and more. All of these considerations are part of the HSI process (Holbrook et al., 2021).

HSI processes occur iteratively throughout the systems engineering process and ensure that design decisions are made with consideration of the human's abilities and limitations (Holbrook et al., 2021; Rochlis and Campbell, 2015). This is particularly important for medical capabilities, as engineers are not trained in nor familiar with the physical application of medical hardware, the space needs, or the nuances of use. Human-in-the-loop testing and rigorous verification and validation of requirements are necessary to ensure that procedural medicine requirements are not only met but also meaningful (Antonsen, 2025b).

The traditional engineering paradigm often emphasizes hardware and software during the design phase, giving less attention to people and the human experience. Integrating humans necessitates understanding how human capabilities and limitations influence hardware and software design, as well as how such designs can impact human performance. Humans should be regarded as subsystems equivalent to hardware and software due to their close interdependencies. Furthermore, the human subsystem should encompass individuals on the ground involved in all aspects of support, including Mission Control, ground support, repair and maintenance, communications—not just those who are fortunate enough to travel to space (Vera and McTigue, 2023; Antonsen, 2025b; Holbrook et al., 2021; Valinia, 2023; Rochlis and Campbell, 2015; Buckland, 2022).

HSI is a multidisciplinary approach that emphasizes system design based on a holistic understanding of human needs, capabilities, and limitations. By integrating human factors into thoughtful design and engineering, one creates spacecraft, tools, and processes that align with astronaut physiology, cognition, behavior, and constraints, aiming to enhance performance and maximize mission success.

In the context of procedural medicine, HSI includes how procedural implements are designed, stored, cleaned, re-used, and more (Antonsen et al., 2022; Valinia, 2022). It also considers the skill retention and ongoing training needs of the astronauts who will be performing procedures when needed during a mission (Blue, 2014; McGuire, 2021; Levin, 2023). In exploration missions where guidance from Earth is not an option, HSI is critical to ensuring that procedural medical capabilities are intuitive and usable regardless of the level of training that the astronaut has received (Mctigue and Parisi, 2021; Parisi, 2024).

Engineers, medical professionals, and mission planners can collaborate and align their efforts early in the design phase to tackle challenges through a shared understanding of human requirements. HSI has been adopted by organizations such as NASA, the Department of Defense, and numerous engineering firms, including commercial space companies. NASA has published a comprehensive guide on its HSI process (Holbrook et al., 2021; Rochlis and Campbell, 2015). By prioritizing human-centered design from the outset, organizations can avoid expensive redesigns and modifications later in the development process. Early integration of HSI helps identify potential usability issues and operational inefficiencies, reducing the risk of project delays and budget overruns (Holbrook et al., 2021). Furthermore, HSI-driven training programs and intuitive interfaces decrease the time required for astronaut preparation and enhance operational efficiency during missions. The synergy created by aligning subsystems with human needs can minimize costs, time, effort, and resources while maximizing the mission's return on investment.

## 3.4 Summary

As physicians and others consider how medical procedures can be performed in space, it is also critically important to stop and consider whether they are needed, and when they will be needed. The benefits of early and iterative procedural learning must be weighed against the amount of risk a medical procedure helps reduce for the astronauts. Processes to help inform these risk–benefit considerations have already been developed in the human spaceflight industry. There is a wealth of organizational knowledge and processes that have been developed at the systems medicine and systems engineering level to guide physicians and others in the process of creating medical systems for spaceflight. Procedural medicine is a key part of that system, and the coming years will see an increase in both the medical conditions that occur in spaceflight as well as a recognition of where and when procedural medical capabilities will be needed. As commercial companies begin to fly people with greater and broader medical risks than the historical government-supported astronaut population, stories of what could have helped in medical situations in space will be shared more broadly.

It is key to remember that the inclusion or exclusion of medical procedural capabilities in a spacecraft system brings both benefits and risk, including risks to the larger system on which all astronauts depend for survival. It is always possible that medical interventions and procedures may cause more problems than they solve. As you read this book, keep in mind the potential system impacts from the medical interventions you seek to include in the space vehicle or habitat.

## References


Anderson, A., Fenbert, J., Easter, B., Lehnhardt, K., 2024. Quantifying medical risk on a long duration lunar mission: a demonstration of NASA's IMPACT tradespace analysis tool.

Anderson, A., Parmar, P., Levin, D., Hilmers, D., 2024. Quantifying the risk of medical evacuation in spaceflight. Acta Astronautica 224, 458–464. Available at https://doi.org/10.1016/j.actaastro.2024.08.026.

Antonsen, E.L., Connell, E., Anton, W., Reynolds, R.J., Buckland, D.M., Van Baalen, M., 2023. Updates to the NASA human system risk management process for space exploration. Microgravity 9 (1). Available at https://doi.org/10.1038/s41526-023-00305-z, https://www.nature.com/npjmgrav/.

Antonsen, E., Hanson, A., Shah, R., Reed, R., Canga, M., 2016. Proceedings of the International Astronautical Congress, IAC 00741795 International Astronautical Federation, IAF United States Conceptual drivers for an exploration medical system. Available at http://conference.researchbib.com/?action=viewEventDetails&eventid=17342&uid=re90b8.

Antonsen, E, 2020. Human System Risk Management Plan.

Antonsen, E., 2025a. Risk and Systems Knowledge in Human Spaceflight Building a Space-Faring Civilization: Advancing the Renaissance of Science, Medicine and Human Performance in Civilian Spaceflight. Elsevier, United States, pp. 181–194. Available at https://www.sciencedirect.com/book/9780443138508, https://doi.org/10.1016/B978-0-443-13850-8.00019-1.

Antonsen, E., 2025b. The Integration of Engineering, Medicine, and Design for Human Spaceflight.

Antonsen, E.L., Myers, J.G., Boley, L., Arellano, J., Kerstman, E., Kadwa, B., et al., 2022. Estimating medical risk in human spaceflight. Microgravity 8 (1). Available at https://doi.org/10.1038/s41526-022-00193-9, https://www.nature.com/npjmgrav/.

Blue, R., Nusbaum, D., Antonsen, E., 2022. Development of an accepted medical condition list for exploration medical capability scoping NASA TM-20190027540. Available at https://ntrs.nasa.gov/citations/20220003136.

Blue, R.S., 2014. Identification of Medical Training Methods for Exploration Missions. 32.

Boyer, R.L., Hamlin, T., 2011. Use of Probabilistic Risk Assessment in Shuttle Decision Making Process.

Buckland, D., 2022. NASA's identified risks of adverse outcomes due to inadequate human systems integration architecture in human spaceflight. Human-Centered Aerospace Systems and Sustainability Applications 26, 1–9.

Canga, M.A., Shah, R., Mindock, J., Antonsen, E., 2016. A strategic approach to medical care for exploration missions. In: 67th International Astronautical Congress, IAC-16.

Conkin, J., 2014. Probabilistic Assessment of Treatment Success for Hypobaric Decompression Sickness. pp. 236.

Engineering, life sciences, and health/medicine synergy in aerospace human systems integration: the rosetta stone project, NASA SP-2017-633, 2018.

Fréville, A., 2004. The multidimensional 0–1 knapsack problem: an overview. European Journal of Operational Research 155 (1), 1–21. Available at https://doi.org/10.1016/s0377-2217(03)00274-1.

Hamlin, T.L., Thigpen, E., Kahn, J., Lo, Y., 2011. AIAA SPACE Conference and Exposition 2011 United States Shuttle risk progression: use of the shuttle Probabilistic Risk Assessment (PRA) to show reliability growth.

Hanson, A., Mindock, J., Okon, S., Hailey, M., McGuire, K., Bardina, J., et al., 2019. IEEE Aerospace Conference Proceedings. Available at https://doi.org/10.1109/AERO.2019.8741864 9781538668542 IEEE Computer Society United States A Model-Based Systems Engineering Approach to Exploration Medical System Development 2019.

Hirshorn, S.R.V., 2017. NASA systems engineering handbook. Available at https://ntrs.nasa.gov/search.jsp?R=20170001761.

Holbrook, J., Novak, B., Rippy, L., Vos, G., DeVasher, J., 2021. NASA human systems integration handbook, NASA/SP-20210010952.

Keenan, A., 2015.The integrated medical model: a probabilistic simulation model predicting in-flight medical risks. In: 45th International Conference on Environmental Systems..

Kim, M.J., 2025. Counting stars and costs: an empirical examination of space launch cost trend at NASA. Acta Astronautica 232, 633–639. Available at https://doi.org/10.1016/j.actaastro.2025.04.011, https://www.sciencedirect.com/science/journal/00945765.

Klaus, D., 2024. An overview of habitat systems supporting human spaceflight. Systems Medicine for Human Spaceflight. World Scientific Publishing Co, United States, pp. 1–20. Available at https://www.worldscientific.com/worldscibooks/10.1142/13713#t=aboutBook, https://doi.org/10.1142/9789811287695_0001.

Knapsack Problem - an overview | ScienceDirect Topics, n.d. Available from: https://www.sciencedirect.com/topics/engineering/knapsack-problem.

Kreykes, A.J., Suresh, R., Levin, D., Hilmers, D.C., 2023. Selecting medical conditions relevant to exploration spaceflight to create the IMPACT 1.0 medical condition list. Aerospace Medical Association, United States Aerospace Medicine and Human Performance 94 (7), 550–557. Available at https://doi.org/10.3357/AMHP.6199.2023, http://ingentaconnect.com/content/asma/amhp;jsessionid=13tlejf2f0ojk.alice.

Levin, D., 2023. A pilot project for quantifying the effect of medical provider knowledge, skills, and abilities on outcomes for spaceflight using a probabilistic risk assessment tool. Available at https://ntrs.nasa.gov/citations/20230011811.

Levin, D.R., McIntyre, L., Steller, J.G., Nelson, A., Zahner, C., Anderson, A., et al., 2025. Predicting exploration crew medical officer training needs: applying evidence-based predictive analytics to space medicine training. SAGE Publications Inc., United States Wilderness and Environmental Medicine 36 (1), 7–23. Available at https://doi.org/10.1177/10806032241292535, https://journals.sagepub.com/home/wem.

Lumpkins, S, 2019. IMPACT Concept of Operations.

Marshall-Goebel, K., Laurie, S.S., Alferova, I.V., Arbeille, P., Auñón-Chancellor, S.M., Douglas, J., et al., 2019. Assessment of jugular venous blood flow stasis and thrombosis during spaceflight. JAMA Network Open 2 (11), e1915011. Available at https://doi.org/10.1001/jamanetworkopen.2019.15011.

Mcginnis, P.J., Harris, B.A., 1998. The re-emergence of space medicine as a distinct discipline. Aviation, Space, and Environmental Medicine 69 (11), 1107–1111.

McGuire, K., 2021. Using systems engineering to develop an integrated crew health and performance system to mitigate risk for human exploration missions. ICES-2021-298 11.

McGuire, K., 2024. Chapter 5 Aerospace Systems Engineering Development Principles for Health Care Providers Systems Medicine for Human Spaceflight. pp. 128–150.

McIntyre, Myers, J., Ong, K., Levin, D., 2024. Computational Modeling and Systems Medicine Systems Medicine for Human Spaceflight. pp. 128–150.

Mctigue, K, Parisi, M, Panontin, T, Wu, S, Vera, A.H., 2021. Extreme Problem Solving: The New Challenges of Deep Space Exploration.

Meadows, D.H., 2008. Thinking in Systems: International Bestseller.

Mindock, J., Lumpkins, S., Rubin, D., Thaxton, S., Lehnhardt, K., 2022. Crew Health and Performance System Overview and Context. Springer Science and Business Media LLC, pp. 1–30. Available at https://doi.org/10.1007/978-3-030-05323-9_5-1.

Mindock, J., Reilly, J., Rubin, D., Urbina, M., Hailey, M., Cerro, J., et al., 2017. AIAA SPACE and astronautics forum and exposition, SPACE. American Institute of Aeronautics and Astronautics Inc, AIAA United States Systems engineering for space exploration medical capabilities. https://doi.org/10.2514/6.2017-5236.

Ocampo, R.P., Klaus, D.M., 2016. A quantitative framework for defining "how safe is safe enough?" in crewed spacecraft. New Space 4 (2), 75–82. Available at https://doi.org/10.1089/space.2015.0040, www.liebertpub.com/overview/new-space/610/.

Parisi, M., 2024. The Impact of Delayed Communication on NASA's Human-Systems Operations: Preliminary Results of a Systematic Review.

Parker, M, Krihak, M, Laing, C, Bardina, J, Anderson, A, 2024. Anderson, Earth-Independent Medical Operations (EIMO) Concept of Operations (ConOps).

Prelich, M.T., 2025. Assessment of Model Outcomes Between the Integrated Medical Model (IMM) and the Medical Extensible Dynamic Probabilistic Risk Assessment Tool (MEDPRAT). Available at https://ntrs.nasa.gov/citations/20240012058.

Rochlis, J, Campbell, P, Miller, M, Witt, E.G., 2015. Human Systems Integration (HSI) Practitioners Guide.

Romero, E., Francisco, D., 2020. The NASA human system risk mitigation process for space exploration. Acta Astronautica 175, 606–615. Available at https://doi.org/10.1016/j.actaastro.2020.04.046.

Shelhamer, M., Antonsen, E., 2023. Systems Medicine for Human Spaceflight. World Scientific. Available at https://doi.org/10.1142/13713.

Valinia, A., 2023. NASA Exploration Systems Maintainability Standards for Artemis and Beyond.

Valinia, A., 2022. Safe Human expeditions beyond Low Earth Orbit (LEO), NASA/TM−20220002905 NESC-RP-20-01589. Available at https://www.nasa.gov/nesc/workshops/safe-human-expeditions-beyond-leo.

Vera, A, McTigue, K, Panontin, T, Parisi, M, Wu, S.C., 2023. Risks from Decreasing Ground Support for Human Spaceflight Missions Beyond Low Earth Orbit.

Walton, M., Antonsen, E.L., 2017. The Integrated Medical Model: A Case Study in Communication Engineering, Life Sciences, and Health/Medicine Synergy in Aerospace Human Systems. Integration: The Rosetta Stone Project, pp. 65–85.

Ward, J., Reynolds, R.J., Connell, E., Anton, W., Kabeel, A., Charvat, J.M., et al., 2024. Levels of evidence for human system risk evaluation. NPJ Microgravity 10 (1). Available at https://doi.org/10.1038/s41526-024-00372-w.

Zahalka, R., Iwry, J., 2024. Risk and Liability in Systems Medicine for Spaceflight. Systems Medicine for Human Spaceflight.

Chapter 4

# Ethics and informed consent

**Paul J. Ford[1], Jon G. Steller[2], and Ruth M. Farrell[3]**
[1]*Neuroethics Program, Cleveland Clinic, Cleveland, OH, United States,* [2]*University of California Irvine, Irvine, CA, United States,* [3]*Obstetrics and Gynecology Institute, Center for Bioethics, Cleveland Clinic, Cleveland, OH, United States*

## 4.1 Introduction

As the science of space medicine advances with increasing numbers of humans in space for increasing durations and distances, there is a growing need to consider the aspects of informed consent (Committee on Ethics Principles and Guidelines for Health Standards for Long Duration and Exploration Spaceflights, 2014; Rajput et al., 2023). Ideally, these complexities would first be discussed during the mission planning stage, and consent reaffirmed and reobtained during spaceflight when necessary. Long-duration exploration missions pose unique medical challenges, necessitating interventions far from Earth. Unlike terrestrial health care, space medicine involves resource limitations, communication delays or blackouts, and ethical complexities regarding autonomy and decision-making. Informed consent practices should ensure ethical standards align with mission demands, balancing individual rights with the operational needs of deep-space exploration. This chapter explores the ethical and legal foundations of informed consent and examines their application in space medicine. It aims to ensure that astronauts can make informed, voluntary decisions regarding medical interventions during long-duration missions beyond Earth while understanding the natural limitations of situations where fully informed consent cannot be obtained.

### 4.1.1 Fundamentals of informed consent

Informed consent is critical to clinical medicine and research, ensuring individuals can make autonomous health and bodily integrity decisions. It is the cornerstone of the ethical and legal conduct of medicine and research (Appelbaum, 2007; Berg et al., 2001; Beauchamp and Childress, 2003; Faden and Beauchamp, 1986). It is a mechanism whereby an individual expresses their autonomous and voluntary decision to undergo a medical procedure or take part in research (Beauchamp and Childress, 2003). Contemporary concepts and practices of informed consent have been articulated in many documents from the Nuremberg Code and, later, the Belmont Report, which emphasizes autonomy, beneficence, nonmaleficence, and justice (Beals et al., 1949; Katz, 1996; National Commission for the Protection of Human Subjects of Biomedical and Behavioral Research, 1979). Informed consent involves providing individuals with comprehensive and comprehensible information about an intervention's nature, risks, benefits, and alternatives, allowing them to make an informed decision free from coercion. Advances in biomedical science and technology, such as precision medicine, genomics, and artificial intelligence (AI) have introduced new ethical complexities, reinforcing the need for rigorous consent processes that balance scientific progress with respect for individual autonomy and ethical integrity (Rigby, 2019; Meslin et al., 1997).

Although informed consent is essential in ethics, competing values shape and counterbalance this concept (Laurijssen et al., 2022; Iserson, 2021; O'Neill, 2003). There are unavoidably coercive situations, times when a person lacks insight or capacity, situations where the public good (or majority good) supersedes the individual, or when an individual bargains away part of their full participation in the process. However, when there is an encroachment on or limitation of informed consent, clear justification and arguments must be presented to support it. For instance, there may not be sufficient time to obtain proper informed consent during an emergency because any delay may increase the risk of mortality or severe morbidity. Fundamental to sound ethics is to identify competing values and to justify why consent or other central ethical values may be limited.

**Procedural Medicine for Spaceflight. DOI:** https://doi.org/10.1016/B978-0-443-27552-4.00007-5

## 4.2 Defining informed consent for procedures performed in space

Informed consent is a mechanism to support an individual's autonomy in their health care or when participating in a study. The informed consent process delineates key differences between these two domains (Levine, 1983). In clinical medicine, the primary goal is to preserve or improve an individual's health using evidence-based, well-established interventions. Informed consent in this context ensures that patients understand the indications, risks, benefits, and alternatives of a given medical procedure, enabling them to make informed choices by weighing the available data.

In contrast, research aims to generate new scientific knowledge by evaluating experimental interventions that may later become standard practice. Informed consent for research emphasizes that complete data on safety and efficacy may be lacking and that participation may not provide direct benefits to the individual but rather to future individuals. These distinctions become increasingly complex when the clinical intervention may benefit others living in close quarters (such as concerns about infecting others or about the patient's debility compromising mission safety). Finally, there may be innovative medical procedures considered with the intention of benefiting the individual but have not been proven effective, and from which rare data may be extracted for future people to benefit.

The lines between clinical decision-making and research consent become even fuzzier in austere environments like space. Spaceflight data and experience to guide medical decision-making are so limited that almost every medical intervention may be both necessary *and* investigational for a crewmember. In spaceflight, even well-established terrestrial medical procedures may be performed under novel and uncertain conditions. In this way, almost all of space medicine includes innovative applications with varying confidence levels in procedural safety and efficacy. While procedural risks and benefits are well characterized in terrestrial settings, significant unknowns arise when adapting these procedures to space.

First, we know that the spaceflight environment alters human physiology, including fluid dynamics, tissue behavior, and wound healing, which may influence surgical techniques and procedural success. However, irrespective of spaceflight-induced pathophysiological considerations, the limited medical system resourcing and spacecraft positioning constraints for performing procedures can impact the feasibility and outcomes of medical interventions. Space missions must prioritize mass and volume restrictions, limiting the types and quantities of medical tools and medications onboard. Such limitations may necessitate modifications to standard procedures, introducing additional uncertainties.

Next, the proficiency of the individual performing the procedure in space may not have the terrestrial standard of training. Research in terrestrial medicine has shown that procedural risk correlates with the practitioner's experience (Kelz et al., 2021). In space, crewmembers may lack extensive, specialized medical training, particularly for complex interventions. While telemedicine and AI-assisted clinical decision support tools may mitigate some of these challenges, the absence of real-time communication with medical experts on the ground adds additional risk considerations. Studies on Earth show that effective communication between healthcare team members is essential for positive outcomes (Hathaway et al., 2023). When communication breakdowns occur, risk and complication rates increase. While such details are thoroughly discussed throughout this book, all of these features must factor into patient-centered counseling and complex decision-making.

For these reasons, the informed consent process for space medicine must ensure that astronauts fully understand the novel context in which a procedure may be performed. This includes acknowledging the lack of precedent, the potential for unforeseen complications, and constraints on available medical support. Unlike on Earth, where advanced medical infrastructure and specialist consultation are readily accessible, astronauts must make decisions with the understanding that medical resources are finite and emergency evacuation may be delayed or even impossible. Furthermore, space medicine presents unique ethical and operational challenges, as informed consent must balance individual autonomy with mission objectives—every decision that affects one crewmember also affects the entire crew and entire mission. Astronauts may feel pressured to accept procedures to maintain crew function and mission success. To uphold ethical integrity, informed consent processes must be robust, transparent, and adaptive, ensuring astronauts can make voluntary, well-informed decisions regarding their health care in the extreme environment of space.

Informed consent for space medicine must account for risks beyond medical complications, particularly ethical concerns about confidentiality. On Earth, privacy laws and institutional policies ensure that individuals undergoing medical procedures or participating in clinical research have their health information protected. However, in the confined setting of a space mission with a small crew, maintaining strict medical confidentiality becomes challenging. Current practices used on the International Space Station (ISS) serve as an example of how potentially sensitive health information may be managed. There are two loops or modes for communication in place on the ISS. One loop allows for open communication of information across the appropriate space and ground crew teams. Another loop allows for private communication between specific individuals on the ISS and the ground. Utilization of a private loop would help support the confidentiality of astronauts with a pending health issue. In addition, when possible, designated spaces on a spacecraft could be used to allow for more private conversations away from the crew. While these types of spaces are available on the ISS, they may not be on smaller spacecraft, calling for specific design considerations for such situations if they were to occur.

Despite these safeguards for privacy and confidentiality, there is the potential for an increasing number of individuals to have access to an astronaut's sensitive health information as both mission control and ground-based medical teams play a role in managing their health during the mission. While astronauts may consent to medical procedures, the nature of spaceflight often requires real-time monitoring and data sharing between crew medical officers (CMOs), ground support, and space agencies. This raises concerns about how much control astronauts have over their medical information and the extent to which they can refuse disclosure of sensitive health data. Despite the steps taken by the institutions that fund and coordinate missions to safeguard privacy, astronauts must acknowledge that their medical conditions and treatments could become more widely known. This is especially significant for conditions involving behavioral health, reproductive/genitourinary function, or other sensitive personal medical concerns, which may have long-term implications for an astronaut's career or future mission eligibility. Thus, informed consent procedures must be adapted for spaceflight, ensuring that astronauts are aware of the potential limitations on medical privacy and can make decisions with a clear understanding of the ethical complexities unique to space medicine.

## 4.3 Preserving the principles of informed consent when conducting medical and surgical procedures in space

Understanding and applying informed consent principles in the evolving field of space medicine is essential to uphold ethical standards while ensuring astronaut health and mission success. This environment presents a series of potential barriers and challenges to urgent or emergent health care. However, several steps can be taken to maintain the integrity of the informed consent process when medical procedures are performed in space. These steps can be found in the principles of informed consent and the experiences of medical fields that must adapt practices in emergent situations or extreme environments (Moskop, 1999; Suah and Angelos, 2018; Starcke and Brand, 2012).

As a first step, it is critical to recognize that the underlying principles of informed consent do not change simply because a procedure is performed in space rather than on Earth. While the core ethical principles—autonomy, beneficence, nonmaleficence, and justice—remain unchanged (Table 4.1), the space environment introduces complexities that require careful consideration (Varkey, 2021). For instance, additional protections may be needed to ensure that an astronaut can make an autonomous, voluntary decision about an intervention without coercion. This includes recognizing potential sources of influence that may undermine these cornerstones of informed consent. It is essential to consider that the operational demands of space missions may affect how an astronaut assesses the risks and benefits of a procedure during flight. Unlike terrestrial healthcare settings, where medical decisions primarily affect the individual, space medicine operates in a highly interdependent environment, where an astronaut's health directly impacts crew performance, mission safety, and overall success. In addition, astronauts are privileged with the rare, highly sought-after opportunity to fly in space, and this may create implicit pressure to comply with any medical recommendations rather than freely exercise their agency and autonomy following counseling. Conversely, there are scenarios in which they may feel pressure not to comply with medical recommendations, such as an astronaut declining a procedure to preserve finite resources they believe may be needed for the care or survival of other crewmembers. These potential sources of undue influence should be discussed with astronauts before flight, and protections should be instituted to ensure astronauts can make voluntary, well-informed medical decisions while seeking medical care in space. In addition, astronauts should understand how this

**TABLE 4.1** The fundamental principles of ethics in medicine.

| The fundamental principles of ethics in medicine | |
|---|---|
| Beneficence | The positive requirement to act for the benefit of the patient |
| Nonmaleficence | The obligation and intention not to harm the patient and relieve suffering |
| Autonomy | All persons have intrinsic and unconditional worth and therefore should be allowed to exercise their capacity for self-determination. |
| Justice | Fair, equitable, and appropriate treatment of persons. |

changes in emergent procedures and situations that may require action without informed consent if an individual is unable to provide it, there is not enough time, or the patient lacks the capacity to consent. In such cases, steps can be taken after the procedure to recognize their autonomy and restore any eroded sense of voluntary decision-making or bodily integrity that may have occurred.

Given the unique complexities of healthcare delivery in space, we recommend that a series of specific approaches to optimize these ethical principles be preserved while medical and surgical procedures are performed hundreds of miles from Earth (Levin et al., 2023a, 2023b). Each crewmember should have preflight visits with their flight surgeons to discuss the diversity of health issues that may arise in spaceflight, in general, and the most likely interventions available to manage them (Levin et al., 2023a, 2023b, 2024). Further, nuanced discussions should also explore the individualized risks to each astronaut based on the personalized health and medical conditions of the astronaut—how spaceflight may exacerbate those conditions or increase the risk of novel complications, how they may impact expected crew mission responsibilities, and specific ways in which those medical problems can be managed within the confines of the limited medical system. That discussion should include potential health outcomes if an urgent or emergent procedure is needed in space and considerations for complications that could arise during flight, launch, or landing. The discussion should also address who may be performing a procedure during flight and whether external supervision or guidance is available, recognizing that crewmembers may carry out a procedure with training that is below the terrestrial standard of care. For instance, a procedure may be performed by a medically trained CMO with expertise in the planned procedure, augmented by clinical decision support tools, and assisted by terrestrial ground support. Alternatively, the CMO may instead be the patient, and an untrained crewmember may find themselves attempting to manage their peer during a communication blackout without appropriate knowledge, skills, abilities, oversight, facilities, or resources.

While these preparatory discussions cannot include all potential risks and exposures, it is intended to set the stage for the type of decision-making needed during flight if a critical medical issue arises. These conversations also provide the opportunity to address medical considerations as well as the ethical and decision-making context in which these decisions may have to be made. Such decisions would occur under acute and stressful situations that would characterize the context of these decisions. Terrestrial studies demonstrate how stressors impact the decision-making process, making it more challenging for an individual to navigate the risks and benefits of a medical decision (Starcke and Brand, 2012). It should be considered how psychological stressors may impact decision-making when a critical choice must be made during flight. Furthermore, there are several ethical considerations to discuss ahead of the flight. Given the interdependence of the crew during spaceflight, an acutely ill astronaut's decision about a procedure will call for weighing the risks of accepting or declining the procedure in the context of the health of the entire crew and mission. While we have discussed situations above in which, after weighing the risks and benefits, an astronaut patient may refuse a proposed intervention, there is also ethical complexity on the side of the caregiver who may refuse to perform a procedure for a myriad of reasons. For instance, the crewmember(s) designated to perform the procedure may express significant moral distress about the possibility of causing undue pain or suffering, or even hastening death for the astronaut by attempting the procedure. There is also the question of medical futility, in which the proposed procedure is unlikely to benefit the ill astronaut, and what steps would be taken if resources had to be allocated to other crewmembers who require medical attention and have a higher chance of survival. That requires considering what alternative interventions or palliative measures could be enacted during the mission in place of the recommended procedure. These are difficult discussions to have when preparing astronauts for the challenges of spaceflight but are critical to ensuring that steps are taken to optimize astronaut autonomy and health and the crew's health.

It is also important to consider that some astronauts may be reluctant to share information about health issues that may impact their ability to participate in spaceflight. In this case, there may be health issues that are undisclosed or underdisclosed to the flight surgeons. That may include information withheld about the existence of the condition, the status of the condition (e.g., whether past or current medical problem), and the severity of the condition when managed under usual conditions on Earth. Significantly, these undisclosed or underdisclosed conditions may manifest or become exacerbated during the unique stressors of spaceflight. If severe enough, these types of medical conditions may warrant medical intervention during the mission. Additionally, there is also the potential that the condition may interfere with the management or proposed intervention for a separate health issue. Thus, there is a need for mechanisms for astronauts to disclose health information before flight and work with a specialist to optimize the management of those conditions without having a punitive consequence for participating in missions. It is also important to rigorously develop medically disqualifying criteria for spaceflight. This process should ensure that when a condition is deemed disqualifying, the reasons regarding the condition and inability to manage risks associated with it during spaceflight are guided by high-quality data, are evidence-based, and are not exclusionary to groups of astronauts with the condition.

Fortunately, most medical conditions are not disqualifying for spaceflight, especially when optimally managed well before flight. With that said, it is understandable that some medical concerns would be kept private by a patient given the complexity of the ethics for a physician working for the government or a private company, as it may cross an ethical line to prioritize hiding a patient's medical condition over the risks to the rest of the crew, the mission, and the institution. A perfect solution for how to best handle this conundrum has not been achieved. However, one consideration may be to establish new roles and processes for astronauts to access a specialized member of the ground crew as an independent medico-ethical ombudsman. This individual would have knowledge of both space and terrestrial medicine and would assemble and lead a specialized panel of experts to provide guidance on the best way to balance autonomy and privacy with integrity and safety in such circumstances. Processes would be needed to ensure that seeking consultation from an independent medico-ethical ombudsman would not raise a "red flag" for a concern but serve as mechanism for all astronauts to discuss their health issues during training and before the mission.

Second, the collection and application of scientific data are pivotal in space medicine, particularly in understanding the risks and benefits of medical procedures. This includes assessing risks for individuals participating in government- or private industry-funded missions, as well as the steps taken before flight to optimize health and performance (Ball et al., 2012; Seylani et al., 2024). To accurately assess these risks, it is crucial to ethically collect comprehensive data that directly reflects the unique needs of astronauts. The development and implementation of research to gather such data must adhere to scientific rigor, ensuring robust and reproducible findings. Those approaches should include conducting research that captures a more comprehensive picture of crew health and performance and considers biological variables like sex, age, anthropomorphic data, and medical comorbidities to provide a more nuanced understanding of how physiological differences impact medical interventions, thereby enabling personalized patient-centered risk–benefit assessments for astronauts.

Third, to ensure ethical and evidence-based medical decision-making in space, it is essential to establish structured processes for data collection, storage, and sharing of data to iterate and adapt. Given the investigational nature of medical procedures in space, encouraging collaboration with government and commercial spaceflight entities to merge datasets from different missions, physiological conditions, and procedural outcomes will be essential for generating meaningful insights while preserving data integrity and minimizing bias. Beyond space missions, collating data from terrestrial analogs like human parabolic flights, HERA (Human Exploration Research Analog), and extreme or austere environments may provide valuable insight into the health risks astronauts face in space. These analogs to space environments offer critical data points on human responses to stressors like microgravity or isolation, contributing to developing evidence-based medical protocols for space missions. Lastly, leveraging advanced computational techniques and modeling systems can simulate potential medical scenarios in space, allowing for real-time adaptations and predictive insights. By making these data more accessible to flight surgeons, astronauts, mission planners, and other key stakeholders, we can refine and unify informed consent processes, ensuring that astronauts make well-informed decisions based on a comprehensive understanding of known and potential risks. This work should be conducted in collaboration with methodologists and epidemiologists who are critical for developing best practices for data analysis and integration with limited sample sizes or using larger Earth-based datasets as a comparator.

Finally, good information, explicit data protections, and voluntariness alone are insufficient. In order to ensure adequate regulatory and ethical oversight of spaceflight medical procedures, additional mechanisms of safeguards and oversight should be required. These need to be implemented in ways that are efficient, effective, and independent. Establishing a specialized oversight body analogous to both an Institutional Review Board and Data Safety Monitoring Boards (DSMB) is a model that could be drawn upon and redeveloped for this specific instance (Evans, 2022; FDA, 2006; NIH, 1998; Fiscella et al., 2020). A body is needed to assess risks, review protections, and provide continuous ethical guidance for medical interventions conducted beyond Earth. That body could provide an independent review of adverse or unexpected events during clinical care or medical research. This would then provide a mechanism to intervene, modify, or stop a procedure if it were an unjustifiable and unconscionable risk. A second resource would be the use of research advocates, a model frequently utilized in clinical research. Specialized research advocates located on Earth could assist astronauts facing the challenging and often time-sensitive decisions about undergoing a procedure in space. Such a mechanism could allow for a two-part consent process—one preflight and then a second confirmed in-flight; similar to consenting for tubal ligations terrestrially for satisfied parity. This resource could also serve as a surrogate if the patient were incapacitated, ensuring the independence of both the advocate and a DSMB-like medical panel in evaluating the risks and benefits of undergoing a procedure in space.

Advances in space medicine, science, and technology make human exploration of space possible. At the same time, these advances raise key questions about the human capacity for long-duration spaceflight, with the unique exposures of space playing a significant role in individual astronaut and crew health and performance. While steps may be taken on Earth to prevent an acute medical issue in space, it is inevitable that as humans travel further into space, there will be a need to perform medical interventions to preserve health in flight. That calls for preparatory steps to preserve the health and well-being of astronauts, not just through the delivery of meticulously planned medical procedures but also through attending to the unique aspects of informed consent. By recognizing these considerations now and developing strategies to support the informed consent process for astronauts, we can ensure that human space exploration can continue.

## References


Appelbaum, P.S., 2007. Clinical practice. Assessment of patients' competence to consent to treatment. The New England Journal of Medicine 357 (18), 1834–1840.

Ball, C.G., Kirkpatrick, A.W., Williams, D.R., Jones, J.A., Polk, J.D., Vanderploeg, J.M., et al., 2012. Prophylactic surgery prior to extended-duration space flight: is the benefit worth the risk? Canadian Journal of Surgery. Journal Canadien de Chirurgie 55 (2), 125–131.

Beals, W.B., Sebring, H.L., Crawford, J.T., 1949. Permissible medical experiments. Trials of War Criminals Before the Nuernberg Military Tribunals. US Government Printing Office, Washington, DC, pp. 1946–1949 vol. 2.

Beauchamp, T.L., Childress, J.F., 2003. Principles of Biomedical Ethics, eighth ed. Oxford University Press, New York.

Berg, J.W., Appelbaum, P.S., Lidz, C.W., Parker, L.S., 2001. Informed Consent: Legal Theory and Clinical Practice. Oxford University Press, New York.

Committee on Ethics Principles and Guidelines for Health Standards for Long Duration and Exploration Spaceflights; Board on Health Sciences Policy; Institute of Medicine, 2014. Health Standards for Long Duration and Exploration Spaceflight: ethics principles, responsibilities, and decision framework.Board on Health Sciences Policy; Institute of Medicine In: Kahn, J., Liverman, C.T., McCoy, M.A. (Eds.), National Academies Press, Washington (DC) Board on Health Sciences Policy; Institute of Medicine.

Evans, S.R., 2022. Independent oversight of clinical trials through data and safety monitoring boards. NEJM Evidence 1 (1) EVIDctw2100005.

Faden, R., Beauchamp, T., 1986. A History and Theory of Informed Consent. Oxford University Press, New York.

Fiscella, K., Sanders, M., Holder, T., Carroll, J.K., Luque, A., Cassells, A., et al., 2020. The role of data and safety monitoring boards in implementation trials: when are they justified? Journal of Clinical and Translational Science 4 (3), 229–232.

Hathaway, J.R., Tarini, B.A., Banerjee, S., Smolkin, C.O., Koos, J.A., Pati, S., 2023. Healthcare team communication training in the United States: a scoping review. Health Communication 38 (9), 1821–1846.

Iserson, K.V., 2021. Bioethical issues in Antarctica. Cambridge Quarterly of Healthcare Ethics: CQ: The International Journal of Healthcare Ethics Committees 30 (1), 136–145. PMID. Available at https://doi.org/10.1017/S0963180120000638, 33371924.

Katz, J., 1996. The nuremberg code and the nuremberg trial: a reappraisal. JAMA: the Journal of the American Medical Association 276 (20), 1662–1666.

Kelz, R.R., Sellers, M.M., Niknam, B.A., Sharpe, J.E., Rosenbaum, P.R., Hill, A.S., et al., 2021. A national comparison of operative outcomes of new and experienced surgeons. Annals of Surgery 273 (2), 280–288.

Laurijssen, S.J., van der Graaf, R., van Dijk, W.B., Schuit, E., Groenwold, R.H., Grobbee DE, et al., 2022. When is it impractical to ask informed consent? A systematic review. Clinical Trials (London, England) 19 (5), 545–560. Available at https://doi.org/10.1177/17407745221103567, 35775421 PMCID: PMC9523816.

Levin, D., Nelson, N.G., McIntyre, L., Anderson, A., Steller, J.G., Hilmers, D.C., 2023a. A Pilot Project for Quantifying the Effect of Medical Provider Knowledge, Skills, and Abilities on Outcomes for Spaceflight Using a Probabilistic Risk Assessment Tool.

Levin, D.R., Steller, J., Anderson, A., Lemery, J., Easter, B., Hilmers, D.C., et al., 2023b. Enabling human space exploration missions through progressively Earth independent medical operations (EIMO). IEEE Open Journal of Engineering in Medicine and Biology 4, 162–167.

Levin, D.R., McIntyre, L., Steller, J.G., Nelson, A., Zahner, C., Anderson, A., et al., 2024. Predicting exploration crew medical officer training needs: applying evidence-based predictive analytics to space medicine training. Wilderness & Environmental Medicine 10806032241292535.

Levine, R.J., 1983. Informed consent in research and practice: similarities and differences. Archives of Internal Medicine 143 (6), 1229–1231.

Meslin, E.M., Thomson, E.J., Boyer, J.T., 1997. The ethical, legal, and social implications research program at the National Human Genome Research Institute. Kennedy Institute of Ethics Journal 7 (3), 291–298. Available at https://doi.org/10.1353/ken.1997.0025.

Moskop, J.C., 1999. Informed consent in the emergency department. Emergency Medicine Clinics of North America 17 (2), 327–340. Available at https://doi.org/10.1016/s0733-8627(05)70062-6 ix-x.

National Commission for the Protection of Human Subjects of Biomedical and Behavioral Research, 1979. The Belmont Report: Ethical Principles and Guidelines for the Protection of Human Subjects of Research. U.S. Department of Health and Human Services.

National Institutes of Health, Policy for data and safety monitoring, 1998. Available at https://grants.nih.gov/grants/guide/notice-files/not98-084.html.

O'Neill, O., 2003. Some limits of informed consent. Journal of Medical Ethics 29, 4–7. Available at https://jme.bmj.com/content/29/1/4.

Rajput, S., Mayor, I., Diamond, M., Rosenberg, M., Cole, V., Bhakare, N., et al., 2023. Medical ethics of long-duration spaceflight. NPJ Microgravity 9 (1), 85.

Rigby, M., 2019. Ethical Dimensions of Using Artificial Intelligence in Health Care. The AMA Journal of Ethic 21 (2), 121–124. Available at https://doi.org/10.1001/amajethics.2019.121.

Seylani, A., Galsinh, A.S., Tasoula, A., I, A.R., Camera, A., Calleja-Agius, J., et al., 2024. Ethical considerations for the age of non-governmental space exploration. Nature Communications 15 (1), 4774.

Starcke, K., Brand, M., 2012. Decision making under stress: a selective review. Neuroscience and Biobehavioral Reviews 36 (4), 1228–1248.

Suah, A., Angelos, P., 2018. How should trauma patients' informed consent or refusal be regarded in a trauma bay or other emergency settings? AMA Journal of Ethics 20 (5), 425–430.

US Federal Food and Drug Administration, 2006. Establishment and operation of clinical trial data monitoring committees: guidance for clinical trial sponsors. Available at https://www.fda.gov/regulatory-information/search-fda-guidance-documents/establishment-and-operation-clinical-trial-data-monitoring-committees.

Varkey, B., 2021. Principles of clinical ethics and their application to practice. Medical Principles and Practice: International Journal of the Kuwait University, Health Science Centre 30 (1), 17–28. Available at https://doi.org/10.1159/000509119 PMID: 32498071; PMCID: PMC7923912.

Section II

# Procedural considerations

Chapter 5

# Vascular access

**Luke Apisa**
*Department of Emergency Medicine, SPEAR Division, Massachusetts General Hospital, Boston, MA, United States*

## 5.1 Introduction

Vascular access, the process of accessing the venous or arterial system for infusion/transfusion, filtration, or interventional procedure, is a mainstay of modern medical care. The first recorded successful IV injection was by Sir Christopher Wren in 1656, using an infusion device made of a pig's bladder attached to the tip of a quill pen (Stegink, 1983). The first IV to be performed in space took place in 1993 on STS-55, when Dr. Bernarned Harris Jr. injected Hans Schlegel with saline. The intervening 300 years are host to a tremendous volume of innovation both in means and types of vascular access.

## 5.2 Types of vascular access

Broadly, vascular access occurs in the venous or arterial aspect of the circulatory system. Access may be transient (lasting only as long as blood is being sampled) or may result in placement for prolonged repeated blood sampling, infusions, and transfusions. Lastly, access may occur peripherally, as in an arm or hand in the superficial orientation, or centrally, as in the anatomically deeper and larger veins found adjacent to known larger arterial sites of the neck or upper femoral areas. Four specific modes of vascular access will be addressed in this chapter: straight needle venipuncture, intravenous (IV) catheter placement, arterial access, and central venous access.

### 5.2.1 Straight needle venipuncture ("straight stick")

At times, persistent vascular access may not be required. Obtaining a blood sample, for example, requires only a single access of the circulatory system, with no need for an indwelling line. In these instances, transient venipuncture is performed with a standard (straight) needle. In these instances, the needle remains in contact with the circulatory system for the duration of sample retrieval and is subsequently removed.

### 5.2.2 Intravenous access

An IV line, or simply "IV," is placed when continuous access to the circulatory system is required, either for repeated blood sampling, installation of therapeutic compounds or treatments, or both. This procedure is similar to straight needle venipuncture in the mode of vascular access, but following the initial venipuncture with a needle, a catheter is advanced into the vein and left in place, providing continuous access to the bloodstream. In modern practice, most IV catheters contain a metal needle with a plastic catheter already prethreaded over the metal needle shaft. Upon confirmation of successful venous access, a spring-loaded mechanism will manually retract the metal needle, leaving only the plastic catheter in place. IVs were historically placed via direct visualization and/or palpation, but modern imaging and other tools can now assist in IV placement. Bedside ultrasound and near-infrared visualizers ("vein finders") are commonly employed examples. IVs are typically inserted into the extremities or superficial vasculature using short catheters and, as such, are categorized as "peripheral access."

### 5.2.3 Central venous line

Central venous lines, in contrast, are long catheters inserted into the neck, chest, or groin, with the distal tip terminating in a large vessel with direct connection to the heart (the vena cava). Central lines often have multiple physically separated

**Procedural Medicine for Spaceflight. DOI:** https://doi.org/10.1016/B978-0-443-27552-4.00008-7

tubes (lumens) within a single catheter to allow for the coadministration of multiple therapeutics. The most common variant has three such tubes, hence a common name for many central lines: triple lumen catheters. Central lines may also have only one or two lumens, such as those used for large volume resuscitation or for dialysis. These lines are more complex to insert and carry a greater risk of complications than peripheral IVs owing to their greater length, longer inserted dwell times, and more proximate positioning to deep internal structures like the lungs and heart. They do, however, allow for the administration of medications that cannot be administered safely or for long durations through peripheral access, such as vasopressors. These lines are a fundamental component of critical care medicine.

### 5.2.4 Arterial access

It should be noted that the venous catheters mentioned only enable access to half of the circulatory system. Accessing an artery, alternatively, may be required for a panoply of reasons: sampling of freshly oxygenated blood, continuous blood pressure monitoring, or catheter guided interventions, to name only a few. This access, however, carries its own distinct risks, which are enumerated below. Arteries may be accessed temporarily, similar to the "straight needle venipuncture" noted above, to obtain measurements of the pH, oxygen, and carbon dioxide content of the arterial circulation ("arterial blood gas"). This is often used in the monitoring of critically ill patients. Indwelling arterial lines may also be placed, usually for the purposes of continuous blood pressure monitoring or for frequent measuring of arterial blood gases. In contrast to serial arm cuff measurements, direct arterial line measurements of blood pressure have the benefit of being continuous in real-time. Arteries may also be accessed in very specific circumstances for the arterial administration of medications, such as intraarterial tPA in the treatment of stroke or frostbite.

### 5.2.5 Intraosseous access

Intraosseous (IO) access is achieved by drilling a needle through the bone of the tibia, humerus, or sternum to directly access the bone marrow space. While the needle tip does not directly enter a blood vessel, the bone marrow space communicates directly with the body's vascular supply, and thus medications or fluids administered into this space will enter the bloodstream. Terrestrially, IO lines are often employed in low-resource environments or in trauma patients where time is of the essence and more delicate IV line placement may be too temporally costly to be feasible. Owing to their logistical simplicity, IO access kits have been routinely flown in human spaceflight and remain an attractive means of vascular access for this specific environment.

### 5.2.6 Other access types

Multiple other types of vascular access exist that are beyond the scope of this chapter. These include dialysis, cardiopulmonary bypass, resuscitative endovascular balloon occlusion of the aorta, extracorporeal membrane oxygenation circuits, etc. These various interventions require a level of logistical complexity and support machinery that far exceed feasible inclusion on space vehicles in the near-to-mid term. The principles of vascular access described in this chapter remain applicable to these more complex operations, but the power, mass, and logistical complexity of these setups extend beyond current and near-future human spaceflight capabilities.

Additionally, for the sake of logistical simplicity, alternatives to direct vascular access may be employed to achieve the same end result. For example, intramuscular (IM) administration of medications obviates the need for time-intensive and complex microgravity IVs. For medications in which a one-time dose is required, IM is frequently favored over IV administration in the spaceflight environment. Similarly, some lab values may be readily obtained by fingerstick. Blood glucose monitoring, for example, is more simply achieved by recurrent fingersticks or a wearable continuous monitor than recurrent blood draws from an IV. Moreover, these methods of blood sampling leverage consumer off-the-shelf techniques and minimize the need for time, mass, and volume-intensive laboratory setups.

Crystalloid administration may also be achieved through alternate means, such as subcutaneous or rectal administration. While the drawbacks of these routes compared with the relative ease of IV administration prevent their routine use as a first line in conventional medicine, they remain worthy considerations in the spaceflight environment, in which IV access carries increased complexity. Similarly, oral rehydration salts provide an excellent means of rehydration for patients with the ability to tolerate substances per os.

## 5.3 Procedural considerations

### 5.3.1 Common elements of vascular access

While the distinct vascular access techniques outlined above are their own individual procedures, many common considerations are shared by each—these will be discussed at length below. The specific common considerations explored below include the following:

- Choice of access and lumen,
- Site identification,
- Site preparation,
- Potential complications.

### 5.3.2 Choice of access and lumen

The indication for vascular access is the first driver of decision-making. If a single blood sample is required—for example, obtaining a single-timepoint research sample—straight needle phlebotomy may be performed. Veins are typically accessed on the hairless, inner surface of the arm. These are found most superficially over the forearm and antecubital fossa, and these are the ideal site for straight needle access.

If frequent venous blood sampling is to be performed or if IV medications must be administered, an IV should be placed. For example, Apollo 13 astronaut Fred Haise developed pyelonephritis in flight during the storied mission. While he ultimately recovered thanks to a timely return, this recovery would be by no means assured on a longer-duration spaceflight, in which IV fluid and antibiotic administration would prove life-saving.

Veins for IV access are routinely accessed over the forearm and antecubital fossa as above, but deeper veins in the upper arm may be used with ultrasound guidance if access in the forearm proves challenging. Leg veins are usuallyavoided due to greater potential for phlebitis, infiltration, and greater risk of dislodgement, though these may also be accessed if necessary. In case of emergent need and poor extremity vein candidates, IVs may also be placed in the external jugular vein of the neck. Peripheral IV flow rates are directly tied to the gauge and length of the catheter inserted, governed by Poiseuille's law. A conventional 14G needle, for instance, can instill up to 240 mL/min with regular flow, whereas a similar length 22G needle can only administer 36 mL/min. Smaller needles are easier and less painful to insert, but they provide more resistance and therefore slower flow rates for sampling blood and administering therapeutics. These considerations must be balanced when choosing the gauge of the needle and catheter to be used. An 18G needle is a standard first choice that balances ease of insertion while minimizing resistance and providing greater flow rates. For large volume resuscitation, a larger 14G or 16G needle might be selected. For a patient with challenging vasculature or smaller veins, 20G or even 22G might be the optimal first choice. The mass and volume of straight needles are roughly comparable to those of IV catheters.

If vasoactive medications for resuscitation must be administered, a central line should be considered, though vasoactive medications can be given through a large bore peripheral IV (i.e., 18G or larger). Central venous catheters have flown in space on multiple missions, including Spacelab Life Sciences 1 and 2 (Buckey et al., 1996). These catheters, however, were placed terrestrially and then flown into space. To date, central venous access has not been performed while in microgravity. Indications for central line placement are for intensive medical care or for research purposes.

Arterial sampling or line placement, similar to central line placement, is also typically performed only in the setting of research or critical care. They are warranted when regular acid-base and respiratory monitoring is required, or when continuous blood pressure monitoring is necessary. Additionally there are rare cases in which they may be utilized for medication administration.

IO access site at the humeral head or tibia will be dictated by the location of patient injury and ease of access of the preferred administration site. Sternal IOs are rarely used due to a greater risk of complications, and the presence (usually) of four alternate sites of access.

### 5.3.3 Site identification

Straight needle venipuncture and arterial access, as well as peripheral IVs, are routinely performed by visualization and palpation. The operator uses anatomic reference points along with direct palpation of the vessel to be accessed to guide needle angulation and depth. Modern visualization methods have emerged to increase first-attempt success. Near-infrared "vein finders" use infrared light to highlight patient vasculature. While these devices do not indicate depth, they show the location of blood vessels through the skin surface.

Dynamic ultrasound guidance has also emerged as the standard of care for placement of central and arterial lines and has become increasingly common in challenging peripheral IV placements (Elshikh et al., 2025). Handheld ultrasounds, which have been manifested on recent spaceflights, have also been studied in the placement of peripheral IVs, with a recent study even demonstrating superior first-attempt success rates compared with high-end ultrasound systems (Gulati et al., 2021).

In the case of central lines, three primary sites exist: femoral, internal jugular, and subclavian. Multiple studies exist comparing the efficacy, outcomes, and complications of these various sites. While all sites provide direct access to the central circulatory system, owing to their varied locations, complications associated with each vary significantly. For example, femoral lines preserve access to the neck when needed for other interventional procedures or airway access but carry with them increased risk of infection. Subclavian lines have the lowest associated rates of infection but also carry with them the risk of inducing a noncompressible hemorrhage in the event of a complication, owing to the deep nature of the accessed vessel to the clavicle (Parienti et al., 2015). Internal jugular lines balance compressibility and risk of infection but may occupy a high-priority location in the event that access to the patient's head and neck are needed for other critical interventions such as intubation or other endovascular techniques. These are discussed at further length in the section on potential complications below.

For arterial lines, radial and femoral artery access points are the most common. Radial artery access is by far the most common. Risks of major hemorrhage are lower, owing to the much smaller caliber and more readily compressible nature of the vessel. Additionally, the more peripheral point of access and the presence of less adjacent tissue minimize the risk of infection and increase the ease of access. Femoral artery access may be necessary, however, if the patient has preexisting conditions limiting ready access to the upper extremities, in cases of vascular upper-extremity injury, or in cases of poor collateral flow to the hands through the ulnar arteries. Additionally, femoral access may be employed for logistical simplicity in critical cases in which the femoral site is already being accessed for the purposes of central-line placement.

For a humeral head IO, insert the needle perpendicular to the bone, 1 cm above the surgical neck of the humerus. For a tibial IO, insert the needle perpendicular to the bone roughly 2 cm distal and 1 cm medial to the tibial tuberosity. Firm, direct pressure should be applied. Manually actuated devices are preferred in the spaceflight environment, though care should be taken by the user to brace themselves as rotational forces of the needle driver will be transmitted to the user.

### 5.3.4 Site preparation

The introduction of any foreign body into the bloodstream carries with it the attendant risk of infection. Even with proper precautions, rates of bloodstream infection resulting from peripheral catheter placement are roughly 0.1% (Maki et al., 2006). This is discussed at greater length below. Cleaning the intended catheter site is the standard of care. For IVs, this is routinely accomplished terrestrially with an alcohol swab. For IOs, arterial lines, and central lines, chlorhexidine is more commonly employed. However, both alcohol and chlorhexidine (frequently formulated mixed with alcohol) are flammable and volatile. Additionally, the volatile resultant components of both alcohol and chlorhexidine must be scrubbed from the enclosed environment utilizing trace contaminant filtration systems presenting further challenges in the spaceflight environment. Benzalkonium chloride (BZK) and povidone iodine are nonflammable antiseptic alternatives that are routinely used for skin surface cleaning and may be an effective alternative for the placement of lines in spaceflight.

Straight sticks, IOs, and IV access are clean procedures. Owing to the direct access to the central circulatory system, arterial lines and central lines are sterile procedures. As such, more extensive personal protective equipment, including sterile gowns and gloves, are required for these procedures. Additionally, sterile drapes are routinely employed to prevent contamination of sterile equipment by contact with nonsterile surfaces.

## 5.4 Potential complications

### 5.4.1 Accidental arterial puncture

In the case of unintended arterial puncture during intended venous access, pressure must immediately be held at the access site. Owing to the high pressure of the arterial system, blood loss may be rapid. While peripheral bleeding from an arterial site may often be swiftly controlled with constant, direct pressure, arterial bleeding during attempted central access may rapidly lead to life-threatening hemorrhage. This blood may also be liberated into the cabin, posing infection control risks discussed at more length below under "Spaceflight-Specific Considerations." Additionally, the known lower body circulation and particularly deep femoral vasculature tendency toward fluid shift and collapse in spaceflight may increase the risk of accidental arterial puncture. Hematoma formation may also occur secondary to venous oozing alone, and while steps are taken to mitigate this risk, the results are less potentially catastrophic than uncontrolled arterial flow.

### 5.4.2 Nerve injury

Veins and arteries commonly run in bundles that also contain an adjacent nerve. Irritation or transection of the nerve sheath may lead to sharp, shooting pain. In the most severe cases, transient or lasting nerve damage may be inflicted. In rare cases, the nerve may be fully transected, leading to lasting disability or neuroma formation (Thrush and Belsole, 1995). Real-time ultrasound guidance helps minimize the risk of this complication.

### 5.4.3 Thrombophelebitis

With peripheral straight needle and IV access, the introduction of a foreign body into small-caliber peripheral vasculature increases the risk of small clots forming in these superficial veins. Clot formation with associated vein irritation is called thrombophlebitis. This complication typically resolves with warm compresses and ibuprofen therapy. While the incidence of thrombophlebitis is typically low and the pathology itself generally is self-limited, it is not yet known whether the hypercoagulable state induced by human spaceflight may increase these risks.

### 5.4.4 Deep vein thrombosis

A known and concerning clotting issue seen in central line access (and deep IV and midline IV access) would be the materialization of a DVT. There have been several documented incidences of DVT formation in prolonged spaceflight, mainly in the upper body circulation, even without the introduction of a foreign body catheter (Marshall-Goebel et al., 2019). Terrestrially, DVT is a known complication of particularly deep vein access. It stands to reason that the potential for DVT formation in spaceflight should be accounted for in-risk assessment when determining the need for vascular access and monitoring. This is discussed at more length below under the section on spaceflight-specific considerations.

### 5.4.5 Bloodstream infection

While all vascular access should be performed with sterile equipment and at a thoroughly cleaned site, postline infections occur with nearly all forms of vascular access. These infectious risks increase with more central forms of access. For example, a systematic review found that bloodstream infection rates with peripheral catheter placement are around 0.1%. For arterial lines, this figure rose to 0.8%. For central catheters, the bloodstream infection risk was higher still at 2.4% (Maki et al., 2006). Terrestrially, minimizing bloodstream infections has been shown effective with several implemented steps in checklist fashion. Use of antiseptic impregnated catheters, antiseptic based caps, and antimicrobial impregnated foam discs surrounding the line at the point of skin penetration are all tools to aid in minimizing attendant infection risk (O'Grady, 2023). Similarly, frequent line checks to rapidly identify early signs of infection minimize the risk of progression to bloodstream infections and sepsis (Buetti et al., 2022).

For the sole purpose of reducing infection risk, the subclavian vein is the preferred site owing to its lower risk of infection. The femoral vein has shown the highest infection risk in studies (Bell and O'Grady, 2017). In spaceflight, however, site selection should also be balanced with known anatomic and physiologic changes of organ shift as the placement of a subclavian line also has the highest risk for pneumothorax development, potentially even higher in the spaceflight environment. Additional checklist components include vigorous hand washing prior to line placement and handling of supplies, as well as sterile insertion with full barrier compliance, including sterile drape, mask, cap, gown, and gloves. Though utilized terrestrially with evidence showing reduced central line infection rates, the cumbersome nature of more material implements listed in the low and zero gravity environment also needs consideration when weighing access options. Utilizing ultrasound guidance when possible to reduce the number of attempts, thereby reducing the number of entries from the skin, helps to avoid repetitive introduction of bacteria from skin surfaces. With each medication administration or blood draw through an accessed site comes the opportunity for contamination. Caps should be disinfected before and after each use and prior to recapping. Needless securement devices such as suturing should be avoided in favor of locking sutureless securement devices coupled with a disinfected impregnated dressing to stave off bacterial load collection at the site of insertion of any line placement, but especially central access. Lastly, frequent evaluation of the chosen access for necessity and possible removal, at least daily, should be performed (Bell and O'Grady, 2017).

## 5.5 Spaceflight specific considerations

### 5.5.1 Microgravity materials handling

Vascular access is an inherently complex, fine-motor procedure requiring an array of supplies. Unlike terrestrial care environments where space can easily be obtained for equipment like procedure trays and large ultrasound machines, vascular access in space will necessarily be performed in cramped and shifting environments. Moreover, the current packaging solutions for vascular access kits do not routinely secure components, making the many small and frequently sharp components a hazard in microgravity.

For routine procedures using similar equipment, such as IM injections, astronauts and commercial space travelers have described using looped tape stuck to a surface as a portable workbench. Components, as they are removed from their packaging, are stuck to the strip of tape. When used, the components are individually removed from the tape workbench and then returned to the tape when no longer in use. NASA has baselined Velcro binders for their medical kit with some similar functionality.

The medication, fluid, or even blood products themselves infused through all lines also experience complications in low- and zero-gravity environments. Pumps and back pressure will be required in these environments to prevent backflow of blood into the placed line and even further into the medication administration receptacle itself.

Sterile technique is aided tremendously by the gravity-bound nature of substances in the Earth's gravity well. A sterile drape rests effortlessly on a patient. Airborne particles on the other end of a trauma bay fall to the ground before reaching a sterile procedure site. Sterile technique in microgravity will require novel approaches to circumvent infection risk from free-floating particles. Proof-of-concept investigations will be required to develop these protocols for future long-duration spaceflight.

As more complex kits, such as central and arterial access kits, are flown in the future, custom packaging and operational solutions must be employed to facilitate their more streamlined use in microgravity.

### 5.5.2 Aerosolized blood

Contact with ambient bacteria during a procedure, in addition to infectious diseases in patient body fluids already present, is a persistent concern when obtaining vascular access. The microgravity environment greatly heightens said risks. Transmissible pathogens are spread by contact with mucous membranes (eyes and mouth) or breaks in the skin. Personal protective equipment, such as gloves, gown, and facemask, are therefore standard precautions with terrestrial blood sampling to protect both the patient and, secondarily, the proceduralist. In spaceflight, liberated body fluids are no longer bound by gravity. As such, they pose a greater chance of contact with mucous membranes that would otherwise be above the field of work during Earth-based procedures. These fluids pose additional risks to other crewmembers in the cabin and to important filter mechanisms, as most body fluids will begin as a liquid before congealing or coagulating to solid form. Historically, spaceflight participants have been heavily screened. As such, risk of blood-borne pathogen exposure has been minimal in part due to excellent screening. A risk-trade exists between robust preflight screening and the need for extensive blood-borne pathogen exposure protocols in the spaceflight medical environment.

Multiple solutions for spaceflight can assist in mitigating these risks. When vascular access is obtained in a 1 g environment, the ends of the inserted catheters are left open to the air and blood return confirms proper placement. Most needles, however, come equipped with a "flash chamber," so named for the flash of blood that appears upon accessing the bloodstream, which allows successful placement to be confirmed without withdrawing the needle. While a brief time is still required to switch the catheter end over to IV, upstream compression at the placement site can prevent most leakage. Likewise, many catheters now contain backstop valves or manual locks to prevent blood leakage at placement. Directly applying an absorbent sterile material such as a gauze product or sterile towel is also an option. Lastly, suction could be used during this time to scavenge any local blood liberated during the transfer. Alternatively, use of a hood with gentle negative pressure at one end could allow an operator to obtain vascular access through a transparent panel and ensure that any liberated blood is safely drawn into a containment vessel.

### 5.5.3 Air–fluid interactions

Many elements of medication administration terrestrially rely on the 1 g for the separation of air from fluid. An IV bag hanging from a pole uses gravity not only to pull the solution downwards for administration to the patient but also serves to separate air towards the top of drip chambers and other devices, minimizing the risk of air embolus. While the administration of an air embolus into the circulatory system could be life-threatening, the amount required to initiate cardiopulmonary vascular complications is typically larger than can be found with line placement alone when lines are properly flushed. A greater risk is that of large air collections in medication and IV fluid prefilled or filled-on-demand bags or large syringes.

Solutions include pressure-flushing of lines prior to use, the use of a pressure bag during medication or fluid administration, mechanical or digital pumps and flow monitors, and three-way stopcocks allowing intermittent syringe flushing to obviate the need for a gravity vector for administration. Gravity-independent air-trapping filters have also been discussed in the literature for use in-line to minimize the risk of bubble transmission (Formanek et al., 2024).

### 5.5.4 Clot formation

Insertion of any foreign body into the bloodstream increases the risk of clot formation. In the case of peripheral IVs, this can result in thrombophlebitis. Central lines can also serve as a nidus for thrombus formation. Current data strongly suggest that spaceflight is a prothrombotic environment (Kim et al., 2021). This increased predisposition to clot formation is of particular clinical salience when considering central lines, which are primarily placed in cephalad vessels in which clots are more likely to occur in spaceflight due to headward fluid shifts. Further study will be required to determine the extent of these risks. Mitigation strategies could include frequent ultrasound surveillance, a bias towards femoral central access, or gentle heparinization.

### 5.5.5 Mass and volume considerations

Minimizing the mass and volume to orbit extends beyond simply being judicious with the medical payload manifest. Rather than simply sending less of a substance, certain alternative formulations are much more attractive for the spaceflight environment. For example, sending sterile IV fluids to orbit is quite costly. The NASA IVGen device was created to solve this problem by creating sterile IV fluids on orbit from reclaimed water from the station along with requisite component electrolytes. Similarly, sending preconstituted medications to orbit is significantly more mass- and volume-intensive than sending lyophilized medications that can be reconstituted once on orbit. Similarly, factory packaging for medications often does not optimize for minimal mass and volume. Alternative packaging solutions have been employed both by NASA and the commercial space sector to minimize the space and volume consumed by medications.

## 5.6 Procedural technique and materials for spaceflight

### 5.6.1 Straight stick

1. Don nitrile gloves.
2. Attach an elastic tourniquet proximal to site of vascular access.
3. Identify the vein or artery to be accessed.
4. Prep overlying skin with BZK wipe.
5. Unsheathe the butterfly needle and insert the needle into the vein until a flash of blood is observed in the flash well.
6. Attach required blood vacuum tubes (or syringe and pull plunger).
7. Remove the butterfly needle and cover sharp end with needle trap.
8. Cover the site with an adhesive bandage or gauze and tape dressing.
9. Dispose of the needle in the sharps box and further waste in the designated waste area.
10. Apply pressure to the site for 60 seconds to reduce the risk of hematoma formation.
11. Ensure dressing is tight to the skin to avoid aerosolization of blood products.

### 5.6.2 Material, mass, volume table: straight stick

Table 5.1 presents the estimated mass, volume, and common applications of straight stick equipment.

### 5.6.3 Intravenous Line

1. Don nitrile gloves.
2. Attach drip chamber and IV tubing to IV fluid bag.
3. Apply pressure bag over IV fluids and pressurize to remove air from attached IV tubing (may also invert bag to remove residual air from IV bag in 1g or partial g). In a microgravity environment ensure that a gravity independent air filter is attached to the IV tubing prior to where it attaches to the patient.

**TABLE 5.1** Estimated mass, volume, and uses of straight stick equipment.

| Item | Mass | Dimensions ($L \times W \times H$) | Indications |
|---|---|---|---|
| 2 × 2 Gauze | 1.5 g | 5 cm × 5 cm × 0.5 cm | Debris/fluid removal |
| 4 × 4 Gauze | 6 g | 10 cm × 10 cm × 0.5 cm | Debris/fluid removal |
| Butterfly needle | 5 g | 2 cm × 2.5 cm × 0.5 cm | Vascular access |
| BZK prep pad | 2 g | 6 cm × 6 cm × 0.25 cm | Surface cleaning |
| Cold compress | 140 g | 13 cm × 10 cm × 2 cm | Hematoma management |
| Medical tape | 20 g | 7 cm × 5 cm × 2 cm | Securing of lines; microgravity workspace |
| NSAID (ibuprofen, 400 mg) | 1 g | 0.4 cm × 0.4 cm × 0.2 cm | Procedural pain management |
| Sharps box | 80 g | 10 cm × 10 cm × 15 cm | Sharps storage |
| Tourniquet | 5 g | 45 cm × 2.5 cm × 0.25 cm | Facilitation of IV access |
| Tegaderm | 2 g | 6 cm × 7 cm × 0.1 cm | Clean adhesive covering |
| Vacutainer tube | 5 g | 1.5 cm × 1.5 cm × 7.5 cm | Sample collection |

*BZK*, Benzalkonium chloride; *H*, height; *IV*, intravenous; *L*, length; *NSAID*, nonsteroidal antiinflammatory drug; *W*, width.

4. Cap end, engage roll clip, or clip or kink tubing to stop fluid flow.
5. Attach an elastic tourniquet proximal to the site of vascular access.
6. Identify the vein or artery to be accessed.
7. Prep overlying skin with BZK wipe.
8. Insert the needle into the vein until a flash of blood is observed.
9. Attach the required blood vacuum tubes if drawing labs.
10. Remove the needle and dispose of it in a sharps container.
11. Cover the site with an adhesive bandage or gauze and tape dressing.
12. Attach IV catheter to the prepped tubing.
13. Dress with clear adhesive dressing to cover the access site and prevent mobilization of further free blood droplets.
14. Ensure dressing is tight to the skin to avoid aerosolization of any blood product. Add a self-adherent elastic bandage wrapping if needed for further site security.
15. Ensure flushing IV site twice daily (5cc NS) if not using but kept in place, or after each medication administration prior to locking if being used.

### 5.6.4 Material, mass, volume table: intravenous access

Table 5.2 presents the estimated mass, volume, and clinical uses of IV access equipment.

### 5.6.5 Central venous catheter

1. Don nitrile gloves, disposable procedure cap, and mask.
2. Identify site of vascular access using anatomic OR ultrasound guidance.
3. Prep overlying skin with BZK wipe utilizing a pattern of starting centrally and making overlying circles outward to create a space at least 5 cm in diameter. BZK will need to dry for at least 30 seconds.
4. Open vascular access kit to establish initial sterile field.
5. Open and preposition sterile gown and gloves.
6. Don sterile gown and gloves utilizing sterile handling technique.
7. Apply sterile drape to patient utilizing sterile technique with the window in the sterile drape encompassing the prepared area (ensure the prepared area circumference is larger than the window in the sterile drape prior to applying. If it is not, utilize additional BZK as previous).

**TABLE 5.2** **Estimated mass, volume, and uses of IV access equipment.**

| Item | Mass | Dimensions ($L \times W \times H$) | Indications |
|---|---|---|---|
| 2×2 Gauze | 2 g | 5 cm × 5 cm × 0.5 cm | Debris/fluid removal |
| 4×4 Gauze | 7 g | 10 cm × 10 cm × 0.5 cm | Debris/fluid removal |
| 10cc Syringe | 7 g | 12 cm × 2 cm × 2 cm | Medication/fluid administration |
| 18G Needle | 5 g | 12 cm × 0.3 cm × 0.3 cm | Access |
| 22G Needle | 4 g | 12 cm × 0.25 cm × 0.25 cm | Access |
| BZK prep pad | 2 g | 6 cm × 6 cm × 0.25 cm | Surface cleaning |
| Cold compress | 140 g | 13 cm × 10 cm × 2 cm | Hematoma management |
| IV fluids | 1000 g | 20 cm × 10 cm × 6 cm | Volume repletion, medication reconstitution, sterile ultrasound medium |
| IV tubing | 50 g | 200 cm × 0.5 cm × 0.5 cm | Medication and fluid administration |
| Medical tape | 20 g | 7 cm × 5 cm × 2 cm | Securing of lines; microgravity workspace |
| NSAID (ibuprofen, 400 mg) | 1 g | 0.4 cm × 0.4 cm × 0.2 cm | Procedural pain management |
| Pressure bag | 150 g | 25 cm × 20 cm × 5 cm | Large volume resuscitation |
| Sharps box | 80 g | 10 cm × 10 cm × 15 cm | Sharps storage |
| Sterile flush | 15 g | 12 cm × 2 cm × 2 cm | Sterile fluid for flushing lines, ultrasound medium |
| Tourniquet | 5 g | 45 cm × 2.5 cm × 0.25 cm | Facilitation of IV access |
| Tegaderm | 2 g | 6 cm × 7 cm × 0.1 cm | Clean adhesive covering |
| Ultrasound | 300 g | 16 cm × 6 cm × 4 cm | Access site identification |

*BZK*, Benzalkonium chloride; *G*, gauge; *H*, height; *IV*, intravenous; *L*, length; *NSAID*, nonsteroidal antiinflammatory drug; *W*, width.

8. In the sterile tray/field, remove the cap from each central line catheter port and flush with included sterile saline flush. Once filled, occlude each line with the tethered clamp prior to disengaging the sterile saline flush syringe, then immediately recap each port.
9. If using dynamic ultrasound guidance, apply sterile probe cover to ultrasound.
10. Apply sterile water to site of access to act as ultrasound medium (this will stay in place via surface tension).
11. Identify desired vascular landmark with ultrasound.
12. Gently swing lidocaine vial to collect fluid in one region of the vial.
13. Draw lidocaine from vial into 5 or 10cc syringe using 18G needle.
14. Remove 18G needle and replace with 22G needle.
15. Inject a wheal of lidocaine at the site of intended access immediately under the skin.
16. Advance the finder needle at the insertion site until return of venous blood.
17. Insert the guidewire through the finder needle.
18. Remove finder needle over guidewire, maintaining constant control of guidewire to prevent accidental dislodgement.
19. Use a scalpel to perform a skin nick overlying the guidewire.
20. Advance the dilator over the guidewire.
21. Remove the dilator and replace it with a CVC catheter.
22. Once the CVC catheter is inserted to the desired depth (see Table 5.2), remove the guidewire.
23. Withdraw blood from each port to ensure good blood return and the presence of dark, venous blood.
24. Flush each port with sterile saline as before.
25. Suture CVC in place.
26. Place a sterile dressing over CVC site.
27. Dispose of all needles and sharps in the sharps box and further waste in the designated waste area.

28. Ensure dressing is tight to the skin to avoid aerosolization of blood product.
29. Perform regular access site checks to assess for signs of irritation or infection.
30. Ensure flushing IV site twice daily (5cc NS) if not using but kept in place, or after each medication administration before locking if actively being used. All ports should remain flushed, locked, and capped when not in use.
31. Sterile transparent dressings should be changed every 5–7 days in sterile fashion or if the dressing becomes wet, dirty, or loose.
32. Gauze and sterile dressing should be changed every 48 hours in sterile fashion or if it becomes wet, dirty, or loose.

## 5.6.6 Material, mass, volume table: central venous catheter

Table 5.3 presents the estimated mass, volume, and clinical uses of central venous catheter equipment.

**TABLE 5.3** Estimated mass, volume, and uses of central venous catheter equipment.

| Item | Mass | Dimensions (*L* × *W* × *H*) | Indications |
|---|---|---|---|
| 2×2 Gauze | 2 g | 5 cm × 5 cm × 0.5 cm | Debris/fluid removal |
| 4×4 Gauze | 7 g | 10 cm × 10 cm × 0.5 cm | Debris/fluid removal |
| 10cc Syringe | 7 g | 12 cm × 2 cm × 2 cm | Medication/fluid administration |
| 18G Needle | 5 g | 12 cm × 0.3 cm × 0.3 cm | Access |
| 22G Needle | 4 g | 12 cm × 0.25 cm × 0.25 cm | Access |
| BZK prep pad | 2 g | 6 cm × 6 cm × 0.25 cm | Surface cleaning |
| Cold compress | 140 g | 13 cm × 10 cm × 2 cm | Hematoma management |
| Dilator | 10 g | 12 cm × 0.5 cm × 0.5 cm | Facilitate CVC catheter passage |
| Finder needle | 3 g | 7 cm × 0.5 cm × 0.5 cm | Guide initial vascular access |
| Guidewire (coiled) | 7 g | 6 cm × 6 cm × 0.5 cm | Catheter guide |
| Iris scissors | 25 g | 11 cm × 5 cm × 0.5 cm | Cutting suture ends |
| IV fluids | 1000 g | 20 cm × 10 cm × 6 cm | Volume repletion, medication reconstitution, sterile ultrasound medium |
| IV tubing | 50 g | 200 cm × 0.5 cm × 0.5 cm | Medication and fluid administration |
| Medical tape | 20 g | 7 cm × 5 cm × 2 cm | Securing of lines; microgravity workspace |
| Lidocaine (10 mL vial) | 20 g | 5 cm × 2 cm × 2 cm | Local analgesia |
| Needle driver | 40 g | 11 cm × 5 cm × 0.5 cm | Securing of access |
| NSAID (ibuprofen, 400 mg) | 1 g | 0.4 cm × 0.4 cm × 0.2 cm | Procedural pain management |
| Pressure bag | 150 g | 25 cm × 20 cm × 5 cm | Large volume resuscitation |
| Sharps box | 80 g | 10 cm × 10 cm × 15 cm | Sharps storage |
| Sterile flush | 15 g | 12 cm × 2 cm × 2 cm | Sterile fluid for flushing lines, ultrasound medium |
| Sterile probe cover (folded) | 15 g | 15 cm × 8 cm × 1 cm | Dynamic ultrasound guidance for complex access |
| Scalpel | 25 g | 18 cm × 3 cm × 1 cm | General cutting implement (e.g., skin nick) |
| Suture (3-0 nylon, single package) | 4 g | 8 cm × 4 cm × 0.5 cm | Securing of access |
| Tegaderm | 2 g | 6 cm × 7 cm × 0.1 cm | Clean adhesive covering |

*(Continued)*

**TABLE 5.3 (Continued)**

| Item | Mass | Dimensions (*L* × *W* × *H*) | Indications |
|---|---|---|---|
| Triple lumen catheter (coiled) | 20 g | 12 cm × 8 cm × 1 cm | See section above |
| Triple lumen catheter caps (each) | 1 g | 1 cm × 1 cm × 0.5 cm | Preventing blood backflow |
| Ultrasound | 300 g | 16 cm × 6 cm × 4 cm | Access site identification |

*BZK*, Benzalkonium chloride; *CVC*, central venous catheter; *G*, gauge; *H*, height; *IV*, intravenous; *L*, length; *NSAID*, nonsteroidal antiinflammatory drug; *W*, width.

## 5.6.7 Arterial line

1. Don nitrile gloves, facemask, and procedure cap.
2. Identify the site of vascular access with anatomic OR ultrasound guidance (e.g., radial artery, femoral artery).
3. Prep overlying skin with BZK wipe.
4. Flush the pressure transducer and connect to the pressure-bagged fluids, in a 1g or partial environment. In a microgravity environment ensure that a gravity independent air filter is attached to the IV tubing prior to where it attaches to the patient.
5. Cap or kink the pressure transducer line to prevent liberation of free sterile fluid.
6. Open an a-line kit in a sterile field.
7. Don sterile gloves.
8. If using dynamic ultrasound guidance, apply sterile probe cover to ultrasound.
9. Apply sterile water to site of access to act as ultrasound medium (this will stay in place via surface tension).
10. Identify desired vascular landmark with ultrasound.
11. Draw lidocaine from the vial into a 5 or 10cc syringe using an 18G needle.
12. Remove the 18G needle and replace with a 22G needle.
13. Deposit a wheal of lidocaine at the site of intended access, immediately under the skin.
14. Advance the finder needle at the insertion site until return of arterial blood.
15. Advance the guidewire through the finder needle.
16. Remove the finder needle over the guidewire, maintaining constant control of the guidewire to prevent accidental dislodgement.
17. Advance the a-line catheter over the wire and remove the wire.
18. Once a line catheter is inserted to the desired depth (see Table 5.2), remove guidewire.
19. Ensure the presence of pulsatile, arterial blood.
20. Connect a-line pressure tubing.
21. Suture the a-line in place.
22. Place a sterile dressing over the a-line site.
23. Zero the a-line pressure gauge and ensure the return of a good waveform with a visible dicrotic notch.
24. Dispose of all needles and sharps in the sharps box and further waste in a designated waste area.

## 5.6.8 Material, mass, volume table: arterial access

Table 5.4 presents the estimated mass, volume, and associated hazards of arterial access equipment.

## 5.6.9 Intraosseous access

1. Attach drip chamber and IV tubing to IV fluid bag.
2. Apply pressure bag over IV fluids and pressurize to remove air from attached IV tubing (may also invert bag to remove residual air from IV bag) in a 1g or partial environment. In a microgravity environment ensure that a gravity independent air filter is attached to the IV tubing prior to where it attaches to the patient.
3. Cap end, engage roll clip, or clip or kink tubing to stop fluid flow.
4. Attach IO needle (25 mm for proximal humerus, 45 mm for tibia) to IO needle driver.
5. Identify site of intended IO access (humeral head or proximal tibia).
6. Prep overlying skin with BZK wipe.
7. Draw lidocaine from the vial into a 5 or 10cc syringe using an 18G needle.

**TABLE 5.4** Estimated mass, volume, and hazards of arterial access equipment.

| Item | Mass | Dimensions (*L* × *W* × *H*) | Indications |
|---|---|---|---|
| 2×2 Gauze | 2 g | 5 cm × 5 cm × 0.5 cm | Debris/fluid removal |
| 4×4 Gauze | 7 g | 10 cm × 10 cm × 0.5 cm | Debris/fluid removal |
| 10cc Syringe | 7 g | 12 cm × 2 cm × 2 cm | Medication/fluid administration |
| 18G Needle | 5 g | 12 cm × 0.3 cm × 0.3 cm | Access |
| 22G Needle | 4 g | 12 cm × 0.25 cm × 0.25 cm | Access |
| Arterial line catheter | 10 g | 8 cm × 0.5 cm × 0.5 cm | Arterial access line |
| Arterial line transducer | 100 g | 10 cm × 6 cm × 3 cm | Real-time blood pressure monitoring |
| BZK prep pad | 2 g | 6 cm × 6 cm × 0.25 cm | Surface cleaning |
| Cold compress | 140 g | 13 cm × 10 cm × 2 cm | Hematoma management |
| Finder needle | 3 g | 7 cm × 0.5 cm × 0.5 cm | Guide initial vascular access |
| Guidewire (coiled) | 7 g | 6 cm × 6 cm × 0.5 cm | Catheter guide |
| Iris scissors | 25 g | 11 cm × 5 cm × 0.5 cm | Cutting suture ends |
| IV fluids | 1000 g | 20 cm × 10 cm × 6 cm | Volume repletion, medication reconstitution, sterile ultrasound medium |
| IV tubing | 50 g | 200 cm × 0.5 cm × 0.5 cm | Medication and fluid administration |
| Medical tape | 20 g | 7 cm × 5 cm × 2 cm | Securing of lines; microgravity workspace |
| Lidocaine (10 mL vial) | 20 g | 5 cm × 2 cm × 2 cm | Local analgesia |
| Needle driver | 40 g | 11 cm × 5 cm × 0.5 cm | Securing of access |
| NSAID (ibuprofen, 400 mg) | 1 g | 0.4 cm × 0.4 cm × 0.2 cm | Procedural pain management |
| Pressure bag | 150 g | 25 cm × 20 cm × 5 cm | Large volume resuscitation |
| Scalpel | 25 g | 18 cm × 3 cm × 1 cm | General cutting implement (e.g., skin nick) |
| Sharps box | 80 g | 10 cm × 10 cm × 15 cm | Sharps storage |
| Sterile flush | 15 g | 12 cm × 2 cm × 2 cm | Sterile fluid for flushing lines, ultrasound medium |
| Sterile probe cover (folded) | 15 g | 15 cm × 8 cm × 1 cm | Dynamic ultrasound guidance for complex access |
| Suture (3-0 nylon, single package) | 4 g | 8 cm × 4 cm × 0.5 cm | Securing of access |
| Tegaderm | 2 g | 6 cm × 7 cm × 0.1 cm | Clean adhesive covering |
| Ultrasound | 300 g | 16 cm × 6 cm × 4 cm | Access site identification |

*BZK*, Benzalkonium chloride; *G*, gauge; *H*, height; *IV*, intravenous; *L*, length; *NSAID*, nonsteroidal antiinflammatory drug; *W*, width.

8. Remove the 18G needle and replace with a 22G needle.
9. Deposit a wheal of lidocaine at the site of intended access, immediately under the skin.
10. Brace against a fixed surface, position the needle and driver at a 90 degree angle to the bone, apply firm pressure, and squeeze the driver actuator.
11. Apply consistent pressure and continue to squeeze the actuator until the needle hub is almost flush with the skin.
12. Remove needle driver from needle hub.
13. Remove the IO needle stylet and dispose in a sharps container.
14. Apply IO stabilizer to skin around IO.
15. Attach saline flush and push to ensure good flow through the IO.
16. If good flow, attach IO to the prepped tubing.

**TABLE 5.5** Estimated mass, volume, and hazards of IO access equipment.

| Item | Mass | Dimensions (*L* × *W* × *H*) | Indications |
|---|---|---|---|
| 2×2 Gauze | 2 g | 5 cm × 5 cm × 0.5 cm | Debris/fluid removal |
| 4×4 Gauze | 7 g | 10 cm × 10 cm × 0.5 cm | Debris/fluid removal |
| 10cc Syringe | 7 g | 12 cm × 2 cm × 2 cm | Medication/fluid administration |
| 18G Needle | 5 g | 12 cm × 0.3 cm × 0.3 cm | Access |
| 22G Needle | 4 g | 12 cm × 0.25 cm × 0.25 cm | Access |
| BZK prep pad | 2 g | 6 cm × 6 cm × 0.25 cm | Surface cleaning |
| Cold compress | 140 g | 13 cm × 10 cm × 2 cm | Hematoma management |
| IO driver | 400 g | 16 cm × 6 cm × 4 cm | IO insertion |
| IO needle | 10 g | 8 cm × 2 cm × 2 cm | IO insertion |
| IO stabilizer | 10 g | 7 cm × 7 cm × 1 cm | IO insertion |
| IV fluids | 1000 g | 20 cm × 10 cm × 6 cm | Volume repletion, medication reconstitution, sterile ultrasound medium |
| IV tubing | 50 g | 200 cm × 0.5 cm × 0.5 cm | Medication and fluid administration |
| Lidocaine (10 mL vial) | 20 g | 5 cm × 2 cm × 2 cm | Local analgesia |
| Medical tape | 20 g | 7 cm × 5 cm × 2 cm | Securing of lines; microgravity workspace |
| NSAID (ibuprofen, 400 mg) | 1 g | 0.4 cm × 0.4 cm × 0.2 cm | Procedural pain management |
| Pressure bag | 150 g | 25 cm × 20 cm × 5 cm | Large volume resuscitation |
| Sharps box | 80 g | 10 cm × 10 cm × 15 cm | Sharps storage |
| Sterile flush | 15 g | 12 cm × 2 cm × 2 cm | Sterile fluid for flushing lines, ultrasound medium |

Abbreviations: *BZK*, Benzalkonium chloride; *H*, height; *IO*, intraosseous; *IV*, intravenous; *L*, length; *NSAID*, nonsteroidal antiinflammatory drug; *W*, width.

### 5.6.10 Material, mass, volume table: intraosseous access

Table 5.5 presents the estimated mass, volume, and associated hazards of IO access equipment.

## 5.7 Summary, recommendations, and considerations for the future

At present, vascular access in space is an uncommon occurrence. However, as the spacefaring population becomes more heterogeneous, as time in low-Earth orbit becomes more common, and as long-duration missions become the primary focus of national space agencies, these procedures will become increasingly necessary. Current use of vascular access in space is primarily relegated to research studies occurring in controlled situations. Over the next one to two decades, these procedures may become more routine and more necessary.

Primary selection of an access site and method is determined by clinical indication. Arterial access will remain uncommon in spaceflight, as most indications are for intensive care and critical care infrastructure is costly, heavy, and infrequently needed.

The type of venous access will be determined by the clinical objective. For one-time blood draws, straight needle venipuncture is the procedure of choice. For repeat blood sampling or IV instillation of therapeutics, placement of an IV is indicated. For critically ill patients or central monitoring, a central line will be required.

Insertion site cleaning is a mainstay of infection mitigation. In spaceflight, the choice of agent is limited by flammability and volatility considerations. Additionally, material handling is made more logistically complex by the microgravity environment. Finally, droplet precautions remain a concern throughout all procedures of vascular access, as any blood that escapes the procedure site will be freed into the cabin where it may enter eyes or other mucous membranes, posing an increased risk of communicable disease.

From a spaceflight operations perspective, it is imperative to consider the mission duration and profile when planning the manifesting of vascular access supplies. For most trips to low-Earth orbit, ICU-level care will be beyond the scope of expected medical capabilities, and arterial and central venous access will almost certainly be unwarranted. Many IV medications can be substituted for IM alternatives, and peripheral IVs may even be beyond the scope of medical care for such flights.

For long-duration missions to the Moon and Mars, however, the breadth of medical capability must be expanded. With a minimum Mars mission duration of over two and a half years, a host of medical capabilities never previously manifested in human spaceflight will be required, possibly including those described here.

## References

Bell, T., O'Grady, N.P., 2017. Prevention of central line-associated bloodstream infections. Infectious Disease Clinics of North America 31 (3), 551–559. Available at https://doi.org/10.1016/j.idc.2017.05.007. Epub 2017 Jul 5. PMID: 28687213; PMCID: PMC5666696.

Buckey, J.C.J., Gaffney, F.A., Lane, L.D., Levine, B.D., Watenpaugh, D.E., Wright, S.J., et al., 1996. Central venous pressure in space. Journal of Applied Physiology (1985) 81 (1), 19–25. Available at https://doi.org/10.1152/jappl.1996.81.1.19 PMID: 8828643.

Buetti, N., Marschall, J., Drees, M., Fakih, M.G., Hadaway, L., Maragakis, L.L., et al., 2022. Strategies to prevent central line-associated bloodstream infections in acute-care hospitals: 2022 update. Infection Control and Hospital Epidemiology the Official Journal of the Society of Hospital Epidemiologists of America 43 (5), 553–569. Available at https://doi.org/10.1017/ice.2022.87. Epub 2022 Apr 19. PMID: 35437133; PMCID: PMC9096710.

Elshikh, S.T., Thomas, B., Taha, M., Shaikh, A.M., Alkhazendar, A.H., Awan, M., et al., 2025. The accuracy of point-of-care ultrasound for peripheral intravenous cannulation in emergency and trauma patients: a systematic review. Cureus 17 (5), e83625. Available at https://doi.org/10.7759/cureus.83625. PMID: 40486335; PMCID: PMC12142272.

Formanek, A., Townsend, J., Ottensmeyer, M.P., Kamine, T.H., 2024. A novel 3D-printed gravity-independent air-eliminating filter for rapid intravenous infusions. Aerospace Medicine and Human Performance 95 (6), 327–332. Available at https://doi.org/10.3357/AMHP.6376.2024. PMID: 38790129.

O'Grady, N.P., 2023. Prevention of central line-associated bloodstream infections. The New England Journal of Medicine 389 (12), 1121–1131. Available at https://doi.org/10.1056/NEJMra2213296 PMID: 37733310.

Gulati, U., Ray, K., Dasgupta, S., Jerusik, B., 2021. Comparison of first-pass peripheral intravenous cannulation using a handheld ultrasound device to using a traditional high-end ultrasound system: a randomized controlled trial. Annals of Emergency Medicine 78 (4).

Kim, David S., Vaquer, Sergi, Mazzolai, Lucia, Roberts, Lara N., Pavela, James, Watanabe, Manabu, Weerts, Guillaume, Green, David A., 2021. The effect of microgravity on the human venous system and blood coagulation: a systematic review. Experimental Physiology 106, 1149–1158. Available at https://doi.org/10.1113/ep089409 Epub 2021 Mar 25. PMID: 33704837.

Maki, D.G., Kluger, D.M., Crnich, C.J., 2006. The risk of bloodstream infection in adults with different intravascular devices: a systematic review of 200 published prospective studies. Mayo Clinic Proceedings 81 (9), 1159–1171. Available at https://doi.org/10.4065/81.9.1159. PMID: 16970212.

Marshall-Goebel, K., Laurie, S.S., Alferova, I.V., Arbeille, P., Auñón-Chancellor, S.M., et al., 2019. Assessment of jugular venous blood flow stasis and thrombosis during spaceflight. JAMA Network Open 2 (11), e1915011. Available at https://doi.org/10.1001/jamanetworkopen.2019. 15011. Erratum in: JAMA Netw Open. 2020 Jan 3;3(1):e1920195. https://doi.org/10.1001/jamanetworkopen.2019.20195. PMID: 31722025; PMCID: PMC6902784.

Parienti, J.J., Mongardon, N., Mégarbane, B., Mira, J.P., Kalfon, P., Gros, A., et al., 2015. 3SITES study group. Intravascular complications of central venous catheterization by insertion site. The New England Journal of Medicine 373 (13), 1220–1229. Available at https://doi.org/10.1056/NEJMoa1500964. PMID: 26398070.

Stegink, L.D., 1983. Amino acids in pediatric parenteral nutrition. Solutions infused–lessons learned. American Journal of Diseases of Children 137 (10), 1008–1016. Available at https://doi.org/10.1001/archpedi.1983.02140360068020. PMID: 6412539.

Thrush, D.N., Belsole, R., 1995. Radial nerve injury after routine peripheral vein cannulation. Journal of Clinical Anesthesia 7 (2), 160–162. Available at https://doi.org/10.1016/0952-8180(94)00029-4. PMID: 7598927.

Chapter 6

# Hemorrhage control

**Peter Alexandrov**[1] **and Amanda N. Prine**[2]
[1]*University of Florida, Gainesville, FL, United States,* [2]*Department of Emergency Medicine, University of Florida, Gainesville, FL, United States*

## 6.1 Introduction

Hemorrhage control and wound repair in spaceflight present immense technical challenges due to the microgravity environment, where the normal gravity that we use to assist us in control and repair techniques is noticeably absent. However, it is still possible, and indeed necessary, to include this capability on board a spacecraft to appropriately care for astronauts. In this chapter, we will review the foundation of bleeding and hemostasis and provide a procedural guide to equip the reader with the knowledge to stop hemorrhage and repair the underlying injury. Certain types of hemorrhage are outside of the scope of this text, namely internal bleeding that requires a surgical or interventional radiology approach to repair. While invasive surgical approaches will undoubtedly become available eventually, the next 10–15 years of spaceflight are not likely to be able to take advantage of them.

## 6.2 Indications and efficacy

### 6.2.1 What is a hemorrhage, anyway?

Hemorrhage is bleeding that is life-threatening. The recognition and management of hemorrhage is often based on three attributes:

1. The rate of bleeding,
2. The location of bleeding,
3. The procedural environment.

The environment will provide a critical context: is the patient in an austere environment, such as climbing a mountain alone? Are they onboard a large spacecraft with an advanced sick bay? Are they exploring a valley on Mars with only first aid provisions or no provisions at all? Considerations like these often determine the capabilities and instruments available for hemorrhage control and repair. Fortunately, the fundamental approach to hemorrhage control is the same:

1. Stop the bleeding expeditiously,
2. Repair the wound if possible,
3. Minimize risk to the spacecraft and crewmembers (Figs. 6.1 and 6.2).

Different types of bleeding demand different responses. A small, stagnant pool of blood in a wound bed that is not increasing in size should elicit a different response than a large bleeding wound or one with pulsatile blood flow. In the former case, the body has adequately stopped bleeding, but in the latter, the body is unlikely to stop the bleeding by its own and often requires medical intervention to ensure the patient's survival. The location of bleeding also evokes a sense of urgency. Bleeding from the distal forearm or ankle is much less likely to cause a life-threatening hemorrhage than a proximal arm or thigh injury, which can kill within minutes if a major artery or vein is lacerated. Ultimately, in spaceflight, the single best definition of hemorrhage is bleeding that will not stop without immediate action by the rescuer.

Hemorrhage is an immediate life threat and must be controlled right away. The initial steps cannot wait for ground support, particularly in situations with communication delays longer than a few seconds, such as on a Mars mission, or even in low Earth orbit flights. Interplanetary spaceflight and future settlements on Mars represent a completely different communication paradigm for medical telemetry; Martian two-way communication would take 8–40 minutes, depending

**Procedural Medicine for Spaceflight. DOI:** https://doi.org/10.1016/B978-0-443-27552-4.00010-5

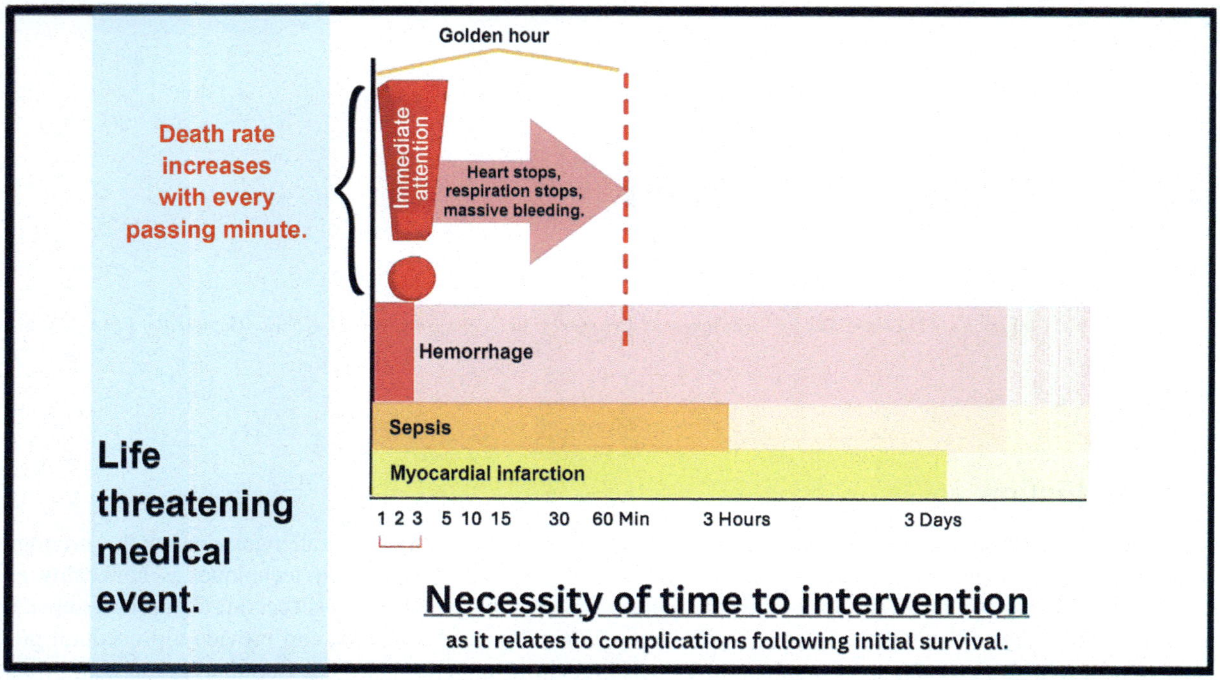


FIGURE 6.1 **Hemorrhage control timeline.** This chart illustrates the time sensitivity of treating hemorrhage. While other conditions must be treated within hours, hemorrhage can be lethal within seconds to minutes.

| Condition | Time to intervention | Key actions |
|---|---|---|
| Hemorrhage | Immediate | Direct pressure, tourniquet, hemostasis |
| Sepsis (with shock) | Within 1 hour | Antibiotics and IV fluids |
| Sepsis (without shock) | Within 3 hours | Assessment and antibiotics |
| Myocardial infarction (STEMI) | Within 90 minutes | Percutaneous Intervention (PCI) if available |
| Myocardial infarction (STEMI) | Within 120 minutes | If PCI is not available, initiate fibrinolytics |

FIGURE 6.2 **Time to intervention per condition.** This chart reinforces the immediacy in treating hemorrhage in relation to other emergent conditions.

on the position of Earth and Mars. As a condition requiring hyperacute management, hemorrhage control capability is directly determined by the immediately available crew and the medical kit. For these reasons the necessary skills should be part of a curriculum taught to all crew members and practiced regularly.

### 6.2.2 Hemostasis

The process of hemostasis occurs naturally but is reliant upon a delicate balance of proteins, enzymes, and platelets, including those that comprise the coagulation cascade. The ultimate purpose of this "cascade" is to stabilize the initial platelet plug that forms in primary hemostasis after an injury. The hematologic system must maintain a balance of coagulation and fibrinolysis, with disruptions leading to a "prothrombotic" or "prohemorrhagic" state. Making things more complex, the two states can coexist, such as in severe infections or trauma.

Some injuries exceed the body's capability to achieve hemostasis. In spaceflight, the majority of life-threatening trauma is likely to be blunt in nature based on historical events, with a reduced likelihood of injuries resulting from penetrating weapons (4). However, the threat of trauma from penetrating injury caused by micrometeorite, pressure-vessel explosion,

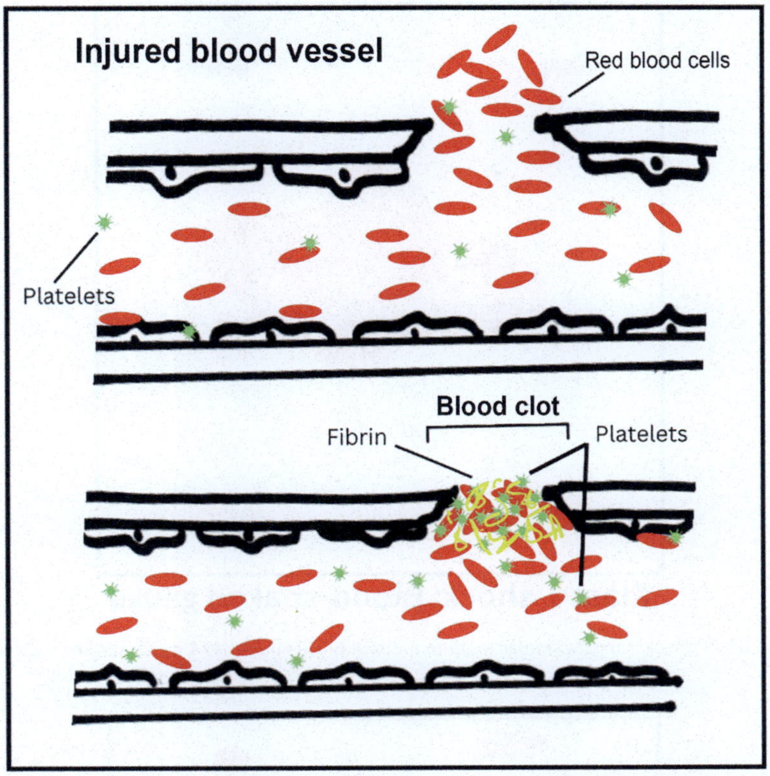


FIGURE 6.3 **Injured blood vessel.** Traumatic injury to blood vessels with an example of vessel spasm and clot composed of platelets and fibrin.

or accidental lacerations, cannot be discounted. Furthermore, less severe injuries resulting in minor trauma will likely be as common as on Earth.

When a laceration occurs, several things happen. First, tissue, including a blood vessel, is damaged. This exposes the various layers of the blood vessel, and it results in vessel spasm (see Fig. 6.3). This acts to limit the rate of blood flow and exposes a layer called the subendothelial matrix. Platelets adhere to this site and activate the coagulation cascade, ultimately forming a fibrin matrix that traps blood cells and acts to help seal the injury. The end result in a normally functioning system is a spasmed blood vessel with a fibrin and platelet plug that stops blood flow from the wound. Some of the tools in our repertoire of hemorrhage control involve taking advantage of this coagulation cascade by embedding specific chemicals into wound-packing and bandage materials that promote the formation of a fibrin and platelet mixture we call a blood clot.

In some cases, the damage is substantial enough that this process is insufficient, and hemorrhage ensues.

## 6.2.3 Classification of hemorrhage

### *6.2.3.1 Degrees of hemorrhage*

The severity of hemorrhage on Earth's 1G gravitational environment is categorized by classes of shock, manifested by changes in vital signs that arise from alterations in circulating blood volume. These scaling symptoms are difficult to apply to the space-traveler in a microgravity environment due to changes in how the body is able to perceive acute blood volume loss in space. Even the appearance of blood differs in altered gravitational environments. While in normogravity, blood will pool dependently; in microgravity, it aggregates and floats, presenting a hazard throughout the cabin to people and spacecraft systems. Even so, knowledge of the classical stages of shock will provide a known context for any hemorrhage encountered in space.

Class 1 shock is defined as loss of less than 750 ml of blood, or 15% of circulating blood volume. This volume of hemorrhage usually does not result in changes in vital signs; the person's heart rate is typically less than 100 beats per minute (bpm) and they have a normal blood pressure (Fig. 6.4).

Class 2 shock is 750–1500 ml of blood, or up to 15%–30% of circulating blood volume. The body begins to compensate, usually with increased heart rate, decreased urine output, and signs of anxiety, but blood pressure typically

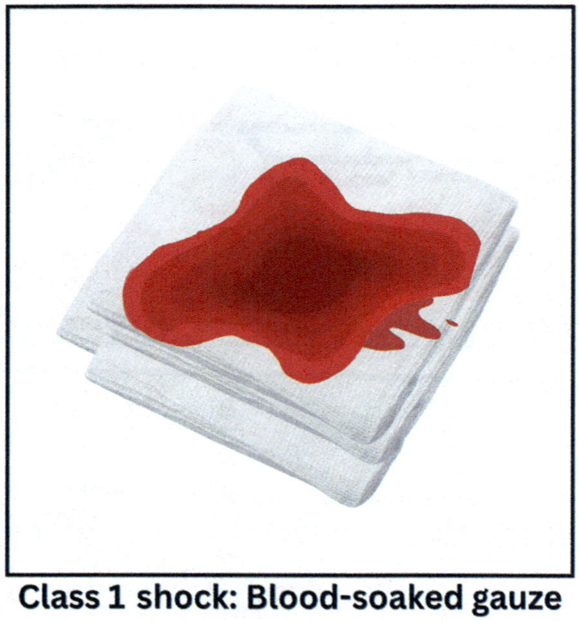


FIGURE 6.4 **Class 1 shock.** Class 1 shock: amount similar to up to a dozen blood-soaked 4×4 gauze stacked together.

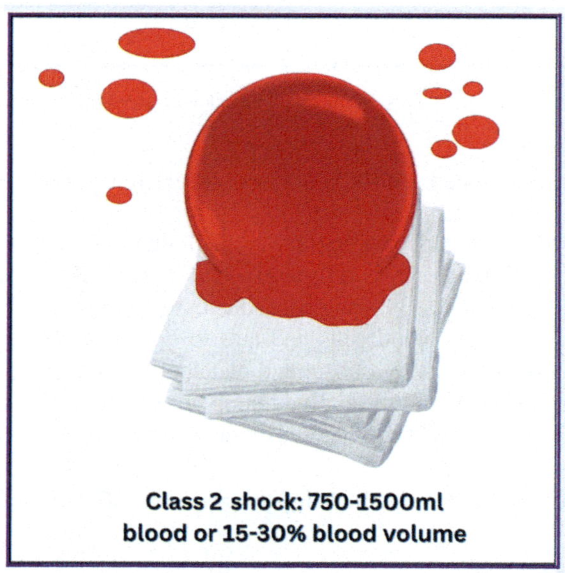


FIGURE 6.5 **Class 2 shock.** Class 2 shock: Similar in amount to a few dozen blood-soaked stacked 4×4 gauze, or a sphere 13 cm in diameter.

remains normal. A patient with this amount of blood loss does not need a blood transfusion unless there is ongoing or anticipated blood loss or other special considerations. Class 2 shock can be visualized by several stacks of bloody gauze or a sphere with a diameter of roughly 13 cm (4.6 inches) (Fig. 6.5).

Class 3 shock is 1,500–2,500 ml, or 30%–40% of circulating blood volume. The patient has a high heart rate (unless masked by drugs) and low blood pressure, and exhibits more profound mental status changes, such as confusion due to cerebral hypoperfusion. Class 3 shock needs immediate blood product administration to prevent or reverse end-organ damage.

Class 4 shock involves greater than 2500 ml or 40% of circulating blood volume loss. This degree of shock may result in nearly irreversible changes in the coagulation cascade that lead to hemodynamic collapse and death even with prompt blood volume resuscitation.

## 6.3 Incidence

### 6.3.1 Injury risk and condition incidence for hemorrhage

Abrasion and lacerations: The injury risk and incidence of minor lacerations requiring repair will likely change as more people travel to space and new mission paradigms develop. A review by Scheuring et al. (9) shows a minor abrasion and laceration incidence so far in spaceflight of approximately 3.11 events/person-year in spaceflight. The majority of these injuries occur on the hand and vary by activity, from shaving to moving through the spacecraft to stowage of gear. They are more likely to occur during specific higher-risk events, such as egress from the vehicle or during extravehicular activities, which mostly imparts a risk of abrasion, notably on the hands. Apollo 12 Astronaut Alan Bean famously suffered a head laceration when a 16 mm camera came loose and struck him on his head during the jolting ride of reentry. This caused a minor laceration that required sutures at a hospital after capsule egress. No abrasions or lacerations to date have caused a loss of mission or death.

Hemorrhage: It is likely that a major hemorrhagic event will occur in spaceflight eventually if given enough time. There are no documented cases of hemorrhage in space thus far. This will change as the number of people flying increases and mission durations grow longer. Modeling of a four-person, 923-day "standard Mars mission," Antonsen et al. (10) showed a probability of a loss-of-crewmember life in 1 of every 935 missions due to hemorrhage shock.

## 6.4 Spaceflight-specific considerations

### 6.4.1 Challenges of hemorrhage control in spaceflight

As noted above, spaceflight renders the standard terrestrial classification of shock inappropriate, although it remains a helpful reference in correlating volume of blood loss with likelihood of end-organ damage. Physiological changes due to microgravity make the classic clinical presentation of shock unreliable for predicting the amount of blood lost. This is due to a myriad of physiological changes that occur in spaceflight in response to the loss of a gravity vector and the closed-loop environment of the spacecraft itself. This closed-loop environment causes an increased $CO_2$ concentration that in conjunction with microgravity may exacerbate hemorrhage through changes in vascular tone. Ultimately, this impairment along with vessel wall changes and long-term exposure to ionizing radiation may alter the relationship between volume of hemorrhage, vital signs, and symptomatology

Many spaceflight induced physiological changes are already characterized in healthy individuals; in those who have chronic illnesses, the challenges in managing medical emergencies, including lacerations and hemorrhage, are expected to become more difficult, increasing the relative benefits from future technologies, such as artificial gravity stations. (Fig. 6.6).

**FIGURE 6.6 Artificial gravity space station.** Difficulties in managing hemorrhage in future space stations will be reduced when artificial gravity is introduced.

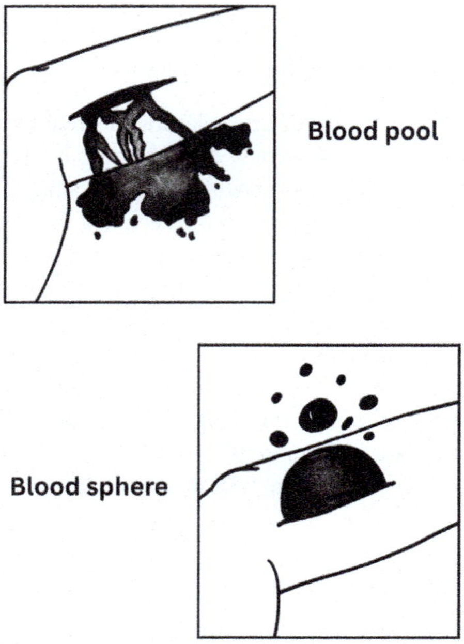


FIGURE 6.7 **Appearance of bleeding on Earth vs in spaceflight.** A bleeding wound in a gravitational environment will pool but in microgravity may dome and spray into the cabin. Bleeding in microgravity can be an immediate threat to other crewmember's health and to spacecraft systems.

### 6.4.2 How injury characteristics change these considerations

The vital signs of a person in hemorrhagic shock in spaceflight will likely not reflect the stage of shock as reliably as on Earth. Moreover, the operational constraints of the spaceflight environment will likely decrease the survival rate due to two major factors:

1. Procedural difficulty in microgravity,
2. Current lack of blood transfusion capabilities.

On Earth, we use gravity to help us treat shock. We can use gravity to help compress wound or elevate the legs to increase blood return to the thorax. In microgravity, due to Newton's Third Law, correct orientation, immobilization, and cantilever action is needed to ensure pressure can be maintained on a wound or to ensure a procedure, such as a laceration, staple repair, or tourniquet tightening, can be correctly performed.

Hemorrhage in spaceflight also creates a risk of blood products entering the cabin environment uncontrolled in microgravity, potentially causing inhalational injury, eye injury, infection exposure, electronic short-circuiting, or environment control and life support system malfunction due to liquid ingestion and clogged filtration. While on Earth blood will tends to pool dependently, in spaceflight, with pressure behind it, blood is more likely float into the cabin or pool into a dome on the wound. A sufficiently powerful and mobile suction device or a sufficient quantity of absorbent materials (e.g., a towel or shirt) should be made immediately available to allow the treating crewmember to contain blood products (Fig. 6.7).

Besides these operational considerations, laceration repair and hemorrhage control are largely the same as in terrestrial medicine. The general approach to hemorrhage control remains the same, with emphasis on immediate hemostasis and blood containment.

## 6.5 Decision making and procedures

The following section details decision-making and methodology for stopping hemorrhage. All of these techniques have been used successfully to stop hemorrhage. They all act to slow or stop the rate of bleeding to give the body's own pro-thrombotic mechanisms time to develop a stable clot. This is achieved through techniques of pressure application, the use of dressings (with or without pro-coagulation chemicals in it), or stopping blood flow to the injury by applying a tourniquet. While the success of many techniques is aided by experience, the application of a tourniquet, wound packing, basic epistaxis control, and procedures such as simple sutures and stapling can be performed with even minimal training.

In the event of a crewmember experiencing hemorrhage, first ensure vehicle safety, then:

1. Call out to crewmembers for help and ask them to bring necessary equipment and to notify ground support as soon as practical.
2. Apply one of the hemorrhage control techniques below with one hand behind the crewmember as support.
3. Instruct other crewmembers to obtain necessary equipment as you maintain the hemorrhage control technique or help translate the rescuer and the affected crewmember to an area where the rescuer and the patient can be secured within a restraint system
4. Begin fluid control via absorbent material or (suction device), or request an assistant do so.
5. Strategize on wound repair technique.

### 6.5.1 One-finger method for hemorrhage control

Indication for wound finger hemorrhage control: A wound with a single point of hemorrhage (Fig. 6.8).

The one-finger method of hemorrhage control can be used in a small or large wound bed with a point hemorrhage that may be arterial or venous in origin. The critical point in the successful use of one-finger hemorrhage control is to avoid using several fingers or the palm of their hand, as this is very often inadequate and the tissue deficit may not be easily covered. One finger will suffice using light to medium pressure. If the wound bed is slippery, fine stabilization of the finger will be necessary. The one-finger method can help buy time to promote clot formation, to signal for additional help from a crew member, and to formulate a plan for the next steps in control.

Procedure: Place one finger over the area of continued bleeding with one hand for support behind the crewmember. Do not remove finger from the area for at least 5–10 minutes after cessation of active bleeding. Call for help in obtaining medical supplies and attempt to translate the patient to an appropriate location in the ship from which to render additional treatment (Fig. 6.9).

### 6.5.2 Wound packing

Indication: Any deep wound hemorrhage. These often result from a penetrating injury such as a deep laceration or a projectile.

Packing a wound may be necessary to help stop bleeding and allow clots to form at the vascular bed. Packing can be done on the extremities but is more commonly done in the groin, buttocks, armpits, shoulders, and neck (Fig. 6.10 There are many different types of packing materials available, each with its own physical characteristics and special properties. They are all lightweight and relatively low volume. Packing a wound with materials (see Fig. 6.11), whether it be with

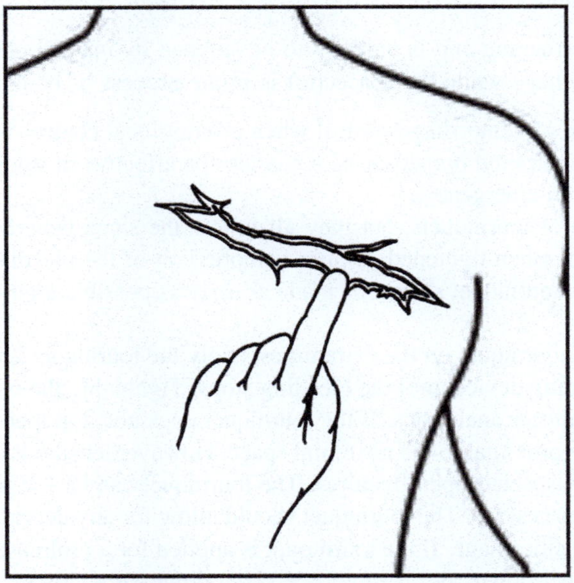

FIGURE 6.8 **Point hemorrhage control.** One finger method of point hemorrhage control involves applying gentle and direct pressure with the pad of a finger on the source of bleeding, while using the other hand to anchor the rescuer to the crewmember.

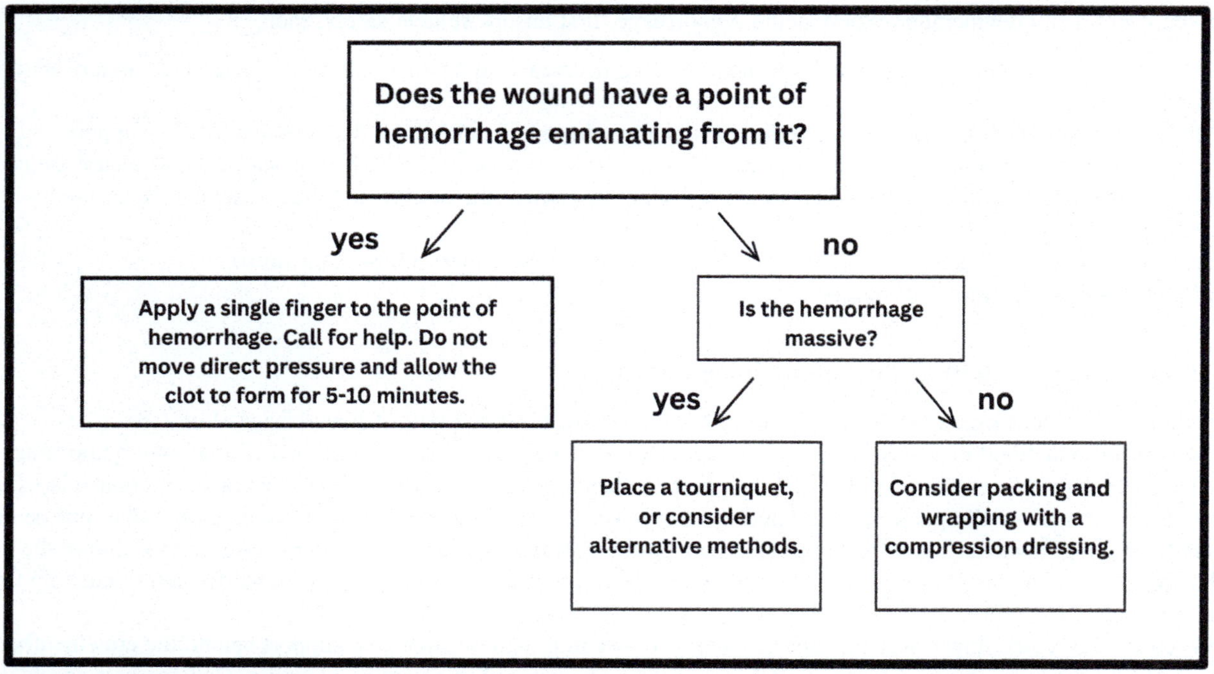


FIGURE 6.9 Point hemorrhage decision making flowchart.

sterile gauze pads or a gauze roll, or with hemostatic bandages or sponge-gel injection fillers (such as X-stat), will always follow the same principle: the hemostatic material is applied from the deepest point of the wound first to avoid formation of a large clot that can cause rebleeding later and serve as a nidus for infection (Figs. 6.12 and 6.13; Table 6.1).

Procedure: Take a finger and the end of the packing material and place it deep into the wound, repeating until the packing is at the top of the wound. Follow with an overlying 4×4 gauze or similar pad and wrap with a bandage. If using a gel such as X-Stat, ensure that the gel or foam is applied first at the bottom of the wound.

### 6.5.3 Tourniquets

Indications for tourniquet placement:

1. Any deep wound that is hemorrhaging and is unlikely to be stopped by manual compression or packing.
2. Any wound that causes hemorrhage while the spacecraft is in the process of being safeguarded in any emergency.

Tourniquets are the cornerstone of hemorrhage control when prompt local control by finger point method, packing, or wrapping is unlikely to provide the desired homeostasis. It can also be effective in stopping bleeding to allow time for the spacecraft to be safed in case of an emergency.

There are many manufacturers of tourniquets, but they all follow the same general concept: it is a band of material wrapped around an extremity that, when tightened, applies compression of the bleeding artery or vein against bone (see Fig. 6.14). The decision to place a tourniquet should be made as early as possible after the injury requiring a tourniquet is identified (Fig. 6.15).

Procedure: For application of a tourniquet on the extremities, apply the tourniquet 2 inches (5 cm) above the injury site and tighten the windlass or tightening device until the bleeding stops. Tightening the windlass will cause severe pain, and it will likely be necessary to administer analgesics. If the tourniquet does not stop bleeding, put on a second tourniquet 2 inches above the first, if necessary proximal to the next joint space. However, ensure it is not placed over a joint space and *never* over any broken bones or suspected open fractures. The tourniquet can be left tightened for 2 hours and extended longer to 6 hours if operationally necessary. The tourniquet should allow for any debridement of the more distal wound or ligation of vessels producing the hemorrhage. If the tourniquet is applied for a prolonged time (>2 hours), great care must be taken when taking the tourniquet down, and it should be taken down very slowly, over the course of many minutes. Taking down the tourniquet could reinitiate bleeding and allow metabolic byproducts of anaerobic respiration and cellular death, such as a large volume of potassium, to circulate back to the heart. If the capabilities of the onboard medical system

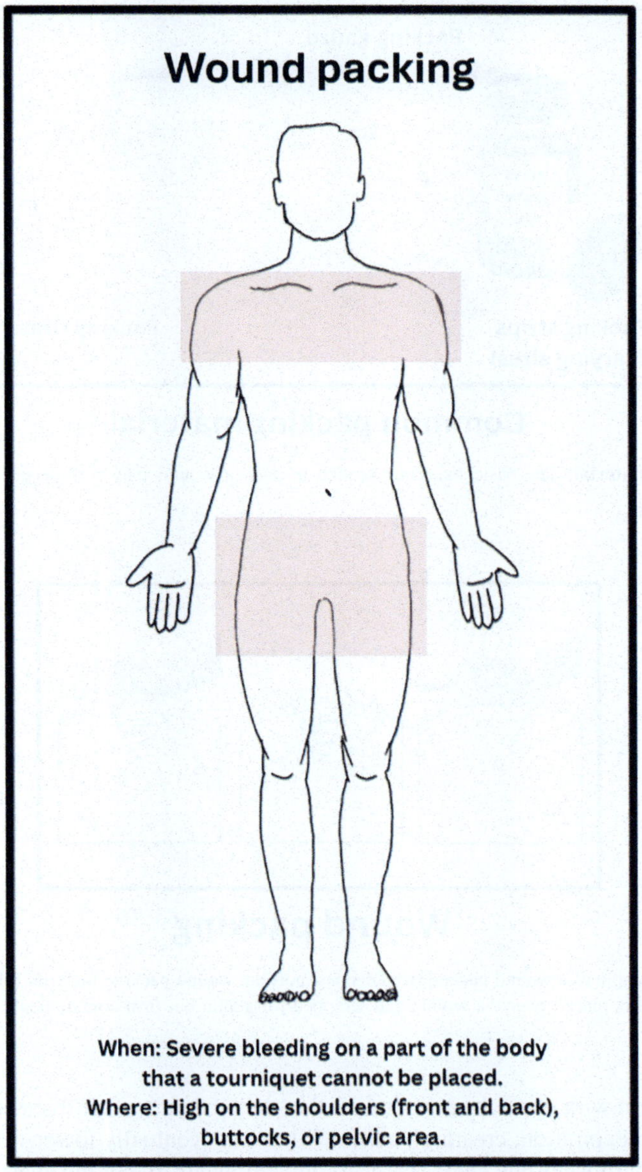


FIGURE 6.10 **Wound packing.** These highlighted areas of the body show where wound packing may be of greatest benefit, as a tourniquet is unlikely to be successfully applied these areas.

allow, the patient should be on the cardiac monitor with peripheral venous access, and medications made immediately available that help stabilize the heart and manage blood potassium levels, including sodium bicarbonate, calcium gluconate, insulin, and normal saline with dextrose. In *some* cases, this much more advanced medical system may not be available, *but in all cases*, a tourniquet is the most appropriate and best treatment for a large hemorrhage from an extremity (Fig. 6.16; Table 6.2). In a situation where definitive care of the injury is not available within 6-12 hours, limb loss should be expected and the tourniquet left up until an amputation can be performed.

### 6.5.4 Chest seals

Indication for chest seal placement: Any wound on the thorax that has a chance of violating the pleural space. More advanced management of thoracic injuries are discussed in Chapter 11.

Chest seals should be employed when there is a chest wound with suspected continuity with the pleural space. Typically, these wounds bubble with blood or serous fluids and may make a hissing or sucking sound. The absence of these sounds however does not mean it is not an open chest wound, and so unless obviously superficial, the wound should

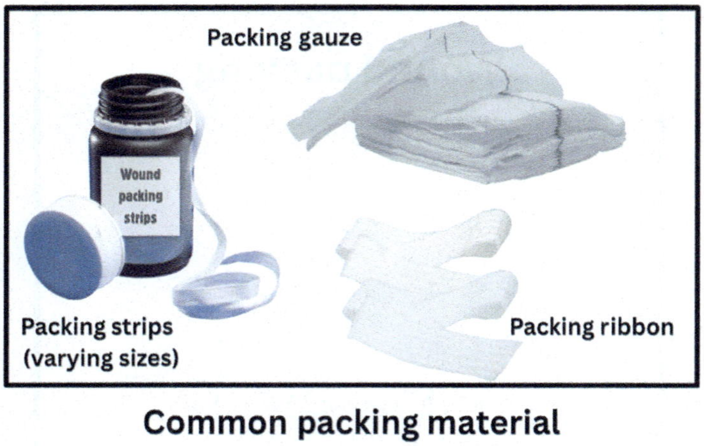


FIGURE 6.11 **Common packing material.** The basic materials needed to pack a wound may include packing strips, gauze, ribbons, or special injectable fillers such as X-Stat.

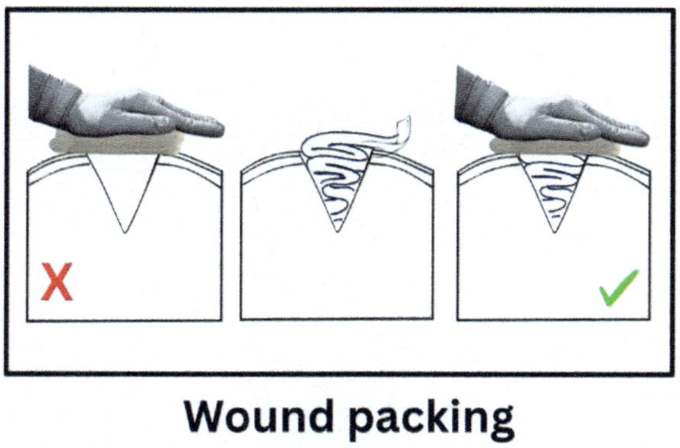


FIGURE 6.12 **Wound packing.** Example of a wound packed from deepest portion, wound packing material folded over, and the wound bed filled up. Then, a 4×4 gauze or similar material is placed over the wound and wrapped for gentle but firm containment.

be covered. The clinical problem with open chest wounds is that they may allow air to enter the chest cavity on inspiration but prevent air from exiting on expiration, creating "tension" that can eventually lower venous return and cardiac output, leading to cardiac arrest. The purpose of the chest seal is to only allow air to exit the chest cavity through the seal. Recall that the diaphragm is curved and may go as low as the umbilicus on the crewmember's sides. As a result, what may look like an abdominal wound may also be a chest wound (Fig. 6.17

Place the seal over a clean and dry chest wound and cover on three corners. This will allow it to serve as a flutter valve, allowing air to come out on expiration and seal on inspiration. If there is no alternative, a petroleum-soaked gauze or similar occlusive dressing may be applied (although not as ideal). There are multiple commercial products available.

Procedure: Identify chest wall defect. Dry the wound edges, then immediately place a chest seal (Fig. 6.18; Table 6.3).

### 6.5.5 Persistent wound bleeding

Indication for using gauze and bandages with pro-coagulant chemicals: Any wound with continued oozing of blood for longer than 5 minutes after applied pressure.

Some wounds are located in specific areas where many small capillary, arteriolar, and venular beds cause persistent bleeding. Sometimes, attempts at hemorrhage control of the wound bed via pressure are unsuccessful. Persistent light to moderate pressure, with the help of gauze and bandages that are impregnated with hemostatic medications and compounds, are very useful in these situations (Table 6.4

Tranexamic Acid, or TXA, may also be applied to a normal nonhemostatic gauze and can stop hemorrhage as well. It comes in a bottle and can be drawn up and soaked onto the gauze.

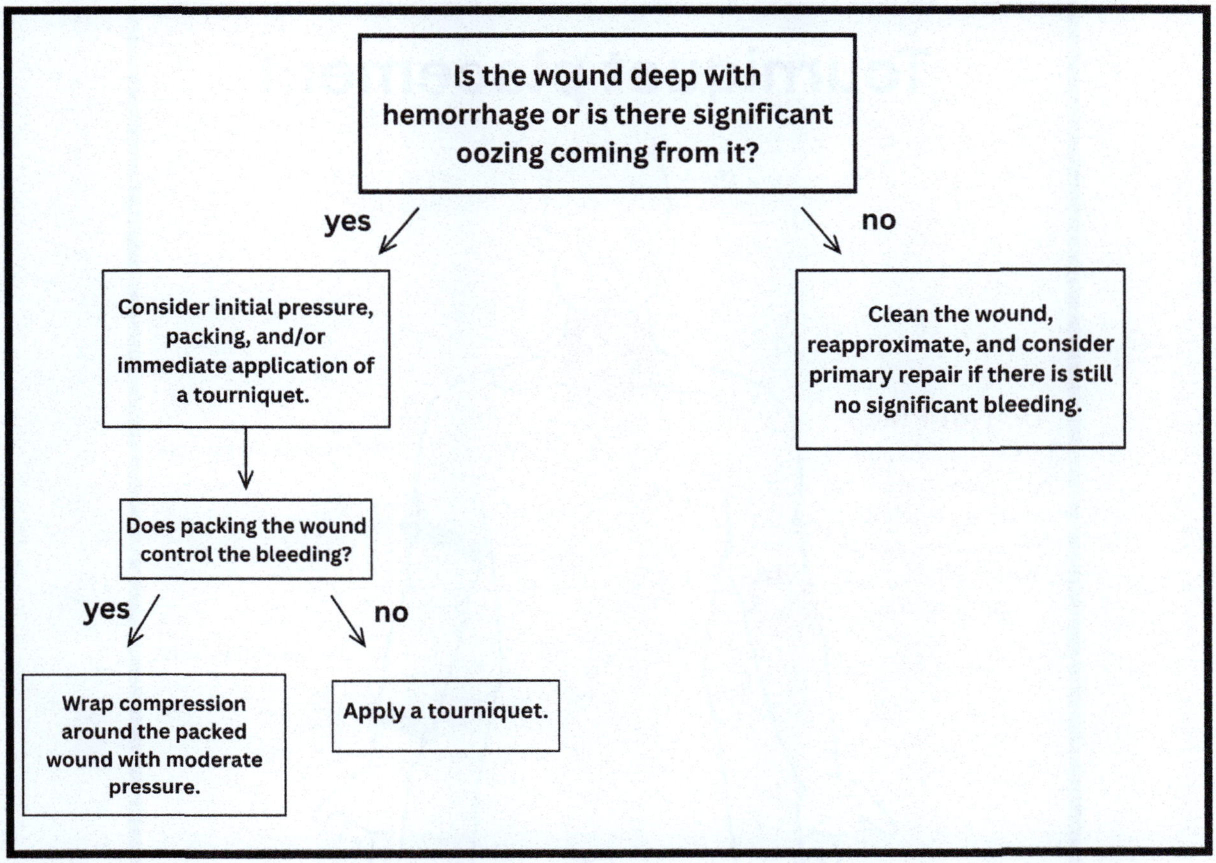


FIGURE 6.13 Wound packing decision-making flowchart.

**TABLE 6.1** Material, mass, volume of basic hemostatic materials for wound packing.

| Item | Mass | Volume | Indication | Alternatives (if any) |
|---|---|---|---|---|
| 4×4 Gauze | 7 g | 16 cm × 14cm × 0.2 cm | | Any other absorbent material (if needed) |
| Hemostatic Bandage (QuikClot Control+®) | 30 g | 19 cm × 16 cm × 2 cm | | Quik clot hemostatic dressing |
| Gauze-Roll | 38 g | 9 cm × 16 cm × 6 cm | | Multiple types of gauze rolls |
| X-Stat® | 130 g | 28 cm × 17 cm × 12 cm | | Multiple types of similar fillers |
| Elastic Bandage | 35 g | 8 cm × 11 cm × 5 cm | | Multiple types of elastic bandages |

Another procedure that may stop bleeding from a wound bed is the injection of lidocaine with epinephrine. Epinephrine constricts blood vessels and may allow hemostasis of the wound. Care should always be taken to avoid injecting lidocaine with epinephrine in the fingers, toes, or ears, as prolonged exposure to epinephrine in these areas may cause tissue damage. If performed, however, care must be taken to keep the affected part warm, with particularly careful consideration to the temperature of a warming pad placed on an anesthetic region of the body so that burns do not occur.

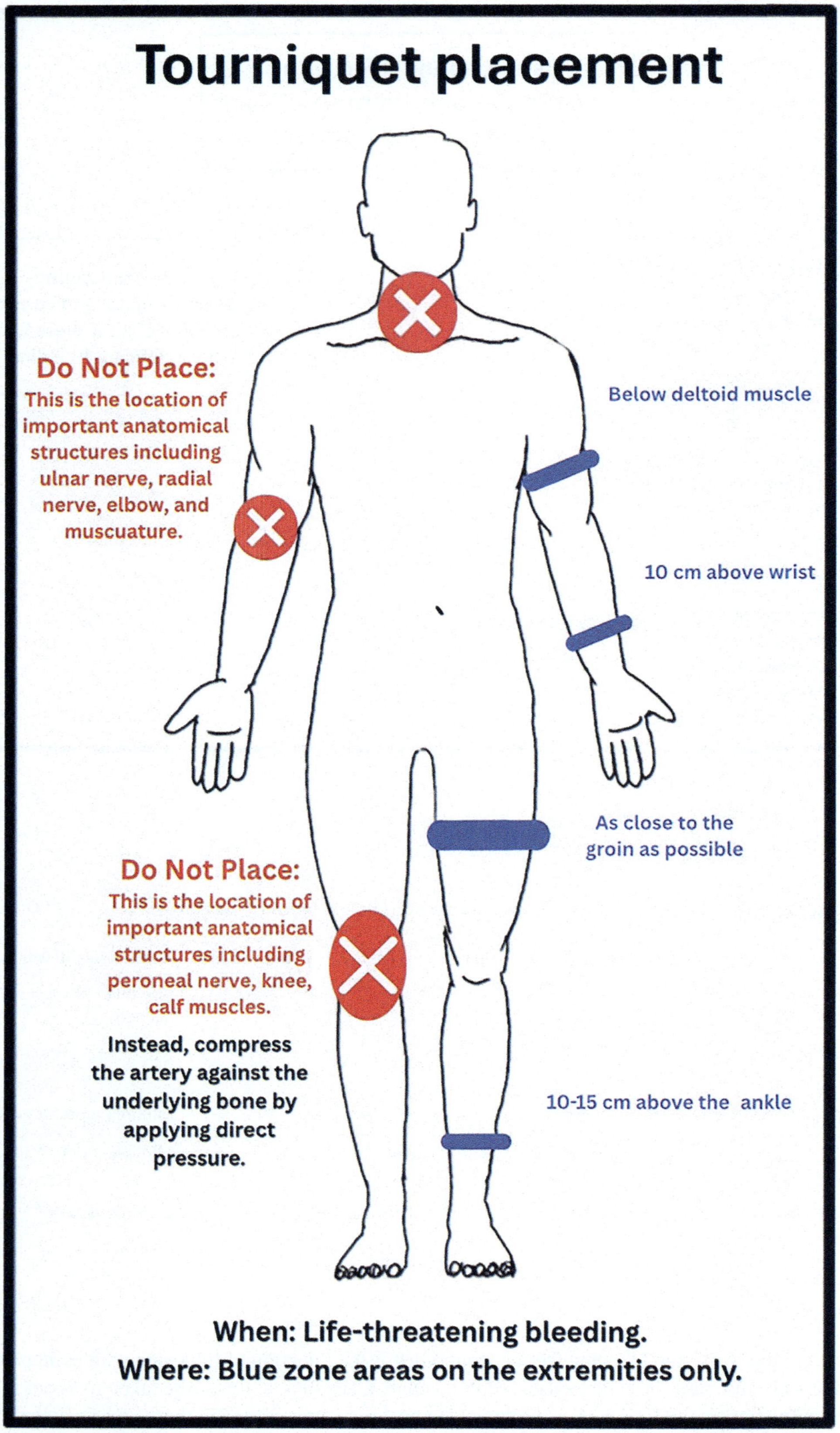


FIGURE 6.14 **Tourniquet placement.** Several locations on the body are ideal for tourniquet application, including the arm, forearm, thigh, and calf. The distal extremities, however, may be less ideal for tourniquet placement due to arteries/veins that run in between the two long bones rather than a single long bone. *Always* avoid placing them over joints and over the wound itself.

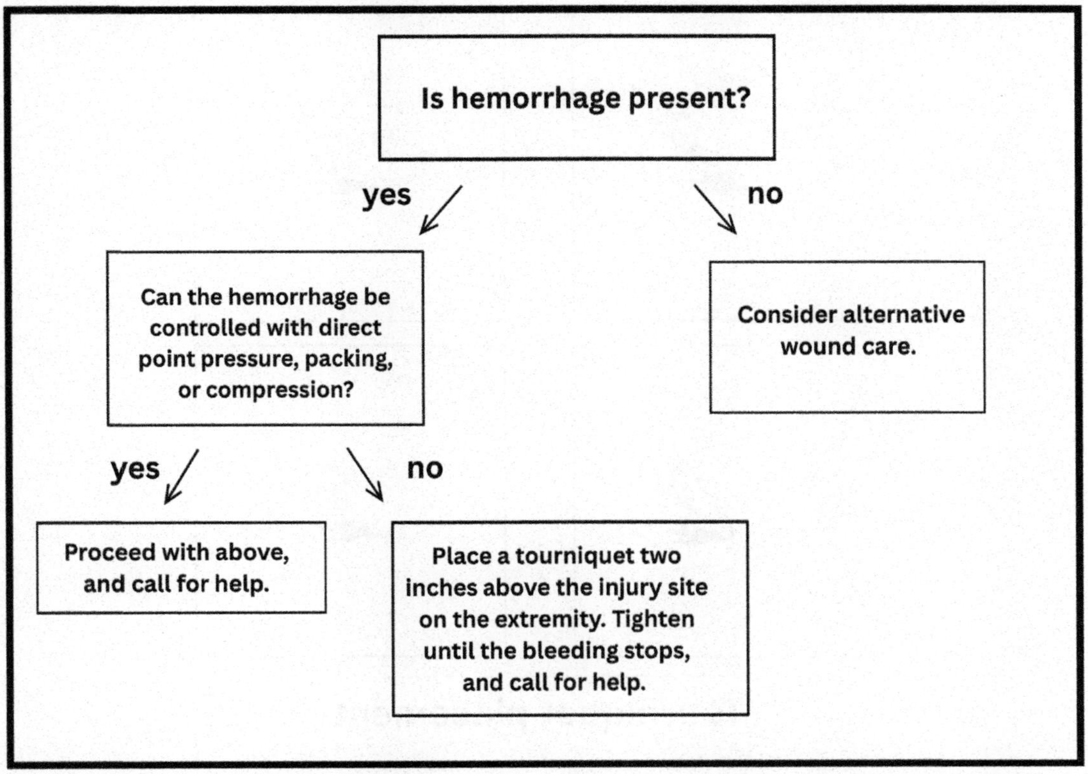


FIGURE 6.15 Tourniquet decision-making flowchart.

## 6.6 Laceration repairs

Indication: Any laceration that is amenable to repair to help stop bleeding, limit the potential for reinjury, and promote healing (Fig. 6.19)

### 6.6.1 Steri-Strips

Lacerations present in all shapes and sizes. Some lacerations can approximate easily, as the wound edges will come together easily without tension. In these cases, lacerations can be repaired by Steri-Strips, which are adhesive strips that keep the wound approximated for several days, a time long enough for fibrin and collagen cross-links to form and begin the healing process.

### 6.6.2 Dermabond

Dermabond can be used for well-approximated lacerations that are smaller in size and are not under tension or overlying joints. Dermabond takes several minutes to seat fully, and immobilization of the wound edges must be maintained as the glue solidifies for at least 60 seconds. If there is any bleeding or wound edge tension after application of the Dermabond, the Dermabond typically will fail to seat and must be removed with reapplication attempted.

### 6.6.3 Staples

Staples represent another means to reoppose lacerations or flaps expeditiously and effectively. They should not be used on the face, hand, distal extremities, or other sensitive areas of the body, as they are relatively imprecise. They can be removed with a specific staple-removal device in 10–14 days. If a staple-removal instrument is not available, the staple can be bent in the center with a forceps so help angle the sides out of the skin. Most commonly, staples are used for scalp lacerations, which may bleed profusely but which is controlled promptly with a quick staple repair (Figs. 6.20 and 6.21; Table 6.5).

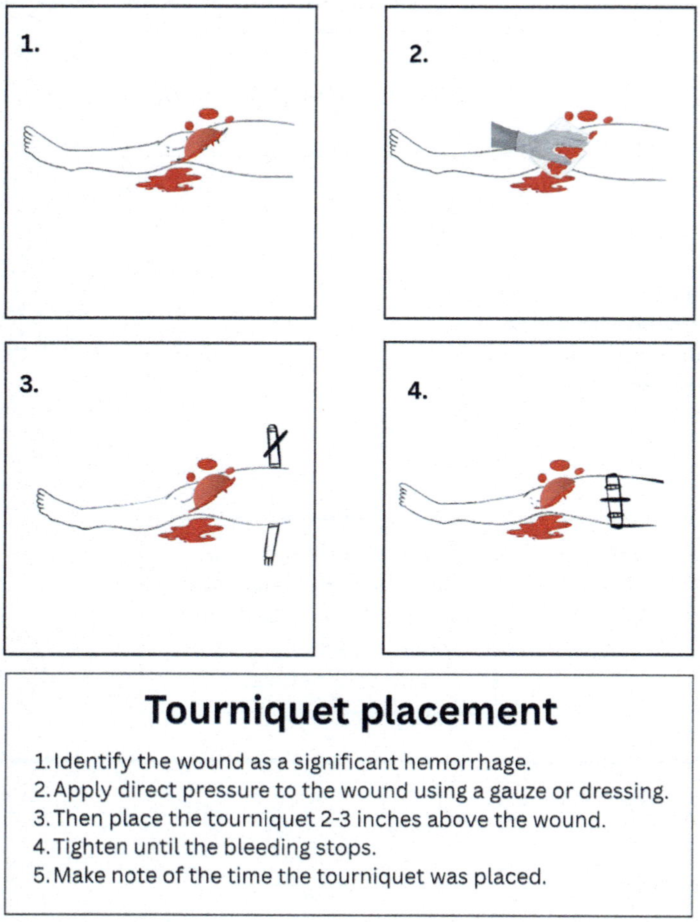


**FIGURE 6.16** Tourniquet placement.

**TABLE 6.2** Material, mass, volume of CAT tourniquet.

| Item | Mass | Volume | Indication | Alternatives (if any) |
|---|---|---|---|---|
| C-A-T® | 88 g | 5 cm × 28 cm × 4 cm | Hemorrhage | Pressure dressing |
| Any availability wide cord, cloth, or belt | N/A | N/A | Hemorrhage | No availability of CAT |

### 6.6.4 Sutures

There are many types of suture materials and suture techniques. The goal of the suture is to secure and oppose tissue layers (such as epidermis, dermis, etc.) and keep them in contact to promote healing. In order to suture, you will need specific instruments that either come as a pack or must be obtained individually. Fig. 6.22 shows the ideal instruments for general laceration repair. Thicker sutures, such as 2-0 and 3-0, can be used on large wounds under tension, as the suture material is thicker and stronger. Thinner sutures, such as 5-0 and 6-0, mean the suture material is smaller in diameter and can be used for delicate tissues such as facial tissue, hands, ears, etc. Permanent monofilament sutures such as polypropylene or nylon have good tensile strength and are easy to tie but must be removed. Still others are made of synthetic absorbable materials (polyglactin or polyglycolic acid which is braided, or poliglecaprone which is monofilament) that can be absorbed by the

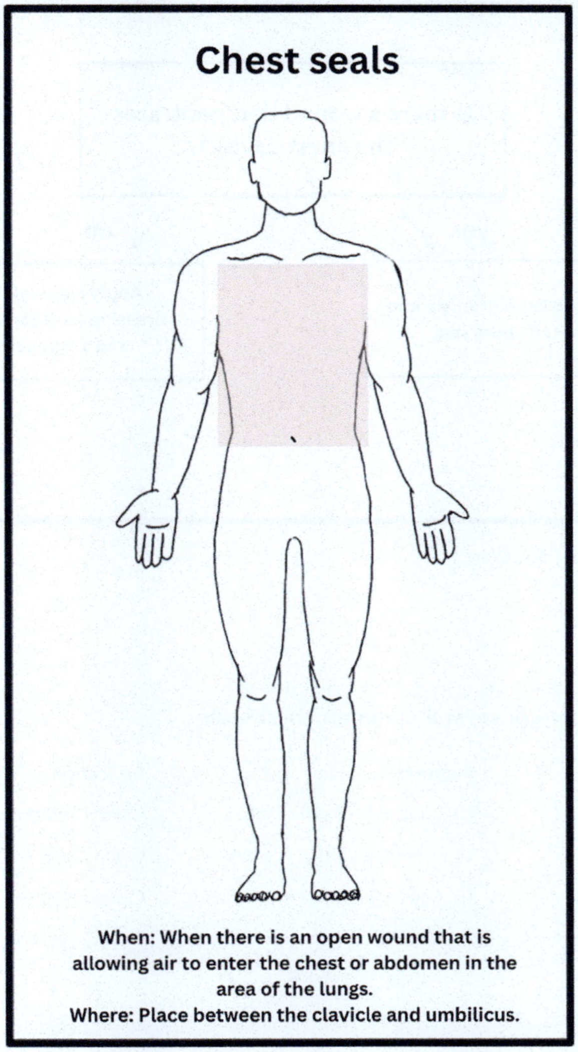


FIGURE 6.17 Chest seals.

body over time and are placed in deeper tissue layers. Suture packaging generally comes in recognizable forms with clearly displayed attributes (Fig. 6.23)

Laceration anesthetization: Prior to repairing a laceration with sutures, a wound is typically anesthetized with lidocaine. Lidocaine may come in 1% or 2%, which means there is 10 mg/ml vs 20 mg/ml of lidocaine. This is important, because the toxic dose of lidocaine is between 200 and 400 mg per person. More precisely, it is 4–7 mg/kg, depending on whether the lidocaine came premixed with epinephrine. A good rule of thumb is that a standard 5–10 ml vial of lidocaine will likely fall below the toxic range for adults. And, when using lidocaine with epinephrine, the body will tolerate higher doses of lidocaine because the epinephrine constricts blood vessels, acting dually as a hemostatic agent and also helps keep lidocaine local, preventing it from being released all at once into the systemic circulation. Thus, a person who receives 2% lidocaine without epinephrine would reach a toxic dose after a much smaller injection than if they were given lidocaine with epinephrine. Read the expiration date on the bottle; however, epinephrine degrades over a period of a few short years and although harmless, likely won't confer any benefits after the bottle has turned pink to brown from epinephrine's degradation product.

Simple interrupted sutures are excellent for more minor, smaller lacerations, while horizontal mattress sutures can be used for more complex lacerations that are under tension. Other suture techniques, such as a running suture (continuous sutures), can be used on longer lacerations and can be done relatively quickly by the experienced operator (Fig. 6.24)

All of these suture techniques have one goal: to oppose the opposite tissue layers of the skin in order to maximize healing efficiency and hemostasis: epidermis to epidermis and dermis to dermis. It is difficult to describe in text exactly

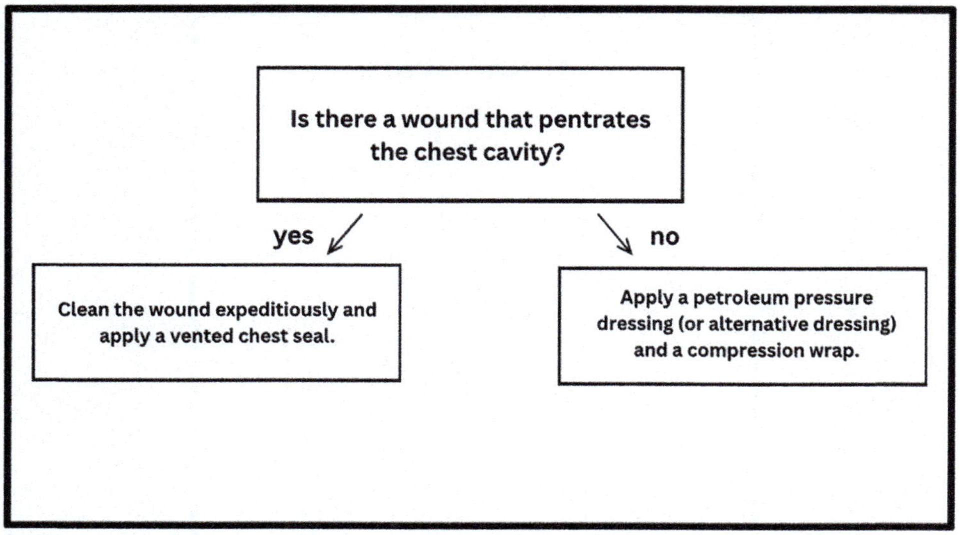


**FIGURE 6.18** Chest seal decision-making flowchart.

**TABLE 6.3** Material, mass, volume of several common chest seals

| Item | Mass | Volume | Indication | Alternatives (if any) |
|---|---|---|---|---|
| Sentinel™ Chest Seal | 28 g | 10.2 cm × 17.8 cm × 0.6 | Open chest wound | Petroleum gauze or similar |
| HALO™ Chest Seal | 57 g | 14 cm × 16.5 cm × 0.6 cm | Open chest wound | Petroleum gauze or similar |
| Hyfin® Chest Seal | 45 g | 10.2 cm × 17.8 cm × 0.6 cm | Open chest wound | Petroleum gauze or similar |

**TABLE 6.4** Common hemostatic agents.

| Product name | Hemostatic agent in the material | Description of active compound |
|---|---|---|
| QuikClot® | Kaolin | Inorganic clay mineral |
| CELOX™ | Chitosan | Polysaccharide derived from shrimp shells |
| X-Stat® | Chitosan in cellulose sponge | Cellulose sponge expansion with chitosan |
| NuStat® | Silica nanofibers and chitosan | Silica-based and cellulose fibers along with chitosan |

which type of laceration pattern would require which suture technique, and even experienced clinicians might differ in their approach to primary closure of a wound. After judging a wound's size, geometry, and location, first choose the type of suture. Then, assess the wound to see how well it reapproximates to determine where to begin and what style of suturing to use (Table 6.6).

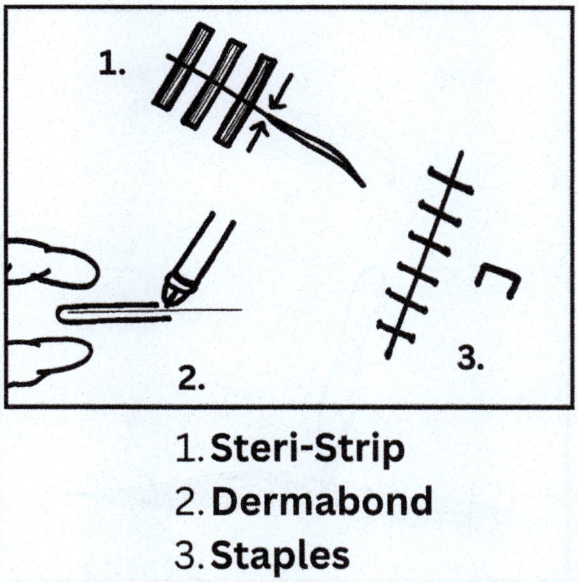


FIGURE 6.19 **Simple laceration repair methods.** Methods 1–3 demonstrate various techniques to repair a simple laceration. A simple laceration is one in which wound edges reapproximate easily and there are no areas under tension.

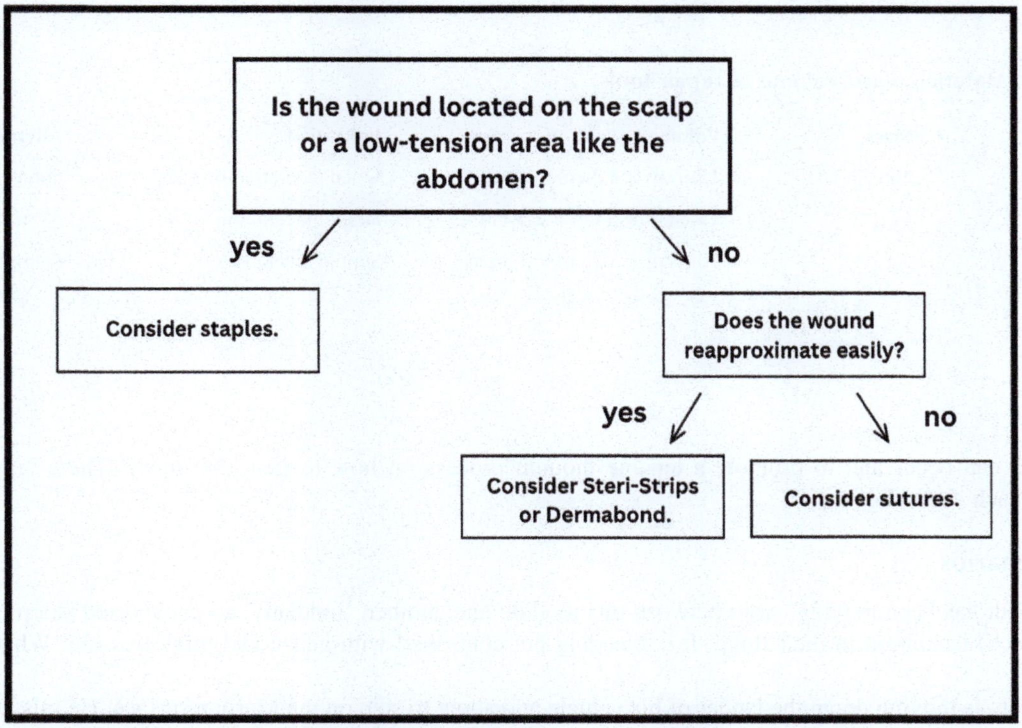


FIGURE 6.20 Sutures, staples, and adhesives decision-making flowchart.

## 6.7 Scenarios and solutions

Hemorrhage is a dynamic process with variable causes. There are many ways to stop a hemorrhage, even though the underlying methods are the same. The following scenarios and solutions are designed to encourage thinking of various

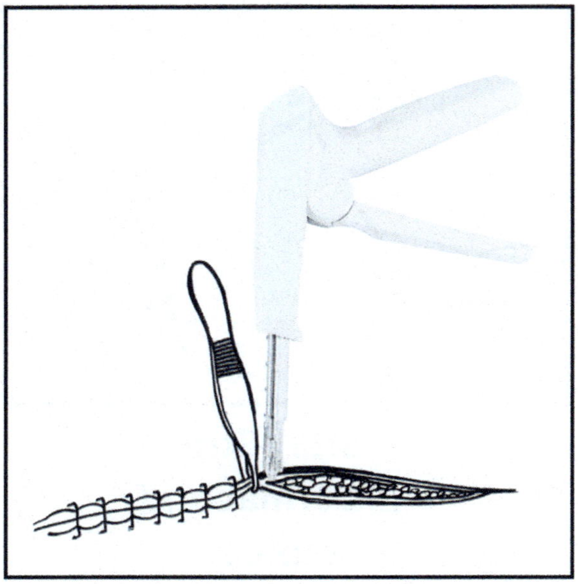

Laceration repair with skin stapler

FIGURE 6.21 Skin stapler.

**TABLE 6.5** Material, mass, volume of repair tools.

| Item | Mass | Volume | Indication | Alternatives (if any) |
|---|---|---|---|---|
| Stapler | 28 g | 15.6 cm × 3.3 cm × 2.9 cm | Open laceration on scalp | Suture |
| Dermabond™ | 12 g | 15.6 cm × 5 cm × 2.5 cm | Simple laceration | Steri-Strip™, Suture |
| Steri-Strip™ | 9 g | 10.2 cm × 5.1 cm × 2.5 cm | Simple laceration | Dermabond™ |

ways injuries can occur and to promote a tenable thought process on how to treat the injury. There may be multiple solutions to each scenario.

## 6.7.1 Scenarios

1. An astronaut has been moving cargo between one module and another. Suddenly, an unexpected sharp edge creates a 4 cm (2 inch) laceration in their thigh. It is bleeding but controlled with gauze and mild pressure. What are the next steps?
2. An astronaut is moving down the ladder of his vehicle and about to step on the Martian surface. He misses the last step and, in doing so, falls 5 feet onto his right leg. He is quickly brought back into the vehicle. After repressurizing, his suit is taken down, and he is found to have an ankle deformity with tissue laceration that is oozing blood. What are the next steps?
3. An explosion on a tank farm sends shrapnel into a member of the groundteam. One of the pieces punctures a crew member's left thigh. Luckily, an airlock is nearby, and the patient is quickly moved inside. A significant amount of blood is coming out of the wound despite pressure. What are the next steps?
4. A particularly tall crewmember hits his head on a cabinet drawer while on duty in Utopia Planitia. There is significant bleeding from the scalp, but it seems to be controlled with pressure. How should the wound be repaired?

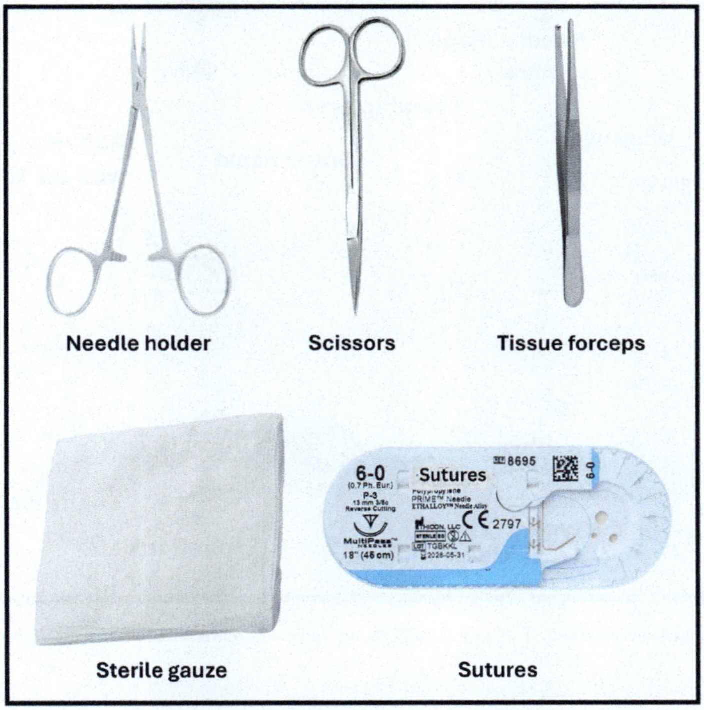


**Suture kit essentials**

FIGURE 6.22 **Suture kit essentials.** There are many suture kits available. The basic components include a needle holder, scissors, forceps, gauze, and suture.

5. An astronaut is working in the kitchen preparing the first tomatoes grown on Mars. He suffers a laceration on a finger. How should the wound be repaired?

### 6.7.2 Solutions

1. In this case, the bleeding on the thigh appears to be controlled with gauze and minor pressure. Keep pressure on the wound for 5–10 minutes and then slowly peel back the gauze to expose the wound for inspection. Clean the wound, remove any debris, and reapply new gauze as needed to control bleeding. After inspection and cleaning, attempt to reapproximate the edges of wound to assess how well it would close. If it reapproximates well without being held and is not under tension, consider using Steri-Strips™ or tape. If it requires some continued force to close but does not widen with movement of the leg, consider 3-0 or 4-0 sutures (likely nonabsorbable) using an interrupted pattern. If it requires substantial force to keep closed particularly with movement of the leg, the wound will be under tension and may require specialized stitched such as a horizontal mattress.
2. An ankle deformity with overlying laceration implies an open fracture. Palpate for pulses and inspect the wound to look for any visualized bone. If none visualized, a mild washout or sponge bath with sterile water or chlorhexidine (if available and permitted) can be applied, and the wound closed with 3-0 or 4-0 nonabsorbable suture.
3. This person is suffering from a hemorrhagic wound from shrapnel damaging major blood vessels. Time is of the essence to prevent shock and continued blood loss. Immediate pressure should be applied to the wound to slow the hemorrhage, if possible. The astronaut should be brought to an airlock and repressurized as quickly as is safely possible. According to the manufacturer's specifications, the suit should be taken down in a rapid manner, all while someone continues to put pressure on the wound. If there is continued uncontrollable bleeding, a tourniquet should be placed at least two inches above the level of the wound. If bleeding stops after application, the wound can be inspected for debris and packed as necessary to further control bleeding. As this is done, another individual should assess for distal extremity pulses (such as popliteal or dorsalis pedis, depending on location). If bleeding returns, the tourniquet should be tightened until bleeding ceases. At this point, a serious traumatic injury has occurred. Now that the hemorrhage has stopped, expert surgical consultation (on-station) should guide the next course of action (the tourniquet

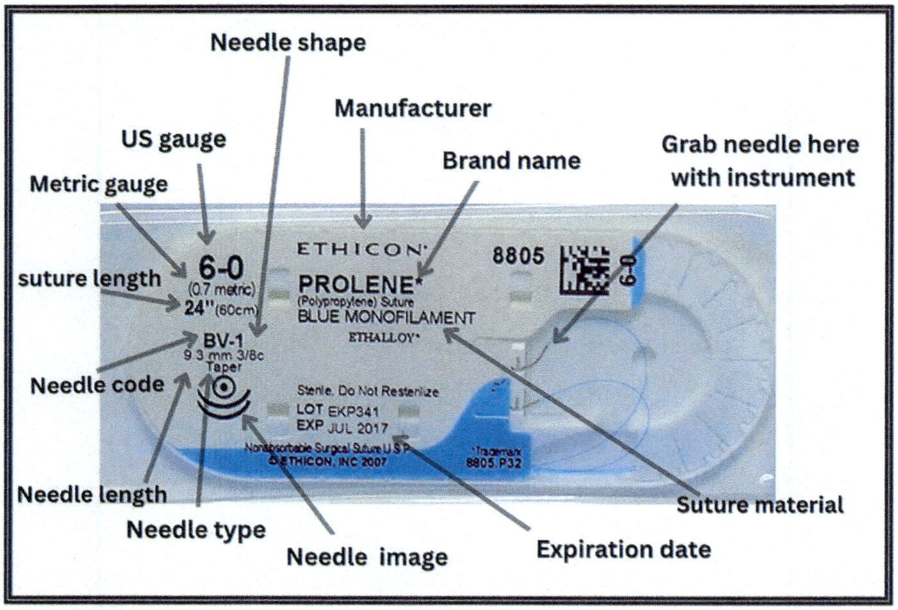


FIGURE 6.23 **Common suture package and key.** This key highlights the important parts of a suture and its packaging of a suture Ethilon™.

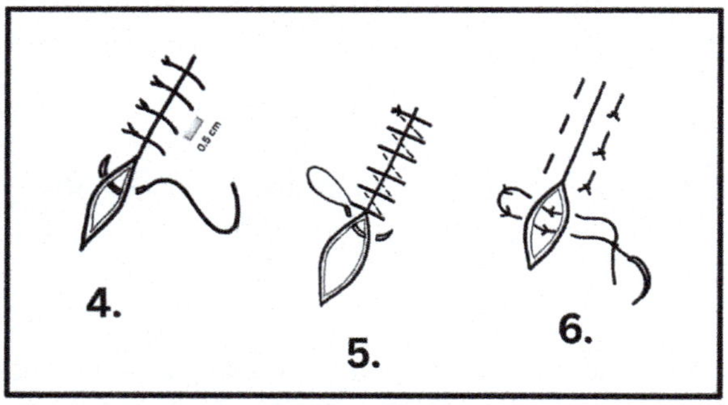


**4. Simple interrupted suture**
**5. Running suture**
**6. Horizontal mattress suture**

FIGURE 6.24 Advanced laceration repair methods.

can be kept up for up to 6 hours in order to reach more advanced medical care–more than that risks limb loss). If there is no way to seek further consultation, the wound can be inspected and any shrapnel removed. Heavy bleeding requiring the use of a tourniquet usually implies injury to a large vessel. If equipped and skill allows, the vessel can be repaired either with simple nonabsorbable sutures and the wound then closed, with deeper layers being closed by absorbable sutures and superficial layers by nonabsorbable sutures or even staples, if the area allows. Managing complex vascular injuries is outside of the scope of this text. Fig. 6.21: Tourniquet placement on affected extremity with step-by-step guide.

4. Scalp lacerations can bleed heavily but usually respond well to pressure. Sometimes, the only way to effectively stop their bleeding is to close them quickly with Staples and then apply gentle pressure. Ensure that no bone is visible. If there is any suspicion of an underlying fracture, do not place a finger into it to explore the wound. Fig. 6.22: Scalp laceration with staples.
5. Finger lacerations are complicated and come in a variety of sizes and types as unique as the hand itself. Some types of lacerations are indeed actually avulsions of skin or amputations of the distal finger tip. Unless bone is exposed, those

**TABLE 6.6** Material, mass, and volume of common components of laceration repair kits compared to all-in-one suture tray.

| Item | Mass | Volume | Indication | Alternatives (if any) |
|---|---|---|---|---|
| Needle driver | 20–80 g | 14 cm × 4 cm ×.02 cm (wide ranging) | Laceration | N/A |
| Scissors | 28 g | 16 cm × 4 cm × 0.2 cm (wide ranging) | Laceration | N/A |
| Tissue forceps | 12 g | 15 cm × 2 cm × 0.4 cm (wide ranging) | Laceration | N/A |
| Anesthetic injection needle | 2 g | 10 × 2 × 0.8 | Laceration | ,N/A |
| 3 ml syringe | 7 g | 10.5 cm × 2.2 cm × 2 cm | Laceration | 10 ml syringe or 1 ml syringe. |
| Suture | 4 g | 6 × 13 × 0.1 | Laceration | Multiple sutures available |
| All-in-one laceration repair kit (including gauze, needle holder) | 184 g | 25 cm × 16 cm × 6 cm | Laceration | Individual components |
| Lidocaine 2% | 24 g | 2.6 cm × 2.6 cm × 5.3 cm | Laceration | Alternate local anesthetic with or without epinephrine |

heal naturally over time and do not usually require primary repair, unless a portion of the flap is still attached. The key to any finger laceration is hemostasis. This can be done by "tying" a tourniquet on the proximal finger tight enough to occlude blood flow. This will be painful; prior to application, inject 1 ml of 2% lidocaine with a 32-gauge "insulin" needle on each side of the finger as a nerve "block," being careful not to inject into joints or into vessels. After hemostasis is achieved, with or without the finger tourniquet, begin by reapproximating the tissue flap (if it is complex) and then repair, taking care to note any tendon or nerve involvement. Usually, these types of repairs require 5-0 or 6-0 Prolene due to the delicacy of these types of lacerations.

## 6.8 Summary recommendations and considerations for the future

This procedural text in hemorrhage control scratches the surface of what will be possible in the future of human spaceflight. The difficulties of procedure performance in microgravity will eventually abate with the emergence of artificial gravity space stations and other craft. Hopefully in short time, more advanced procedures that can control major hemorrhage will be developed as the launch vehicles and interplanetary spacecraft become larger and more capable.

To date, there has never been a fatality in spaceflight due solely to blood loss from trauma; however, there have been 21 fatalities involving injury from blunt trauma or depressurization. Spaceflight is dangerous, and as we endeavor to reach for the stars, we must carry with us the medical capabilities to manage contingencies as much as possible in order to maximize our chances of staying permanently. Eventually, an injury resulting in hemorrhagic shock in space is likely to occur and it behooves us to be prepared for when it does. When we begin colonizing Mars and beyond, we will be bringing our humanity with us, including our susceptibilities and strengths. This brings opportunities to educate, prepare, and engineer for medical emergencies as we endeavor to create human-centered spacecraft.

## Further reading

Antonsen, E., Van Baalen, M., 2021. Comparison of health and performance risk for accelerated Mars mission scenarios. Available from: https://ntrs.nasa.gov/citations/20210009779.

Convertino, V.A., 1996. Clinical aspects of the control of plasma volume at microgravity and during return to one gravity. Medicine and Science in Sports and Exercise 28 (10 Suppl), S45–52. Available at https://doi.org/10.1097/00005768-199610000-00033 PMID: 8897404.

Kirkpatrick, A.W., Campbell, M.R., Jones, J.A., Broderick, T.J., Ball, C.G., McBeth, P.B., et al., 2005. Extraterrestrial hemorrhage control: terrestrial developments in technique, technology, and philosophy with applicability to traumatic hemorrhage control in long-duration spaceflight. Journal of the American College of Surgeons 200 (1), 64–76. Available at https://doi.org/10.1016/j.jamcollsurg.2004.08.028 PMID: 15631922.

Kunz, H., Quiriarte, H., Simpson, R.J., Ploutz-Snyder, R., McMonigal, K., Sams, C., et al., 2017. Alterations in hematologic indices during long-duration spaceflight. BMC Hematology 17, 12. Available at https://doi.org/10.1186/s12878-017-0083-y PMID: 28904800; PMCID: PMC5590186.

Nicogossian, A.E., Sawin, C.F., Huntoon, C.L., 1993. Overall physiologic response to spaceflight. In: Nicogossian, A.E., Hunton, C.L., Pool, S.L. (Eds.), Space Physiology and Medicine. Williams & Wilkins, pp. 213–227 third ed.

Scheuring, R.A., Mathers, C.H., Jones, J.A., Wear, M.L., 2009. Musculoskeletal injuries and minor trauma in space: incidence and injury mechanisms in U.S. astronauts. Aviation, Space, and Environmental Medicine 80 (2), 117–124. Available at https://doi.org/10.3357/ASEM.2270.2009.

Thoolen, S.J.J., Kuypers, M.I., 2023. External hemorrhage control techniques for human space exploration: lessons from the battlefield. Wilderness & Environmental Medicine 34 (2), 231–242. Available at https://doi.org/10.1016/j.wem.2023.01.006.

Udden, M.M., Driscoll, T.B., Pickett, M.H., Leach-Huntoon, C.S., Alfrey, C.P., 1995. Decreased production of red blood cells in human subjects exposed to microgravity. Journal of Laboratory and Clinical Medicine 125 (4), 442–449 PMID: 7706898.

Welling, D.R., McKay, P.L., Rasmussen, T.E., Rich, N.M., 2012. A brief history of the tourniquet. Journal of Vascular Surgery 55 (1), 286–290. Available at https://doi.org/10.1016/j.jvs.2011.10.085.

Chapter 7

# Resuscitation and critical care

**Austin Almand[1,2], Tovy Haber Kamine[3], and Dana R. Levin[4,5,6]**

[1]*Department of Surgery, MedStar Georgetown University Hospital – Washington Hospital Center, Washington, DC, United States,* [2]*123d Special Tactics Squadron, Air Force Special Operations Command, Louisville, KY, United States,* [3]*Trauma, Acute Care Surgery, and Surgical Critical Care, University of Massachusetts Chan Medical School –Baystate, Springfield, MA, United States,* [4]*Space Medicine Operations, Mission Operations Vast Inc., Long Beach, CA, United States,* [5]*Department of Emergency Medicine, University of California, Los Angeles, CA, United States,* [6]*Division of Space, Ecological, Arctic, and Resource-Limited Medicine, Massachusets General Hospital, Boston, MA, United States*

## 7.1 Introduction

Given enough time, space missions to low Earth orbit, cis-lunar space, the lunar surface, and eventually Mars will inevitably face medical emergencies that require resuscitation and critical care interventions (Robertson et al., 2020). As person-years in each of these environments increase, so too does the probability of a catastrophic medical event.

While there are many medical conditions that can cause an individual's body to fail, most will take weeks or months to progress, providing ample time for a thorough diagnostic process and development of a definitive care plan. These are not considered medical emergencies. A true medical emergency is a condition that carries a substantial risk of imminent, permanent, and irreversible consequence (e.g., death or disability). In spaceflight, as with other operational environments, this is often expanded to include medical events that risk an irreversible loss of mission objectives even if the crew is later able to recover.

True medical emergencies will progress to catastrophic consequences in seconds, minutes, or hours, severely limiting the time for providers to recognize what is happening and direct appropriate interventions. This is why terrestrial medical systems train specialized personnel and devote a large amount of infrastructure to place medical experts in front of patients in crisis as fast as possible. For true emergencies, providers must operate with a high degree of uncertainty, taking action before the complete picture is known and rapidly adjusting to new information as it becomes known.

Fortunately, emergency medical conditions tend to follow a limited set of endpoints that, once stabilized, buy the provider time to better assess the situation. These are conditions like a blocked airway, inability to breathe, stopped heart, or severe bleeding.

The common theme in these conditions is an interruption in the supply of oxygen to vital organs. Cells rely on oxygen as the final step in generating energy for their processes. Since human cells have very limited oxygen reserves, cutting off the supply rapidly stalls energy generation and halts cellular processes. The most metabolically active cells, such as those in the brain, will stall out first and die within 4–6 minutes, setting a functional time limit for providers to respond. After the first few minutes, the longer it takes to restore oxygen supply to the brain, the more brain functions are permanently lost and the lower the chances of neurologically intact survival. Table 7.1 contains examples of common medical emergencies labeled with the initial level of diagnostic certainty required, the time to respond before consequences, and the immediate actions a provider should consider.

Emergency medicine treats this lack of time as an asset. Rather than approaching a case as potentially needing all the knowledge and interventional capabilities medicine has to offer, the lack of time forces a provider to focus on the most immediate needs. The response is split into four main stages, detection/recognition, triage, resuscitation, and critical care, with the first two stages geared towards assessing the amount of time and the primary targets for intervention, followed by initial stabilizing interventions to stabilize the patient and buy time, and finally ending with an ongoing cycle of observation and rapid intervention aimed at maintaining stability and preventing decompensation until definitive care can be rendered or the patient recovers and the emergency has passed.

**Procedural Medicine for Spaceflight. DOI:** https://doi.org/10.1016/B978-0-443-27552-4.00009-9

**TABLE 7.1** Examples of medical emergencies, time to respond, and required initial interventions.

| Emergent event | Example relevant to spaceflight | Time to respond | Immediate interventions |
|---|---|---|---|
| Loss of airway | Choking | 1–2 minutes | Heimlich maneuver |
| Breathing stopped | Suffocation | 4–6 minutes | Artificial respirations (e.g., rescue breathing, bag-valve mask, or ventilator) |
| Heart stopped | Ventricular TachycardiaVentricular Fibrillation | 4–6 minutes | Chest compressions/CPR Defibrillation |
| Severe bleeding | Trauma | <30 minutes | Hold direct pressure. Apply tourniquet |
| Difficulty breathing | Injury from smoke inhalation | <30 minutes | Provide oxygen, bronchodilators. Monitor blood oxygen levels. Monitor for fatigue. Consider noninvasive positive-pressure ventilation |
| Unconsciousness | $CO_2$ narcosisDrug overdoseHypoglycemia (low blood sugar) | 30–45 minutes | Protect airway. Monitor vital signs and response to stimuli. Consider oxygen, IV fluids, and further interventions as necessary |
| Heart attack | ST elevation myocardial infarction | <1 hour | Administer thrombolytic medications (e.g., TPA). Consider evacuation for cardiac catheterization if achievable within 24 hours |
| Stroke | Ischemic cerebrovascular accident | <4.5 hours | Administer thrombolytic medications (e.g., TPA). Consider evacuation for clot retrieval if consistent with a large vessel proximal occlusion and intervention is achievable within 72 hours |

*CPR*, Cardiopulmonary resuscitation; *IV*, intravenous; *TPA*, tissue plasminogen activator.

### 7.1.1 Detection

The first stage is always to recognize that an emergency is occurring. This may be obvious, as when an accident causes a visible injury with bleeding, or subtle, as when low blood sugar causes a patient to quietly lose consciousness. Detection of an emergency is often a multifactorial process that involves passive monitors (e.g., smoke detectors) as well as active systems (e.g., crew activating an alarm or calling a flight surgeon). Depending on needs, a future station may also employ passive vital sign tracking or motion tracking to monitor crew and arrange early alerts for elevated, depressed, or absent heart beats. Technology for such monitoring is already being developed (Canepa, Levin and Padaki, 2025; Padaki et al., 2025).

### 7.1.2 Triage

Triage is the early stage emergency version of a differential diagnosis. The aim is to identify the most immediate threats and assess how much time a provider has to intervene. This can be applied to a single patient or a group of patients if a mass-casualty event has occurred. There are multiple systems employed for this with common strategies, each leveraging simple rating scales based on easily assessed metrics intended to predict those most at risk for immediate decompensation. The table below provides examples of commonly used methods. Table 7.2 compares common triage methods.

Fig. 7.1 contains a modified version of the Emergency Severity Index (ESI) intended to support triage in spaceflight settings. This Medical Spaceflight Triage Acuity Rating (Med-STAR) is meant to account for illness severity as well as resource and time demands to facilitate the assessment of patients in flight. Med-STAR 1 and 2 are high acuity cases, distinguished by the time left to respond and are most directly applicable to the contents of this chapter. Med-STAR 3, 4, and 5 are lower acuity cases distinguished primarily by the number of resources or crew time required to manage the patient's condition. These would likely leave sufficient time to follow a nonemergent diagnostic process. It should also be noted that this is only one potential tool and that provider judgment plays a significant role in assessing the severity and identifying the immediate life threats. If a provider is concerned for any reason, it is often advisable to treat the case as if it were a higher acuity than triage schemas might suggest. Patient acuity level will also change as more information is gained

**TABLE 7.2** Triage method comparison.

| System | Scale | Intended use | Metrics assessed |
|---|---|---|---|
| ESI (emergency severity index) | 5 levels (1 = most urgent, 5 = least urgent) | Hospital emergency departments. National standard in the United States | Immediate life threat<br>Vital signs (HR, BP, RR, temp, $O_2$ sat)<br>Level of consciousness/mental status<br>Stability of airway, breathing, circulation<br>Anticipated resource utilization (labs, imaging, consults) |
| CTAS (Canadian Triage and acuity scale) | 5 levels (resuscitation → nonurgent) | Hospital emergency departments. National standard in Canada, used internationally | Presenting complaint (chief symptom)<br>Vital signs (HR, BP, RR, temp, $O_2$ sat)<br>Level of consciousness<br>Pain severity<br>Mechanism of injury & risk factors |
| SALT (sort, assess, life-saving interventions, treatment/ transport) | 4–5 categories (immediate, delayed, minimal, expectant, dead) | Mass-casualty incidents; disaster response; austere environments | Ability to follow commands / ambulate<br>Breathing (presence, rate, distress)<br>Circulation (major hemorrhage)<br>Mental status<br>Response to brief life-saving interventions (airway, bleeding control) |
| START (simple triage and rapid treatment) | 4 categories (red, yellow, green, black) | Mass-casualty incidents; disaster response; austere environments | Respiratory rate/presence of breathing<br>Perfusion (capillary refill or pulse)<br>Ability to follow simple commands<br>Walking ability (ambulatory = green) |

*HR*, Heart rate; *BP*, blood pressure; *RR*, respiratory rate; *temp*, temperature; *$O_2$ sat*, oxygen saturation.

or the condition progresses. By tradition, during the acute phase of a condition, once a triage level is assigned, it may be increased but not decreased. This helps prevent complacency and maintain an appropriate level of vigilance. For these reasons, triage schemas are useful for rapidly assessing the risk to an individual patient in the clinical setting. They should not be used to characterize conditions or medical risk outside of triage settings (e.g. these re not optimized for mission assurance consequence tables and hazard reports.

While the Med-STAR or similar schema will help stratify patients for communication with other members of the care team and with mission contro , the practical method by which these scores are assessed is referred to as the primary survey. This often follows a treat-as-you-go pattern where as soon as a problem is encountered, the provider must stop and manage it before moving on to the next phase. Here too there are different algorithms employed such as ABCDEs (Airway, Breathing, Circulation, Disability, Exposure/Environment) or MARCH (Massive Hemorrhage, Airway, Respiration, Circulation, Hypothermia/Head injury). Both are effective and the specific algorithm used often depends on provider preference and base training. The key is to become familiar with a method and develop confidence in applying it. MARCH was developed for use with military environments where trauma, and therefore bleeding, is the most likely

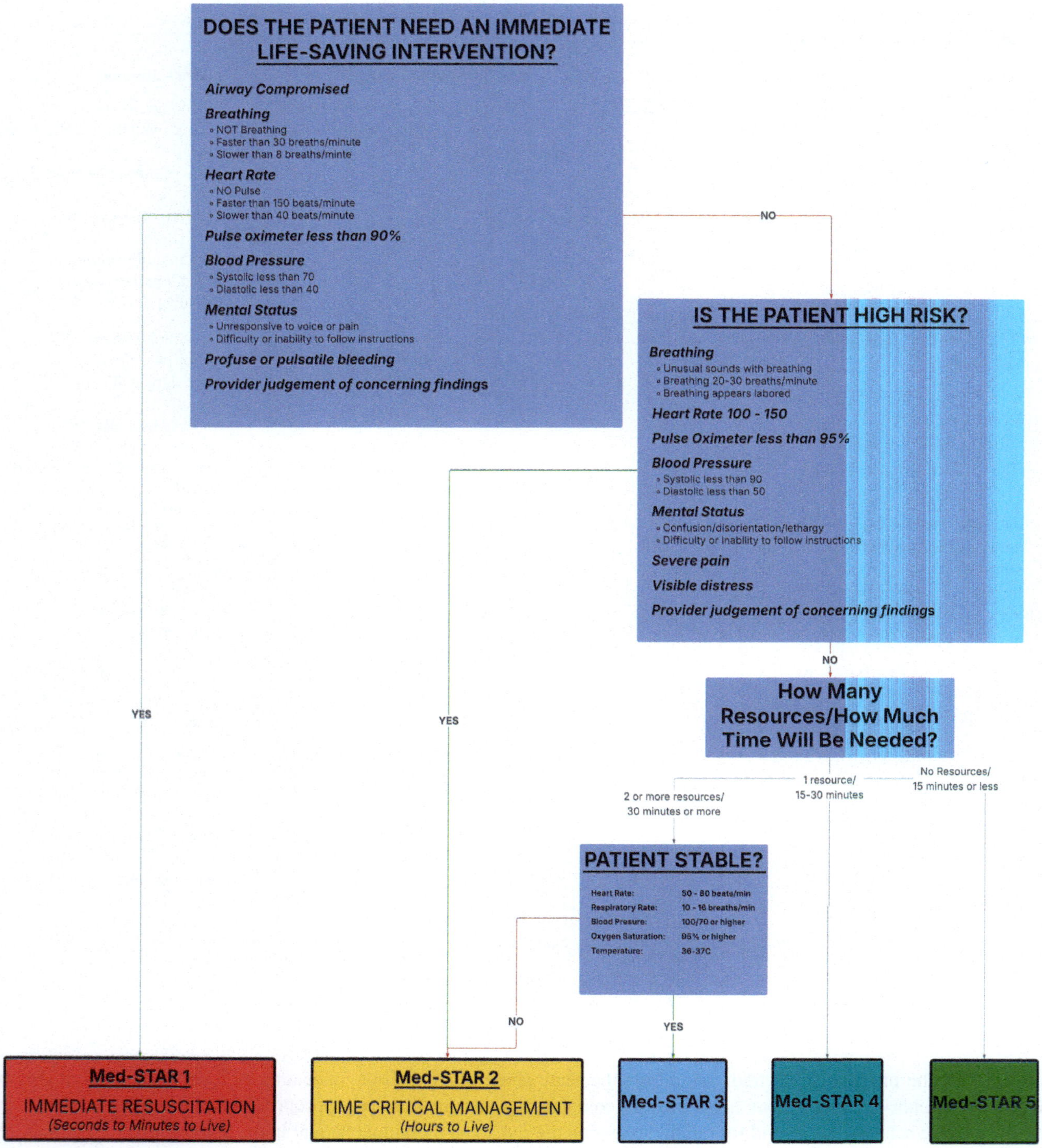


**FIGURE 7.1 Medical spaceflight triage acuity rating.** **Triage scheme created by Dr. Dana Levin, 2025.**

immediate life threat. Since the near-term future of spaceflight is more likely to involve medical events than traumatic ones, we will focus this chapter on the more general ABCDE approach. However, providers and planners should recognize that other models are just as valid and may be more applicable to their conops.

### 7.1.3 Resuscitation

Resuscitation is a general term for the stabilization of unstable patients. Unstable patients are at imminent risk of death or permanent disability; thus, action must be taken rapidly and decisively, even in the setting of incomplete information. This means resuscitative techniques focus on fast, short-term interventions to buy time for later definitive care, and diagnosis is limited to identifying life threats rather than specific causes. In general, resuscitation is an active, hands-on process of rapid assessment and treat-as-you-go interventions meant to rapidly interrupt life-threatening processes and stabilize a patient to buy more time to gather information and determine a more definitive care plan.

Resuscitation in this chapter is defined as the treatments, steps, maneuvers, and interventions needed in the acute phase of an illness or injury to restore vital functions (e.g., breathing, oxygenation, circulation). In practice, it will almost always occur in tandem with the triage step and represents a careful and nuanced balance of assessment and intervention. Providers must learn to gather just enough information to enable them to act, as delays in interventions at this stage can be life threatening. However, acting with insufficient data may also cause problems. Resuscitation algorithms such as the primary survey (ABCDE) and secondary survey approach such as those used in Advanced Trauma Life Support or Prehospital Trauma Life Support are designed to maximize this balance for inexperienced providers. In general, during a medical emergency, it is better to act than not (e.g., a patient without a pulse will not become more dead from an injury sustained during CPR). It is also important to note that resuscitation is not typically an end in itself since it is primarily meant to buy time for a more complete diagnostic workup and more definitive care. Thus, a stabilized patient is later handed off to other specialized teams for continued monitoring and treatment.

### 7.1.4 Critical care

Critical care is an intensive period of monitoring and managing a patient once they have been stabilized through their initial resuscitation. Newly resuscitated patients are tenuous and likely to decompensate rapidly with little warning. Critical care providers keep a close watch for early signs of decompensation and seek to intervene to keep a patient stable during the acute phase of their recovery. It is the continued life-sustaining treatment required to keep a severely injured or ill patient alive until recovery or transfer to a definitive care facility. There is much overlap between critical care and resuscitation, and often a patient undergoing critical care can require acute interventions due to decompensation, recurrence of an acute illness, or complications of a previous injury.

In the spaceflight environment, what may be routine in a modern Earth-based healthcare facility becomes more complicated and restrained. The microgravity environment not only alters how the human body functions but also affects the effectiveness of resuscitation and critical care tools and techniques. Furthermore, spaceflight places a premium on mass, power, volume, data, and bandwidth (MPVDB), with hard limits that must be shared between all mission aspects (i.e., a liter of normal saline fluid can come at the cost of a liter of rocket fuel). The limited crew and constrained resources, combined with the long times for transit from space to a definitive care facility, make the choice to undergo any medical intervention a complicated balance between mission, vehicle, crew, patient, resources, and time.

For example, while cardiopulmonary resuscitation (CPR) may be possible in space, the routine critical care often required after the return of spontaneous circulation in cardiac arrest may not be feasible. Especially during dynamic phases of flight (launch, reentry, landing, and recovery). The volume of fluids, variety of medications, the mass of the machines and monitors, and the diagnostic and interventional techniques required to provide the current standard of care are simply too heavy, too large, too power hungry, too untested, or require substantial engineering to be directly translated to spaceflight. The mission risk reduction gained by manifesting such equipment or tasking engineers must be weighed against the likelihood of occurrence, the consequences of the outcome, and other mission risks that might be introduced by directing resources to critical care medicine. That is not to say that some level of critical care, or more accurately prolonged field care, is impossible. In this chapter, we will take a broad look at the concepts of resuscitation, the associated critical care, and discuss how these concepts, techniques, diagnostics, interventions, and treatments can be modified, adjusted, forgone, or supplanted in spaceflight, given the constraints discussed.

Additionally, the medical team is likely to be small, perhaps limited to 1–2 individuals with the necessary training, as routine functions to maintain the ship cannot be neglected for long. This means that a medical event requiring critical care activities will place a substantial strain on the mental resources of the crew in addition to the medical resources available for the mission. Mission planners must keep this in mind when planning for critical care support and account for periods of rest as well as ensuring the remaining crewmembers can properly maintain the vehicle so as not to turn a single patient emergency into a mass-casualty event.

### 7.1.5 The ABCs of space medicine

Resuscitation and critical care in space is more akin to care in resource limited environments like ocean vessels, remote towns, research station , and wilderness research camps. Some conditions that may be relatively simple to treat on Earth may be significantly more challenging or even impossible to treat in space. Resuscitation and stabilization for space missions over the next 10–15 years must focus on techniques that can be adapted for a single skilled provider and focus on stabilizing and maintaining patients for extended transport times and/or survival through the higher G forces of reentry.

A final consideration is that these are, by definition, time sensitive processes dependent on immediate responses to rapid and potentially unexpected changes in a patient's status. While real-time telemedical guidance from Earth may be possible in Earth orbit, missions to the Moon, Mars, or even certain missions in Earth orbit may involve delayed or absent communications. Due to the time-sensitive nature of these procedures, it is likely that providers on board the vehicle will need to be empowered and capable of responding independent of ground support. Mission planners must account for these risks on each mission and carefully weigh the intense training and resource demands for resuscitation and critical care with the risks that such events may occur on a given mission.

This is where the ABCDEs assessment is most useful. These are high stress and high task load moments, so an algorithmic approach is an excellent tool to maintain focus and provide guidance even when the provider feels overwhelmed. ABCDE is designed to be treat-as-you-go, requiring the provider to pause at each identified issue and fix it before moving on. Each step is brief and often feels incomplete because they are not designed to be definitive. They are designed only to buy time and allow the provider to evaluate the next step. However, even a few seconds or minutes can turn an emergency into a stable scenario where the patient can heal. This is a core priciple of emergency medicine and emergency medical systems.

For example, the A—airway step asks only if there is a path for air to get into the lungs. If not the provider must open it. Everything else, including breathing is secondary, but if the airway is not open, rescue breaths cannot get into the patient's lungs, so step 1 is always to make sure there is a pathway for air to flow. B—breathing answers only whether a patient is breathing on their own and, if not, asks the provider to breath for them. Once breathing is established, C—circulation asks two questions; is the patient's heart beating AND is the patient bleeding? if there is no heartbeat, a provider must attempt CPR (Section 4); if there is bleeding, the provider must achieve hemostasis (Chapter 6). D—disability is an assessment of mental status and obvious injury. These issues are not immediate life threats but may become so if not addressed. In general, if A, B, and C are absent, the patient is Med-STAR 1, if they are present but abnormal, the patient is Med-STAR 2. When the provider can move on to step D , the case is most likely Med-STAR 3 or lower. Already, just with simple interventions and following a step-by-step approach, the provider has bought their patient hours to days of time and saved their life with minimal cognitive load. It all begins with A. E refers to environment and exposure. In this context, it is a quick assessment of threats to life from the environment such as hypothermia or unstable surfaces and leads into the secondary survey with the reminder to expose all of the patient in preparation for a thorough examination.

## 7.2 Airway

### 7.2.1 Anatomy and description

The upper airway consists of the nose, nasal cavity, nasopharynx, mouth, oral cavity, oropharynx, laryngopharynx (including the epiglottis, vocal cords, and glottis). The lower airway includes the trachea, main bronchi, and bronchioles to the alveoli. To enable oxygenation and ventilation, air must be able to travel from the outside environment, exchange oxygen for $CO_2$ in the alveoli, and return back to the outside environment. These are a series of progressively smaller tubes to draw air in under negative pressure developed by the muscles of the chest and diaphragm and allow it to flow out under passive and/or active contraction of the chest muscles.

The most common airway challenges are obstructions, and these are most commonly food or other ingested objects in the large airways (choking), swelling and/or mucous plugs in the smaller airways (bronchitis/bronchiolitis/pneumonia), or irritant/muscular constriction and inflammation of too many bronchioles (bronchiolitis/reactive airway disease). Swelling of the higher structures such as the epiglottis, larynx, and pharyngeal tissue may also occur, typically in the setting of severe smoke inhalation or overwhelming infection. Issues often present suddenly, are terrifying for the patient causing severe agitation, and must be managed immediately or the patient will lose consciousness and die.

The first indication may be new sounds audible in the airway such a wheezing or snoring (stridor) or an absence of sound indicating the patient has stopped breathing or ibreating ineefectively. Both are highly concerning if audible without an adunct such as a stethoscope and should be considered an emergency until proven otherwise. Table 7.3 illustrates some examples of sounds, their likely meanings, and a suggested initial intervention/important context for the provider.

**TABLE 7.3** Examples of common airway sounds and what they might mean.

| Airway sound | Phase of breathing | Medical name | Anatomy most likely responsible | Important clues | Interventions |
|---|---|---|---|---|---|
| High-pitched harsh sound, louder on inspiration | Inspiration (breathing in) | Stridor | Upper airway (larynx, trachea, epiglottis) | *Sudden onset*. Recent ingestion (choking or allergic exposure). Recent trauma*Gradual onset*. Fever (infection—croup, epiglottitis) | Airway first: sit upright, give oxygen, avoid agitation; consider nebulized epinephrine, steroids; prepare for intubation (Chapter 7) and/or surgical airway (Chapter 9) |
| High-pitched musical sound, mostly expiratory | Expiration (breathing out) | Wheezing | Lower airways (bronchi, bronchioles) | History of asthma, chronic obstructive pulmonary disease, smoking, allergic trigger, viral illness | Bronchodilators (albuterol), oxygen, steroids if indicated |
| Crackling/ bubbling | Either or both | Crackles/rales | Fluid in the alveoli (pulmonary edema, pneumonia, bronchitis, heart failure) | History of asthma or heart failure. Recent exposure to smoke. Recent illness | If hypoxic—provide oxygen, consider positive pressure if needed. Treat underlying condition. Infection—antibiotics. Heart failure—diuretics |
| Diminished or absent sounds | Either or both | Absent or decreased breath sounds | Complete obstruction of the airway, collapsed lung, or liquid in the lung (effusion) | Is the patient breathing? Do they have a pulse? Injury to the ribs? Recently eating | Oxygen, Evaluate for choking—Heimlich. Chest imaging with pneumothorax or pleural effusion—consider chest tube |
| Wet gurgling | Either or both | Gurgling | Fluid (vomit, blood, secretions) in the throat | Is the mental status intact?Secretions or vomit. Recent trauma—blood | Suction/remove excess fluid from the airway |

### 7.2.2 Triage

*Airway* is the first step in the ABCDE primary survey. Obstructions in the airway prevent oxygen flow and will kill a patient imminently. Fortunately, it usually takes only a few seconds to gather the necessary context and formulate a plan for this step.

#### *7.2.2.1 Assessing the airway*

Evaluating the airway is often as simple as asking the patient a question such as "hey, can you hear me?" If the patient can understand the question and respond in complete sentences, they are able to move air in and out of their lungs, across their vocal chords, and out into the environment (A), there is adequate gas exchange (B), and there is sufficient blood flow to the brain (C). Since cognition is intact in this scenario D is also covered and all that's left is an analysis of the environment and a reminder to expose all relevant parts of the patient to thoroughly evaluate them for hidden injuries or signs of illness. A normal response to this kind of question means the provider has completed nearly the entire primary survey in less than 5 seconds and is often able to assigne the patient to at least Med-STAR of 2 if not an even less acute number since no immediate life threats exist.

However, in the rare case of a true airway obstruction, providers will often arrive after the patient has lost or is losing consciousness and must proceed with the algorithmic approach.

- Step 1: Pause and assess the scene
  - Yourself: are you in a position to respond? Is the patient in a place you can get to without much risk?
  - The environment: was it an accident that caused a traumatic injury? Was the patient recently eating? Is there a fire and/or smoke? Are there obvious signs to explain what happened?
  - Your Teammates: is anyone else in harm's way? Is anyone else in a position to assist? Call for assistance if possible. You will likely need help.

If the scene is safe, and you are able to respond with limited risk, proceed. If not, avoid putting yourself at risk since this both adds a new patient AND removes a potential responder.

- Step 2: Assess the patient look and listen
  - Can the patient move and respond to instructions? Do they appear conscious?
  - Are there unusual or absent sounds?
  - Was this a sudden, unexpected event?
  - Can they talk? How many words can they speak before pausing for breath (1, 2, 3, 4, 5?)
  - What does their voice sound like
- Step 3: If easily done, assign a mental number for the acuity of the case (Fig. 7.1). This step can be skipped if necessary, but the number will help rapidly communicate acuity level to others. If all else fails, call it red/emergent (seconds or minutes to live), yellow/time critical (less than 12 hours to live), or green/delayed (more than 12 hours to live).

This step often takes less than 1 minute. You may feel it is not complete and you do not have enough information; however, in nearly every case, if you have been observing for 30 seconds or longer, you have sufficient information to proceed to resuscitative airway steps.

### 7.2.3 Resuscitation

Resuscitation at this stage has one goal. Secure and establish the airway. Everything else is a secondary concern and can wait until the airway is secure.

Initial interventions use what is immediately available, typically only the provider's own hands and whatever tools your colleagues are able to bring.

#### *7.2.3.1 Manual maneuvers*

These are temporary measures intended to help clear an airway that is momentarily blocked, often by the patient's own pharyngeal tissues. It is worth noting that these maneuvers will have limited utility in a reduced or microgravity setting. In microgravity, obstruction by the tongue and other oropharyngeal tissues is less likely since gravity is not present to collapse them. However, it is not clear how significant these effects may be in reduced gravity settings such as the moon or Mars. Additionally this will become important on artificial gravity stations. Each of these maneuvers is simple, fast, and requires no specialized equipment.

#### 7.2.3.1.1 Recovery position

Place the patient on their side, with their head resting on their arm and the upper knee bent, to help drain any secretions or emesis. This is intended to allow an unconscious patient, who is breathing on their own to recover in settings where a single provider cannot provide them constant attention, for example, mass-casualty events, limited resources where the provider must also tend to ship duties and/or prepare other medical interventions without assistance. It is most useful in mass-casualty events and/or drug overdoses and largely prevents a patient from aspirating their own vomit or collapsing their oropharynx. It is heavily dependent on gravity and thus has limited utility in microgravity settings. Table 7.4 includes indications, a step-by-step description, and modifications recommended for spaceflight. Fig. 7.2 illustrates the position in a terrestrial setting.

#### 7.2.3.1.2 Head tilt/chin lift

This involves moving the patient's chin “up” and away from the chest, forcing the neck to extend. This moves the tongue off of the patient’s oropharynx and clears a way for air to move past the oropharynx to the lungs. This maneuver also relies on gravity and is likely to be of limited use in microgravity. It may still be valuable in certain settings and in reduced gravity environments. Table 7.4 includes indications, a step-by-step description, and modifications recommended for spaceflight. Fig. 7.3 illustrates this maneuver.

#### 7.2.3.1.3 Jaw thrust

This involves forcing the patient’s mandible forward/anterior to pull the tongue away from the back of the oropharynx. This is the preferred method if there is concern for cervical/neck trauma in the patient and is less reliant on gravity so it may be more effective, if not frequently necessary, in microgravity. It is accomplished by applying pressure behind both angles of the mandible and pressing forward. In the microgravity environment, the provider should anchor their hands to the patients face/temples and rotate their hands to provide the pressure behind the mandible with their fingers and counter torque on the temple/face with the thumbs when approaching the patient from the front (opposite if approaching from the rear). This maneuver would be successful in microgravity as it physically moves the tongue forward. Table 7.4 includes indications, a step-by-step description, and modifications recommended for spaceflight. Fig. 7.4 illustrates this maneuver.

### *7.2.3.2 Basic airway adjuncts*

These are devices inserted into the upper airway to stent open the tissues present there. They function similarly to the basic maneuvers but do not require a provider to continuously hold them. Again they may have limited utility in the microgravity or reduced gravity environment. However, there is one scenario where these adjuncts may be particularly useful in spaceflight. Reentry exposes individuals to higher than normal g forces, often in the $+G_x$ (chest to back) direction. Evacuating a patient from orbit will require a reentry and a provider will not be able to hold basic airway maneuvers during this time. The hypergravity of this phase may exacerbate airway collapse in unconscious or semiconscious individuals, particularly in the presence of swelling or injury. Another scenario may be to prevent an airway collapse in individuals prone to obstructive sleep apnea during these high G, dynamic flight phases.

- Nasopharyngeal Airway (NPA): this is a temporary airway that involves placing a flexible tube, often made of silicon or rubber, into the nostril and through to the nasopharynx. It is indicated for patients who are breathing on their own but have some occlusion in the oropharynx that is not easily cleared through positioning (e.g., angioedema, intraoral masses or trauma, cervical spine instability, trismus). This is usually moderately well tolerated by patients and would have few modifications needed for spaceflight. In general when a patient is conscious enough to notice the discomfort and remove the adjunct, they likely no longer need it. Table 7.5 includes indications, a step-by-step description, and modifications recommended for spaceflight. Fig. 7.5 shows the typical appearance of an NPA.
- NPAs need to be sized appropriately for each patient; however, in spaceflight each potential patient will be known ahead of time which may allow fewer sizes to be flown. A typical set contains six NPAs, has a mass of 0.11–0.17 kg, and requires 0.35–0.6 L of volume.
- Oral airways: This is a temporary airway that involves placing a rigid plastic tool into the mouth that is shaped in a way to depress and retract the tongue in the oral cavity while a flanged end protrudes between the jaws resting on the patient's lips to keep the mouth open. It is used for patients who are able to breathe on their own but may have trouble keeping their mouth or oropharynx open. It can be inserted into the airway sideways and then reoriented with the aid of a tongue depressor. It is often an adjunct to manual maneuvers and would have few modifications needed for spaceflight. It is

**TABLE 7.4** Manual maneuver descriptions.

| Procedure | Indications | Step-by-step instructions (Earth-based standard) | Modifications/considerations in space (microgravity) |
|---|---|---|---|
| **Recovery position** | • Unconscious but breathing normally<br>• Risk of vomiting or aspiration<br>• No suspected spinal injury | **1.** Kneel beside patient<br>**2.** Straighten the patient's legs<br>**3.** Place the arm nearest you at a right angle to the patient's body<br>**4.** Place the hand farthest from you against the patient's cheek<br>**5.** Bend the patient's knee furthest from you to 90 degrees<br>**6.** Roll patient toward you onto their side, The angled arm should now be in front of them with the cheek supported by their other hand, palm down to the ground to prevent rolling forward. The bent knee will serve a similar purpose. An alternative is to position the angled arm behind the patient to provide a degree of protection for them rolling backwards.<br>**7.** Tilt head back slightly to maintain airway<br>**8.** Adjust the position of the patient's limbs as needed to achieve stability<br>**9.** Do not leave them unattended for more than a 5 minutes at a time. | "Side-lying" has no meaning in microgravity. The goal is to prevent vomit and secretions from blocking the airway, and control the position of the tongue and other tissues to prevent collapse of a patient's airway.In a microgravity setting,<br>• Secure patient with restraints to a fixed location<br>• Position head lower than torso relative to airflow/vent direction if possible<br>• If available, place active low suction into the patient's mouth to catch fluids.<br>• If suction is not available, place an absorbent pad in the corner of the patient's mouth and secure it with tape or a strap, so as not to obstruct the airway<br>• Do not leave them unattended for more than 5 minutes at a time. |
| **Head tilt/ chin lift** | • Unconscious patient<br>• No concern for cervical spine injury<br>• Suspected tongue obstruction | **1.** Place one hand on the patient's forehead<br>**2.** Place your fingers under bony part of the patient's chin<br>**3.** Gently tilt the patient's head backward and lift the chin upward to open airway<br>**4.** Avoid pressure on soft tissues under chin<br>**5.** Maintain this position to prevent recollapse of the airway or, if the provider needs their hands, limit the time not holding this maneuver in place to less than 1 minute at a time. | • In microgravity, the tongue and oropharyngeal tissues are less likely to collapse the throat due to the absence of a gravitational vector, and the neck muscles may somewhat extend the neck without assistance making this maneuver less relevant in microgravity. It is still possible for a patient to be positioned in a way that compresses these structures, and reduced gravity settings may also need this procedure.<br>In a reduced or microgravity setting,<br>• Rescuer and patient must both be anchored (foot loops, restraints) to avoid pushing away from each other<br>• Head tends to float back excessively, minimal tilt may be required<br>• Head straps or bulky padding may be able to maintain position once airway is opened |

(Continued)

**TABLE 7.4** (Continued)

| Procedure | Indications | Step-by-step instructions (Earth-based standard) | Modifications/considerations in space (microgravity) |
|---|---|---|---|
| **Jaw thrust** | • Unconscious patient<br>• Suspected or confirmed cervical spine injury<br>• Trauma patients | **1.** Position provider at patient's head<br>**2.** Place fingers behind angle of the patient's mandible<br>**3.** Provider's thumbs rest on their cheeks or chin<br>**4.** Lift jaw forward by pressing the mandible angle towards the provider's thumbs and without tilting head<br>**5.** Maintain this position to prevent recollapse of the airway or, if the provider needs their hands, limit the time not holding this maneuver in place to less than 1 minute at a time. | • In microgravity, the tongue and oropharyngeal tissues are less likely to collapse the throat due to the absence of a gravitational vector, and the neck muscles may somewhat extend the neck without assistance making this maneuver less relevant in microgravity. It is still possible for a patient to be positioned in a way that compresses these structures, and reduced gravity settings may also need this procedure.<br>In a reduced or microgravity setting,<br>• This maneuver braces the pulling force against the patient, limiting the need for additional restraints and thus may be preferable to the head tilt/chin life in microgravity settings. However, it may be easier for the provider to be facing the patient rather than positioned above their head. Rescuer and patient anchors (foot loops, restraints) are still likely to be beneficial. Avoid rotational forces that could move cervical spine |

FIGURE 7.2 **Recovery position.**


often quite uncomfortable for patients and will precipitate a gag reflex in most patients. Table 7.5 includes indications, a step-by-step description, and modifications recommended for spaceflight. Fig. 7.5 shows the typical appearance of an OPA.

- OPAs need to be sized appropriately for each patient; however, in spaceflight, each potential patient will be known ahead of time which may allow fewer sizes to be flown. A typical set contains six OPAs, has a mass of 0.1–0.12 kg, and requires 0.25–035 L of volume.

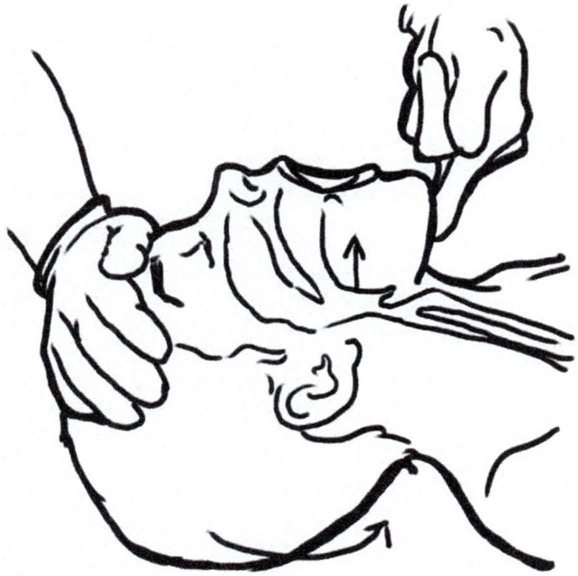

FIGURE 7.3 Head tilt, chin lift maneuver.

FIGURE 7.4 Jaw thrust maneuver.


### *7.2.3.3 Supraglottic airways*

Supraglottic airways: this is a temporary airway that involves placing a tube through the mouth and oropharynx to the epiglottis where it stents open the epiglottis and communicates with the opening to the trachea, masking the rest of the oropharynx and esophagus either through soft rubber occlusion or inflation of a balloon. These were designed as bridges to a definitive airway (discussed in the next section) intended to be used in a patient with a difficult or failed airway or by providers with less technical skill to stabilize them for transport to a higher level of care. They have a high rate of first pass success, even by nonexpert providers, and have been tested in microgravity and other space analog environments via parabolic flight and underwater studies with good results (Hinkelbein et al., 2021; Warnecke et al., 2019, 2021). They have also been used successfully in surgical procedures and in a few cases as bridge therapy to a definitive airway for up to 2 days. However, cases that require this level of airway support tend to be quite severe with injuries or illnesses that often require several days to weeks on a ventilator. These tubes are placed deep in the oropharynx and are poorly tolerated by conscious patients, requiring continuous sedation. Continuous sedation is a significant limitation in the spaceflight

**TABLE 7.5** Basic airway adjuncts.

| Airway adjunct | Estimated mass | Estimated volume | Indications | Contraindications/ cautions | Insertion—step-by-step (Earth standard) | Spaceflight / microgravity modifications |
|---|---|---|---|---|---|---|
| Oropharyngeal airway (OPA) | Single: ~12–20 g<br>Packaged Set of 6:<br>~0.10–0.12 kg | Single:<br>~35–50 $cm^3$<br>Packaged Set of 6:<br>~250–350 $cm^3$ | • Unconscious patient<br>• No gag reflex<br>• Tongue obstruction<br>• Need to facilitate bag-valve-mask ventilation or reentry while unconscious | • Conscious or semiconscious patient<br>• Intact gag reflex<br>• Oral trauma | **1.** Measure from corner of mouth → angle of mandible<br>**2.** Open mouth with "scissor" technique<br>**3.** Insert with tip toward hard palate<br>**4.** Advance until soft palate reached<br>**5.** Rotate 180° (adults)<br>**6.** Flange rests on lips<br>**7.** Secure OPA with tape or strap once placed | • Patient and rescuer must be restrained to prevent recoil during insertion<br>• Debris control is critical—loose secretions float and may be pushed back into the throat |
| Nasopharyngeal airway (NPA) | Single: ~18–30 g<br>Packaged Set of 6: 110–170 g | Single:<br>~70–100 $cm^3$<br>Packaged. Set of 6:<br>~350–600 $cm^3$ | • Altered mental status<br>• Intact gag reflex<br>• Trismus or oral obstruction<br>• Need for airway support with minimal stimulation | • Suspected basilar skull fracture<br>• Severe midface trauma | **1.** Measure from nostril → earlobe or angle of mandible<br>**2.** Diameter ≈ patient's small finger<br>**3.** Lubricate NPA generously with surgical lubricant<br>**4.** Insert into larger nostril if known<br>**5.** Bevel toward septum<br>**6.** Advance gently, following nasal floor<br>**7.** Stop if resistance is met and reposition or try the other nostril<br>**8.** Flange rests at nostril | • Blood and mucus do not drain as effectively in microgravity, suction, or clear the nostril before and after insertion<br>• Advance slowly to avoid floating recoil |

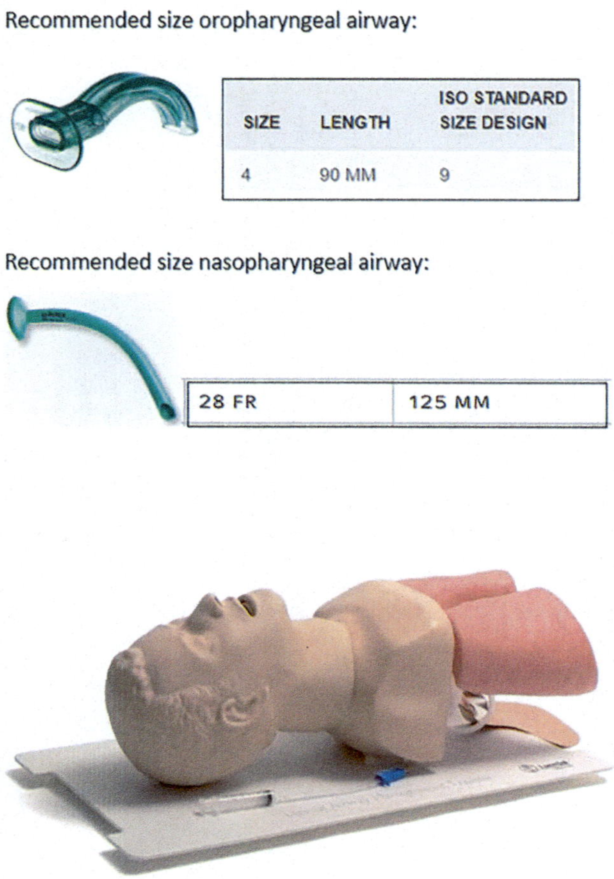


FIGURE 7.5 Basic airway adjuncts.
https://laerdal.my.site.com/HelpCenter/s/article/What-are-the-recommended-sizes-of-oropharyngeal-airway-and-nasopharyngeal-airway-for.

environment, requiring substantial resources and high levels of expertise. It may be unavoidable in some cases but will need to be considered carefully within the limitations other spaceflight environment.

More concerning for the supraglottic airway devices in spaceflight is that prolonged use is associated with a high rate of complications, including aspiration and/or swelling in the upper airway, that may make further airway management and/or weaning them off of the device difficult (Naaz et al., 2020; Subramanian et al., 2016; Sandefur et al., 2020; Staudinger et al., 1993). This is not a problem if they are intended for short-term use, as they might be for airway protection during a surgical procedure or as a bridge therapy to a definitive airway. However, procedures involving prolonged general anesthesia are not likely to be done in space in the near-term future and emergencies requiring ventilatory support are likely to need support for more than 1 day. Mission planners should be cautious when considering these tools because the use case is likely to be narrow and, despite the tempting ease of use, the risks associated with using them for prolonged periods may not help solve the problems for which they are manifested.

For these reasons, it may be best to spend the time and resources to train CMOs in definitive airway techniques, particularly since endotracheal intubation success rates for trained novices under emergency conditions are close to 80% with direct laryngoscopy alone and 90% with video laryngoscopy techniques (Garcia et al., 2023).

### 7.2.3.4 Definitive airways

Definitive airways are essentially stents with an inflatable barier at the end positioned below the larynx to keep a patient's airway open, protect the airway from aspiration of secretions or vomit, or to bypass an obstruction. The intention is to be able to support the patient for an extended period of time through mechanical ventilation. Definitive airway typically refers to either an endotracheal tube, a cricothyroidotomy, or a tracheostomy. The latter two are surgical techniques discussed in Chapter 9 and will not be covered here. For the purposes of this chapter, we will focus on endotracheal intubation as a temporary, nonsurgical procedure to secure a patient's airway for extended periods of time.

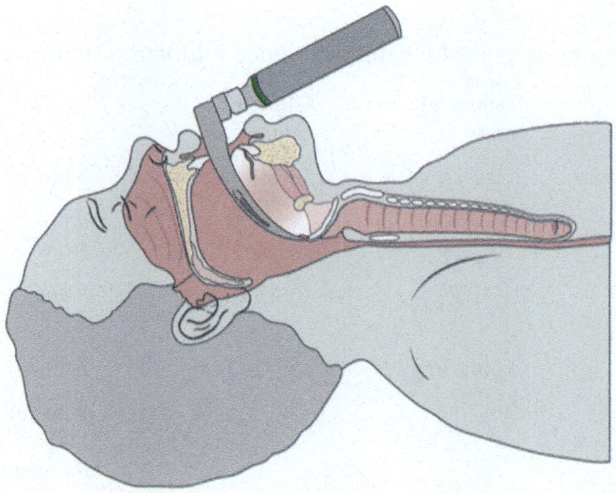

FIGURE 7.6 **Endotracheal intubation.** In this cross section, a laryngoscope is inserted through the mouth into the oropharynx to enable direct visualization of the vocal cords and intubation.
Courtesy Open Critical Care & World Health Organization via Creative Commons.

Endotracheal intubation (Fig. 7.6 involves inserting an endotracheal tube (ET tube) through the mouth and vocal cords until the tip rests just above that carina (where the trachea splits into the left and right main stem bronchi). A balloon or "cuff" located near the end of the tube is inflated in the trachea distal to the vocal cords to block secretions and vomit from entering the lungs. The tube itself is then secured to the patient's face to prevent it from becoming dislodged and connected to a mechanism for ventilation and oxygenating the patient (e.g., a bag-valve-mask [BVM] or a ventilator). Intubation is considered the "gold standard" for securing an airway during resuscitation. See Chapter 9 for discussion of surgical airways.

The standard of care for emergency intubation is called rapid sequence intubation (RSI). The equipment necessary includes the tube itself, the laryngoscope, paralytic and sedative medications, a mechanism for mechanical ventilation, suction, a rescue device, methods for confirming placement of the tube, a method for securing the tube in place, a syringe to inflate the cuff, and physical restraints/foot holds for both the provider and the patient. Table 7.6 contains a representative example of the equipment needed for RSI. This does not include the equipment needed to start and maintain vascular access nor for sedating and monitoring the patient beyond the minimum listed in the table. Both of these are essential for successful RSI and intubation management. These are discussed in Chapter 5 and Chapter 19, respectively.

In addition to these materials, adjunct mediations should be considered to manage the common complications that may occur, such as IV push doses of phenylephrine (50–200 mcg/bolus) and epinephrine (5–20 mcg/bolus) for hypotension, as well as additional sedation and paralytic agents as needed. There are many more medications and considerations for this type of procedure, but it is beyond the scope of this chapter to cover all possible contingencies. The general principle is that anticipating what can go wrong and preparing accordingly helps ensure uncomplicated success, and there is always time to prepare, even in emergencies.

Table 7.7 describes a representative procedure for rapid sequence intubation and some suggested modifications for the spaceflight environment.

Once the tube is in place, ongoing sedation, assessment, and management are critical. Patients occasionally attempt to pull the tube out and it may be necessary to restrain their limbs to prevent this and avoid turning a stabilized situation into a dangerous and chaotic one.

#### 7.2.3.4.1 Common contingency responses

Can't see cords? → Optimize once → **abort attempt**

Failed attempt? → Can you ventilate?

- **YES** → Reoxygenate → change technique→ try again
- **NO** → Use a fescue airway technique (e.g. bougie or smaller tube) immediately

Can't ventilate + can't intubate? → **surgical airway**; see Chapter 9.

**TABLE 7.6 Equipment and materials needed for rapid sequence intubation (RSI).**

| Equipment | Estimated mass | Estimated volume | Estimated power requirement |
|---|---|---|---|
| Two endotracheal tubes (ETT) half a size difference apart (e.g., 7.5 and 8Fr) | 20–30 g each | 30–50 $cm^3$ each | None |
| Stylet (ETT stylet) | 10–15 g | 10–20 $cm^3$ | None |
| A 10 ml syringe filled with saline or sterile water for cuff inflation | 5–10 g | 10–15 $cm^3$ | None |
| Laryngoscope handle (video) | 200–350 g | 200–300 $cm^3$ | Battery powered (≈1–3 W during use) |
| Video laryngoscope blade (single-use) | 40–70 g | 60–100 $cm^3$ | Powered via rechargeable battery. The Mcgrath uses a 3.6 V battery with ~800 milliamp hours |
| *Alternative: direct laryngoscope blade* | *60–100 g* | *80–120 $cm^3$* | *Alkaline AA or AAA battery powered handle for the tip light* |
| Bag-valve-mask (adult) | 400–600 g | 1.5–2.5 L | None |
| Oxygen source (portable regulator only) | 200–400 g | 300–600 $cm^3$ | None |
| Suction device (portable, powered) | 1000–2500 g | 1000–1500 $cm^3$ | 10–30 W (intermittent) |
| Yankauer suction catheter | 10–20 g | 20–40 $cm^3$ | None |
| End-tidal $CO_2$ detector (colorimetric) | 15–30 g | 20–40 $cm^3$ | None |
| Pulse oximeter (finger unit) | 40–70 g | 30–60 $cm^3$ | Battery powered (≈0.1–0.5 W) |
| Blood pressure cuff | | | |
| Tape or tube-securing device | 10–30 g | 20–50 $cm^3$ | None |
| Personal protective equipment (gloves, mask) | 20–40 g | 50–100 $cm^3$ | None |
| Lubricant (single-use packet) | 3–5 g | 5–10 $cm^3$ | None |
| Backup Bougie stylet | 30–60 g | 150–300 $cm^3$ | None |
| Paralytic medication:<br>• Rocuronium (~100 mg or 1.2 mg/kg) | 25–40 g | 40–80 $cm^3$ | None |
| Induction sedative medication:<br>• Etomidate, 10 ml vial (~20 mg or 0.3 mg/kg)<br>• Ketamine, 10 ml vial (~150 mg or 2 mg/kg) | ~25–40 g | ~40–80 $cm^3$ | None |
| Syringe 5 mL (wrapped) | 8–12 g | 20–35 $cm^3$ | |
| Syringe 10 mL (wrapped) | 10–15 g | 25–45 $cm^3$ | |
| BZK wipes | 2–4 g | 6–12 $cm^3$ | |
| Label(s) + small zip bag | 2–5 g | 30–80 $cm^3$ | |

Hypotension? → Push-dose vasopressor (e.g. 1-2mg of phenlephrine every 2-3 minutes) → fluids → reduce sedation

#### 7.2.3.4.2 Rapid sequence intubation checklist

One potential tool to increase the probability of success is the use of checklists. This section contains an example of a checklist that may be useful for mission planners.

**TABLE 7.7** Endotracheal intubation steps.

| Phase | Step | Standard RSI action | Typical timing | Spaceflight/microgravity modifications |
|---|---|---|---|---|
| **Preparation** | 1 | Examine the patient's airway and assess for potential difficulties. | 1.5 min | Lower tolerance for failure.Decide early. |
| | 2 | Confirm decision to proceed with RSI | 30 s | Confirm with ground teams if possible |
| | 3 | Assign roles, call for help | 30–60 s | Limited crew, likely 1 provider and 1 assistant. Role assignments must be explicit and clear. |
| | 4 | Prepare equipment & backup | 2–3 min | All items tethered and within arm's reach. |
| | 5 | Check suction & oxygen, connect to bag-valve mask and position near operator | 30–60 s | Critical—secretions/vomit do not drain in microgravity. |
| **Positioning** | 6 | Position patient (sniffing/ramped) | 30–60 s | "Sniffing" replaced by head–neck alignment via restraints; padding and straps required.Consider alternative positions to the head of bed, based on provider familiarity, and best view. However, ice-pick, free floating, or head between knee approaches, along with anterior approaches, have not been shown to enhance success over more traditional head-of-bed positioning (Groemer et al., 2005; Warnecke et al., 2019; Thierry et al., 2023). |
| | 7 | Secure patient & operator | 30–60 s | Both must be physically anchored to avoid unintentional movements. |
| | | Assistant in position | | Assistant should be ready and positioned with suction, to provide cricoid pressure, and to hand materials to the provider. |
| **Preoxygenation** | 8 | High-flow $O_2$ (>15 L/min) | 2–5 min | To be done while preparing.Aim to ensure patient oxygenation saturation is 100% or stable for at least 2 min.A minimum of 90 s is needed to determine oxygen's effect on blood oxygen saturation.This step is important but should not delay insertion of the airway. |
| **Monitoring** | 10 | Attach monitors ($SpO_2$, BP, ECG) | 30–60 s | Chapter 19 has more details on monitoring.Assistant should monitor vitals with careful attention to pulse oximetry so operator can focus on procedure.A good rule of thumb is to abort the airway attempt when pulse oxygenation reaches 90%. |
| **Induction** | 11 | Administer induction agent | 10–20 s | |
| | 12 | Immediately administer paralytic and flush line | 5–10 s | ***Paralysis commits you; the patient cannot breathe on their own until the paralytic wears off.*** |
| **Onset/apnea** | 13 | Wait for paralysis | 30–60 s | If ventilation is needed: two-person seal preferred, beware of patient/provider drift from forces applied |

*(Continued)*

**TABLE 7.7** (Continued)

| Phase | Step | Standard RSI action | Typical timing | Spaceflight/microgravity modifications |
|---|---|---|---|---|
| **Laryngoscopy** | 14 | Insert laryngoscope, video laryngoscopy preferred | ≤30 s | Operator must be anchored to avoid pushing away from patient. Video laryngoscopy strongly preferred due to higher success rates for inexpert providers (Starck et al., 2020). |
| | 15 | Visualize Vocal cords and/or arytenoids | ≤15 s | Cricoid pressure applied by the assistant may be necessary. |
| **Tube placement** | 16 | Pass ET tube through vocal cords | ≤15 s | |
| | 17 | Advance ET tube to proper depth (21–23 cm markings positioned at patient's teeth) | 5 s | |
| **Securing airway** | 18 | Inflate cuff using 10 ml syringe | 5–10 s | Use saline rather than air if possible to avoid changes in cuff pressure in the event of atmospheric pressure changes. Excessive pressure can lead to necrosis of surrounding tissue. |
| | 19 | Remove stylet | 5 s | Brace tube on patient's face to hold in place while removing stylet, care must be taken not to withdraw the tube with the stylet. |
| **Confirmation** | 20 | Attach BVM/ventilator | 5 s | |
| | 21 | Attach colorimetric or digital $ETCO_2$ monitor if available | Within 5–10 s | Necessary to confirm placement over 4–6 breaths |
| | 22 | Observe chest rise | 10–20 s | If X ray available, confirmation with X ray will be optimal to confirm placement. |
| **Postintubation** | 23 | Secure tube | 30–60 s | Tape is sufficient, securing device preferred |
| | 24 | Sedation & analgesia | Immediately | Paralysis ≠ sedation, depending on combination used, sedative may wear off before paralytic. Patient should not be left paralyzed but awake. |
| | 25 | Ventilator setup, arrange ventilator according to settings described in later sections | 1–2 min | |
| | 26 | Ongoing reassessment | Continuous | Limited crew, criteria-based alerts should be set to allow inexpert crew or automated system to alert CMO. |
| | | Total time | < 22 min | |

*BP*, Blood pressure; *BVM*, bag-valve mask; *CMO*, crew medical officer; *ECG*, electrocardiogram; *ET*, endotracheal; *$ETCO_2$*, end-tidal carbon dioxide; *$O_2$*, oxygen; *RSI*, rapid sequence intubation; *$SpO_2$*, peripheral oxygen saturation.

Preintubation:

*Preparation:*

☐ Airway assessment complete,
☐ RSI decision confirmed,
☐ Roles assigned (operator, assistant, meds, monitor),
☐ Primary + backup airway tools immediately reachable,
☐ Suction on and tested,
☐ Oxygen source confirmed and connected to bag-valve mask,
☐ All equipment secured/tethered,
☐ Confirm IV in pace and functional,
☐ Operator and patient anchored.

*Preoxygenate:*

☐ $O_2$ to patient applied,
☐ Preoxygenate 2–5 minutes (longer if possible),
☐ $SpO_2$ 100% or stable for at least 2 minutes.

*Monitor:*

☐ Pulse oximetry,
☐ Blood pressure,
☐ ECG,
☐ $ETCO_2$ ready for immediate use.

*Medications ready:*

☐ Induction agent correct dose for weight,
☐ Paralytic correct dose for weight,
☐ Push-dose vasopressor available,
☐ Sedation/analgesia for postintubation prepared,
☐ Suction catheter, bag-valve mask, and intubation tools within reach.

Postintubation:

Confirm tube placement:

1. ☐ ETT passed through cords,
2. ☐ Depth 21–23 cm as measured by markings at patient's teeth,
3. ☐ Cuff inflated,
4. ☐ Stylet removed,
5. ☐ Ventilate,
6. ☐ $ETCO_2$ color change or waveform present,
7. ☐ Bilateral chest rise
8. ☐ X-Ray for placement (if available),
9. ☐ Tube secured in place.

Postintubation:

1. ☐ Sedation started,
2. ☐ Ventilator settings established,
3. ☐ Reassessment schedule in place,
4. ☐ Criteria based intervention plan in place.

#### 7.2.3.4.3 Surgical airway

If intubation is unsuccessful after two attempts and the patient continues to decompensate (e.g., cannot intubate and BVM ventilation is ineffective), providers must proceed to a surgical airway (see Chapter 9).

## 7.2.4 Critical care

Airway critical care generally involves maintenance of the tube and attempts to diagnose or definitively repair the airway. There are limited interventions that will be possible in spaceflight over the next 10–20 years, and the approach will be governed by each specific case. Some general considerations are presented here.

#### 7.2.4.1 Diagnostics

Laryngoscope: This tool is used for intubation but can also be used to inspect the posterior oropharynx, epiglottis, and vocal chords. These can be used for direct visualization or can include a fiberoptic video device for easier visualization. This is employed on a sedated patient since it is poorly tolerated while awake and can be useful if there is an upper airway obstruction or concern for trauma or swelling and edema. Flexible fiberoptic laryngoscope was recently tested successfully in space on the SpaceX Polaris Dawn Mission in 2024 (Polaris Program Science and Research, 2025). Chapter 9 discusses care of the upper airway and associated procedures in more detail.

Bronchoscope: A bronchoscope is a flexible fiber-optic camera with an articulating tip and light that can be used to direct the scope as well as provide different angles for viewing. Bronchoscopes often also include the ability to suction from the tip, with larger ones occasionally also including a working channel for threading tools for further diagnostics or procedures. Bronchoscopes are used either for placement of airways in difficult patients with oropharyngeal trauma, swelling, or injury or more commonly to evaluate the lower airways in the trachea and bronchi. Chapter 11 discusses bronchoscopy in more detail.

#### 7.2.4.2 Monitoring, maintenance, and treatment

During a prolonged period of critical care, it is important to be able to monitor the airway, maintain its patency, and provide any treatment. The primary ongoing care for the airway is appropriate sedation and pain control to aid the patient in tolerating whatever airway adjuncts or definitive care is in place.

### 7.2.5 Unique spaceflight considerations

#### 7.2.5.1 Launch and reentry

Launch and reentry present significant challenges to spaceflight medical care. The forces experienced make movement within the spacecraft extremely limited and likely prohibit any active interventions. These same forces also pose risks to a patient's stability and may dislodge support equipment or limit the effectiveness of any interventions or treatments. During these phases, securing the airway will be critical and may even make wearing a pressure suite during entry, descent, and landing impossible. Additionally, pump systems will need to be developed to ensure highly accurate dose delivery in spite of changing gravity. Syringe pumps that use a threaded rod as a linear actuator may present one solution to this difficulty.

#### 7.2.5.2 Sedation

Sedation is a critical element to both establish the airway as well as maintain it. An inappropriately sedated patient with a definitive airway is unable to speak, cough, or move without significant assistance. Often the airway adjunct causes significant discomfort and patients are not able to tolerate the presence or insertion of an endotracheal tube or the surgical insertion of a cricothyroidotomy without some form of analgesia and/or sedation. Inadequately sedated patients often become confused and agitated. This in itself can become an emergency if they begin to interfere with their own equipment (e.g., pull out their endotracheal tube or begin flailing about in a panic). Table 7.8 describes some common intravenous medications, common doses, volume needed, and some advantages and disadvantages of each. Chapter 19 describes sedation in space in more detail.

One solution used in prolonged field care settings can be periodic dosing of ketamine. Ketamine can be attractive as a sedative because it has less cardio suppressive side effects and minimal respiratory depression. Bolus dosing is often preferred in a resource limited setting as it allows some degree of titration of medication and prevents over-sedation if a lone provider becomes task saturated and loses track of the medication (Zietlow et al., 2019). With some medications, bolus doses can even be administered intramuscularly, sparing the need for an IV line and pump set up altogether. However, this is not practical over long periods of time. Even bolus dosing will run into spaceflight-specific concerns regarding limited supply as well as difficulty in drawing up and mixing drugs in microgravity and shelf stability (Daniels and Williams, 2023). Further, ketamine may not provide adequate sedation for maintenance of an endotracheal tube without similarly large volumes.

General concepts for sedative medications include sedating only to the level necessary; prioritizing pain control first, with sedation second; and considering scheduled breaks from sedation to avoid delirium and/or adverse reactions from prolonged medication use. It is also worth noting that lightly sedated patients, even when intubated, are still able to hear and potentially interact. As with any care plan, talking to a patient, reminding them they are safe, orienting them to the circumstances, and reassuring them can often make a substantial difference in management by reducing sedation needs, reducing delirium, and potentially speeding recovery.

**TABLE 7.8** Common intravenous sedative medications, volume needed, advantages, and disadvantages for a 75 kg individual.

| Medication | Induction dose | Maintenance dose (ICU) | Volume/day | Volume/ 3 days | Advantages | Disadvantages |
|---|---|---|---|---|---|---|
| Ketamine (10 mg/ml) | 1–2 mg/kg IV bolus | 0.5–2 mg/kg/h (used: 1 mg/kg/h) | ~180 mL/day | ~540 mL | Analgesia + sedation; bronchodilation; preserves BP | Hallucinations; ↑ secretions; ↑ HR/BP |
| Midazolam | 0.05–0.1 mg/kg IV | 0.02–0.1 mg/kg/h (used: 0.05 mg/kg/h) | ~90 mL/day (1 mg/mL) | ~270 mL | Familiar; amnestic; cheap | Delirium; accumulation; prolonged ventilation |
| Fentanyl *(analgesic sedative adjunct often used with midazolam)* | 1–2 mcg/kg IV | 1–3 mcg/kg/h (used: 2 mcg/kg/h) | ~36 mL/day (50 mcg/mL) | ~108 mL | Strong analgesia; hemodynamically stable | Chest wall rigidity; resp depression |
| Dexmedetomidine (4 mcg/ml) | *No bolus usually* | 0.2–0.7 mcg/kg/h (used: 0.5 mcg/kg/h) | ~225 mL/day | ~675 mL | Cooperative sedation; minimal resp depression | Bradycardia; hypotension; limited depth |
| Propofol (10 mg/ml) | 1–2 mg/kg IV bolus | 5–50 mcg/kg/min (used: 20 mcg/kg/min) | ~216 mL/day | ~648 mL | Rapid on/off; predictable; antiemetic | Hypotension; respiratory depression; PRIS risk |
| Etomidate (2 mg/ml) | 0.3 mg/kg IV bolus | *Not recommended for maintenance sedation due to risk of adrenal suppression when used for extended periods* | | | Hemodynamically neutral induction | Adrenal suppression; no analgesia |

*BP*, Blood pressure; *HR*, heart rate; *ICU*, intensive care unit; *IV*, intravenous; *PRIS*, propofol-related infusion syndrome; *resp*, respiratory.

### 7.2.5.3 Suctioning

Managing the secretions of a patient with an invasive airway, particularly one that is sedated and unable to do so themselves through coughing, requires frequent monitoring and intervention. This is usually done through suctioning via a handheld suctioning device or through an "in-line" suctioning device to clear secretions and condensation from an airway tube and prevent blockage or drainage into the lungs. Although this is a low-skill activity, it does require attention and manual direction of the suction catheter as well as suction capability engineered into the spacecraft. While portable suction units do exist, their mass, power, and volume requirements may be prohibitive for many spacecraft. Manual suction through squeeze bulb systems, expelled into a cloth between uses or hand driven pumps similar to those used in the 1800s, may be a good solution when intermittent suction is acceptable.

Furthermore, given the dominance of surface tension in spaceflight, it is not clear how oropharyngeal and airway secretions would inhibit a sedated patient given that the common terrestrial gravity gradient drawing the fluids down into the airway and lungs would be less of a concern. However, spaceflight is known to interfere with mucociliary clearance and causes cephalad fluid shifts, often increasing airway secretions, making the true need for airway suction in a spaceflight critical care scenario unclear. Similarly, as secretions are not likely to fall deeper into the lungs, they may be more likely to inspissate in the larger airways causing obstruction in spaceflight. Of note, medical suction itself is a critical component of both resuscitation and critical care and is discussed repeatedly through this chapter. To date, no clear solution for medical suction exists for spaceflight, although many have been proposed. Ultimately, a solution will need to incorporate spacecraft suction capability combined with an in-line fluid collection system (Houtchens, 1993; McCuaig and Houtchens, 1992).

### 7.2.5.4 Suits

Suits designed to help astronauts withstand the various spaceflight environments present unique challenges to medical care (see Fig. 7.7). While many current and future designs include telemetry and vital sign monitoring devices, interventions within a suit, including securing an airway to a ventilator and running continuous sedation, would be difficult. For launch and reentry, it may be reasonable to forgo a patient's suit and accept the elevated but small risk of decompression to allow life-saving interventions to continue.

### 7.2.5.5 Limited resources

Of all the components of resuscitation and critical care, airways may present the lowest burden on spacecraft budgets for MPVDB. Most of the equipment used terrestrially for the airway is lightweight plastic or rubber. Laryngoscopes, which are critical for intubation, are purpose made and not likely to serve any dual purposes; however, the video component for a video laryngoscope can likely be used on a multiuse tablet or screen. As noted previously, the securing the airway is not the end of the procedure and the MPVDB considerations for prolonged sedation and critical care in an intubated patient

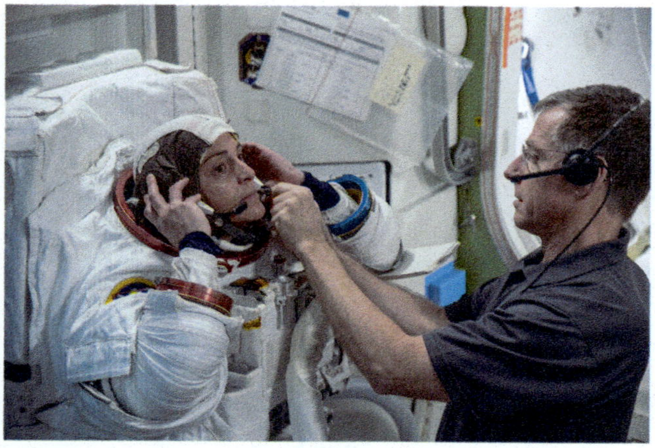

FIGURE 7.7 **Astronaut donning an extravehicular mobility unit (EMU) for an extravehicular activity (EVA) or spacewalk.** This photo demonstrates some of the limitations of neck mobility and interventional access that spacesuits can present.
Courtesy NASA.

TABLE 7.9 Additional airway equipment mass, power, and volume data.

| Item | Mass (g) | Volume ($cm^3$) |
|---|---|---|
| Bag-valve mask | 493 g | 1,049 $cm^3$ |
| Ambu bag kit | 318 g | 4759 $cm^3$ |
| ETCO2 | 15–30 g | 30–60 $cm^3$ |
| In-line suction catheter system for an ET tube | 80–160 g | 400–1000 $cm^3$ |
| Surgical airway kit (tube, hemostat, scalpel, syringe, hook) | 85 g | 591 $cm^3$ |
| Bulb suction with bag | 207 g | 1803 $cm^3$ |

must be considered. Bronchoscopes and suction equipment on the other hand are likely to be sacrificed in a risk–benefit analysis when mission planners evaluate the low likelihood of a prolonged medical airway situation and the associated costs and likely outcomes. Even more costly in terms of MPVDB is the associated sedation and ventilation equipment and resources, both of which will be discussed in further depth below. This likelihood-resource-cost-outcome analysis is really at the core of any medical planning for spaceflight. MPVDB considerations for additional airway equipment is demonstrated in Table 7.9. This decision can be somewhat objectified with new prediction and model tools and will be discussed further at the end of this chapter.

### 7.2.6 Decision-making

The decision to perform a maneuver, procedure, or intervention for a patient's airway is somewhat straightforward in most standard resuscitation and critical care scenarios. If a patient is or may shortly start having trouble protecting their airway or moving air to their lungs, then an airway should be established. However, like all interventions in medicine, it should be bridging to an additional or definitive treatment or intervention. Care after a resuscitation or intervention is usually as simple as transporting to a definitive care facility where prolonged care can be accomplished in a presumably resource-rich environment. However, in spaceflight, the decision to intervene is complicated by the limited ability and resources to provide definitive care for a prolonged period which may be necessary to transport the patient to a definitive care facility.

The list of any and all medical emergencies that could occur to a human (even if highly selected and screened) is extremely long and it is not possible to plan and equip a crew for every contingency. Fortunately, within this long list, there are many insults or emergencies that are either too minor to warrant dedicated planning/equipment (e.g., a sore throat).On the other end of the spectrum, some that are so devastating and the likelihood of survival so low (e.g., completely transecting a patient's airway and major neck blood vessels) that equipping for treatment would be moot. Between these ends lies the decision space for optimizing a medical support system that treats the most important things while maintaining the requisite limits on MPVDB. This space becomes somewhat strained when we consider extended treatment timelines required for critically ill patients to survive an extended transit to definitive medical care from an exploration mission, adding not only the capacity to treat but the capacity to *continue* to treat over time to consideration. An example of these concept and considerations applied to intubation can be seen below in Fig. 7.8.

## 7.3 Breathing

### 7.3.1 Anatomy and description

Breathing is the movement of gas into and out of the lungs enabling the exchange of oxygen and carbon dioxide between the atmosphere and the blood. Broadly breathing serves two functions.

- Oxygenation: transfer of $O_2$ into the blood.
- Ventilation: transfer of $CO_2$ out of the blood.

This exchange depends on two physiological functions:

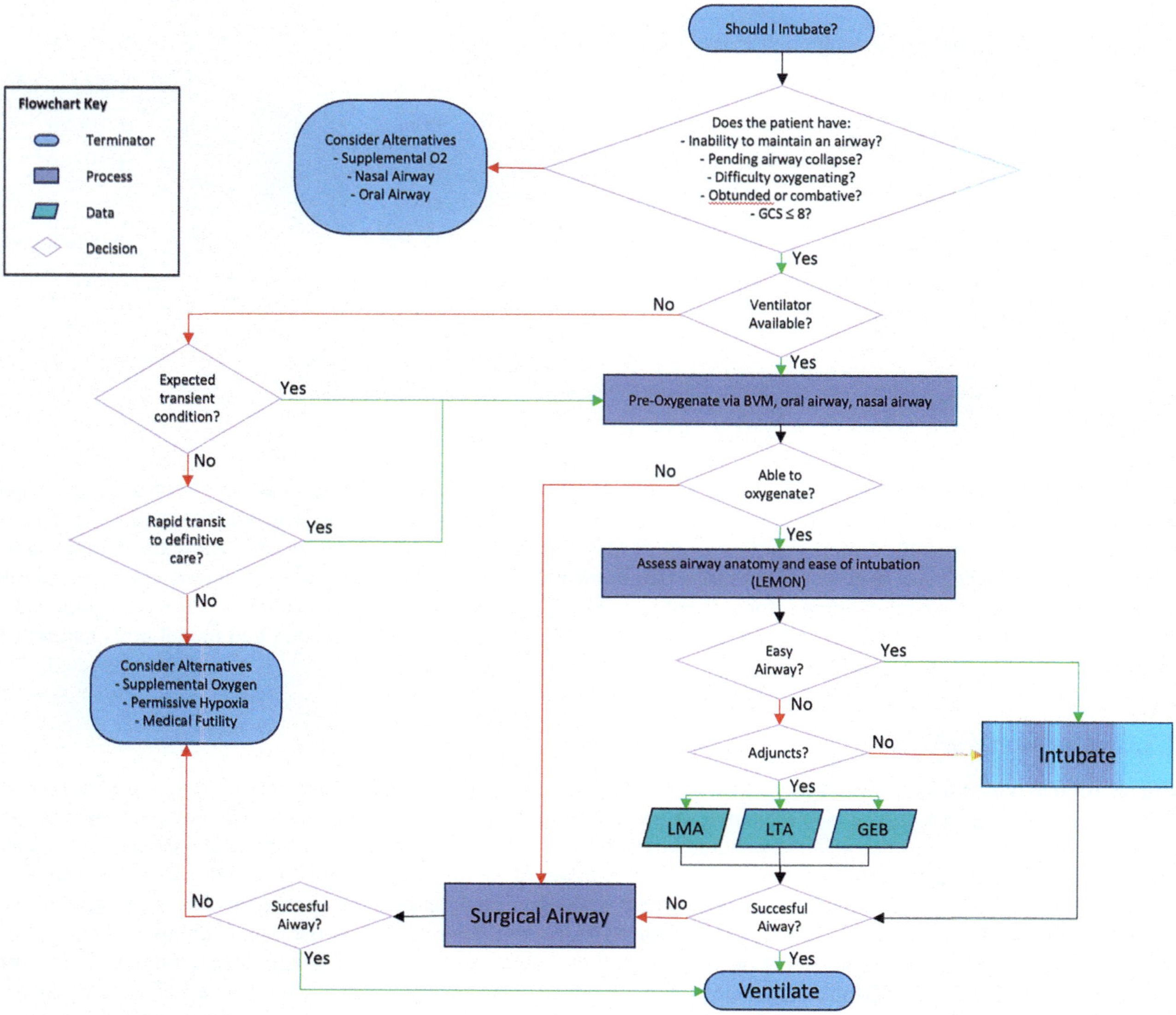


**FIGURE 7.8** Intubation decision tree.
Here, the decision to intubate is broken into steps that consider the local capabilities, evacuation options, and long-term prognosis of the patient. *GCS*, Glasgow Coma Scale; *GEB*, gum elastic bougie; *LMA*, laryngeal mask airway; *LTA*, laryngeal tube airway

- Movement of air in and out of the lungs, which involves pressure gradients generated by muscle-driven expansion and contraction of the chest cavity, involving the lungs, visceral and parietal pleura, ribs, chest wall muscles, and diaphragm (Fig. 7.9).
- Gas exchange, which takes place primarily in the alveolar sacks of the lungs and their associated capillary beds.

## 7.3.2 Triage

Breathing is the second step in the primary assessment, as it depends on an open airway. Again, if the patient is able to respond to a question verbally, then they are breathing. However, if they appear out of breath while at rest or are unable to speak a full sentence due to a need to pause to breath in between words, their breathing may be at imminent risk and they are likely a Med-STAR 1 or 2. The classic assessment for breathing is described as look, listen, and feel.

- Look
  - for chest rise/fall or cyclical fogging of a mask or airway tube,

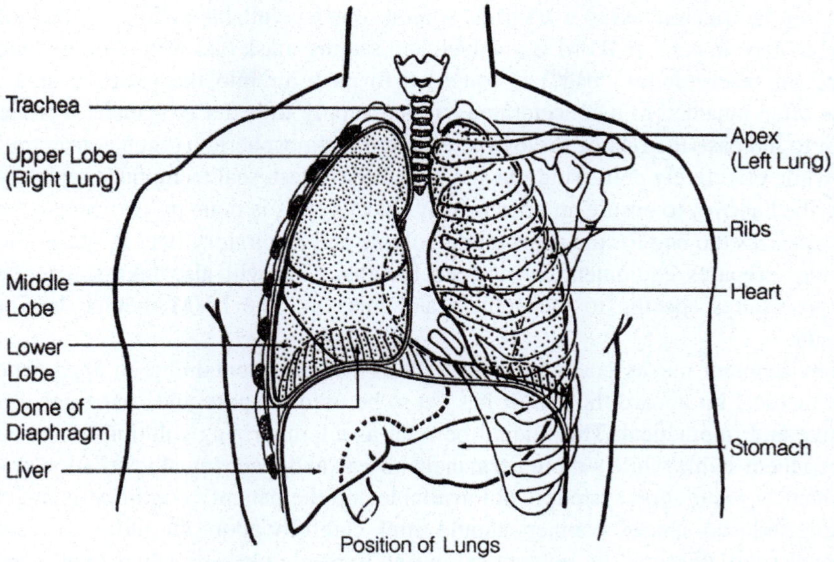


FIGURE 7.9 Basic anatomy for breathing movement.
This image was released by the National Cancer Institute, an agency part of the National Institutes of Health, with the ID 1774.

- is the rate fast or slow? This step is meant to be qualitative. While a resting breathing rate should be roughly between 10 and 16 breaths per minute, stopping to count breaths at this stage is an unnecessary delay.
- Are they straining with nasal flaring or use of accessory chest wall and neck muscles?
- is there any sign of injury that could interfere with chest movement?
- is there evidence for cyanosis (blue lips, nail beds, etc.)?

- Listen
  - Can you hear breath sounds?
- Feel
  - Can you feel air movement in and out of the patient's mouth and/or nose?

If breathing is not detected, provisions for rescue breaths must be initiated before moving on to assess circulation. In a spaceflight environment, visual and auditory cues may be difficult to ascertain, given the often-noisy environment and the possible restrictions of a spacesuit or other safety equipment (e.g., harness, straps, etc.).

## 7.3.3 Resuscitation

### *7.3.3.1 Manual ventilation*

Manual Ventilation or rescue breathing is the first step. If the patient is not able to breathe on their own, a provider must push air into their lungs for oxygenation and allow time for passive exhalation to wash out the $CO_2$. This is called positive-pressure ventilation since it does not rely on the normal *negative* pressure gradient developed by expansion of the chest wall. This should be delivered initially at a rate of 6–10 breathes per minute. Providers are often tempted to push air faster, but it is important to recognize that time is needed to permit passive exhalation and avoid building up pressure in the chest cavity. This may be even more important in space since the collapse of the anterior chest wall will not be assisted by gravity. Providers should consider counting 6 seconds in between delivering breaths.

The simplest version of rescue breathing is "mouth-to-mouth" where the provider pinches the patient's nose, closes their mouth over the patient's, and exhales directly into the patient's airway. It is critical to block the patient's nose to direct air into the lungs and prevent it escaping through the lower resistance path. Ideally a barrier with a one way valve should be used between the provider and patient to prevent secretions or vomit from entering the mouth of the provider, though this is not an absolute necessity in the early stages, particularly with small space crews since crewmembers are screened before flight and unknown infections are less of a risk. The oxygen concentration delivered through this method is typically 16%–17%, which is more than enough to sustain a patient in this state.

Mouth to mouth should be transitioned to a BVM as soon as one is available to free a provider's attention and enable preparations for the next steps in care. A BVM is a valved and sealing mask that is held to the face of the patient while a flexible self-expanding air reservoir (or "bag") is squeezed to push air into the patient's lungs, followed by passive exhalation. The bag is often attached to a concentrated oxygen supply to better oxygenate a patient. However, in space, care must be taken not to increase the risk of fire by increasing the atmospheric oxygen concentration. Practice is needed to ensure a tight seal while effectively delivering oxygen. The BVM mask seal techniques typically work well with a jaw thrust maneuver (described above) to ensure an open airway. Ideally, this is done by two people, with one focused on the mask while the other squeezes the bag for the appropriate volume and respiratory rate. In space, the assistant can hold the mask while the provider prepares equipment in between breaths. This will also take some practice and will require planners to develop procedures specific to the vehicle and mission. The BVM method has been shown to present difficulties in spaceflight.

The microgravity environment makes positioning difficult as well as maintaining an appropriate seal with the mask. While included in the medical kit aboard the ISS, it has yet to be used in space and "bagging" a patient for an extended period is labor-intensive and impractical. This should be seen as a temporizing solution for an acutely decompensating patient and used to keep them stable while a more permanent airway and mechanical ventilation system are established. In other words, if a mechanical ventilation system is not available and the patient's condition is not rapidly reversible (e.g., electrocution or choking induced apnea), planners should think carefully about the utility of manifesting these devices, particularly since manual ventilation will be impractical during dynamic phases such as launch and reentry due to the G forces and need for occupant restraints.

In general, if a patient requires a BVM, they are highly likely to also require a definitive airway (see previous section). Studies have demonstrated that BVM ventilation can be feasibly accomplished and consistent for up to 6 hours; however, this is also labor-intensive and impractical to maintain during dynamic phases of flight (Halpern et al., 2019).

#### *7.3.3.2 Monitoring during resuscitation*

In many cases, a patient who is unable to breathe on their own will require prolonged management with artificial ventilation. This means a provider will need tools to enable them to respond to changes in physiological state such as hypoxia (low $O_2$) and hypercapnia (high $CO_2$). As with everything in resuscitation, the balance of time, complexity, and need for information is challenging, and determining the "minimum necessary equipment" is highly situation-dependent and difficult to predict ahead of time. Tools such as probabilistic risk assessment, described in Chapter 2, may be useful. A good place to start is having the ability to assess vital signs, heart rate, respiratory rate, blood oxygen saturation, blood pressure, and temperature, though not all of these are needed continuously. A representative set of individual small sensors for resuscitation monitoring is described in Table 7.10 with examples of integrated monitoring systems included in Table 7.11. Planners will need to consider the use case for these devices. It is likely that for longer duration missions, or those where the risk analysis leans towards including resuscitative equipment, integrated monitoring systems will be valuable. This is because of increased capabilities for continuous monitoring and, in some cases, the ability to include interventional features such as defibrillation or pacing. Chapter 19 has more detailed discussion of monitoring needs.

#### *7.3.3.3 Supplemental oxygen and associated risk*

Often the early stages of stabilizing a patient struggling to breathe include providing supplemental oxygen. In the resuscitation stage, it is important to have the capability to deliver 100% oxygen at a flow rate of at least 15 liters per minute. There are several methods used terrestrially depending on the level of support needed. Each delivery device provides some mixing with ambient air that limits the actual percentage that the patient receives. For example, a nasal cannula delivers ~30%, a nonrebreather mask delivers ~50%, and rebreather masks deliver ~70%. The specific system is less important than the ability to titrate the percentage of oxygen received. Given the time sensitivity, it is a good idea to start with 100% oxygen. However, once the patient's oxygen saturation has stabilized, the percent should be reduced. This is both to conserve limited supplies as well reduce health risks associated with prolonged use of high oxygen levels.

In a closed environment, this is problematic, since humans only absorb 3%–5% of the oxygen we inhale. If 100% supplemental oxygen is used, the exhaled 95% oxygen will raise the cabin oxygen concentration and increase the risk of fire. While the exact risk of this will be vehicle specific, increasing fire risk is a problem in the space environment. Solutions to this problem include closed-loop systems, where exhaled gas is captured and disposed of or recycled without exposing the rest of the cabin or oxygen concentrators. Closed loop systems will require a seal against the patient's face which may make interventions difficult but might be effective for mechanical ventilation through a definitive airway or through noninvasive positive-pressure ventilation (discussed below). Oxygen concentrator systems are likely better for

**TABLE 7.10** Examples of minimum monitoring needed in resuscitation.

| | Parameter | Tool | Target physiological values | Mass | Volume | Power |
|---|---|---|---|---|---|---|
| Continuous | Blood oxygenation | Noninvasive pulse plethysmography ("pulse oximeter") | >95% saturation | 0.10–0.14 kg | 250–400 cm³ | Battery powered (~0.2–0.5 W) |
| | Heart rate | Noninvasive pulse plethysmography ("pulse oximeter") | 60–100 beats/minute | | | |
| | Respiratory rate | Timing device or impedance monitoring | ~12 breaths/minute | 0.20–0.40 kg | 500–1000 cm³ | Battery powered (~0.5–2 W) |
| Intermittent | Blood pressure | Automatic blood pressure cuff | Systolic >100 mmHgDiastolic >50 mmHgMean arterial pressure ~65 mmHg | 0.40–0.80 kg | 1.0–2.0 L | Battery powered (~5–15 W intermittent) |
| | Temperature | Electronic thermometer | ~37°C | 0.05–0.12 kg | 100–250 cm³ | Battery powered (~0.05–0.2 W) |

supplemental oxygen since these concentrate oxygen from the existing atmosphere. These systems increase oxygen to the patient by locally depleting the atmosphere. The net neutral exchange results in only a transient and local increase in oxygen concentration upon the patient's exhalation that rapidly diffuses into the cabin atmosphere and significantly limits fire risk. Mission planners must assess the risk of temporary oxygen-dependent conditions occurring in flight and weigh these risks against the complexity, mass, volume, and training limitations for their specific mission (Lange et al., 2005, Blakeman et al., 2016).

## 7.3.4 Critical care

With regard to supporting breathing, the line between resuscitation and critical care is often blurry. For the purposes of this chapter, we treat resuscitation as interrupting the initial process and critical care as interventions that maintain stability. Other texts may select different dividing lines which are equally appropriate. This section will cover noninvasive positive-pressure ventilation, mechanical ventilation, and touch on diagnostic considerations that may be useful in this phase. Monitoring is even more critical in this step with the integrated systems described in Table 7.11 being preferred from a clinical perspective.

### *7.3.4.1 Mechanical ventilation*

For conditions anticipated to require more than 1 hour of ventilatory support and where a definitive airway has been established, mechanical ventilation is necessary in order to free up resources and enable more precise control. A mechanical ventilator is functionally a pump with a triggering mechanism to deliver a breath, along with controls that can vary the volume of gas delivered, the rate of delivery, the oxygen concentration, the maximum pressure delivered, and the minimum pressure allowed. Triggers are typically by time or by patient-initiated breathing. Control of ventilators is a complex and nuanced discipline which requires a textbook in and of itself. This chapter will only cover a few basic concepts intended for basic understanding and intervention. We encourage interested parties, or those intending to manage patients seek professional level training.

For the purposes of this chapter, we use the medically standard units of $cmH_2O$ (1 $cmH_2O$ = 0.7356 mmHg = 0.09807 kPa = 0.01422 psi). Table 7.12 contains examples of portable ventilators available on the market with mass, power, and volume estimates.

The overall guiding principal is that mechanical ventilation is intended to as follows:

- Actively deliver oxygen into the patient's lungs because they are unable to do so on their own.
- Passively allow $CO_2$ to be expelled by the elastic contraction of the chest.

Important baseline considerations:

Mechanical ventilation is typically well tolerated and complications are relatively rare. However, prevention, preparation, and anticipation are powerful tools in the emergency management toolkit. In general, providers should aim to use the lowest oxygen, lowest pressures, and longest expiration possible while still supporting the patient. The rationale to keep in mind is that as follows:

- **Oxygen is corrosive:**
  - It strips surfactant from the lungs, decreasing lung compliance and increasing the work of breathing/the pressure required to maintain open lower airways.
  - It may also injure some tissues if administered for more than 24 hours. However, if a patient is severely hypoxic (<85%), high levels of oxygen are worth the risks.
- **Positive-pressure ventilation is a mechanical stress:**
  - Pressure in the chest can rupture lung tissue, allowing air to escape into the pleural cavity; this is called a pneumothorax. If not managed promptly, pressure may continue to build, compressing structures such as the heart and the remaining intact lung, compromising blood flow and resulting in a tension pneumothorax—another life-threatening emergency.
  - It can also compress veins, decreasing blood return to the heart. At high pressures (>10 $cmH_2O$), this can artificially lower blood pressure and potentially cause mild-to-moderate fluid retention, and pressure increases in the head.

### *7.3.4.2 Definitions*

- *Respiration:* $O_2$ in/$CO_2$ out.
- *Oxygenation:* Delivery of oxygen to tissues.

**TABLE 7.11** Examples of integrated monitoring systems.

| Device | Monitoring capabilities | Estimated mass | Estimated volume | Power/battery |
|---|---|---|---|---|
| **LIFEPAK 15** (physio-control) | ElectrocardiogramNoninvasive pulse PlethysmographyNoninvasive positive-pressure ventilationEnd-tidal $CO_2$Respiratory rateTemperatureDefibrillationPacing | 7.0–7.6 kg | 12–16 L | ~25–40 W active; ~4–6 h battery |
| **LIFEPAK 20e** | ElectrocardiogramNoninvasive pulse PlethysmographyNoninvasive positive-pressure ventilationEnd-tidal $CO_2$Respiratory rateTemperatureDefibrillation | **6.9–7.5 kg** | 12–16 L | ~25–40 W active; ~4–6 h battery |
| **Tempus Pro** (Philips/ Remote Diagnostic Technologies) | ElectrocardiogramNoninvasive pulse PlethysmographyNoninvasive positive-pressure ventilationEnd-tidal $CO_2$Respiratory rateTemperatureOptional ultrasound | 3.0–3.5 kg | 4–6 L | ~10–20 W; ~8 h battery |
| **ZOLL X Series** | ElectrocardiogramNoninvasive pulse PlethysmographyNoninvasive positive-pressure ventilationEnd-tidal $CO_2$Respiratory rateTemperatureDefibrillationPacing | **5.2–5.7 kg** | **9–12 L** | ~20–35 W; ~4–5 h battery |
| **Philips MRx** | ElectrocardiogramNoninvasive pulse PlethysmographyNoninvasive positive-pressure ventilationEnd-tidal $CO_2$Respiratory rateTemperatureDefibrillationPacing | **6.0–6.5 kg** | **10–14 L** | ~20–35 W |
| **Corpuls3** (modular) | ElectrocardiogramNoninvasive pulse PlethysmographyNoninvasive positive-pressure ventilationEnd-tidal $CO_2$Respiratory rateTemperatureDefibrillation | ≈6.0 kg (total) | 10–14 L | ~20–35 W |
| **Propaq M/Propaq MD** | ElectrocardiogramNoninvasive pulse PlethysmographyNoninvasive positive-pressure ventilationEnd-tidal $CO_2$Respiratory rateTemperatureDefibrillation (MD model) | 2.7–3.5 kg | 4–6 L | ~8–15 W |
| **Tempus ALS** | ElectrocardiogramNoninvasive pulse PlethysmographyNoninvasive positive-pressure ventilationEnd-tidal $CO_2$Respiratory rateTemperature | 2.5–3.0 kg | 3.5–5 L | ~8–15 W |

**TABLE 7.12** Examples of portable ventilators available on the market.

| Device | Mass (kg) | Volume (L) | Typical power (W) | Maximum power (W) | Battery (Wh) | Advantages | Disadvantages | Sources |
|---|---|---|---|---|---|---|---|---|
| Hamilton Medical HAMILTON-T1 | 6.5 | 22.63 | 50 | 150 | 72 | ICU-capable modes; internal blower (no compressed air needed); internal battery; robust transport design; optional $CO_2$/$SpO_2$ integrations. | Heavier and higher peak power; humidity/condensation management in microgravity; needs secure mounting and filter/spares; high $FiO_2$ still requires $O_2$ source. | Hamilton T1 tech specs PDF (weight/dims/power/battery): https://www.hamilton-medical.com/(HAMILTON-T1_tech-specs_v3.1.x_en_usa_10103182.04.pdf) |
| Dräger Oxylog 3000 plus | 5.8 | 9.34 | 30 | 60 | | Compact/robust; optional integrated capnography; works from $O_2$ cylinder/pipeline; internal battery supports transport. | Requires high-pressure medical $O_2$ for operation; $O_2$ logistics burden for long-duration missions; limited by $O_2$ supply; battery Wh not specified in datasheet. | Dräger Oxylog 3000 plus datasheet PDF: https://www.draeger.com/Content/Documents/Products/oxylog-3000-plus-ds-9066915-en-na.pdf |

| | | | | | | | | |
|---|---|---|---|---|---|---|---|---|
| ZOLL EMV+ 731 Series | 4.4 | 7.36 | 10 | 101 | 97 | Integrated compressor (ambient air); long battery life; altitude compensation; rugged transport focus; can operate without external $O_2$ (but supports $O_2$). | Higher peak power when running/charging; compressor noise/vibration; consumables (filters) and condensation management; still needs $O_2$ for high $FiO_2$ targets. | EMV+ 731 spec sheet PDF + operator guide excerpt: https://sema.dps.mo.gov/documents/emv-731-spec-sheet.pdf; https://manualmachine.com/zoll/emv/6508044-user-manual/ |
| Philips Respironics Trilogy Evo | 5.8 | 11.56 | 40 | 78 | | Broad patient range; long nominal runtime with internal + detachable batteries; supports multiple modes; accessory monitoring options ($SpO_2$/$EtCO_2$). | Battery Wh not clearly specified; larger volume; humidification/condensation and circuit management in microgravity; may need oxygen blender/attachments depending on $FiO_2$ needs. | Trilogy Evo spec sheet PDF: https://www.atlanticrespiratory.com/uploads/userfiles/files/documents/products/trilogy%20evo%20spec%20sheet.pdf |

(*Continued*)

**TABLE 7.12** (Continued)

| Device | Mass (kg) | Volume (L) | Typical power (W) | Maximum power (W) | Battery (Wh) | Advantages | Disadvantages | Sources |
|---|---|---|---|---|---|---|---|---|
| ResMed Astral 150 | 3.2 | 5.70 | 12 | | 95 | Very good power-to-weight; low average power draw; internal battery (95 Wh) and external battery options; compact form factor. | Max power not specified in brochure; condensation/humidification management; limited to device's intended environments/operating constraints; may require external $O_2$ for higher $FiO_2$. | ResMed Astral brochure (weight/dims/battery): https://document.resmed.com/documents/products/machine/astral-series/product-brochure/1017216_astral_brochure_hospital_apac_eng.pdf |
| Medtronic Puritan Bennett 560 | 4.5 | 11.40 | 30 | 180 | 111 | Compact; long internal battery runtime; AC or DC input; broad modes for chronic ventilation; widely available in home/clinical settings. | Older platform; high max VA; humidification/condensation management; oxygen enrichment requires external supply and monitoring kit; consumables and infection-control burdens. | PB560 brochure/spec sheet PDF: https://www.azcuenaga.ar/wp-content/uploads/2015/10/pb560.pdf |

| | | | | | | | | |
|---|---|---|---|---|---|---|---|---|
| WEINMANN MEDUMAT Standard² | 2.5 | 3.70 | 10 | 30 | | Low mass and volume; designed for emergency/transport; max power consumption listed (30 W); robust environmental qualifications in transport contexts. | Uses medical oxygen (operating gas) with high-pressure supply requirements; long-duration oxygen logistics; limited compared with ICU ventilators; battery capacity not specified in excerpt. | MEDUMAT Standard² technical data (Manualzz): https://manualzz.com/doc/o/2c0nie/weinmann-ventilation—medumat-standard%C2%B2-ventilator-instr.-technical-data |
| Vortran GO2VENT (single-use pneumatic ventilator) | 0.1 | 1.80 | 0 | 0 | 0 | Extremely low mass/volume; no electrical power; simple, disposable. | Requires continuous gas flow source of at least 10 L per min; limited monitoring/ventilation sophistication; not suitable for prolonged ICU ventilation; high gas-use risk in closed habitat. | GO2VENT product listing w/ size/weight: https://www.epsimed.com/single-use-ventilator-go2vent/ |

*AC*, Alternating current; *DC*, direct current; *EtCO₂*, end-tidal carbon dioxide; *FiO₂*, fraction of inspired oxygen; *ICU*, intensive care unit; *SpO₂*, peripheral oxygen saturation.

- *Positive end-expiratory pressure (PEEP):* Minimum pressure allowed in the airway circuit. Functionally a stent to keep small airways open and recruit more lung tissue.
- *Inspiratory/expiratory ratio (I:E ratio):* The ratio of time spent by the vent delivering pressure versus allowing passive exhalation; for example, a 1:2 ratio with a rate of 10 breaths per minute (1 breath every 6 seconds) would be 2 seconds for the inspiratory phase and 4 seconds for the expiratory phase.
- *Ventilation:* Removal of $CO_2$ from the body.
- *Rate:* Number of breaths per minute.
- *Tidal volume ($V_T$):* Volume inspired/expired in a single breath.
- *Hypoxia:* Blood oxygen less than 95% saturation.

### 7.3.4.3 Mechanics and settings of the ventilator

Ventilators are air pumps connected to a trigger mechanism to start and stop the pump.

#### 7.3.4.3.1 Triggers to turn the pump on

- Patient breath (best when the patient is breathing on their own)
- Time (best when the patient is truly apneic)
- Hybrid (best all-purpose setting)

#### 7.3.4.3.2 Triggers to turn pump off

- Time (not ideal because it can cause overpressure and lung damage)
- Pressure maximum (prevents barotrauma, might deliver lower volumes)
- Volume target (delivers fixed volume, might cause barotrauma)
- Hybrid (volume target with a max pressure cutoff, compromise solution)

#### 7.3.4.3.3 Oxygenation

Oxygenation is a function of the partial pressure in the delivered gas (percent of oxygen × mean airway pressure) and the amount of lung tissue available for transferring oxygen into the blood. Ventilators have mechanisms to manage both. Percentage in the mixture increases the oxygen concentration delivered to the patient, PEEP increases the mean airway pressure, and stents open more alveoli providing more surface area for gas exchange.

***Percentage*** Terrestrially, the minimum value is typically 21% because this is the percentage of oxygen in the atmosphere. In a spacecraft, the atmosphere may be nonstandard. Planners and providers will need to consider oxygenation in terms of the partial pressure actually being delivered based on the atmosphere designed for the spacecraft. We recommend that planners build procedures that account for any nonstandard atmospheres used in the vehicle.

***Strategy*** Start high (100% $FiO_2$) to stabilize the patient, titrate down as quickly as possible, often over several hours monitoring the patient's oxygen saturations.

- Reasons for high $O_2$
  - Patients with a blood oxygen saturation problem (fluid in lungs, fibrosis, PE, etc.)
- Problems
  - High $O_2$ increases free radicals which increases oxidative stress on tissues. The first clinically significant manifestation is stripping the surfactant from the lungs which takes several hours to develop which can make it difficult to keep alveoli open and add to the work of breathing and require higher saturations or higher pressures. This is not often catastrophic, but it does create a cycle that can prolong a patient's need to remain on a ventilator. Lower $O_2$ as much as possible as early as possible, balancing oxygen percent against end-expiratory pressure (PEEP).
  - In terrestrial ICUs, providers may use invasive arterial lines to gather samples of blood for testing and monitoring arterial gases (see Chapter 5). This is resource- and expertise-intensive, making it less palatable for spaceflight. For this reason, we focus primarily on noninvasive saturation monitoring. However, prolonged ventilation is best adjusted via blood gas measurements and relying on noninvasive measures alone is suboptimal.

***Positive end-expiratory pressure*** PEEP is used to recruit more lung surface area for gas exchange and is the second lever commonly employed to preserve oxygenation saturations in a patient. Physiologically, the human body maintains a PEEP of ~5 $cmH_2O$.

***Strategy*** Higher PEEP will recruit more alveoli but also risks injuring the tissue and reducing blood flow back to the heart by increasing the intrathoracic pressure. Since 5 $cmH_2O$ (3.7 mmHg) is the physiological level, it is best to keep the PEEP at or above this value. In most cases, a PEEP of 5–10 $cmH_2O$ is sufficient to enable lowering oxygen percentage. Above this level, complications are more likely with a practical upper limit of 25 $cmH_2O$. This will significantly reduce venous return to the heart and, in spaceflight, may induce significant hypotension. Providers will need to be alert for this risk and should consider leaving oxygen percentage elevated or accepting a lower saturation percentage if a PEEP >10 $cmH_2O$ is required. In cases where oxygen saturation is consistently <90%, while on 100% $O_2$, the provider should consider pathology such as acute respiratory distress syndrome or a physiological shunt such as a pulmonary embolism.

#### 7.3.4.3.4 Ventilation

Ventilation is the removal of $CO_2$. This is largely a passive process and requires air to be pushed out of the lungs by the elasticity of the chest wall and lungs. The ventilator levers that affect this are levers that allow for more frequent or longer expirations, vent rate and tidal volume, that is, those that effect minute ventilation or the total liters of air moved in and out of the lungs in one minute.

Again, this is typically managed by sampling blood gases and adjusting the ventilator accordingly. In space, this will most likely not be possible and a provider must use noninvasive methods such as end-tidal $CO_2$. This means that if blood gas analysis is not available, $ETCO_2$ is a mandatory capability on board spacecraft enabled for critical care since $CO_2$ can induce an acidosis and carries significant pathological potential.

**Vent rate and delivered volume**

*Strategy*

In general, a provider should monitor four aspects of the clinical picture to manage $CO_2$ levels and ventilation concerns:

- $ETCO_2$ value,
- Waveform shape,
- Respiratory rate,
- $SpO_2$ and hemodynamics.

Start with the same rate of breathing the patient had before being placed on a ventilator. If no rate is known, a rate of 12–14 breaths per minute is reasonable. A tidal volume of 6–8 mL/kg (~450 ml for a 75 kg patient) is also a good place to start. In the setting of acute respiratory distress syndrome (ARDS), that is, a patient who is markedly hypoxic from noncardiac pulmonary edema, a low-tidal volume ventilation (4–6 mL/kg) is preferred as it is lung-protective. Titrating a ventilator by $ETCO_2$ alone is not ideal. Ideally, an initial blood gas is taken used to calibrate the $ETCO_2$ and establish the gradient difference, $PaCO_2$–$ETCO_2$. Arterial gas sampling is preferred over venous but venous can be used if necessary. The typical gradient is 2–5 mmHg. Following this calibration, the ventilator can be titrated based on changes in the $ETCO_2$ targeting; 35–45 mmHg for most cases, and 45–55 mmHg if lung-protective ventilation is needed for ARDS or after an ebullism (explosive decompression) event. In cases where elevated intracranial pressures are a concern, a lower range of 30–35 mmHg may be targeted. Obtain a repeat blood gas if there is a change in the patient's clinical stability. If no blood gas is available, assume a gradient of 2–5 mmHg exists and titrate accordingly.

Using these target ranges, providers should focus on the trends in $ETCO_2$. Change the rate of ventilation first and only adjust the tidal volume if the rate change is not sufficient.

- Increasing $ETCO_2$: increase respiration rate (if not sufficient, consider increasing $V_T$ up to a maximum of 10 ml/kg, except in ARDS when the maximum $V_T$ should be 6–8 mL/kg)
- Decreasing $ETCO_2$: decrease respiration rate (if not sufficient, consider decreasing $V_T$ down to a minimum of 4 ml/kg)

Providers should aim to make small changes and wait to observe the effect. Resuscitation is about speed and efficiency; critical care is about details and diligence.

#### 7.3.4.3.5 Inspiratory/expiratory ratio

I:E ratio is a method for adjusting the time over which the ventilator pump delivers gas to the patient. Generally this is set to 1:2. For conditions where air is likely to be trapped in the lungs (asthma or emphysema), it is critical to enable sufficient time for expiration to prevent breath stacking or autoPEEP. This occurs when gas is not allowed sufficient time to exit the lungs and builds up within the chest cavity. This can interfere with ventilation and increase the risk of pneumothorax and/or hypotension. While unlikely, reduced or microgravity may contribute to breath stacking in an intubated patient. This is because the reduction in gravity may mean slightly reduced forces acting to push air out of the lungs. Since no one has yet been intubated in space, providers should be alert for this possibility and consider setting longer I:E ratios to compensate.

However, since the majority of expiratory force comes from the elasticity of the chest wall and lungs, we still recommend starting with a standard ration of 1:2 until evidence suggests otherwise.

### *7.3.4.4 Common indications for ventilation*

While there are many potential settings for ventilators, in a resuscitation scenario, it is often helpful to have a simple starting point from which it can be optimized. A typical initial ventilator setting is as follows:

- Assist control—volume control,
- Tidal volume of 6–8 mL/kg (450–600 mL for a 75 kg person),
- Rate of 12–16 breaths per minute,
- 100% $FiO_2$,
- Positive end-expiratory pressure of 5 $cmH_2O$.

From this starting point, the onboard providers can work with ground teams to optimize the settings for the specific case. Since the crew is known ahead of time, mission planners may also include some baseline settings for tidal volume and critical weight base medications that are specific to each crewmember in the electronic medical record to simplify the calculations for the CMO. Table 7.13 contains common conditions, suggested initial vent settings, and key considerations that may be useful to consider including on board. Table 7.14 describes common ventilator modes, when they are indicated, and pitfalls.

### *7.3.4.5 Trouble shooting steps*

Mechanical ventilation and support of patients requiring it requires substantial attention and monitoring. There are many aspects of the patient's condition, the ventilator circuit, and the monitoring equipment that can change and create complications or require adjustments. This is why monitoring critical care patients requires a high level of expertise and training. Table 7.15 illustrates some common complications and suggested actions should they occur.

### *7.3.4.6 Noninvasive positive-pressure ventilation*

In many cases, patients may benefit from temporary support with noninvasive methods. In general, these are best used when the patient is conscious and needs short-term support while other treatments take effect (e.g., asthma or pulmonary edema) or for when a patient is at risk of fatigue induced respiratory failure (e.g. viral pneumonitis). In general, there are three types of NIPPV.

#### 7.3.4.6.1 Continuous positive airway pressure

This is similar to the oxygen masks used in fighter aircraft and requires a tight fitting mask or helmet built to deliver higher than ambient pressures into the patient's airway. CPAP is designed to enhance physiological PEEP and improve oxygenation. As with invasive ventilation this is typically set between 5 and 15 $cmH_2O$. CPAP can deliver variable fractions of inspired oxygen ($FiO_2$) up to 100%

#### 7.3.4.6.2 Bi-level positive airway pressure

BiPAP is similar to CPAP except that the machine is able to vary the pressure in response to the patient's breathing, a timed signal, or both. The goal is to maintain PEEP above physiological levels and also assist the patient by lowering the work of breathing. BiPAP can deliver variable fractions of inspired oxygen ($FiO_2$) up to 100%.

#### 7.3.4.6.3 High-flow nasal cannula

This is a specialized nasal cannula designed to deliver up to 60 L/min of gas at up to 100% fraction of inspired oxygen ($FiO_2$). These units often require a stand-alone pressure, heating, and humidification unit, which may be prohibitive from a MPVDB standpoint and 60 L/min of oxygen could rapidly deplete oxygen reserves as well as raise cabin oxygen levels if not accounted for in design.

**TABLE 7.13** Examples of common conditions and ballpark initial ventilatory settings.

| Condition | Mode (typical) | Tidal volume | Target $ETCO_2$ | Respiratory rate | PEEP | $FiO_2$ | I:E ratio | Primary goals | Criteria and parameters to monitor |
|---|---|---|---|---|---|---|---|---|---|
| **ARDS/ ebullism** | Volume control (AC/VC) | 4–6 mL/ kg IBW | 45–55 mmHg | 18–30 bpm | 8–15 $cmH_2O$ | Start 100%, titrate ↓ | 1:1–1:2 | Oxygenation, lung protection | Permissive hypercapnia acceptable; monitor plateau pressure |
| **Sepsis** | Volume control or pressure control | 6–8 mL/ kg IBW | 35–45 mmHg | 16–24 bpm | 5–10 $cmH_2O$ | Start 100%, titrate ↓ | 1:2 | Avoid hypoxia & hypotension | Watch for ARDS evolution; hemodynamic instability common |
| **Trauma** | Volume control | 6–8 mL/ kg IBW | 35 -45 mmHg30–35 mmHg (if concerned for elevated ICP) | 14–20 bpm | 5–8 $cmH_2O$ | 100% initially | 1:2 | Oxygenation, CO2 control | Avoid hyperventilation unless TBI with herniation |
| **Pneumonia** | Volume control | 6–8 mL/ kg IBW | 35–45 mmHg | 14–24 bpm | 5–10 $cmH_2O$ | Start 100%, titrate ↓ | 1:2 | Gas exchange | May progress to ARDS; monitor compliance |
| **Asthma/ obstructive** | Pressure control | Low $V_T$ (allow variable) | 45–55 mmHg | 8–12 bpm | 5 $cmH_2O$ | 100% initially | 1:3–1:5 | Prevent air trapping | Allow permissive hypercapnia; long expiratory time |
| **Procedural/ short-term** | Volume control | 6–8 mL/ kg IBW | 35–45 mmHg | 12–16 bpm | 5 $cmH_2O$ | 30–50% | 1:2 | Comfort & safety | Minimal sedation; reassess frequently |

**TABLE 7.14** Common ventilator modes, indications, and complications.

| Ventilator mode | What It means (plain language) | Good for/typical uses | Cautions/pitfalls |
|---|---|---|---|
| **Volume control (VC/ AC-VC)** | Ventilator delivers a set tidal **volume** each breath | Default ICU mode; ARDS (with low $V_T$); predictable minute ventilation | Airway pressures can rise if compliance worsens; watch plateau pressure |
| **Pressure control (PC/ AC-PC)** | Ventilator delivers breaths to a set **pressure** | ARDS, poor compliance, limiting barotrauma | Tidal volume varies—can drop with worsening compliance |
| **Assist control (AC)** | Ventilator has a set minimum rate, but the patient can trigger breaths and reset the timing. Every breath, whether initiated be the patient or the machine breath, is fully supported | Sick or paralyzed patients; early ventilation | Risk of hyperventilation if patient overbreathes |
| **Synchronized intermittent mandatory ventilation (SIMV)** | Mandatory breaths + spontaneous breaths. Similar to assist control but with less support. | Weaning (less common now) | Can increase work of breathing; inferior to PSV for weaning |
| **Pressure support ventilation (PSV)** | Patient triggers all breaths; ventilator adds pressure. **If the patient is not breathing spontaneously, no air will be delivered** | Optimal for weaning, spontaneous breathing trials | Not for apneic or heavily sedated patients |
| **APRV** | Continuous high pressure with brief releases for ventilation | Severe ARDS, refractory hypoxemia. Potentially useful for ebullism events | Complex; risk of volutrauma; requires expertise |
| **High-frequency ventilation** | Very small $V_T$ at very high rates | Refractory ARDS. Potentially useful for ebullism events | Specialized equipment; difficult monitoring |
| **Inverse ratio ventilation (IRV)** | Inspiration longer than expiration. E.g., an I:E ratio of 1:2 | Severe hypoxemia unresponsive to standard modes | High sedation/paralysis needs; hemodynamic compromise |

## 7.3.5 Unique spaceflight considerations

### *7.3.5.1 Barotrauma*

Astronauts experience multiple pressure changes during spaceflight. Currently protocols for extravehicular activities (EVAs) are gradual decompressions with little chance for abrupt changes that would pose a threat of lung barotrauma. However, systems can fail, potentially placing astronauts at risk. An unplanned rapid transition from a low-pressure environment (like a suit) to a high pressure environment (like the ISS) poses little risk of over expansion of the lungs. However, sudden transitions from a high pressure environemnt to a low pressure one risk expansion of trapped gasses in the body and can cause significant damage to the involved tissues.

proposed (albeit never tested) treatment for air embolism and decompression sickness (known risks of EVAs) aboard the station would be to pressurize the patient using a combination of the station's ambient pressure, an airlock, and a pressure suit (Conkin et al., 2014). A failure in this protocol would potentially see a rapid transition from a high-pressure state to a low-pressure state, which could put astronauts at risk for parenchymal lung injury due to over inflation of the high-pressure intralung gas, potentially causing a pneumothorax. Similarly, a rapid decompression from a momentary suit failure or of a surface-based habitat could also present significant risk for barotrauma (assuming an astronaut could survive what is likely a catastrophic failure). While operational protocols and engineering safeguards are the primary means of preventing these injuries, the ability to treat the injuries of mishaps while working in an environment with extreme pressures is worth the consideration.

**TABLE 7.15** Common complications and responses.

| Complication | Common causes | Immediate checks | Ventilator adjustments (step-by-step) | Adjunct/supportive actions |
|---|---|---|---|---|
| **Low oxygen (hypoxemia)** | DOPES; tube **displacement:** → reconnectMucous plug or other **obstruction** → suction**Patient** malfunction<br>• pneumothorax → chest tube<br>• atelectasis → increase PEEP<br>• pneumonia → antibiotics, increase $O_2$<br>• ARDS → permissive hypoxia, increase PEEP<br>**Equipment** malfunction/check<br>• Low $FiO_2$ → increase $FIO_2$<br>• low PEEP → increase PEEP<br>• low power → alternate power source, BVM<br>**Stacking** breaths → disconnect vent, squeeze chest | $SpO_2$<br>PEEP<br>$ETCO_2$<br>waveform<br>tube position/<br>depth<br>chest rise<br>Ventilator<br>power state | **1.** ↑ $FiO_2$<br>**2.** ↑ PEEP (1-2 $cmH_2O$ at a time)<br>**3.** Ensure adequate $V_T$ (4-8 ml/kg)<br>**4.** In-line suction<br>**5.** Consider disconnecting patient from Vent and using BVM | Suction, recruit lung, check for pneumothorax, prone if ARDS |
| **High $CO_2$ (hypercapnia)** | Low minute ventilation,obstruction,breath stacking,increased dead space | $ETCO_2$<br>Rate<br>$V_T$<br>$ETCO_2$<br>waveform | **1.** ↑ RR<br>**2.** ↑ $V_T$ (maximum of 8 ml/kg)<br>**3.** Reduce dead space in ventilator circuit | Bronchodilators, suction, accept permissive hypercapnia if lung-protective ventilation employed |
| **Low $CO_2$ (hypocapnia)** | Excess minute ventilation, pain/anxiety, over-assist | $ETCO_2$ trend<br>Rate | **1.** ↓ RR<br>**2.** ↓ $V_T$<br>**3.** Reduce pressure support | Optimize sedation, avoid hyperventilation (esp. TBI unless indicated) |
| **Hypotension** | High PEEP, excessive mean airway pressure, hypovolemia, sedation | Blood pressure<br>Heart rate<br>PEEP<br>Mean airway<br>Pressure | **1.** ↓ PEEP if possible<br>**2.** ↓ $V_T$<br>**3.** ↓ inspiratory time | Fluids, vasopressors, reduce sedatives |
| **High plateau pressure (>30 $cmH_2O$)** | Low lung compliance (prolonged $O_2$ use), ARDS, excessive $V_T$, pneumothorax | Measure plateau pressure | **1.** ↓ $V_T$ (4–6 mL/kg IBW)<br>**2.** Optimize PEEP | Accept hypercapnia, check for pneumothorax |
| **Breath stacking/ autoPEEP** | Obstruction, high rate, short expiratory time | Flow curve not returning to baseline | **1.** ↓ RR<br>**2.** ↑ expiratory time (I:E 1:3–1:5) | Bronchodilators, suction, deep sedation |

*(Continued)*

**TABLE 7.15 (Continued)**

| Complication | Common causes | Immediate checks | Ventilator adjustments (step-by-step) | Adjunct/supportive actions |
|---|---|---|---|---|
| **Vent dyssynchrony** | Inadequate sedation, poor trigger settings, inappropriate mode | $ETCO_2$ waveform, patient effort | **1.** Adjust trigger sensitivity<br>**2.** Change mode (PC ↔ VC)<br>**3.** Consider increased sedation/analgesia | Optimize sedation/analgesia; consider neuromuscular blockade |
| **High peak pressure, normal plateau** | Airway resistance (secretions, kink, bronchospasm) | Compare peak vs plateau | **1.** Treat obstruction, suction<br>**2.** Maintain $V_T$ | Suction, bronchodilators, check tubing |
| **Sudden loss of $ETCO_2$** | Disconnection, tube dislodgement, cardiac arrest | Circuit integrity, Patient pulse | **1.** Reconnect circuit<br>**2.** Verify tube position | CPR if indicated; immediate airway reassessment |

### 7.3.5.2 Lack of convection

In the microgravity environment, many of the normal fluid dynamics that we experience in the background of everyday life change. Namely, natural convection, or movement of air along a gravity gradient due to pressure differences, does not occur. Similar to how a bubble moves up underwater, low density gas (usually due to heat) doesn't move "up" in free-fall, since there is no "up." Rather, gases move through simple diffusion, or random motion. Practically for astronauts, this means there are no natural currents in a spacecraft, and exhaled gases will remain near the astronaut until forced away. This can result in increased $CO_2$ inhalation, as exhaled $CO_2$ often doesn't make it very far from an astronaut's airway. This increased $CO_2$ level, as mentioned previously, is important to keep in consideration if using measurements such as $EtCO_2$ and VBGs to guide a patient's resuscitation and treatment (James and Ryder, 2020).

### 7.3.5.3 Inhalation

When transiting from the EVA environment back into a spacecraft, or following docking procedures, astronauts potentially expose the closed environment and atmosphere of the spacecraft to external contaminants. In the spaceflight environment, the main contaminants of concern are propulsion chemicals and extraterrestrial regolith that could be brought into the cabin on a spacesuit exposed to them on EVA (see Fig. 7.10). Propulsion chemicals are particularly toxic if inhaled with the potential to cause mucosal irritation, cough, dyspnea, headache, and even pulmonary edema and hypoxemia. Regolith, on the other hand, is to likely cause relatively mild acute inflammation but is may cause a more chronic irritation akin to silicosis (James and Ryder, 2020). The closed environment with a lack of natural convection can mean that these materials can potentially remain in the environment for extended periods of time, turning a potentially acute issue in large amounts to continued chronic concern. Injuries associated with the inhalation of these materials can potentially lead to airway and respiratory compromise, requiring resuscitation and critical care support.

### 7.3.5.4 Limited resources

Like all areas of resuscitation and critical care in space, the materials and practice of diagnostics, interventions, and treatments are challenged by the severe resource limitations of spaceflight. Breathing for a patient could be critical to keeping them alive; however, manually ventilating a patient with a BVM, while likely lower on the MPVDB scale, is a significant burden to the crew as discussed previously. Mechanical ventilation, depending on form factor, will almost certainly be a higher cost in terms of MPVDB but would save valuable crew time and effort. Furthermore, the treatments required in the resuscitation and critical care of a patient with breathing issues can require significant consumable resources. Even something as simple as enriched oxygen can potentially be in short supply, to say nothing of the sedative drugs needed for a patient to tolerate invasive airways. Medications used to treat airway disease, as well as antibiotics and antiseptics to protect patients from infection during invasive procedures, come at a cost—not only in terms of launch and

FIGURE 7.10 Astronaut with protective mask.
An astronaut wears a protective breathing apparatus that would be used in the unlikely event of a fire or hazardous chemical leak inside the closed pressurized air volume of the International Space Station.
Courtesy NASA.

storage, but also because once they are used on one patient, they are no longer available for future treatments and interventions.

Ultimately, spaceflight presents some unique risks that can cause issues with breathing, and the resource costs needed to safely and effectively care for an injured astronaut must be weighed carefully against the likelihood of injury and the possibility of successfully reaching definitive care.

## 7.4 Circulation

### 7.4.1 Anatomy and description

The chief organs of the circulatory system relevant to resuscitation and critical care are the heart, lungs, arteries, arterioles, capillaries, and veins. The circulatory system function in this context is to deliver the oxygen pulled in through the airway and exchanged through breathing to the organs that sustain life. It also functions to pick up the $CO_2$ waste gas at these end organs and transport this to the lungs to be removed. This movement is driven by a pressure gradient between the arteries and the veins maintained by the muscular contractions of the heart. The pressure is maintained by four elements:

- The rate of heart muscle contractions (heart rate or chronotropy),
- The strength of heart muscle contractions (inotropy),
- The distributed volume controlled by muscle tone in the arterioles and open capillary beds (peripheral vascular tone),
- The overall volume of blood (~55–70 mg/kg or ~4–5 L for a 75 kg individual).

### 7.4.2 Triage

There are two steps to the assessment of circulation. The first is assessing the function of the heart with the second focused on assessing the integrity of the blood vessels and blood volume. For this chapter, we will consider restoration of an absent pulse to be resuscitation with follow on maintenance of the heart rhythm and circulatory system critical care. Hemorrhage control is discussed in Chapter 5.

#### *7.4.2.1 Heart function*

The primary assessment of heart function is the presence or absence of a pulse. If there is no pulse, a provider must make an immediate judgment of the probability the patient will survive with meaningful neurological activity.

#### *7.4.2.2 Hemorrhage and circulatory assessment*

If a pulse is present, the next assessment is to determine the presence or absence of massive, life-threatening external hemorrhage and capillary refill, which is a measure of the volume of blood under pressure. Hemorrhage is typically assessed visually, including assessing the patient's clothing and back for signs of blood loss and uncontrolled bleeding. Chapter 5 discusses the management steps for this. Capillary refill should be assessed by briefly squeezing the patient's warm fingertip and nailbed or compressing the skin of the forehead to assess the time it takes for blood to return once the pressure is released. Normal capillary refill should be <3 seconds, prolonged refill is a sign of inadequate circulation or shock (Brun, 2018). Delayed refill can signal poor distal perfusion either due to vasoconstriction or poor oxygenation. Of note, trauma to the hands and fingertips is a well-documented injury in spaceflight caused by the prolonged griping and holding associated with EVAs and could impair the utility of this assessment (Viegas et al., 2004). Table 7.16 highlights the examples of common types of shock.

### 7.4.3 Resuscitation and critical care

#### *7.4.3.1 Absent pulse*

##### 7.4.3.1.1 Chest compressions

When a patient's heart is no longer beating effectively, oxygen is not delivered to tissues and the patient will suffer irreversible brain damage within 4–6 minutes. Chest compressions are a temporizing mechanism intended to buy time until the heart rhythm can be restored. The provider attempts to establish a degree of circulation to the brain by manually compressing and releasing the heart through the patient's chest. When combined with rescue breathing, the procedure is commonly referred to as CPR.

CPR is primarily intended for scenarios where an otherwise healthy individual suffers an injury that causes the heart muscle to enter a nonperfusing rhythm, specifically ventricular fibrillation or ventricular tachycardia. This is usually the

**TABLE 7.16** Types of shock.

| Type of shock | Primary problem | Skin findings | Heart rate/BP | Lung exam | Other key physical exam clues |
|---|---|---|---|---|---|
| **Hypovolemic** | Low intravascular volume (bleeding, dehydration) | Cool, pale, clammy | Tachycardia, hypotension, narrow pulse pressure | Clear | Dry mucous membranes, delayed capillary refill, low urine output |
| **Cardiogenic** | Pump failure (MI, cardiomyopathy) | Cool, clammy, mottled | Tachycardia, hypotension, narrow pulse pressure | Crackles/ pulmonary edema | S3 gallop, weak pulses, chest pain history |
| **Distributive—septic** | Vasodilation + capillary leak | **Warm early**, flushed → cool late | Tachycardia, hypotension, **wide pulse pressure early** | Variable | Fever or hypothermia, bounding pulses early, altered mental status |
| **Distributive—anaphylactic** | Vasodilation + capillary leak | Warm, flushed, urticaria | Tachycardia, hypotension | Wheezing, stridor | Angioedema, rash, GI symptoms |
| **Distributive—neurogenic** | Loss of sympathetic tone | Warm, dry | **Bradycardia**, hypotension | Clear | Spinal injury, loss of temperature regulation |
| **Obstructive—pulmonary embolism** | RV outflow obstruction | Cool, clammy | Tachycardia, hypotension | Often clear | Sudden dyspnea, pleuritic chest pain, hypoxia |
| **Obstructive—tension pneumothorax** | Impaired venous return | Cool, clammy | Tachycardia, hypotension | **Unilateral absent breath sounds** | Tracheal deviation (late), hyperresonance |
| **Obstructive—cardiac tamponade** | Impaired cardiac filling | Cool, clammy | Tachycardia, hypotension | Clear | Muffled heart sounds, pulsus paradoxus |

result of an electric shock, suffocation, or an interruption in blood supply to the heart such as occurs in some myocardial infarctions or with conditions like hypertrophic cardiomyopathy. Since CPR is not a definitive treatment in itself, it is used terrestrially to buy time until rescuers can arrive with the tools and expertise to restore a reversible cause of cardiac arrest or until the patient can be transported to a facility where these experts and tools exist. It is rarely effective after 30–40 minutes (Okubo et al., 2024).

In spaceflight, with limited providers, reduced or absent gravity, and limited volume, CPR may be of limited use. Time may be better spent setting up definitive care tools than securing the patient and provider to initiate CPR. However, if sufficient assistance is available to retrieve and arrange supplies AND perform CPR, modified techniques for chest compressions that do not rely on gravity could have been evaluated. The three main one-person methods are as follows:

- Handstand (HS) method, where the rescuer braces their feet on a ceiling or wall and "kicks" downward with arms extended (Fig. 7.11).
- Evetts–Russomano (ER) method, where the rescuer straddles the supine patient, hooking their legs around the patient's torso and shoulder to anchor themselves and compress the chest by rocking their hips (Fig. 7.12).
- Reverse Bear Hug (RBH), where the rescuer positions behind the patient and squeezes the chest between their arms (Fig. 7.13).

Studies indicate that *none* of these manual methods achieve terrestrial-quality CPR, but they can generate some circulation. The HS method produces the greatest compression depth (~40–50 mm achieved on a manikin), approaching but still slightly below the 50–60 mm guideline target. It also is the least fatiguing method since it uses the rescuer's large leg muscles (Braunecker et al., 2015). However, HS requires sufficient space for the patient to be laid flat and a surface to brace the providers feet against that is close enough for the provider to place their hands on the patient's chest and may be impossible in a large volume module or if the rescuer's height doesn't match the cabin dimensions.

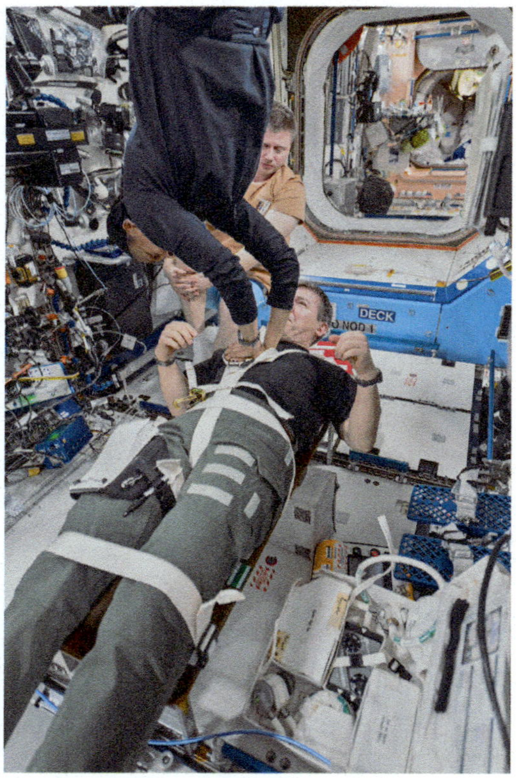


**FIGURE 7.11** Handstand cardiopulmonary resuscitation (CPR) compressions.
An astronaut practices handstand chest compressions aboard the International Space Station on a crewmember restrained on the Crew Medical Restraint System (CMRS).
Courtesy NASA.

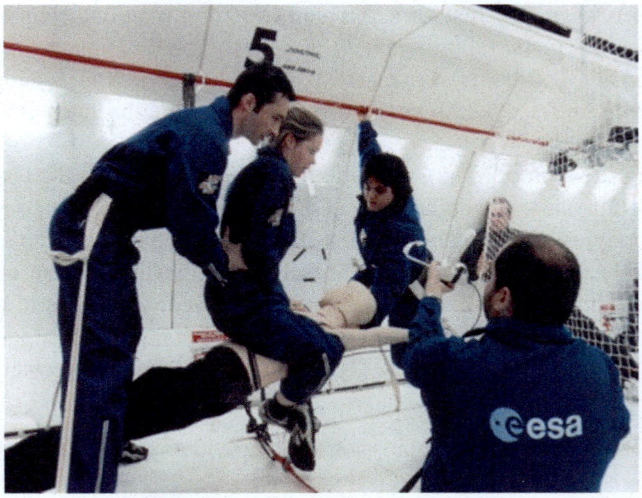


FIGURE 7.12 Evetts–Russomano cardiopulmonary resuscitation (CPR) compressions.
Here, a study participant demonstrates the Evetts–Russomano method of compressions in a parabolic flight study.
Courtesy ESA in the public domain.

FIGURE 7.13 Reverse-bear-hug cardiopulmonary resuscitation (CPR) compressions.
Here, a study participant demonstrates the reverse-bear-hug method of compressions in a parabolic flight study.
Courtesy NASA.

The ER method can be performed immediately at the collapse site without any equipment or wall support, making it versatile in application. Its compression depth has varied widely in tests (as high as ~46 mm in some trials, but as low as ~27 mm in others). The ER tends to allow a high compression rate (often >100 /min) but may be more fatiguing to the rescuer.

The RBH method is simple and requires minimal setup but generally yields the shallowest compressions (<42 mm) and can strain the rescuer (Rehnberg et al., 2014). Current expert consensus is to start with the ER method for immediate response, then once the patient can be secured on a firm surface (e.g., a crew medical restraint system), transition to the HS method for more effective compressions. If ER fails (e.g., rescuer cannot get adequate depth), the rescuer may switch to RBH as a stopgap.

Mechanical CPR devices offer an appealing potential solution to microgravity chest compressions. A 2024 parabolic flight study compared manual compressions to three automated chest compression devices (LUCAS 3, AutoPulse, EasyPulse) during weightlessness. The LUCAS 3 consistently achieved guideline-compliant depths (~53 mm median depth at 101 compressions/min) suggesting that an automated CPR unit may outperform manual CPR in microgravity. However, there is a considerable mass (approximately 8.5 kg), volume (approximately 50 L), and power cost (approximately 3.3 Amp-hours, 86 Watt-hours) associated with the Lucas 3. Additionally, the setup time will likely be better spent retrieving and setting up for a definitive treatment such as a defibrillator or surgical interventions (Reynette et al., 2024).

In partial gravity environments (Moon or Mars), CPR might be somewhat easier than in 0 g since rescuers and patients will have weight, providing friction and some ability to lean. However, a Moon rescuer weighs only ~1/6 of Earth weight, so they cannot rely purely on body weight for compressions. They will need to push harder and possibly strap themselves to the floor or patient. No published studies to date have measured CPR quality in lunar or Martian gravity analogs, but it is reasonable to expect compression depth will still be suboptimal. Anecdotal reports of two expert acute care providers performing simulated chest compressions on a lunar gravity parabolic flight are suggestive that despite the reduced gravity, it was neither difficult nor fatiguing to repeatedly generate substantial downward force (Levin, 2025).

Procedure for CPR:

- CPR is performed in cycles. Every 30 chest compressions, the providers deliver two breaths, and every 2 minutes, the providers check for a pulse.
- Since it takes time to build up pressure to circulate blood, for a single rescuer, it is often advisable to skip breaths all together and continue compressions/pulse checks until a defibrillator is in place.
- If two rescuers are available, one delivers 30 compressions and pauses while the other delivers breaths.
- Compressions should be delivered at a rate of 100 per minute, and breaths should be spaced about 3–4 seconds apart.
- If a timer is available, pulse checks should occur every 2–3 minutes, or roughly every 5-6 cycles if a timer is not available
- The provider should time their compressions to a well-known melody at 100 beats per minute such as Gloria Gaynor's "I Will Survive" by or "Another one Bites the Dust" by Queen.
- If a definitive airway is in place, providers do not need to pause chest compressions, and breaths can be delivered continuously.
- CPR is a temporizing method and should not interfere with or delay placement of a defibrillator and/or surgical decompression of the chest or heart.
- Consider terminating CPR when:
  - The patient is not found to be in ventricular tachycardia or ventricular fibrillation and no reversible cause for the arrest is found (Table 7.17).
  - There are obvious signs of death such as a nonsurvivable injury or rigor mortis is present.
  - CPR has been ongoing for more than 40 minutes (Ishii et al., 2024).

#### 7.4.3.1.2 Advanced cardiac life support medications

Standard of care for Advanced Cardiac Life Support (ACLS) by the American Heart Association recommends the use of vasopressors like epinephrine and antiarrhythmics such as Amiodarone or Lidocaine. These medications may improve the chance of restoring a pulse; however, current data does not support that these medications improve the chances of neurologically intact survival (Finn et al., 2019). Planners may wish to consider the cost/benefit of including ACLS medications in light of supporting evidence. Lidocaine in particular may be better utilized as a local anesthetic in spaceflight when quantities are limited (see Chapter 20). However, as of the latest updates to ACLS, the American Heart Association still maintains as recommendation to administer epinephrine. Table 7.18 contains the doses, intervals, and additional information for ACLS medications.

#### 7.4.3.1.3 Definitive treatments for cardiac arrest

Table 7.17 describes the limited set of reversible causes of cardiac arrest. These are often referred to as the H's and T's. Two additional causes are added: electrocution and ventricular tachycardia/fibrillation. Electrocution induces lethal arrythmias which can often be reversed with a second electric shock (Zoll et al., 1956). In fact, this was the historical inspiration for developing the capabilities that later became standard of practice (Kouwenhoven et al., 1960). Ventricular tachycardia and ventricular fibrillation are the only nonperfusing rhythms that are reversible with present resuscitative practices, and neurologically intact survival after defibrillation approaches 75% for all causes (Valenzuela et al., 2000). It is likely that spaceflight survival might be even higher, given that crews are likely to be a healthier cohort and response times in the small spacecrafts volume will be nearly instantaneous.

**TABLE 7.17** Reversible causes for cardiac arrest "H's and T's."

| Cause | Mechanism | Key clues | Immediate stabilization steps |
|---|---|---|---|
| **Hypoxia** | Inadequate oxygen delivery | Cyanosis, low $SpO_2$, airway issues | Ensure airway patency100% oxygenVentilate with BVM or ETT |
| **Hypovolemia** | Inadequate preload | Trauma, bleeding, | Rapid IV/IO access (Chapter 5)Aggressive fluids and blood productsImmediate hemorrhage control (Chapter 6) |
| **Hydrogen ion (acidosis)** | Depressed myocardial function | Prolonged arrest, DKA, renal failure | Effective ventilation >10 breaths per minuteRapid IV/IO access (Chapter 5)Consider sodium bicarbonate (selective) |
| **Hypo-/hyperkalemia** | Lethal arrhythmias | Renal failure, dialysis, ECG changes | Rapid IV/IO access (Chapter 5)2 g IV calcium chloride (stabilize membrane)10 units IV insulin + 25 g IV glucoseNebulized albuterol (if available) |
| **Hypothermia** | Electrical instability | Cold exposure, low core temp | Active rewarmingContinue CPR until patient is >32°C (90°F) |
| **Tension pneumothorax** | Obstructed venous return | Unilateral breath sounds, JVD | Immediate needle decompression (Chapter 11)Chest tube (Chapter 11) |
| **Tamponade (cardiac)** | Impaired cardiac filling | JVD, muffled heart sounds | Pericardiocentesis (Chapter 11) |
| **Toxins (overdose)** | Myocardial depression or arrhythmia | History, empty bottles | Airway/ventilation4 mg intranasal Naloxone (or 0.4 mg IV) if narcotics are suspectedAntidotes if known |
| **Thrombosis—coronary** | Acute MI | Chest pain prior, ECG changes | CPR + defibrillation if indicatedemergent thrombolysis |
| **Thrombosis—pulmonary (PE)** | RV outflow obstruction | Sudden collapse, hypoxia | CPR thrombolysis (if massive PE suspected) |
| **Ventricular tachycardia** | Lethal arrhythmia | Sudden collapse, electrocution, ECG | CPR + defibrillation |
| **Ventricular fibrillation** | Lethal arrhythmia | Sudden collapse, electrocution, ECG | CPR + defibrillation |

Abbreviations: *BVM*, Bag-valve mask; *CPR*, cardiopulmonary resuscitation; *DKA*, diabetic ketoacidosis; *ECG*, electrocardiogram; *ETT*, endotracheal tube; *IO*, intraosseous; *IV*, intravenous; *JVD*, jugular venous distention; *MI*, myocardial infarction; *PE*, pulmonary embolism; *RV*, right ventricle; *$SpO_2$*, peripheral oxygen saturation.

The converse is equally important. Neurologically intact survival for nonventricular tachycardia or ventricular fibrillation cardiac arrests are on the order of 0.5% (Dwivedi et al., 2025; Ishii et al., 2024). This means that patients found to be in ventricular tachycardia or ventricular fibrillation cardiac arrest have a high chance of surviving, and the interventions described in Table 7.19 are applicable. Patients found in asystole or pulseless electrical activity that are not suffering a mechanical obstruction (e.g., tension pneumothorax or cardiac tamponade) are unlikely to survive and practitioners should consider conserving limited medical resources.

### *7.4.3.2 Defibrillation*

Defibrillation is an unsynchronized cardioversion intended to depolarize misfiring ventricular cells and allow the heart's natural pacemaker to restart a normal rhythm. Defibrillators have varying degrees of automation to assist providers of varying levels. Simple models will recognize shockable rhythms and instruct providers to deliver a shock if indicated. Models intended for more expert providers typically have more functions for managing different types of arrythmias (described in the next section) and often are combined with monitoring capabilities (Table 7.11).

**TABLE 7.18 Cardiac arrest advanced cardiovascular life support (ACLS) medications.**

| Drug | Recommended dosage | Notes |
|---|---|---|
| **Epinephrine** | 1 mg (1:10,000) intravenous (IV) every 4 minutes | Available in 1:1,000 and 1:10,000 concentrations, be aware of which concentration is being usedA good timing method is to alternate epinephrine and defibrillation (if indicated) with each other pulse<br>• 2 minutes → pulse check/epinephrine → 2 minutes → pulse check and shock, etc.<br>Lacks evidence for neurointact survival |
| **Amiodarone** | First dose: 300 mg rapid bolusSecond dose: 150 mgMaximum of 2.2 g over 24 hours | Lacks evidence for neurointact survival |
| **Lidocaine** | First dose: 1–1.5 mg/kg IV bolusSecond dose: 0.5–0.75 mg/kg IV bolusThird dose: 0.5–0.75 mg/kg IV bolusInfusion: 1–4 mg per minute | Lacks evidence for neurointact survival |

An Automated External Defibrillator in spaceflight has been included as part of the crew medical system since the Space Shuttle era and is preferred since it can remove the need for a crewmember to be trained and practiced in interpreting EKG tracings. Currently, an AED's use is recommended in the event of a cardiac arrest in spaceflight, with special care taken to ensure the patient is restrained onto an electrically isolated surface to prevent damaging electrical currents from being conducted to the rescuer or other critical spacecraft systems (Hinkelbein et al., 2020).

#### 7.4.3.2.1 Procedure for an automated external defibrillator

**A. Power on the AED and follow voice and visual prompts immediately.**

**B. Attach pads**

1. Plug pads into the AED if required.
2. Confirm correct placement as shown on the pads/device (Fig. 7.14).
   a. Standard placement
      - **Right (sternal) pad:** Below right clavicle, to the right of the sternum.
      - **Left (apical) pad:** Left lateral chest, mid-axillary line, below the nipple.

   b. Alternative if unable to use standard (e.g., implanted devices or severe obesity)
      - Sternal pad (right pad) on patient's **chest**
      - Apical pad (left pad) on patient's **back**

   c. Confirm pads are firmly adhered with full skin contact.

**C. Allow rhythm analysis**

1. Loudly state: "Clear for analysis."
2. Ensure **no one touches the patient.**

**D. If shock is advised**

1. Ensure patient is positioned on a grounded and electrically isolated surface
2. Confirm oxygen is turned off and/or removed from the patient
3. State clearly: *"Clear!"* (check head-to-toe).
4. Visually confirm no one is touching the patient

**E. Press the shock button** when prompted (or allow autoshock if applicable).

**F. Immediately resume CPR** (Do *not* check pulse or rhythm unless prompted by the AED).

**G. Continue AED-guided cycles**

1. Follow prompts for CPR, analysis, and additional shocks as indicated.
2. Pulse checks every 2 minutes

#### 7.4.3.2.2 Procedure for an manual defibrillator

**A. Power on and prepare the defibrillator**

**TABLE 7.19** Interventions, equipment, mass, volume, and key steps for reversible causes of cardiac arrest.

| Reversible cause | Intervention | Materials needed | Estimated mass | Estimated volume | Key steps to perform |
|---|---|---|---|---|---|
| **Hypoxia** | Oxygenation & ventilation | BVM, mask, $O_2$ source | 0.6–1.2 kg | 2–4 L | Open airway → seal mask → ventilate with 100% $O_2$ |
| | Endotracheal intubation | Laryngoscope, ETT, stylet, syringe, $ETCO_2$ | 1.5–3 kg | 5–10 L | Visualize cords → insert ETT → inflate cuff → confirm with $ETCO_2$ |
| **Hypovolemia** | IV/IO fluid resuscitation | IV/IO kit, crystalloid (1 L) | ~1.1 kg | ~1 L | Establish access → rapid bolus → reassess perfusion |
| | Hemorrhage control | Tourniquet, pressure dressing | 0.1–0.3 kg | 0.3–0.6 L | Direct pressure → tourniquet if uncontrolled |
| **Hydrogen Ions (acidosis)** | Ventilation optimization | Ventilator or BVM | — | — | Increase minute ventilation |
| | Sodium bicarbonate (selective) | 50 mEq syringe | ~60 g | ~60 $cm^3$ | IV push if indicated |
| **Hyperkalemia** | Calcium administration | $CaCl_2$ or Ca-gluconate syringe | ~40 g | 20–30 $cm^3$ | IV push to stabilize myocardium |
| | Potassium shift | Insulin + D50 | ~80 g | 60–80 $cm^3$ | Give insulin → dextrose |
| | Potassium shift | 50 mEq syringe sodium bicarbonate | ~60 g | ~60 $cm^3$ | IV push if indicated |
| | Adjunct bronchodilation | Albuterol + neb | ~150 g | ~0.5 L | Nebulize to shift potassium intracellularly |
| **Hypothermia** | Active rewarming | Blankets, warming device | 0.5–2 kg | 2–6 L | Remove wet clothes → warm core → continue CPR |
| **Tension pneumothorax** | Needle decompression | 14 g catheter/needle | ~10 g | ~10 $cm^3$ | Insert at 5th ICS AAL or 2nd ICS MCL |
| | Chest tube placement | Tube, scalpel, clamp | 0.3–0.6 kg | 0.8–1.5 L | Insert tube → connect to seal |
| **Cardiac tamponade** | Pericardiocentesis | Needle, syringe, ECG lead | ~150 g | ~300 $cm^3$ | Subxiphoid approach → aspirate pericardial blood |
| **Toxins (e.g., opioids)** | Naloxone | Naloxone syringe | ~20 g | ~10 $cm^3$ | IV or IM administration |
| | Supportive airway care | BVM/ETT | — | — | Oxygenate, ventilate, support circulation |

*(Continued)*

**TABLE 7.19** (Continued)

| Reversible cause | Intervention | Materials needed | Estimated mass | Estimated volume | Key steps to perform |
|---|---|---|---|---|---|
| **Coronary thrombosis (MI)** | CPR + defibrillation | Monitor/defibrillator | 3–7 kg | 6–15 L | High-quality CPR → defibrillate VF/VT |
| | Epinephrine | Epi syringe | ~20 g | ~10 $cm^3$ | IV/IO per ACLS |
| **Pulmonary embolism (massive)** | Thrombolysis | tPA vial/syringe | 50–100 g | 50–100 $cm^3$ | Administer during CPR if indicated |
| | Ventilatory & pressor support | Ventilator, vasopressors | — | — | Support RV function and oxygenation |
| **Ventricular tachycardia** | Defibrillation | Defibrillator | See Table 7.10 | See Table 7.10 | Remove patient clothing→ secure patient to a grounded, electrically isolated location→ apply defibrillator pads → charge defibrillator → administer shock |
| **Ventricular fibrillation** | Defibrillation | Defibrillator | See Table 7.10 | See Table 7.10 | Remove patient clothing→ secure patient to a grounded, electrically isolated location→ apply defibrillator pads → charge defibrillator → administer shock |

*AAL*, Anterior axillary line; *ACLS*, advanced cardiovascular life support; *BVM*, bag-valve mask; *$CaCl_2$*, calcium chloride; *Ca-gluconate*, calcium gluconate; *CPR*, cardiopulmonary resuscitation; *D50*, 50% dextrose; *ECG*, electrocardiogram; *Epi*, epinephrine; *$ETCO_2$*, end-tidal carbon dioxide; *ETT*, endotracheal tube; *ICS*, intercostal space; *IM*, intramuscular; *IO*, intraosseous; *IV*, intravenous; *MCL*, midclavicular line; *MI*, myocardial infarction; *RV*, right ventricle; *tPA*, tissue plasminogen activator; *VF*, ventricular fibrillation; *VT*, ventricular tachycardia.

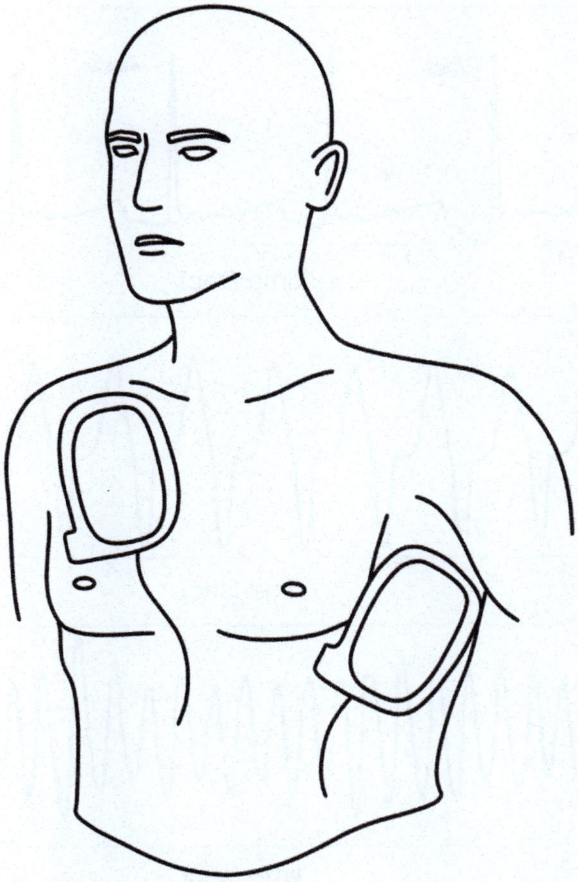

FIGURE 7.14 Defibrillator pad locations.
Description: illustration of AED electrode placement, February 25, 2020, Author Baedr-9439. This file is made available under the Creative Commons CC0 1.0 Universal Public Domain Dedication. The person who associated a work with this deed has dedicated the work to the public domain by waiving all of their rights to the work worldwide under copyright law, including all related and neighboring rights, to the extent allowed by law. You can copy, modify, distribute, and perform the work, even for commercial purposes, all without asking permission.

1. Confirm biphasic mode (default on modern units).
2. Attach defibrillation pads and connect to the monitor.

**B. Attach pads**

1. Plug pads into the AED if required.
2. Confirm correct placement as shown on the pads/device (Fig. 7.14).
   a. Standard placement
      - **Right (sternal) pad:** Below right clavicle, to the right of the sternum.
      - **Left (apical) pad:** Left lateral chest, midaxillary line, below the nipple.
   b. Alternative if unable to use standard (e.g., implanted devices or severe obesity)
      - Sternal pad (right pad) on patient's **chest**
      - Apical pad (left pad) on patient's **back**
   c. Confirm pads are firmly adhered with full skin contact.

**C. Assess rhythm**

**D. If rhythm is ventricular tachycardia or ventricular defibrillation, charge the defibrillator** (Fig. 7.15).

1. Select energy level
   a. First shock: 200 Joules
   b. Second shock: 300 Joules
   c. Third shock: 360 Joules

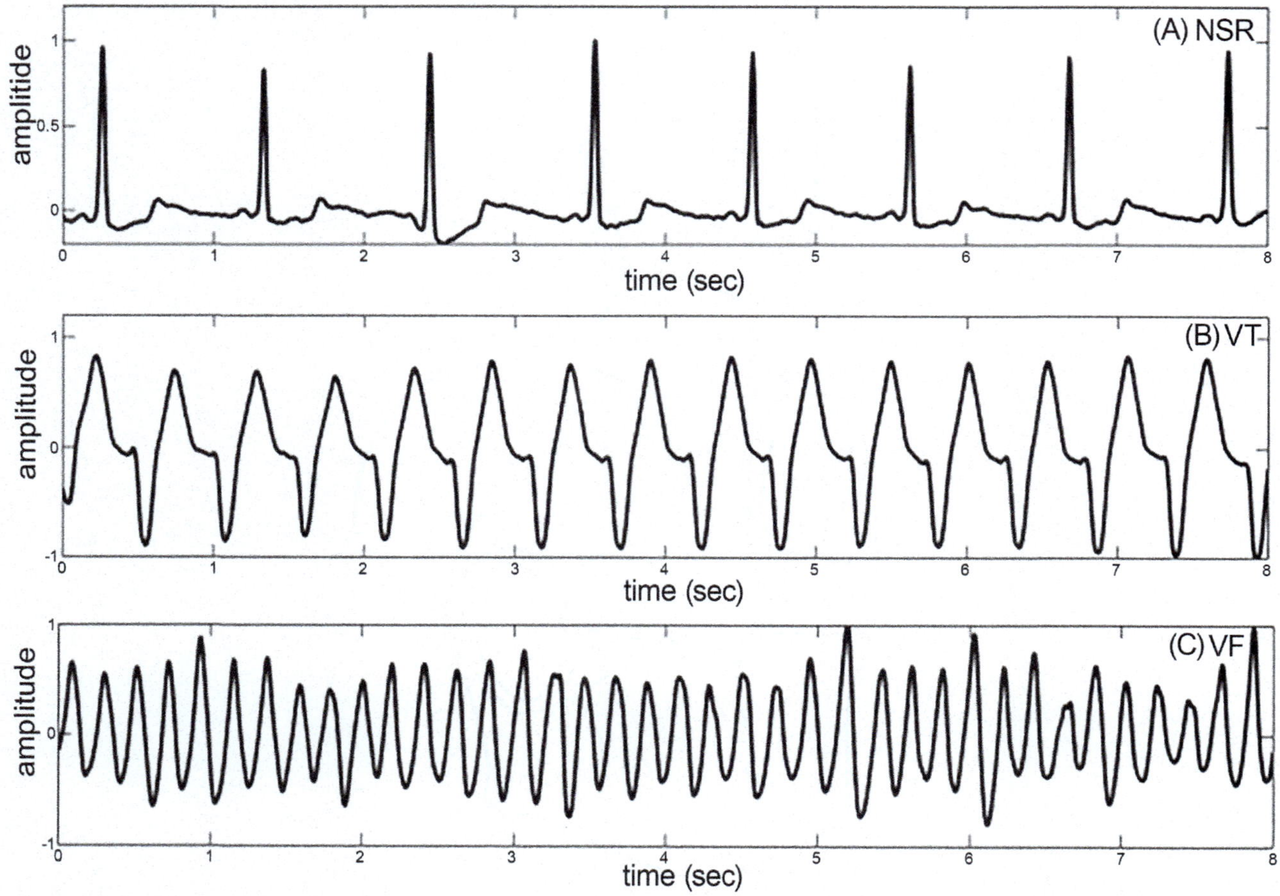


FIGURE 7.15 Illustration of normal sinus rhythm (A), ventricular tachycardia (B), and ventricular fibrillation (C).
Creative Commons Attribution 2.0 Generic TY—JOUR AU—Abu Anas, E.M., Lee, S.Y. & Hasan, M.K., 2010. Sequential algorithm for life-threatening cardiac pathologies detection based on mean signal strength and EMD functions. BioMedical Engineering Online, 9:43. https://doi.org/10.1186/1475-925X-9-43.

2. Wait for signal that the defibrillator is charged

**E. Clear the patient**

1. Ensure patient is positioned on a grounded and electrically isolated surface
2. Confirm oxygen is turned off and/or removed from the patient
3. State clearly: *"Clear!"* (check head-to-toe).
4. Visually confirm no one is touching the patient

**F. Press the shock button**

**G. Immediately resume CPR**

1. Begin chest compressions without delay.
2. Do not pause for rhythm or pulse checks until the next planned assessment.

**H. Check for a pulse at next pulse check**

### *7.4.3.3 Other arrhythmias*

Other disturbances of cardiac rhythm may cause a patient to become unstable and put them at imminent risk for cardiac arrest. In this case, rapid interventions with synchronized electrical cardioversion and/or medications are effective and necessary. Arrhythmias can be broadly divided into two categories: bradyarrhythmias (slow heartbeat, <50 beats per minute) and tachyarrhythmias (fast heartbeat, >100 beats per minute). Tachyarrhythmias can be further divided into wide complex and narrow complex arrythmias based on their appearance on an ECG.

The common workup is to take the patient's vital signs and obtain an electrocardiogram (ECG) to evaluate the rhythm.

#### 7.4.3.3.1 Bradyarrhythmia

Slow heart rhythms may cause instability when the heartbeat is too infrequent to generate adequate pressure for perfusing body tissues. While the criteria is often a heartbeat less than 50 beats per minute, if a patient is asymptomatic and has stable blood pressure, a heartbeat above 40 beats per minute may be physiologically normal for them, particularly in highly conditioned athletes. Rates less than 40 beats per minute are rarely normal. It is important to first confirm the patient's stability by taking vital signs, and then check previous records to see if this is a change from their baseline. It is important to recognize and manage the route cause for these rhythms once the patient is stable. Table 7.20 contains initial interventions for bradyarrhythmia.

#### 7.4.3.3.2 Tachyarrhythmia

Heartbeats faster than 100 beats per minute are considered tachycardia and should raise suspicion for an impending or active emergency medical process. In most cases, a regular heartbeat between 100 and 150 is a normal sinus rhythm. This indicates the patient is under physiological or psychological stress, but the issue may not be primarily cardiac. While a normal sinus rhythm faster than 150 is possible, a resting patient with a heart rate this fast should raise suspicion for an arrhythmia. Providers will need to check the patient's vital signs and obtain ECG to determine the rhythm.

A wide complex tachycardia should be considered ventricular tachycardia until proven otherwise. If the patient is unstable, this should be considered a cardiac arrest and treated accordingly. If they are stable the provider should prepare for a synchronized cardioversion. Table 7.21 contains examples of interventions for wide complex tachyarrhythmias.

Narrow complex: Narrow complex rhythms may also cause instability but are often less immediately problematic, particularly in a microgravity setting where low blood pressure may not be as immediately impactful as in a terrestrial setting. Table 7.22 describes common narrow complex tachyarrhythmias and their associated interventions. However, if a patient is unstable immediate synchronized cardioversion is indicated.

### *7.4.3.4 Shock*

Shock is a generic term for any condition that results in inadequate perfusion and oxygen delivery to tissues with end-organ ischemia. There are many causes for shock, but the common theme is that providers must support the patient's blood pressure. This will require establishing venous access (described in Chapter 5) and preparing IV fluids or blood products as appropriate. Patients may need substantial amounts. In some cases, this may also require medications to support blood pressure (vasopressors). Table 7.23 illustrates the common types of shock, provider priorities, and the initial interventions to meet them.

In many cases, the exact cause of shock will not be clear and a provider will need to take action before having a complete diagnosis. As with everything discussed so far, the goal is to buy time by interrupting the most immediately life-threatening processes, leaving time for a more definitive diagnosis and intervention later. The principles of triage, primary assessment, and secondary assessment are the same. Assuming that the airway is intact, and the patient is breathing, the provider should secure IV or IO access, administer a 1 L bolus of fluid, and set up physiological monitors. Following this, the provider should assess the context clues available to them. Since near-term astronauts will all have thorough preflight medical screenings and be healthy and stable at baseline, the most likely causes of shock in near-term space travel will be

**TABLE 7.20** Interventions for bradyarrhythmia.

| Arrhythmia | Primary intervention | Medications (dose & interval) | Key notes |
|---|---|---|---|
| **Symptomatic bradycardia** | Atropine/ epinephrine → pacing | **Push dose epinephrine:** 1–2 ml of 10 mcg/ ml**Epinephrine infusion:** 2–10 mcg/ min**Atropine:** 1 mg IV q3–5 min (max 3 mg) | Atropine is often ineffectiveConsult with cardiology |
| **High-grade AV block** | Electrical Pacing | Dopamine 5–20 mcg/kg/min (bridge) | Prepare for pacing earlyConsult with cardiology |

*AV*, Atrioventricular; *IV*, intravenous.

**TABLE 7.21 Interventions for wide complex tachyarrhythmias.**

| Arrhythmia | Primary intervention | Medications | Key notes |
|---|---|---|---|
| **Unstable ventricular tachycardia (with pulse)** | Synchronized cardioversion | Sedation if time allows (e.g., **ketamine 0.5–1 mg/kg IV**) | Hypotension or altered mental status = unstableConsult with cardiology |
| **Unstable ventricular tachycardia (without a pulse)** | Defibrillation | ACLS cardiac arrest (epinephrine 1 mg, 1:10,000) | Hypotension or altered mental status = unstableConsult with cardiology |
| **Torsades de pointes** | MagnesiumSynchronized Cardioversion | **Magnesium sulfate** 2 g IV over 1–2 min | Avoid QT-prolonging drugs |
| **Ventricular tachycardia (stable)** | Antiarrhythmic MedicationsSynchronized Cardioversion | **Procainamide** 20–50 mg/min (max 17 mg/kg) OR **Amiodarone** 150 mg IV over 10 min × 2 | Avoid if hypotensiveMonitor patient for signs of instabilityConsult with cardiology |

*ACLS*, Advanced cardiovascular life support; *IV*, intravenous.

**TABLE 7.22 Interventions for narrow complex tachyarrhythmias.**

| Arrhythmia | Primary intervention | Medications (dose & interval) | Key notes |
|---|---|---|---|
| **Normal sinus rhythm** | Fluids, identify underlying cause | IV fluids 2–4 L | Consider potential causes of shock. |
| **Supra-ventricular tachycardia (AVNRT/AVRT)** | Vagal maneuvers → medications | **Adenosine:** First dose: 6 mg IV push. Second dose 12 mg IV once | Rapid push + flush required.Monitor patient for instabilityIf unstable, perform immediate cardioversion.Consult with cardiology. |
| **Atrial fibrillation/ flutter (known issue for longer than 48 hours)** | Medications, synchronized Cardioversion | **Procainamide** 20–50 mg/min (max 17 mg/kg) OR **Amiodarone** 150 mg IV over 10 min × 2 | Monitor patient for instability.If unstable perform immediate cardioversion.Consult with cardiology.If longer than 48 hours, anticoagulation should be considered to prevent stroke. |
| **Atrial fibrillation/ flutter (new onset)** | Rate control | **Metoprolol** 5 mg IV q5 min (max 15 mg) OR **Diltiazem** 0.25 mg/kg IV bolusIf needed, a second dose, 0.35 mg/kg | Metoprolol is less likely to cause hypotension and may be better in spaceflight.Oral medications may also be effective if patient is stable, they just take longer to act.If IV medications are effective, oral medications should follow to maintain effect.Consult with cardiology. |

*ACLS*, Advanced cardiovascular life support; *AVNRT*, atrioventricular nodal reentrant tachycardia; *AVRT*, atrioventricular reentrant tachycardia; *IV*, intravenous.

anaphylaxis, sepsis, or trauma. Another possibility is a pulmonary embolism from a theoretically increased risk of DVT. The incidence of this is likely to be low, but it is not yet known and will be difficult to diagnose. Therefore provider must maintain a high level of suspicion. Traumatic causes for shock will likely be the result of an injury or obstructive shock and should be relatively easy to diagnose from the context. For management of obstructive shock, see Chapter 11. For the remaining two, anaphylaxis is a rapid onset and typically associated with an allergic style rash, wheezing, and facial swelling. Sepsis is more gradual, typically associated with a fever, and may follow a recent or active infection, most

**TABLE 7.23** Early interventions for various types of shock.

| Type of shock | Context clues | Primary problem | Immediate priorities | Initial interventions (first line) | Key cautions |
|---|---|---|---|---|---|
| **Distributive—anaphylactic** | **History/context**<br>• Recent exposure (food, medication, insect sting, contrast)<br>• Rapid onset (minutes)<br>**Physical exam clues**<br>• Warm, flushed skin<br>• Urticaria, rash, itching<br>• Swelling of the lips, tongue, face (angioedema)<br>• Wheezing/ stridor<br>• Hypotension with tachycardia | Severe allergic reaction causing vasodilation, Reactive airways, and capillary leak | Reverse allergic reaction | **Epinephrine 0.3 mg of 1:1000 administered intramuscularly (IM)**Antihistamines (Cetirizine and Famotidine) and corticosteroids are potentially valuable adjuncts**IV crystalloids** (2 L initial bolus) OxygenAirway protection from swelling | Epinephrine is lifesaving. Everything else in this context is secondary. |
| **Distributive—septic** | **History/context**<br>• Recent infection (pneumonia, UTI, abdominal)<br>• Fever or hypothermia<br>• Indwelling lines, recent surgery<br>**Physical exam clues**<br>• Warm, flushed skin early-→ | Vasodilation + capillary leak | Restore vascular tone & volumeTreat Infection | **IV crystalloids** (typically 2–4 L, 30 ml/kg)Broad spectrum IV antibiotics (e.g., Ceftriaxone or Piperacillin/ Tazobactam) **Vasopressors** (norepinephrine) if persistently hypotensive after 3–4 LOxygen if hypoxic | Hypotension may persist despite fluids. |

*(Continued)*

**TABLE 7.23** (Continued)

| Type of shock | Context clues | Primary problem | Immediate priorities | Initial interventions (first line) | Key cautions |
|---|---|---|---|---|---|
| | cool and clammy later<br>• Bounding pulses, wide pulse pressure (early)<br>• Hypotension despite fluids<br>• Altered mental status | | | | |
| **Hypovolemic** (hemorrhage, dehydration) | **History/context**<br>• Trauma, bleeding, surgery<br>• GI losses (vomiting, diarrhea)<br>• Burns, dehydration, heat exposure<br>• Long EVA with poor water intake<br>• Use of diuretics<br>**Physical exam clues**<br>• Cool, pale, clammy skin<br>• Flat neck veins (may not be present in microgravity)<br>• Tachycardia,<br>• Narrow pulse pressure<br>• Delayed capillary refill<br>• Dry mucous membranes<br>• low urine output | Low circulating blood volume, often caused by trauma or severe dehy-dration | Restore preload & perfusion | Control bleeding**IV crystalloids** (2 L initial bolus)**Type matched blood products** if hemorrhagicProvide oxygen if hypoxic | Consult with emergency medicine and trauma surgeon.Establish control of bleeding.Consider permissive hypotension. |

| | | | | | |
|---|---|---|---|---|---|
| **Obstructive—pulmonary Embolism** | **History/context**<br>• Sudden collapse or dyspnea<br>• Recent/active deep vein thrombosis<br>• Use of estrogen based hormonal supplements (birth control)<br>• Prolonged exposure to spaceflight (>30 days)<br>**Physical exam clues**<br>• Hypoxia out of proportion to lung exam<br>• Bulging neck veins (may be difficult to assess in microgravity)<br>• Clear lungs<br>• Tachycardia, hypotension<br>• Pleuritic chest pain (if conscious) | RV outflow obstruction | Reduce obstruction & support RV | Oxygen**Carefully monitored IV crystalloids** (1 L initial bolus) VasopressorsThrombolysis | Excess fluids can worsen RV failure |
| **Obstructive—tension pneumothorax** | **History/context**<br>• Trauma<br>• Receiving positive-pressure ventilation, particularly with high PEEP<br>• Sudden deterioration while on ventilator | Impaired venous return | Relieve pressure immediately | **Needle decompression** (Chapter 11)Chest tube (Chapter 11)Oxygen | Do not delay for imaging |

*(Continued)*

**TABLE 7.23** (Continued)

| Type of shock | Context clues | Primary problem | Immediate priorities | Initial interventions (first line) | Key cautions |
|---|---|---|---|---|---|
| | **Physical exam clues**<br>• Unilateral absent breath sounds<br>• Hyperresonance on affected side<br>• Bulging neck veins (may be difficult to assess in microgravity)<br>• Tracheal deviation (late sign)<br>• Rapid hypotension | | | | |
| **Obstructive—cardiac tamponade** | **History/context**<br>• Penetrating chest trauma<br>• Recent cardiac procedure<br>• Pericarditis<br>• Malignancy<br>**Physical exam clues**<br>• Bulging neck veins (may be difficult to assess in microgravity)<br>• Muffled heart sounds<br>• Hypotension<br>• Pulsus paradoxus<br>• Electrical Alternans<br>• Clear lungs | Impaired filling | Relieve pericardial pressure | **Pericardiocentesis** (Chapter 11)**IV crystalloids** (2 L initial bolus)Oxygen | Positive-pressure ventilation may worsen shock |

| | | | | | |
|---|---|---|---|---|---|
| **Cardiogenic** (myocardial infarction, drug overdose) | **History/context**<br>• Known coronary artery disease<br>• Recent or ongoing MI<br>• Heart failure, cardiomyopathy<br>• Arrhythmias (VT, AF with RVR)<br>• Recent large fluid load causing shortness of breath<br>**Physical exam clues**<br>• Cool, clammy skin<br>• Bulging neck veins (may be difficult to assess in microgravity)<br>• Pulmonary crackles<br>• Hypoxia<br>• Weak pulses<br>• Chest pain<br>• ECG changes | Poor cardiac output | Improve perfusion without fluid overload | **Vasopressors/inotropes** (e.g., norepinephrine, dobutamine)Diuretics (e.g., furosemide, budesonide) if congestedOxygen if hypoxic | Avoid large fluid boluses, Monitor patient during fluid resuscitationConsult with Cardiology |
| **Distributive—neurogenic** | **History/context**<br>• Spinal cord injury (especially cervical)<br>• Recent trauma with neurologic deficit<br>• High spinal anesthesia | Loss of sympathetic tone, often due to spinal trauma | Restore vascular tone | **Vasopressors** (norepinephrine, phenylephrine)**IV crystalloids** (2 L initial bolus)Spinal immobilization | Bradycardia is common, fluids alone insufficient |

(*Continued*)

**TABLE 7.23** (Continued)

| Type of shock | Context clues | Primary problem | Immediate priorities | Initial interventions (first line) | Key cautions |
|---|---|---|---|---|---|
| | **Physical exam clues**<br>• Warm, dry skin<br>• Bradycardia with hypotension<br>• Flat neck veins (may not be present in microgravity)<br>• Lack of peripheral vasocon-striction<br>• Temperature dysregulation | | | | |

*AF*, Atrial fibrillation; *EVA*, extravehicular activity; *GI*, gastrointestinal; *IM*, intramuscular; *IV*, intravenous; *MI*, myocardial infarction; *PEEP*, positive end-expiratory pressure; *RV*, right ventricle; *RVR*, rapid ventricular response; *UTI*, urinary tract infection; *VT*, ventricular tachycardia.

commonly urinary or skin. If in doubt, a provider should feel comfortable administering IM epinephrine and/or antibiotics, since there is little risk of harm and the potential to save the patient's life.

#### 7.4.3.4.1 Rectal-fluid administration

In spaceflight, limitation on spacecraft systems may mean that large volumes of sterile IV fluids, or the tools to start an intravenous or intraosseous line are unavailable. In these circumstances, fluid resuscitation should be performed orally using an oral rehydration solution. This can be manufactured by adding 25 g of sugar and 3 g of salt to 1 liter of water. If a patient is awake and able to take in fluids by mouth, this is the preferred method (WHO/FCH/CAH/06.1 World Health Organization, 2006).

However, a patient in shock may be unable to take in fluids by mouth. In this case, providers must rapidly infuse fluids using proctoclysis. This method was first described in 1909 by Murphy (1909) and was widely used to great effect prior to the wide availability of IV lines. It lends itself well to improvised fluid resuscitation with readily available supplies and does not require sterile fluids (Grocott et al. (2005)). The basic concept is to place a tube into the rectum, connect this to a drink bag containing a rehydration solution, and squeeze the solution into the patient's colon. The rectum must remain blocked to avoid a reflux of the infused water back into the cabin.

If a purpose built rectal tube is not available, a Foley catheter is an effective substitute. All spacecraft should carry foley catheters to manage urinary retention, as described in Chapter 14. Using this method, we suggest the following procedure:

1. *Preparation*
   - Explain procedure to patient (reduces sphincter tone)
   - Position patient
   - Microgravity: patient restrained with pelvis stabilized
   - Reduced gravity: supine or slight lateral
   - Prime tubing with fluid to remove air
2. *Catheter insertion*
   - Generously lubricate a 14Fr Foley catheter (or whatever size is available)
   - Gently insert 10–15 cm into rectum
   - Inflate balloon with 10–15 mL sterile water
   - Goal: retention, not occlusion
   - Gently retract until balloon rests against rectal wall
   - Secure catheter externally
   - Overinflation may cause cramping and leakage
3. *Fluid administration*
   - Connect catheter to a drink bag with an oral rehydration solution or similarly hydrating beverage. In a pinch, fruit juices will work.
   - Fluids should be warm if possible
   - Gently apply pressure to the drink bag to force the fluid into the rectum.
   - Start slow and increase pressure to effect.
   - Likely to attain a rate of ~1 L/hour or ~20 ml/min.
   - Faster rates dislodge the tube, cause cramping, or cause fluid leakage into the cabin
4. *Monitoring and adjustment*
   - Monitor patient's heart rate, blood pressure, and mental status
   - Slow or stop fluid administration if:
   - Patient develops abdominal discomfort or cramping
   - There is leakage around catheter
5. *Ongoing use*
   - Aim to deliver 2–3 liters in 3 hours
   - Reassess perfusion frequently
   - Prepare for evacuation.

## 7.5 Disability, Exposure, and Secondary Survey

The primary assessment covers immediate life threats and focuses on the ABCs. However, once immediate threats are mitigated or ruled out, providers must conduct a rapid but thorough evaluation to look for injuries or illness that may become a threat later. In the ABC paradigm, this stage begins with D for disability and E for exposure and environmental concerns, and is completed with the secondary survey, a full head to toe physical exam to evaluate for additional injuries.

## 7.5.1 Disability

### *7.5.1.1 Definition and description*

Disability in the context of resuscitation and critical care refers to either mechanical or neurological deficits and includes assessments of mental state, orientation, motor movement, pain response, eye movement, and any signs suspicious for an underlying neurologic or vascular injury (e.g., head or spinal trauma or grossly deformed limbs). In terrestrial-based critical care and resuscitation, the focus is on the identification of potential injuries, prevention of further harm, diagnosis, and advanced treatment or transfer to a facility with advanced treatment capability. Treatment interventions for neurological and vascular injuries can be highly specialized and resource-intensive, potentially requiring neurological or orthopedic surgery, or complex endovascular interventions with the aid of real-time neuroradiological imaging. For these reasons, many of the interventions are well beyond the scope of what can be provided in a spaceflight context, and the discussion here will focus on identification of potential injuries, stabilization, and some temporizing adjuncts with an aim to evacuate the patient to definitive care as soon as possible. More information on procedures to care for these potential issues may be found in Chapter 16.

### *7.5.1.2 Identification and assessment*

Rapid identification and assessment of potential neurological injury is key in the immediate efforts of resuscitation. Several tools and scoring systems exist to help identify potential issues and concerns.

#### 7.5.1.2.1 Mechanism

It is important to consider the mechanism of injury and immediately take steps to prevent further damage if a neurological injury exists. In the case of high velocity or high impact trauma, cervical immobilization and spinal precautions are usually applied until an injury to this area can be ruled out. Cervical immobilization is usually accomplished through applying a cervical collar to stabilize the neck, and spinal precautions usually entail careful movement of the patient via a flat spinal board and using "log roll" techniques when moving them to inspect their backside. These techniques together are known as spinal motion restriction. Movement of a patient in microgravity or reduced gravity will be greatly simplified, but it is important to ensure that the patient's body and head move together and no unnecessary leverage is applied that might cause them to move in opposition.

In a microgravity or reduced gravity setting without the typical terrestrial causes of trauma (high speed motor vehicles, guns, knives, etc.), severe trauma is unlikely outside of the dynamic phases of flight (e.g., launch and landing). Based on prior experience, injuries sustained during these phases are unlikely to be survivable (Rogers et al., 1986; Barry et al., no date; NTSB/AAR-15/02 National Transportation Safety Board Washington D.C., 2014). Interestingly, while the American College of Surgeons (ACS) recommends the use of a spinal board and log-roll techniques, the Wilderness Medical Society (WMS) recommends against the use of rigid backboards and cervical collars in austere environments, citing the lack of evidence for benefit and the increased logistical complexity (Brun, 2018; Hawkins et al., 2024). In the light of these somewhat contrasting recommendations between the ACS and WMS, future exploration missions will need to weigh both arguments against the MPVDB restrictions of spaceflight. In any case, whether on a backboard or utilizing existing restraint systems, immobilization will be key to prevent further injury and enable a multitude of interventions through patient and provider restraint as discussed elsewhere.

#### 7.5.1.2.2 Mental status

There are two commonly used components for assessing a patient's mental status. AVPU is a quick and rapid assessment of mental status focused primarily on the level of responsiveness a patient has.

- A—Alert
  - The patient is awake and interactive.
  - The follow-up is to ask three questions to determine how oriented the person is.
  - Can you tell me your name?
  - Can you tell me where you are?
  - Can you tell me the date/year

- If a person can answer all three questions, they are alert and oriented × 3 (A&O×3). If they answer only two correctly they would be described as A&O×2 with the incorrect answer noted in parentheses (e.g., time).
- Some systems choose to add a 4th criteria "situation." If the patient is able to describe what happened, they may be described as A&O×4.
- Some judgment must be exercised in this criteria since it may not always be reasonable to expect the patient to remember the exact date.

- V—Verbal
  - This means the patient responds to verbal cues but is otherwise not alert. For example, a patient who opens their eyes and/or moans when you call their name is considered responsive to verbal stimuli.
- P—Pain
  - This means, the patient will not respond unless painful stimuli are applied.
  - A person who withdraws from pain is likely to have a better mental status and prognosis than one who twitches or moves towards the pain.
- U—Unresponsive
  - If the patient responds to nothing, not even painful or shouted stimuli. they are considered unresponsive.

The second common assessment is the Glasgow Coma Scale (GCS). This is a simple scale originally designed to determine which patients would likely need to be transferred to a hospital with neurosurgical capability. It has been adopted as a common scoring system for mental status by trauma services around the world that allows for rapid assessment of a patient's overall neurological status and provides a basis for rapid decision-making based on the GCS score. The revised scale is composed of three components: eye opening, verbal response, and best motor response with points scored for each component (Table 7.24). Table 7.25 describes a schema for interpreting the GCS. It is important to remember that GCS was intended for trauma and has limited applicability in other circumstances.

**TABLE 7.24** Glasgow coma score.

| Component | Score | Response |
|---|---|---|
| **Eye opening (E)** | 4 | Spontaneous |
| | 3 | To speech |
| | 2 | To pain |
| | 1 | No eye opening |
| **Verbal response (V)** | 5 | Oriented |
| | 4 | Confused conversation |
| | 3 | Inappropriate words |
| | 2 | Incomprehensible sounds |
| | 1 | No verbal response |
| | T | Unable to assess |
| **Motor response (M)** | 6 | Obeys commands |
| | 5 | Localizes pain |
| | 4 | Withdraws from pain |
| | 3 | Abnormal flexion (decorticate) |
| | 2 | Abnormal extension (decerebrate) |
| | 1 | No motor response |

**TABLE 7.25** Interpretation of the Glasgow Coma Scale (GCS).

| GCS score | Level of consciousness | Primary risks | Recommended actions |
|---|---|---|---|
| **15** | Fully alert | None acute | Routine monitoring; serial neurologic exams |
| **13–14** | Mild impairment | Subtle brain injury, intoxication, early deterioration | Frequent reassessment, consider imaging |
| **10–12** | Moderate impairment | Loss of protective reflexes possible; deterioration | Continuous monitoring, imaging if available |
| **9** | Borderline severe | Impaired airway reflexes | Close observation, imaging if available |
| **≤8** | Severe coma | **Unable to protect airway; aspiration risk** | **Definitive airway (intubation)**; mechanical ventilation |
| **6–7** | Deep coma | Hypoventilation; aspiration; ↑ intracranial pressure | |
| **3–5** | Profound coma | Brainstem dysfunction; respiratory failure | |

### 7.5.1.2.3 Neurological assessment

In the immediate resuscitation and evaluation setting, neurologic assessment consists of attaining a GCS, pupillary exam, and focal neurological exam.

Pupillary exam: This is accomplished through shining a light into a patient's eyes and noting the pupillary response. A normal direct response would be pupillary constriction when exposed to light with relaxation after the light is removed. Consensual response is when the contralateral pupil constricts to light being shined in the other eye. Unequal pupillary reaction or fixed dilated pupils can indicate a brain injury or herniation. Overly constricted or dilated eyes with minimal response can indicate a potential cranial injury or perhaps an intoxication or substance/drug reaction.

Focal neurological exam: This includes an examination of the extremities for movement, strength, and sensation. A deficit in any of these should be further investigated to help determine if there is potentially a focal neurologic deficit that could correspond to a cranial or spinal injury. Dermatome and nerve mapping of different areas of sensation and motor are helpful here to help evaluate what level a potential injury could be.

Spinal assessment: A key portion of the initial evaluation of a patient is the palpation of their vertebral column and assessment of sphincter tone. In the terrestrial environment, this is done by log-rolling the patient to one side and systematically firmly pressing on each vertebra from the skull down to the coccyx and repeatedly asking the patient if they are feeling any pain while noting any palpated step-offs or mobile fractures or injuries. This assessment is concluded with an assessment of sphincter tone to evaluate for cauda equina or conus medullaris (damage or impingement of the terminal nerves from the spinal cord). If the patient is awake and oriented, they are asked to "squeeze their cheeks" as if they are defecating. If they are not alert and oriented, a quick digital exam of the rectum to evaluate for tone is done. Lack of squeeze or rectal tone can be concerning for a distal spinal injury (Brun, 2018). In microgravity, this may require some adjustment and counter pressure or restraint to enable the appropriate level of pressure to evaluate for pain and injury without the patient and provider simply floating apart.

### 7.5.1.2.4 Further injuries that may be discovered during disability exam

The initial secondary assessment focuses on injuries that may disrupt circulation to various body parts. Such as a severely displaced fracture or dislocation or compartment syndrome.

Displaced fracture/dislocation: These are generally managed by attempting to place the injured body part back into alignment to restore a kinked off blood vessel. Providers should attempt to reposition and splint injured limbs after ABCs are stabilized. Providers should check for a pulse. If the limb is missing its pulse, providers should reposition it in anatomical location and recheck. Chapter 15 covers this in more detail.

Compartment syndrome: After a crush impact injury or a period of prolonged ischemia, it is critical to evaluate the patient for the development of compartment syndrome. This occurs when swelling in the limb causes sufficient pressure to cut off circulation. Signs of this in an injured and swelling limb include the following:

- Pain that is increasing over time and appears out of proportion with the injury.
- Parasthesias. The feeling of pins and needles is often a sign of decreased blood supply.
- Pale skin. Loss of blood supply may cause pallor.
- Tightness/fullness. Severe swelling will cause a noticeable tightness in the overlying skin.
- Pulselessness. At the end stages, the pressure may be high enough to overcome the pressures in the main arteries supplying the limb. This is a late finding but is highly concerning.

Compartment syndrome is treated by incising the tissue compartments and leaving the wound open to relieve the pressure while it heals. While this is beyond the scope of this chapter, this procedure is one that can be performed in a spacecraft with a scalpel and wound dressings. The main concern is to know which compartments to incise. Given that there is time to prepare for this procedure, we recommend that on board providers consult with a surgeon to prepare, even when a substantial communication delay exists. See Chapter 15 for more detail.

### 7.5.2 Exposure and environmental considerations

#### *7.5.2.1 Definition and description*

Finally, expose the patient as fully as practical and perform a head-to-toe exam, including the back, to ensure no injuries or wounds are missed. If the patient suffered a burn or chemical irritation, it is important to remove any contaminated clothing or garments quickly and to ensure other crewmembers are protected from exposure in the process. All these considerations are even more important in a spaceflight context. The operational environment of spaceflight can lead to bulky suits and equipment that can complicate medical assessment and treatment. Furthermore, the spaceflight environment has a unique mixture of exposures and intoxicant risks. The enclosed environment and atmosphere of a spacecraft can mean even small amounts of particulates or gases can become a major hazard to the crew. Movement in and out of the spacecraft also exposes the crew to various exotic and hazardous materials such as fuels, exhausts, and extraterrestrial regolith and soil and risks of electrical injury from exposed equipment outside the craft. Mission planners and space medicine practitioners must make sure that the risk of harmful exposures are limited to the greatest extent possible and prevent unnecessary harmful materials from being manifested on the spacecraft.

Full exposure of a patient often involves completely removing all of the patient's clothing. Following this, care must be taken to cover the patient back up to protect them from environmental concerns like cold. In the spaceflight environment, it may not be possible to fully expose a patient given the hostility of the environment outside of a spacecraft. In these circumstances, the patient must be brought to a location where a full exam can be performed as quickly as possible without exposing the surviving crewmembers to unnecessary risk.

#### *7.5.2.2 Assessment*

Scene safety: the most important consideration when dealing with any injured patient is safety for the rescuers. The worst thing a provider can do for a patient is become a patient themselves. In the spaceflight context, this means that any treatment likely needs to happen in a delayed fashion after the patient is transported back to a controlled environment (i.e., back inside a spacecraft) and has any contaminated or hazardous equipment or clothing removed.

While this may seem simple, it will likely take discipline and training on the part of the entire crew. Astronauts are carefully chosen, often with teamwork as a key desirable trait, and crews are often assembled years ahead of a mission and work closely day in and day out over years (Landon et al., 2018). This can lead to a strong bond and an automatic desire to help as soon as possible. While there have been few examples of these conflicts in spaceflight, combat medicine has a relatively large body of teaching and protocols that instruct medics and teammates to prioritize safety first with “care under fire” only including immediate life-saving interventions that do not put the provider at risk (Montgomery et al., 2017). Civilian emergency first-aid guidelines similarly reflect this primary consideration of the care giver or responder over the patient (Hewett Brumberg et al., 2024). While it may be difficult and complex in the spaceflight setting with close knit teams, it is critical that the provider and crew are safe before they begin treating an injured crewmember.

#### 7.5.2.3 *Treatment and critical care*

There are many potential injuries a crewmember may sustain that will require medical attention. It is beyond the scope of this chapter to address the management for all of them. At this stage of the secondary assessment, immediate life threats are identified and mitigated, and the immediate time sensitivity of the emergency has passed. Injuries and illnesses identified at this stage typically do not require immediate interventions. While they may require procedural interventions, these are likely to be covered in other chapters of this book.

## 7.6 Prolonged critical care

In an austere medical scenario such as human spaceflight, immediate interventions will be key to saving and stabilizing a patient. However, continued care while transporting the patient to a definitive care facility may be just as important to ensure survival. In a spaceflight scenario, this transport may take from days (in a cis-lunar mission) to possibly months (in the case of a Mars mission). This critical care will likely be severely limited by the amount of equipment, capabilities, and resources aboard (to include crew time and skills). Nevertheless, some considerations are necessary to help provide an injured crewmember the best chance at survival and the maximum recovery possible.

Follow-up critical care in many cases is aimed at supporting the body systems that support the ABCs and allowing the body to heal and/or the care teams time to develop a definitive diagnosis and care plan. Care is intensive but often centers on monitoring the patient's stabilized airway, breathing, and circulation. In this sense, critical care is essentially the same as resuscitation, just slower and with a lot more information. Primarily it involves monitoring trends and changes, interpreting when these changes are not in a favorable direction and intervening to nudge the patient back on track. This is an area where checklists and scheduled assessments may be helpful. These help ensure providers keep track of all necessary parameters and are able to identify trends early. One example of such a tool is described below. In addition, telemedical support for critical care will be required in this setting.

### 7.6.1 Patient rounding checklist

#### 7.6.1.1 *Airway*

☐ Airway patent
☐ ETT/trach position confirmed (if present)
☐ Cuff pressure acceptable

**Normal/target**

- ETT depth stable; no air leak
- Cuff pressure **20–30 cm$H_2O$**
- No signs of aspiration

#### 7.6.1.2 *Breathing/ventilation*

☐ $SpO_2$ acceptable
☐ Ventilator settings reviewed
☐ Plateau/peak pressures acceptable
☐ Vent synchrony

**Normal/target**

- $SpO_2$ **92%–96%**
- VT **6–8 mL/kg** IBW (4–6 in ARDS)
- Plateau pressure **≤30 cm$H_2O$**
- $ETCO_2$ **35–45 mmHg** (unless permissive hypercapnia)

#### 7.6.1.3 *Circulation/perfusion*

☐ MAP adequate
☐ HR appropriate
☐ Vasopressors reviewed
☐ Perfusion signs assessed

**Normal/target**

- MAP **≥65 mmHg** (patient-specific)
- HR **60–100 bpm** (context-dependent)
- Cap refill **<2 seconds**
- Lactate stable or decreasing, if lab capability exists
- $ETCO_2$ stable **35–45 mmHg** (unless permissive hypercapnia)

### *7.6.1.4 Neurologic status*

☐ GCS or Richmond Agitation–Sedation Scale documented
☐ Pupils checked
☐ Delirium assessed

**Normal/target**

- GCS stable or improving
- Sedation score **–1 to 0** (unless deep sedation needed; Table 7.26
- Pupils equal and reactive
- Delirium assessment (if RASS > –4; Table 7.27)

**TABLE 7.26** Richmond agitation–sedation scale (RASS).

| RASS score | Category | Description |
|---|---|---|
| +4 | Combative | Overtly combative, violent, immediate danger to staff |
| +3 | Very agitated | Pulls or removes tubes/lines; aggressive behavior |
| +2 | Agitated | Frequent nonpurposeful movement; fights ventilator |
| +1 | Restless | Anxious or apprehensive; movements not aggressive |
| 0 | Alert and calm | Awake, calm, makes eye contact |
| –1 | Drowsy | Not fully alert; sustained eye contact to voice (>10 s) |
| –2 | Light sedation | Brief eye contact to voice (<10 s), then drifts off |
| –3 | Moderate sedation | Movement or eye opening to voice; no eye contact |
| –4 | Deep sedation | No response to voice; movement or eye opening to physical stimulation |
| –5 | Unarousable | No response to voice or physical stimulation |

**TABLE 7.27** Confusion assessment method for the intensive care unit.

| Feature | What to assess | Positive if... |
|---|---|---|
| **Feature 1** | **Acute onset or fluctuating course** | Acute change from baseline **OR** mental status fluctuates during the day |
| **Feature 2** | **Inattention** | Difficulty focusing; easily distracted; misses >2 items on attention test (e.g., letter "A" test) |
| **Feature 3** | **Altered level of consciousness** | Anything other than alert and calm (**RASS ≠ 0**) |
| **Feature 4** | **Disorganized thinking** | Illogical answers, incoherent speech, or inability to follow simple commands |

CAM-ICU is considered positive if 3 of the 4 features are present. Feature 1 AND 2 are present + EITHER Feature 3 OR Feature 4.

#### 7.6.1.5 Sedation and analgesia

☐ Pain assessed
☐ Sedation goal defined
☐ Daily sedation interruption considered
**Normal/target**

- Pain score **≤3/10** (or equivalent)
- Sedation matched to indication
- No unnecessary continuous paralytics

#### 7.6.1.6 Hemodynamics and fluids

☐ Volume status assessed
☐ Daily fluid balance reviewed
**Normal/target**

- Clinically euvolemic
- Net balance **neutral to mildly negative**
- No signs of fluid overload

#### 7.6.1.7 Renal function

☐ Urine output adequate
☐ Creatinine trend reviewed (if lab tests are available)
**Normal/target**

- UOP **≥0.5 mL/kg/h**
- Creatinine stable or improving
- Electrolytes in safe range

#### 7.6.1.8 Electrolytes and acid base

☐ $Na^+$, $K^+$, $Ca^{2+}$, $Mg^{2+}$ reviewed (if lab tests are available)
☐ Acid–base status reviewed (if lab tests are available)
**Normal/target**

- $Na^+$ **135–145 mEq/L**
- $K^+$ 3.5–5.0 mEq/L
- $Mg^{2+}$ **≥2.0 mg/dL** (often targeted)
- pH **7.35–7.45** (or acceptable deviation)

#### 7.6.1.9 Cardiac rhythm

☐ ECG reviewed
☐ Rate control adequate
**Normal/target**

- Sinus rhythm or controlled atrial fibrillation or atrial flutter
- No untreated malignant arrhythmias
- QTc monitored if relevant drugs used

#### 7.6.1.10 Infection/sepsis

☐ Source identified/considered
☐ Antibiotics reviewed
☐ Cultures reviewed (if culture or PCR tests are available)
**Normal/target**

- Appropriate antibiotics started within **1 hour**
- De-escalation considered daily

- Afebrile or improving fever curve

#### 7.6.1.11 Nutrition and glucose

☐ Nutrition route confirmed
☐ Caloric adequacy reviewed
☐ Glucose controlled (if lab tests are available)

**Normal/target**

- Enteral feeding within **24–48 hours** if possible
- Glucose **140–200 mg/dL**
- Tight glucose control has been associated with higher rates of complications in the acute phase of an illness and while in critical care
- No recurrent hypoglycemia

#### 7.6.1.12 Hematology and coagulation

☐ Hemoglobin reviewed (if lab tests are available)
☐ Platelets reviewed (if lab tests are available)
☐ Coagulation status assessed (if lab tests are available)

**Normal/target**

- Hgb **≥7 g/dL** (unless otherwise indicated)
- Platelets adequate for procedures
- INR/PTT acceptable or intentionally managed

#### 7.6.1.13 Prophylaxis

☐ Consider need for DVT prophylaxis
☐ Stress ulcer prophylaxis
☐ VAP prevention

**Normal/target**

- Pharmacologic or mechanical DVT prophylaxis is used terrestrially for prolonged immobility during a stay in the ICU. In space, DVT risk is likely elevated but no critical illness DVT prophylaxis protocols have been developed. Consider and enact on an as needed basis.
- If in gravity, HOB elevated **30°–45°** in microgravity, consider Famotidine as indicated
- Oral care performed

#### 7.6.1.14 Lines, tubes, devices

☐ All lines reviewed
☐ Indication still present

**Normal/target**

- No unnecessary central lines
- Foley removed unless indicated
- Line days documented and removed as soon as practicable

#### 7.6.1.15 Skin and mobility

☐ Skin inspected
☐ Pressure injury prevention
☐ Mobility assessed

**Normal/target**

- No new pressure injuries
- Repositioning q2h
- Mobility at highest tolerated level

#### 7.6.1.16 *Trajectory and goals*

☐ Improving/stable/worsening
☐ Anticipated problems
☐ Goals of care documented
**Normal/target**

- Clear daily plan
- Code status documented
- Family updated as appropriate

## 7.7 Additional considerations

Many of the diagnostic, resuscitative, and treatment options used on Earth are technically possible in spaceflight with some adjustments to account for the environment. The main issues surround the limitations of MPVDB budgets for the vehicle. Medical systems must be matched to the likelihood and consequence of each eventuality. Resuscitation and critical care on Earth are characterized by an "all-hands-on-deck" approach with many people, many resources, and many tools used to help save and stabilize a patient. In spaceflight, there are few people, finite resources, and limited tools. This difficulty is further exacerbated by the length of mission, the distance from Earth, and the time to evacuate to definitive care. Medical systems must necessarily tend more towards independent medical capabilities as these factors increase, but the risks and benefits of the system must still be traded with all other vehicle systems to avoid lowering medical risk at the expense of increasing overall mission risk (Levin et al., 2023).

Once a medical system is optimized to support a crew successfully across an acceptable range of possible medical scenarios, tough decisions remain that are somewhat unique to critical care such as;

- While stabilization of a critically ill patient may be possible, is it the best use of time and resources?
- A critical care intervention may be possible, but can it be sustained until the patient arrives at a definitive care location?
- If not, what risk does continued care and an evacuation pose to the remaining crew?
- What happens if the patient's clinical course does not improve? What are the thresholds for medical futility in space, and when should an astronaut be allowed to?

These decisions regarding care and evacuation will need careful consideration, and the use of established frameworks may help provide objective assessments on what will ultimately be subjective decisions (see Chapter 29; Almand et al., 2023).

Ultimately, the decisions underlying the design of an exploration mission medical system will likely be unique for each mission. What can be treated, what should be treated, and what must be treated will depend heavily on the specifics of MPVDB limitations of the vehicle, the training and capability of the crew, and the hard objective distances and times associated with the mission. Ideally the information presented in this chapter combined with PRA design tools and decision-making frameworks can provide a basis to design capable and quality medical support for future crews.

## References

Almand, A., Ko, S.Y., Anderson, A., Keller, R.J., Zero, M., Anderson, A.P., et al., 2023. A qualitative investigation of space exploration medical evacuation risks. Aerospace Medicine and Human Performance 94 (12), 875–886. Available at https://doi.org/10.3357/AMHP.6262.2023. Available from: http://ingentaconnect.com/content/asma/amhp;jsessionid=13tlejf2f0ojk.alice.

Barry, J.L., Deal Duane, W., Hallock James, N., Hess Kenneth, W., Hubbard, G., Scott Logsdon John, M., et al., n.d. NASA National Aeronautics and Space Administration Washington D.C. Columbia Accident Investigation Board Report Volume 1.

Sandefur, B.J., Driver, B.E., Brown, C.A., Reardon, R.F., 2020. Definitive airway management of patients with a King laryngeal tube TM in place in the COVID-19 pandemic. The western journal of emergency medicine 21 (3), 542–545. Available at https://doi.org/10.5811/westjem.2020.4.47462.t.

Blakeman, T.C., Rodriquez, D., Britton, T.J., Johannigman, J.A., Petro, M.C., Branson, R.D., 2016. Evaluation of oxygen concentrators and chemical oxygen generators at altitude and temperature extremes. Military Medicine 181 (5), 160–168. Available at https://doi.org/10.7205/MILMED-D-15-00130. Available from: http://publications.amsus.org/doi/pdf/10.7205/MILMED-D-15-00130.

Braunecker, S., Douglas, B., Hinkelbein, J., 2015. Comparison of different techniques for in microgravity—a simple mathematic estimation of cardiopulmonary resuscitation quality for space environment. The American Journal of Emergency Medicine 33 (7), 920–924. Available at https://doi.org/10.1016/j.ajem.2015.04.018.

Brun, S., 2018. ATLS advanced trauma life support 10th edition student course manual. American College of Surgeons. Available from: https://research.bond.edu.au/en/publications/atls-advanced-trauma-life-support-10th-edition-student-course-man/.

Canepa, C.A., Levin, D.R., Padaki, A.S., 2025. Comparison of camera-acquired vital signs to conventional vital signs in a space-analog environment. Wilderness & Environmental Medicine 36 (1_suppl). Available at https://doi.org/10.1177/10806032241291994.

Conkin, J., Abercromby, A.F.J., Dervay, J.P., Feiveson, A.H., Gernhardt, M.L., Norcross, J.R., et al., 2014. Probabilistic assessment of treatment success for hypobaric decompression sickness. Available from: http://www.sti.nasa.gov.

Daniels, V.R., Williams, E.S., 2023. Exploring the complexities of drug formulation selection, storage, and shelf-life for exploration spaceflight. British Journal of Clinical Pharmacology. Available at https://doi.org/10.1111/bcp.15957. Available from: http://onlinelibrary.wiley.com/journal/10.1111/(ISSN)1365-2125.

Dwivedi, D.B., Ball, J., Smith, K., Nehme, Z., 2025. Incidence and outcomes of out-of-hospital cardiac arrest from initial asystole: a systematic review and meta-analysis. Resuscitation 212.

Finn, J., Jacobs, I., Williams, T.A., Gates, S., Perkins, G.D., 2019. Adrenaline and vasopressin for cardiac arrest. The Cochrane Database of Systematic Reviews 1 (1). Available at https://doi.org/10.1002/14651858.CD003179.pub2.

Garcia, S.I., Sandefur, B.J., Campbell, R.L., Driver, B.E., April, M.D., Carlson, J.N., et al., 2023. First-attempt intubation success among emergency medicine trainees by laryngoscopic device and training year: a national emergency airway registry study. Annals of Emergency Medicine 81 (6), 649–657.

Grocott, M.P.W., McCorkell, S., Cox, M.L., 2005. Resuscitation from hemorrhagic shock using rectally administered fluids in a wilderness environment. Wilderness & Environmental Medicine 16 (4), 209–211.

Groemer, G.E., Brimacombe, J., Haas, T., de Negueruela, C., Soucek, A., Thomsen, M., et al., 2005. The feasibility of laryngoscope-guided tracheal intubation in microgravity during parabolic flight: a comparison of two techniques. Anesthesia & Analgesia 101 (5), 1533–1535. Available at https://doi.org/10.1213/01.ane.0000181001.25777.53.

Halpern, P., Dang, T., Epstein, Y., Van Stijn-Bringas Dimitriades, D., Koenig, K.L., 2019. Six hours of manual ventilation with a bag-valve-mask device is feasible and clinically consistent. Critical Care Medicine 47 (3), e222. Available at https://doi.org/10.1097/CCM.0000000000003632. Available from: http://journals.lww.com/ccmjournal/pages/default.aspx.

Hawkins, S.C., Williams, J., Bennett, B.L., Islas, A., Quinn, R., 2024. Wilderness medical society clinical practice guidelines for spinal cord protection: 2024 update. Wilderness & Environmental Medicine 35 (1_suppl), 78S. Available at https://doi.org/10.1177/10806032241227232.

Hewett Brumberg, E.K., Douma, M.J., Alibertis, K., Charlton, N.P., Goldman, M.P., Harper-Kirksey, K., et al., 2024. American Heart Association and American Red Cross guidelines for first aid. Circulation 150 (24), e519. Available at https://doi.org/10.1161/CIR.0000000000001281. Available from: http://circ.ahajournals.org.

Hinkelbein, J., Ahlbäck, A., Antwerber, C., Dauth, L., DuCanto, J., Fleischhammer, E., et al., 2021. Using supraglottic airways by paramedics for airway management in analogue microgravity increases speed and success of ventilation. Scientific Reports 11 (1). Available at https://doi.org/10.1038/s41598-021-88008-x.

Hinkelbein, J., Kerkhoff, S., Adler, C., Ahlbäck, A., Braunecker, S., Burgard, D., et al., 2020. Cardiopulmonary resuscitation (CPR) during spaceflight - a guideline for CPR in microgravity from the German Society of Aerospace Medicine (DGLRM) and the European Society of Aerospace Medicine Space Medicine Group (ESAM-SMG). Scandinavian Journal of Trauma, Resuscitation and Emergency Medicine 28 (1). Available at https://doi.org/10.1186/s13049-020-00793-y. Available from: http://www.sjtrem.com/.

Houtchens, B.A., 1993. Medical-care systems for long-duration space missions. Clinical Chemistry 39 (1), 13–21. Available at https://doi.org/10.1093/clinchem/39.1.13.

Ishii, J., Nishikimi, M., Kikutani, K., Ohki, S., Ota, K., Anzai, T., et al., 2024. Resuscitation attempt and outcomes in patients with asystole out-of-hospital cardiac arrest. JAMA Network Open 7 (11). Available at https://doi.org/10.1001/jamanetworkopen.2024.45543.

James, J.T., Ryder, V.E., 2020. Health effects of atmospheric contamination. Principles of Clinical Medicine for Space Flight. Springer New York, United States, pp. 153–174. Available at http://doi.org/10.1007/978-1-4939-9889-0, https://doi.org/10.1007/978-1-4939-9889-0_4 Available from:.

Kouwenhoven, W.B., Jude, J.R., Knickerbocker, G.G., 1960. Closed-chest cardiac massage. JAMA: the Journal of the American Medical Association 173, 1064–1067.

Landon, L.B., Slack, K.J., Barrett, J.D., 2018. Teamwork and collaboration in long-duration space missions: going to extremes. American Psychologist 73 (4), 563–575. Available at https://doi.org/10.1037/amp0000260.

Lange, K.E., Perka, A.T., Duffield, B.E., Jeng, F.F., Sverdrup, J., 2005. Bounding the spacecraft atmosphere design space for future exploration missions. Available from: http://www.sti.nasa.gov.

Levin, D., 2025. 1/6th G Parabolic Flight Bordeaux, FR Personal Conversations.

Levin, D.R., Steller, J., Anderson, A., Lemery, J., Easter, B., Hilmers, D.C., et al., 2023. Enabling human space exploration missions through progressively earth independent medical operations (EIMO). IEEE Open journal of Engineering in Medicine and Biology 4, 162–167. Available at https://doi.org/10.1109/OJEMB.2023.3255513.

McCuaig, K.E., Houtchens, B.A., 1992. Management of trauma and emergency surgery in space. Journal of Trauma and Acute Care Surgery 33 (4). Available at https://journals.lww.com/jtrauma/fulltext/1992/10000/management_of_trauma_and_emergency_surgery_in.18.aspx.

Montgomery, H.R., Butler, F.K., Kerr, W., Conklin, C.C., Morissette, D.M., Remley, M.A., et al., 2017. TCCC guidelines comprehensive review and update: TCCC guidelines change 16-03. Journal of Special Operations Medicine 17 (2), 21. Available at https://doi.org/10.55460/zgaf-inzu.

Murphy, 1909. Proctoclysis in the treatment of peritonitis. Journal of the American Medical Association LII (16), 1248–1250. Available at https://doi.org/10.1001/jama.1909.25420420028003a.

Naaz, S., N.Sahay, R., Kumar, A., Kumar, A., Asghar, 2020. Prolonged use of supraglottic airway device for mechanical ventilation in the intensive care unit. Indian Journal of Critical Care Medicine: Peer-Reviewed, Official Publication of Indian Society of Critical Care Medicine 24 (10), 1001–1002. Available at https://doi.org/10.5005/jp-journals-10071-23630.

NTSB/AAR-15/02 National Transportation Safety Board Washington D.C., 2014. In-Flight Breakup During Test Flight Scaled Composites SpaceShipTwo, N339SS Near Koehn Dry Lake, California.

Okubo, M., Komukai, S., Andersen, L.W., Berg, R.A., Kurz, M.C., Morrison, L.J., et al., 2024. Duration of cardiopulmonary resuscitation and outcomes for adults with in-hospital cardiac arrest: retrospective cohort study. BMJ (Clinical Research Ed.) 384. Available at https://doi.org/10.1136/bmj-2023-076019.

Padaki, A., Zarzour, A.L., Keene, K.R., Canepa, C.A., Levin, D.R., Antonsen, E.L., 2025. Clinical validation of non-contact vital signs in an emergency department setting. Frontiers in Medical Technology.

Polaris Program Science and Research, 2025. Available from: https://polarisprogram.com/polaris-dawn-crew-meets-with-principal-investigators-of-science-and-research-experiments-to-discuss-initial-findings/#:~:text=University%20of%20Texas%20Medical%20Branch:%20The%20Airway, transnasal%20endoscopy%20and%20airway%20ultrasound%20by%20non

Rehnberg, L., Ashcroft, A., Baers, J.H., Campos, F., Cardoso, R.B., Velho, R., et al., 2014. Three methods of manual external chest compressions during microgravity simulation. Aviation, Space, and Environmental Medicine 85 (7), 687–693. Available from: Available at https://doi.org/10.3357/ASEM.3854.2014, http://docserver.ingentaconnect.com/deliver/connect/asma/00956562/v85n7/s2.pdf.

Reynette, N., Sagnières, L., Pequignot, B., Levy, B., Zuily, S., Chenuel, B., et al., 2024. Manual versus automatic chest compression devices for cardiopulmonary resuscitation under zero gravity (The MACCC - 0G STUDY). Resuscitation 203, 110385. Available at https://doi.org/10.1016/j.resuscitation.2024.110385.

Robertson, J.M., Dias, R.D., Gupta, A., Marshburn, T., Lipsitz, S.R., Pozner, C.N., et al., 2020. Medical event management for future deep space exploration missions to Mars. Journal of Surgical Research 246, 305–314. Available from: Available at https://doi.org/10.1016/j.jss.2019.09.065, http://www.elsevier.com/inca/publications/store/6/2/2/9/0/1/index.htt.

Rogers, W.P., Acheson, D.C., Feynman, R.P., Kutyna, D.J., Rummel, R.W., Walker Jr., A.B.C., et al., 1986. Rogers Commission Report to the President on the Space Shuttle Challenger Accident. National Aeronautics and Space Administration, Washington, D.C..

Starck, C., Thierry, S., Bernard, C.I., Morineau, T., Jaulin, F., Chapelain, P., et al., 2020. Tracheal intubation in microgravity: a simulation study comparing direct laryngoscopy and videolaryngoscopy† British Journal of Anaesthesia 125 (1), e47. Available at https://doi.org/10.1016/j.bja.2019.11.029. Available from: https://www.journals.elsevier.com/british-journal-of-anaesthesia.

Staudinger, T., Brugger, S., Watschinger, B., Röggla, M., Dielacher, C., Löbl, T., et al., 1993. Emergency intubation with the Combitube: comparison with the endotracheal airway. Annals of Emergency Medicine 22 (10), 1573–1575.

Subramanian, A., Garcia-Marcinkiewicz, A.G., Brown, D.R., Brown, M.J., Diedrich, D.A., 2016. Definitive airway management of patients presenting with a pre-hospital inserted King LT(S)-D laryngeal tube airway: a historical cohort study. Canadian Journal of Anaesthesia = Journal Canadien d'anesthesie 63 (3), 275–282. Available at https://doi.org/10.1007/s12630-015-0493-x.

Thierry, S., Jaulin, F., Starck, C., Ariès, P., Schmitz, J., Kerkhoff, S., et al., 2023. Evaluation of free-floating tracheal intubation in weightlessness via ice-pick position with a direct laryngoscopy and classic approach with indirect videolaryngoscopy. Nature Research, France npj Microgravity 9 (1). Available at https://doi.org/10.1038/s41526-023-00314-y. Available from: https://www.nature.com/npjmgrav/.

Valenzuela, T.D., Roe, D.J., Nichol, G., Clark, L.L., Spaite, D.W., Hardman, R.G., 2000. Outcomes of rapid defibrillation by security officers after cardiac arrest in casinos. The New England Journal of Medicine 343 (17), 1206–1209.

Viegas, S.F., Williams, D., Jones, J., Strauss, S., Clark, J., 2004. Physical demands and injuries to the upper extremity associated with the space program. Journal of Hand Surgery 29 (3), 359–366. Available at https://doi.org/10.1016/j.jhsa.2004.01.015. Available from: http://www.elsevier.com/inca/publications/store/6/2/3/1/4/5/index.htt.

Warnecke, T., Dauth, L., Ahlbäck, A., DuCanto, J., Fleischhammer, E., Glatz, C., et al., 2021. Time to ventilation and success rate of airway devices in microgravity: a randomized crossover manikin-trial using an underwater setting. Acta Anaesthesiologica Scandinavica 65 (5), 681–687. Available at https://doi.org/10.1111/aas.13780.

Warnecke, T., Tochtermann, F., Kerkhoff, S., Komorowski, M., Neuhaus, C., Hinkelbein, J., 2019. Airway management in microgravity: a systematic review. Acta Anaesthesiologica Scandinavica 63 (1), 2–7. Available at https://doi.org/10.1111/aas.13251.

WHO/FCH/CAH/06.1 World Health Organization, 2006. Oral rehydration salts, production of the new ORS.

Zietlow, J., Berns, K., Jenkins, D., Zietlow, S., 2019. Prehospital use of ketamine: effectiveness in critically ill and injured patients. Military Medicine 184 (Supplement_1), 542–544. Available at https://doi.org/10.1093/milmed/usy422.

Zoll, P.M., Linenthal, A.J., Gibson, W., Paul, M.H., Norman, L.R., 1956. Termination of ventricular fibrillation in man by externally applied electric countershock. The New England Journal of Medicine 254 (16), 727–732.

Chapter 8

# Irrigation

Dana R. Levin[1,2,3]

[1]*Space Medicine Operations, Mission Operations Vast Inc., Long Beach, CA, United States,* [2]*Department of Emergency Medicine, University of California, Los Angeles, CA, United States,* [3]*Division of Space, Ecological, Arctic, and Resource-Limited Medicine, Massachusets General Hospital, Boston, MA, United States*

## 8.1 Introduction

Irrigation is the act of applying a stream of liquid to a body part or wound. It is used to remove debris, minimize tissue damage from reactive chemicals, reduce the risk of infection, and facilitate examination and wound repair. While it is not a technically difficult procedure, several variables influence its effectiveness and must be considered in the spaceflight environment. These include characteristics of the injury, characteristics of the procedure, management of consumables, and waste disposal.

## 8.2 Does it work?

There are four reasons to irrigate an injury: reduce infection risk, facilitate wound healing, remove debris and reactive compounds, and facilitate examination and wound repair.

### 8.2.1 Facilitate examination and wound repair

Irrigation for improved physical examination of a wound is a common use case well supported by clinical experience. Blood, tissue, and debris can obscure visualization, and the adequacy of wound repair depends on an accurate exam, with some techniques requiring clean, dry surfaces (e.g., adhesive strips).

For these reasons, any wound with visible contamination that impairs a provider's ability to fully examine the wound or properly repair it should be cleaned. However, the extent of this cleaning may not be substantial. Some wounds may require only patting the area dry with a clean absorbent material such as gauze or rinsing off dried blood or other material. Others may require the washing out of debris trapped under the surface to better visualize the extent of the injury. Irrigating contaminated wounds, deep wounds, and wounds that carry a high risk of infection in sensitive areas (e.g., open fractures and intraperitoneal injuries) will be discussed under the infection control section.

However, for the purposes of examining and facilitating repair, irrigation is recommended by all major terrestrial guidelines, and evidence supports this use. The caveat is that the volume, pressure, and type of solution, as well as other variables, are specific to the type of wound and often do not require research trials to demonstrate the value and efficacy of cleaning a surface for visual inspection and the use of adhesives.

### 8.2.2 Removal of debris and reactive or bioactive compounds

Removal of debris can also be important to limit ongoing tissue damage. Debris can act as a vehicle for microorganisms to seed tissue. Reactive chemical compounds can cause direct tissue damage or even systemic effects through their interaction with tissues, cells, and enzymes. A classic example of this is chemical exposure to the eye or chemical burns to the skin.

While little research has been performed comparing irrigation of ocular chemical exposures to a nonirrigated control, the risks of chemical burns causing scarring are well established (VanHoy et al., 2025). What research has been performed supports the idea that copious amounts of water is the most effective solution.

**Procedural Medicine for Spaceflight. DOI:** https://doi.org/10.1016/B978-0-443-27552-4.00012-9

The principle is that a highly reactive compound, such as an acid, base, or other highly caustic chemical, will react with the molecules of the exposed tissue. These reactions change the chemical and physical properties of the tissue, rendering it ineffective, such as the dissolution of skin or eye tissue. This renders the tissue dysfunctional at best and permanently damaged at worst. In some cases, this can result in loss of function for a critical organ, such as blindness, or destruction of sufficient skin and underlying structures to cause loss of function, loss of limb, or even death (Patek et al., 2025). By irrigating the offending chemical, it is possible to wash off any remaining material and/or neutralize the reaction, thus halting any ongoing damage. The shorter the time until treatment, the less damage is done. This is the principle behind having eyewash stations in every lab and instructions to wash skin under running water for every caustic household chemical (OSHA, 2013).

This is a strong argument for the availability of copious irrigation fluid on any space station where crew may encounter caustic chemicals, even in the absence of empirical studies. While the best strategy is prevention, through strategies including avoiding the use of such chemicals within spacecraft, only flying them in multi-barrier leak-proof containers, or only storing them external to the vehicle, no system of exposure prevention will ever be 100% effective. It is important for medical system designers to know which chemicals will be on board to enable them to plan accordingly for the volume of irrigation solution.

With regard to irrigation solutions, water will be sufficient in almost every case (Tan and Wong, 2015). This is because most damaging chemicals are acids or bases and are water-soluble. Even in cases where oils are the primary offender, such as with capsaicin-based sprays, water appears to be equivalent or superior to emulsifier solutions (e.g., shampoo, milk; Stopyra et al., 2018; Barry et al., 2009).

The advantage of this is that water will be present on all crewed spacecraft at baseline. The disadvantage is that water is a limited resource that is needed for a variety of life critical purposes and irrigating a caustic wound turns potable water into a biohazardous waste product which may not be recyclable. The risk of a chemical exposure occurrence, and the potential damage it may cause, must be traded against the amount of water required to safely irrigate, the risk of consuming those supplies poses, and the risks of damaging or consuming contaminant filters.

The other advantage is that the chemicals and quantities on board the sealed environment of a spacecraft can be selected, controlled, and measured. Since water is the best overall choice, barring any specific needs for specific chemicals, the main questions for a spaceflight physician are "how much to carry?" and "is the on-board supply sufficient or does a separate supply of water for irrigation need to be part of the manifest?". The spaceflight physician must, at a minimum, consider the following items.

1. What chemicals will be present on board?
2. What other uses does the station have for water?
3. Can contaminated water be recycled?
4. If disposed of, what are the available volumes and protections for waste containment?
5. How much mass/volume is available on board for medical use, and what resources will need to be excluded if a specific medical supply is manifested?
6. How much mass/volume does the equipment for irrigation, droplet protection, and waste collection require?

If the risk is high enough and the water cannot be recycled, the medical system may need to provide its own supply of water for irrigation purposes.

### 8.2.3 Wound healing

Another often-cited theoretical benefit of irrigation is that it may help speed healing. The theory behind this approach is that removing microscopic debris and microorganisms can reduce impediments to healing and accelerate wound closure. Unfortunately, several high-quality systematic reviews have shown no significant changes in healing time associated with saline, tap water, distilled water, or antiseptic solutions as compared to no irrigation. This is a strong argument against the efficacy of irrigation as a wound-healing accelerant.

### 8.2.4 Infection reduction

Irrigation of acute traumatic wounds is widely recommended to reduce the risk of infection. This is based on the theory that debris and microorganisms may lodge under the breach in skin, and the mechanical pressure of the irrigation solution can remove them before they result in infection. However, evidence supporting this is mixed and of low quality. In general, infection rates from acute traumatic wounds range from 3% to 9%, and studies suggest that the level and type of debris or contamination may be more significant for infection risk than whether it was irrigated. However, infection reduction may also be influenced by the type of solution, the volume of solution used, the irrigation pressure, and the time elapsed before intervention.

#### *8.2.4.1 Choice of solution*

A Cochrane review from 2022 examining studies comparing tap water, normal saline, and no irrigation found no difference in infection rates (8.6% and 8.3%, respectively), speed of wound healing, wound size reduction, patient satisfaction, pain scores, or cost. Similarly, a 2015 systematic review comparing six randomized controlled trials also found tap water to be as effective as sterile saline (Huang et al., 2015). Distilled water has also been compared to normal saline in open fractures, with no significant difference in infection rates or wound healing outcomes (Olufemi and Adeyeye, 2017) These findings suggest that nonsterile tap water is a safe and effective alternative to sterile saline solutions for wound irrigation, particularly in low-resource settings (Mitha et al., 2024).

However, comparison of sterile saline irrigation to no irrigation after abdominal surgery adds uncertainty to Mitha's 2024 conclusion. Studies suggest that irrigation with sterile saline is likely equivalent to no irrigation at all (Ambe and Rombey, 2020). Given the constraints of the spacecraft, this would make a strong argument against manifesting the mass and volume required for irrigation solutions for these indications.

Conversely, a systematic review (Soeslo et al., 2022) found that antiseptic solutions, such as povidone iodine, significantly reduced infection rates in acute traumatic wounds and surgical site infections, respectively. Other studies evaluating polyhexanide and betadine solutions found rapid inactivation of bacteria and reduced bacterial loads when compared to saline (Watson et al., 2024). Findings are similar to postsurgery antiseptic irrigation. A systematic review and network meta-analysis from 2024 found that antiseptic solutions significantly reduced surgical site infections compared to no irrigation, while saline irrigation showed no statistically significant difference. Similar findings are seen with solutions infused with antibiotic medications; however, concerns about antimicrobial resistance have led to recommendations against routine use (de Jonge et al., 2017).

The findings are summarized in Table 8.1. Evidence suggests that the antiseptic properties of the solution may be more effective than the mechanical pressure from irrigation for reducing infection rates. Since the same studies show an incidence of 3%–9% for infections after acute traumatic wounds, it is worth considering that small volumes of antiseptic solutions (or their lyophilized equivalent) mixed with potable water may be a worthwhile trade against carrying additional antibiotic courses to manage post-wound infections on space missions. These studies also imply that, from an infection perspective, designing and flying complex mechanisms for generating high pressure while managing stray spray and droplets is not worth the mass trade.

For these reasons, the optimal resource inventory for space-based irrigation to reduce infections after traumatic wounds may be lyophilized antiseptic powder, which can be rehydrated with potable water from the spacecraft supply in a small volume (500–1000 ml) fluid bag, and an absorbent pad or towel to remove waste and excess water.

**TABLE 8.1 Summary of solution type and infection-related outcomes.**

| Solution | Outcomes |
|---|---|
| No irrigation | No significant difference in bacterial load or infection rates when compared with sterile saline after abdominal surgery (Ambe and Rombey, 2020) |
| Tap water | Tap water was equivalent to saline in reducing infection rates and promoting wound healing (Huang et al., 2015; Resende et al., 2016). |
| Distilled water | No significant difference in infection rates or wound healing outcomes compared to sterile saline (Olufemi and Adeyeye, 2017) |
| Sterile saline | No significant difference in bacterial load or infection rates when compared with no irrigation after abdominal surgery (Ambe and Rombey, 2020) |
| Antiseptic solutions | Povidone iodine and polyhexanide reduced bacterial load and infection rates more effectively than sterile saline in acute traumatic wounds (Soeslo et al., 2022; Nyoman et al., 2023; Watson et al., 2024; Fu et al., 2022). |

#### 8.2.4.2 Volume of irrigation

Another variable is the volume of solution used. While there is substantially less literature surrounding this question, there are plenty of guidelines, and the available evidence suggests that large volumes are more effective than smaller volumes. One study evaluating the use of 3, 6, and 9 liters of sterile saline found that 9 liters was the most effective in reducing bacterial colonization after an open fracture. Another study comparing polyhexanide and betaine solutions to sterile saline found that the antiseptic solutions reduced bacterial load more effectively. While not conclusive, these data suggest that antiseptic solutions may achieve similar levels of infection reduction with substantially lower volumes of fluid. Again, this implies that the antibiotic properties of the fluid are more effective than the mechanical flush at reducing infections.

#### 8.2.4.3 Pressure of irrigation

Schottel (2016) found no difference in reoperation rates after open fracture repairs with irrigation pressures ranging from 1 to ~20 psi (Schottel, 2016). A second, much smaller randomized controlled trial also found the same result. Additionally, a 2007 clinical trial revealed no difference between irrigation at 40 psi and a high-pressure parallel waterjet at between 5025 and 7360 psi (Granick et al., 2007).

Perhaps most tellingly, a 2024 systematic review and meta-analysis of 41 randomized controlled trials (RCTs) found that antiseptic solutions, rather than irrigation pressure, were associated with a reduction in surgical site infections (RR, 0.60; 95% CI, 0.44–0.81). As with the findings around choice of solution and volume, the results of this review suggest that the mechanical forces of irrigation are less effective at preventing infections than the chemical properties of an antiseptic solution.

#### 8.2.4.4 Time until intervention

There is little evidence on time until intervention and the use of irrigation to reduce infection. One study, by Bhandari et al. (2000) found that high-pressure irrigation (70 psi) was more effective than low-pressure irrigation (1–2 psi) when treatment for open fractures was delayed by more than 4 hours. This implies that irrigation may have some effect on reducing infection rates with delayed wound closure; however, the data is limited and insufficient to drive practice changes.

## 8.3 Procedural considerations for irrigation

### 8.3.1 Injury risk and incidence

The risk of injuries that may require irrigation in space is difficult to assess, as it is highly dependent on mission plans and vehicle design. Terrestrial activities involve gravity, which is a significant contributor to injuries, and research to date has not focused on separating gravity and non-gravity-dependent injuries. To date, only minor lacerations have been publicly reported in space operations. A review of musculoskeletal and minor trauma injuries from 2009 reports a total of 219 injuries over 231,725 person-hours of spaceflight, equating to an overall incidence of 0.023 injuries/crew-day. Of these, most were musculoskeletal injuries, but 98 were described as abrasions (70) or lacerations (28). This translates to an incidence rate of 0.01 dermatologically compromising injuries per crew-day (1 in 100 days). The majority of injuries (approximately ⅓) were reported to have occurred on the hands, though the data does not distinguish whether these were skin-compromising injuries or other musculoskeletal injuries. While all but one were described as minor, this is the closest approximation of the incidence rate for in-space acute traumatic injuries that may require irrigation (Scheuring et al., 2009).

For eye injuries that may require irrigation, a 2023 review by Meer et al. (2023) found four incidents of ocular debris and six chemical eye burns that occurred in spaceflight through Expedition 13 in 2006. This is the same time frame as the aforementioned Scheuring et al. study and yields an incidence of 0.0004 debris incidents per person-day (1 in 2500 days) and 0.0006 chemical eye burns per person-day (1 in 1667 days). This yields a combined incidence for eye injuries that may require irrigation at 0.001 injuries/person-day of 1 every 1000 days. All incidents were minor.

One other consideration is that Lunar and Martian dust will likely act as an irritant. All Apollo astronauts reported eye irritation on the Lunar surface. This was managed with lubricating eye drops, but copious amounts were recommended for future missions. While not strictly irrigation, this is something for a spaceflight physician to consider for missions involving surface operations.

Chemical skin burns have not been reported in published literature for spaceflight. This is likely due to rigorous protections around the use of caustic chemicals in flight. However, the risk continues to exist and will likely increase as more missions are flown and more experiments are conducted. The incidence of chemical burns in the eye may be a valid proxy.

## 8.3.2 Spaceflight specific considerations

The main challenges to irrigation in spaceflight are limited storage for the necessary volume of irrigation fluid, altered or absent gravity, and managing the resulting, now biohazardous, waste fluid.

### *8.3.2.1 Volume*

The volume of irrigation fluid is typically not specified in guidelines. Most medical guidelines for irrigating the eye discuss rinsing under a faucet or eyewash station until the pH of the eye surface normalizes or a certain time has elapsed. For skin burns, a minimum rinse time is often given rather than a volume (e.g., 15–20 minutes). For this reason, and given that the volume required depends on the amount and specific type of chemical, it is difficult to specify a specific volume. However, if a time of 20 minutes using a standard U.S. faucet with a flow rate of 1.5 gallons per minute is selected for reference, irrigation for a chemical burn would require approximately 91 liters, weighing 91 kg (~200lbs). This is not likely feasible on-board spacecraft for the foreseeable future.

Smaller volumes may be more practical and possible if the chemical exposures are limited to a small area. For example, one paper recommends using between 1 and 10 L of fluid to normalize the pH for ocular chemical burns sustained during spaceflight (Meer et al., 2023). If applied judiciously and in conjunction with rigorous controls, this is a reasonable ballpark for the volume of irrigation solution needed per event.

### *8.3.2.2 Gravity*

Terrestrially, gravity is helpful for generating pressure (as in a down-facing faucet), removing waste flow, and containing any droplets that fly off from the stream. This will be possible for habitats placed in a gravity well (such as the moon or Mars). With lower gravity, a system to add more pressure may be desirable, but lower flow rates will also limit the formation of droplet spray.

In a microgravity environment, the pressure, droplet containment, and waste flow must be design elements of the systems. Systems to produce pressure, contain droplets, and remove waste in microgravity may require substantial system design elements, including an inflow line and pump, a body part containment "rinsing" vessel to limit droplet spray and enable water contact time with the affected body part, a waste outflow line, and a biohazardous waste containment vessel (Fig. 8.1).

A more straightforward solution may be to wrap the affected body area in an absorbent fabric "towel" and spray water from a syringe or flexible drink bag held near the area to limit droplet spray. This will have to be done carefully, with the recognition that some droplets are likely to escape absorption by the towel. For this reason, irrigation with this method will require the treating providers and the patient to wear personal protective equipment such as gloves, goggles, and a facemask to avoid contaminating themselves as well. Additionally, the procedure should be performed near an airflow vent that can "vacuum" up any stray droplets. This method may put substantial strain on the contaminant filters and still carries some risk of exposing other crew members to the chemical.

For a spinning artificial gravity station, flowing water will be possible, but Coriolis forces will cause the stream to appear curved. While this will require operational considerations about where to place the affected body part so that water flows over it and can still be caught by the waste receptacle, the rest of the irrigation process will likely be much the same as on Earth. In many cases, this means that sink faucets or pressure syringes are placed or held off to the side, rather than above the drain hole.

### *8.3.2.3 Reusability*

Captured water will either need to be treated and recycled or disposed of. Irrigation wastewater will contain traces of the irritant chemical that will need to be inactivated or filtered out before this water can be used for other purposes, such as drinking or rehydrating food. While a closed-loop recycling system for water is preferable for many reasons, it adds substantial complexity to the water system. It may involve consumable components such as filters or phase separation centrifuges.

Disposed water cannot be reused and must be contained in a sealed container to prevent release or dumped overboard. In the former scenario, sufficient volume must be present in the waste container to manage the anticipated wastewater for the expected mission duration. In the latter, a complex series of valves must be constructed to enable waste to be pushed

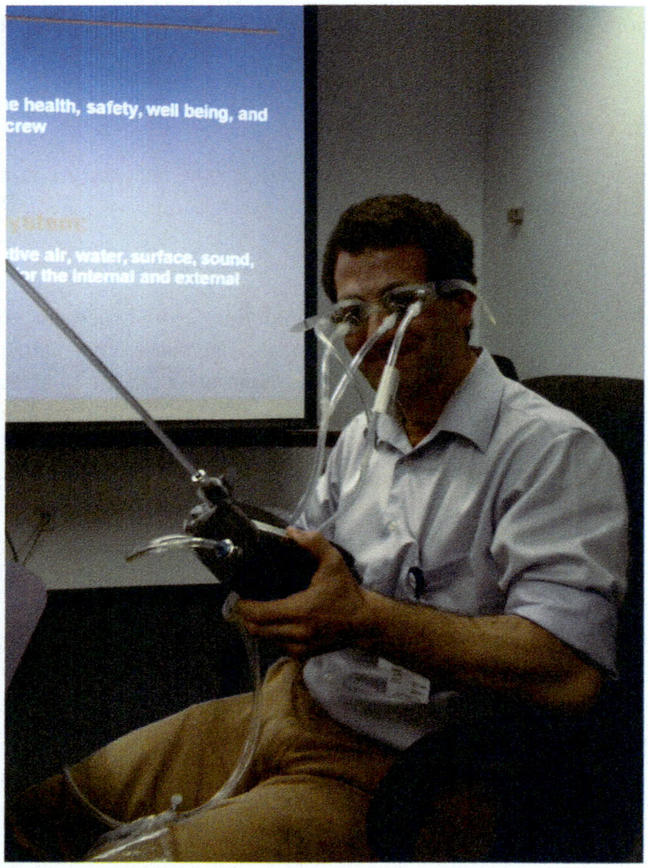


FIGURE 8.1 **NASA's International Space Station eye irrigation system.** This is an example of an individual wearing the ocular irrigation system developed by NASA for the International Space Station.

overboard without compromising hull integrity or creating a system that can accidentally expose the cabin to the vacuum of space if mishandled.

## 8.4 Injury characteristics

### 8.4.1 Location

The location of the injury often dictates the risks associated with it. The locations requiring unique treatment include eye, trunk wounds, head and neck wounds, and extremity wounds. Injuries and chemical exposures to the cornea are particularly high risk for loss of function since scarring and destruction of corneal tissue can cause loss of function of the eye. Trunk, extremity, and head and neck injuries are generally lower risk for loss of function, as the healing process is more robust and more tolerant to delays in treatment.

Each category of location also possesses unique considerations for irrigation technique. For example, it is relatively easy to irrigate an extremity in a controlled manner by placing the extremity inside a bag or box, while irrigating a head or trunk wound may require a more complex method for capturing wastewater and droplet spray without compromising an individual's ability to breathe or spreading contamination to other parts of the body.

### 8.4.2 Complexity of the injury

The complexity of a wound also plays into the decision to irrigate. Complex wounds are those that violate multiple layers of tissue (deep penetrating wounds), involve loss of a significant portion of tissue (avulsions or large abrasions/burns), involve injuries to multiple tissue types (e.g., compound fractures), injuries involving chemical compounds (e.g., with strong acids or bases), injuries that risk loss of function in the affected body part, and injuries with substantial retained debris (Nefihancoro and Putra, 2021; Nyoman et al., 2023).

In contrast, a simple wound is one with none of these characteristics. Generally, these are superficial, clean, do not involve chemicals or debris, do not involve multiple tissue layers, and are not in an area likely to cause loss of function. Some complex injuries may benefit more from irrigation than simple wounds, though evidence is mixed and inconclusive (Bansal et al., 2002; Groenen et al., 2024).

### 8.4.3 Level of contamination

Debris can serve as a delivery vehicle for microorganisms, and reactive chemical compounds can directly damage tissue. In both cases, the threshold for irrigating to remove these compounds should be set very low, particularly in the event of a chemical exposure (Abid et al., 2019). This is likely the strongest case for high-volume irrigation.

### 8.4.4 Type of contamination

The type of contamination should also be considered, since biologically inert materials like retained glass are mechanically irritating but likely pose little to no increased risk for infection, wound healing, or loss of function. However, depending on the location and shape of the debris, it may cause mechanical damage to body structures, such as a sharp piece of glass near a major blood vessel or embedded in a tendon or ligament. Chemical compounds that can damage human tissue can pose a substantial risk and should be neutralized and removed as fast as possible to minimize the damage done.

## 8.5 Decision making

Based on available evidence, irrigation may be necessary in spaceflight to manage infection risk from acute traumatic injuries, to facilitate examination and/or repair procedures, and to remove chemically reactive compounds, biologically active fomites (e.g., organic material), or mechanically damaging debris (e.g., glass shards). For infection prevention, evidence suggests using small volumes of antiseptic solutions administered under low pressure. For mechanically removing biologically active or mechanically dangerous debris, larger volumes of 1–5 L and higher pressures may be considered. Finally, for neutralizing chemically reactive material, volumes of 10 L or more may be needed. The optimal irrigation solution for spaceflight is likely to be potable water. This will be abundantly available on the spacecraft and has multiple uses. Potable water will also not likely pose a substantial infection risk in itself. Evidence does not support the need for solutions to be fully sterile at this time.

### 8.5.1 Description

The decision to irrigate is based on the following factors:

1. What is the goal of irrigation?
    a. Infection control, removal of debris, or removal/neutralizing caustic compounds
2. Are there sufficient vehicle/habitation supplies of irrigation solution?
    a. If there is an insufficient supply of potable water, the benefits of irrigation may not outweigh the risks.
3. Can the vehicle environmental contaminant systems contain the wastewater?
    a. If there is no safe way to contain the wastewater, the benefits of irrigation may not outweigh the risks.
4. Is the wound superficial or deep?
    a. Superficial wounds may not need irrigation
    b. Deep or complex wounds may benefit from irrigation
5. Is the wound located in an area at high risk for loss of function or dangerous scarring (e.g., eye, across a joint, near the airway, etc.)?

### 8.5.2 Flowchart

(Fig. 8.2).

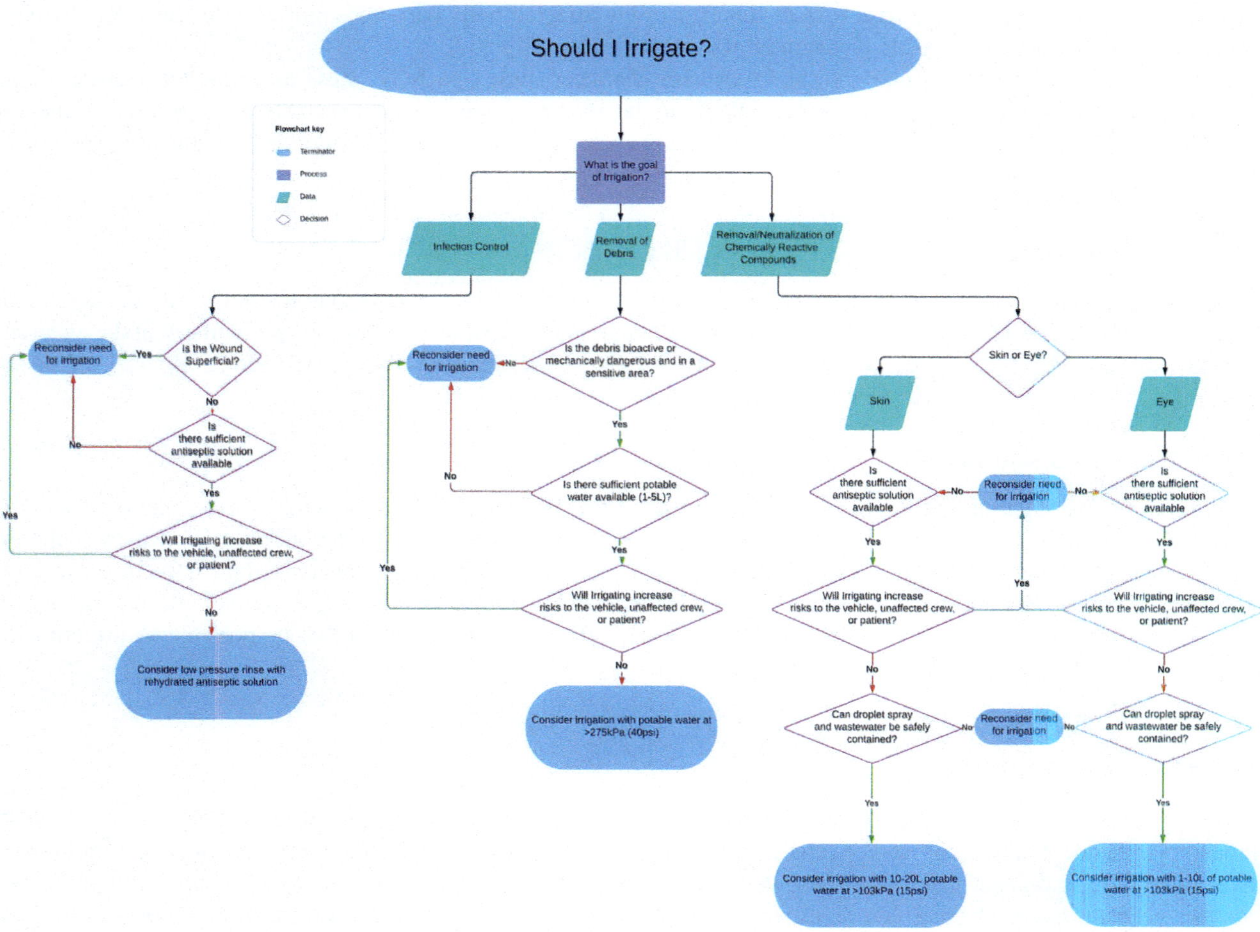


FIGURE 8.2 **Irrigation decision flowchart.** This flowchart depicts an example decision tree for irrigation as a procedure in spaceflight.

## 8.6 Procedural technique and materials for spaceflight

### 8.6.1 Material, Mass, Volume Table

(Table 8.2).

**Step by Step**

1. Ensure the scene is safe for the provider to approach
2. Direct one crew member to secure any loose sharps or chemical spills
3. Remove the patient to the treatment area
4. Consider pain medications, subcutaneous, topical, or oral, depending on the injury.
5. Examine the affected area
   a. **Eyes:**
      i. Don provider personal protective equipment
      ii. Consider tetracaine eye drops if pain is severe but weigh their use against the potential for interaction with the chemical
      iii. Place irrigation goggles on patient's eyes
      iv. Secure pump assembly or irrigation syringe
      v. Secure wastewater bags
      vi. Begin irrigation with 1–10 L of potable water
      vii. After each liter, check pH with an indicator strip

**TABLE 8.2** Estimated mass and volume of irrigation equipment. This does not include pain control supplies or equipment, as these are described in other chapters.

| Item | Mass | Volume | Indication |
|---|---|---|---|
| Potable water | 1 g/ml | 1–20 L | Debris removal, neutralization/removal of chemically reactive compounds |
| Polyhexanide powder | ~2400 g/mol | 0.04 g/100 ml of water (0.04% aqueous solution) | Antiseptic solution |
| Povidone iodine powder | 365 g/mol | 5 g/100 ml of water (5% aqueous solution) | Antiseptic solution |
| 60cc tumi syringe | ~60 g | ~250 $cm^3$ | Manual irrigation |
| Absorbent cotton towel | 400 g | ~900 $cm^3$ | Manual catching of wastewater |
| Eyewash goggles and pump system | ~1 kg | ~1500 $cm^3$ | Irrigation of the eyes for chemical injury |
| 5 L plastic wastewater containment bag | 10 g | ~500 $cm^3$ | Collection of wastewater from irrigation of a limb |
| Nitrile gloves (1 pair) | 12 g | ~50 $cm^3$ | Provider personal protective equipment |
| Facemask | 4 g | ~100 $cm^3$ | Provider personal protective equipment |
| Protective goggles | 150 g | ~900 $cm^3$ | Provider personal protective equipment |
| Tetracaine 0.5% (bottle) | ~18 g | ~25 $cm^3$ | Eye pain control |
| pH indicator strips (set of 100) | ~20 g | ~25 $cm^3$ | Measuring eye pH after a chemical burn |

**viii.** Stop irrigation after pH returns to 7

**b. Skin on a limb (chemical burn or debris)**

**i.** Don provider personal protective equipment
**ii.** Place the affected limb in a wastewater bag
**iii.** Obtain an irrigation bag with potable water
**iv.** Spray antiseptic solution directly at the injury within the bag

**c. Skin on a limb (wound)**

**i.** Don provider personal protective equipment
**ii.** Place the affected limb in a wastewater bag
**iii.** Obtain an irrigation bag with antiseptic powder
**iv.** Rehydrate with potable water and mix
**v.** Carefully and with low pressure, spray 250–1000 mL of antiseptic solution directly at the injury within the bag

**d. Skin on trunk (chemical burn or debris)**

**i.** Instruct a second crew member to assist with catching droplet spray and wastewater.
**ii.** Don provider personal protective equipment
**iii.** Obtain absorbent towels
**iv.** Obtain an irrigation bag with potable water
**v.** Position one towel wrapped around the end of the syringe to create a loose cone, which will contact the skin around the injury
**vi.** Position a second towel directly behind the area if possible or have a second provider hold the towel around the edges of the first towel to catch droplet spray
**vii.** Begin irrigation
**viii.** Stop and replace towels when saturated
**ix.** Place used towels in a biohazard containment bag or waste container.

- **x.** Consider irrigation with up to 20 L of potable water, depending on the extent of the wound and the caustic chemical

**e. Skin on trunk (wound)**

- **i.** Instruct a second crew member to assist with catching droplet spray and wastewater.
- **ii.** Don provider personal protective equipment
- **iii.** Obtain absorbent towels
- **iv.** Obtain an irrigation bag with antiseptic powder
- **v.** Rehydrate with potable water and mix
- **vi.** Position one towel wrapped around the end of the syringe to create a loose cone, which will contact the skin around the injury
- **vii.** Position a second towel directly behind the area if possible or have a second provider hold the towel around the edges of the first towel to catch droplet spray
- **viii.** Carefully and with low pressure, spray 250–1000 mL of antiseptic solution directly at the injury
- **ix.** Clean the area thoroughly with a towel when irrigation is complete
- **x.** Place used towels in a biohazard containment bag or waste container.

**f. Skin on face/head/neck (chemical burn or debris)**

- **i.** Instruct a second crew member to assist with catching droplet spray and wastewater.
- **ii.** Don provider personal protective equipment
- **iii.** Obtain absorbent towels
- **iv.** Obtain an irrigation bag with potable water
- **v.** Position one towel wrapped around the end of the syringe to create a loose cone, which will contact the skin around the injury
- **vi.** Position a second towel directly behind the area if possible, or have a second provider hold the towel around the edges of the first towel to catch droplet spray
- **vii.** Place a third towel over any sensitive areas, such as eyes, ears, nose, mouth, and ensure the patient's airway is not obstructed
- **viii.** Place protective goggles on the patient's eyes and instruct them to close their mouth. Take particular care when irrigating around the ears and nose
- **ix.** Begin irrigation
- **x.** Stop and replace towels when saturated.
- **xi.** Place used towels in a biohazard containment bag or waste container
- **xii.** Consider irrigation with up to 20 L of potable water, depending on the extent of the wound and the caustic chemical

**g. Skin on face/head/neck (wound)**

- **i.** Instruct a second crew member to assist with catching droplet spray and wastewater.
- **ii.** Don provider personal protective equipment
- **iii.** Obtain absorbent towels
- **iv.** Obtain an irrigation bag with antiseptic powder
- **v.** Rehydrate with potable water and mix
- **vi.** Position one towel wrapped around the end of the syringe to create a loose cone, which will contact the skin around the injury
- **vii.** Position a second towel directly behind the area if possible or have a second provider hold a towel around the edges of the first towel to catch droplet spray
- **viii.** Place a third towel over any sensitive areas, such as eyes, ears, nose, and mouth, and ensure the patient's airway is not obstructed
- **ix.** Place protective goggles on the patient's eyes and instruct them to close their mouth. Take particular care when irrigating around the ears and nose.
- **x.** Carefully and with low pressure, spray 250–1000 mL of antiseptic solution directly at the injury
- **xi.** Clean the area thoroughly with a towel when irrigation is complete
- **xii.** Place used towels in a biohazard containment bag or waste container.

## 8.7 Summary recommendations and considerations for the future

In the present era of spaceflight, and likely for the next 10–15 years, irrigation will be confined to cleaning injuries rather than as part of invasive surgery. The main indications for irrigation are to facilitate an exam or procedure, to reduce infection after an acute traumatic injury that breaches the skin, and to neutralize and wash out reactive chemicals or debris. For exam purposes and chemical/debris removal, potable water is a sufficient and effective chemical. For exams, use only the amount needed to clean the area sufficiently to perform the exam or procedure. For cleaning debris off the skin or eye or for chemical neutralization, large amounts of potable water are typically needed, in some cases up to 20 liters. An optimal specific pressure is not easily defined, but approximately 103–275 kPa is likely sufficient. For chemical burns of the eye, irrigation should continue until pH indicator strips show the eye surface has returned to 7.

The secondary considerations when performing irrigation are protecting the vehicle and other crew members from droplet spray and biohazardous waste. It is always important to ensure the area is safe to approach before retrieving a patient for treatment. The danger may persist and injure a second responding crew member, which removes an assistant or provider and adds a patient, particularly with chemical exposures.

For infection control after an acute traumatic injury, evidence suggests that the antiseptic properties of a solution are more effective than mechanical flow in reducing infection rates. While injuries may be contaminated and a larger volume of irrigation may be needed to remove contaminants, irrigation of a clean wound may only require a small amount of low-pressure antiseptic solution to reduce the baseline 3%–9% post-injury infection rate by as much as 50% (Fernandez et al., 2022).

From a design perspective, it is important to consider the occurrence rates of potential consequences associated with not irrigating and trade these against the risks introduced by opting to include the necessary mass and volume for the irrigation solution at the expense of other supplies. This is particularly true when considering the need for large volume irrigation. In most cases, prevention is the best strategy, namely, by avoiding caustic chemicals on the spacecraft or using strict and extensive engineering controls to contain them. Additionally, short-duration missions with limited supplies may not pose a sufficient risk of consequences to justify the allocation of mass and volume for irrigation. In addition, the need to contain and dispose of or treat wastewater should be considered. The volume available on board for waste products will be limited, as will the ability of the filters and contaminant removal systems to remove hazardous waste. This is true both for bulk waste collected while irrigating as well as droplet spray waste that inevitably arises from irrigation in microgravity. Future missions to planetary surfaces or with artificial gravity stations will have a significantly easier time with collecting and containing wastewater and droplet spray, but it will remain an issue even in these environments.

As vehicles become more capable and missions more complex, it is likely that we will see an increased risk for injuries requiring irrigation. In the far future, this may even be needed as part of on-station invasive surgical procedures. As these progress, the optimal solution will be to care for patients in an environment with artificial or real gravity, closed-loop water recycling systems with mechanisms to treat and remove any toxic biological or chemical compounds, and mechanisms to produce or recycle antiseptic solution chemicals. While these are not presently scoped for near future missions, the technology to support them is within reach, and when they are needed, little development will be necessary.

## References


Abid, M., Syed, A., Haider, A., Sidra, A., Ahmed, K., Ahmed, S., et al., 2019. Comparison of irrigation with povidone-iodine versus normal saline on wound infection after open appendectomy. Annals Abbasi Shaheed Hospital & Karachi Medical & Dental College 24 (3), 265–271.

Ambe, P., Rombey, T., Rembe, J.-D., Dorner, J., Zirngibl, H., Peiper, D., 2020. The role of saline irrigation prior to wound closure in the reduction of surgical site infection: a systematic review and meta-analysis. Patient Safety in Surgery 14 (1).

Bansal, B.C., Wiebe, R., Perkins, S., Abramo, T., 2002. Tap water for irrigation of lacerations. American Journal of Emergency Medicine 20 (5), 469–472.

Barry, J., Hennessy, R., McManus, J., 2009. A randomized controlled trial comparing treatment regimens for acute pain for topical oleoresin capsaicin (pepper spray) exposure in adult volunteers. Prehospital Emergency Care 12 (4), 432–437.

Bhandari, M., Thompson, K., Adili, A., Shaughnessy, S., 2000. High and low pressure irrigation in contaminated wounds with exposed bone. International Journal of Surgical Investigations 2 (3), 179–182.

Fernandez, R., Green, H., Griffiths, R., Atkinson, R., Ellwood, L., 2022. Water for wound cleansing. Cochrane Database of Systematic Reviews.

Fu, C., Linghong, M., Miguo, M., Na, L., Jingcen, Z., 2022. Effect of wound irrigation on the prevention of surgical site infections: A meta-analysis. International Wound Journal 19 (7), 1878–1886. Available at https://doi.org/10.1111/iwj.13794.

Granick, M., Tenenhaus, M., Knox, K., Ulm, J., 2007. Comparison of wound irrigation and tangential hydrodissection in bacterial clearance of contaminated wounds: results of a randomized, controlled clinical study. Ostomy Wound Management 53 (4), 64–66.

Groenen, H., Bontekoning, nathan, Jalalzadeh, H., Buis, D., Dreissen, Y., Goosen, J., et al., 2024. Or, incisional wound irrigation for the prevention of surgical site infection: a systematic review and network meta-analysis. JAMA Surgery 159 (7), 792–800.

Huang, C.-Y., Choong, M.-Y., Li, T.-S., 2015. Cleansing of wounds by tap water? An evidence-based systemic analysis. International Wound Journal 12 (4), 493–494.

de Jonge, S., Boldingh, Q., Solomkin, J., Allegranzi, B., Egger, M., Dellinger, E., et al., 2017. Systematic review and meta-analysis of randomized controlled trials evaluating prophylactic intra-operative wound irrigation for the prevention of surgical site infections. Surgical Infections 18 (4), 508–519.

Meer, E., Grob, S., Antonsen, E., Sawyer, A., 2023. Ocular conditions and injuries, detection and management in spaceflight. NPJ Microgravity 9 (37). Available at https://doi.org/10.1038/s41526-023-00279-y.

Mitha, S., Rasul, T., Bergholz, D., Weinstock, K., Akhtar, S., Sultan, U., et al., 2024. Wound irrigation using normal saline 0.9% versus tap water: a review for street medicine and low-resource providers. Journal of Dermatology Research 5 (1), 1–12.

Nefihancoro, U., Putra, M., 2021. The effect of irrigation to the amount of bacterial colony formation in open fracture. Journal of Advances in Medicine and Medical Research 33 (172), 214–220.

Nyoman, I., Raditya, Y., Rida, K., Nengah, N., Fatmawati, D., Suyasa, I., et al., 2023. Differences in decreasing bacteria colonization wound irrigation using 0.9% Nacl compared with wound irrigation using active ingredients polyhexanide 0.1% and 0.1% betaine in open fractures of long bone. Open Access Macedonian Journal of Medical Sciences 514–518.

Olufemi, O., Adeyeye, A., 2017. Irrigation solutions in open fractures of the lower extremities: evaluation of isotonic saline and distilled water. Société Internationale de Chirurgie Orthopédique et de Traumatologie 3 (7). Available at https://doi.org/10.1051/sicotj/2016031.

OSHA, 2013. Washington, DC United States Department of Labor OSHA Brief: Hazard Communication Standard: Labels and Pictograms.

Patek, G., Amanda, B., Zanaboni, A., 2025. StatPearls: Ocular Burns. StatPearls Publishing.

Resende, M.M., Rocha, C.A., Corrêa, N.F., Veiga, R.R., Passos, S.J., Novo, N.F., et al., 2016. Tap water versus sterile saline solution in the colonisation of skin wounds. Int Wound J, 13, (4), 526–530. Available at https://doi.org/10.1111/iwj.12470.

Scheuring, R., Mathers, C., Jones, J., Wear, M., 2009. Musculoskeletal injuries and minor trauma in space: incidence and injury mechanisms in U.S. astronauts. Aviation, Space, and Environmental Medicine 80 (2), 117–124.

Schottel, P., 2016. Wound irrigation pressure did not affect reoperation rate after open fracture repair, but normal saline solution was better than Castile soap. Journal of Bone and Joint Surgery 98 (10).

Soeslo, D., Yolanda, R., Zita, M., Theresia, S., Astiarani, Y., Santi, B., 2022. Antiseptic versus non-antiseptic solutions for preventing infection in acute traumatic wounds: a systematic review. Journal of Wound Care 31 (2), 162–169. Available at https://doi.org/10.12968/jowc.2022.31.2.162.

Stopyra, J., Winslow III, J., Johnson III, J., Hill, K., Bozeman, W., 2018. Baby shampoo to relieve the discomfort of tear gas and pepper spray exposure: a randomized controlled trial. Western Journal of Emergency Medicine 19 (2), 294–300.

Tan, T., Wong, D., 2015. Chemical burns revisited: what is the most appropriate method of decontamination? Burns: Journal of the International Society for Burn Injuries 41 (4), 761–763.

VanHoy, T., Metheny, H., Patel, B., 2025. StatPearls: Chemical Burns. StatPearls Publishing.

Watson, F., Chen, R., Saint Bezard, J., Percival, S., 2024. Comparison of antimicrobial efficacy and therapeutic index properties for common wound cleansing solutions, focusing on solutions containing PHMB. 16 (19).

Chapter 9

# Head and neck procedures

Kaleigh Stabenau[1,2] and Khushi Bhatt[3]
[1]Department of Otolaryngology-Head and Neck Surgery, University of California Irvine, Orange, CA, United States, [2]Arizona Program for Exploration (APEX) Medicine and Surgery, University of Arizona, Phoenix, AZ, United States, [3]University of California Irvine, Orange, CA, United States

## 9.1 Introduction

The head and neck, while encompassing a relatively small portion of the human as a whole, represent the intersection of multiple organ systems. Pathology in any one system can affect the others due to their close proximity, and pathology in this area can be rapidly life-threatening (e.g., upper airway obstruction, spread of infection through a valveless venous system, and even exsanguination from difficult-to-reach arterial bleeding). Thus, any medical officer in a spacecraft will require knowledge of these systems and their potential pathology, as well as diagnostic and management strategies.

Simple procedures of the ear, nose, and throat lend themselves well to the spaceflight environment, as they can be performed without general anesthesia and in an in-office setting in terrestrial environments. Many common diagnostic and therapeutic procedures can be performed under local or regional analgesia, or in the absence of analgesia entirely. These simple procedures are generally not resource-intensive and can be feasibly taught to any crewmember, regardless of prior medical experience.

When building a medical system for a specific mission, many factors must be considered. Mass, power, and volume are limited, and devoting more of any one of these resources to a single system requires sacrificing resources for another system in a trade. Thus, medical systems for spaceflight are designed to address anticipated medical contingencies based on terrestrial data and/or modeling. Medical contingencies are assessed in a risk and consequence matrix, measuring impacts to crew and mission health as quantified by task time lost, need for removal to definitive care (i.e., evacuation), and loss of life. Because human spaceflight experience is limited, these medical models help elucidate the medical conditions that are

1. likely to occur during spaceflight,
2. have the potential to significantly impact crew and mission,
3. are within the scope of technical feasibility in the spaceflight environment.

Once these conditions are addressed, we can determine which require procedural interventions and begin planning for mass/power/volume allocation in the context of the wider spaceflight systems design.

In this chapter, we will cover procedures of the ears, nose, and throat that help diagnose or treat problems that can occur during spaceflight. These procedures have been specifically chosen to address medical conditions identified by the Informing Mission Planning via Analysis of Complex Tradespaces (IMPACT) model, which baselines lunar missions for a NASA astronaut population, as well as common problems that we see on Earth that are applicable to commercial or private crew. It is important to note that while we do not have a medical model for a mission to Mars yet, all of the procedures covered in this chapter would apply to a crewed mission to Mars as well. While some procedures may require adaptation for the microgravity environment (see Table 9.1), most do not require significant technical innovation to be feasible in spaceflight. However, areas where technical innovation would be helpful are highlighted.

## 9.2 Otologic procedures

### 9.2.1 Cerumen removal

Cerumen, or ear wax, is naturally produced by the sebaceous glands within the hair-bearing skin of the cartilaginous external auditory canal (EAC). It serves a protective function and prevents debris from reaching the delicate tissues of the

**Procedural Medicine for Spaceflight. DOI:** https://doi.org/10.1016/B978-0-443-27552-4.00011-7

**TABLE 9.1** Mass and volume parameters for the minimal supplies needed for all head and neck procedures described.

| Supplies and equipment | Volume ($cm^3$)[a] | Mass (g)[b] |
|---|---|---|
| 10 cc syringe | 46.2 | 9 |
| 18G needle | 2.9 | 1 |
| 25G needle | 2.2 | 1 |
| 18G angiocath | 24.1 | 5 |
| Frazier suction (3 Fr) | 27.0 | 4 |
| Frazier suction (5 Fr) | 27.0 | 8 |
| Frazier suction (7 Fr) | 27.0 | 8 |
| Frazier suction (10 Fr) | 27.0 | 14 |
| Frazier suction (12 Fr) | 27.0 | 14 |
| 11-blade scalpel | 4.5 | 8 |
| 15-blade scalpel | 4.5 | 8 |
| Needle driver | 43.8 | 46 |
| Scissors | 15.2 | 21 |
| Boies elevator | 42.8 | 75 |
| Bayonet forceps | 12.6 | 27 |
| Nasal speculum | 243.0 | 88 |
| Tympanostomy tube | 1.12 | 5 |
| Curved hemostat (e.g., Kelly) | 236.8 | 39 |
| Myringotomy knife | 34.0 | 6 |
| Flexible laryngoscope | 468.0 | 500 |
| Otoscope | 508.4 | 322 |
| Suction bulb | 396.0 | 36 |
| Foley catheter | 55.0 | 27 |
| 88% phenol and angled applicator | 74.4 | 11 |
| Oxymetazoline 0.05% nasal spray | 126.3 | 46 |
| Local anesthesia (e.g., 1% lidocaine with 1:100,000 epinephrine) | 32.8 | 32 |
| Povidone iodine | 108.9 | 66 |
| Surgicel | 8.1 | 4 |
| Surgiflo Kit | 1596.0 | 187 |
| 2-O nylon suture | 8.8 | 4 |
| Tongue blades | 2.4 | 3 |
| Vaseline gauze | 18.9 | 21 |
| Dental roll | 5.8 | 6 |
| Xeroform | 20.0 | 11 |
| ¼″ iodoform or strip gauze | 168.4 | 27 |
| Steri-Strips (¼″) | 5.2 | 3 |

*(Continued)*

**TABLE 9.1 (Continued)**

| Supplies and equipment | Volume ($cm^3$)[a] | Mass (g)[b] |
|---|---|---|
| Merocel | 7.15 | 3 |
| Denver or thermoplastic splint | 4.3 | 8 |
| RhinoRocket (10 cm) | 26.4 | 15 |
| Otowick | 0.09 | 2 |
| Gauze (200 pack) | 1843.0 | 233 |
| Bougie | 13.75 | 17 |

[a] *Some measurements are from prepackaged or bulk packaged materials. Measurements may vary depending on manufacturer and if items are repackaged for spaceflight.*
[b] *Some measurements are from prepackaged or bulk packaged materials. Measurements may vary depending on manufacturer and if items are repackaged for spaceflight.*

membranous canal and tympanic membrane. The cartilaginous EAC is generally self-cleaning; however, anatomic obstruction or mechanical impaction of cerumen into the medial canal can lead to impairment or failure of this mechanism and accumulation of cerumen within the EAC.

Terrestrial Indications: Cerumen impaction is the most common otologic complaint presented by patients, prompting 12 million scheduled clinic visits and 8 million cerumen removal procedures annually within the United States (Sevy et al., 2023). Approximately 10% of all children and 57% of all elderly experience cerumen impaction (Michaudet and Malaty, 2018). Cerumen removal can be indicated if impacted cerumen interferes with otoscopic examination or produces symptoms of ear fullness, hearing loss, tinnitus, otalgia, dizziness, or pruritus. Removal of cerumen may also facilitate audiometric examination or surveillance of chronic otologic conditions (i.e., cholesteatoma).

Efficacy: Studies of patients undergoing cerumen removal for otologic or vestibular symptoms show a 76.8% success rate with manual removal techniques and a 99% satisfaction rate (Lian et al., 2025). If symptoms persist after cerumen removal, alternative diagnoses should be considered.

Alternatives and Consequences: Alternatives to mechanical cerumen removal include chemical cerumen removal with hydrogen peroxide or carbamide peroxide otic drops. These agents act to soften the cerumen to allow for the natural cleaning mechanisms of the EAC to be more effective, or to allow for removal via irrigation. Chemical cerumen removal is contraindicated in the presence of a perforated tympanic membrane or tympanostomy tube. In general, chemical cerumen removal should be pursued as a first-line treatment for cerumen impaction if no contraindications exist, as it presents less risk than instrumentation. Additionally, it can make mechanical removal easier by softening the impacted cerumen.

Performance in Austere Environments: Both otoscopy and mechanical cerumen removal have been conducted on Mir and International Space Station by Russian cosmonauts. In all cases, otoscopy could successfully identify cerumen impaction and mechanical cerumen removal (either by oneself or another cosmonaut) was successfully conducted and relieved symptoms of EAC discomfort, moderate hearing loss, and autophony (Bogomolov et al., 2021).

Indication for Spaceflight: Impacted cerumen that interferes with hearing, communication, or use of communication and hearing protection devices (i.e., fitted ear plugs)

Relevant Anatomy:

- Pinna
- Tragus
- External auditory canal (EAC)
- Cartilaginous EAC
- Membranous EAC
- Tympanic membrane
- Ossicular chain

Preprocedural Care: Cerumen impaction can be identified based on symptoms and direct visualization on otoscopic examination. Cerumen can be softened with carbamide peroxide drops prior to instrumentation. In fact, chemical cerumen removal should be first-line prior to considering instrumentation, due to the inherent risks of injury to the EAC or tympanic membrane with instrumented removal.

**TABLE 9.2 Optimal and minimal supplies and equipment for cerumen impaction.**

| Optimal | Minimum |
|---|---|
| Otologic microscope | Curette (with light and camera) |
| Ear specula | Viewing device (PC or tablet/phone) |
| Curette | Gauze |
| Alligator forceps | Otoscope (optional) |
| Frazier suction (7 Fr) | |
| Suction and tubing | |
| Gauze | |

Supplies and Equipment: See Table 9.2

There are several methods for performing physical cerumen removal. In the terrestrial setting, this procedure is generally accomplished with the use of irrigation or with a combination of instruments, including curettes, forceps, and suction, under microscopic visualization. Irrigation is messy even on Earth and would present a significant challenge to the cleanliness of the cabin environment in a microgravity setting. Thus, instrumented cerumen removal is more suited to the spaceflight environment. At minimum, the crew would require an instrument (e.g., a curette), a method of magnification, and restraint of the patient and operator. The large mass and volume of an otologic microscope preclude its use in space. Fortunately, curettes with integrated lights and cameras are available that meet this requirement with a more reasonable footprint.

Instructions:

1. Confirm presence of cerumen impaction via direct or indirect visualization (otoscope vs curette with camera).
2. Obtain all supplies that will be used during the procedure if not used in Step 1.
3. Power on the viewing device and connect the curette. Ensure that all equipment is working properly prior to restraining the patient. Place the viewing device in a secured location that doesn't require turning your head away from the patient to see it.
4. Restrain the patient in their seat or a dedicated medical working area.
5. Restrain yourself in a position that allows you the necessary degrees of freedom to perform the procedure safely and effectively.
6. Using your nondominant hand, place your index finger over the tragus and gently pinch the posterior pinna with your thumb and middle finger, then gently spread the EAC os to allow for better visualization.
7. Insert the curette in the EAC.
8. Remove cerumen under endoscopic guidance. Potential maneuvers include gently sliding the curette between the cerumen and the EAC to release any adherence, and placing the tip of the curette behind the cerumen to "scoop" it out of the canal. The most effective maneuver will depend on the amount and consistency of the cerumen.
9. Withdraw the curette. Using gauze, remove cerumen from the curette. TIP: It's helpful if someone else wipes the curette for you, as it will allow you to avoid having to reset your nondominant hand position.
10. Repeat steps 7 through 9 until cerumen is clear, there is no space behind the cerumen (e.g., cerumen is against the tympanic membrane), the patient can no longer tolerate the procedure, or bleeding is encountered.
11. When the procedure is finished, clean supplies (if they are to be reused) and restow or dispose of single-use supplies in the designated location.

Postprocedural Care: In general, there is no specific postprocedural care following cerumen removal. The patient should be counseled on preventing cerumen impaction in the future, including avoidance of EAC instrumentation (e.g., with Q-tips) and only wearing communication devices and in-ear hearing protection when absolutely necessary.

Pearls and Pitfalls: Though it seems like a straightforward procedure, there is a risk of great harm if cerumen removal is performed incorrectly. Cerumen removal via instrumentation should only be performed if the cerumen is in the outer 1/3 of the EAC. This is because the outer 1/3 of the EAC is comprised of skin and subcutaneous soft tissue. The medial 2/3 of the canal is comprised of a thin layer of skin directly adherent to periosteum. This portion of the canal is exquisitely sensitive and easily injured. Circumferential or near-circumferential injury of the EAC can result in scarring and stenosis. Without an adequate understanding of the typical length and course of the adult EAC, the risk of injury to the medial canal or tympanic membrane increases significantly. Tympanic membrane perforations result in intense pain and conductive hearing loss. While the vast majority of these will heal on their own, a minority will persist and require surgical correction. With enough force, a curette can even disarticulate the ossicular chain, resulting in a greater degree of conductive hearing loss that may be permanent without surgery. Marking the curette with a "do not pass" depth can reduce this risk, especially for a medical officer who is not a healthcare professional. Cerumen medial to the outer canal should be treated with carbamide peroxide drops or similar, which, as previously stated, should be first-line prior to any instrumentation.

If significant injury occurs to the EAC, placement of an otowick may be necessary to prevent stenosis. Any injury resulting in bleeding should be treated with otologic drops (ofloxacin or similar) to prevent occlusive scabbing and re-impaction. Tympanic membrane perforations that are clean and without significant bleeding can be monitored; drops are not indicated.

### 9.2.2 External auditory canal debridement

Otitis externa is a bacterial infection of the outer ear, with *Pseudomonas aeruginosa* being the most common causative organism. It is characterized by edema and purulence of the EAC and subsequent epithelial maceration and desquamation. While ototopical antibiotics are the first-line treatment, this swelling and debris accumulation can interfere with the efficacy of topical therapies, necessitating debridement.

Terrestrial Indications: EAC debridement is indicated in acute cases of otitis externa causing considerable debris, mucous, or desquamated epithelium accumulation. Annually, over 2.4 million cases of otitis externa are seen by healthcare providers, with 95% of cases being acute and 6% of all diagnoses occurring in those 20 years and older (CDC, 2011).

Efficacy: EAC debridement facilitates otitis externa treatment by clearing accumulated debris to increase topical therapy efficacy. Studies have demonstrated a 94% complete recovery or improvement rate amongst patients with otitis externa who have been treated with both surgical debridement and an antibiotic regimen (Parize et al., 2009). While debridement reduces inflammation and the bacterial load, antibiotics should be concurrently used to resolve the underlying bacterial infection.

Alternatives and Consequences: The alternatives to EAC debridement in the setting of moderate to severe otitis externa include combination ototopicals and oral antibiotics and/or placement of an otowick to assist with ototopical distribution. However, there is a higher risk of failure or prolonged symptoms with these methods in the absence of debridement.

Performance in Austere Environments: Studies of otitis externa in the deserts of western Australia (Chidlow and van Bockxmeer, 2019) and in rural combat zones (Brennan et al., 2015) mention EAC debridement as one of the most effective interventions regularly used to clear EAC debris. EAC debridement is a first-line procedure in such austere environments and has been shown to facilitate otowick placement and expedite otitis media treatment.

Indication for Spaceflight: Otitis externa resulting in significant debris accumulation that interferes with the efficacy of ototopical treatment and/or use of communication or hearing protection devices.

Relevant anatomy:

- Pinna
- Tragus
- External auditory canal (EAC)
- Cartilaginous EAC
- Membranous EAC

Preprocedural Care: Otitis externa is a clinical diagnosis with symptoms including pain and decreased hearing and signs including otorrhea and swelling of the EAC. Pain control should be achieved primarily with antiinflammatories (e.g., NSAIDs) and acetaminophen +/− opioid medications for severe cases.

Supplies and Equipment: See Table 9.3

Instructions:

1. Confirm diagnosis via history and physical with direct or indirect visualization (otoscope vs curette with camera).
2. Obtain all supplies that will be used during the procedure if not used in Step 1.
3. Power on the viewing device and connect the curette. Ensure that all equipment is working properly prior to restraining the patient. Place the viewing device in a secured location that doesn't require turning your head away from the patient to see it.
4. Restrain the patient in their seat or dedicated medical working area.
5. Restrain yourself in a position that allows you the necessary degrees of freedom to perform the procedure safely and effectively.

**TABLE 9.3 Optimal and minimal supplies and equipment for external auditory canal debridement.**

| Optimal | Minimum |
|---|---|
| Otologic microscope | Curette (with light and camera) |
| Ear specula | Viewing device (PC or tablet/phone) |
| Frazier suction (5 and 7 Fr) | Gauze |
| Suction and tubing | Otowick (for moderate to severe cases) |
| Gauze | Forceps (alligator or other) (to place otowick) |
| Otowick | Otoscope (optional) |
| Forceps (alligator or other) | |

6. Using your nondominant hand, place your index finger over the tragus and gently pinch the posterior pinna with your thumb and middle finger, then gently spread the EAC os to allow for better visualization.
7. Insert the curette in the EAC.
8. Remove debris under endoscopic guidance. Potential maneuvers include gently sliding the curette between the debris and the EAC to release any adherence, and placing the tip of the curette behind the debris to "scoop" it out of the canal. The most effective maneuver will depend on the amount and consistency of the debris.
9. Withdraw the curette. Using gauze, remove debris from the curette. TIP: it's helpful if someone else wipes the curette for you, as it will allow you to avoid having to reset your non-dominant hand position.
10. Repeat steps 7 through 9 until debris is clear, there is no space behind the debris (e.g., debris is against the tympanic membrane), patient can no longer tolerate the procedure, or bleeding is encountered.
11. FOR SEVERE CASES: If the tympanic membrane cannot be visualized before or after debridement, place an otowick into the EAC following debridement using forceps and apply topical antibiotic and steroid drops.
12. When the procedure is finished, clean supplies (if they are to be reused) and restow or dispose of single-use supplies in the designated location.

Postprocedural Care: Following debridement, the placement of an otowick can ensure that the EAC remains open and assist with ototopical medication administration and efficacy by ensuring adequate distribution of the medication within the EAC. An otowick should be placed if the tympanic membrane cannot be visualized due to edema. This should be removed within 3–5 days and can be replaced if prolonged treatment is required.

Pearls and Pitfalls: Otitis externa is painful. Terrestrial debridement relies on gentle suction to clear the EAC, enabling effective use of ototopical treatment, which is first-line for mild to moderate cases and adjunctive in severe cases. Possible terrestrial complications of EAC debridement include pain and discomfort during the procedure, bleeding (due to trauma or inflammation), or infection due to the introduction of novel bacteria. However, installing and operating a suction device in microgravity is a challenge. While not insurmountable, there are currently no commercially available medical suction devices available for use in a microgravity environment. Teams are working to develop suction devices for spaceflight, and prototypes have been demonstrated during parabolic flight. Therefore, this EAC debridement procedure has been modified to be similar to cerumen removal, in that it relies on instrumentation with a lighted curette and video monitoring. As this makes use of equipment already used for another procedure, it does not add any additional mass or volume (unless curettes are disposable and require additional supplies to be flown to account for this use case). This method is also very similar to cerumen removal, reducing the training burden for future crew medical officers. However, this method is less effective and more painful than suctioning, and thus will require additional pain medication (ideally multimodal pain management) and a lower threshold to treat with topical and oral antibiotics and steroids.

## 9.2.3 Myringotomy +/− tympanostomy tube placement

Terrestrial Indications: In cases of severe acute otitis media or otitis media causing intracranial complications or impending mastoiditis, a myringotomy is indicated. There are approximately 391.3 million global cases of otitis media annually (Huang et al., 2024), and 7.1% of cases occur amongst those 15–39 years old (Rijk et al., 2021). If diagnosed with recurrent acute otitis media (more than three episodes over 6 months), tympanostomy tubes should be placed. For patients with otitis media with effusion and hearing loss of more than 30 dB, antibiotic-resistant otitis media, or otitis media with effusion lasting more than 3 months, a myringotomy with or without tympanostomy tubes can be performed. It is estimated that over 667,000 tympanostomy tubes are placed annually, with more than 8% of children having one set before the age of three and 20% of individuals having two sets in their lifetime (Spaw et al., 2024). Aside from otitis media, a myringotomy and/or tympanostomy tube placement may be indicated for patients with craniofacial abnormalities (i.e., cleft palate) or barotitis media in urgent situations.

Efficacy: A myringotomy and/or tympanostomy tube placement provides a drainage pathway, culture material, and otic drop entry in cases of otitis media. Studies show decreased occurrences of recurrent otitis media after tympanostomy tube placement (Steele et al., 2017). Lower Eustachian Tube Dysfunction Questionnaire (ETDQ-7) scores are also reported after myringotomy and tympanostomy tube placement in patients with otitis media with effusion or otitis media that is unresponsive to antibiotics (Yu et al., 2025). In cases of otitis media causing intracranial complications or mastoiditis, a myringotomy is needed to resolve infection in 90% of cases (Mattos et al., 2014). Myringotomy alone cannot resolve the underlying bacterial infection in otitis media; however, and antibiotics should be concurrently used. In situations of dysfunctional or blocked Eustachian tube function, as in urgent barotitis media or craniofacial abnormalities, myringotomy and/or tympanostomy tube placement allows for equalization of middle and external ear pressures (Llewellyn et al., 2014).

Performance in Austere Environments: Studies have assessed the efficacy of myringostomy with tympanostomy tube placement amongst civilian (Konishi et al., 2012) and military (Fouts and Mortimer, 2018) airplane passengers and pilots. In all cases, flying was concluded to have no deleterious impact on graft healing rate and otitis media resolution rates (Raju et al., 2020).

Indication for Spaceflight: While there are several indications for myringotomy with or without tympanostomy tube placement, including recurrent acute otitis media, chronic otitis media with effusion, acute mastoiditis, etc., perhaps the most relevant indication for spaceflight is Eustachian tube dysfunction that interferes with the ability to equalize the middle ear space with atmospheric pressure changes. This is particularly important when transitioning to an exploration atmosphere or prebreathe state in preparation for extravehicular activity (EVA), which occurs at low pressures (around 4–5 psi) and high oxygen concentrations (100%). The tympanic membrane ruptures at 5 psi, which can easily occur during routine spaceflight if equalization does not occur. Pressure differentials across the tympanic membrane can also occur following the EVA after return to atmospheric or exploration pressures, as the high concentration of metabolically active oxygen is absorbed from the middle ear space, leading to a negative pressure differential if inert gases (such as nitrogen) are not reintroduced. This is known as oxygen ear and is the reason why astronauts must participate in active ear clearing and are not allowed to sleep for several hours following an EVA.

Relevant anatomy:

- Pinna
- Tragus
- External auditory canal (EAC)
- Cartilaginous EAC
- Membranous EAC
- Tympanic membrane
- Ossicular chain

Preprocedural Care: Premedicating with pain medication can be helpful, as the topical anesthetic required for this procedure (phenol) burns during application.

Supplies and Equipment: See Table 9.4

Instructions:

This procedure requires an assistant.

1. Confirm diagnosis (via fluid in the middle ear space, retraction of the tympanic membrane, or immobile tympanic membrane with Valsalva) via direct or indirect visualization (otoscope vs curette with camera).
2. Obtain all supplies that will be used during the procedure. If cerumen is present, proceed to steps 3 through 9. If not, skip to step 10.
3. Power on the viewing device and connect the curette. Ensure that all equipment is working properly prior to restraining the patient. Place the viewing device in a secured location that doesn't require turning your head away from the patient to see it.
4. Restrain the patient in their seat or a dedicated medical working area.

**TABLE 9.4 Optimal and minimal supplies and equipment for myringotomy +/− tympanostomy tube placement.**

| Optimal | Minimum |
|---|---|
| Otologic microscope | Endoscopic camera or magnified instrumenting otoscope |
| Ear specula | Ear specula |
| Curette | Curette |
| Alligator forceps | Alligator forceps |
| Frazier suction (3, 5, and 7 Fr) | 88% phenol and angled applicator |
| Suction and tubing | Myringotomy knife |
| 88% phenol for topical application and angled applicator | Tympanostomy tube |
| Myringotomy knife | |
| Tympanostomy tube (e.g., Armstrong) | |

5. Restrain yourself in a position that allows you the necessary degrees of freedom to perform the procedure safely and effectively.
6. Using your nondominant hand, place your index finger over the tragus and gently pinch the posterior pinna with your thumb and middle finger, then gently spread the EAC os to allow for better visualization.
7. Insert the curette in the EAC.
8. Remove cerumen under endoscopic guidance until the tympanic membrane is fully visualized.
9. Withdraw and disconnect the curette from the camera. Using gauze, remove cerumen from the curette. TIP: It's helpful if someone else wipes the curette for you, as it will allow you to avoid having to reset your nondominant hand position.
10. Insert the ear speculum, angling anteriorly and inferiorly.
11. Gently insert the endoscope through the speculum and visualize the anterior inferior quadrant of the tympanic membrane. Do not pass the endoscope beyond the distal end of the ear speculum.
12. Instruct your assistant to obtain the phenol and angled applicator. Prior to opening the phenol, centrifuge the tube to ensure the phenol is collected at the bottom of the tube. CAUTION: Phenol is caustic and can cause injuries to skin.
13. Instruct the assistant to dip the applicator into the phenol, then push the tip of the applicator against the side of the tube to remove the excess.
14. Take the applicator from your assistant. Pinch at the angle between your thumb and index/middle fingers. Use your ring and pinky fingers to brace your hand against the patient's head to improve steadiness.
15. Under DIRECT visualization, insert the tip of the applicator into the ear speculum. Back out the camera until it is behind the applicator tip, then follow the applicator in with the camera.
16. Gently press the tip of the applicator to the anterior inferior quadrant for 3–5 seconds. The tympanic membrane will blanch. Lift and reapply until the entire quadrant is blanched. NOTE: This portion of the procedure is uncomfortable for the patient and burns.
17. Remove the applicator, with care taken not to touch it to the exposed portion of the EAC.
18. Take the myringotomy knife from your assistant. Pinch at the angle between your thumb and index/middle fingers. Use your ring and pinky fingers to brace your hand against the patient's head to improve steadiness.
19. Under DIRECT visualization, insert the tip of the knife into the ear speculum. Back out the camera until it is behind the knife tip, then follow the knife in with the camera.
20. Make a radial incision in the anterior inferior quadrant, taking care not to incise near the middle of or at the edge of the tympanic membrane.
21. Remove the knife from the EAC.
22. If placing a tympanostomy tube, proceed to steps 23 through 28. If not, skip to step 29.
23. Instruct your assistant to load the tympanostomy tube into the alligator forceps. The way this is accomplished will vary based on the specific tube. For an Armstrong tube, this is loaded by gripping the inferior portion of the anterior flange, with the beveled portion of the tube facing away with the bevel pointed inferiorly (see photo). IMPORTANT: Do not grip the tube too tightly, or else it will stick to the forceps when attempting to release it.

24. Take the tube and forceps from your assistant, ensuring that the forceps remain closed.
25. Under DIRECT visualization, insert the forceps and tube into the ear speculum. Back out the camera until it is behind the tube, then follow the tube in with the camera.
26. Insert the superior edge of the beveled end into the myringotomy incision. Then, in a single swooping motion, move your wrist forwards and downwards until the inferior edge of the beveled end is engaged with the tympanic membrane.
27. Release the forceps from the tube. If the tube is only partially seated, a small curette can be used to manipulate it into place. CAUTION: Overmanipulation or too large of a myringotomy incision can lead to medialization of the tube into the middle ear space. If this occurs, leave the tube. Do not attempt to retrieve it.
28. Remove the forceps, endoscope, and ear speculum. Apply five drops of topical antibiotic to the EAC and pump the tragus to ensure these reach the tube.
29. When the procedure is finished, clean supplies (if they are to be reused) and restow or dispose of single-use supplies in the designated location.

Postprocedural Care: If a tympanostomy tube is placed, then it is important to follow the procedure with a short course of ototopical drops (e.g., ofloxacin) to prevent any bleeding or fluid from clogging the tube and rendering it ineffective. If any bleeding is encountered, including bleeding from inadvertent injury to the EAC, then ototopical drops should also be used to prevent occlusion of the EAC by blood and to promote healing. Myringotomy incisions without tympanostomy tubes typically heal within 48–72 hours, while tympanostomy tubes can remain in place for months to years depending on the type of tube and placement.

Pearls and Pitfalls: Myringotomy with or without tympanostomy tube placement is a deceptively challenging procedure, made even more so in the spaceflight environment without optimal resources. It requires a steady hand and knowledge of external and middle ear anatomy in order to avoid injury and morbidity, including hearing loss, which could be permanent. Possible terrestrial complications of this procedure include tympanosclerosis, hearing loss, EAC damage, permanent tympanic membrane perforation, and continuous otorrhea. Currently, difficulty in equalizing the middle ear space in spaceflight environments is treated with topical and systemic decongestants, such as oxymetazoline and pseudoephedrine. However, if these were to be ineffective on an exploration-class mission to the Moon or Mars, the affected crewmember would be exempt from future EVAs. Given the current architectures for these missions involve multiple EVAs in close proximity, this would jeopardize mission objectives.

One of the most challenging aspects of this procedure is visualization. Magnification is required to be able to adequately see landmarks and ensure proper placement of the incision and tympanostomy tube. This is traditionally achieved via the use of a microscope, though it can be performed endoscopically (Kanotra et al.), which is likely how it would be performed in the spaceflight environment due to the high mass and volume of an instrumenting microscope. Instruments must be used in a way that they do not block the view of the tympanic membrane. While topical phenol provides analgesia and hemostasis to the tympanic membrane, injury to the skin of the EAC can lead to bleeding, which quickly obscures the tympanic membrane, a situation that may lead to aborting the procedure in the absence of suction capabilities. Overall, this is a procedure that would lend itself well to future innovations. For example, optical instruments would allow for more precise performance of the procedure with minimal mass and volume requirements. Practically speaking, crew would likely rely on myringotomy only, as placing a tympanostomy tube into the myringotomy incision is the most technically challenging portion of the procedure.

### 9.2.4 Auricular hematoma incision, drainage, and bolstering

Auricular hematoma is a collection of blood between the perichondrium and the cartilage, usually as a result of direct trauma to the auricle or pinna. The cartilage does not have a dedicated blood supply and relies on diffusion of nutrients from the vascular perichondrium. Bleeding between these tissues physically separates the perichondrium from the cartilage, increasing diffusion distance and leading to a nutrient-poor environment and cartilage death.

Terrestrial Indications: Approximately 40% of auricular hematomas are a consequence of contact sports (i.e., wrestling), whereas 31% are spontaneous (Dalal et al., 2020). All auricular hematomas should be drained as soon as possible to prevent auricular deformation, but indications for the method of drainage vary. For newer lesions with a diameter less than two centimeters, a needle aspiration may be attempted. Incision and drainage are indicated for older, larger, or recurrent hematomas. Hematomas that reappear after being drained via needle aspiration should subsequently be managed via incision and drainage.

Efficacy: There is a 25% chance of a recurrent auricular hematoma after drainage (Dalal et al., 2020). Both needle aspiration & incision and drainage are equally effective at draining auricular hematomas and preventing recurrent cases.

Bolstering is associated with a 14% recurrence rate of auricular hematomas, whereas there is a 60% recurrence rate without bolstering (Dalal et al., 2020).

Alternatives and Consequences: Conservative management includes observation; however, this carries the risk of prolonged cartilage ischemia resulting in cartilage death and scarring of the pinna (cauliflower ear). Additionally, any fluid collection is at risk of secondary infection, which can result in further deformity, functional deficits, and systemic complications (e.g., sepsis).

Performance in Austere Environments: NASA's undersea experiments at the National Extreme Environment Medical Operations habitat have tested aquanauts' ability to perform abscess incision and drainage in the presence and absence of premission training, telemonitoring, or a procedure manual. In all cases, aquanauts successfully performed the procedure with a success rate of 90% (Thirsk et al., 2007). Studies on the efficacy of incision and drainage in austere surgical environments have also shown no major complications (Lucha et al., 2000).

Indication for Spaceflight: Auricular hematoma that interferes with the use of communication or hearing protection devices

Relevant Anatomy:

- Pinna
- Perichondrium
- Cartilage

Preprocedural Care: Avoiding medication that can increase the risk of bleeding (e.g., NSAIDs) can be helpful in the peri-procedural period to reduce bleeding during the procedure and reduce the risk of reaccumulation following drainage.

Supplies and Equipment: See Table 9.5

The ideal drainage method is dependent upon the age of the hematoma. Subacute injuries are often filled with thick, clotted blood, requiring an incision and use of a hemostat to mechanically disrupt clots to facilitate drainage. Acute, older injuries, or re-accumulated fluid are often less viscous and therefore amenable to needle aspiration, which has the benefit of being less technically challenging and less resource-intensive. Thus, it would be recommended to perform this procedure in the acute period prior to blood clotting.

While it is traditionally taught that local anesthetics containing epinephrine should not be used in acral areas such as the ears and nose due to the risk of ischemia, several studies have demonstrated no ischemic complications with epinephrine-containing local anesthetics in the ears and nose. Epinephrine has the additional benefit of reducing blood loss from the procedure and extending the duration of local anesthesia.

Instructions:

1. Confirm diagnosis via history and physical examination with direct visualization.
2. Obtain all supplies that will be used during the procedure.

**TABLE 9.5 Optimal and minimal supplies and equipment for incision, drainage, and bolstering of an auricular hematoma.**

| Optimal | Minimum |
|---|---|
| Otologic microscope | Endoscopic camera or magnified instrumenting otoscope |
| Ear specula | Ear specula |
| Curette | Curette |
| Alligator forceps | Alligator forceps |
| Frazier suction (3, 5, and 7 Fr) | 88% phenol and angled applicator |
| Suction and tubing | Myringotomy knife |
| 88% phenol for topical application and angled applicator | Tympanostomy tube |
| Myringotomy knife | |
| Tympanostomy tube (e.g., Armstrong) | |

3. Restrain the patient in their seat or dedicated medical working area.
4. Restrain yourself in a position that allows you the necessary degrees of freedom to perform the procedure safely and effectively.
5. Thoroughly clean the area to be drained with the cleaning solution available and allow the solution to fully dry.
6. Attach the 18G needle to the 10 cc syringe and use this to draw up the local anesthetic (no more than 3 cc).
7. Cap the 18G needle and remove from the syringe, taking care not to depress the plunger.
8. Attach the 25G needle.
9. Inject the local anesthetic into the area to be drained and bolstered. Ensure that you are injecting the anterior and posterior aspects of the pinna.
10. Cap the 25G needle and remove from the syringe. Waste the remaining local anesthetic into a gauze or bag.
11. Reattach the 18G needle and uncap. Ensure the plunger is fully depressed.
12. Insert the needle into the fullest part of the hematoma and begin to slowly pull the plunger of the syringe.
13. When no more fluid is draining into the syringe, remove the needle and syringe and cap the needle.
14. Apply gauze over the puncture site and massage the hematoma to express any residual blood.
15. Apply bolster material to the anterior and posterior surface of the drainage site. The configuration of the bolster is dependent on the specific bolster material. A single piece of Xeroform can be molded to fit the anterior and posterior surfaces, while two dental rolls must be used to achieve the same coverage.
16. Open the outer packaging of the suture. Grasp the needle of the nylon suture with the needle driver approximately 75% of the length from the tip and remove the suture fully from its remaining packaging.
17. Insert the needle through the bolster on the anterior surface first. Pierce the pinna and then insert the needle through the bolster on the posterior surface. TIP: This does not need to be accomplished in a single throw and can be achieved stepwise.
18. Pull the suture about halfway through, leaving a tail long enough to tie to on the anterior surface.
19. Insert the needle through the bolster on the posterior surface approximately 1 to 1.5 cm from the exit point. Pierce the pinna and then insert the needle through the bolster on the anterior surface, again 1 to 1.5 cm from the entry point. TIP: This does not need to be accomplished in a single throw and can be achieved stepwise.
20. Pull the suture through until it is tight against the posterior bolster.
21. Cut and remove the needle.
22. Hand tie the suture to affix the bolster to the pinna, ensuring that it is tightly flush against the bolster. Throw at least 6 knots. TIP: After throwing the first knot, you can instruct your assistant to gently clamp the knot with the needle driver until you can throw a second knot. This helps to ensure the suture remains flush against the bolster and appropriately tight.
23. Cut the excess suture away, leaving at least 1 cm "tails" from the final knot.
24. When the procedure is finished, clean supplies (if they are to be reused) and restow or dispose of single-use supplies in the designated location.

Postprocedural Care: It is essential to keep the bolster dry to minimize the risk of infection or tissue maceration. Additionally, use of a cartilage-penetrating oral antibiotic such as ciprofloxacin will help prevent infection while the bolster is in place. In general, the bolster should remain in place for at least 5 days to allow for the perichondrium to re-anneal to the cartilage to reduce the risk of reaccumulation.

Pearls and Pitfalls: Possible complications of this procedure include a greater risk of chronic deformity (i.e., cauliflower ear), necrosis of the aural cartilage, perichondritis, persistent infection, or recurrence of the hematoma. This proposed method of auricular hematoma management in spaceflight relies on infiltration of local anesthesia in the area of drainage and bolstering. Some practitioners advocate for the use of auricular blocks instead. This is accomplished by infiltrating local anesthesia around the base of the pinna in a diamond shape to reduce sensation in the auriculotemporal nerve, lesser occipital nerve, great auricular nerve, auricular branch of the vagus nerve, and nervus intermedius. However, this method often leads to residual sensation and uses a larger amount of medication compared to local infiltration.

Not applying a bolster or removing the bolster too early can lead to re-accumulation and the need for repeat drainage, which uses more resources. Depending on the bulk of the bolster, this can interfere with use of communication or hearing protection devices and could potentially interfere with suit entry and use based on the design of the suit and helmet. Therefore, it would be important to have a plan for bolster management or removal in the event of emergent suit donning (e.g., in the event of cabin depressurization or fire response). Low bulk and deformable bolsters such as Xeroform would help to minimize this risk (compared to dental rolls or similar).

## 9.3 Nasal procedures

### 9.3.1 Closed nasal bone reduction

Nasal fractures are the most common type of facial fracture in adults and result from blunt force trauma to the bony structures of the nose, which include the nasal bones, frontal processes of the maxilla, and bony nasal septum, comprised of the vomer and perpendicular plate of the ethmoid, all of which can be involved in fractures of the nose in descending frequency. Occasionally, the cartilage of the nasal septum will also fracture. This is associated with septal hematoma, abscess, and ultimately nasal septal perforation if left untreated.

Terrestrial Indications: Comprising 40%–50% of all facial fractures, nasal bone fractures are the most common facial fracture (Atighechi and Karimi, 2009; Bartkiw et al., 1995). Closed nasal bone reduction should be conducted for simple nasal bone fractures that are less than two weeks old and not severely comminuted or open.

Efficacy: Closed nasal bone reduction has success rates of 60%–90% for simple, noncomminuted fractures (Staffel, 2002). Osteotomies may increase closed reduction efficacy by facilitating bone mobilization if reduction has not been accomplished 10–14 days after fracture onset.

Alternatives and Consequences: The alternative to closed nasal bone reduction following displaced nasal bone fracture is observation, which can result in functional deficits due to obstruction and cosmetic deformity with the need for corrective open septorhinoplasty in the future. While cosmesis is not the chief concern in the spaceflight environment, inability to breathe through one or both sides of the nose can lead to performance decrements and decreased ability to equalize the middle ear space during ambient pressure changes, such as in preparation for an EVA.

Performance in Austere Environments: Analog studies of the efficacy of closed nasal bone reduction in austere environments have reported 50% of outcomes as "excellent," 30% of outcomes as "fair," 19% of outcomes as "good," and 1% of outcomes as "poor" (Raghvi et al., 2023). Nasal deviations were apparent in 31% of cases, bony irregularities in 44% of cases, and olfactory disturbances in 4% of cases (Raghvi et al., 2023).

Indication for Spaceflight: Closed fracture of one or both nasal bones/frontal processes of the maxilla with mild to moderate comminution and displacement

Relevant Anatomy:

- Nasal bone (bilateral)
- Frontal process of the maxilla (bilateral)
- Nasal septum (cartilage, vomer, perpendicular plate of the ethmoid)
- Nares (singular: naris)

Preprocedural Care: Preprocedural care for nasal fractures hinges on pain management, which can be accomplished with oral and parenteral medications. Note that while nasal fractures are almost invariably associated with epistaxis, closed nasal bone reduction often results in repeat epistaxis, and thus aggressive control prior to the procedure is not indicated, as this will consume resources unnecessarily. Reducing bleeding with oxymetazoline 0.05% nasal spray can be helpful.

Supplies and Equipment: See Table 9.6

Instructions:

1. Confirm diagnosis via history and physical examination.
2. Obtain all supplies that will be used during the procedure.
3. Restrain the patient in their seat or a dedicated medical working area.
4. Restrain yourself in a position that allows you the necessary degrees of freedom to perform the procedure safely and effectively.
5. Liberally apply oxymetazoline nasal spray to the bilateral nasal cavities (3–5 sprays per side).
6. Thoroughly clean the nose and midface with the cleaning solution available and allow the solution to fully dry.
7. Attach the 18G needle to the 10 cc syringe and use this to draw up the local anesthetic (4–5 cc).
8. Cap the 18G needle and remove from the syringe, taking care not to depress the plunger.
9. Attach the 25G needle to the syringe.
10. Palpate the infraorbital foramina, insert the needle perpendicular to the skin until it hits bone, aspirate back, and then inject approximately 1 cc of local anesthetic. Repeat on the contralateral side.
11. Using the nasal speculum, open the naris and expose the lateral nasal wall.
12. From inside the naris, insert the needle subcutaneously along the lateral nasal wall to the level of the nasal bones.
13. Aspirate to ensure that the lateral nasal artery has not been cannulated, then slowly inject approximately 1 cc of local anesthetic while pulling the needle back. Repeat on the contralateral side.

**TABLE 9.6 Optimal and minimal supplies and equipment for a closed nasal bone reduction.**

| Optimal | Minimum |
|---|---|
| Cleansing solution (e.g., alcohol prep pad, Chloraprep, povidone iodine) | Nonflammable cleansing solution (e.g., povidone iodine) |
| Local anesthesia (e.g., 1% lidocaine with 1:100,000 epinephrine) | Local anesthesia (e.g., 1% lidocaine with 1:100,000 epinephrine) |
| 10 cc syringe (for local anesthetic) | 10 cc syringe (for local anesthetic and aspiration) |
| 18G needle or blunt-tipped draw needle (for local anesthetic) | 18G needle (for local anesthetic and aspiration) |
| 25G needle | 25G needle |
| 10 cc syringe (for needle aspiration) | Bolster material (e.g., Xeroform, Vaseline gauze, dental roll) |
| 18G needle (for needle aspiration) | 2-O nylon suture on a curved cutting needle (e.g., FS) |
| 11-blade scalpel | Needle driver |
| Curved hemostat (e.g., Kelly) | Scissors |
| Gauze | Oral antibiotic with cartilage penetration (e.g., ciprofloxacin) |
| Bolster material (e.g., Xeroform, Vaseline gauze, dental roll) | Gauze |
| 2-O nylon suture on a curved cutting needle (e.g., FS) | |
| Needle driver | |
| Scissors | |
| Oral antibiotic with cartilage penetration (e.g., ciprofloxacin) | |

14. Cap the 25G needle and remove from the syringe. Waste the remaining local anesthetic into a gauze or bag. Dispose of the needle and syringe in appropriate containers.
15. Manually reduce the nasal bones using fingers by applying medial pressure to the portion deviated laterally from its anatomic position
16. Obtain the Boies elevator and measure it against the lateral wall of the nose, ensuring the tip does not pass the middle of the nasal bones/frontal processes of the maxilla. Use your fingers to mark this length at the naris. CAUTION: Inserting the Boies elevator beyond this length can cause injury to the medial orbit and/or skull base (Fig. 9.1).
17. Insert the Boise elevator into the nasal cavity to the previously measured depth. Lift up and out while palpating the nasal bones with the contralateral hand (Fig. 9.2). Successful reduction can be confirmed via palpation (it will feel as though the bone "clicks" back into place; Fig. 9.3).
18. Repeat steps 16 and 17 on the contralateral side. This ensures the nasal bones are not collapsed into the nasal cavities.
19. Reapply oxymetazoline nasal spray (2–3 sprays per side).
20. Roll the Surgicel into a cylinder approximately 0.5 × 2 cm.
21. Using the Bayonet or similar long forceps, guide the roll of Surgicel into the nasal cavity beneath the nasal bones for support
22. Apply oxymetazoline nasal spray (1–2 sprays)
23. Repeat steps 20 through 22 on the contralateral side.
24. Cut the Steri-Strips diagonally to create graded strips.
25. Apply the Steri-Strips to the nasal dorsum, going from small to large, taking care to avoid the medial canthus and ensure the ala and tip are covered (Fig. 9.4).
26. With the Velcro attached, prebend the Denver splint to the shape of the nose.
27. Remove the Velcro from the metal splint.
28. Remove the adhesive backing from the Velcro and place it over the Steri-Strips on the nose, taking care to avoid placing the adhesive directly on the skin.

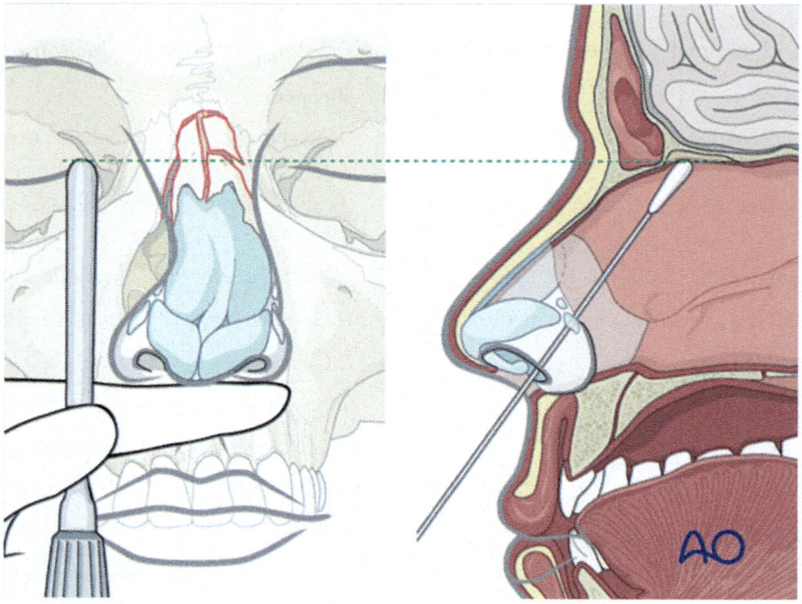


FIGURE 9.1 Prevention of medial orbit/skull base injury during nasal fracture reduction.
Care should be taken to prevent penetration through the cribriform plate by aligning the Boies elevator against the lateral wall of the nose.
**Source:** AO Surgery Reference. https://surgeryreference.aofoundation.org. Copyright by AO Foundation, Switzerland.

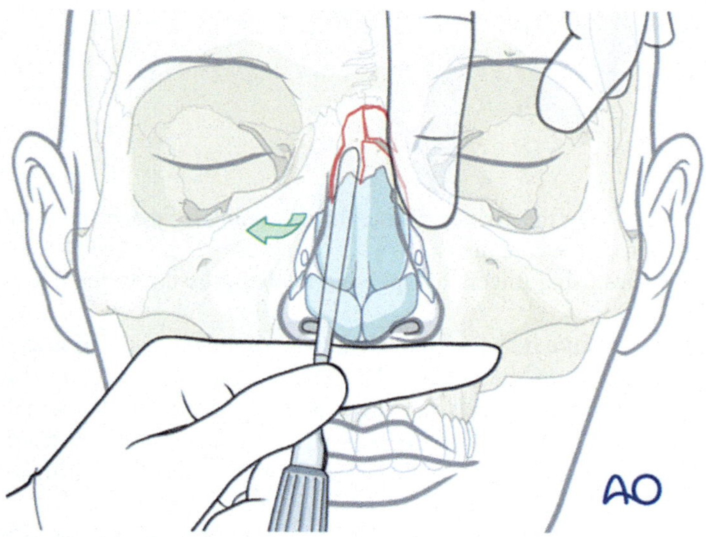


FIGURE 9.2 Reduction of nasal bones.
The Boies elevator should be elevated up and out while the contralateral hand palpates the nasal bones during reduction.
**Source:** AO Surgery Reference. https://surgeryreference.aofoundation.org. Copyright by AO Foundation, Switzerland.

29. Place the metal splint onto the Velcro, gently pinching the sides until it is flush with the contour of the nose, taking care not to overbend it.
30. When the procedure is finished, clean supplies (if they are to be reused) and restow or dispose of single-use supplies in the designated location.

Postprocedural Care: Following closed nasal bone reduction, continued pain management and control of epistaxis take priority. Nasal saline spray is crucial for maintaining mucosal moisture, promoting healing, and assisting with the dissolution of blood products and absorbable packing materials. This author recommends oxymetazoline nasal spray twice daily for three days and nasal saline spray three to four times daily for 2 weeks. Steroids can assist with swelling, and pain

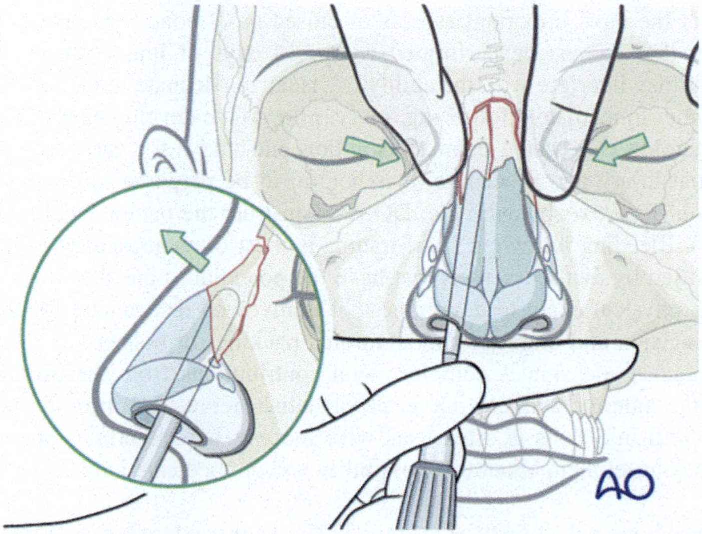


FIGURE 9.3 Completion of fracture reduction.
Palpation and visual inspection of the of the fracture can confirm successful fracture reduction, with the fracture feeling like it "clicks" into place.
**Source:** AO Surgery Reference. https://surgeryreference.aofoundation.org. Copyright by AO Foundation, Switzerland.

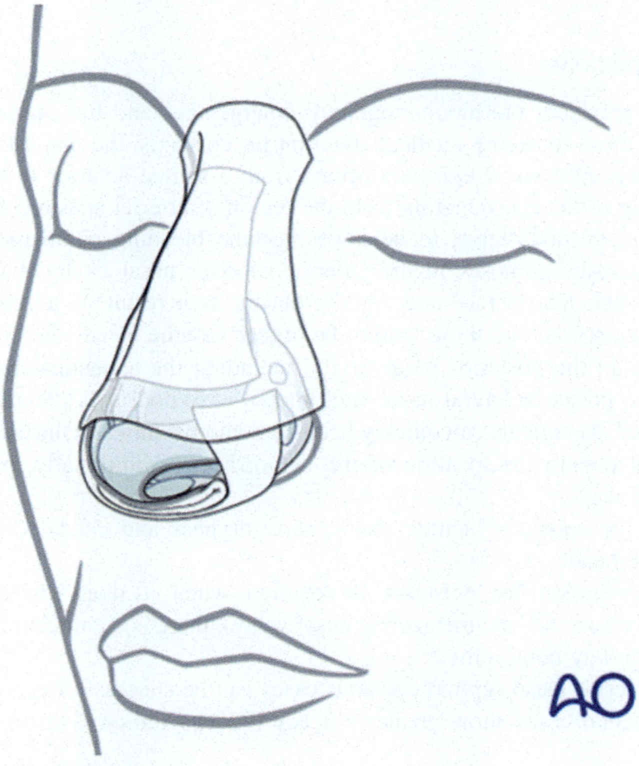


FIGURE 9.4 Splinting of nasal bones post-reduction.
After reduction, adhesive strips (e.g., Steri-Strips) are cut into strips and applied to the nose.
**Source:** AO Surgery Reference. https://surgeryreference.aofoundation.org. Copyright by AO Foundation, Switzerland.

management is typically accomplished with a combination of acetaminophen, ibuprofen, and narcotic pain medication as needed. Ice packs to the nose are also helpful for swelling and pain, though care must be taken to avoid prolonged cooling (we recommend no more than 15 minutes every 1 hour). The Denver splint is generally removed within 5 days, but it often falls off earlier and can be safely left off.

Pearls and Pitfalls: One of the most important aspects of closed nasal bone reduction is the timing of the procedure. While most literature states that it must be performed within 14 days of injury, this is often too long. However, if attempted too soon, swelling may interfere with the ability to visualize adequate reduction. Therefore, it is recommended to perform this procedure either immediately following injury, prior to the development of significant edema, or 3–7 days later. In these instances, steroids can hasten edema resolution and allow for earlier reduction. As bleeding is often encountered during the initial injury and procedure, it is helpful to be prepared to control epistaxis with the available resources (see *Management of Epistaxis* below; note: DO NOT instruct the patient to blow their nose to remove clots if bleeding is due to fractures). Bleeding following nasal trauma is most often secondary to mucosal disruption at fracture sites, but it can rarely be caused by avulsed vessels that have the potential to bleed profusely. It is important to evaluate the nasal septum during the physical exam to determine if it is involved or evaluate for the presence of a nasal septal hematoma, which requires incision and drainage and additional packing for bolster.

Sensation of the nasal mucosa and skin is complex, with contributions from the infratrochlear nerve, external and internal nasal branches of the anterior ethmoid nerve, nasopalatine nerve, and infraorbital nerve. Regional anesthesia utilizing blocks and lateral wall injections is often used with success, but oftentimes patients still feel pain during the procedure. It is important to counsel them about this possibility and premedicate with oral or parenteral analgesics before the procedure.

Finally, while a closed nasal bone reduction is a deceptively simple procedure for a common injury, it can be dangerous if performed incorrectly. Fracture of the cribiform plate and cerebrospinal fluid leaks have been described in the literature, and this would be a difficult-to-manage and potentially life-threatening problem in the spaceflight environment. Other complications associated with this procedure include nasal deformity, septal deviation, olfactory disturbance, nasal obstruction, and rhinorrhea.

### 9.3.2 Management of epistaxis

Epistaxis is one of the most prevalent otorhinolaryngology emergencies and has an estimated lifetime prevalence of 60% (Passali et al., 2020). Of those receiving medical attention for epistaxis, the majority (40%) are more than 50 years old (Passali et al., 2020). Ninety percent of epistaxis occurs from a source anterior to the basal lamella of the middle turbinate, with the vast majority of these originating from the region of arterial anastomoses on the caudal septum called Kiesselbach's plexus. Anterior epistaxis tends to be lower volume bleeding compared to epistaxis from a posterior source (note: there exist arterial anastomoses in the anterior superior nasal cavity that can result in life-threatening hemorrhage when injured, though this is rare and usually occurs as a result of iatrogenic injury). Posterior sources of epistaxis comprise 10% of epistaxis, and these tend to be higher volume bleeds due to the concentration of relatively larger caliber arterial branches in the posterior nasal cavity, including the sphenopalatine artery and its branches, and the venous anastomoses on the posterior lateral nasal wall known as Woodruff's plexus. True posterior epistaxis is an otorhinolaryngologic emergency, as patients can quickly become hemodynamically unstable and conventional methods of control are often unsuccessful due to the location of the hemorrhage. Additionally, posterior epistaxis has a higher potential to compromise the airway.

Microgravity-specific risks for epistaxis include atmospheric dryness and caudal cephalad fluid shifts resulting in greater venous pressures in the head.

Terrestrial Indications: Intervention for epistaxis is required when it does not resolve spontaneously or with conservative measures (e.g., pressure +/– oxymetazoline nasal spray), there is significant blood loss and/or hemodynamic instability, or there is risk of airway compromise.

Efficacy: 78% to 88% of anterior nasal septum epistaxis cases can be successfully controlled by nasal packing (Beck et al., 2018). Posterior nasal septal cases show greater efficacy (97%) with surgical treatment than with nasal packing (62%; Beck et al., 2018).

Alternatives and Consequences: Alternatives include conservative measures such as the application of local pressure and oxymetazoline nasal spray. Except for obviously high-volume bleeds, hemodynamic instability, or airway compromise, these measures should always be the first-line treatment of epistaxis, as instrumentation and packing are traumatizing to the nasal mucosa and may cause additional bleeding. Unresolved epistaxis can result in loss of task time, clinical instability, airway obstruction, and death, depending on the severity. In the microgravity environment, uncontrolled bleeding also represents an atmospheric contaminant and biohazard.

Performance in Austere Environments: Oxymetazoline and nasal packing have been demonstrated to be effective at managing cases of epistaxis aboard the ISS without causing significant complications (Alexander, 2020; Antonsen, 2019).

Even in austere environments, nasal packing and Foley catheters have been cited as effective interventions in managing cases of anterior nasal septum and posterior nasal septum epistaxis, respectively (Breeze and Bryant, 2009).

Indication for Spaceflight: The indications for control of epistaxis in spaceflight are the same as those terrestrially.

Relevant Anatomy:

- Naris (plural: nares)
- Nasal septum
- Nasal cavity
- Nasal vestibule
- Floor of nose
- Kiesselbach's plexus
- Woodruff's plexus
- Sphenopalatine artery
- Posterior septal artery

Preprocedural Care: As previously stated, for epistaxis that is not high-volume in a hemodynamically stable patient without airway compromise, first-line treatment should be conservative measures. The authors recommend gently blowing the nose to remove clots, liberally applying oxymetazoline nasal spray bilaterally (3–5 sprays per side), then applying hard pressure to the soft portion of the nose using fingers (ideally while leaning forward in the presence of a gravity vector). To achieve adequate pressure, pinch the nose until it hurts, then let up slightly. Nasal clamps (including those created with tongue blades) often do not provide an adequate level or a large enough area of pressure and should be avoided. Hold constant pressure for 15 minutes without checking. If bleeding continues despite these measures, repeat once more before moving on to more invasive interventions.

It is important to establish vascular access in any patient with hemodynamic compromise and utilize airway protective mechanisms up to and including intubation if there is a risk of impending airway compromise. For bleeding secondary to trauma, it is NOT recommended to instruct the patient to blow their nose to clear clots, as this can lead to subcutaneous and orbital emphysema and even pneumocephalus depending on other bony involvement.

Supplies and Equipment: See Table 9.7

There are several options when it comes to nasal packing materials, and the choice of material depends on many factors, including the suspected location of the bleed and the volume of the bleed. In general, durable (nonabsorbable) nasal packing should be avoided where possible, as these types of packing invariably cause more trauma and can exacerbate ongoing epistaxis if inserted using poor technique. Additionally, durable nasal packing has a higher risk of iatrogenic

**TABLE 9.7 Optimal and minimal supplies and equipment for the management of epistaxis.**

| Optimal | Minimum |
|---|---|
| Oxymetazoline 0.05% nasal spray | Oxymetazoline 0.05% nasal spray |
| Nasal speculum | Nasal speculum |
| Headlight or headmirror | Headlight |
| Frazier suction (10 and 12 Fr) | Suction bulb |
| Suction canister and tubing | Bayonet forceps (other long forceps, such as DeBakey or Gerald forceps, will also suffice) |
| Suction device | Surgicel mesh |
| Bayonet forceps | Floseal/Surgiflo kit |
| Absorbable packing materials (Nasopore, PosiSep, Floseal/Surgiflo, Surgicel, etc.) | RhinoRocket (10 cm) |
| Durable packing materials (10 cm PVA sponge, Epistat vs Foley catheter for posterior bleeds) | Foley catheter, 10 cc syringe, tape |
| Vitals monitor | Vitals monitor |

injury to superior arterial anastomoses if inserted along an incorrect trajectory, which can result in life-threatening hemorrhage that is difficult to control with packing (bleeds in these locations generally require control in an operative setting).

Absorbable packing materials include polyurethane foam (Nasopore), chitosan-containing foam (PosiSep), gelatin foam (Gelfoam), thrombin-containing gelatin matrix (Floseal, Surgiflo), and oxidized regenerated cellulose mesh or powder (Surgicel). Choice of absorbable nasal packing material is based on material stability (e.g., Nasopore requires refrigeration and is therefore unlikely to be available in spaceflight), patient allergies (e.g., PosiSep contains chitin, which is contraindicated in patients with a shellfish allergy), and ability to reconstitute (e.g., Floseal/Surgiflo requires reconstitution of thrombin and then mixing of thrombin and gelatin granules). Of all the choices listed, oxidized regenerated cellulose mesh and thrombin-containing gelatin matrices are likely the best options in the spaceflight environment, despite challenges with reconstitution in a microgravity environment. Cellulose mesh can be laid flat over areas of bleeding or rolled and shaped as needed to apply pressure, and thrombin-gelatin foams have the ability to fill the numerous nooks and crannies of the nasal cavity with hemostatic material to assist with coagulation.

Durable packing materials include hydroxylated polyvinyl acetate (PVA) sponges (Merocel, RhinoRocket) and balloon-based nasal packing devices with or without carboxymethylcellulose mesh (RapidRhino). Options available for posterior nasal cavity and nasopharyngeal packing include elongated PVA (10 cm) sponges and balloons (7.5 cm), dual-balloon systems (Epistat), Foley catheters (10 cc balloon), and gauze rolls secured through the nose and mouth with strip gauze (last resort, requires catheter for placement and has a high risk of parasympathetic stimulation and bradycardia). The best options for the spaceflight environment are the RhinoRocket (PVA sponge in an applicator for streamlined insertion) and a Foley catheter (which will already be available for use in urinary retention). The most important thing to ensure regarding durable nasal packing besides appropriate insertion technique is the inclusion of a string or other retrieval method for removal.

Instructions:

1. Confirm diagnosis via history and physical examination. Attempt control with oxymetazoline nasal spray and pressure before proceeding.
2. Obtain all supplies that will be used during the procedure.
3. Restrain the patient in their seat or a dedicated medical working area.
4. Connect the patient to the vitals monitor.
5. Don headlight.
6. Restrain yourself in a position that allows you the necessary degrees of freedom to perform the procedure safely and effectively.
7. Instruct the patient to gently blow their nose to remove clots.
8. Liberally apply oxymetazoline nasal spray to the bilateral nasal cavities (3–5 sprays per side)
9. Using the nasal speculum, open the naris and attempt to visualize the location of the bleeding. Use the suction bulb as needed to clear additional clots. If bleeding is anterior and/or low volume, proceed to step 9. If bleeding is posterior and/or high volume, proceed to step 15.
   a. **NOTE**: In the microgravity environment, blood will tend to pool around its source due to surface tension being the predominant force. High-pressure (e.g., larger caliber arterial) bleeding may result in spray. Recall that arterial bleeding is bright red while venous bleeding is darker; this may aid in source identification and material selection for packing.
10. Cut a portion of the Surgicel into postage-stamp-sized (approximately 1.5 × 2 cm) rectangles. Using the Bayonet or similar long forceps, gently lay the Surgicel flat over the area of bleeding.
11. Roll additional Surgicel into a cylinder of the appropriate shape and size to be able to apply pressure against the area of bleeding. Gently insert the Surgicel roll into the nasal cavity using the Bayonet forceps, taking care not to disturb the previously placed Surgicel.
12. Reconstitute the Floseal/Surgiflo per the instructions.
   a. **NOTE**: It is acceptable for air to be drawn into the syringe following reconstitution, as this medication is intended for topical use and not for intravenous injection.
13. Generously apply the Floseal/Surgiflo into the nasal cavity, focusing on the suspected location of bleeding.
14. Apply oxymetazoline nasal spray (2–3 sprays).
15. **IF POSTERIOR OR HIGH VOLUME**: In the event of a low-volume bleed with a suspect posterior source, refer to steps 12 and 13 for application of Floseal/Surgiflo. For high-volume posterior bleeds, skip to step 18.
16. Insert the RhinoRocket into the nasal cavity along the floor of the nose (i.e., straight back, NOT up or at an angle) by opening the naris with the nasal speculum, aligning the opening of the applicator with the nasal vestibule perpendicular to the coronal plane of the face, and depressing the plunger of the applicator in a single, smooth motion.

a. **CAUTION**: This is extremely painful. Premedication with narcotic pain medication can be helpful, but it is helpful to warn the patient and ensure they are restrained in a way that they cannot move their head posteriorly ("run away") or reflexively grab the operator's arms. If topical lidocaine is available, this can be applied with mixed success.

b. **CAUTION**: Stimulation of the walls of the nasal cavity can result in parasympathetic overdrive, bradycardia, and hypotension. In severe cases, it can result in asystole.

17. Apply oxymetazoline nasal spray (3–5 sprays)
18. In the event of a high-volume posterior bleed, open the Foley catheter and test the balloon with AIR to ensure it is functional. Fully deflate the balloon after testing.
19. Insert the Foley catheter into the nasal cavity along the floor of the nose (i.e., straight back, NOT up or at an angle) by opening the naris with the nasal speculum, aligning the tip of the Foley with the nasal vestibule perpendicular to the coronal plane of the face, and sliding it along the nasal floor until it is just visible in the oropharynx.
20. Inflate the balloon with 10 cc of AIR while visualizing the tip of the catheter in the oropharynx.
21. Gently pull the catheter back until you meet resistance.
22. Secure the Foley catheter to the cheek with tape.
23. Refer to steps 12 and 13 for application of Floseal/Surgiflo.
    a. **CAUTION**: Stimulation of the nasopharynx can result in parasympathetic overdrive, bradycardia, and hypotension. In severe cases, it can result in asystole.
24. Prior to cleaning up and stowing supplies, monitor the patient for at least 30 minutes to ensure bleeding is slowing or resolved.
    a. NOTE: Patients with bilateral posterior nasal or nasopharyngeal packing require continuous vitals monitoring while packing is in place.
25. When the procedure is finished, clean supplies (if they are to be reused) and restow or dispose of single-use supplies in the designated location.

Postprocedural Care: Following control of epistaxis, it is essential to initiate a nasal hygiene regimen to promote healing of the nasal mucosa, dissolve absorbable packing materials, and prevent recurrent bleeding. This is generally accomplished with nasal saline spray at least twice daily, ideally three to four times daily. Nasal saline gel sprays are longer-lasting and require less frequent application. Saline gels placed in the nares before bed help moisturize throughout the day and overnight. If the bleeding source is the anterior septum at Kiesselbach's plexus, mupirocin ointment for 1–2 weeks can aid in healing. Vaseline and other petroleum-based products should be avoided, as they can lead to paradoxical mucosal irritation and weakness due to cytotoxic effects and, if inadvertently inhaled, rarely result in lipoid pneumonia. Ensuring that the environment is appropriately humidified is another supportive measure. The authors also recommend scheduled oxymetazoline nasal spray twice daily for 3 days, as it promotes vasoconstriction and reduces the risk of recurrent bleeding. Rhinitis medicamentosa can develop with prolonged use of oxymetazoline; therefore, durations over 5–7 days should be avoided.

The patient should be instructed to avoid blowing their nose or partaking in any activity that increases blood pressure in the head (e.g., Valsalva, heavy exercise) for 2 weeks to reduce the risk of recurrence. EVAs are therefore contraindicated while recovering from moderate to severe episodes of epistaxis.

Durable packing materials should be left in place for 5 days. Saline and oxymetazoline should be applied on and around the packing as above to keep it moist, which helps to mitigate trauma from removal. While it was historically taught that antibiotic prophylaxis with *Staphylococcus* coverage for indwelling sponges is required to prevent toxic shock syndrome, this has become controversial in recent years. In the spaceflight environment where removal to definitive care is significantly delayed or impossible, antibiotic prophylaxis is recommended (cephalexin or similar) while the packing is in place.

Patients with bilateral posterior or nasopharyngeal packing require continuous vitals monitoring while the packing is in place due to the risk of parasympathetic overstimulation.

Pearls and Pitfalls: Epistaxis is a common medical condition with the potential for severe sequelae, including death. Despite the well-intentioned interventions of healthcare providers, a significant number of injuries occur due to attempted control of epistaxis by inexperienced operators, typically secondary to poor technique or repeated insertion and removal of nasal packing materials. Potential complications of this procedure include the development of septal hematomas, necrosis secondary to tight packing, abscesses, neurogenic syncope during nasal packing, sinusitis, and toxic shock syndrome. Improper use of a RhinoRocket or Foley catheter in cases of posterior nasal septal epistaxis or high-volume bleeding may lead to septal perforations, airway obstruction, or infection. It is important not to take epistaxis for granted and have a healthy respect for the anatomy and vascularity of the nose. Patience is also important when addressing any bleeding, but certainly when addressing epistaxis. As long as the patient remains stable, the process should be slow and meticulous in an effort to "get it right" the first time.

Tranexamic acid (TXA) applied topically is another option for control of epistaxis. This can be applied to absorbable and durable packing, and some studies have shown it to be superior to oxymetazoline (Whitworth et al., 2020). However, while oxymetazoline is a commonly used medication in spaceflight and thus likely to be available in future medical kits, TXA may not be available. If TXA is included, it can be applied concomitantly with oxymetazoline, as they have different mechanisms of action.

Preventing epistaxis in spaceflight is better than treating it. This can be achieved by adopting a nasal hygiene regimen, controlling environmental humidity, avoiding intranasal medications that can be drying or irritating to the nasal mucosa, and avoiding nasal instrumentation.

Nasal cautery is not discussed in this section as chemical cautery (e.g., with silver nitrate) is generally not effective in the acute setting, and electrocautery is unlikely to be available (even if it is, nasal electrocautery often requires specialized bipolars, endoscopic cameras, and significant training). Silver nitrate cautery can help prevent further episodes of nuisance bleeding from Kiesselbach's plexus after the acute bleed has resolved, especially if a target vessel can be identified. Silver nitrate cautery should NEVER be used bilaterally in the same location, as this can result in septal necrosis and perforation. Silver nitrate becomes less useful for bleeds posterior to Kiesselbach's plexus, as visualization is more challenging and the risk of iatrogenic injury from runoff increases.

## 9.4 Oral/throat/neck procedures

### 9.4.1 Flexible laryngoscopy

Flexible laryngoscopy is the workhorse of otolaryngology and can be used to visualize the structures of the nasal cavities, nasopharynx, oropharynx, hypopharynx, and larynx. With adequate topical anesthesia, the scope can even be guided through the true vocal folds to examine the proximal trachea and used for nasal or oral fiberoptic intubation. Flexible laryngoscopes have also been used to visualize structures in the oral cavity and EACs.

While most terrestrial otolaryngology practices utilize reusable flexible laryngoscopes, disposable flexible laryngoscopes became popular during the COVID pandemic and continue to be used today. While marketed as disposable, these can safely be reused multiple times in the same patient with diagnostic image quality. Disposable scopes generally require an electronic viewing device, which can be used to record and store picture and video, whereas reusable scopes can either have an eyepiece for visualization or require a separate viewing device. Disposable scopes also tend to be lighter than reusable scopes with a similar volume. Given the decreased mass, presumed desire for photo and video recording for transmission to ground-based operators, and avoidance of complicated cleaning methods, this author recommends disposable scopes for spaceflight.

Terrestrial Indications: Flexible laryngoscopy is used to diagnose functional, infectious, allergic, and malignant diseases of the nasal cavities, nasopharynx, oropharynx, hypopharynx, and larynx. Signs and symptoms prompting referral to ENT for laryngoscopy include chronic throat pain, caustic inhalations, smoke inhalation, foreign body sensation, dysphonia, and stridor. There is an estimated annual incidence of 3.47 cases of chronic laryngitis per 1000 individuals (Stein and Noordzij, 2013). Laryngoscopy should also be conducted in those experiencing acute airway obstruction, vocal cord palsy, or suspected laryngeal cancer. Laryngeal cancer makes up 0.6% of all cancer cases, and there are approximately 88,075 cases of laryngeal cancer within the United States (*Cancer Stat Facts: Laryngeal Cancer*, 2025).

Efficacy: Laryngoscopy has a sensitivity and specificity of 95% and 96%, respectively, for diagnosing laryngeal cancer (Du et al., 2025). For cases of vocal cord palsy, laryngoscopy has a sensitivity of 100% and specificity of 80.7% (Omokanye et al., 2021). A flexible laryngoscope is effective at visualizing the erythema and edema of different stages of acute laryngitis, and it is especially useful for confirmatory diagnoses in cases of laryngitis refractory to treatment (Gupta, 2022). In cases of chronic laryngitis, however, flexible laryngoscopy is less effective since symptoms are nonspecific and may originate from many conditions (Zhukhovitskaya and Verma, 2019). For airway management, studies report that flexible laryngoscopy is associated with easier airway intubations and shorter intubation times compared to other visualization methods (Xu et al., 2022).

Alternatives and Consequences: Alternatives to flexible laryngoscopy include mirror laryngoscopy and direct laryngoscopy, which is often utilized when a biopsy is required. Mirror laryngoscopy can be challenging for the inexperienced operator, relies on having a patient without a strong gag reflex, and cannot be recorded. Direct laryngoscopy must be performed in the operative setting and generally is preceded by diagnostic flexible laryngoscopy. Flexible bronchoscopy can also be used to visualize the same structures, though these scopes are generally larger caliber and more uncomfortable for awake patients. See Chapter 11: Thoracic Procedures. The consequences for the inability to perform diagnostic flexible laryngoscopy depend on the reason for the procedure. For dysphonia without airway obstruction or signs and symptoms consistent with malignancy, the consequences are generally minor and involve quality of life. However, for acute airway obstruction due to anaphylaxis, the inability to perform diagnostic flexible laryngoscopy can be life-threatening.

Performance in Austere Environments: Studies investigating the efficacy of tele-flexible laryngoscopy by non-otolaryngologists in rural, low-resource environments demonstrate success in visualizing the larynx and in transmitting the image for review (Rustad et al., 2025). Designs of flexible laryngoscopies for resource-limited areas have reported mean System Usability Systems scores of 88.93/100, indicating excellent usability in austere environments (Srinivasan et al., 2024).

Indication for Spaceflight: Signs and symptoms of nasal, nasopharyngeal, oropharyngeal, hypopharyngeal, or laryngeal pathology that require visualization for diagnosis, including nasal airway obstruction, chronic throat pain, dysphagia, odynophagia, dysphonia, stridor, bleeding, or signs and symptoms of a systemic allergic reaction with head and neck manifestations

Relevant Anatomy:

- Naris (plural: nares)
- Nasal cavity
- Floor of the nose
- Inferior turbinate
- Middle turbinate
- Eustachian tube orifice
- Nasopharynx
- Velum/soft palate
- Velopharyngeal opening
- Oropharynx
- Tonsils
- Base of tongue
- Valleculae
- Supraglottis
- Epiglottis
- Arytenoid complexes
- Aryepiglottic folds
- Piriform sinuses
- Post-cricoid region
- False vocal folds
- Glottis
- True vocal folds
- Subglottis
- Trachea

Preprocedural Care: While flexible laryngoscopy can be performed without any medication or lubricant, it is generally kinder to premedicate the nasal cavities and lubricate the scope. This also helps decrease the risk of iatrogenic injury. The nasal cavities are typically prepared with a 50:50 mixture of topical decongestant such as oxymetazoline or neosynephrine and 4% topical lidocaine. Lidocaine jelly can be used as an alternative. Water-based lubricants are applied to the distal 2–3 cm of the scope. It is important not to over-lubricate the scope as this can make handling it more difficult. If a dedicated defogging solution is available, it is helpful to place one or two drops on the tip of the scope. If not, as is likely in the spaceflight environment, isopropyl alcohol can be used as an alternative. Fogging will decrease as the scope warms to body temperature, and can be cleared by gently tapping it against mucosa.

Supplies and Equipment: See Table 9.8

Instructions:

1. Confirm procedure is necessary for diagnosis or intervention based on history and other physical examination findings
2. Obtain all supplies that will be used during the procedure.
3. Restrain the patient in their seat or dedicated medical working area.
   a. **NOTE**: Ensure the head is in a neutral position.

**TABLE 9.8 Optimal and minimal supplies and equipment for a flexible laryngoscopy.**

| Optimal | Minimum |
|---|---|
| Flexible laryngoscope | Flexible laryngoscope |
| Viewing/recording device | Viewing/recording device |
| Oxymetazoline 0.05% nasal spray | |
| 4% topical lidocaine | |
| Syringe with atomizer | |
| Water-based lubricant | |
| Anti-fog solution or isopropyl alcohol pad | |

4. Restrain yourself in a position that allows you the necessary degrees of freedom to perform the procedure safely and effectively.
5. If medications are available, proceed to step 6. Otherwise, skip to step 10.
6. Apply 2 sprays of oxymetazoline nasal spray in each naris, aiming posteriorly and laterally.
7. Draw up approximately 2 cc of 4% topical lidocaine using the syringe.
8. Attach atomizer tip to the syringe.
9. Apply 2 sprays of lidocaine in each naris, aiming posteriorly and laterally.
   a. **NOTE**: Topical lidocaine has a very bitter taste. Despite this, encourage your patient to swallow it as it will help reduce the gag reflex. Numbness lasts for 20–30 minutes.
10. Connect the flexible laryngoscope to the viewing device and ensure the viewing device is working and the picture is in focus.
11. Apply a small amount of water-based lubricant to the distal 2 to 3 cm of the scope, taking care to avoid applying lubricant to the camera.
12. Open isopropyl alcohol pad and gently swirl over the camera tip. Keep this close to reapply as needed.
13. Position your dominant hand over the proximal scope controls and place your thumb over the driving lever.
14. Pinch the scope approximately 4 to 5 cm from the tip between the thumb and index finger of your nondominant hand
15. Insert the tip of the scope into each naris to determine which side of the nose is more patent. This can be influenced by anatomic variations such as septal deviation and physiologic variations such as the nasal cycle.
16. Once a side is chosen, insert the scope and slide it along the floor of the nose. Keep the fingers of the nondominant hand at the opening of the naris, releasing and replacing pressure in a gentle pinching motion to allow for scope motion.
   a. **TIP**: If the scope begins to torque, you can pinch and roll it using the fingers of your nondominant hand.
17. Upon reaching the nasopharynx, use the lever to turn the tip of the scope inferiorly and advance the scope to follow the nasopharynx into the oropharynx through the velopharyngeal opening.
   a. **TIP**: Instruct the patient to breathe through their nose to open the velopharynx.
18. Once the scope has reached the level of the larynx, and using a combination of finger torque and lever motion, use the scope to sequentially evaluate the structures of concern.
   a. **TIP**: If the scope is cloudy or covered with secretions, instruct the patient to swallow. This will clear the scope. Repeat as needed.
   b. **NOTE**: Avoid touching the tip of the scope to laryngeal structures, including the epiglottis or the true vocal folds. This may cause the patient to gag, cough, or laryngospasm.
19. Upon completion of the exam, slowly withdraw the scope, using the lever to bring the tip into a neutral position to pass through the nose.
20. When the procedure is finished, clean supplies (if they are to be reused) and restow or dispose of single-use supplies in the designated location.

Postprocedural Care: There is no specific postprocedural care following flexible laryngoscopy unless iatrogenic injury (e.g., epistaxis) occurs.

Pearls and Pitfalls: Flexible laryngoscopy is a very versatile procedure and can be instrumental in the diagnosis of pathology in the nose and throat. While the steps above focus on using flexible laryngoscopy in its traditional sense, that is to evaluate the larynx as an endoscopic camera, it can be used to evaluate anything between the nares and the true vocal folds with ease. With practice, this procedure can take between 30 seconds and 1 minute.

Possible procedure complications when performing flexible laryngoscopy include mucosal tearing, bleeding secondary to trauma, hoarseness, tongue soreness, gagging, laryngospasm, and an adverse reaction to preprocedural anesthetics. One of the largest pitfalls commonly observed in trainees learning how to perform flexible laryngoscopy is navigating the scope through the nasal cavity, especially in patients with anatomic variations such as septal deviation or septal spurs. Because this is the narrowest area that the scope will pass through, hesitating and repeated retreating and advancing can be very uncomfortable for the patient and increase the risk of epistaxis. As noted in the *Management of Epistaxis* section above, knowledge of the nasal anatomy is instrumental to developing a good technique.

While fiberoptic intubation is out of the scope of this chapter, a flexible laryngoscope can also be used as a tool to accomplish this, though generally a flexible bronchoscope is preferred due to the increased length.

## 9.4.2 Incision and drainage—oral and neck abscesses

While some abscesses of the head and neck are too deep to drain outside of the operating room, the vast majority of abscesses within this region are drained in the ambulatory setting. This section will focus on abscesses that are amenable to bedside management, including odontogenic subperiosteal abscesses, sublingual and submandibular abscesses, and peritonsillar abscesses (Fig. 9.5).

Terrestrial Indications: Patients with oral or neck abscesses greater than 2 cm in diameter should present for incision and drainage, as abscesses of this size are unlikely to resolve with antibiotic treatment alone (Urban et al., 2022). The estimated incidence of deep neck abscesses is 10/100,000 individuals per year, with the primary causes being bacterial tonsillitis (32%) and odontogenic infection (24%; Brito et al., 2017).

Efficacy: Incision and drainage leads to oral abscess clearance 7 to 10 days after drainage, with there being a 28% chance of recurrent abscess formation (Ju and Koyfman, 2015). For neck abscesses, studies show up to 40% need a secondary incision and drainage after initial drainage (Ho et al., 2022). Antibiotics may increase incision and drainage efficacy by resolving the underlying bacterial infection. Delays in abscess incision and drainage are associated with a 2.38-fold increase in morbidity and mortality for a delay greater than two days (Cramer et al., 2016).

Alternatives and Consequences: Sole medical management of smaller abscesses is a reasonable first-line treatment, with the potential to avoid a procedure and the risks and resources associated with it. However, sole medical management of

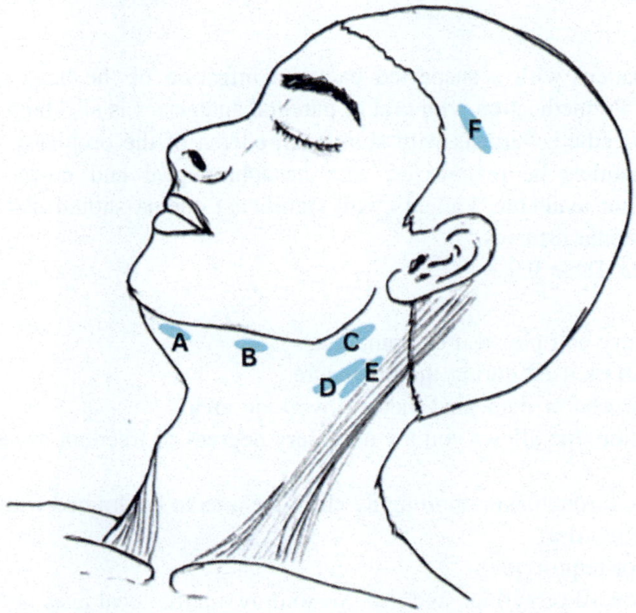


**FIGURE 9.5** Common incision and drainage sites for deep fascial space infections.
A: Submental; B: submandibular; C: submandibular, submasseteric, and pterygomandibular; D: lateral pharyngeal; E: retropharyngeal; F: temporal.
**Source:** Created by Amanda N Prine, MPAS, PA-C, ATC.

larger oral and neck abscesses often requires prolonged courses of antibiotics and has a higher rate of treatment failure. Untreated or poorly managed abscesses of the sublingual and submandibular spaces and oropharynx can result in upper airway obstruction requiring intubation, which may or may not be an available medical capability in spaceflight.

Terrestrially, needle aspiration of an abscess is often classified as a separate procedure when compared to incision and drainage. Needle aspiration results in less pain and bleeding than traditional incision and drainage, but it removes the ability to irrigate or pack the abscess cavity. For the purposes of performing this procedure in spaceflight, needle aspiration will be combined with incision and drainage in a way that reduces free-floating fluids and can avoid the need for powered suctioning devices.

Performance in Austere Environments: Uses of incision and drainage to resolve oral abscesses amongst astronauts aboard SkyLab have reported 100% efficacy rates without abscess recurrence (Johnston and Dietlein, 1977). Similarly, reports of incision and drainage efficacy within the British Polaris submarine patrol have documented successful resolution of oral and neck abscesses without significant complications in such austere environments (Glover and Taylor, 1981; Tansey and Wilson, 1979).

Indication for Spaceflight: Abscesses of the head and neck ≥2 cm that are in a location easily reachable by an 18G needle

Relevant Anatomy:

- For oral and sublingual abscesses:
- Maxillary and mandibular alveolus
- Gingiva
- Teeth
- Tongue
- Floor of mouth
- Submandibular gland ducts and ostia (Wharton's ducts)
- For peritonsillar abscesses:
- Oropharynx
- Tonsils
- Tonsillar pillars
- Soft palate
- Uvula
- For submandibular space abscesses:
- Border of the mandible
- Angle of the mandible
- Marginal mandibular nerve

Preprocedural Care: Any patient with a suspected bacterial infection of the head and neck should be started on antibiotics as soon as possible. Premedication with oral or parental analgesics is also helpful, though local anesthesia will still be required during the procedure. Patients with significant edema of the oropharynx or submandibular region and airway compromise can sometimes be temporized with nasopharyngeal and oropharyngeal airways, respectively, especially when intubation is not available. Patients with significant edema should also receive high-dose steroids for at least 24 hours or until the edema resolves.

Supplies and Equipment: See Table 9.9

Instructions:

1. Confirm diagnosis via history and physical examination.
2. Obtain all supplies that will be used during the procedure.
3. Restrain the patient in their seat or dedicated medical working area.
4. Restrain yourself in a position that allows you the necessary degrees of freedom to perform the procedure safely and effectively.
5. If performing the procedure through skin, thoroughly clean the area to be drained with the cleaning solution available and allow the solution to fully dry.
   a. **NOTE**: Mucosa does not require prep.
6. Attach the 18G needle to the 10 cc syringe and use this to draw up the local anesthetic (generally 2–4 cc is required depending on abscess location and size).
7. Cap the 18G needle and remove from the syringe, taking care not to depress the plunger.
8. Attach the 25G needle.

**TABLE 9.9** Optimal and minimal supplies and equipment for the incision and drainage of oral and neck abscesses.

| Optimal | Minimum |
|---|---|
| Cleansing solution (e.g., alcohol prep pad, Chloraprep, povidone iodine) (skin only) | Nonflammable cleansing solution (e.g., povidone iodine; skin only) |
| Local anesthesia (e.g., 1% lidocaine with 1:100,000 epinephrine) | Local anesthesia (e.g., 1% lidocaine with 1:100,000 epinephrine) |
| 10 cc syringe (for local anesthetic) | 10 cc syringe (for local anesthetic and aspiration) |
| 18G needle or blunt-tipped draw needle (for local anesthetic) | 18G needle (for local anesthetic and aspiration) |
| 25G needle | 25G needle |
| 10 cc syringe (for needle aspiration) | 11-blade scalpel |
| 18G needle (for needle aspiration) | Curved hemostat (e.g., Kelly) |
| 11-blade scalpel | Gauze |
| Curved hemostat (e.g., Kelly) | ¼″ iodoform or strip gauze |
| Culture swab | Tongue blades (oral or oropharyngeal abscesses) |
| Sterile saline | |
| 10 cc syringe | |
| 18G angiocath | |
| Gauze | |
| ¼″ iodoform or strip gauze | |
| Tongue blades (oral or oropharyngeal abscesses) | |
| Yankauer suction | |
| Suction and suction tubing | |

9. Palpate the most fluctuant area of the abscess.
   a. **TIP**: For abscesses that are difficult to palpate due to location or depth, use an ultrasound probe to locate the fluid collection and decide the best site for drainage.
   b. **TIP**: For peritonsillar abscesses, three standard drainage sites are attempted in the following order: superior pole, middle pole, and inferior pole.
   c. **TIP**: For intraoral abscesses, use tongue blades to retract the tongue or oral commissure.
10. Inject the local anesthetic into the area to be drained. Ensure that you are injecting a wide enough area to encompass all potential drainage sites and the 1–1.5 cm incision that will be created later.
11. Cap the 25G needle and remove from the syringe. Waste the remaining local anesthetic into a gauze or bag.
12. Reattach the 18G needle and uncap. Ensure the plunger is fully depressed.
13. Insert the needle into the selected drainage site and begin to slowly pull the plunger of the syringe.
    a. **TIP**: If no purulence is encountered, gently advance or withdraw the needle while changing directions slightly and applying constant backwards pressure to the plunger.
14. When no more fluid is draining into the syringe, remove the needle and syringe and cap the needle.
15. Obtain the scalpel and retract the guard.
16. Use the scalpel to create a 1–1.5 cm incision that includes the initial aspiration site. For skin, incise through the epidermis and dermis into the subcutaneous tissues. For the mucosa, incise the mucosal layer.
17. Insert the curved hemostat, tips closed, into the incision, and apply gentle pressure to bluntly enter the abscess cavity.
18. Use the hemostat to break up loculations by inserting it into the cavity with tips closed, spreading, then coming out with tips open. Repeat this in several directions.
    a. **CAUTION**: Never close the tips of the hemostat while it's in the abscess pocket, as it can inadvertently close over neurovascular structures and cause injury or bleeding.

19. Apply gauze over the drainage site and massage the site to express any residual purulence (skin only).
20. If irrigation capabilities are available, proceed with irrigation of the abscess pocket per Chapter 8: Irrigation.
21. If the abscess is >2 cm, use the hemostat to pack the drained abscess pocket with strip gauze, leaving a tail of at least 3 cm outside of the wound (skin only). Note the length of packing inserted in centimeters.
22. When the procedure is finished, clean supplies (if they are to be reused) and restow or dispose of single-use supplies in the designated location.

Postprocedural Care: Antibiotics are required following incision and drainage of any abscess. Intraoral and neck abscesses tend to be multiorganismal, so it is important to utilize broad-spectrum antibiotics that cover normal oral flora, including aerobes, facultative anaerobes, and anaerobes. Amoxicillin-clavulanic acid is a good first-line antibiotic, with ampicillin-sulbactam as an intravenous alternative. Pseudomonal coverage is generally not required. While uncommon, methicillin-resistant *Staphylococcus aureus* is a possible causative organism. In these cases, doxycycline is a great alternative to amoxicillin-clavulanic acid as it is a broad-spectrum antibiotic with activity against MRSA and normal oral flora. Trimethoprim-sulfamexazole is another option, though it lacks robust Gram-negative coverage and generally must be combined with another antibiotic in this setting. Duration for antibiotic treatment should be 7–10 days, depending on the size and location of the abscess.

Packing the abscess cavity with packing material such as strip or Iodoform gauze allows for the drainage site to remain open to prevent reaccumulation of purulence, as well as assist in the contraction and healing of the space. The packing should be removed gradually, 1–5 cm per day depending on the amount that was inserted, with a goal of at least 72 hours up to 5 days for complete removal.

Pearls and Pitfalls: Diagnosis of an abscess can be challenging. The natural history of soft tissue infections generally begins with a cellulitis before progressing to phlegmon followed by abscess. In the context of the oral cavity, many abscesses are odontogenic in origin, progressing from periapical abscess into subperiosteal or deep neck space abscesses after direct spread via bony erosion. In the spaceflight environment where procedures are difficult and resources are limited, it will be incredibly important to avoid unnecessary procedures, including attempted incision and drainage of phlegmon that has not yet matured into a formal abscess, or small abscess (<2 cm) that is likely to resolve with conservative management. While history and physical exam, including identifying overlying skin or mucosal changes and palpating for fluctuance, is the crux of diagnosis, ultrasound can be used as an adjunct when the diagnosis or need for a procedure is uncertain. Ultrasound is also useful for the aspiration or incision and drainage of deeper abscesses, especially when there is significant soft tissue inflammation and induration without a clear window into the abscess.

Pain is the most common complication resulting from this procedure. Other possible complications include bleeding, abscess recurrence, damage to surrounding anatomical structures, and scarring. Inadequately treated abscesses may continue to expand, spread to additional spaces, or result in sepsis.

### 9.4.3 Cricothyrotomy/tracheostomy

Surgical airways are relatively common in the terrestrial setting, but are not anticipated to be a necessary or included capability for spaceflight in the short-term. However, we have included this section to review emergent surgical airway techniques that can be adapted for future exploration-class missions. This section will review the two main types of surgical airways before focusing on the anatomical knowledge and technique required to perform a cricothyrotomy. While there are different methods of performing cricothyrotomy, including percutaneous techniques that utilize kits, this chapter will focus on open and percutaneous needle methods to avoid reliance on specialized equipment.

Of note, while reading this chapter, it is important to pay attention to the terminology used to describe the procedure versus the result. When discussing the procedure, the suffix "-otomy" is used to denote the creation of a tract between two surfaces, that tract then being referred to with the suffix "-ostomy." For example, one performs a tracheotomy, which results in the creation of a tracheostomy.

Terrestrial Indications: In cases of oral, oropharyngeal, hypopharyngeal, or laryngeal obstruction, cervical spine trauma, orofacial anatomic abnormalities, or patients with oral, maxillofacial, or neck trauma where urgent airway management is needed but endotracheal intubation has failed or is not possible, a cricothyrotomy is indicated to establish a patent airway. In prehospital care settings, cricothyrotomy rates range from approximately 0.06% to 0.72%, whereas rates range from 0.14% to 1.4% in the emergency department (McMahon et al., 2025). A cricothyrotomy is often converted to a more secure tracheostomy, both for airway security and to avoid tracheal stenosis within 24–48 hours. However, in the spaceflight environment, a cricothyrotomy may be used for a prolonged period of time (up to a few days) to avoid a tracheostomy if a patient is likely to return to Earth in that time frame. In an emergent situation with acute loss of airway, a cricothyrotomy is preferred because the cricothyroid membrane is relatively fixed and is just underneath the skin with little tissue intervening, as opposed to the trachea farther down, which has intervening thyroid and muscle tissue requiring careful dissection. A tracheotomy may be conducted either as an emergent or elective surgery. An emergent tracheotomy should also be performed in the case of penetrating laryngeal trauma. A patient may undergo an elective tracheotomy if they have experienced prolonged ventilator dependence, chronic aspiration, or obstructive sleep apnea unresponsive to other interventions. An elective tracheotomy may also be conducted prophylactically for head and neck cancer or in patients with neuromuscular disease. More than 100,000 tracheotomies are performed annually (Raimonde et al., 2025).

Efficacy: Cricothyrotomies used in situations requiring emergent airway management have been associated with success rates up to 83.3% (DeVore et al., 2019). In patients undergoing a tracheotomy for acute respiratory failure, post-cricothyrotomy, or in cases of neuromuscular weakness, studies report a mortality rate of 19%, of which only 57% of survivors are freed from mechanical ventilation. This is mostly due to the underlying pathophysiology associated with needing a tracheostomy, rather than the tracheostomy itself; however, a tracheotomy is potentially a complicated and morbid procedure. In situations of penetrating laryngeal trauma, a tracheostomy is associated with shorter ICU stays and better airway edema management to prevent additional injury. Longer mean intubation times and bleeding in 11.1% of cases is found in situations where tracheotomy is used for LeFort fractures. In cases of obstructive sleep apnea, tracheotomy leads to significant decreases in mean apnea index, apnea-hypopnea index, and oxygen desaturation index scores with an overall reduction in mortality (1.7%) compared to those not treated with tracheotomy (13.8%). A significant reduction in secretion burden and the volume of aspiration is also seen in patients who undergo a tracheotomy for purposes of relieving chronic aspiration (Camacho et al., 2014). And, amongst head and neck cancer patients who also underwent a tracheotomy, a 3.2% mortality rate has been reported with a 9% readmission rate.

Alternatives and Consequences: Assuming every other method of establishing a definitive airway has been attempted and failed, cricothyrotomy or tracheotomy is the last resort to secure the airway. The alternative for someone with an acute upper airway obstruction is death. However, there are complications associated with cricothyrotomy/tracheotomy, including bleeding, infection, failure to secure the airway, iatrogenic injury to other structures in the neck, including the vocal cords, nerves, carotid artery, and internal jugular vein.

Performance in Austere Environments: Emergent cricothyrotomy use in military battlefields reveals an overall 62% success rate, of which 33% were incorrectly placed (Hessert and Bennett, 2013). Case reports of awake tracheotomies in austere settings also present the procedure as a safe and overall successful alternative in establishing an open airway in cases of severe facial trauma (Choi, 2016).

Indication for Spaceflight: Acute upper airway obstruction when intubation from above is not available or not possible (a "can't intubate, can't oxygenate" or CICO condition)

Relevant Anatomy, see Fig. 9.6:

- Hyoid bone
- Thyroid cartilage
- Thyroid cartilage notch
- Cricoid cartilage
- Cricothyroid membrane
- Trachea
- Sternal notch
- Anterior jugular veins
- Strap muscles
- Thyroid gland

Preprocedural Care: Cricothyrotomy is generally performed in an emergent setting, thus there may be little to no preprocedural care. If time allows, IV access with at least one, ideally two, large-bore (18G or larger) should be established, and the skin of the neck should be prepped with an antiseptic solution. Patients in acute respiratory distress also benefit from supplemental oxygen, especially if administered prior to significant obstruction, as breathing 100% oxygen preoxygenates the patient and allows more time to complete the procedure.

Supplies and Equipment:

Supplies and Equipment (Open): Table 9.10

Instructions (Open):

**NOTE**: If intubation capability is available, attempt intubation first (See: Chapter 7: Resuscitation and Critical Care). Only proceed with cricothyrotomy if intubation fails or is not possible.

1. Obtain all supplies that will be used during the procedure.
2. Restrain the patient in their seat or a dedicated medical working area.

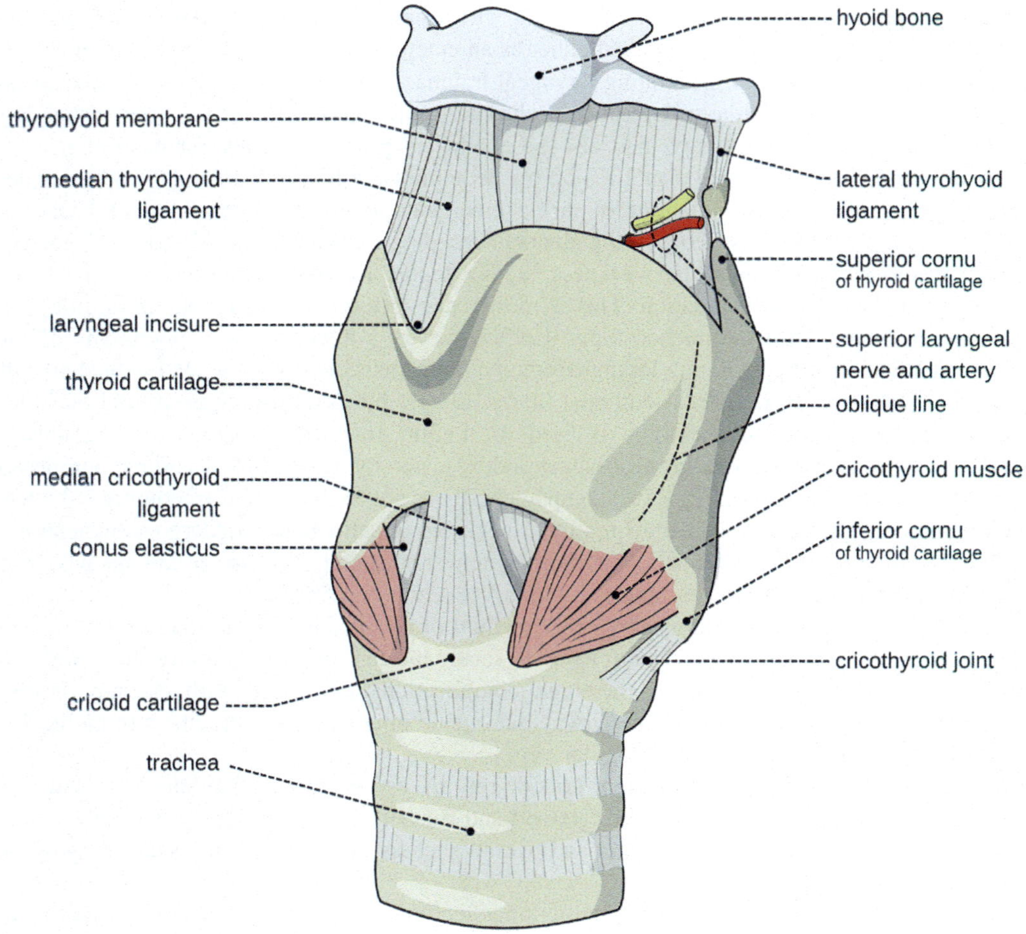


**FIGURE 9.6** Relevant anatomy for cricothyrotomy.
The cricothyroid ligament is between the thyroid and cricoid cartilage.
**Source:** Olek Remesz (wiki-pl: Orem, commons: Orem)—Larynx external en.svg.

3. Restrain yourself in a position that allows you the necessary degrees of freedom to perform the procedure safely and effectively.
4. Don gloves.
5. Prep skin using the available cleaning solution.
6. Palate the relevant landmarks, including the thyroid cartilage, thyroid cartilage notch, cricoid, and cricothyroid membrane.
   a. **TIP**: If laryngeal landmarks are not easily palpable, ultrasound can be used to assist with the identification of the cricothyroid membrane.
7. If time allows, proceed with the injection of local anesthesia. Otherwise, skip to step 13.
   a. **NOTE**: If anticipating a cricothyrotomy, it can be helpful to inject the local anesthetic during an intubation attempt. Otherwise, we generally proceed without it in an emergency scenario.
8. Attach the 18G needle to the 10 cc syringe and use this to draw up the local anesthetic.
9. Cap the 18G needle and remove from the syringe, taking care not to depress the plunger.
10. Attach the 25G needle.
11. Inject the local anesthetic over the area of the cricothyroid membrane. Ensure that you are injecting a wide enough area to encompass a 3 to 5 cm vertical incision.
12. Cap the 25G needle and syringe.
    a. **TIP**: Keep the needles and syringe nearby. A fluid-filled syringe can be used to locate the airway if there is difficulty with direct visualization.
13. Obtain the scalpel and retract the guard.

**TABLE 9.10 Supplies and equipment for open cricothyrotomy.**

| Optimal | Minimum |
|---|---|
| Cleansing solution (e.g., alcohol prep pad, Chloraprep, povidone iodine) (skin only) | Nonflammable cleansing solution (e.g., povidone iodine; skin only) |
| Local anesthesia (e.g., 1% lidocaine with 1:100,000 epinephrine) | Local anesthesia (e.g., 1% lidocaine with 1:100,000 epinephrine) |
| 10 cc syringe (for local anesthetic and aspiration) | 10 cc syringe (for local anesthetic and aspiration) |
| 18G needle or blunt-tipped draw needle (for local anesthetic and airway location) | 18G needle (for local anesthetic and airway location) |
| 25G needle | 25G needle |
| 15-blade scalpel | 15-blade scalpel |
| Cricoid/tracheal hook | 5.5 endotracheal tube |
| Trousseau dilator | Gauze |
| Curved or straight hemostat (e.g., Kelly) | Bougie |
| Cricothyrotomy tube or 5.0 endotracheal tube | |
| Sterile saline | |
| Gauze | |
| Yankauer suction | |
| Suction and suction tubing | |
| Bougie | |

14. Use the scalpel to create a 3–5 cm vertical incision through the skin and subcutaneous tissues over the cricothyroid membrane. This incision can be lengthened as needed for access.
15. Identify the cricothyroid membrane. This can be identified by feel as the cricothyroid membrane will feel like a divot between the fixed and hard thyroid and cricoid cartilage.
    a. **TIP**: Aspiration into a saline or other fluid-filled syringe can be used to locate the airway in the event of bleeding, soft tissue obstruction, or distorted anatomy. Place an 18G needle onto a 10 cc syringe that is half full (5 cc) with fluid (saline or lidocaine). Insert the needle into the suspected airway and pull back on the plunger. The presence of bubbles indicates that the needle is in the airway.
16. Use the scalpel to create a 1 cm horizontal incision through the cricothyroid membrane. Close the scalpel guard.
17. Insert the bougie into the incision in the membrane, passing it down into the trachea.
    a. **NOTE**: If a bougie is unavailable, a hemostat can be used instead, with the tube inserted between the blades of the hemostat.
18. Place the tube over the bougie and advance the tube over the bougie into the trachea.
    a. **CAUTION**: Care is needed to ensure the tube is deep enough that the balloon and approximately 3–5 cm of tube above the balloon are in the trachea. Advancing the tube too far can result in mainstem intubation.
19. Inflate the cuff of the tube using an empty 10 cc syringe, then remove the bougie, careful ot to move the tube.
    a. **TIP**: The syringe used for the local anesthetic can be emptied and reused for this purpose.
20. Connect the tube to a ventilatory circuit or bag.
21. Confirm tube position via observation of chest rise, auscultation of breath sounds, and end tidal $CO_2$ if available.
22. Once the position is confirmed, carefully remove the hemostats while applying counterpressure to the tube.
23. Secure the tube around the neck with soft ties.
    a. **NOTE**: The ties should be tightened so that the operator can comfortably fit two fingers beneath them (“two fingerbreadths”).
24. Continue resuscitative efforts.

**25.** When the procedure is finished and the patient is stable, dispose of sharps per sharps disposal procedure, clean supplies (if they are to be reused), and restow or dispose of single-use supplies in the designated location(s).

Supplies and Equipment (Needle) Table 9.11:
Instructions (Needle)

**NOTE**: If intubation capability is available, attempt intubation first. Only proceed with cricothyrotomy if intubation fails or is not possible.

**CAUTION**: A needle cricothyrotomy is nearly always insufficient to ventilate the patient and can be used for oxygenation only for less than 20 minutes while a more secure airway is placed.

**1.** Obtain all supplies that will be used during the procedure.
**2.** Restrain the patient in their seat or dedicated medical working area.
**3.** Restrain yourself in a position that allows you the necessary degrees of freedom to perform the procedure safely and effectively.
**4.** Don gloves.
**5.** If time allows, proceed with skin preparation using the available cleaning solution.
**6.** Draw up 5 cc of sterile saline using the 10 cc syringe.
**7.** Attach the 14G angiocatheter with needle to the 10 cc syringe.
**8.** Palpate the relevant landmarks, including the thyroid cartilage, thyroid cartilage notch, cricoid, and cricothyroid membrane.
  **a. TIP**: If laryngeal landmarks are not easily palpable, ultrasound can be used to assist with the identification of the cricothyroid membrane.
**9.** Insert the 14G angiocatheter with needle into the airway through the cricothyroid membrane.
  **a. NOTE**: Ensure the needle is pointing towards the feet and inserted at a 30–45 degree angle from the skin.
**10.** Aspirate by pulling back on the plunger as the needle is advanced. The presence of bubbles indicates that the needle is in the airway.
**11.** Advance the catheter over the needle until the hub is flush with the skin.
**12.** Remove the needle and syringe.
**13.** Cap the 14G needle and syringe.
  **a. NOTE**: If oxygen tubing is available, proceed to step 14.
  **b. NOTE**: If 3 cc syringe, 7.0 endotracheal tube connector, and ventilation bag are available, skip to step 15.
**14.** Connect the oxygen tubing to the angiocatheter via a luer-lock connector and begin oxygenating the patient. Secure the angiocatheter with tape or other adhesive material.
**15.** Remove the plunger from the 3 cc syringe.
**16.** Connect the 7.0 endotracheal tube connector to the flanged end of the 3 cc syringe.

**TABLE 9.11 Supplies and equipment for needle cricothyrotomy.**

| Optimal | Minimum |
|---|---|
| Cleansing solution (e.g., alcohol prep pad, Chloraprep, povidone iodine; skin only) | Nonflammable cleansing solution (e.g., povidone iodine; skin only) |
| 14G angiocatheter | 14G angiocatheter |
| 10 cc syringe | 10 cc syringe |
| 3 cc syringe | 3 cc syringe (optional) |
| 7.0 endotracheal tube connector | 7.0 endotracheal tube connector (optional) |
| Ventilation bag | Ventilation bag (optional) |
| Oxygen tubing | Oxygen tubing |
| Sterile saline | Sterile saline |
| Gauze | Gauze |

17. Connect the angiocatheter to the luer-lock end of the 3 cc syringe.
18. Connect the ventilation bag to the 7.0 endotracheal tube connector and ventilate the patient.
19. Continue resuscitative efforts.
20. When the procedure is finished and the patient is stable, dispose of sharps per sharps disposal procedure, clean supplies (if they are to be reused), and restow or dispose of single-use supplies in the designated location(s).

Postprocedural Care: A cricothyrostomy is not considered a secure airway and is meant to act as a temporizing measure to the more definitive tracheostomy. Needle cricothryostomies in particular are generally only used for short-term oxygenation, as carbon dioxide elimination is significantly limited and should be replaced by a surgical cricothyrotomy as soon as possible. If a patient is unconscious, then continual ventilation is required, either by hand via a ventilation bag or through the use of a formal ventilator if available. The tube should be secured to the neck to prevent accidental dislodgement, and the underlying cause of the upper airway obstruction should be treated.

Pearls and Pitfalls:

The laryngeal airway can feel deceptively superficial on examination and can become difficult to locate as the dissection proceeds, especially if significant bleeding is encountered. Neck extension via placement of a shoulder roll or dropping the head of the bed with relation to the rest of the table are techniques used terrestrially to bring the airway more anteriorly. In a microgravity environment, this could be accomplished via placement of a shoulder roll with securement of the head and chest to force neck extension. A cricoid or tracheal hook, while not mandatory, would likely be very helpful in this environment, specifically given the challenges with positioning. The above procedures were written assuming a cricoid hook would be unavailable; however, in any mission where cricothyrotomy is flown as a capability, it would be wise to include one.

The most common complication of a cricothyrotomy is bleeding. Severe hemorrhage occurs in up to 50% of cases, which poses risks of aspiration or asphyxiation. This risk is mitigated by utilizing midline vertical incisions through the skin and soft tissues to avoid the anterior jugular veins. If the tube is advanced too far or deviates from the tracheal lumen, endobronchial intubation or false tract formation may result. Other cricothyrotomy complications include fracturing local structures (i.e., thyroid and cricoid cartilage) or unintentional tracheostomy, which may result from misplaced incisions. Poor technique during a cricothyrotomy may also lead to posterior tracheal injury or hypoxia. Late complications of a cricothyrotomy include possible infection, scarring, subglottic stenosis, dysphagia, fistula formation, and voice changes.

Tracheostomy complications can be grouped into operative, early postoperative, and late postoperative complications. The most common operative complication is bleeding. Other outcomes include an airway fire (caused by the electrocautery unit's proximity to high volumes of oxygen in the anesthetic tubing), pneumothorax, or pneumomediastinum (caused by misplacement of the tracheostomy tube). Early postoperative complications following a tracheostomy include infection or obstruction of the tracheostomy tube by blood or mucus. Acute obstruction may also result if the tube is dislodged. Late complications post-tracheostomy include pressure necrosis, subglottic stenosis, and tracheal stenosis. Though rare, other possible and devastating complications include tracheoesophageal fistulas and tracheoinnominate fistulas. A persistent tracheocutaneous fistula may also develop if granulation tissue persists after removal of the tracheostomy tube.

## References


Alexander, D.J., 2020. Illnesses seen in spaceflight. In: Young, L.R., Sutton, J.P. (Eds.), Handbook of Bioastronautics. Springer International Publishing, pp. 1–15. Available at https://doi.org/10.1007/978-3-319-10152-1_133-1.

Antonsen, E, 2019. Human System Risk in Spaceflight. NASA Johnson Space Center.

Atighechi, S., Karimi, G., 2009. Serial nasal bone reduction: a new approach to the management of nasal bone fracture. Journal of Craniofacial Surgery 20 (1), 49–52. Available at https://doi.org/10.1097/SCS.0b013e318190def5, http://journals.lww.com/jcraniofacialsurgery.

Bartkiw, T.P., Pynn, B.R., Brown, D.H., 1995. Diagnosis and management of nasal fractures. International Journal of Trauma Nursing 1 (1), 11–18. Available at https://doi.org/10.1016/S1079-2104(05)80405-6.

Beck, R., Sorge, M., Schneider, A., Dietz, A., 2018. Current approaches to epistaxis treatment in primary and secondary care. Deutsches Arzteblatt International 115 (1-2), 12–22. Available at https://doi.org/10.3238/arztebl.2018.0012, https://www.aerzteblatt.de/archiv/195549/Therapiekonzepte-der-Epistaxis-in-Praxis-und-Klinik.

Bogomolov, V.V., Polyakov, A.V., Matsnev, E.I., Popova, I.I., Kovachevich, I.V., Alferova, I.V., et al., 2021. Diagnosis and treatment of ENT diseases in russian cosmonauts during missions to the orbital station MIR and ISS. Human Physiology 47 (7), 810–814. Available at https://doi.org/10.1134/s0362119721070021.

Breeze, J., Bryant, D., 2009. Current concepts in the epidemiology and management of battlefield head, face and neck trauma. Journal of the Royal Army Medical Corps 155 (4), 274–278. Available at https://doi.org/10.1136/jramc-155-04-07.

Brennan, J.A., Holt, G.R., Thomas, R.W., 2015. Otolaryngology/head and Neck Surgery Combat Casualty Care in Operation Iraqi Freedom and Operation Enduring Freedom. Office of the Surgeon General, United States Army.

Brito, T.P., Hazboun, I.M., Fernandes, F.L., Bento, L.R., Zappelini, C.E.M., Chone, C.T., et al., 2017. Abscessos cervicais profundos: estudo de 101 casos. Brazilian Journal of Otorhinolaryngology 83 (3), 341–348. Available at https://doi.org/10.1016/j.bjorl.2016.04.004, http://www.journals.elsevier.com/brazilian-journal-of-otorhinolaryngology/.

Camacho, M., Certal, V., Brietzke, S.E., Holty, J.-E.C., Guilleminault, C., Capasso, R., 2014. Tracheostomy as treatment for adult obstructive sleep apnea: a systematic review and meta-analysis. The Laryngoscope 124 (3), 803–811. Available at https://doi.org/10.1002/lary.24433.

CDC, 2011. Estimated burden of acute otitis externa—United States, 2003–2007. Morbidity and Mortality Weekly Report 60 (19), 605–609. Available at https://www.cdc.gov/mmwr/preview/mmwrhtml/mm6019a2.htm.

Chidlow, C., van Bockxmeer, J., 2019. Otitis externa-climatic associations and evidence-based management strategies for Australian practice. Australian Journal of Rural Health 27 (3), 251–256. Available at https://doi.org/10.1111/ajr.12516.

Choi, Y.U., 2016. Awake tracheostomy in an austere setting. In: Lim, C. (Ed.), Surgery During Natural Disasters, Combat, Terrorist Attacks, and Crisis Situations. Springer, Cham, pp. 73–76. Available at https://doi.org/10.1007/978-3-319-23718-3_7.

Cramer, J.D., Purkey, M.R., Smith, S.S., Schroeder, J.W., 2016. The impact of delayed surgical drainage of deep neck abscesses in adult and pediatric populations. Laryngoscope 126 (8), 1753–1760. Available at https://doi.org/10.1002/lary.25835, http://onlinelibrary.wiley.com/journal/10.1002/(ISSN)1531-4995.

Dalal, P.J., Purkey, M.R., Price, C.P.E., Sidle, D.M., 2020. Risk factors for auricular hematoma and recurrence after drainage. Laryngoscope 130 (3), 628–631. Available at https://doi.org/10.1002/lary.28310, http://onlinelibrary.wiley.com/journal/10.1002/(ISSN)1531-4995.

DeVore, E.K., Redmann, A., Howell, R., Khosla, S., 2019. Best practices for emergency surgical airway: a systematic review. Laryngoscope investigative otolaryngology 4 (6), 602–608. Available at https://doi.org/10.1002/lio2.314.

Du, Guo, Huang, Liu, Zhang, Lu, 2025. Diagnostic accuracy of deep learning-based algorithms in laryngoscopy: a systematic review and meta-analysis. European Archives of Oto-Rhino-Laryngology: Official Journal of the European Federation of Oto-Rhino-Laryngological Societies (EUFOS): Affiliated with the German Society for Oto-Rhino-Laryngology - Head and Neck Surgery 282 (1), 351–360. Available at https://doi.org/10.1007/s00405-024-09049-2.

Fouts, B., Mortimer, D., 2018. Air Force Research Laboratories Stresses of Flight during Aeromedical Transport: An Integrated Review.

Glover, S.D., Taylor, E.W., 1981. Surgical problems presenting at sea during 100 British Polaris submarine patrols. Journal of The Royal Naval Medical Service 67 (2), 65–69. Available at https://doi.org/10.1136/jrnms-67-65.

Gupta, G., 2022. NIH Acute Laryngitis.

Hessert, M.J., Bennett, B.L., 2013. Optimizing emergent surgical cricothyrotomy for use in austere environments. Wilderness & Environmental Medicine 24, 53–66. Available at https://doi.org/10.1016/j.wem.2012.07.003.

Ho, C.Y., Wang, Y.C., Chin, S.C., Chen, S.L., 2022. Factors creating a need for repeated drainage of deep neck infections. Diagnostics 12 (4). Available at https://doi.org/10.3390/diagnostics12040940, https://www.mdpi.com/2075-4418/12/4/940/pdf.

Huang, G.J., Lin, B.R., Li, P.S., Tang, N., Fan, Z.J., Lu, B.Q., 2024. The global burden of otitis media in 204 countries and territories from 1992 to 2021: a systematic analysis for the Global Burden of Disease study 2021. Frontiers in Public Health 12. Available at https://doi.org/10.3389/fpubh.2024.1519623, http://journal.frontiersin.org/journal/public-health/section/public-health-education-and-promotion#archive.

Johnston, R.S., Dietlein, L.F., 1977. Biomedical Results from Skylab. NASA Lyndon B. Johnson Space Center.

Ju, N.R., Koyfman, A., 2015. Do oral antibiotics after incision and drainage of simple abscesses improve cure rates? Annals of Emergency Medicine 65 (1), 112–113. Available at https://doi.org/10.1016/j.annemergmed.2014.08.007, http://www.elsevier.com/inca/publications/store/6/2/3/2/7/8/index.htt.

Konishi, M., Sivalingam, S., Shin, S.H., Vitullo, F., Falcioni, M., 2012. Effects of early commercial air travel on graft healing rates after tympanoplasty. Annals of Otology, Rhinology and Laryngology 121 (2), 110–112. Available at https://doi.org/10.1177/000348941212100207, http://www.annals.com/toc/auto_article_process.php?year=2012&page=110&id=15816&sn=0.

Lian, Y., Yuan, K., Nie, L., Xiao, L., Ding, L., 2025. Evaluation of the safety and efficacy of manual removal and irrigation in cerumen removal in children. Science Progress 108 (2). Available at https://doi.org/10.1177/00368504251345886.

Llewellyn, A., Norman, G., Harden, M., Coatesworth, A., Kimberling, D., Schilder, A., et al., 2014. Interventions for adult Eustachian tube dysfunction: asystematic review. Health Technology Assessment 18 (46). Available at https://doi.org/10.3310/hta18460.

Lucha, P.A., Wallace, D., Pasque, C., Brickhouse, N., Olsen, D., Styk, S., et al., 2000. Surgical wound morbidity in an austere surgical environment. Military Medicine 165 (1), 13–17. Available at https://doi.org/10.1093/milmed/165.1.13, https://academic.oup.com/milmed/issue.

Mattos, J.L., Colman, K.L., Casselbrant, M.L., Chi, D.H., 2014. Intratemporal and intracranial complications of acute otitis media in a pediatric population. International Journal of Pediatric Otorhinolaryngology 78 (12), 2161–2164. Available at https://doi.org/10.1016/j.ijporl.2014.09.032, www.elsevier.com/locate/ijporl.

McMahon, K., Tariq, A., Morley, E., 2025. Cricothyroidotomy. StatPearls Publishing, Treasure Island (FL). Available at https://www.ncbi.nlm.nih.gov/books/NBK537350/.

Michaudet, C., Malaty, J., 2018. Cerumen impaction: diagnosis and management. American Family Physician 98 (8), 525–529. Available at https://www.aafp.org/pubs/afp/issues/2018/1015/p525.html#afp20181015p525-b1.

NIH - National Cancer Institute Cancer Stat Facts: Laryngeal Cancer, 2025. Available at https://seer.cancer.gov/statfacts/html/laryn.html

Omokanye, H.K., Alabi, S.B., Idris', S.O., Ayodele, S.O., Nasir, AA, Salaudeen, GA, et al., 2021. Diagnostic accuracy of flexible fiberoptic laryngoscopy: experience from a tertiary health institution in Nigeria. European Archives of Oto-Rhino-Laryngology: Official Journal of the European Federation of Oto-Rhino-Laryngological Societies (EUFOS): Affiliated with the German Society for Oto-Rhino-Laryngology - Head and Neck Surgery 278 (8), 2937–2942. Available at https://doi.org/10.1007/s00405-021-06824-3.

Parize, P., Chandesris, M.O., Lanternier, F., Poirée, S., Viard, J.P., Bienvenu, B., et al., 2009. Antifungal therapy of aspergillus invasive otitis externa: efficacy of voriconazole and review. Antimicrobial Agents and Chemotherapy 53 (3), 1048–1053. Available at https://doi.org/10.1128/AAC.01220-08France, http://aac.asm.org/cgi/reprint/53/3/1048.

Passali, D., Damiani, V., Passali, F.M., Tosca, M.A., Motta, G., Ciprandi, G., et al., 2020. An international survey on the pragmatic management of epistaxis. Acta Biomedica 91, 5–10. Available at https://doi.org/10.23750/abm.v91i1-S.9241, https://www.mattioli1885journals.com/index.php/actabiomedica/article/download/9241/8572.

Raghvi, A., Priya, K., Rajasekaran, S., Prabakaran, S., Navin, R.B.N., Balaji, D., et al., 2023. Evaluating the outcomes of closed reduction of different types of nasal bone fractures. Indian Journal of Otolaryngology and Head and Neck Surgery 75 (4), 2998–3006. Available at https://doi.org/10.1007/s12070-023-03894-z, https://www.springer.com/journal/12070.

Raimonde, A., Gaston, S., Wang, C., 2025. StatPearls [Internet]. Tracheostomy. StatPearls Publishing, Treasure Island (FL).

Raju, A.V.K., Saxena, S., Mohapatra, S.S., 2020. In situ grommet and fitness for flying. Indian Journal of Aerospace Medicine 64 (2), 100–104.

Rijk, M.H., Hullegie, S., Schilder, A.G.M., Kortekaas, M.F., Damoiseaux, R.A.M.J., Verheij, T.J.M., et al., 2021. Incidence and management of acute otitis media in adults: a primary care-based cohort study. Family Practice 38 (4), 448–453. Available at https://doi.org/10.1093/fampra/cmaa150, http://fampra.oxfordjournals.org/.

Rustad, J.L., Lipsky, A., Herrington, H., Weismann, A., 2025. Tele-flexible laryngoscopy enables remote specialty airway evaluation in rural emergency department. Laryngoscope. Available at https://doi.org/10.1002/lary.32425, http://onlinelibrary.wiley.com/journal/10.1002/(ISSN)1531-4995.

Sevy, J.O., Hohman, M.H., Singh, A., 2023. Cerumen Impaction Removal. NIH.

Spaw, M, Nikki, A, Macario, C, 2024. NIH Tympanostomy Tube Insertion.

Srinivasan, Y., Shamritsky, D., Bhatta, A., Chou, E., Pham, T., Sanghvi, Y., et al., 2024. Design and usability of an open-source, low-cost flexible laryngoscope for resource-limited settings. JAMA Otolaryngology - Head and Neck Surgery 150 (4), 342–348. Available at https://doi.org/10.1001/jamaoto.2024.0063, http://archotol.jamanetwork.com/issues.aspx.

Staffel, J.G., 2002. Optimizing treatment of nasal fractures. The Laryngoscope 112 (10), 1709–1719. Available at https://doi.org/10.1097/00005537-200210000-00001.

Steele, D.W., Adam, G.P., Di, M., Halladay, C.H., Balk, E.M., Trikalinos, T.A., 2017. Effectiveness of tympanostomy tubes for otitis media: a meta-analysis. Pediatrics 139 (6). Available at https://doi.org/10.1542/peds.2017-0125, http://pediatrics.aappublications.org/content/pediatrics/139/6/e20170125.full.pdf.

Stein, D.J., Noordzij, J.P., 2013. Incidence of chronic laryngitis. Annals of Otology, Rhinology and Laryngology 122 (12), 771–774. Available at https://doi.org/10.1177/000348941312201207, http://www.annals.com/toc/auto_article_process.php?year=2013&page=771&id=16085&sn=0.

Tansey, W.A., Wilson, J.M., Schaefer, K.E., 1979. Analysis of health data from 10 years of Polaris submarine patrols. Undersea Biomedical Research 6 Suppl.

Thirsk, R., Williams, D., Anvari, M., 2007. 4 IAA Humans in Space Symposium, Graz, Austria, pp. 512–517. NEEMO 7 undersea mission 60.

Urban, M.J., Masliah, J., Heyd, C., Patel, T.R., Nielsen, T., 2022. Peritonsillar abscess size as a predictor of medical therapy success. Annals of Otology, Rhinology and Laryngology 131 (2), 211–218. Available at https://doi.org/10.1177/00034894211015590, aor.sagepub.com/.

Whitworth, K., Johnson, J., Wisniewski, S., Schrader, M., 2020. Comparative effectiveness of topically administered tranexamic acid versus topical oxymetazoline spray for achieving hemostasis in epistaxis. The Journal of Emergency Medicine 58 (2), 211–216. Available at https://doi.org/10.1016/j.jemermed.2019.11.038.

Xu, F., Liu, C., Zhou, Y., Li, M., Guo, X., 2022. A test prototype of a novel flexible video laryngoscope and preliminary verification in a difficult airway management simulator. BioMedical Engineering OnLine 21 (1). Available at https://doi.org/10.1186/s12938-022-01043-1.

Yu, Q., Zhang, Y., Lu, H., Zhao, Y., Zhou, X., 2025. Incidence and prognosis of otitis media with effusion in COVID-19 patients a comparative case-control analysis. American Journal of Otolaryngology 46 (4), 104649. Available at https://doi.org/10.1016/j.amjoto.2025.104649.

Zhukhovitskaya, A., Verma, S.P., 2019. Identification and management of chronic laryngitis. Otolaryngologic Clinics of North America 52 (4), 607–616. Available at https://doi.org/10.1016/j.otc.2019.03.004, http://www.elsevier.com/inca/publications/store/6/2/3/1/7/0/index.htt.

Chapter 10

# Ophthalmic procedures

**Carlo Andre Canepa**[1] **and Elana Meer**[2]
[1]*Department of Emergency Medicine, Alaska Native Tribal Health Consortium, Anchorage, AK, United States,* [2]*Department of Ophthalmology, University of California San Francisco, San Francisco, CA, United States*

## 10.1 Introduction

Ocular injuries are common in spaceflight but are often more irritating than disabling. In a postflight symptom dataset drawing on Space Shuttle and ISS astronauts (N≈257), eye irritation was reported by ~34%, foreign body sensation by ~21%, and dry eyes by ~14% of astronauts. This is often a function of dust particles remaining suspended in the microgravity environment rather than settling on surfaces under the influence of gravity (Ong et al., 2025). In this setting, the most likely injury is a corneal abrasion, though chemical burns and ocular infections have also been reported (Meer et al., 2023). Traumatic injuries sustained from impacting a surface with the open or closed eye are also possible but are controlled mainly through engineering and design to avoid sharp, protruding corners. Within this context, the following procedures will be discussed as the most likely to be needed in the near-term future of spaceflight to manage ophthalmic injuries and illness:

1. Ocular irrigation (discussed in Chapter 8)
2. Fluorescein staining
3. Antibiotic/antifungal drop/ointment application
4. Tonometry
5. Corneal foreign object removal/rust ring removal
6. Lateral canthotomy and cantholysis
7. Eyelid laceration repair

One additional condition worth mentioning is Spaceflight-Associated Neuro-ocular Syndrome (SANS). SANS is a constellation of posterior-segment and neuro-ophthalmic findings, most often including optic disc edema, globe flattening, choroidal folds, and hyperopic refractive shifts that have been documented in astronauts during and after long-duration missions. Up to 70% of long-duration crew members exhibit one or more signs consistent with SANS, but the mechanisms and clinical significance of this condition remain uncertain. While some astronauts do report changes in visual acuity, likely from globe flattening, SANS has not yet shown a need for procedural intervention and is most often managed with corrective lenses (Nguyen et al., 2025).

## 10.2 Procedures

### 10.2.1 Fluorescein staining

Fluorescein staining is performed as part of the assessment for corneal injuries, infections, foreign objects, and globe injuries. It is a fluorescent yellow dye, often stored on dry paper strips, that is activated when the strip comes into contact with water, most often the tear film on the corneal surface. This is performed when an individual complains of significant eye pain without apparent injury, after encounters with an ocular foreign body, after ocular trauma, or after contact with a caustic chemical in the eye. Fluorescein strips are a routine part of spacecraft medical kits, and this procedure has been successfully performed in flight (Suh et al., 2025). Table 10.1 contains the necessary equipment for fluorescein staining.

**Procedural Medicine for Spaceflight. DOI:** https://doi.org/10.1016/B978-0-443-27552-4.00018-X

**TABLE 10.1** Minimum and optimal fluorescein stain supplies, mass, volume, and power requirements for the described supplies/equipment, as well as any other anticipated difficulties in their use (e.g., related to device communication, microgravity effects on the functioning of a device, etc.).

| Item | Mass | Volume | Power |
|---|---|---|---|
| ***Minimum necessary equipment*** | | | |
| Fluorescein strip/dye | 0.0008 kg | 60 mm × 7 mm | N/A |
| Cobalt blue light pen light | 0.0143 kg | 130 mm × 12 mm | 2 AAA battery |
| ***Best practice additions*** | | | |
| Topical ophthalmic anesthetic solution 4 mL | 0.02 kg | 40 mm × 15 mm | N/A |
| Handheld anterior segment camera (Volk Viva nonmydriatic handheld retinal camera) | 0.600 kg | 184.15 mm × 82.55 mm × 222.25 mm | 9 V lithium polymer battery |
| Portable slit lamp (microclear hyperion digital portable slit lamp) | 0.85 kg | 140 mm × 118 mm × 225 mm | 3.7 V lithium battery |
| Preservative-free artificial tears solution vial 10 mL | 0.02 kg | 66 mm × 20 mm | N/A |
| Balanced salt solution ophthalmic vial 15 mL | 0.005 kg | 66 m × 23.4 mm | N/A |

#### *10.2.1.1 Procedure*

1. Remove the fluorescein strip from the protective sheath.
2. Apply either 1–2 drops of topical anesthetic solution (such as Tetracaine) or, if unavailable, 1–2 drops of water or artificial tears directly onto the fluorescein strip tip. This should cause the drop of fluid to color. Tetracaine will cause a stinging sensation for a few seconds after administration that rapidly dissipates and will leave the surface numb for 10–30 min.
3. Gently apply the tip of the fluorescein strip to the lower, inner lid of the eye (the space between the lower eyelid and the globe). Avoid touching the cornea; sufficient staining will occur by touching fluorescein to the inferior fornix.
4. Have the patient blink a couple of times, and a yellow-green solution should spread across the cornea.
5. The eye can now be assessed for any corneal defects by shining a Wood's lamp or any cobalt blue light source over the cornea.
6. Any areas of more intense blue staining are indicative of a corneal injury.

#### *10.2.1.2 Postprocedural care specific to the procedure*

1. The fluorescein will briefly discolor the surface of the cornea and tears, but this should wash out within an hour. The contents can stain clothing or other soft goods so care should be taken to avoid contact with materials that should not be discolored.

#### *10.2.1.3 Common or important procedural complications and how they can be managed in the spaceflight environment*

1. There is an exceedingly rare but theoretically possible risk of causing a corneal abrasion by incorrectly applying the fluorescein strip to the surface of the eye. If this occurs, it can be treated the same as any other corneal abrasion with antibiotic eye drops and pain control. Since the point of using Tetracaine in many cases is to evaluate for eye pain and injury, and the risk is small (limited evidence to provide hard numbers but reported more often as a theoretical risk than an actual one), the risk is most often worth the potential benefit.
2. If the patient requires additional pain medications after the procedure, it is worth considering alternatives to Tetracaine anesthetics. While some studies do report safe and efficacious use of topical ocular anesthetics for up to 24 hours (Waldman et al., 2014), a thorough review of literature performed in 2023 found insufficient evidence to evaluate them

for effective pain control, complication rates, nor delayed healing effects.(Leslie et al., 2025). However, there is some evidence to suggest that corneal opacities and delayed healing may be increased with topical anesthetic use for longer than 72 hours (Leslie et al., 2025). So while it is clear that topical anesthetics used as ocular procedure pain control is safe and effective, use beyond the perioperative period may increase healing time and rates of corneal opacity. We recommend using it primarily for anesthesia during procedures and sparingly for pain control. Avoid regular use beyond 24 hours.

3. Another potential complication is that numbing the corneal surface will blunt the blink reflex, leaving the eye more susceptible to injury while the anesthesia remains active. For this reason, it is important for the patient to be particularly careful to avoid additional eye injuries (e.g., floating debris, self-induced from fingers or materials placed in the eye, etc.).

#### *10.2.1.4 Interpretation*

Fig. 10.1 illustrates examples of common fluorescein stain patterns in cobalt blue light.

### 10.2.2 Eye drop/ointment application (antibiotics, antifungal, corneal anesthesia)

Antibiotics are used in the eye for the same reason as they are used anywhere else in the body to manage or prevent infection. This is often done for the following indications:

1. Bacterial conjunctivitis
2. Foreign object embedded in and/or removed from the eye
3. Corneal abrasion
4. Corneal ulceration
5. Fungal keratitis

Topical ophthalmic antibiotic drops and ointment have been used to treat suspected corneal abrasions and corneal ulcerations in space (Waisberg et al., 2024). Ophthalmic drops have also been used to manage chronic glaucoma and elevated ocular hypertension, with some astronauts reporting difficulty with the sterile administration of eye drops in space (Dalal et al., 2021). In general, drops are easier to use than ointments because gravity does not prevent the more viscous ointment from separating, whereas aqueous solutions separate more easily when in contact with the tear film. This is not an absolute contraindication, especially when in gravity wells, but it is worth considering from an ease-of-use perspective. Table 10.2 contains a description of the minimum necessary materials for performing this procedure, along with estimated mass and volume. No electrical power is required for this procedure, though certain medications may require cold storage. In general, we recommend avoiding medications that require cold storage to limit power needs.

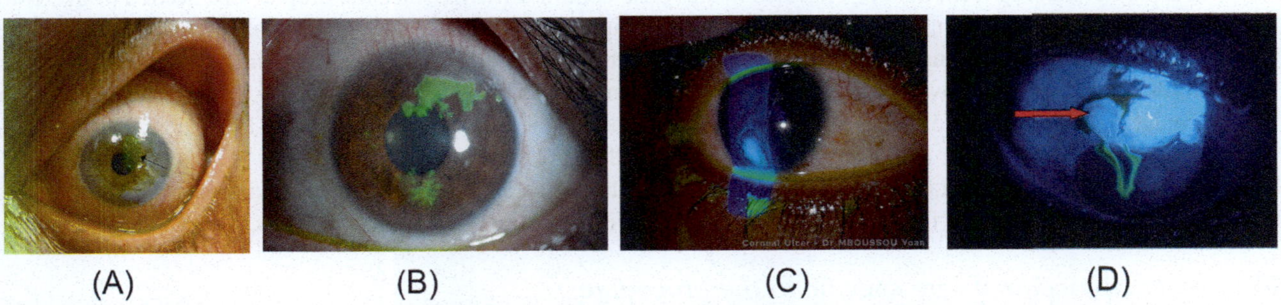


FIGURE 10.1 Examples of common fluorescein stain patterns cobalt blue light. (A) This file is licensed under the Creative Commons Attribution-Share Alike 3.0 Unported license. From James Heilman, MD via Wikimedia.org. (B) Prakash G., Avadhani K., Srivastava D. The three faces of herpes simplex epithelial keratitis: a steroid-induced situation. BMJ Case Reports. 2015 Apr 2;2015:bcr2014209197. https://doi.org/10.1136/bcr-2014-209197. PMID: 25837655; PMCID: PMC4401969. (C) This file is licensed under the Creative Commons Attribution-Share Alike 4.0 International license. From Yoanmb via Wikimedia.org. (D) Management of complications in glaucoma surgery. Vijaya L, Manish P, Ronnie G, Shantha B - Indian journal of ophthalmology (2011).

**TABLE 10.2 Antibiotic/antifungal drop/ointment supplies.**

| Item | Mass | Volume | Power |
|---|---|---|---|
| ***Minimum necessary equipment*** | | | |
| Ophthalmic antibiotic drops (polymyxin B and trimethoprim 10 mL vial) | 0.0015 kg | 22 mm × 55 mm | N/A |
| Ophthalmic antibiotic ointment (erythromycin 3.5 g tube) | 0.0035 kg | 90 mm × 13 mm | N/A |
| Antifungal drops (natamycin 5% 15 mL bottle) | 0.0035 kg | 25 mm × 75 mm | N/A |

#### *10.2.2.1 Procedure*

1. After diagnosing a corneal abrasion or bacterial conjunctivitis, apply 2–4 drops of antibiotic drops or a 1 cm strip of antibiotic ointment or antifungal ointment according to the dose schedule of the medication. This is often used every 2–4 hours because tear film quickly washes antibiotics off the surface of the eye.
2. Position the fluid in the bottle:
   a. Fluids within vials in microgravity tend to adhere to the walls of the vial due to surface tension. Swinging the arm with the bottle held by the tip, with the tip facing out, provides gentle centrifugal force and brings the fluid to the tip of the dropper.
3. Position the bottle near the patient's eye:
   a. Fluid droplets will easily fall off the tip with even gentle movement, creating a glob of medication debris. To avoid this, a provider must limit movement to the extent possible by positioning the bottle next to the patient's eye before expressing the droplet from the tip.
   b. Due to fluid adherence to the walls, expressing the droplet may require higher external pressure than typical on the ground. Providers should start with slight pressure and gradually increase until the desired droplet size is reached.
4. Titrate the droplet size by varying the squeeze applied to the bottle. More squeezing will create a larger droplet; less squeezing will cause some of the medication to reenter the bottle, making the droplet smaller.
5. Instruct the patient to look medially (towards their nose).
6. The expressed droplet should be applied to the lateral canthus (cheek side of the eye). Gently move the droplet until it contacts the lateral surface of the eye and begins to spread into the tear film.
7. Maintain a constant squeeze on the bottle while moving the bottle laterally (towards the patient's cheek). This should allow the droplet to catch on the inner surface of the eyelid and make separation easier.
   a. Since gravity will not be present to help the droplet separate from the tip, the provider must rely on surface tension to separate the drop from the tip. This also makes dose control difficult, since the droplet size needed to overcome the surface tension holding it to the bottle tip varies with available gravity. This will make inadvertent overdoses much more likely. Fortunately, excessive antibiotic in the eye has little consequence aside from using up limited stores of medication. Providers must be mindful of the number of drops released and make a visual estimate of the droplet size needed while avoiding adding extra pressure to the bottle (Dalal et al., 2021).
8. Avoid contamination of the eye dropper with contact of the patient's eyelid or lashes.
9. Once used, the droplet container should only be used with that patient to avoid cross-contamination.
10. Application of an ointment is similar, but the viscosity will make separation more difficult.

#### *10.2.2.2 Postprocedural care specific to the procedure*

Ensure no medication drops or ointment continues to float within the cabin, as they can become new foreign body material for the crew or damage sensitive electronic equipment.

#### *10.2.2.3 Common or important procedural complications and how they can be managed in the spaceflight environment*

Monitor for appropriate improvement in the infection. If worsening or spreading after 24 hours, the infection may not be sensitive to the selected antibiotic. Consider changing antibiotics, an alternative diagnosis, or even adding oral antibiotics, though these are less effective at reaching therapeutic concentrations in the cornea due to limited circulation. The duration of therapy should be directed by experts and will vary depending on the infection type, antibiotic used, and severity. As a rule of thumb, 5–10 days of treatment are required to clear most infections.

### 10.2.3 Tonometry

#### *10.2.3.1 Purpose*

The purpose of tonometry is to measure intraocular pressure (IOP). Normal IOP is less than 20 mmHg, with most readings centering around 15–16 mmHg. In space, a transient rise of 4–7 mmHg is expected in the first 24 hours, which typically resolves after a few days. This generally is not concerning unless the pressure is substantially higher than 20 mmHg and/or associated with symptoms or visual changes.

1. Suspicion of elevated IOP
2. Ocular trauma (without concern for a ruptured globe)
3. Concern for iritis (inflammation of the iris)

Table 10.3 contains a representative set of equipment for measuring intraocular pressure.

#### *10.2.3.2 Performance of the procedure in space or analog environments*

Self and provider assessed tonometry has been performed in space many times and documented in published literature from a German Spacelab mission in March 1992 (Draeger et al., 1995). A tonometer, along with a battery and tip covers, is part of the current ISS medical kit and was previously flown on the Space Shuttle (Dalal et al., 2021; Fig. 10.2) There do not appear to be significant deviations from terrestrial procedures. Still, available data suggest that IOP transiently increases by as much as 4–7 mmHg after initial arrival in microgravity, but gradually normalizes within a few days

**TABLE 10.3** Tonometry supplies.

| Item | Mass | Volume | Power |
|---|---|---|---|
| ***Minimum necessary*** | | | |
| Portable intraocular pressure tonometer (Tono-Pen XL) | 0.071 kg | 160 mm × 20 mm × 44 mm | 3 V lithium battery |
| Sterile tonometer tip cover | 0.0008 kg | 19 mm × 6 mm | N/A |
| ***Supplies for managing elevated intraocular pressure*** | | | |
| Latanoprost 2.5 mL ophthalmic solution bottle | 0.02 kg | 45 mm × 20 mm | N/A |
| Timolol 0.5% 5 mL ophthalmic solution bottle | 0.01 kg | 75 mm × 23 mm | N/A |
| Brimonidine 0.2% 5 mL ophthalmic solution bottle | 0.01 kg | 83 mm × 33 mm | N/A |
| Dorzolamide 2% 10 mL ophthalmic solution bottle | 0.016 kg | 60 mm × 21 mm | N/A |
| Acetazolamide (systemic oral carbonic anhydrase inhibitor) | 0.0008 kg | 17 mm × 6 mm | N/A |
| ***Optimal equipment*** | | | |
| Goldmann applanation tonometer (Haag-Streit AT-900-R) | 0.85 kg | 90 mm × 47 mm × 33 mm | N/A |
| Cosopt (timolol and drozolamide) 10 mL ophthalmic solution bottle | 0.018 kg | 25 mm × 80 mm | N/A |
| Combigan (timolol and brimonidine) 5 mL ophthalmic solution bottle | 0.05 kg | 34 mm × 83 mm | N/A |
| Topical ophthalmic anesthetic solution 4 mL | 0.02 kg | 40 mm × 15 mm | N/A |

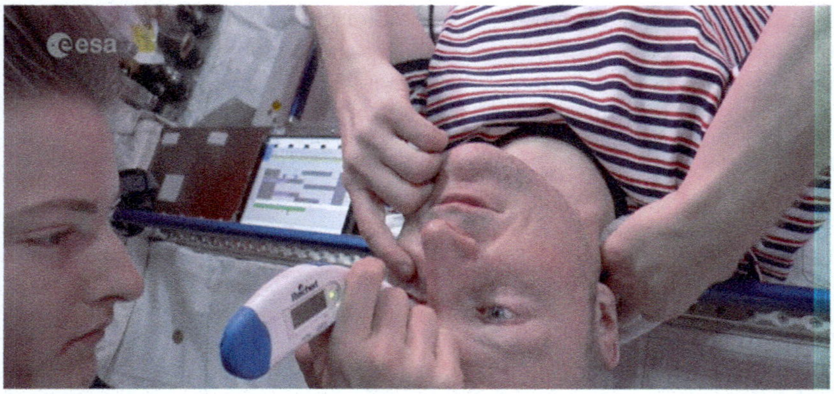


**FIGURE 10.2 Tonometry performed on the International Space Station.** A NASA astronaut Kayla Barron performs a tonometry examination on ESA astronaut Matthias Maurer aboard the ISS as part of the regular medical testing. https://www.esa.int/ESA_Multimedia/Videos/2022/02/Eye_exam_in_orbit_Cosmic_Kiss. (n.d.).

(Masalkhi et al., 2023). This pattern is similar to the cephalad fluid shifts seen in other body systems, and likely does not have long-term consequences for most astronauts. However, individuals with known glaucoma may need to be monitored more closely during the space adaptation period.

#### *10.2.3.3 Procedure*

1. The patient and operator should both be in comfortable positions, with the patient and operator positioned, so they can remain in close proximity. Ideally, the position allows both of the provider's hands to be free.
2. Apply anesthetic drops (if available) to reduce procedural discomfort.
3. Turn on the tonometer and apply a sterile cover to the tip.
4. Hold the tonometer between the thumb and the middle finger of the dominant hand, leaving the index finger free to activate any buttons.
5. Positioning of the patient and provider is primarily a function of patient comfort. Terrestrially, the patient is often seated and facing the provider. In microgravity, an analogous position will also work; however, alternative orientations are also possible, such as that illustrated in Fig. 10.2.
6. Brace the pinky or base of the hand holding the tonometer against the patient's face to stabilize the hand and avoid unintentional injury when pressing the tonometer against the patient's eye.
7. Raise the patient's upper eyelid with the nondominant hand and hold it in place to prevent blinking. The patient should maintain a steady forward gaze throughout the procedure.
8. Ensure the tonometer tip is centered on the cornea and *gently* tap the probe tip to the cornea. Use the lightest and briefest motion that will obtain a reading. Do not push into the eye, as this may artificially raise pressure and yield erroneous readings.
9. Most portable tonometers require averaging several readings. Tap the probe tip against the cornea repeatedly until the tonometer displays a final reading, often with an audible beep and the pressure number.
10. If a reading appears unusually high (>20 mmHg), repeat the reading to ensure accuracy.

#### *10.2.3.4 Postprocedural care specific to the procedure*

Anesthetic drops will numb the surface of the eye. Avoid any contact with the eye until the anesthesia wears off, as one can more easily injure the eye when anesthetized.

#### *10.2.3.5 Common or important procedural complications and how they can be managed in the spaceflight environment*

Avoid tonometry if there is concern for globe laceration or injury, as further damage, like vitreous extrusion, could be caused if excessive pressure is placed on the globe. Excessive pressure or a scraping motion during the procedure may also result in a corneal abrasion. Finally, using the tonometer without a sterile tip may transfer bacteria to the corneal surface, leading to an infection. Using a sterile tip, light pressure, and bracing one's hand on the patient's face further reduces these already small risks.

### 10.2.4 Corneal foreign object removal/rust ring removal

Floating debris or items ejected from an activity or damaged equipment may impact the surface of the eye. While most cause little to no damage, some may result in corneal abrasions, and a few may embed themselves in the cornea and impair vision or lead to more significant damage or ulceration of the cornea. Fig. 10.3 is an example of a rust ring left behind by a foreign body impacting the eye. The ring itself is discolored from debris and rarely needs to be removed. However, the appearance of a foreign body in the center of such a ring should be addressed. Table 10.4 contains a representative set of

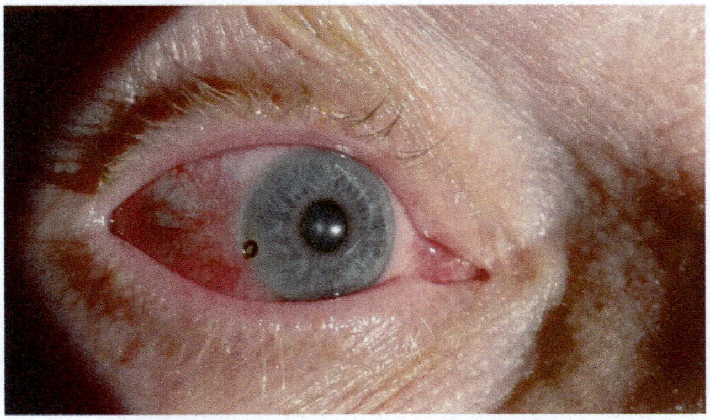

FIGURE 10.3 Corneal rust ring from a superficial foreign body.


**TABLE 10.4** Corneal foreign object removal/rust ring removal supplies.

| Item | Mass | Volume | Power |
|---|---|---|---|
| ***Minimum necessary equipment*** | | | |
| Topical ophthalmic anesthetic solution 4 mL | 0.02 kg | 40 mm × 15 mm | N/A |
| Sterile cotton-tipped applicator 6 in | 0.0004 kg | 153 mm × 4.78 mm | N/A |
| Ophthalmic antibiotic drops (polymyxin B and trimethoprim 10 mL vial) | 0.0015 kg | 22 mm × 55 mm | N/A |
| Normal saline irrigation (5 mL prefilled flush syringe) | 0.005 kg | 87 mm × 14 mm | N/A |
| Sterile golf club spud | 0.04 kg | 114 mm | N/A |
| Ophthalmic Betadine solution (5% 30 mL vial) | 0.04 kg | 63 mm × 34 mm | N/A |
| Portable slit lamp (microclear hyperion digital portable slit lamp) | 0.85 kg | 140 mm × 118 mm × 225 mm | 3.7 V lithium battery |
| ***Alternative supplies*** | | | |
| 30 gauge needle | 0.0009 kg | 12.7 mm × 0.312 mm | N/A |
| 1cc syringe | 0.003 kg | 7.6 mm × 100 mm | N/A |
| Artificial Tears solution vial 10 mL | 0.02 kg | 66 mm × 20 mm | N/A |

equipment for managing retained superficial ocular foreign bodies. It should be noted that penetrating injuries of the eye are far more severe than those discussed here and often require much more involved care. This requires ophthalmological expertise and substantial equipment to manage, which is likely beyond the scope of what is and will be possible in spaceflight over the next decade and consequently beyond the scope of this chapter. These injuries are not common, but when they occur they are often associated with high velocity impacts from shrapnel kicked off during specific activities like grinding metal. For now, limiting activities that are likely to generate high velocity debris, containing equipment or activities in protective containers, and requiring eye protection are the best strategies for preventing these injuries. Should a globe-penetrating injury occur in flight, the medical team will need to consult with and involve an ophthalmologist as soon as possible and plan management in real time.

### *10.2.4.1 Performance of the procedure in space or analog environments*

A review article by Meer et al. documented 55 ocular foreign bodies during spaceflight, which included 48 during the shuttle era and seven during the ISS era. This was the second most common ocular complaint, accounting for 22.7% of all ocular complaints. This can be attributed to the high number of particulates suspended in the vehicle air, including sweat, antifog treatment, food particles, dust, lint, spacesuit irritants, salt crystals, lithium hydroxide dust, and other materials (Meer et al., 2024). None of these injuries was severe, and most were managed with minimal interventions.

The absence of gravity in orbit or transplanetary space makes foreign body impacts common. While most is likely to be dust or tiny particles, which are often easily cleared by the eye's own self-washing processes, larger objects may be retained on the surface and lead to corneal abrasions or ulcerations if not treated.

### *10.2.4.2 Procedure*

1. Perform a fluorescein stain to facilitate identifying the foreign body, confirm no signs of globe rupture, and identify any long streaking abrasions that may suggest a foreign body hidden under the eyelids.
2. Topical anesthesia is necessary for both the exam and procedure to reduce patient discomfort, minimize the changes of involuntary movement on the part of the patient, and maximize the chances of procedure success.
3. Attempt to remove the foreign body by flushing the eye with saline (see Ocular Irrigation in chapter 8). This is the least complicated and least risky method for removing a foreign body.
4. If irrigation fails, attempt to dislodge the foreign body with a moistened cotton-tipped applicator or a cotton ball rolled into a fine tip.
   a. Methodically and gently slide the moistened cotton tip over the foreign body to create sufficient friction to attract it to the tip.
   b. A small circular motion or gentle side-to-side movement may dislodge the foreign body.
   c. If the object is dislodged and becomes mobile but remains on the corneal surface, return to irrigation to flush the object out of the eye.
5. If the cotton-tipped applicator is unsuccessful, the object is likely partially embedded in the cornea. The provider should attempt to dislodge with a sterile golf club spud under portable slit lamp guidance.
   a. A golf club spud is a long, thin ophthalmic instrument with a golf club-shaped scoop on one end, designed for precise control, like removing ocular foreign bodies.
   b. Brace the heel of the provider's hand against the patient's face on the side closest to the injured eye. Use the portable slit lamp or other magnifying lens to aid in visualizing the injury. Fig. 10.4 illustrates this process.
   c. Make sure cabin lighting is sufficiently bright and neutral white. Local task lighting with a fixed medical lamp, head lamp, portable slit lamp, or assistant-held light may be necessary. If these items are not part of the ship's manifest, consider adding them to the medical kit.
6. Alternatively, a 30-gauge needle attached to a 1cc syringe can be used to pick off the foreign body. This is also useful if the golf club spud fails, but it is a more technically challenging procedure and carries greater risk, as a needle can penetrate the eye.
   a. Ophthalmic Betadine solution should be applied to the cornea, since this is likely to leave a small corneal ulceration, and the risk of infection is high.
   b. Both the patient and the operator should be restrained in a way that leaves both of the providers' hands free. Fine motor coordination is required to prevent any further injury.
   c. Position the needle parallel to the patient's eye with the bevel facing out (Fig. 10.5).

1. Ensure that no thruster firings, exercise, crew activities, or other potential sources of unplanned motion that may be transmitted to the provider are halted, and that none are scheduled nor likely to occur during the procedure.

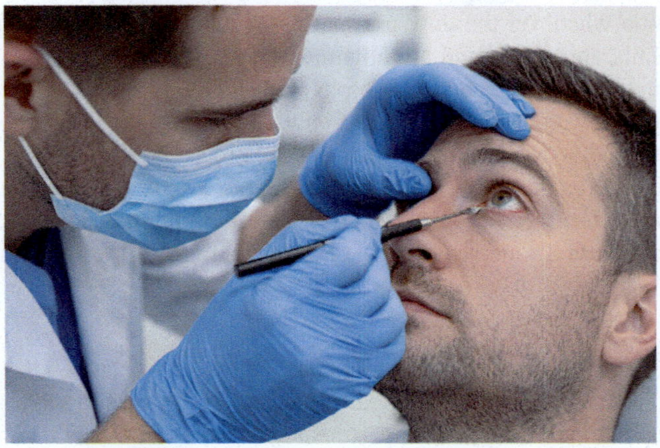

FIGURE 10.4 Illustration of a golf club spud.
Image generated using AI.

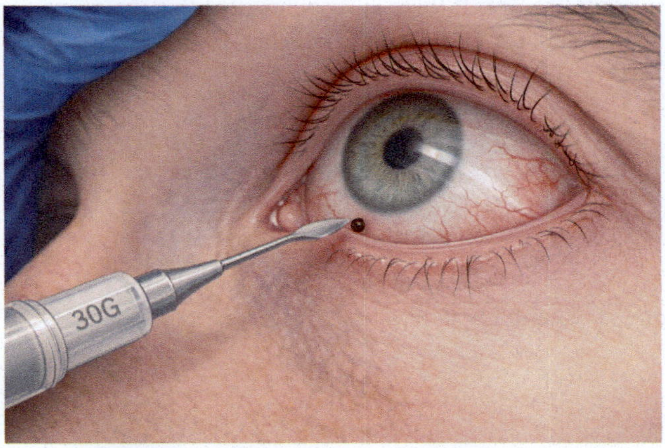


FIGURE 10.5 Illustration of a 30 gauge needle positioned to remove an ocular foreign body.
This image was generated using AI.

2. Paying close attention to avoid penetrating the eye surface, gently use the needle tip and bevel as a scoop to wedge under the object and nudge it away from the eye surface.
   a. Terrestrially, this is done with a gentle flicking motion. In space, the object is likely to be propelled away and into the cabin atmosphere. This will pose a low risk of another crew member encountering it, so if it is possible to direct the object towards an air filter intake, this should be done. If not, positioning a piece of gauze or clothing in the likely path of the object may help catch it. However, this should not interfere with the movements or vision of the provider.

#### 10.2.4.3 Postprocedural care specific to the procedure

Antibiotic drops such as moxifloxacin may be administered three times daily for 1 week and should be applied after the procedure is performed to prevent any introduced bacteria from spreading into an infection. This is necessary if the object was difficult to remove, caused a corneal abrasion, or required a needle. For easily removed objects, antibiotics should be left up to the provider's discretion based on the specific circumstances of the injury.

#### 10.2.4.4 Common or important procedural complications and how they can be managed in the spaceflight environment

1. A corneal abrasion often accompanies an injury related to a foreign body in the eye. This may come from the object itself or be created during its removal. Treat the corneal abrasion with pain control and topical ophthalmic antibiotics.

2. Conjunctivitis is also common whenever the thin, transparent membrane that protects the eye is irritated. This can be treated with topical ophthalmic antibiotics and antiinflammatories.
3. Open globe (or globe rupture) is a rare but dangerous full-thickness corneal laceration. This is an ophthalmic emergency that requires emergency consultation with an ophthalmologist and often operative intervention to repair the globe. Given the complexity of this injury and the fact that treatment is often highly dependent on the context, it is presently beyond the scope of this chapter. If the provider suspects that a globe-penetrating injury may be present, consult the flight surgeon immediately and prepare to perform a follow-up fluorescein stain. Do not press on the eye and avoid any activities that may cause an increase IOP. Follow the advice of the flight surgeon.

## 10.2.5 Lateral canthotomy and cantholysis

A lateral canthotomy and cantholysis (LCC) is a surgical procedure performed to rapidly relieve pressure on the globe from a mass or fluid building up behind it (orbital compartment syndrome). The lateral canthus is the corner of the eyelid, and the lateral canthal tendon is a band of tissue that holds the eyelid in place, creating a restraining ring to hold the eyeball in the orbit. A lateral canthotomy is an incision that opens the skin over this tendon, and a cantholysis severs the tendon. This provides more space for the eye and relieves pressure, potentially restoring blood supply and preserving a patient's vision. Orbital compartment syndrome is typically diagnosed when IOP is found to be acutely above 40 mmHg after the patient experiences trauma to the eye. If there is suspicion that this injury may have ruptured the globe, a lateral canthotomy is not indicated and should not be performed. Table 10.5 contains a representative list of supplies and equipment for performing an LCC.

### *10.2.5.1 Performance of the procedure in space or analog environments*

Orbital compartment syndrome has never been diagnosed in space nor has a LCC ever been performed.

**TABLE 10.5** Lateral canthotomy and cantholysis supplies.

| Item | Mass | Volume | Power |
|---|---|---|---|
| ***Minimum necessary equipment*** | | | |
| Pen light | 0.0143 kg | 130 mm × 12 mm | 2 AAA battery |
| Portable intraocular pressure tonometer (Tono-Pen XL) | 0.071 kg | 160 mm × 20 mm × 44 mm | 3 V Lithium battery |
| Sterile gloves (size 7 in sterile package) | 0.02 kg | 152 mm × 229 mm | N/A |
| Straight hemostat (5 in mosquito) | 0.028 kg | 127 mm × 15 mm | N/A |
| Sterile suture or iris scissors | 0.057 kg | 65 mm × 32.5 mm × 5 mm | N/A |
| Forceps with teeth | 0.028 kg | 10 mm × 4 mm × 140 mm | N/A |
| ***Optimal equipment*** | | | |
| Local anesthesia (Lidocaine 1%–2% with epinephrine 20 mL vial) | 0.100 kg | 58 mm × 32 mm | N/A |
| 30 gauge needle | 0.0009 kg | 12.7 mm × 0.312 mm | N/A |
| 5cc syringe | 0.011 kg | 78 mm × 15 mm × 13 mm | N/A |
| Topical antiseptic solution (Betadine, chlorhexidine) | 0.0002 kg | 67 mm × 30 mm | N/A |
| Sterile drape (18 in × 26 in) | 0.023 kg | 457 mm × 660 mm | N/A |
| Gauze (4 in × 4 in) | 0.0003 kg | 100 mm × 100 mm | N/A |
| Lubricating ointment (erythromycin 3.5 g tube) | 0.0035 kg | 90 mm × 13 mm | N/A |

### 10.2.5.2 Procedure

LCC requires a sterile surgical field, topical antibiotic preparation, and sterile gloves, as well as sterile surgical instruments. The risk of infection without these materials is high, as normally intact tissue is often violated and left open during this process. Fig. 10.6 Shows the anatomy relevant to this procedure. Understanding the location of the targeted tissue and the tissues to be avoided is key to success without complications.

1. Confirm IOP is above 40 mmHg and confirm which eye will require the procedure. It is also worth asking the patient about visual changes if they remain conscious. In many cases, elevated pressures like this will severely decrease or eliminate vision. Conduct a rapid vision assessment to identify any obvious deficits and establish a preprocedure baseline. However, since this is a time-sensitive issue, and delays may cause permanent blindness, this assessment should be quick and not delay the setup or performance of the procedure.
2. If communication is possible, confirm with the flight surgeon or senior medical officer that this procedure is needed.
3. If video is possible, position cameras to enable transmission of images
4. Both the patient and the operator should be securely restrained before the procedure begins, with both of the providers' hands free.
5. Ensure that no thruster firings, exercise, crew activities, or other potential sources of unplanned motion that may be transmitted to the provider are halted, and that none are scheduled nor likely to occur during the procedure.
6. Prepare all necessary instruments near the patient's head and within reach on a sterile "tray."
   a. The "tray" will need to be designed to hold equipment even in the absence of gravity. This may be a sterile bag, or a sterile surface with velcro or small magnets to secure equipment and hold it in place for use. This may be positioned on the vehicle or as a "leg board" on the provider, as long as it can be maintained as a sterile surface.
7. Perform initial skin prep around the eye with a single antiseptic wipe such as BZK.
   a. Clean the area and allow it to air dry.
   b. Avoid contact with the patient's eye.
8. Anesthetize the incision site.
   a. Use the 5cc syringe with a 25-gauge or smaller needle.

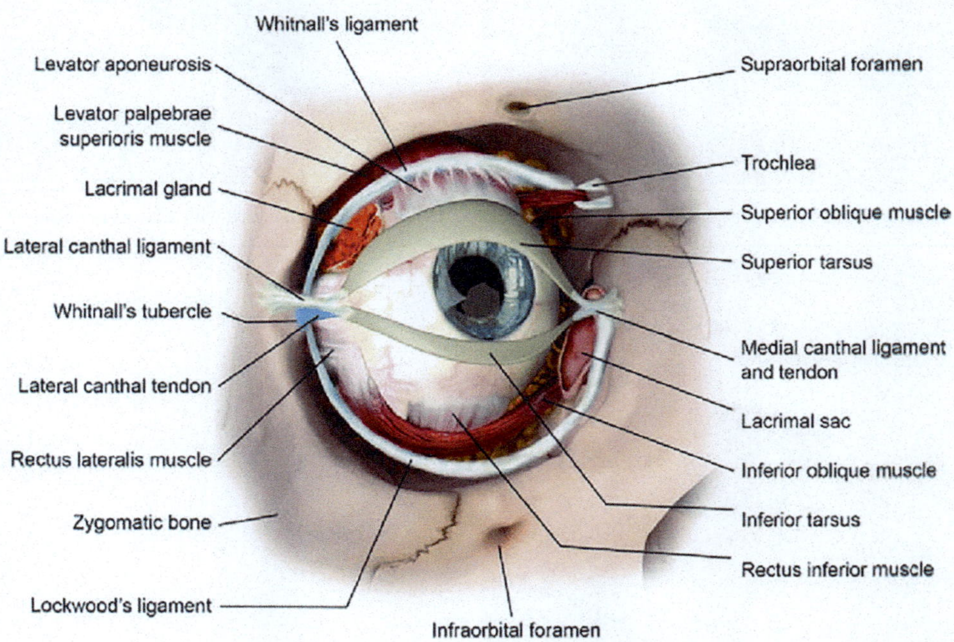


FIGURE 10.6 Anatomy of the lateral canthus.
Botti, Giovanni, Botti, Chiara, Rossati, Leonardo, Gualdi, Alessandro, Nocini, Pierfrancesco, Nocini, Riccardo, Bertossi, Dario. (2019). "Dynamic Canthopexy" drill hole canthal repositioning. Aesthetic Surgery Journal. 39. https://doi.org/10.1093/asj/sjz077.

   b. With the nondominant hand, place one finger on the upper eyelid and the other on the lower lid.
   c. Gently but firmly pull the skin medially (towards the nose) to provide tension to counter the needle pressure and hold the skin in place
   d. Aim the needle away from the globe of the eye to avoid damaging it.
   e. Insert the needle into the skin at the lateral (cheek side) corner of the eye. Make sure the tip of the needle stays buried within the canthus and does not poke through the eyelid skin
   f. Stop when the needle contacts the bone of the orbit.
   g. Inject 1–2 mL of local anesthetic solution, such as 1% lidocaine with epinephrine, into the lateral canthal incision site. Slowly withdrawing the needle and leaving the anesthetic solution in the entirety of the needle's track.
   h. Once the injection is complete, secure the needle tip with a needle locking mechanism to prevent unintentional needle sticks, and set it aside
9. Prepare the patient's skin around the eye with additional antiseptic wipes such as BZK.
   a. Clean the area and allow it to air dry two more times.
   b. Avoid contact with the patient's eye.
   c. If available, place a sterile drape around the eye.
10. Use a new disinfectant wipe to wash the provider's hands and apply sterile gloves without touching the outer surface of the gloves to preserve sterility to the extent possible.
11. Lateral canthotomy and cantholysis (Fig. 10.7)
   a. Use a straight hemostat to crush the lateral canthus to the rim of the orbit.

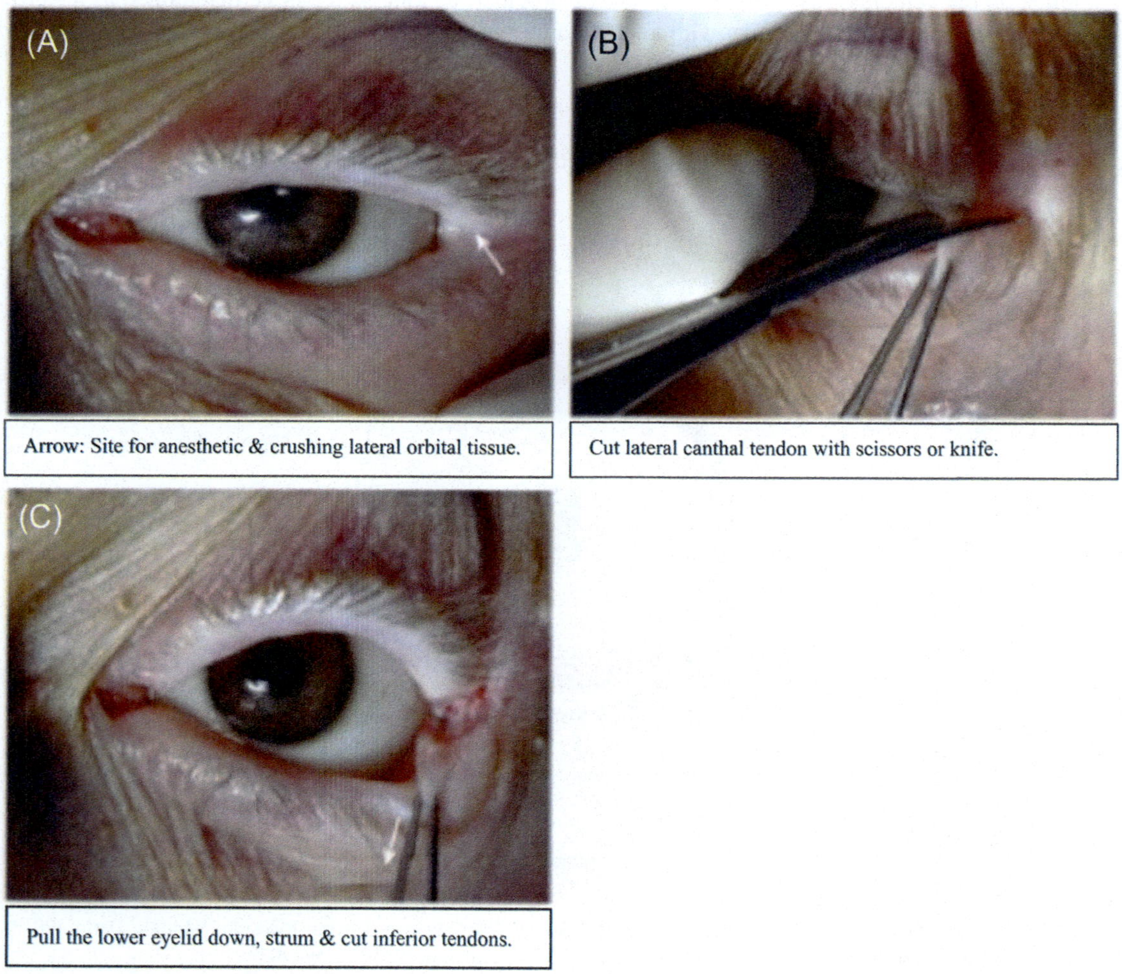


**FIGURE 10.7** Lateral canthotomy and cantholysis.
Iserson, K.V., Luke-Blyden, Z., Clemans, S. (2016). Orbital compartment syndrome: alternative tools to perform a lateral canthotomy and cantholysis. Wilderness & Environmental Medicine. 27(1):85–91. doi:10.1016/j.wem.2015.09.002.

   **i.** Unlock the hemostat and slightly open the tip.
   **ii.** Slide the tip over the lateral canthus so that one end is in the orbit, and one end is outside with the skin of the canthus in between the two jaws, with the tip contacting the rim of the orbit bone (about 1–2 cm).
   **iii.** Close the jaws to pinch the skin firmly between them, then release them. This is done to reduce bleeding, speed up the procedure, and reduce pain. When done correctly, the skin should be blanched with a clearly visible imprint of the hemostat tip.
   **iv.** Withdraw the hemostat and set it aside
 **b.** Use the sterile scissors to cut the lateral canthus:
   **i.** Position the scissors in the same line as the hemostat all the way down to the rim of the orbit bone (about 1–2 cm).
   **ii.** With the scissors pointing away from the globe, cut the lateral canthal tendon and overlying skin.
   **iii.** Pull the lower eyelid down to expose the inferior crus of the lateral canthal tendon.
   **iv.** If it is still intact, cut it as well.

**12.** Recheck the IOP using tonometry to confirm a successful procedure. The pressure should immediately be decreased to baseline levels.

### *10.2.5.3 Postprocedural care specific to the procedure*

1. Reevaluate the globe for injury after the procedure.
2. Check visual acuity and visual field after the procedure and at regular intervals for the first few hours to assess for damage.
3. Lubricating ointment, antibiotic drops, or ointment like erythromycin, and a sterile dressing should be applied over the procedure areas without pressure on the globe.
4. Lateral canthotomy incisions are not sutured at the time of the canthotomy and are typically allowed to heal on their own (by secondary intention). The tendon heals as well.
5. If IOP remains elevated despite LCC, it is possible that there is an additional injury to the intraocular circulation system. Consider adding topical medications or systemic therapy to lower the IOP.

### *10.2.5.4 Common or important procedural complications and how they can be managed in the spaceflight environment*

LCC is an invasive procedure in a small space with delicate, tightly packed structures. While it is not difficult to perform this procedure, the provider must have expertise and practice to ensure success. The provider must be familiar with the anatomy of the area and practice the movements and techniques to ensure the targeted structure (the lateral canthal ligament) is reached. In contrast, other important structures nearby are preserved. Because of this, many of the injuries that can result are not common but can carry severe and/or permanent disability in vision. For these reasons, the procedure must be done only when necessary, confirmed with a medical expert before beginning, and practiced before it is needed. Incisions should not be made until the risk of orbital compartment syndrome (loss of vision) outweighs the risks of the procedure. In most cases, complications of this procedure will require immediate consultation with an expert in ocular injuries. Some injuries may require rapid return to definitive care, and others may not be salvageable at all. That said, this procedure can be safely performed consistently and with success by inexperienced practitioners with appropriate training and practice.

1. Pain at the site of incision
   **a.** This is mitigated with local anesthetics and oral pain medications. Bandages applied to the surgical site with sterile surgical lube may also decrease pain.
2. Incomplete cantholysis
   **a.** This is common with inexperienced providers since there is a tendency to cut less than is needed.
   **b.** Examine the remaining tissue around the incision site to identify the pale white tendon and repeat the cutting step as needed.
3. Iatrogenic injury to the globe
   **a.** This is an ocular emergency, and steps should be taken to minimize the risk, such as aiming sharps away from the globe, minimizing the chance of unintentional bumps to the provider during the procedure, and providing adequate lighting. Ultimately, the best prevention is practice. If this procedure is deemed necessary by medical system designers and risk assessments, then the crew medical officer will need to practice this procedure terrestrially to ensure competence.

b. Injury to the globe can be confirmed with fluorescein staining. Avoid applying pressure to any part of the globe, consult with a flight surgeon or on-board medical authority, and request an emergent ophthalmology consult.

4. Iatrogenic injury to the lateral rectus muscle
   a. The lateral rectus muscle is located on the side of the eye and tracks back into the orbit (Fig. 10.8).
   b. Severing this muscle will result in the patient being unable to pull the eye laterally (lateral gaze palsy) and induce double vision whenever the patient attempts to do so, since one eye will look full in that direction while the other does not.
   c. This can be a significantly debilitating injury during a mission, and the severity depends on the amount of muscle tissue severed. A complete transection would cause an inability to pull the eye laterally and would require a high level of surgical expertise to repair. A partial tear would cause pain and weakness but may preserve the ability to gaze laterally with the eye. In either case, if a lateral rectus injury is suspected, consult the flight surgeon as soon as practical to develop a care plan.
5. Ptosis (caused by injury to the levator aponeurosis)
   a. The levator aponeurosis is a connective tissue scaffold under the upper lid that helps raise the eyelid out of the field of view (Fig. 10.9).

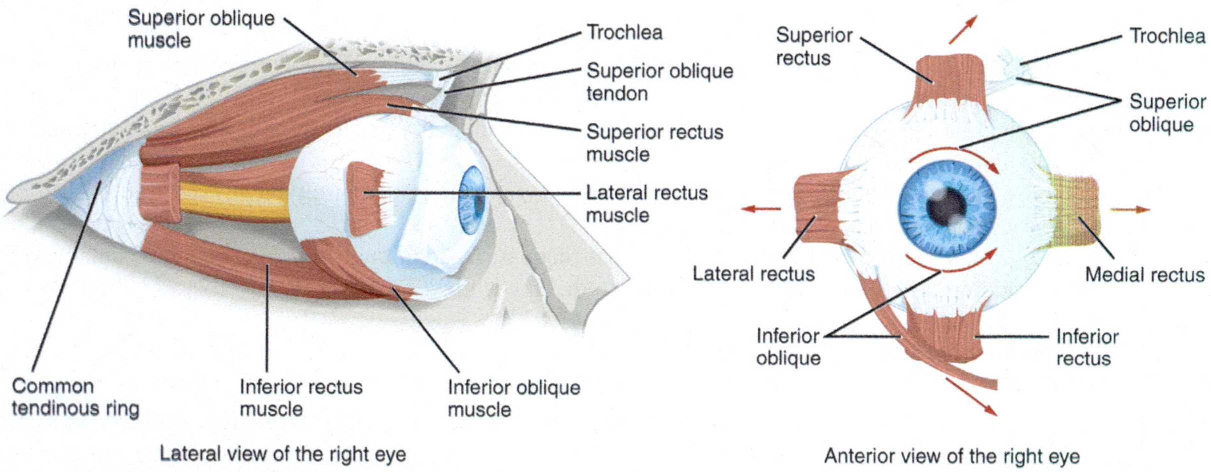


FIGURE 10.8 Location of the lateral rectus muscle.


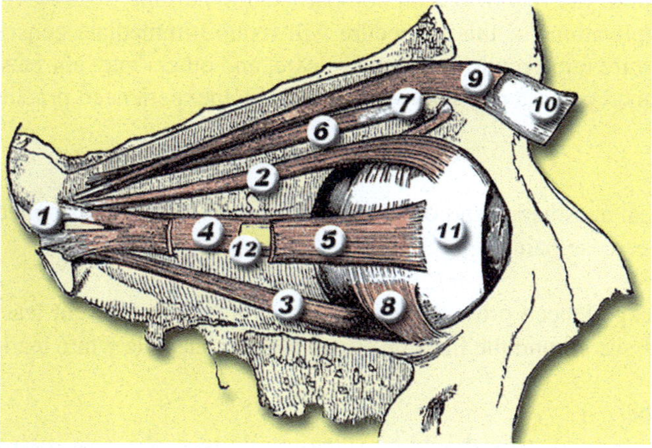


FIGURE 10.9 **Location of the levator palpebrae superioris muscle and the levator aponeurosis.** Rectus muscles: 2=superior, 3=inferior, 4=medial, 5=lateral, Oblique muscles: 6=superior, 8=inferior, Other muscle: 9=levator palpebrae superioris, Other structures: 1=annulus of zinn, 7=trochlea, 10=superior tarsus, 11=sclera, 12=optic nerve.

   **b.** If this is injured, it will cause a droop in the eyelid called ptosis. Understanding the location of this connective tissue is critical to avoiding the injury. Fortunately, it is located above the eye and is not in the operative field laterally.
   **c.** This most often requires surgical repair but can be deferred until after a return to Earth.
**6.** Injury to the lacrimal gland and the lacrimal artery
   **a.** The lacrimal gland is located superior to the eye and superolateral to the canthus (Fig. 10.10). Its blood supply comes from behind the gland and traces back into the orbit.
   **b.** An injury to the gland is likely to cause excessive tearing if mild, or destruction of the gland and a lack of tears if severe. The latter injury will require permanent supplementation with artificial tears.
   **c.** A compromised blood supply will also destroy the gland but may also result in substantial bleeding. In this event, the bleeding must be stopped by holding pressure on the area. This can be difficult given the location, and compression of the eye may cause other complications. However, the general principles of hemostasis outlined in Chapter 6 still hold. In general, identifying the source of the bleeding and applying pressure there will stem the flow. With regard to the eye, avoid putting pressure on the globe since this can induce a reflex that slows the heart, and the eye is already under excess pressure, which is the reason to perform the LCC in the first place. Pressing the eyelid superolaterally and against the orbital ridge is a good first step while contacting the flight surgeon or medical authority for further guidance.
**7.** Ectropion, outward turning of the eyelid margin (Fig. 10.11)
   **a.** This is a cosmetic, mild functional deficit resulting from the intended injury to the canthal tendon. It is often simply a cosmetic defect, but it may interfere with tear drainage and lead to dry eye. In these cases, it is most often managed with artificial tears. Surgical repair is possible but is best deferred until after the mission has concluded.
**8.** Bleeding form the incision site.
   **a.** A small amount of bleeding from the incision site is to be expected. However, much of this can be mitigated with the use of lidocaine with epinephrine and the hemostat to crush the tissue and push the blood out of the area. If the incision covers tissue that is not pretreated, or if epinephrine is not available, then bleeding is likely to occur. This will most likely be capillary bleeding, easily controlled with direct pressure held until bleeding stops.
   **b.** One of the common causes for orbital compartment syndrome is a traumatic injury inducing bleeding behind the eye. In this case, if the source of that bleeding is not well controlled, or if a large amount of blood has pooled behind the eye, this may be released as the procedure is performed. Observe the flow of blood and determine whether it is controlled and slowing or continuing and requiring pressure to achieve hemostasis. If basic hemostatic techniques are not successful, you will need to consult with the flight surgeon or the mission's medical authority to determine the next steps.

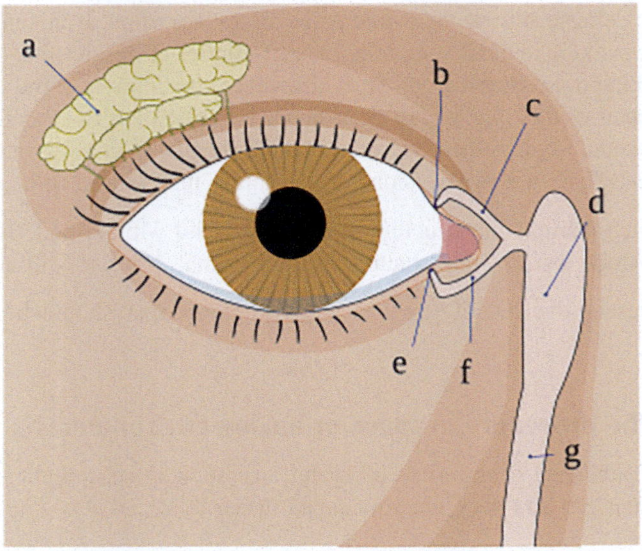


**FIGURE 10.10 Location of the lacrimal gland.** a. tear gland/lacrimal gland, b. superior lacrimal punctum, c. superior lacrimal canal, d. tear sac/lacrimal sac, e. inferior lacrimal punctum, f. inferior lacrimal canal, g. nasolacrimal canal.

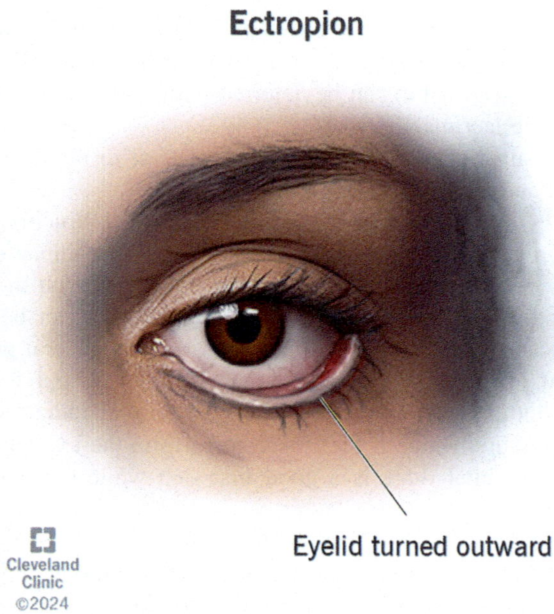


**FIGURE 10.11** Ectropion.
https://my.clevelandclinic.org/health/diseases/24420-ectropion.

9. Infection at the incision site.
   a. Infection is a known risk for every surgical procedure, as disruption of the skin may introduce bacteria. Using sterile technique as described in the procedure will help limit this complication, but cannot completely eliminate it. The use of postprocedure antibiotic drops will further help mitigate this risk. However, an infection can still occur, and the patient should be monitored for signs of infection such as erythema, swelling, and/or increased pain. Should these occur, the patient will need oral antibiotics.

## 10.2.6 Eyelid laceration repair

Trauma to the eye can tear the delicate skin of the eyelid and cause a laceration. If this injury is superficial, there is little need for an intervention, and it is likely to heal on its own. Deeper injuries or injuries that tear the eyelid itself may require repair to maintain function and improve cosmetic outcomes. However, the thinness of the skin and the densely packed tiny structures in the orbit make repair of this type of injury difficult. It is often left to specialists; even emergency medicine physicians often defer this procedure unless there are no specialists available in a reasonable time frame, and no way to transfer a patient to reach them. There are two versions of this injury that will be addressed here.

1. Marginal lid laceration, which disrupts the structure of the general lid without underlying structure injury
2. Canalicular lid laceration, which disrupts the structure of the lacrimal duct and will affect tear drainage.

Table 10.6 contains a representative list of supplies and equipment that should be manifested to perform eyelid laceration repairs.

## 10.2.7 Performance of the procedure in space or analog environments

There have been at least two published occurrences of eyelid laceration in spaceflight in the available literature. Both occurred on the ISS, but it is unclear how they were managed (Meer et al., 2024).

### *10.2.7.1 Procedure*

Evaluate for nasolacrimal drainage system (Fig. 10.10), levator aponeurosis/superior rectus involvement (Fig. 10.9), violation of orbital septum, visible orbital fat, canthal avulsion, and extensive tissue loss—as these injuries are likely beyond the capabilities and training that will be available on board near-term space missions and thus indications for

**TABLE 10.6** Eyelid laceration repair supplies.

| Item | Mass | Volume | Power |
|---|---|---|---|
| Topical ophthalmic anesthetic solution 4 mL | 0.02 kg | 40 mm × 15 mm | N/A |
| Needle driver (6 in) | 0.057 kg | 152 mm × 102 mm × 25 mm | N/A |
| Forceps | 0.028 kg | 10 mm × 4 mm × 140 mm | N/A |
| Sutures (6-0 plain gut) | 0.01 kg | 128 mm × 97 mm × 30 mm | N/A |
| Gloves | 0.02 kg | 152 mm × 229 mm | N/A |
| Local anesthesia (Lidocaine 1%–2% with epinephrine 20 mL vial) | 0.100 kg | 58 mm × 32 mm | N/A |
| 30 gauge needle | 0.0009 kg | 12.7 mm × 0.312 mm | N/A |
| 5cc syringe | 0.011 kg | 78 mm × 15 mm × 13 mm | N/A |
| Topical antiseptic solution (Betadine, chlorhexidine) | 0.0002 kg | 67 mm × 30 mm | N/A |
| Sterile drape (18 in × 26 in) | 0.023 kg | 457 mm × 660 mm | N/A |
| Gauze (4 in × 4 in) | 0.0003 kg | 100 mm × 100 mm | N/A |
| Lubricating drops or ointment (e.g., erythromycin 3.5 g tube) | 0.0035 kg | 90 mm × 13 mm | N/A |

evacuation to definitive care. If these are present, discuss the next steps in management with the flight surgeon or medical authority to develop an appropriate care plan (Cochran and Czyz, 2022)

1. Thoroughly examine the injury and identify the extent of the wound and any underlying structures or other areas that may be injured.
2. If communication is possible, confirm with the flight surgeon or senior medical officer that this procedure is needed.
3. If video is possible, position cameras to enable transmission of images.
4. Both the patient and the operator should be securely restrained before the procedure begins, with both of the providers' hands free.
5. Ensure that no thruster firings, exercise, crew activities, or other potential sources of unplanned motion that may be transmitted to the provider are halted, and that none are scheduled nor likely to occur during the procedure.
6. Prepare all necessary instruments near the patient's head and within reach on a sterile "tray."
   a. The "tray" will need to be designed to hold equipment even in the absence of gravity. This may be a sterile bag, or a sterile surface with velcro or small magnets to secure equipment and hold it in place for use. This may be positioned on the vehicle or as a "leg board" on the provider, as long as it can be maintained as a sterile surface.
7. Apply a topical anesthetic solution (e.g., Tetracaine) to the affected eye.
8. Ensure any foreign body or particulate matter is removed from the affected area. Irrigate the surrounding skin using an adaptation of the procedure in Chapter 8 for ocular irrigation and clean the area with a topical antiseptic solution (if available). The skin should be gently prepped with a BZK wipe or betadine, careful to avoid getting it in the eye.
9. Pat the area dry.
10. Using the 5cc syringe, inject 1–2 mL of local anesthetic solution containing epinephrine into the laceration site, using the minimal amount necessary. If the patient is significantly uncomfortable during the exam/irrigation steps, anesthesia may be administered earlier.
11. Examine the wound to identify critical structures:
    a. Lid margins (the edges of the eyelids where the lashes attach)
    b. The tarsus (connective tissue plate, below the skin of the upper and lower lids, just behind the margins)
    c. Gray line, Fig. 10.12, the line separating the anterior and posterior parts of the eyelid.
    d. Anterior lash line, Fig. 10.12
    e. And the mucocutaneous junction, Fig. 10.12 where the mucosal (interior) surface of the lid transitions to the epidermal surface.

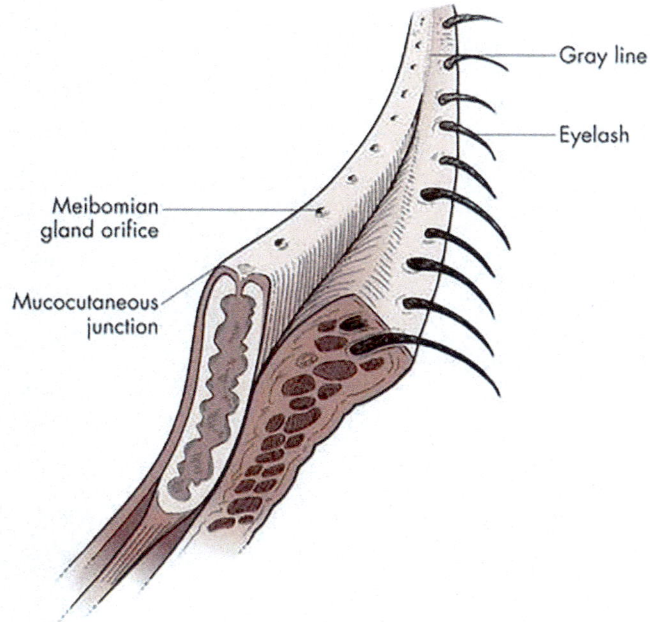


FIGURE 10.12 Gray line and mucocutaneous junction.
http://plasticsurgerykey.com (available online), used for educational purposes.

f. Fundamentally, the eyelid is just thin skin, and the techniques outlined in Chapter 6 are applicable to this area. However, the location makes unintentional injury to delicate underlying structures like the eye, the extraocular muscles, the lacrimal gland, and the nerves highly likely if precautions are not taken. It is also easy to misalign tiny structures, leading to a noticeable cosmetic defect. For these reasons, examination and identification of these structures is critical.

1. With these structures identified, attempt to align them as closely as possible to their original positions. Aligning each of these key structures will ensure a good cosmetic and functional outcome.
   a. With these held in temporary alignment, assess the tension on the wound. If it is too high, a lateral canthotomy may be necessary to reduce tension and permit healing. This should only be considered in conjunction with an expert consultant.
2. Suture the wound
   a. Begin by aligning the tarsus, Fig. 10.13, along its vertical axis.
   b. Evert (pick up and turn out) the edges of the wound and place a 6-0 sutures through the gray line, Fig. 10.12, on either side of the wound approximately 2 mm from the wound edge.
   c. Avoid placing sutures through the mucosal ocular surface of the lid since this will scratch the eye with blinking and lead to a corneal abrasion. Placing sutures in the epidermal surface and remaining above the mucosal surface will also avoid damaging the lid glands.
      i. Experts will often use a vertical mattress suture to hold these in place, if the provider is familiar with this technique it is encouraged since this reduces wound tension and leads to a better cosmetic and functional outcome.
      ii. For less experienced practitioners, a simple interrupted suture technique is often easier to use and likely sufficient in most cases.
   d. Tie the simple interrupted suture in place, cut the suture material, and begin the next suture approximately 2–3 mm from the initial one.
   e. Using the same technique of everting the wound edge and placing the suture through the gray line, add another interrupted suture.
   f. Continue this process until the wound is held together in anatomical alignment.
   g. Use the minimum number of sutures necessary to hold the eyelid in position.

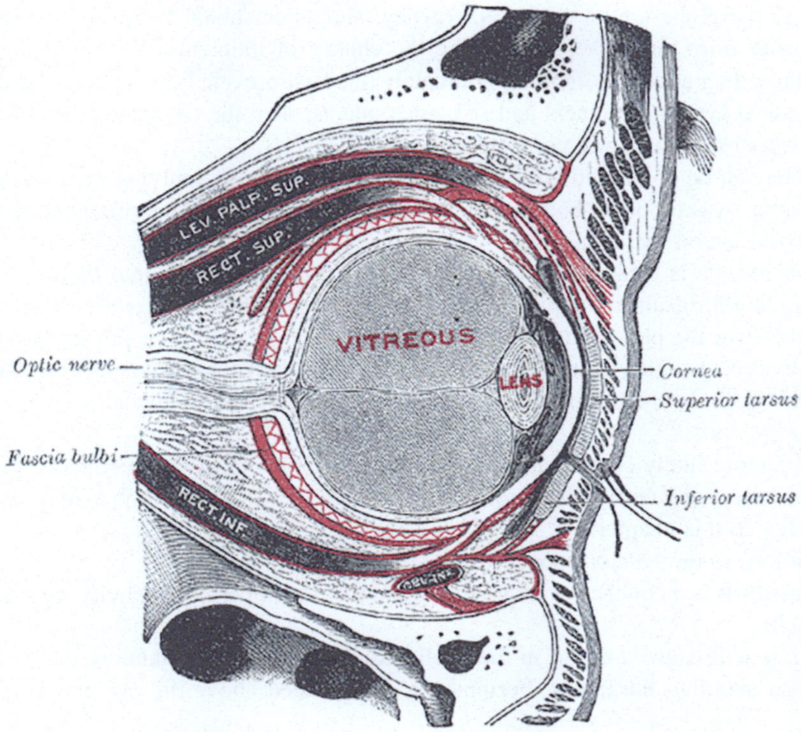


FIGURE 10.13 Tarsus of the eyelid.
This work is in the public domain in the United States because it was published (or registered with the U.S. Copyright Office) before January 1, 1930.

### *10.2.7.2 Postprocedural care specific to the procedure*

1. Monitor the wound for healing, alignment changes, and signs of infection.
2. The timing of suture removal is a balance between the tensile strength of the healing skin and the formation of scar tissue. In general, the longer the sutures are in place, the more scar tissue forms. This strengthens the healing wound but also adds functional and cosmetic defects.
   a. Lid sutures can be removed in 5–7 days.
   b. Sutures in the periorbital skin or on the lid margins should be kept in place for 10 days.
   c. Absorbable sutures will often fall out on their own but may also need to be removed if they are present longer than 7–10 days, depending on their location.
   d. Assess the wound for sufficient strength before removing the sutures.
3. While the sutures are present, apply a thin film of a lubricating jelly (such as a petroleum jelly) with or without an antibiotic to the wound as often as needed but at least 3–4 times per day to facilitate healing and lubricate the sutures.
4. The majority of healing will be complete in 2 weeks, but full healing may take as long as 6–12 months.

### *10.2.7.3 Common or important procedural complications and how they can be managed in the spaceflight environment*

1. Pain at the site of repair:
   a. This is mitigated with local anesthetics and oral pain medications. Repairing the wound will often decrease pain.
   b. Swelling at the site will also increase pain and the absence of gravity may cause greater than terrestrial levels of swelling. If this occurs, use of a cold pack may help reduce swelling and decrease pain.
2. Iatrogenic injury to the globe:
   a. In many cases, this will be a superficial injury caused by unintentionally contacting the globe with the tip of the suture needle. Superficial globe injuries may be treated similarly to a corneal abrasion and monitored closely for signs of infection.

   b. Deeper injuries to the globe are an ocular emergency, and steps should be taken to minimize this risk, such as aiming sharps away from the globe, minimizing the chance of unintentional bumps to the provider during the procedure, and providing adequate lighting. Ultimately, the best prevention is practice. If this procedure is deemed necessary by medical system designers and risk assessments, then the crew medical officer will need to practice this procedure terrestrially to ensure competence.
   c. Injury to the globe can be confirmed with fluorescein staining. Avoid applying pressure to any part of the globe, consult with a flight surgeon or on-board medical authority, and request an emergent ophthalmology consult.
3. Iatrogenic injury to the lateral rectus muscle:
   a. The lateral rectus muscle is located on the side of the eye and tracks back into the orbit (Fig. 10.8).
   b. Injuring this muscle will result in the patient being unable to pull the eye laterally (lateral gaze palsy) and induce double vision whenever the patient attempts to do so since one eye will look full in that direction while the other does not. If a lateral rectus injury is suspected, consult the flight surgeon as soon as practical to develop a care plan. In most cases, this will also be a minor injury since the suture needle is a small tool.
4. Iatrogenic injury to the cornea:
   a. This is one of the more likely complications given the manipulation of tissue so close to the cornea.
   b. These injuries are essentially corneal abrasions and should be treated in the same manner as other corneal abrasions as described earlier in this chapter.
5. Ptosis (caused by injury to the levator aponeurosis):
   a. The levator aponeurosis is a connective tissue scaffold under the upper lid that helps raise the eyelid out of the field of view (Fig. 10.9).
   b. If this is injured, it will cause a droop in the eyelid called ptosis. Understanding the location of this connective tissue is critical to avoiding the injury. Fortunately, it is located above the eye and is not in the operative field laterally.
   c. This most often requires surgical repair but can be deferred until after a return to Earth.
6. Injury to the lacrimal gland and the lacrimal artery:
   a. The lacrimal gland is located superior to the eye and lateral but above the canthus (Fig. 10.10). It's blood supply comes from behind the gland and traces back into the orbit.
   b. An injury to the gland is likely to cause excessive tearing if mild, or destruction of the gland and a lack of tears if severe. The latter injury will require permanent supplementation with artificial tears.
   c. A compromised blood supply will also damage or destroy the gland and may also result in substantial bleeding. In this event, the bleeding must be stopped by holding pressure on the area. This can be difficult given the location, and compression of the eye may cause other complications. However, the general principles of hemostasis, as outlined in Chapter 6, still hold. In general, identifying the source of the bleeding and applying pressure there will stem the flow. With regard to the eye, avoid putting pressure on the globe since this can induce a reflex that slows the heart. Pressing the eyelid superiolaterally and against the orbital ridge is a good first step while contacting the flight surgeon or medical authority for further guidance.
   d. Lacrimal injury from suturing is likely to be relatively minor, but care must be taken to avoid placing a suture through the gland if possible.
7. Ectropion, outward turning of the eyelid margin (Fig. 10.11):
   a. This is a cosmetic and mild functional deficit caused by the intended injury to the canthal tendon. It is often simply a cosmetic defect, but it may interfere with tear drainage and lead to dry eye. In these cases, it is most often managed with artificial tears. Surgical repair is possible but is best deferred until after the mission has concluded.
8. Bleeding from the site:
   a. A small amount of bleeding from the incision site is to be expected. This will most likely be capillary bleeding that is easily controlled with direct pressure held until the bleeding stops.
   b. If an arteriole is damaged, then a steady but small stream of pulsatile blood may be seen. Applying pressure will temporarily stem the flow, but these "pumpers" often require additional interventions, such as a small suture to tie them off to stem the flow of blood.
9. Infection at the incision site:
   a. Infection is a known risk for every surgical procedure, as disruption of the skin may introduce bacteria. Using sterile technique as described in the procedure will help limit this complication, but cannot completely eliminate it. The use of postprocedure antibiotic drops will further help mitigate this risk. However, an infection can still occur, and the patient should be monitored for signs of infection such as erythema, swelling, and/or increased pain. Should these occur, the patient will need oral antibiotics.

10. Irregular eyelid contour and notching:
    a. These are common cosmetic issues that may occur when the wound is repaired while it is misaligned. This will likely need to be corrected after the mission has concluded.
11. Wound dehiscence:
    a. Sutures under tension may not successfully hold the wound in place sufficiently to facilitate healing. If the wound is under significant tension, discuss with the flight surgeon or medical authority and an ophthalmologist or plastic surgeon consultant to determine the best path forward.
    b. Should the wound dehisce after repair, healing and management may be quite complex and varied depending on the specifics of the wound. This is best handled in conjunction with the flight surgeon or medical authority and an ophthalmologist or plastic surgeon consultant.

## 10.3 Conclusion

In the uniquely demanding environment of space, protecting and preserving crew vision is essential to maintaining crew health, performance, and mission readiness. The procedures listed in this chapter demonstrate how a handful of terrestrial medical skills, modified and adapted for microgravity, can have an outsized impact on mission success. By understanding the changes that occur in spaceflight, the operational constraints of spacecraft, and the limited resources available to crew, mission planners can better prepare for ocular emergencies in space.

## References

Cochran, M.L., Czyz, C.N., 2022. Eyelid laceration. Available from: https://www.ncbi.nlm.nih.gov/books/NBK470367/.

Dalal, S.R., Ramachandran, V., Khalid, R., Manuel, F.K., Knowles, J.R., Jones, J.A., 2021. Increased intraocular pressure in glaucomatous, ocular hypertensive, and normotensive space shuttle crew. Aerospace Medicine and Human Performance 92 (9), 728–733. Available at https://doi.org/10.3357/AMHP.5455.2021, http://ingentaconnect.com/content/asma/amhp;jsessionid=13tlejf2f0ojk.alice.

Draeger, J., Schwartz, R., Groenhoff, S., Stern, C., 1995. Self-tonometry under microgravity conditions. Aviation, Space, and Environmental Medicine 66 (6), 568–570.

Leslie, L., Liu, S.-H., Kuo, I.C., 2025. Topical ophthalmic anesthetics for corneal abrasions: findings from a Cochrane systematic review and meta-analysis. Communications medicine 5 (1). Available at https://doi.org/10.1038/s43856-025-00979-4.

Masalkhi, M., Ong, J., Waisberg, E., Berdahl, J., Lee, A.G., 2023. Intraocular pressure during spaceflight and risk of glaucomatous damage in prolonged microgravity. Encyclopedia 3 (4), 1187–1196. Available at https://doi.org/10.3390/encyclopedia3040086, https://www.mdpi.com/journal/encyclopedia.

Meer, E., Grob, S., Antonsen, E.L., Sawyer, A., 2023. Ocular conditions and injuries, detection and management in spaceflight. NPJ Microgravity 9 (1). Available at https://doi.org/10.1038/s41526-023-00279-y.

Meer, E., Grob, S.R., Lehnhardt, K., Sawyer, A., 2024. Ocular complaints and diagnoses in spaceflight. NPJ Microgravity 10 (1). Available at https://doi.org/10.1038/s41526-023-00335-7, https://www.nature.com/npjmgrav/.

Nguyen, T., Ong, J., Brunstetter, T., Gibson, C.R., Macias, B.R., Laurie, S., et al., 2025. Spaceflight Associated Neuro-ocular Syndrome (SANS) and its countermeasures. Progress in Retinal and Eye Research 106.

Ong, J., Mader, T., Gibson, C.R., Suh, A., Panzo, N., Memon, H., et al., 2025. The ocular surface during spaceflight: post-mission symptom report, extraterrestrial risks, and in-flight therapeutics. Life Sciences in Space Research 46, 169–186.

Suh, A., Ditelberg, S., Szeto, J.J., Kumar, D., Ong, J., Robert Gibson, C., et al., 2025. Safety protocols, precautions, and countermeasures aboard the International Space Station to prevent ocular injury. Survey of Ophthalmology 70 (5), 1003–1012. Available at https://doi.org/10.1016/j.survophthal.2024.08.005.

Waisberg, E., Ong, J., Lee, A.G., 2024. Corneal abrasions in space: current therapeutics and future directions. Eye (Basingstoke) 38 (7), 1238–1239. Available at https://doi.org/10.1038/s41433-023-02911-3, https://www.nature.com/eye/.

Waldman, N., Densie, I.K., Herbison, P., 2014. Topical tetracaine used for 24 hours is safe and rated highly effective by patients for the treatment of pain caused by corneal abrasions: a double-blind, randomized clinical trial. Academic Emergency Medicine: Official Journal of the Society for Academic Emergency Medicine 21 (4), 374–382. Available at https://doi.org/10.1111/acem.12346.

Chapter 11

# Thoracic procedures

**Matthew Marc Cooper**
*Peak Aerospace Medicine & Cardiovascular Surgery, Lake Elmo, MN, United States*

## 11.1 Introduction

Thoracic emergencies or urgent issues requiring timely intervention have traumatic or nontraumatic causes. Traumatic injuries may be penetrating or blunt in nature. Nontraumatic etiologies include inflammatory processes, infection, previously unidentified malignancy, and cardiovascular disease, including coronary ischemia. Regardless of etiology, the presented common thoracic procedures treat a finite set of clinical manifestations.

The procedures included in this chapter are those considered within the realm of safe performance by operators with some training, though without the mastery of a surgical specialist in an extraterrestrial environment. As such, therapeutic approaches involving complicated thoracoscopic, robotic, or other minimally invasive techniques are not included. While describing all conceivable procedures is impractical, the procedures presented are designed to flexibly encompass the spectrum of clinical scenarios most likely to occur. All of the procedures described may be performed by an unassisted operator. Time spent in preparation, setup of the procedural field, organization of required instrumentation, etc., will critically impact the ease, efficiency, and success of the intervention. Instances are called out in which a second set of hands from an assistant will particularly facilitate performance and mitigate both anticipated and unexpected procedural challenges. By intent, a minimalist, adequate, and safe approach has been used to describe required equipment, preparation, and procedural technique.

There exists a set of general capabilities and equipment that is optimal and another considered to be the minimum required for both effective decision-making and therapeutic intervention. These items are categorized and described initially. The procedures described subsequently will draw as needed from this common supply list. Preprepared dedicated kits will accelerate the setup and performance of the described procedures.

As of this writing, there have been no documented cases of these procedures being done on humans in space or parabolic flight. Thoracotomies were, however, performed in rats as part of the STS-90 Neurolab Shuttle Mission. Consideration for gravity-dependent aspects of the maladies, their manifestations, and the potential interventions are addressed where appropriate.

## 11.2 Resource needs

### 11.2.1 Diagnostic imaging

Radiographic imaging, including chest x-ray with multiple views, is ideal for the diagnosis of intrathoracic issues, especially rib fractures and pneumothorax. Portable capability currently used in large animal veterinary practices and resource-limited medical settings is available. The first on-orbit radiograph was performed of the hand as part of the Fram2 SpaceX mission on March 31, 2025 (Dinner, 2025).

Computed tomography (CT) and fluoroscopy would be optimal, though size and weight requirements given current technology preclude their inclusion in existing craft. Static CT is a potential solution for this challenge; further discussion is found in the chapter dedicated to Diagnostic Tools.

Point-of-care handheld ultrasound is currently available from multiple vendors and has already been deployed in flight. Its use may be both diagnostic and procedural. Image interpretation based on artifactual shadows has increased diagnostic application to the lungs and pleura. Artificial intelligence to mitigate dependence on operator expertise for obtaining and interpreting usable images is under development. Sterile probe covers may be required for some procedures.

In the following procedures, when imaging is unnecessary or unavailable, performance via anatomic landmarks is described.

**Procedural Medicine for Spaceflight. DOI:** https://doi.org/10.1016/B978-0-443-27552-4.00031-2

### 11.2.2 Procedural imaging

- Laptop, tablet, or other device with a monitor and processing unit able to communicate with a wireless uniport scope (without cable) and also able to transmit images to a remote guide/observer.

### 11.2.3 Scopes

- Video fiberoptic (flexible) bronchoscope: 3.8–6.2 mm external diameter and 180 degree superior deflection/130 degree inferior deflection.
- Thoracoscope: 5 or 10 mm high resolution fiberoptic, 0 degree lens preferred.
- Flexible upper endoscope: 10 mm adult.

Single-use disposable and reusable scopes are available. The limited instrumentation needed for procedures during spaceflight lends itself to single-use scopes and avoids cleaning and sterilization considerations. Alternatively, reusable scopes could be considered, with chemical cleansing and sterilization most readily achieved.

- Stethoscope: Modern stethoscopes have noise canceling, sound amplification, and AI-enhanced interpretation that work even in noisy environments.

### 11.2.4 Supplemental lighting

- Headlight or overhead light(s): surgeon-worn headlights may also contain cameras.

### 11.2.5 Anesthesia and medications

- Airway management and anesthesia are discussed in Chapters 7, 9, 20, and 21.
- Supplemental oxygen should be available to maintain oxygen saturation of at least 90%. Given the potential contribution to ambient flammability risk, this should be delivered in a closed circuit via rebreather mask or endotracheal tube. Use of an oxygen concentrator may mitigate this risk by concentrating oxygen from the atmosphere (by depleting it somewhere else). The local increase in an individual's lungs is offset by the oxygen removed from the cabin by the concentrator. This greatly reduces the fire risk as it is a net neutral exchange with only a transient increase, which is not much different from typical mismatches related to air revitalization systems. Fans and ducts manage the local increase in oxygen and decrease it rapidly.

### 11.2.6 Patient preparation

- Platform for patient positioning and fixation (Velcro/hook and loop).
- An assortment of wedge-shaped cushions (30, 60, and 45 degrees) that may be used to optimize patient positioning in combination with straps.

### 11.2.7 Procedural site preparation

- Nonflammable antiseptic prep solutions are recommended for use in confined space environments. Therefore, water-based rather than alcohol-based products are preferred. Chlorhexidine Gluconate (CHG), as used in the operating room, is mixed with isopropyl alcohol and is potentially flammable until drying and dispersal time are allowed. However, when water-based, CHG would be an acceptable prep. Povidone-iodine solutions are nonflammable and do not require a specified time to mitigate the risk of ignition, as do alcohol-based preps. Iodine was used as a disinfectant for the potable water system on the Skylab missions. Hydrogen peroxide is nonflammable and breaks down into water and oxygen. Products containing accelerated hydrogen peroxide technology have been specifically approved for use aboard the International Space Station (ISS).

### 11.2.8 Surgeon/operator tether

- Stable fixation of the surgeon's feet to a surface will facilitate normalization of procedures learned on earth by presenting a familiar site picture and patient access. Mechanisms of such fixation may include Velcro, bicycle style clip-in pedal mechanisms, or equivalent. As an alternative to anchoring the feet, potentially preferable design concepts have been explored that incorporate the body's tendency to assume a semi-seated position in reduced gravity. In such a case, tethering at the level of the thighs should allow unimpeded hand and arm movement for task performance (Howard et al., no date). If a procedural assistant is required and available, similar tethering may be used. Limited accommodations may achieve the required anchoring, so that a dedicated procedure care space is neither required nor must be included in vehicle design.

### 11.2.9 Irrigation

- Please see Chapter 8: Irrigation for a description of accessible irrigation equipment and techniques.

### 11.2.10 Suction and related equipment

- Suction for use on the operative field may be provided by relatively robust portable closed canister systems that are powered by rechargeable batteries (e.g., Laerdal Medical).

Following thoracic procedures, a dry-suction, dry-seal closed chest drainage system such as the Pleur-evac™ Sahara Chest Drainage System may be considered (Sahara Chest Drainage System. Teleflex). However, this still requires demonstration of reliable function in the space operational environment. In particular, the water seal portion of a closed chest drainage system relies on gravity. A Heimlich valve may be used to accomplish the required creation of a one-way valve that will allow air and fluid out of the pleural space on expiration without allowing air into the pleural space on inspiration. Preparation should include the availability of portable mechanical hand pumps that have been proven effective in intermittent evacuation of pleural contents via indwelling thoracostomy tubes (Fig. 11.1). Jackson-Pratt bulbs/suction grenades may also be used postoperatively to prevent the accumulation of blood, serous fluid, or pus.

### 11.2.11 Air insufflation

- Used to fill the esophagus and stomach to facilitate endoscopy.

### 11.2.12 Cautery

- Ignition and ambient flammability risk are elevated in the closed environment. As such, a battery-powered handheld "ophthalmic" pen-type cautery with a 5″ (12.7 cm) extended tip may be considered at lower risk than electrocautery (Fig. 11.2). Its required use is expected to be low; however, it can still start a fire. Therefore, it may reasonably be deleted from a minimum capability/equipment list; consideration of a more complex reduced ignition cautery, such as a harmonic/ultrasonic scalpel, is also unnecessary.

### 11.2.13 Point-of-care laboratory testing

- This is discussed in Chapter 27.

### 11.2.14 Blood and products for trans-/infusion

- Considerations for prolonged or exploration-class missions must include the current shelf-life limitations of refrigerated blood. As such, developments in the use of freeze-dried red blood cells in combination with infused factors are important (Elder et al., 2024).

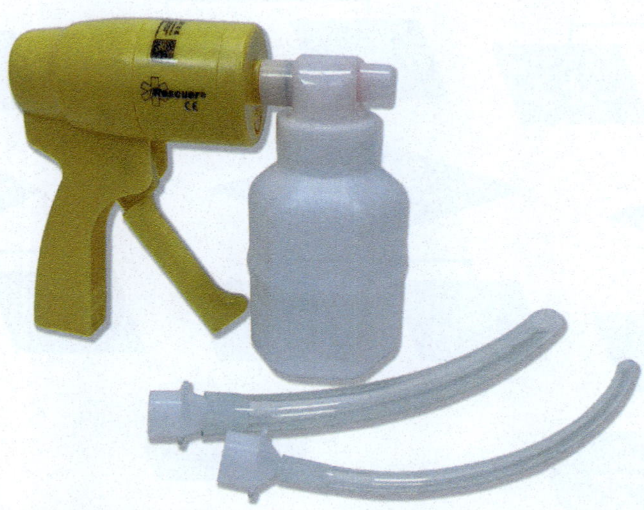


**FIGURE 11.1 Rescuer manual pump.** May be used intermittently for evacuation of pleural contents. Mass 190 g, volume 392 $cm^3$. Rescuer® Manual Vacuum Pump with adult and child catheter. From Rescuer Products.

FIGURE 11.2 **Battery powered handheld cautery (Bovie) with 5" (12.7 cm) tip.** As necessary, it may be associated with reduced risk of ignition. Bovie website. From Aspen Surgical (Bovie). https://www.aspensurgical.com/.

### 11.2.15 Management of sharps and biohazards

- Sharps disposal containers with a seal to prevent escape.

## 11.3 Procedures, conditions, and indications

### 11.3.1 Bronchoscopy (flexible fiberoptic)

#### *11.3.1.1 Decision making*

Bronchoscopy in the setting of spaceflight is predominantly therapeutic rather than diagnostic; it has not reportedly been performed in space. Its reach is at the level of the large airways (trachea, mainstem, segmental, and subsegmental bronchi). Indications include aspiration of foreign body or of other materials that is insufficiently cleared with coughing, percussion, or Heimlich maneuver. Food aspiration, in part related to gastroesophageal reflux, is a more consistent concern in microgravity. Mucus plugging of airways or significant hemoptysis may similarly obstruct airways. Potential manifestations of airway obstruction include persistent cough, shortness of breath, clinically significant postobstructive airway atelectasis or lung collapse, pneumonia, hypoxemia, and sepsis. Auscultation with the stethoscope may detect decreased breath sounds in certain areas, a manifestation of airspaces consolidated with secretions. Further diagnosis of occlusion of the large airways may be accomplished by x-ray showing lobar or segmental collapse with abrupt cut-off of a bronchus (Fig. 11.3).

During spaceflight, use of aerosolized compounds requires assurance of relative containment to the operative field, focused delivery, and atmospheric filtering.

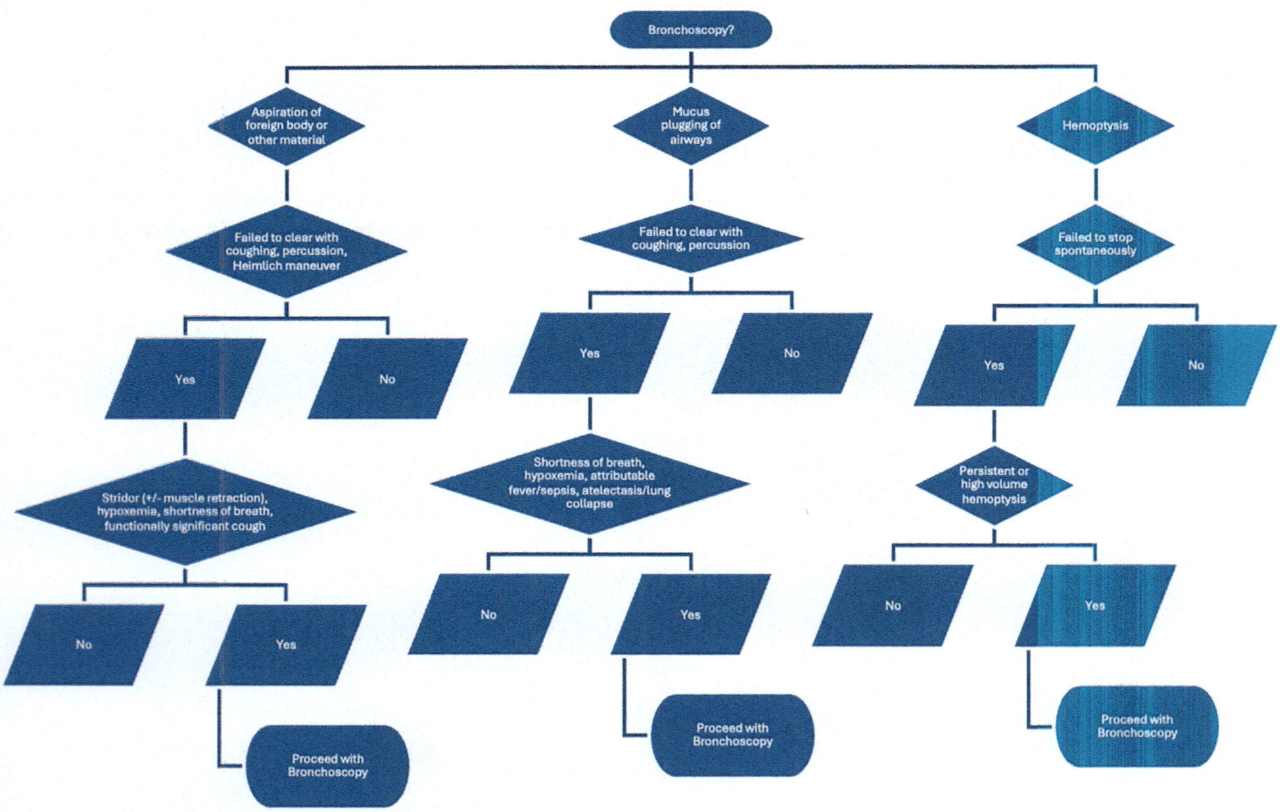


FIGURE 11.3 Bronchoscopy decision flowchart.

### 11.3.1.2 Description

Severity and/or persistence of attributable symptoms and/or functional impairment should prompt consideration of the potential benefit of bronchoscopy. Fever, green sputum production, elevated white blood cell count, persistent or high volume hemoptysis, or indication of persistent lung parenchymal collapse on chest x-ray or ultrasound will influence the decision whether to proceed. Diagnostic bronchoscopy for Gram stain and culture is not likely to be useful in the space environment. Terrestrial bronchoscopy is often performed with conscious sedation. However, in the space environment, in order to minimize mass and volume constraints, as well as to avoid utilizing anesthetic resources, it may be performed awake with premedication with oral benzodiazepine for anxiolysis.

### 11.3.1.3 Procedural technique and materials for spaceflight

Materials, mass, volume table (Table 11.1).

Step by step (Daughtry, no date):

1. Prepare video-assisted bronchoscope (including attachment to suction tubing; turn on suction).
2. Prepare the patient with prebreathing of 100% oxygen.
3. Prefill several saline syringes for irrigation.
4. Administer aerosolized lidocaine to the posterior oropharynx. You may use a tongue depressor or a 4×4 to depress the tongue.
5. Don a surgical mask and gloves.
6. Place the bite block and pass the scope orally. If intubated for other reasons, pass through the endotracheal tube.
7. Once the vocal cords are visible, inject lidocaine through the bronchoscope so that it sprays the vocal cords. Following this, have the patient pant, take a deep breath, or say "Ahhh" to abduct the cords. Pass scope. The carina should be the next visible landmark.
8. Systematically begin clearing the right mainstem bronchus and its branches. The right side is more vertical and is more likely, thereby, to contain significant material. Then clear the left system similarly.
9. Use irrigation with partial or full volume from 10 ml Luer tip saline-filled syringes and suction to clear inspissated secretions, blood, etc.
10. If the indication for bronchoscopy is the retrieval of a foreign body, the first priority is to identify it. Having done so, introduce biopsy forceps via the operating port of the bronchoscope. Open the forceps once outside the end of the scope, close on the object, and hold it closed tightly while you remove the entire scope and biopsy forceps apparatus. Do not attempt to withdraw the biopsy forceps back into the scope once the object is caught. If the biopsy forceps is unable to grasp, the snare may be introduced. The snare is first passed downstream of the foreign body and then used to grasp the object as it is drawn back. Once ensnared, the entire scope is removed while holding the foreign body tightly in the snare.
11. The retrieved foreign body should be inspected once out of the body to assess whether it has been completely removed. If this is unclear, the bronchoscope may be reintroduced to examine the airway.
12. During the procedure, monitor closely for oxygen desaturation. When saturation drops into the 80 s, pause to allow the patient to recover before recommencing.
13. Use the stethoscope to listen for symmetric breath sounds in all distributions. A chest radiograph may be performed, if available.
14. Following the procedure, oxygen is weaned as tolerated.

**TABLE 11.1 Minimum materials, mass, volume for bronchoscopy.**

| Item | Mass | Volume | Purpose |
|---|---|---|---|
| Stethoscope | 186 g | 1,192 $cm^3$ | Evaluate breath sounds. |
| Conscious sedation (including monitor) | 5 kg | 13,000 $cm^3$ | May be considered. However, awake bronchoscopy can be easily, if somewhat uncomfortably, performed with lidocaine injected through the scope at the vocal cords. |
| Pulse oximetry | 50 g | 63 $cm^3$ | Monitoring. |
| Patient positioning restraints | 6 kg | <10,000 $cm^3$ | Fix patient location to facilitate procedure. |
| Operator positioning restraints | Variable based on type | Variable based on type | Fix Operator location to facilitate procedure. |
| Flexible fiberoptic video-assisted bronchoscope | <1 kg | 14 $cm^3$ | Critical to perform procedure. |
| Viewing monitor | 8 kg | 40,000 $cm^3$ | Optimized viewing for Operator, team, and remote. |
| Suction setup (portable self-contained device) | 4 kg | <10,000 $cm^3$ | Required to clear view of secretions/materials. |
| Irrigation (10 ml saline-filled syringes—Luer slip tip) | 27 g | 10 $cm^3$ | Required to clear view of secretions/materials. |
| Lidocaine 2% liquid (25 ml) | 70 g | 35 $cm^3$ | Inject through the scope at the level of the vocal cords to anesthetize the cords and allow awake procedure. |
| Nebulized lidocaine | 18 g | 9 $cm^3$ | Spray posterior pharynx and vocal cords prior to instrumentation—*optional. Challenge of containing nebulized delivery supports avoidance of use.* |
| Surgical mask | 3 g | <100 $cm^3$ | Asepsis and personal protection, Operator. |
| Sterile gloves | 14 g | <100 $cm^3$ | Asepsis and personal protection, Operator. |
| Bite block | <3 g | <100 $cm^3$ | For access by oral route. |
| Oxygen (E size cylinder)<br>Oxygen concentrator | 11 kg<br>9 kg | 15,000 $cm^3$<br>17,500 $cm^3$ | Oxygenate patient before, during, and following procedure. |
| Biopsy (grasping) forceps | <100 g | 500 $cm^3$ | Used for lodged foreign body extraction. |
| Snare | <100 g | 500 $cm^3$ | Used for lodged foreign body extraction. |
| Chest radiography (machine) | 10 kg | 17,500 $cm^3$ | Post procedure imaging—*optional.* |
| Handheld ultrasound | 307 g | 100 $cm^3$ | Post procedure imaging—*optional.* |

## 11.3.2 Upper endoscopy (flexible fiberoptic)

### *11.3.2.1 Decision making*

Somewhat in contrast to bronchoscopy performed during spaceflight, upper gastrointestinal endoscopy is both therapeutic and diagnostic to guide further therapy. Its span includes the posterior pharynx, esophagus, stomach, and proximal duodenum. Indications include esophageal obstruction from a food bolus, hematemesis, or dark tarry stools that test positive for blood. Causes of upper gastrointestinal bleeding are stress ulceration of the stomach or duodenum and that associated with significant esophageal injury due to forceful or prolonged vomiting, retching, or straining (Mallory-Weiss tear: mucosal tear of esophagus; Boerhaave's Syndrome: transmural rupture of the esophagus; Fig 11.4)

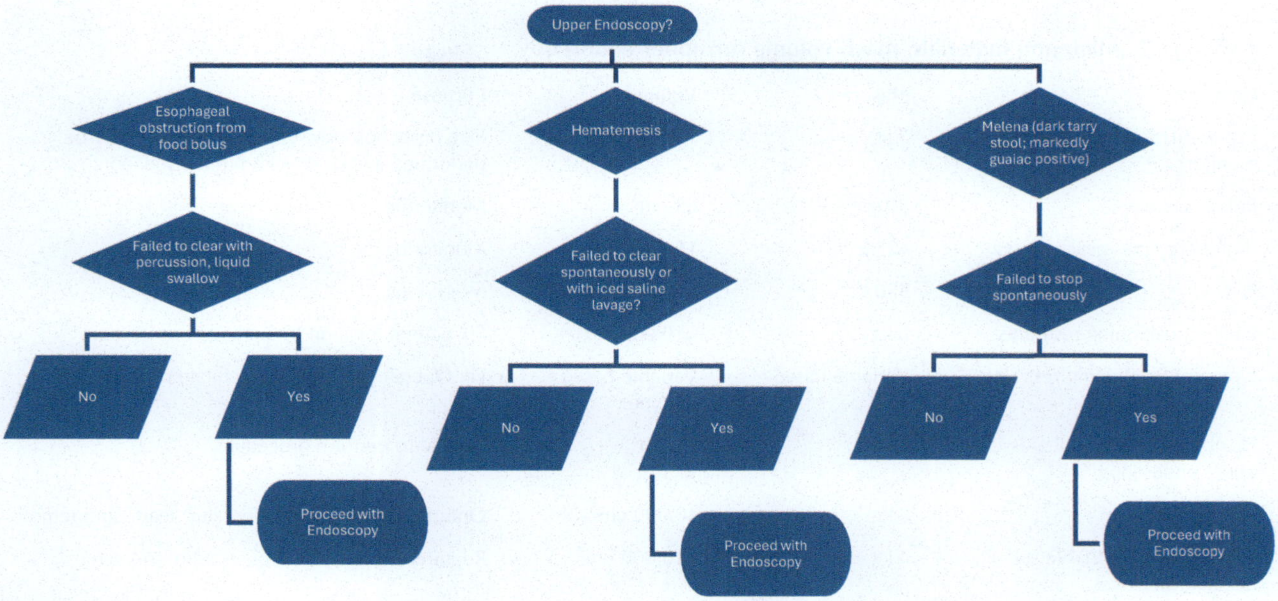


**FIGURE 11.4** Upper endoscopy decision flowchart.

### *11.3.2.2 Description*

Obstruction due to a food bolus should be a rare phenomenon. If the astronaut is symptomatically stable, oral papain treatment may be initiated first. Endoscopy would be considered for acute or persistent upper gastrointestinal bleeding of a significant volume. Such bleeding may ultimately be treated by iced saline lavage through a nasogastric tube. However, details about specific etiology gleaned from endoscopy can significantly impact the therapeutic plan. For example, bleeding from a severe gastritis or ulceration may be primarily treated with H2 blocker medication.

### *11.3.2.3 Procedural technique and materials for spaceflight*

Materials, mass, volume table (Table 11.2).

Step by step:

1. Performance on an empty stomach or after a period of fasting is optimal.
2. Parenteral sedation should be considered.
3. Initiate pulse oximetry and blood pressure monitoring; ECG may be added.
4. A bite block is placed to keep the mouth open and prevent biting of the endoscope.
5. Attach suction to the endoscope.
6. Fill multiple 10 ml Luer slip-tipped syringes with saline.
7. Don a surgical mask and gloves.
8. The endoscope is inserted in the mouth and guided down the esophagus. It is assisted by the patient's swallowing.
9. Gentle air insufflation may be used while passing the scope. The scope should never be passed without direct vision of the lumen ahead or forced.
10. Areas of bleeding or otherwise of interest may be irrigated with a stream of saline delivered from the 10 ml Luer-tipped syringes.
11. Foreign body extraction begins with the introduction of biopsy forceps through the endoscope's operating port. Open the forceps once outside the end of the scope, close on the object, and hold it closed tightly while you remove the entire scope and biopsy forceps apparatus. Do not attempt to withdraw the biopsy forceps back into the scope once the object is caught. If the biopsy forceps are unable to grasp, the snare may be introduced. The snare is first passed downstream of the foreign body and then used to grasp the object as it is drawn back. Once ensnared, the entire scope is removed while holding the foreign body tightly in the snare.
12. The retrieved foreign body should be inspected once out of the body to assess if it has been completely removed. If this is unclear, the endoscope may be reintroduced.
13. Suction is applied continuously as the scope is removed to remove accumulated air.
14. Following the procedure, if sedation was employed, the patient is monitored until the effects abate.

**TABLE 11.2** Minimum materials, mass, volume for upper endoscopy.

| Item | Mass | Volume | Purpose |
|---|---|---|---|
| Conscious sedation(including monitor) | 5 kg | 13,000 $cm^3$ | Like bronchoscopy, upper endoscopy may be performed awake or with limited sedation. |
| Pulse oximetry | 50 g | 63 $cm^3$ | Monitoring. |
| Blood pressure measurement | 425 g | 1600 $cm^3$ | Monitoring. |
| ECG | 550 g | <100 $cm^3$ | Monitoring |
| Patient positioning restraints | 6 kg | <10,000 $cm^3$ | Fix patient location to facilitate procedure. |
| Operator positioning restraints | Variable based on type | Variable based on type | Fix Operator location to facilitate procedure. |
| Flexible fiberoptic video-assisted upper endoscope | <1 kg | 14 $cm^3$ | Critical to perform procedure. |
| Viewing monitor | 8 kg | 40,000 $cm^3$ | Optimized viewing for Operator, team, and remote. |
| Suction setup (portable self-contained device) | 4 kg | <10,000 $cm^3$ | Required to clear view of secretions/materials. |
| Air insufflation | 9 kg | 22,500 $cm^3$ | Visualization. It is necessary to insufflate the stomach and esophagus to appropriately visualize the structures. |
| Irrigation (10 ml saline-filled syringes—Luer slip tip) | 27 g | 10 $cm^3$ | Required to clear view of secretions/materials. |
| Bite block | <3 g | <100 $cm^3$ | To prevent damage to scope. |
| Oxygen (E size cylinder)<br>Oxygen concentrator | 11 kg<br>9 kg | 15,000 $cm^3$<br>17,500 $cm^3$ | Oxygenate patient before, during, and following procedure. |
| Surgical mask | 3 g | <100 $cm^3$ | Asepsis and personal protection, Operator. |
| Sterile gloves | 14 g | <100 $cm^3$ | Asepsis and personal protection, Operator. Sterile gloves are ideal but not required |
| Biopsy (grasping) forceps | <100 g | 500 $cm^3$ | Used for lodged foreign body extraction. |
| Snare | <100 g | 500 $cm^3$ | Used for bolus extraction. |

## 11.3.3 Chest wall trauma: minor procedures

### *11.3.3.1 Decision making*

Blunt and superficial penetrating trauma may be expected to occur particularly during extravehicular activity. Even superficial body penetration, however, may involve perforation of the spacesuit with its attendant required actions. The gold standard of care for rib fractures is pain control and respiratory protection. Pain control may be provided by enteral or parenteral medications or by intercostal block (below). Respiratory protection is facilitated by using incentive spirometry.

### *11.3.3.2 Description*

Injuries may be clearly nonpenetrating. Where there is penetration of the skin, exploration is required to determine the depth of and whether the pleural space has been entered (Fig. 11.5).

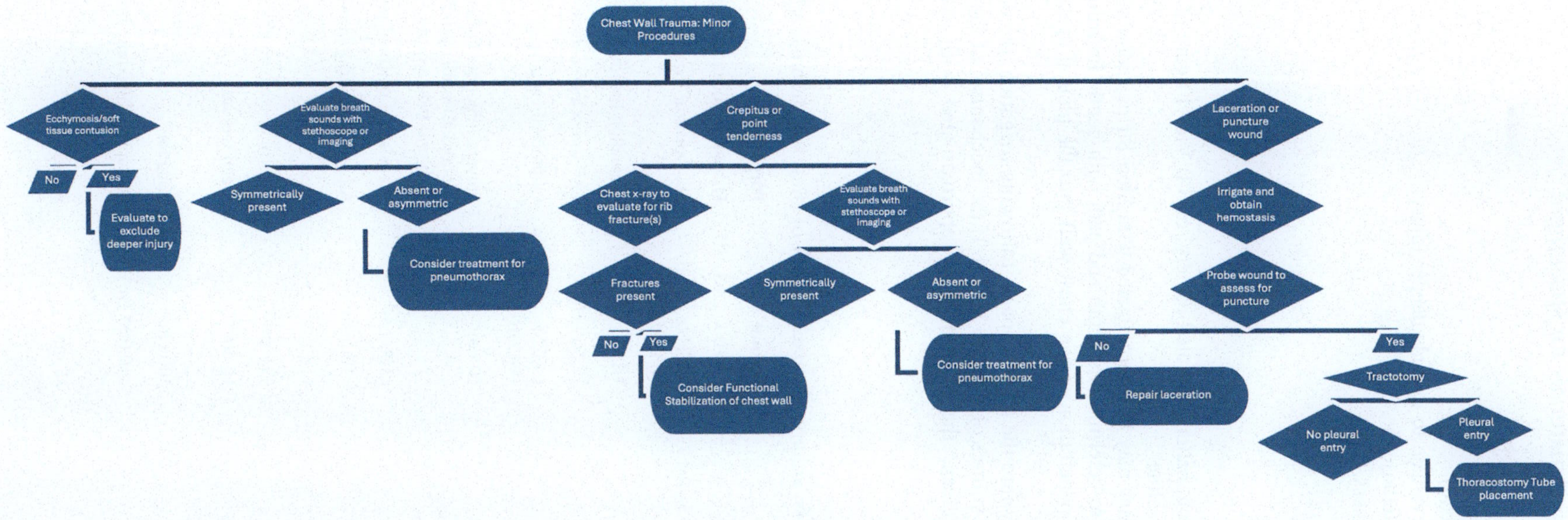


**FIGURE 11.5** Chest wall injury: minor procedures decision flowchart.

### 11.3.3.3 Procedural technique and materials for spaceflight

Materials, mass, volume table (Table 11.3).

Step by step:

1. Analgesia as indicated (oral/parenteral).
2. Ecchymosis/soft tissue contusion: evaluate to exclude deeper injury.
   a. Evaluate breath sounds with a stethoscope.
   b. Feel for crepitus or point tenderness.
3. Laceration or puncture wound.
   a. Don a surgical mask and gloves.
   b. Prep the operative site.
   c. Drape the operative site.
   d. Prophylactic antibiotic.
   e. Achieve hemostasis—compression, battery-operated cautery, or 3-0 Vicryl figure of 8 stitch
   f. Irrigate the wound with normal saline.
   g. Small, clean, superficial lacerations may be approximated with Steri-Strips and/or skin glue.
   h. Explore the tract of the wound to determine its depth and path with gentle insertion and probing with a curved Kelly clamp. If there has been penetration of the chest wall into the pleural space, tube thoracostomy will be required.
   i. If there is no indication of entry into the pleural space, following irrigation, the deep soft tissue layers may be approximated using interrupted sutures of 0 or 3-0 Vicryl. If the mechanism of the injury and/or the appearance of the wound suggest it is clean, the skin may be approximated with staples and followed closely postoperatively.

**TABLE 11.3** Minimum materials, mass, volume for management of chest wall trauma: minor procedures.

| Item | Mass | Volume | Purpose |
|---|---|---|---|
| Patient positioning restraints | 6 kg | <10,000 cm3 | Fix patient location to facilitate procedure. |
| Operator positioning restraints | Variable based on type | Variable based on type | Fix Operator location to facilitate procedure. |
| Stethoscope | 186 g | 1,192 $cm^3$ | Evaluation breath sounds. |
| Chest radiography (machine) | 10 kg | 17,500 $cm^3$ | CXR to exclude deeper injury. |
| Headlight +/- camera<br>Battery pack/light source | 304 g<br>260 g | <4,620 $cm^3$<br>194 $cm^3$ | Optimized visualization. |
| Prophylactic antibiotic | 31 g | 10 $cm^3$ | Prevent infection. Cefazolin (Clindamycin for penicillin allergic). Consider prophylactic therapy for maximum 72 h and then adjust per clinical findings. |
| Surgical mask | 3 g | <100 $cm^3$ | Asepsis and personal protection, Operator. |
| Sterile gloves | 14 g | <100 $cm^3$ | Asepsis and personal protection, Operator. |
| Prep sponge | 113 g | 118 $cm^3$ | Prep. |
| Self-adhesive drapes | 90 g | 431 $cm^3$ | Control operative field. |
| Irrigation (10 ml saline filled syringe) | 27 g | 10 $cm^3$ | Clean wound. |
| Bulb syringe (60 ml saline) | 59 g | 300 $cm^3$ | Irrigation. |
| Suction setup (portable self-contained device) | 4 kg | <10,000 $cm^3$ | To capture bleeding, irrigation fluid, and debris. |
| Local anesthetic (25 ml) | 70 g | 35 $cm^3$ | Analgesia. |

*(Continued)*

**TABLE 11.3 (Continued)**

| Item | Mass | Volume | Purpose |
|---|---|---|---|
| Kelly clamp—curved | 36 g | 70 $cm^3$ | For exploring extent/depth of tract. |
| 10 ml syringe (empty) | 20 g | 8 $cm^3$ | Local infiltration of anesthetic. |
| # 22 gauge needle | <1 g | 21 $cm^3$ | Local infiltration of anesthetic. |
| Toothed forceps | 57 g | <10 $cm^3$ | Tissue retraction for wound exploration. |
| Needle holder | 70 g | 8 $cm^3$ | Wound closure. |
| Mayo scissors | 85 g | 11 $cm^3$ | Cutting sutures. |
| 3-0 Vicryl suture | <5 g | 0.6 $cm^3$ | Soft tissue approximation or hemostasis. |
| 0 Vicryl suture | <5 g | 0.6 $cm^3$ | Soft tissue approximation. |
| Handheld cautery (battery-powered) | 62 g | 10 $cm^3$ | Hemostasis. |
| Skin glue (e.g., Dermabond) | 0.7 g | 0.7 $cm^3$ | Skin/superficial wound approximation. |
| 1″ Steri-Strips | 15 g | 41 $cm^3$ | Skin approximation. |
| Skin stapler | 92 g | 550 $cm^3$ | Skin/wound approximation. |
| Absorbent dressings | 86 g | 524 $cm^3$ | To absorb predominantly serous ooze. |
| "Dressing" Tape | 19 g | 50 $cm^3$ | For dressing fixation. |
| Incentive spirometer | 215 g | 1,721 $cm^3$ | Lung expansion, prevent atelectasis, enhance recovery. |

## 11.3.4 Chest wall trauma: intercostal nerve block

### *11.3.4.1 Decision making*

Blunt and penetrating chest trauma may be associated with persistent moderate to severe pain in the setting of rib fracture and/or that which compromises respiration and/or function. In addition, dosing of narcotic analgesics sufficient to provide desired relief could result in unacceptable functional impairment (Fig. 11.6).

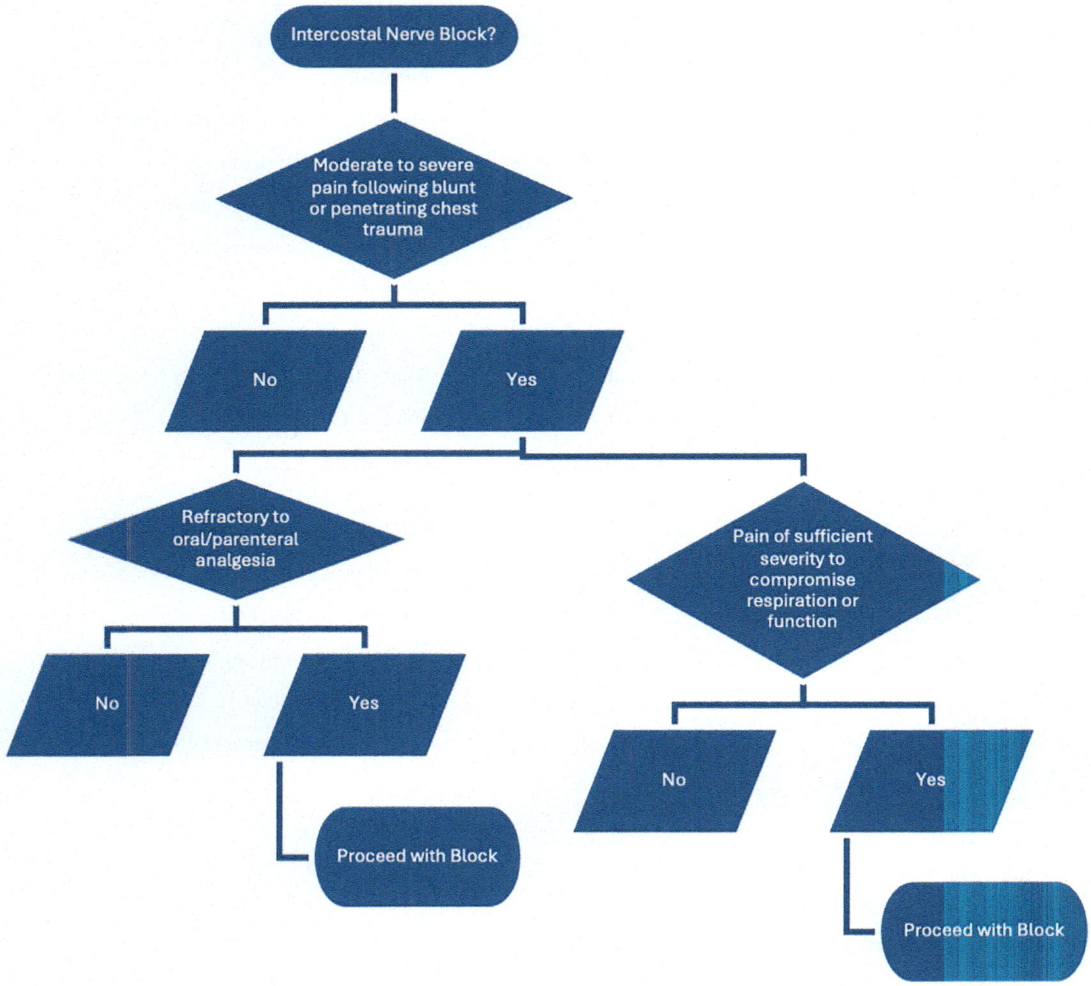


FIGURE 11.6 Intercostal nerve block decision flowchart.

### *11.3.4.2 Description*

Consider this procedure to supplement other analgesic modalities; an intercostal block may be administered at several confluent levels based on the extent of the injury to provide dermatomal symptomatic relief.

### *11.3.4.3 Procedural technique and materials for spaceflight*

Materials, mass, volume table (Table 11.4).

Step by step (Lopez-Rincon et al., no date):

1. The procedure may be performed using ultrasound guidance or anatomical landmarks. The former is preferred to potentially increase effectiveness and decrease complications.
2. Place the patient upright, lateral decubitus, or prone (assuming a partial gravity environment).
3. Palpate the target rib spaces; mark the skin just forward of the junction of the affected rib with the bony spine.
4. Prep the skin and administer local anesthetic.
5. If ultrasound is used, guide the needle to the neurovascular bundle at the inferior aspect of the target rib. If by anatomic landmark, the needle is angled ~20 degrees cephalad and walked off the inferior aspect of the target rib (whose interspace is to be blocked).
6. Advance the spinal needle just underneath the inferior aspect of the target rib. Aspirate the syringe to ensure no return of blood to prevent an intravascular injection.
7. Inject the long-acting local anesthetic (~3–5 ml per interspace of 0.25% bupivacaine).
8. Repeat for each involved interspace.

**TABLE 11.4** Minimum materials, mass, volume for intercostal block.

| Item | Mass | Volume | Purpose |
|---|---|---|---|
| Patient positioning restraints | 6 kg | <10,000 $cm^3$ | Fix patient location to facilitate procedure. |
| Operator positioning restraints | Variable based on type | Variable based on type | Fix Operator location to facilitate procedure. |
| Handheld ultrasound | 307 g | 100 $cm^3$ | Guidance. |
| Sterile cover for ultrasound probe | <14 g | 35 $cm^3$ | Asepsis. |
| Surgical mask | 3 g | <100 $cm^3$ | Asepsis and personal protection, Operator. |
| Sterile gloves | 14 g | <100 $cm^3$ | Asepsis and personal protection, Operator. |
| 10 ml syringe (empty) | 20 g | 8 $cm^3$ | Administer block. |
| #22 g spinal needle | <10 g | 0.13 $cm^3$ | Administer block. |
| Skin marker | 12 g | 18 $cm^3$ | Guidance. |
| Prep kits, including sponge(s) impregnated with prep solution | 136 g | 6,500 $cm^3$ | Prep. |
| Bupivacaine (Marcaine) 0.25% (25 ml) | 70 g | 35 $cm^3$ | Analgesia/block. |
| Band-Aid | <1 g | 19 $cm^3$ | For skin puncture site. |

## 11.3.5 Chest wall trauma—initial treatment open pneumothorax (sucking chest wound)

### *11.3.5.1 Decision making*

A penetrating and apparently "sucking" chest wound is an emergency and must be occluded as quickly as possible to prevent air entry through the wound as this will rapidly create a large pneumothorax. When such a wound is in excess of two-thirds the diameter of the trachea, air may preferentially enter the chest cavity through the wound rather than through the trachea, and the patient could suffocate with insufficient air entering the lungs. The wound may be acutely stabilized by applying a dressing that allows air and blood to exit the pleural space on expiration and prevents air from entering the pleural space on inspiration. Some degree of respiratory improvement is intended, and development of tension pneumothorax is unlikely (Fig. 11.7).

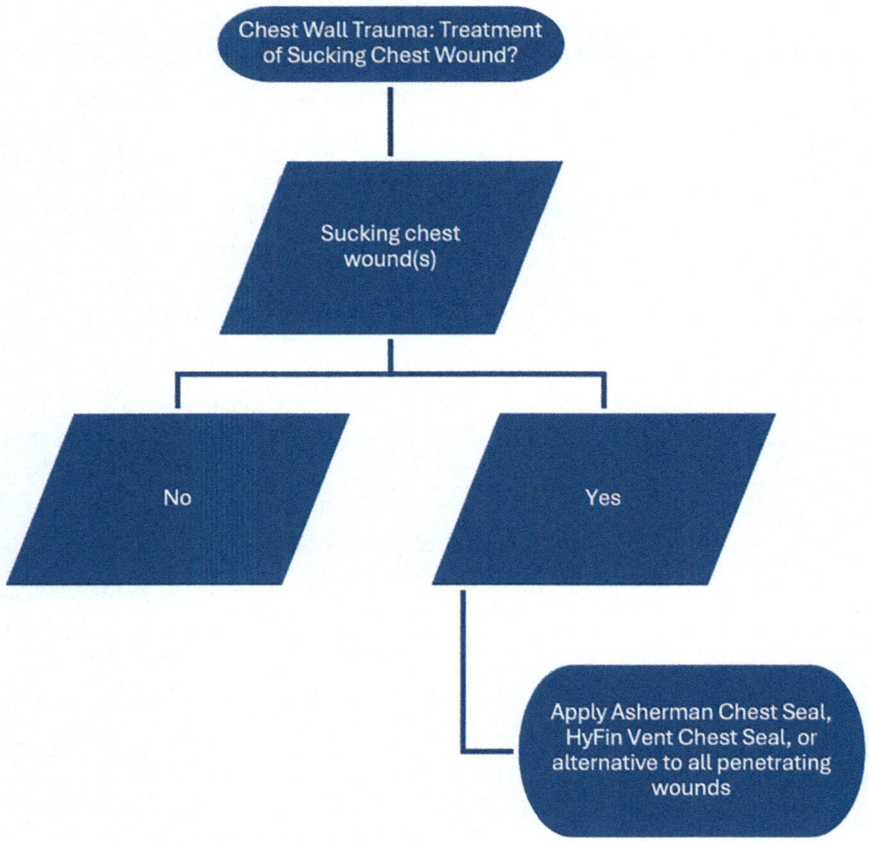


**FIGURE 11.7** Chest wall trauma open pneumothorax decision flowchart.

### *11.3.5.2 Description*

Application of either of the two modified dressings described below benefits from rapid application. In the absence of purpose-built dressings that are available terrestrially, a one-way valve partially occluding dressing can be approximated using Vaseline gauze and an adhesive plastic dressing. However, all of these dressings are temporary and are to be combined with tube thoracostomy as well as other potentially needed thoracic interventions as soon as feasible.

### *11.3.5.3 Procedural technique and materials for spaceflight*

Materials, mass, volume table (Table 11.5).

Step by step:

### *11.3.5.4 Asherman Chest Seal*

1. Dry around the wound.
2. Remove liner from dressing.
3. Center over wound and apply at end expiration. Smooth out the adhesive circumferentially. Reinforce adhesive borders with cloth tape as needed.
4. A second or additional wound is treated similarly.

### *11.3.5.5 HyFin Vent Chest Seal Twin Pack (North American Rescue)*

1. Applied as per the Ascherman at the end of expiration.
2. A second or additional wound is treated similarly.

#### 11.3.5.6 Alternative dressing

1. Clean and dry the area around the open wound
2. Place Vaseline gauze over the open wound
3. Place a plastic adhesive dressing around the Vaseline gauze, covering it and securing it on three sides so that air can escape from the fourth side.

Following dressing placement, place the thoracostomy tube (see below).

**TABLE 11.5** Minimum materials, mass, volume for management of open pneumothorax.

| Item | Mass | Volume | Purpose |
|---|---|---|---|
| Patient positioning restraints | 6 kg | <10,000 $cm^3$ | Fix patient location to facilitate procedure. |
| Operator positioning restraints | Variable based on type | Variable based on type | Fix Operator location to facilitate procedure. |
| Prophylactic antibiotic | 31 g | 10 $cm^3$ | Prevent infection. Cefazolin (Clindamycin or Vancomycin for penicillin allergic). Prophylactic therapy for maximum 72 h and then adjust per clinical findings. |
| Absorbent towel (s) | 25 g | 36 $cm^3$ | Dry application site. |
| Asherman Chest Seal | 24 g | 430 $cm^3$ | Rapid initial control of open pneumothorax. |
| HyFin Vent Chest Seal Twin Pack | 39 g | 557 $cm^3$ | Rapid initial control of open pneumothorax. |
| Vaseline gauze | 17 g | 44 $cm^3$ | Alternative one-way dressing; Vaseline gauze will not allow air into the wound but will allow air out. |
| Plastic adhesive dressing (like Tegaderm ®) | 5 g | 41 $cm^3$ | To secure the Vaseline gauze |
| Sterile gloves | 14 g | <100 $cm^3$ | Asepsis and personal protection, Operator. Sterile gloves are ideal but not required |
| Cloth tape | 19 g | 50 $cm^3$ | To bolster Asherman, HyFin, or Tegaderm adhesive. |

### 11.3.6 Needle decompression for pneumothorax

#### 11.3.6.1 Decision making

Tension pneumothorax may be associated with respiratory distress and/or hemodynamic compromise. As such, needle decompression can be done quickly while setting up for tube thoracostomy. Further, the catheter used for needle decompression may be used to advance a wire to place a thoracostomy tube by Seldinger technique (Fig. 11.8).

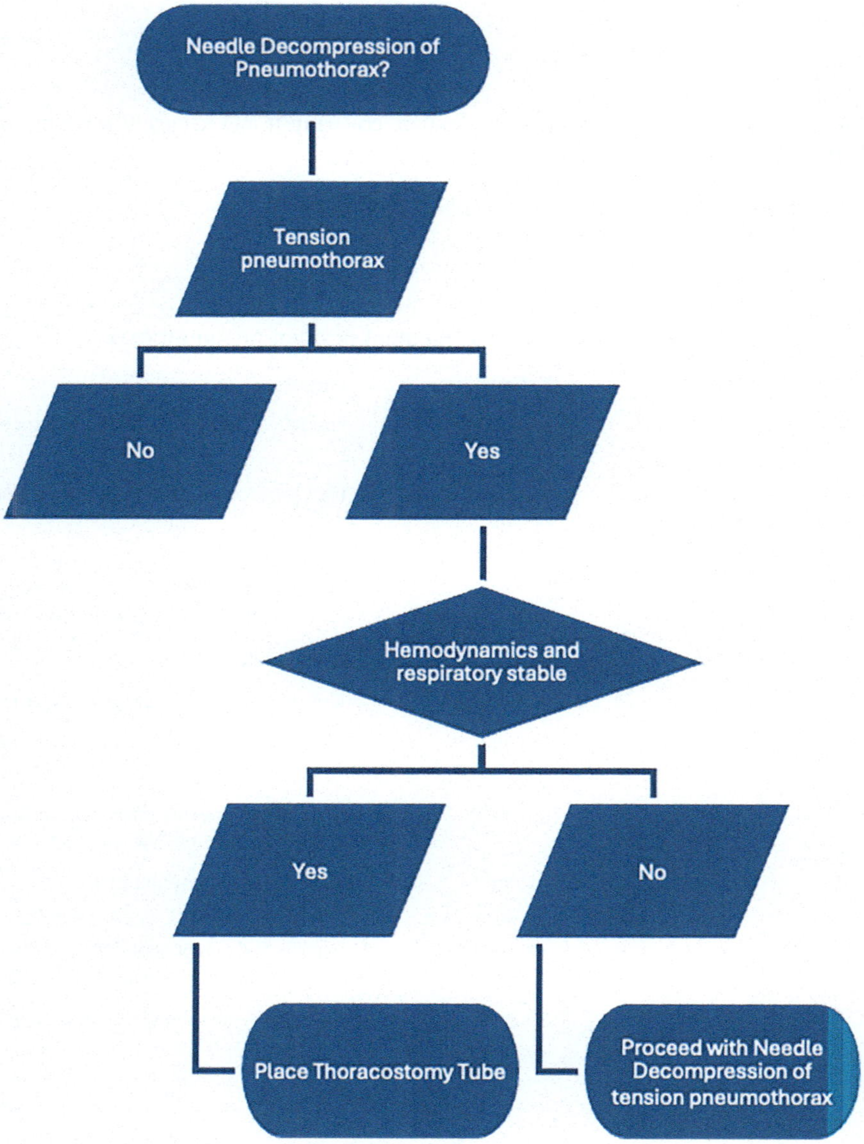


**FIGURE 11.8** Needle decompression for tension pneumothorax decision flowchart.

### *11.3.6.2 Description*

The area should be prepped if possible, especially if it will be used for the placement of a tube thoracostomy. Guided by anatomic landmarks, the placement of the angiocatheter is accompanied by a rush of air released from its pressurized state. Both the $2^{nd}$ intercostal space in the midclavicular line (ICS2-MCL) and the $4^{th}/5^{th}$ intercostal space in the anterior axillary line (ICS 4/5-AAL) are preferred locations for needle decompression. Midclavicular placement may be preferred in those with a thicker chest wall. This is temporizing, and the catheter should be replaced by a thoracostomy tube as soon as possible.

### *11.3.6.3 Procedural technique and materials for spaceflight*

Materials, mass, volume table (Table 11.6).

Step by step (Azizi et al., 2021):

1. Prep the target site.
2. Identify the midclavicular line by tracing a line up from the nipple.

3. The second rib may be identified as corresponding to the palpable Angle of Louis at the manubrial-sternal junction or as the first rib whose superior and inferior surfaces may be palpated inferior to the clavicle.
4. After identifying the second rib, find the superior aspect of the third rib. Push in the angiocatheter in such a manner as to follow the top surface of the third rib. Controlled pressure will be required to pop into the pleural space, which is then usually accompanied by a rush of air.
5. Following this, a tube thoracostomy must be placed, either by Seldinger technique through the angiocatheter or by open technique (see below).

**TABLE 11.6** Minimum materials, mass, volume for needle decompression of tension pneumothorax. .

| Item | Mass | Volume | Purpose |
|---|---|---|---|
| Patient positioning restraints | 6 kg | <10,000 $cm^3$ | Fix patient location to facilitate procedure. |
| Operator positioning restraints | Variable based on type | Variable based on type | Fix Operator location to facilitate procedure. |
| Sterile gloves | 14 g | <100 $cm^3$ | Asepsis and personal protection, Operator. |
| Prep sponge | 113 g | 118 $cm^3$ | Asepsis |
| 10 ml Luer lock syringe | 20 g | 8 $cm^3$ | Local anesthetic. |
| #14 or 16 gauge angiocatheter | 14 g | 177 $cm^3$ | Rapid initial decompression of tension pneumothorax. |

## 11.3.7 Tube thoracostomy for pneumothorax, pleural effusion, hemothorax, and empyema

### *11.3.7.1 Decision making*

Thoracostomy is the standard of care for treatment of pneumothorax to reexpand the collapsed lung (Fig. 11.9). Pneumothoraces that are minimally symptomatic and stable may be serially observed. Small pneumothoraces may also be treated with 100% oxygen therapy since cellular metabolism will use up oxygen in the cavity and create a relative vacuum that will reinflate the lung over several hours. This process is too slow and creates too little vacuum to manage pneumothoraxes greater than 30% of the lung volume.

Primary treatment for large pneumothoraces (>30%) is thoracostomy tube placement. Otherwise, thoracostomy tube placement may follow emergent decompression of a tension pneumothorax with an angiocatheter as described above. Placement of a thoracostomy tube, especially in the space environment, should be reserved for pneumothoraces that are assessed as moderate to large on chest x-ray or other imaging.

Tube thoracostomy is also appropriate for drainage of fluid, blood, or pus in the pleural space (Fig. 11.10). Several factors guide a decision to drain accumulated material in this space. Such considerations include volume of fluid sufficient to cause significant atelectasis or lobar collapse, attributable symptoms, sanguinous nature of fluid that would be expected to organize and entrap the lung (prevent full expansion and impair ventilation/gas exchange), or postpneumonic effusion/empyema associated with infection and potential sepsis. Ultrasound-guided tube thoracostomy placement theoretically affords the least invasive and most precise placement of a drainage tube, potentially for drainage of a loculated rather than free-flowing effusion. If this is the intent, the operator should first be prompted to consider the clinical significance of such a localized collection (Anderson et al., 2022).

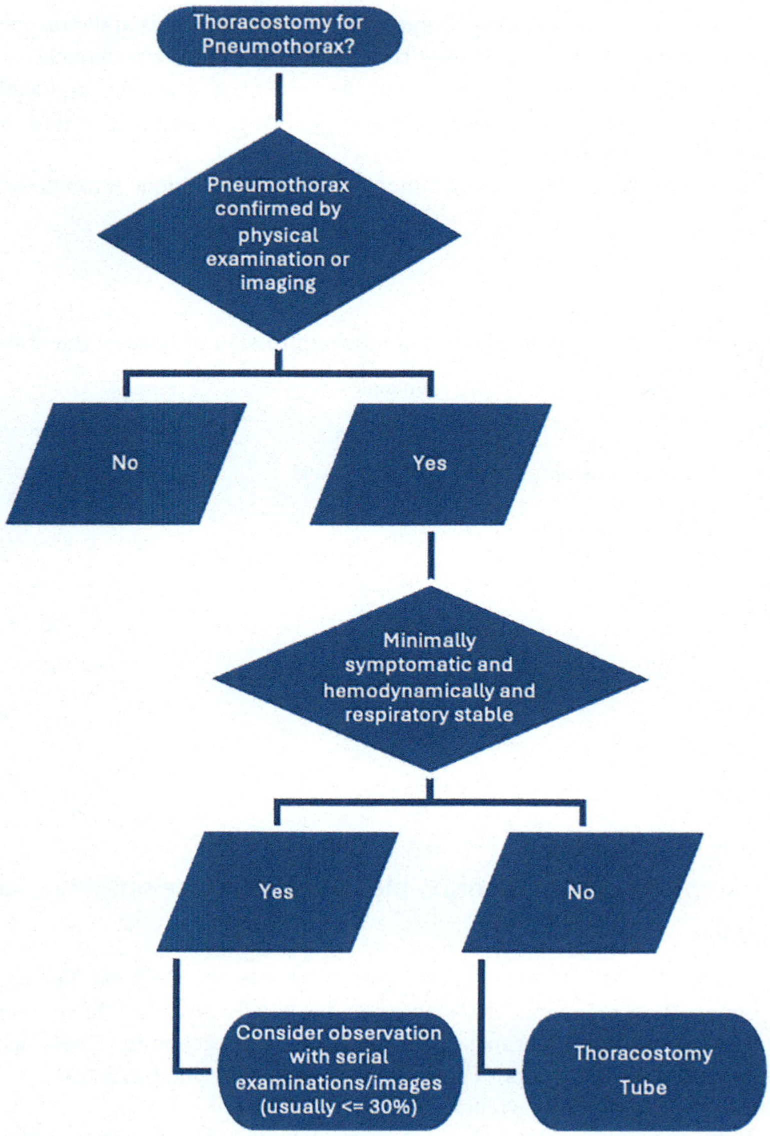


**FIGURE 11.9** Thoracostomy for pneumothorax.

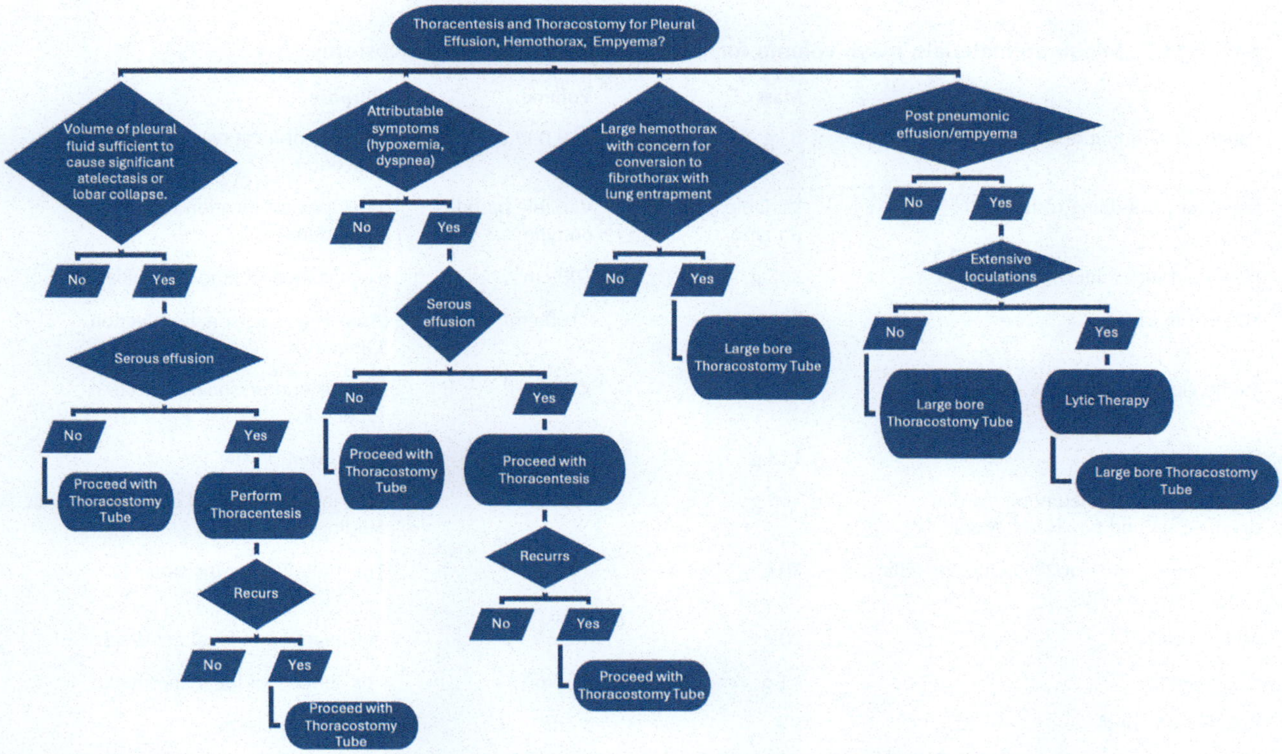


**FIGURE 11.10** Thoracentesis and thoracostomy for pleural effusion, hemothorax, empyema decision flowchart.

### *11.3.7.2 Description*

Thoracostomy tube placement may use a relatively small pigtail catheter placed via Seldinger technique or "open" placement of a larger tube via a small incision. Decision-making for placement of the tube should be based on the familiarity of the Operator with the technique and available resources. Small catheters can be placed by Seldinger technique anteriorly in the 2nd interspace, midclavicular line, or laterally in the 4th interspace, anterior axillary line for treatment of pneumothorax. Open chest tubes should only be placed laterally.

On earth, with standard +Gz directed gravity, auscultation and percussion may be used to estimate the height of fluid in the pleural space. In the presence of a decreased or absent gravity vector, this may not be possible. Standard terrestrial placement for fluid drainage is the $5^{th}$ or $6^{th}$ ICS directed posteriorly and cephalad. However, given reduced gravity, the presence or absence of a free communicating pleural space will impact this. Therefore, ultrasound guidance may be used for a coarse localization of the greatest dimension of the fluid collection to be drained. The viscosity of the material to be drained has implications for both the size of the drainage catheter/tube employed as well as the choice of thoracentesis vs placement of an indwelling tube for a defined period of time. The size of the tube should be selected to ensure adequate drainage based on the volume to be drained and the anticipated viscosity. Blood and purulence usually drain more expeditiously with larger bore catheters or tubes. Hemothorax in various stages of clot formation and empyema (pus) usually require at least a size 32Fr thoracostomy tube.

### *11.3.7.3 Procedural technique and materials for spaceflight*

Materials, mass, volume table (Table 11.7).

Step by step:

1. If placing the chest tube for treatment of pneumothorax, standard placement is the 4th ICS at the anterior axillary line. Chest tubes can be placed via the Seldinger technique through the anterior chest location in the 2nd interspace at the midclavicular line as described above for needle decompression.
2. Identify the $4^{th}$ and $5^{th}$ ribs and the corresponding fourth intercostal space. It is roughly just below the level of the nipple. Place the patient's arm over their head to stretch out the intercostal spaces and better identify the location to perform thoracostomy.

**TABLE 11.7 Minimum materials, mass, volume for thoracentesis and tube thoracostomy.**

| Item | Mass | Volume | Purpose |
|---|---|---|---|
| Patient positioning restraints | 6 kg | <10,000 $cm^3$ | Fix patient location to facilitate procedure. |
| Operator positioning restraints | Variable based on type | Variable based on type | Fix Operator location to facilitate procedure. |
| Handheld ultrasound | 307 g | 100 $cm^3$ | Localization of fluid/air; guidance. |
| Surgical mask | 3 g | <100 $cm^3$ | Asepsis and personal protection, Operator. |
| Sterile gloves | 14 g | <100 $cm^3$ | Asepsis and personal protection, Operator. |
| Prep sponge | 113 g | 118 $cm^3$ | Prep insertion site. |
| Fuhrman pleural/pneumopericardial drainage set and tray (Cook Medical) | 400 g | 190 $cm^3$ | Less invasive method of placement via Seldinger technique. |
| Local anesthetic (1 or 2% Lidocaine with epinephrine; 25 ml) | 70 g | 35 $cm^3$ | For infiltration of incision site. |
| 10 ml syringe | 20 g | 8 $cm^3$ | For injection of local anesthetic. |
| #22 g needle | <1 g | 21 $cm^3$ | For delivery of local anesthetic. |
| #10 or 15 Scalpel | 8 g | 3 $cm^3$ | To create incision. |
| Chest tube 20 Fr (with trocar) | 41 g | 34 $cm^3$ | Drainage of pleural space (pneumothorax). |
| Chest tube 32 Fr or 36 Fr (with trocar) | <100 g | 61 $cm^3$ | Drainage of pleural space (fluid, blood, or purulence). |
| Needle holder | 70 g | 8 $cm^3$ | Used to fix thoracostomy tube to skin. |
| Mayo scissor | 85 g | 11 $cm^3$ | To cut suture. |
| 0 Silk sutures ×3 | <5 g (each) | 0.6 $cm^3$ (each) | On curved (cutting) needle. |
| Vaseline gauze | 17 g | 44 $cm^3$ | To effect improved seal around chest tube. |
| 4x4 gauze dressing | 5 g | 0.1 $cm^3$ | To collect low volume ooze around tube. |
| Male to male connector sized to chest tube and tubing | <10 g | 15 $cm^3$ | Connection between thoracostomy tube and drainage tubing. |
| Thoracostomy drainage (tubing and Peur-evac™) | 0.7 kg | 11,400 | Directs air and drainage from thoracostomy tube. |
| Heimlich valve | 23 g | 13 $cm^3$ | Allows one-way exit of air /fluid from pleural space. |
| Tube or Kelly clamp | 36 g | 70 $cm^3$ | Thoracostomy tube occlusion prior to connection to drainage system. |
| Cloth tape | 19 g | 500 $cm^3$ | For tube and dressing fixation. |

**3.** Identify (and mark) this level at the anterior axillary line.
**4.** Use of a prepared insertion kit is preferred for placement via the Seldinger technique.
**5.** Prep the site and apply self-adhesive encircling drapes.
**6.** Administer local anesthetic.
**7.** Seldinger technique:

a. For placement of a small kit catheter for treatment of pneumothorax, an angiocatheter is used to "pop" into the pleural space by traveling across the top of the 5th rib. The trocar from the catheter is removed, a guidewire passed, and the angiocatheter is then exchanged for the pleural catheter, always controlling the end of the guidewire. The guidewire is then removed, and the newly placed chest tube is attached to tubing connecting to the Heimlich valve or Pleur-evac™. Always tape the chest tube connection to the tubing.
b. For drainage of fluid or other material, the placement site is the 5th or 6th ICS with posterior direction of the catheter. This means that entry into the pleural space is similar to the above across the top of the 6th or 7th rib, respectively. If the pleural effusion is expected to be transudative (serous) and the underlying etiologic process has or is being treated, thoracentesis may be considered in which the catheter is removed after acute withdrawal of the pleural contents.
c. The catheter is then sewn to the skin with 2-0 Silk suture.

8. "Open" Tube Placement
  a. Make a 2–3 cm horizontal incision through the skin with #15 scalpel blade at the 4th interspace at the anterior axillary line for treatment of pneumothorax.
    i. You may then place the chest tube in one of two ways.
      i. Using a Kelly clamp, with the tip directed toward the patient's foot, dissect through skin and soft tissue, and then use the clamp to glide over the upper aspect of the lower rib in the target interspace of entry. A "pop" should be felt on entering the pleural space. Spread the size of the opening with the clamp.
      ii. Then, pass your index finger into the hole in the pleural space to palpate and clear tissue in all directions.
      iii. Guide the selected chest tube anteriorly and cephalad by grasping the tip of the tube with a curved clamp. After the final side hole is well inside the patient, sew the tube to the chest wall with silk suture and connect it to the Heimlich valve or Pleur-evac™. Always tape the chest tube to the tubing connection.
    ii. An alternative technique employs a chest tube that has an integral trocar. A gentle curve is introduced to the trocar and tube. After making the skin incision in the chest wall, push the combined tube/trocar into the pleural space with the concavity of the bend in the trocar directed toward the feet. Loss of resistance signals entry, at which point you must terminate forward motion. Once in the pleural space, the curvature in the trocar is rotated 180 degrees with the concavity now directed cephalad. This technique requires finesse so that a deep parenchymal injury does not occur. The tube is then directed anteriorly and cephalad. Remove the trocar and suture the tube to the skin. Connect to the drainage tubing as described.
    iii. As per placement of a thoracostomy tube for treatment of pneumothorax, for drainage of fluid or other material, entry into the pleural space may be achieved reliably in a controlled fashion by using a curved clamp to glide over the upper aspect of the lower rib in the target interspace of entry. Alternatively, a trocar tube may be utilized. Note the recommended larger size of the selected tube, 32 or 36 Fr, for drainage of fluid. Once introduced into the pleural space, the tube is directed posteriorly and cephalad. Suture the tube to the skin. .
9. A purse-string suture may now be placed around the inserted tube and knotted to itself to be used at the time of tube removal.
10. Wrap Vaseline gauze around the tube at the insertion site and then place an absorbent dressing of dry 4×4 and cloth tape.
11. Chest tube removal
  a. Chest tubes must be removed in such a way as to avoid the patient inspiring and sucking air into the pleural cavity, which could result in reaccumulation of the pneumothorax.
  b. Prepare an occlusive bandage with Vaseline gauze, 4×4, and tape. Strips of tape already torn to the appropriate length are adhered to an adjacent surface, so that they may be grasped and applied at the appropriate time with one hand.
  c. Don surgical gloves.
  d. Remove the chest tube dressing. If a purse-string suture had been placed at the time of tube placement, cut the knot off the end and prepare the two ends to be tied.
  e. Cut the tube's tethering suture(s) while holding the chest tube in place to prevent premature dislodgment.
  f. Have the patient take a maximal inspiration; then have them make a maximal expiration. At the end of expiration, pull the tube out completely. If there is a purse string to be tied, the nondominant index finger is used to occlude the hole while the purse string suture ends are tightened and tied. Alternatively, the prepared Vaseline gauze occlusive dressing is applied as the tube is withdrawn and held firmly in place with the nondominant hand. This dressing is now taped occlusively in place using the pretorn strips of tape.

## 11.3.8 Pleural space washout via thoracostomy

### *11.3.8.1 Decision making*

Washout through tube thoracostomy would be expected to be performed infrequently (Fig. 11.11). In general, it would be indicated for significant respiratory impairment or attributable systemic signs of infection/sepsis with presumed retained infected material (empyema). It may be performed to evacuate retained/incompletely drained hemothorax that significantly prevents lung reexpansion or loculated pleural effusion of significant size. Clinically significant improvement may be difficult or impossible to achieve in a microgravity environment. As such, this procedure should be reserved for situations where the patient is far from evacuation home and is suffering from decreased oxygen saturation requiring oxygen despite tube thoracostomy drainage.

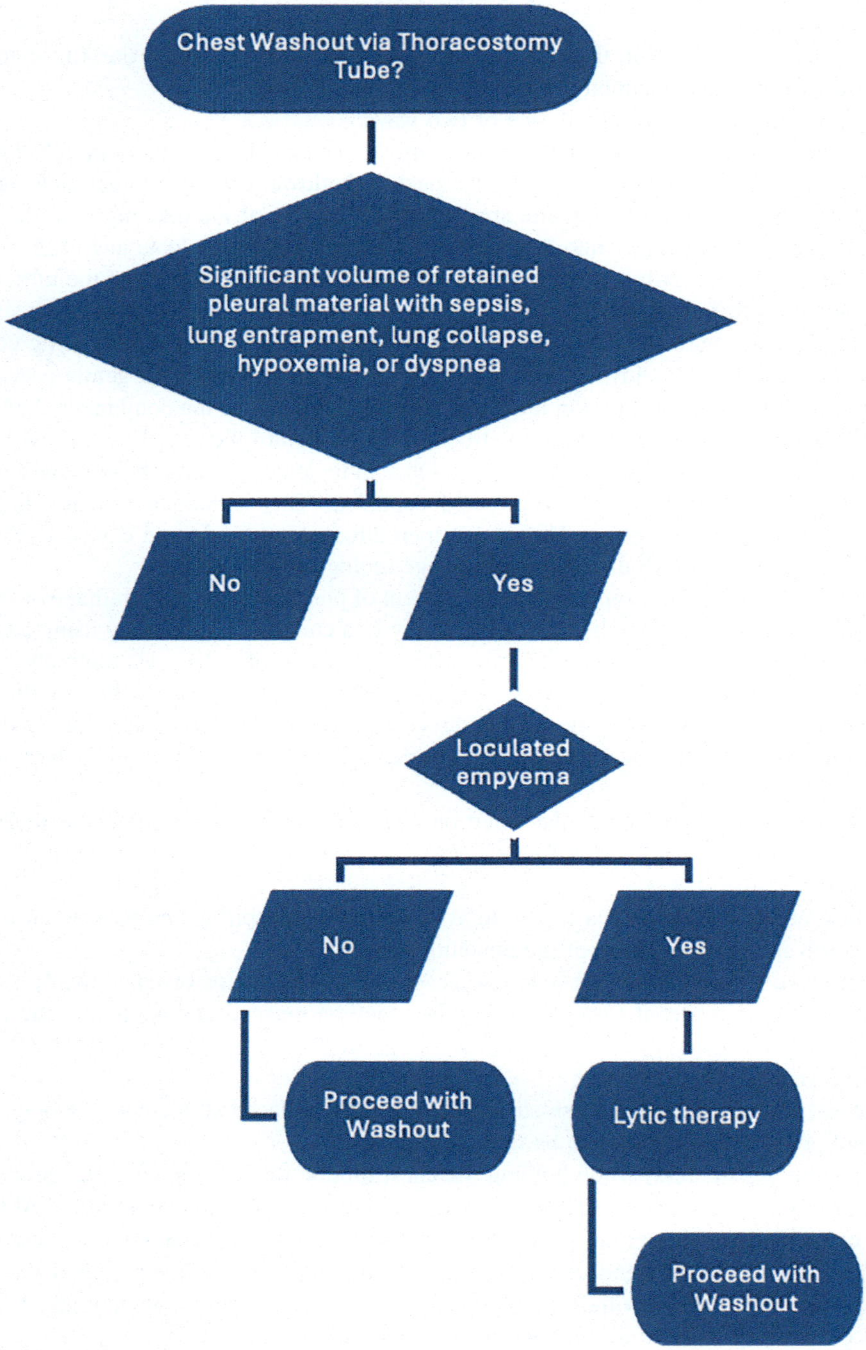


FIGURE 11.11 Chest washout via thoracostomy tube decision flowchart.

#### 11.3.8.2 Description

The consistency/viscosity of the intrapleural material to be drained will determine the utility of this procedure as compared with a more invasive thoracoscopy or thoracotomy. Further, instillation of an active lytic agent such as alteplase (rt-PA), urokinase, or tenecteplase (TNK) can be used to break down loculations and fibrinous adhesions to facilitate drainage (Heimes et al., 2017). Use of lytic agents should precede the washout procedure.

#### 11.3.8.3 Procedural technique and materials for spaceflight

Materials, mass, volume table (Table 11.8).

Step by step:

1. Position the patient with the affected side facing the Operator.
2. If a lytic agent is to be used, mix 6 mg of alteplase, for example, in 50 ml of saline and administer through the pleural tube. Clamp the tube for 4 hours. This may be repeated daily for 3 days as guided by auscultation, ultrasound, or radiographic imaging.
3. When initiating the washout procedure, clamp the indwelling chest tube and separate it from the drainage system. Manipulation of the chest tube connector should be aseptic at all times.
4. Use a bulb syringe to introduce 100–200 ml of warm saline into the pleural space at a time. In normo- or partial gravity, the fluid may be allowed to drain passively. In microgravity, active aspiration may be required by using the bulb syringe or placing the chest tube back to suction. Alternatively, the manual hand pump may be used to evacuate the pleural space.
5. Repeat the above cycle multiple times until there is a significant decrease in the yield of purulence or debris dislodged and removed by the lavage.
6. Finally, reconnect the chest tube to its drainage system and remove all clamps.

**TABLE 11.8** Materials, mass volume for pleural space washout via thoracostomy.

| Item | Mass | Volume | Purpose |
|---|---|---|---|
| Patient positioning restraints | 6 kg | <10,000 cm$^3$ | Fix patient location to facilitate procedure. |
| Operator positioning restraints | Variable based on type | Variable based on type | Fix Operator location to facilitate procedure. |
| Surgical mask | 3 g | <100 cm$^3$ | Asepsis and personal protection, Operator. |
| Sterile gloves | 14 g | <100 cm$^3$ | Asepsis and personal protection, Operator. |
| 20 ml Syringe | 23 g | 125 cm$^3$ | For drawing up active ingredient. |
| #22 gauge needle | <1 g | 21 cm$^3$ | For drawing up active ingredient. |
| 1 liter bottle Sterile Saline | 1.1 kg | 1200 cm$^3$ | Irrigant/instillate. Bottle with controlled aperture sized to fit with bulb syringe. |
| Bulb syringe (60 ml saline) | 59 g | 300 cm$^3$ | Irrigation. |
| Lytic agent such as alteplase (rt-PA), urokinase, or tenecteplase (TNK). | 100 g | 125 cm$^3$ | Used to break down loculations and fibrinous adhesions to facilitate drainage |
| Tube or Kelly clamp ×2 | 36 g (each) | 70 cm$^3$ (each) | Thoracostomy tube occlusion prior to connection to drainage system. |

### 11.3.9 Pericardiocentesis

#### 11.3.9.1 Decision making

Indications include the relief of impending or actual tamponade as a result of blunt or penetrating trauma or infectious/inflammatory process. The drainage of even a relatively small volume of fluid can have a significant hemodynamic impact. It is important to note that in the setting of cardiac trauma with pericardial effusion, pericardiocentesis is not indicated and is not consistent with the principles of Advanced Trauma Life Support. In the setting of cardiac injury with pericardial effusion, the appropriate diagnostic and therapeutic procedure is pericardial window (below) or anterolateral thoracotomy. These more advanced procedures are unlikely to be successful when performed in microgravity by an unskilled operator and should probably not be performed. If a trained surgeon is not available, pericardiocentesis with drain placement may be considered. However, the likelihood of success may be low, and there is a risk of complicating the cardiac injury with exsanguination (Fig. 11.12).

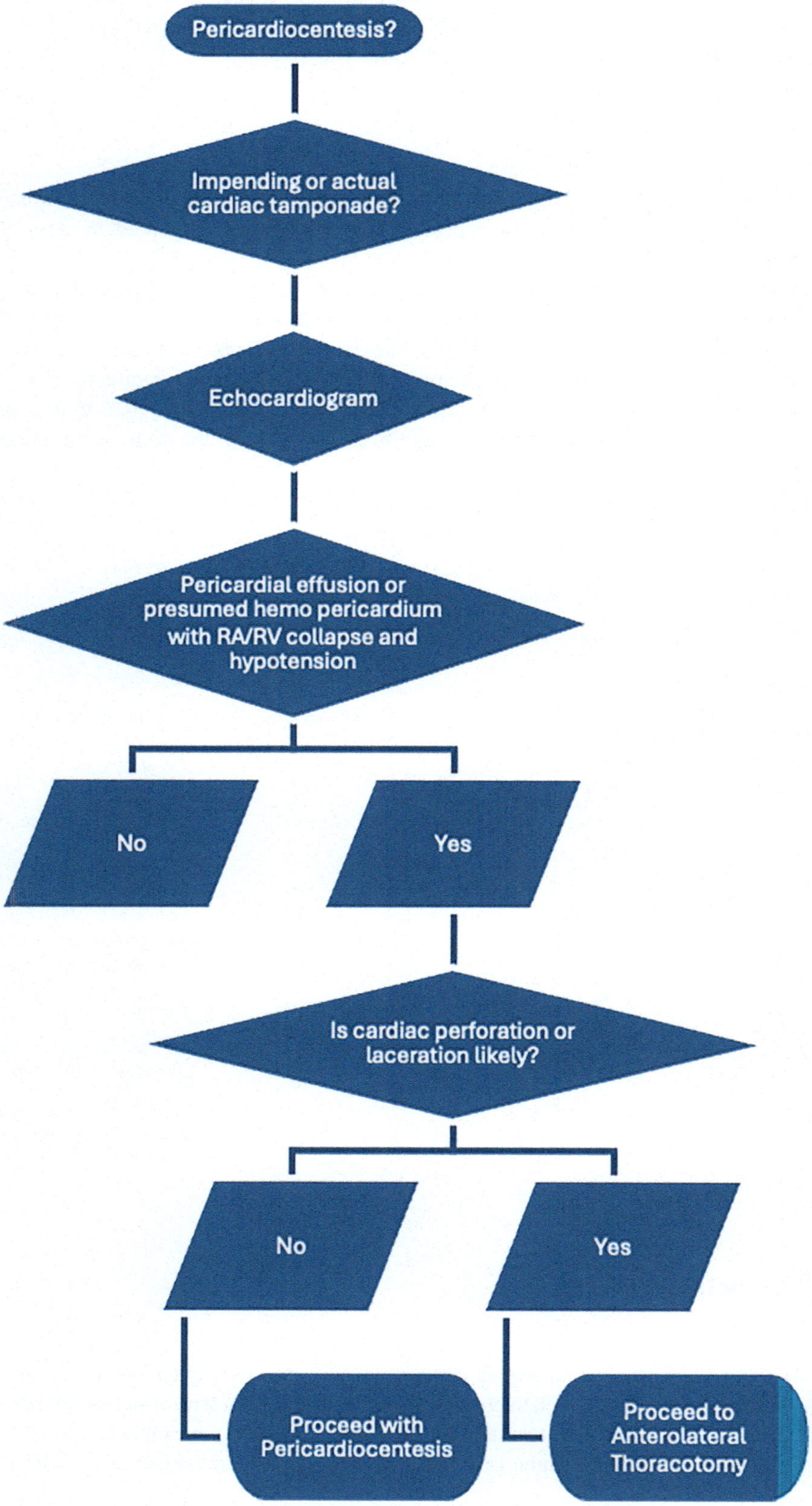


FIGURE 11.12 Pericardiocentesis decision flowchart.

### 11.3.9.2 Description

This procedure is optimally performed under ultrasound guidance.

### 11.3.9.3 Procedural technique and materials for spaceflight

Materials, mass, volume table (Table 11.9).

Step by step (Willner et al., no date; Fig. 11.13):

1. Patient monitoring should include blood pressure, pulse oximetry, and ECG; reliable IV access should be assured.
2. The patient may be relatively supine or propped 30–45 degrees "head up." If hemodynamically compromised, he/she should be supine if in partial or full gravity.
3. A subcostal approach is favored for the less experienced as it is easier to visualize the pericardium, the heart, and the track of the needle by ultrasound from this perspective.
4. Use of a prepared kit is preferred.
5. Prep the subcostal region around the xiphoid process, then drape and infiltrate with local anesthetic.
6. Insert a sheathed 16 g or 18 g 9 cm (spinal) needle attached to a 12cc syringe from the subxiphoid/subcostal location at a 45-degree angle directed towards the tip of the left scapula. Aspiration (negative pressure) should be maintained constantly on the syringe as the needle is advanced slowly until fluid is withdrawn. Ultrasound is used to confirm the location of the needle.

**TABLE 11.9** Materials, mass, volume for pericardiocentesis.

| Item | Mass | Volume | Purpose |
|---|---|---|---|
| Patient positioning restraints | 6 kg | <10,000 $cm^3$ | Fix patient location to facilitate procedure. |
| Operator positioning restraints | Variable based on type | Variable based on type | Fix Operator location to facilitate procedure. |
| Handheld ultrasound | 307 g | 100 $cm^3$ | Guidance. |
| Surgical mask | 3 g | <100 $cm^3$ | Asepsis and personal protection, Operator. |
| Sterile gloves | 14 g | <100 $cm^3$ | Asepsis and personal protection, Operator. |
| Fuhrman pleural/pneumopericardial drainage set and tray | 400 g | 190 $cm^3$ | Preprepared kit incorporating prep, local, needle, syringe, and catheters are available and recommended. |
| 12Fr Pigtail catheter | 25 g | 325 $cm^3$ | For continued drainage of pericardial space. |
| Grenade for pericardial catheter | 70 g | 175 $cm^3$ | Collect drainage. |

7. A guidewire may now be introduced through the needle and exchanged for an appropriate catheter via the Seldinger exchange technique. A 6–8 Fr pigtail catheter may be sufficient for drainage of serous fluid. However, it may require upsizing for hemorrhagic or purulent content (available 12 Fr). This may require the passage of a dilator along the created tract before the larger drainage catheter can be passed.
8. Gently aspirate all of the fluid from the pericardial space with repeated syringe aspirations. Watch the blood pressure. If the systolic pressure decreases with continued drainage, cease aspiration of fluid. It is critical to assess the potential for cardiac laceration or perforation that is partly tamponaded by the pericardial contents. If this is the case, consider proceeding to thoracotomy. Once aspiration is completed, attach the catheter to the tubing leading to a drainage grenade.
9. Suture/tether the catheter to the skin with 2-0 Silk suture.
10. The nature of the fluid, serous, purulent, or hemorrhagic, may have implications for further intervention, as may the mechanism of the injury (Fig. 11.12).
11. ***Risks:** Puncture of the ventricle is the most significant complication. Left ventricular wounds may become hemostatic while right ventricular are less likely to do so. Differentiating hemorrhagic pericardial fluid from frank intracardiac blood may be challenging. Intravascular/cardiac chamber blood typically forms a clot. Bradycardia and hemodynamic instability have been reported after rapid drainage of large volumes of pericardial fluid.

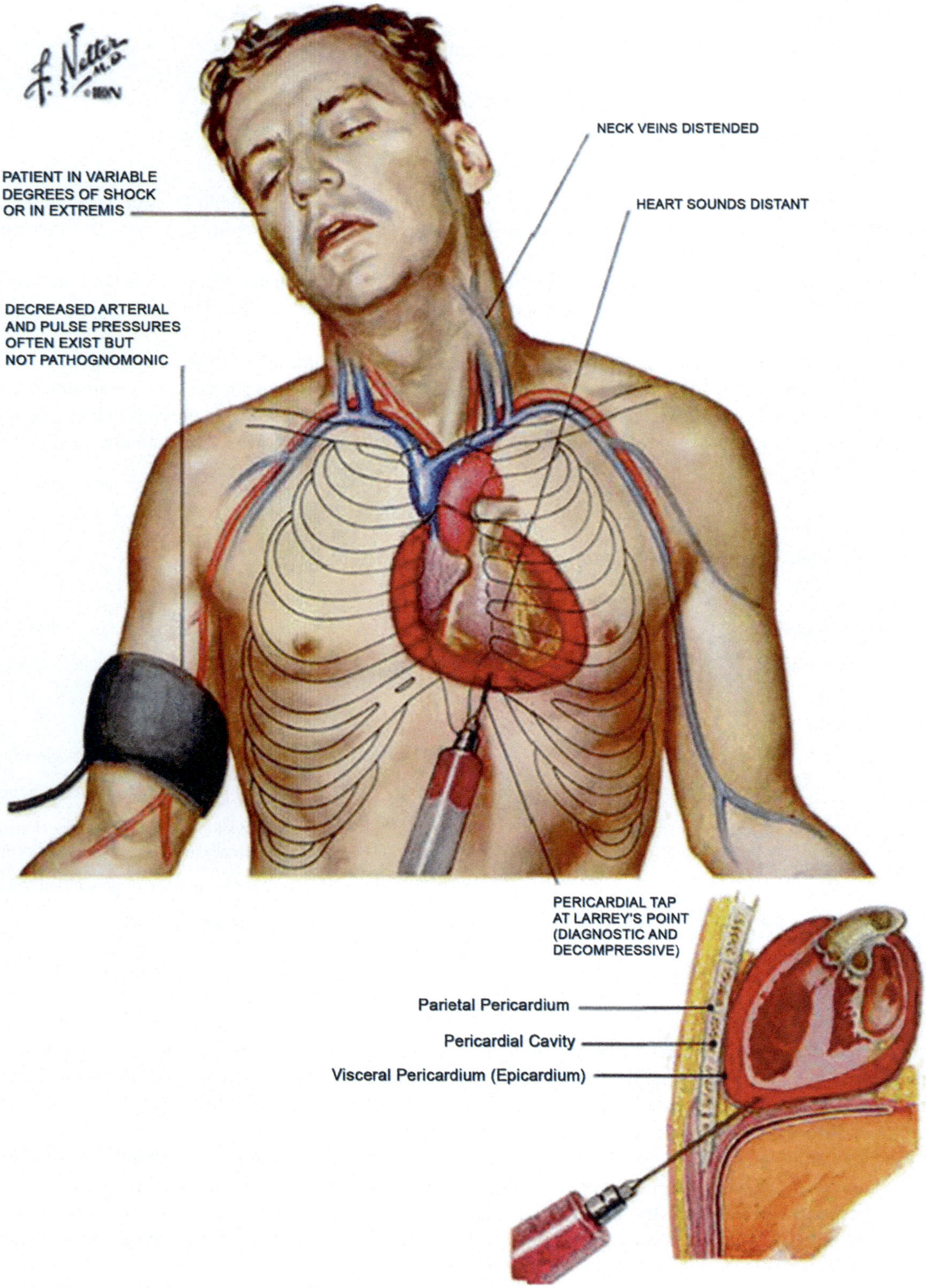


FIGURE 11.13 Pericardiocentesis. The substernal approach is preferred and described.
Reproduced from Vergen, J.A., Zipf, J.J., Richardson, E.S., Hill, A.J., Iaizzo, P.A, 2024. The Pericardium. In: Iaizzo, P.A. (eds) Handbook of Cardiac Anatomy, Physiology, and Devices. Springer, Cham. https://doi.org/10.1007/978-3-031-72581-4_13, with permission.

## 11.3.10 Pericardial window, video-assisted thoracoscopic surgery (VATS), anterolateral thoracotomy

The remaining three procedures in this chapter are major surgical operations that are unlikely to be successful if performed by an untrained operator in the space environment. While they can be lifesaving when performed by trained surgeons in the terrestrial setting, they are highly complex and risk causing major injury or death. However, if, and *only* if, a trained surgeon is available, they are reasonable to consider in the setting of thoracic trauma where observation or the minor procedures above have been unsuccessful, and the patient is at imminent risk of death. These procedures may be performed by a single operator. However, if available, a Surgical Assistant would facilitate. Further, VATS and thoracotomy require general anesthesia and are preferentially performed with a double-lumen endotracheal tube or bronchial blocker. Thus, a trained anesthesia provider is also a requirement for at least these two procedures. As such, in the absence of a surgeon and anesthesia provider, these procedures should not be attempted.

### *11.3.10.1 Pericardial window*

#### 11.3.10.1.1 Decision making

Indications include the relief of impending or actual tamponade as a result of blunt or penetrating trauma or infectious/inflammatory process. A window would be required, rather than pericardiocentesis, if the material to be drained is of high viscosity, such as already clotted blood, etc. (Fig. 11.14). It is reasonable in some cases to consider the creation of a pericardial window only if pericardiocentesis has failed to achieve the necessary therapeutic effect. In the context of cardiac perforation or laceration, evacuation of the pericardial space could relieve a tamponade that is preventing exsanguination. Therefore, it is critical to assess the likelihood of significant cardiac injury. If this is considered likely, then anterolateral thoracotomy with direct cardiac repair could be required and is the only course consistent with survival. Pericardial window can be performed under local anesthesia if necessary with parenteral anxiolytics.

#### 11.3.10.1.2 Description

A plastic Yankauer suction instrument may be used to gently break up and extract pericardial content.

#### 11.3.10.1.3 Procedural technique and materials for spaceflight

Materials, mass, volume table (Table 11.10).

Step by step (Fig. 11.15):

1. Don headlight or direct overhead light.
2. Monitoring should include blood pressure, oxygen saturation, and ECG; reliable IV access should be assured.
3. Administer prophylactic antibiotic.
4. Conscious sedation or general anesthesia is recommended; this procedure can be performed with local anesthesia.
5. Position the patient supine.
6. Prep the operative site.
7. Square off the operative site with self-adhesive sterile drapes.
8. Make a vertical midline incision over the palpable xiphoid process, 10–15 cm.
9. Cauterize significant bleeding as needed.
10. Incise the linea alba in the midline.
11. Dissect down to and below the xiphoid process with Metzenbaum scissors.
12. Use an Army-Navy or small Richardson retractor to lift up on the xiphoid. If available, an Assistant may continue to do so.
13. Excision of the xiphoid process to enhance exposure may be required. It may be cut out with the Mayo scissor.
14. Place the self-retaining Weitlaner retractor to facilitate exposure.
15. Continue dissection superiorly, retro-xiphoid and retro-sternal. A smooth visible surface/plane is the appearance of the pericardial sac. Use toothed forceps to elevate the pericardium and Metzenbaum scissors to nick/open to relieve underlying pericardial pressure.
16. Regrasp and elevate the edge of the pericardium with an Allis clamp and excise a window of tissue of approximately 5 cm diameter.
17. Attach an Allis clamp to the edge of the pericardium. While elevating the edge, you may use the plastic Yankauer sucker to break up any adhesions/loculations and extract pericardial contents. When wielding the Yankauer, push out against the inner surface of the pericardium to avoid cardiac injury.

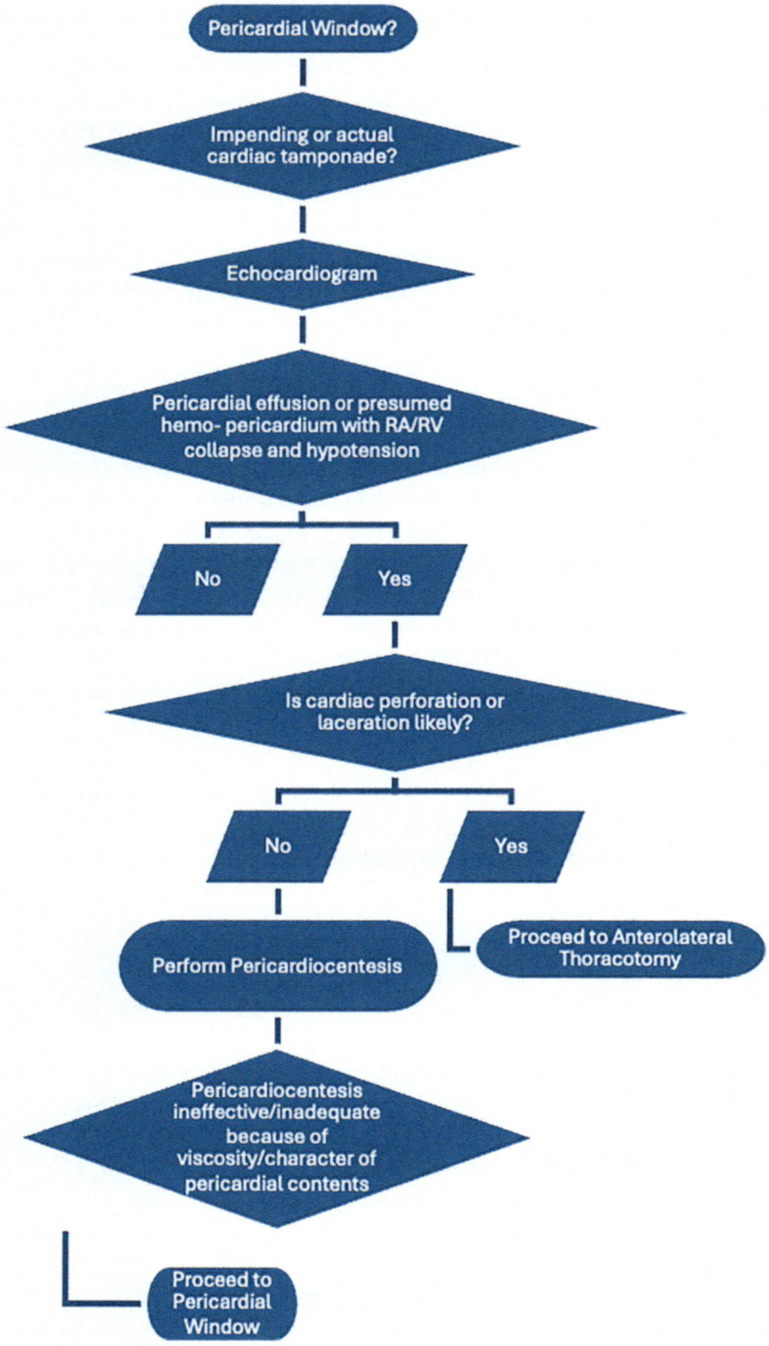


FIGURE 11.14 Pericardial window decision flowchart.

18. Deliver the connection tube ends of the Jackson-Pratt drains via separate stab wounds made with the #15 scalpel blade with a Tonsil clamp and connect to their respective grenades—do not yet apply suction. Use DeBakey forceps to guide and then feed the two 10 mm flat drains into the pericardial space. One is directed anterior and superior; the other posteriorly on the diaphragmatic surface. Tether the drain(s) to the skin with 2-0 Silk suture.
19. Irrigate the operate field with saline.
20. Approximate the deep soft tissue with 0 Vicryl.
21. Approximate the subcutaneous tissue with 3-0 Vicryl.
22. Activate the Jackson-Pratt grenades.
23. Approximate the skin with staples.
24. Apply absorbent dressings over the incision and drain sites.

**TABLE 11.10 Materials, mass, volume for pericardial window.**

| Item | Mass | Volume | Purpose |
|---|---|---|---|
| Patient positioning restraints | 6 kg | <10,000 $cm^3$ | Fix patient location to facilitate procedure. |
| Operator positioning restraints | Variable based on type | Variable based on type | Fix Operator location to facilitate procedure. |
| Prophylactic antibiotic | 31 g | 10 $cm^3$ | Prevent infection. Cefazolin (Clindamycin or Vancomycin for penicillin allergic). Prophylactic therapy limited to perioperative period. |
| Handheld ultrasound | 307 g | 100 $cm^3$ | Guidance. |
| Headlight +/- camera<br>Battery pack/light source | 304 g<br>260 g | <4,620 $cm^3$<br>194 $m^3$ | Optimized visibility. |
| Mask | 3 g | <100 $cm^3$ | Asepsis and personal protection, Operator. |
| Surgical gown | 173 g | 405 $cm^3$ | Asepsis and personal protection, Operator. |
| Sterile gloves | 14 g | <100 $cm^3$ | Asepsis and personal protection, Operator. |
| Prep sponge | 113 g | 118 $cm^3$ | Prep. |
| Sterile self-adhesive drape | 90 g | 431 $cm^3$ | Asepsis. |
| #15 Scalpel | 8 g | 3 $cm^3$ | Incision. |
| Handheld cautery (battery-powered) | 62 g | 10$cm^3$ | Hemostasis. |
| Army-Navy or Small Richardson Retractor | 126 g | 45 $cm^3$ | Exposure. |
| Plastic Yankauer suction tip | 21 g | 18 $cm^3$ | To assist in breaking up and extracting pericardial contents. |
| Suction setup (portable self-contained) | 4 kg | <10,000 $cm^3$ | On operative field. |
| Irrigation (10 ml saline-filled syringes) | 17 g | 10 $cm^3$ | Irrigation of pericardial space (if required) and soft tissue layers prior to closure. |
| Bulb syringe (60 ml saline-filled) | 59 g | 300 $cm^3$ | Irrigation. |
| Large forceps with teeth | 100 g | 30 $cm^3$ | Exposure. |
| DeBakey Forceps | 90 g | 30 $cm^3$ | Exposure. |
| Metzenbaum scissors | 70 g | 9 $cm^3$ | Dissection, incision of pericardial sac, and excision of pericardial window. |
| Mayo scissors (curved) | 85 g | 11 $cm^3$ | Removing xiphoid. |
| Tonsil clamp | 36 g | 70 $cm^3$ | Elevate pericardium |
| Allis clamp × 2 | 45 g (each) | 139 $cm^3$ | Pericardial retraction. |
| Weitlaner retractor (blunt) | 400 g | 168 $cm^3$ | Self-retaining soft tissue retraction. |
| Needle holder | 70 g | 8 $cm^3$ | For use with curved needles |
| 0 Vicryl suture | <5 | 0.6 $cm^3$ | Deep closure |
| 3-0 Vicryl suture | <5 | 0.6 $cm^3$ | Superficial closure. |
| Skin stapler | 92 g | 550 $cm^3$ | Skin closure. |
| 2-0 Silk suture | <5 g | 0.6 $cm^3$ | Fix the pericardial drain(s) to the skin. |
| 10 mm flat Jackson-Pratt drain × 2 | 20 g (each) | 20 $cm^3$ (each) | Pericardial drainage. |
| Grenade for pericardial drain | 70 g | 175 $cm^3$ | Collect drainage. |
| Adhesive and absorbent dressing. | 86 g | 524 $cm^3$ | Dressing. |

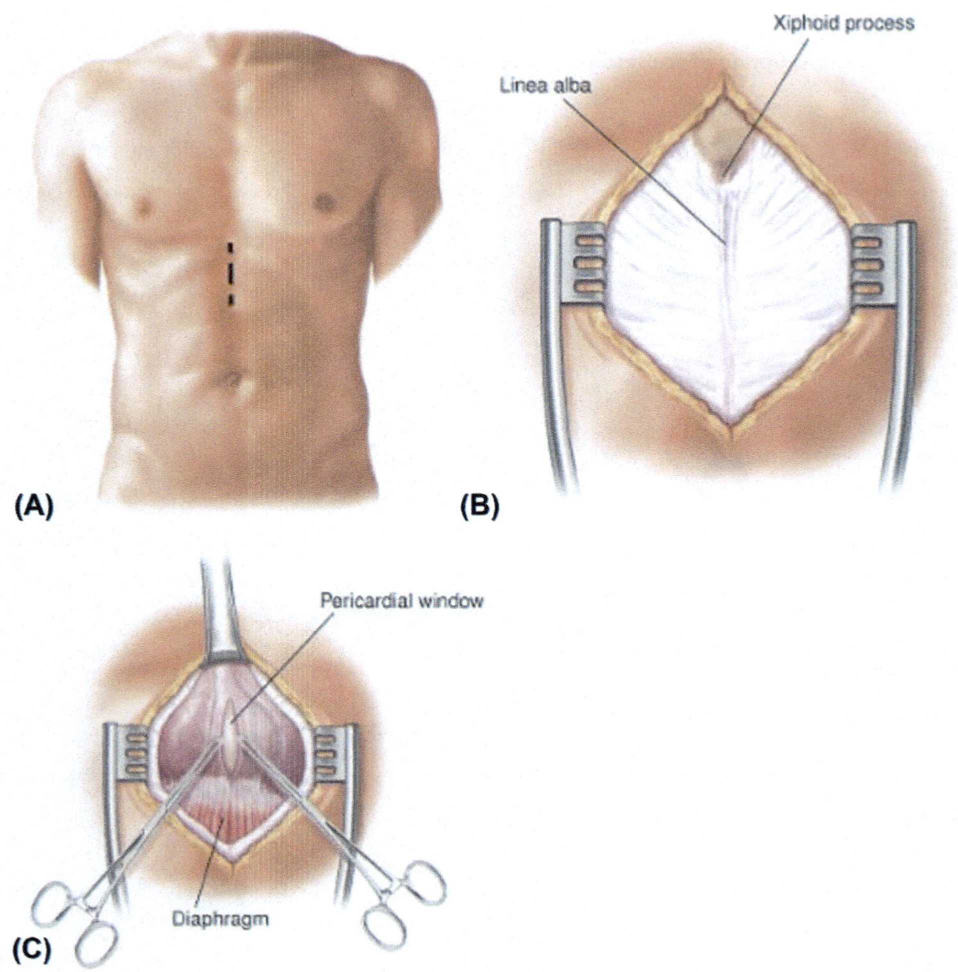


FIGURE 11.15 Pericardial window, subxiphoid approach.
Stasik C.N., Calhoon J.H., 2016. Repair of penetrating and blunt cardiac trauma. Thoracic Key.

### *11.3.10.2 Video-assisted thoracoscopic surgery (VATS)*

#### 11.3.10.2.1 Decision making

VATS is indicated for visually guided and potentially definitive treatment of complications of thoracic trauma or infection (Mehrotra et al., no date). Significant bleeding will impair visualization and suggest proceeding to anterolateral thoracotomy (Decision Flowcharts: Figs. 11.16 and 11.18). However, decortication and/or breakup and drainage of loculated pleural processes (hemorrhagic or infectious/postpneumonic) to effect near complete pleural drainage, achieve reexpansion of entrapped lung, and decrease infectious load are appropriate for this technical approach (Fig. 11.16).

#### 11.3.10.2.2 Description

This is optimally performed under general anesthesia using a double lumen endotracheal tube that allows single lung ventilation with selective collapse of the lung on the operative side during the procedure. Second choice would be single lumen endotracheal intubation with use of a bronchial blocker. These require a higher level of intubation and airway management expertise. Of note, there is a growing experience with awake nonintubated thoracoscopy. While this is of significant consideration in performance in an austere setting, it would significantly impede and likely impair the effectiveness of the procedure for this group of patients (Yanik, 2023). One camera port and at least one and preferably two working ports are recommended.

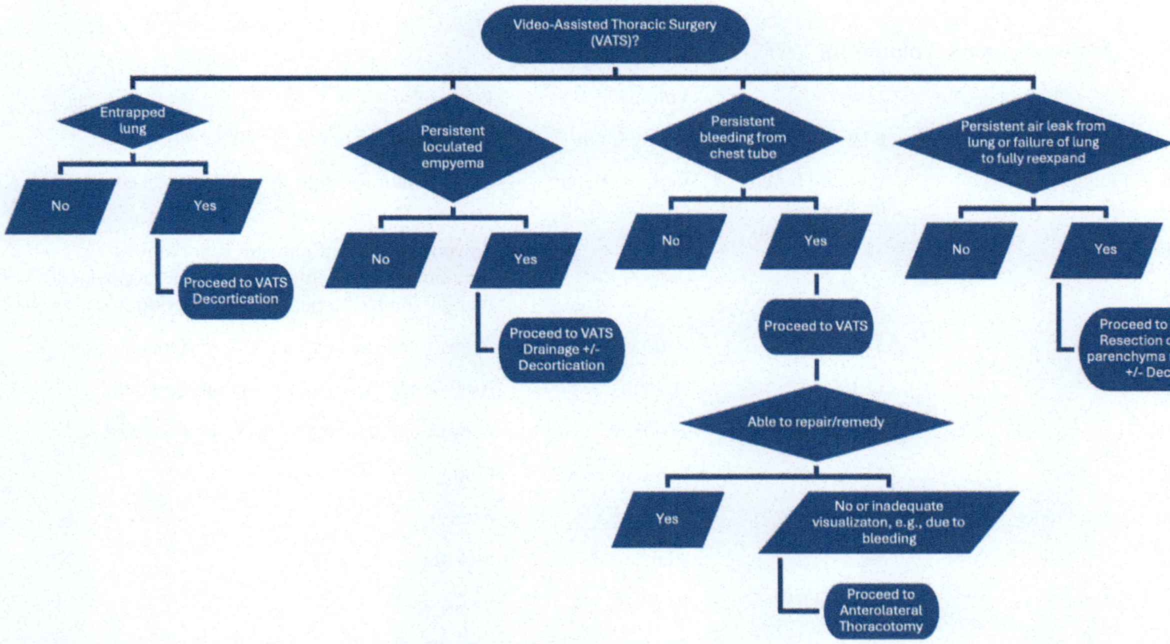


**FIGURE 11.16** Video-assisted thoracoscopic surgery (VATS) decision flowchart.

#### 11.3.10.2.3 Procedural technique and materials for spaceflight

Materials, mass, volume table (Table 11.11).

Step by step (Fig. 11.17):

1. Set up scope and video monitor(s).
2. Attach monitoring: blood pressure, oxygen saturation, ECG; reliable IV access should be assured.
3. Administer general anesthesia.
4. Administer prophylactic antibiotics.
5. Position the patient in a 45-degree or lateral decubitus position with the operative side up.
6. Plan and mark triangulated port sites × 3.
7. Don a mask.
8. Don a surgical gown and gloves.
9. Prep the chest; place sterile drapes.
10. Collapse the lung on the operative side, if possible.
11. Port site incisions are horizontal, 3 cm in length. The first incision should be $7^{th}$ ICS midaxillary line. Similar to open placement of a thoracostomy tube, use a curved Kelly clamp to enter into the pleural space. Then insert the first thoracoscopy port with its trocar. The camera is introduced here and now used to look superiorly and anteriorly. This allows direct visualization of creation and placement of the second port in a similar fashion, $4^{th}$ ICS anterior axillary line. The third port is then placed similarly as observed from the pleural camera in the $5^{th}$ ICS posteriorly, approximately below the scapula.
12. Introduce and manipulate operative instruments initially through the second and third ports. Use the thoracoscopic suction tip to break up the clot and suction out. Irrigation may facilitate this.
13. Following completion of the procedure, irrigate the operative site with saline and suction all fluid.
14. A posterior chest tube and an anterior chest tube, both directed apically, are now passed via two of the port sites. Placement direction is facilitated by the use of curved ring forceps or a chest tube passer. Suture the tubes to the skin with 0 Silk and connect via tubing to the Pleur-evac™.
15. The lung on the operative side is now ventilated and expanded under direct vision.
16. Infiltrate the port sites with 0.25% bupivacaine.
17. Use 0 Vicryl to approximate deep soft tissue and 3-0 Vicryl to approximate subcutaneous tissue at each port site. Skin is closed with staples.
18. Apply dry occlusive dressings to all sites.
19. ***N.B.:** If at any time visualization is inadequate or bleeding ensues that obscures vision, the operation should be converted to an open anterolateral thoracotomy as described below.

**TABLE 11.11 Materials, mass, volume for VATS.**

| Item | Mass | Volume | Purpose |
|---|---|---|---|
| Patient positioning restraints | 6 kg | <10,000 $cm^3$ | Fix patient location to facilitate procedure. |
| Operator positioning restraints | Variable based on type | Variable based on type | Fix Operator location to facilitate procedure. |
| Prophylactic antibiotic | 31 g | 10 $cm^3$ | Prevent infection. Cefazolin (Clindamycin or Vancomycin for penicillin allergic). Prophylactic therapy limited to perioperative period. |
| Mask | 3 g | <100 $cm^3$ | Asepsis and personal protection, Operator. |
| Surgical gown | 173 g | 405 $cm^3$ | Asepsis and personal protection, Operator. |
| Sterile gloves | 14 g | <100 $cm^3$ | Asepsis and personal protection, Operator. |
| Prep sponge | 113 g | 118 $cm^3$ | Prep. |
| Sterile self-adhesive drape | 90 g | 431 $cm^3$ | Asepsis. |
| Scalpel #10 | 8 g | 3 $cm^3$ | Incision |
| Thoracoscopic port × 3 | 30 g each | 1 $cm^3$ | Access. |
| Thoracoscope | 200 g | 40 $cm^3$ | Visualization. |
| Thoracoscopic linear stapler with lung parenchymal and vascular loads | 430 g | 180 $cm^3$ | For resection or repair. Loads sizes should include 30 mm, 45 mm, and 60 mm. |
| Thoracoscopic suction tip | 60 g | 16 $cm^3$ | Procedural. |
| Suction (for thoracoscopic instrument; portable self-contained) | 4 kg | <10,000 $cm^3$ | On operative field. |
| Handheld cautery (battery-powered) | 62 g | 10 $cm^3$ | Hemostasis. |
| Irrigation (10 ml saline filled syringes) | 17 g | 10 $cm^3$ | Irrigation of space and layers prior to closure. |
| Bulb syringe (60 ml saline) | 59 g | 300 $cm^3$ | Irrigation. |
| Large forceps with teeth | 100 g | 10 $cm^3$ | Exposure. |
| Metzenbaum scissors | 70 g | 9 $cm^3$ | Dissection. |
| Long curved ring forceps × 2 | 250 g (each) | 16 $cm^3$ (each) | Thoracoscopy length; for decortication |
| Long curved chest tube passer | 225 g | 16 $cm^3$ | |
| Needle holder | 70 g | 8 $cm^3$ | For use with curved needles |
| Marcaine 1% with epinephrine (25 ml) | 70 g | 35 $cm^3$ | Intercostal block and incisional injection. |
| #22 spinal needle | <10 g | 0.13 $cm^3$ | Intercostal block; postoperative analgesia. |
| 20ml syringe | 23 g | 125 $cm^3$ | Intercostal block. |
| 0 Vicryl suture | <5 g | 0.6 $cm^3$ | Deep closure |
| 3-0 Vicryl suture | <5 g | 0.6 $cm^3$ | Superficial closure. |
| Skin stapler | 92 g | 550 $cm^3$ | Skin closure. |
| 28Fr chest tube | 75 g | 51 $cm^3$ | Reexpansion of lung and drainage. |
| Thoracostomy drainage (tubing and Pleur-evac™) | 700 g | 11400 $cm^3$ | Collect drainage. |
| Heimlich valve | 23 g | 13 $cm^3$ | |
| Adhesive and absorbent dressing | 86 g | 524 $cm^3$ | Dressing. |
| Incentive spirometer | 215 g | 1721 $cm^3$ | Lung expansion, prevent atelectasis, enhance recovery. |

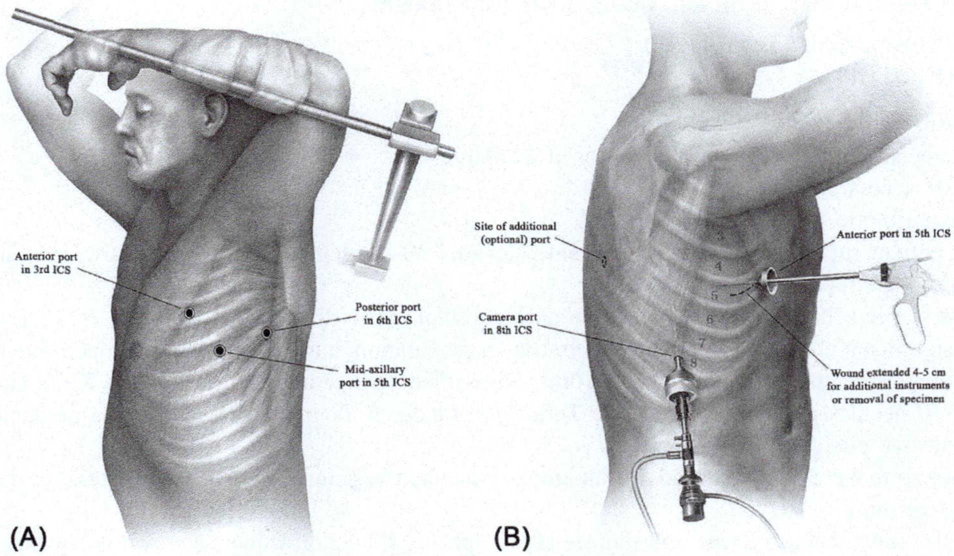


FIGURE 11.17 Triangulated video-assisted thoracoscopic surgery port sites. (A) Left. (B) Right. The choice of interspace may be modified according to the target of the intervention and surgeon preference.
From: Hirji, S.A. et al., 2022. Video-assisted thoracic surgery technique for chest wall resection. Operative Techniques in Thoracic and Cardiovascular Surgery 27 (3) 345-358.

### *11.3.10.3 Anterolateral thoracotomy*

#### 11.3.10.3.1 Decision making

This approach on the left side offers exposure for definitive repair of cardiac or left-sided vascular or pulmonary injuries (Hood RM et al. Thoracic Trauma. 1989. WB Saunders Company., no date). Open cardiac massage is readily achieved via this incision. A right-sided incision may be used for right-sided vascular or thoracic injuries. This operation should only be performed by a trained surgeon or experienced emergency/trauma physician on a patient in extremis. This is a potentially lifesaving intervention but requires significant surgical knowledge and ability to localize and repair injuries. Even if successful, such an operation requires significant surgical, anesthetic, and postoperative critical care resources. There will likely be significant contamination of the cabin with blood and bodily fluids. An untrained operator should not attempt this. As such, it is unlikely to be performed in the space environment in the near future. It is listed here for reference and completeness only (Fig. 11.18).

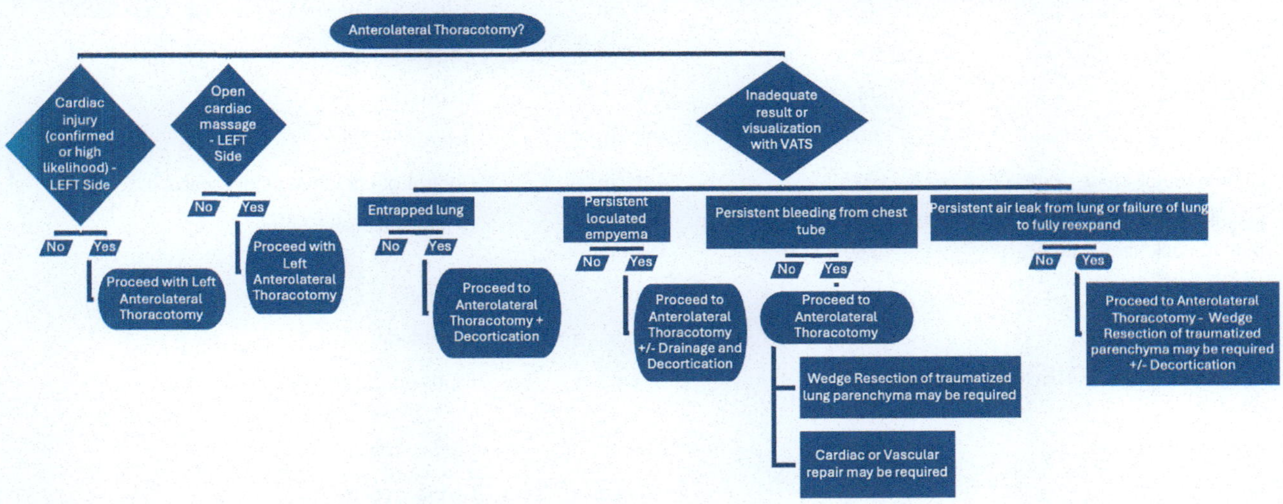


FIGURE 11.18 Anterolateral thoracotomy decision flowchart.

#### 11.3.10.3.2 Description

This approach may be performed in an emergency or urgently.

#### 11.3.10.3.3 Procedural technique and materials for spaceflight

Materials, mass, volume table (Table 11.12).

Step by step (Fig. 11.19):

1. Full monitoring
2. General anesthesia with single lung ventilation if available.
3. Large-bore IV access.
4. Prophylactic antibiotic.
5. Position the patient supine with the operative side elevated 30–45 degrees and the operative side arm abducted.
6. Fasten positioning straps.
7. Prep, place a sterile self-adhesive drape and an antimicrobial incise drape.
8. If there is a significant chance of a cardiac penetration or perforation, have prepared on the operative field: size 12, 16, and 20 Fr Foley catheters each attached to a 10ml saline-filled syringe and clamped with a Kelly clamp; 2-0 Tevdek suture on T-10 needles passed through large Teflon felt pledgets. Four of these should be prepared as well as the corresponding free pledgets.
9. Use #10 blade to make a 15 cm curved, inframammary incision beginning medially just lateral to the junction of the ribs with the sternum.
10. Division of the subcutaneous tissue and muscle layers (pectoralis major, minor, and serratus anterior) may be done with the same scalpel or Metzenbaum scissors. Because of the presumed patient extremis, do not belabor and focus on complete hemostasis at this time.

**TABLE 11.12 Materials, mass, volume for anterolateral thoracotomy.**

| Item | Mass | Volume | Purpose |
|---|---|---|---|
| Patient positioning restraints | 6 kg | <10,000 $cm^3$ | Fix patient location to facilitate procedure. |
| Operator positioning restraints | Variable based on type | Variable based on type | Fix Operator location to facilitate procedure. |
| Prophylactic antibiotic | 31 g | 10 $cm^3$ | Prevent infection. Cefazolin (Clindamycin or Vancomycin for penicillin allergic). Prophylactic therapy limited to perioperative period. |
| Mask | 3 g | <100 $cm^3$ | Asepsis and personal protection, Operator. |
| Surgical gown | 173 g | 405 $cm^3$ | Asepsis and personal protection, Operator. |
| Sterile gloves | 14 g | <100 $cm^3$ | Asepsis and personal protection, Operator. |
| Prep sponge | 113 g | 118 $cm^3$ | Prep. |
| Sterile self-adhesive drape | 90 g | 431 $cm^3$ | Asepsis. |
| Antimicrobial incise drape | 90 g | 431 $cm^3$ | Control of operative field; asepsis. |
| Headlight +/- camera<br>Battery pack/light source | 304 g<br>260 g | <4.620 $cm^3$<br>194 $cm^3$ | Optimized visualization. |
| Scalpel #10 blade | 8 g | 3 $cm^3$ | Incision |
| Triangle lung clamp | 295 g | 51 $cm^3$ | Procedural. |
| Egg beater (DeBakey–Dietrich) lung retractor | 350 g | 51 $cm^3$ | Exposure. |
| 4-0 Prolene suture | <5 g | 0.6 $cm^3$ | Cardiac repair. |
| Teflon pledgets | <1 g | 0.05 $cm^3$ | Cardiac repair. |
| Plastic Yankauer suction tip | 21 g | 18 $cm^3$ | Suction. |
| Satinsky vascular clamps (curved for partial occlusion) | 115 g | <100 $cm^3$ | Hemostasis to allow vascular repair. |

(Continued)

**TABLE 11.12 (Continued)**

| Item | Mass | Volume | Purpose |
|---|---|---|---|
| Suction (for Yankauer; portable self-contained) | 4 kg | <10,000 $cm^3$ | On operative field. |
| Handheld cautery (battery-powered) | 61 g | 10 $cm^3$ | Hemostasis. |
| 12, 16, 20 Fr Foley catheters | 23 g (each) | 84 $cm^3$ (each) | Repair cardiac perforation. |
| Irrigation (10 ml saline filled syringe) | 17 g | 10 $cm^3$ | Irrigation of space and layers prior to closure. |
| Bulb syringe (60 ml saline) | 59 g | 300 $cm^3$ | Irrigation. |
| Large forceps with teeth | 100 g | 30 $cm^3$ | Exposure. |
| DeBakey forceps | 90 g | 30 $cm^3$ | Procedural |
| Rib spreader (intercostal retractor; Finochietto) × 2 | 2 kg (each) | 250 $cm^3$ (each) | Exposure. |
| Metzenbaum scissors | 70 g | 9 $cm^3$ | Dissection. |
| Long curved ring forceps × 2 | 250 g (each) | 16 $cm^3$ (each) | Thoracoscopy length; for decortication. |
| Endo linear stapler with lung parenchymal and vascular loads | 430 g | 180 $cm^3$ | For resection or repair. Loads sizes should include 30 mm, 45 mm, and 60 mm. |
| Long curved chest tube passer | 115 g | 16 $cm^3$ | Precise positioning of chest tube. |
| Needle holder | 70 g | 8 $cm^3$ | For use with curved needles |
| Marcaine 1% with epinephrine (25 ml) | 70 g | 35 $cm^3$ | Intercostal block and incisional injection. |
| #22 spinal needle | <10 g | 0.13 $cm^3$ | Intercostal block; postoperative analgesia. |
| 20 ml syringe | 23 g | 125 $cm^3$ | Intercostal block. |
| 0 Silk sutures | <5 g | 0.6 $cm^3$ | Pericardial stay sutures. |
| #1 Vicryl sutures | <5 g | 0.6 $cm^3$ | Pericostal approximation. |
| 0 Vicryl suture | <5 g | 0.6 $cm^3$ | Deep closure. |
| 3-0 Vicryl suture | <5 g | 0.6 $cm^3$ | Superficial closure. |
| Skin stapler | 92 g | 550 $cm^3$ | Skin closure. |
| 28Fr chest tube (nontrocar) | 75 g | 51 $cm^3$ | Reexpansion of lung and drainage. |
| Thoracostomy drainage (tubing and Pleur-evac™) | 700 g | 11,400 $cm^3$ | Collect drainage. |
| Heimlich valve | 23 g | 13 $cm^3$ | For one-way drainage of air/fluid. |
| Adhesive and absorbent dressing | 86 g | 524 $cm^3$ | Dressing. |
| Incentive spirometer | 215 g | 1,721 $cm^3$ | Lung expansion, prevent atelectasis, enhance recovery. |

11. Identify the 4th ICS and then similarly divide the intercostal muscles superior to the edge of the 5th rib.
12. You may now collapse (stop ventilating) the lung on the operative side.
13. Enter into the pleural space.
14. Insert the rib spreader/retractor and open widely. A second rib spreader may be placed 90 degrees rotated from the first to expand the exposure. If needed, the incision may be extended even across the midline (sternum) or posteriorly to further improve access.

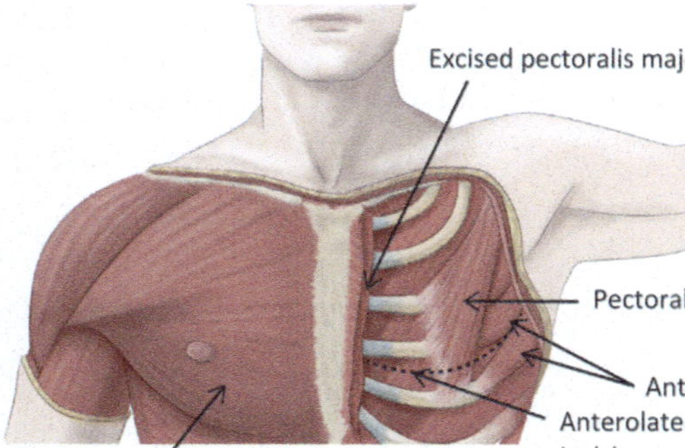


FIGURE 11.19 Anatomic layers of anterolateral thoracotomy.
Demetriades, D., et al., 2020. Emergency room resuscitative thoracotomy. Anesthesia Key.

15. Note the phrenic nerve traveling vertically on the outside of the pericardium and avoid it. The pericardium is now opened vertically anterior to the phrenic nerve using forceps to elevate and Metzenbaum scissors to nick and then open. Lift up with the lower blade of the scissors while cutting to avoid injury to the heart.
16. Tack up the "4 corners of the pericardium" to the skin using 0 Silk suture. The net effect is to elevate the pericardium and its contents into the surgical field.
17. Repair of a cardiac laceration may be performed with 4-0 Prolene pledgeted horizontal mattress sutures. When tying, gently follow the knot down and focus carefully so as not to lift up when tying or tighten sufficiently to tear myocardium.
18. If the heart is acutely bleeding from an apparently transmural cardiac perforation, insert one of the prepared Foley catheters with a Kelly clamp across its lumen, inflate the balloon with saline, and apply very gentle traction; this will allow placement of a purse-string hemostatic suture using deep bites with the prepared pledgeted Tevdek sutures. As noted, when tying, gently follow the knot down and focus carefully so as not to lift up when tying or tighten sufficiently to tear myocardium. After the first throw, the Foley balloon may be deflated and the catheter removed to complete tying with at least six squared throws.
19. Following cardiac repair, the pericardial stay sutures are cut.
20. The operative site should be irrigated with saline and all fluid suctioned.
21. Two apically directed chest tubes should be placed, one directed anteriorly and one posteriorly.
22. Load #1 Vicryl sutures. These pericostal sutures are placed by introducing the needle, skimming the top of the rib above, and then immediately directing the tip of the needle to stay superficial behind the rib. Follow the curve of the needle to exit inferior to the lower edge of the rib below. Several such interrupted sutures should then be similarly placed.
23. Once these sutures are placed, the central one may be crossed and held tightly by a gloved Assistant, if available, and the other sutures tied. The sutures should be snug against the superior aspect of the rib above and the inferior aspect of the rib below. At least 6 squared throws should be used to tie each suture.
24. Prior to closure, an intercostal block may be performed at the posterior-most aspect of the exposed ribs. This should include the intercostal levels/rib spaces one above and one below the incision.
25. Irrigate with saline prior to closure of each layer.
26. Use 0 Vicryl running or several figure-of-8 interrupted sutures to approximate the muscle layers. Tighten to approximate the tissues without tearing or causing ischemia.
27. Approximate the subcutaneous layer with 3-0 Vicryl similarly.
28. Approximate the skin with staples.

## 11.4 Summary, recommendations, and considerations for the future

Experience in space, parabolic flight, and simulations all suggest that Earth-based procedures can reasonably be expected to be performed in microgravity, requiring acclimation of the Operator to the environment and additional considerations. Resource, mass, and volume constraints will continue to dominate. Critical future developments include the design and operationalization of an isolation/containment capability that insulates the operative field from the closed-loop vehicular atmosphere. Exploration-class missions will require medical autonomy without the fallback of timely evacuation. As such, related crew selection, medical expertise, training, and a cushion of redundancy will be essential.

## Biography

Matthew Marc Cooper MD, MBA, FACS, FAsMA, FCAMA, CPL, CFI is Principal and CEO at Peak Aerospace Medicine & Cardiovascular Surgery. He previously served as Vice President Medical Affairs, Corporate Health, & Safety for 3M Health Care, including Chief Medical Officer Medical Solutions. He is a Diplomat of the American Boards of Thoracic Surgery and Surgery, and a Fellow of the American College of Surgeons, the Aerospace Medical Association, and the Civil Aviation Medical Association. He is a graduate of Franklin & Marshall College, New York University School of Medicine, and Spears School of Business, Oklahoma State University. In addition to Residency and Chief Residency in General Surgery (New York University Medical Center, The University of Iowa Hospitals & Clinics) and Thoracic Surgery (Columbia University), he also completed Medical Staff Fellowship in the Surgery Branch of the National Heart, Lung & Blood Institute of the National Institutes of Health and Fellowship in Pediatric Cardiothoracic Surgery and Transplantation at The Hospital for Sick Children, Great Ormond Street, United Kingdom. He has also studied Health Policy at the John F. Kennedy School of Government, and Comprehensive Medical Simulation, both at Harvard University.

His research has included pioneering methods of immunosuppression for primate cardiac xenograft transplantation and adaptation of those with abnormal cardiovascular physiology to austere conditions such as increased altitude and flight.

He is also an expert in Aviation Physiology, a Special Consultant to the Federal Air Surgeon, and a Senior Aviation Medical Examiner. He is a commercial and aerobatic pilot, flight instructor, and has been a leader in the application of aviation-based safety culture to health care.

## AI disclosure

During the preparation of this work, the author used Google Search AI Mode in order to assist in estimating and calculating mass and volume of required equipment for each procedure. After using this tool/service, the author reviewed and edited the content as needed and takes full responsibility for the content of the publication.

## References


Anderson, D., Chen, S.A., Godoy, L.A., Brown, L.M., Cooke, D.T., 2022. Comprehensive review of chest tube management: a review. JAMA Surgery 157 (3), 269–274. Available at https://doi.org/10.1001/jamasurg.2021.7050, http://archsurg.jamanetwork.com/issues.aspx.

Azizi, N., ter Avest, E., Hoek, A.E., Admiraal-van de Pas, Y., Buizert, P.J., Peijs, D.R., et al., 2021. Optimal anatomical location for needle chest decompression for tension pneumothorax: a multicenter prospective cohort study. Injury 52 (2), 213–218. Available at https://doi.org/10.1016/j.injury.2020.10.068.

Daughtry, B., n.d. Bronchoscopy. TSRA Primer- Thoracic. AATS 2025.

Dinner, J., n.d. SpaceX's Fram2 mission captures 1st human X-ray in Earth orbit. Space.com. 2025.

Elder, C.A., Moore, J., Janis, B.R., Shacklette, S., Jones, C., Cantrell, R., et al., 2024. Rehydration outcomes for freeze-dried red blood cells in reduced gravity. Acta Astronautica 214, 64–71. Available at https://doi.org/10.1016/j.actaastro.2023.10.006.

Heimes, J., Copeland, H., Lulla, A., Duldulao, M., Bahjri, K., Zaheer, S., et al., 2017. The use of thrombolytics in the management of complex pleural fluid collections. Journal of Thoracic Disease 9 (5), 1310–1316. Available at https://doi.org/10.21037/jtd.2017.04.56.

Hood, R.M., et al., n.d. Thoracic Trauma. 1989. WB Saunders Company.

Howard, R.L., et al., n.d. A Crew Seat for Human Exploration in Multiple Gravity Environments. Johnson Space Center. Document ID 20230014002. Nasa.gov.

Lopez-Rincon, R.M., et al., n.d. Ultrasound-Guided Intercostal Nerve Block. StatPearls [Internet]. July 19, 2023.

Mehrotra, M., et al., n.d. Video-Assisted Thoracoscopy. StatPearls [Internet]. May 1, 2024.

Willner, D.A., et al., n.d. Pericardiocentesis. StatPearls [Internet]. 2025.

Yanik, F., 2023. Current overview of awake, non-intubated, video-assisted thoracic surgery. Wideochirurgia I Inne Techniki Maloinwazyjne 18 (3), 445–452. Available at https://doi.org/10.5114/wiitm.2023.128070, https://www.termedia.pl/Current-overview-of-awake-non-intubated-video-assisted-r-nthoracic-surgery,42,50833,0,1.html.


## Further reading


Joseph, C.T., Appanavar, V.S., 2025. Bronchoscopy in space: overcoming medical and engineering challenges in a zero-gravity environment. International Journal of Advanced Research 13 (1), 594–597.

Chapter 12

# Abdominal procedures

**Margaret Siu**
*Fellow of Surgical Critical Care, R Adams Cowley Shock Trauma Center, University of Maryland School of Medicine, Baltimore, MD, United States*

## 12.1 Introduction

Half of the population will experience intraabdominal surgical pathology in their lifetime (Nunoo-Mensah et al., 2009), and approximately 300 million surgeries are performed globally every year (Weiser et al., 2008). As spaceflight becomes more accessible to civilians and the number of humans in space increases, so too does the likelihood of a surgical emergency. Currently, there is limited literature on intraabdominal procedures during spaceflight (Kamine et al., 2022; Siu et al., 2022; Kamine et al., 2024; Lazow et al., 2024; Levin et al., 2022). This chapter reviews the procedures for management of common intraabdominal pathology that may occur during spaceflight.

## 12.2 Appendicitis

### 12.2.1 Background

Appendicitis is the inflammation and subsequent infection of the appendix. Symptoms include abdominal pain, nausea, vomiting, anorexia, and sometimes diarrhea. The diagnosis of appendicitis can be accomplished by a handheld ultrasound, demonstrating an tubular structure in the right lower quardant with fluid stranding around the appendix and potential solid lesion at the base, known as a fecalith. The average lifetime probability of developing acute appendicitis is 6.6% with one-third of cases presenting as perforated appendicitis (Ferris et al., 2017). Thus, it is important to review the treatment of nonperforated and perforated appendicitis.

### 12.2.2 Nonperforated appendicitis

Uncomplicated appendicitis, that is, appendicitis without perforation, may be treated with broad-spectrum antibiotics or with an appendectomy. If antibiotics are utilized as mono-therapy, broad-spectrum antibiotics such as a third-generation Cephalosporin with concomitant anaerobic coverage, such as with Metronidazole, are generally recommended for 2–4 weeks. However, appendectomy remains the gold standard for treatment of uncomplicated appendicitis (Doleman et al., 2022). If surgical capabilities are available, the authors believe that an appendectomy is preferred over medical treatment. Additionally, if antibiotic use does not improve symptoms, surgery will be needed.

### 12.2.3 Appendectomy

In this section, we discuss open appendectomy, as rendering pneumoperitoneum in microgravity requires laparoscopic equipment that is both too big and too heavy for most space missions. Ideally, two operators are present, one for administering anesthesia and the other for performing the surgery. However, this operation has been described with a single operator and performed with local anesthesia injected into the skin, subcutaneous tissue, and fascia of the abdominal wall. Table 12.1 summarizes the optimal supplies needed.

Procedure of appendectomy:

1. The patient is positioned supine with the abdominal wall exposed. The entire body should be secured to a flat surface with straps, holding the extremities in place alongside the torso.

**Procedural Medicine for Spaceflight. DOI:** https://doi.org/10.1016/B978-0-443-27552-4.00033-6

**TABLE 12.1** Materials needed for appendectomy in nonperforated appendicitis.

| Optimal supplies (within spacecraft limitations) | Level of need | Alternatives/improvised substitutes | Mass (g) | Volume (ml) | Power required | Modifications | Hazards |
|---|---|---|---|---|---|---|---|
| Sterile gloves | Required | | 33.2 | 110.5 | No | | |
| Scalpel, 10 blade or 15 blade | Required | | 8.4 | 58.5 | No | | Sharp to be disposed |
| Pickup instrument (i.e., DeBakey forceps) | Preferred | Any pickup instrument | 45.2 | 40 | No | | |
| Self-retaining retractor (Weitlander retractors) | Preferred | Handheld retractors | 94.6 | 432 | No | | |
| Kelly clamp | Required | | 40.3 | 76.5 | No | | |
| Nontraumatic bowel clamp | Preferred | Any clamp | 50.4 | 85.5 | No | | |
| Suction capability | Preferred | Abdominal sponges | 137.8 | 600 | Yes | | |
| 0-0 Silk ties | Preferred | Any suture material with needle removed | 3.6 | 13 | No | | |
| 1-0 Polydioxanone suture | Preferred | Any suture | 3.8 | 13 | No | | Needle to be disposed |
| 3-0 Vicryl suture | Preferred | Any suture | 4.4 | 13 | No | | Needle to be disposed |
| 4-0 Monocryl suture | Preferred | Skin glue or skin staples | 5.3 | 13 | No | | Needle to be disposed |
| Gauze packet | Required | | 21.3 | 188.5 | No | | |
| Skin tape | Required | | 20.4 | 50.3 | No | | |

2. In an optimal environment, the patient would be under general anesthesia with a protected airway; however, local anesthesia on the abdominal wall in the right lower quadrant alone remains an option.
3. The performer should maximize sterility by washing hands with betadine and saline solution, and donning sterile gloves.
4. The right anterior superior iliac spine and umbilicus should be identified. Midway between these landmarks lies McBurney's point, the location most commonly overlying the appendix.
5. The scalpel is used to make a skin incision approximately 3–5 cm in size obliquely over McBurney's point, running right (superolateral) to left (inferomedial). Alternatively, the incision can be made over the point of maximum tenderness as this indicates the location of most severe peritoneal inflammation and likely location of the appendix.
6. The subcutaneous tissue is then dissected down either using a blade or, if available, using electrocautery. The fascia and peritoneum are to be entered sharply. If there is abdominal fluid, the fluid could be suctionedto maximize exposure. Fig. 12.1 shows commonly used suction and electrocautery equipment.
7. A self-retaining retractor should be set up to provide exposure.
8. The appendix is identified as a blind tubular structure that pouches off the beginning of the colon. It should be secured with pickups. The base of the appendix is then identified to enter directly into the colon.
9. A window is created using the blunt Kelly clamp at the base of the appendix and mesoappendix.
10. At this point, the bowel should only be manipulated using the nontraumatic bowel clamp to avoid injury to healthy parts of the bowel.
11. At the base of the appendix at the cecum, the appendix is ligated using two separate 0-0 silk ties (nonabsorbable suture ties) approximately 5 mm apart, and then transected sharply between the ties. Fig. 12.2 displays the common ties and sutures needed.
12. The mesoappendix is ligated using 0-0 silk ties (nonabsorbable suture ties) and then subsequently divided distal to the ligation.
13. The appendix is then disposed of in a sealed container.
14. Hemostasis should be obtained either using electrocautery or Vicryl (braided absorbable) sutures. Fig. 12.1 depicts electrocautery equipment.

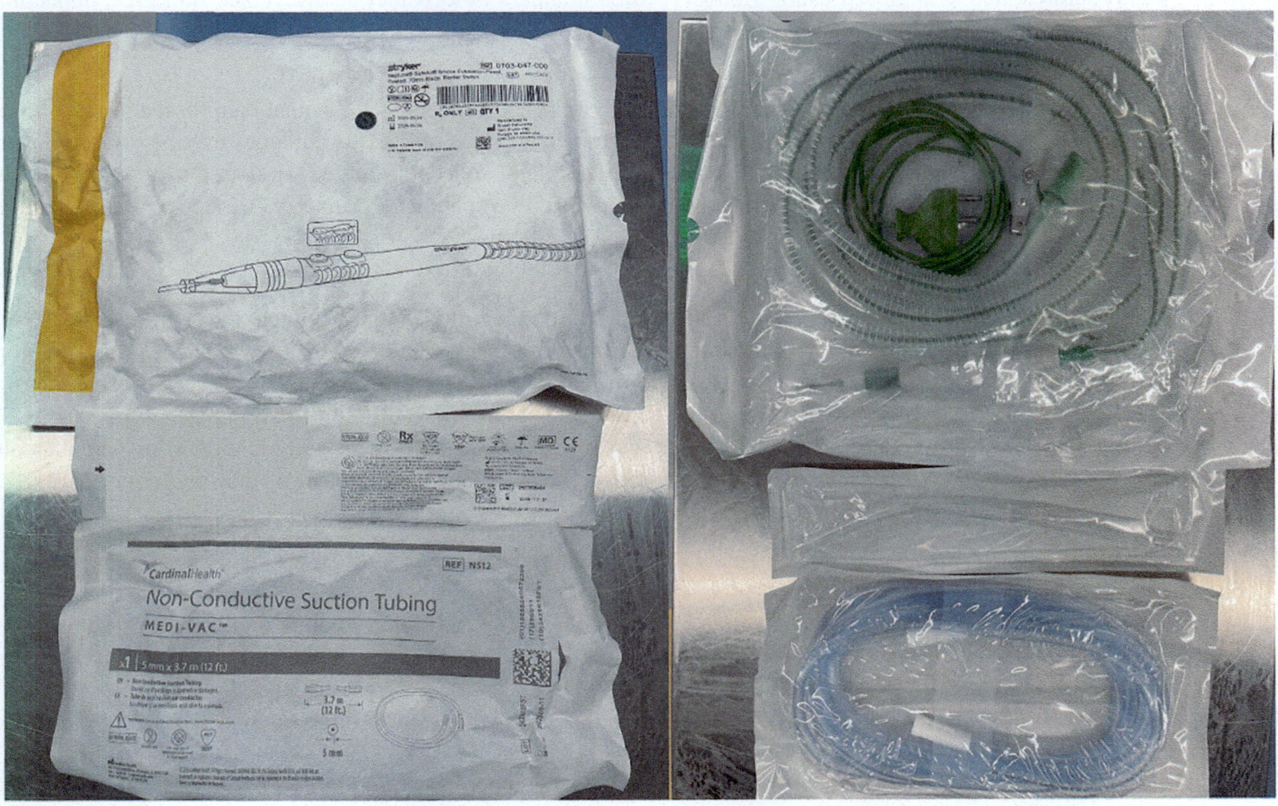


FIGURE 12.1 Suction and electrocautery.
Suction and tubing on bottom and electrocautery set up on top.

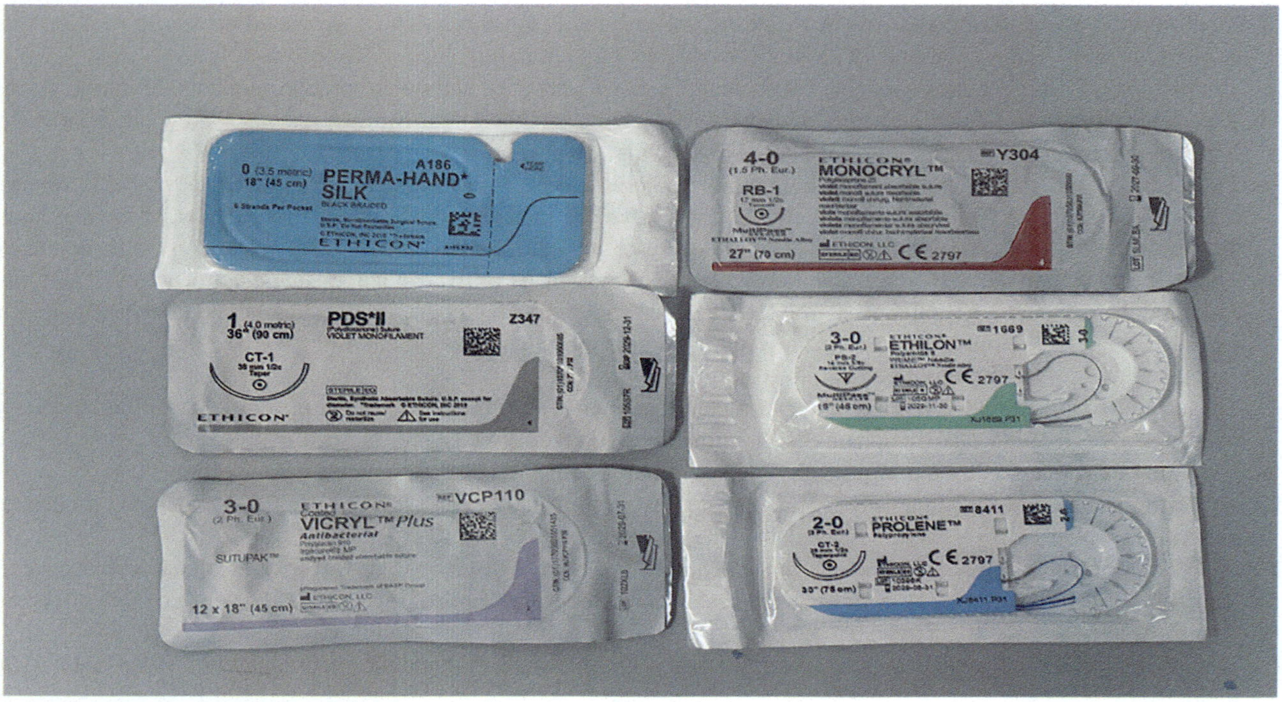


FIGURE 12.2 Variety of sutures.
Commonly used ties and sutures in the described abdominal procedures, listed in Tables 12.1 through 12.4.

15. Fascia should be closed using polydioxanone (absorbable) sutures in a running fashion, subcutaneous tissue closed using Vicryl sutures, and skin closed using Monocryl (monofilament absorbable) suture versus skin glue, depending on which is available.
16. Dry sterile dressing is then applied.

The patient should be monitored for bleeding, drainage, fevers, chills, and worsening abdominal pain for the next 72 hours. Diets may be resumed once the patient has a return of bowel function in the form of flatus. Antibiotics may be stopped if there was no spillage of gastrointestinal contents or if there were no fluid collections noted around the appendix. If the aforementioned spillage or fluid collections are present, antibiotics should be continued for four days following the procedure.

## 12.2.4 Perforated appendicitis

In the setting of multiple days of abdominal pain with a fluid collection identified on ultrasound in the right lower quadrant adjacent to the appendix, the diagnosis of perforated appendicitis with abscess can be confidently made. In this situation, an appendectomy is not the optimal treatment. The patient will require antibiotics. Intravenously infused broad-spectrum antibiotics such as piperacillin/tazobactam are preferred; however, they are more difficult to administer in microgravity. Thus, oral antibiotics such as amoxicillin/clavulanate for two weeks would suffice, in combination with percutaneous drainage of fluid collection. Table 12.2 lists the supplies needed for percutaneous drainage of the fluid collection.

Percutaneous drainage for perforated appendicitis:

1. To successfully perform percutaneous drainage of intraabdominal abscesses, imaging guidance is favored. In spaceflight, ultrasound guidance alone can be used.
2. The performer should maximize sterility by washing hands with betadine and saline solution and donning sterile gloves.
3. Under ultrasound guidance, an 18-gauge or larger needle attached to a syringe is directed into the abscess cavity to first allow drainage of purulent fluid.
4. The syringe is then removed, leaving the needle in place, and a wire is threaded through the needle into the collection (Fig. 12.3), followed by removal of the needle.
5. The skin around the puncture site should be incised to enlarge the puncture wound to approximately 5 mm in length using a scalpel blade.
6. Then, a drainage catheter is placed over the wire and directed into the fluid collection.

**TABLE 12.2** Materials needed for percutaneous drainage for perforated appendicitis.

| Optimal supplies (within spacecraft limitations) | Level of need | Alternatives/improvised substitutes | Mass (g) | Volume (ml) | Power required | Modifications | Hazards |
|---|---|---|---|---|---|---|---|
| Sterile gloves | Required | | 33.2 | 110.5 | No | | |
| Portable ultrasound | Required | | | | Yes | | |
| 18-gauge needle | Preferred | May use any large-bore needle | 3.3 | 67.5 | No | | Needle to be disposed |
| 60 mL syringe | Preferred | May use any size syringe | 46.5 | 1250 | No | | |
| Guidewire | Required | | 15.4 | 100 | No | | Wire to be disposed |
| Scalpel, 10 blade or 15 blade | Required | | 8.4 | 58.5 | No | | Sharp to be disposed |
| Bulb suction drain, includes drain catheter and bulb | Required | | 105.7 | 4312 | No | | |
| 3-0 nylon suture | Preferred | Any suture | 4.8 | 21.4 | No | | Needle to be disposed |

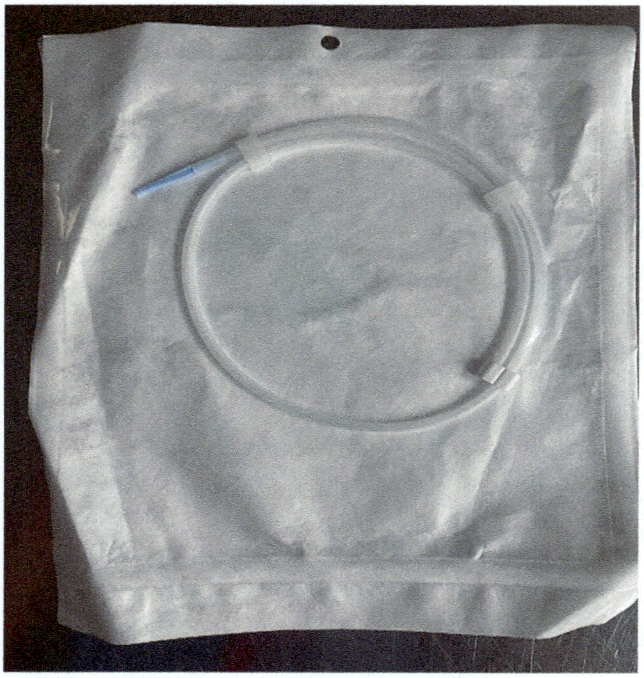

FIGURE 12.3 Guidewire.
Thin, flexible guidewire to be thread into needle and guide catheter placement.

7. The wire is removed, and the drainage catheter is connected to negative-pressure drainage to bulb suction (Fig. 12.4).
8. Following placement of the drain, the drain is sutured in with nylon or any nonabsorbable sutures.

Oral amoxicillin/clavulanate should be continued for 4 days after source control with a drain. Patients should be monitored for drain output content and volume, bleeding from the drain site, worsening abdominal pain, fever, and erythema around the drain insertion site. If the patient is improving clinically and if the drain output is clear and less than 30 mL daily for three consecutive days, the drain may be removed. If the patient does not improve following drain placement, he/she will need to return to Earth for washout, as a standard appendectomy will no longer suffice.

## 12.3 Cholecystitis

### 12.3.1 Background

Cholecystitis is the inflammation and subsequent infection of the gallbladder. In the United States, approximately 200,000 individuals are diagnosed with cholecystitis each year (Gallaher and Charles, 2022). Gallbladder disease may be secondary to gallstones obstructing the cystic duct, leading to infection, or it may be acalculous in nature. In healthy patients, the definitive treatment for cholecystitis is a cholecystectomy (Wang et al., 2014). Recent literature has described the limited utility of cholecystostomy during spaceflight (Lazow et al., 2024). Indications to perform a cholecystectomy in spaceflight include sepsis secondary to cholecystitis and lack of improvement of cholecystitis with antibiotics. In situations where the patient improves with antibiotic use alone, we recommend delaying cholecystectomy until after return as the procedure itself may entail morbidity in nonideal environments (Wang et al., 2014). Antibiotics should be initiated with every cholecystitis diagnosis. We recommend two weeks of broad-spectrum antibiotics such as piperacillin/tazobactam if intravenous infusions are possible, or oral amoxicillin/clavulanate if infusions are not possible.

### 12.3.2 Diagnosis and cholecystectomy for cholecystitis

Cholecystitis usually presents with abdominal pain, commonly localized to the right upper quadrant, with associated symptoms including fever, nausea, and vomiting. While a full workup, including a complete blood count, would be beneficial, an ultrasound examination of the right upper quadrant is more useful. Handheld ultrasound would suffice, showing a thickened gallbladder wall, peri-cholecystic fluid, dilated cystic duct, and multiple gallstones.

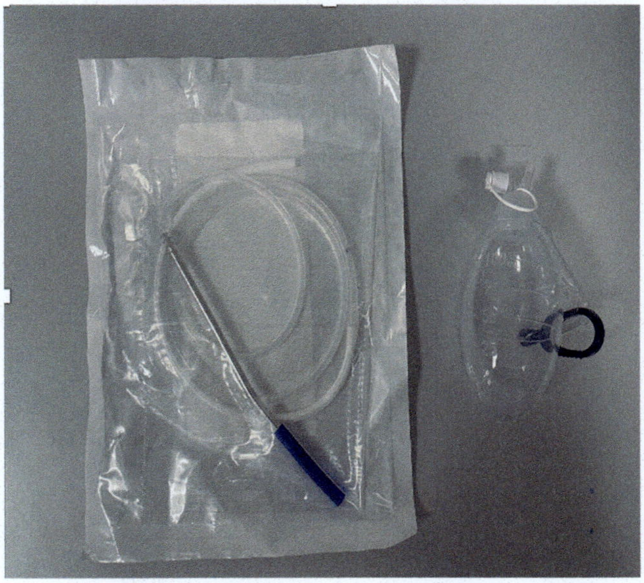

FIGURE 12.4 Bulb suction drainage.
Drain with bulb suction placed into an intraabdominal fluid collection.

Materials needed to perform a cholecystectomy are listed in Table 12.3. Optimally, three crewmembers would be involved in the procedure: one for anesthesia, one to perform the cholecystectomy, and one to retract. Without the assistance of a crew member to retract or the availability of a self-retaining retractor, the case may be quite difficult. Additionally, general anesthesia is preferred for complete paralysis and pain control. However, an open cholecystectomy can be performed with local anesthesia to the abdominal wall at the right upper quadrant through all the abdominal wall layers.

Procedure of open cholecystectomy:

1. The patient is positioned supine and secured to a flat surface, and the abdominal wall is exposed. As with other abdominal procedures, the entire body should be secured to a flat surface with straps, holding the extremities in place alongside the torso.
2. The performer should maximize sterility by washing hands with betadine and saline solution and donning sterile gloves.
3. A right subcostal oblique skin incision is made with a scalpel two fingerbreadths inferior to the most inferior palpated rib.
4. Using electrocautery (Fig. 12.1), subcutaneous tissue is dissected down, and each layer of the muscle is divided using electrocautery, with extreme care taken to ensure hemostasis as each layer of muscle is encountered.
5. The assistant then retracts the abdominal wall on both sides of the incision using handheld retractors.
6. Three to four laparotomy pads are used to lift the gallbladder and liver anteriorly while pushing the hollow viscus organ inferiorly.
7. The gallbladder is visualized and grasped with a Kelly clamp. Using electrocautery, the gallbladder peritoneum is divided laterally and medially, and the gallbladder is subsequently divided off the liver. This dissection of the gallbladder off the liver is continued until the infundibulum and the cystic duct are encountered.
8. Using manual palpation, the cystic duct and the cystic artery are encircled and clamped with a right-angle clamp.
9. The gallbladder itself is sharply transected distal to the right-angle clamp with a scalpel.
10. The cystic duct and cystic artery are ligated with two separate 0-0 silk ties (nonabsorbable ties) and the right-angle clamp removed.
11. Hemostasis on the cystic plate is confirmed, and the laparotomy pads are removed.
12. The anterior abdominal wall is closed in multiple layers using 1–0 polydioxanone (absorbable) sutures on the fascial level, followed by Vicryl (absorbable) sutures on the deep dermis layer, and then followed by skin staples versus Monocryl (absorbable) sutures on the skin.
13. The skin is dressed with dry sterile dressing.

Antibiotics may be discontinued after cholecystectomy. Diet may be resumed on postoperative day one. The patient is then monitored for fevers, drainage, and skin infection. If these symptoms do develop, antibiotics should continue for four

**TABLE 12.3** Materials needed for cholecystectomy.

| Optimal supplies (within spacecraft limitations) | Level of need | Alternatives/improvised substitutes | Mass (g) | Volume (ml) | Power required | Modifications | Hazards |
|---|---|---|---|---|---|---|---|
| Sterile gloves | Required | | 33.2 | 110.5 | No | | |
| Scalpel, 10 blade or 15 blade | Required | | 8.4 | 58.5 | No | | Sharp to be disposed |
| Pickup instrument (i.e. DeBakey forceps) | Preferred | Any pickup instrument | 45.2 | 40 | No | | |
| Self-retaining retractor (Weitlander retractors) | Preferred | Handheld retractors | 94.6 | 432 | No | | |
| Large handheld retractor | Preferred | Laparotomy pads | 135.6 | 626 | No | | |
| Laparotomy pads | Required | | 113.5 | 1632 | No | | |
| Kelly clamp | Required | | 40.3 | 76.5 | No | | |
| Right-angle clamp | Required | | 51.7 | 171 | No | | |
| Suction capability | Preferred | Laparotomy pads | 137.8 | 600 | Yes | | |
| Cautery | Required | | 231.0 | 4144 | Yes | | |
| 0-0 Silk ties | Preferred | Any suture material with needle removed | 3.6 | 13 | No | | |
| 1-0 Polydioxanone suture | Preferred | Any suture | 3.8 | 13 | No | | Needle to be disposed |
| 3-0 Vicryl suture | Preferred | Any suture | 4.4 | 13 | No | | Needle to be disposed |
| 4-0 Monocryl suture | Preferred | Skin glue or skin staples | 5.3 | 13 | No | | Needle to be disposed |
| Gauze packet | Required | | 21.3 | 188.5 | No | | |
| Skin tape | Required | | 20.4 | 50.3 | No | | |

additional days. If there remains no improvement or if the patient becomes hypotensive, develops bleeding, or worsening abdominal pain, an emergent return to Earth is warranted for surgical exploration.

### 12.3.3 Cholecystostomy tube placement

There are situations where cholecystostomy tube placement, that is, the drainage of the gallbladder, is preferred over cholecystectomy. On earth, patients who are too ill to undergo operative intervention would usually undergo a cholecystostomy tube placement. During space travel, instances where individuals are uncomfortable performing the procedure or if equipment is not available, cholecystostomy tube may temporize cholecystitis while bridging patient to return to Earth for cholecystectomy. Table 12.4 lists the materials needed to place a cholecystostomy tube.

Performing a cholecystostomy:

1. The performer first uses a handheld ultrasound to identify the gallbladder in the right upper quadrant of the abdomen.
2. The performer should maximize sterility by washing hands with betadine and saline solution and donning sterile gloves.
3. Using a large-bore needle, preferably 18-gauge or above, attached to a syringe, the gallbladder is accessed under ultrasound guidance.
4. Once the needle is noted to be inside the gallbladder, bile should be aspirated.
5. The syringe is then removed, leaving the needle in place, and a wire is threaded through the needle, followed by the removal of the needle.
6. The skin is incised to enlarge the puncture wound to approximately 5 mm using a scalpel blade.
7. Then a drainage catheter is then placed over the wire and directly into the gallbladder.
8. The wire is removed, and the drainage catheter will be connected to negative-pressure drainage to bulb suction.
9. Following placement of the drain, the drain is sutured in with nylon or any nonabsorbable sutures.

Following cholecystostomy, antibiotics should be continued for 4 days after drain placement, or source control. If at the four-day mark, symptoms of right upper quadrant pain and fever do not improve, antibiotics may be continued. This cholecystostomy tube will be left in place until the individual returns to Earth for a definitive cholecystectomy. Patients can usually be conservatively managed with the cholecystostomy tube and antibiotics. On Earth, cholecystectomy takes place 6–8 weeks following cholecystostomy tube placement. If there is leakage of bile around the tube or erythema and drainage from the tube site, he or she should return earlier for operative intervention and definitive removal of the gallbladder.

## 12.4 Hernias

### 12.4.1 Background on ventral hernia

A ventral hernia is a defect of the strength layer, or the fascia, of the anterior abdominal wall. Ventral hernias include, but are not limited to, umbilical hernias, incisional hernias, and/or Spigelian hernias. Ventral hernia repair is one of the most common surgeries performed in the United States, occurring approximately 350,000 times per year (Bhardwaj et al., 2024). Thus, it is important to be able to perform ventral hernia repairs in spaceflight.

Ventral hernias may be reducible, meaning the abdominal contents in the hernia can be manually pushed back into the abdominal cavity; or incarcerated, meaning the hernia contents cannot be manually returned to the abdominal cavity. These hernias may also be strangulated, which occurs when hernia tissue is necrosing secondary to vascular compromise due to being trapped in the fascial defect. Ventral hernias that are incarcerated lead to pain and discomfort. Strangulated hernias require urgent management to prevent further necrosis, which ultimately results in infection and subsequent morbidity and mortality. Systematic reviews and meta-analyses have assessed multiple types of ventral hernia repairs utilizing either mesh or other techniques to yield lower rates of recurrence (Siddiqui et al., 2023). However, in limited resource environments such as space travel, a primary repair with permanent sutures is appropriate given its simplicity and its ability to prevent progression to strangulation and sepsis.

### 12.4.2 Ventral hernia repair

This section focuses on procedural steps to repair a ventral hernia in spaceflight, a topic not yet described in the literature. Table 12.5 lists the optimal equipment needed for the hernia repair.

Procedural steps of ventral hernia repair:

1. The patient is placed supine with the anterior abdominal wall exposed, with the extremities and torso secured to a flat surface using safety belts.

**TABLE 12.4** Materials needed for percutaneous cholecystostomy tube placement.

| Optimal supplies (within spacecraft limitations) | Level of need | Alternatives/improvised substitutes | Mass (g) | Volume (ml) | Power required | Modifications | Hazards |
|---|---|---|---|---|---|---|---|
| Sterile gloves | Required | | 33.2 | 110.5 | No | | |
| Portable ultrasound | Required | | | | Yes | | |
| 18-gauge needle | Preferred | May use any large-bore needle | 3.3 | 67.5 | No | | Needle to be disposed |
| 60 mL syringe | Preferred | May use any size syringe | 46.5 | 1250 | No | | |
| Guidewire | Required | | 15.4 | 100 | No | | Wire to be disposed |
| Scalpel, 10 blade or 15 blade | Required | | 8.4 | 58.5 | No | | Sharp to be disposed |
| Bulb suction drain, includes drain catheter and bulb | Required | | 105.7 | 4312 | No | | |
| 3-0 nylon suture | Preferred | Any suture | 4.8 | 21.4 | No | | Needle to be disposed |

**TABLE 12.5** Ventral hernia repair materials.

| Optimal supplies (within spacecraft limitations) | Level of need | Alternatives/improvised substitutes | Mass (g) | Volume (ml) | Power required | Modifications | Hazards |
|---|---|---|---|---|---|---|---|
| Sterile gloves | Required | | 33.2 | 110.5 | No | | |
| Scalpel, 10 blade or 15 blade | Required | | 8.4 | 58.5 | No | | Sharp to be disposed |
| Pickup instrument (i.e., DeBakey forceps) | Preferred | Any pickup instrument | 45.2 | 40 | No | | |
| Self-retaining retractor (Weitlander retractors) | Preferred | Handheld retractors | 94.6 | 432 | No | | |
| Large-bore permanent suture | Required | | 4.9 | 21.4 | No | | Needle to be disposed |
| Suction capability | Preferred | Laparotomy pads | 137.8 | 600 | Yes | | |
| Cautery | Preferred | Scalpel, and sponges/sutures for hemostasis | 231.0 | 4144 | Yes | | |
| 4-0 Monocryl suture | Preferred | Skin glue or skin staples | 5.3 | 13 | No | | Needle to be disposed |
| Gauze packet | Required | | 21.3 | 188.5 | No | | |
| Skin tape | Required | | 20.4 | 50.3 | No | | |

2. The performer should maximize sterility by washing hands with betadine and saline solution and donning sterile gloves.
3. The skin overlying the hernia is first anesthetized with local anesthetic.
4. A scalpel is used to make a skin incision overlying the hernia along the entire hernia length, which is identified using manual palpation. The initial incision should be done carefully and only penetrate the skin, as bowel may lie immediately under the skin.
5. Using electrocautery, the subcutaneous tissue is dissected down to the level of the fascial defect.
6. Self-retaining retractors and suction, with at least 3 L canisters available for collection, are used to provide optimal exposure of the fascia.
7. Once the anterior surface of the fascia is entirely clear, the deeper posterior surface of the fascia is cleared bluntly to ensure that no viscera from the abdominal cavity is adhered to the hernia defect.
8. Nonabsorbable sutures are used to approximate the fascia without any tension, either in a running fashion or with interrupted sutures.
9. Skin is then closed using 4-0 monocryl sutures or absorbable suture, and the wound covered with a dry sterile dressing.

### 12.4.3 Additional considerations

If optimal equipment is not available, such as no electrocautery or suction capabilities, then dissection could also be carried down using sharp dissection. However, providers would need to take extra care to ensure hemostasis at the completion of the case. Following the procedure, the patient should be monitored for skin or deep tissue infection, bleeding from the site, or drainage from the site. Such complications should prompt initiation of antibiotics. While antibiotics such as vancomycin and piperacillin/tazobactam are preferred, amoxicillin/clavulanate may be used in microgravity; and patient is to return to Earth for reexploration of the surgical site. With the lack of mesh, recurrence may be common, warranting close monitoring and evaluation by a surgeon once returned from space travel. If recurrence takes place in spaceflight and there are symptoms of bowel obstruction, that is, nausea, vomiting, and obstipation, there would be concern for bowel incarceration or strangulation secondary to complication of the hernia repair. This complication would warrant immediate return to Earth for complex repair.

### 12.4.4 Inguinal hernias

Inguinal hernias are hernias that develop in the groin due to weakness or defect of the abdominal wall; they are one of the most common hernias in adults (Ruhl and Everhart, 2007). These usually present as a mass felt in the groin that enlarges upon an increase in intraabdominal pressure or with standing. If the hernia content consists of bowel, obstruction may occur. Subsequently, the patient may experience nausea, vomiting, or obstipation. A majority of the time however, inguinal hernias are reducible (Jenkins and O'Dwyer, 2008). This latter point is reassuring, as inguinal hernia repair, whether an open approach or a laparoscopic approach, is quite difficult.

In the event an individual experiences an acute development of an inguinal hernia, certain key aspects should be assessed. Specifically, the individual should be monitored for bowel function, as a lack of bowel function may indicate obstruction. Moreover, the patient should be assessed for overlying skin changes above the mass, as erythema may indicate bowel compromise in the inguinal hernia defect. Lastly, the individual needs to be examined for whether or not the hernia contents are reducible. To do so, the patient should lie flat, be relaxed, and preferably ice the area over a 10-minute interval. Another individual should then attempt to reduce the hernia into the defect by using gentle pressure with two hands overlying the mass, gently pushing the mass into the hernia defect.

We advise against operative intervention for inguinal hernia in space travel as this operative procedure carries high-level complexity, requires mesh placement, and at times bowel resection and anastomosis. The safest approach to inguinal hernias during space travel would be to attempt reduction. If the hernia remains incarcerated or potentially strangulated, meaning there is an inability to reduce the hernia or there is concern for loss of blood flow, respectively, we advise a return to Earth for definitive repair.

## 12.5 Diverticulitis

### 12.5.1 Background on diverticulitis

Diverticulitis is one of the most common colon pathologies middle-aged individuals develop (Strate and Morris, 2019). Presenting symptoms include abdominal pain, nausea, vomiting, and at times fevers, chills, and significant abdominal pain. Diverticulitis most commonly occurs in the sigmoid colon; however, it can occur anywhere along the gastrointestinal tract (Floch and Bina, 2004). Uncomplicated diverticulitis is described as inflammation of diverticula. Complicated diverticulitis involves the development of perforation of the diverticula and the development of fluid collections and abscesses. With a diagnosis of uncomplicated diverticulitis, 2 weeks of amoxicillin/clavulanate should be initiated. In complicated diverticulitis, the patient would need 2 weeks of antibiotics and fluid collections associated with complicated diverticulitis require drainage.

## 12.6 Percutaneous drainage for perforated diverticulitis

Ideally, diagnoses require computed tomography (CT) scan enhanced by intravenous contrast. However, this equipment would require a large volume and mass and is not feasible in space travel. Portable ultrasound may detect loculated fluid in the abdomen if perforated diverticulitis is suspected. Ultrasound would capture rim-enhancing fluid collections located in the pelvis secondary to perforated sigmoid diverticulitis, or around other colonic segments with perforated diverticula. If imaging demonstrates accessibility with a needle, these collections may be drainable under ultrasound guidance. Table 12.6 demonstrates the supplies needed to drain fluid collections secondary to perforated diverticulitis.

Ultrasound-guided drainage for perforated diverticulitis with abscess:

1. The performer first uses a hand-held ultrasound to identify the fluid collection surrounding the perforated diverticulitis. If ultrasound is not sufficient, we advise against further procedural intervention.
2. The performer should maximize sterility by washing hands with betadine and saline solution and donning sterile gloves.
3. Using a large-bore needle, preferably 18-gauge or above, attached to a syringe, the fluid collection would be accessed under ultrasound guidance.
4. Once the needle is noted to be inside the collection, fluid should be aspirated.
5. The syringe is then removed and the needle left in place. A wire (Fig. 12.3) is threaded through the needle, followed by the removal of the needle.
6. The skin should be incised to enlarge the puncture wound to approximately 5 mm using a scalpel blade.
7. Then a drainage catheter is then placed over the wire directly into the collection.
8. The wire is then removed, and the drainage catheter is connected to a negative-pressure drainage via bulb suction.
9. Following placement of the drain, the drain should be sutured in with nonabsorbable sutures.

The drain should be maintained until output is less than 30 mL for three consecutive days and the output content is serous in nature. Following drain placement, antibiotics should be continued for 4 days after source control. If at the 4-day mark symptoms of pain and fever do not improve, antibiotics may be continued. Diet may be advanced based on patient's tolerance. If there is drainage of murky fluid around the tube or erythema and drainage from the tube site, he or she should return to Earth for imaging workup, further intravenous antibiotics treatment, and potentially washout of infection source. If the patient initially presents with unrelenting abdominal pain, rigid abdomen, involuntary abdominal guarding, and unstable vital signs such as hypotension and tachycardia, immediate return to Earth for complex abdominal exploration or bowel resection is warranted.

## 12.7 Small bowel obstruction

### 12.7.1 Background and treatment for small bowel obstruction

Another common emergency general surgery diagnosis is small bowel obstruction. Small bowel obstruction is defined as a mechanical obstruction along any portion of the intestines, excluding the colon. It accounts for approximately 12%–16% of emergency general surgery admissions and 20% of emergency surgical procedures (Liang et al., 2024). The patient usually presents with nausea, vomiting, and constipation. Risk factors for small bowel obstruction include previous surgery, hernias, and malignancy (Barmparas et al., 2010). For diagnosis, imaging such as plain film usually reveals dilated loops of bowel with air-fluid levels in normogravity. On CT imaging, dilated loops of small bowel are usually present. On imaging enhanced by oral contrast, the contrast would abruptly stop at the point of obstruction. During space travel, x-ray and CT imaging may not be available. However, based on the patient's medical history, clinical symptoms, and physical exam, small bowel obstruction may be diagnosed without definitive imaging. For instance, individuals with previous abdominal surgery or known hernia who present with these symptoms have a high likelihood of having a small bowel obstruction.

**TABLE 12.6** Materials needed for ultrasound-guided drainage of perforated diverticulitis with abscess.

| Optimal supplies (within spacecraft limitations) | Level of need | Alternatives/improvised substitutes | Mass (g) | Volume (ml) | Power required | Modifications | Hazards |
|---|---|---|---|---|---|---|---|
| Sterile gloves | Required | | 33.2 | 110.5 | No | | |
| Portable ultrasound | Required | | | | Yes | | |
| 18-gauge needle | Preferred | May use any large-bore needle | 3.3 | 67.5 | No | | Needle to be disposed |
| 60 mL syringe | Preferred | May use any size syringe | 46.5 | 1250 | No | | |
| Guidewire | Required | | 15.4 | 100 | No | | Wire to be disposed |
| Scalpel, 10 blade or 15 blade | Required | | 8.4 | 58.5 | No | | Sharp to be disposed |
| Bulb suction drain, includes drain catheter and bulb | Required | | 105.7 | 4312 | No | | |
| 3-0 nylon suture | Preferred | Any suture | 4.8 | 21.4 | No | | Needle to be disposed |

**TABLE 12.7** Materials needed for gastric decompression in small bowel obstruction.

| Optimal supplies (within spacecraft limitations) | Level of need | Alternatives/ improvised substitutes | Mass (g) | Volume (ml) | Power required | Modifications | Hazards |
|---|---|---|---|---|---|---|---|
| Nasogastric tube | Required | | 227 | 803 | No | | |
| Suction capability | Required | | 137.8 | 600 | Yes | | |
| Skin tape | Preferred | | 20.4 | 50.3 | No | | |

For patients who are hemodynamically stable, such as those with normal blood pressure, normal heart rate, and tolerable abdominal pain, decompression of the stomach and bowel rest (no oral intake) remains the standard of care. Specifically, the stomach should be decompressed with a nasogastric tube connected to a suction canister and applied with intermittent low suction, which is less than −125 mmHg. Table 12.7 lists the supplies needed.

Gastric decompression:

1. Measure how deep the nasogastric tube needs to be inserted by lining the tube from either earlobe to the patient's epigastrium (about 5 cm above the umbilicus).
2. The nasogastric tube will be inserted into either naris and directly into the stomach, ending at the mark previously measured. It is then secured with skin tape.
3. The tube is then connected to intermittent low suction of approximately −125 mmHg.
4. Once there is return of bowel function in the form of flatus and/or stool output, and once the nasogastric tube output is below one liter daily, the nasogastric tube may be removed.
5. The patient's diet can then be slowly advanced.

In situations where the patient is hemodynamically unstable, such as hypotension and tachycardia, and has significant abdominal pain with rigidity or involuntary guarding, we recommend immediate return to Earth for exploration of the abdomen and identification of the point of obstruction, potential bowel resection, and potential anastomosis. We advise against exploratory surgery in space due to the complexity of the surgery itself and the tremendous amount of equipment and instruments needed.

## 12.8 Conclusion

Intraabdominal surgical emergencies are common. Certain procedures, such as appendectomy, percutaneous drain placement, cholecystectomy, and ventral hernia repair, may be eventually required during space travel. This chapter provides a brief summary of the necessary equipment and procedural steps to perform the aforementioned operations. To optimize outcomes, further research is warranted to test necessary adaptations that translate these procedures to the spaceflight environment.

## References


Barmparas, G., Branco, B.C., Schnüriger, B., Lam, L., Inaba, K., Demetriades, D., 2010. The incidence and risk factors of post-laparotomy adhesive small bowel obstruction. Journal of Gastrointestinal Surgery 14 (10), 1619–1628. Available at https://doi.org/10.1007/s11605-010-1189-8.

Bhardwaj, P., Huayllani, M.T., Olson, M.A., Janis, J.E., 2024. Year-over-year ventral hernia recurrence rates and risk factors. JAMA Surgery 159 (6), 651–658. Available at https://doi.org/10.1001/jamasurg.2024.0233, http://archsurg.jamanetwork.com/issues.aspx.

Doleman, B., Fonnes, S., Lund, J.N., Boyd-Carson, H., Javanmard-Emamghissi, H., Moug, S., Williams, J.P., 2022. Appendectomy versus antibiotic treatment for acute appendicitis. Cochrane Database of Systematic Reviews.

Ferris, M., Quan, S., Kaplan, B.S., Molodecky, N., Ball, C.G., Chernoff, G.W., Bhala, N., Kaplan, G.G., et al., 2017. The global incidence of appendicitis: a systematic review of population-based studies. Annals of Surgery. 266, 237–241.

Floch, M.H., Bina, I., 2004. The natural history of diverticulitis: fact and theory. Journal of Clinical Gastroenterology. 38 (Suppl. 1), S2–S7. https://doi.org/10.1097/01.mcg.0000124003.07433.ee.

Gallaher, J.R., Charles, A., 2022. Acute cholecystitis: a review. JAMA 327, 965–975.

Jenkins, J.T., O'Dwyer, P.J., 2008. Inguinal hernias. BMJ 336 (7638), 269–272. Available at https://doi.org/10.1136/bmj.39450.428275.ad.

Kamine, T.H., Siu, M., Stegemann, S., Formanek, A., Levin, D., 2024. Long round-trip time delay effects on performance of a simulated appendectomy task. Aerospace Medicine and Human Performance 95 (9), 703–708. Available at https://doi.org/10.3357/AMHP.6367.2024, http://ingentaconnect.com/content/asma/amhp;jsessionid=13tlejf2f0ojk.alice.

Kamine, T.H., Siu, M., Kramer, K., Kelly, E., Alouidor, R., Fernandez, G., Levin, D., 2022. Spatial volume necessary to perform open appendectomy in a spacecraft. Aerospace Medicine and Human Performance 93 (10), 760–763. Available at https://doi.org/10.3357/AMHP.6062.2022, http://ingentaconnect.com/content/asma/amhp;jsessionid=13tlejf2f0ojk.alice.

Lazow, S.P., Siu, M., Brown, L., Kamine, T.H., 2024. Percutaneous cholecystostomy for acute cholecystitis during spaceflight. Aerospace Medicine and Human Performance 95 (10), 771–776. Available at https://doi.org/10.3357/AMHP.6352.2024, https://asma.kglmeridian.com/view/journals/amhp/amhp-overview.xml.

Levin, D.R., Siu, M., Kramer, K., Kelly, E., Alouidor, R., Fernandez, G., Kamine, T., 2022. Time cost of provider skill: a pilot study of medical officer occupied time by knowledge, skill, and ability level. Aerospace Medicine and Human Performance 93 (11), 816–821. Available at https://doi.org/10.3357/AMHP.6093.2022, http://ingentaconnect.com/content/asma/amhp;jsessionid=13tlejf2f0ojk.alice.

Liang, J.-T., Liao, Y.-T., Chen, T.-C., Huang, J., Hung, J.-S., 2024. Changing patterns and surgical outcomes of small bowel obstruction in the era of minimally invasive surgery for colorectal cancer. International Journal of Surgery 110 (3), 1577–1585. Available at https://doi.org/10.1097/js9.0000000000000980.

Nunoo-Mensah, J.W., Rosen, M., Chan, L.S., Wasserberg, N., Beart, R.W., 2009. Prevalence of intra-abdominal surgery: what is an individualg's lifetime risk? Southern Medical Journal 102 (1), 25–29. Available at https://doi.org/10.1097/SMJ.0b013e318182575b.

Ruhl, C.E., Everhart, J.E., 2007. Risk factors for inguinal hernia among adults in the US population. American Journal of Epidemiology 165 (10), 1154–1161. Available at https://doi.org/10.1093/aje/kwm011.

Siddiqui, A., Lyons, N.B., Anwoju, O., Cohen, B.L., Ramsey, W.A., O'Neil, C.F., Ali, Z., Liang, M.K., 2023. Mesh type with ventral hernia repair: a systematic review and meta-analysis of randomized trials. Journal of Surgical Research 291, 603–610. Available at https://doi.org/10.1016/j.jss.2023.07.003, http://www.elsevier.com/inca/publications/store/6/2/2/9/0/1/index.htt.

Siu, M., Levin, D., Christiansen, R., Kelly, E., Alouidor, R., Kamine, T.H., 2022. Prophylactic splenectomy and hyposplenism in spaceflight. Aerospace Medicine and Human Performance 93 (12), 877–881. Available at https://doi.org/10.3357/amhp.6079.2022.

Strate, L.L., Morris, A.M., 2019. Epidemiology, pathophysiology, and treatment of diverticulitis. Gastroenterology 156 (5), 1282. Available at https://doi.org/10.1053/j.gastro.2018.12.033, http://www.journals.elsevier.com/gastroenterology/.

Wang, C.-H., Chou, H.-C., Liu, K.-L., Lien, W.-C., Wang, H.-P., Wu, Y.-M., 2014. Long-term outcome of patients with acute cholecystitis receiving antibiotic treatment: a retrospective cohort study. World Journal of Surgery 38 (2), 347–354. Available at https://doi.org/10.1007/s00268-013-2311-3.

Weiser, T.G., Regenbogen, S.E., Thompson, K.D., Haynes, A.B., Lipsitz, S.R., Berry, W.R., Gawande, A.A., 2008. An estimation of the global volume of surgery: a modelling strategy based on available data. The Lancet 372 (9633), 139–144. Available at https://doi.org/10.1016/S0140-6736(08)60878-8, http://www.journals.elsevier.com/the-lancet/.

Chapter 13

# Gynecologic evaluation and procedures

**Ruth M. Farrell**[1] **and Jon G. Steller**[2]

[1]*Obstetrics and Gynecology Institute, Center for Bioethics, Cleveland Clinic, Cleveland, OH, United States,* [2]*University of California Irvine, Irvine, CA, United States*

## 13.1 Overview

Since the first flight of Valentina Tershkova in 1963 followed by Svetlana Savitskaya and Sally Ride in 1982 and 1983, women have had an increasing presence in both governmental and commercial spaceflight. Valentina was also the youngest female to ever fly at 26 years old, and Wally Funk flew a sub-orbital flight with Blue Origin at the age of 82 in 2021. Just like other astronaut anthropomorphics, age and sex are two human characteristics that we design around for spaceflight and this chapter introduces the diagnosis and management of gynecologic conditions in space throughout the reproductive to the menopausal period.

Fortunately, gynecologic morbidity requiring procedural medicine has not occurred in space to date. This lies partly in the lower number of total female days spent in space, partly in optimal preflight mitigation of gynecologic pathology, and partly in the overall low incidence of such complications. However, gynecologic and reproductive conditions have arisen during analogs such as military deployment, in Antarctic research stations, and in spaceflight analog simulations (Moraes et al., 2020; Jennings and Baker, 2000; Braun et al., 2016; Hughes-Fulford et al., 2024). Our understanding of gynecologic pathology in space is further limited because such events may not be perfectly tracked or reported given privacy concerns and/or the desire to prevent affecting the chances for future personal flight or flight for women at large (Farrell et al., 2025). Thus, it is still prudent for astronauts, crew medical officers, and flight surgeons to be aware of what conditions may arise, what diagnostic and therapeutic procedural interventions are pragmatic for spaceflight in the present, and what procedures may become a reality as mass, volume, and power limitations are eased for long-duration spaceflight and colonization within a Lunar or Martian base.

In space, there are a number of interrelated gynecologic considerations, including the following (Fig. 13.1):

1. the volume, duration, and timing of menses and the possibility of abnormal uterine bleeding (AUB);
2. maintenance of the vaginal microbiome and risk for vulvovaginal candidiasis or bacterial vaginosis;
3. bone mineral density and the risk of osteopenia, osteoporosis, or future fractures;
4. ovarian cyst production and risk of ovarian torsion;
5. the risk of gynecologic or breast cancer;
6. the risk of thrombosis/venous thromboembolism;
7. and the risk of pregnancy (Strock, Rivas and Goebel, 2023; Steller et al., 2020).

A set of hormonal modalities can be used to prevent and treat many of these conditions. However, they can also alter the risk profiles of other conditions (Jain et al., 2020; Zwart et al., 2022). Additionally, it is not clear how the pharmacokinetics and other factors may be affected by spaceflight (Blue et al., 2019). Thus, there is complexity in understanding how all of these gynecologic and reproductive considerations are related and nuance in the counseling of female astronauts going to space with or without the usage of hormonal modalities.

With respect to the tradespace of mass, volume, and power within present and future medical systems, and the limited time for astronaut preflight training, this chapter does not seek to compel mission planners to include the resourcing and training outlined below. Rather, it proposes resources for consideration as mission planners and subject matter experts conduct risk assessments.

**Procedural Medicine for Spaceflight. DOI:** https://doi.org/10.1016/B978-0-443-27552-4.00016-6

FIGURE 13.1 **Common gynecologic considerations in space.** These common gynecologic conditions are closely related to one another, as well as to hormonal modalities (birth control methods), and are important when considering how to optimize the health and performance of women in space. *BMD*, Bone mineral density; *BCM*, birth control methods; *VTE*, venous thromboembolism.

### 13.1.1 Gynecologic procedural technique and materials for spaceflight

The recommended equipment proposed in this chapter is listed in Table 13.1. Bolded items are not currently available in the ISS medical equipment kit (NASA, 2025; TRISH Orbit Rendezvous Teleconference: Artemis Medical System Overview, 2025). Mass and volume estimations for these supplies are described in Table 13.2.

This does not include pain control supplies or equipment, as these are described in other chapters.

## 13.2 Menstrual management in spaceflight

### 13.2.1 Introduction

From menarche to menopause, the ovaries and the uterus respond to hormonal alterations from the pituitary gland by way of the hypothalamus in a cycle known as the "menstrual cycle." The menstrual cycle lasts 21–35 days (average 28 days) and can vary based on one's age, diet, exercise frequency, stress levels, health conditions, and medications (Fraser and Critchley, 2007; Woolcock et al., 2008). During the *proliferative* phase, estrogen causes the lining of the uterus to thicken creating the ideal conditions for implanting an embryo. If that embryo does not arrive, the uterine lining is shed during the *secretory* phase, leading to bleeding we call menses or a "period." While the effects of spaceflight on the menstrual cycle are not completely understood, we do know that female astronauts who do not use hormonal contraception for menstrual suppression continue to have cycles during spaceflight (Coelho et al., 2025; Puukila et al., 2024).

### 13.2.2 Hormonal modalities

Hormonal contraceptives were originally developed to provide a highly effective method of birth control. Combined hormonal contraceptives (CHCs), which contain estrogen and progesterone, prevent pregnancy by suppressing pituitary gonadotropin release through negative feedback, thereby inhibiting ovulation (Jensen and Creinin, 2019; US Medical Eligibility Criteria US MEC for Contraceptive Use, 2024). When exogenous CHCs are discontinued, pituitary suppression is lifted, resulting in a withdrawal bleed. For patients who use CHCs for 3 weeks followed by a 1-week placebo interval, this bleeding may feel analogous to menstruation; however, it reflects repeated cycles of hormonal suppression and withdrawal rather than a true physiologic menses.

This suppression is highly effective at preventing pregnancy, and the withdrawal bleed has historically served as reassurance that ovulation, fertilization, and implantation have not occurred, as pregnancy-associated endogenous hormone elevations would prevent such bleeding. Beyond this reassurance, however, there is no intrinsic health benefit to experiencing a withdrawal bleed. Rather than incorporating a placebo week, CHCs can be taken continuously by initiating a new pack without interruption. Continuous, or "suppressive," CHC use reduces both the volume and duration of bleeding (American College of Obstetricians and Gynecologists, 2013; American College of Obstetricians and Gynecologists' Committee on Practice Bulletins-Gynecology, 2012; Anderson et al., 2003; Creinin, 2022), and

**TABLE 13.1** Gynecologic procedural technique and materials for spaceflight.

| Analgesia | Antiseptic | General materials and equipment | Medications | Medical devices |
|---|---|---|---|---|
| • Lidocaine 1% or 2% or OR Bupivacaine 0.25% or 0.5% (1–20 cc) | • Iodine swab stick/ BZK wipe × 2<br>• Chlorhexidine prep stick<br>• Sterile gloves<br>• Sterile drapes<br>• Biohazard bag | • Crew medical restraint system<br>• Speculum<br>• Tenaculum<br>• Scalpel<br>• Needle holder<br>• Scissors<br>• Forceps<br>• Hemostats<br>• Hysterscopy<br>• Gauze<br>• Adhesive bandage<br>• 2-0 to 3-0 delayed absorbable suture<br>• Absorbent pads<br>• Measuring tape<br>• Surgical marker<br>• Steri-Strips<br>• Pressure bandage | • Antibiotics<br>• IV hydration (i.e., normal saline)<br>• Sterile saline for hysteroscopy or vaginoscopy<br>• Hormonal agents (including contraceptives and agents to manage abnormal uterine bleeding)<br>• Tranexamic acid<br>• NSAIDs<br>• Opioids | • IUD (LNG-IUS)<br>• Contraceptive implant |
| Needles | Syringes | Catheters | | |
| • 18- to 20-gauge needle | • 10 mL syringe | When needed to provide fluid monitoring in conjunction with a GYN procedure:<br>• Foley catheter and collection system<br>For uterine tamponade:<br>• 30 mL Foley catheter | | |
| **Medical imaging** | | | | |
| • Portable ultrasound machine<br>• Ultrasound gel (sterileand nonsterile)<br>• Sterile probe cover | | | | |

60%–88% of women using CHCs continuously for 12 months may achieve complete amenorrhea. Thus, in addition to pregnancy prevention, hormonal contraceptive modalities may offer substantial operational and quality-of-life benefits for women, both on Earth and in unique environments such as space.

Historically, the most commonly used hormonal modalities during spaceflight have been combined oral contraceptives and the levonorgestrel intrauterine device (LNG IUD; Steller et al., 2020, 2021). A CHC vaginal ring and progesterone-only subdermal implants have also been used in space. In contrast, the CHC transdermal patch and intramuscular depot medroxyprogesterone acetate have not been utilized during spaceflight for contraception or menstrual suppression.

Progesterone-only modalities, such as the subdermal implant and the LNG IUD, do not rely on pituitary suppression as their primary mechanism of pregnancy prevention; however, they can substantially alter bleeding patterns. These effects result from continuous progesterone exposure to the endometrium, which limits endometrial proliferation and prevents the cyclic buildup and shedding that occurs in the absence of hormonal modulation (Guttinger and Critchley, 2007; Jensen, 2005).

**TABLE 13.2** Material, mass, volume table.

| Item | Mass | Volume |
|---|---|---|
| *Sterile drapes* | ~300 g | ~3 L |
| *5 L plastic wastewater containment bag* | 10 g | ~500 cm$^3$ |
| *Nitrile gloves (1 pair)* | 12 g | ~50 cm$^3$ |
| *20 mL bottle of lidocaine or bupivacaine* | 50–70 g | |
| *Iodine swab stick (balances sterilization capability with avoiding trace contamination and fire risks associated with other sterilization solutions like alcohol or chlorhexidine)* | Single swab package: 3–4 g3 swab package: 10–12 g | Single swab package: 2.5 ml3 swab package: 7 ml |
| *Biohazard bag* | 15 g per bag | |
| *Speculum* | 40 g (plastic) | |
| *Tenaculum* | ~100–150 g | |
| *Scalpel* | ~10–20 g (disposable plastic) | |
| *Needle holder* | 70–150 g | |
| *Forceps* | 100 to 150 g | |
| *Hemostat* | 60–120 g | |
| *Hysteroscope* | ~145 g (disposable) | |
| *Gauze* | 4.5 inches × 4.1 yards: ~25–40 g | |
| *2-0 to 3-0 delayed absorbable suture* | 5–10 g/pack | |
| *Absorbent pads* | 8–12 g | |
| *Measuring tape* | ~2–5 g | |
| *Surgical marker* | 8–15 g | |
| *Steri-Strips* | 2–6 g | |
| *Pressure bandage* | 50–100 g | |
| *IUD (LNG-IUS)* | With inserter and sterile packaging: ~25–30 g | |
| *Contraceptive implant* | With sterile applicator and packaging: ~4–6 g | |
| *18- to 20-gauge needle* | With plastic hub and sterile packaging: ~2–4 g | |
| *10 mL syringe* | Plastic syringe without needle: ~20–30 g | |
| *Sterile gel and probe cover* | ~27 g | ~6 cm$^3$ |

Although a comprehensive discussion of all available options is beyond the scope of this chapter, a wide range of hormonal contraceptives are available for consideration by astronauts. When thoughtfully selected, these modalities can serve as powerful tools for minimizing menses, managing AUB from a variety of etiologies, and modifying other gynecologic risks, as summarized in Fig. 13.1.

IUDs that have been used for menstrual management in addition to contraception include the levonorgestrel and copper containing IUDs. Copper containing IUDs may be associated with heavier or prolonged menstrual bleeding and, thus, are not readily indicated for menstrual management for astronauts. Levonorgestrel IUDs are available in different hormonal doses and are used for varying durations (e.g., 3–8 years) (American College of Obstetricians and Gynecologists' Committee on Practice Bulletins-Gynecology, Long-Acting Reversible Contraception Work Group, 2017). To date, these hormone-containing IUDs have been used by astronauts during flight without any known documented complications. In these cases, the IUDs were all placed appropriately at the uterine fundus preflight, confirmed to be in the appropriate location sonographically, and did not have any known complications (e.g., malposition) prior to flight. Amongst a large terrestrial cohort of 25–44 year old females with a LNG IUD, the rate of mechanical complication within 1 year was 1.2% (Berenson et al., 2013). However, this risk may be partially mitigated for astronauts by placing the IUD under ultrasound guidance and following expectantly for 6–12 months preflight for any complication. For long duration spaceflight, it is possible that a previously-placed LNG IUD may be due to be replaced or a new gynecologic condition may arise in space for which the LNG IUD is a reasonable management consideration. Recent advances in point-of-care ultrasound have paved the way for evaluation of an IUD during spaceflight if a potential complication (e.g., malposition of the IUD, abnormal bleeding, nonpalpable IUD strings) was suspected or occurred (Fischetti et al., 2025; Mathyk et al., 2025). While it is unlikely that a LNG IUD will be placed in the present, it is possible that this may be a consideration in the near future.

Similarly, the leading subdermal contraceptive is a small (68 mg) rod of ethylene vinyl acetate that contains etonogestrel (ACOG, 2017). The device was designed to slowly elute progesterone to provide 3 years of contraception. The rod is placed subdermally on the medial upper arm, specifically 8–10 cm from the medial epicondyle of the humerus and 3–5 cm posterior to the groove between the triceps and biceps (American College of Obstetricians and Gynecologists' Committee on Practice Bulletins-Gynecology, Long-Acting Reversible Contraception Work Group, 2017). When placed at the appropriate location and depth, the implant will remain in this location for the duration of use. However, there is a small chance of implant migration. When this occurs, it most often entails migration proximal to the site of insertion. There have been rare cases of more significant degrees of migration from the upper extremity and other anatomical sites (Barlow-Evans et al., 2017; Hindy et al., 2020). Individuals who elect to use a subdermal implant should be able to palpate the device in the upper arm for reassurance of its location. Though mechanical complications from both of these foreign bodies are low, there is a small chance that these could need to be triaged during spaceflight with unplanned removal. Furthermore, as we move towards long-duration spaceflight, there is even the chance of needing to remove the device at its hormonal end of service and consider replacing it with a new device. Below, we discuss resourcing required to complete these procedures.

## 13.2.3 Procedural considerations of hormonal modalities in spaceflight

The following procedures have been modified to reflect some of the unique challenges of performing these procedures in spaceflight (American College of Obstetricians and Gynecologists' Committee on Practice Bulletins-Gynecology, Long-Acting Reversible Contraception Work Group, 2017; Speroff and Darney, 2011).

### *13.2.3.1 Subdermal implant removal and/or replacement*

**Removal of subdermal implant**

| **Procedure** | **Removal of subdermal implant** |
|---|---|
| **Purpose** | Remove subdermal implant |
| **Indications** | • Mechanical complication, migration, or pain<br>• Device lifespan (depleted hormone)<br>• Side effect of medication (i.e., abnormal uterine bleeding, mood changes) |
| **Contraindications** | N/A |
| **Equipment and supplies** | |
| **Operators** | One minimum; likely the Crew Medical Officer |
| **Antiseptic** | • Iodine swab stick, Chlorhexadine swab stick, or Benzalkonium Chloride (BZK) Antiseptic Moist Towelette × 2 (if only option available)<br>• Sterile gloves<br>• Biohazard bag |

| | |
|---|---|
| **Analgesia** | • Local anesthetic (20cc)<br>• Lidocaine 1% or 2%<br>• OR Bupivacaine 0.25% or 0.5% |
| **Materials** | • Crew medical restraint system<br>• 5 mL syringe<br>• 18- & 22-gauge needles<br>• Surgical marker<br>• Scalpel (ideally a #15 blade)<br>• Forceps (straight and/or curved mosquito)<br>• Portable ultrasound machine<br>• Ultrasound gel (sterile; or water if in microgravity)<br>• Ultrasound probe cover (sterile)<br>• Gauze<br>• Measuring tape<br>• Biohazard bag<br>• Steri-Strips<br>• Pressure bandage |

**Steps**

1. Consent. Risks include bleeding, infection, discomfort during removal, hematoma, paresthesias, scarring, and incomplete removal of implant.
2. Positioning. Restrain the patient supine with arm flexed at the elbow and externally rotated so that her hand is behind her head.
3. Locate the implant by palpation. Push down the end of the implant closest to the shoulder to stabilize it; a bulge should appear indicating the tip of the implant that is closest to the elbow.
4. Mark the distal end with a surgical marker (this is where you will make the stab incision).
5. Time out.
6. Clean the site with an antiseptic solution.
7. Anesthetize the site with 1 mL 1%–2% lidocaine where the incision will be made. Be sure to inject the lidocaine under the implant to keep the implant close to the skin's surface.
8. Push down the proximal end of the implant to stabilize it throughout the procedure. Starting over the distal tip, make a stab incision which can be carried longitudinally (parallel to the implant) ~2 mm toward the elbow.
9. Guide the distal tip of the implant to pop through the incision.
10. Confirm the entire implant, which is 4 cm long, has been removed by measuring it.
11. Close the incision with a Steri-Strip.
12. Apply a sterile gauze and pressure bandage to minimize bruising.

**Postprocedural care**

- Analgesia as needed
- The patient may remove the pressure bandage in 24 hours and the Steri-Strip in 3–7 days.

**Potential challenges, complications, and management approach**

| Potential challenge or complication | Management approach |
|---|---|
| Cannot palpate the implant | 1. Utilize POCUS to locate the implant. It will be linear, 4 cm long, hyperechoic, and will display acoustic shadowing |
| Can palpate the implant but cannot palpate the distal tip. | 1. *Locate the implant by palpation. Push down the end of the implant closest to the shoulder to stabilize it; a bulge should appear indicating the tip of the implant that is closest to the elbow.*<br>a. If the tip does not pop up, removal of the implant should be performed by a provider experienced with removing deeper implants (terrestrial Family Planning specialists). |
| If the distal tip of the implant is not popping through the incision. | 1. *Guide the distal tip of the implant to pop through the incision.*<br>a. If it does not, gently push the implant towards the incision until the tip is visible and grab it with the forceps.<br>b. If needed, gently remove adherent tissue from the tip of the implant using blunt dissection.<br>c. If that does not expose the implant tip, make an incision into the tissue sheath and then remove the implant with the forceps. |
| Excessive bleeding | 1. If there is excessive bleeding, hold pressure with gauze for 60 seconds and reevaluate. If still bleeding, hold pressure for another 60 seconds. |

### 13.2.3.2 Intrauterine device removal

**Removal of intrauterine device**

| Procedure | Removal of intrauterine device |
|---|---|
| **Purpose** | Removal of intrauterine device |
| **Indications** | • Mechanical complication, malposition, or pain<br>• Device lifespan (depleted hormone)<br>• Side effect of medication (i.e., abnormal uterine bleeding, mood changes) |
| **Contraindications** | Cannot visualize or palpate IUD strings coming through the cervix |
| **Equipment and supplies** | |
| **Operators** | One minimum; Subject can perform themselves, or likely would be the Crew Medical Officer |
| **Antiseptic** | • N/A<br>• Sterile gloves<br>• Biohazard bag |
| **Analgesia** | • NSAID 30–60 minutes prior<br>• Heating pad |
| **Materials** | • Crew medical restraint system<br>• Speculum (if available)<br>• Tenaculum<br>• Uterine sound<br>• Scissors<br>• Portable ultrasound machine (if needed)<br>• Ultrasound gel (sterile; or water if in microgravity) |

**Steps**

Removal

1. Consent. Risks include bleeding, infection, discomfort during removal, and incomplete removal of implant.
2. Positioning. Restrain patient and flex and abduct hips to lithotomy-like position.
3. Locate the IUD strings by digital palpation with index finger moving in a circular pattern in the back of the vagina where the cervix is (if self-removal). Locate IUD strings by visualization with speculum (if nonself-removal).
4. For self-removal, grab the strings using your pointer/thumb or pointer/middle fingers and pull with gentle traction. If nonself-removal, after visualizing the strings, grasp the strings with forceps and pull with gentle traction. If not making steady descent, abort the procedure.

**Postprocedural care**

- Analgesia as needed
- Heating pad

**Potential challenges, complications, and management approach**

| Potential challenge or complication | Management approach |
|---|---|
| Cannot locate the IUD strings | 1. Use transabdominal POCUS to visualize if the IUD is within the uterine cavity. It will be a hyperechoic linear structure sitting within the endometrium. |
| Excessive bleeding | 1. If there is excessive bleeding immediately following removal, if a speculum is present, attempt to visualize if there are any lacerations of the cervix or vagina.<br>2. Whether or not this can be visualized, gauze packing can be temporarily placed within the vagina for pressure. |

### 13.2.3.3 Intrauterine device placement

**Placement of intrauterine device**

| Procedure | Placement of intrauterine device |
|---|---|
| **Purpose** | Placement of intrauterine device |
| **Indications** | • Device lifespan (depleted hormone) of previous device has been exhausted and replacement is necessary |

| | |
|---|---|
| | • Treatment for abnormal uterine bleeding |
| **Contraindications** | An IUD is already in place. Knowledge, skills, abilities, or medical resourcing is not accessible to safely place. |
| **Equipment and supplies** | |
| **Operators** | One minimum; Would be the Crew Medical Officer |
| **Antiseptic** | • Vaginal povidone iodine or chlorhexidine<br>• Sterile gloves<br>• Biohazard bag |
| **Analgesia** | • Placing a cervical or pundendal block are beyond the capabilities of spaceflight; thus, analgesia would be with oral medication.<br>• Heating pack<br>• (Optional) misoprostol 24–48 hours prior to insertion |
| **Materials** | • Crew medical restraint system<br>• Speculum<br>• Uterine sound<br>• Device applicator<br>• Tenaculum (if present)<br>• Long scissors (if present)<br>• Portable ultrasound machine<br>• Ultrasound gel (sterile) (or water if in microgravity) |

**Steps**

1. Consent. Risks include bleeding, infection, discomfort during placement, uterine perforation.
2. Positioning. Restrain patient and flex and abduct hips to lithotomy-like position.
3. Bimanual exam to assess uterine position, size, and mobility.
4. Place speculum in vagina in order to visualize the cervix.
5. Apply tenaculum (if present) to anterior lip of cervix to straighten the canal.
6. Measure depth of the uterus with a uterine sound (if available) under abdominal ultrasound guidance.
7. Load the LNG IUD into the inserter per manufacturer instructions.
8. Set the flange to the measured uterine depth from the sound (should be 6–10 cm).
9. Gently insert the loaded LNG IUD inserter through the cervix into the uterine cavity until the flange reaches the cervix. You may also feel the inserted gently touch the uterine fundal wall.
10. Pull back less than a centimeter and release the arms of the device under US guidance.
11. Advance back rostrally to fundus, then withdraw the inserter, slowly wiggling it side to side to release the LNG IUD fully in the fundus of the uterus under US guidance.
12. Cut strings to 2–3 cm from external os.
13. Remove tenaculum, control bleeding if necessary.
14. Remove speculum.

**Postprocedural care**

• Analgesia as needed

**Potential challenges, complications, and management approach**

| Potential challenge or complication | Management approach |
|---|---|
| Uterine sound or inserter perforating uterine fundal wall | 1. Use transabdominal POCUS to visualize the uterine sound and/or the LNG IUD inserter moving toward the fundus of the uterus throughout the procedure to prevent this complication. Management of this complication would require major abdominal surgery. |
| Excessive bleeding | 1. If there is excessive bleeding during or immediately following placement and if a speculum is present, attempt to visualize if there are any lacerations of the cervix or vagina.<br>2. Consider removal of the device by pulling on the IUD strings and following the instructions above.<br>3. Gauze packing can be temporarily placed within the vagina for pressure. |

## 13.3 Management of abnormal uterine bleeding in spaceflight

### 13.3.1 Introduction

AUB is defined as vaginal bleeding that occurs outside of the expected menstrual cycle. This can include irregular duration between menses (i.e., <21 days or >35 days), intermenstrual spotting/bleeding, or an abnormal volume or duration of menses (i.e., >7 days) (American College of Obstetricians and Gynecologists, 2013; American College of Obstetricians and Gynecologists' Committee on Practice Bulletins-Gynecology, 2012). AUB encompasses a range of bleeding patterns, from a small amount of bleeding that may present as spotting or light bleeding to a large volume which may present with or without clots. This information plays an important role in determining the etiology of the bleeding, approach to evaluation, and appropriate management.

### 13.3.2 Etiologies

AUB may result from several different conditions (Munro and Critchley, 2018; Wheeler and Hemingway, 2024). Bleeding may result from an anatomical cause, such as an endometrial polyp, adenomyosis, leiomyoma (also known as a fibroid), or endometrial hyperplasia/malignancy. There may also be nonanatomical causes, more often associated with anovulation (i.e., polycystic ovarian syndrome, hypothalamic hypogonadism, premature ovarian insufficiency) or other cause that may lead to the disruption in the regular ovulatory cycle (e.g., perimenopause). It may also be due to a coagulopathy, endometrial dysfunction, or iatrogenic as caused by medications and hormonal agents for contraception or menstrual control. A helpful pneumonic to determine the source of bleeding is known as PALM-COEIN (Munro and Critchley, 2011) as noted in Fig. 13.2:

Current NASA guidelines require comprehensive gynecologic screening for astronauts during initial candidacy, as well as periodic evaluations during annual recertification and/or preflight medical examinations (NASA Office of the Chief Health and Medical Officers, 2024, 2025; Steller et al., 2020). These assessments include pelvic and speculum examinations, routine human papillomavirus screening, sexually transmitted infection testing, and transvaginal ultrasound. Collectively, these evaluations identify the majority of conditions associated with AUB. In particular, anatomic etiologies classified under the PALM framework can be detected and subsequently managed with medical or surgical interventions as appropriate, while most nonanatomic causes can be medically optimized prior to flight. While this process may rarely identify any major gynecologic conditions that may preclude safe participation in spaceflight, it also more pragmatically identifies stable conditions at risk of evolving over time and leading to future risk of AUB. For example, a submucosal fibroid with characteristics of increased risk of prolapse or a uterine/cervical polyp may be identified leading to the decision to undergo definitive surgical management before a flight even when not causing any AUB at present.

Given the preflight opportunity for evaluation and management of AUB, there are three remaining causes that are more likely to present in spaceflight: ovulatory dysfunction, endometrial factor, and iatrogenic. Ovulatory dysfunction results from hormonal dysregulation of the hypothalamic–pituitary–ovarian axis that results in altered hormone secretion and response in the ovaries and uterine endometrium (American College of Obstetricians and Gynecologists, 2013). Ovarian

**FIGURE 13.2 Causes of abnormal uterine bleeding (AUB).** Etiologies of AUB can be recalled using the acronym PALM-COEIN (P: endometrial or endocervical polyp, A: adenomyosis, L: leiomyomata, M: malignancy or premalignancy, C: coagulopathy, O: ovulary dysfunction, E: endometrial factor, I: Iatrogenic, N: not yet classified).

Adapted from Munro M.G., Critchley H.O., Broder M.S., Fraser I.S. FIGO Working Group on Menstrual Disorders. FIGO classification system (PALM-COEIN) for causes of abnormal uterine bleeding in nongravid women of reproductive age. International Journal of Gynecology & Obstetrics. 2011 Apr;113(1):3–13. https://doi.org/10.1016/j.ijgo.2010.11.011. Epub 2011 Feb 22. PMID: 21345435. (n.d.).

dysfunction is most often medically managed with the same hormonal modalities discussed above that are used for contraception or other suppressive agents (e.g., GnRH agonists or antagonists; American College of Obstetricians and Gynecologists' Committee on Practice Bulletins-Gynecology, 2012). Endometrial etiologies include endometritis resulting from infection or inflammation of the lining of the uterus. Endometrial causes are most often treated medically with hormonal agents, antibiotics, or antiinflammatories as determined by the nature of the etiology. Iatrogenic causes may be due to medications including hormonal contraceptives themselves (e.g., missed or delayed pills, or use of triphasic formulations for continuous cycling to achieve menstrual suppression). Bleeding may also result when other medications are used in conjunction with hormonal contraceptive agents which may alter the pharmacokinetics of the contraceptive agent. It may also include anticoagulants or other medications which may interfere with endometrial homeostasis and lead to bleeding.

One of the challenges to consider is that astronauts, as part of private or commercial efforts, may reflect a different population of individuals with different demographic characteristics and different underlying medical risk factors. Further, there may be variability in preflight screening, preflight management, and astronaut autonomy regarding the types of hormonal modalities they may want to consider (i.e., using CHCs in a suppressive fashion, vs progesterone-only modalities, vs no modalities at all).

## 13.3.3 Approach to evaluation

If an acute abnormal bleeding event occurs during flight, the first step is a thorough review of the menstrual history and an examination (Bradley, 2011; Marnach and Laughlin-Tommaso, 2019). A review of menstrual history should be performed to determine whether the bleeding appears to be from ovulatory dysfunction or a possible anatomical lesion. If the cycle length has become irregular, this may indicate ovulatory dysfunction. If the timing of menses are regular, but the volume or duration are heavier, this may be an anatomical finding. Whether long-duration spaceflight may introduce novel endometrial or hemostatic alterations that can increase the risk of AUB is yet unknown; however, thus far, we do not have evidence of short-duration spaceflight increasing the risk of AUB.

As a next step, a pelvic and/or speculum exam (if available) is needed to determine the origin of the bleeding, whether from the vulva, vagina, cervix, or uterus. This will allow for identifying potential lacerations, erosions, or lesions that may cause bleeding (e.g., polyp). Ideally a transvaginal ultrasound would then be performed (Kelekci et al., 2005; Farquhar et al., 2003). However, with the understanding that an intracavitary probe is unlikely to be included in near-future medical systems, a transabdominal ultrasound of the pelvis will also provide additional insight into the uterus and endometrium, potentially identifying an intracavitary lesion that may be causing the bleeding. Information gained through physical and sonographic evaluation will help determine if management should be medical or surgical. In most cases during spaceflight, etiologies which would usually be managed surgically will call for a temporizing medical management approach during the mission. However, there may be circumstances in which an urgent procedural approach during flight is required because of the nature of the condition or if it is refractory to medical management.

## 13.3.4 Procedural considerations

### Intrauterine tamponade

| Procedure | Intrauterine tamponade |
|---|---|
| Purpose | To control uterine bleeding until a sufficient medical or surgical approach can be taken. |
| Indications | Acute vaginal or uterine bleeding of unknown etiology not responsive to other pharmaceutical or procedural approaches. |
| Contraindications | Mullerian anomaly that does not allow insertion or deployment of an intrauterine balloon or packing/gauze. |
| **Equipment and supplies** | |
| Operators | Minimum 1, Ideally 2 |
| Antiseptic | • Povidone or 4% chlorhexidine gluconate solution on an applicator or swab (selection based on availability and potential allergy) |
| Analgesia | • Paracervical block with 2% lidocaine, 1.5% mepivacaine, 0.25% bupivacaine, or equivalent anesthetic solution for paracervical injection<br>• Additional PO anxiolytics as needed<br>• Nonsteroidal antiinflammatory medications PO 60–90 minutes prior to the procedure<br>• Ongoing analgesia while the balloon or gauze is in place |

| | |
|---|---|
| **Materials** | • Crew medical restraint system<br>• Close drape system adhesive to the perineum for management of blood<br>• Sterile speculum (if available)<br>• Single tooth tenaculum (if available)<br>• Sterile gloves<br>• Sterile saline<br>• Antiseptic solution with applicators<br>• 10 mL syringe<br>• 18- to 20-gauge needle<br>• 30 mL Foley catheter to function as an intrauterine balloon or a larger balloon device if available (e.g., Bakri Cook balloon with a 500 mL capacity) as needed based on response and uterine size<br>• Kerlix sterile gauze or equivalent gauze (with or without added hemostatic agent) with similar thickness, width, and length (small squares or individual pieces of gauze should not be used because of the risk of retention after packing removal). Thrombin solution (if available)<br>• Long forceps (e.g., Kelly, ring, or uterine packing) |

**Steps**

**Intrauterine balloon placement**

1. Follow procedures for Foley catheter placement to assist in monitoring I/Os during the acute bleeding episode.
2. Prepare sterile field for supplies to be readily accessible.
3. Insert the speculum into the vagina to obtain visualization of the cervix and cervical os (if available)
4. Apply antiseptic solution directly to the cervix two times.
5. Don sterile gloves.
6. Place a single tooth tenaculum on the cervix to stabilize the cervix and uterus during the procedure.
7. Perform a paracervical block (if resourcing available)
8. Grasp the Foley/balloon with the forceps or fingers with a sterile glove. Gently guide the material through the cervical os and into the endometrial cavity. If using forceps, reference preflight ultrasound uterine measurements or use an endometrial sound to determine the length of the endometrial cavity as steps to prevent perforation while inserting the material into the uterus.
9. Once the catheter/balloon is adequately placed in the endometrium, fill the balloon to the desired amount using a syringe filled with normal saline until tampanode occurs. The total amount of the fluid will depend on the size of the uterus, response to insufflation, and patient tolerance.

**Intrauterine gauze placement**

**Follow steps 1–7 for intrauterine balloon placement as described above:**

1. Though unlikely, if thrombin is available, dilute 5000 units of thrombin in 5 mL of sterile saline in an enclosed container large enough to accommodate the gauze. Saturate the gauze in this solution.
2. Grasp the gauze with the forceps or fingers. Gently guide the material through the cervical os and into the endometrial cavity. Reference preflight ultrasound uterine measurements or use an endometrial sound to determine the length of the endometrial cavity as steps to prevent perforation while inserting the material into the uterus. The total amount of the gauze will depend on the size of the uterus, response to insufflation, and patient tolerance.

**Balloon removal**:

Based on clinical use in Earth-based medicine, the balloon should be removed after 24 hours.

1. Insert the speculum into the vagina to obtain visualization of the cervix and cervical os.
2. Use a syringe to remove the fluid and deflate the balloon. Deflation should be performed gradually and at a slow rate (e.g., 10–20 mL/hour) to prevent disruption of the clotting processes.
3. Inspect the balloon to ensure all components have been removed and not retained in the uterus. A vaginal exam and vaginal sweep performed at the end of the procedure to be certain all materials and instruments are removed at the close of the procedure.

**Gauze removal**

Based on clinical use in terrestrial medicine, the gauze should be removed after 24 hours.

1. Insert the speculum (if available) into the vagina to obtain visualization of the cervix and cervical os.
2. Gauze removal should take place over several hours (e.g., 3–5 cm/hour) to prevent disruption of the clotting processes.
3. Inspect the gauze to ensure all components have been removed and not retained in the uterus. A vaginal exam and vaginal sweep performed at the end of the procedure to be certain all materials and instruments are removed at the close of the procedure.

**Postprocedural care**

- Analgesia as needed
- I/O monitoring
- Peripad to quantify ongoing blood loss

**Potential challenges, complications, and management approach**

| Potential challenge or complication | Management approach |
|---|---|
| *Balloon or gauze cannot be inserted secondary to pain* | Do not force the material into the uterus because of the risks of uterine perforation. Pause the procedure. Reevaluate pain management approaches prior to the procedure. Consider additional NSAIDs and anxiolytics as needed. |
| *Balloon or gauze cannot be removed secondary to pain* | Pause the procedure. Reevaluate pain management approaches prior to the procedure. Consider additional NSAIDS and anxiolytics as needed. |

## Myomectomy of a prolapsed fibroid

| | |
|---|---|
| **Procedure** | **Myomectomy of a prolapsed fibroid** |
| **Purpose** | To control uterine bleeding caused by a prolapsed fibroid. |
| **Indications** | Acute uterine bleeding secondary to a prolapsed fibroid. ***The procedure should only be performed if the prolapsed fibroid has been identified as the cause of bleeding and medical management to control bleeding has not been successful.*** |
| **Diagnosis** | A mass observed at the external os separate from the cervix. |
| **Contraindications** | The indications and contraindications have been determined in a terrestrial environment with adequate procedure or operating room resources. Studies show that a small fibroid with a narrow base and graspable stalk may be removed in an office-based procedure. ***Fibroids that do not meet these criteria should not be removed.*** Specific **contraindications** include a fibroid that is:<br>**1.** Larger than 4 cm in diameter,<br>**2.** the pedicle cannot be visualized or palpated<br>**3.** or the fibroid has a pedicle >2 cm in diameter. |
| **Supplies** | |
| **Operators** | Minimum 1, Ideally 2 |
| **Antiseptic** | • Swab stick with povidone or 4% chlorhexidine gluconate solution<br>• Sterile gloves |
| **Analgesia** | • Paracervical block with 1%–2% lidocaine, 1.5% mepivacaine, 0.25% bupivacaine, or equivalent anesthetic solution for paracervical injection<br>• Additional PO anxiolytics as needed<br>• Nonsteroidal antiinflammatory medications PO 60–90 minutes prior to the procedure |
| **Materials** | • Crew medical restraint system<br>• Close drape system adhesive to the perineum for management of blood<br>• Sterile speculum (if available)<br>• Single tooth tenaculum (if available)<br>• Sterile gloves<br>• Gauze pads<br>• 10 mL syringe<br>• 18- to 20-gauge needle<br>• 2-0 to 3-0 delayed absorbable suture<br>• Mayo scissors or similar heavy tissue scissors |

**Steps**

1. Follow procedures for Foley catheter placement to assist in monitoring I/Os during the acute bleeding episode.
2. Prepare sterile field for supplies to be readily accessible.
3. Insert the speculum into the vagina to obtain visualization of the cervix and cervical os if available. Otherwise, attempt to palpate the mass and stalk digitally.
4. Apply antiseptic solution directly to the cervix two times.
5. Don sterile gloves.
6. Place a single tooth tenaculum (if available) on the cervix to stabilize the cervix and uterus during the procedure.
7. Perform a paracervical block.
8. Grasp the fibroid with a tenaculum and, with gentle traction, bring the fibroid and stalk into the mid to lower vagina.
9. Using a Kelly forceps or similar device, grasp the pedicle, ensuring the whole base is included in the clamp. If possible, place a second forceps in parallel to the first to secure the base.
10. The pedicle should be circled with a 2-0 to 3-0 delayed absorbable suture and secured with a surgical knot. This process should be repeated. The sutures should not be cut at this stage.

11. Once sutures are securely placed, Mayo scissors or equivalent heavy tissue scissor may be used to remove the fibroid from the base. A scalpel may be used but only with caution given the risks of incidental laceration of the surrounding vaginal or cervical tissue.
12. The fibroid stalk will usually retract into the uterine cavity after the fibroid is removed from the pedicle. For this reason, the pedicle should be stabilized using the forceps or remaining suture material to ensure hemostasis. If continued bleeding is observed, additional sutures may be placed at the base. Topical agents, such as Monsel's solution, silver nitrate, or other topical clotting agent, may be applied. Once the area is hemostatic, the sutures can be cut.
13. A vaginal exam and vaginal sweep performed at the end of the procedure to be certain all materials and instruments are removed at the close of the procedure.

**Postprocedural care**

- Analgesia as needed
- I/O monitoring
- Peripad to quantify ongoing blood loss

**Potential challenges, complications, and management approach**

| Potential challenge or complication | Management approach |
|---|---|
| *The fibroid stalk cannot be identified.* | Do not attempt the procedure. Optimize medical management approaches. |
| *Uterine bleeding continues after the procedure.* | Perform a uterine tamponade procedure. |

## 13.4 Management of Vulvar/Bartholin's Gland Cyst

### 13.4.1 Introduction

Female astronauts may be at risk of developing vulvar masses during long-duration spaceflight. While many of these lesions may resolve spontaneously with conservative management, including analgesia and antibiotic therapy, there may be cases in which procedural intervention is necessary during the mission to alleviate symptoms and prevent complications that could become more challenging to manage at a later stage.

One of the most common vulvar masses requiring intervention is a Bartholin's cyst or abscess (Illingworth et al., 2020; Kroese et al., 2017; Wechter et al., 2009). The Bartholin's glands, located bilaterally at the vaginal vestibule, can become obstructed, leading to cyst formation. If the cyst enlarges beyond 3 cm or progresses to an abscess, it can cause significant pain and increase the risk of further complications. In such cases, an incision and drainage procedure is recommended to provide symptom relief and prevent progression.

**Bartholin's cyst or abscess cyst incision and drainage**

| **Procedure** | **Bartholin's cyst or abscess cyst incision and drainage** |
|---|---|
| **Purpose** | To symptomatically manage a vulvar cyst causing significant pair or negatively impacting function. |
| **Diagnosis** | A physical exam is required to diagnose: Bartholin cyst: This presents as a soft, nontender mass at the site of the Bartholin duct and gland. Bartholin abscess: This presents as a soft, tender, and frequently fluctuant mass at the site of the Bartholin duct and gland. There may be surrounding edema, erythema, and a focal point where the abscess may begin to drain spontaneously. |
| **Indications** | The main indication for the procedure is a Bartholin's cyst or abscess or other large vulvar cyst causing pain or affecting function. This procedure will help relieve the symptomatic aspects of the mass and support healing. It is not a definitive treatment for the cyst, as it may recur without additional management (e.g., cyst excision or marsupialization). This procedure should only be undertaken if steps to manage pain with NSAIDS and other medications have been unsuccessful. |
| **Contraindications** | There are no absolute contraindications, with the exception of the use of a catheter in an individual who has a Latex allergy. |
| **Supplies** | |
| **Operators** | Minimum 1 operator, ideally 2 |
| **Antiseptic** | • Povidone or 4% chlorhexidine gluconate solution on an applicator or swab (selection based on availability and potential allergy) |
| **Analgesia** | • Subcutaneous injection of local anesthetic with 1% lidocaine or equivalent<br>• Additional PO anxiolytics as needed<br>• Nonsteroidal antiinflammatory medications PO 60–90 minutes prior to the procedure |

| | |
|---|---|
| **Materials** | • Crew medical restraint system (See Image 1)<br>• Close drape system adhesive to the perineum for management of blood<br>• Sterile speculum<br>• Sterile gloves<br>• Antiseptic solution<br>• Gauze pads<br>• Suction/irrigation device (optional)<br>• 3-5 mL syringe with 25- or 30-gauge needle for local anesthetic administration<br>• 1 or 2% lidocaine without epinephrine, 1.5% mepivacaine, 0.25% bupivacaine, or equivalent anesthetic solution for paracervical injection<br>• Small or medium forceps<br>• Small or medium hemostat<br>• Scalpel with a #11 blade<br>• Word catheter<br>• 3–5 mL syringe for filling Word catheter with saline<br>• Sterile saline |

**Steps**

**Incision and drainage with placement of a Word catheter or gauze**

1. Prepare equipment and sterile field
2. Apply an antiseptic to the procedural area, with focus on the intended site of the incision (e.g., at the site overriding the cyst behind the hymenal ring or where the abscess may develop spontaneous drainage)
3. Inject 1–3 mL of the selected local anesthetic agent to the intended incisions site.
4. If the cyst is fluctuant, stabilize the intended incision area either with a small/medium forceps or between the thumb and forefinger.
5. Make a stab incision using the scalpel. An initial superficial stab may be sufficient to allow effluent material to be expressed from the cyst. If this initial stab incision does not produce drainage, then make a deeper stab incision at the same site as the first incision.
6. Extend the incision to 5 mm to allow for expulsion of the material and subsequent placement of the Word catheter.
7. Using pressure, express all of the material from the cyst. A hemostat may be needed to disrupt loculations within the cyst to ensure adequate drainage. A suction device placed within the cavity or at the incision site may be used to facilitate this process. Drainage is complete once additional pressure does not produce additional effluent material.
8. The cyst should be opened to allow additional drainage. The preferred approach is to place a small catheter (e.g., Word catheter) within the cyst walls. Alternatively, if a Word catheter is not available or cannot be secured in place, an alternative approach may involve the use of a narrow continuous piece of gauze (e.g., Nu Gauze). If a Word catheter will be used, test the catheter to confirm there are no leaks when the balloon is filled. Insert the catheter to the full depth of the cyst. Grasp one of the cyst walls with a hemostat or forceps to ensure the catheter is situated within the cyst cavity. Once in place, insufflate the catheter balloon with 1–3 mL saline (depending upon cavity size) using the 3–5 mL syringe until the catheter is secured within the cyst wall. Remove the syringe once the balloon is full and is secured in place. The end of the catheter can be tucked into the site or into the superficial aspect of the vagina. If gauze will be used, then use a hemostat to introduce the gauze into the cyst cavity. Leave of 1–2 cm tail of gauze from the incision site to allow for wicking and to facilitate gauze removal.
9. A vaginal exam and vaginal sweep performed at the end of the procedure to be certain all materials and instruments are removed at the close of the procedure.

**Removal of the Word catheter**: The Word catheter may stay in place 2–4 weeks as needed to ensure the cyst cavity is empty of contents and allow for a new entry point to form.

1. At the end of the selected period of time, a small syringe can be attached to the end of the catheter and the balloon deflated. The catheter should be easily removed from the cavity. A small hemostat may be needed in the unlikely case the wall of the catheter has become adhered to the wall.
2. A vaginal exam and vaginal sweep performed at the end of the procedure to be certain all materials and instruments are removed at the close of the procedure.

**Removal of the gauze:** Gauze should be removed within 24–48 hours.

1. The gauze is removed by grasping the tail and, with gentle traction, gently remove the full length of the gauze from the cyst.
2. A vaginal exam and vaginal sweep performed at the end of the procedure to be certain all materials and instruments are removed at the close of the procedure.

**Postprocedural care**

- Analgesia as needed
- I/O monitoring
- Peripad to quantify additional effluent and bleeding
- Wound care
- Observation to ensure the abscess does not recur

**Potential challenges, complications, and management approach**

| Potential challenge or complication | Management approach |
|---|---|
| *The Word catheter will not remain in place* | To prevent this issue, maintain a 5 mm or smaller incision site so that the catheter is not supported in place by the surrounding tissues. If the catheter will not remain in place, ensure that the balloon is fully inflated. If these steps do not solve the problem, then use gauze to maintain an open cyst cavity during the healing process. |

## 13.5 Management of vulvar/vaginal injuries

### 13.5.1 Introduction

A vulvar or vaginal injury may occur if an individual experiences a saddle injury or other traumatic process to the perineum (Lopez et al., 2018). While the microgravity environment of spaceflight may reduce the incidence of vulvar and vaginal lacerations and hematomas, sudden changes in position, force, or gravity may still result in a hematoma-inducing injury. An injury may initially present as a laceration. However, an injury may result in a hematoma without evidence of a laceration and may have a more delayed onset compared to a laceration.

Vulvar and vaginal injuries can be particularly complex to assess and manage given the nature of the tissue planes in the perineum. Because of the fascial planes, a laceration may lead to occult bleeding and hematomas. Thus, a complete physical exam with or without the assistance of ultrasound is required to determine the nature and extent of an injury.

Approaches to the vaginal and vulvar injuries are described below (Lopez, Focseneanu and Merritt, 2018).

**Vaginal or vulvar hematoma evacuation**

| Procedure | Vaginal or vulvar hematoma evacuation |
|---|---|
| **Purpose** | The purpose is to achieve hemostasis if there is evidence of active bleeding within the hematoma. |
| **Indications** | This procedure is for hematomas that are actively expanding, leading to hemostatic instability, or compromising the function of adjacent anatomical structures. The first line of management of a vulvar hematoma is to observe and allow it to resolve on its own. It is strongly advised to avoid a surgical procedure, if possible, given the vascular nature of the vulva and vaginal in addition to the unique anatomical considerations of these areas. Careful and regular observation is recommended. Surgical intervention may be necessary if there is expansion of the hematoma that is impinging upon the urethra or evidence of expansion of the hematoma across the vulvar or vaginal tissue planes causing acute blood loss or hemodynamic instability. |
| **Diagnosis** | Diagnosis is made through a pelvic exam. A speculum examination may be required if the hematoma extends into the vagina. A hematoma may appear as a mass with ecchymosis and/or violaceous color with surrounding tenderness. There may be labia asymmetry if the hematoma is within the vagina. However, the indications for intervention are based on the individual vital signs, symptoms, and indicators of hemodynamic instability. |
| **Contraindications** | If the hematoma appears stable and/or is not impinging on proximal structures. |
| **Supplies** | |
| **Operators** | 2 operators are required |
| **Antiseptic** | • Povidone or 4% chlorhexidine gluconate solution on an applicator or swab (selection based on availability and potential allergy) |
| **Analgesia** | • Topical anesthetic applied to the procedural site prior to subQ injection of local anesthetic<br>• SubQ injection of local anesthetic with 1% or 2% lidocaine, 1.5% mepivacaine, 0.25% bupivacaine, or equivalent anesthetic solution for subQ injection<br>• Additional PO anxiolytics as needed<br>• Nonsteroidal antiinflammatory medications PO 60–90 minutes prior to the procedure |

| | |
|---|---|
| **Materials** | • Crew medical restraint system<br>• Close drape system adhesive to the perineum for management of blood<br>• Sterile speculum<br>• Sterile gloves<br>• Gauze<br>• Suction/irrigation device<br>• Cautery device<br>• Topical anesthetic<br>• Forceps<br>• Small to medium hemostat<br>• Needle driver<br>• Suture scissors<br>• Scalpel with #10 blade<br>• 2-0 or similar delayed absorbable suture<br>• Vaginal packing with povidone iodine |

**Steps**

1. Place a Foley catheter as needed to allow for bladder drainage and facilitate identification of the urethra if the hematoma is in close approximation to the urethral meatus.
2. If topical anesthetic is available, apply 60 minutes prior to the intended site of the injected local anesthetic.
3. Prepare equipment and sterile field.
4. Apply an antiseptic to the procedural area, with focus on the intended incision site.
5. Using a scalpel, make an incision over the hematoma to allow visualization of any vessels that may be bleeding. A suction/irrigation device may assist in visualization.
6. Vessels noted with active bleeding should be cauterized. If this does not control the bleeding, then the vessel base can be suture ligated using a 2-0 or similar delayed absorbable suture. Alternatively, a hemostatic agent may be used to facilitate hemostasis as needed.
7. Once hemostasis is achieved, the incision site may be closed with monofilament absorbable suture. After closure, vaginal packing with povidone iodine may be used if the additional pressure will assist with maintaining hemostasis.

**Postprocedural care**

- Analgesia as needed
- I/O monitoring
- Peripad to quantify ongoing blood loss

**Potential challenges, complications, and management approach**

| Potential challenge or complication | Management approach |
|---|---|
| *The hematoma continues to expand and/or there is increasing signs of hemodynamic stability* | The repaired area may be reopened, irrigated, and inspected for bleeding vessels. This area may be cauterized, suture ligated, or managed with a hemostatic agent. |

# 13.6 Approach to pelvic pain

## 13.6.1 Introduction

An essential component for flight preparedness is the ability to evaluate and manage acute pelvic pain of unknown etiology. In female astronauts, acute pelvic pain may result from a gynecologic origin. This may include conditions related to the uterus, ovary and adnexal structures, including an ovarian cyst, ovarian torsion, pelvic inflammatory disease (PID), and tubo-ovarian abscess (TOA; Juganavar and Joshi, 2022). In rare instances for the astronaut who has undergone appropriate preflight management and evaluation, pain may indicate an abnormal early pregnancy, including an ectopic pregnancy or miscarriage. While other gynecologic conditions may cause pain, most of these would have been diagnosed and managed prior to flight (e.g., endometriosis, adenomyosis) and have a very low risk of presenting as acute pain during the mission. It is important to recognize that acute pain presenting in the pelvic region may reflect a urinary issue, such as a urinary tract infection or cystolithiasis. It may also represent a gastrointestinal or other etiology, including acute appendicitis and diverticulitis. To date, there are no data indicating increased rates of ovarian pathology among female astronauts during flight. As the number of female astronauts and the duration of flights increase, it will be important to have adequate preparedness to rule out and manage an ovarian or other gynecologic cause of acute pain during the mission.

The initial step in management is a thorough clinical evaluation and accurate diagnosis, which determines whether the condition can be managed in flight or requires emergency intervention. For example, acute pelvic pain may be caused by an ovarian cyst (Mohamadian, Bayani and Shakki Katouli, 2023). In many cases, these are functional cysts arising from normal ovarian physiology, including follicular and corpus luteum cysts, which most often resolve spontaneously (American College of Obstetricians and Gynecologists' Committee on Practice Bulletins-Gynecology, 2016). These cysts can enlarge and, when adjacent to vascular structures, may become hemorrhagic. Hemorrhagic cysts typically present with acute-onset pain that subsequently stabilizes, and management is generally supportive, focusing on appropriate analgesia.

In rare cases, a hemorrhagic cyst may rupture or lead to hemoperitoneum, presenting with worsening abdominal pain, dizziness, vomiting, and signs of hemodynamic instability. Under these circumstances, emergency evaluation is required. Although terrestrial management may include computed tomography or ultrasound-guided cyst drainage, such interventions are not currently feasible during spaceflight. Future advances in laparoscopic or minimally invasive techniques may eventually allow similar management approaches to be implemented in the spaceflight environment.

Ovarian cysts may also result in ovarian torsion. This is a condition in which the ovary shifts or twists around it supporting pedicle (Liu, Shih and Yang, 2008; Argenta et al., 2000; Houry and Abbott, 2001). As these ligaments contain the ovarian vessels, the torsion may compromise the blood supply to the ovary. Torsion may present with sudden-onset severe pain, which is often accompanied by nausea, vomiting, and fever. In some cases, the ovary will spontaneously untwist, resulting in improvement of the pain. The torsion may or may not recur within a short interval of time. However, in other cases, surgical management is required to manually untwist the ovary or to perform a cystectomy or oophorectomy. What is remarkable is that the ovary can withstand torsion of the vessels and with reverse of the torsion (either spontaneously or surgically) the ovary will return to normal function in some cases.

It is unlikely that nonfunctional cysts, such as endometriomas, teratomas, cystadenomas, or ovarian cancer, which do not resolve spontaneously, will be the cause. In general, these ovarian cysts would be identified as part of preflight sonographic diagnosis and, if at risk for torsion in-flight, will likely have been surgically managed prior to flight.

The differential diagnosis of acute pelvic pain of gynecologic etiology may also include PID and TOA (Jenkins and Vadakekut, 2025; Taira et al., 2022). PID is an infection of the uterus (most often limited to the endometrium), fallopian tubes, and ovaries. TOA is a complication of PID and describes a condition in which the infection develops into an abscess involving the ovary and/or the fallopian tubes. Both PID and TOA may present with acute-onset pelvic and abdominal pain which may also include nausea, vomiting, and fever. The causing infectious agent is most often the result of an untreated chlamydia and gonorrhea infection (Algeri et al., 2025). Thus, these conditions have low likelihood of occurring during flight following appropriate preflight evaluation. With that said, such screening tests are imperfect, and PID and TOAs may result from other infectious etiologies which may not be detectable through routine sexually transmitted infection testing (Jenkins and Vadakekut, 2025; Taira, Broussard and Bugg, 2022).

Despite preflight screening, it is important to consider these etiologies in the evaluation of acute pelvic pain. PID may be treated with antibiotics, which may be feasible in-flight. Most often, these infections are polymicrobial with multiple antibiotic agents required for management. Additional management for pain, nausea, and fever may be needed. However, because of the nature of the abscess wall, TOAs typically require surgical management. Antibiotics may be used as a temporizing measure until evacuation can be performed. However, they will not sufficiently treat the TOA and TOA symptomatology may worsen until surgical management is complete.

### 13.6.2 Evaluation

The physical examination is the initial step in the diagnostic evaluation (As-Sanie et al., 2026). Ideally, this assessment includes both a digital pelvic examination and a sterile speculum examination. The presence of purulent discharge or material emanating from the cervical os is highly suggestive of infection, while cervical motion tenderness on digital examination is considered pathognomonic for PID. However, a normal or negative examination does not exclude PID or TOA, as infection may remain localized to the upper genital tract or present atypically in the microgravity environment.

Transvaginal or transabdominal ultrasound is an essential tool in determining the source of pelvic pain, particularly when gynecologic in origin (Boccatonda et al., 2025). While transvaginal ultrasound is the gold standard for visualizing the uterus and adnexa, it is unlikely this will be available in near-future spaceflight. Thus, transabdominal views may still have diagnostic value. The sonographic exam should begin with a general inspection of the pelvic anatomy. Any sonographic images obtained prior to flight may be used to compare the real-time ultrasound findings with prior anatomical findings. The scan should be able to delineate the major landmarks of the pelvis, including the uterus, ovaries, and adnexal structures. An initial inspection should also evaluate for any masses or free fluid in the pelvis. This may appear as an echogenic (fluid or blood) or heterogenous area (if clot is present). The uterus can be evaluated for echotexture of the myometrium and endometrium. While PID and TOA commonly present with adnexal findings, diminished visualization of the uterine border or landmarks may indicate pelvic infection.

The adnexa should be inspected to identify the ovaries and surrounding structures. Indicators of PID include loss of the normal adnexal architecture, including dilated or fluid filled fallopian tubes (e.g., hydrosalpinx) which may also have additional cystic findings. A TOA may be characterized by a complex adnexal mass with a heterogeneous appearance that may include septations and internal debris. There may also be free fluid in the pelvis, with localization in the cul-de-sac. A proximal infectious process, like appendicitis or abscess associated with diverticulitis, may present in a similar fashion given the nature and extent of the infectious process. Thus, the sonographic evaluation should also inspect for additional findings with the pelvis and abdomen that may suggest these nongynecologic conditions.

Functional ovarian cysts will have a different appearance. Most often the surrounding anatomical borders are normal. The ovary may be enlarged with a simple appearing thin and smooth walled anechoic structure representing the cystic area. A hemorrhagic cyst may have a similar appearance but with an internal echotexture indicating blood. This may appear as internal debris or appear in a reticular pattern. If the ovary is torsed, it may appear enlarged with a cystic appearance. Doppler ultrasound may indicate absent or reduced flow in the surrounding vascular structures, indicating the effect of twisted vessels leading to and from the ovary.

The absence of obvious sonographic findings does not rule out a gynecologic condition. For example, in the case of a ruptured functional ovarian cyst, the ovary may have a normal appearance following the spontaneous rupture of the cyst and release of the fluid into the pelvis. Yet, pain may persist after the rupture and in the context of an otherwise normal appearing adnexa. Another example includes early stage PID, which may be initially localized with the mid to lower genital tract but presents with pain from the ascending infection. In these cases, a thorough history should be collected to help determine the nature of the pain.

## 13.7 Management of early pregnancy complications

### 13.7.1 Introduction

The best birth control modalities, emergency contraceptives, and human intentions are imperfect, and thus the risk for pregnancy immediately preflight and in spaceflight is nonzero (Santaguida and Dubé, 2023). The effects of spaceflight on fertilization, implantation, and early pregnancy are yet unknown as a mammalian pregnancy has never occurred in space (Steller et al., 2018). If and when a pregnancy does happen, the risks of an early pregnancy loss (EPL) or ectopic pregnancy are at least as likely to occur in space as on Earth, if not potentially higher. Terrestrially, a pregnancy is often detected when menses is delayed, and a urine pregnancy test is performed. However, a urine or serum test for β-HCG can be performed typically as early as 10–14 days after ovulation depending upon the assay.

In the near-term, we are unlikely to have serum β-HCG capabilities. However, as the human presence grows in space, this may become more of a reality, at which time further discussion of how a β-HCG discriminatory zone or even level trend can assist with understanding whether an early pregnancy is developing properly or abnormally (Barnhart et al., 2011). Importantly, the other critical modality for monitoring whether an early pregnancy is indeed intrauterine and consistent with dating by the last menstrual period is transvaginal ultrasound (ACOG, 2017). A pregnancy that is <8 weeks gestation is very difficult to visualize with high fidelity on a handheld portable transabdominal ultrasound. As above, given mass and volume considerations, a transcavitary probe is not currently being considered for any current or near-future medical systems. This may complicate our ability to triage any complications of early pregnancy if it occurs (ACOG, 2018). For example, when some of the common ultrasound features diagnostic of an EPL or ectopic pregnancy are outlined, the assessment of mean sac diameter, visualization of a yolk sac, and the "ring of fire" sign around a tubal ectopic pregnancy would still be limited with transabdominal ultrasound (see Fig. 13.3 Barnhart et al., 2011; American College of Obstetricians and Gynecologists, 2017; Steller, Alberts and Ronca, 2018).

If a pregnancy is intrauterine, it may develop appropriately, or end in an EPL. Occasionally, an EPL occurs without the symptomatology of a spontaneous abortion such as bleeding or contractions (i.e., a missed abortion). The early pregnancy simply stops growing and the human body has not recognized it yet. In this setting, expectant management can be a reasonably safe option for a limited period of time (e.g., a few weeks). Other times, the presenting features of the miscarriage are bleeding and cramping. Bleeding may be heavier than normal menses but generally >2 soaked maxi pads/hour × 2 hours is concerning and requires immediate management (ACOG, 2018). Most EPLs are completed within 7–10 days; however, misoprostol or mifepristone can be offered to patients with EPLs <84 days of gestation who want to expedite completion and recovery from their EPL if either of these are included in the medical system. The regimen with

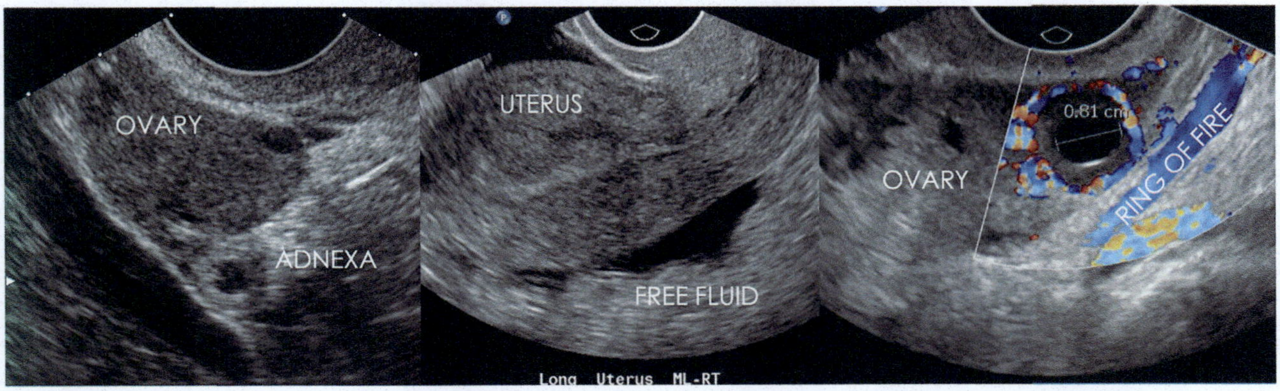


**FIGURE 13.3 Sonographic features of ectopic pregnancy.** Three concerning signs consistent with ectopic pregnancy on transvaginal ultrasound.

the highest success includes 200 mg of oral mifepristone (a progesterone receptor antagonist) followed by 800 mcg of misoprostol (a prostaglandin E1analogue) administered buccally or vaginally. If utilizing misoprostol only, the dosing remains the same; however, a 2nd misoprostol dose can be provided after 3–48 hours if needed to also increase efficacy. Procedural management of an EPL would include manual vacuum aspiration (MVA) or dilation & curettage (D&C). Both of these procedures are beyond the scope of this chapter of near-future procedural care; however, they are introduced in the future considerations section below. Importantly, if the pregnancy itself or EPL has not already led to end of mission with plans for evacuation, continued bleeding from a miscarriage heavier than two pads/hour would warrant more invasive management and returning to definitive care immediately. In this setting, it is likely that the continued bleeding is stemming from retained products of conception that need to be removed. Without the ability to perform a curettage, a temporizing management strategy would be to try and place a foley balloon into the uterus and fill it with 40–60cc in order to achieve uterine tampanode. Gauze packing can be placed in the vagina in an attempt to support the balloon's placement. This procedure has been discussed above for AUB and would follow the same steps and would only be temporizing in this setting.

If a pregnancy is extrauterine, this is known as an ectopic pregnancy (ACOG, 2018). When an ectopic pregnancy occurs in space, it will be a medical emergency as it can cause life-threatening intraabdominal bleeding leading to loss of life. An ectopic pregnancy that is found early enough and is unruptured can be treated with methotrexate 50 mg/m$^2$ IM once if the patient does not have contraindications and this medication is included in the medical system. However, given how rare the combined risks are—that a pregnancy occurs in space, that it is ectopic, and that the ectopic pregnancy can be diagnosed with a handheld portable ultrasound before it becomes large enough to rupture—it is unlikely that this medication should be included in near-future systems. If an ectopic pregnancy is ruptured, definitive management can only be surgical, and this is a dire situation. If not on Earth with full surgical capabilities and a blood bank, this will likely lead to crew mortality.

## 13.8 Future considerations

When the human presence in space expands and is sustained beyond low Earth orbit (LEO), the inability to evacuate rapidly will necessitate a much more robust and autonomous medical system capable of triaging and stabilizing most medical conditions. It is at that point that the mass, volume, and power needs of invasive equipment will become worth the cost to launch them and build them into a Lunar or Martian medical bay. In the gynecologic setting, such procedural considerations will expand to include the possibility of hysteroscopic, laparoscopic, and robotic surgical modalities. Other nearer-term life-saving procedures would include the possibility of D&C and MVA. Below are some of the procedures our authors feel are the next procedures to be considered in spaceflight after the above.

## Diagnostic vaginoscopy and hysteroscopy

| Procedure | Diagnostic vaginoscopy and hysteroscopy |
|---|---|
| Purpose | Direct visualization of the vagina and endometrial cavity |
| Indications | • Management of acute vaginal or uterine bleeding of unknown etiology |
| Contraindications | • Evidence of pyometra or other acute infectious process<br>• Evidence of ulcerative vaginal or vulvar lesion suggestive of including genital herpes infection<br>• Positive HCG (whether pregnancy is of known or unknown location) |
| **Supplies** | |
| Operators | Minimum 1, ideally 2 |
| Antiseptic | • Swab stick with povidone or 4% chlorhexidine gluconate solution<br>• Sterile gloves<br>• Biohazard bag |
| Analgesia | Diagnostic hysteroscopy with a small and/or flexible device is most often performed without anesthesia. The components of the procedure most likely to cause pain include cervical placement of a tenaculum, cervical dilation, insertion of the hysteroscope, uterine distention with hysteroscopic fluid, and endometrial sampling. A diagnostic hysteroscopy can be performed without many of these steps, based on available visualization, anatomical considerations, and procedural findings. The selection of the analgesic agent:<br>• Nonsteroidal antiinflammatory medications PO 60–90 minutes prior to the procedure<br>• Additional PO anxiolytics as needed<br>• Optional: paracervical block with 1% lidocaine or equivalent |
| Materials | • Crew medical restraint system (see Image 1)<br>• Close drape system adhesive to the perineum for management of hysteroscopic fluids<br>• Sterile speculum<br>• Sterile gloves<br>• Hysteroscope (3 mm) with a sterile cannula<br>• For operative hysteroscopy: appropriate hysteroscopic sheaths and graspers<br>• Compatible hysteroscopic video display module with associated USB cables<br>• Normal saline for uterine distention<br>• 60 cc syringes with tubbing for hysteroscopic fluid medium instillation<br>• Cervical dilators<br>• Monsel's solution or silver nitrate sticks<br>• Equipment for a paracervical block (as needed), including 10 mL syringe with an 18- to 20-gauge needle<br>• 2% lidocaine, 1.5% mepivacaine, 0.25% bupivacaine, or equivalent anesthetic solution for paracervical injection |

**Steps**

**Diagnostic hysteroscopy**

1. Prepare sterile field for supplies to be readily accessible.
2. Prepare the hysteroscope in a sterile fashion according to manufacture guidelines with respect to appropriate sheaths, fluid instillation, and video monitor connection. Optimize visualization with the appropriate white balance and set focus.
3. Insert the speculum into the vagina to obtain visualization of the cervix.
4. Inspect the vaginal walls for lesions that may be the source of bleeding
5. Apply antiseptic solution directly to the cervix two times.
6. Don sterile gloves.
7. Preparation for uterine distention: Prepare the hysteroscopic fluid for instillation by filling the 60 cc sterile syringe with normal saline. Attach tubing to connect the syringe to the inflow port on the hysteroscope.
8. If a paracervical block will be performed, draw up the selected anesthetic agent in the syringe. The local anesthetic is then may be injected in a two-point (at 4 and 8 o'clock only) fashion, with additional injection at 8 and 10 o'clock as needed. Withdrawal before injection at each site is recommended to prevent intravascular injection.

Note: While Earth-based procedures may use vasopressin or epinephrine to assist in hemostatic control, research is needed regarding the efficacy and safety of this approach with procedures performed in flight given the vascular nature of the cervix and surrounding stroma.

1. The primary operator should insert the hysteroscope through the cervix and into the endometrial cavity with direct visualization at all points in time. The assistant should slowly apply pressure to the syringe plunger to instill the hysteroscopic fluid into the uterine cavity. This process should be done slowly to prevent pain and/or uterine muscle response that may prevent successful fluid instillation.
2. Visualize the endocervix, general endometrial cavity, and ostia. If the patient has a didelphic uterus with two cervixes, perform hysteroscopy in the standard fashion by inserting the hysteroscope through each cervix and evaluating the cavities separately.
3. If a polyp is noted, proceed with operative hysteroscopy as described below.
4. End the procedure once a complete inspection of the cavity is performed or if adequate visualization cannot be achieved.
5. Remove the hysteroscope from the uterus and vagina. Inspect the endocervix to ensure there is no active bleeding. If a tenaculum was used, remove the tenaculum. It is normal for there to be a small amount of initial bleeding from the tenaculum site which will most often resolve on its own. Apply Monsel's solution or silver nitrate to the tenaculum site if bleeding continues after 3–5 minutes of observation.
6. A vaginal exam and vaginal sweep performed at the end of the procedure to be certain all materials and instruments are removed at the close of the procedure.

**Operative hysteroscopy**

Follow steps 1–7 as described for diagnostic hysteroscopy.

1. Insert the hysteroscopic grasper through the operating channel of the hysteroscope. Bring the edges of the grasper to the base of the polyp. Close the grasper at the base of the polyp and, using gentle traction on the grasper, avulse the polyp from its base. Repeat the procedure until the polyp is removed in entirety.
2. Inspect the area where the lesion as removed for hemostasis. This area may initially ooze for a few minutes but should resolve on its own without intervention.
3. Remove the hysteroscope from the uterus and vagina. Inspect the endocervix to ensure there is no active bleeding. If a tenaculum was used, remove the tenaculum. It is normal for there to be a small amount of initial bleeding from the tenaculum site which will most often resolve on its own. Apply Monsel's solution or silver nitrate to the tenaculum site if bleeding continues after 3–5 minutes of observation.
4. A vaginal exam and vaginal sweep performed at the end of the procedure to be certain all materials and instruments are removed at the close of the procedure.

**Postprocedural care**

- Analgesia as needed
- Use peripads to collect hysteroscopic fluid

**Potential challenges, complications, and management approach**

| Potential challenge or complication | Management approach |
|---|---|
| *Hysteroscope cannot be inserted secondary to pain* | Apply a tenaculum to the cervix and apply gentle traction to align the hysteroscopic approach with the endocervical canal. |
| | If the endocervical canal is narrow, gentle dilation can be performed to a diameter that allows for the hysteroscope to pass. The diagnostic hysteroscope will be 3–5 mm, so only minimal dilation is needed. Avoid overdilation as this will prevent adequate distention of the uterus. |
| *Hysteroscopy cannot be performed secondary to pain* | Perform a paracervical block as described above. |
| | Pause or slow instillation of the hysteroscopic distention medium. |
| | Terminate the procedure if adequate analgesia cannot be achieved. |
| *Complete visualization of the endometrial cavity cannot be obtained* | Slow or pause hysteroscopic fluid instillation to allow the endometrial cavity to relax, which may then allow for successful instillation of fluid on a repeat attempt. |
| | Slowly introduce additional hysteroscopic fluid into the endometrial cavity to flush blood and clots as needed. |
| *Uterine perforation suspected or confirmed* | Perforation should be suspected if direct visualization demonstrates the hysteroscope is in the pelvic cavity (e.g. visualization of bowel, bladder, change in appearance of tissue). End the procedure. Withdraw the hysteroscope and inspect the cervix for several minutes to ensure good hemostasis is achieved. |
| *Hemostasis at the lesion site cannot be achieved* | Proceed with uterine tamponade (see uterine tamponade) |

## Dilation and curretage

| Procedure | Dilation and curretage |
|---|---|
| **Purpose** | Abnormal uterine bleeding |
| **Indications** | • Removal of products of conception/management of missed abortion<br>• Hemorrhage from retained products of conception after a spontaneous abortion<br>• Interruption of early pregnancy<br>• Uterine/cervical polyp<br>• Pedunculated submucosal fibroid<br>• Endometrial hyperplasia<br>• Abnormal bleeding of unknown etiology in space |
| **Contraindications** | Knowledge, skills, abilities, or medical resourcing is not accessible to perform. |
| **Equipment and supplies** | |
| **Operators** | One minimum; would be the Crew Medical Officer |
| **Antiseptic** | • Vaginal povidone iodine or chlorhexidine<br>• Sterile gloves<br>• Biohazard bag |
| **Analgesia** | • Cervical block, pundendal block, or spinal anesthesia<br>• Oral/IV anesthesia |
| **Materials** | • Crew medical restraint system<br>• Speculum<br>• Tenaculum (if available)<br>• Vaginal dilators (if available)<br>• Uterine sound<br>• Sharp curet<br>• Portable ultrasound machine<br>• Ultrasound gel (sterile; or water if in microgravity) |

**Steps**

1. Consent. Risks include bleeding, infection, discomfort, uterine perforation.
2. Positioning. Restrain patient and flex and abduct hips to lithotomy-like position.
3. Bimanual exam to assess uterine position, size, and mobility.
4. Place speculum in vagina order to visualize the cervix.
5. Apply tenaculum (if present) to anterior lip of cervix to straighten the canal.
6. Gradually dilate cervix with dilators, starting with smallest size needed for cannula passage (if available)
7. Measure depth of the uterus with a uterine sound (if available) under abdominal ultrasound guidance.
8. Insert sharp curet gently through the cervix into the uterine cavity up to the fundus.
9. Scrape the endometrial surface with firm but controlled strokes. Work in a systematic manner, usually starting at the fundus and moving along the anterior, posterior, and lateral walls.
10. Collect curetted tissue for pathology or diagnostic evaluation.
11. Continue until a gritty sensation (contact with endometrium) is felt and aspirated material appears scant and clear. Bleeding should be minimal; if persistent, reinspect and curette lightly.
12. Withdraw curet.
13. Control bleeding as needed.
14. Remove instruments and speculum.

**Postprocedural care**

• Analgesia as needed

**Potential challenges, complications, and management approach**

| Potential challenge or complication | Management approach |
|---|---|
| Uterine sound or inserter perforating uterine fundal wall | 1. Use transabdominal POCUS to visualize the uterine sound and/or the LNG IUD inserter moving toward the fundus of the uterus throughout the procedure to prevent this complication. |
| Excessive bleeding | 1. If there is excessive bleeding during or immediately following procedure and if a speculum is present, attempt to visualize if there are any lacerations of the cervix or vagina.<br>2. Consider performing transabdominal ultrasound to evaluate the intrauterine contents and if the lining is still thick and therefore concerning for retained products, consider repeat D&C. |

| | |
|---|---|
| | 3. If the uterine lining is thin, a Foley catheter can temporarily be placed in the uterus for tamponade as discussed above and gauze packing can be temporarily placed within the vagina for pressure.<br>4. If there is bleeding on the cervix from the tenaculum sites, consider application of topical silver nitrate (if available). |

## Manual vacuum aspiration

| Procedure | Manual vacuum aspiration |
|---|---|
| **Purpose** | Removal of products of conception. |
| **Indications** | • Management of missed abortion<br>• Hemorrhage from retained products of conception after a spontaneous abortion<br>• Interruption of early pregnancy |
| **Contraindications** | Desired pregnancy. Knowledge, skills, abilities, or medical resourcing is not accessible to perform. |
| **Equipment and supplies** | |
| **Operators** | One minimum; would be the Crew Medical Officer |
| **Antiseptic** | • Vaginal povidone iodine or chlorhexidine<br>• Sterile gloves<br>• Biohazard bag |
| **Analgesia** | • Performing a cervical block, pundendal block, or spinal anesthesia are beyond the capabilities of spaceflight, thus analgesia would be with oral or IV medications. |
| **Materials** | • Crew medical restraint system<br>• Speculum<br>• Tenaculum (if available)<br>• Vaginal dilators (if available)<br>• Uterine sound<br>• Manual vacuum aspirator<br>• Portable ultrasound machine<br>• Ultrasound gel (sterile; or water if in microgravity) |

**Steps**

1. Consent. Risks include bleeding, infection, discomfort, and uterine perforation.
2. Positioning. Restrain the patient and flex and abduct hips to lithotomy-like position.
3. Bimanual exam to assess uterine position, size, and mobility.
4. Place speculum in vagina order to visualize the cervix.
5. Apply tenaculum (if present) to anterior lip of cervix to straighten the canal.
6. Gradually dilate cervix with dilators, starting with smallest size needed for cannula passage (if available)
7. Measure depth of the uterus with a uterine sound (if available) under abdominal ultrasound guidance.
8. Attach the cannula to the MVA syringe (prepped by creating vacuum with syringe).
9. Insert cannula gently through the cervix into the uterine cavity up to the fundus.
10. Release vacuum and aspirate contents by rotating and moving cannula within uterine cavity.
11. Continue until a gritty sensation (contact with endometrium) is felt and aspirated material appears scant and clear.
12. Withdraw cannula and inspect aspirated tissue for completeness (look for villi/gestational sac).
13. Control bleeding as needed.
14. Remove instruments and speculum.

**Postprocedural care**

• Analgesia as needed

**Potential challenges, complications, and management approach**

| Potential challenge or complication | Management approach |
|---|---|
| Uterine sound or inserter perforating uterine fundal wall | 1. Use transabdominal POCUS to visualize the uterine sound and/or the LNG IUD inserter moving toward the fundus of the uterus throughout the procedure to prevent this complication. |

| | |
|---|---|
| Excessive bleeding | 1. If there is excessive bleeding during or immediately following procedure and if a speculum is present, attempt to visualize if there are any lacerations of the cervix or vagina.<br>2. Consider performing transabdominal ultrasound to evaluate the intrauterine contents and if the lining is still thick and therefore concerning for retained products, consider repeat MVA.<br>3. If the uterine lining is thin, a Foley catheter can temporarily be placed in the uterus for tamponade as discussed above, and gauze packing can be temporarily placed within the vagina for pressure.<br>4. If there is bleeding on the cervix from the tenaculum sites, consider application of topical silver nitrate (if available). |

**Manual vacuum aspiration**

An example of the device is shown in Fig. 13.4.

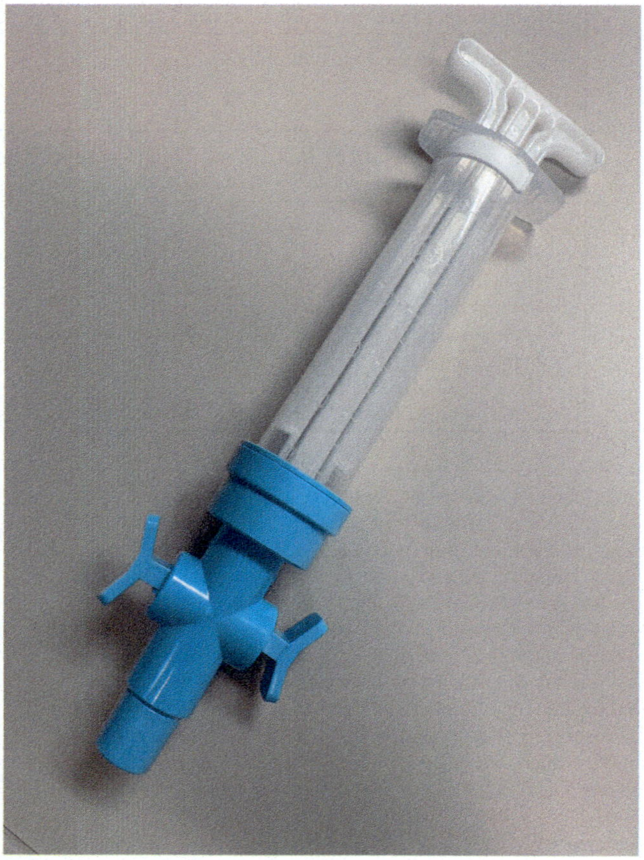

FIGURE 13.4 **Manual vacuum aspirator.** An example of a manual vacuum aspirator. https://upload.wikimedia.org/wikipedia/commons/6/60/Manual_vacuum_aspirator.jpg. (n.d.).

## References

Algeri, P., Spazzini, M.D., Pinna, N., Garbo, S., Villa, A., 2025. The challenge of diagnosing tubo-ovarian abscess and the necessity for aggressive management. Journal of the Turkish-German Gynecological Association 26 (1), 68–69. Available at https://doi.org/10.4274/jtgga.galenos.2025.2024-9-11.

American College of Obstetricians and Gynecologists' Committee on Practice Bulletins-Gynecology, 2018. ACOG Practice Bulletin No. 200: Early pregnancy loss. Obstetrics and Gynecology 132 (5), e197–e207. Available at https://doi.org/10.1097/AOG.0000000000002899. PMID: 30157093.

American College of Obstetricians and Gynecologists' Committee on Practice Bulletins-Gynecology, Long-Acting Reversible Contraception Work Group, 2017. Practice Bulletin No. 186: long-acting reversible contraception: implants and intrauterine devices. Obstetrics and Gynecology 130 (5), e251–e269. Available at https://doi.org/10.1097/AOG.0000000000002400. PMID: 29064972.

American College of Obstetricians and Gynecologists' Committee on Practice Bulletins-Gynecology, 2016. Practice Bulletin No. 174: evaluation and management of adnexal masses. Obstetrics and Gynecology 128 (5), e210–e226. Available at https://doi.org/10.1097/AOG.0000000000001768. PMID: 27776072.

American College of Obstetricians and Gynecologists' Committee on Practice Bulletins-Gynecology, 2012. Practice nulletin No. 128: diagnosis of abnormal uterine bleeding in reproductive-aged women. Obstetrics and Gynecology 120 (1), 197–206. Available at https://doi.org/10.1097/AOG.0b013e318262e320. PMID: 22914421.

American College of Obstetricians and Gynecologists, 2017. Committee opinion No 700: methods for estimating the due date. Obstetrics and Gynecology 129 (5), e150–e154. PMID: Available at https://doi.org/10.1097/AOG.0000000000002046, 28426621.

American College of Obstetricians and Gynecologists, 2013. Practice bulletin No. 136: management of abnormal uterine bleeding associated with ovulatory dysfunction. Obstetrics and Gynecology 122 (1), 176–185. Available at https://doi.org/10.1097/01.AOG.0000431815.52679.bb. PMID: 23787936.

Anderson, F.D., Hait, H., Baldwin, D., Kerber, I., Moffet, A., Simon, J.A., et al., 2003. A multicenter, randomized study of an extended cycle oral contraceptive. Contraception 68 (2), 89–96. Available at https://doi.org/10.1016/S0010-7824(03)00141-0, www.elsevier.com/locate/contraception.

Argenta, P.A., Yeagley, T.J., Ott, G., Sondheimer, S.J., 2000. Torsion of the uterine adnexa: pathologic correlations and current management trends. Journal of Reproductive Medicine for the Obstetrician and Gynecologist 45 (10), 831–836.

As-Sanie, S., Ross, W.T., Till, S.R., 2026. Evaluation and treatment of chronic pelvic pain. Obstetrics and Gynecology 147 (1), 21–43. Available at https://doi.org/10.1097/AOG.0000000000006123. Epub 2025 Nov 20. PMID: 41264919; PMCID: PMC12704687.

Barlow-Evans, R., Jaffer, K., Balogun, M., 2017. Migration of a Nexplanon contraceptive implant to the pulmonary artery. BMJ Case Rep 2017. Available at https://doi.org/10.1136/bcr-2017-219259. PMID: 28500111; PMCID: PMC5612561.

Barnhart, K., Van Mello, N.M., Bourne, T., Kirk, E., Van Calster, B., Bottomley, C., et al., 2011. Pregnancy of unknown location: a consensus statement of nomenclature, definitions, and outcome. Fertility and Sterility 95 (3), 857–866. Available at https://doi.org/10.1016/j.fertnstert.2010.09.006.

Berenson, A.B., Tan, A., Hirth, J.M., Wilkinson, G.S., 2013. Complications and continuation of intrauterine device use among commercially insured teenagers. Obstetrics and Gynecology 121 (5), 951–958. Available at https://doi.org/10.1097/AOG.0b013e31828b63a0.

Blue, R.S., Bayuse, T.M., Daniels, V.R., Wotring, V.E., Suresh, R., Mulcahy, R.A., et al., 2019. Supplying a pharmacy for NASA exploration spaceflight: challenges and current understanding. npj Microgravity 5 (1). Available at https://doi.org/10.1038/s41526-019-0075-2, https://www.nature.com/npjmgrav/.

Boccatonda, A., Tiraferri, V., De Meis, L., Fiorini, G., Serra, C., 2025. Role of ultrasound in the evaluation of patients with acute pelvic pain. Internal and Emergency Medicine 20 (5), 1339–1351. Available at https://doi.org/10.1007/s11739-025-03982-8, https://www.springer.com/journal/11739.

Bradley, L.D., 2011. Diagnosis of abnormal uterine bleeding with biopsy or hysteroscopy. *Menopause* 18 (4), 425–433. https://doi.org/10.1097/gme.0b013e31821255cc.

Braun, L.A., Kennedy, H.P., Womack, J.A., Wilson, C., 2016. Integrative literature review: U.S. military women's genitourinary and reproductive health. Military Medicine 181 (1), 35–49. Available at https://doi.org/10.7205/MILMED-D-15-00242, http://publications.amsus.org/doi/pdf/10.7205/MILMED-D-15-00242.

Coelho, L.F., Miranda, C., Canas, J., Morgado, M., Nunes, D., Henriques, A.F., et al., 2025. One giant leap for womankind: first menstrual cups tested in spaceflight conditions. NPJ Women's Health 3 (1). Available at https://doi.org/10.1038/s44294-025-00112-9.

Creinin, M.D., 2022. ACOG Clinical Consensus No. 3: general approaches to medical management of menstrual suppression. Obstetrics and Gynecology 140 (6), 1079–1080. Available at https://doi.org/10.1097/AOG.0000000000005007, http://journals.lww.com/greenjournal.

Farquhar, C., Ekeroma, A., Furness, S., Arroll, B., 2003. A systematic review of transvaginal ultrasonography, sonohysterography and hysteroscopy for the investigation of abnormal uterine bleeding in premenopausal women. Acta Obstetricia et Gynecologica Scandinavica 82 (6), 493–504. Available at https://doi.org/10.1034/j.1600-0412.2003.00191.x.

Farrell, R.M., Fogarty, J.A., Covington, M., Van Baalen, M., Aunon-Chancellor, S., Steller, J.G., 2025. A moonshot for female astronaut health, a win for all astronauts and space exploration. NPJ Microgravity 11 (1). Available at https://doi.org/10.1038/s41526-025-00495-8, https://www.nature.com/npjmgrav/.

Fischetti, C., Rajagopal, A.B., Frisch, E.H., Steller, J., Pissaris, A., Clark, K., et al., 2025. SPOCUS©: spaceflight point-of-care ultrasound training curricula for crew medical officers. Wilderness and Environmental Medicine 36 (1), 3–6. Available at https://doi.org/10.1177/10806032251330504, https://journals.sagepub.com/home/wem.

Fraser, I.S., Critchley, H.O., Munro, M.G., Broder, M., Writing Group for this Menstrual Agreement Process., 2007. A process designed to lead to international agreement on terminologies and definitions used to describe abnormalities of menstrual bleeding. Fertility and Sterility 87 (3), 466–476. Available at https://doi.org/10.1016/j.fertnstert.2007.01.023. 2007 Aug;88(2):538. PMID: 17362717.

Guttinger, A., Critchley, H.O.D., 2007. Endometrial effects of intrauterine levonorgestrel. Contraception 75 (6), S93. Available at https://doi.org/10.1016/j.contraception.2007.01.015.

Hindy, J.-R., Souaid, T., Larus, C.T., Glanville, J., Aboujaoude, R., 2020. Nexplanon migration into a subsegmental branch of the pulmonary artery. Medicine 99 (4), e18881. Available at https://doi.org/10.1097/md.0000000000018881.

Houry, D., Abbott, J.T., 2001. Ovarian torsion: a fifteen-year review. Annals of Emergency Medicine 38 (2), 156–159. Available at https://doi.org/10.1067/mem.2001.114303, http://www.elsevier.com/inca/publications/store/6/2/3/2/7/8/index.htt.

Hughes-Fulford, M., Carroll, D.J., Allaway, H.C.M., Dunbar, B.J., Sawyer, A.J., 2024. Women in space: a review of known physiological adaptations and health perspectives. Experimental Physiology. Available at https://doi.org/10.1113/EP091527, http://onlinelibrary.wiley.com/journal/10.1111/(ISSN)1469-445X.

Illingworth, B.J.G., Stocking, K., Showell, M., Kirk, E., Duffy, J.M.N., 2020. Evaluation of treatments for Bartholin's cyst or abscess: a systematic review. BJOG: An International Journal of Obstetrics and Gynaecology 127 (6), 671–678. Available at https://doi.org/10.1111/1471-0528.16079, http://obgyn.onlinelibrary.wiley.com/hub/journal/10.1111/(ISSN)1471-0528/.

Jain, V., Ploutz-Snyder, R., Young, M., Charvat, J.M., Wotring, V.E., 2020. Potential venous thromboembolism risk in female astronauts. Aerospace Medicine and Human Performance 91 (5), 432–439. Available at https://doi.org/10.3357/AMHP.5458.2020, http://docserver.ingentaconnect.com/deliver/connect/asma/23756314/v91n5/s9.pdf?expires=1592548286&id=0000&titleid=72010555&checksum=2DEE00CDB8380685527D56A5567CA778.

Jenkins, S.M., Vadakekut, E.S., 2025. Pelvic inflammatory disease. StatPearls [Internet]. StatPearls Publishing, Treasure Island (FL). PMID: 29763134.

Jennings, R.T., Baker, E.S., 2000. Gynecological and reproductive issues for women in space: a review. Obstetrical and Gynecological Survey 55 (2), 109–116. Available at https://doi.org/10.1097/00006254-200002000-00025.

Jensen, J.T., 2005. Contraceptive and therapeutic effects of the levonorgestrel intrauterine system: an overview. Obstetrical and Gynecological Survey 60 (9), 604–612. Available at https://doi.org/10.1097/01.ogx.0000175805.90122.af.

Jensen, J.T., Creinin, M.D., 2019. Speroff & Darney's Clinical Guide to Contraception, 6th ed. Wolters Kluwer Health, Philadelphia.

Juganavar, A., Joshi, K.S., 2022. Chronic pelvic pain: a comprehensive review. Cureus 14 (10), e30691. Available at https://doi.org/10.7759/cureus.30691. PMID: 36465795; PMCID: PMC9709590.

Kelekci, S., Kaya, E., Alan, M., Alan, Y., Bilge, U., Mollamahmutoglu, L., 2005. Comparison of transvaginal sonography, saline infusion sonography, and office hysteroscopy in reproductive-aged women with or without abnormal uterine bleeding. Fertility and Sterility 84 (3), 682–686. Available at https://doi.org/10.1016/j.fertnstert.2005.03.036.

Kroese, J.A., van der Velde, M., Morssink, L.P., Zafarmand, M.H., Geomini, P., van Kesteren, P.J.M., et al., 2017. Word catheter and marsupialisation in women with a cyst or abscess of the Bartholin gland (WoMan-trial): a randomised clinical trial. BJOG: An International Journal of Obstetrics and Gynaecology 124 (2), 243–249. Available at https://doi.org/10.1111/1471-0528.14281, http://obgyn.onlinelibrary.wiley.com/hub/journal/10.1111/(ISSN)1471-0528/.

Liu, Y.P., Shih, S.L., Yang, F.S., 2008. Sudden onset of right lower quadrant pain after heavy exercise. American Family Physician 78 (3), 379–380. Available at http://www.aafp.org/afp/20080801/photo.html.

Lopez, H.N., Focseneanu, M.A., Merritt, D.F., 2018. Genital injuries acute evaluation and management. Best Practice and Research: Clinical Obstetrics and Gynaecology 48, 28–39. Available at https://doi.org/10.1016/j.bpobgyn.2017.09.009, http://www.elsevier.com/inca/publications/store/6/2/3/0/0/4/index.htt.

Marnach, M.L., Laughlin-Tommaso, S.K., 2019. Evaluation and management of abnormal uterine bleeding. Mayo Clinic Proceedings 94 (2), 326–335. Available at https://doi.org/10.1016/j.mayocp.2018.12.012, http://www.journals.elsevier.com/mayo-clinic-proceedings.

Mathyk, Pandya, Wright Beatty, Anderson, Kohut, 2025. Handheld point-of-care ultrasonography for gynecology: insights into space travel through parabolic flight. Obstetrics and Gynecology 146 (5), 751–757. Available at https://doi.org/10.1097/AOG.0000000000006009.

Mohamadian, A., Bayani, L., Shakki Katouli, F., 2023. A simplified approach to ovarian lesions based on the O-RADS US risk stratification and management system. Ultrasonography 42 (1), 165–171. Available at https://doi.org/10.14366/usg.22037.

Moraes, M.M., Bruzzi, R.S., Martins, Y.A.T., Mendes, T.T., Maluf, C.B., Ladeira, R.V.P., et al., 2020. Hormonal, autonomic cardiac and mood states changes during an Antarctic expedition: from ship travel to camping in Snow Island. Physiology and Behavior 224. Available at https://doi.org/10.1016/j.physbeh.2020.113069, www.elsevier.com/locate/physbeh.

Munro, M.G., Critchley, H.O., Broder, M.S., Fraser, I.S., FIGO Working Group on Menstrual Disorders, 2011. FIGO classification system (PALM-COEIN) for causes of abnormal uterine bleeding in nongravid women of reproductive age. International Journal of Gynaecology and Obstetrics: The Official Organ of the International Federation of Gynaecology and Obstetrics 113 (1), 3–13. Available at https://doi.org/10.1016/j.ijgo.2010.11.011 Epub 2011 Feb 22. PMID: 21345435.

Munro, M.G., Critchley, H.O.D., Fraser, I.S., FIGO Menstrual Disorders Committee, 2018. The two FIGO systems for normal and abnormal uterine bleeding symptoms and classification of causes of abnormal uterine bleeding in the reproductive years: 2018 revisions. International Journal of Gynaecology and Obstetrics: the Official Organ of the International Federation of Gynaecology and Obstetrics 143 (3), 393–408. Available at https://doi.org/10.1002/ijgo.12666. Epub 2018 Oct 10. Erratum in: Int J Gynaecol Obstet. 2019 Feb;144(2):237. https://doi.org/10.1002/ijgo.12709. PMID: 30198563.

NASA. Medical kit contents and reference. https://www.nasa.gov/wp-content/uploads/2015/03/medical_kit_checklist_-_full_release.pdf (accessed 01.11.25).

NASA Office of the Chief Health and Medical Officers. 2024. NASA space flight medical selection, recertification and mission evaluation standards. http://nasa.gov/wp-content/uploads/2023/04/ochmo-std-100.1a.pdf (accessed 09.09.25).

Puukila, S., Alwood, J.S., Christenson, L.K., Ronca, A.E., Steller, J.G., 2024. Women's Health in Spaceflight Precision Medicine for Long and Safe Permanence of Humans in Space. Elsevier, United States, pp. 137–150. Available at http://doi.org/10.1016/B978-0-443-22259-7.00038-2, https://www.sciencedirect.com/book/9780443222597.

Santaguida, M., Dubé, S., 2023. Sexual health in space: a 5-year scoping review. Current Sexual Health Reports 15 (3), 148–179. Available at https://doi.org/10.1007/s11930-023-00368-9, http://www.springer.com/journal/11930.

Speroff, L., Darney, P.D., 2011. A Clinical Guide for. Contraception Lippincott Williams & Wilkins.

Steller, J.G., Alberts, J.R., Ronca, A.E., 2018. Oxidative stress as cause, consequence, or biomarker of altered female reproduction and development in the space environment. International Journal of Molecular Sciences 19 (12). Available at https://doi.org/10.3390/ijms19123729, https://www.mdpi.com/1422-0067/19/12/3729/pdf.

Steller, J.G., Blue, R.S., Burns, R., Bayuse, T.M., Antonsen, E.L., Jain, V., et al., 2020. Gynecologic risk mitigation considerations for long-duration spaceflight. Aerospace Medicine and Human Performance 91 (7), 543–564. Available at https://doi.org/10.3357/AMHP.5538.2020, http://docserver.ingentaconnect.com/deliver/connect/asma/23756314/v91n7/s3.pdf.

Steller, J.G., Blue, R.S., Zahner, C., Frisch, E.H., Bayuse, T., Aunon-Chancellor, S., et al., 2021. Menstrual Management considerations in the space environment. Reviews in Human Space Exploration. Available at https://doi.org/10.1016/j.reach.2021.100044.

Strock, N., Rivas, E., Goebel, K.M., 2023. The effects of space flight and microgravity exposure on female astronaut health and performance. In: Proceedings of the IEEE Aerospace Conference 2023. IEEE Computer Society. https://doi.org/10.1109/AERO55745.2023.10115765.

Taira, T., Broussard, N., Bugg, C., 2022. Pelvic inflammatory disease: diagnosis and treatment in the emergency department. Emergency Medicine Practice 24 (12), 1–24.

TRISH Orbit Rendezvous Teleconference: Artemis Medical System Overview. https://ntrs.nasa.gov/citations/20210021709 (accessed 01.11.25).

US Medical Eligibility Criteria (US MEC) for Contraceptive Use, 2024. https://www-cdc-gov.ccmain.ohionet.org/mmwr/volumes/73/rr/rr7304a1.htm (accessed 01.11.25).

Wechter, M.E., Wu, J.M., Marzano, D., Haefner, H., 2009. Management of bartholin duct cysts and abscesses. Obstetrical and Gynecological Survey 64 (6), 395–404. Available at https://doi.org/10.1097/OGX.0b013e31819f9c76.

Wheeler, A.P., Hemingway, C.O., 2024. Quantifying menorrhagia and overview of nonsurgical management of heavy menstrual bleeding. Hematology (Amsterdam, Netherlands) 2024 (1), 367–375. Available at https://doi.org/10.1182/hematology.2024000661.

Woolcock, J.G., Critchley, H.O.D., Munro, M.G., Broder, M.S., Fraser, I.S., 2008. Review of the confusion in current and historical terminology and definitions for disturbances of menstrual bleeding. Fertility and Sterility 90 (6), 2269–2280. Available at https://doi.org/10.1016/j.fertnstert.2007.10.060.

Zwart, S.R., Auñón-Chancellor, S.M., Heer, M., Melin, M.M., Smith, S.M., 2022. Albumin, oral contraceptives, and venous thromboembolism risk in astronauts. Journal of Applied Physiology 132 (5), 1232–1239. Available at https://doi.org/10.1152/japplphysiol.00024.2022, https://journals.physiology.org/doi/full/10.1152/japplphysiol.00024.2022.

Chapter 14

# Genitourinary procedures

Laurence Tang Hou[1], Jon G. Steller[2], and Katie Harris[3]
[1]*Hackensack University Medical Center, Hackensack, NJ, United States,* [2]*University of California Irvine, Irvine, CA, United States,* [3]*University of Toronto, Toronto, ON, Canada*

## 14.1 Overview

Spaceflight presents distinct challenges to human physiology, particularly in the field of urology. The history of urological issues in spaceflight dates to the early days of human space exploration. Urinary retention was first identified as a concern due to microgravity reducing pressure on the bladder exit and potentially exacerbated by the use of antiemetic medications like promethazine ("Risk of Urinary Retention", 2025). Nephrolithiasis has also emerged as a significant potential risk attributed to altered calcium metabolism and fluid shifts in microgravity. Urinary tract infections (UTIs) have also occured on several occaisions during flight, most notably from a drug resistant strain of Pseudomonas causing urosepsis in Apollo 13 astronaut Fred Haise (Cole et al., 2016).

This chapter examines the procedural approaches to managing urinary retention, nephrolithiasis, urosepsis, testicular torsion, and urogynecological considerations in the spaceflight environment. By addressing these conditions, this chapter seeks to equip medical practitioners with the tools and knowledge needed to safeguard astronaut urogenital health during current and future missions.

This chapter will delve into the practical management of these conditions. It will cover the placement and management of straight catheters, Foley catheters, and suprapubic catheters as well as the strategies for hematuria management. For nephrolithiasis, the management will focus on diagnosis and treatment with ultrasound-guided percutaneous nephrostomy (PCN) tubes to address obstructing ureteral stones while highlighting the limitations of nephrolithiasis management in space. Additionally, the theoretical risks of testicular torsion will be reviewed, emphasizing the constraints of current medical capabilities for surgical intervention during near-Earth missions. Finally, urogynecological considerations will provide a broader perspective on astronaut health and medical preparedness.

### 14.1.1 Procedural technique and materials for spaceflight

The recommended equipment proposed in this chapter is listed in Table 14.1 (NASA 3.001 Medical kit—contents and reference, 2015). The estimated mass and volume of suggested additional equipment is listed in Table 14.2.

## 14.2 Urinary catheterization in spaceflight

### 14.2.1 Introduction

Urinary catheterization allows for drainage of the bladder when voluntary voiding is not possible. In spaceflight, urinary retention can affect crew health and performance by causing pain and discomfort and increasing the risk of infection or kidney injury. Risk factors include age, sex, spaceflight adaptation, pharmacologic side effects (e.g., anticholinergic use), anatomical obstruction, psychosocial factors, and infection. Urethral (straight or Foley) and suprapubic catheterization routes are both effective, and each has its own advantages and disadvantages. In this section, we will review how to assess bladder volume and how to perform straight catheterization, Foley catheter placement, and suprapubic catheter placement.

**Procedural Medicine for Spaceflight. DOI:** https://doi.org/10.1016/B978-0-443-27552-4.00026-9

**TABLE 14.1** Recommended equipment for urinary procedures in spaceflight.

| Analgesia | Antiseptic | General materials | Medications |
|---|---|---|---|
| • GLYDO lidocaine HCl USP jelly, 2% gel<br>• Lidocaine 1% or 2% | • Iodine swab stick/ benzalkonium chloride (BZK) wipe × 2<br>• Chlorhexidine prep stick<br>• Sterile gloves<br>• **Sterile drapes**<br>• Biohazard bag<br>• Antiseptic solution | • Crew medical restraint system<br>• Needle holder<br>• Scissors<br>• Forceps<br>• Gauze<br>• Adhesive bandage<br>• Razor and shaving cream<br>• Absorbent pads | • Antibiotics<br>• IV hydration (i.e., normal saline)<br>• Alpha-blockers (i.e., tamsulosin)<br>• Nonsteroidal antiinflammatory drugs (NSAIDS)<br>• Opioids |
| **Needles** | **Syringes** | **Catheters** | **Stents/tubes** |
| • 0.035-inch guidewire<br>• Hydrophilic coating guidewire<br>• 18G needle<br>• 22G needle | • 60 mL catheter-tip syringe<br>• 10 mL syringe | • 8 Fr catheter<br>• 14G pigtail teflon fatheter<br>• Suprapubic catheter adapter clamp<br>• 5 Fr open-ended ureteral catheter<br>• 16 Fr or 18 Fr councill tip catheter<br>• Foley catheter and collection system<br>• 16 or 18 Fr coude catheter | • Amplatz dilators |
| **Medical imaging** | | | |
| • Portable ultrasound machine<br>• Ultrasound gel or water depending on anticipated gravity environment (sterile and nonsterile)<br>• Sterile probe cover | | | |

Bolded items are not currently included in the International Space Station medical equipment toolkit.

**TABLE 14.2** Estimated mass and volume of suggested additional equipment. This does not include pain control supplies or equipment, as these are described in other chapters.

| Item | Mass | Volume |
|---|---|---|
| Sterile drapes | ~300 g | ~3000 $cm^3$ |
| 60 mL catheter-tip syringe | ~60 g | ~250 $cm^3$ |
| 0.035-inch guidewire | ~8 g | ~1 $cm^3$ |
| 0.035-inch hydrophilic coating guidewire | ~8 g | ~1 $cm^3$ |
| 5 L plastic wastewater containment bag | 10 g | ~500 $cm^3$ |
| Nitrile gloves (1 pair) | 12 g | ~50 $cm^3$ |
| 18G needle | 1 g | ~1 $cm^3$ |
| 8 Fr catheter | ~25 g | ~40 $cm^3$ |
| Suprapubic catheter adapter clamp | ~10 g | ~3 $cm^3$ |
| 14 Fr pigtail teflon catheter | ~60 g | ~40 $cm^3$ |
| 5 Fr open-ended ureteral catheter | ~6 g | ~3 $cm^3$ |
| 16 Fr councill tip catheter | ~80 g | ~50 $cm^3$ |
| Amplatz dilator | ~80 g | ~20 $cm^3$ |
| Sterile gel and probe cover | ~27 g | ~6 $cm^3$ |

### 14.2.2 Does it work?

Urinary catheterization has been successfully performed multiple times in low Earth orbit (LEO). The standard International Space Station (ISS) medical kit contains supplies for both indwelling and straight catheterization ("3.001 Medical kit—contents and reference," 2015). While procedural principles are similar as on Earth, microgravity requires additional steps to secure equipment, control fluids, and maintain sterility.

Suprapubic catheterization has not been performed on humans in spaceflight, but it has been successfully demonstrated in anesthetized pigs during parabolic flight (Jones et al., 2007). This remains a viable contingency option if urethral access is unsuccessful.

The unique skillsets for both self-catheterization and catheterization of others are also important, given the possible risk of hematuria in microgravity due to kidney stones, history of previous pelvic radiation, theoretically altered clotting pathways, and potential for traumatic injuries to the bladder and kidney. As mission durations increase, the possibility of encountering hematuria increases due to UTIs, nephrolithiasis, radiation exposure, traumatic catheterization, and traumatic injuries to the abdomen and pelvis. It is critical to address hematuria using catheterization and efficient irrigation given the limited medical capabilities and resourcing available during spaceflight.

### 14.2.3 Pharmacologic adjuncts to treat urinary retention

Two drug classes may help maximize the chance of successfully removing a catheter: α-blockers and muscarinic agonists. The archetypal α-blocker, Tamsulosin (Flomax), relaxes the bladder neck and prostate, reducing outlet resistance, and is recommended terrestrially by the American Urologic Association (AUA) Guidelines (Sandhu et al., 2024) after catheterization for suspected bladder outlet obstruction (BOO). It is taken orally once daily and requires no additional monitoring. Terrestrially, this medication is used to treat lower urinary tract symptoms associated with benign prostatic hyperplasia (BPH) or manage postcatheter urinary retention. It is typically continued for at least 3 days prior to performing a trial without a catheter (TWOC). Given its safety profile and once-daily dosing, it may also be considered for female astronauts with suspected outlet obstruction, though the evidence is limited and based on small terrestrial studies.

Bethanecol, a cholinergic agonist, is a theoretical option to treat detrusor hypocontractility without obstruction; however, diagnosis requires urodynamic testing, which has never been performed in space. Bethanechol also carries a significant adverse event profile, including drowsiness, nausea, abdominal pain, headaches, bronchospasm, and hypotension, which could impair mission performance and should be held at least 48–72 hours before landing or mission-critical activities.

### 14.2.4 Procedural considerations for urinary catheterization

#### *14.2.4.1 Injury risk and incidence*

From 1981 to 2019, National Aeronautics and Space Administration (NASA) records documented 16 in-flight urinary retention events, an incidence of 0.018 events per person-flight, with similar rates for Shuttle (0.019) and ISS (0.021) missions (Cole et al., 2016). However, incidence may be as high as 2 in 10 crewmembers based on discussions with subject matter experts. This is most likely due to altered fluid pressure gradients in the setting of microgravity, though motion sickness drugs such as promethazine and/or scopolamine may also be a contributing factor ("Risk of Urinary Retention," 2025). Notably, female astronauts appear more likely to develop urinary retention compared to males which is reversed from the typical terrestrial pattern; however, this may be an artifact of smaller numbers of female astronauts have flown.

The true incidence of urinary catheterization in space is not well documented, though there are no publicly reported in-flight complications from the procedure. NASA medical literature does confirm that acute urinary retention (AUR) has occurred during both Shuttle and ISS missions, and that cases have been successfully managed with catheterization (Stepaniak et al., 2007; "Urinary health (OCHMO-MTB-003)," 2025). UTIs, sometimes associated with catheter use, have also been documented during spaceflight, with most cases occurring on missions lasting more than one week (Stepaniak et al., 2007; "Urinary health (OCHMO-MTB-003)," 2025). While catheter-related trauma (e.g., urethral mucosal injury, false passage) is a recognized terrestrial complication, intermittant self catheterization is routinely performed by many terrestrial patients with few complications and no such events have been reported to date from space missions. However, it is worth noting that the microgravity environment complicates patient positioning, visualization, and fluid containment, each of which can increase the risk for less-experienced operators. Given the limited publicly available data as of march 2026 , it is difficult to draw accurate conclustions on incidence and complications. However, given reported incidence as high as 20% of crew members it is clear that AUR is a common occurance and that catheters are an effective countermeasure. Crewmembers and crew medical officers should be provided with sufficient training and supplies to manage these events (Prieto et al., 2021).

### 14.2.5 Assessment of bladder volume

When urinary retention was suspected, historical evaluation included physical exam and catheterization as both a diagnostic and therapeutic management option. As point-of-care ultrasound (POCUS) devices have become increasingly portable and are now included in most spaceflight medical systems, the bladder volume can be estimated sonographically.

To perform:

1. Secure the crewmember. Place the ultrasound probe above the symphysis pubic in the transverse midline position with the probe indicator facing the crewmember's right side.
2. Bring the bladder into view, stopping at the widest dimension.
3. Freeze the image and measure the transverse (width) and anterior–posterior (depth) dimensions.
4. Unfreeze the image and rotate the probe 90 degrees clockwise. The bladder should still be centered in the image.
5. Measure from left to right on the screen to reflect the craniocaudal dimension.
6. Calculate the bladder volume using the ellipsoid formula of depth × width × height × 0.52 (correction factor); however, this can be auto-calculated using some ultrasound software (Karakala et al., 2021).

Many POCUS devices with beam-forming capabilities now include artificial intelligence (AI) auto-capture functionality to obtain bladder volume by placing the probe in one position in the midline over the bladder, which makes assessment of bladder volume more feasible for those with minimal training (Hurd et al., 2025). Notably, normal bladder findings include a smooth wall measuring 3–5 mm, which may decrease to 2–3 mm when filled. A healthy adult's full bladder volume is 300–500 mL and does not vary significantly between men and women. If calculating a bladder volume after attempted complete urination, a normal postvoid residual volume in patients under 65 years old is <50 mL and <100 mL in those over the age of 65. A postvoid residual of >100 mL raises suspicion for urinary retention on Earth; however, there is no firm cut-off for spaceflight, in which postvoid residuals can be as low as 79 mL in the setting of retention or as high as 150 mL ("Risk of Urinary Retention"," 2025). We recommend that medical teams develop algorithms with logically determined cutoff points to help guide crews in decision making. A good rule of thumb is that a post void residual by ultrasound of >100 mL or perceived symptoms of urinary retention should be a trigger to contact flight surgeon for further guidance.

### 14.2.6 Gross hematuria in space

Gross hematuria refers to visible blood in urine and represents a clinically significant finding in any environment. In spaceflight, its occurrence is of particular concern due to limited diagnostic capabilities, constrained resources, and the potential for urinary obstruction from clot formation. Potential causes of hematuria include iatrogenic injury associated with urinary catheterization, nephrolithiasis, UTI, blunt genitourinary injury, pelvic radiation, and malignancy. To date, no confirmed cases of gross hematuria have been reported during spaceflight; however, an instance of hematuria and clot retention in an astronaut 2 weeks after mission completion required hospital admission and subsequent indwelling catheter placement with irrigation of blood clots (Stepaniak et al., 2007). The etiology of this incident was likely multifactorial, involving traumatic catheter insertion as a result of multiple catheter attempts and a UTI. As mission duration increases, the risk of hematuria and its associated complications is also expected to rise. In spaceflight, if gross hematuria results in bladder outlet obstruction from clot formation, prompt catheterization is required, with manual aspiration of blood clots and bladder irrigation as needed to restore drainage and prevent complications such as urinary retention, UTI, bladder perforation, and postrenal acute kidney injury.

### 14.2.7 Spaceflight specific considerations

#### *14.2.7.1 Volume*

For catheter maintenance, about 100 mL of sterile water or normal saline is typically sufficient to clear temporary obstructions caused by debris, sediment, or small clots. In the setting of gross hematuria, irrigation needs may increase substantially: smaller amounts (around 250 mL) are often adequate to keep the catheter draining, but more severe bleeding or visible clots may require 500 mL or more, potentially exceeding the available onboard sterile fluid supplies. In such cases, alternative drainage strategies such as suprapubic or PCN tube placement may be required. If urinary flow cannot be maintained despite these measures, medical evacuation should be considered, as persistent obstruction in space poses an immediate risk to crew health and mission safety.

#### 14.2.7.2 Reusability

A 2021 Cochrane systematic review concluded that there is low-certainty evidence that there is no meaningful difference in symptomatic UTI rates between sterile single-use and appropriately cleaned reusable catheters (Prieto et al., 2021). Catheter reuse has also been described historically in spaceflight contexts, demonstrating feasibility when resources are constrained. Nevertheless, given the operational severity of even minor infections in spaceflight, sterile single-use catheters are favored as standard practice when resource availability allows. Reuse may be considered only in emergencies, though it introduces distinct microgravity-specific challenges such as ensuring effective cleaning and disinfection, mitigating fluid aerosolization during handling, and maintaining sterile storage of the reused device. The ongoing MultICath randomized trial, expected to report in 2025, may further inform best practices for catheter reuse under constrained conditions (Fader et al., 2024).

#### 14.2.7.3 Clean intermittent catheterization as an alternative to indwelling catheter

Clean intermittent catheterization (CIC) is often preferable to indwelling catheters for long-duration bladder management, as it avoids risks associated with continuous drainage, such as urethral trauma, encrustation, and stone formation (Neumeier et al., 2023). Although UTI rates may not differ significantly between CIC and indwelling catheters, many guidelines, including those from the United States Center for Disease Control (CDC), favor CIC to reduce long-term complications (Healthcare Infection Control Practices Advisory Committee, 2009). CIC is typically performed at regular intervals, often every 4–6 hours, with each voiding episode targeting bladder volumes of approximately 400–500 mL (Neumeier et al., 2023).

Catheter hygiene is foundational to infection prevention and device functionality. On Earth, best practices include cleansing the insertion site with soap and water (rather than antiseptics), strict hand hygiene, maintaining a closed drainage system, and disinfecting sampling ports (Healthcare Infection Control Practices Advisory Committee, 2009). These principles remain essential in spaceflight, though they must be adapted to the constraints of microgravity. In orbit, catheters and tubing should be securely anchored to prevent movement that could cause trauma or dislodgement. Because gravity-assisted drainage is absent, urine must instead be emptied at regular intervals using suction or manual expulsion. For long-duration missions, such as Mars transit, frequent drainage and vigilant monitoring will be especially important to prevent stasis and reduce infection risk. For a comprehensive terrestrial framework, crews should refer to the CDC's Guideline for Prevention of Catheter-Associated Urinary Tract Infections (Healthcare Infection Control Practices Advisory Committee, 2009).

#### 14.2.7.4 Retrograde urine flow

In microgravity, urinary drainage systems lack gravitational assistance, increasing the risk of retrograde flow from the collection bag back into the bladder. This can lead to infection, overdistension, or upper tract compromise. Closed drainage systems should incorporate one-way valves, antireflux chambers, or negative-pressure suction mechanisms to minimize reflux. Tubing must be arranged to avoid kinking and always be anchored to the body, as unsecured lines in the spacecraft environment can snag on equipment or other crewmembers. Such accidental traction could result in catheter dislodgement, urethral trauma, or gross hematuria from prostatic or bladder injury. Safe practice, therefore, requires regular inspection of the drainage system, timely emptying into sealed waste containers, and engineering designs that ensure valve function independent of orientation during long-duration missions.

#### 14.2.7.5 Urinary tract infections

The presence of an indwelling urinary catheter, whether a Foley or a suprapubic catheter, significantly increases the risk of UTIs due to both bacterial introduction and biofilm formation. The risk of bacteriuria with an indwelling catheter rises by approximately 3%–10% per day, with colonization becoming nearly universal after several weeks ("Urinary health (OCHMO-MTB-003)," 2025). Terrestrially, catheter-associated UTIs (CAUTIs) are responsible for 75%–80% of complicated UTIs. In spaceflight, urosepsis poses a critical threat and can lead to medical evacuation, making prevention and rapid treatment imperative for mission success.

In terms of causative pathogens, CAUTIs are most frequently due to Gram-negative organisms, with *Escherichia coli* (24%) and *Candida* spp. (24%) representing the most common isolates, followed by *Enterococcus* spp. (13%), *Pseudomonas aeruginosa* (10%), and *Klebsiella* spp. (10%) in large multicenter surveillance data (Weiner et al., 2016).

Notably, *Pseudomonas* was the causative organism in a severe UTI bordering on urosepsis in Apollo 13 astronaut Fred Haise, ("Urinary health (OCHMO-MTB-003)," 2025). This event was likely triggered by prolonged use of a condom catheter in clearly off nominal circumstances on board the crippled spacecraft. However, underscores the importance of considering atypical infections when manifesting medical resources for spaceflight.

The Infectious Diseases Society of America 2025 clinical practice guideline on complicated UTIs provides the most current terrestrial framework, which can be adapted for space medicine (Trautner et al., 2025). The guideline defines complicated UTI pragmatically as an infection in the presence of risk factors apparent at the point-of-care (e.g., catheterization, obstruction, systemic illness). For patients with sepsis or systemic features, empiric intravenous therapy is recommended, with options including a third- or fourth-generation cephalosporin, piperacillin-tazobactam, a carbapenem, or a fluoroquinolone. For nonseptic complicated UTI, empiric therapy may begin with a third- or fourth-generation cephalosporin, piperacillin-tazobactam, or fluoroquinolone, reserving carbapenems for patients at risk for resistant organisms such as extended spectrum beta-lactamase producers. Once culture and susceptibility results are available, therapy should be deescalated to the narrowest effective agent (Trautner et al., 2025). Given the substantial additional mass of IV antibiotics and the impracticality of IV antibiotic therapy during an active mission, a complicated UTI requiring IV treatment is grounds for crewmember evacuation.

#### *14.2.7.6 Role of prophylactic antibiotics*

The benefit of routine prophylactic antibiotics for urinary catheterization in spaceflight is unclear. In terrestrial environments, routine prophylactic antibiotics are not recommended for simple outpatient cystoscopy, urodynamic procedures, catheterization, or catheter changes per the AUA Guidelines (Lightner et al., 2020). The terrestrial data have not shown decreased infection rates as a result of prophylactic antibiotic use, and prolonged antibiotic use increases the risk of antibiotic resistance.

### 14.2.8 Considerations for evacuation

All crewmembers should undergo a trial without catheter (TWOC) before evacuation, extravehicular activity (EVA), or returning to ground if there is a clinically reasonable chance that urinary retention has resolved, and they currently have an indwelling Foley catheter. A successful TWOC reduces the risks associated with traveling or performing strenuous activity while catheterized. However, if the original catheterization was technically difficult, or if a suprapubic catheter is in place, the risks of attempting removal may outweigh the benefits. In these cases, failed TWOC could necessitate urgent reinsertion of a difficult Foley or lead to complications from disruption of a suprapubic tract, both of which could be more dangerous than maintaining the catheter. The risks must be carefully weighed against other operational risks such as cabin depressurization on reentry, or failure to meet mission objectives.

To perform a TWOC in spaceflight, the indwelling catheter should be removed during a period of clinical observation, ideally during daytime hours. The crewmember should be allowed to void spontaneously, with adequate hydration encouraged, and urine output monitored over a 4–8 hour interval. Postvoid residual volume (PVR) should be assessed using bladder ultrasound. A trial is generally considered successful when spontaneous voiding occurs without significant discomfort and residual urine volume remains acceptably low (PVR $\leq$ 100 mL). Failed trials, defined by inability to void, progressive bladder distension, or elevated PVR >100 mL, warrant prompt recatheterization and reassessment.

If the catheter is to remain in place, best practices should be followed to minimize complications and discomfort under the purpose-made intravehicular activity suits. After draining the collection bag, the catheter should be clamped, with gentle suprapubic pressure applied beforehand to ensure full drainage. The catheter must be well secured to the leg or lower abdomen to prevent traction, leakage, or accidental dislodgement during movement or high workload. Attention should also be given to infection risk and comfort, as even minor complications can become significant in the operational environment of spaceflight. Please see the decision tree below for a summary of the above (Fig. 14.1).

In high-G scenarios such as during launch, reentry, or abrupt maneuvering, urinary catheter systems face significant functional challenges. While the catheter itself remains mechanically intact, drainage problems such as backflow or leakage occur if antireflux mechanisms are absent. Historical evidence from NASA confirms that later Gemini missions upgraded their urine collection systems by adding check valves to the inflatable cuff to prevent backflow, signaling the operational necessity of one-way valves in dynamic fluid environments (Stapleton and Baccus, 2013). Modern aviation has taken this further; devices like the Advanced Mission Extender Device and the U.S. Air Force's Skydrate and AIRUS systems actively function to provide secure, leak-free relief under high-G loads (Adams and Cox, 2014; Nieberg, 2025).

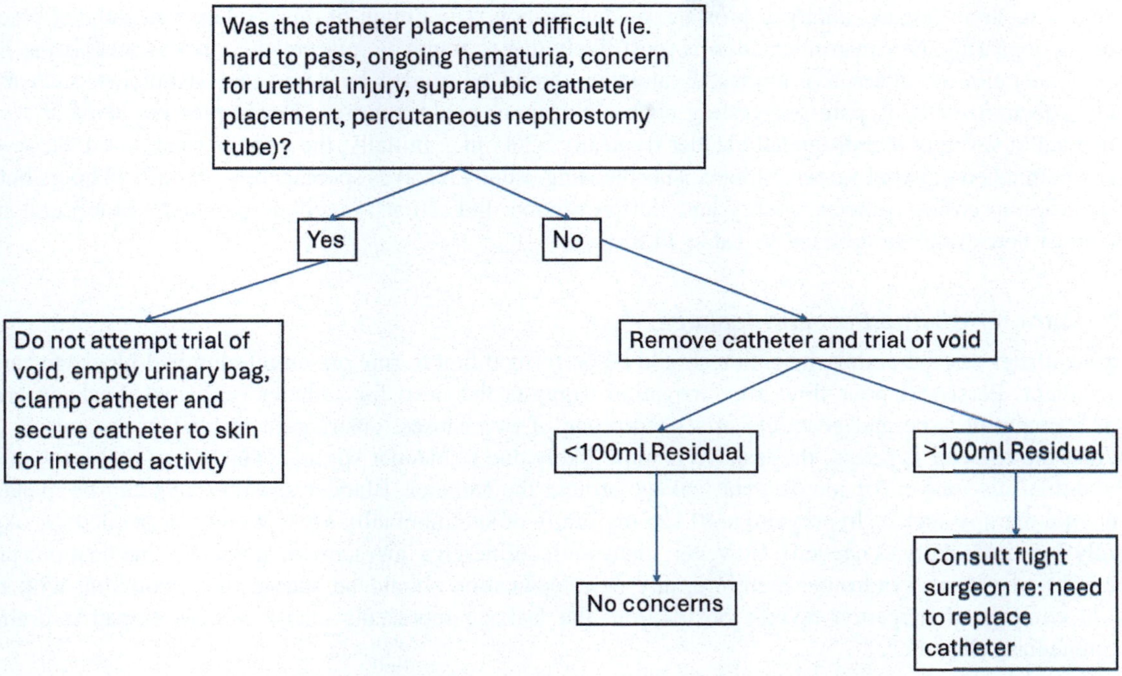


FIGURE 14.1 Decision flow tree for attempting trial of void before extravehicular activity/return/evacuation.

These insights support the integration of lightweight antireflux valves or active pumping mechanisms in space catheter designs, in addition to standard practices like predraining, clamping before G-transitions, and ensuring secure tubing anchoring.

### 14.2.9 Injury characteristics and complications

#### *14.2.9.1 False passage and prostatic trauma*

The most frequent injuries during Foley or intermittent catheter placement are the creation of a false passage and prostatic trauma. A false passage often presents as the catheter "bouncing back," indicating the catheter has created a new, nonphysiologic channel, typically through the prostate. Traumatic catheter manipulation resulting in urothelial injury may produce blood at the urethral meatus. If the catheter is placed correctly in the bladder (verified through in-and-out irrigation with a syringe), minor prostatic bleeding usually self-resolves within 48 hours as the catheter tamponades the site of prostatic trauma, and any false passage may heal spontaneously in that period.

#### *14.2.9.2 Enlarged prostate*

An enlarged prostate is another frequent source of difficulty for placing urethral catheters, especially in older men. The term refers to increased prostate size, most often due to BPH. Still, enlargement can also result from prostatitis or malignancy (more likely to be detected preflight). In catheterization, the enlarged prostate can produce resistance or "bounce-back" at the prostatic urethra. The safest approach when encountering resistance is to withdraw the catheter a few centimeters and gently reattempt placement one time, rather than forcing the catheter multiple times, which increases the risk of creating a false passage or traumatizing the prostate. A coude catheter, with its curved tip designed to navigate the prostatic urethra more effectively, is strongly preferred in these cases and should be considered in medical kits, especially if astronauts with known BPH are flying. Coude catheters can also be used interchangeably with normal straight, Foley, or indwelling suprapubic catheters in male and female patients.

#### *14.2.9.3 Hematuria*

Hematuria following catheterization requires careful characterization. Bright red or "fruit punch" urine is more concerning for active bleeding with clot risk, whereas darker maroon, brown, or tea-colored urine usually represents older blood or upper tract sources and is less likely to form significant new clots once irrigated clear. Possible etiologies for hematuria include infection, stone disease, enlarged prostate, or malignancy. The timing of onset relative to catheter placement is critical: blood appearing after instrumentation is most likely due to traumatic placement, whereas preexisting hematuria warrants consideration of alternative causes. Evaluation should be guided by associated symptoms, such as dysuria (suggesting infection) or flank pain (suggesting nephrolithiasis or pyelonephritis). Decompressive hematuria can occur after drainage of a severely overdistended bladder (typically >300 mL). Initially, the urine is clear, but it becomes blood-tinged as overstretched mucosal vessels rupture. This bleeding usually resolves spontaneously within 48 hours but requires close observation to ensure catheter patency and clot prevention. Indications, required equipment, procedural steps, and management of hematuria are outlined in Table 14.3.

#### 14.2.9.4 *Catheter obstruction and leakage*

If a urinary catheter stops draining, irrigation should be performed first to rule out an intraluminal blockage caused by a clot or sediment. Persistent poor flow after irrigation suggests the need for catheter exchange. Leakage around the catheter is a common issue and generally arises from one of two causes: obstruction or bladder spasms. If irrigation confirms that the catheter is patent, the leakage is more likely due to bladder spasms, which may present with urgency, lower abdominal discomfort, or intermittent leakage around the catheter. Bladder spasms can often be managed with anticholinergic therapy such as hyoscyamine (0.125 mg orally or sublinguinally every 4 hours as needed) or oxybutynin (5 mg orally 2–3 times daily as needed). However, many anticholinergics interact with antinausea medications prescribed for spaceflight adaptation syndrome; therefore, any new medication should be started in consultation with the flight surgeon. In contrast, if irrigation reveals obstruction, the leakage represents overflow from impaired drainage and requires immediate clearance.

#### 14.2.9.5 *Postobstructive diuresis*

A notable complication after relieving urinary retention is postobstructive diuresis. No data are currently available regarding the risk of this complication in spaceflight, and terrestrial data are variable, suggesting that 0.5%–52% of patients may experience it (Nyman et al., 1997). When the bladder has been chronically distended, tubular concentrating mechanisms may be impaired, leading to excessive urine production once the obstruction is relieved. Clinically, it is defined as urine output exceeding 200 mL per hour for two consecutive hours or more than 3 L over 24 hours. This diuresis can precipitate significant hypovolemia and electrolyte derangements if not carefully managed.

Management centers on close monitoring of urine output and fluid balance. When possible, patients should be encouraged to drink freely to match losses. If oral intake is insufficient or urine output is excessive, intravenous replacement with 0.9% or 0.45% saline at roughly half the rate of ongoing losses (approximately 0.5 mL IV for each 1 mL of urine output) can be considered; however, oral intake is likely to be sufficient in spaceflight (Halbgewachs and Domes, 2015).

#### 14.2.9.6 *Pelvic trauma*

Pelvic fractures carry a high-risk of urethral injury, most often signaled by blood at the urethral meatus. Another classic finding is a high-riding prostate, identified on digital rectal examination when the prostate is absent or poorly palpable due to displacement from a pelvic hematoma. In terrestrial settings, retrograde urethrography is used to confirm urethral integrity, but this is not feasible in spaceflight. If the practitioner has prior experience with catheter placement, one careful attempt at urethral catheterization with a Foley may be considered in either scenario. However, if resistance is encountered, further attempts should be avoided to prevent false passage, and suprapubic catheterization should be performed to ensure bladder drainage and allow the urethra to heal, followed by immediate medical evacuation.

#### 14.2.9.7 *Risks of suprapubic catheter placement*

While suprapubic catheter placement is generally safe, bowel injury remains the most feared complication, particularly in patients with prior abdominal surgery or adhesions. Preprocedural assessment should include a detailed surgical history and, when available, a POCUS to confirm bladder distension and identify any intervening bowel loops anterior to the bladder dome. If ultrasound suggests bowel interposition, a proper window should be identified, and suprapubic catheter placement should be attempted with great care. Even with reassuring imaging, meticulous technique is essential: insertion

**TABLE 14.3** Hematuria management.

| **Procedure** | **Catheter irrigation** |
|---|---|
| Purpose | To manage and prevent urinary obstruction due to clots or blood, ensuring the patency of the urinary catheter and bladder drainage. |
| Indications | • Gross hematuria (often caused by renal, bladder, or urethral trauma, renal stones, cystitis, rhabdomyolysis, radiation cystitis, bladder or upper tract tumors on Earth, but most likely renal stones in space)<br>• Postsurgical bladder irrigation<br>• Obstructed catheter due to blood clots or debris |
| Contraindications | • None |
| **Supplies** | |
| Operators | Minimum 1, ideally 2 operators |
| Antiseptic | • Antiseptic/alcohol wipes (optional) |
| Analgesia | • Lidocaine hydrochloride USP jelly 2% |
| Materials | • 60 mL catheter-tip syringe<br>• Sterile irrigation solution (i.e., saline or sterile water)<br>• Biohazard container for irrigant<br>• Sterile gloves<br>• Ultrasound (if available) to assess fluid retention in the bladder |
| Estimated equipment mass/volume (not including ultrasound) | ~720 g/1300 $cm^3$ |
| **Steps** | |
| **1.** Position and properly restrain the patient crewmember.<br>**2.** Prepare a sterile field with all supplies.<br>**3.** Clean the irrigation port with antiseptic.<br>**4.** Attach the irrigation syringe to the port.<br>**5.** Assess the current suspected fluid volume in the bladder.<br>**a.** If a full bladder is suspected, slowly instill 50 mL sterile irrigation fluid into the bladder using a 60 mL catheter-tip syringe and pull back syringe to aspirate fluid/clot. It is acceptable to reuse irrigant in a syringe repeatedly unless a clot is aspirated, in which case the effluent should be disposed of and fresh sterile irrigation should be instilled.<br>**b.** Allow the fluid to drain, monitoring for clots or obstruction.<br>**6.** *Microgravity adaptation: Capture expelled fluid and clots into a biohazard bag immediately to prevent floating debris.<br>**7.** Repeat until hematuria color clears or until the effluent is free of clot. As a general rule, if clot is encountered, continue to irrigate 3–5 additional times until no clot is encountered.<br>**8.** Secure the catheter and dispose of used materials properly. | |
| **Postprocedure care** | |
| • Monitor for recurrent obstruction or systemic signs of infection.<br>• Record urinary output and characteristics.<br>• Ensure catheter stability or secure in place to prevent displacement. | |
| **Potential complications and solutions** | |
| Complication | Solution |
| Ineffective irrigation | • Catheter may not be in bladder: deflate balloon, push catheter all the way to the level of the foley port bifurcation, and inflate the balloon. Then, irrigate.<br>• Catheter-tip may be stuck on bladder wall: move the catheter around the patient's body in a 360 degree manner while irrigating to change the catheter-tip positioning. Another potential problem is that the bladder may be empty; in this case, instill 100 mL of sterile irrigant and begin irrigation<br>• If above techniques have been tried, the catheter may be completely obstructed, in which case it should be deflated, removed, and a new catheter should be placed. |
| Catheter dislodgement | Replace catheter using procedures detailed above. |
| Persistent hematuria | Consider systemic interventions (i.e., clotting factor replacement, if available). |
| Infection | Treat with empiric antibiotics. |

should be performed in the midline, two finger-breadths above the pubic symphysis, where the risk of bowel interposition is lowest. Continuous monitoring for peritonitis or sepsis postprocedure is required, as delayed recognition of bowel injury can be catastrophic.

## 14.3 Decision making

The decision to place a urinary catheter in spaceflight must weigh the benefits of immediate bladder decompression against the risks of infection, urethral trauma, and finite supply use. On Earth, catheterization is often routine, but in orbit it represents a high-stakes intervention that should be reserved for well-defined indications: AUR, BOO, the need for precise urine output measurement in critically ill crew, or urinary diversion when urethral catheterization is not possible due to trauma.

Operationally, the choice between urethral and suprapubic access depends on the clinical context. Straight or Foley catheterization is appropriate for most cases of urinary retention, but if resistance is encountered or urethral injury is suspected, further attempts risk creating a false passage. In these situations, suprapubic catheterization is the safer option. Each technique carries distinct hazards, such as urethral trauma with repeated passage attempts or bowel injury with suprapubic insertion, both of which are magnified in microgravity. Careful preprocedural evaluation, secure device anchoring, and meticulous fluid management are therefore essential for safe catheterization during spaceflight.

## 14.4 Description

The decision to place a urinary catheter is based on the following factors:

1. What is the indication?
   a. Urinary retention, accurate urine monitoring, hematuria with clot retention, or trauma.
2. Can the indication be addressed nonprocedurally?
   a. Some causes of retention may be reversible.
   b. Constipation can be treated with a bowel regimen.
   c. Medication-induced retention (e.g., from anticholinergics such as oxybutynin or hyoscyamine, or sympathomimetics) may resolve with discontinuation.
   d. Outlet obstruction from prostatic enlargement may improve with an alpha-blocker (likely known from preflight screening).
   e. Timed voiding, suprapubic pressure, or valsalva maneuvers (forced expiration against a closed airway) may also be effective.
3. What is the ideal catheterization modality given the clinical scenario?
   a. Straight, Foley, or suprapubic (see below for procedure-specific indications).
4. Is urethral access safe?
   a. Signs of urethral injury, such as blood at the meatus, a high-riding prostate, or associated pelvic trauma, make urethral catheterization unsafe; suprapubic catheterization is favored in these scenarios.
5. Has resistance been encountered?
   a. Difficulty advancing the catheter may indicate an enlarged prostate or urethral obstruction. Repeated attempts should be avoided. A coude catheter should be used in these cases.
6. What resources are available?
   a. Supplies must be sufficient for ongoing management, including irrigation fluids, sterile catheters, and antibiotics if infection risk is high.
7. Is medical evacuation necessary?
   a. Uncontrolled hematuria, persistent urinary retention despite urinary diversion, and severe infection progressing to sepsis require immediate evacuation.

This is summarized in the decision tree below (Fig. 14.2).

Procedural indications, required equipment, and step-wise guidance for straight and Foley urinary catheterization and for suprapubic catheterization are outlined in Tables 14.4 and 14.5, respectively. Guidance for urinary catheter removal is provided in Table 14.6.

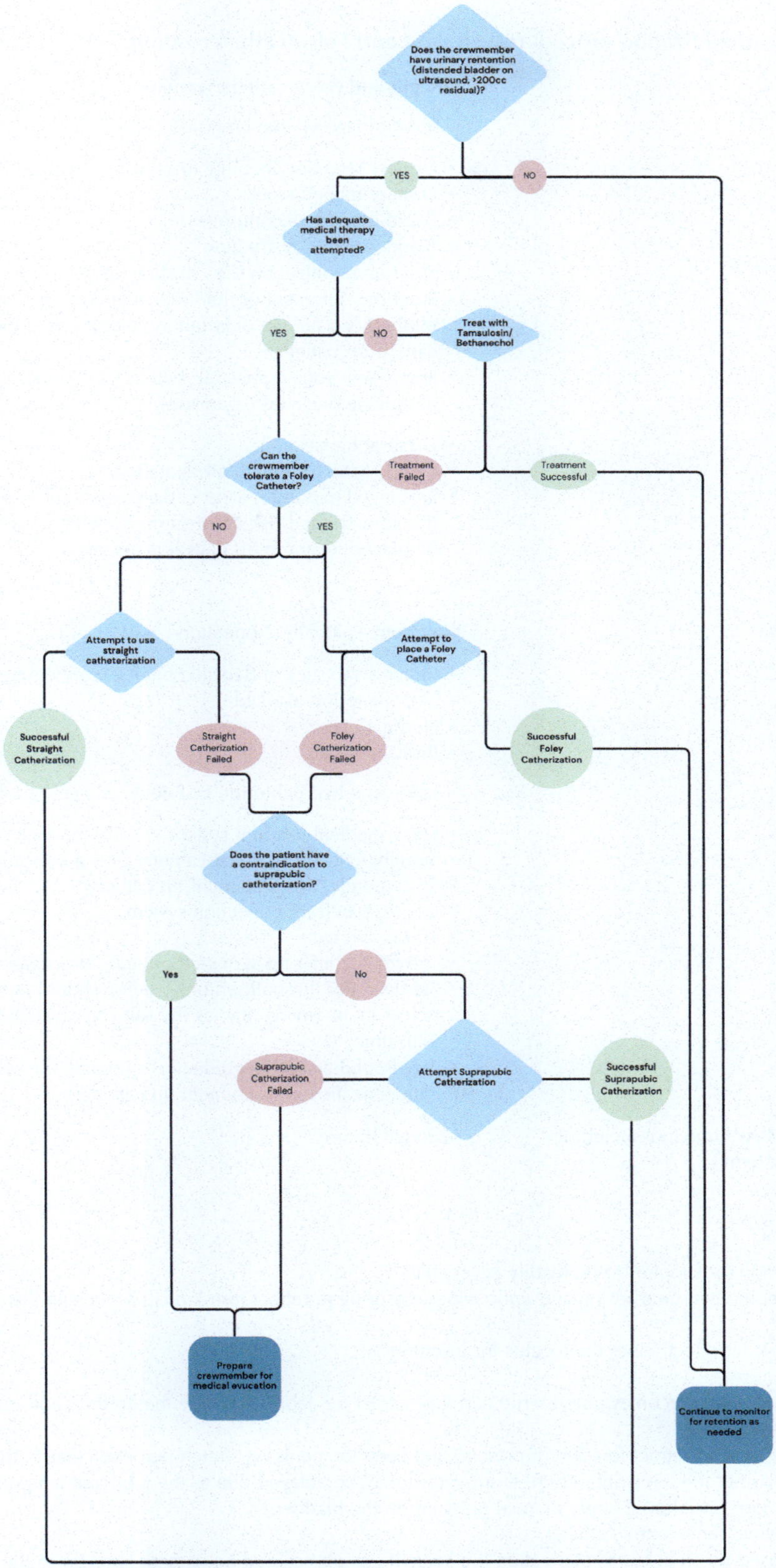


**FIGURE 14.2** Decision flow chart for the described types of catheterization.

**TABLE 14.4 Urinary catheterization procedures: straight and Foley catheterization.**

| **Procedure** | **Straight and Foley catheterization** |
|---|---|
| Purpose | Urine drainage via the urethra |
| Indications | • Urinary retention/inability to void<br>• Urinary incontinence<br>• Rapid bladder decompression<br>• Bladder outlet obstruction<br>• Strict immobilization (i.e., postfracture)<br>• Strict measurement of fluid inputs and outputs (i.e., sepsis, acute kidney injury, heart failure, liver failure, electrolyte imbalances, postoperative care, burn patients)<br>• Emphysematous cystitis (air within the bladder lumen or wall from gas-producing bacterial organisms) |
| Contraindications | • Urethral transection<br>• Urethral stricture (without dilatation)<br>• Prolonged exposure. Most catheters should only be placed for 2–4 days due to increased risk of infection (Cole et al., 2016). However, they can be placed for up to 30 days if necessary. |
| **Supplies** | |
| Operators | Minimum 1, Ideally 2 operators |
| Antiseptic | • Iodine swab stick or Benzalkonium Chloride Antiseptic Moist Towelette × 2 (if only option available)<br>• Sterile gloves<br>• Biohazard bag |
| Analgesia | • Lidocaine hydrochloride USP jelly, 2% gel or lubricant |
| Materials | • Crew medical restraint system<br>• Straight catheter or Foley catheter plus a collection system<br>• 16 Fr coude, 16 Fr councill tip catheter<br>• Leg strap/belt for collection system<br>• 10 mL syringe<br>• *60 mL catheter-tip syringe (optional, for irrigation)<br>• Sterile water (for balloon only; saline should not be used for Foley balloon unless in an emergency, as the salt can crystallize and rupture the balloon with time)<br>• *Hydrophilic coating guidewire (optional, for use during difficult catheterization with councill tip catheter) |
| Estimated equipment mass/volume (not including crew medical restraint system) | ~200 g/830 $cm^3$ |
| **Steps** | |
| Straight catheterization<br>**1.** Prepare a sterile field/tray for supplies to be readily accessible.<br>**2.** Clean the urethral opening two times using antiseptic wipes, ensuring to retract foreskin/labia to clean the area thoroughly.<br>**3.** Don sterile gloves.<br>**4.** Inject 5–10 mL of lubricant directly into the meatus for insertion.<br>**5.** Catheter insertion<br>**a.** For females: use one hand (becomes a nonsterile hand) to retract the labia, visualize the urethra, and insert the catheter until return of urine is visualized.<br>**b.** For males: use one hand to retract foreskin (if present) and keep the penis on stretch perpendicular to the body. Insert the catheter all the way until there is urine return. Urine return may be delayed due to the lubricant temporarily obstructing the catheter, in which case one should apply manual pressure on the bladder.<br>**6.** When urine stops flowing, remove the catheter.<br>*Foley catheters can be used for straight catheterization if a straight catheter cannot be passed. Sometimes the straight catheters are too rigid to pass initially, this occurs more commonly in males.Foley catheterization: | |

*(Continued)*

**TABLE 14.4 (Continued)**

| **Procedure** | **Straight and Foley catheterization** |
|---|---|
| 1. Prepare a sterile field/tray for supplies to be readily accessible.<br>2. Test Foley catheter balloon.<br>a. Fill a 10 mL syringe with sterile water, attach to the balloon port, and inflate the balloon to check for leaks.<br>b. Deflate before insertion, leaving the syringe attached.<br>3. Attach Foley catheter to a urinary collection bag.<br>4. Position and properly restrain the crewmember.<br>5. Clean the urethral opening two times using antiseptic wipes, ensuring to retract the foreskin/labia to clean the area thoroughly.<br>6. Don sterile gloves.<br>7. Inject 5–10 mL of lubricant directly into the meatus for insertion.<br>8. Catheter insertion.<br>a. For females: use one hand (becomes a nonsterile hand) to retract the labia, visualize the urethra, and insert the catheter until the return of urine is visualized, then insert the Foley a few more centimeters and inflate the balloon using a 10 mL syringe.<br>b. For males: use one hand to retract the foreskin and keep the penis on stretch perpendicular to the body. Insert the catheter all the way to the branching point of the catheter and inflate the balloon using a 10 mL syringe. Urine return is ideal but may be delayed due to the lubricant temporarily obstructing the catheter, in which case put manual pressure on the bladder.<br>9. Detach and dispose of the syringe.<br>10. Retract Foley until resistance is felt (i.e., balloon is flush against the bladder neck).<br>11. Secure urine collection system to crewmember's leg.<br>12. Dispose of used materials in biohazard bag.<br>13. To remove the Foley catheter, use a 10 mL syringe, deflate the balloon, and gently withdraw the catheter. | |
| **Postprocedural care** | |
| • Analgesia as needed (NSAIDS or opioids if refractory)<br>• Straight: No other dedicated postprocedural care required<br>• Foley: Clean the urethral opening 1–2× per day<br>• Use frequent lubrication to avoid discomfort<br>• Monitor urine output to ensure Foley is working<br>• Frequently empty urine collection bag to prevent backflow/infection and monitor for signs of infection | |
| **Potential complications and solutions** | |
| Complication | Solution |
| Foley catheter does not inflate/breaks | Reattempt with a new catheter. Use straight catheterization as needed. |
| Urosepsis | Maintain strict sterile protocol (hand-washing, maintain sterile field, minimize urinary backflow). Exchange Foley catheter if active UTI or sepsis is present due to concern for biofilm accumulation. Treat with empiric antibiotics. |
| Foley catheter does not drain | Irrigate (flush) the catheter with 50 mL of normal saline or sterile water using a 10 mL syringe (or 60 mL catheter-tip syringe, if available). |
| Catheter will not pass | If a difficult catheterization secondary to urethral stricture or false passage is suspected, one option is to pass a moistened hydrophilic coated guidewire into the urethra and into the bladder. If resistance is encountered, use a back-and-forth motion as if "picking a lock," until resistance is able to be bypassed with gentle forward advancement. If no resistance is felt after >20 cm of guidewire is inserted then the wire is likely in the bladder. Then, pass a 16 Fr councill tip over the wire while keeping tension on the wire with finger tips. Inflate the balloon with 10 mL. **Caution: the guidewire will be extremely slippery, so use a strong grip or the tip of one's fingernails to pinch the wire in place.* |
| Blood at urethral meatus | • If blood per meatus is noted, a traumatic catheterization may have occurred. The treatment for this complication is to insert and leave a catheter in the proper position, as it will tamponade the bleeding.<br>• If resistance is felt at the distal end of the urethra, it is possible there is a urethral or fossa navicularis stricture. For a fossa navicularis stricture, push the 10 mL syringe tip hard into the tip of the penis to break through the stricture, you will feel a pop and the catheter should be able to be inserted. |

(*Continued*)

**TABLE 14.4** (Continued)

| **Procedure** | **Straight and Foley catheterization** |
|---|---|
| | • If resistance is felt at the mid/bulbar urethra, a false passage may have been created. The solution is to switch to a 16 Fr coude catheter and attempt insertion. If resistance is still felt, rotate the catheter 90 degrees and attempt again gently. If still unable to pass, rotate the catheter until a full 180 degrees from the initial position and attempt again gently. The other possibility is that the patient has an enlarged prostate, and therefore a coude catheter will increase the chance of successful catheterization. |

**TABLE 14.5** **Urinary catheterization procedures: suprapubic catheter placement.**

| **Procedure** | **Suprapubic catheter placement** |
|---|---|
| Purpose | Drain urine from the bladder when urethral drainage is not possible |
| Indications | • In the spaceflight context, a suprapubic catheter should only be attempted if urethral catheterization is unsuccessful.<br>• Unsuccessful urethral catheterization could be due to an enlarged prostate, the creation of a false urethral passage, or a urethral stricture. |
| Contraindications | • Suspected bowel in the path of insertion<br>• Coagulopathy or bleeding disorders (relative contraindication)<br>• Presence of prosthetic devices in the lower abdomen (relative contraindication) |
| **Supplies** | |
| Operators | Minimum 1, Ideally 2 operators |
| Antiseptic | • Iodine swab stick, BZK Antiseptic Moist Towelette × 2, or Chlorhexidine prep stick |
| Analgesia | • Local anesthetic<br>• Lidocaine 1% or 2% OR bupivacaine 0.25% or 0.5% |
| Materials | • Crew medical restraint system<br>• Razor and shaving cream<br>• Sterile water<br>• Small cloth<br>• Antiseptic (i.e., Iodine swab stick/BZK wipe) × 2<br>• Lidocaine 1% or 2% OR bupivacaine 0.25% or 0.5%<br>• 18G, 22G needle<br>• 10 mL syringe × 2<br>• Sterile drape and gloves<br>• Portable ultrasound machine (optional)<br>• Ultrasound gel (sterile)<br>• Ultrasound probe cover (sterile)<br>• 0.035-inch guidewire<br>• Amplatz dilators<br>• Gauze dressing and tape<br>• Balloon catheter<br>• Urine collection system<br>• Leg strap/belt for collection system<br>• Biohazard bag |
| Estimated equipment mass/volume (not including crew medical restraint system or ultrasound machine) | ~730 g/3900 $cm^3$ |

(Continued)

**TABLE 14.5 (Continued)**

| **Procedure** | **Suprapubic catheter placement** |
|---|---|
| **Steps** | |
| **1.** Prepare sterile field/tray for supplies to be readily accessible.<br>**2.** Position the patient<br>**a.** Place and restrain the patient supine with the anterior lower abdomen exposed.<br>**b.** Ensure the bladder is sufficiently distended either by palpation or with ultrasound.<br>**3.** Shave the patient<br>**a.** Prepare the suprapubic area by first wetting the area with water, applying shaving cream, and then shaving a small area in the midline, 2–3 cm (or two finger breadths) above the pubic symphysis.<br>**b.** Clean with cloth and store appropriately.<br>**4.** Prepare the field<br>**a.** Cleanse the field using an antiseptic solution, with the boundaries being superiorly to the level of the umbilicus, inferiorly to the mid-pubic area, and laterally to the anterior superior iliac spine.<br>**b.** Wash hands and don sterile gloves.<br>**c.** Drape the area to maintain sterility.<br>**5.** Ultrasound<br>**a.** Place a small amount of sterile gel within the ultrasound probe cover and/or on the skin.<br>**b.** Have an assistant cover the ultrasound probe with sterile cover.<br>**c.** Localize distended bladder with ultrasound, ensuring no intervening structures such as bowel are anterior to the bladder, then plan approach.<br>**6.** Local anesthesia<br>**a.** Draw up local anesthetic in a 10 mL syringe with an 18G needle, then recap and replace with a 22G needle.<br>**b.** Insert the needle and aspirate to check that a blood vessel was not entered.<br>**c.** Raise a bleb with 1–2 mL of local anesthetic under the entry area. Direct the needle into the subcutaneous tissue and fascia of the insertion site and infiltrate with local anesthetic.<br>**7.** Test catheter balloon<br>**a.** Fill a 10 mL syringe with fluid, attach it to catheter balloon port, and inflate balloon to check for leaks.<br>**b.** Deflate balloon before insertion, leaving the syringe attached.<br>**8.** Insertion<br>**a.** Palpate the bladder or confirm its position using ultrasound guidance.<br>**b.** Insert an 18G needle into the distended bladder under ultrasound guidance.<br>**c.** Attach a 10 mL syringe and attempt to aspirate urine to ensure proper placement.<br>**d.** Feed the guidewire through the needle, and then retract the needle from the skin using the Seldinger technique.<br>**e.** Sequentially dilate the tract to the size of the catheter using the dilators up to 14–16 Fr.<br>**f.** Insert councill tip catheter (or a councillized catheter) over a guidewire until urine appears in the catheter.<br>**g.** To councillize a regular Foley catheter, push a 18G needle through the tip of the catheter into the lumen. Alternatively, use a pair of scissors to cut the tip off the regular catheter until a lumen is visible and can fit a guidewire. **Cutting the tip of the catheter can sometimes violate the balloon channel, so it is important to confirm the balloon has no leaks.*<br>**h.** Remove the guidewire.<br>**i.** Inflate the balloon using 10 mL syringe and detach and dispose of the syringe.<br>**j.** Retract Foley until resistance is felt (i.e., balloon is flush against bladder neck)<br>**9.** Attach the urine drainage bag.<br>**10.** Dispose of used materials in a biohazard bag and sharps in a sharps container. | |
| **Postprocedure care** | |
| • Monitor for complications<br>• Check for bleeding, infection, or signs of improper placement<br>• Ensure urine flows freely into the drainage bag<br>• Check for normal bowel sounds and bowel movements<br>• Wash the area around the catheter daily with a cleansing wipe<br>• The catheter will need to be replaced every 30 days | |
| Potential complications and solutions | |
| Complication | Solution |
| Bleeding | Apply gauze dressings and pressure to the area of bleeding. |
| Bowel injury | Remove the suprapubic catheter and monitor for hemodynamic instability and signs of sepsis. |

*(Continued)*

**TABLE 14.5 (Continued)**

| **Procedure** | **Suprapubic catheter placement** |
|---|---|
| Catheter does not drain | Irrigate the catheter with 50 mL of normal saline or sterile water using a 60 mL catheter-tip syringe (use a 10 mL syringe if not available) |
| Postobstructive diuresis (urine output greater than 200 mL for >2 hours) | For mild postobstructive diuresis, oral hydration is usually sufficient. For moderate to severe cases, IV fluids will be needed to replace losses and maintain electrolyte balance. If oral intake is insufficient or urine output is excessive, intravenous replacement with 0.9% or 0.45% saline at roughly half the rate of ongoing losses (approximately 0.5 mL IV for each 1 mL of urine output) can be considered. Adjust fluid replacement hourly based on ongoing urine output. |
| Infection | Treat with antibiotics, with selection depending on if the infection is related to the surgical site or urinary tract. |

**TABLE 14.6 Urinary catheter removal procedure.**

| **Procedure** | **Catheter removal** |
|---|---|
| Purpose | To remove catheter to replace it or discontinue use |
| Indications | • Routine replacement<br>• Suspected infection (Urosepsis) |
| Contraindications | • Difficult/traumatic catheter placement with continued need for catheterization |
| **Supplies** | |
| Operators | 1 |
| Antiseptic | • Antiseptic/alcohol wipes (optional) |
| Analgesia | • None |
| Materials | • Biohazard bag |
| Estimated equipment mass/volume | ~5 g/50 $cm^3$ |
| **Steps** | |
| **1.** Position and properly restrain the patient crewmember.<br>**2.** Attach a 10 mL syringe to the balloon port and aspirate the full volume<br>**3.** Gently retract the catheter. It should come easily.<br>**4.** Dispose of used materials properly. | |
| **Postprocedure care** | |
| • Continue to monitor for recurrence of urinary retention. | |
| **Potential complications and solutions** | |
| Complication | Solution |
| Catheter balloon will not deflate | Several options are available:<br>• Inject mineral oil into the balloon port to dissolve the balloon<br>• Cut the catheter balloon port, resulting in spontaneous drainage of balloon fluid. Alternatively, can cut the foley catheter; however, make sure there is adequate tubing extending from the urethra so the catheter does not drift back into the urethra.<br>• If cutting the balloon port does not work, gently push a guidewire through the balloon port to pop the balloon or remove the obstruction.<br>If the balloon will not deflate, confirm with ultrasound if available. Consider medical evacuation as the foley catheter can increase risk of infection, especially after 30 days. |
| Catheter breaks | Replace catheter using procedures detailed above. |

## 14.5 Summary, recommendations, and considerations for the future

Urinary catheterization in space should be considered a last resort, reserved for situations in which nonprocedural strategies have failed or are inappropriate. The most common indications include AUR, BOO, trauma requiring diversion, and the need for precise urine output monitoring in critical illness. On Earth, catheterization may be routine; in space, it is a high-stakes procedure that consumes limited resources and carries elevated risk.

Whenever possible, medical management should be attempted first. Alpha-blockers can help relieve obstruction from an enlarged prostate, anticholinergics may help control bladder spasms, and adjustments in hydration or other modifiable factors may prevent retention altogether. When these measures are insufficient, the choice between straight, Foley, or suprapubic catheterization must be guided by the clinical scenario, with careful attention to balancing efficacy and safety. Clean intermittent catheterization offers a promising alternative to indwelling catheters, reducing infection risk and improving crew independence, though it requires training and protocols for safe reuse in resource-limited settings.

Long-duration exploration missions will amplify the importance of conserving supplies and preventing infection. Preventing backflow into drainage bags is a key priority, as it increases the risk of ascending infection. Lightweight, easy-to-use one-way valves should be developed, and complementary technologies such as ultraviolet sterilization or compact sanitation devices could extend catheter reusability without compromising safety. Additional research is also needed to better understand the drivers of UTI and urosepsis in confined environments, including the impact of antimicrobial resistance. Waste containment must be engineered carefully to prevent droplet spread and contamination in microgravity.

As missions progress toward planetary surfaces and artificial gravity habitats, catheterization may become more manageable. Gravity-assisted drainage will simplify waste control, reduce the risk of backflow, and lessen the technical burden on providers. In such environments, reusable catheter systems, onboard sterilization, and closed-loop fluid recycling can be more easily integrated into daily operations. Until then, the best safeguard is prevention through careful crew selection, comprehensive preflight screening, and proactive management of urinary risk factors, so that the need for catheterization in space can be avoided whenever possible.

## 14.6 Procedural management of nephrolithiasis in space

### 14.6.1 Introduction

Nephrolithiasis is a common condition on Earth, affecting up to 10%–12% of men and 5%–7% of women during their lifetime, with recurrence rates approaching 50% within 10 years (Scales et al., 2012). While most kidney stones can be managed terrestrially with imaging, analgesia, and a range of definitive procedures such as extracorporeal shockwave lithotripsy (ESWL), ureteroscopy with laser lithotripsy, percutaneous nephrolithotomy (PCNL), or pyelolithotomy, such options are not feasible in space due to the size, weight, and power requirements of the equipment. In orbit, management thus far has primarily centered on prevention, hydration, and medical therapy such as pain medication, antiemetics, and expulsive agents.

Microgravity itself may contribute to stone risk. Increased bone resorption elevates urinary calcium levels, while fluid shifts and reduced intake lower urine volume. Dietary and environmental factors may compound these effects, producing a urinary environment favorable to stone formation. To date, there has been one suspected in-flight renal stone in a Russian cosmonaut and 43 terrestrial stone events documented in 29 U.S. astronauts since 2016 ("Urinary health (OCHMO-MTB-003)," 2025). For mission planners and flight surgeons, the primary concern is not small, asymptomatic calculi, but the possibility of an obstructing renal or ureteral stone leading to hydronephrosis and infection. In such scenarios, PCN remains the only feasible and potentially life-saving intervention until definitive treatment can be achieved on return to Earth.

### 14.6.2 Does it work?

On Earth, ultrasound-guided PCN is a well-established method for decompressing an obstructed collecting system due to a renal or ureteral stone. The purpose of the procedure is to divert urine from the renal pelvis into an external drainage system, thereby relieving pressure, preserving renal function, and reducing the risk of infection. The terrestrial success rate of PCN tube placement ranges from 84%–99% (Pabon-Ramos et al., 2016).

PCN may be required when obstruction is accompanied by infection or sepsis, when renal function is threatened in the setting of a solitary kidney (likely identified preflight) or rising creatinine, or when pain and oral intake cannot be controlled with medications. On Earth, ureteral stenting or endoscopic approaches would be alternatives, but these are not currently feasible in space given the equipment and disposables required. In the flight environment, PCN remains the only realistic option for definitive decompression once conservative care fails.

Microgravity introduces unique barriers. Drainage will occur, but the absence of gravity increases the risk of backflow from the collection bag, potentially leading to contamination, infection, or overdistension unless anti-reflux valves or suction systems are incorporated. Limited sterile supplies, constrained resources, and the steep learning curve for both ultrasound and procedural technique further complicate implementation.

If PCN cannot be performed, the consequences are significant. Persistent obstruction may result in UTI, sepsis, and irreversible renal injury, all of which can jeopardize crew health and mission success. In such cases, the only fallback is medical evacuation, which may not be possible on long-duration exploration missions.

## 14.7 Procedural considerations

### 14.7.1 Injury risk and incidence

NASA's 2024 *Renal Stone Risk Evaluation Evidence Report* provides the most detailed assessment of stone risk in astronauts to date. To quantify operational risk, NASA developed the *Informing Mission Planning via Analysis of Complex Tradespaces* (IMPACT; now renamed IMM: Exploration) model, an evolution of the earlier Integrated Medical Model (IMM) that was limited to LEO missions. Unlike IMM, IMPACT was specifically designed to account for the unique medical uncertainties of deep-space missions, integrating terrestrial, astronaut, and analog data to model complex outcomes. Using this tool, the mean incidence of nephrolithiasis in astronauts was estimated at 0.00396 events per person-year, approximately four cases per 1000 person-years, higher than in the U.S. general population (Stratton and Lumpkins, 2024). Mission modeling further showed that an untreated obstructing stone carried a probability of impairing 26.8% of mission tasks and a 100% probability of requiring return to definitive care in severe cases. Loss of crew life was modeled at 0%, though this does not account for complications such as sepsis or renal failure.

### 14.7.2 Spaceflight specific considerations

Access to basic laboratory testing may be available depending on the mission; however, given the limited nature of these resources, it is critical to consider the purpose of each test in diagnostic management. For example, in a clinically febrile crewmember with costovertebral angle tenderness and symptoms suggestive of UTI, it should be considered whether obtaining a white blood cell count or urinalysis would meaningfully alter clinical decision-making.

#### *14.7.2.1 Diagnostic laboratory testing for nephrolithiasis*

In spaceflight, laboratory testing cannot be assumed to be routine. Every blood test requires consumables such as needles, vacutainers, reagents, and crew time to collect and process samples. The decision to order a test must therefore be justified by whether the result will meaningfully alter management. For example, leukocytosis is a standard terrestrial marker of infection, but in orbit clinical signs and urinalysis may provide sufficient evidence to begin empiric antibiotics and consider decompression. Likewise, serum creatinine is unlikely to rise appreciably in unilateral obstruction, since the contralateral kidney maintains clearance. Meaningful changes would be expected only in cases of bilateral obstruction or underlying renal disease, both of which would generally be known prior to flight through pre-mission screening.

For reference, in terrestrial practice, laboratory testing is an essential part of nephrolithiasis management. The American Urological Association recommends urinalysis for all patients, with urine culture obtained when infection is suspected based on clinical features or laboratory evidence (Assimos et al., 2016). A basic metabolic panel, including serum electrolytes and creatinine, is also valuable for establishing baseline renal function and monitoring during surveillance. However, this is unlikely to be necessary for an uncomplicated UTI in space if there is sufficient clinical evidence.

Preprocedural laboratory evaluation is particularly important in patients undergoing invasive interventions such as nephrostomy tube placement. According to the Society of Interventional Radiology (SIR) consensus guidelines, nephrostomy tube exchange is considered a low-risk procedure for bleeding, whereas initial tube placement is classified as high-risk. For low-risk procedures, routine coagulation testing is not required in patients without bleeding risk factors, but may be appropriate in those on anticoagulants or with conditions that increase bleeding tendency. When included, suggested laboratory thresholds include an International Normalized Ratio (INR) less than 2.0–3.0, with platelet transfusion considered if platelet count is below $20 \times 10^9$/L.

In contrast, high-risk procedures such as initial nephrostomy tube placement warrant preprocedural coagulation testing. For these, correction of INR to 1.5–1.8 or less is recommended, and platelet transfusion should be considered when platelet count falls below 50 × $10^9$/L (Patel et al., 2019). Fortunately, severe thrombocytopenia or a coagulopathy is unlikely to be present for most cases where PCN is necessary.

Given these constraints, urinalysis should be emphasized as the primary and most informative study. It is relatively simple to perform and provides actionable data on hematuria, pyuria, bacteriuria, and urinary pH. The sensitivity and specificity of isolated blood/RBCs on a urinalysis alone for nephrolithiasis is 77% and 33%, respectively (Alnazari et al., 2024). When urinalysis suggests infection in the presence of obstruction, empiric antibiotics and decompression are indicated without additional laboratory confirmation. Urine culture is recommended by AUA guidelines, but it is not currently feasible in orbit; whether to incorporate culture capability into future missions will depend on careful mass, volume, and frequency-of-use tradeoffs. Complete blood count and serum electrolytes should be reserved for scenarios where results directly influence decision-making, such as suspected sepsis with hemodynamic instability, unexplained anemia, or prior to PCN placement when bleeding risk is a concern.

In summary, while terrestrial guidelines provide a broad framework for laboratory evaluation, spaceflight demands a streamlined, decision-driven approach. Urinalysis should be prioritized as the most resource-efficient and clinically informative test, with other studies used selectively when their impact on management is clear.

### 14.7.2.2 Diagnostic imaging for nephrolithiasis

On Earth, noncontrast computed tomography (CT) is the gold standard for ureteral stone detection, with sensitivities and specificities approaching 97% (Fulgham et al., 2013). In spaceflight, CT is currently unavailable, and imaging is constrained to portable modalities. Ultrasound is currently the only routinely available option: it is compact, radiation-free, and astronauts can be trained in focused renal imaging with teleguidance. Its greatest strength is detecting hydronephrosis, which is present in ~83% of cases, though hydronephrosis alone does not predict the need for intervention. Direct stone visualization is possible depending on stone location, but stone sizing is often imprecise, and the AUA surgical guidelines caution that ultrasound is not reliable for accurate size assessment. Per the panel that drafted the "Clinical Effectiveness Protocols for Imaging in The Management of Ureteral Calculous Disease: AUA Technology Assessment," the median reported sensitivity and specificity for ultrasound compared to CT are 61% and 97%, respectively (Fulgham et al., 2013).

If available, plain abdominal radiography (Kidney, Ureters, Bladder or KUB) can complement ultrasound, particularly for radiopaque stones. Combining KUB with ultrasound improves detection (reported sensitivity 58%–100%, specificity 37%–100%), and performance may be optimized by fasting to reduce bowel gas or adding oblique films to avoid obscured stones (Fulgham et al., 2013). Recently, NASA has begun evaluating miniature X-ray systems such as the Fujifilm Xair (generator mass of 3.5 kg), MinXray Impact (8.5 kg), and Remedi REMEX-KA6 (2.4 kg). Expert consensus modeling found that X-ray outperformed ultrasound for 21 conditions, was equivalent for nephrolithiasis, and could serve as a high-value adjunct in mission medical kits. X-ray has been successfully performed in microgravity as of 2025 on the Fram2 mission.

In many cases, a ureteral stone will not be directly visualized; therefore, management algorithms in space should frame stone-dependent decisions as if the stone were identified. When stone size cannot be confirmed, evacuation thresholds will likely need to be lower, prioritizing crew safety. For now, ultrasound, augmented by radiograph if equipment is flown, remains the most practical diagnostic approach.

### 14.7.2.3 Analgesia for nephrolithiasis

Pain control is central to renal stone management and carries special considerations in spaceflight. On Earth, NSAIDs such as ketorolac are considered first-line therapy for renal colic. Randomized trials have shown that NSAIDs reduce pain scores by approximately 50% within one hour of administration (Holdgate and Pollock, 2005). Their mechanism includes inhibition of prostaglandin-mediated ureteral smooth muscle spasm, which is particularly effective for colic. However, concerns about impaired renal perfusion and gastrointestinal side effects necessitate caution for prolonged use during missions. Intramuscular ketorolac provides rapid analgesia, whereas oral agents such as naproxen may be preferred for sustained pain control when immediate relief is not required. Acetaminophen remains a safe adjunct for mild to moderate pain, while opioids are reserved for refractory cases.

Opioids remain an option for renal colic that does not respond to NSAIDs and acetaminophen, but their use in space must be highly judicious. NASA formulary records confirm that agents such as morphine and meperidine have been carried and administered during missions, though infrequently compared with oral or intramuscular medications (Putcha et al., 1999). The primary advantage of opioids in orbit is their rapid and reliable analgesia, which may be necessary to restore a crewmember's function during a stone event. However, the risks are amplified in microgravity. Sedation and cognitive slowing can impair mission performance, opioid-induced constipation compounds gastrointestinal changes already present in-flight, and respiratory depression is particularly dangerous where advanced monitoring and ventilatory support are limited. Intravenous administration also requires line placement, additional consumables, and crew time. For these reasons, when opioids are used, the preference is for minimal effective dosing, ideally by oral or intramuscular routes, with careful observation of the crewmember's alertness, respiratory status, and ability to continue operational duties.

#### *14.7.2.4 Antiemetics for nephrolithiasis*

On the ISS, the primary antiemetic options are ondansetron, meclizine, and promethazine (Diaz et al., 2024). Ondansetron, a serotonin (5-$HT_3$) antagonist, is highly effective for acute nausea, carries minimal sedation risk, and has a low incidence of urinary retention, making it the most suitable first-line agent for nephrolithiasis in space. Meclizine and promethazine can also control nausea and motion sickness, but both have anticholinergic activity that may theoretically increase the risk of urinary retention and subsequent UTI. Given these considerations, ondansetron should be the preferred antiemetic for nephrolithiasis in spaceflight.

#### *14.7.2.5 Location of the stone*

Stone location strongly influences both urgency and treatment options. On Earth, surgical approaches differ substantially for renal, ureteral, and bladder stones, but in spaceflight, where definitive modalities such as ureteroscopy, ESWL, and PCNL are not available, management simplifies to determining whether a stone is obstructing, infected, and/or causing functional compromise.

Nonobstructing renal stones are usually asymptomatic and rarely require intervention. In both terrestrial and spaceflight practice, they can be managed conservatively with hydration, analgesia, and preventive pharmacologic measures. The AUA guidelines note that observation is appropriate unless stones are symptomatic, enlarging, or associated with infection (Assimos et al., 2016).

When obstruction arises at the level of a calyx or infundibulum, stones are far more difficult to treat. On Earth, ureteroscopy, PCNL, or open/robotic pyelolithotomy may be required, as spontaneous passage is unlikely. In space, however, targeted calyceal access would be prohibitively complex for nonspecialist providers. If such stones cause obstruction and infection, the recommended course is medical stabilization with antibiotics and, when possible, evacuation for definitive management.

Ureteral stones are the most clinically significant inflight, as they are the most likely to cause acute obstruction, hydronephrosis, and infection. Spontaneous passage is more common with distal stones, particularly when ≤10 mm, and α-blockers such as tamsulosin have been shown to improve expulsion rates for distal ureteral stones (Hollingsworth et al., 2006; Assimos et al., 2016). In the presence of severe, uncontrolled pain, refractory vomiting, fever, hemodynamic instability, or UTI, decompression is mandatory. On Earth, either ureteral stenting or PCN may be performed, with equivalent outcomes (Pearle et al., 1998). Since ureteral stenting is not feasible in space due to required technical expertise, specialized equipment, and weight limitations, ultrasound-guided PCN is the only available intervention to relieve obstruction and prevent urosepsis and must be weighed against the feasibility of evacuation in medically meaningful timeframes.

Bladder stones are less common but should also be considered. In terrestrial settings, these are often associated with urinary retention, bladder outlet obstruction, or recurrent UTI. In spaceflight, bladder stones generally do not require intervention unless they cause significant retention or recurrent infections, in which case catheter drainage would be the primary management option until definitive treatment can occur after return to Earth (Assimos et al., 2016).

#### *14.7.2.6 Size of the stone*

Stone size is one of the strongest predictors of whether a stone will pass spontaneously or require intervention. The AUA guidelines reflect this, recommending observation and medical expulsive therapy (MET) for stable patients with ureteral stones ≤10 mm, while larger stones typically require procedural management such as ureteroscopy or PCNL (Assimos et al., 2016). For spaceflight, these thresholds carry special weight: stones <10 mm in the ureter may be managed conservatively if the astronaut remains stable and free of infection, but stones ≥10 mm—particularly if lodged in the renal pelvis or calyces—pose a significant risk of obstruction, sepsis, and impaired mission performance, making them indications for early evacuation given the limited availability of definitive stone procedures in orbit. Bladder stones, in contrast, are less immediately dangerous and may often be tolerated during flight, but if they cause urinary retention or recurrent infections, they too may necessitate intervention. However, the decision to treat conservatively should be made in consultation with the flight surgeon, as a lower threshold to treat may be reasonable depending on the mission duration, medical capabilities, and expected time of return.

#### 14.7.2.7 Medical expulsive therapy in spaceflight

Medical expulsive therapy refers to the use of medications to facilitate the spontaneous passage of ureteral stones. The most commonly used agents are α-adrenergic blockers (such as Tamsulosin), which work by selectively blocking α-1 receptors in the ureteral smooth muscle, thereby promoting stone passage. Evidence from randomized trials and meta-analyses, cited in the AUA guidelines, demonstrates that α-blockers significantly improve expulsion rates for distal ureteral stones >5 mm but offer little additional benefit for stones ≤5 mm, which already have high spontaneous passage rates (Assimos et al., 2016; Furyk et al., 2016) The AUA Panel further noted that most MET trials enrolled patients with distal ureteral stones, while studies including mid and proximal stones showed no clear benefit. As a result, the panel did not recommend MET for stones in these locations (Assimos et al., 2016).

In spaceflight, operational risks must be weighed against benefits. Tamsulosin carries side effects particularly relevant to astronauts, including orthostatic changes (objective in 15%–19%, symptomatic in less than 1%), dizziness (~15%), headache (~20%), and nausea (~3%). These effects may amplify postflight orthostatic intolerance, especially landings where crew may need to self-evacuate. Given Tamsulosin's approximately 12-hour half-life, discontinuing therapy 48–72 hours before landing or evacuation is recommended to minimize postflight hemodynamic risk (Flomax (tamsulosin hydrochloride) prescribing information, 2025). In summary, for the spaceflight environment, MET with α-adrenergic blockers should be reserved for symptomatic distal ureteral stones >5 mm managed conservatively on-orbit, withheld when evacuation is imminent, and avoided in smaller or more proximal stones where risks outweigh uncertain benefits.

Ongoing surveillance is also critical. Natural history studies have shown that ~83% of ureteral stones pass spontaneously, with 95% of those that pass doing so within 6 weeks (Miller and Kane, 1999). The AUA Panel recommends a maximum 6-week trial of MET, based on experimental data showing that complete unilateral obstruction beyond this interval can result in irreversible renal damage (Vaughan and Gillenwater, 1971; Assimos et al., 2016). While not all ureteral stones cause complete obstruction, adopting a 6-week threshold balances safety with the likelihood of spontaneous passage if the patient remains stable. During this surveillance period, consider obtaining a baseline serum creatinine (if available) and then every 1–2 weeks to monitor for evolving obstruction if medical supplies permit. Additionally, repeat ultrasound can be considered every 1–2 weeks to assess for hydronephrosis, or sooner if symptoms worsen. Throughout the episode, all urine should be strained to confirm stone passage, which allows cessation of surveillance imaging and medications. In a spacecraft environment, this could be accomplished by placing a filter or strainer between the urethra if the waste management system can allow for this. Fortunately, since all spaceflight waste management systems operate in vacuum, this capability may still be possible if planned for during design. The stone should be kept in storage and sent for stone analysis once back on Earth.

#### 14.7.2.8 Anesthesia considerations for nephrostomy tube placement

Procedural anesthesia for PCN is more challenging in space. Local infiltration with a local anesthetic such as lidocaine is the foundation of analgesia, but procedural sedation is frequently required. Ketamine is generally the preferred agent in austere or microgravity environments because it preserves airway reflexes and maintains cardiovascular stability. By contrast, propofol is an effective agent but is associated with increased risks of hypotension and respiratory depression in settings lacking advanced critical care support (Barr et al., 2013). Adjunctive antiemetics such as ondansetron are also important, as nausea and vomiting can be exacerbated in microgravity. For astronauts undergoing nephrostomy, the anesthesia protocol should therefore emphasize local anesthesia with cautious use of ketamine-based sedation, consistent with NASA's broader contingency guidance for procedural care in austere environments. 14.7.2.3 provides a more detailed discussion of analgesia in spaceflight for the interested reader.

#### 14.7.2.9 Urinary tract infection

Infection is a common complication of nephrostomy tube placement. Bacterial colonization of the tube is almost universal over time, and in a study by Radecka et al., UTIs were found in approximately 14% of patients following placement (Radecka and Magnusson, 2004). More severe infection is less common but important, with sepsis reported in 1%–10% of cases, particularly when decompressing an already infected system (Pabon-Ramos et al., 2016). The organisms most frequently implicated are gram-negative bacteria, including *Escherichia coli*, *Proteus* species, *Klebsiella* species, and *Enterococcus* species (Hausegger and Portugaller, 2006). Prophylaxis with a first-generation cephalosporin can be considered at the time of placement, while established infections are typically managed with combination regimens such as a cephalosporin plus gentamicin to provide broader gram-negative coverage. In the context of spaceflight, the risks are magnified by limited antibiotic availability, challenges in maintaining sterility, and altered drainage dynamics in microgravity that may promote urinary stasis and backflow. Therefore, the decision to place a PCN should be made with consideration of available antibiotics and mission duration.

### 14.7.2.10 Tube care and maintenance

Nephrostomy tubes require meticulous care to prevent obstruction, infection, and accidental dislodgement. The exit site should be covered with a sterile gauze pad or a transparent occlusive dressing. Dressings should be changed at least twice per week or sooner if they become wet or soiled, and the skin around the site should be gently cleaned with sterile saline or soap and water at each change. The nephrostomy tube must be secured to the skin with adhesive devices or sutures to minimize traction and reduce the risk of dislodgement.

When drainage decreases or obstruction is suspected, the tube can be flushed with 5–10 mL of sterile normal saline using a Luer-lock syringe. The recommended technique is to gently instill saline and then aspirate to check for debris or clots. Forceful flushing should be avoided to prevent renal trauma or overdistension. If flushing fails to restore patency, imaging, such as ultrasound or KUB radiography, should be performed to confirm the nephrostomy tube position and rule out kinking or dislodgement.

On Earth, routine exchange of nephrostomy tubes is performed every 6–12 weeks to reduce the risk of infection. In spaceflight, where exchange capability may not be available, prevention becomes paramount. Daily inspection for erythema, leakage, or decreased output, coupled with strict adherence to sterile dressing changes and judicious flushing, represents the cornerstone of long-term tube maintenance. These measures are critical to sustaining drainage and minimizing complications in a resource-limited environment. It is also crucial to consider catheter fixation during EVA/return versus removal to reduce the risk of dislodgement. This decision should be made in consultation with the flight surgeon, depending on the expected natural history of the cause of obstruction.

### 14.7.2.11 Ureteral stenting in spaceflight

The use of ureteral stents in spaceflight remains purely investigative. A stent removes the need for external tubing, which reduces the risk of accidental traction or dislodgement in a confined spacecraft. Despite this advantage, stents create new challenges. They are prone to encrustation in high-calcium urine and may require frequent exchanges in these settings. In pregnant patients where there is physiologic hypercalciuria, ureteral stents must be exchanged every four weeks (Assimos et al., 2016). Placement is also complex as it requires specialized training, careful patient positioning, a closed-loop fluid management system, and a considerable amount of equipment (including a flexible cystoscope, irrigation fluids, a light source, a monitor, and the stent itself). The additional mass and volume of this equipment make stenting impractical for current missions. Although stents have been placed with minimal sedation or even in awake patients in low-resource and emergency terrestrial settings, concerns about sterility, patient discomfort, and the need for fluoroscopic guidance remain barriers in spaceflight.

In the future, ureteral stents may become a more viable option as medical infrastructure expands on lunar or Martian surfaces. With access to technologies such as laser lithotripsy, they could provide an alternative to nephrostomy tubes for stone management. For now, however, ureteral stenting in space should be regarded as experimental and a subject for further research.

### 14.7.2.12 Considerations for evacuation

Placement of a nephrostomy tube in space represents a high-risk scenario that generally mandates evacuation when feasible. For LEO or lunar missions, nephrostomy placement should be considered an indication for immediate return to Earth given the significant risks of infection and encrustation. The same principles used for Foley catheters during medical evacuations apply: the tube and drainage system must be fully emptied and clamped before transit, the collection bag securely fastened to prevent leakage or traction, and all tubing stabilized to reduce the chance of dislodgement during high workload or high-G events.

For Mars or other exploration-class missions, evacuation is not possible, and sustained nephrostomy management will require the ability to perform routine tube exchanges, maintain sterile supplies, and deliver targeted antibiotics to manage recurrent infection. These requirements highlight the critical operational challenges that nephrostomy tubes present for long-duration exploration missions.

### 14.7.3 Complications of nephrostomy

Complications from PCN tubes are uncommon but important to recognize, especially in the spaceflight environment where resources are limited. One of the most frequent complications is bleeding. Minor hematuria is reported in the vast majority of patients and typically resolves within 48 hours, while perinephric hematomas are identified in roughly 13% of cases. Clinically significant bleeding that requires transfusion occurs in about 1%–4% of patients (Pabon-Ramos et al., 2016). On Earth, persistent arterial hemorrhage can be managed with selective embolization, but this is not feasible in space. Inflight management would therefore be limited to fluids, transfusion if blood products were available, observation, and prompt evacuation.

Infection is another well-described risk, especially in the setting of obstruction or a colonized collecting system. Reported rates of sepsis after nephrostomy range from 1%–10% depending on the patient population and definitions used (Pabon-Ramos et al., 2016). Prevention depends on strict sterile technique and peri-procedural antibiotics. If infection occurs despite these measures, treatment requires continued drainage, broad spectrum antimicrobial therapy, and aggressive supportive care, with evacuation considered if the patient deteriorates.

Nephrostomy tube-related problems, such as dislodgement, malposition, and urine leakage around the insertion site, are among the most frequent postprocedural complications. On Earth, patients may experience persistent urine drainage around the tube, which, if bothersome, is usually managed by upsizing the nephrostomy tube (Dablan et al., 2025). Risk factors include first-time placement, limited operator experience, and a shorter cortex-to-skin distance (Panach-Navarrete et al., 2020; Dablan et al., 2025). Dislodgement may present as persistent obstruction on ultrasound, flank pain, or leakage around the catheter. Management involves confirming tube position with imaging, such as ultrasound or KUB, when available, and then repositioning or exchanging the tube, ideally over a wire using the existing tract. Preventive strategies include secure fixation, regular inspection of the drainage system, and minimizing line tension, which is especially important in microgravity where tubing and collection bags may shift unpredictably.

## 14.8 Decision making

Nephrolithiasis presents a unique challenge in spaceflight, as stones can cause sudden pain, obstruction, or infection in an environment with limited diagnostic and treatment options. On Earth, definitive management can be achieved with procedures such as ESWL, laser lithotripsy, or PCNL, but these are not currently feasible in orbit due to equipment and resource constraints. In space, management must therefore rely on medical therapy, hydration, expulsive agents, and, when infection and obstruction coexist, PCN tube placement as the only available procedural option.

The decision to intervene for nephrolithiasis is based on the following factors:

1. Does the patient appear sick or septic?
   - **a.** Assess for fever, tachycardia, hypotension, altered mental status, and diaphoresis.
   - **b.** Sepsis with obstruction mandates immediate antibiotics and decompression.
2. Is the patient hemodynamically stable?
   - **a.** Stability determines whether conservative management is appropriate.
   - **b.** Unstable patients require urgent decompression if obstruction is present.
3. Is the patient diabetic or immunocompromised?
   - **a.** These conditions increase the risk of rapid progression and lower the threshold for intervention.
4. Is the pain controllable with analgesics?
   - **a.** If unable to control pain and it is interfering with the functioning and/or safety of the crew, decompression should be considered.
5. Is the patient able to tolerate oral nutrition and medication?
   - **a.** If unable to tolerate oral nutrition and medication, decompression should be considered.
6. Is the stone obstructing or nonobstructing?
   - **a.** Use ultrasound to assess for hydronephrosis as a marker of obstruction.
   - **b.** Nonobstructing stones may be managed medically with analgesia, hydration, and expulsive therapy.
7. What do laboratory results show?
   - **a.** Urinalysis and cultures: presence of infection alters management toward urgent intervention.
8. Is intervention warranted?
   - **a.** Obstruction with infection, solitary kidney, or anuria requires decompression.
   - **b.** Otherwise, consider conservative management if pain is controllable and no infection is present.

9. What resources are available?
   a. Ultrasound, nephrostomy kit (defined below), antibiotics, fluids, and analgesics.
10. Can real-time or delayed consultation with a flight surgeon or ground-based expert be obtained?

## 14.9 Diagnosis of nephrolithiasis

The signs and symptoms of obstruction include flank pain or costovertebral angle tenderness, which is acute, severe, and colicky in nature. The crewmember may experience fevers, nausea and/or vomiting.

The initial workup of a crewmember with these features involves vital sign monitoring to assess for fever, tachycardia, or hypotension, which would be suggestive of sepsis. Then, imaging interrogation can be performed using POCUS in conjunction with a KUB radiograph if radiography is available. On ultrasound, findings consistent with a stone include direct visualization of a calculus, absence of ureteral jets, and presence of hydronephrosis. The stone may be directly visualized on radiography if it is radiodense and large enough.

Once nephrolithiasis with suspected obstruction has been diagnosed, subsequent management is guided by the crewmember's clinical stability, evidence of infection, likelihood of spontaneous stone passage, and available medical resources. Particular attention should be paid to signs of sepsis, hemodynamic instability, refractory pain, inability to tolerate oral intake, and patient-specific risk factors such as diabetes or immunocompromised status, all of which lower the threshold for intervention. Stones larger than 10 mm are unlikely to pass spontaneously and, when associated with obstruction or infection, warrant consideration of urinary diversion. Stones smaller than 10 mm that are associated with obstruction and infection also warrant consideration of urinary diversion. In this context, and in the absence of endourologic options, ultrasound-guided PCN represents the primary means of decompression, provided that personnel expertise, equipment, and medical support are available. The indications, technique, and postprocedure considerations for nephrostomy placement are outlined in Table 14.7

See Fig. 14.3 for an illustration of the avascular plane of Brödel.

## 14.10 Procedural management of testicular torsion in space

### 14.10.1 Introduction

Testicular torsion is a urological emergency caused by the twisting of the spermatic cord, which compromises the blood supply to the testis. This condition typically results from a congenital anatomical defect known as the bell clapper deformity, where the tunica vaginalis attaches abnormally high, allowing the testicle to rotate freely within the scrotum. Testicular torsion can occur at any age but is most common in adolescents aged 12–18 years. It is rare in men over 30, accounting for fewer than 10% of cases (Lee et al., 1983). Recent data suggest an incidence rate of 0.52 per 100,000 person-years for males between 30 and 39, and 0.19 for males between 40 and 49 (Choi et al., 2022). Precipitating factors include trauma, rapid growth during puberty, and vigorous physical activity, although it can occur spontaneously.

If untreated, testicular torsion leads to ischemia and infarction, with testicular viability sharply decreasing after 6 hours of torsion. Prompt diagnosis and management are essential to prevent testicular loss and long-term complications such as infertility or chronic pain.

### 14.10.2 Does it work?

While the definitive management is surgical, successful manual detorsion can restore blood flow to the testicle until access to surgical care is regained. The success rate is quoted from 25%–70% of patients (Vasconcelos-Castro et al., 2020; Manual Detorsion of the Testes, 2024). There are no known complications of manual detorsion, though the procedure is uncomfortable for the patient.

### 14.10.3 Procedural considerations

#### *14.10.3.1 Injury risk and incidence*

There are no recorded incidents of testicular torsion during spaceflight, and it remains a theoretical risk as most torsions occur in adolescent populations.

**TABLE 14.7 Comprehensive protocol for ultrasound-guided nephrostomy tube placement.**

| Procedure | Ultrasound-guided nephrostomy tube placement |
|---|---|
| Purpose | To drain urine directly from the renal pelvis |
| Indications | • Hydronephrosis<br>• Nephrolithiasis<br>• Urine diversion (bladder injury)<br>• Pyonephrosis<br>• Obstructive pyelonephritis |
| Contraindications | • No absolute contraindications<br>• Relative contraindications:<br>• Untreated urinary tract infection<br>• Coagulopathy<br>• No hydronephrosis (difficult to assess appropriate calyx) |
| **Supplies** | |
| Operators | Minimum 2, Ideally 3 |
| Antiseptic | • Iodine swab stick, BZK Antiseptic Moist Towelette × 2<br>• If suspected infection, antibiotics to be given preprocedure<br>• Sterile gloves |
| Analgesia | • Local anesthetic (20 mL)<br>• Lidocaine 1% or 2%<br>• OR bupivacaine 0.25% or 0.5% |
| Materials | • Amplatz dilator set<br>• 8 Fr percutaneous nephrostomy catheter<br>• Portable ultrasound machine<br>• Ultrasound gel (sterile; or water if in microgravity)<br>• Ultrasound probe cover (sterile)<br>• Razor and shaving cream<br>• Sterile water<br>• Small cloth<br>• Gauze<br>• Crew medical restraint system<br>• Urine collection system<br>• Leg strap/belt for collection system<br>• Biohazard bag<br>• 5 mL syringe<br>• 18G & 22G needles<br>• Scalpel (if not included in dilator kit) |
| Estimated equipment mass/volume (not including crew restraint system or portable ultrasound) | ~345 g/780 $cm^3$ |
| **Steps** | |

1. Pretreatment with antibiotics if suspected infection.
2. Prepare sterile field/tray for supplies to be readily accessible.
3. Draw up local anesthetic in 5 mL syringe with 18G needle, then recap and replace with 22G needle.
4. Position and restrain the crewmember patient.
5. Shave the posterior lateral upper abdomen, if applicable.
   a. Microgravity: wet area with water, apply shaving cream.
   b. Shave hair.
   c. Clean with cloth and store appropriately.
6. Ultrasound target kidney, identify avascular plane of Brodel (just posterior to the lateral cortex of the kidney, see Figure 14.3). Make sure there are no intervening structures (i.e., bowel). Plan approach.
7. Clean area with sterile wipes × 2.
8. Don sterile gloves.
9. Cover ultrasound probe with sterile cover, placing a small amount of gel within ultrasound probe cover.
10. Raise a bleb with lidocaine under entry area, then under ultrasound guidance anesthetize the tract to the kidney.

(Continued)

**TABLE 14.7** (Continued)

| **Procedure** | **Ultrasound-guided nephrostomy tube placement** |
|---|---|
| **11.** Insert an 18G needle into the renal pelvis under ultrasound guidance. Attach a 10 mL syringe and attempt to aspirate urine to ensure placement within the collecting system.<br>**12.** Feed guidewire through the needle into the pelvis, then retract the needle from skin.<br>**13.** Make a nick in the skin approximately the width of a standard scalpel or pinky fingernail.<br>**14.** Dilate the tract to the size of the catheter using the dilator set.<br>**15.** Insert the 8 Fr catheter over the guidewire until urine appears in the catheter. Remove guidewire.<br>**16.** Pull the attached string to curl the pigtail catheter in the renal pelvis. Secure string.<br>**17.** Place gauze and bandage around catheter entry at the skin.<br>**18.** Attach urine collection system.<br>**19.** Secure urine collection system to crewmember's leg.<br>**20.** Dispose of used materials in biohazard bag. | |
| **Postprocedure care** | |
| • Monitor for complications<br>• Check for bleeding, infection, or signs of improper placement.<br>• Ensure urine flows freely into drainage bag.<br>• Check for normal bowel sounds and bowel movements.<br>• Wash the area around the catheter daily with a cleansing wipe.<br>• The catheter will need to be removed after 6–12 weeks and a new one will need to be placed.<br>• The catheter may stay in place under intravehicular activity suit for reentry if necessary with extra gauze and bandaging around the exposed catheter to limit risk of dislodgement. | |
| **Potential complications and solutions** | |
| Complication | Solution |
| Difficulty localizing the renal pelvis | Delay procedure until sufficiently distended. |
| Breakage of pigtail catheter string | If already coiled in the renal pelvis, secure the pigtail catheter to the skin. When attempting removal, cut the pigtail catheter far away from the skin to prevent retraction, ensure the string is free, and remove the catheter. |
| Leak in urine collection system | Replace the collection system, ensuring the connections are tightly secured. |
| Bleeding | Apply gauze dressings and pressure to the area of bleeding. |
| Hemorrhage (i.e., subcapsular hematoma and pseudoaneurysm) | On ultrasound, if there is an echogenic collection surrounding the kidney, this may represent a hematoma. Alternatively, if the crewmember becomes hemodynamically unstable or appears weak/dizzy/pale/cold, consider the possibility of hemorrhage.Do not remove the catheter. Maintain hemodynamic stability and arrange for crewmember return if possible. |
| Bowel/adjacent organ injury | Remove the catheter and monitor for hemodynamic instability and signs of sepsis. Treat with empiric antibiotics (i.e., broad spectrum, such as metronidazole and cephalosporin). |
| Urinoma | Monitor size with ultrasound every 3 days or as symptoms change, as well as for signs of sepsis (fever and/or hemodynamic instability). If infected urinoma is suspected, begin empiric antibiotics (defined above). If infection persists, consider percutaneous drain placement into urinoma. |
| Catheter does not drain | Irrigate the catheter with either a 10 mL syringe or ideally with a catheter-tip syringe using 50 mL of normal saline or sterile water (see below). |
| Postobstructive diuresis (urine output greater than 200 mL for >2 hours) | For mild postobstructive diuresis, oral hydration is usually sufficient. For moderate to severe cases, IV fluids will be needed to replace losses and maintain electrolyte balance. If oral intake is insufficient or urine output is excessive, intravenous replacement with 0.9% or 0.45% saline at roughly half the rate of ongoing losses (approximately 0.5 mL IV for each 1 mL of urine output) can be considered. Adjust fluid replacement hourly based on ongoing urine output. |
| Infection | Treat with antibiotics with selection depending on if the infection is related to the surgical site (Amox-Clav or Cefalexin) or a urinary tract infection (Ciprofloxacin). |

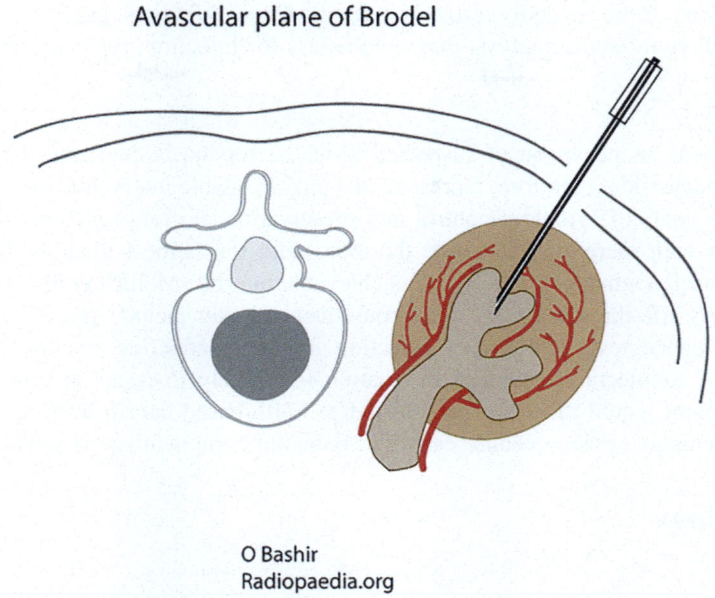


FIGURE 14.3 Avascular plane of Brodel. Credit: Radiopedia.
From Radiopaedia.org, Avascular plane of Brödel, licensed under Creative Commons Attribution-NonCommercial-ShareAlike (CC BY-NC-SA).

## 14.10.4 Spaceflight specific considerations

Definitive treatment is surgical exploration, which involves detorsion and bilateral orchiopexy to prevent recurrence. However, in the unique constraints of the spaceflight environment, such surgical intervention is not practical. Anesthesia and the required surgical expertise are likely outside the scope of astronaut training and require too many crewmembers for active treatment, making surgery infeasible during missions. This underscores the importance of early detection and manual detorsion as critical temporizing measures in space.

## 14.10.5 Decision making

The decision to intervene for testicular torsion is based on the following factors:

1. Is the clinical history consistent with testicular torsion?
   a. Sudden onset of testicular pain, inciting event such as trauma, prior history of torsion, high-riding testicle with horizontal lie, and absence of cremasteric reflex.
2. Are the ultrasound findings consistent with testicular torsion?
   a. Whirlpool sign of twisted spermatic cord.
   b. Absence of arterial/venous flow in the bilateral testicles.
   c. Possible reactive hydrocele.
3. Is there an alternative diagnosis that is more likely?
   a. If there is a fever, leukocytosis and/or hyperemia/abscess on imaging, it is more likely orchitis/epididymitis, and the patient should be treated with antibiotics.

### *14.10.5.1 Diagnosis of torsion*

Testicular torsion typically presents with sudden onset severe unilateral scrotal pain. The affected scrotum may appear swollen and erythematous. On examination, the testicle is often retracted and may lie horizontally. A hallmark clinical feature is the absence of the cremasteric reflex; when the upper inner thigh is stroked, the testicle fails to retract. Patients may also experience nausea and vomiting. Several conditions can mimic torsion and should be considered, including epididymitis, orchitis, torsion of the appendix testis, hydrocele, varicocele, and hernia.

Evaluation begins with vital signs, noting the presence of fever or tachycardia, which may serve as a proxy for pain. A testicular ultrasound is the imaging modality of choice. Key findings include the whirlpool sign indicating a twisted vascular stalk, assessment of arterial and venous flow within the testicle, and evaluation for hydrocele, hyperemia suggestive of infection, torsion of the appendix testis, or a hernia. Clinical findings guide laboratory testing: if the patient is febrile, a white blood cell count and urinalysis may help assess for infection, typically performed after imaging.

#### 14.10.5.2 Management

In the spaceflight environment, management of suspected testicular torsion is constrained by the absence of definitive surgical capability. Manual detorsion, therefore, represents the only available intervention to restore perfusion and reduce ischemia, with the primary goals of symptom control and preservation of crew function. Terrestrial data indicate that testicular viability declines with increasing ischemic duration, with the highest likelihood of salvage within the first 6–8 hours after symptom onset, a window unlikely to be achievable in orbit (Mellick et al., 2019). Although unsuccessful detorsion is not immediately life-threatening, delayed consequences may include persistent or severe pain that could compromise mission performance, testicular atrophy, infection, impaired endocrine function, and decreased spermatogenesis. Given current mission architectures, medical evacuation is unlikely to occur in time for testicular salvage, and definitive surgical management would therefore be deferred to Earth-based care following temporization. A structured overview of indications, technique, postprocedure care, and potential complications is provided in Table 14.8.

## 14.11 Urogynecology

### 14.11.1 Introduction

Female astronauts may be at higher risk of unique sex-specific urological challenges in prolonged microgravity. While the majority of urogynecological conditions are likely to be screened out during the selection process for government astronauts, the increase in commercial astronauts and spaceflight participants increases the likelihood of facing female urinary obstruction, incontinence, and pelvic organ prolapse (POP) during spaceflight.

This section will discuss the basic types of urinary incontinence and the suggested management during spaceflight. Pessary insertion will not be discussed, as in the case of pessary complication and subsequent removal, it is unlikely to be replaced. Furthermore, additional surgical/interventional procedures to address urinary incontinence, such as Botox injections, urethral bulking injections, and neuromodulation, require resources and expertise that are not feasible currently or in the near future.

### 14.11.2 Does it work?

There is very limited data on urogynecological conditions during spaceflight. This is partially secondary to the relatively lower number of female spaceflight participants, as well as less dedicated urogynecological research. While incontinence has not been documented explicitly during microgravity, its prevalence on Earth necessitates the formulation of basic management techniques in microgravity.

Theoretically, pelvic organ prolapsemay be ameliorated in microgravity due to reduced caudal pressure. While pessaries are a first-line treatment on Earth and may be used in space as part of a personal medical kit, a misplaced pessary can be an independent cause of urinary obstruction.

### 14.11.3 Female urinary incontinence management and spaceflight considerations

Urinary incontinence is the involuntary loss of bladder control, which can significantly impact quality of life. It is categorized into three primary types: stress urinary incontinence (SUI), urge urinary incontinence (UUI), and overflow incontinence. In spaceflight, moistness in the genital region after urination may be misattributed to one of these forms of incontinence; however, this is less likely to be secondary to incontinence of any kind and more likely secondary to the effect of surface tension in microgravity.

**Stress urinary incontinence (SUI):** SUI occurs when physical activities such as coughing, sneezing, or exercising increase abdominal pressure, leading to urine leakage. This is due to weakened pelvic floor muscles or urethral sphincter dysfunction. In space, where microgravity alters fluid distribution and muscle conditioning, SUI may present differently but remains theoretically possible.

**Management in space:** Conservative approaches, such as the use of absorbent pads, are practical and low-resource solutions. Optimal management includes preflight mitigation with pelvic floor exercises. Positioning for more sophisticated exercises beyond Kegels might be more challenging to implement effectively during flight in microgravity. Interventional options like urethral bulking agents or surgeries are not feasible in the near-term due to mass, volume, and power constraints of the equipment and technical skills of the crew medical officer.

**TABLE 14.8** Manual detorsion of testicular torsion: indications, technique, and postprocedure considerations.

| **Procedure** | **Manual detorsion of testicles** |
|---|---|
| Purpose | Temporizing measure to restore blood flow and alleviate ischemia. |
| Indications | • Suspected testicular torsion when surgical intervention is delayed.<br>• Sudden onset of severe scrotal pain.<br>• Swelling and erythema of the scrotum.<br>• A high-riding testicle with a horizontal lie.<br>• Absence of the cremasteric reflex.<br>• Nausea and vomiting. |
| Contraindications | • None |
| **Supplies** | |
| Operators | 1 person |
| Antiseptic | • None |
| Analgesia | • NSAIDS or opioids as needed |
| Materials | • Ultrasound machine<br>• Probe cover<br>• Gel/Saline/Water |
| Estimated equipment mass/volume (not including portable ultrasound) | ~30 g/6 $cm^3$ |
| **Steps** | |
| **1.** Using ultrasound, determine if there is absence of flow and which testicle is affected. Attempt to identify a whirlpool sign of the twisted spermatic cord.<br>**2.** Determine the direction of torsion, which is typically medial-to-lateral.<br>**3.** Begin by rotating the affected testicle laterally ("opening the book" by rotating clockwise for right testis from perspective of patient or counter-clockwise for the left testis) in 180-degree increments.<br>**4.** After each rotation, reassess for pain relief and return of normal testicular position. Since testicular torsion is not usually bilateral, symmetry is the primary goal. Repeat up to two full rotations (720 degrees) if necessary.<br>**5.** If pain worsens or does not improve after two full rotations, reverse the rotation direction as some torsions are lateral-to-medial.<br>**a.** Successful detorsion often results in immediate pain relief and improved testicular position.<br>**6.** If available, repeat Doppler ultrasound to confirm restored blood flow. | |
| Postprocedure care | |
| • Monitor for pain recurrence<br>• Prepare patient for repatriation to medical facilities as soon as possible (no specific measures for securing the affected testis) | |
| Potential complications and solutions | |
| Complication | Solution |
| Persistent ischemia | Continue manual detorsion attempts, palliate with pain medication. |
| Nonviable testis | Provide pain medication as needed. Perform orchiectomy once surgical treatment is available. |
| Infection | Pain medication as needed. If febrile, begin broad spectrum antibiotics for a 5–7 day course. |
| Abscess | Pain medication as needed. If febrile, begin broad spectrum antibiotics for a 5–7 day course. Consider incision and drainage of abscess if abscess is >2 cm and not improving with antibiotics. |

**Urge urinary incontinence (UUI):** UUI, or "overactive bladder," involves a sudden, intense urge to urinate, often resulting in leakage. It is associated with detrusor muscle overactivity and may be exacerbated by microgravity-related alterations in autonomic bladder control.

**Management in space:** Absorbent pads are again the primary option. Pharmacological interventions include anticholinergics or beta-3 agonists, which could have adverse effects in the space environment. These may include dry mouth and cognitive impairment, which can affect mission-critical tasks. Advanced treatments like Botox injections or neuromodulation are also not feasible due to their procedural and equipment demands.

**Overflow incontinence:** This type of incontinence results from incomplete bladder emptying, often due to obstruction or detrusor underactivity. It may manifest as continuous dribbling or urinary retention.

**Management in space:** Overflow incontinence can be managed effectively with CIC or in prolonged cases, an indwelling Foley catheter (as discussed above).

### 14.11.4 Considerations for the future

As human exploration extends to the Moon, Mars, and beyond, the need for advanced medical capabilities to manage urological conditions becomes increasingly critical. Addressing these challenges will require integrating innovative technologies and tailoring procedures to the constraints of deep-space missions.

### 14.11.5 Robotic telesurgery

Robotic telesurgery systems, such as Intuitive Surgical's da Vinci Surgical System hold significant potential for long-duration missions, particularly as communication latency and resource constraints challenge traditional models of care delivery. These systems could enable remote surgical intervention, allowing expert surgeons on Earth to manage complex abdominal and urologic conditions across mission distances. In addition, robotic platforms may mitigate skill gaps by reducing dependence on highly specialized onboard medical personnel. The integration of artificial intelligence with surgical platforms may further enchance these capabilities by supporting surgical decision making, augmenting image guidance, and automating procedural steps. Importantly, robotic systems facilitate minimally invasive approaches, which may reduce physiologic stress, limit wound related complications, and accelerate recovery. These advantages are especially valuable in environments with constrained medical resources and limited capacity for postoperative care. However, substantial barriers remain, including significant power, volume, and mass requirements of current robotic systems, which complicate integration into spaceflight architectures, latency effects from the time delay inherent to communications outside of low Earth orbit, and the use of $CO_2$ to create pneumoperitoneum amongh other issues, making robotic surgery infeasible for the foreseeable future. To address these challenges, progressive validation in analog enviornments will be essential to evaluate system performance under conditions of signal delay, microgravity constraints, and remote assistance workflows.

### 14.11.6 Miniaturized and portable devices

Longer missions necessitate compact, low-resource diagnostic and therapeutic tools that can be reliably deployed in constrained environments. Burst Wave Lithotripsy represents a noninvasive and portable approach for stone fragmentation, making it particularly well suited for medical system miniaturization in space (Harper et al., 2025). As human exploration extends to lunar and Martian habitats, it is anticipated that a spaceflight compatible system will be developed that integrates the control unit, electronic pulser, and display into a single compact platform suitable for deployment within base medical infrastructure.

In parallel, advances in imaging technology will be essential, with the development of compact systems capable of precise diagnostics and visualization. These may include portable ultrasound platforms as well as emerging cone beam and mobile CT architectures, which could provide critical diagnostic capability while maintaining feasibility within the mass, volume, and power constraints of long duration missions.

### 14.11.7 Artificial intelligence and automation

Artifical intelligence (AI) systems have the potential to enhance urologic care in space by supporting surgical planning, real time clinical decision making, and postprocedural monitoring. During urologic emergencies, AI-assisted tools may improve diagnostic accuracy through integration with portable imaging platforms and assisting with image acquisition and interpretation. In settings where direct human surgical assistance is unavailable, AI-based systems may also provide procedural guidance to support autonomous and/or semi-autonomous medical interventions.

### 14.11.8 Prolonged mission considerations

For missions extending beyond one year, additional considerations arise related to the durability and sustainability of medical capabilities. Long duration care may require robust stent and nephrostomy systems constructed from durable materials that are designed to reduce complications during extended use. A self-sustaining medical ecosystem will also be necessary, incorporating sterilization, maintenance, and repair capabilities for surgical and diagnostic equipment. In parallel, reliable telemedicine infrastructure will remain essential to facilitate communication with Earth-based medical teams and support remote triage, consultation, and procedural guidance throughout prolonged missions.

The convergence of urology, robotics, and prolonged space exploration underscores the need for innovative approaches to delivering complex medical care in extreme and resource-limited environments. As human spaceflight extends beyond LEO and mission durations increase, urologic conditions that are routinely managed on Earth will require novel diagnostic, procedural, and technological solutions to preserve crew health and mission integrity. Advances in miniaturized instrumentation, image-guided interventions, AI, and autonomous medical systems hold the potential to redefine the standard of care in space. Importantly, the lessons learned from developing these capabilities will extend beyond spaceflight, informing the delivery of urologic and surgical care in remote, austere, and underserved settings on Earth.

## References


Adams, MS, Cox, JM, 2014. U.S. Army Aeromedical Research Laboratory Integration Evaluation of the Advanced Mission Extender Device Max.

Alnazari, M., Bakhsh, A., Shaqroon, H.A., Rajih, E.S., Al-Nakshabandi, N.A., Rabah, D.M., 2024. Hematuria: is it useful in predicting renal or ureteral stones in patient presenting to emergency department with flank pain? Urology Annals 16 (1), 71–74. Available at https://doi.org/10.4103/ua.ua_66_23.

Assimos, D., Krambeck, A., Miller, N.L., Monga, M., Murad, M.H., Nelson, C.P., et al., 2016. Surgical management of stones: American Urological Association/Endourological Society guideline, part I. Journal of Urology 196 (4), 1153–1160. Available from: Available at https://doi.org/10.1016/j.juro.2016.05.090, http://www.sciencedirect.com/science?_ob=Publication.

Barr, J., Fraser, G.L., Puntillo, K., Ely, E.W., Gélinas, C., Dasta, J.F., et al., 2013. Clinical practice guidelines for the management of pain, agitation, and delirium in adult patients in the intensive care unit. Critical Care Medicine 41 (1), 263–306. Available from. Available at https://doi.org/10.1097/CCM.0b013e3182783b72, http://journals.lww.com/ccmjournal/pages/default.aspx.

Choi, J.B., Han, K.H., Lee, Y., Ha, U.S., Cho, K.J., Kim, J.C., et al., 2022. The incidence of testicular torsion and testicular salvage rate in Korea over 10 years: a nationwide population-based study. Investigative and Clinical Urology 63 (4), 448–454. Available at https://doi.org/10.4111/icu.20220122. Available from: https://www.icurology.org/pdf/10.4111/icu.20220122.

Cole, R.L., Jennifer, M., Sara, Y.M., 2016. JSC-CN-34668 NASA Technical Report Server NASA Astronaut Urinary Conditions Associated With Spaceflight.

Dablan, A., Okur, Z., Cingöz, M., Erdim, Ç., Arslan, M.F., Türksayar, O., et al., 2025. Predictors of nephrostomy catheter dislodgement: insights from a retrospective analysis. Cardiovascular and Interventional Radiology 48 (5), 633–642. Available at https://doi.org/10.1007/s00270-025-04020-y.

Diaz, T.E., Ives, E.C., Lazare, D.I., Buckland, D.M., 2024. Expiration analysis of the International Space Station formulary for exploration mission planning. NPJ Microgravity 10 (1). Available at https://doi.org/10.1038/s41526-024-00414-3.

Fader, M., Macaulay, M., Wilson, N., Goudie, N., Chadwick, T.J., Abouhajar, A., et al., 2024. Trial to compare mixed-use (multi-use and single-use) intermittent catheter management with single-use management over 12 months (The MultICath Trial): protocol for a noninferiority randomised controlled trial. BMJ Open 14 (8). Available at https://doi.org/10.1136/bmjopen-2024-088483. Available from: http://bmjopen.bmj.com/content/early/by/section.

Flomax (tamsulosin hydrochloride) prescribing information. Available from: https://dailymed.nlm.nih.gov/dailymed/drugInfo.cfm?setid=6771ad8e-ac92-4aec-b484-5d8350a353f8., 2025.

Fulgham, P.F., Assimos, D.G., Pearle, M.S., Preminger, G.M., 2013. Clinical effectiveness protocols for imaging in the management of ureteral calculous disease: AUA technology assessment. Journal of Urology 189 (4), 1203–1213. Available at https://doi.org/10.1016/j.juro.2012.10.031.

Furyk, J.S., Chu, K., Banks, C., Greenslade, J., Keijzers, G., Thom, O., et al., 2016. Distal ureteric stones and tamsulosin: a double-blind, placebo-controlled, randomized, multicenter trial. Annals of Emergency Medicine 67 (1), 86. Available from. Available at https://doi.org/10.1016/j.annemergmed.2015.06.001, http://www.elsevier.com/inca/publications/store/6/2/3/2/7/8/index.htt.

Halbgewachs, C., Domes, T., 2015. Postobstructive diuresis: pay close attention to urinary retention. Canadian family physician Medecin de famille canadien 61 (2), 137–142. Available from. Available at http://www.cfp.ca/content/61/2/137.full.pdf+html.

Hausegger, K.A., Portugaller, H.R., 2006. Percutaneous nephrostomy and antegrade ureteral stenting: technique-indications-complications. European Radiology 16 (9), 2016–2030.

Harper, J.D., Dunmire, B., Thiel, J., Burke, B.H., Wang, Y.N., Totten, S., et al., 2025. Facilitated clearance of small, asymptomatic renal stones with burst wave lithotripsy and ultrasonic propulsion. Journal of Urology 214 (1), 41–47. Available at https://doi.org/10.1097/JU.0000000000004533.

Healthcare Infection Control Practices Advisory Committee 2009 Centers for Disease Control and Prevention Guideline for Prevention of Catheter-Associated Urinary Tract infections, 2009.

Holdgate, A., Pollock, T., 2005. Nonsteroidal anti-inflammatory drugs (NSAIDs) versus opioids for acute renal colic. The Cochrane Database of Systematic Reviews 2004 (2).

Hollingsworth, J.M., Rogers, M.A.M., Kaufman, S.R., Bradford, T.J., Saint, S., Wei, J.T., et al., 2006. Medical therapy to facilitate urinary stone passage: a meta-analysis. Lancet (London, England) 368 (9542), 1171–1179.

Hurd, V.S., Valle, M.D., Kravets, V.G., Anderson, A.L., Le, N.N., Kendall, J.L., et al., 2025. Artificial intelligence assistance in point-of-care ultrasound skill retention for novice users in space medicine scenarios. Wilderness and Environmental Medicine 36 (1), 24–33. Available at https://doi.org/10.1177/10806032241304441. Available from: https://journals.sagepub.com/home/wem.

Jones, J.A., Kirkpatrick, A.W., Hamilton, D.R., Sargsyan, A.E., Campbell, M., Melton, S., et al., 2007. Percutaneous bladder catheterization in microgravity. The Canadian Journal of Urology 14 (2), 3493–3498.

Karakala, N., Córdoba, D., Chandrashekar, K., Lopez-Ruiz, A., Juncos, L.A., 2021. Point-of-care ultrasound in acute care nephrology. Advances in Chronic Kidney Disease 28 (1), 83–90. Available at https://doi.org/10.1053/j.ackd.2021.06.003. Available from: http://www.elsevier.com/wps/find/journaldescription.cws_home/703170/description#description.

Lee, L.M., Wright, J.E., McLoughlin, M.G., 1983. Testicular torsion in the adult. Journal of Urology 130 (1), 93–94. Available at https://doi.org/10.1016/S0022-5347(17)50975-1.

Lightner, D.J., Wymer, K., Sanchez, J., Kavoussi, L., 2020. Best practice statement on urologic procedures and antimicrobial prophylaxis. Journal of Urology 203 (2), 351–356. Available at https://doi.org/10.1097/JU.0000000000000509. Available from: https://www.auajournals.org/.

Manual Detorsion of the Testes, 2024. Available from: https://emedicine.medscape.com/article/1413565-overview#a1?form=fpf.

Mellick, L.B., Sinex, J.E., Gibson, R.W., Mears, K., 2019. A systematic review of testicle survival time after a torsion event. Pediatric Emergency Care 35 (12), 821–825. Available at https://doi.org/10.1097/PEC.0000000000001287.

Miller, O.F., Kane, C.J., 1999. Time to stone passage for observed ureteral calculi: a guide for patient education. The Journal of Urology 162 (3 Pt 1).

NASA 3.001 Medical kit—contents and reference, 2015.

Neumeier, V., Stangl, F.P., Borer, J., Anderson, C.E., Birkhäuser, V., Chemych, O., et al., 2023. Indwelling catheter vs intermittent catheterization: is there a difference in UTI susceptibility? BMC Infectious Diseases 23 (1). Available at https://doi.org/10.1186/s12879-023-08475-7.

Nieberg, P., 2025. Air Force pilots get a new way to pee at 30,000 feet. Available from: https://www.yahoo.com/news/air-force-pilots-way-pee-144532467.html.

Nyman, M.A., Schwenk, N.M., Silverstein, M.D., 1997. Management of urinary retention: rapid versus gradual decompression and risk of complications. Mayo Clinic Proceedings 72 (10), 951–956. Available at https://doi.org/10.4065/72.10.951. Available from: https://www.sciencedirect.com/science/journal/00256196.

Pabon-Ramos, W.M., Dariushnia, S.R., Walker, T.G., Janne d'Othée, B., Ganguli, S., Midia, M., et al., 2016. Quality improvement guidelines for percutaneous nephrostomy. Journal of Vascular and Interventional Radiology: JVIR 27 (3), 410–414.

Panach-Navarrete, J., Tonazzi-Zorrilla, R., Martínez-Jabaloyas, J.M., 2020. Dislodgement in long-term patients with nephrostomy tube: risk factors and comparative analysis between two catheter designs. Journal of Endourology 34 (2), 227–232. Available at https://doi.org/10.1089/end.2019.0655.

Patel, I.J., Rahim, S., Davidson, J.C., Hanks, S.E., Tam, A.L., Walker, T.G., et al., 2019. Society of Interventional Radiology Consensus guidelines for the periprocedural management of thrombotic and bleeding risk in patients undergoing percutaneous image-guided interventions-part II: recommendations: endorsed by the Canadian Association for Interventional Radiology and the Cardiovascular and Interventional Radiological Society of Europe. Journal of Vascular and Interventional Radiology: JVIR 30 (8).

Pearle, M.S., Pierce, H.L., Miller, G.L., Summa, J.A., Mutz, J.M., Petty, B.A., et al., 1998. Optimal method of urgent decompression of the collecting system for obstruction and infection due to ureteral calculi. The Journal of Urology 160 (4), 1260–1264.

Prieto, J.A., Murphy, C.L., Stewart, F., Fader, M., 2021. Intermittent catheter techniques, strategies and designs for managing long-term bladder conditions. The Cochrane Database of Systematic Reviews 10 (10). Available at https://doi.org/10.1002/14651858.CD006008.pub5.

Putcha, L., Berens, K.L., Marshburn, T.H., Ortega, H.J., Billica, R.D., 1999. Pharmaceutical use by U.S. astronauts on space shuttle missions. Aviation, Space, and Environmental Medicine 70 (7), 705–708.

Radecka, E., Magnusson, A., 2004. Complications associated with percutaneous nephrostomies. A retrospective study. Acta Radiologica (Stockholm, Sweden: 1987) 45 (2), 184–188.

Risk of Urinary Retention, 2025.

Sandhu, J.S., Bixler, B.R., Dahm, P., Goueli, R., Kirkby, E., Stoffel, J.T., et al., 2024. Management of lower urinary tract symptoms attributed to benign prostatic hyperplasia (BPH): AUA guideline amendment 2023. The Journal of Urology 211 (1), 11–19. Available at https://doi.org/10.1097/JU.0000000000003698.

Scales, C.D., Smith, A.C., Hanley, J.M., Saigal, C.S., 2012. Prevalence of kidney stones in the United States. European Urology 62 (1), 160–165. Available at https://doi.org/10.1016/j.eururo.2012.03.052.

Stapleton, T.J., Baccus, S., 2013. International Conference on Environmental Systems (ICES) Vail, CO, USA National Aeronautics and Space Administration Development of a Universal Waste Management System.

Stepaniak, P.C., Ramchandani, S.R., Jones, J.A., 2007. Acute urinary retention among astronauts. Aviation, Space, and Environmental Medicine 78 (4 Suppl).

Stratton, E, Lumpkins, S, Antonsen, E, 2024. National Aeronautics and Space Administration Risk of Renal Stone Formation.

Trautner, B, Nicolas, C.-P., Kalpana, G, Elizabeth, H, Molly, H, Gregory, M, et al., 2025. Infectious Diseases Society of America Complicated Urinary Tract Infections (cUTI): Clinical Guidelines for Treatment and Management.

Urinary health (OCHMO-MTB-003), 2025.

Vasconcelos-Castro, S., Flor-de-Lima, B., Campos, J.M., Soares-Oliveira, M., 2020. Manual detorsion in testicular torsion: 5 years of experience at a single center. Journal of Pediatric Surgery 55 (12), 2728–2731.

Vaughan, E.D., Gillenwater, J.Y., 1971. Recovery following complete chronic unilateral ureteral occlusion: functional, radiographic and pathologic alterations. The Journal of Urology 106 (1), 27–35.

Weiner, L.M., Webb, A.K., Limbago, B., Dudeck, M.A., Patel, J., Kallen, A.J., et al., 2016. Antimicrobial-resistant pathogens associated with healthcare-associated infections: summary of data reported to the National Healthcare Safety Network at the Centers for Disease Control and Prevention, 2011–2014. Infection Control and Hospital Epidemiology 37 (11), 1288–1301.

Chapter 15

# Upper and lower extremity procedures

**Shawna Pandya[1,2,]*, Prashant Parmar[3,]*, and Richard Meehan[4]**

[1]*Space Medicine Group, International Institute for Astronautical Sciences, Boulder, CO, United States,* [2]*Faculty of Medicine & Dentistry, University of Alberta, Edmonton, AB, Canada,* [3]*Rheumatology, Veros Health, Denver, CO, United States,* [4]*Rheumatology, National Jewish Health, Denver, CO, United States*

## 15.1 Introduction

Treating injuries in spaceflight requires consideration of medical practice in an austere environment, owing to the resource-limited and remote nature of space, as well as the unique environmental challenges of spaceflight, including microgravity, spaceflight-induced physiological changes, sarcopenia and osteopenia, fluid shifts, and immunosuppression. In this chapter, we will discuss considerations of the spaceflight environment as they pertain to extremity injuries that may be encountered, with a special emphasis on physiological adaptations that may increase the risk of injury or impede healing, resource constraints, and the logistics involved with performing procedures in micro- and partial gravity environments that may require modifications from earth-based procedures.

Musculoskeletal conditions are expected to be some of the most common pathologies that will be addressed by a Crew Medical Officer (CMO) in spaceflight. A CMO is thus expected to be familiar with routine diagnosis and standard management of musculoskeletal pathology. As future spacecraft, stations, and even potential lunar and planetary explorations evolve, the environment has the potential to change in such a way that acute pathologies may become more prominent, and a CMO would be expected to be able to manage such conditions as they arise.

The procedures discussed in this section focus on pathologies that are likely to be encountered in spaceflight through trauma, occupational/overuse injuries, and/or activities specific to the spaceflight environment, such as extravehicular activities (EVAs). These include :

- Local blocks: hematoma and digital ring blocks
- Tourniquet application
- Blister drainage
- Splinting
- Fractures, dislocations, and reductions
- Subungual hematoma drainage
- Finger and toenail removal
- Extensor tendon repair
- Fasciotomy
- Joint and bursa injections

Other common procedures that often accompany these injuries, including laceration repair and suturing, wound irrigation, debridement and packing, intravenous (IV) access and placement, abscess drainage, hemorrhage control, and vessel ligation, are discussed elsewhere in the text.

* Lead authors.

**Procedural Medicine for Spaceflight. DOI:** https://doi.org/10.1016/B978-0-443-27552-4.00035-X

## 15.2 Special considerations of the spaceflight environment

### 15.2.1 Trauma and physiologic adaptations in the spaceflight environment

Spaceflight data to date demonstrate osteopenia, particularly in weight-bearing bones, occurring at a rate of 1%–2% per month (Genah et al., 2021). While trauma impacts are ostensibly decreased in microgravity, astronauts may be more susceptible to fractures, particularly at lower-than-expected impacts, owing to this osteopenia (Genah et al., 2021). Similarly, the combination of immunosuppression, increased germ pathogenicity, and delayed wound healing that has been previously noted in spaceflight may theoretically increase the risk of infection in wounds and blisters and impair astronaut performance depending on the location of the injury; however, this has not been seen for abrasions and EVA-induced blisters in spaceflight (Jacob et al., 2023).

### 15.2.2 Spaceflight specific activities: considerations

In addition to spaceflight-induced physiological changes, the nature of work and activities performed in the spaceflight environment will further inform the incidence and specificity of particular injury patterns. Microgravity EVAs, for example, are known to cause pressure injuries, strain injuries, subungual hemorrhages, and blistering (Thompson, 2025). While the data on injury and strains sustained during planetary EVAs is limited to the eleven EVAs performed during the Apollo program, it is reasonable to imagine that the combination of steep rocky terrain, osteopenia and sarcopenia from microgravity exposure, and EVA suit ergonomics may predispose astronauts to falls and potentially fractures in EVA on the Moon, Mars, and beyond (Belobrajdic et al., 2021).

### 15.2.3 Resource considerations, limitations, and specialist support

While most procedures can be performed in a multitude of ways—there are at least eight ways to reduce an anterior shoulder dislocation, for example—for the sake of brevity, the procedures put forth in this chapter will focus on one or two commonly known methods of execution that also optimize procedure completion in the context of the known challenges of the spaceflight environment, namely, mass, volume, power, and shelf-life viability. This is not to say that the guidance given on procedures within this chapter is absolute—the principles are more important—for example, using a large-bore needle for subungual hematoma drainage rather than specifying a blunt needle versus an 18 g needle. An additional focus will be placed on using clinical acumen to overcome existing limitations in spaceflight. For example, despite a promising demonstration of radiography during a recent commercial orbital spaceflight (First X-rays taken in space mimic historic first radiograph captured on earth, 2025), the availability of radiography is not a standard feature of current spaceflight missions. The availability of onsite portable or handheld ultrasonography is a reasonable assumption, given its long-standing presence on the International Space Station (ISS), and recent demonstrations of point-of-care ultrasound (PoCUS) on recent short-duration commercial orbital missions have demonstrated the feasibility of newer commercial off-the-shelf devices in spaceflight, such as during the Inspiration4 and Polaris Dawn all-civilian missions. At this time, computerized tomography and magnetic resonance imagery (MRI) are assumed to be unavailable in spaceflight, although some institutions have proposed portable MRI capture in the spaceflight environment (Jullienne et al., 2024).

Where able, we recommend full adherence to the same standard of care recommended by terrestrial guidelines, including specialist consultation where indicated. The mission profile will determine the feasibility and timing of the consultation: Martian explorers, for example, will have to rely on asynchronous communications for specialist guidance. For example, the healing of a significant burn can be monitored through daily image review and guidance by a plastic surgeon back on Earth. Conversely, an astronaut with a similar injury in Low Earth Orbit can receive real-time guidance regarding wound debridement and dressing, and a flyer in suborbital spaceflight could simply dress the wound until landing, given the 11–90 minute flight profile on a suborbital flight.

### 15.2.4 Procedures in altered gravity environments

Procedures don't inherently change in their steps between terrestrial and partial gravity environments, but they do change in their execution, firstly due to resource availability, as noted above, and also due to the logistics of execution in partial gravity—particularly microgravity. As a guiding principle, the authors advise planning to undertake the following steps prior to initiating any procedure:

- Restrain the patient and identify the relevant anatomy
- Stabilize the proceduralist to avoid movements that may impede or even contribute to further injury

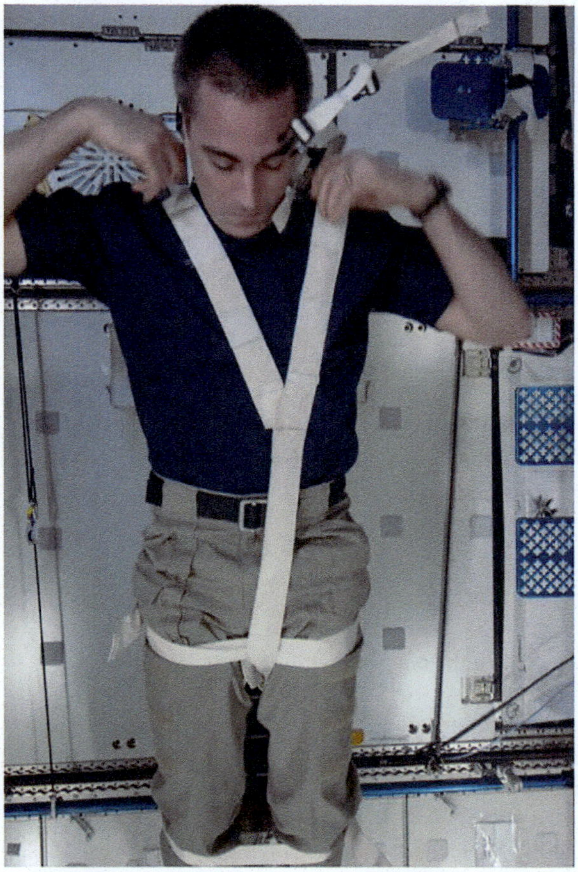

FIGURE 15.1 Crew medical restraint system. An astronaut demonstrates the use of the crew medical restraint system in the microgravity environment aboard the International Space Station.
From: Astronaut Chris Cassidy Checks Out Medical Gear, 2013. NASA.

- Manage/immobilize instruments that will be used during the procedure
- Delineate a plan for waste, biohazard, and sharp containment and disposal ahead of time.

For example, an anterior shoulder dislocation may be elegantly reduced in terrestrial or 1 g conditions by employing the Spaso technique—that is, flexing the shoulder to 90 degrees anteriorly, and then applying vertical distraction while simultaneously externally rotating the shoulder until the dislocated joint is reduced. In microgravity settings, the patient should be restrained ahead of time using an existing device, such as the Crew Medical Restraint System (CMRS) or a similar method of restraint (Fig. 15.1). The proceduralist can similarly stabilize themselves through the use of tethers, restraints, bungee cords, makeshift fixtures for immobilization, or even by leveraging themselves against the patient, as has been suggested previously for cardiopulmonary resuscitation and/or the Heimlich maneuver in microgravity (Hinkelbein et al., 2020).

### 15.2.5 Time to definitive care

As each mission profile is different, procedures in space need to be adapted based on the mission profile and the time to definitive medical care. For example, a spaceflight participant who sustains a painful shoulder impact upon the transition from microgravity to gravity during suborbital spaceflight will likely do just as well holding their shoulder in place to immobilize it during descent and landing, while a crew on Mars must take time to properly sedate, reduce, and immobilize an anterior shoulder dislocation, while taking time to check in, asynchronously, with an orthopedist and physiotherapist to ensure the best possible recovery.

### 15.2.6 Proceduralist experience

The procedures described in this chapter are described at an attending primary care or emergency physician's level, and certain baseline knowledge, such as the risks, benefits, side effects, and alternatives involved with consenting a patient for each procedure, is assumed. With this assumption in mind, we advise the proceduralist to adhere to their comfort level and scope of practice. As such, a CMO with no medical background and minimal experience is advised to wait for further specialist guidance from Mission Control.

## 15.3 Options for local analgesia

Patient comfort is critical for injury management, and in this section, we discuss local anesthesia, specifically the digital ring block and the hematoma block. Regional and systemic analgesia options, such as regional nerve blocks and intravenous (IV) sedation, are discussed elsewhere.

### 15.3.1 Digital ring block

#### *15.3.1.1 Indications*

The digital ring block is a simple, elegant technique for providing anesthesia to the distal finger or toe when repairing a laceration, removing a nail from the nailbed, reducing a dislocated finger, or reducing a displaced fracture. The items necessary for the digital ring block can be found in Table 15.1 below.

#### *15.3.1.2 Technique*

1. Obtain the patient's consent for the procedure after explaining the risks, benefits, side effects, and alternatives of the proposed procedure.
2. Position and immobilize the affected digit using a restraint.
3. Don nitrile gloves.
4. The proceduralist is to affix or tether themselves to stabilize.
5. Landmark the site of injection of the digital nerves at the proximal base of the affected digit at the webspace.
6. Clean the intended injection sites on the medial and lateral aspects of the affected digit using BZK wipes. Let the site dry before proceeding.
7. Using a dorsal approach, take a 10cc lidocaine-filled syringe loaded with a 22–25 g needle and insert the needle just distal to the webspace with the needle directed in the palmar direction.
   a. Ensure that the needle does not pierce through the skin on the palmar aspect of the digit.
   b. Draw backward on the syringe to ensure that the needle has not entered a blood vessel.
   c. Slowly infiltrate 2–4cc of anesthetic while withdrawing the needle dorsally so as to infiltrate the palmar and dorsal digital nerves.
   d. Repeat the same step on the opposite side of the digit.
   e. Alternative methods to the dorsal approach:
      i. The proceduralist may opt to pursue a lateral approach, entering the digit laterally 1 cm distal to the webspace with the needle angled toward the bone, and after following steps 7a and b, can slowly fan the needle out dorsally and volarly/plantarly to infiltrate in a distributed fashion; OR

**TABLE 15.1** Estimated mass and volume of equipment required for a digital ring block.

| Item | Mass | Volume | Indication |
|---|---|---|---|
| *10cc syringe* | 8 g | 10 cm$^3$ | To hold the local anesthetic |
| *10cc 1 or 2% lidocaine without epinephrine* | 14 g | 10 cm$^3$ | To provide anesthesia |
| *22–25 g 1-1 ½" needle* | 2 g | 8 cm$^3$ | To infiltrate the digital nerves with local anesthetic |
| *BZK wipe* | 1.5 g | 4 cm$^3$ | To clean the areas of injection prior to injection |
| *4x4 gauze* | 2 g | 15 cm$^3$ | To staunch any bleeding post infiltration |
| *Nitrile gloves (1 pair)* | 12 g | ~50 cm$^3$ | Provider personal protective equipment |

**ii.** The proceduralist may opt to pursue a webspace approach, entering the digit with the needle angled toward the bone, and after following steps 7a and b, can slowly fan the needle out dorsally and volarly/plantarly to infiltrate the digital nerve in a distributed fashion.

**f.** Dab away any blood at the injection site using gauze.

**g.** Allow 2–3 minutes for the lidocaine to take effect prior to commencing the relevant procedure.

The dorsal approach is depicted in Fig. 15.2 below.

### *15.3.1.3 Contraindications*

Do not proceed if there is:

- A history of allergy to local anesthetics
- Severe coagulopathy
- An active infection or overlying cellulitis
- There is neurovascular compromise

### *15.3.1.4 Complications*

The proceduralist should be aware of rare but potential complications associated with the digital nerve block, including, but not limited to as follows:

- Poor uptake necessitating reinfiltration
- Nerve damage
- Bleeding
- Infection

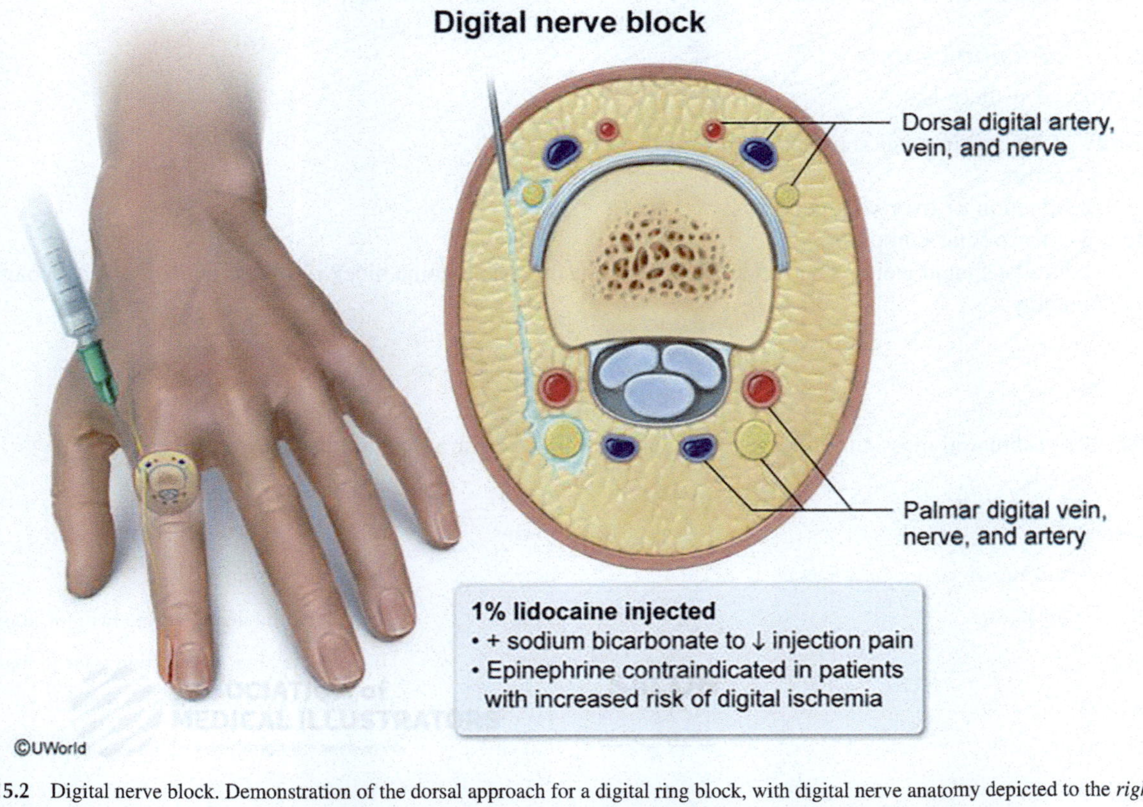


FIGURE 15.2 Digital nerve block. Demonstration of the dorsal approach for a digital ring block, with digital nerve anatomy depicted to the *right* of the image.
From: Local Anesthetic Agents, (n.d.). https://medical.uworld.com/library/pharmacology/792/local-anesthetic-agents/.

## 15.3.2 Hematoma block

### *15.3.2.1 Indications*

The hematoma block is an option for providing local pain control when reducing a fracture and when the fracture site is easily accessible. It is most commonly used in distal radius fractures and is performed by injecting a local anesthetic at the fracture site. The items necessary for the hematoma block can be found in Table 15.2 below.

### *15.3.2.2 Technique*

1. Obtain the patient's consent for the procedure after explaining the risks, benefits, side effects, and alternatives of the proposed procedure.
2. Position and immobilize the affected anatomy using a restraint while taking care not to jostle the fracture site.
3. Don nitrile gloves.
4. The proceduralist is to affix or tether themselves to stabilize.
5. Landmark the site of injection over the fracture site, identified by deformity, a step-off on palpation, or point-of-care ultrasound.
6. Clean the intended injection site using BZK wipes, then let it dry.
7. Using a 10cc lidocaine-filled syringe loaded with a 20 g needle, infiltrate a superficial wheal of 1cc of local anesthetic to minimize the discomfort of the injection. **Note:** Typically, the proceduralist will use a smaller gauge needle (e.g., 27 g) for superficial anesthesia, then switch to a larger (e.g., 18 g) needle for the deeper infiltration. However, to minimize the resources required for this procedure, it is reasonable to use a midgauge (e.g., 20 g) needle for both.
8. After 1 minute, insert the needle through the wheal until bone is encountered.
9. Confirm placement by aspirating blood.
10. Proceed to alternately infiltrate and aspirate blood until the entire volume of anesthetic has been infiltrated along the periosteum.
11. Apply gauze to the infiltration site to staunch any bleeding.
12. Allow at least 10 minutes for the anesthetic to take effect before proceeding with the reduction.

The hematoma block is depicted in Fig. 15.3 below (Graham and Reichman, 2018).

### *15.3.2.3 Contraindications*

Do not proceed if there is:

- A history of allergy to local anesthetics
- An open fracture
- An active infection or overlying cellulitis
- There is neurovascular compromise
- The proceduralist should weigh the risks and benefits of using a hematoma block if there is a history of anticoagulant use or coagulopathy

**TABLE 15.2** Estimated mass and volume of equipment required for a hematoma block.

| Item | Mass | Volume | Indication |
|---|---|---|---|
| *10cc syringe* | 8 g | 10 $cm^3$ | To hold the local anesthetic |
| *10cc 1 or 2% lidocaine without epinephrine* | 14 g | 10 $cm^3$ | To provide anesthesia |
| *22–25 g 1-1 ½" needle* | 2 g | 8 $cm^3$ | To infiltrate the digital nerves with local anesthetic |
| *BZK wipe* | 1.5 g | 4 $cm^3$ | To clean the areas of injection prior to injection |
| *4×4 gauze* | 2 g | 15 $cm^3$ | To staunch any bleeding post infiltration |
| *Nitrile gloves (1 pair)* | 12 g | ~50 $cm^3$ | Provider personal protective equipment |

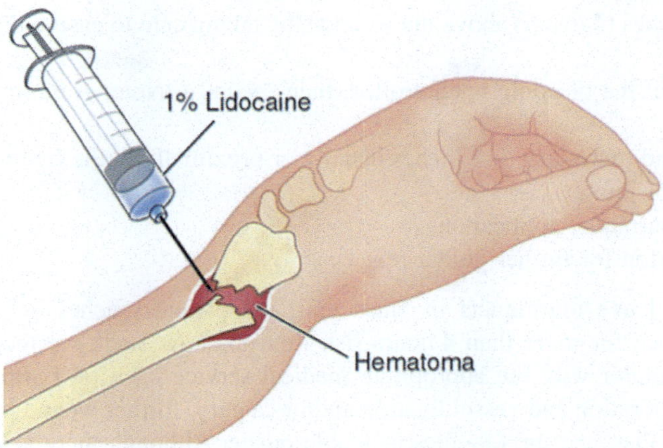


**FIGURE 15.3** Hematoma block. A schematic of the application of the hematoma block.
From: Graham, T.P., Reichman, E.F., 2018. Hematoma Blocks. In: Reichman's Emergency Medicine Procedures, third ed. McGraw-Hill Education. http://accessemergencymedicine.mhmedical.com/content.aspx?aid=1159805309.

**TABLE 15.3** Estimated mass and volume of equipment required for tourniquet application.

| Item | Mass | Volume | Indication |
|---|---|---|---|
| *1–2 Tourniquets* | 80–160 g | 100–200 $cm^3$ | To provide hemostasis |
| *Nitrile gloves (1 pair)* | 12 g | ~50 $cm^3$ | Provider personal protective equipment |

### *15.3.2.4 Complications*

The proceduralist should be aware of rare but potential complications associated with the digital nerve block, including, but not limited to as follows:

- Poor uptake necessitating reinfiltration
- Nerve damage
- Bleeding
- Infection

## 15.4 Tourniquet application

### 15.4.1 Indications

A tourniquet should be applied to an extremity when there is a need for hemorrhage control and can be used alone or as an adjunct to direct wound pressure and/or wound packing. A tourniquet cannot be used at junctional sites, that is, in the groin or axilla, over joints, or over nonextremity areas, such as the head, neck, torso, abdomen, or pelvis. A second tourniquet may be applied either proximally or distally to the first tourniquet if hemorrhage control is not adequate. The material required for this procedure is summarized in Table 15.3 below.

### 15.4.2 Technique

1. Obtain the patient's consent for the procedure after explaining the risks, benefits, side effects, and alternatives of the proposed procedure.
2. Immobilize the patient using a restraint or crew restraint system while taking care not to jostle the bleeding site.
3. Don nitrile gloves.
4. The proceduralist is to affix or tether themselves to stabilize.
5. Identify the site of bleeding.

6. Apply a tourniquet 2–3 inches (5–6 cm) above the wound site, taking care to ensure the tourniquet is not placed over a wound or joint.
7. Tighten the tourniquet until the bleeding is controlled, then secure it. Note: Appropriately tightened tourniquets are quite painful.
8. If necessary, apply a second tourniquet 2–3 inches distally or proximally to the first tourniquet, again taking care not to place it over a wound or joint.
9. Mark down the time of tourniquet application.
10. Initiate specialist consultation for further guidance.

The typical teaching is to leave tourniquets in place until the patient reaches definitive medical care. However, tourniquets that are left in place for more than 4 hours risk limb loss. As such, we recommend undertaking specialist consultation with the CMO, and/or with the appropriate medical service back on Earth, to help determine next steps, which may include wound exploration and vessel ligation and/or cautery, further wound packing, for example, depending on the skillset of the proceduralist, and the facilities available onsite. Although there is some controversy regarding the efficacy of makeshift and nonapproved tourniquets, the literature from battlefield, austere environment, and wilderness medicine settings generally indicates a positive benefit with the proper application of even makeshift tourniquets (Kragh and Dubick, 2017).

Tourniquet application and makeshift tourniquets are depicted in Fig. 15.4 below.

### 15.4.3 Contraindications

Do not proceed if the hemorrhage is (SAM Splint - Tacmed Australia, 2025; Mhaskar and Maheshwari, 2015):

- Junctional (e.g., in the groin or axilla)
- Over a joint
- In a nonextremity (e.g., head, neck, torso, abdomen, pelvis).=

### 15.4.4 Complications

There is a very real risk of loss of limb due to limb ischemia and prolonged tourniquet time, and the proceduralist should carefully assess the risks of "life over limb" in the context of the severity of bleeding, and resources available for definitive hemorrhage control, including specialist consultation, surgical capabilities, and evacuation potential. Additional complications associated with prolonged tourniquet time over 4–6 hours include tissue necrosis, nerve injury, rhabdomyolysis, thromboembolic events, and reperfusion injury (Dayan et al., 2008).

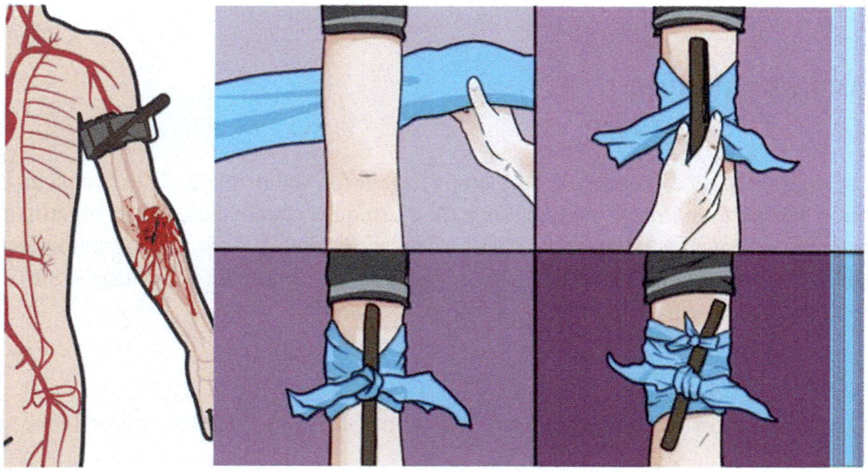

FIGURE 15.4 Tourniquet application techniques. *Left*: Demonstration of the correct placement of a tourniquet 2–3 inches (5–6 cm) above a nonjunctional, nonjoint wound. *Right*: Demonstration of the application of makeshift tourniquet application.
From *(left)*: DeployedMedicine - Tourniquets in TCCC, 2024. https://books.allogy.com/web/tenant/8/books/a30c619d-7270-4bfe-be4f-eb4d27adc783/#id3155bc0e-e45c-4815–8782-165517a34a57; From *(right)*: Lee, L. (Ed.), 2025. How to Decide to Use a Tourniquet (Home Remedy). WikiHow. https://www.wikihow.com/Decide-to-Use-a-Tourniquet-%28Home-Remedy%29

## 15.5 Blister drainage

### 15.5.1 Indications

Blisters generally occur as a result of repeated friction, causing skin shearing and separation within the epidermis. When nondisabling or asymptomatic, blisters are generally left alone; however, drainage can be indicated when the location and/or size of the blister impedes the astronaut's function owing to position and/or pain (e.g., a large blister over a friction point where an EVA or IVA suit may be worn). Drainage may also be indicated if the blister is infected, as evidenced by blister contents that may be purulent-looking in appearance, and/or cellulitic changes to the surrounding skin (e.g., redness, swelling, warmth to touch, and/or tenderness). It is generally recommended to aspirate any blisters while maintaining the integrity of the blister roof, to minimize both patient discomfort and the risk of secondary infection (John, 2025). The material required for this procedure is summarized in Table 15.4 below.

### 15.5.2 Technique

1. Obtain the patient's consent for the procedure after explaining the risks, benefits, side effects, and alternatives of the proposed procedure.
2. Immobilize the affected appendage using a restraint or crew restraint system while taking care not to jostle the bleeding site.
3. Don nitrile gloves.
4. The proceduralist is to affix or tether themselves to stabilize.
5. Clean the roof of the blister with a BZK wipe, taking care not to rupture it.
6. Identify an aspiration site at the edge of the blister to keep the blister roof intact.
7. With a 20 or 22 g needle locked to a 10cc syringe, introduce the needle at the identified aspiration site while gently but steadily drawing back on the syringe plunger while advancing the needle parallel to the blister interface to aspirate the blister contents. Light pressure may be applied to the roof of the blister to aid with drainage.
8. Once aspiration is complete, inspect the blister to ensure complete or near-complete drainage. Large blisters may be best drained by introducing multiple aspiration sites.
9. Leave the blister roof intact as a biological dressing and consider applying a dressing or bandage to the blister after drainage to protect the site from further friction or shearing trauma.

Blister drainage is depicted in Fig. 15.5 below.

### 15.5.3 Contraindications

Do not proceed with aspiration if the blister is asymptomatic, painless, and/or nondisabling, as this can introduce a risk of secondary infection

**TABLE 15.4** Estimated mass and volume of equipment required for blister aspiration.

| Item | Mass | Volume | Indication |
|---|---|---|---|
| *10cc syringe* | 8 g | 10 $cm^3$ | To form a receptacle in which to drain the blister contents |
| *Nitrile gloves (1 pair)* | 12 g | ~50 $cm^3$ | Provider personal protective equipment |
| *20–22 1-1 ½" needle* | 2 g | 8 $cm^3$ | To puncture and aspirate the blister contents |
| *BZK wipe* | 1.5 g | 4 $cm^3$ | To clean the areas of injection prior to aspiration |
| *4×4 gauze* | 2 g | 15 $cm^3$ | To clean any fluid post aspiration |
| *Dressing or bandaid* | 1 g | 5 $cm^3$ | To protect the blister site post drainage |

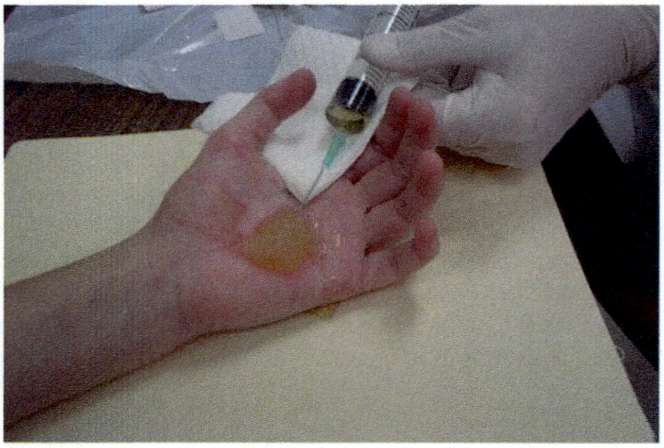

FIGURE 15.5 Blister drainage. Demonstration of needle aspiration of a disabling blister, with the needle introduced at the edge of the blister and kept relatively parallel to the blister floor while simultaneously advancing the needle and aspirating the plunger.
From: Minor Burn Management, 2021. https://litfl.com/tag/minor-burn/.

### 15.5.4 Complications and additional considerations

Ruptured blisters that have been de-roofed should be cleansed, monitored, and dressed daily to prevent further trauma and to monitor for healing and signs of developing infection. Antibiotics are not typically indicated unless signs or symptoms of infection are present, such as spreading erythema in the surrounding soft tissues and/or purulent drainage from the blister site. In these cases, topical or oral antibiotics may be indicated, according to the clinician's assessment. Where the proceduralist lacks the clinical experience to make this pronouncement, Earth-based consultation should be initiated. In these cases, de-roofing of the blister is indicated using forceps and either scissors or a scalpel, so as to prevent bacterial sequestration.

## 15.6 Immobilization and splinting: upper and lower extremity, with or without reduction

### 15.6.1 Indications

Musculoskeletal injuries can occur from overuse or trauma. Ideally, diagnostic imaging can help assess the severity of the injury. PoCUS, for example, is currently available in spaceflight and can help aid in the diagnosis of swelling, sprains, tears, and/or fractures. Portable x-ray modalities, while not currently widely available, have been deployed experimentally in spaceflight and can help in the diagnosis of fractures, dislocations, and fracture-dislocations (First X-rays taken in space mimic historic first radiograph captured on earth, 2025). For overuse, low-grade strain, and sprain injuries, wrapping and/or taping can provide temporary relief and prevent further injury. High-grade strains and sprains, fractures, dislocations, and/or fracture-dislocations diagnosed clinically and/or by point-of-care imaging should be splinted and immobilized. The duration of immobilization depends on the location, nature, and severity of injury, and specialist consultation should be sought for specific guidance. The materials required for wrapping, taping, and splinting procedures are summarized in Table 15.5 below.

SAM splints, or similar clinical-grade splinting materials, can make for helpful immobilization devices and are used for reference; however, when not available, any type of rigid material that can be adequately molded and immobilize the affected injury, including the joints above and below, may be sufficient. As the indications and types of splinting can vary widely based on the location and type of injury, this chapter will address the general principles of splinting for upper and lower extremity splinting, with a few select examples given. In general, however, low-grade strains and sprains and repetitive motion injuries will benefit from wrapping, taping, and rest, while high-grade sprains and tears, fractures, and dislocations should be immobilized and splinted. Specialist consultation should be sought for nonhealing repetitive motion injuries, high-grade sprains and strains, complex fractures, dislocations, and where the crew member's inexperience or lack of medical training prevents them from making an informed clinical judgment as to the nature of the injury.

**TABLE 15.5** Estimated mass and volume of equipment required for splinting, taping, and/or immobilization of an injured or fractured joint.

| Item | Mass | Volume | Indication |
|---|---|---|---|
| *SAM splint* | 110 g | 450 $cm^3$ | Immobilization of an injured or fractured extremity |
| *Gauze wrap* | 35 g | 120 $cm^3$ | To provide padding and protection for the skin |
| *Tensor bandage* | 50 g | 180 $cm^3$ | To secure the SAM splint in place |
| *Fastening clip, pin, or tape* | 5 g | 10 $cm^3$ | To secure the tensor bandage in place |
| *Athletic tape* | 60 g | 90 $cm^3$ | For wrapping and supporting injured joints |
| *Sling* | 45 g | 200 $cm^3$ | To provide support for an injured arm. |

### 15.6.2 Technique

1. The proceduralist should obtain a thorough history as to the mechanism, duration, severity, and quality of injury to inform their formulation of a reasonable differential diagnosis.
2. Obtain the patient's consent for the procedure after explaining the risks, benefits, side effects, and alternatives of the proposed procedure.
3. Immobilize the affected appendage using a restraint or crew restraint system while taking care not to jostle the affected site.
4. The proceduralist is to affix or tether themselves to stabilize.
5. Assess the most distal part of the patient's affected appendage for neurovascular integrity and open wounds. Altered, decreased, or absent sensation, poor or absent capillary refill, or a pale, cold limb are all indicators of neurovascular compromise and can be limb-threatening, and should be considered a medical emergency. Open wounds with a concomitant fracture comprise a compound fracture and are considered a surgical emergency. In these cases, after initial assessment and management, specialist consultation should be sought immediately. In the setting of open fracture or concern for open fracture, IV antibiotics should be given.
6. If available, and if the proceduralist's skillset permits, PoCUS and/or x-ray imaging should be obtained to assess for the presence of local swelling, joint effusions, tendinous and/or ligamentous tears, and/or fractures. Fractures should be assessed for displacement, angulation, and comminution. The acquisition and interpretation of PoCUS and/or portable x-ray is beyond the scope of this chapter; where point-of-care imaging is suggested, it is assumed that the practitioner has previously trained in this skillset.
7. Based on their clinical assessment of the injury, the practitioner may decide to wrap, tape, or splint the injury. The general principles of each are described below; however, specific techniques will depend on the type and location of the injury involved, and the proceduralist should have a low threshold for initiating specialist consultation.
   a. Wrapping. Minor sprains, strains, and repetitive motion injuries may benefit from support via taping or wrapping with an elastic-type bandage, as well as rest, analgesic medication, and an individualized rehabilitation or physiotherapy plan. Additional elements of recovery for simple Earth-based musculoskeletal injuries, such as ice and elevation, may not be possible in the microgravity and resource-limited environment of a space mission.
      i. Position the affected joint in a fairly anatomical, neutral position. Ankle sprains, for example, must be maintained in a way that avoids excessive dorsi- or plantar-flexion.
      ii. Create an anchor point either proximally or distally to the injury and start wrapping in the opposite direction (e.g., if anchoring proximally, start wrapping distally, and vice-versa), maintaining even pressure and with 50% bandage overlap with each wrap. Wrapping can proceed more controlled if the roll is held facing outward toward the proceduralist rather than inward toward the patient.
      iii. Joints and/or smaller parts (e.g., over the back of the hand) may benefit from figure-of-8 or twisting wraps for better support.
      iv. Ensure that fingers or toes are left unwrapped for ease of neurovascular assessment.
      v. Secure the elastic bandage with tape, pins, or clips.

vi. Reassess the neurovascular status of the limb post wrapping to ensure the wrap is not too tight. Neurovascular status should be assessed at regular intervals throughout the duration of the wrap. If a patient is neurologically and cognitively well, this can be achieved by asking the patient to report any changes in sensation, pain, or circulation in the affected area.

vii. An example of wrist wrapping is depicted in Fig. 15.6 below.

1. **Taping**. Joint taping can be used for both injury prevention and treatment, to reduce load on joints, to prevent pressure on pain points, to provide support to joints where a previous injury has occurred, or for pain relief from minor sprains, strains, and repetitive motion injuries (Wilkerson, 2002). Taping may be particularly effective for EVAs, where the joints of an EVA suit can add extra pressure points on joints. Principles of joint taping will vary on the joint, but as a basic principle, the joint should be positioned in an anatomical, neutral position, underwrap should be applied first to protect the skin underneath, the taping should not be excessively tight, the taping strategy should focus on stabilizing/ supporting a specific joint or muscle group, and distal appendages, like fingers and toes, should remain open to allow for neurovascular assessments and mobility. Ankle taping is described here as an example.
   a. Position the ankle in a neutral position, taking care not to overplantar- or dorsi-flex the joint.
   b. Protect the skin by applying underwrap using the wrapping technique described in Fig. 15.7.
   c. Establish anchor points at the top and bottom of the wrap as depicted in Fig. 15.7.
   d. Proceed to establish three stirrups from the top medial anchor point to the lateral aspect of the anchor, with 1–2 cm of overlap on each strip, as depicted in Fig. 15.7. You may choose to apply 2–3 anchoring strips over the stirrups at the proximal anchor point.
   e. Establish 2–3 figure-of-8 overlaps, starting proximally and laterally, descending inferiorly to crossover the anterior ankle to arrive at the midfoot medially, passing under the lateral midfoot, and passing supero-medially to again cross the anterior of the ankle, returning to the medial proximal ankle, and rejoining the lateral starting point to finish the figure-of-8 anchors from medial to lateral.
   f. Lock the heel by establishing 2–3 J-shaped strips starting medially, wrapping under the calcaneus, and ending back at the medial starting point, and then repeating this from lateral to medial.
   g. Repeat the figure-of-8 pattern (see Fig. 15.7).
   h. Close any gaps in the ankle taping with additional strips.
   i. Verify that the taping isn't too tight or uncomfortable for the patient, and the circulation isn't compromised to the distal digits, and that neurovascular status remains intact.

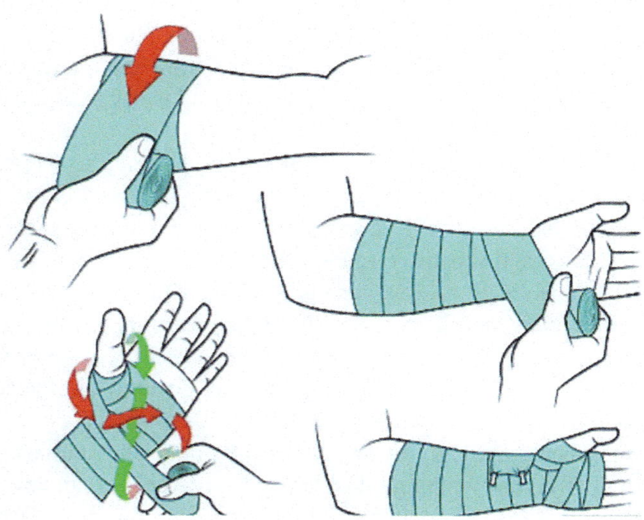

FIGURE 15.6 Wrist wrapping. Wrapping techniques should start by establishing an anchor point either distally or proximally to the injury, then proceeding in the opposite direction with controlled, even pressure and 50% of overlap of the previous wrap, until the wrap provides support and stabilization proximally, distally, and directly over the image. The wrap should be secured using clips, pins or tape, and fingers or toes should be left open to allow for neurovascular assessment. While wrapping with the roll facing inwards as shown is not wrong, wrapping with the roll facing the practitioner allows for more control and a decreased chance of dropping and unraveling the bandage.
How to Wrap a Wrist, 2025. https://www.wikihow.com/Wrap-a-Wrist.

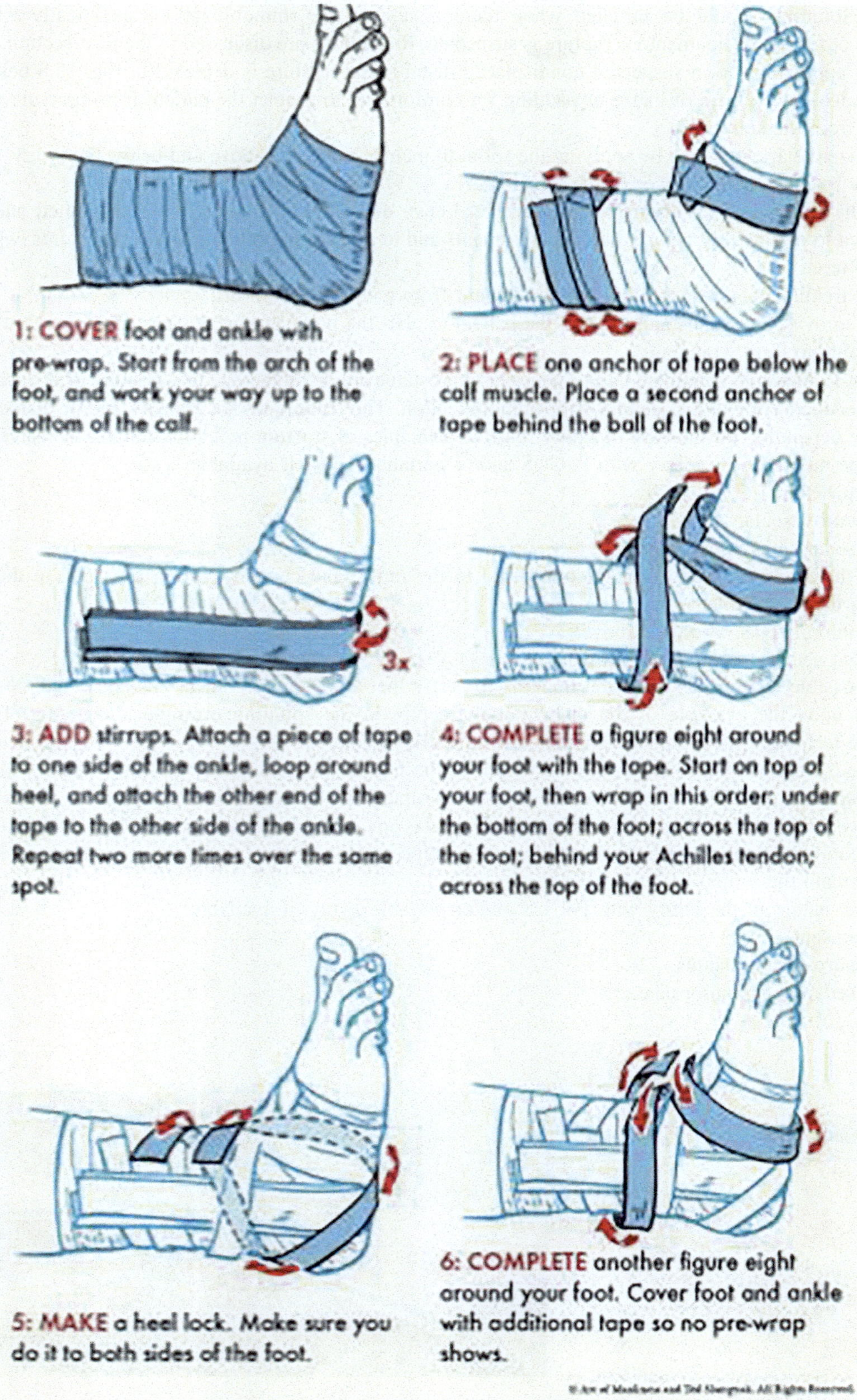


**FIGURE 15.7** Ankle taping. Taping techniques should start with applying underwrapping to protect the skin underneath (step 1), then establishing anchor points with tape distally and proximally (step 2). Next, tape should be applied in a stirrup fashion (step 3), and completed with a figure-of-8 (steps 4 and 6), alternating with a heel lock (step 5), before proceeding to fill in any remaining blank spaces with controlled, even pressure and 50% of overlap of the previous wrap, until the wrapping provides support and stabilization proximally, distally, and directly over the injury. The wrap should be secured using clips, pins or tape, and fingers or toes should be left open to allow for neurovascular assessment.
Modified from Ted Slampysk, "How to Tape a Sprained Ankle." https://www.reddit.com/r/selfreliance/comments/lnb2zx/how_to_tape_an_ankle_an_illustrated_guide/#lightbox.

2. **Splinting**. Splinting should be initiated when there is a need for immobilization, specifically when a fracture, dislocation, or tendon or ligamentous rupture is suspected. Reductions are discussed in the next section. An example of SAM splint application for a suspected nondisplaced distal radial fracture is depicted in Fig. 15.8 below.
   **a.** Apply at least two layers of gauze or padding for comfort and to protect the patient from pressure points that may develop from the splint.
   **b.** Immobilize the affected area by applying the splint to include the joints above and below the injury. Mold the splint to follow the anatomical curves of the affected area.
   **c.** Secure the splint with an elastic wrap or gauze. Leave distal digits (fingers, toes) unsplinted and unwrapped if unaffected to permit easy neurovascular assessments and to avoid immobilizing unaffected joints, which can lead to joint stiffness.
   **d.** Upper extremity fractures may additionally benefit from a sling for comfort.
   **e.** Verify neurovascular status and counsel the patient to alert the proceduralist if signs of neurovascular compromise develop due to tight wrapping.
3. **Reductions**. If anatomic alignment and/or neurovascular status are disrupted due to a fracture or dislocation, the injury should be reduced prior to proceeding with immobilization. The techniques for fracture and/or dislocation reduction vary greatly depending on the type of injury, but the principles of fracture reduction remain the same:
   **a.** Verify the nature of the injury with PoCUS and/or portable X-ray, if available.
   **b.** Obtain consent.
   **c.** Verify neurovascular status.
   **d.** Provide sedation as appropriate.
   **e.** Distract the fracture by applying traction distal to the injury and counter-traction proximal to the injury, before returning the injury to anatomic alignment.
   **f.** Splint/immobilize as described above.
   **g.** Verify that neurovascular status remains intact.
   **h.** Where possible, obtain postreduction imaging to verify that anatomic alignment has been restored.
   Building upon the example of the distal radial fracture in the splinting example above, Fig. 15.9 depicts the reduction technique for a dorsally angulated distal radius fracture.
4. **Dislocations**. A dislocation refers to the disruption of the normal joint congruity and may occur with or without an accompanying fracture. When neurovascular status is disrupted due to a dislocation, *this can be limb-threatening, and specialist consultation should be sought immediately, and evacuation considered*, if feasible. Like fractures, techniques for dislocation reduction vary greatly depending on the type and locations of dislocation, but the principles of fracture reduction remain the same:
   **a.** Verify the nature of the injury with PoCUS and/or portable X-ray, if available.
   **b.** Obtain consent.
   **c.** Verify neurovascular status.
   **d.** Provide sedation as appropriate.

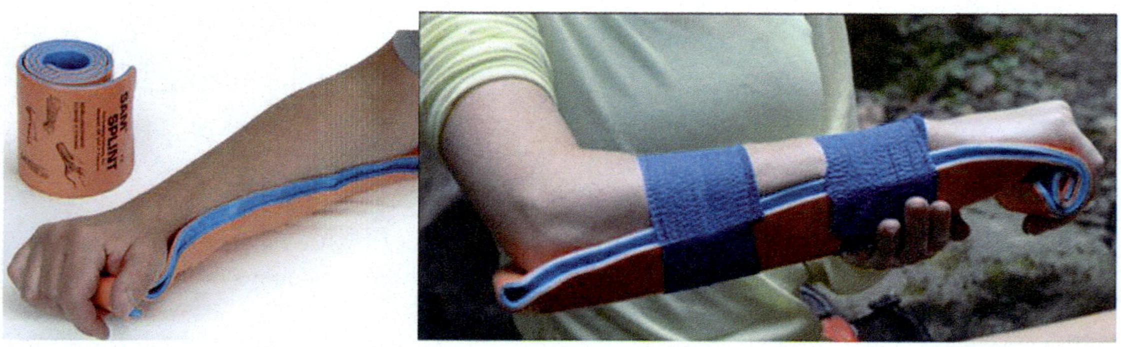

**FIGURE 15.8** SAM splint application. SAM splint application for a suspected nondisplaced distal radius fracture *(left)*. Where possible, the skin should be wrapped with a protective layer such as gauze for comfort. Note how the splint is molded to support a neutral anatomic position. The splint should then be secured with gauze or elastic wrap.
From SAM Splint - Tacmed Australia, 2025. https://tacmedaustralia.com.au/products/sam-splint-original?srsltid=AfmBOor5gzwx4zL_k965Hv3PeDsxIDg1i_LBaOf79kVrV1u88Y6iZ2dj.

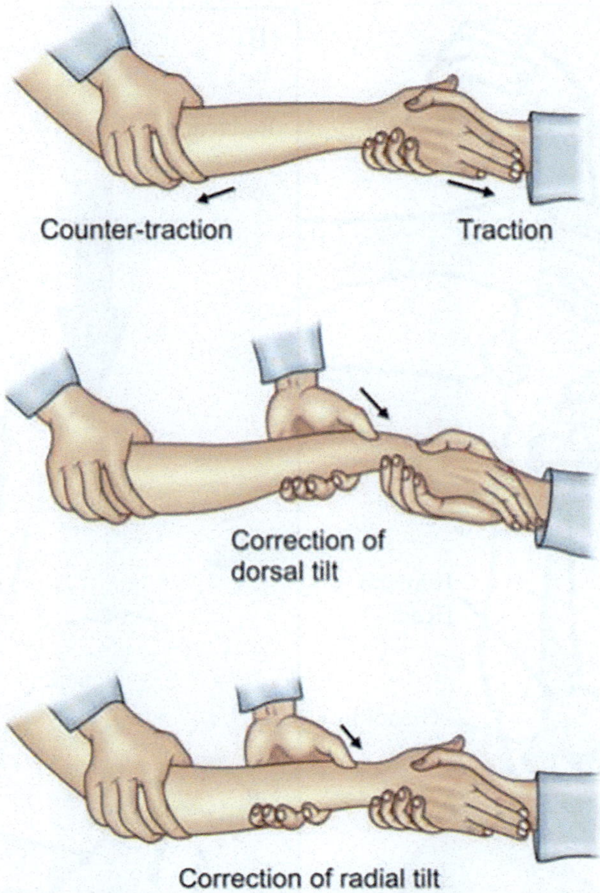


FIGURE 15.9 Reduction technique of a dorsally displaced distal radius (Colles) fracture. The fracture is first distracted through traction and counter-traction *(top)*, and the anatomic alignment restored in both the dorsal and radial planes *(middle, bottom)*, after which a splint is applied, postreduction imaging obtained, and neurovascular status reestablished.
From Mhaskar, V., Maheshwari, J., 2015. Essential Orthopaedics (Including Clinical Methods), fifth th ed. Jaypee Brothers Medical Publishers. https://doi.org/10.5005/jp/books/12526_16

**e.** Many dislocations will necessitate the application of gentle traction distal to the injury and counter-traction proximal to the injury, before restoring the joint to normal anatomic alignment. **Note:** Note that this is a general principle only, and specific techniques may apply to different joints.
**f.** Splint/immobilize as described above.
**g.** Verify that neurovascular status remains intact.
**h.** Obtain postreduction imaging to verify that anatomic alignment has been restored.

The reduction of an anterior shoulder dislocation is depicted in Fig. 15.10

Similarly, lower extremity dislocations follow the same principles of reduction, but the technique is wholly dependent on the location and mechanism of injury. A lateral patellar dislocation may be reduced on the spot through gentle lateral replacement if the mechanism was not the result of a high-energy trauma (Fig. 15.11).

### *15.6.2.1 Emergent cases: neurovascular compromise, compound fractures, pelvic fractures, and femur fractures*

Special attention is paid to cases considered life-or-limb-threatening and necessarily mission-ending.

#### 15.6.2.1.1 Neurovascular compromise and compartment syndromes

Trauma, injury, or rhabdomyolysis has the potential to compromise the neurovascular status of the affected limb. The proceduralist should sound the alarm if they identify decreased, altered, or absent sensation, sluggish or no capillary refill,

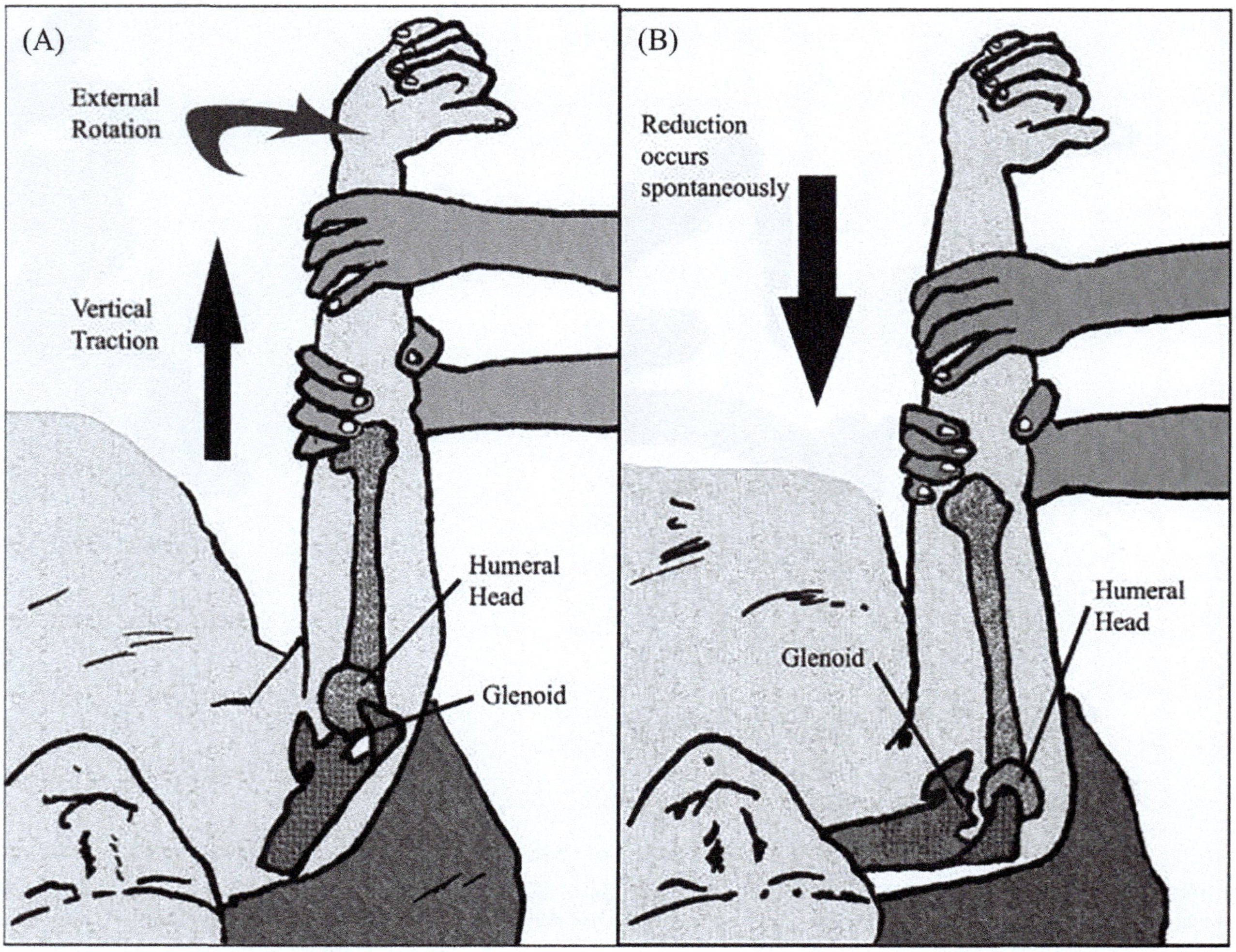


**FIGURE 15.10** Reduction of a shoulder dislocation using the Spaso technique. Reduction technique of an anterior shoulder dislocation using the Spaso technique. The upper extremity is held vertically with gentle traction upward vertical traction to open up the joint space, while the weight of the patient's torso acts as counter-traction *(left)*. The upper extremity is then gently externally rotated until the humeral head is reduced back into the glenohumeral joint *(right)*.
From: Reduction of anterior shoulder dislocations by spaso technique: clinical results, 2008. The Journal of Emergency Medicine, 34 (4), 383–387. https://doi.org/10.1016/j.jemermed.2007.07.026.

and/or a cold, pale, pulseless limb. If caused by the injury itself, the proceduralist may have success in reduction and should assess the neurovascular status of the affected limb prereduction and postreduction. Where caused by the reduction itself, the proceduralist should immediately seek consultation with a specialist for further consultation; however, such an injury is typically limb-threatening without surgical intervention, and where possible, immediate plans for evacuation should be initiated.

In cases where compartment syndrome is suspected, often due to a crush injury, closed fracture, or burn, the affected limb will also be quite tense, due to the pressures of the fascial compartment exceeding the perfusion pressures of the same. Diagnosis can be challenging in the unconscious patient and should be assessed based on mechanism of injury, clinical exam, and index of suspicion (Compartment Syndrome, 2013). Compartment syndromes may present as pale or purple in color, pulses may be weak or absent, the patient may complain of altered or absent sensation, paralysis or paresis, and palpation of the limb may cause pain out of proportion to the magnitude of touch. However, early compartment syndrome may present only as severe pain and swelling in the affected limb. On physical exam, the presence of severe pain with passive stretch should raise the alarm for the presence of a compartment syndrome. **Note that the presence of a distal pulse does not preclude the presence of a compartment syndrome** (Ardolino et al., 2010). Without intervention, disruption of neurovascular supply can progress to limb ischemia, and ultimately, loss of the limb. If available, a pressure

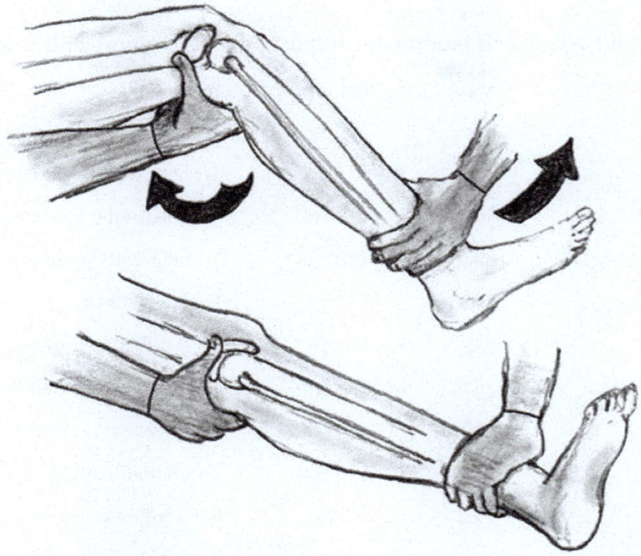

FIGURE 15.11 Technique for reduction of a laterally dislocated patella. Reduction technique of a lateral patellar displacement. The lower extremity is stabilized distally, while the proceduralist palpates for the lateral edge of the displaced patella, stabilizing their hand by placing their fingers in the popliteal fossa *(top)*. The patellar is gently translated medially, while the knee is straightened, facilitating the reduction of the patella *(bottom)*. From: Reducing a dislocated patella I Wilderness Medicine Newsletter, n.d. https://wildernessmedicinenewsletter.wordpress.com/2006/11/11/reducing-a-dislocated-patella/

monitor such as a handheld manometer should be used to confirm the compartment pressure. In general, a compartment pressure greater than 30 mmHg is sufficient to warrant a fasciotomy; however, fasciotomy should not be delayed to obtain compartment pressure if compartment syndrome is clinically diagnosed (Ardolino et al., 2010). In cases where a compartment syndrome is suspected, the proceduralist should carefully weigh the risks and benefits of a fasciotomy, taking into account:

- Their own skillset and training
- The resources they have at their disposal
- The possibility and timing of evacuation
- The time to definitive medical care

Regardless, surgical consultation should always be sought, where available, given the complexity of undertaking such a procedure in a resource-limited, reduced-gravity environment.

## 15.7 Fasciotomies

The principles for a fasciotomy are generally the same, regardless of anatomical location, although the anatomical landmarks will depend on the compartment and limb that are affected. A thorough discussion of fasciotomy procedures, different approaches, and critical anatomical structures of every upper and lower limb compartment is beyond the scope of this chapter. An upper extremity superficial and deep volar compartment fasciotomy of the forearm and lower extremity fasciotomy of the are reviewed here as examples. The materials needed are reviewed in Table 15.6 below.

### 15.7.1 Forearm fasciotomy

1. Obtain consent.
2. Verify neurovascular status and obtain quantitative compartment pressures, where possible.
3. Cleanse the site with an antiseptic solution.
4. Administer 2 g of IV cefazolin, if available.
5. Provide local anesthesia and/or sedation as appropriate. Ideally, this is a regional block; however, if necessary, it can be done with local infiltration within the line of the incision.

**TABLE 15.6 Estimated mass and volume of equipment required for fasciotomy and fasciotomy aftercare.**

| Item | Mass | Volume | Indication |
|---|---|---|---|
| *Scalpel* | 6 g | 20 $cm^3$ | To make an incision to access the relevant compartments |
| *Forceps* | 25 g | 60 $cm^3$ | To help with tissue manipulation |
| *Scissors* | 50 g | 120 $cm^3$ | To release the fascia containing a compartment |
| *Electrocautery (suggested not required)* | 80 g | 150 $cm^3$ | To help with hemostasis |
| *Local anesthetic (e.g., 2% Lidocaine)* | 14 g | 15 $cm^3$ | For pain control |
| *BZK wipes (4)* | 6 g | 16 $cm^3$ | To sterilize the tissue prior to starting the procedure |
| *Dressing (gauze, wrap)* | 25 g | 200 $cm^3$ | To dress the affected limb postprocedure |
| *Saline (250cc)* | 270 g | 280 $cm^3$ | To irrigate and cleanse the affected area |
| *Splinting material* | 110 g | 450 $cm^3$ | To immobilize the affected area postprocedure |
| *Cefazolin* | 25 g | 30 $cm^3$ | For antibiotic prophylaxis |
| *Intravenous needle & catheter (18–22 g)* | 5 g | 15 $cm^3$ | For intravenous access |

6. The provider should be aware of critical anatomical structures in the superficial and deep volar compartments, namely (Compartment Syndrome, 2013):
   a. Superficial volar compartment:
      i. Muscle groups: pronator teres, flexor carpi radialis, flexor carpi ulnaris, flexor digitorum superficialis, palmaris longus
      ii. Vessels: ulnar artery
      iii. Nerves: median and ulnar nerves
   b. Deep volar compartment:
      i. Muscle groups: flexor digitorum superficialis, pronator quadratus, flexor digitorum profundus, flexor pollicis longus
      ii. Vessels: anterior interosseous artery
      iii. Nerves: anterior interosseous nerves
7. Proceed with a backwards "S" skin incision over the volar forearm, starting 1 cm proximal and 2 cm lateral to the medial epicondyle, extending obliquely over the antecubital fossa, arcing around to the midvolar forearm, extending distally down to the wrist (Fig. 15.12).
8. Once the fascia is exposed, open up the superficial volar compartment by incising the fascia with scissors.
9. Dissect down over the palmaris longus to expose the deep volar compartment.
10. Incise the deep volar compartment over the flexor digitorum superficialis using scissors.
11. Use the electrocautery for hemostasis where necessary.
12. ***Postprocedure care***. Associated fractures of the upper extremity should be reduced to protect the limb from further injury. Position the upper extremity in a neutral or functional position with gentle elevation, with the hand positioned in the position of safety, or "intrinsic plus" position: the metacarpal-phalangeal (MP) joint positioned at 70–90 degrees; the proximal interphalangeal (PIP) and distal interphalangeal (DIP) joints in extension; and the wrist in slight extension to prevent extension contractures (Fig. 15.12). The wound should be packed with wet-to-dry or bulky dressings, as it is assumed that negative pressure wound therapy is not available in the spaceflight environment. The extremity should be monitored hourly, and the wound assessed for debridement and closure at 24–48 hours (Compartment Syndrome, 2013).
13. ***Complications***. Excessive edema can delay primary closure to 7–10 days or even require a partial-thickness skin graft. Delayed fasciotomies performed at 12 hours or beyond carry a higher risk of reperfusion injury or infection and carry a two-fold amputation risk and a three-fold mortality risk. Even early fasciotomies don't guarantee adequate reperfusion. Additional complications include the risk of neurovascular injury (especially to the medial antebrachial nerve proximally and the palmar branch of the median nerve distally), rhabdomyolysis, and acute renal failure from

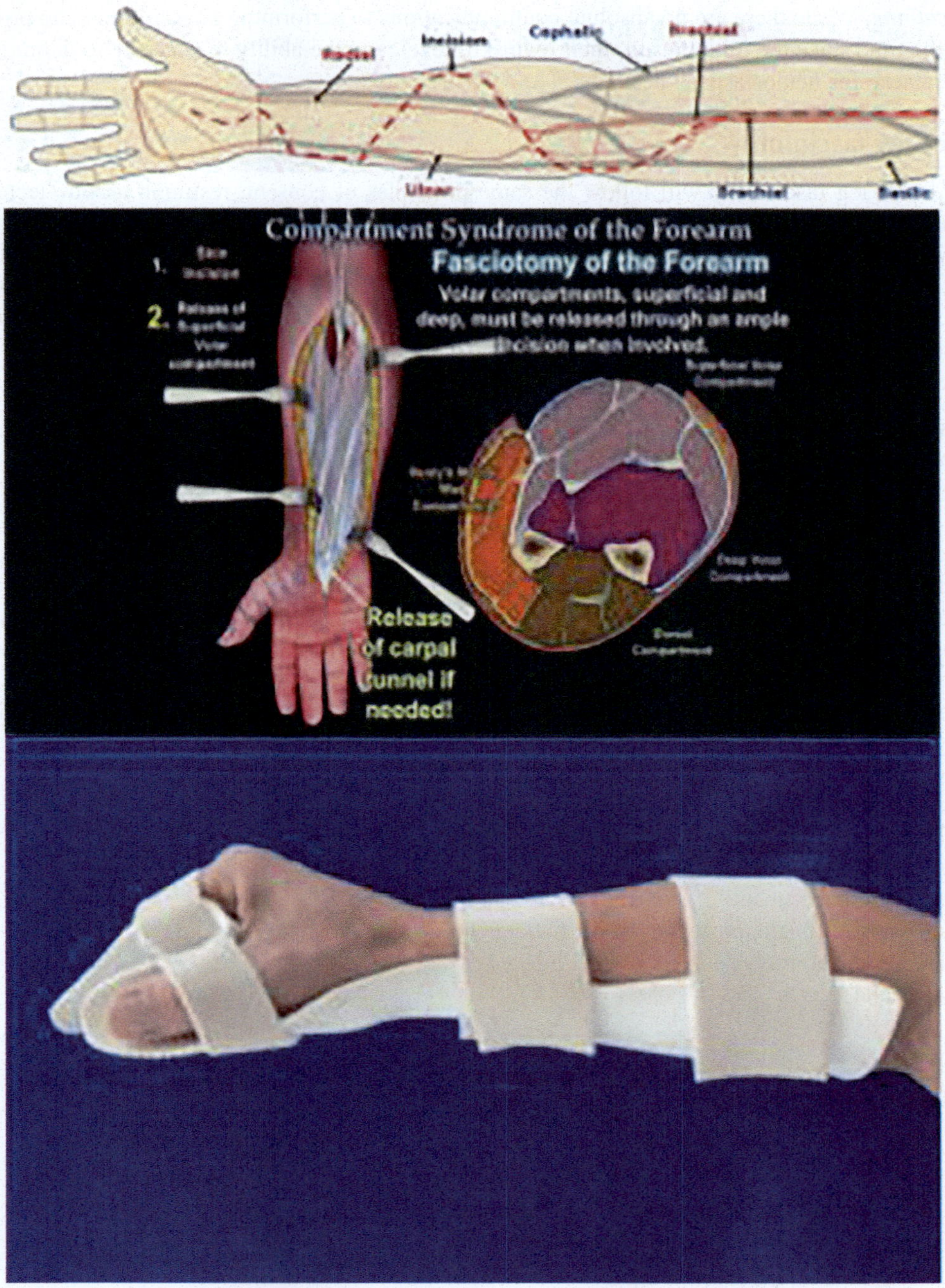


**FIGURE 15.12** Relevant landmarking, anatomy and splinting for a forearm compartment syndrome release. Landmarking and trajectory for the "S"-shaped incision needed for a superficial and deep volar forearm compartment release *(top)*, the relevant muscle groups, nerves, vessels, and relationships of the compartments to one another *(middle)*. Postprocedure, the upper extremity should be extended and the hand, wrist, and MP, PIP, and DIP joints positioned in the position of safety, or the "instrinsic plus" position *(bottom)*.
*Top image*: From Compartment Syndrome, 2013. AO Surgery Reference. https://surgeryreference.aofoundation.org/orthopedic-trauma/adult-trauma/forearm-shaft/further-reading/compartment-syndrome#diagnosis; *Middle image*: From Compartment Syndrome of the Arm – Everything You Need to Know, 2012. Dr. Nabil Ebraheim. https://www.youtube.com/watch?v=Q5Rrch-0TBA. *Bottom image*: From Bersch, I., Krebs, J., Fridén, J., 2022. A Prediction Model for Various Treatment Pathways of Upper Extremity in Tetraplegia. Frontiers in Rehabilitation Sciences, 3. https://doi.org/10.3389/fresc.2022.889577.

muscle necrosis, gangrene, contractures, venous thromboembolisms (VTEs), complex regional pain syndrome, or even death (Ormiston and Marappa-Ganeshan, 2023). Additionally, incomplete fasciotomies may necessitate revision fasciotomy to extend the initial incision or open up additional compartments, the risks of which increase with anatomy distortion, for example, from severe trauma or prior scarring (Compartment Syndrome - Lower Limb, 2024).

14. ***Contraindications***. While there are no absolute contraindications to performing a fasciotomy, the proceduralist needs to consider the time since the initial insult, their own skill level, and the ability to evacuate to definitive medical care. Use electrocautery for hemostasis as needed.

### 15.7.2 Lower leg fasciotomy

The steps for a lower leg fasciotomy will follow the same principles of consent, restraint, aseptic technique, antibiotic prophylaxis, analgesia, and sedation as outlined in steps 1–6 for superficial and deep compartment volar forearm fasciotomy above but will differ in their positioning, anatomical landmarking, and after care. All four compartments of the lower leg must be fully decompressed, and decompression of the lower leg is achieved through two incisions: an anterolateral incision to decompress the anterior and lateral compartments and a separate medial incision to decompress the deep and superficial posterior compartments. The steps for a lower leg fasciotomy are outlined below, and the materials used are the same as outlined for forearm fasciotomy in Table 15.6 above.

1. Obtain consent.
2. Verify neurovascular status and obtain quantitative compartment pressures, where possible.
3. Cleanse the site with an antiseptic solution.
4. Administer 2 g of IV cefazolin, if available.
5. Provide local anesthesia and/or sedation as appropriate. As above, regional anesthesia is preferred, but it can be done with local infiltration
6. The provider should be aware of the four compartments of the lower leg, and the critical structures within each, as named below, and pictured in Fig. 15.13 namely (Compartment Syndrome - Lower Limb, 2024):
   a. Anterior compartment:
      i. Muscle groups: tibialis anterior, extensor hallucis, longus, extensor hallucis longus, extensor digitorum longus
      ii. Vessels: anterior tibial artery
      iii. Nerves: deep peroneal nerve
   b. Lateral compartment:
      i. Muscle groups: peroneus longus, peroneus brevis
      ii. Nerves: superficial peroneal nerve
   c. Deep posterior compartment:
      i. Muscle groups: tibialis posterior, flexor hallucis longus, flexor digitorum longus
      ii. Vessels: peroneal artery, posterior tibial artery
      iii. Nerves: tibial nerve

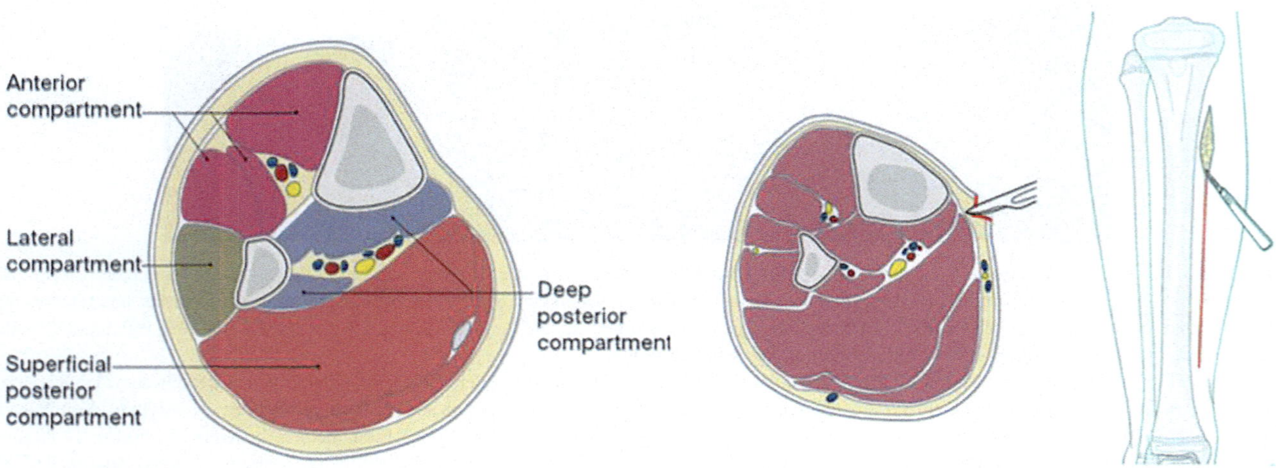


FIGURE 15.13 Critical structures, landmarks and anatomy for the lower extremity fasciotomy. The proceduralist should be aware of the four compartments of the lower extremity, and the critical structures contained therein. Here, a coronal section demonstrates the spatial relationship of the anterior, lateral, deep posterior, and superficial posterior compartments of the lower extremity *(left)*. A single vertical posteromedial incision is used to expose the deep and superficial posterior compartments, beginning 1 cm posterior to the palpated edge of the posteromedial tibia, extending 2 cm superior to the medial malleolus *(right)*. The anterolateral incision is not pictured.
From Compartment Syndrome - Lower Limb, 2024. https://surgeryreference.aofoundation.org/orthopedic-trauma/pediatric-trauma/tibial-shaft/further-reading/compartment-syndrome.

   **d.** Superficial posterior compartment:
      **i.** Muscle groups: gastrocnemius, soleus
      **ii.** Nerves: sural nerve

**7.** Medial incision for posteromedial dissection and release. The proceduralist must take care to protect the superficial peroneal nerve laterally, and the saphenous nerve and long saphenous vein medially while making initial incisions, and also protecting the posterior tibial nerve and artery during deep dissection. A single vertical posteromedial incision is used to expose the deep and superficial posterior compartments, beginning 1 cm posterior to the palpated edge of the posteromedial tibia, extending 2 cm superior to the medial malleolus (Fig. 15.13). If the saphenous vein is injured on incision, it should be ligated with silk suture or a clip. Do not attempt to control it with electrocautery.

**8.** After dissecting down to the level of the fascia of the superficial posterior compartment, the proceduralist will dissect anteriorly toward the posterior tibial edge to spare the long saphenous vein and saphenous nerve. This both spares these critical neurovascular structures and also meets the deep posterior compartment fascia at its most superior and accessible point. From here, the proceduralist should proceed to open the deep posterior fascia with scissors, taking care to **completely** release the fascia distally and proximally under the soleus muscle, **while taking care to protect the posterior tibial neurovascular bundle**.

**9.** The proceduralist should then proceed to open the fascia of the superior posterior compartment completely, both proximally and distally, by moving their scissors just posterior and parallel to the fasciotomy of the deep posterior compartment.

**10.** Lateral incision for anterolateral dissection and release. A single 20 cm vertical incision is made to access both the anterior and lateral compartments, starting 1 cm lateral to the anterolateral edge of the tibia to protect the deep perforators, and extending longitudinally to expose the length of both compartments. Scissors are used to open the fascia of the anterior compartment both distally and proximally, following the same line as the skin incision. The proceduralist should take care to spare visible superficial nerves, which will take on fine, shiny, sinewy hair-like strands in appearance.

**11.** To reach the lateral compartment, the proceduralist is advised to reflect the extensor muscles medially using retractors to expose the lateral intermuscular septum of the lateral compartment, as it borders on the anterior compartment. This lateral fascia is then released using scissors along the length of the fascia.

**12.** Electrocautery should be used for hemostasis where necessary.

**13.** ***Postprocedure care***. Associated fractures should be reduced, and the foot and ankle immobilized with a splint or backslab in a neutral or dorsiflexed position to mitigate contracture formation, while leaving the toes unencumbered. The fasciotomy wounds should be left open and packed with wet-to-dry or bulky dressings, as it is assumed that negative pressure wound therapy is not available in the spaceflight environment. The extremity should be monitored hourly, and the wound assessed for repeat debridement 48–96 hours, and again for wound closure at 5–7 days (Leg Compartment Release - Two Incision Approach, 2017). While split skin grafts may be required for soft-tissue closure, this is likely not an option in the spaceflight environment, and delayed skin closure may be a more realistic option; however, it should only be pursued if wound closure can be performed without inflicting excessive skin tension. Some sources recommend a vessel loop shoelace stitch to mitigate tension (Ormiston and Marappa-Ganeshan, 2023). Note that where critical neurovascular and periosteal structures are exposed, such as after the debridement of necrotic tissues, a full-thickness flap may be indicated, and a specialist should be consulted to review (Compartment Syndrome - Lower Limb, 2024).

**14.** ***Complications***. Excessive edema can delay primary closure to 7–10 days or even require a partial-thickness skin graft. Delayed fasciotomies performed at 12 hours or beyond carry a higher risk of reperfusion injury or infection and are associated with a twofold increase in amputation risk and a three-fold increase in mortality. In contrast, early fasciotomies performed within 6 hours typically result in near-complete functional recovery of the affected limb, compared to 68% at 6–12 hours and 8% beyond 12 hours. However, some sources dispute these findings (Ormiston and Marappa-Ganeshan, 2023), and early fasciotomy does not guarantee adequate reperfusion. Additional complications include the risk of neurovascular injury, rhabdomyolysis, and acute renal failure from muscle necrosis, gangrene, contractures, VTEs, complex regional pain syndrome, or even death (Compartment Syndrome - Lower Limb, 2024). Additionally, incomplete fasciotomies may necessitate revision fasciotomy to extend the initial incision or open up additional compartments, the risks of which increase with anatomy distortion, for example, from severe trauma or prior scarring (Compartment Syndrome - Lower Limb, 2024).

**15.** ***Contraindications***. While there are no absolute contraindications to performing a fasciotomy, the proceduralist needs to consider the time since the initial insult, their own skill level, the ability to evacuate to definitive medical care, and the resources at their disposal. Specialist consultation should be sought, if such a capability exists. The operator should also consider the time to presentation, as delayed fasciotomies beyond 12 hours increase the risk of reperfusion injury, as discussed above.

## 15.8 Compound and pelvic fractures

Like compartment syndromes, compound fractures are limb-threatening and potentially life-threatening and are considered a surgical emergency. A compound fracture is defined as any wound creating congruity between the fracture site and the surrounding environment. These fractures carry a higher risk of neurovascular compromise, infection, functional impairment, and mortality and should be identified and treated immediately.

1. Verify the nature of the injury with PoCUS and/or portable X-ray, if available.
2. Obtain consent.
3. Verify neurovascular status.
4. Identify and tend to uncontrolled hemorrhage through packing, pressure, ligation and/or tourniquet application (please refer to the appropriate chapter on hemorrhage control).
5. Provide local anesthesia and/or sedation as appropriate.
6. Cleanse the site with an antiseptic solution.
7. Administer 2 g of IV cefazolin or other antibiotic with Gram-positive and Gram-negative coverage.
8. Administer a tetanus booster intramuscularly if it has been more than 5 years since their last booster.
9. Copiously irrigate the wound with saline or sterile water and a syringe to washout the wound and visible debris.
10. Dress the wound.
11. Reduce and splint the fracture as previously described.
12. Obtain postreduction PoCUS and/or portable X-ray, if available.
13. Promptly consult an orthopedic surgeon and evaluate for medical evaluation, if available.

### 15.8.1 Contraindications

There are no contraindications to washout and reduction of open fractures.

### 15.8.2 Complications and additional considerations

Complications include failing to appropriately reduce the fracture, causing the fracture to heal inappropriately. In addition, failure to adequately irrigate the wound and provide antibiotic coverage can lead to infection, osteomyelitis, bone loss, and/or death.

#### *15.8.2.1 Closed femur midshaft fractures and traction splinting*

Closed, displaced midfemoral shaft fractures are the result of a high-energy mechanism and can be considered **life-threatening**, owing to the risk of hemorrhage of up to 1–1.5 L into the potential space, although new reviews dispute this. Hemorrhage may be the result of soft-tissue injury, major vessel injury, or bone marrow bleeding. As such, immediate recognition through history, mechanism, physical exam—including the presence of pain, swelling, deformity—and/or diagnostic imaging should spur the proceduralist to close the potential space through traction splinting. Critically, the suspicion of the presence of a femoral shaft fracture should prompt the proceduralist to perform a complete trauma survey and obtain imaging of the hip and knee to look for other injuries in nearby joints. The goals of treatment are pain control and transfer to definitive care. Orthopedic specialist consultation should be sought in all cases of suspected or confirmed femur fracture. Note that some sources suggest simply immobilizing the extremity in the position it was found if there is no neurovascular compromise; however, reduction should be pursued when there is neurovascular compromise. The use of traction splinting, although recommended by some orthopedic surgeons, lacks major clinical evidence, with those arguing in favor of traction splinting cite alleviation of patient discomfort, better fracture alignment, and potentially improved arterial flow, with some additional evidence demonstrated a decreased incidence of blood transfusions and pulmonary embolism with traction splinting (Gammons et al., 2025). Traction splints cannot be left on indefinitely, as they can cause pressure ulcers where they contact the skin. Note that patients with open midshaft femur fractures require additional open fracture care, including antibiotics and tetanus, as outlined in the section on compound fractures. The materials needed are summarized in Table 15.7 below.

1. Verify the nature of the injury with PoCUS and/or portable X-ray, if available.
2. Obtain consent.
3. Verify neurovascular status.
4. Provide sedation as appropriate. An ultrasound-guided femoral nerve block may also be considered, depending on practitioner comfort and experience.

**TABLE 15.7** Estimated mass and volume of equipment required for traction splinting of a midshaft femur fracture.

| Item | Mass | Volume | Indication |
|---|---|---|---|
| *Hare or Thomas traction splint* | 2.5–3.0 kg | ~8000 $cm^3$ | A traction splint is indicated for displaced midshaft femur fractures. |

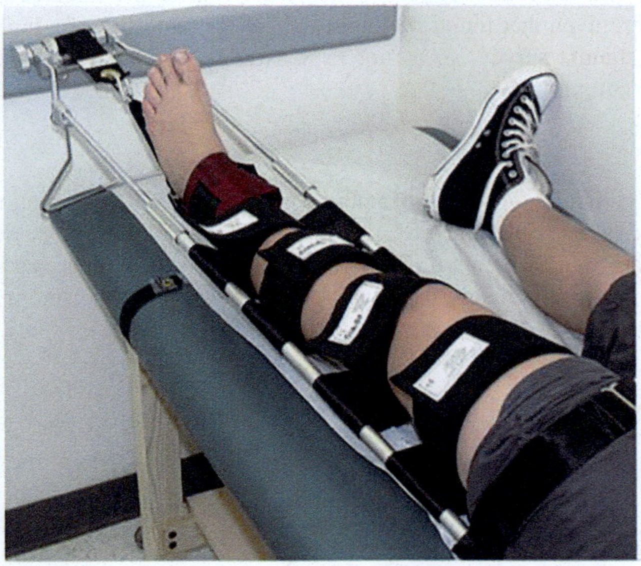

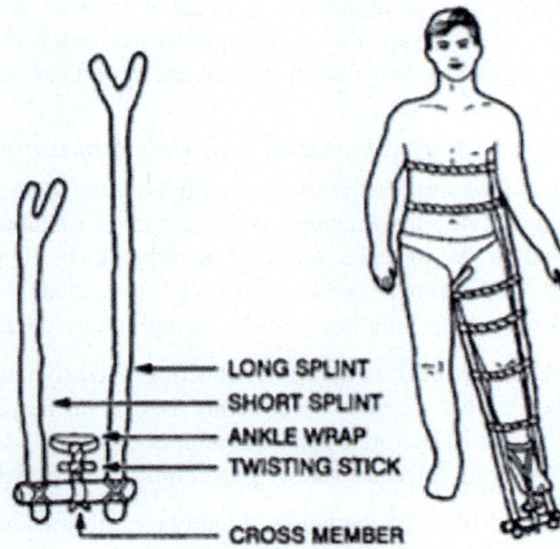


FIGURE 15.14 Traction splinting for midshaft femur fractures. A commercial Hare traction splint has been applied, securing the traction splint to the ankle distally and to pelvis proximally *(left)*, with traction then applied by pulling on the ankle distally to distracting the fracture components, until the affected limb is approximately the same length as the unaffected limb. In the absence of having a formal splint, the proceduralist and treating team may consider fashioning a traction splint by padding the lower extremity, then using components at their disposal that replicate the long rods of the splint, and securing them in place with triangular bandages or similar, securing the top of the splint around the pelvis, and creating a loop at the bottom of the splint, where traction can be applied by securing a small rod (or similar), and twisting to apply traction for fracture distraction *(right)*.
Image credit *(left)*: From Gammons, M., Aspulund, C., Mezzanotte, T., Grayzel, J. (Eds.), 2025. Mid-shaft Femur Fractures in Adults. UpToDate. https://www.uptodate.com/contents/midshaft-femur-fractures-in-adults; Image credit *(right)*: From Femoral Traction Splinting: Evolution, Engineering and Systems, 2017. https://rescue-essentials.com/rescue-essentials-blog/femoral-traction-evolution/?srsltid=AfmBOoo44lGPE-WDfV9kpumLe_2-UQM6lVwCdAG0VWiv7Pm5H-AMw7bi.

5. Apply a Hare or Thomas traction splint as shown in Fig. 15.14 below. Secure the traction splint to the ankle distally and the pelvis proximally, then apply traction by pulling on the ankle distally as the proximal portion of the splint braces and immobilizes the pelvis, distracting the fracture components. In the absence of having a formal splint, the proceduralist and treating team may consider fashioning a traction splint by padding the lower extremity, then using components at their disposal that replicate the long rods of the splint, and securing them in place with triangular bandages or similar, securing the top of the splint around the pelvis, and creating a loop at the bottom of the splint, where traction can be applied by securing a small rod (or similar), and twisting, as depicted in Fig. 15.14.
6. Verify that neurovascular status remains intact.
7. Obtain postreduction imaging to verify that anatomic alignment has been restored.
8. Initiate specialist consultation.

### 15.8.3 Contraindications

Hemodynamic instability is considered an absolute contraindication to traction splinting of midshaft femur fractures, as it can worsen the injury and cause further blood loss. Additionally, hemodynamic instability should be treated immediately with fluid and blood resuscitation, hemostasis, and pressors, if appropriate. While traction splints can be applied to compound midshaft femur fractures, some sources caution that a traction splint could theoretically allow for the retraction of contaminated bone fragments back into the wound-bed. However, ultimately, the proceduralist should prioritize hemorrhage control, restoration of neurovascular integrity, fracture stabilization, and the prevention of further soft-tissue injury over theoretical contamination. The presence of fracture or fracture-dislocation injuries at the hip, knee, and/or ankle is considered a relative contraindication to traction splinting and should be assessed on a case-by-case basis.

## 15.8.4 Complications and additional considerations

While midshaft fractures tend to have a low rate of complications, they still include infection, malunion, chronic pain, hemorrhage, neurovascular injury, compartment syndrome, and refracture. Rarely, patients may suffer life-threatening complications from pulmonary or fat embolisms. Importantly, given that femur fractures are usually the result of a high-injury mechanism, the provider must perform a complete trauma survey, so as not to miss additional major injuries causing hemorrhage, organ failure, or respiratory or airway compromise.

### *15.8.4.1 Pelvic fractures and pelvic binding*

Pelvic fractures are the result of a high-energy trauma and encompass a spectrum of fracture patterns, including disruption of the pelvic ring, fractures of the sacrum or acetabulum, and/or avulsion injuries, and are classified as stable or unstable, based on the presence of two or more breaks in the pelvic ring. A classification of pelvic ring fractures is summarized in Fig. 15.15 below. The clinician should immediately be suspicious of a history including a high-velocity mechanism, or a physical exam pointing to such, including any of the following:

- Flank, perineal, or scrotal contusions or ecchymoses;
- Tenderness over any of the bony aspects of the pelvis—particularly over the sacrum or sacroiliac joints;
- Hematuria and/or vaginal or rectal hemorrhage;
- Abnormal lower extremity positioning, focal lower extremity weakness, or diminished sensation.

Examination of the pelvis should be performed carefully and only once, so as to avoid further disruption of fractures that may, in turn, cause further hemorrhage. Any fracture pattern has the potential to cause significant hemorrhage. Pelvic ring disruptions and open book pelvic fractures, in particular, can create a potential space in the pelvis, in which up to 3–4 L of blood can accumulate, creating a risk of hemorrhagic shock leading to hemodynamic instability and death (Bulger et al., 2024).

Additionally, given the type of high-energy mechanisms involved with this type of fracture pattern, the clinician should be wary of the potential for concurrent major injuries, including intraabdominal bleeding, and a point-of-care ultrasound assessment for free fluid of the abdomen, as well as a comprehensive primary and secondary trauma assessment should accompany any pelvic fracture assessment. Patients with pelvic fractures often have accompanying life-threatening injuries, including brain injuries, torn thoracic aortas, and bladder, penile, urethral, rectal, vaginal, vascular, and/or spinal injuries. Assessment may be particularly challenging in the altered or comatose patient, and clinicians should have a low threshold for proceeding with pelvic binding based on the history and physical. The goals of treatment are pelvic

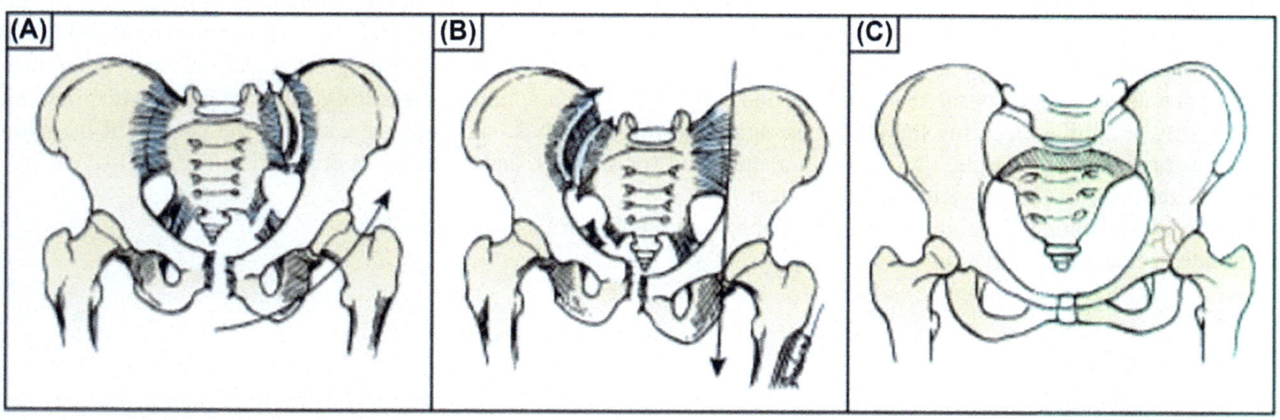


FIGURE 15.15 Patterns of pelvic fractures. Unstable pelvic fracture patterns will typically be due to rotational (A) or vertical stress (B), or may be undisplaced (C).
From Pelvis Fracture, n.d. https://www.brainkart.com/article/Pelvis—Fracture_32605/.

stabilization and, where possible, transfer to definitive care. Pelvic stabilization decreases the pelvic volume, promotes tamponade from bleeding of venous origin, and prevents fractures from shifting further, limiting any secondary hemorrhage.

Surgical consultation should be sought in all cases of suspected or confirmed pelvic fracture. It should go without saying that any unstable patients should immediately be fluid resuscitated and undergo available mass transfusion protocols as able, in tandem with pelvic binding. The materials needed are summarized in Table 15.8 below.

1. Restrain the patient while avoiding jostling the pelvis.
2. Two proceduralists are needed to provide adequate traction when binding the pelvis. The proceduralists should stabilize themselves.
3. Verify the nature of the injury with PoCUS and/or portable X-ray, if available, taking care to perform a focused assessment with sonography in trauma (FAST) of the abdomen to look for a concurrent intraabdominal bleed, if available.
4. Obtain consent.
5. Provide sedation as appropriate.
6. Initiate pelvic stabilization with either a bedsheet or commercially available pelvic binder, wrapped tightly around the circumference of the pelvis at the level of the greater trochanters, with the aim of stabilizing the injuries, while taking care not to over-reduce by applying a sheet or stabilizer too tightly. The two proceduralists should maintain steady, even tension on the sheet, crossing it across the midline of the pubic symphysis, twisting it, tightening it, and repeating these motions until the sheet is tight and the pelvis is stable. The binding should then be secured in place, using clamps if using a bedsheet, or using the closure of the commercial binder. Application methods for pelvic binders and sheets are demonstrated in Fig. 15.16 below.

**TABLE 15.8** Estimated mass and volume of equipment required for pelvic binding.

| Item | Mass | Volume | Indication |
|---|---|---|---|
| *Pelvic binder or sheet* | 300 g | 1200 cm$^3$ | A pelvic immobilizer is indicated for the stabilization of a pelvic fracture. |
| *Clamps × 6* | 90 g | 120 cm$^3$ | Clamps are necessary to secure the sheet/immobilizer in place |

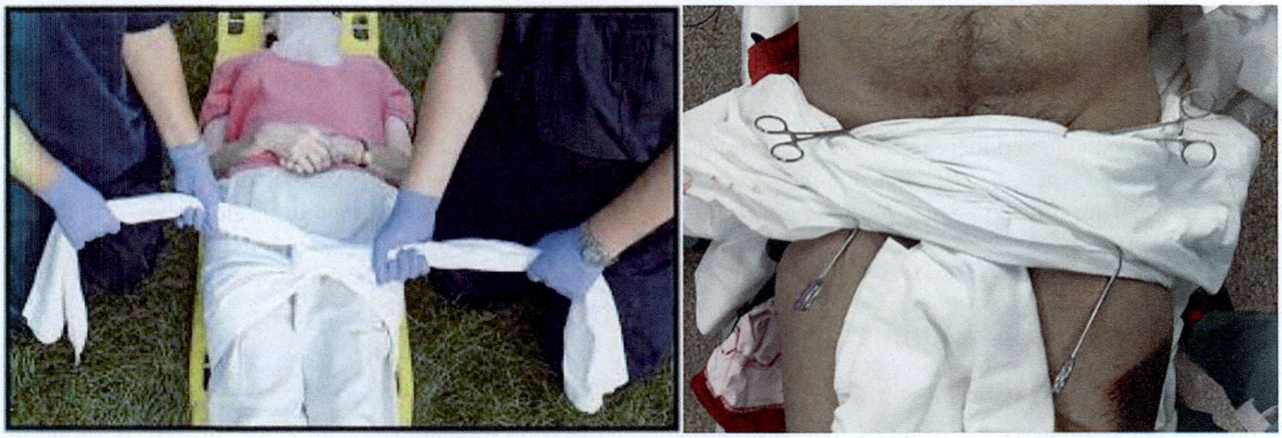

FIGURE 15.16 Pelvic binding technique. Pelvic binding using a bedsheet. Note that the top of the sheet reaches the level of the tops of the greater trochanters bilaterally. The sheet is crossed over while each operator maintains even tension on each end, then crosses the sheet over, twists, pulls, and repeats until the unstable pelvis is secured *(left)*. The sheet is then secured in place with clamps, as depicted on the *right*.
*Left image*: From McHenry Western Lake County EMS System Paramedic, EMT-B and PHRN Optional Continuing Education Pelvic Fractures, 2019. McHenry Western Lake County EMS System Paramedic, EMT-B and PHRN Optional Continuing Education. *Right image*: From Bulger, E., Khurana, B., Collins, K., Burlew, C. & Jenson, W. (Eds.), 2024. Severe Pelvic Fracture in the Adult Trauma Patient. UpToDate. https://www.uptodate.com/contents/severe-pelvic-fracture-in-the-adult-trauma-patient.

**7.** Verify that neurovascular status remains intact postapplication of the binder or sheet.
**8.** Initiate specialist consultation.

### 15.8.5 Considerations and complications

As pelvic fractures are the result of a high-injury mechanism, it is critical that the provider perform a complete trauma survey and maintain a low index of suspicion for additional life-threatening injuries, including traumatic brain injury, intrathoracic injury, including aortic injury, and intraabdominal hemorrhage. Because the risk of hemorrhagic shock and hemodynamic instability is high in pelvic fracture traumas, pelvic binding can be a life-saving measure. As there is a risk of worsening instability, the pelvis with a suspected pelvic ring fracture should only be examined once. While a commercially available pelvic binder or bedsheet is typically recommended, the proceduralist is encouraged to improvise to find a solution when neither option is available.

Taking the whole picture, the standard of care for a suspected pelvic fracture includes access to a mass transfusion protocol, trauma service, intensive care unit, pelvic arteriography, resuscitative endovascular balloon occlusion of the aorta (REBOA) capabilities, medical evacuation capabilities, and an orthopedic specialist consultant. Depending on the space mission profile, most existing and future medical capabilities and mission profiles lack most or all of these. As such, the proceduralist may be forced to simply stabilize the pelvis and resuscitate the patient with IV fluid resuscitation.

The proceduralist should be careful to monitor the duration and condition of pelvic binding, as both can lead to skin necrosis, breakdown, and ulceration as early as 3 hours after application. The literature also recommends that binders be removed within 24 hours of application, particularly since polytrauma patients may have a higher risk of soft-tissue compromise due to systemic factors in the postinjury period (Mitchnik et al., 2022). Further, it must be noted that the pelvic binder is only a **temporary solution** for a few hours at most. Any patient with a pelvic fracture requiring a pelvic binder is likely to exsanguinate from associated injuries in short order without surgical fixation, angioembolization of bleeding vessels, or other necessary surgical procedures. Pelvic binders are also only appropriate for the use in "open book" pelvic fractures secondary to anterior–posterior compression with widening of the pubic symphysis. Vertical shear injuries of the pelvis can be worsened by the placement of a pelvic binder, leading to death.

Additionally, severe pelvic fractures present a real risk for VTE, and patients should receive early thromboprophylaxis, consisting of pneumatic compression (if available) until bleeding is controlled. After this time, pharmacologic VTE prophylaxis should be initiated. For patients in whom pharmacological VTE prophylaxis is contraindicated, an inferior vena cava filter is an option on Earth but is unlikely to be available in the spaceflight environment, and consultation with an Earth-based specialist should be sought (Pelvic Trauma: Initial Evaluation and Management, 2025).

### 15.8.6 Contraindications

There is some criticism about the application of a pelvic binder before having obtained adequate diagnostic imaging, as the determination of the presence of a pelvic ring injury versus an isolated hip fracture based on physical exam alone is unreliable. Additionally, while some studies have suggested that a pelvic binder may do more harm than good for lateral compression fractures due to their unstable nature, trauma protocols still advocate for pelvic binding when pelvic ring imaging is unavailable and there is clinical concern for pelvic ring. Regardless, the application of a pelvic immobilizer is still recommended when there is clinical concern of a pelvic ring injury, although it is contraindicated for injuries other than a pelvic ring fracture, such as a hip or femur fracture, or vertical shear injury of the pelvis (Khaliq and Rodham, 2024).

## 15.9 Subungual hematoma drainage

### 15.9.1 Indications

Subungual hematomas are the result of trauma or sustained pressure to the nailbed of the finger or toes, resulting in a contusion within the closed space of the soft-tissues under the nailbed. Sustained and repeated mechanical pressure from EVA suit gloves is one known such cause and can eventually lead to onycholysis. Many subungual hematomas will be asymptomatic or nondisabling and will resolve on their own without intervention; however, larger contusions can become quite painful as blood breakdown products increase in volume with time, interfering with astronaut performance and function and will require drainage. Drainage can also be indicated if the subungual collection becomes infected, as evidenced by a collection that is purulent-looking in appearance, and/or cellulitic changes to the surrounding skin (e.g., redness, swelling, warmth to touch, and/or tenderness), in which case antibiotics may also be indicated. The material required for this procedure is summarized in Table 15.9 below.

### 15.9.2 Technique

1. Obtain the patient's consent for the procedure after explaining the risks, benefits, side effects, and alternatives of the proposed procedure.
2. Immobilize the affected finger or toe appendage using a restraint or crew restraint system.
3. Don nitrile gloves.
4. The proceduralist is to affix or tether themselves to stabilize.
5. Clean the affected nail with a BZK wipe.
6. Identify an appropriate drilling site within the subungual hematoma.
7. Use an 18 g needle to trephine into the nail over the greatest density of hematoma collection, rolling the needle slowly and gently clockwise and counterclockwise until the full-thickness of the nail has been pierced, and blood is visible.
8. Squeeze periungual soft-tissue to "milk" the blood out of the drilled hole and use gauze to wick away the blood as it is squeezed out. The subungual collection should recede as this is done, revealing the normal nailbed underneath.

Subungual hematoma drainage is depicted in Fig. 15.17 below.

**TABLE 15.9** Estimated mass and volume of equipment required for the drainage of a subungual hematoma.

| Item | Mass | Volume | Indication |
|---|---|---|---|
| *Nitrile gloves (1 pair)* | 12 g | 50 $cm^3$ | Provider personal protective equipment |
| *18 g 1" needle* | 2 g | 8 $cm^3$ | To drill through the nail and release the subungual hematoma |
| Electrocautery (optional instead of 18 g) | 80 g | 150 $cm^3$ | To drill through the nail and release the subungual hematoma (instead of an 18 g needle) |
| *BZK wipe* | 1.5 g | 4 $cm^3$ | To clean the areas of injection prior to aspiration |
| *4x4 gauze* | 2 g | 15 $cm^3$ | To clean any fluid post aspiration |

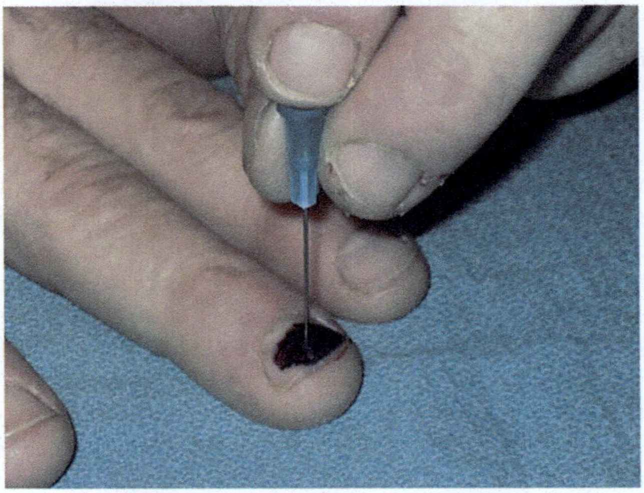

**FIGURE 15.17** Demonstration of subungual hematoma drainage using a large-gauge hollow-bore needle for trephination.
From FAST Track: Subungual Hematomas - Brown Emergency Medicine Blog, 2024. https://brownemblog.com/blogposts/2024/2/26/subungual-hematomas.

### 15.9.3 Contraindications

Do not proceed with trephination for asymptomatic subungual hemorrhages and nontraumatic subungual presentations, such as splinter hemorrhages and subungual nevi, the distinction of which will be apparent by history and patient report of pain. Nailbed injuries where blood is leaking out around the nailbed edges are already considered to be self-draining and do not require trephination. Nailbed removal is indicated when a laceration threatens the viability of the nail or digit and is covered in the next section.

### 15.9.4 Additional considerations

Electrocautery may also be used in place of a large-gauge hollow-bore needle, if available. As previously stated, asymptomatic or small collections do not need intervention. The proceduralist should consider imaging to assess for a fracture for large hematomas that occupy over 50% of the nailbed space, and/or where there is a history of a traumatic mechanism. Analgesia and digital ring blocks are typically not needed, as nail trephination is a relatively painless procedure, unless the patient has sustained a concurrent laceration, distal phalanx fracture, or both. Some sources only advocate for drainage within the first 48 hours of injury, after which time the collection may be too clotted for easy drainage. In the author's experience, timing is less important than patient self-report of pain at any time, and older collections may become more uncomfortable with time as blood clots then degrade, and still tend to be amenable to drainage. Specialist consultation with a hand surgeon is indicated for displaced fractures, intraarticular fractures, and extensive nailbed injuries. The presence of infection and cellulitic changes in the periungual soft-tissues should prompt the proceduralist to assess the need for antibiotics. Lacerations and wounds should also prompt for the assessment of tetanus status, although ostensibly this would have been updated prespaceflight (Fastle et al., 2025).

### 15.9.5 Complications

The patient should be counseled regarding potential complications, including pain and discomfort associated with the procedure, discoloration of the nail (usually resolving within 4 weeks), potential deformity, or loss of the nail from the initial injury, clotting of the hematoma, leading of reaccumulation of the subungual collection, and secondary infection of the nailbed, surrounding tissues, and rarely, deeper structures (Fastle et al., 2025).

### 15.9.6 Follow-up care

After trephination is complete, the patient should keep the affected nailbed clean and dry. The patient should be counseled to return to care if the hematoma reaccumulates to the point of causing pain and impeding activities, or if signs and symptoms of infection, including purulent drainage, soft-tissue swelling, redness, tenderness, and warmth, develop, although this is rare.

## 15.10 Joint injection

### 15.10.1 Indications

Terrestrially, both acute and chronic conditions that might benefit from localized inflammatory suppression are the traditional indications for a joint injection with corticosteroids (alternate injection medications are beyond the scope of this text, though the techniques are roughly similar no matter which medication is used). The most common chronic indication for corticosteroid joint injection terrestrially is osteoarthritis of the affected joint.

In a spaceflight environment, which is expected to be time-limited, only acute pathologies would be expected to arise. These pathologies can include acute musculoskeletal sprains and strains, acute bursitis that is adjacent or overlying a joint area, and acute impingement syndromes. Thus, common microgravity scenarios may include movement in an unnatural plane, causing unusual strain on the joint, EVA-related joint injury, or exacerbations of underlying conditions such as osteoarthritis or crystalline arthritis due to the demands of the spaceflight environment. In a spaceflight participant with a preexisting autoimmune condition, a corticosteroid injection may also be helpful for a localized flare.

Spaceflight also presents unique stressors on the joint spaces, which are, at the time of writing, incompletely characterized. These stressors include lower mechanical stimulus of the joint, as well as potential radiation and inflammatory characteristics that may create an environment where joint degradation is more likely. It is currently unclear how a localized corticosteroid injection will interact with this unique environment within the joint space.

Acute pathologies also include rapid fluid accumulation within a joint cavity, which can cause a painful mass effect. In such cases, drainage of the affected joint can facilitate pain relief and can either be done alone or combined with a corticosteroid injection after fluid has been drained. This would most commonly occur due to a flare of an underlying chronic condition, such as gout or osteoarthritis. If a rapid fluid accumulation occurs as a result of joint trauma with high concern for a hemarthrosis, available diagnostic tools should first be used to determine as accurately as possible if a hemarthrosis has occurred. If so, this is a relative contraindication to fluid drainage but still may be appropriate in some circumstances. This will be further discussed below.

In general, more precise injection techniques of the soft-tissue which would require drug delivery into a tendon sheath (such as for carpal tunnel syndrome) require more specialized terrestrial training given the higher rate of potential adverse events with unpracticed technique.

### 15.10.2 Medication

Corticosteroid dosing for a joint injection is not standardized, and wide variation exists amongst practitioners. In the interest of minimizing the physical amounts of medication to be considered within a ship's medical system, a typical recommendation will be given. In general, for a large joint (the shoulders, knees, and large bursae), 40 mg of triamcinolone or 40 mg of methylprednisolone is appropriate. Standard concentration for these medications is 40 mg/1 ml. For medium joints (the ankle, wrist, elbow), 20 mg of triamcinolone or 20 mg of methylprednisolone is appropriate. Solutions should be mixed 1:1 with an anesthetic agent, for example, 1 ml of lidocaine for 1 ml of corticosteroid.

The antiinflammatory properties of corticosteroids generally take between 24 and 72 hours for an effect to begin, and there is usually a rapid onset to peak efficacy. Pain relief immediately following the injection is likely either from the mixed anesthetic agent (traditionally lidocaine) that was used in the injection or, if an aspiration has been performed, from a relief of the mass effect (Dickson et al., 2018; Cushman et al., 2019).

Systemic absorption of corticosteroids does take place after the medication is injected into the intraarticular space, though the systemic effects of steroids are generally expected to be transient. As an example, while triamcinolone was found in the serum for up to 12 weeks after injection, adrenal suppression only occurred for 4 days after injection, and similar blood pressure and blood sugar effects were transient. Corticosteroid injections should not be repeated in a joint more often than every 12 weeks, and ideally no more often than every 16 weeks (Hepper et al., 2009).

### 15.10.3 Contraindications

Absolute contraindications to joint aspiration include an active infection over or near the injection site or a systemic bacteremia. Absolute contraindications to a corticosteroid injection include septic arthritis, local malignancy, allergy to medications, or a nearby fracture. Relative contraindications include uncontrolled systemic chronic diseases that may be influenced by transient systemic corticosteroids, such as diabetes. Uncontrolled coagulopathy should be considered with any procedure that requires breaking the skin barrier. Aspiration or injection of a prosthetic joint should not be done except in consultation with an orthopedic specialist (either a specially trained CMO or with ground support), as the risk of a prosthetic joint infection is substantially higher than a native joint. The medications to be used should be evaluated for any particular concerning interactions with the spaceflight participant's physiology or with any medications they are currently taking (e.g., ritonavir and triamcinolone; Kamel et al., 2024).

### 15.10.4 Complications and follow-up care

An injected joint may have a slight increase in pain and swelling symptoms for the 24–48 hours after injection until the antiinflammatory effect of the corticosteroid becomes more pronounced. Substantial and rapid pain, warmth, and swelling after injection should prompt reevaluation of the joint for complications, procedural trauma, or alternate considerations such as native or iatrogenic septic introduction to the joint. Care should be taken to ensure that the initial injection has the corticosteroid agent mixed with an anesthetic, as a "crystalline reaction" can take place when corticosteroid alone is injected. If a postprocedural joint has mild discomfort and swelling that is not concerning for rapid swelling with severe, out-of-proportion symptoms, rest and ice are the mainstay of care consistent with standard musculoskeletal sprain/strain treatment, with expected symptom improvement in 24–48 hours.

In general, expert consensus is that a postprocedural joint should not engage in heavy activity (e.g., "nothing heavier than the equivalent of lifting a milk jug in 1 G") for 48–72 hours because of the hypothetical risk of tendon rupture with an acute corticosteroid infusion near the tendon insertion point (Weitoft and Forsberg, 2010; Jüni et al., 2015). This is an expert recommendation, and there is no well-researched evidence base to support a recommendation either way.

Nevertheless, given the unique physiologic adaptations to microgravity, there is a possibility of increased tendon vulnerability or sprains/strains in unusual planes of force. Within each subsection, specific recommendations are given about limiting activity.

### *15.10.4.1 Modifications for aspiration*

In general, larger gauge needles are more conducive to smooth aspiration and are less likely to become clogged with synovial debris, depending on the differential diagnosis. A minimum of a 22 g needle should be used for aspiration, and preferably a 20 g or 18 g needle. If a larger gauge needle is used, it is recommended to use an additional anesthetic agent to numb the surface and the insertion tract before aspiration for patient comfort and safety (to prevent patient movement that may cause trauma and needle dislodging). If septic arthritis is of great concern, then an anesthetic is not recommended as it may have bactericidal properties that will hamper diagnostic sensitivity.

Appropriate containers for immediate (inter-procedure) and long-term disposal or storage (for surface analysis) of joint fluid should be prepared prior to the procedure and within ready access of the proceduralist. If laboratory analysis is available and desired, the appropriate sample tubes should be prepared.

When appropriate, injection should be performed after aspiration to prevent aspirating the medication itself, as well as enhancing the efficacy of a corticosteroid injection, as a fluid-filled joint can impair corticosteroid efficacy.

### *15.10.4.2 Item list*

The items in this table are not all mandatory, but a medical system should instead be customized for the expected medical conditions that are of concern (Table 15.10).

### *15.10.4.3 A note on small joint injections*

As a general principle, unless the CMO has specific training in small joint injections or there is a substantial medical need specifically for an aspiration or injection (such as for septic arthritis or a gout flare that is compromising operations), small joint injections (the MCPs, PIPs, MTPs, and related joint sizes) should be avoided when possible. There is a higher risk of infection with smaller joint spaces that have relatively less protection, and there is a higher risk of not successfully penetrating the joint space with a smaller joint, thus causing infiltration of corticosteroid into a relatively low-fat and low-muscle area (such as the dorsum of the hand).

### *15.10.4.4 A note on septic joint and hemarthrosis*

#### 15.10.4.4.1 Septic joint

Terrestrially, a septic joint is considered a medical emergency as there can be rapid, permanent destruction of a joint with an untreated septic arthritis. The standard of care for septic joint management is urgent joint surgery for source control. This will likely not be feasible in a spaceflight environment. Thus, when possible, management of a suspected or confirmed septic joint should be in consultation with orthopedic support. Confirmation or high suspicion of septic joint should prompt serious consideration of medical evacuation, if feasible.

Risk factors for a septic joint include a feasible route of entry (such as systemic bacteremia or trauma), uncontrolled diabetes, prior immunocompromise (in addition to the immune effects of spaceflight), and the presence of a prosthesis.

A septic joint traditionally presents with an isolated, overtly swollen, warm joint with substantial reduction of BOTH active AND passive range of motion, with severe pain from any manipulation. Systemic symptoms for a septic joint are similar to those of bacteremia and include fever, diaphoresis, and SIRS-like vital sign changes. However, these are not always present as the joint space may partially shield the rest of the body. If laboratory analysis is available, a septic joint can yield results similar to those of bacteremia, with leukocytosis and elevated inflammatory markers. If joint fluid analysis is available, a cell count of 50,000 or greater white cells is considered at high risk for a septic joint, while 10,000–50,000 would be regarded as inflammatory, though a septic joint cannot be excluded based on cell count alone. A bacterial Gram stain or another method for detecting bacterial signatures, if available, can confirm the diagnosis. Terrestrially, cell culture is also often performed, though it is unlikely to be feasible under the constraints of spaceflight. A microscope, if available, can assist in a differential diagnosis, as crystalline arthritis can have a nearly identical presentation, especially if the patient has a history of crystalline arthritis. It should be noted that crystalline arthritis and a septic joint can coexist in the same joint.

While surgical intervention is considered standard of care for a septic joint, an alternate treatment approach includes serial drainage of the joint with initial and each effusion recurrence combined with IV (preferred) or oral (if absorption

**TABLE 15.10** Mass and volume table for joint injections.

| Name | Rationale | Mass/volume |
|---|---|---|
| Joint immobility device | This can be something as simple as straps and a rigid, immobile surface. In a spaceflight environment, stability in multiple planes is important when any procedure is indicated. Thus, immobility of the relevant area should be achieved as thoroughly as possible. | Variable and can potentially be improvised from surroundings/personnel |
| 3 ml syringe | The typical volume of an injection will be 1–2 ml | 4.55 g/7.23 $cm^3$ (individually packaged) |
| 5 ml/10 ml syringe | Larger syringes can be utilized for drainage, though a small syringe with a secure disposal container is a reasonable alternative | 5.58 g/17.32 $cm^3$ (individually packaged) |
| Biohazard waste bag | Sealed waste bag for drained bodily fluids if aspiration is done | 5.00 g/18.88 $cm^3$ |
| 18–25 g needles, 1.5″ | In general, drawing up a crystalline suspension is easiest with an 18 g needle. Aspirations should be done with an 18–22 g needle. Injections can usually be accomplished with a 22–25 g needle. For large and medium joints, a needle length of 1.5″ is appropriate. | 1.10 g/1.7 $cm^3$ (individually packaged) |
| Surgical gloves | Sterile vs nonsterile | 3.44 g/7.98 $cm^3$ individual700 g/2388.20 $cm^3$ box of 200 |
| Needle disposal device | For securing midprocedure and postprocedure | Variable depending on deviceE.g., small cylindrical sharps container:29.68 g/256.59 $cm^3$ |
| Surgical marking pen | For identification and visualization of palpated landmarks and injection site | 3.30 g/5.79 $cm^3$ |
| Antiseptic | Alcohol, chlorhexidine, or iodine. All three suggested antiseptics require evaporation, so this should be done next to an area of automated or manual air movement | Alcohol swab:1.16 g/2.54 $cm^3$ individual131.34 g/383.04 $cm^3$ box of 100Chlorhexidine applicator:11.32 g/118.03 $cm^3$ |
| Corticosteroid bottle | Triamcinolone or methylprednisolone are the two most commonly used | Variable depending on medication and vial sizeE.g., Single dose 1 ml triamcinolone bottle7.51 g/8.24 $cm^3$E.g., Multi dose 5 ml triamcinolone bottle 13.52 g/12.88 $cm^3$ |
| Anesthetic agent | Lidocaine or bupivacaine without epinephrine. It is important to mix a corticosteroid suspension with the anesthetic agent as a "crystal arthritis-like" reaction can occur when corticosteroid is injected alone | Variable depending on medication and vial sizeE.g., Single dose 2 ml lidocaine 1% bottle7.12 g/5.39 $cm^3$E.g., Multi dose 5 ml lidocaine 1% bottle 13.96 g/12.88 $cm^3$ |
| Gauze | For control when breaking the skin barrier | 2.00 g/5.20 $cm^3$Individually wrapped |
| Fabric bandage | Postprocedure | 0.89 g/1.66 $cm^3$ Individually wrappedBox size variable pending storage sizeE.g.,10.71 g/47.04 $cm^3$ pack of 8 |
| Vacuum tube ("red-top"; optional) | If lab analysis is available, needed for bacteria stain and analysis | 7.14 g/18.91 $cm^3$ |
| EDTA Vacuum tube ("purple-top"; optional) | If lab analysis is available, needed for cell count and differential | 11.65 g/14.31 $cm^3$ |
| Ultrasound device (optional) with gel | Procedures can be "landmark-guided" or ultrasound can be utilized for optimizing procedural approach | Variable depending on probe type and gel containerE.g., Butterfly probe309 g/320 $cm^3$250 ml nonsterile gel bottle250.04 g/688.26 $cm^3$ |
| Microscope (optional) | For crystal analysis | Variable depending on microscope type.E.g., portable microscope379.40 g/2203.32 $cm^3$ |

will be similar) antibiotics. Intraarticular antibiotics have also been proposed for treatment, though the evidence for and against this is beyond the scope of this document and should be undertaken in consultation with a specialist when possible (and certain antibiotics are highly chondrotoxic; Hassan et al., 2017).

#### 15.10.4.4.2 Hemarthrosis

Hemarthrosis in a spaceflight environment should be suspected when a plausible traumatic mechanism involving the joint is present. Spontaneous hemarthrosis is rare and usually only presents in patients with an underlying coagulopathy. A hemarthrosis will present with rapid swelling of the joint with warmth, tenderness to palpation, and active restriction of range of motion.

When possible, aspiration and drainage of a frank hemarthrosis should be avoided to minimize the risk of infection and rebleeding and promote the "tamponade" effect that will control further bleeding into the joint space. Indications for drainage of a hemarthrosis include a "tense" joint with substantial functional compromise, when rapid (potentially temporary) recovery is needed for operational purposes, and when conservative management (including rest, ice, and compression) has failed (Ravi et al., 2019).

### *15.10.4.5 A note on ultrasound-guided procedure*

Ultrasound is a valuable adjunct to joint injection procedures, both diagnostically and therapeutically, and is an alternative to "landmark-guided" injection techniques.

Diagnostically, ultrasound can visualize a joint space of concern and determine whether fluid is present within the joint space itself or overlying soft-tissue swelling, potentially negating the need for an aspiration. Debris within the fluid space can also be visualized for assistance in differential diagnosis, as well as grayscale evaluation if hemarthrosis is on the differential. If power Doppler is available, a trained practitioner can also assess if a joint effusion is concerning for an inflammatory pathology. Musculoskeletal ultrasound imaging can also be helpful in coordination with ground specialty support.

Therapeutically, the most common application of ultrasound is guidance for joint aspiration and injection. Both indirect visualization and direct visualization techniques are possible. With indirect visualization, the ultrasound is used to determine the optimal approach, including angle, to access the joint space in preparation for a successful injection. If resources to perform sterile ultrasound technique are available or the ultrasound can be positioned outside the sterile injection field, then direct visualization can be performed where the ultrasound will be used to visualize the pathway of the needle into the joint space and confirm aspiration and injection to ensure targeted delivery of the corticosteroid.

### *15.10.4.6 Upper extremity joint injections*

#### 15.10.4.6.1 Large joint injection (shoulders)

**Glenohumeral joint (posterior):**

1. Traditionally, 40 mg of triamcinolone or methylprednisolone is mixed with 1 ml of anesthetic.
2. The glenohumeral joint is the most mobile joint in the body, stabilized by multiple soft-tissue attachments, and the combination of these two characteristics makes it a joint particularly vulnerable to acute pathology.
3. Injection of the glenohumeral joint aims to access the posterior glenohumeral joint space. In preparation for this access, one possible technique to aid stabilization and maximize access in the microgravity environment is for the patient to place their ipsilateral hand on their contralateral shoulder, which opens the posterior space.
4. Ensure that adequate immobilization has occurred. This includes immobilization of both the patient and the shoulder joint, accounting for the three-dimensional nature of spaceflight.
5. Identify all relevant items needed for the procedure and ensure an appropriate plan for capped and uncapped needle management before, during, and after the procedure to ensure safety.
6. The head of the humerus, the acromion, and the coracoid process (on the patient's front) should be identified. If a surgical marking pen is available, it can be used to identify and demarcate the lower horizontal border of the acromion and the vertical line where the edge of the humerus head curves inward.
7. Identify the posterolateral corner of the acromion process. The injection site will be 2–3 cm (or approximately two finger widths) below this posterolateral corner. Ensure that the joint space can be palpated in this area.
8. After appropriate local site sterilization with BZK and appropriate preparation, including personal protective equipment and positioning, the appropriately selected needle and syringe can be inserted at the marked site. The needle should be angled towards the previously identified ipsilateral coracoid process (on the front of the patient). The needle may be hubbed if the joint space is deep enough.

9. Slightly aspirate to ensure that the needle is within the joint space and is not within a blood vessel. A small bubble may appear in the fluid as the joint's negative space is aspirated. Apply gentle but firm pressure to inject the medication at a steady rate. If resistance is encountered, slightly withdraw and reposition the needle rather than pressing past it.
10. Withdraw and secure the needle and utilize appropriate postprocedure techniques, such as any bleeding control and protection of the surgical site with a bandage.
11. Heavy activity should be avoided around the joint for the next 48–72 hours, though this is based on clinical expertise rather than trial data. Conversely, joint immobilization is not needed postprocedure.

**Glenohumeral joint (anterior):**

1. Injection of the glenohumeral joint with an anterior approach aims to access the anterior glenohumeral joint space. In preparation for this access, one possible technique to aid stabilization and maximize access in the microgravity environment is for the patient to place their ipsilateral hand on their ipsilateral thigh or knee and relax the musculature, which provides access to the anterior space.
2. Ensure that adequate immobilization has occurred. This includes immobilization of both the patient and the shoulder joint, accounting for the three-dimensional nature of spaceflight.
3. Identify all relevant items needed for the procedure and ensure that there is an appropriate plan for capped and uncapped needle management before, during, and after the procedure to ensure safety.
4. The medial head of the humerus and the coracoid process (on the patient's front) should be identified. If a surgical marking pen is available, it can be used to identify and demarcate the vertical line where the edge of the humerus head curves inward, as well as the coracoid process and the anterior course of the collarbone.
5. Identify the inward curve of the medial side of the humerus and identify the coracoid process. The injection site is approximately 1–2 cm (or 1–2 finger widths) lateral to the coracoid process within the groove medial to the humeral head.
6. After appropriate local site sterilization with an antiseptic (with appropriate evaporation time), and appropriate preparation including personal protective equipment and positioning, the appropriately selected needle and syringe can be inserted at the marked site. The needle should be angled with a slight upward and lateral tilt. The needle may be hubbed if the joint space is deep enough.
7. Slightly aspirate to ensure that the needle is within the joint space and is not within a blood vessel. A small bubble may appear in the fluid as the joint's negative space is aspirated. Apply gentle but firm pressure to inject the medication at a steady rate. If resistance is met, slightly withdraw and reposition the needle as opposed to pressing past resistance.
8. Withdraw and secure the needle and utilize appropriate postprocedure techniques, such as any bleeding control and protection of the surgical site with a bandage.
9. Heavy activity should be avoided with the joint for the next 48–72 hours, though this is considered based on clinical expertise and not trial data. Conversely, joint immobilization is not needed postprocedure (Fig. 15.18).

**Subacromial bursa:**

1. Injection of the subacromial bursa aims to access the space underneath the acromion to relieve acute bursitis. In preparation for this access, one possible technique to aid in stabilization as well as maximize access in the microgravity environment is for the patient to allow their arm to fully straighten out in a neutral posture (without locking the elbow) straight down. It may help to secure the wrist to the patient's thigh, either if the patient is in a seated position or with a straight leg.
2. Ensure that adequate immobilization has occurred. This includes immobilization of both the patient and the shoulder joint, accounting for the three-dimensional nature of spaceflight.
3. Identify all relevant items needed for the procedure and ensure that there is an appropriate plan for capped and uncapped needle management before, during, and after the procedure to ensure safety.
4. The lateral course of the acromion should be identified. If a surgical marking pen is available, it can be used to identify and demarcate the lower border of the acromion.
5. The approach will be lateral to the patient, perpendicular to the shoulder. Identify the space below the acromion. The injection site will be 1–2 cm (or approximately 1–2 finger widths) below the acromion in the middle of the lateral shoulder space and will usually be at the point of maximal tenderness. Ensure that the joint space can be palpated in this area.
6. After appropriate local site sterilization with BZK and appropriate preparation including personal protective equipment and positioning, the appropriately selected needle and syringe can be inserted at the marked site. The needle should be angled directly inwards.

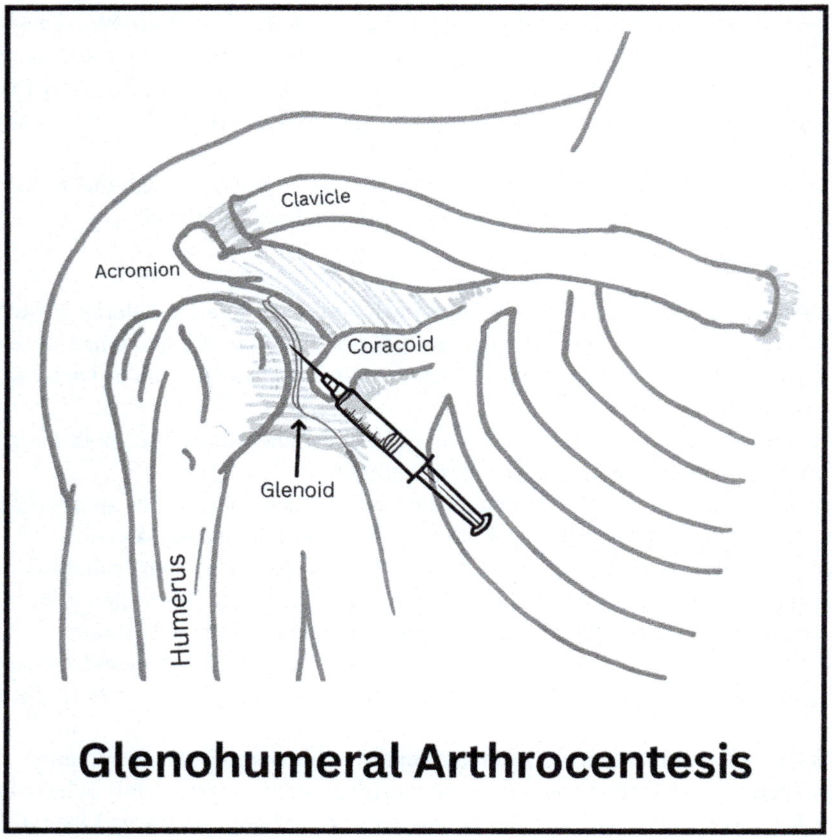


FIGURE 15.18 Glenohumeral anterior approach.
Credit: Amanda Prine.

7. Slightly aspirate to ensure that the needle is within the space and is not within a blood vessel. Apply gentle but firm pressure to inject the medication at a steady rate. If resistance is met, slightly withdraw and reposition the needle as opposed to pressing past resistance. Note that the subacromial space in certain patients may be more restricted, and the full 2 ml of fluid may not be fully injected.
8. Withdraw and secure the needle and utilize appropriate postprocedure techniques such as any bleeding control and protection of the surgical site with a bandage (Fig. 15.19).

### 15.10.4.6.2 Modifications for aspiration

1. Aspiration of the glenohumeral joint can be accomplished with a needle 22 g or greater, and ideally, an 18 g or 20 g needle will be available. If a larger gauge of needle is used, it is important to first anesthetize the needle tract with 2–3 ml of lidocaine or an equivalent anesthetic UNLESS a septic arthritis is suspected as the anesthetic may lower diagnostic yield. This should be done after the surgical site is sterilized and before the aspiration needle is placed.
2. If therapeutic aspiration is being considered, ensure there is a proper immediate and long-term plan for the fluid waste, including whether it is to be stored in syringes or other collection containers, and whether it needs to be packaged for surface analysis, further research, or disposed of with proper biohazard precautions. If laboratory analysis is available, ensure that there is a plan for appropriate fluid transfer to the relevant equipment. If microscopy is planned for crystal analysis, note that the fluid should be analyzed within a few hours of collection for optimal yield.
3. After successfully inserting the needle and syringe intended for aspiration, withdraw the syringe plunger slowly; excessive negative force can cause the needle to become clogged. Note the color and consistency of the fluid that is being removed.
4. Once fluid has been removed, if a corticosteroid injection is desired, ensure the syringe containing the corticosteroid and anesthetic mixture is readily available. Unscrew the syringe from the hubbed needle in the patient in a counterclockwise direction. Secure the first syringe. Attach the syringe containing the medication to the hubbed needle and inject following the instructions above.

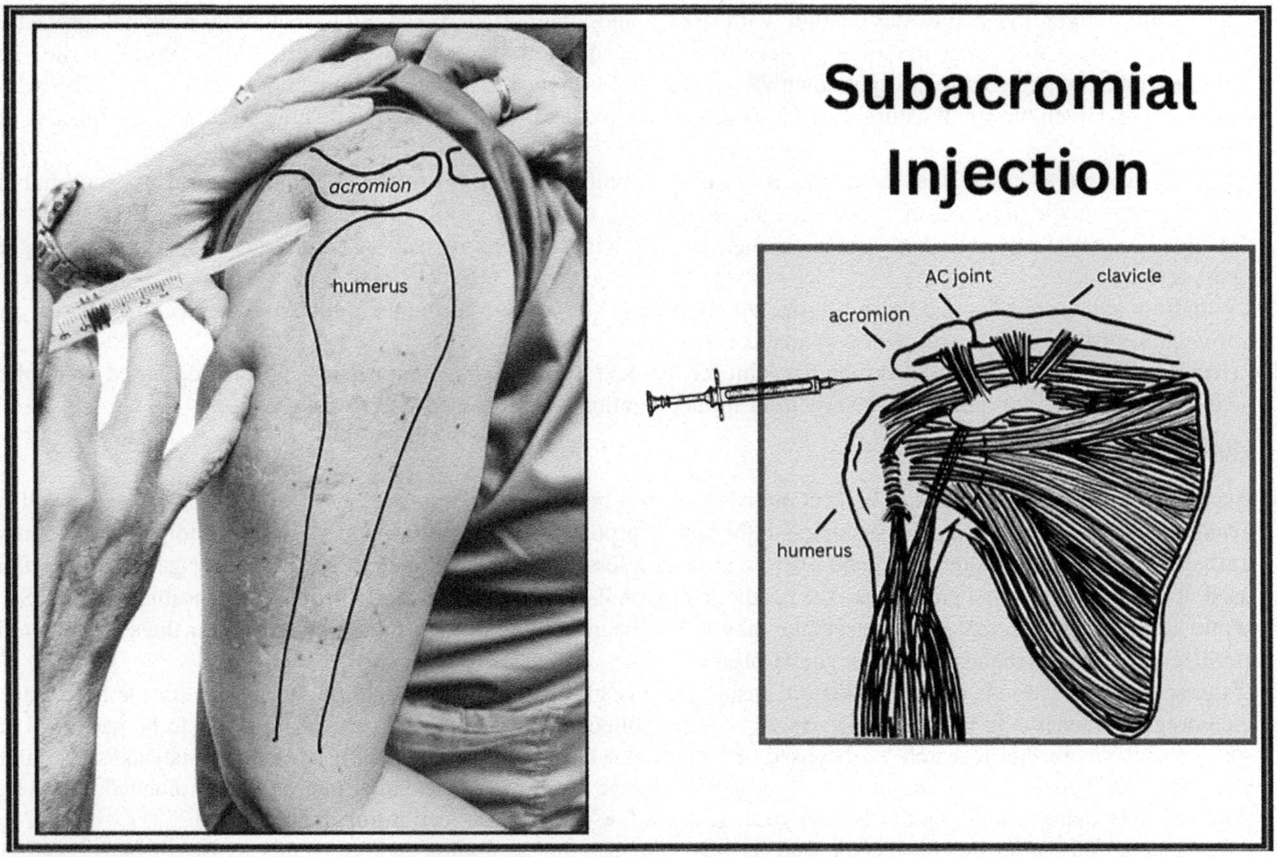


FIGURE 15.19 Subacromial approach.
Credit: Amanda Prine.

### *15.10.4.7 Medium joint injection*

#### 15.10.4.7.1 Elbow joint

**Technique (with and without ultrasound):**

1. Traditionally, 20 mg of triamcinolone or methylprednisolone is mixed with 0.5 ml of anesthetic. If available, a 1″ or ⅝″ needle may provide additional ease and safety for this procedure, though a 1.5″ needle can be used.
2. The elbow joint is a versatile medium joint prone to overuse injuries and operational trauma, especially with repetitive use. Care should be taken when assessing an elbow injury for amenability to relief with injection that the elbow joint itself is affected and that symptoms are not being caused either by epicondylitis or olecranon bursitis.
3. In preparation for this access, one possible technique to aid stabilization and maximize access in the microgravity environment is for the patient to place their arm, elbow bent at 90 degrees, on a stabilizing surface, allowing an approach perpendicular to the forearm and parallel to the upper arm. Given the nature of the microgravity environment, it may be necessary to secure the arm in this position to prevent drift in multiple directions. One approach for this would be to use a pillow or other soft object between the bent forearm and the patient's chest to allow the patient's own body to stabilize the angle.
4. Ensure that adequate immobilization has occurred. This includes immobilization of both the patient and the elbow joint, accounting for the three-dimensional nature of spaceflight.
5. Identify all relevant items needed for the procedure and ensure an appropriate plan for capped and uncapped needle management before, during, and after the procedure to ensure safety.
6. The head of the radius, the end of the lateral epicondyle, and the olecranon process should be identified. These three landmarks form a triangle. If a surgical marking pen is available, demarcate these three points and draw lines between them to form the triangle visually. The injection site is the soft area at the center of the triangle.

7. After appropriate local site sterilization with BZK, and appropriate preparation including personal protective equipment and positioning, the appropriately selected needle and syringe can be inserted at the marked site. The needle should be angled with a slight downward and lateral tilt towards the medial epicondyle, which should be below the triangle. Given the small nature of the joint space, the needle may not be able to be hubbed if the joint space is not deep enough.
8. Slightly aspirate to ensure that the needle is within the joint space and is not within a blood vessel. A small bubble may appear in the fluid as the joint's negative space is aspirated. Apply gentle but firm pressure to inject the medication at a steady rate. If resistance is met, slightly withdraw and reposition the needle as opposed to pressing past resistance.
9. Withdraw and secure the needle and utilize appropriate postprocedure techniques, such as any bleeding control and protection of the surgical site with a bandage.
10. Heavy activity should be avoided with the joint for the next 48–72 hours, though this is considered based on clinical expertise and not trial data. Conversely, joint immobilization is not needed postprocedure.

**Modification for aspiration:**

1. Aspiration of the elbow joint can be accomplished with a needle 22 g or greater, and ideally, a 20 g needle will be available. An 18 g needle can also accomplish the same purpose but may not be desired given its additional discomfort in the relatively smaller joint space and over an area with less fascial and fat protection. If a larger gauge of needle is used, it is important to first anesthetize the needle tract with 2–3 ml of lidocaine or an equivalent anesthetic UNLESS a septic arthritis is suspected, as the anesthetic may lower diagnostic yield. This should be done after the surgical site is sterilized and before the aspiration needle is placed.
2. If therapeutic aspiration is being considered, ensure there is a proper immediate and long-term plan for the fluid waste, including whether it is to be stored in syringes or other collection containers, and whether it needs to be packaged for surface analysis, further research, or disposed of with proper biohazard precautions. If laboratory analysis is available, ensure that there is a plan for appropriate fluid transfer to the relevant equipment. If microscopy is planned for crystal analysis, note that the fluid should be analyzed within a few hours of collection for optimal yield.
3. After successfully inserting the needle and syringe intended for aspiration, withdraw the syringe plunger slowly; excessive negative force can cause the needle to become clogged. Note the color and consistency of the fluid that is being removed.
4. Once fluid has been removed, if a corticosteroid injection is desired, then ensure that the syringe containing the corticosteroid and anesthetic mix is readily available. Unscrew the syringe from the hubbed needle in the patient in a counterclockwise direction. Secure the first syringe. Attach the syringe containing the medication to the hubbed needle and inject following the instructions above.

### 15.10.4.7.2 Lateral/medial epicondyle

Epicondylitis is a common result of repetitive motion or overuse injury. If a physical exam is suggestive of epicondylitis and the patient is refractory to conservative treatments such as rest, ice, and NSAIDs, then a corticosteroid injection may provide relief and return to operational status (Figs. 15.20 and 15.21).

1. A 20 mg of triamcinolone or methylprednisolone and 0.5 ml of anesthetic is usually sufficient for antiinflammatory pain relief.
2. The preparation and procedural course is similar to that described above.
3. For lateral epicondylitis, the elbow is bent at a 45-degree angle with a pronated wrist. The injection site is the point of maximal tenderness, usually located approximately 1 cm anterior to the lateral epicondyle.
4. For medial epicondylitis, the arm can be abducted with the wrist supinated. The injection site is the point of maximal tenderness, usually located approximately 1 cm anterior to the medial epicondyle.
5. After appropriate local site sterilization with BZK, and appropriate preparation including personal protective equipment and positioning, the appropriately selected needle and syringe can be inserted at the marked site. Insert the needle perpendicularly to the level of the bone, then slightly withdraw the needle and inject the medication.
6. Apply gentle but firm pressure to inject the medication at a steady rate. If resistance is met, slightly withdraw and reposition the needle as opposed to pressing past resistance.
7. Withdraw and secure the needle and utilize appropriate postprocedure techniques, such as any bleeding control and protection of the surgical site with a bandage.

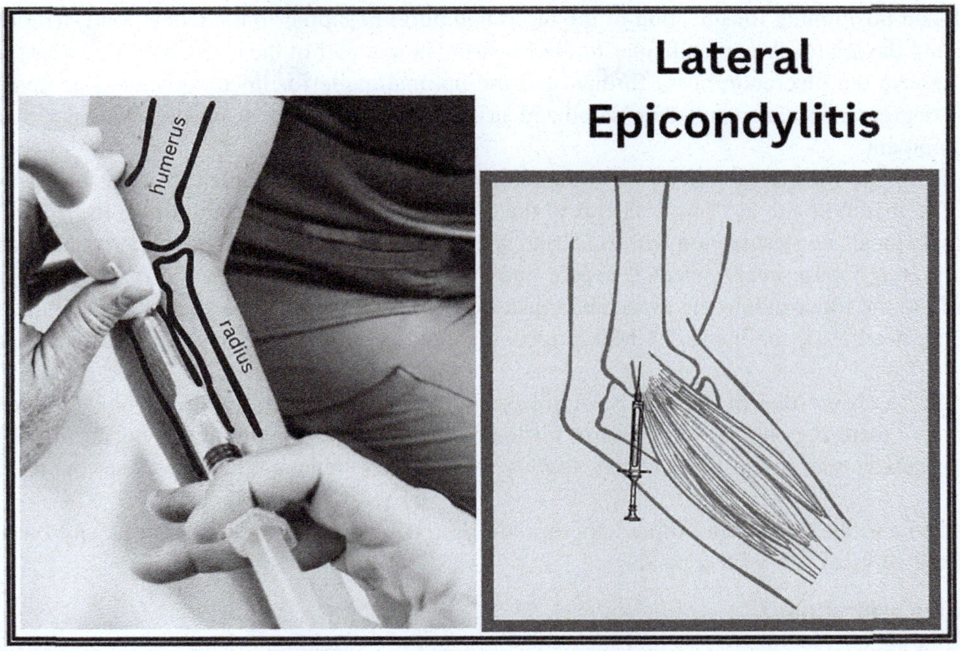


FIGURE 15.20 Lateral epicondyle approach.
Credit: Amanda Prine.

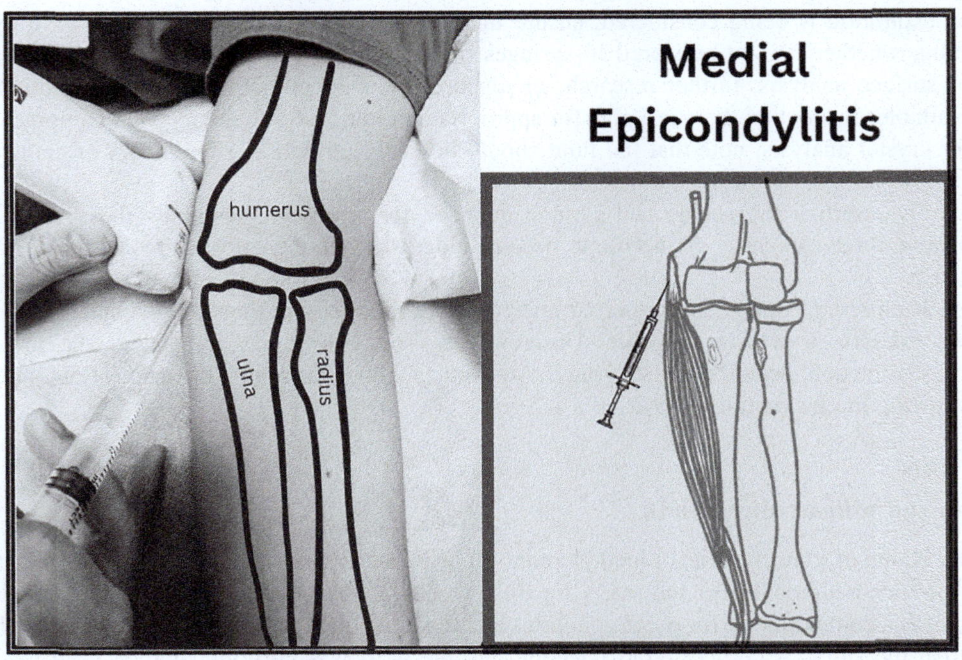


FIGURE 15.21 Medial epicondyle approach.
Credit: Amanda Prine.

### 15.10.4.7.3 Olecranon bursa

**Technique (with and without ultrasound):**

The olecranon bursa is another common site prone to overuse injury, repetitive motion injury, or trauma. Injury to the olecranon bursa is readily identified with swelling and palpation of a "fluid-filled sac" at the tip of the elbow.

Note that the olecranon bursa can be a frequent site of infection, so if an infection is suspected, then corticosteroid should not be injected. An infected bursa raises the risk for, but is not diagnostic of, a concurrent septic arthritis.

1. The approach and positioning for injection of the olecranon bursa is similar to the elbow joint, with the elbow at a 90-degree angle and the approach perpendicular to the forearm and parallel to the upper arm. If desired, additional elbow flexion may expose the olecranon bursa further, and the optimal angle for injection should be determined with arm flexion to 90 degrees or greater and then stabilized in that position, either using the patient's own body or using additional equipment.
2. Identify the tip of the olecranon process. If a surgical marking pen is available, the edge of the olecranon can be demarcated. The injection site is slightly lateral to the olecranon in the swollen fluid area itself.
3. After appropriate local site sterilization with BZK and appropriate preparation, including personal protective equipment and positioning, the appropriately selected needle and syringe can be inserted at the marked site. Insert the needle perpendicularly to the forearm into the space and inject the medication. Given the small nature of the bursal space, it is unlikely that the needle will be hubbed. CMO judgment should be relied upon to determine if excess penetration has occurred.
4. Slightly aspirate to ensure that the needle is within the joint space and is not within a blood vessel. Note that with traumatic bursitis, there is potential for blood to fill the olecranon bursa. Apply gentle but firm pressure to inject the medication at a steady rate. If resistance is met, slightly withdraw and reposition the needle as opposed to pressing past resistance.
5. Withdraw and secure the needle and utilize appropriate postprocedure techniques, such as any bleeding control and protection of the surgical site with a bandage.

**Modification for aspiration:**

1. Aspiration of the olecranon bursa can be accomplished with a needle 22 g or greater. An 18 g needle can also accomplish the same purpose but may not be desired given its additional discomfort in the relatively smaller space; the higher gauge may also create "leakage" around the insertion point during or after the procedure is performed. Note that anesthetizing the overlying skin is not practical or feasible for an olecranon bursa aspiration.
2. If therapeutic aspiration is being considered, ensure there is a proper immediate and long-term plan for the fluid waste, including whether it is to be stored in syringes or other collection containers, and whether it needs to be packaged for surface analysis, further research, or disposed of with proper biohazard precautions. If laboratory analysis is available, ensure that there is a plan for appropriate fluid transfer to the relevant equipment. If microscopy is planned for crystal analysis, note that the fluid should be analyzed within a few hours of collection for optimal yield.
3. After successfully inserting the needle and syringe intended for aspiration, withdraw the syringe plunger slowly; excessive negative force can cause the needle to become clogged. Note the color and consistency of the fluid that is being removed.
4. Once fluid has been removed, if a corticosteroid injection is desired, ensure the syringe containing the corticosteroid and anesthetic mixture is readily available. Unscrew the syringe from the hubbed needle in the patient in a counterclockwise direction. Secure the first syringe. Attach the syringe containing the medication to the hubbed needle and inject following the instructions above.

#### 15.10.4.7.4 Wrists

**Technique (with and without ultrasound):**

1. Traditionally, 20 mg of triamcinolone or methylprednisolone is mixed with 0.5 ml of anesthetic. If available, a 1 or ⅝″ needle may provide additional ease and safety for this procedure, though a 1.5″ needle can be used.
2. The wrist joint is a common joint for overuse injuries and repetitive motion injuries, but frequently, this is responsive to conservative measures. A corticosteroid injection into the wrist is a fairly uncommon need for acute pathology except for a flare of an underlying inflammatory arthritis (which a spaceflight participant would presumably have been screened for and thus a medical system should be adequately prepared for).
3. The wrist joint has three compartments: the radiocarpal joint, the ulnocarpal joint, and the ulnar styloid joint. The most common acute injuries will usually respond to a radiocarpal injection. While the joints do not traditionally communicate, it has been demonstrated that contrast injected into the radiocarpal joint will have unidirectional communication with the other joint spaces, potentially due to pressure inside the joint space.
4. In preparation for this access, one possible technique to aid in stabilization as well as maximize access in the microgravity environment is for the patient to place their hand and arm on a flat surface to facilitate a superior approach. The wrist joint can be more easily secured on a flat, hard surface.

5. Ensure that adequate immobilization has occurred. This includes immobilization of both the patient and the wrist joint, accounting for the three-dimensional nature of spaceflight. With a superior approach and the wrist secured to a flat surface, there is a relatively smaller risk of the patient's bodily movement compromising the procedure.
6. Identify all relevant items needed for the procedure and ensure that there is an appropriate plan for capped and uncapped needle management before, during, and after the procedure to ensure safety.
7. The end of the radius and the relevant extensor tendons as they cross the wrist should be identified. Tendons can be identified by the patient extending each relevant finger, allowing palpation of that individual tendon. For a radiocarpal joint, the injection site is the soft area distal to the radius and between either the second and third extensor tendons or between the first and second extensor tendons. If a surgical marking pen is available, demarcate the first, second, and third extensor tendons and draw a horizontal line over the end of the radius. For an ulnocarpal joint, the injection site is the soft area distal and medial to the ulnar styloid and between either the third and fourth extensor tendons or the fourth and fifth extensor tendons. If a surgical marking pen is available, demarcate the third, fourth, and fifth extensor tendons and draw a horizontal line over the end of the ulna.
8. After appropriate local site sterilization with BZK and appropriate preparation, including personal protective equipment and positioning, the appropriately selected needle and syringe can be inserted at the marked site. The needle should be angled perpendicular to the skin. Given the small nature of the joint space, the needle may not be able to be hubbed if the joint space is not deep enough, especially if a longer needle is used.
9. Once the needle has been placed into the joint, have the patient extend the relevant fingers to allow engagement of the tendons. If the needle is within the joint space, then it should not move excessively with movement of the tendons. If the needle is within a tendon sheath or tendon, the needle will move with the motion of the tendons and likely cause discomfort.
10. Slightly aspirate to ensure that the needle is within the joint space and is not within a blood vessel. A small bubble may appear in the fluid as the joint's negative space is aspirated. Apply gentle but firm pressure to inject the medication at a steady rate. If resistance is met, slightly withdraw and reposition the needle as opposed to pressing past resistance. Given the small joint space, it may not be possible to inject all the fluid.
11. Withdraw and secure the needle and utilize appropriate postprocedure techniques, such as any bleeding control and protection of the surgical site with a bandage.
12. Heavy activity should be avoided with the joint for the next 48–72 hours, though this is considered based on clinical expertise and not trial data. Conversely, joint immobilization is not needed postprocedure.

**Modification for aspiration:**

1. Aspiration of the wrist joint can be accomplished with a needle 22 g or greater, and ideally, a 20 g needle will be available. An 18 g needle can also accomplish the same purpose but may not be desired given its additional discomfort in the relatively smaller joint space and over an area with less fascial and fat protection. If a larger gauge of needle is used, it is important to first anesthetize the needle tract with 1–2 ml of lidocaine or an equivalent anesthetic UNLESS a septic arthritis is suspected as the anesthetic may lower diagnostic yield. This should be done after the surgical site is sterilized and before the aspiration needle is placed.
2. If therapeutic aspiration is being considered, ensure there is a proper immediate and long-term plan for the fluid waste, including whether it is to be stored in syringes or other collection containers, and whether it needs to be packaged for surface analysis, further research, or disposed of with proper biohazard precautions. If laboratory analysis is available, ensure that there is a plan for appropriate fluid transfer to the relevant equipment. If microscopy is planned for crystal analysis, note that the fluid should be analyzed within a few hours of collection for optimal yield.
3. After successfully inserting the needle and syringe intended for aspiration, withdraw the syringe plunger slowly; excessive rapidity of withdrawal can cause the needle to become clogged. Note the color and consistency of the fluid that is being removed.
4. Once fluid has been removed, if a corticosteroid injection is desired, then ensure that the syringe containing the corticosteroid and anesthetic mix is readily available. Unscrew the syringe from the hubbed needle in the patient in a counterclockwise direction. Secure the first syringe. Attach the syringe containing the medication to the hubbed needle and inject following the instructions above.

### *15.10.4.8 Stenosing flexor tenosynovitis ("trigger finger") injections*

#### 15.10.4.8.1 Technique

Stenosing flexor tenosynovitis occurs when the flexor tendon does not have a free range of motion through the pulley system, usually at the A1 level over the MCP joint area. A stenosed sheath in an operational environment would be expected to be commonly caused by a buildup of debris and scar tissue from overuse, repetitive trauma to the area, or strain, resulting in a finger that can become "locked" in a flexed or extended position that requires manual movement to "unlock." If conservative measures such as manual massage are ineffective, a glucocorticoid injection in the area of the stenosed sheath can provide relief.

1. If available, a shorter needle can be used for ease and safety, and if a smaller gauge (e.g., 27 g) is available, that will likely be sufficient. Traditionally, 10–20 mg of triamcinolone or methylprednisolone mixed 1:1 with 0.25–0.5 ml of lidocaine will be sufficient for the small space.
2. In preparation for this access, one possible technique to aid in stabilization as well as maximize access in the microgravity environment is for the patient to place their hand and arm on a flat surface to facilitate a superior approach. The wrist joint can be more easily secured on a flat, hard surface, and the hand should be supinated to access the palm.
3. Ensure that adequate immobilization has occurred. This includes immobilization of both the patient and the wrist joint, accounting for the three-dimensional nature of spaceflight. With a superior approach and the wrist secured to a flat surface, there is a relatively smaller risk of the patient's bodily movement compromising the procedure.
4. Identify all relevant items needed for the procedure and ensure that there is an appropriate plan for capped and uncapped needle management before, during, and after the procedure to ensure safety.
5. The A1 area of the pulley system should be identified on the palmar aspect. This is commonly approximately 1 cm medial to the crease where the finger and the palm meet. Often, there will be pain with palpation of that area, as well as potentially a small palpable buildup of scar tissue. If a surgical marking pen is available, demarcate the point of maximal tenderness or the area of scar tissue buildup. The approach will be perpendicular to the skin, though a 30-degree angle angled medially can also be utilized.
6. After appropriate local site sterilization with BZK and appropriate preparation, including personal protective equipment and positioning, the appropriately selected needle and syringe can be inserted at the marked site. The needle should be angled perpendicular to the skin. Given the small nature of the joint space, the needle may not be able to be hubbed if the joint space is not deep enough, especially if a longer needle is used.
7. Once the needle has been placed into the joint, have the patient flex the relevant fingers to allow engagement of the tendons. If the needle is within the correct area and not positioned on/inside the tendon, then it should not move excessively with movement of the tendons. If the needle is within a tendon sheath or tendon, it will move with the tendons' motion and likely cause discomfort.
8. Slightly aspirate to ensure that the needle is within the joint space and is not within a blood vessel. A small bubble may appear in the fluid as the joint's negative space is aspirated. Apply gentle but firm pressure to inject the medication at a steady rate. If resistance is met, slightly withdraw and reposition the needle as opposed to pressing past resistance. Given the small joint space, it may not be possible to inject all the fluid.
9. Withdraw and secure the needle and utilize appropriate postprocedure techniques, such as any bleeding control and protection of the surgical site with a bandage. Additional pressure may be needed to prevent extravasation. A gentle massaging technique with the applied pressure can aid in diffusion of the medication.
10. Heavy activity should be avoided with the finger/hand for the next 24–48 hours, though this is considered based on clinical expertise and not trial data. Conversely, joint immobilization is not needed postprocedure.

## 15.11 Lower extremity joint injections

### 15.11.1 Joint and bursal Injections/aspiration

It is notable that while many common joint injections are done for the lower extremities, spaceflight may change this expectation, such that a lower extremity joint injection may be relatively uncommon in a microgravity environment. For exploration-class missions that have a partial gravity surface component, it is theorized that joint-stabilizing and supporting structures, such as muscle-tendon, may atrophy due to disuse in microgravity, which could predispose to injury due to nonroutine cyclic compressive loading when force is reapplied or upon resuming some weight-bearing activity on Mars or the lunar surface. At the time of writing, however, data informing surface operations are currently nonexistent and remain theoretical, and the clinical decision to administer a joint injection and/or aspiration will require the expertise of the CMO.

Spaceflight also presents unique stressors on the joint spaces, which are, at the time of writing, incompletely characterized. These stressors include altered lower mechanical stimulus of the joint, as well as potential radiation and inflammatory characteristics that may create an environment where joint degradation is more likely. It is currently unclear how a localized corticosteroid injection will interact with this unique environment within the joint space.

In general, more precise injection techniques of the soft tissue, which would require drug delivery into a tendon sheath (such as for tendonitis) or into a very small compartment (carpal tunnel syndrome), are best performed by a clinician with terrestrial experience in the specific procedure and with the aid of ultrasound for optimum needle visualization and guidance.

This section concerns itself with some of the more common and easily accessible injection sites. Due to the need for a specialized curvilinear ultrasound probe, proximity to neurovascular structures, and required expertise, hip joint injections and aspirations are not included.

## 15.11.2 Large joint injection (knees)

Illustration demonstrating approach to knee arthrocentesis (Fig. 15.22).

### *15.11.2.1 Technique (with and without ultrasound)*

The knee joint is a deceptively complex structure. It is the combination of two functional joints (the femoral-patellar and femoral-tibial joints) as well as the floating mobile kneecap, which is stabilized by multiple ligaments and adjacent tendons. There is also a communicating suprapatellar bursal space. The knee is designed to absorb magnified force from the foot and to stabilize and cushion descending force from the upper body. As such, it is often prone to sprains and strains. The relatively large joint space lends itself to effusions. This combination of factors makes the knee joint the most commonly aspirated and injected joint (Jüni et al., 2015).

There are multiple windows into the knee joint, and as such, there are multiple approach techniques. Two of the more common approaches (the superomedial and superolateral approaches) will be highlighted here.

The superomedial approach is preferred for injections as it minimizes pain, given the relative lack of fat pads and nerve endings compared to other knee windows. The goal of the injection is to infiltrate the suprapatellar bursa, which will communicate the medication within the intraarticular joint compartment. In contrast, the superolateral approach is preferred for aspiration to access fluid buildup behind the patella and inside the joint.

Note that both injection and aspiration approaches will be heavily informed by ultrasound guidance, if available, to determine if the largest fluid is identified in the medial or lateral region.

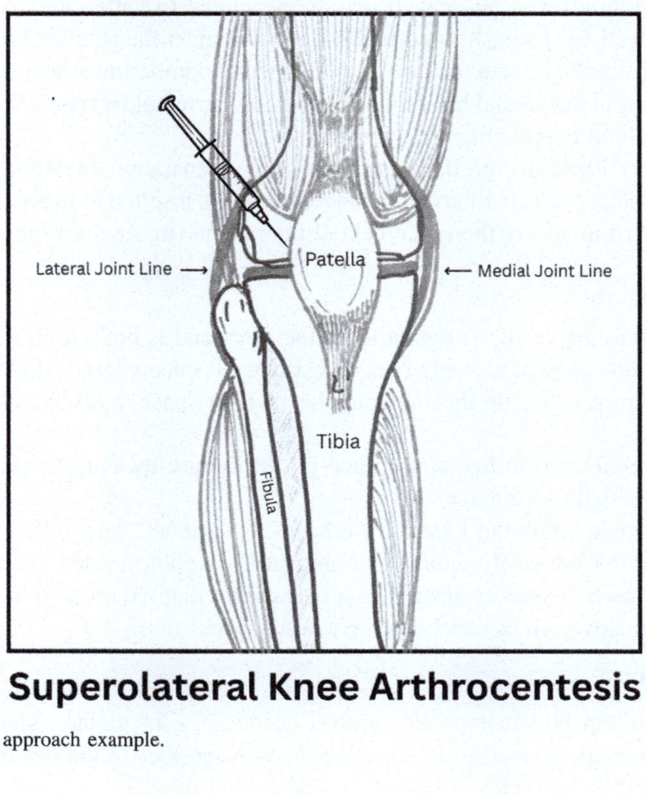


**FIGURE 15.22** Knee arthrocentesis approach example.
Credit: Amanda Prine.

In preparation for this access, one possible technique to aid in stabilization as well as maximize access in the microgravity environment is for the knee to be immobilized at an approximately 30-degree bend, with the physician perpendicular to the leg so that the patella is oriented "up" in the three-dimensional microgravity space. This can be accomplished with the leg placed on a flat surface and a bulky flexible item placed under the knee to achieve the 30-degree bend, with immobilization above and below the joint. Securing the patient's upper body may aid in preventing an unusually angled force to the hip joint during the technique.

Identify all relevant items needed for the procedure and ensure that there is an appropriate plan for capped and uncapped needle management before, during, and after the procedure to ensure safety. Traditionally, 40 mg of triamcinolone or methylprednisolone is mixed with 1.0 ml of anesthetic. A 1.5″ needle is preferred to access the joint space successfully.

### *15.11.2.2 Superomedial approach*

1. The top (proximal) half of the medial border of the patella and the lower border of the femur as it projects near the patellar space. Ultrasound can more accurately identify the optimum site for needle insertion based on the actual location of the synovial fluid, which is anechoic or black on the ultrasound imaging screen. If a surgical marking pen is available, use it to demarcate these lines. The injection site is the soft area under the border of the patella at an approximately 45-degree angle from the midline (Fig. 15.22).
2. After appropriate local site sterilization with BZK, and proper preparation including personal protective equipment and positioning, the appropriately selected needle and syringe can be inserted into the marked site. The needle should be angled with a proximal tilt towards the suprapatellar bursa above the patella. An alternative approach is to angle the needle straight down to access the space directly behind the patella. The needle will often be hubbed to access the space, depending upon the size of the individual knee joint. If the needle is advanced in the incorrect position, then striking the pain-sensitive bone (patella or femur) will cause immediate discomfort, and the needle must be withdrawn slightly and advanced using a different angle during insertion.

### *15.11.2.3 Superolateral approach*

1. The top (proximal) half of the lateral border of the patella and the lower border of the femur as it projects near the patellar space. If a surgical marking pen is available, use it to demarcate these lines. As above, an ultrasound can easily identify the optimal site for needle insertion. The injection site is the soft area under the border of the patella at an approximately 45-degree angle from the midline. If an effusion is present, this space will often feel more "open" when compared to a joint without an effusion present. It may be beneficial to flatten the knee entirely and remove the 30-degree bend, as this may effectively "squish" an effusion out from under the patella. Another technique that identifies a knee effusion is to compress the fluid from the lateral to the medial compartment toward the lateral compartment, then rapidly "milk" the effusion from the medial back to the lateral compartment by rapidly compressing the undersurface of the patella to observe the lateral compartment bulging.
2. After appropriate local site sterilization with BZK and appropriate preparation, including personal protective equipment and positioning, the appropriately selected needle and syringe can be inserted at the marked site. The needle should be angled straight and downward to access the space behind the patella (figure number).

### *15.11.2.4 For both approaches*

1. Slightly aspirate to ensure that the needle is within the joint space and is not within a blood vessel. Apply gentle but firm pressure to inject the medication at a steady rate. If resistance is encountered, slightly withdraw and reposition the needle rather than pressing past it. Inside the suprapatellar or joint space, resistance should feel minimal even if an effusion is not present.
2. Withdraw and secure the needle and utilize appropriate postprocedure techniques such as any bleeding control and protection of the surgical site with a bandage.
3. Heavy activity should be avoided with the joint for the next 24–48 hours, though this is considered based on clinical expertise and not trial data. Conversely, joint immobilization is not needed postprocedure. In a microgravity environment, encountering lower extremity activity at a stress level that would be relatively contraindicated would be unusual short of deliberate loads such as exercise or EVA activities.

**Modification for aspiration:**

1. Aspiration of the knee joint can be accomplished with a needle 22 g or greater, and ideally a 18 g needle will be available. If a larger gauge needle is used, it is important to first anesthetize the needle tract with 2–3 ml of 1%–2%

lidocaine without epinephrine or an equivalent anesthetic. This should be done after the surgical site is sterilized and before the aspiration needle is placed.

2. If therapeutic aspiration is being considered, ensure there is a proper immediate and long-term plan for the fluid waste, including whether it is to be stored in syringes or other collection containers, and whether it needs to be packaged for surface analysis, further research, or disposed of with proper biohazard precautions. If laboratory analysis is available, ensure that there is a plan for appropriate fluid transfer to the relevant equipment. If microscopy is planned for crystal analysis, collect the fluid into a heparin tube and analyze it within a few hours of collection for optimal yield.
3. After successful insertion of the needle and syringe intended for aspiration, withdraw the syringe plunger slowly, as excessively rapid opposing force can cause the needle to become clogged, have the needle bevel against the joint capsule, or fail to remove high viscosity fluid of noninflammatory joint fluid. Note the color and consistency of the fluid that is being removed.
4. Once fluid has been removed, if a corticosteroid injection is desired, ensure the syringe containing the corticosteroid and anesthetic mixture is readily available. Unscrew the syringe from the hubbed needle in the patient in a counterclockwise direction. Secure the first syringe. Attach the syringe containing the medication to the hubbed needle and inject following the instructions above.

## 15.11.3 Medium joint injection

### *15.11.3.1 Ankles*

Illustration demonstrating ankle arthrocentesis approach (Fig. 15.23).

#### 15.11.3.1.1 Technique (with and without US)

1. Traditionally, 20 mg of triamcinolone or methylprednisolone is mixed with 0.5 ml of anesthetic. A 1″ or 1.5″ needle is preferred to access the joint space.
2. The ankle joint is a relatively well-protected and stabilized joint since it must be able to transfer the force of impact from walking or running through the joint and into the upper leg. The joint itself is where the tibia, fibula, and talus meet, and the many attachments originating from this joint space and the surrounding bones provide flexion and extension to the foot and the toes. As such, care must be taken with procedures involving the ankle to avoid tendon injury.

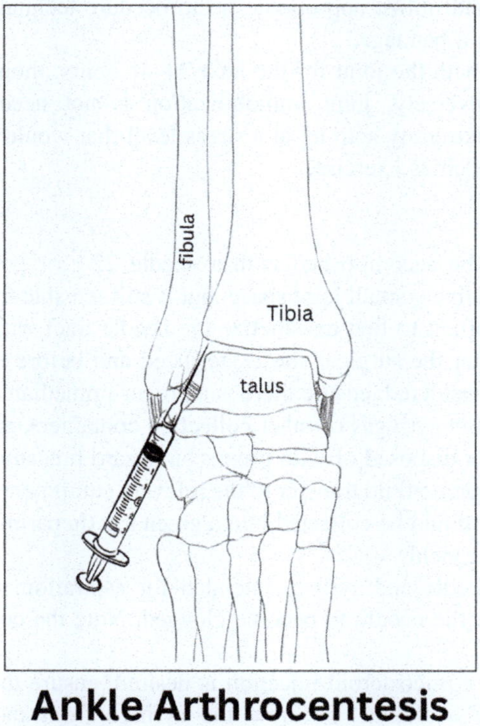


FIGURE 15.23 Ankle arthrocentesis approach.
Credit: Amanda Prine.

3. The primary approach to the ankle joint is perpendicular from the anterior side of the joint. In preparation for this access, one possible technique to aid stabilization and maximize access in the microgravity environment is to immobilize the ankle at a neutral angle; in microgravity, this will likely form a slight obtuse angle relative to the leg. Ensure that the patient's body is secured as well to ensure that torque from unusual angles will not be applied to the patient's body during the procedure (e.g., place the patient in a chair).
4. Identify the proximal border of the talus as well as the tibialis anterior tendon as it courses over the ankle crease. These areas can be more easily palpated with flexion of the ankle joint to more easily raise the tibialis anterior tendon. The injection site is most commonly lateral to the tibialis anterior tendon, though medial to the tendon is also a valid approach. Flex and extend the ankle during palpation until a soft spot overlying the joint space can be identified. If there is no overlying object, the place will not change in feel during either flexion or extension. When the injection site is located, immobilize the joint in a neutral position to maintain an open joint space. The joint angle can be adjusted to maximize open space. The multiple tendons anterior to the ankle joint often make it difficult to identify an effusion. If ultrasound is available, the site of optimal needle placement can easily be identified on ultrasound imaging using a longitudinal orientation of the probe to identify and avoid the adjacent tendons, as well as identify and mark on the skin surface the very small tibial-talus joint space, which is usually only 3–4 mms.
5. Identify all relevant items needed for the procedure and ensure that there is an appropriate plan for capped and uncapped needle management before, during, and after the procedure to ensure safety.
6. After appropriate local site sterilization with BZK and appropriate preparation, including personal protective equipment and positioning, the appropriately selected needle and syringe can be inserted into the marked site. If the needle is inadvertently placed into a tendon, there is immediate pain (unless the track was previously anesthetized with lidocaine) and the needle must be immediately withdrawn and repositioned. The needle should be angled downward and perpendicular to the joint to access it. It may be beneficial for the patient to move their toes at this point to ensure that the needle is not within any tendon sheath. If the needle does not move when the tendons are activated, then it is likely not within a tendon.
7. Slightly aspirate to ensure that the needle is within the joint space and is not within a blood vessel. Apply gentle but firm pressure to inject the medication at a steady rate. If resistance is encountered, slightly withdraw and reposition the needle rather than pressing past it. Inside the joint space, resistance should feel minimal during the majority of the injection, even if an effusion is present. A slight resistance may be present at the end of the injection as the joint may not be able to hold the full amount of fluid; if that is the case, inject as much medication as is reasonable and comfortable and do not force the full amount.
8. Withdraw and secure the needle and utilize appropriate postprocedure techniques, such as any bleeding control and protection of the surgical site with a bandage.
9. Heavy activity should be avoided with the joint for the next 24–48 hours, though this is considered based on clinical expertise and not trial data. Conversely, joint immobilization is not needed postprocedure. In a microgravity environment, encountering lower extremity activity at a stress level that would be relatively contraindicated would be unusual short of deliberate loads such as exercise.

**Modification for aspiration:**

1. Aspiration of the ankle joint can be accomplished with a needle 22 g or greater. An 18-gauge needle may cause excessive discomfort, given the relatively small joint space, but is still feasible for the procedure if necessary. If a larger gauge of needle is used, it is important to first anesthetize the needle tract with 2–3 ml of lidocaine or an equivalent anesthetic. This should be done after the surgical site is sterilized and before the aspiration needle is placed.
2. If therapeutic aspiration is being considered, ensure there is a proper immediate and long-term plan for the fluid waste, including whether it is to be stored in syringes or other collection containers, and whether it needs to be packaged for surface analysis, further research, or disposed of with proper biohazard precautions. If laboratory analysis is available, ensure that there is a plan for appropriate fluid transfer to the relevant equipment. If polarized microscopy is planned for crystal analysis, note that the fluid should be collected into a green top (heparin-containing) tube and analyzed within a few hours of collection for optimal yield.
3. After successfully inserting the needle and syringe intended for aspiration, withdraw the syringe plunger slowly; excessive negative force can cause the needle to become clogged. Note the color and consistency of the fluid that is being removed (Table 15.11).
4. Once fluid has been removed, if a corticosteroid injection is desired, ensure the syringe containing the corticosteroid and anesthetic mixture is readily available. Unscrew the syringe from the hubbed needle in the patient in a counterclockwise direction. Secure the first syringe. Attach the syringe containing the medication to the hubbed needle and inject following the instructions above.

**TABLE 15.11** Joint fluid appearance.

| | Color | Clarity | Viscosity | Interpretation concern |
|---|---|---|---|---|
| Physiologic | Colorless | Clear | | No pathology |
| Noninflammatory/osteoarthritic | Yellow-tinged | Clear | High | Osteoarthritic flare |
| Inflammatory | Yellow | Cloudy | Low | Inflammatory arthritic flare/acute gout attack |
| Infectious | Dark yellow/green | Cloudy | Low | Developing septic joint |
| Septic/gout | White | Opaque/pus | High | Septic joint and/or acute gout attack |
| Traumatic | Pink | Clear/cloudy | Low | Trauma without bleeding into joint capsule |
| Frank hemarthrosis | Red | Opaque | Low | Bleeding into joint capsule |

## 15.12 Bursa injections

### 15.12.1 Greater trochanteric bursa

#### *15.12.1.1 Technique (with and without US)*

1. Traditionally, 40 mg of triamcinolone or methylprednisolone is mixed with 1.0 ml of anesthetic. A 1.5″ needle is preferred to access the space successfully.
2. The trochanteric bursa is a large bursa that overlies the lateral prominence of the greater trochanter. As such, it is particularly vulnerable to chronic pressure and repeated trauma that can cause inflammation, leading to pain. Multiple tendons that facilitate hip abduction, including the gluteal tendons, meet under the bursal areas. Microtears in these tendons can lead to bursal trauma. In a microgravity environment, spacesuit pressure or repetitive exercise motions would be expected to be the primary mechanisms of trauma to this area. Bursal injections generally offer an easier approach and a lower risk of infection than joint injections, since the bursal space is often better protected than the inner joint space.
3. The primary approach to the bursa is perpendicular, lateral to the patient. One recommended approach is a mirror of the terrestrial approach, with the patient angled supine and "laying" on one side such that the bursal area is pointed "upward" towards the proceduralist's head. Immobilization can be achieved with a table or flat surface to which the patient can be secured so that one side of the patient is contacting the surface and the relevant side for the procedure is accessible.
4. Identify the bony prominence on the lateral aspect of the hip. The injection site will be the point of maximal tenderness, which is usually at the furthest point on the bony prominence. Reproducing the patient's pain by direct pressure confirms that the bursa is the origin of the patient's pain, and it is not from the hip (whereas pain upon gentle passive internal and external rotation of the femur suggests pain is originating from the hip joint). While ultrasound can easily identify the greater trochanter, it is rare for the bursae to be distended on ultrasound in this condition.
5. Identify all relevant items needed for the procedure and ensure an appropriate plan for capped and uncapped needle management before, during, and after the procedure to ensure safety.
6. After appropriate local site sterilization with BZK and appropriate preparation, including personal protective equipment and positioning, the appropriately selected needle and syringe can be inserted at the marked site. The needle should be angled perpendicular and downward until the osseous surface is touched by the tip of the needle. Slightly withdraw the needle and inject the medication so that it is delivered slightly above the osseous surface. If the bursa is particularly inflamed, there may be some increased discomfort both with the initial contact and the pressure of the injection.
7. Withdraw and secure the needle and utilize appropriate postprocedure techniques, such as any bleeding control and protection of the surgical site with a bandage.

### 15.12.2 Pes anserine bursa

#### *15.12.2.1 Technique (with and without US)*

1. Traditionally, 40 mg of triamcinolone or methylprednisolone is mixed with 1.0 ml of anesthetic. A 1.5″ needle is preferred to access the space successfully.

2. The pes anserine bursa is the other lower extremity bursa that may be vulnerable to repeated trauma in a microgravity environment with either spacesuit pressure or repeated traumatic exercise. The pes anserine bursa is located where the sartorius, gracilis, and semitendinosus tendons contact the shin bone and is posterior to the medial collateral ligament. Bursal injections generally offer an easier approach and a lower risk of infection than joint injections, since the bursal space is often better protected than the inner joint space.
3. The primary approach to the bursa is perpendicular to the proceduralist position to have access to the medial side of the knee and leg. A similar positioning and immobilization to a knee injection can be done, with the knee flexed at an approximately 30-degree angle.
4. Identify the medial collateral ligament, the tibial tubercle, and, if palpable, the course of the three relevant muscles as they course over the joint medially and meet together. The injection site will be the point of maximal tenderness over a localized area, which is usually slightly distal and medial to the tibial tubercle (approximately 2–3 finger widths). Reproducing the patient's pain by direct pressure in this area confirms that the pain is originating from the pes anserine bursae and not from the knee joint (intraarticular knee pain would more likely be aggravated by passive flexion and extension of the knee). The bursa is rarely distended on ultrasound evaluation.
5. Identify all relevant items needed for the procedure and ensure that there is an appropriate plan for capped and uncapped needle management before, during, and after the procedure to ensure safety.
6. After appropriate local site sterilization with BZK and appropriate preparation, including personal protective equipment and positioning, the appropriately selected needle and syringe can be inserted at the marked site. The needle should be angled downward and perpendicular to the osseous surface until the tip of the needle contacts the osseous surface. Slightly withdraw the needle and inject the medication so that it is delivered slightly above the osseous surface. If the bursa is particularly inflamed, there may be increased comfort with initial contact and with the pressure of the injection.
7. Withdraw and secure the needle and utilize appropriate postprocedure techniques, such as any bleeding control and protection of the surgical site with a bandage.

## Acknowledgments

The authors wish to thank Amanda Prine for her work in creating select images in this chapter.

## Abbreviations

| | |
|---|---|
| **CMRS** | Crew medical restraint system |
| **COTS** | Commercial off-the-shelf |
| **CT** | Computerized tomography |
| **DIP** | Distal interphalangeal |
| **EVA** | Extravehicular activity |
| **FAST** | Focused abdominal sonography for trauma |
| **ISS** | International Space Station |
| **IV** | intravenous |
| **IVC** | Inferior vena cava |
| **LEO** | Low earth orbit |
| **MP** | metacarpophalangeal |
| **MRI** | Magnetic resonance imaging |
| **PIP** | Proximal interphalangeal |
| **PoCUS** | Point-of-care ultrasound |
| **REBOA** | Resuscitative endovascular balloon occlusion of the aorta |
| **VTE** | Venous thromboembolism |

## References

Ardolino, A., Zeineh, N., O'Connor, D., 2010. Experimental study of forearm compartmental pressures. The Journal of Hand Surgery 35 (10), 1620–1625. Available at https://doi.org/10.1016/j.jhsa.2010.06.017.

Belobrajdic, B., Melone, K., Diaz-Artiles, A., 2021. Planetary extravehicular activity (EVA) risk mitigation strategies for long-duration space missions. NPJ Microgravity 7 (1). Available at https://doi.org/10.1038/s41526-021-00144-w.

Bulger, E., Khurana, B., Collins, K., Burlew ,C., Jenson W., 2024. Severe pelvic fracture in the adult trauma patient. UpToDate. Available at https://www.uptodate.com/contents/severe-pelvic-fracture-in-the-adult-trauma-patient.

Compartment Syndrome, 2013. AO surgery reference. Available at https://surgeryreference.aofoundation.org/orthopedic-trauma/adult-trauma/forearm-shaft/further-reading/compartment-syndrome#diagnosis.

Compartment Syndrome - Lower Limb. 2024. Available at https://surgeryreference.aofoundation.org/orthopedic-trauma/pediatric-trauma/tibial-shaft/further-reading/compartment-syndrome.

Cushman, D.M., Christiansen, J., Clements, N.D., Cunningham, S., Teramoto, M., McCormick, Z.L., 2019. Infections after large joint or bursa injection a survey of 554 sports medicine physicians. American Journal of Physical Medicine and Rehabilitation 98 (12), 1106–1109. Available at https://doi.org/10.1097/PHM.0000000000001255, http://journals.lww.com/ajpmr/pages/default.aspx.

Dayan, L., Zinmann, C., Stahl, S., Norman, D., 2008. Complications associated with prolonged tourniquet application on the battlefield. Military Medicine 173 (1), 63–66.

Dickson, R.R., Reid, J.M., Nicholson, W.T., Lamer, T.J., Hooten, W.M., 2018. Corticosteroid and cortisol serum levels following intra-articular triamcinolone acetonide lumbar facet joint injections. Pain Practice 18 (7), 864–870. Available at https://doi.org/10.1111/papr.12686, http://onlinelibrary.wiley.com/journal/10.1111/(ISSN)1533-2500.

Fastle, R., Bothner, J., Stack, A., Wolfson, A., Ganetsky, M., 2025. Subungual hematoma. UpToDate. Available at https://www.uptodate.com/contents/subungual-hematoma.

First X-rays taken in space mimic historic first radiograph captured on earth, 2025. Radiology Business. Available at https://radiologybusiness.com/topics/clinical/clinical-research/first-x-rays-taken-space-mimic-historic-first-radiograph-captured-earth.

Gammons, M., Aspulund, C., Mezzanotte, T., Grayzel, J., 2025. Midshaft femur fractures in adults. UpToDate. Available at https://www.uptodate.com/contents/midshaft-femur-fractures-in-adults.

Genah, S., Monici, M., Morbidelli, L., 2021. The effect of space travel on bone metabolism: considerations on today's major challenges and advances in pharmacology. International Journal of Medical Sciences 22 (9). Available at https://pmc.ncbi.nlm.nih.gov/articles/PMC8123809/.

Graham, T.P., Reichman, E.F., 2018. Hematoma Blocks Reichman's Emergency Medicine Procedures, third ed. McGraw-Hill Education, New York, NY, http://accessemergencymedicine.mhmedical.com/content.aspx?aid=1159805309.

Hassan, A.S., Rao, A., Manadan, A.M., Block, J.A., 2017. Peripheral bacterial septic arthritis: review of diagnosis and management. Journal of Clinical Rheumatology 23 (8), 435–442. Available at https://doi.org/10.1097/RHU.0000000000000588, http://journals.lww.com/jclinrheum/pages/default.aspx.

Hepper, C.T., Halvorson, J.J., Duncan, S.T., Gregory, A.J.M., Dunn, W.R., Spindler, K.P., 2009. The efficacy and duration of intra-articular corticosteroid injection for knee osteoarthritis: a systematic review of level I studies. Journal of the American Academy of Orthopaedic Surgeons 17 (10), 638–646. Available at https://doi.org/10.5435/00124635-200910000-00006, http://www.jaaos.org/archive/.

Hinkelbein, J., Kerkhoff, S., Adler, C., Ahlbäck, A., Braunecker, S., Burgard, D., et al., 2020. Cardiopulmonary resuscitation (CPR) during spaceflight - a guideline for CPR in microgravity from the German Society of Aerospace Medicine (DGLRM) and the European Society of Aerospace Medicine Space Medicine Group (ESAM-SMG). Scandinavian Journal of Trauma, Resuscitation and Emergency Medicine 28 (1). Available at https://doi.org/10.1186/s13049-020-00793-y.

Jacob, P., Oertlin, C., Baselet, B., Westerberg, L.S., Frippiat, J.-P., Baatout, S., 2023. Next generation of astronauts or ESA astronaut 2.0 concept and spotlight on immunity. NPJ Microgravity 9 (1).

John, Z., 2025. Friction blisters. UpToDate. Available at https://www.uptodate.com/contents/friction-blisters.

Jullienne, A., Malo, M., Shaw, K., Zheng, Y., Johnston, J.D., Kontulainen, S., et al., 2024. Musculoskeletal perturbations of deep space radiation: assessment using a Gateway MRI. Life Sciences in Space Research 42, 74–83.

Jüni, P., Hari, R., Rutjes, A.W.S., Fischer, R., Silletta, M.G., Reichenbach, S., et al., 2015. Intra-articular corticosteroid for knee osteoarthritis. Cochrane Database of Systematic Reviews 2015 (10). Available at https://doi.org/10.1002/14651858.CD005328.pub3, https://www.cochranelibrary.com/cdsr/table-of-contents.

Kamel, S.I., Rosas, H.G., Gorbachova, T., 2024. Local and systemic side effects of corticosteroid injections for musculoskeletal indications. American Journal of Roentgenology 222 (3). Available at https://doi.org/10.2214/AJR.23.30458, http://www.ajronline.org/.

Khaliq, F., Rodham, P., 2024. StatPearlsEMS Pelvic Binders. StatPearls Publishing, Treasure Island, FL. Available at https://www.ncbi.nlm.nih.gov/books/NBK598968/.

Kragh, J.F., Dubick, M.A., 2017. Bleeding control with limb tourniquet use in the wilderness setting: review of science. Wilderness & Environmental Medicine 28 (2S).

Leg compartment release - two incision approach. 2017. OrthoBullets. Available at https://www.orthobullets.com/approaches/12236/leg-compartment-release–two-incision-approach.

Mhaskar, V., Maheshwari, J., 2015. Essential Orthopaedics (Including Clinical Methods). Jaypee Brothers Medical Publishers. Available at https://doi.org/10.5005/jp/books/12526_16 Vol. 5.

Mitchnik, I.Y., Talmy, T., Radomislensky, I., Chechik, Y., Shlaifer, A., Almog, O., et al., 2022. Femur fractures and hemorrhagic shock: implications for point of injury treatment. Injury 53, 3416–3422. Available at https://doi.org/10.1016/j.injury.2022.08.053.

Ormiston, R., Marappa-Ganeshan, R., 2023. Fasciotomy. StatPearls - NCBI Bookshelf. StatPearls Publishing, Treasure Island, FL. Available at https://www.ncbi.nlm.nih.gov/books/NBK556153/.

Pelvic trauma: initial evaluation and management, 2025. Available at https://www.uptodate.com/contents/pelvic-trauma-initial-evaluation-and-management.

Ravi, B., Hosack, L., Backstein, D., Spangehl, M., 2019. Recurrent hemarthrosis after total knee arthroplasty: evaluation and treatment. Journal of the American Academy of Orthopaedic Surgeons 27 (17), 652–658. Available at https://doi.org/10.5435/jaaos-d-18-00182.

SAM splint - tacmed Australia, 2025. Available at https://tacmedaustralia.com.au/products/sam-splint-original?srsltid=AfmBOor5gzwx4zL_k965Hv3PeDsxIDg1i_LBaOf79kVrV1u88Y6iZ2dj.

Thompson, R., 2025. Physiological and musculoskeletal injury considerations pertaining to Extravehicular Activities (EVA). KBR-NASA Johnson Space Centre. American Physical Therapy Association Combined Sections Meeting. Houston, Texas. NASA.

Weitoft, T., Forsberg, C., 2010. Importance of immobilization after intraarticular glucocorticoid treatment for elbow synovitis: a randomized controlled study. Arthritis Care & Research 62 (5), 735–737. Available at https://doi.org/10.1002/acr.20039.

Wilkerson, G.B., 2002. Biomechanical and neuromuscular effects of ankle taping and bracing. Journal of Athletic Training 37 (4), 436–445.

Chapter 16

# Spinal and central nervous system procedures

David Mampre[1], Connor Rupp[1], Amanda N. Pine[1], and Terje Saehle[2]
[1]*Department of Neurosurgery, University of Florida, Gainesville, FL, United States,* [2]*Civil Aviation Authority, Bodø, Norway*

## 16.1 Introduction

Neurosurgical emergencies are statistically rare, but the consequences of delayed or inadequate treatment in the isolated environment of space may potentially be catastrophic. Even noncritical neurosurgical conditions can significantly impact crew survival and mission success. Such conditions may present with a wide array of symptoms, ranging from subtle cognitive impairment, personality changes, or headaches to severe neurological deficits, vision loss, or loss of consciousness. In some scenarios, elevated intracranial pressure (ICP) may result in cerebral herniation and compression of critical brainstem centers that regulate respiration and cardiovascular function. Without emergent intervention, this would invariably result in death.

The most likely insult astronauts may encounter in a long-duration mission is trauma to the head or spine through manual labor, extravehicular activity (EVA), and/or operation of large equipment. On Earth, head injuries are responsible for 50%–90% of prehospital trauma-related deaths, and probabilistic risk assessments have indicated that the risk of head injuries on the ISS should not be neglected (Brasel, 2013; Chowdhury et al., 2014; Weaver et al., 2013). During future exploration-class missions beyond low earth orbit, this risk is expected to rise further due to more frequent EVAs, including surface exploration on planetary bodies with partial gravity and variable terrain, building habitats, and the use of rovers for cargo movement. Fatigue and dynamic phases of launch or landing would also pose additional risks for trauma. Moreover, environmental factors and mission profile may increase the risk of fatigue, cognitive load, and decreased situational awareness.

Cephalad fluid shifts under decreased gravitational force may result in decreased cerebral venous outflow, leading to venous stagnation and risk of thrombosis in the internal jugular veins or intracranial venous sinuses. Both ground-based analog studies and in-flight ultrasound investigations have demonstrated increased cross-sectional areas of jugular veins and jugular venous pressure, with stagnant or reversed jugular venous blood flow in more than half of crewmembers (Marshall-Goebel et al., 2019). The incidental detection of jugular vein thrombosis in astronauts has, in particular, heightened concerns, as such thromboses can lead to life-threatening complications, including cerebral venous infarction and pulmonary embolism. Cranial venous stasis may also involve the ocular structures, such as the choroid, and play a role in the manifestation of Spaceflight-Associated Neuro-Ocular Syndrome (SANS). Signs of SANS have been identified in a majority of astronauts and include hyperopic refractive error shift, optic disc edema, and globe flattening, among other ocular abnormalities. While there are many theories on the etiology of SANS, including venous overload of the choroid, the contributions of ICP are unknown, and the continued progression of SANS during exploration-class missions raises serious concerns about potential visual loss or neurological complications without mitigation. SANS has therefore been classified as one of the few remaining "red risk" areas for exploration-class missions, according to NASA's Human Research Program (Patel et al., 2020; Human Research Roadmap. Risk of Spaceflight Associated Neuro-ocular Syndrome SANS, 2024).

Another neurologic concern is ventricular enlargement after long-duration missions, likely secondary to cephalad fluid redistribution that includes shifting of the cerebrospinal fluid (CSF) spaces (Roberts and Petersen, 2019). Additionally, MRI studies have revealed increased volumes of perivascular spaces, suggesting disruptions in the brain's glymphatic system, which clears waste from the central nervous system. This hypothesis is also supported by studies demonstrating an

**Procedural Medicine for Spaceflight. DOI:** https://doi.org/10.1016/B978-0-443-27552-4.00025-7

increase in blood markers for amyloid-beta proteins linked with neurodegenerative disease and brain injury from pre- to post-spaceflight. Other changes to the brain seen in association with microgravity include white matter changes, cognitive changes, motion sickness, and neurovestibular changes, though these are likely of low surgical significance (Lee et al., 2019; Roberts et al., 2019; Hupfeld et al., 2021; Pechenkova et al., 2019; Seidler et al., 2024).

During current missions to the International Space Station (ISS), the capability to perform neurosurgical procedures is virtually nonexistent for several reasons. The ISS lacks the necessary surgical equipment, advanced imaging modalities, sterile operating environments, and even the equipment and procedures to effectively restrain the surgeon and the patient. Astronauts receive medical training, but they are not equipped with specialized neurosurgical skills. Furthermore, the microgravity environment introduces significant challenges; fluids like blood and CSF behave differently, complicating efforts to control bleeding or maintain a clear operative field. Long-duration missions will also include additional risk factors for neurosurgical emergencies, as well as communication delays and complete isolation from Earth-based medical facilities. This necessitates a reevaluation of neurosurgical preparedness and capabilities, in particular due to the potential catastrophic consequences of serious neurosurgical events. In addition to enhancing neurosurgical and medical preparedness, neurosurgical procedures may aid spaceflight research to improve our understanding of the anatomical and physiological changes that occur during sustained microgravity exposure. Conditions such as SANS, ventricular enlargement, cerebral venous stagnation, and potential effects of microgravity on glymphatic circulation present unique challenges that require in-depth investigation. For example, several astronauts have undergone postflight lumbar puncture for direct ICP measurement to understand the relation between ICP and SANS. The eight astronauts who have had direct postflight lumbar puncture (LP) days to years later and were found to have normal or only mildly elevated ICP (Fig. 16.1; Mader et al., 2021, 2011, 2017, 2013). However, reliable inflight data on ICP dynamics remains limited and an important area for future study.

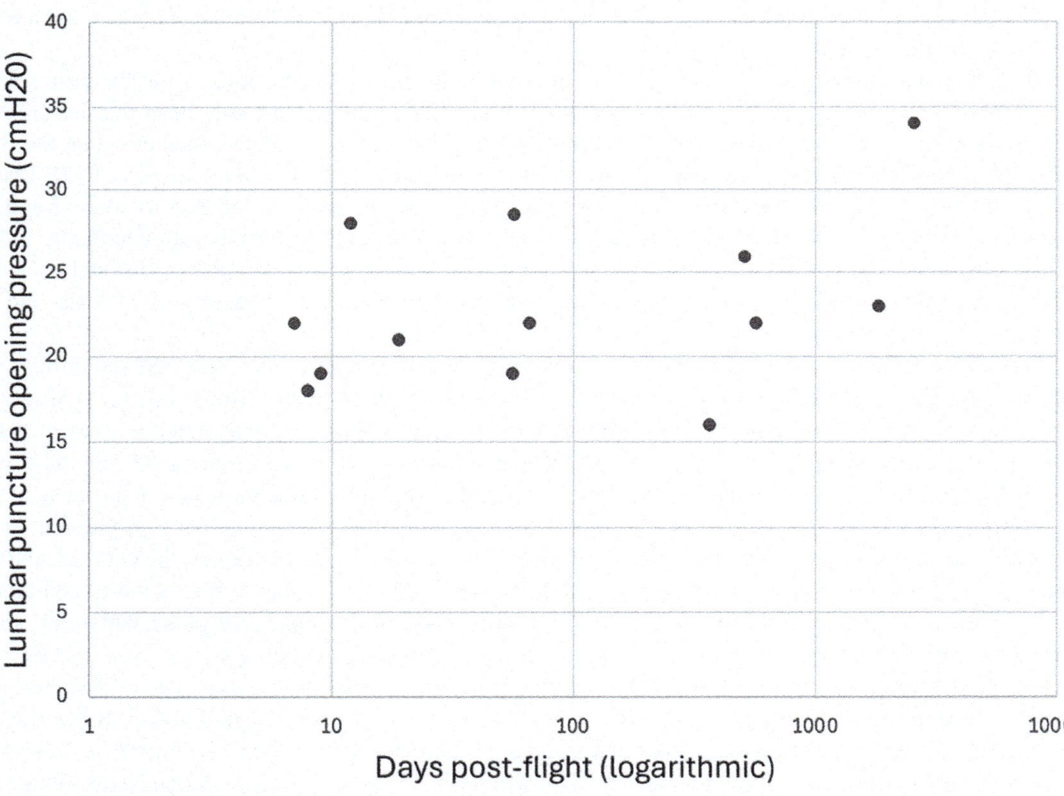


**FIGURE 16.1 Lumbar puncture opening pressures measured in astronauts by days postflight (Mampre et al., 2025).** Lumbar puncture opening pressure vs days postflight.

From Mampre, David, Spaide, Richard, Mason, Sara, Van Baalen, Mary, Gibson, C. Robert, Mader, Thomas H., Wostyn, Peter, Briggs, John, Brown, David, Lee, Andrew G., Patel, Nimesh, Tarver, William & Brunstetter, Tyson. (2025). Spaceflight associated neuro-ocular syndrome as a potential variant of venous overload choroidopathy. Aerospace Medicine and Human Performance, 96(6), 496–508. https://doi.org/10.3357/amhp.6602.2025.

## 16.2 The role of diagnostic imaging

The current era of neurosurgery relies largely on CT and/or MRI for preintervention diagnosis, modalities that may not be available in spaceflight due to mass/volume/power constraints. Some companies are developing smaller portable CT and MRI scanners, but adaptation for flight has yet to be performed (Neurocritical care, 2024). As such, neurosurgery in near-term long-duration spaceflight may be akin to an era prior to the invention of CT in 1971. During the decades of practice until the diagnostic imaging revolution of the 1970s, neurosurgeons practiced on the basis of history, neurologic exam, and X-ray findings. Plain X-ray films harbored clues of intracranial pathology, such as midline shift of pineal calcifications, as well as evaluation of spinal fractures. Adding contrast modalities to X-ray provided more information, such as injecting intrathecal air to create pneumoencephalography imaging of the ventricular system, which contains information about mass lesions and hydrocephalus. After injecting intra-arterial contrast, angiograms using X-ray cranial imaging can reveal vascular pathologies such as stroke and vascular anomalies, as well as indications of extra-axial mass lesions such as subdural hematomas. Intrathecal contrast can also provide imaging of severe spinal stenosis requiring decompression. While CT and MRI modalities may be in the distant future for use in spaceflight, portable X-ray is also in development and could provide diagnostic capabilities similar to the first seven decades of neurosurgical practice (KA Imaging. A Historic Image: The First X-ray in Space Is Here press release, 2025; Portable x-ray providers, 2020). Since its installation aboard the ISS in 2014, bone densitometry has successfully been performed for over 150 inflight scans on mice, demonstrating that radiographic imaging is both technically feasible and operationally sustainable in microgravity conditions. If employed in microgravity settings, several imaging characteristics could differ from terrestrial findings. Terrestrial imaging characteristics can be gravity-dependent, however, for example, fluid-fluid levels and the location of pneumocephalus.

## 16.3 Safety of commercial spaceflight with neurosurgical intervention

The United States Congress has issued a legislative moratorium preventing the Federal Aviation Administration from issuing regulations on the health and safety of human spaceflight participants. Screening and clearance are typically at the discretion of the commercial entity. With historical astronauts being heavily screened for medical conditions, there is limited data for generalizing the safety of spaceflight to the broader population. Neurosurgical patients have previously flown in space postoperatively after anterior cervical discectomy and fusion (Menger et al., 2019). While some neurosurgical diagnoses and postoperative interventions may be safe for spaceflight, other neurosurgical conditions and procedures may be exclusionary, and there is a role for neurosurgeon involvement as the commercial space develops guidelines and practices for screening commercial spaceflight participants.

### 16.3.1 Current procedural capabilities

This section provides a list of various practical inflight procedures that are already used or feasible to implement in the absence of further research or innovation.

## 16.4 Management of head trauma and elevated intracranial pressure

1. Assessment and management of head injuries should follow the American College of Surgeons guidelines and require careful evaluation due to the potential for intracranial bleeding or increased ICP (Best Practices Guidelines: The Management of Traumatic Brain Injury guideline, 2024). Obtain detailed information about the circumstances of the injury, including the mechanism and any immediate symptoms such as loss of consciousness. Look for external signs of injury, such as lacerations, contusions, or skull deformities. Assess for signs of basal skull fracture, like periorbital ecchymosis (raccoon eyes) or mastoid ecchymosis (Battle's sign). Use the Glasgow Coma Scale (GCS) to assess the level of consciousness and monitor for changes over time (Table 16.1). Monitor for nausea, vomiting, seizures, amnesia, and focal neurological deficits.
2. In cases of reduced consciousness, ensure the patient's airway is clear, breathing is adequate, and circulation is maintained (ABCs of trauma care). While it is plausible that microgravity could reduce the tendency of the tongue or pharyngeal tissue to collapse into the airway compared to the supine position in terrestrial environments, factors like fluid redistribution and possible oropharyngeal swelling may still complicate airway maintenance in space. Thus, even if airway patency might be somewhat easier to maintain, it is still recommended to follow standard protocols to ensure

**TABLE 16.1** Glascow coma scale.

| Exam | Response |
|---|---|
| Eye opening response | 4=Spontaneously<br>3=To speech<br>2=To pain<br>1=No response |
| Verbal response | 5=Oriented to person, place, and time<br>4=Confused<br>3=Inappropriate words<br>2=Incomprehensible sounds<br>1=No response |
| Motor response | 6=Obeys command<br>5=Localizes pain<br>4=Withdraw from pain<br>3=Abnormal flexion (decorticate)<br>2=Abnormal extension (decerebrate)<br>1=No response |

a clear airway. Continuously observe for signs of neurological deterioration, such as decreasing GCS score or new neurological deficits. Recognizing increased ICP is critical, and diagnosis relies on identifying clinical signs and utilizing diagnostic tools adapted for the microgravity environment. Papilledema may be observed via fundoscopy and potentially be supported by ONSD measurement, transcranial doppler (TCD), and pupillometry in combination, although with limited accuracy.

3. Managing increased ICP during spaceflight presents unique challenges due to the microgravity environment, limited medical resources, and absence of immediate surgical options. Traditional terrestrial measures, such as elevating the head of the bed, are ineffective in microgravity without the gravitational effect on venous drainage. Therefore, alternative strategies must be employed. Osmotherapy using Mannitol (osmotic diuretic) or hypertonic saline may reduce ICP, but its use in spaceflight is limited by the availability of supplies, need for careful monitoring, and risk of side effects such as electrolyte imbalance and rebound effect with increased ICP (Panesar et al., 2020, 2019).
4. Lower Body Negative Pressure (LBNP) devices create a negative pressure environment around the lower body, simulating the effects of gravity by pooling blood in the lower extremities and reducing cephalad fluid shifts that contribute to increased ICP. The astronaut should be placed within the LBNP chamber, which seals at the waist. Negative pressure is applied incrementally while monitoring physiological responses. However, some studies suggest that LBNP might inadvertently worsen venous outflow by inducing collapse of the jugular veins, thereby potentially exacerbating intracranial venous stasis (Arbeille et al., 2021). In the terrestrial ICU setting, LBNP has shown significant decreases in ICP (Petersen et al., 2019). The role of LBNP in management of elevated ICP has not yet been validated in microgravity, and any use should be approached with caution and accompanied by careful monitoring of cardiovascular and cerebrovascular effects.
5. Elevated $CO_2$ levels cause cerebral vasodilation, increasing cerebral blood flow and ICP. Intentional hyperventilation to lower $CO_2$ is impractical in a closed spacecraft environment or for sustained control of ICP, but the patient should be kept away from $CO_2$ pockets if possible. Given the risk of hypercapnia in spaceflight from limited scrubbing resources, ventilation, and local elevations, there may be a risk of worsened ICP compared to a terrestrial environment.
6. Euvolemia should be maintained as both dehydration and overhydration can adversely affect ICP. Oxygenation should be optimized, and supplemental oxygen should be administered to maintain $SpO_2$ levels above 95%. Also, maintaining optimal mean arterial pressure (MAP) is essential to ensure adequate cerebral perfusion pressure (CPP). Although CPP is often described by the approximate relationship CPP=MAP – ICP, in practice this equation is a simplified conceptual tool. Real-world cerebral perfusion depends on additional factors, including variations in intravascular resistance, autoregulatory mechanisms, and intracranial compliance, all of which fluctuate continuously and may be affected by the microgravity environment. Both hypertension and hypotension can negatively impact cerebral blood flow and ICP. Terrestrially, cerebral blood pressure is dependent on head-of-bed positioning. The head of the bed can be elevated in cases where decreased cerebral blood pressure is desired, such as hemorrhage, and lowered in cases where elevated

pressure is desired, such as cerebral ischemia. The neurosurgeon uses gravity as therapy in managing ICP, where elevating the head of the bed decreases arterial blood flow and increases venous drainage; however, this would be ineffective in microgravity.

### 16.4.1 Quantitative pupillometry

Pupillometry involves the quantitative assessment of pupil size and reactivity, providing information about autonomic nervous system function and potential neurological impairment. Quantitative pupillometry has been used as a supplementary tool for ICP assessment in selected patients at hospitals on Earth. Unfortunately, the predictive value is limited in cases of gradual or moderate ICP changes. The validity of pupillometry in a microgravity environment is also uncertain due to the potential effects fluid shifts may have on the autonomic nervous system. Proper calibration and consistent environmental lighting conditions are necessary to obtain the most accurate results.

1. Equipment: Utilize a portable, handheld infrared pupillometer designed for clinical use.
2. Baseline Measurement: In a controlled lighting environment, assess the baseline pupil size.
3. Light Reflex Test: Deliver a standardized light stimulus using the device and record the pupil's constriction and subsequent dilation response.
4. Data Analysis: Evaluate parameters such as latency of response, constriction velocity, dilation velocity, and amplitude of pupillary change.

## 16.5 Spine immobilization

In the setting of microgravity, pathological changes to the spine may increase the risk of spinal injury both during and after spaceflight. In the initial phase of spaceflight, space adaptation back and neck pain is experienced by up to 86% of astronauts, and more than 40% continue to experience such back pain after returning to Earth (Sauer et al., 2023). There are a number of proposed mechanisms of space adaptation back pain, including changes in spine morphology with spinal elongation due to straightening of the lumbar and thoracic curvature, weakening of paraspinal support muscles, ligament strain, and nerve root tension (Swinney and Allison, 2018). While there has not been an observed change in spine disc height, there is evidence of muscle atrophy and weakening of the paraspinal and other spinal support muscles after spaceflight (Chang et al., 2016). Additionally, vertebral bone degradation in microgravity may increase the risk of osteoporotic fractures, such as a compression fracture, on return to a load-bearing environment. Due to a number of postflight cervical and lumbar intervertebral disc herniations diagnosed in astronauts, it was initially thought that unloading of the spine in microgravity may predispose to a higher incidence of disc damage on reloading. However, postflight astronaut surveillance has suggested that preflight risk factors are more responsible for the disc herniations than spaceflight exposure, and as such NASA has inactivated the risk of intervertebral disc damage from the Human Research Roadmap (Risk: Risks to Human Health and Performance. Human Research Roadmap, no date).

Prompt recognition and assessment of spinal injuries is critical. Evaluation includes obtaining a detailed account of how the injury occurred and assessing clinical signs such as localized spinal pain and neurological deficits. Loss of bladder or bowel control suggests possible spinal cord involvement. In a spaceflight setting, this exam finding may be confounded by the presence of urinary retention secondary to microgravity or anticholinergic medication that may be administered for Space Motion Sickness. In the setting of suspected injury, a thorough neurological examination should be performed, including motor and sensory function, reflexes, and proprioception. Providers should be aware that an increase in musculotendinous stiffness and tendon stretch reflexes has been observed in astronauts during and after spaceflight, despite a lack of pathology.

Spine immobilization is indicated when there is suspicion of blunt spinal injury, identified by altered mental status, midline spinal pain or tenderness, focal neurologic deficits, anatomic deformity, severe mechanism, or injuries that distract the patient's ability to provide a reliable examination (Best Practice Guidelines: Spine Injury, no date). In these scenarios, initiate the spinal motion restriction protocol as recommended by the American College of Surgeons Committee on Trauma, American College of Emergency Physicians, and the National Association of EMS Physicians (Best Practice Guidelines: Spine Injury, no date; Hawkins et al., 2024, 2019; Fischer et al., 2018). Spine immobilization should be implemented immediately on the scene of injury prior to any transport, as any movement can result in further compression of the spinal cord and permanent deficit. Extrication from vehicles or other enclosed spaces must be done while restoring and maintaining neutral alignment of the spine. Avoid the use of rigid tools during extrication, as these may cause resistance, increase pain, or precipitate new or worsening neurological deficits (Hawkins et al., 2024, 2019).

Perform neurological examinations before and after any patient movement or transfer to detect any changes in neurological status. Axial loading force is the force of gravity on the spine terrestrially in the upright position, which can cause further alignment distraction after fracture. For this reason, terrestrial patients are typically maintained in a flat position after suspected spine fracture. Axial loading force is absent in microgravity; however, spine flexion, extension, coronal bending, and rotation remain force vectors that could contribute to further spine compromise. Spine immobilization in microgravity seeks to maintain the spine in a straight position and limit these movement forces. Continuous monitoring of vital signs should also be performed for signs of neurogenic shock, characterized by hypotension and bradycardia due to disruption of autonomic pathways.

## 16.6 Cervical collar placement

1. Manually stabilize the head and neck using the trap squeeze method to maintain neutral, in-line spine alignment throughout the procedure.
2. To perform the trap squeeze method, the rescuer should place both forearms on either side of the patient's head while gripping the trapezius/upper shoulders with their hands (Fig. 16.2). While doing so, use the forearms to stabilize the head/neck in strict, neutral alignment. Neutral alignment refers to the head, neck, torso, hips, and legs all being in strict, straight alignment.

1. While one crewmember stabilizes the head and neck via the trap squeeze method, another crewmember may select devices to immobilize the spine. This may include a cervical collar, but in resource-limited settings could also include other stabilizing equipment such as splints, IV bags, rolled towels, tape, or other available tools (Eisner et al., 2022). A properly sized cervical stabilization device should fit snugly without causing airway compromise.
2. Avoid any manipulation or forceful movement of the cervical spine that elicits pain during collar placement.

### 16.6.1 Thoracolumbar immobilization

1. Once the cervical spine is immobilized with a rigid collar, immobilize the rest of the spine using an appropriate device, such as a hard backboard, scoop stretcher, or vacuum splint mattress, without impeding necessary medical access (Best Practice Guidelines: Spine Injury, no date; Hawkins et al., 2024, 2019).
2. Patient should remain on the floor (or be secured to the floor with tape in microgravity) while supplies are gathered. A backboard would be utilized to transfer the patient to an alternative location.
3. Terrestrially, teams of 4–6 people are often utilized for transfers onto immobilization devices, though these numbers may be limited in a spaceflight setting (Fig. 16.3). Fortunately, the absence of gravity may facilitate fewer team members applying a backboard without the need to lift the patient against gravity.

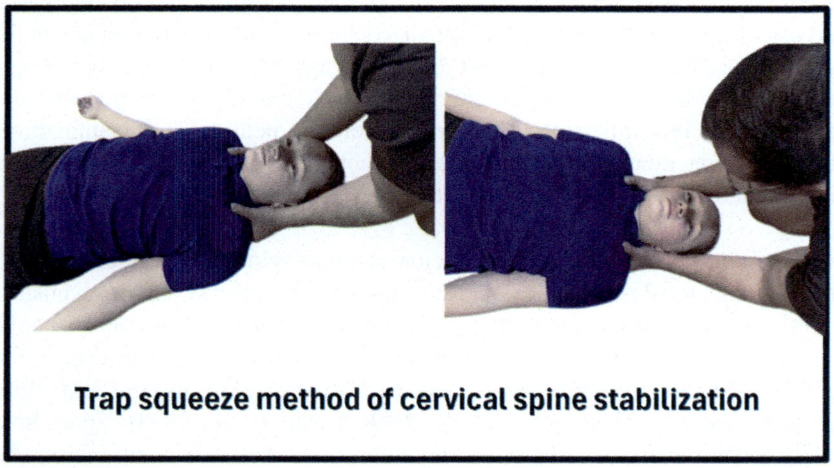


FIGURE 16.2 **Trap squeeze method of cervical stabilization.** To perform the trap squeeze method, the rescuer should place both forearms on either side of the patient's head while gripping the trapezius/upper shoulders with their hands. While doing so, use the forearms to stabilize the head/neck in strict, neutral alignment. Neutral alignment refers to the head, neck, torso, hips, and legs all being in strict, straight alignment.

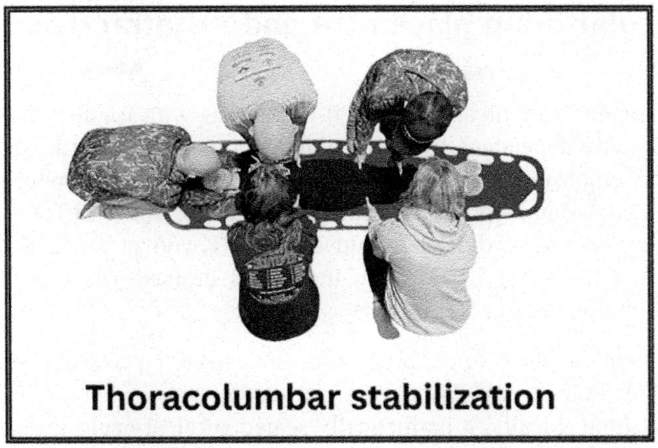


FIGURE 16.3 **Thoracolumbar spine immobilization.** One person typically manages the patient's head to maintain cervical spine neutral alignment using the trap squeeze method with the C-collar remaining in place, while other team members manage the torso, hips, and legs.

4. One person typically manages the patient's head to maintain cervical spine neutral alignment using the trap squeeze method with the C-collar remaining in place, while other team members manage the torso, hips, and legs.
5. The patient should be secured to the immobilization device using straps.

## 16.7 Orthotic removal

Spine immobilization would be incompatible with spacesuits for return/landing. If there is a high suspicion for continued spine instability at the time of return, flight surgeons may need to consider returning without donning an intravehicular suit or securing the patient to the cabin in a location other than the seat. For clearance of orthotic devices, the patient can be reevaluated for pain and symptoms on a daily basis. Radiography would have high utility if available. If there is resolution of pain, neurologic symptoms, and potentially an XR confirming spinal alignment, orthoses may be cleared by first assessing for pain with palpation, pain with passive range of motion, and pain with active range of motion (Best Practice Guidelines: Spine Injury, no date; Ham et al., 2014; Hewett Brumberg et al., 2024).

### 16.7.1 Future procedural capabilities

Animal studies aboard the Space Shuttle Neurolab have previously demonstrated the feasibility of performing craniotomy and laminectomy on rats during microgravity (Campbell et al., 1996, 2001, 2005). However, no neurosurgical procedures have ever been performed on humans in a spaceflight environment, nor on directly comparable large-animal or human analog models. Thus, these neurosurgical capabilities remain in the distant future for human healthcare. As such, this section will discuss the procedural considerations for less invasive interventions, including external ventricular drain, ICP monitor, subdural evacuating port, burr holes, lumbar puncture, and lumbar drain. The descriptions below represent a theoretical and conceptual framework, extrapolated from terrestrial neurosurgical principles, experiences from analog environments, as well as known microgravity physics and physiology. Any potential future implementation of such procedures on humans should be preceded by extensive testing and validation in high-fidelity environments and validated models.

Microgravity presents challenges for all surgical fields, neurosurgery included, such as hemostasis, sterility management, tool management, and mass/volume/power constraints. Both patient and surgeon must be adequately restrained to prevent unintended movements, as the application of force leads to equal and opposite reactions (Newton's Third Law). This could involve adjustable harnesses, body straps, or specialized medical pods designed for space applications. In addition to these common surgical challenges, microgravity may also impose challenges unique to the neurosurgical field. Increased infection risk postulated in spaceflight (Cowen, Zhang and Komorowski, 2024; Taylor, 2015) would be of higher consequence after neurosurgery. Additionally, impaired wound healing has been described and could pose a risk for other postoperative complications, such as CSF leaks.

## 16.8 External ventricular drain placement and/or intracranial pressure monitor placement

EVD and/or ICP monitor placement are typically indicated in patients with traumatic brain injury and GCS <9. While terrestrial EVDs depend on gravity-dependent hydrostatic fluid columns for setting a drainage pressure, alternative mechanisms would need to be employed for CSF drainage in a microgravity environment. For example, there are electronic pressure-controlled and volume-controlled drainage systems, which differ from conventional gravity-based collection chambers that depend on the hydrostatic fluid column (Kwon et al., 2016). Other mechanisms could be developed; for example, the first historical EVDs from the 1800s drained via a wicking mechanism by implanting horsehair into the ventricle (Srinivasan et al., 2014).

1. Secure the patient and stabilize the head for the procedure, such as with a cervical collar. The operator should also be properly restrained to provide counter-stability during instrumentation.
2. Prep and drape in sterile fashion. Ideally, a hermetically sealed surgical enclosure should be used.
3. Infiltrate local anesthetic around the insertion site.
4. Make a linear incision at Kocher's point (Fig. 16.4).
5. Reflect the pericranium from the skull surface.
6. Perform skull trepanation using a twist drill or power drill with depth stop (C). Control any free-floating bone dust, blood droplets, or CSF aerosols within the sealed field. Bone dust may be controlled by applying an adherent gel (such as ultrasound gel) to the skull prior to drilling, which traps the escape of surgical dust. The adherent gel (or sterile water in microgravity) may also act to contain blood and CSF from escaping the burr hole. Other methods of controlling surgical fluids like blood and CSF have yet to be identified but may include surgical canopy enclosures that incorporate suction, airflow, and irrigation capabilities. In microgravity, drilling generates equal and opposite torque on the operator and the patient. Therefore, effective restraint and use of a low-torque or mechanically stabilized drill with torque-limiting and reaction-arm systems are recommended to prevent inadvertent motion or deviation.
7. Incise the dura and insert an EVD and/or ICP monitor into the brain. Minimize risk of air entrainment into the cranial cavity by using a catheter prefilled with sterile saline. In microgravity, buoyancy does not separate air and fluid, so even tiny bubbles will tend to persist and adhere to surfaces.
8. Drain/wire is tunneled and secured to the scalp with suture/staples. Unlike in a terrestrial environment, the catheter should be immediately connected fluid-to-fluid to a preprimed drainage system. The drainage system is preprimed by injecting sterile saline through the tubing to eliminate air within the tubing system. This is to maintain a sealed, air-free fluid column.
9. Close the incision with suture/staples and apply an occlusive dressing.
10. All fluids and disposables must be captured and sealed.

The drain can be weaned by raising the pressure required for output on a daily basis until a clamp trial can be performed. Drain should be clamped during patient transport or during dynamic phases of flight, when vehicle acceleration could alter drain output.

## 16.9 Emergency trepanation (burr hole)

Burr holes can be used to evacuate extra-axial hematomas causing mass effect. This intervention is most commonly used for chronic subdural hematomas terrestrially; however, it may also be applied to emergency burr holes for other pathologies such as acute subdural or epidural hematomas.

Life-threatening epidural hematoma (EDH) represents the most plausible indication for emergency cranial trepanation during a spaceflight mission. The typical mechanism is a focal blunt impact to the head with arterial bleeding from the middle meningeal artery, and this can occur during intra-ventricular or EVAs, even in pressurized helmets or when struck by floating equipment. In contrast to subdural or intraparenchymal hemorrhage, which usually require more extensive surgery and postoperative monitoring, EDH can often be temporized or treated definitively by prompt decompression through a single burr hole.

Clinical suspicion must be based entirely on history and examination, since neuroimaging will be unavailable. A classic presentation includes:

- Head trauma with a brief loss of consciousness, followed by a lucid interval of apparent recovery, and subsequent progressive neurological deterioration (decreasing GCS)
- Unilateral fixed and dilated pupil (suggesting ipsilateral mass effect and uncal herniation)

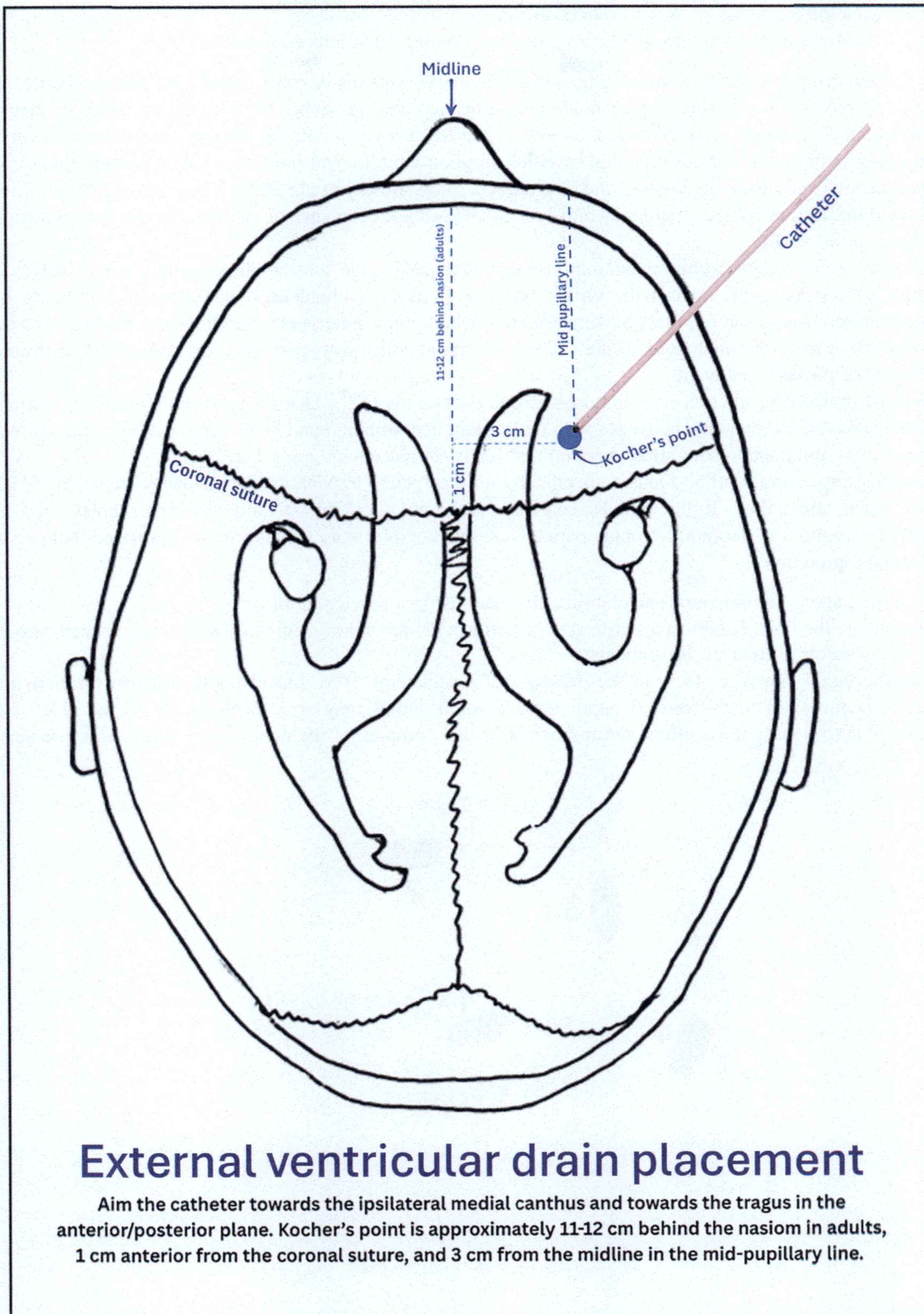


FIGURE 16.4 **While there are multiple cranial entry points for ventricular access, Kocher's point is the most commonly utilized. Kocher's point is located approximately 11 cm from the nasion (1 cm anterior to coronal suture) and 3 cm lateral (midpupillary line). External ventricular drain trajectory aims medially towards the medial canthus of the ipsilateral eye and towards the tragus in the anterior/posterior plane.**

- Contralateral motor weakness or decerebrate posturing
- Progressive bradycardia, hypertension, and irregular respiration, consistent with rising ICP.

In such a scenario, particularly if lateralizing signs and a clear trauma side exist, failure to intervene carries an almost 100% mortality risk, while emergency burr hole evacuation may be life-saving even if performed by nonneurosurgical crew under telemedical supervision. Historical terrestrial precedents exist for successful exploratory trepanation in austere or prehospital environments without imaging, resulting in good neurological outcomes when diagnosis was correct and decompression achieved promptly (Bishop and Drummond, 2006; Bosley et al., 2025; Wilson et al., 2012). Prior authors have suggested indications for the emergency burr hole as GCS<8 with evidence of an extra-dural hematoma with unequal pupils (Wilson et al., 2012)

Ethically, the decision to perform trepanation in space must balance the substantial procedural risks (such as infection, vessel injury, or misdiagnosis) against the virtual certainty of death without decompression. In a scenario of rapidly deteriorating consciousness and pupillary asymmetry after head trauma, emergency burr holes can therefore be justified as a heroic, last-resort measure, analogous to pre-CT era trauma neurosurgery, provided that the system, instruments, and crew are prepared for such an event.

As described previously, microgravity imposes unique constraints like a lack of hydrostatic drainage, torque-induced motion of the operator and patient, challenges with the containment of blood and bone dust, and heightened infection risk. The same restraint and containment strategies outlined for EVD placement apply here.

The following procedural outline focuses specifically on emergency trepanation for suspected acute EDH/SDH under the already listed constraints. If the initial trepanation does not yield blood, additional burr holes may be placed sequentially following the temporal–frontal–parietal–occipital exploratory sequence as illustrated below, mirroring terrestrial trauma protocols.

1. Restrain the patient and operator and stabilize the head using a cervical collar.
2. Prep and drape the field. Establish a sterile area or perform the procedure within a sealed surgical enclosure to contain fluids and aerosols. Tether all instruments.
3. Identify the target site (Fig. 16.5. In the presence of a unilateral fixed dilated pupil, perform the first trepanation ipsilateral (same side) to the dilated pupil. Only a small subset may arise contralateral to the side of pupillary dilatation due to a large mass effect and midline shift that compresses the contralateral oculomotor nerve.

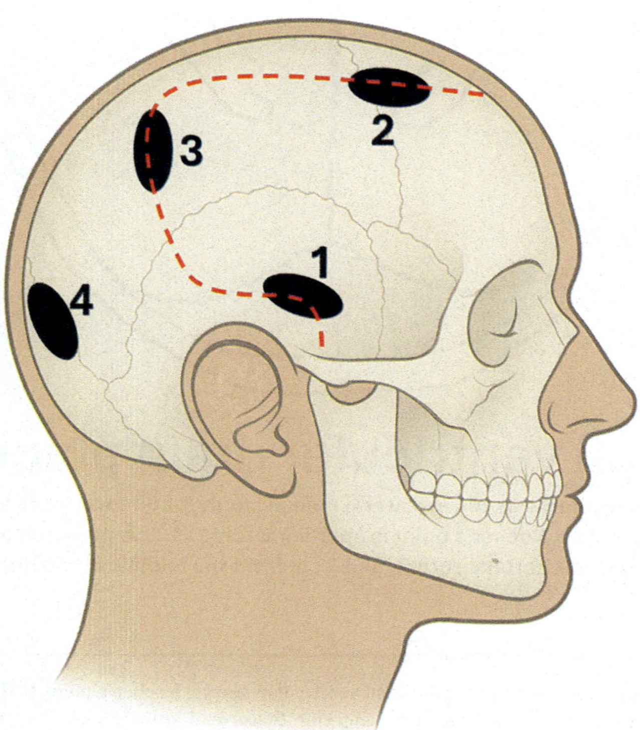


**FIGURE 16.5** **Burr hole incision and targets.** 1: Temporal (key burr hole), 2: frontal, 3: parietal, 4: occipital.

a. The temporal key point remains the most likely site for an epidural hematoma. For practical orientation, this spot lies slightly above the top of the ear, on the flat part of the skull. For a more detailed description: The optimal drilling site is located approximately 3–4 cm (about two finger breadths) above the midpoint of the zygomatic arch and 2.5–3 cm (about one finger breadth) posterior to the frontal process of the zygoma (the front-most part of the cheekbone). Avoid going below the zygomatic arch, too close to the orbit, behind the external auditory canal, or far above the temporal line.
b. If the temporal burr hole does not yield blood or the patient fails to improve, proceed to the frontal burr hole, 2–3 cm anterior to the temporal key point and 2–3 cm above the lateral orbital rim or the frontotemporal hairline. Avoid drilling directly over the frontal sinus. This location may be the primary attempt if trauma occurred to the forehead or superior orbital ridge.
c. The parietal burr hole should be located 5–6 cm above the external auditory canal and 4–5 cm posterior to the temporal key point. Avoid crossing the superior temporal line too far posteriorly, where the sagittal sinus or lambdoid suture region becomes a risk zone.
d. The occipital burr hole is reserved for cases where the injury or swelling suggests posterior fossa or occipital EDH—rare but catastrophic if not relieved. This one is located 3 cm above the inion (the external occipital protuberance, or the small bony bump at the back of the head) and 3 cm lateral to the midline. Avoid too inferior placement or midline drilling below the transverse sinus (risk of venous sinus injury).

4. Infiltrate local anesthetic. Maintain airway control and mild hyperventilation if possible to lower ICP.
5. Either make a continuous question mark incision as shown in the illustration above, or preferably make a 2–3 cm incision over the chosen site. Incisions are typically made several centimeters from midline in frontal and/or parietal areas.
6. After incision, obtain hemostasis with pressure or bipolar cautery.
7. Expose the bone by reflecting the pericranium.
8. Skull trephination is performed using a low-torque twist drill (A) or a power drill with irrigation. If possible, use a mechanically stabilized drill with a depth stop. Bore through both tables of the skull until a sudden loss of resistance indicates entry into the epidural space (C). In microgravity, ensure that both the patient and operator remain motionless.
9. If SDH is suspected, incise dura.
10. Blood may passively escape through the burr hole if under pressure. Hematoma can also be actively evacuated with a combination of irrigation and gentle aspiration with a syringe or low-flow suction. Maintain containment. Decompression may be indicated by renewed brain pulsation, improved vital signs, or pupillary constriction. Avoid incision of the dura or deep exploration that increases the risks.
11. If active bleeding is visualized, apply bone wax or sterile pledgets. Persistent arterial spurting likely indicates a torn middle meningeal branch; tamponade rather than attempt dissection.
12. Optional drainage: a soft catheter may be placed in the epidural or subgaleal cavity and connected to a closed, preprimed, air-free drainage system with low, electronically regulated suction, such as described under EVD principles. Do not use open or gravity-based drains. Wound is closed by suturing galea followed by superficial closure with sutures/staples. Secure any tubing along restraint tie-downs to prevent motion.
13. Contain and seal all fluids and disposables.

## 16.10 Lumbar puncture and drain

The most likely indications for lumbar puncture in spaceflight are CSF diversion for traumatic brain injury or in research to understand CSF dynamics in microgravity. Understanding and monitoring CSF pressure dynamics during spaceflight may be essential due to the unique physiological changes induced by microgravity, as well as concerns related to SANS and ventricular enlargement. Noninvasive methods for estimating ICP, such as optic nerve sheath diameter ultrasound, tympanic membrane displacement, pupillometry, scalp pulsations, and transcranial Doppler ultrasonography are currently available but have very limited predictive values and lack validation in the spaceflight environment (Frigieri et al., 2021). These methods may be influenced by individual variability as well as other physiological changes in microgravity. Also, they may not reliably detect subtle but clinically significant changes in ICP. Invasive measurements, on the other hand, provide direct and accurate assessments of ICP and allow for the evaluation of additional variables of importance. Traditional methods of measuring ICP rely on hydrostatic pressure gradients, which are absent in microgravity environments. Therefore, utilizing sensors that do not depend on hydrostatic gradients, such as piezoresistive and piezoelectric sensors, fiber optic, or microelectromechanical systems, is crucial for accurate ICP monitoring during space missions. Lumbar access allows for the placement of sensors that continuously monitor CSF dynamics, providing data critical for both clinical management and research purposes. LP also allows for the collection of CSF for analysis and therapeutic drainage in cases of elevated ICP. Thus, several researchers have advocated for the use of LP during spaceflight to assess the influence of microgravity on CSF dynamics. Specifically, one team demonstrated a successful LP on an anthropomorphic radiological phantom model of an adult lumbar spine and a subsequent cadaver using portable radiography (Lerner et al., 2016). However, due to continued concerns about procedural safety, infection risk, and the logistics of performing LP in microgravity, the procedure has yet to be conducted in-flight on human patients.

Concerning risks of LP in space, microgravity causes cephalad fluid shifts that may alter CSF dynamics, potentially inducing postprocedural complications or altering ICP readings in such environments (Jasien et al., 2022; Seidler et al., 2024). Additionally, paraspinal muscle atrophy, spinal elongation, and loss of lumbar lordosis during long-duration spaceflight missions complicate anatomical landmark identification and needle trajectory during traditional landmark-based LP (Bailey et al., 2018; Green and Scott, 2018; Lazzari, Aria and Menger, 2021), increasing the risk for LP procedural complications such as nerve damage or downward herniation. To mitigate these risks, researchers developed an innovative AI-powered program that utilizes neural networks, ultrasound images, and a needle-guidance system to identify anatomical landmarks and changes in ICP noninvasively, offering detection and measurement of CSF dynamics in real-time (Saragoça, 2024). One meta-analysis found ultrasound-guided LPs to be associated with higher success rates (OR 2.22), fewer traumatic LPs (OR 0.28), shorter procedural times, fewer needle passes, and lower patient pain scores when compared to traditional landmark-based LPs, although this was not in the setting of microgravity (Gottlieb et al., 2019). Further, use of remote-guided ultrasound, where licensed physicians instruct crewmembers through telecommunications, has been proven successful for the application of ultrasound-guided LP (Lerner et al., 2019). Because ultrasound is readily available during spaceflight missions, like on the ISS, crewmembers can be trained or guided remotely. This approach offers a practical, minimally invasive solution for inflight diagnostics and expands the scope of medical autonomy and research capability in long-duration missions.

1. Terrestrial positioning for lumbar puncture usually involves lateral positioning to equilibrate cranial and lumbar CSF pressures. Due to homogenous CSF pressure in microgravity with the absence of hydrostatic pressure, the patient could assume any body orientation with the spine vertically oriented and the lumbar spine flexed to promote access through the interspinous spaces. The patient may be positioned by "hugging" their knees, with their arms restrained to their shins/ankles, and then secured to a cabin surface.
2. Identify the bony ridges of the iliac crests on both sides (the superior edges of the pelvis on each side). Draw a straight parallel intercristal line to connect the highest points of the iliac crests and mark the midline intersection point at the palpated intersected spinous process. This is the approximate level of L4 (Fig. 16.6). Palpate the adjacent interspinal space directly below the marked point to identify the L4–5 interspinal space and mark as needle insertion point.
3. Prep and drape in sterile fashion.
4. Inject local anesthetic.
5. With a slight cephalad angle toward the umbilicus, slowly insert the spinal needle in the interspinous space at the L4–5 level (Fig. 16.7). You may feel a "pop" or sudden decrease in resistance as the needle enters the subarachnoid space. Check for CSF fluid at this point. Prior authors have recommended using a 20–22G nontraumatic spinal needle for spaceflight application (Barr, 2014).
6. Remove stylet and collect CSF.
7. Similar to measuring CSF pressure via EVD, the absence of hydrostatic pressure requires an alternative method to the terrestrial manometers, such as electronic pressure transducers.

If a lumbar drain were to be inserted, a larger trocar would be used in place of the spinal needle. Once the inner cannula is removed, the drain is inserted, after which the trocar is withdrawn and the drain secured in place with a nylon suture.

Due to elongation of the spine in microgravity and flattening of the lumbar curvature, access through interspinous spaces may be easier in spaceflight. However, this elongation is thought to put traction on the brainstem, where the release of CSF from the lumbar space could exacerbate downward herniation (Kershner and Binhammer, 2004). Venous engorgement in spaceflight could increase the risk of bleeding complications, though the propensity for bleeding in CSF spaces during spaceflight is unknown. Post-procedure, post-LP headache may be less significant in microgravity due to less tension on the dura and a lack of downward gravitational vector forces on the brain (Barr, 2014).

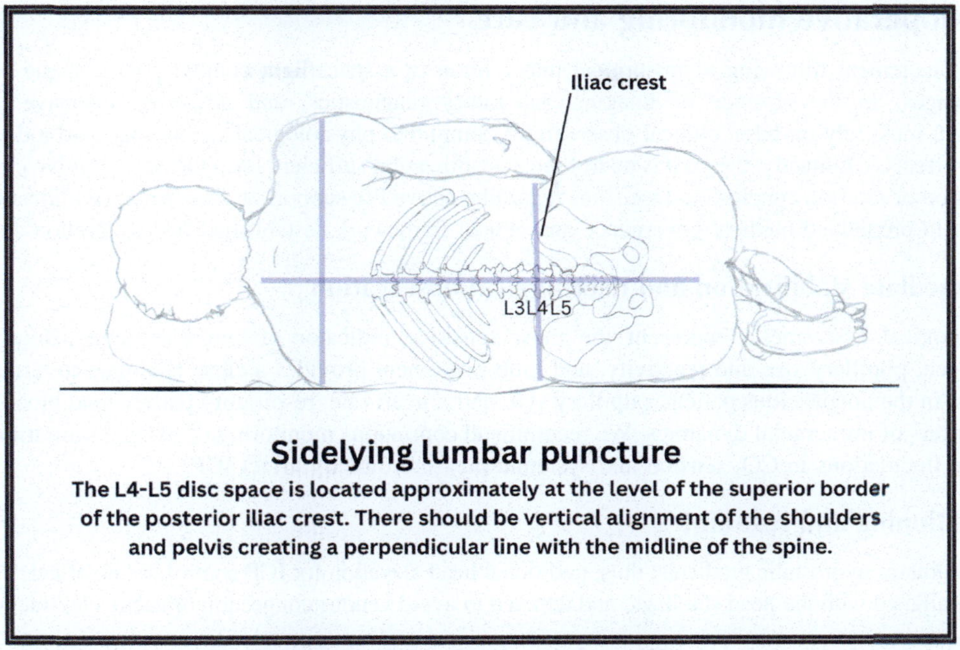


FIGURE 16.6 The L4–5 disc space is identified at the level of the iliac crest.

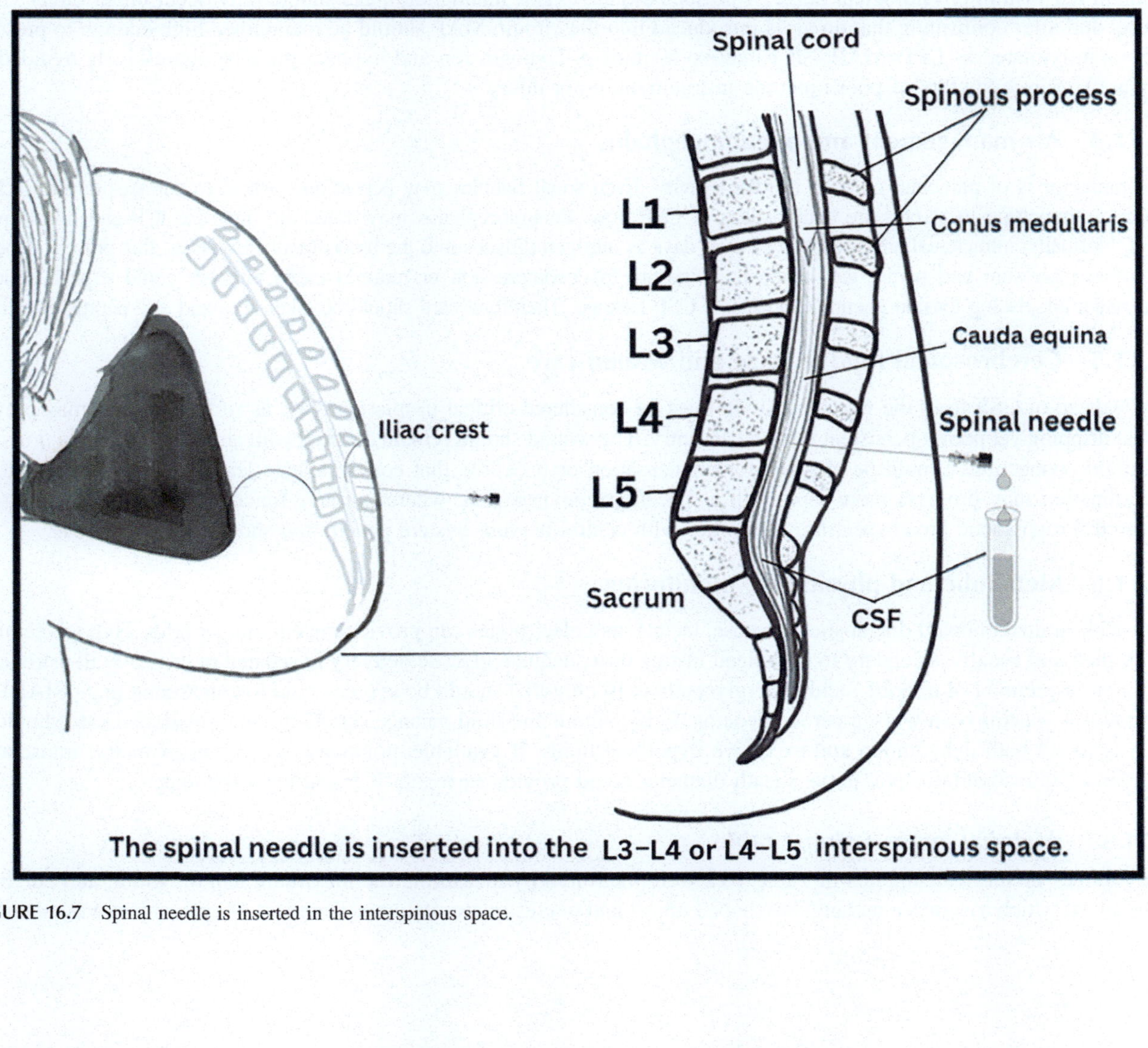


FIGURE 16.7 Spinal needle is inserted in the interspinous space.

## 16.11 Postoperative monitoring and care

Postoperative management following neurosurgical intervention in a spaceflight context poses unique physiological and logistical challenges. In the absence of imaging, laboratory diagnostics, and advanced intensive care technology, postoperative care must rely on basic clinical observations, simplified physiological monitoring, and a deep understanding of microgravity effects. Ultimately, the goal is neurological stabilization sufficient for return to definitive care, either through medical evacuation or mission conclusion. Decisions regarding active life support or crew resource allocation must be made in accordance with predefined medical governance and ethical frameworks established for exploration-class missions.

### 16.11.1 Immediate stabilization and neurological observation

Frequent neurological assessments represent the most sensitive indicator of cerebral status. Hourly evaluation of orientation, speech, pupillary size and reactivity, and limb movement provides a close analogue to terrestrial neuro-ICU scoring systems. In the unconscious patient, pupillary symmetry, heart rate, respiratory pattern, and blood pressure remain the primary markers of intracranial dynamics. We recommend continuous monitoring of oxygen saturation, end-tidal $CO_2$, and MAP, since fluctuations in $CO_2$ tension and systemic pressure directly affect ICP.

### 16.11.2 Positioning and restraint

Microgravity eliminates hydrostatic gradients; thus, traditional head elevation for ICP control has no effect. The patient should remain neutrally aligned with the head stabilized and tethered to avoid venous congestion or accidental traction on any drains.

### 16.11.3 Intracranial pressure and hemodynamic control

Without gravity-based drainage, closed-loop electronic or pressure-controlled drainage systems may be required to maintain ICP within physiological range. If no such system exists, manual drainage should only occur under direct visual control and after confirming that no air is introduced into the circuit. MAP should be maintained high enough to preserve CPP (approximated as CPP=MAP – ICP) above 60 mmHg. Hypotension and hypoxia must be aggressively avoided, as they are strong predictors of poor outcome in traumatic brain injury.

### 16.11.4 Air management and pneumocephalus

Intracranial air is of particular concern in microgravity. Even small bubbles may persist and adhere to surfaces due to lack of buoyancy, impairing local pressure equilibration or CSF flow. Pneumocephalus may transiently increase ICP and mimic mass effect. Normally, intracranial air is reabsorbed over days as nitrogen diffuses into the bloodstream. However, this process depends on ambient pressure and partial gas gradients. Exposure to decompression or reduced cabin pressure could expand residual intracranial air, risking tension pneumocephalus or CSF leakage. Therefore, care should be taken to avoid any depressurization.

### 16.11.5 Cerebrospinal fluid leakage and wound care

Any leakage of CSF from the wound or drain must be considered critical in microgravity, as surface tension may prevent visible dripping yet permit persistent loss and air entry. The wound should remain covered with an occlusive sterile dressing, and drain connections should be checked for condensation or backflow that could indicate leakage. Low-grade bacterial contamination may progress more rapidly in a spaceflight environment, where immune function and wound healing are impaired. Prophylactic broad-spectrum antibiotics with central nervous system penetration should be considered.

### 16.11.6 Metabolic and physiologic monitoring

Even with limited onboard diagnostics, glucose, lactate, and electrolytes can provide meaningful insights. Hypoglycemia is uncommon and usually secondary to prolonged fasting or inadequate glucose delivery in sedated or critically ill astronauts. Both hypoglycemia (<4 mmol/L) and hyperglycemia (>10 mmol/L) should be avoided, and hyponatremia (e.g., SIADH or cerebral salt wasting) can worsen cerebral edema. Urine output and fluid balance should be monitored, and care should be taken to avoid both dehydration and excessive hypotonic fluids. If available, noninvasive cerebral oximetry, transcranial Doppler, or ultrasound of optic nerve sheath diameter could provide surrogate ICP and perfusion data.

#### *16.11.6.1 Material, mass, volume table*

The volumes of surgical supplies in Table 16.2 were measured by measuring the maximum length, width, and height of each object, which assumes a rectangular-shaped object and overestimates volume. The mass of each object was evaluated

**TABLE 16.2** Mass and volume of neurosurgical supplies by different tiers of capabilities.

| Procedure | Equipment | Mass (g) | Volume (cc) | Total mass (g) | Total volume (cc) |
|---|---|---|---|---|---|
| Quantitative pupillometry | Handheld pupillometer | 370 | 1458 | 320 | 1458 |
| Immobilization | Backboard and straps | 6803 | 33,036 | 9918 | 53,332 |
| | C-collar | 186 | 9195 | | |
| | (OR) SAM splint | 115 | 160 | | |
| | (OR) Rolled towels | 600 | 5242 | | |
| | (OR) IV bags, ×2 * | 2094 | 5416 | | |
| | (AND/OR) Tape | 120 | 283 | | |
| External ventricular drain | Surgical prep, ChloraPrep * | 72 | 921 | 1250 | 21,644 |
| | Razor, disposable | 8 | 119 | | |
| | Local anesthetic, 20 mL vial | 53 | 59 | | |
| | Syringe, 20cc * | 17 | 78 | | |
| | Hypodermic needle, 23G * | 3 | 3 | | |
| | Hypodermic needle, 19G * | 2 | 2 | | |
| | Marking pen | 4 | 5 | | |
| | C-collar | 186 | 9195 | | |
| | Scalpel, disposable * | 10 | 58 | | |
| | Heis retractor | 16 | 21 | | |
| | Hemostat | 24 | 38 | | |
| | Gauze sponges, ×10 * | 37 | 614 | | |
| | Hand drill w/ drill bits | 256 | 1334 | | |
| | Ventricular drain + trocar * | 54 | 639 | | |
| | CSF external drainage collection system kit * | 393 | 8431 | | |
| | Needle driver | 51 | 35 | | |
| | Adson tissue forceps, 4 ¾″ | 22 | 22 | | |
| | Nylon suture ×3 | 15 | 38 | | |
| | Bone wax * | 6 | 16 | | |
| | Iris scissors | 21 | 16 | | |
| Intracranial pressure monitor | EVD supplies | 1250 | 21,644 | 1361 | 23,976 |
| | Microsensor kit (includes ICP wire and 14G Tuohy needle with stylet) * | 113 | 2333 | | |
| Lumbar puncture | Local anesthetic, 20 mL vial | 53 | 59 | 151 | 1069 |
| | Syringe, 20cc * | 17 | 78 | | |
| | Hypodermic needle, 23G * | 3 | 3 | | |
| | Hypodermic needle, 19G * | 2 | 2 | | |
| | Surgical prep * | 72 | 922 | | |
| | Spinal needle, 18G * | 4 | 5 | | |

*(Continued)*

**TABLE 16.2 (Continued)**

| Procedure | Equipment | Mass (g) | Volume (cc) | Total mass (g) | Total volume (cc) |
|---|---|---|---|---|---|
| | Lumbar drainage catheter kit (includes 80 cm closed tip catheter, 100 cm guide wire, 14G Tuohy needle, female luer lock and luer lock cap) * | 87 | 1758 | 536 | 10,237 |
| Lumbar drain | CSF external drainage collection system kit * | 393 | 8431 | | |
| | Needle driver | 51 | 35 | | |
| | Nylon suture * | 5 | 13 | | |
| Burr hole | Surgical prep, ChloraPrep * | 72 | 922 | 1984 | 5976 |
| | Razor, disposable | 8 | 119 | | |
| | Local anesthetic, 20 mL vial | 53 | 59 | | |
| | Syringe, 20cc * | 17 | 78 | | |
| | Hypodermic needle, 23 G * | 3 | 3 | | |
| | Hypodermic needle, 19 G * | 2 | 2 | | |
| | Marking pen | 4 | 5 | | |
| | Scalpel, disposable * | 10 | 58 | | |
| | Weitlaner retractor | 108 | 355 | | |
| | Hemostat | 24 | 38 | | |
| | Gauze sponges, ×10 * | 37 | 614 | | |
| | Power drill | 962 | 873 | | |
| | Needle driver | 51 | 35 | | |
| | Adson tissue forceps (4 ¾″) | 22 | 22 | | |
| | Iris scissors | 21 | 16 | | |
| | Irrigation bulb, 3 oz * | 63 | 737 | | |
| | Vicryl Suture, 2-0 CT-2 ×32 * | 56 | 167 | | |
| | Skin stapler with staples * | 94 | 903 | | |
| | Cord for Bovie or Bayonet cautery * | 63 | 51 | | |
| | Bovie * | 81 | 229 | | |
| | Bovie tip * | 6 | 2 | | |
| | Bayonet bipolar forceps | 37 | 76 | | |
| | Bone wax * | 6 | 16 | | |
| | Woodson | 35 | 3 | | |
| | Collection bulb for drain, 100cc * | 57 | 415 | | |
| | Flat Drain, 7 mm | 35 | 118 | | |
| | 14 Fr. suction tube | 30 | 30 | | |

*Item was measured in sterile packaging.*

with a digital scale. Items were kept in sterile packaging where possible, with the sterile packaging reduced to its smallest possible size for measurement.

The total estimated mass and volume of supplies were 15,520 grams and 117,692 cm³. The surgical tools assembled in Table 16.1 account for procedural needs only and do not account for mass/volume requirements due to imaging modalities, personal protective equipment, visual aids (such as loupes), anesthesia needs (such as IV lines and respiratory equipment), and patient positioning needs (such as operating table and safety padding). Additionally, all items considered were commercial off-the-shelf items without considering custom-made, multipurpose, or 3D printed tools. Neurosurgical specialty capabilities not listed in this table include functional, endoscopic, endovascular, and complex spine due to training requirements and mass/volume constraints. While endovascular surgery may be of high utility in spaceflight as a minimally invasive approach that can aid in diagnosis, this intervention requires a level of trained skill likely not attainable by the astronaut population in the near future.

### *16.11.6.2 Areas for research and innovation*

Areas of focus include portable diagnostic tools and microgravity-compatible surgical equipment. In cases of internal jugular vein thrombosis, the development of a safe and effective technique for jugular vein catheterization in microgravity would enable both diagnostic and therapeutic interventions. Catheter-directed thrombolysis or mechanical thrombectomy could be life-saving procedures in the event of thrombosis leading to severe cerebral venous outflow obstruction. Establishing transjugular venous access in microgravity poses technical challenges due to altered fluid dynamics and the need for precise vascular navigation without gravity-assisted blood flow. Overcoming these challenges through research and innovation would not only provide treatment options for life-threatening conditions but also facilitate the assessment of intravenous pressure dynamics and improve understanding of cerebral hemodynamics during exposure to microgravity.

Sterile surgical enclosures for spaceflight have yet to be determined, which could range from the surgical canopy used on STS to other solutions like an aqueous immersion surgical enclosure that has previously been proposed for cranial surgery in spaceflight (Fig. 16.8; Hayden et al., 2013). Environmental controls, such as isolation curtains and localized air filtration systems equipped with high-efficiency particulate air filters, help maintain a controlled operative environment by preventing the spread of contaminants and ensuring air quality.

Augmented reality (AR) technology can provide invaluable assistance during both training and actual procedures. Wearable AR devices integrated with neuronavigation systems can overlay virtual instructions, anatomical landmarks, and guidance directly onto the operator's field of view in real-time. This technology can compensate for the lack of neurosurgical expertise among crewmembers, enhancing procedural accuracy and safety. The AR system should be designed to function offline and be fully self-contained, ensuring reliability even when communication with Earth-based support is delayed or unavailable. Artificial Intelligence (AI) can be integrated into decision support systems and assist in

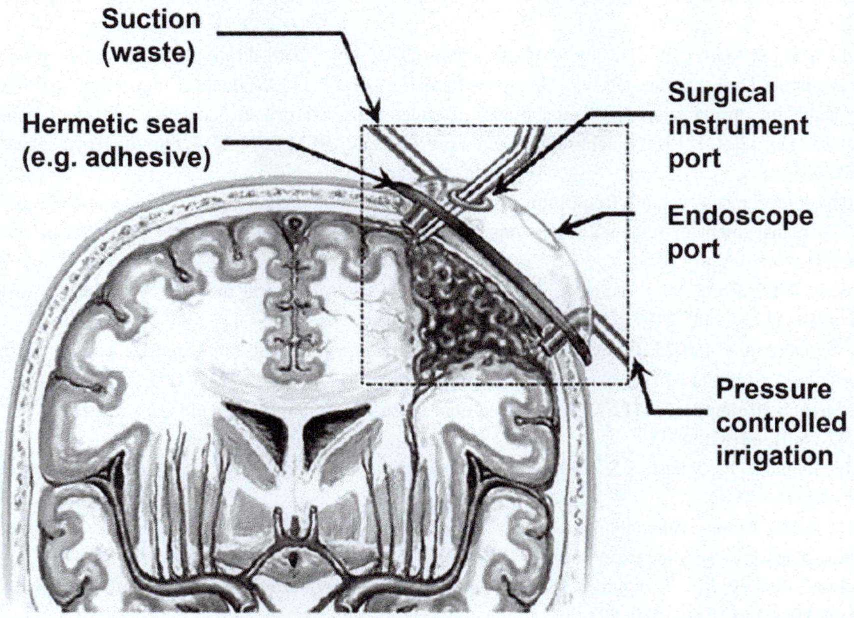


FIGURE 16.8 **Illustration of the aqueous immersion surgical enclosure system that has been proposed for spaceflight.** The transparent enclosure hermetically seals the surgical environment and has numerous ports that serve as access points for the surgeon, much like endoscopic procedures. From Hayden, J.A., Pantalos, G.M., Burgess, J.E. & Antaki, J.F. (2013). A hermetically sealed, fluid-filled surgical enclosure for microgravity. Aviation Space and Environmental Medicine, 84(12), 1298–1303. https://doi.org/10.3357/ASEM.3751.2013.

both diagnosis and treatment planning. During procedures, AI can also offer real-time guidance, integrated with AR to display step-by-step instructions.

Adapting equipment for spaceflight involves designing multifunctional, compact tools that can operate effectively in microgravity. 3D printing technology offers a supplementary solution, enabling the on-demand fabrication of surgical instruments using biocompatible materials. In-situ resource utilization strategies could further reduce the need to carry extensive supplies by utilizing local materials on planetary surfaces.

## 16.12 Summary

Although neurosurgical procedures are statistically rare, inadequate treatment for such pathologies in spaceflight risks loss of crew and mission objectives. The most likely insult to the brain and spinal cord requiring procedural intervention in spaceflight is trauma, which is most often treated with decompression and stabilization. Neurosurgeons may play a role in both procedural care and medical management, such as management of ICP. An understanding of basic neurosurgical procedures is important both for procedures within current capabilities, including transcranial doppler, pupillometry, optic nerve sheath diameter, the medical management of head and spine trauma, and for potential future neurosurgical interventions ranging from lumbar puncture to trepanation. Mass and volume of supplies for a robust suite of neurosurgical tools to manage traumatic conditions were measured to be in the range of 15,520 grams and 117,692 $cm^3$.

Common across all surgical specialties, ongoing research and development are essential to improve surgical equipment, sterility, imaging, training, management of communication delays, and handling of body fluids. In neurosurgery, specific areas for development in the microgravity environment include manometry, CSF diversion, portable diagnostics, neurosurgical tools, and management of cephalad fluid shifts and venous drainage. Besides treating neurological disorders on Earth, neurosurgical diagnosis and procedures may also contribute to research and countermeasures for space-related conditions such as SANS, ventricular enlargement, thrombosis, and space adaptation back pain. Preparing neurosurgery for space missions is a complex, multidisciplinary effort. Although current capabilities are limited, the risks of long-duration missions to distant locations make it crucial to proactively improve neurosurgical readiness.

## References


Arbeille, P., Zuj, K.A., Macias, B.R., Ebert, D.J., Laurie, S.S., Sargsyan, A.E., et al., 2021. Lower body negative pressure reduces jugular and portal vein volumes and counteracts the elevation of middle cerebral vein velocity during long-duration spaceflight. Journal of Applied Physiology 131 (3), 1080–1087. Available at https://doi.org/10.1152/japplphysiol.00231.2021, https://journals.physiology.org/doi/full/10.1152/japplphysiol.00231.2021.

Bailey, J.F., Miller, S.L., Khieu, K., O'Neill, C.W., Healey, R.M., Coughlin, D.G., et al., 2018. From the international space station to the clinic: how prolonged unloading may disrupt lumbar spine stability. The Spine Journal 18 (1), 7–14. Available at https://doi.org/10.1016/j.spinee.2017.08.261.

Barr, Y.R., 2014. Lumbar puncture during spaceflight: operational considerations, constraints, concerns, and limitations. Aviation, Space, and Environmental Medicine 85 (12), 1209–1213. Available at https://doi.org/10.3357/ASEM.3674.2014, http://ingentaconnect.com/content/asma/asem/2014/00000085/00000012/art00008.

Best practice guidelines: Spine injury, n.d. Available from: https://www.facs.org/media/k45gikqv/spine_injury_guidelines.pdf.

Best practices guidelines: The management of traumatic brain injury [guideline], 2024. Available from: https://www.facs.org/media/vgfgjpfk/best-practices-guidelines-traumatic-brain-injury.pdf.

Bishop, C.V., Drummond, K.J., 2006. Rural neurotrauma in Australia: implications for surgical training. ANZ Journal of Surgery 76 (1-2), 53–59. Available at https://doi.org/10.1111/j.1445-2197.2006.03642.x.

Bosley, L., Gibbs, C., Joo, E., Dobson, G., 2025. Can non-neurosurgeons operate on traumatic brain injuries in non-metropolitan areas? A scoping review. Emergency Medicine Australasia 37 (3). Available at https://doi.org/10.1111/1742-6723.70055.

Brasel, K.J., 2013. Advanced trauma life support (ATLS®): the ninth edition. Journal of Trauma and Acute Care Surgery 74 (5), 1363–1366. Available at https://doi.org/10.1097/TA.0b013e31828b82f5.

Campbell, M.R., Billica, R.D., Jennings, R., Johnston, S., 1996. Laparoscopic surgery in weightlessness. Surgical Endoscopy 10 (2), 111–117. Available at https://doi.org/10.1007/bf00188354.

Campbell, M.R., Kirkpatrick, A.W., Billica, R.D., Johnston, S.L., Jennings, R., Short, D., et al., 2001. Endoscopic surgery in weightlessness: the investigation of basic principles for surgery in space. Surgical Endoscopy 15 (12), 1413–1418. Available at https://doi.org/10.1007/s004640080178.

Campbell, M.R., Williams, D.R., Buckey, J.C., Kirkpatrick, A.W., 2005. Animal surgery during spaceflight on the neurolab shuttle mission. Aviation, Space, and Environmental Medicine 76 (6), 589–593.

Chang, D.G., Healey, R.M., Snyder, A.J., Sayson, J.V., Macias, B.R., Coughlin, D.G., et al., 2016. Lumbar spine paraspinal muscle and intervertebral disc height changes in astronauts after long-duration spaceflight on the International Space Station. Spine (Philadelphia, Pa.: 1986) 41 (24), 1917–1924. Available at https://doi.org/10.1097/BRS.0000000000001873, http://journals.lww.com/spinejournal.

Chowdhury, T., Kowalski, S., Arabi, Y., Dash, H.H., 2014. Pre-hospital and initial management of head injury patients: an update. Saudi Journal of Anaesthesia 8 (1), 114–120. Available at https://doi.org/10.4103/1658-354X.125971.

Cowen, D., Zhang, R., Komorowski, M., 2024. Infections in long-duration space missions. The Lancet Microbe 5 (9), 100875. Available at https://doi.org/10.1016/s2666-5247(24)00098-3.

Eisner, Z.J., Delaney, P.G., Pine, H., Yeh, K., Aleem, I.S., Raghavendran, K., et al., 2022. Evaluating a novel, low-cost technique for cervical-spine immobilization for application in resource-limited LMICs: a non-inferiority trial. Spinal Cord: The Official Journal of the International Medical Society of Paraplegia 60 (8), 726–732. Available at https://doi.org/10.1038/s41393-022-00764-3, http://www.nature.com/sc/index.html.

Fischer, P.E., Perina, D.G., Delbridge, T.R., Fallat, M.E., Salomone, J.P., Dodd, J., et al., 2018. Spinal motion restriction in the trauma patient–a joint position statement. Prehospital Emergency Care 22 (6), 659–661. Available at https://doi.org/10.1080/10903127.2018.1481476, http://www.tandfonline.com/loi/ipec20.

Frigieri, G., Rabelo, N.N., De Sena Barbosa, M.G., Cagle, Y., Pelligra, R., 2021. Letter: neurosurgery and manned spaceflight. Neurosurgery 89 (1). Available at https://doi.org/10.1093/neuros/nyab133.

Gottlieb, M., Holladay, D., Peksa, G.D., 2019. Ultrasound-assisted lumbar punctures: a systematic review and meta-analysis. Academic Emergency Medicine 26 (1), 85–96. Available at https://doi.org/10.1111/acem.13558, http://onlinelibrary.wiley.com/journal/10.1111/(ISSN)1553-2712.

Green, D.A., Scott, J.P.R., 2018. Spinal health during unloading and reloading associated with spaceflight. Frontiers in Physiology 8. Available at https://doi.org/10.3389/fphys.2017.01126, https://www.frontiersin.org/articles/10.3389/fphys.2017.01126/full.

Ham, W., Schoonhoven, L., Schuurmans, M.J., Leenen, L.P.H., 2014. Pressure ulcers from spinal immobilization in trauma patients: a systematic review. Journal of Trauma and Acute Care Surgery 76 (4), 1131–1141. Available at https://doi.org/10.1097/TA.0000000000000153, http://journals.lww.com/jtrauma/pages/default.aspx.

Hawkins, S.C., Williams, J., Bennett, B.L., Islas, A., Kayser, D.W., Quinn, R., 2019. Wilderness medical society clinical practice guidelines for spinal cord protection. Wilderness & Environmental Medicine 30 (4_suppl), S87. Available at https://doi.org/10.1016/j.wem.2019.08.001.

Hawkins, S.C., Williams, J., Bennett, B.L., Islas, A., Quinn, R., 2024. Wilderness medical society clinical practice guidelines for spinal cord protection: 2024 update. Wilderness & Environmental Medicine 35 (1_suppl), 78S. Available at https://doi.org/10.1177/10806032241227232.

Hayden, J.A., Pantalos, G.M., Burgess, J.E., Antaki, J.F., 2013. A hermetically sealed, fluid-filled surgical enclosure for microgravity. Aviation, Space, and Environmental Medicine 84 (12), 1298–1303. Available at https://doi.org/10.3357/ASEM.3751.2013UnitedStates, http://docserver.ingentaconnect.com/deliver/connect/asma/00956562/v84n12/s14.pdf.

Hewett Brumberg, E.K., Douma, M.J., Alibertis, K., Charlton, N.P., Goldman, M.P., Harper-Kirksey, K., et al., 2024. 2024 American Heart Association and American Red Cross Guidelines for First Aid. Circulation 150 (24), e519. Available at https://doi.org/10.1161/CIR.0000000000001281, http://circ.ahajournals.org.

Human Research Roadmap. Risk of Spaceflight Associated Neuro-ocular Syndrome (SANS). Available from: https://humanresearchroadmap.nasa.gov/Risks/risk.aspx?i=105, 2024.

Hupfeld, K.E., McGregor, H.R., Reuter-Lorenz, P.A., Seidler, R.D., 2021. Microgravity effects on the human brain and behavior: dysfunction and adaptive plasticity. Neuroscience & Biobehavioral Reviews 122, 176–189. Available at https://doi.org/10.1016/j.neubiorev.2020.11.017.

Jasien, J.V., Laurie, S.S., Lee, S.M.C., Martin, D.S., Kemp, D.T., Ebert, D.J., et al., 2022. Noninvasive indicators of intracranial pressure before, during, and after long-duration spaceflight. Journal of Applied Physiology 133 (3), 721–731. Available at https://doi.org/10.1152/JAPPLPHYSIOL.00625.2021, https://journals.physiology.org/doi/10.1152/japplphysiol.00625.2021.

KA Imaging. A historic image: the first X-ray in space is here [press release], 2025. Available from: https://kaimaging.com/wp-content/uploads/2025/04/2025-04-02-SpaceXray-first-image-PR.pdf.

Kershner, D., Binhammer, R., 2004. Intrathecal ligaments and nerve root tension: possible sources of lumbar pain during spaceflight. Aviation, Space, and Environmental Medicine 75 (4), 354–358.

Kwon, Y.S., Lee, Y.H., Cho, J.M., 2016. Early experience of automated intraventricular type intracranial pressure monitoring (LiquoGuard®) for severe traumatic brain injury patients. Korean Journal of Neurotrauma 12 (1), 28. Available at https://doi.org/10.13004/kjnt.2016.12.1.28.

Lazzari, Z.T., Aria, K.M., Menger, R., 2021. Neurosurgery and spinal adaptations in spaceflight: a literature review. Clinical Neurology and Neurosurgery 207, 106755. Available at https://doi.org/10.1016/j.clineuro.2021.106755.

Lee, J.K., Koppelmans, V., Riascos, R.F., Hasan, K.M., Pasternak, O., Mulavara, A.P., et al., 2019. Spaceflight-associated brain white matter microstructural changes and intracranial fluid redistribution. JAMA Neurology. American Medical Association, United States, pp. 412–419. Available at 10.1001/jamaneurol.2018.4882, http://archneur.jamanetwork.com/issues.aspx 76.

Lerner, D.J., Parmet, A.J., Don, S., Shimony, J.S., Goyal, M.S., 2016. Technique for performing lumbar puncture in microgravity using portable radiography. Aerospace Medicine and Human Performance 87 (8), 745–747. Available at https://doi.org/10.3357/AMHP.4601.2016, http://docserver.ingentaconnect.com/deliver/connect/asma/23756314/v87n8/s12.pdf.

Lerner, D.J., Chima, R.S., Patel, K., Parmet, A.J., 2019. Ultrasound guided lumbar puncture and remote guidance for potential in-flight evaluation of VIIP/SANS. Aerospace Medicine and Human Performance 90 (1), 58–62. Available at https://doi.org/10.3357/AMHP.5170.2019, http://docserver.ingentaconnect.com/deliver/connect/asma/23756314/v90n1/s13.pdf?expires=1547290373&id=0000&titleid=72010555&checksum=0E17C63C5D484C70749B2C8DDFF38EAE.

Mader, T.H., Gibson, C.R., Pass, A.F., Kramer, L.A., Lee, A.G., Fogarty, J., et al., 2011. Optic disc edema, globe flattening, choroidal folds, and hyperopic shifts observed in astronauts after long-duration space flight. Ophthalmology 118 (10), 2058–2069. Available at https://doi.org/10.1016/j.ophtha.2011.06.021.

Mader, T.H., Gibson, C.R., Pass, A.F., Lee, A.G., Killer, H.E., Hansen, H.C., et al., 2013. Optic disc edema in an astronaut after repeat long-duration space flight. Journal of Neuro-Ophthalmology. 33 (3), 249–255. Available at https://doi.org/10.1097/WNO.0b013e31829b41a6.

Mader, T.H., Gibson, C.R., Otto, C.A., Sargsyan, A.E., Miller, N.R., Subramanian, P.S., et al., 2017. Persistent asymmetric optic disc swelling after long-duration space flight: implications for pathogenesis. Journal of Neuro-Ophthalmology. 37 (2), 133–139. Available at https://doi.org/10.1097/WNO.0000000000000467, http://journals.lww.com/jneuro-ophthalmology.

Mader, T.H., Gibson, C.R., Barratt, M.R., Miller, N.R., Subramanian, P.S., Killer, H.E., et al., 2021. Persistent globe flattening in astronauts following long-duration spaceflight. Neuro-Ophthalmology 45 (1), 29–35. Available at https://doi.org/10.1080/01658107.2020.1791189, http://www.tandfonline.com/loi/ioph20.

Mampre, D., Spaide, R., Mason, S., Van Baalen, M., Gibson, C.R., Brunstetter, T., et al., 2025. Spaceflight associated neuro-ocular syndrome as a potential variant of venous overload choroidopathy. Aerospace Medicine and Human Performance 96 (6), 496–508. Available at https://doi.org/10.3357/amhp.6602.2025.

Marshall-Goebel, K., Laurie, S.S., Alferova, I.V., Arbeille, P., Auñón-Chancellor, S.M., Ebert, D.J., et al., 2019. Assessment of jugular venous blood flow stasis and thrombosis during spaceflight. JAMA Network Open 2 (11). Available at https://doi.org/10.1001/jamanetworkopen.2019.15011, https://jamanetwork.com/journals/jamanetworkopen.

Menger, R., Wolf, M., Thakur, J.D., Nanda, A., Martino, A., 2019. Astronaut Michael Collins, Apollo 8, and the anterior cervical fusion that changed the history of human spaceflight. Journal of Neurosurgery: Spine 31 (1), 87–92. Available at https://doi.org/10.3171/2018.11.spine18629.

Neurocritical care, 2024. Available from: https://hyperfine.io/specialties/neurocritical-care.

Panesar, S.S., Fernandez-Miranda, J.C., Kliot, M., Ashkan, K., 2019. In reply: neurosurgery and manned spaceflight. Neurosurgery 85 (1), E159. Available at https://doi.org/10.1093/neuros/nyz107.

Panesar, S.S., Fernandez-Miranda, J.C., Kliot, M., Ashkan, K., 2020. Neurosurgery and manned spaceflight. Neurosurgery 86 (3), 317–324. Available at https://doi.org/10.1093/neuros/nyy531.

Patel, Z.S., Brunstetter, T.J., Tarver, W.J., Whitmire, A.M., Zwart, S.R., Smith, S.M., et al., 2020. Red risks for a journey to the red planet: the highest priority human health risks for a mission to Mars. npj Microgravity 6 (1). Available at https://doi.org/10.1038/S41526-020-00124-6, https://www.nature.com/npjmgrav/.

Pechenkova, E., Nosikova, I., Rumshiskaya, A., Litvinova, L., Rukavishnikov, I., Mershina, E., et al., 2019. Alterations of functional brain connectivity after long-duration spaceflight as revealed by fMRI. Frontiers in Physiology 10 (May). Available at https://doi.org/10.3389/fphys.2019.00761.

Petersen, L.G., Lawley, J.S., Lilja-Cyron, A., Petersen, J.C.G., Howden, E.J., Sarma, S., et al., 2019. Lower body negative pressure to safely reduce intracranial pressure. Journal of Physiology 597 (1), 237–248. Available at https://doi.org/10.1113/JP276557.

Portable x-ray providers, 2020. Available from: https://www.minxray.com/portable-x-ray-providers.

Risk: risks to human health and performance. Human Research Roadmap, n.d. Available from: https://humanresearchroadmap.nasa.gov/Risks/risk.aspx?i=78.

Roberts, D.R., Petersen, L.G., 2019. Studies of hydrocephalus associated with long-term spaceflight may provide new insights into cerebrospinal fluid flow dynamics here on Earth. JAMA Neurology 76 (4), 391–392. Available at https://doi.org/10.1001/jamaneurol.2018.4891, http://archneur.jamanetwork.com/issues.aspx.

Roberts, D.R., Asemani, D., Nietert, P.J., Eckert, M.A., Inglesby, D.C., Bloomberg, J.J., et al., 2019. Prolonged microgravity affects human brain structure and function. Journal of Neuroradiology 40 (11), 1878–1885. Available at https://doi.org/10.3174/ajnr.A6249, http://www.ajnr.org/content/ajnr/40/11/1878.full.pdf.

Saragoça, Bd.S.P.G., 2024. Ultrasound-guided lumbar puncture with AI-based image labeling for intracranial pressure assessment in spaceflight.

Sauer, A., Vigouroux, M., Dougherty, P., Cata, J., Ingelmo, P., 2023. Pain experience and sensory changes in astronauts during and after short-lasting commercial spaceflight: a proof-of-concept study. Journal of Pain Research 16, 4253–4266. Available at https://doi.org/10.2147/jpr.s440630.

Seidler, R.D., Mao, X.W., Tays, G.D., Wang, T., zu Eulenburg, P., 2024. Effects of spaceflight on the brain. The Lancet Neurology 23 (8), 826–835. Available at https://doi.org/10.1016/S1474-4422(24)00224-2, https://www.sciencedirect.com/science/journal/14744422.

Srinivasan, V.M., O'Neill, B.R., Jho, D., Whiting, D.M., Oh, M.Y., 2014. The history of external ventricular drainage. Journal of Neurosurgery 120 (1), 228–236. Available at https://doi.org/10.3171/2013.6.jns121577.

Swinney, C.C., Allison, Z., 2018. Spaceflight and neurosurgery: a comprehensive review of the relevant literature. World Neurosurgery 109, 444–448. Available at https://doi.org/10.1016/j.wneu.2017.10.062, http://www.elsevier.com/wps/find/journaldescription.cws_home/722082/description#description.

Taylor, P.W., 2015. Impact of space flight on bacterial virulence and antibiotic susceptibility. Infection and Drug Resistance 8, 249–262. Available at https://doi.org/10.2147/IDR.S67275, http://www.dovepress.com/getfile.php?fileID=26227.

Weaver, A.S., Zakrajsek, A.D., Lewandowski, B.E., Brooker, J.E., Myers, J.G., 2013. Predicting head injury risk during international space station increments. Aviation, Space, and Environmental Medicine 84 (1), 38–46. Available at https://doi.org/10.3357/ASEM.3429.2013, http://docserver.ingentaconnect.com/deliver/connect/asma/00956562/v84n1/s9.pdf?.

Wilson, M.H., Wise, D., Davies, G., Lockey, D., 2012. Emergency burr holes: "how to do it" Scandinavian Journal of Trauma, Resuscitation and Emergency Medicine 20 (1), 24. Available at https://doi.org/10.1186/1757-7241-20-24.

Chapter 17

# Dental procedures

Michael Hodapp[1,*], Shawna Pandya[2,3], James Yacyshyn[3], Shahnawaz Khijmatgar[4,*], Saranjeev Lalh[3], and Ravjot Kaur[5,*]

[1]*General Dentist, Private Practice, Houston, TX, United States,* [2]*Space Medicine Group, International Institute for Astronautical Sciences, Boulder, CO, United States,* [3]*Faculty of Medicine & Dentistry, University of Alberta, Edmonton, AB, Canada,* [4]*School of Dentistry, Royal College of Surgeons in Ireland, Dublin, Ireland,* [5]*Ohio State University College of Dentistry, Columbus, OH, United States*

## 17.1 Introduction

The field of oral health remains one of the least developed fields of spaceflight medicine. The US National Institute of Dental and Craniofacial Research states that dental caries is one of the most prevalent chronic diseases. Nearly 90% of adults between the ages of 20 and 64 years exhibit some form of tooth decay in their permanent teeth, with more than one in four adults having untreated dental caries (NIHICDR, 2025). Teeth with a history of dental caries carry a higher likelihood of developing future complications such as pulpitis, fracture, or infection since prior decay weakens tooth structure and may have bacterial colonies present in a compromised pulp. Although appropriate treatment can significantly reduce these risks, it cannot eliminate them entirely, making dental intervention more likely.

Although most spaceflight dental events would likely result in a slight inconvenience to a crewmember, such as loss of crown, fractured cusp, or a chipped tooth, We must be prepared in the unlikely event that a dental emergency occurs. Facial trauma, tooth fracture that exposes the pulp, an infected tooth with apical periodontitis, acute symptomatic irreversible pulpitis, and acute apical abscess, are all dental emergencies that could occur during spaceflight, especially in a heavily restored multirooted tooth. The first line of defense for most of these would be antibiotics and analgesics. However, with dental infections, especially acute apical abscess where the tooth is raised above the occlusal plane and the patient becomes febrile, antibiotics are only a temporary measure. When a tooth becomes necrotic or infected, the blood supply to the pulp chamber is compromised and the antibiotics will not reach the multiplying bacterial colonies within the pulp chamber. Some form of definitive care, such as root canal or extraction, will be necessary. Since current astronaut training and available supplies for dental emergencies are limited on the International Space Station (ISS), an early return to Earth may be necessary in these situations.

The main incapacitating symptom is often pain, and pain management will play a role enabling capability for most dental procedures. However, symptomatic irreversible pulpitis or infected teeth may be very difficult to anesthetize. Inflammation and infection lower tissue pH in the region surrounding the tooth. The reduction in pH will favor the formation of a greater proportion of the cation form (RNH+) of the anesthetic relative to the base form (RN), thus inhibiting penetration of the anesthetic into the sodium channels of the nerve.

The primary aim of this chapter is to outline evidence-based dental care strategies for space crews that may be needed over the next 10 years, collate the relevant evidence on dental procedures in microgravity, and establish best-practice approaches for managing common dental complications during space missions. While the risks are usually not catastrophic, they must be weighed and considered by planners when determining the quantities and types of antibiotics, analgesics, and the type of dental supplies and equipment for each mission. This is particularly true as we move into an era of longer and more distant destinations.

## 17.2 Historical context

Dental issues have occurred throughout the history of human spaceflight. In 1978, cosmonaut Yuri Romanenko suffered excruciating dental pain during the last 2 weeks of his mission aboard Salyut 6. The only available dental care supplies

* Lead authors.

**Procedural Medicine for Spaceflight. DOI:** https://doi.org/10.1016/B978-0-443-27552-4.00015-4

were pain medications and mouthwash which failed to provide effective relief for the remainder of his mission (Harika et al., 2023). During the period between March 1995 and June 1998, an incidence of dental caries requiring management with a temporary filling was recorded on the Mir space station, comprising 1% of the medical events reported during that period (Gontcharov et al., 2005). During the Apollo program, five of the thirty-three astronauts (15%) required treatment within a 90-day period around their flights. Of these, there was one preflight occurrence and one postflight occurrence of pulpitis that, according to the report, would have affected crewmember performance if it had occurred during their mission (Johnston et al., 1975). During the Shuttle program, there were two cases that required endodontic therapy, one of which occurred 2 weeks prior to launch (Hodapp and Jeske, 2020). There is also a published dental incident aboard the ISS where American astronaut Mark Kelly recemented the crown of his cosmonaut colleague (Kelly, 2017).

To aid with risk development and resource planning, NASA developed the Integrated Medical Model (IMM). IMM is a probabilistic risk assessment tool used for mission planning that uses the anticipated medical conditions, and the consequences associated with treating or not treating them to identify which medical resources will have the greatest chance of enhancing mission success. IMM includes dental cavities, abscesses, pulpal exposure, tooth avulsion, dental crown replacement, and dental filling replacement as potential dental conditions (Harika et al., 2023; Myers et al., 2015). An IMM risk assessment of ISS missions conducted in 2012 found that a dental abscess was the condition most likely to cause flight surgeons to consider evacuating an astronaut from the ISS (Menon, 2012).

Based on this, the ISS does include a compact dental suite capable of covering most low-impact dental contingencies with the assumption that if the onboard capabilities are exceeded, the crew can return to Earth within a few hours focused on operations where an early evacuation back to definitive care is possible within a few hours (Hodapp and Jeske, 2020).

The dental suite includes the following:

- a mouth mirror,
- combination explorer/Williams probe,
- small spoon excavator,
- small dental carver/file instrument,
- an orange wood stick,
- syringe with both long and short 27-gauge needles,
- local anesthetic,
- Red Cross toothache kit (eugenol/sesame liquid, tweezers, and cotton pellets),
- temporary crown luting agent,
- Cavit temporary filling material,
- Small and large dental elevators,
- Upper and lower posterior forceps,
- and lower universal forceps for upper and lower premolar and anterior tooth removal.

Crew Medical Officers (CMOs) on the ISS receive 1 hour of dental training for most missions as part of their 12–40-hour CMO course. This is largely focused on the use of onboard dental materials and procedures, including how to coordinate with ground-based specialists when needed. Because CMOs are not required to be physicians, they may be engineers, pilots, biologists, or astrophysicists; therefore, their skills may be limited to what is provided through NASA's CMO program. The fact that these interventions must be performed in a microgravity environment adds to the complexity, though recent investigations of common dental procedures on a zero G flight demonstrate that microgravity did not affect the capabilities of moderately experienced practitioners (Sefic et al., 2026). This approach relies on the mission's proximity to Earth which enables real-time telemedical guidance and the potential for timely evacuation. It has thus far been sufficient for managing the in-flight dental events that have occurred. Missions further from Earth will need to function with more autonomy and include sufficient training and supplies to enable more robust stabilization or perhaps a wider range of definitive care techniques.

## 17.3 A brief review of space-analog dental literature

Two studies on men living for 1–2 weeks in space-analog environments revealed changes in their oral microbiome. Specifically, Sun et al. (2022) found that 16 young men under 6-degree head-down bed rest conditions exhibited increases in counts of salivary bacteria associated with dental infections at the expense of their normal commensal bacteria population. Another study found that 12 young men at the Mars Desert Research Station in Utah, United States, showed elevated streptococcal mutans activity relative to their baselines (Rai and Kaur, 2013). It is worth noting that neither study used a space food diet and both had participants change their diets and dental habits from baseline.

A 2011 head-down bed rest study (Rai et al., 2011) on 20 male volunteers for 60 days found significant changes in serum and salivary antioxidants but noted no significant periodontal changes at 2 months. This study noted that oxidative stress changes may have been a function of prolonged inactivity rather than microgravity. Finally, one 2010 study on bone mineral density BMD in the mandible and alveolar bone in 10 male and 10 female volunteers undergoing 3 weeks of head-down bed rest demonstrated a decrease in mandibular bone mineral density (Rai et al., 2010).

Taken together, these findings are intriguing targets for future research, although they do not support a need for changing present-day operations. It is difficult to draw conclusions given the study design, limited generalizability to actual spaceflight, and small numbers of participants, though they do suggest that changes in diet, task loads, oral hygiene, and environment might alter the oral microbiome, oxidative stress responses, and bone health, even over relatively short periods.

A recent study tested how microgravity affected senior dental students performing common dental procedures. The researchers found that they had little difficulty performing the skills. They noted that kneeling added stability and that positioning a right-handed provider on the left side of the patient's head (the 9 o'clock position) helped facilitate the procedure. The researchers also recommended avoiding water sprays for cleaning, since this could scatter biohazardous debris, and instead rely on wet sponges or water gels to clean the tooth area. Gauze or suction were identified as options for managing saliva (Sefic et al., 2026).

## 17.4 Missions beyond low Earth orbit

Exploration class missions will not have the rapid return capability that we currently take for granted on ISS. In 2023, NASA and its collaborators established the IMPACT 1.0 Medical Condition List as part of a program to build a medical risk assessment tool like IMM for exploration class missions. This list also highlights the IMM dental conditions, dental abscess, pulpitis, tooth fracture, exposed pulp, loss of filling, and crown displacement, as important considerations (Kreykes et al., 2023). Medical probabilistic risk assessment is a powerful tool to help planners compare the risk reduction benefits of treating specific conditions or including specific supplies. This helps medical system designers to determine which resources are truly necessary versus nice-to-haves and further enables trades to reduce overall mission risk rather than optimizing risk reduction from medical events alone at the expense of other ship systems such as life support, propulsion, communications, etc. The concept is further described in Chapter 2 of this text which covers the concepts of systems medicine. While the tool itself has not been operationalized as of 2026, the IMPACT 1.0 conditions list emphasizes that dental conditions remain an important consideration for medical mission planning.

A condition worth considering for exploration missions is an acute apical abscess. These infections can progress rapidly and cause intense pain. These infections often raise the tooth above the occlusal plane, and when this occurs, the crewmember will have trouble swallowing, eating, sleeping, or focusing attention. Mission planners should consider manifesting a dental handpiece to reduce the occlusion on an affected tooth and have sufficient instrumentation for extraction.

One study estimated the incidence of dental infections requiring hospital admission to be 1 in 2600 person-years, which is a reasonable estimation for how often such a condition may need to be evacuated from a space mission (Wang et al., 2005). While this number is reasonably low, it does not account for the number of milder infections. The CDC's National Center for Healthcare Statistics estimates dental visits to the ER at more than 15 times higher than admission rate (1 in 168 person-years (Wang et al., 2005)). For a 6-person, 3-year mission to Mars, this means a dental infection that would cause someone to visit an emergency room may occur on roughly 1 out of every 10 missions.

These infections may be unlikely to require an evacuation but can still cause significant disability and interfere with mission performance. Antibiotics will stabilize a dental infection for a short period of time; however, definitive treatment will be necessary to remove the source of infection, making a reasonably strong argument for manifesting equipment to manage these conditions. Although root canal is considered the standard of care for dental infections or symptomatic irreversible pulpitis, it is a complex procedure and the difficulties of performing it in spaceflight are currently prohibitive. In the event where a crewmember requires definitive care for a dental emergency in spaceflight without the possibility of early return, extraction of the compromised tooth is the most reasonable approach.

To move beyond Earth's orbit, we need to continue research to enhance our dental management capabilities in spaceflight. Since exploration class missions will have longer evacuation times, the exploration class dental system will need additional resources to reduce the probability of an evacuation triggering condition as well as stabilize or definitively manage dental issues that do occur. Most likely this will provide for some degree of examination and dental hygiene, but less temporary solutions for dental caries and fractures may also be worth considering, such as bonding materials, glass ionomers, and flowable composites.

Enhanced capabilities will also mean additional training; therefore, in addition to trades of mass, power, and volume, crew training time and in-flight training resources will also need to be considered against the likelihood of mission-relevant consequences for each dental condition. Training may include hands-on patient care to develop skillsets of diagnostic and dental risk reasoning skills. These may be performed on Earth before flight, but just-in-time treatment videos, simulations, and practice exams/procedures may also be conducted in flight to prevent skill erosion, reduce terrestrial training times, and offset any communication delays.

In cases where evacuation is not possible, but the acuity of a dental issue is high and cannot wait for definitive care on Earth, crewmembers may need to extract a tooth. For exploration missions, mission planners should consider including surgical instruments such as root tip forceps, luxators or proximators, periosteal elevators, surgical spoon excavators, suture instrumentation, and suture material. A special note: there are resources and references for production of printed dental instrumentation for spaceflight. 3D printing may be applicable for procedures that require little force such as an instrument for temporary filling placement (Bean Quincy et al., 2015). However, surgical and hygienic dental instrumentation requires the use of high-grade stainless-steel that must be run through a high-temperature 1000+ degree heat-treatment hardening process, a quenching procedure, a low temperature tempering process, and finally a sharpening process to be effective in dental procedures. With current technology, the 3D printing of hygiene instrumentation and oral surgery applications may result in instrument failure and iatrogenic injury.

## 17.5 Procedures and guidance

The remainder of this chapter will describe dental emergency methods, including those currently used by NASA and those anticipated to be useful for managing dental issues during spaceflight over the next 10–15 years. These procedures are not definitive treatment. Rather they are intended to stabilize a crewmember and temporize the issue until intervention by a dental professional can be obtained. Special consideration should be given to performing procedures in the microgravity environment. Notably, both the patient and provider should be restrained or stabilized during the procedure. Similarly, instruments should be secured using hook and loop tape or similar, and all medical systems need a waste management plan and appropriate resources manifested in order to manage body fluids, debris, and biohazard materials generated during the procedure. To prepare our space program for deep space exploration class missions, additional research and development will likely be needed to aid inexperienced crewmembers in managing conditions relating to dental trauma, and/or oral, dental, and craniofacial disease when an early return to seek professional care will be substantially delayed or not possible.

### 17.5.1 Diagnosis: "the key to proceed"

A "toothache" can be defined as pain originating from a tooth or its surrounding structures, and it typically arises from two main causes: disease or trauma. Some traumatic injuries are obvious and straightforward to manage, while others such as temporomandibular disorders, myofascial pain dysfunction, and sleep-related issues linked to occlusal abnormalities and maxillofacial discrepancies may be far more complex to diagnose. Microtrauma can also occur in the form of noncarious lesions where excessive occlusal forces stress the enamel and dentin near the gumline, leading to thermal sensitivity and in severe cases pulp exposure. Dental diseases can be equally challenging to identify. Periodontal disease and dental caries often progress silently, producing little more than occasional bleeding or thermal sensitivity. Due to the absence of pain, these conditions may progress undetected without regular exams, underscoring the importance of routine dental assessments to catch disease early while management remains relatively simple and not resource intensive. Table 17.1 contains a basic description of symptoms for dental diagnosis and initial management strategies.

**TABLE 17.1 Diagnosing dental conditions.**

| Condition | Signs/symptoms | Initial treatments | Alternative care |
|---|---|---|---|
| Pulpitis (mild to moderate) | • Painful stimulus to heat and/or cold<br>• May have mild sensitivity to bite pressure | • Avoid cold/hot foods<br>• Use of nonsteroidal antiinflammatory medications (NSAIDs) | • None |
| Pulpitis (severe) | • Spontaneous paroxysms of pain<br>• Intense pain to thermal stimulus<br>• Sometimes relieved by hot or cold stimulus | • nonsteroidal antiinflammatory drugs (NSAIDS)<br>• NSAIDS/narcotic combo | • Early return<br>• Extraction |
| Cracked tooth | • Sharp stabbing pain when chewing, biting together, or upon release<br>• Unusual cold sensitivity | • Avoid tooth | • Early return<br>• Extraction |

*(Continued)*

**TABLE 17.1** (Continued)

| Condition | Signs/symptoms | Initial treatments | Alternative care |
|---|---|---|---|
| Dental infection (symptomatic) | • Intense dull pain/intensifies when biting together or chewing | • Antibiotics with coverage for Gram (+) and anerobic bacteria | • Early return<br>• Extraction |
| Dental infection (asymptomatic) | • Occasional unprovoked dull pain or when chewing<br>• Small pimple head next to affected tooth | • No treatment<br>• Monitor daily | |
| Acute apical abscess | • Intense pain<br>• Tooth raised above occlusal table<br>• Patient may become febrile | • Antibiotics<br>• Early return event ASAP | • Early return ASAP<br>• Extraction |
| Temporomandibular disorder or myofascial pain dysfunction | • C/O headaches with dull ache<br>• Pain or tenderness during palpation of temporomandibular joints.<br>• Pain or tenderness during palpation of oral musculature | NSAIDS | |
| Trigeminal neuralgia | • Tic or shock type pain with gentle touch on portion of face | • Avoid touching trigger area of face | • Carbamazepine or other neuralgic pain medications |
| Local gingival inflammation | • Local red swollen area around gum or in between teeth | • Floss, flush with sterile saline and curved tip syringe | |

## 17.5.2 Anesthetic delivery

Anesthetic delivery is one of the most crucial parts of providing any type of dental treatment. Successful anesthesia of the teeth requires an understanding of the type of anesthetic used for dental procedures and a working knowledge of the anatomical landmarks for proper delivery of the anesthetic. For example, the mandibular or inferior alveolar nerve (IAN) block is necessary to numb the lower teeth and requires a bolus of anesthetic (1.8 mL) with epinephrine next to the inferior alveolar nerve where it inserts into the mandibular ramus. The epinephrine is necessary to maintain the anesthetic long enough for it to absorb into the large nerve trunk (Figs. 17.1–17.3).

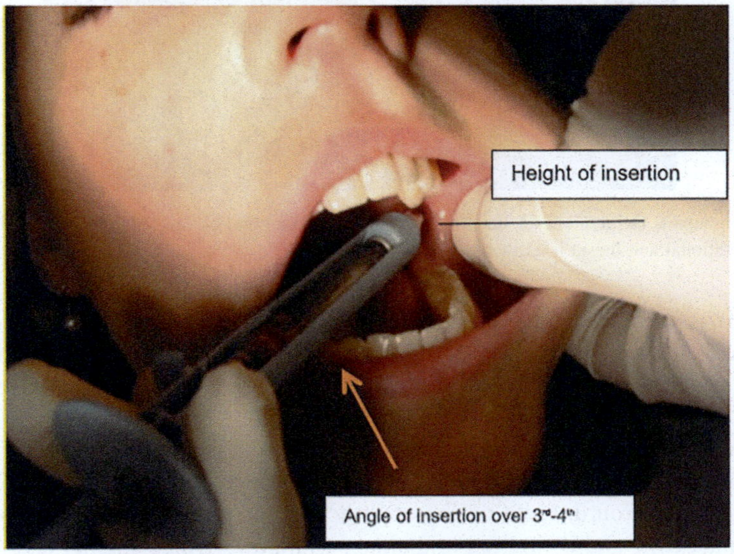


**FIGURE 17.1** Mandibular block procedure. Use an imaginary line drawn through the deepest portion of the coronoid notch to the commissure of the pterygomandibular raphe where it turns upwards towards the maxilla. Insert the needle 2/3rds the distance posteriorly along this line, keeping the barrel of the syringe over the third and fourth tooth on the opposite side of the arch.

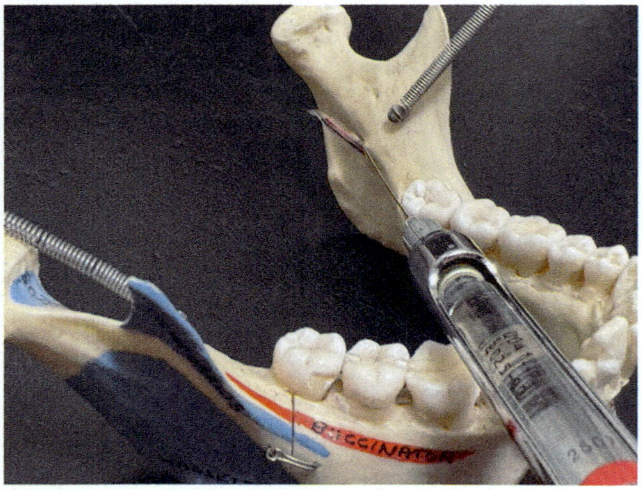

FIGURE 17.2 The image above depicts the target area. The target area is slightly above the inferior alveolar nerve where it inserts into the mandible. Note: if the anesthetic is delivered below the target area, the teeth will not be anesthetized.

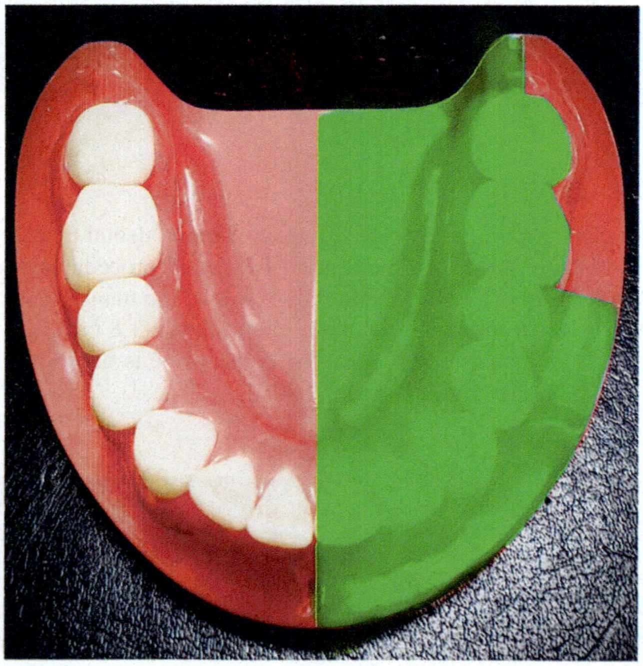

FIGURE 17.3 The image above depicts the area anesthetized during the inferior alveolar block (IAN). Note: the buccal mucosa next to the first and second molars would require additional anesthesia if extraction were necessary.

A long needle (31 mm) is used for this procedure. With the patient's mouth open wide, place the thumb of the nondominant hand into the deepest portion of the coronoid notch of the mandibular ramus. Using an imaginary line drawn from the deepest portion of the coronoid notch to the angle of the pterygomandibular raphe as it turns up toward the maxilla, insert the needle at 2/3 the distance posteriorly along this line. Keep the barrel of syringe of the dominant hand against the teeth on the opposite side and over the premolar area. Insert the needle until bone is contacted. Approximately 5 mm of the needle should be exposed. If more than 5 mm of needle is exposed, withdraw needle slightly and insert needle with syringe barrel over third and fourth tooth on the opposite side of the arch until the appropriate depth is achieved. If the needle does not contact bone and is inserted to the hub, withdraw syringe slightly and position over 4–5th to until proper depth is achieved and bone is contacted. The needle should contact bone and have approximately 5 mm of needle exposed prior to administering anesthetic. Administer approximately 1.8 mL of anesthetic, withdraw syringe 2 mm and administer additional 2 mL of anesthetic solution to anesthetize the lingual nerve.

The target nerve (inferior alveolar nerve) is housed deep inside the buccal mucosa and will anesthetize the lower (mandibular) teeth on the side of the jaw where the injection is delivered. The height and depth of insertion is critical to achieve profound anesthesia of the lower teeth. The buccal mucosa next to the molars would require anesthesia as well if extraction were necessary (see Fig. 17.3).

### 17.5.3 Local anesthetic with vasoconstrictor

A 2% lidocaine with 1:100,000 epinephrine is one of the preferred anesthetics for dental procedures and is what has been available on the ISS. The concentration is sufficient to allow enough of the base form to transfer across sodium channels of a large nerve trunk such as the IAN. The addition of 1:100,000 epinephrine, the equivalent of 18 μg per each 2 mL of lidocaine, is present to keep the anesthetic in the vicinity of the IAN nerve or cortical plate of bone long enough to absorb and anesthetize the teeth. Without epinephrine, the teeth will still have sensation; however, the surrounding tissue will remain numb for a period of 60–90 minutes. The epinephrine performs an additional safety function by limiting the amount of lidocaine absorbed into the bloodstream, thus reducing the chance of affecting cardiac inhibition. A study performed by Phero et al. (2017) was able to demonstrate that the concentration of lidocaine was able to be reduced to a 1% concentration for infiltration of the maxilla. This was accomplished by buffering with sodium bicarbonate immediately prior to injecting, thereby increasing the amount of base form available to cross sodium channels. A 1:100,000 epinephrine concentration (18 μg per 2 mL) was still required to anesthetize the teeth. Recent evidence suggests that 4% articaine with 1:100,000 epinephrine has the ability to anesthetize the mandibular first molar area via infiltration, due to its 4% concentration and lower pKa of 7.8, which increases the base form available to cross sodium channels (Majid, 2025; Luo et al., 2022; Tables 17.2–17.7).

**TABLE 17.2** Mandibular block anesthetic delivery.

| Item | Mass | Volume |
|---|---|---|
| Nitrile gloves | 6–12 g | 40 mL |
| Syringe (5–6 mL) | 20–30 g | 40–55 mL |
| 27 g long needle (31 mm) | 0.5–1 g | 0.5–1 mL |
| 2% lidocaine/1:100,000 epi | 2–6 g | 7–9 mL |
| Schematic anesthetic delivery (laminated card) | 1–15 g | 5–15 mL |
| **Segment total** | **29.5–64 g** | **92.5–120 mL** |

| | |
|---|---|
| **Procedure** | **Mandibular block anesthetic delivery** |
| **Purpose** | Anesthetize lower teeth with block anesthesia. |
| **Indications** | • Uncontrolled dental pain<br>• Dental extractions<br>• Acute dental trauma<br>• Dental infections causing severe discomfort<br>• Acute apical abscess<br>• Dental procedures requiring anesthetic<br>• Acute symptomatic irreversible pulpitis<br>• Anesthesia testing for diffuse pain |
| **Contraindications** | • Lack of pain<br>• Allergy |
| **Supplies** | |
| **Operators** | Minimum 1, ideally 2 operators |

(Continued)

**TABLE 17.2 (Continued)**

| | |
|---|---|
| **Antiseptic** | • Gloves |
| **Analgesia** | • 2% lidocaine with 1:100,000 (18 μg) epinephrine |
| **Materials** | • Crew medical restraint system<br>• 5–6 mL dental syringe<br>• Long 31 mm 27-gauge needle<br>• 2% lidocaine with 1:100,000 epinephrine (2 mL)<br>• Schematic to perform proper delivery<br>• Digital or hard copy diagram or video to facilitate proper delivery |

**Steps**

1. Locate dental syringe with thumb aspiration ring. If syringe with aspirating ring is unavailable, fabricate an aspiration ring out of tape on the plunger portion of the syringe. When delivering anesthetic around large blood vessels it is important to aspirate prior to injection to prevent inadvertently delivering anesthetic into the bloodstream. Place 27-gauge long (31 mm) needle onto syringe body. Load the syringe with 1.8 mL of 2% lidocaine with 1:100,000 epinephrine.
2. Review the anesthetic block delivery diagram before proceeding further.
3. With the patient's mouth opened wide, use the nondominant hand to palpate the deepest portion of the coronoid notch of the ramus on the side the anesthetic is to be injected.
   a. Use an imaginary line drawn from the deepest portion of the coronoid notch of the ramus to the commissure of the pterygomandibular raphe to determine the height of insertion for the 27-gauge long needle and the insertion point for the needle 2/3rds the way back on this imaginary line.
   b. The barrel of the syringe should remain over the third and fourth tooth on the opposite side of the arch.
   c. Slowly insert the needle. If the angulation and position is held correctly, the needle should hit bone with about 5 mm of the needle still remaining outside the tissue. If the needle hits bone and the needle has more than 5 mm of needle exposed, pull needle back slightly and reinsert with the hub of the of the needle over the second to third tooth. If the needle does not hit bone and is inserted to the hub, withdraw needle slightly and reinsert with the hub of the syringe over the fourth and fifth tooth.
   d. Aspirate the syringe to verify the needle is not in a blood vessel. If there is no aspirant (blood drawn into cartridge), slowly inject 1.8 mL of anesthetic over a period of about 1 minute.
4. Once the anesthetic is delivered, wait 3–5 minutes for anesthetic to take effect.
5. Test for mandibular anesthetic effectiveness by gently poking the gingival tissue below the mandibular canine on the side where the anesthetic was delivered.

**Postprocedural care**

Once procedure is completed,
- Wait for anesthetic to wear off completely before eating. Fluids are allowed.
- Monitor patient closely for any postop care needs.

**Potential complications and solutions**

| **Complication** | **Solution** |
|---|---|
| Lack of numbness | 1. Check that correct anesthetic with epinephrine was used.<br>2. Review description and diagram and deliver additional dose of 1.8 mL anesthetic.<br>3. Wait 3–5 minutes and test the target area again. |
| The patient felt a shock during needle penetration. | The needle contacted the lingual or inferior alveolar nerve during penetration. Pull back slightly and continue penetration at a slightly different angle.<br>Then deliver anesthetic as described in the procedures above. |
| Blood enters syringe during aspiration | Pull needle back slightly and then penetrate at a slightly different angle.<br>Aspirate again. If there is no aspirant present, slowly deliver contents of syringe. |
| Patient feels heart racing after anesthetic delivery | Anesthetic was delivered into blood a vessel. The little epinephrine in the anesthetic triggered the release of epinephrine from the adrenal gland. Wait a few minutes and the heart will slow again. |
| Syncope reaction | Monitor patient and wait a few minutes and patient will awaken. |

**TABLE 17.3 Infiltration anesthetic delivery.**

| Item | Mass | Volume |
|---|---|---|
| Nitrile gloves | 6–12 g | 40 mL |
| Syringe (5–6 mL) | 20–30 g | 40–55 mL |
| 27 g short needle (24 mm) | 0.5–1 g | 0.5–1 mL |
| 2% lidocaine/1:100,000 epinephrine | 2–6 g | 7–9 mL |
| Schematic anesthetic delivery (laminated card) | 1–15 g | 5–15 mL |
| **Segment total** | **29.5–64 g** | **92.5–120 mL** |

| | |
|---|---|
| **Procedure** | **Infiltration anesthetic delivery** |
| **Purpose** | Anesthetize maxillary teeth, lower buccal mucosa, and all palatal tissue. |
| **Indications** | • Uncontrolled dental pain<br>• Dental extractions<br>• Acute dental trauma<br>• Dental infections causing severe discomfort<br>• Acute apical abscess<br>• Dental procedures requiring anesthetic<br>• Acute symptomatic irreversible pulpitis<br>• Anesthesia testing for diffuse pain |
| **Contraindications** | • Lack of pain<br>• Allergy |
| **Supplies** | |
| **Operators** | • Minimum 1, ideally 2 operators |
| **Antiseptic** | • Sterile gloves |
| **Analgesia** | • 2% lidocaine with 1:100,000 (18 μg) epinephrine |
| **Materials** | • Crew medical restraint system<br>• 5–6 mL syringe.<br>• Short 24 mm 27-gauge needle<br>• 2% lidocaine with 1:100,000 epinephrine<br>• Digital or hard copy diagram or video to facilitate proper delivery |

**Steps**

1. Locate dental syringe with thumb aspiration ring. If syringe with aspirating ring is unavailable, fabricate an aspiration ring out of tape on the plunger portion of the syringe. When delivering anesthetic around large blood vessels, it is important to aspirate prior to injection to prevent inadvertently delivering anesthetic into the bloodstream. Place 27-gauge short needle onto syringe body. Load the syringe with 1.0 mL of 2% lidocaine with 1:100,000 epinephrine.
2. Place the needle at the height of the muccobuccal fold over the tooth to be anesthetized. Insert needle and slowly advance the needle into the tissue toward the bone, at about the height of the root tip (approximately 4× the crown length of the tooth).
3. Aspirate the syringe to verify the needle is not in a blood vessel. If there is no aspirant (blood drawn into cartridge), slowly inject about 0.6 mL of the anesthetic over a period of about 1 minute.
4. Once the anesthetic is delivered, wait 3–5 minutes for anesthetic to take effect.
5. Test for anesthetic effectiveness by gently poking the gingival tissue next to the tooth where the anesthetic was delivered.
6. If there is a need for an extraction, the lingual tissue will require anesthesia as well.

**Postprocedural care**

Once procedure is completed,
- Wait for anesthetic to wear off completely before eating. Fluids are allowed.
- Monitor patient closely for any postop care needs.

*(Continued)*

**TABLE 17.3 (Continued)**

| Potential complications and solutions | |
|---|---|
| **Complication** | **Solution** |
| Lack of numbness | **1.** Check that the correct anesthetic with epinephrine was used.<br>**2.** Review description and diagram and deliver a second dose of 0.6 mL anesthetic.<br>**3.** Wait 3–5 minutes and test the target area again. |
| The patient felt a shock during needle penetration. | The needle contacted portion of V-2 branch of trigeminal nerve during penetration. Pull back slightly and reposition needle at a slightly different angle.<br>Then deliver anesthetic as described in the procedures above. |
| Blood enters syringe during aspiration | Pull needle back slightly and then penetrate at a slightly different angle.<br>Aspirate again. If there is no aspirant present, slowly deliver 1/3 of syringe. |
| Patient feels heart racing after anesthetic delivery | Anesthetic was delivered into blood vessel. The little epinephrine in the anesthetic triggered the release of epinephrine from the adrenal gland. Wait a few minutes and the heart will slow again. |
| Patient passes out during anesthetic delivery | Syncope reaction. Wait a few minutes and patient will awaken. |

**TABLE 17.4** Cementation of displaced crown.

| Item | Mass | Volume |
|---|---|---|
| Nitrile gloves | 6–12 g | 40 mL |
| Syringe (5–6 mL) | 20–30 g | 40–55 mL |
| 27 g short or long needle (24–31 mm) | 0.5–1 g | 0.5–1 mL |
| 2% lidocaine/1:100,000 epi | 2–6 g | 7–9 mL |
| Schematic anesthetic delivery (laminated card) | 1–15 g | 5–15 mL |
| Carver | 15–25 g | 10–15 mL |
| Explorer/probe | 12–22 g | 8–14 mL |
| Temporary crown cement with mixing tips | 35–60 g | 30–50 mL |
| 2×2 gauze pads (×4) | 7–10 g | 20–30 mL |
| Wooden tongue depressor | 1.5–2 g | 4–5 mL |
| Cotton swab (×1) | 0.3–0.5 g | 1–2 mL |
| Dental floss | 10–15 g | 15–25 mL |
| **Segment total** | **110.3–198.5 g** | **180.5–261 mL** |

| | |
|---|---|
| **Procedure** | **Cementation of displaced crown** |
| **Purpose** | To restore function and cover sensitive dentin. |
| **Indications** | • There must be sound tooth structure, the crown is intact, and retention without cement should be present to consider recementation.<br>• Warning: If there is no dentin sensitivity and the loss of crown does not affect function, it is best to stow the crown away. This is temporary cement, and the risk of the crown dislodging again is high, especially on long duration missions. There is an inherent risk of additional tooth breakage when chewing or even aspiration if crown were to dislodge again. |

(*Continued*)

**TABLE 17.4 (Continued)**

| | |
|---|---|
| | • The exposed dentin is sensitive.<br>• The patient is unable to masticate effectively. |
| **Contraindications** | • Tooth structure is missing due to caries or fracture causing a lack of retention.<br>• Crown has a post, and root is cracked or broken.<br>• Portion of tooth of filling is still in crown and retention is poor.<br>• Extensive decay present.<br>• Exposed pulp from tooth structure fracture. |
| **Supplies** | |
| **Operators** | Minimum 1, ideally 2 operators |
| **Antiseptic** | • Sterile gloves |
| **Analgesia** | • 2% lidocaine with 1:100,000 epinephrine |
| **Materials** | • Crew medical restraint system<br>• Dental syringe<br>• 27-gauge needle (short for upper procedures, long for lower procedures)<br>• 2% lidocaine with 1:100,000 epinephrine<br>• Schematic to perform proper anesthetic delivery<br>• Carver<br>• Explorer/probe<br>• Temporary crown cement with mixing tips<br>• 2×2 cotton gauze pads (4)<br>• Tongue depressor (1)<br>• Cotton swab (1)<br>• Dental floss |

**Steps**

1. Anesthetize patient as described in local anesthetic procedure above.
2. Remove residual cement with carver from crown with dental carver. Cement may be difficult to see and remove on tooth-colored restorations. (look closely)
3. Remove residual cement from tooth structure with dental carver. Be careful not to disturb gingival tissue. If bleeding occurs, delay cementation until bleeding stops.
4. Check the fit of crown without cement. There should be retention present.
5. Remove cap from temporary cement and extrude the base/catalyst onto gauze before placing mixing tip.
6. Dry crown.
7. Dry and isolate tooth.
8. Squeeze enough temporary filling material into base of crown to fill it.
9. Place the crown onto the dry tooth and push it into place with positive rotational force.
10. Place cotton tip applicator over crown and have patient bite crown fully into place.
11. Have patient open, remove cotton tip applicator, and then bite together to verify crown is fully seated. If crown is not fully seated, remove crown immediately and proceed again with step 2.
12. If crown is fully seated, place cotton tipped applicator over crown, and have patient continue mild bite pressure for 5 minutes.
13. Once cement is set, gently clean cement around tooth with a dental carver and floss. Use gauze to retain cement particles.
14. When flossing the tooth, the floss must be pulled out the side and not back up through contact or the crown may be dislodged.
15. Stubborn interproximal cement can be removed with an explorer by chipping into small pieces and gently tease to the side of the tooth for removal.

**Postprocedural care**

Once procedure is completed,

- Wait for anesthetic to wear off completely before eating. Fluids are allowed.
- Monitor patient closely for any postop care needs.
- The patient should avoid using the crowned tooth when eating and should chew food on the opposite side of the mouth.
- If crown loosens again, consider stowing or recement following the same procedures.

(*Continued*)

**TABLE 17.4** (Continued)

| **Potential complications and solutions** | |
|---|---|
| **Complication** | **Solution** |
| Gross caries is present under crown | It is best to stow crown in these situations, so long as the patient can function without it. If the crown must be recemented, then all caries must be removed down to hard dentinal structure. If a pulp exposure occurs during the caries removal process, do not cement crown, and treat "exposed pulp" only. |
| A portion of tooth is inside the crown | It is best to stow crown in these situations, so long as the patient can function without it. If the crown must be recemented, then leave the retained filling or tooth structure in the crown and verify retention without cement present. If satisfactory, proceed with recementation of crown. |
| Crown dislodges during the cement removal process | Either the retention was originally not present or tooth/crown was wet. Either stow crown or proceed with recement crown procedures starting with step 2. |
| Recemented crown is too high and causing discomfort | Gently pry up crown around margin and either stow crown or go through recementation procedures and verify that there is not hidden cement retained on crown or tooth structure. |
| Recemented crown dislodges while eating or sleeping | Consider stowing crown or proceed with recement crown procedures carefully. Caution on the side of stowing since aspiration and breaking another tooth may occur. |

**TABLE 17.5** Care for a lost restoration.

| **Item** | **Mass** | **Volume** |
|---|---|---|
| Nitrile gloves | 6–12 g | 40 mL |
| Syringe (5–6 mL) | 20–30 g | 40–55 mL |
| 27 g short or long needle (24–31 mm) | 0.5–1 g | 0.5–1 mL |
| 2% lidocaine/1:100,000 epi | 2–6 g | 7–9 mL |
| Schematic anesthetic delivery (laminated card) | 1–15 g | 5–15 mL |
| Carver | 15–25 g | 10–15 mL |
| Spoon excavator | 12–20 g | 8–14 mL |
| Explorer/probe | 12–22 g | 8–14 mL |
| Temporary filling material Cavit (7 g tube) | 10–12 g | 1012 mL |
| 2×2 gauze pads (×4) | 7–10 g | 20–30 mL |
| Wooden tongue depressor | 1.5–2 g | 4–5 mL |
| **Segment total** | **87–155 g** | **162.5–210 mL** |

| | |
|---|---|
| **Procedure** | **Temporary filling placement** |
| **Purpose** | To restore function and cover sensitive dentin. |

*(Continued)*

**TABLE 17.5 (Continued)**

| | |
|---|---|
| **Indications** | There should be sound tooth structure and works best if the missing portion<br>of tooth has some retentive areas. |
| **Contraindications** | • Exposed pulp<br>• Extensive decay present (must be removed first)<br>• Exposed pulp occurs during decay removal process<br>• Exposed pulp from tooth structure fracture |
| **Supplies** | |
| **Operators** | Minimum 1, ideally 2 operators |
| **Antiseptic** | • Sterile gloves |
| **Analgesia** | • 2% lidocaine with 1:100,000 epinephrine |
| **Materials** | • Crew medical restraint system<br>• Dental syringe<br>• 27-gauge needle (short for upper teeth, long for lower procedures)<br>• 2% lidocaine with 1:100,000 epinephrine<br>• Schematic to perform proper anesthetic delivery<br>• Carver<br>• Spoon excavator<br>• Explorer/probe<br>• Temporary filling material (Cavit)<br>• 2×2 cotton gauze pads (4)<br>• Tongue depressor (1)<br>• Dental floss |
| **Steps** | |
| **1.** Anesthetize patient as described in local anesthetic procedure above.<br>**2.** Check tooth to verify remaining structure is intact.<br>**3.** If decay is present, use spoon excavator to remove caries.<br>**4.** Check dentinal area to verify if exposed pulp is present. Open hole in dentin.<br>  **a.** Bleeding will be present if it is a vital tooth.<br>  **b.** No bleeding and an empty deep hole will be present if it is a nonvital tooth.<br>  **c.** If exposed pulp, do not place filling and treat with exposed pulp procedure.<br>**5.** If no exposed pulp present, dry tooth with gauze or blow dry with large empty syringe.<br>**6.** Expel enough filling material onto gloved hand to form a 1/4″ ball.<br>**7.** Place into open cavity<br>**8.** Press into walls of cavity with spoon excavator or carver file until cavity is full.<br>**9.** Wipe off excess with gauze.<br>**10.** Check to make sure filling is not high and filling material is not pushed into gingival tissue.<br>**11.** Clean up excess with explorer, carver, and dental floss. When using floss, push through contact then pull through side of contact. Do not lift up through contact or it may dislodge the filling. | |
| **Postprocedural care** | |
| Once procedure is completed,<br>• Wait for anesthetic to wear off completely before eating. Fluids are allowed.<br>• Monitor patient closely for any postop care needs.<br>• The patient should avoid the tooth when eating and chew all food on opposite side of mouth.<br>• If filling comes out, it can be replaced by following procedures above. | |
| **Potential complications and solutions** | |
| **Complication** | **Solution** |
| Gross caries is present. | Caries should be removed down to solid dentin prior to filling tooth. |

(Continued)

| TABLE 17.5 (Continued) | |
|---|---|
| Pulp exposure occurs during filling removal process. | Do not place filling. Go to exposed pulp section. |
| Small portion of temporary filling dislodges while eating. | Can either leave out if it does not affect the patients eating or daily functions.<br>Can usually add additional filling material without anesthetizing tooth. |
| Temporary filling material too high and causing discomfort. | Reduce filling material with carver. |
| Gum tissue irritated where filling was placed. | Check gingival tissue for impaction of temporary filling material and remove carefully. |
| Patient face starts to swell and has pain from tooth after a few days. | May have covered an exposed pulp and/or tooth is infected. Start antibiotics.<br>May need to remove filling and check for exposed pulp. |

**TABLE 17.6** Treatment of an exposed pulp.

| Item | Mass | Volume |
|---|---|---|
| Nitrile gloves | 6–12 g | 40 mL |
| Syringe (5–6 mL) | 20–30 g | 40–55 mL |
| 27 g short or long needle (24–31 mm) | 0.5–1 g | 0.5–1 mL |
| 2% lidocaine/1:100,000 epi | 2–6 g | 7–9 mL |
| Schematic anesthetic delivery (laminated card) | 1–15 g | 5–15 mL |
| Explorer/probe | 12–22 g | 8–14 mL |
| Red Cross toothache kit (eugenol/sesame oil, tweezers, cotton pellets) | 20–35 g | 30–45 mL |
| 2×2 gauze pads (×4) | 7–10 g | 20–30 mL |
| Goggles | 60–90 g | 150–250 mL |
| **Segment total** | **177.5–221 g** | **175.5–246 mL** |

| | |
|---|---|
| **Procedure** | **Treatment of exposed pulp** |
| **Purpose** | To prevent swelling and pain (short term). |
| **Indications** | • To temporarily stabilize a patient that has experienced a pulpal exposure either through facial trauma or the fracturing of teeth.<br>• It is important to note that pulp exposures can be present without bleeding from the tooth, and these must be treated the same as if blood was present.<br>• Warning: This is a temporary method only (2–4-week period) and is used only to stabilize a patient until they can be seen by a dental professional.<br>• The tooth will eventually necrose and finally become infected. |

(Continued)

**TABLE 17.6 (Continued)**

| | |
|---|---|
| **Contraindications** | Tooth is infected and is pushed up above the occlusal table. |
| **Supplies** | |
| **Operators** | Minimum 1, ideally 2 operators |
| **Antiseptic** | • Sterile gloves |
| **Analgesia** | • 2% lidocaine with 1:100,000 epinephrine |
| **Materials** | • Crew medical restraint system<br>• Dental syringe<br>• 27-gauge needle (short for upper procedures, long for lower procedures)<br>• 2% lidocaine with 1:100,000 epinephrine<br>• Schematic to perform proper anesthetic delivery<br>• Explorer/probe<br>• Red Cross Medication Kit<br>  ◦ Eugenol/Sesame oil<br>  ◦ Tweezers<br>  ◦ Cotton pellets<br>• 2×2 cotton gauze pads (4)<br>• Goggles |
| **Steps** | |
| 1. Anesthetize patient as described in local anesthetic procedure above.<br>2. Don gloves and protective goggles<br>3. If exposed pulp is a vital pulp due to trauma, use spoon excavator to remove as much tissue as accessible.<br>4. Isolate exposed pulp area with rolled cotton gauze.<br>5. Place one cotton pellet in tweezers.<br>6. Carefully open eugenol/sesame oil mixture being careful to keep mixture retained in bottle. Warning: solution is an eye irritant. Handle with caution.<br>7. Dip cotton pellet on tweezers into solution and place into a cotton gauze to blot dry.<br>8. Place the cotton pellet next to pulp opening and use probe to push a portion of the cotton into the pulp chamber. Leave cotton pellet in pulp chamber as a vent.<br>9. If patient were to lose the cotton pellet, it may be replaced usually without the need for anesthetic. | |
| **Postprocedural care** | |
| Once procedure is completed,<br>• Wait for anesthetic to wear off completely before eating. Fluids are allowed.<br>• Monitor patient closely for any postop care needs.<br>• Prepare for an early return event.<br>• Tooth will become necrotic and become infected. | |
| **Potential complications and solutions** | |
| **Complication** | **Solution** |
| Pain | Treat with NSAIDS |
| Cotton pellet comes out while eating | Proceed to replace cotton pellet following the instructions above. May not require anesthetic when replacing cotton pellet. |
| Pain with swelling | Start patient on antibiotics ASAP. Schedule an early return. |
| Patient experiences a white burning irritation on cheek next to cotton pellet | Cotton pellet was not blotted well before placement. Blot cotton pellet well before placing into pulpal exposure. |
| Solution gets out of bottle during procedure. | Bottle was not closed immediately after wetting cotton pellet. Stop procedure. Cap bottle and use goggles, gloves, and 4×4 gauze to capture any loose Red Cross fluid and place in a<br>sealed bag and dispose of in the appropriate manner. |

**TABLE 17.7 Incision and draining of dental abscess.**

| Item | Mass | Volume |
|---|---|---|
| Nitrile gloves | 6–12 g | 40 mL |
| Syringe (5–6 mL) | 20–30 g | 40–55 mL |
| 27 g short or long needle (24–31 mm) | 0.5–1 g | 0.5–1 mL |
| 2% lidocaine/1:100,000 epi | 2–6 g | 7–9 mL |
| Schematic anesthetic delivery (laminated card) | 1–15 g | 5–15 mL |
| Sterile scalpel with a #15 blade (disposable) | 15–25 g | 15–25 mL |
| Large syringe (20 mL) with 18–20 g needle | 16–22 g | 40–60 mL |
| Saline solution | Unspecified qty | Unspecified qty |
| 2×2 gauze pads (×8) | 14–20 g | 40–60 mL |
| **Segment total** | **151–212 g** | **175–245 mL** |

| | |
|---|---|
| **Procedure** | **Incision and drainage of an abscess** |
| **Purpose** | To temporarily stabilize an oral infection. |
| **Indications** | • Infection of a tooth with swelling of the soft tissues.<br>• Reduce pain and swelling of the oral structures.<br>• Fever or systemic symptoms associated with spreading infection. |
| **Contraindications** | • No exudate present upon aspiration with syringe, indicating cellulitis.<br>• Dental infection without soft tissue involvement.<br>• Cellulitis. |
| **Supplies** | |
| **Operators** | Minimum 1, ideally 2 operators |
| **Antiseptic** | • Sterile gloves |
| **Analgesia** | • 2% lidocaine with 1:100,000 epinephrine |
| **Materials** | • Crew medical restraint system<br>• Dental syringe<br>• 27-gauge needle (short for upper procedures, long for lower procedures)<br>• 2% lidocaine with 1:100,000 epinephrine<br>• Schematic to perform proper anesthetic delivery<br>• Sterile scalpel with #15 blade.<br>• Large empty syringe with 18–20-gauge needle.<br>• Saline solution<br>• 2×2 cotton gauze pads (8) |
| **Steps** | |

1. Stabilize patient using harness.
2. Administer local anesthetic according to steps in local anesthetic delivery.
3. Once tissue is fully anesthetized.
   a. Pick up large syringe with 18–20-gauge needle.
   b. Puncture the abscess with needle and gently draw back on plunger to check for aspirant.
   c. If there is aspiration of exudate (pus and/or pus and blood), incision and drainage can proceed. If there is no aspiration, reposition needle and try again. Once aspiration is positive, incise and drain that area.
   d. If there is no aspiration, there may be cellulitis only (do not incise and drain).

*(Continued)*

**TABLE 17.7 (Continued)**

| | |
|---|---|
| **4.** With sterile scalpel, incise tissue where exudate was aspirated.<br>**5.** If incision easily accessed, remove 18–20-gauge needle and use syringe to aspirate. If it is hard to access area, carefully use syringe with needle, or milk exudate with gloved finger into gauze.<br>**6.** To allow continuous draining of wound, it may be beneficial to leave a small portion of gauze in wound to prevent closure. | |
| **Postprocedural care** | |
| Once procedure is completed,<br>• Administer appropriate antibiotics.<br>• Follow patient closely.<br>• Schedule an early return event or consider extraction of infected tooth if early return is unavailable. | |
| **Potential complications and solutions** | |
| **Complication** | **Solution** |
| No aspirant present when aspirating syringe | Cellulitis. Do not incise and drain. Consider early return event or extraction of infected tooth. |
| Continuous pain | Treat with appropriate antibiotics and NSAIDS. Schedule early return event, if possible.<br>May need to consider extraction as a last resort. |
| Swelling under tongue area | Antibiotics may not be sufficient due to anaerobes and facultative bacteria. Consider switching antibiotics or adding Flagyl. Extraction and/or early return event may be required. |
| Fluctuant facial swelling near infraorbital foramen remains after incising and drain | Caution patient to not put pressure over swelling. Antibiotics may not be sufficient due to anaerobes and facultative bacteria. Consider switching antibiotics or adding Flagyl. Extraction and/or early return event may be required. |

## 17.5.4 Necrosis and infections

The following procedures are all related to dealing with necrosis and infections of dental origin. Necrosis and infections of teeth cannot be treated with antibiotics alone. There must be some form of definitive care such as extraction or root canal to remove the infection. The reason for this is that a necrotic or dead tooth is devoid of a blood supply and thus harbors the bacteria in the pulp chamber. The bacteria will continue to multiply and extrude though the open apex, forming active colonies in the alveolar bone below. As the colonies grow, they can elevate a tooth above the occlusal plane, as well as erode through the cortical plate and spread to vital areas in the head and neck region of the body. The cases listed below are likely to result in an early return event. Currently, the only method to avoid an early return event is an extraction. The decision to extract must be made with extreme caution since the risk is high for fracture of root tips, jaw fracture, and several other complications in the hands of an inexperienced provider. However, for exploration class missions without the option for an early return event, this may be the only and best option available. If the option for an early return event is available for the following procedures, it should be a priority in the planning process.

### *17.5.4.1 Dental extraction*

A dental extraction in spaceflight is reserved as a last resort. The risk of root fracture or other complications is moderately high in inexperienced hands, especially while working in a microgravity environment. The technique described below is one of the safest methods and uses the natural anatomy of the tooth to aid in the extraction process. The key is to slowly stretch the periodontal ligaments that retain the tooth in bone, with the understanding that if excessive force is applied, a tooth can break since the bone is dense, especially around lower molars. This procedure can serve as a definitive form of treatment if done carefully. However, there are unforeseen situations like curved roots, bulbous root, and multiple rooted teeth in dense bone that can make extraction especially difficult. It must be understood that antibiotics for dental-related infections are only a temporary measure, and time is of the essence. If the option for early return is available, it should be scheduled the moment that the diagnosis of dental infection is determined. If early return is not feasible, then extraction may be the best option for the safety of the patient (Table 17.8).

**TABLE 17.8** Dental extraction.

| Item | Mass | Volume |
|---|---|---|
| Nitrile gloves | 6–12 g | 40 mL |
| Syringe (5–6 mL) | 20–30 g | 40–55 mL |
| 27 g short or long needle (24–31 mm) | 0.5–1 g | 0.5–1 mL |
| 2% lidocaine/1:100,000 epinephrine | 2–6 g | 7–9 mL |
| Schematic anesthetic delivery (laminated card) | 1–15 g | 5–15 mL |
| Curved tip irrigation syringe, empty (×1) | 8–15 g | 15–25 mL |
| Curved tip irrigation syringe w/ saline 12 mL (×1) | 20–27 g | 27–37 mL |
| Explorer/probe | 12–22 g | 8–14 mL |
| Saline solution | Unspecified qty | Unspecified qty |
| 2×2 gauze pads (×8) | 14–20 g | 40–60 mL |
| Large elevator (#34) | 25–40 g | 15–25 mL |
| Small elevator (#301) | 20–30 g | 12–18 mL |
| Extraction forceps (region specific) | 120–160 g | 50–70 mL |
| **Segment total** | **248.5–378 g** | **259.5–369 mL** |

| **Procedure** | **Dental extraction** |
|---|---|
| **Purpose** | To remove the source of an oral infection. |
| **Indications** | • Infection of a tooth with swelling of the soft tissues<br>• Tooth raised above occlusal plane, thereby affecting function, mastication, and sleep.<br>• Fever or systemic symptoms associated with spreading infection |
| **Contraindications** | • End of mission within 2–3 weeks (antibiotics may control swelling until then)<br>• Early return determined to be best option.<br>• Lower molars or heavily curved roots. |
| **Supplies** | |
| **Operators** | Minimum 1, ideally 2 operators |
| **Antiseptic** | • Sterile gloves |
| **Analgesia** | • 2% lidocaine with 1:100,000 epinephrine |
| **Materials** | • Crew medical restraint system<br>• Dental syringe<br>• 27-gauge needle (short for upper procedures, long for lower procedures)<br>• 2% lidocaine with 1:100,000 epinephrine<br>• Schematic to perform proper anesthetic delivery<br>• Two curved tip syringes<br>  ◦ One empty<br>  ◦ One with sterile saline<br>• Saline solution<br>• 2×2 cotton gauze pads (8)<br>• Large elevator (34)<br>• Small elevator (301)<br>• Appropriate forceps<br>• Explorer/probe |

(Continued)

**TABLE 17.8 (Continued)**

**Steps**

1. Stabilize patient using harness.
2. Administer local anesthetic according to steps in local anesthetic delivery.
   a. For mandibular molar extractions, the buccal tissue must be anesthetized as well.
   b. For Maxillary extractions, the palatal tissue must be anesthetized too.
3. Once tissue is fully anesthetized.
4. Place explorer into the gingival crevice of the tooth to be extracted with the tip angled toward the root and contact bone.
   a. Pull the explorer slowly and deliberately through the crevice to separate the attached tissue from the tooth.
   b. Continue this through the interproximal area of the tooth as well.
5. With the appropriate forceps (upper molar #10S, lower molar #17, all other teeth #151A), place the forceps over the tooth, and with the thumb or finger of the nondominant hand, make sure the forceps are seated on the tooth as deep into the socket as possible.
6. For premolars and all other anterior teeth: start with very gentle and very small rotational forces. Do not use excessive force! Note: a tooth has a natural conical taper. This taper creates a natural force for the tooth to erupt vertically, and by using the very gentle rotational forces, you will be using this taper to your advantage. Too much force and there is a risk of breaking the tooth. Be patient! Perform this for a period of 1–2 minutes. If the tooth is starting to feel as if it is loosening, continue with the gentle rotations. If your hand becomes tired, stop and rest. During this time, inflammation will set in around the root and aid in the release of the periodontal membrane around the tooth. Begin again with gentle rotations, continue this until the tooth slowly elevates and releases from the socket. For molars: the addition of side-to-side forces will need to be used as well since there are multiple roots involved and rotational forces will be limited. The key is to be patient and use gentle forces. When using the side-to-side forces, it is best to hold the lateral force for a period of 30 seconds to 1 minute in each direction. When performing lower molar extractions, hold the lower jaw with the nondominant hand to prevent stress on the temporomandibular joints. If the tooth does not feel as if there is loosening after 5–10 minutes of gentle rotations, use the elevators as described below.
7. Elevator use. Start with the small elevator (#301). The "C" shaped end of the elevator is placed interproximal between two teeth with bottom edge of the "C" engaging the tooth to be extracted. The elevator is rotated so that the bottom edge of the elevator produces a lifting force on the tooth to be extracted. The force should be mild to moderate and held for a period of 30 seconds to a minute. Release the force and then move to the interproximal surface on the other side of the tooth to be extracted and repeat. Keep alternating until the tooth starts to raise up, or the gap between the teeth becomes too great for the small elevator. If the space becomes too large for the small elevator, switch to the large elevator (#34) and continue with the mild to gentle force. Alternate between the rotational and side-to-side forces for molar until the tooth is extracted.

**Postprocedural care**

Once procedure is completed,
- Administer appropriate antibiotics, if determined to be necessary.
- Follow patient closely

**Potential complications and solutions**

| Complication | Solution |
|---|---|
| Pain | Treat with NSAIDS |
| Continuous pain with open socket with foul odor or taste. | Dry socket. Anesthetize area. Clean any food debris out of socket with syringe and sterile saline, and/or schedule an early return event. |
| Swelling under tongue area | Antibiotics and early return. |
| Fluctuant facial swelling near infraorbital foramen remains after incising and drain | Caution the patient to not put pressure over swollen area. Antibiotics and early return. |
| Jaw breakage | Soft food diet, NSAIDS, early return. |
| Broken root tip in extraction site | Monitor patient carefully. Antibiotics and NSAIDS as needed. Schedule for early return. |
| Can't close mouth due to dislocated jaw | Stabilize patient and so as to be facing each other. With gloved hands, place the thumbs of your right and left hands on the buccal aspect of the gum tissue next to the molars. (Do not place thumbs over teeth.) Gently push down at the molar area to release temporomandibular joints. If unsuccessful, the patient may require a muscle relaxer. Wait the allotted time to take effect and attempt again. |
| Extraction site continues to bleed | Bite on folded 2×2 gauze for 20 minutes. If bleeding continues, continue biting another 2×2 gauze for 20 minutes. If hemostatic cellulose is available, it can be placed into socket to prevent food impaction. |

## Acknowledgments

The authors would like to thank Dr. Kris Lehnhardt for his input and guidance in drafting this chapter.

## References

Bean Quincy, A., Cooper, K.G., Edmundson, J.E., Johnston, M.M., Wekheiser, M.J., 2015. NASA Technical Report Server 20150016234 NASA Headquarters, Department of the Army, Department of the Air Force, Department of the Navy Spacecraft Propulsion Meeting Nashville, TN, Marshal Spaceflight Center International Space Station (ISS) 3D printer performance and material characterization methodology.

Gontcharov, I.B., Kovachevich, I.V., Pool, S.L., 2005. In-Flight medical incidents in the NASA-MIR program. Aviation, Space, and Environmental Medicine 76 (7), 692–696.

Harika, P.D., Mehta, K.H., Pulluri, S.S., Rana, P., Rajani, H., Aiman, A., 2023. Oral health in zero gravity: a comprehensive review of orofacial effects and countermeasures in spaceflights. Cureus 15 (11). Available at https://doi.org/10.7759/cureus.49035.

Hodapp, M.H., Jeske, A.H., 2020. Dental considerations. Principles of Clinical Medicine for Space Flight. Springer New York, United States, pp. 861–884. Available from: https://doi.org/10.1007/978-1-4939-9889-0.

Wang, J., Ahani, A., Pogrel, M.A., 2005. A five-year retrospective study of odontogenic maxillofacial infections in a large urban public hospital. International Journal of Oral and Maxillofacial Surgery 34 (6), 646–649.

Johnston, R.S., Dietlein, L.F., Berry, C.A., James, F., Parker, J., West, V., 1975. Biomedical results of Apollo. Available from: https://ntrs.nasa.gov/citations/19760005580.

Kelly, M., 2017. Endurance: A Year in Space, A lifetime of Discovery. Alfred A Knopf. Viking Press ISBN 978-1-5347-3159-5.

Kreykes, A.J., Suresh, R., Levin, D., Hilmers, D.C., 2023. Selecting medical conditions relevant to exploration spaceflight to create the IMPACT 1.0 medical condition list. Aerospace Medicine and Human Performance 94 (7), 550–557. Available at https://doi.org/10.3357/AMHP.6199.2023. Available from http://ingentaconnect.com/content/asma/amhp;jsessionid=13tlejf2f0ojk.alice.

Luo, W., Zheng, K., Kuang, H., Li, Z., Wang, J., Mei, J., 2022. The potential of Articaine as new generation of local anesthesia in dental clinics: a review. Medicine 101 (48). Available at https://doi.org/10.1097/MD.0000000000032089.

Majid, O.W., 2025. Articaine buccal infiltration for mandibular first molars with irreversible pulpitis: is the evidence enough? Evidence-Based Dentistry 26 (2), 89–90. Available at https://doi.org/10.1038/s41432-025-01153-4.

Menon, A., 2012. Review of spaceflight dental emergencies. NASA/TM–2012-217368.

Myers, J., Boley, L., Foy, M., Goodenow, D., Griffin, D., Keenan, E. et al., 2015. Integrated medical model overview. Available from: https://ntrs.nasa.gov/citations/20150002713.

National Institute of Health Institute of Dental and Craniofacial Research Dental Caries (Tooth Decay) in adults (ages 20 to 64 years), 2025. Available from: https://www.nidcr.nih.gov/research/data-statistics/dental-caries/adults#:~:text=Nearly%2090%25%20of%20adults%20ages,between%20the%20two%20NHANES%20cycles.

Phero, J.A., Warren, V.T., Fisher, A.G., Reside, G., Phillips, C., White, R.P., 2017. Buffered 1% lidocaine with epinephrine can be as effective as 2% lidocaine with epinephrine for maxillary field block. Journal of Oral and Maxillofacial Surgery: Official Journal of the American Association of Oral and Maxillofacial Surgeons 75, 2071–2075.

Rai, B., Kaur, J., 2013. Periodontal status, salivary immunoglobulin, and microbial counts after short exposure to an isolated environment. Journal of Oral Science 55 (2), 139–143. Available at https://doi.org/10.2334/josnusd.55.139.

Rai, B., Kaur, J., Catalina, M., 2010. Bone mineral density, bone mineral content, gingival crevicular fluid (matrix metalloproteinases, cathepsin K, osteocalcin), and salivary and serum osteocalcin levels in human mandible and alveolar bone under conditions of simulated microgravity. Journal of Oral Science 52 (3), 385–390.

Rai, B., Kaur, J., Catalina, M., Anand, S.C., Jacobs, R., Teughels, W., 2011. Effect of simulated microgravity on salivary and serum oxidants, antioxidants, and periodontal status. Journal of Periodontology 82 (10), 1478–1482. Available at https://doi.org/10.1902/jop.2011.100711.

Sefic et al., 2026. Effect of microgravity on the feasibility and the accuracy of dental procedures. Available at https://doi.org/10.1038/s41526-025-00552-2.

Sun, H., Zhou, Q., Qiao, P., Zhu, D., Xin, B., Wu, B., et al., 2022. Short-term head-down bed rest microgravity simulation alters salivary microbiome in young healthy men. Frontiers in Microbiology 13. Available at https://doi.org/10.3389/fmicb.2022.1056637.

Section III

# Anesthesia and pain management

Chapter 18

# Pharmaceutical considerations

Michelle S. Hong[1] and Tina Bayuse[2]

[1]*McGovern Medical School, The University of Texas Health Science Center at Houston, Houston, TX, United States,* [2]*Pharmacy Operations, KBR Inc., NASA Johnson Space Center, League City, TX, United States*

## 18.1 Introduction

The 21st century marks the beginning of a new era in spaceflight characterized by increased accessibility to space due in part to a growth in commercial space as the industry shifts away from government-led space programs. With expanding accessibility, the need to transport, store, use, and safely discard pharmaceuticals becomes increasingly important. Medications have been flown into space since the beginning of US human spaceflight. These medical kits have evolved from minimal, emergency-only medications during the early Mercury missions (Berry, 1965; Link, 1965) to comprehensive, modular systems supporting long-duration missions (see Table 18.1). The evolution of pharmaceuticals used has varied based on mission duration and lessons learned from prior missions. In general, the complexity and capability of onboard medical care increased in parallel with increasing mission duration. Current International Space Station (ISS) medical systems support long-duration semi-autonomous medical care that spans the gamut of drug administration routes and include over 190 unique medications (NASA Office of the Chief Health and Medical Officer

**TABLE 18.1** Pharmaceutical capabilities by space program.

| Program | Mission duration and crew size | Approx. number of medications | Key innovations |
|---|---|---|---|
| Mercury (1958–1963) | <34 hours, 1 astronaut | ~4–7 | Focused on emergency response; developed administration techniques that considered in-suit drug administration |
| Gemini (1965–1966) | <14 days, 2 astronauts | ~10–12 | Expanded medication formulary to support longer missions and dual-crew operations; included an antibiotic |
| Apollo (1965– 1972) | <13 days, 3 astronauts | ~25–30 + LM kit | Customized medication selection based on mission history and risk assessments; supported dual-vehicle operations with a separate lunar module (LM) kit |
| Skylab (1973– 1974) | Up to 84 days, 3 astronauts | ~60+ | Shift to sustained pharmacological care for extended orbital stays; added dental kit |
| STS (1981– 2011) | 2–17 days, 5–7 astronauts | ~100–120 | Adopted modular medication packs; added redundancy; Crew Medical Officer assigned to each mission |
| NASA-Mir (1994–1998) | Up to 437 days, 3–6 astronauts | ~120–150 | Joint US–Russian formulary and resupply capability |
| ISS (1993–est. 2030) | 6+ months, 7+ astronauts | >190 | Comprehensive formulary; included personal medication kits |

**Procedural Medicine for Spaceflight. DOI:** https://doi.org/10.1016/B978-0-443-27552-4.00029-4

(OCHMO), 2023). Looking ahead, exploration-class missions to the moon and beyond will require even greater medical self-sufficiency, extended drug stability, and potentially in-situ pharmaceutical production.

As space exploration and missions continue to push beyond Earth's orbit, the need for reliable and effective healthcare in space becomes increasingly critical. Pharmaceuticals play a vital role in maintaining the health and well-being of astronauts during spaceflight, but their physical and chemical composition are impacted by multiple facets of the spaceflight environment. Transportation, storage, and stability concerns are a major hurdle, as medications must withstand the extreme conditions of launch, microgravity, and potentially extended storage periods. Thermal cycling, radiation exposure, and mechanical stressors, such as vibration during launch, can all compromise pharmaceutical integrity as well. Beyond physical stability, the pharmacokinetics and pharmacodynamics of medications may be significantly altered due to changes in human physiology in the spaceflight environment, such as altered fluid distribution, cardiac output, hepatic blood flow, and renal function. These perturbations can impact drug absorption, distribution, metabolism, and elimination, potentially necessitating dosage adjustments for medications with narrow therapeutic indices. Operational considerations further complicate medication use in space, where mass and volume limitations restrict medication quantity and impact packaging solutions. Altered fluid dynamics in microgravity transform simple procedures like reconstituting powdered medications or managing injectable solutions into complex tasks requiring specialized techniques. Environmental concerns include the potential impact of pharmaceutical waste on closed-loop life support systems, drug fragmentation debris posing inhalation risks, or even packaging debris damaging elements of the spacecraft. Lastly, administering medications during launch and reentry, extravehicular activities (EVAs), or lunar surface exploration where astronauts are confined within spacesuits requires advanced operational planning. This chapter will delve into the considerations for the use of pharmaceuticals in spaceflight, providing an overview of the challenges and potential solutions.

## 18.2 Transportation, storage, and stability

**Practical tip**

Visually inspect medications prior to administration. Look for particulates, color changes, and any deviation of the expected consistency of the drug product, such as a gelatinous or frozen appearance when the product should be in solution.

One of the primary challenges in the use of pharmaceuticals in space is the transportation of medications to their intended destination. In developing mission-specific medical kits, a key decision is whether to prestage medications at the destination, such as aboard a spacecraft or planetary surface, or to transport them with the crew. Prestaging partially mitigates some mass/volume considerations, particularly when they are not needed during transit. However, this approach requires medications to be stored well in advance, which introduces challenges since pharmaceutical products in space exhibit a shorter shelf life than their terrestrial counterparts (Jaworske and Myers, 2016). In contrast, sending medications with the crew minimizes lead time and helps preserve drug potency by maximizing the usable shelf life but at the cost of additional mass/volume in crewed spacecraft.

Regardless of the transport strategy, careful consideration must be given to storage conditions, as environmental factors can significantly impact pharmaceutical stability. Variables such as temperature, humidity, and light exposure are known to influence drug stability (Veronica et al., 2022; Quality of Biotechnological Products, 1995; United States Pharmacopeia. (n.d.); World Health Organization WHO, 2018) in both terrestrial and space environments. In spaceflight, additional stressors, including vibration during launch and reentry, potential vacuum conditions, and exposure to ionizing radiation, pose unique risks to medication integrity (NASA Office of the Chief Health and Medical Officer (OCHMO), 2023; Mehta and Bhayani, 2017). Without appropriate handling and environmental controls, pharmaceutical degradation may occur, which can reduce therapeutic efficacy and increase the risk of harmful degradation byproducts. Table 18.2 outlines key characteristics commonly associated with drug degradation that crewmembers should be trained to assess prior to use.

### 18.2.1 Light

Exposure to light, both visible and ultraviolet (UV), poses another significant challenge to medication stability (Elste et al., 2023; Ioele et al., 2017; Perkins et al., 2020). UV light consists of high-energy wavelengths that are capable of breaking chemical bonds within drug molecules, leading to photodegradation, photolysis, or isomerization. This can result in discoloration, changes in odor, and reduced potency (Tønnesen, 2001; Janga et al., 2018). Visible light, while lower in

**TABLE 18.2** Common physical signs of drug degradation by dosage form.

| Dosage form | Common physical signs of drug degradation |
|---|---|
| Solid dosage forms (general) | Clumping, fog or moisture inside container, crystal deposition, odor change |
| Hard and soft gelatin capsules | Shell hardening or softening, distortion, gas release (e.g., distended seals) |
| Uncoated tablets | Crumbling, cracks, chips, swelling, mottling, discoloration, fusion, foreign crystals |
| Coated tablets | Cracks in coating, mottling, tackiness, clumping |
| Dry powders and granules | Caking, hard masses, discoloration |
| Reconstitutable powders/granules | Caked appearance, internal fog or droplets, objectionable odor |
| Effervescent products | Swelling, internal gas formation, premature reaction indicators |
| Solutions, elixirs, syrups | Precipitation, cloudiness, gas formation |
| Emulsions | Phase separation (breaking), not easily redispersible |
| Suspensions | Caking (not resuspendable), large particle growth |
| Tinctures/fluidextracts | Precipitation |
| Sterile liquids (including ophthalmic/parenteral) | Haze, discoloration, cloudiness, surface film, flocculent matter, gas, broken seal integrity |
| Creams | Emulsion breakage (phase separation), crystal growth, shrinkage, discoloration, microbial growth |
| Ointments | Consistency change, bleeding (liquid separation), granule formation |
| Suppositories | Softening, drying, hardening, shriveling, oil stains on packaging |

**Source:** From United States Pharmacopeia. (n.d.). Stability considerations in dispensing practice. In General Chapter. United States Pharmacopeia. https://doi.org/10.31003/USPNF_M99910_03_01.

**TABLE 18.3** Light-sensitive oral medications.

| | |
|---|---|
| Acetazolamide | Meclizine |
| Amoxicillin | Metronidazole |
| Aripiprazole | Promethazine |
| Doxycycline | Sulfamethoxazole/trimethoprim |
| Levofloxacin | Valacyclovir |

**Source:** Adapted from Perkins, S., Evans, A., King, A., 2020. Updated list of light-sensitive oral medications. Hospital Pharmacy. 55 (6), 349–365. https://doi.org/10.1177/0018578719844699.

energy, can also affect medication stability due to photosensitization, free radical damage, and color sensitivity. Photosensitization occurs when drugs absorb visible light and become energized, transferring this energy to oxygen molecules to create reactive oxygen species (ROS) that degrade the drug. Free radicals from ROS can lead to oxidative degradation, reducing the effectiveness of the drug (Tønnesen, 2001; Janga et al., 2018). Certain drugs are more vulnerable to specific wavelengths of visible light (i.e., color sensitive), which can also impact stability (Tønnesen, 2001).

To mitigate these effects, light-sensitive medications should be stored in amber packaging or other opaque materials such as aluminum foil (Perkins et al., 2020; Tønnesen, 2001; Janga et al., 2018). These protective measures help reduce light exposure, preserving the medication's efficacy and safety for use during space missions. A list of light-sensitive medications found in previously flown medical kits is shown in Table 18.3.

### 18.2.2 Humidity

**Pearl**

**Practical tip**

Avoid colocating medications near areas that may transiently contain more moisture such as near exercise equipment.

Excess humidity can promote microbial contamination and trigger chemical or physical changes in drug products. Relative humidity (RH) is an important metric in maintaining pharmaceutical stability. NASA's human spaceflight standards recommend a RH range of 25%–75% (NASA, 2025), which differs from international pharmaceutical guidelines that suggest maintaining RH at 60%–65% ± 5% (World Health Organization WHO, 2018; United States Pharmacopeia, General Chapter, 2023). The United States Pharmacopeia (USP), a nonprofit organization that publishes pharmaceutical standards, states that medications requiring dry storage should be maintained in a location that does not exceed 40% average RH at 20°C or the equivalent water vapor pressure at other temperatures (U.S. Pharmacopeial Convention, 2020).

Thoughtful selection of products can reduce the impact of humidity on drug stability. Aqueous-based products packaged in semi-permeable containers are particularly susceptible to water loss, which can impact their stability. For moisture-sensitive drugs, choosing solid dosage forms with excipients (i.e., inactive substances in a pharmaceutical) that are independent of RH can mitigate the influence of environmental conditions (Veronica et al., 2022, 2023).

### 18.2.3 Thermal

Appropriate packaging can effectively limit the effects of light and humidity on stability, but temperature constraints require more planning and coordination. While crewed vehicles are designed to maintain a stable and habitable environment, cargo vehicles may not have the same level of climate control, potentially exposing medications to a wider temperature range.

Temperature has a profound effect on the stability of medications. In space, the method of heat exposure, whether constant or cyclical, can significantly influence the degree and chemical composition of drug degradation (Brown et al., 2004). For instance, additives in medications, such as preservatives and antioxidants, can alter their stability under different thermal conditions, and thus temperature variability should be avoided. (see Table 18.4).

**Pearl**

**Practical tip**

Avoid colocating medications near potentially heat-generating equipment (Table 18.4).

**TABLE 18.4** Spacecraft cabin locations to avoid for thermal-sensitive medication storage.

| Cabin location or surface | Rationale |
|---|---|
| Near avionics racks or electronic panels | These areas dissipate heat during operation and can exceed ambient temperatures |
| Adjacent to walls facing the Sun (Sun-facing structure) | External surfaces exposed to solar radiation may conduct heat inward, warming internal storage areas |
| Behind panels or in stowage near thermal control ducts | Heat from life support or environmental control system ducts can cause localized hot spots |
| Near lighting elements or powered hardware | When on or in use, these items may increase the local ambient temperature |
| Unventilated stowage lockers | Poor airflow can trap heat and lead to temperature cycling or localized hot spots |
| Areas near air returns or filters | Temperature fluctuation possible due to air movement variability and proximity to hardware |

### 18.2.4 Radiation

While pharmaceuticals on Earth are rarely exposed to meaningful levels of ionizing radiation, the ambient radiation in low Earth orbit (LEO) is significantly higher than terrestrial background levels. Over long-duration space missions, chronic exposure to this elevated radiation environment has the potential to affect the chemical and physical stability of pharmaceuticals, raising concerns about loss of potency, degradation byproducts, and altered therapeutics.

Support for these concerns comes from terrestrial data obtained during radiation drug sterilization, where high-dose ionizing radiation (typically 25–50 kilogray) is used to eliminate microbial contamination in lieu of aseptic technique. These doses have been shown to degrade certain APIs and excipients, especially in aqueous or semi-solid formulations (Hasanain et al., 2014; Sarcan and Ozer, 2020). While solid oral dosage forms tend to be more resistant, degradation and alterations in drug performance have still been observed (Maquille et al., 2008).

Radiation in space differs not only in intensity but also in quality. This difference is typically described in terms of linear energy transfer (LET), which refers to the amount of energy that ionizing radiation deposits per unit length as it travels through matter. Pharmaceutical sterilization employs low-LET radiation such as gamma rays, X-rays, or electrons, while space radiation consists of a complex mixture of low-LET protons and electrons as well as high-LET radiation in the form of alpha particles and heavy ions. Daily doses in LEO only average ~0.3–0.5 milligray, but cumulative exposure over months or years, particularly outside Earth's protective magnetic field, raises concerns for radiodegradation of pharmaceutical products (Schimmerling, no date; Blue et al., 2019). Differences in radiation quality can lead to varied degradation pathways, suggesting that the high-LET radiation encountered in LEO may cause more complex damage compared to typical sterilization procedures (Du et al., 2011).

Several terrestrially based studies have simulated space radiation conditions to examine the potential for degradation of pharmaceuticals during long-duration missions. These studies showed that most solid oral medications, including acetaminophen, amoxicillin, and promethazine, maintained chemical stability and potency after exposure to simulated galactic cosmic rays and solar particle events (Bhayani et al., 2022; Plante et al., 2025). In contrast, liquid and moisture-sensitive formulations demonstrated greater susceptibility to radiation-induced degradation and were consistent with results from the pharmaceutical sterilization studies. This is likely due to the presence of water, which enhances free radical formation under radiation conditions and thereby accelerates degradation (Du et al., 2011). A separate study focusing on amlodipine besylate found various levels of degradation in both a pure API solid state and in consumer ready tablet forms postexposure, although tablets still fared better overall (Bhayani et al., 2019).

Another study evaluated the stability of multiple solid-state oral medications, including acetaminophen, amoxicillin, and promethazine, after exposure to simulated intravehicular space radiation over 3 years. The results showed no clinically significant degradation, and the majority of the APIs retained over 90% of their labeled potency (Plante et al., 2025). These findings align with earlier retrospective analyses at NASA, which concluded that radiation is not the primary driver of instability for most solid-state pharmaceuticals during space missions (Daniels et al., 2017).

Although spaceflight data is limited, available results largely support these same observations. In an analysis of pharmaceuticals stored aboard the ISS for 550 days, eight out of nine analyzed drug products met USP requirements without the generation of unusual byproducts 2–8 months past their respective expiration dates (Wotring, 2016). While the radiation environment in space poses a potential risk to drug stability, current evidence suggests that solid-state pharmaceuticals are generally robust under both simulated and real spaceflight conditions. Continued investigation is warranted, especially for liquid formulations and longer missions beyond LEO where cumulative radiation and high-LET exposures will increase. In general, proper formulation strategies and protective packaging can effectively mitigate radiation-induced instability in many pharmaceutical products (Du et al., 2011). Retrospective NASA analyses also point to storage conditions like temperature and humidity as more significant drivers of instability than radiation for most medications (Daniels et al., 2017).

### 18.2.5 Vibration

Mechanical stressors such as shaking and agitation can impact the physical and chemical stability of medications, particularly suspensions and protein-based formulations (Staniforth and Rees, 1982; Wang, 1999; Wang et al., 2007; Fleischman et al., 2017). In some cases, physical agitation may have beneficial effects, such as redistributing active ingredients uniformly in suspensions, which are often labeled with instructions to "shake well before use." However, mechanical stress can also lead to undesirable outcomes, including particle aggregation, protein denaturation, oxidation, and other degradative processes (Wang et al., 2007; Fleischman et al., 2017). While regulatory standards are well established for environmental conditions such as temperature control and humidity, the effects of mechanical vibration, especially during transport, are less clearly defined (Menzen, 2024; Theobald et al., 2023).

Studies on drone, helicopter, pneumatic tube, and rough-terrain transport have investigated the potential for vibrational stress to affect drug quality with varying results (Theobald et al., 2023; Oakey et al., 2021; Beck et al., 2020; De Winter et al., 2016; Coliat et al., 2023; Guo et al., 2016; Güngören et al., 2024). Ground-based analogs offer valuable insights into the type and magnitude of vibrational forces that pharmaceuticals may experience during launch and orbital operations, but it is important to note that the frequency of vibration and duration of exposure may differ significantly between different modes of transport, making direct comparisons difficult (Menzen, 2024; Oakey et al., 2021; De Winter et al., 2016). In fixed-wing and rotary-wing drone studies, amplitude and frequency spectra varied widely depending on airframe design, payload placement, and flight profile. Postflight insulin formulations have been shown to maintain acceptable turbidity, a measure of opacity of an aqueous pharmaceutical that is associated with proteins coming out of solution (Oakey et al., 2021). However, monoclonal antibodies exhibited increased aggregation and altered bioactivity after prolonged or repeated drone flights, especially without adequate packaging (Güngören et al., 2024). Similarly, pneumatic tube systems (which may be more comparable to rocket launch and landing profiles, albeit with higher peak G-forces; Güngören et al., 2024) have been shown to induce conformational changes in sensitive protein drugs, with increased turbidity and reduced functional activity observed in monoclonal antibody formulations (Coliat et al., 2023; Linkuvienė et al., 2022).

Space vehicles impose mechanical stress on pharmaceutical payloads during several mission phases (Pachiyappan et al., 2024). The most intense vibrational loads typically occur during launch and stage separation (Arenas and Margasahayam, 2006). Reboost maneuvers aboard the ISS, intended to maintain orbital altitude, can also impart mechanical vibration, though at a much lower magnitude than during launch. These perturbations may be sufficient to agitate loosely packed materials or contribute to the gradual degradation of sensitive drug products over time. While generally considered a low-risk contributor to pharmaceutical instability, cumulative exposure to vibrations could potentiate degradation pathways during extended storage (Menzen, 2024). Properly engineered packaging may mitigate some of these mechanical stresses (Waters et al., 2024) and therefore has the potential to serve as a countermeasure.

### 18.2.6 Reduced atmospheric pressure and vacuum

Exposure to reduced atmospheric pressure or vacuum can accelerate physical and chemical alterations in drug formulations. Early NASA programs deliberately flew at pressures lower than sea-level, and although this is no longer typical for spaceflight cabins, altered atmospheric pressures are still encountered at various stages of a mission such as during emergency depressurizing procedures to remove atmospheric contaminants, airlock usage, or on surface vehicles (Friello, 2024; Shennara, 2025).

Altered atmospheric pressure can cause physical changes to packaging (Friello, 2024), potentially compromising its integrity and ability to maintain pharmaceutical stability. Vacuum and low-pressure conditions have been shown to accelerate off-gassing (i.e., release of gases from solid materials) through sublimation or evaporation of volatile components (Stabenau et al., 2024; Back and Thomas, 1970; Fleischer, 2021), resulting in the potential for altered drug properties, exposure risk to crew, and an additional challenge for environmental control and life support systems (ECLSS). In some cases, packaging materials themselves may off-gas volatile compounds under low-pressure conditions, contributing to cabin air contamination. Pressurized drug delivery systems may also behave unpredictably under fluctuating atmospheric conditions. For example, inaccurate dosage delivery from insulin pumps has been reported due to changes in pressure associated with simulated and standard commercial flight profiles, highlighting the need for caution when using such systems in low-pressure environments (King et al., 2011; Class 2 Device Recall Paradigm, 2025). Furthermore, reduced barometric pressures may influence pharmacokinetic and pharmacodynamic properties, resulting in altered drug response (Hawkins et al., 1971).

Susceptibility to reduced-pressure environments may also be formulation-dependent. A longitudinal analysis of nineteen medications relevant to Artemis missions found minimal differences (<5%) in API potency between control and vacuum-exposed medication after eight hours of simulated spaceflight altitude equivalent to 120,000 feet (Shennara, 2025). However, certain medications fell outside standard reference ranges (95%–105%), warranting further investigation into specific formulations. The same study also showed that tablets and capsules removed from manufacturer packaging showed different responses compared to liquid medications that remained within manufacturer packaging. It is important to note that this study did not examine dissolution rates of oral medications. Changes to the way the medication leaves the matrix (i.e., the biological material, such as blood, in which the pharmaceutical is present) may be such that the medication doesn't get to the tissues regardless of how much API is intact. Prefilled syringes present their own unique challenges under reduced atmospheric pressure. Plunger-stopper displacement increases exponentially as atmospheric pressure decreases, following

Boyle's law of gas expansion. Air bubble size in prefilled syringes should be minimized when possible to mitigate sterility and container-closure integrity risks (Mehta et al., 2022).

**Key point**

**Formulation selection**

- **Solid over liquid dosage forms:** Solid formulations are favored over liquids due to a generally improved shelf life and reduced mass and volume.
- **Lyophilized preparations:** Lyophilization, a technique used in space food systems, can be applied to some pharmaceutical products to enable room-temperature storage and avert cold-chain logistics (Menzen, 2024). However, maintaining lyophilized products in cool, low-humidity conditions may further prolong their shelf life (Samyuktha et al., 2025)
- **Dual-use preference for ocular drops:** When designing multi-use medications, ocular drops are preferred over otic drops, as eye drops can safely be administered in the ear when needed, whereas the reverse is not recommended.

## 18.3 Mitigation strategies

### 18.3.1 Ground testing

Terrestrial testing of medications for spaceflight is a foundational component of pharmaceutical risk mitigation and mission readiness. Preflight evaluation focuses on drug stability, potency, and container-closure integrity when exposed to space-relevant stressors outlined in this chapter. Testing protocols can also assess physicochemical changes that may affect performance, such as altered dissolution, sedimentation, or phase separation, as well as compatibility with delivery systems (e.g., syringes, autoinjectors, infusion devices). Ground-based studies are also increasingly incorporating human analogs and modeling approaches to inform dosing and administration strategies to better characterize effects on pharmacokinetics that result from spaceflight-induced changes in human physiology.

Medication integrity in spaceflight can also be tested through stability studies comparing pharmaceuticals stored on spacecraft (e.g., the ISS) to ground controls. Analysis is typically performed via high-performance liquid chromatography and gas chromatography coupled with mass spectrometry, which quantify API content and identify degradation products or impurities after return to Earth against terrestrially age and lot-number-matched controls. Together, these investigations aim to minimize uncertainty and ensure that medications used in spaceflight remain safe and operationally reliable across the full mission lifecycle.

Lastly, ground testing of medications intended for astronaut use can be conducted to evaluate the safety and tolerability of individual crewmembers before in-orbit administration. The spaceflight environment can alter human physiology and drug response, introducing both clinical and operational risks. Supervised ground-based evaluations provide flight surgeons with critical data to reduce the likelihood of adverse or operationally consequential reactions during missions. For medications that affect the central nervous system, testing can include assessment of cognitive performance under mission-relevant conditions, such as abrupt awakenings, to ensure that crewmembers retain the capacity to respond appropriately to time-critical events.

## 18.4 Medication use in spacesuits

As the distance from Earth increases, the time in a spacesuit conducting EVAs or intravehicular activities also increases. Given that these activities are often long-duration, suit design should allow for medical treatment capabilities. A known challenge in provision is that inactive ingredients (i.e., excipients) of topical medications may affect the integrity of the suit and lead to suit failure. Excipients serve multiple purposes in topical medications, including enhancing drug solubility, ensuring stability, improving drug release, and providing desirable sensory properties. A common inactive ingredient in topical medications is benzyl alcohol, which may damage coatings in the helmet and weaken suit bladders, the airtight layer of the pressure garment system. Additionally, when cooling unit fibers are exposed to surfactants present in topical medications, they may lose their hydrophobicity, resulting in water leakage. This can be highly problematic in microgravity, especially if free liquid is present in large quantities in a suit or helmet. Lastly, medications that possess certain properties, such as highly flammable compounds like hydrocarbons, or those with high off-gassing such as siloxanes, carry additional risk when used in-suit.

Unfortunately, there are no studies to help identify problem compounds or the volume of a compound that reaches a risk threshold. The current mitigation strategy in spaceflight operations is to distance the time of application of the medication from the donning of the suit by eight hours. However, this strategy may not be feasible in exploration-class missions. Complicating the issue is that a medication with different formulations or different manufacturers of the same API may have different excipients, or inactive contents.

### 18.4.1 Biomanufacturing of pharmaceuticals

In the field of pharmacy, biomanufacturing refers to the production of pharmaceuticals using living systems such as microorganisms, plants, or animal cells. The concept of producing medications in situ during a mission is not new. McNulty et al. (2021a) introduced the term molecular pharming, describing how terrestrial successes in this area could translate into the potential for plants grown in space to provide pharmaceuticals for space travelers. However, this approach presents several challenges. Ensuring consistent drug content across production batches, verifying appropriate dosing through reliable testing, and addressing regulatory considerations related to biomanufacturing all remain areas requiring further development (McNulty et al., 2021b).

## 18.5 Formulation specifics

### 18.5.1 Eye drops in microgravity

The administration of ocular medications in space requires modifications to standard terrestrial techniques due to altered fluid dynamics in microgravity. On Earth, gravity determines droplet size, facilitates drainage, and affects drug absorption by influencing contact time with the ocular surface. In microgravity, these effects are absent, and surface tension becomes the dominant force shaping droplet behavior (Langbein, 1990; Memon et al., 2024; Lee et al., 2025). As a result, droplets do not fall naturally and can remain suspended at the dropper tip, requiring deliberate contact with the eye for instillation. Additionally, without gravity to regulate drop detachment, excessive fluid can accumulate at the tip, leading to overmedication if not properly controlled (Memon et al., 2024; Van Santvliet and Ludwig, 2004). These changes necessitate careful technique to ensure accurate dosing and prevent contamination.

#### *18.5.1.1 Contact lens precautions*

Before administering any ocular medication, contact lenses should be removed unless the drops are specifically labeled for use with contact lenses or identified as rewetting drops. Most medicated eye drops, including those for infection, inflammation, or allergy, can interact with contact lens materials by either reducing drug efficacy or damaging the lenses. Similarly, vasoconstrictor drops and many lubricating drops for dry eye may leave deposits or cause clouding of the contact lens. After drop administration, wait at least 15 minutes before reinserting lenses to allow the medication to absorb and reduce the risk of chemical interaction with lens material.

#### *18.5.1.2 Recommended procedure for instilling eye drops in microgravity*

##### 18.5.1.2.1 Preparation

1. Inspect the medication for discoloration, cloudiness, particulate matter, or residue on the tip of the dropper. Confirm the expiration date. Discard any medication that appears compromised. Note: While single-use formulations are ideal for maintaining sterility, they are not always feasible due to mass and volume constraints.
2. Clean hands thoroughly and remove contact lenses if present.
3. Clean the tip of the dropper with antiseptic (typically a benzalkonium chloride wipe).
4. Assume a stable position and consider using foot restraints, as free-floating can make precise hand-eye coordination more difficult.

##### 18.5.1.2.2 Administration

1. Slowly squeeze to allow a single droplet to form at the tip. Note: Squeezing too slowly may result in abnormally large droplets due to prolonged surface tension, potentially delivering up to six times the intended volume (Memon et al., 2024).
2. Using one hand, gently pull the upper and lower eyelids apart to create a wider exposed area of the ocular surface. This can provide a larger, more stable target for the droplet to adhere to.

3. Have the patient look upward to expose the inferior conjunctival surface. This creates a pocket for the droplet to promote absorption (Bains et al., 2023) and minimizes the risk of abrading the central part of the cornea, which is critical for vision.
4. With the droplet suspended at the tip, carefully move the eye toward the drop, allowing it to make contact with the ocular surface without touching the dropper tip; or, if needed, lightly touch the droplet to the eye to transfer it by capillary action.

#### 18.5.1.2.3 Optimize drug absorption

1. Gently close your eyelid and keep it closed without blinking. Repeated blinking promotes drainage and reduces drug absorption. In contrast, gentle movement of the eye while keeping the eyelid closed may improve drug absorption (Hedaya, 2024).
2. Apply light pressure to the inner canthus (over the nasolacrimal duct) for 1–2 minutes to minimize drainage (Bains et al., 2023). Note: Drainage increases systemic drug absorption, potentially leading to unwanted side effects (Van Santvliet and Ludwig, 2004; Bains et al., 2023; Šklubalová and Zatloukal, 2005). This may be even more critical in microgravity, where altered tissue pressures and changes in ocular and circulatory physiology may further influence local drug kinetics (Memon et al., 2024; Lee et al., 2025; Kast et al., 2017).
3. With the eye still closed, gently blot away any excess fluid from the eyelids with gauze or a cloth.
4. Wait 3–5 minutes between administrations if instilling more than one drop or using multiple medications in the same eye to prevent dilution or washout.

**Key point**

**Why ear drops should not be used in the eye**

While many ophthalmic drops can be safely used in the ear, particularly antibiotic and glucocorticoid preparations, otic drops should never be used in the eye unless explicitly labeled for ophthalmic use. Most ear drops (otic preparations) are nonsterile, unless specifically intended for use in a perforated eardrum or before surgery. This is because sterility is not required for routine external application to the ear canal. In contrast, the eye is much more sensitive, and otic drops may contain preservatives, excipients, or have a pH/osmolarity that can lead to eye irritation, chemical injury, or infection (Hedaya, 2024).

## 18.5.2 Ear drops in microgravity

As with ocular drops, otic drops are bound by the same fluid dynamics of spaceflight and the loss of passive gravity increases difficulty in administration. Alternative strategies to ensure that the medication reaches and remains in the external auditory canal include using the tragal pumping technique alone or with the aid of an ear wick.

**Key point**

**Perforated eardrum and otic drops**

Exercise caution before administering otic drops in patients with a known or suspected tympanic membrane (TM) perforation. Many otic formulations can be ototoxic if they enter the middle ear. Only drops explicitly labeled as safe for use with perforated TMs should be used (typically sterile, preservative-free fluoroquinolone antibiotic drops).

### *18.5.2.1 Recommended procedure for instilling ear drops in microgravity*

#### 18.5.2.1.1 Preparation

1. Inspect the medication for discoloration, cloudiness, particulate matter, or residue on the tip of the dropper. Confirm the expiration date. Discard any medication that appears compromised. Pro tip: Warm the medication to body temperature to help minimize vestibular stimulation and discomfort during administration. This can be accomplished by holding the container in your hand for a few minutes prior to use.
2. Clean hands thoroughly.
3. Clean the tip of the bottle with an antiseptic (typically a benzalkonium chloride wipe).
4. Assume a stable position and consider using foot restraints, as free-floating can make precise hand-eye coordination more difficult.

#### 18.5.2.1.2 Administration

1. Straighten the ear canal by gently grasping the auricle and pulling the outer ear up and back.
2. While maintaining this position, slowly squeeze the bottle to allow a droplet to form at the tip. Note: Unlike with ocular administration, larger droplet sizes are generally not a concern, aside from good stewardship of resources, since typical otic dosages range from 4 to 5 drops, though this may vary by drug and indication.
3. Bring the ear toward the droplet or gently touch the drop to the entrance of the ear canal. Avoid contact between the dropper tip and skin to minimize microbial contamination.
4. Use tragal pumping by applying gentle inward pressure on the tragus 3–5 times to create a pressure change and draw the medication deeper into the canal.

#### 18.5.2.1.3 Ear wick (optional or for refractory cases)

*An ear wick acts like a dehydrated sponge and expands rapidly when moistened. It can be used to promote extended drug contact and potentially deeper drug penetration into the ear canal.*

1. If using an ear wick, complete steps 1 through 4 as above.
2. Unwrap and remove the ear wick from the packaging.
3. Straighten the ear canal by gently grasping the auricle and pulling the outer ear up and back.
4. While maintaining this position, gently insert a dry ear wick into the ear canal. The wick should sit securely in the external ear canal with the tip of the wick visually exposed to an outside observer. Do not force or bend the wick.
5. Prime the wick by instilling six drops of the otic medication onto the wick to allow it to fully expand and conform to the canal (Bola et al., 2017).
6. Wait until the wick softens and expands, ensuring good contact with the canal wall.
7. Now apply additional medication until the wick is fully saturated.
8. Use tragal pumping by applying gentle inward pressure on the tragus 3–5 times to create a pressure change and draw the medication deeper into the canal.
9. Keep the wick saturated by readministering the drops three to four times a day.

**Key point**

**Wick and string technique** (Badran and Jani, 2006)

To improve hygiene and further promote self-care in microgravity, an alternative method uses a string attached to an ear wick.

- Attach a suture or string by threading and tying it at the midpoint of the ear wick or a 14 cm ribbon gauze.
- Insert the wick into the ear canal while the string is taped to the skin in front of (or behind) the auricle.
- Prime and administer the medication as above.
- This eliminates the need to leave a wick tail exposed, reducing the risk of premature dislodgement while providing secure and hygienic access for later removal.

### 18.5.3 Intranasal medications

To date, there are no published studies evaluating the pharmacokinetics of intranasal medications in spaceflight despite the fact that topical nasal decongestants are often used (e.g., oxymetazoline). Concerns that are not well characterized include the vasoconstriction activity of oxymetazoline causing a change in blood flow rate or decreased absorption in the spaceflight environment, which would increase clearance of the medication by mucosal cilia from the site of action (Sardu and Poggesi, 2024).

Terrestrial studies of intranasal medications used with oxymetazoline have shown mixed results. One study in which oxymetazoline was administered one hour prior to esketamine had no effect on esketamine pharmacokinetics (Zannikos et al., 2023), but another small study showed treatment with oxymetazoline one hour prior decreased the rate, but not absolute dose absorbed of ketamine (Carr et al., 2009). A third study showed that oxymetazoline decreased both the rate and absolute dose absorption of intranasal fentanyl (Kaessner et al., 2012).

Due to the inconclusiveness of the current knowledge, further study is needed regarding intranasal medication pharmacokinetics in spaceflight as novel treatment therapies become standard clinical care. Dosing adjustments or changes to the frequency of doses may be warranted.

### 18.5.4 Use of parenteral products

Administering parenteral medications in microgravity presents some of the most formidable pharmaceutical challenges in spaceflight. On Earth, parenteral drug preparation is reliant on gravity to allow liquids to settle and bubbles to rise such that the medication can be drawn up from a vial and air bubbles expelled from the syringe. In space, these processes are far more complicated and labor-intensive. Liquids adhere to container surfaces, fail to pool predictably, and remain intermixed with air, making it extremely difficult to draw up accurate doses, prevent the inadvertent injection of air, and maintain aseptic technique (see Fig. 18.1).

Ready-to-use injectable medications in the form of prefilled syringes are a partial solution in some situations, particularly for drugs intended for time-critical emergencies where immediate availability is required. These preparations eliminate the need for reconstitution or dose transfer, significantly reducing the likelihood of contamination, dosing error, or procedural delay. The expulsion of air from the syringe prior to administration is still a concern and many drugs are not available in prefilled, ready-to-use formulations. Furthermore, prefilled syringes tend to be more expensive per unit, add packaging weight, and may have shorter shelf lives. The added expense of the medication itself may be counteracted by the need for less crew time and supplies to prepare the medication for use; however, all of these trade-offs need to be considered when developing a formulary for operational use in the space environment.

This strategy of prioritizing medications that are easier to use with less room for human error reflects a broader risk-mitigation approach. In the context of human spaceflight, where medical interventions must be executed by crewmembers with variable medical training and in suboptimal conditions, formulation design becomes a critical factor in ensuring proper drug use and crew safety. The reliance on prefilled syringes or other ready-made formulations, such as autoinjectors, is not merely a convenience but may be a necessity dictated by the physiological and operational constraints of space.

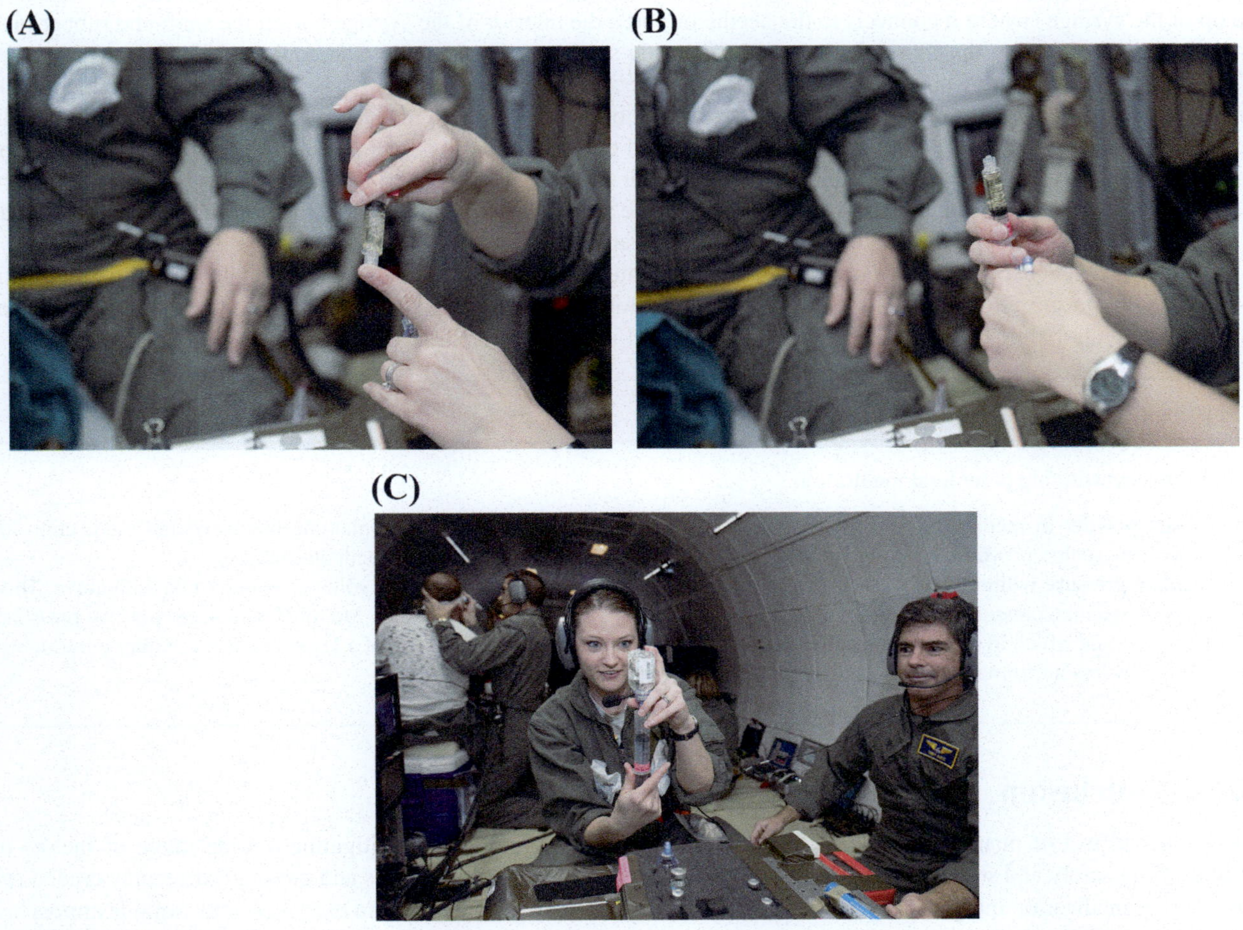

FIGURE 18.1 **Bubble studies on C9 microgravity flight.** Bubble formation in parenteral (injectable) medication vial and prepared syringe during parabolic flight, highlighting a potential operational risk in spaceflight care.
Courtesy Tina Bayuse.

In nonemergency settings where reconstitution or dose preparation is unavoidable (such as certain antibiotics, biologics, or products requiring lyophilized formulations for shelf stability), specially designed, sealed reconstitution systems or containment devices that limit free fluid handling may help address containment concerns and the ability to maintain aseptic conditions. However, these systems still face the same challenges with air–fluid separation and demand additional time and training to ensure safe use.

#### *18.5.4.1 Reconstituting medication*

The absence of gravitational settling in space introduces significant challenges to the safe and effective reconstitution of parenteral medications. Notably, separating air from liquid becomes much more difficult, increasing the risk of air embolism during administration of the prepared product. Incomplete mixing of drug and diluent is also a concern. The behavior of fluids in microgravity, compounded by adhesion forces of parenteral medications interacting within syringes and vials, further complicates drug preparation. While there is no single validated method to reconstitute and prepare parenteral medications in microgravity, several best practices adapted from terrestrial pharmaceutical practice can be applied. The goal is to minimize the risk of microbial contamination while increasing the confidence that the parenteral preparation contains the accurate dose and has minimal risk for introducing air bubbles into the bloodstream during administration. Of note, the human lung is a great natural air filter for smaller bubbles, but there is still a risk for air emboli, particularly in individuals with undiagnosed paradoxical shunts, such as a patent foramen ovale, that permit air bubbles to pass into body areas of greater clinical concern.

Cleanroom environments are not currently feasible in space, but aseptic technique and thorough hand hygiene should be performed prior to handling any materials involved in reconstitution. Surfaces such as vial stoppers and injection ports must be disinfected and allowed to dry completely before being accessed with a needle or spike. Care must be taken to avoid nonsterile contact with any sterile components of the injection system, including the needle bevel, shaft, and hub; the tip of the syringe (where it connects to the needle or port); the interior of the syringe barrel; the shaft and rubber piston of the plunger (excluding the finger flange); and the inside surfaces of any medication containers or fluid reservoirs.

Incomplete dissolution of powdered formulations is a recognized risk terrestrially; however, this risk is increased in microgravity, where visual confirmation of mixing can be more difficult due to the lack of sedimentation or fluid layering. Gentle swirling and leveraging centripetal force is preferred over vigorous shaking, as the latter may generate air bubbles that are hard to remove. To avoid generating additional air bubbles when preparing parenteral products, remember the adage "slow is smooth and smooth is fast." After mixing, the solution should be carefully inspected for visible particulates or evidence of incomplete reconstitution prior to administration.

Finally, when reconstituted drugs are not used immediately, syringes should be clearly labeled with the medication name, dose or concentration, and the time of preparation. Storage conditions must align with known stability profiles, and unnecessary delay before administration should be avoided unless validated data support extended use.

**Key point**

**Tips for reconstituting parenteral medication**

- **Loosen powder in medication vial.** Before adding the diluent, gently tap the vial containing the powdered medication to loosen any settled powder and powder that has adhered to the side of the vial. This facilitates mixing.
- **Equalize pressure within the vial.** Draw an amount of air into the syringe equal to the volume of fluid to be withdrawn. This helps to equalize pressure and makes withdrawing the diluent easier. However, adding too much air causes positive pressure within the vial which in turn can aerosolize contents that will enter the ECLSS. Instead of a 1:1 exchange, consider introducing 10%–20% less air to maintain a slightly negative pressure within the vial.

## 18.6 Conclusion

The successful use of pharmaceuticals in space depends on understanding and mitigating a wide range of factors that influence drug safety and stability in the extraterrestrial environment. Stability studies and ground testing play crucial roles in predicting medication integrity under conditions of radiation, vibration, temperature extremes, and variable atmospheric pressures. Beyond stability concerns, altered human physiology in microgravity introduces additional uncertainty in drug absorption, distribution, metabolism, and elimination, making dosage and route optimization a continuing area of investigation.

Operational limitations such as restricted storage capacity, waste management challenges, and design for crewmembers without pharmaceutical expertise further highlight the importance of practical pharmaceutical design. As exploration missions extend beyond LEO, innovative approaches such as in-situ biomanufacturing and molecular pharming, that is, the use of genetically modified plants to produce recombinant pharmaceutical proteins, may offer sustainable pathways for medication supply. However, significant regulatory, analytical, and consistency challenges remain.

Despite incremental progress, data on pharmaceutical performance, degradation, and clinical efficacy in space remain limited. Lack of published data requires that mission planners and other stakeholders consult with space pharmaceutical experts, whose specialized knowledge can help bridge the current evidence gap and guide future research priorities. Continued collaboration among pharmacists, space medicine clinicians, scientists, and engineers will be critical to ensure that future astronauts have access to safe, effective, and reliable medications.

## References


Arenas, J.P., Margasahayam, R.N., 2006. Noise and vibration of spacecraft structures. Ingeniare. Revista Chilena de Ingeniería 14 (3), 251–264.

Back, K.C., Thomas, A.A., 1970. Aerospace problems in pharmacology and toxicology. Annual Review of Pharmacology 10 (1), 395–412. Available at https://doi.org/10.1146/annurev.pa.10.040170.002143.

Badran, K., Jani, P., 2006. How we do it: a new "one stop" ear wick. Clinical Otolaryngology: Official Journal of ENT-UK and the Official Journal of the Netherlands Society for Oto-Rhino-Laryngology and Cervico-Facial Surgery 31 (2), 151–152. Available at https://doi.org/10.1111/j.1749-4486.2006.01128.x.

Bains, U.K., Syed, Z.A., Henderer, J.D., Rapuano, C.J., 2023. Ocular pharmacology. In: Brunton, L.L., Knollmann, B.C. (Eds.), Goodman & Gilman's: The Pharmacological Basis of Therapeutics, 14th ed. McGraw-Hill Education Accessed August 1, 2025. Available at accessmedicine.mhmedical.com/content.aspx?aid=1193243178.

Beck, S., Bui, T., Davies, A., et al., 2020. An evaluation of the drone delivery of adrenaline auto-injectors for anaphylaxis: pharmacists' perceptions, acceptance, and concerns. Drones 4 (4), 66. Available at https://doi.org/10.3390/drones4040066.

Berry, C., 1965. Pharmaceuticals in space medicine. Journal of the American Pharmaceutical Association 5 (7), 358–379. Available at https://doi.org/10.1016/S0003-0465(15)33906-9.

Bhayani, D., Mehta, P., Patel, M., Naik, H., Nathaniel, T.N., Khan, S., 2022. Ground-based selected ionizing space radiation effects on stability of APIs and their formulations. Journal of Pharmaceutical and Biomedical Analysis 220, 115019. Available at https://doi.org/10.1016/j.jpba.2022.115019.

Bhayani, D., Naik, H., Nathaniel, T.N., Khan, S., Mehta, P., 2019. Simulated space radiation: investigating ionizing radiation effects on the stability of amlodipine besylate API and tablets. European Journal of Pharmaceutical Sciences: Official Journal of the European Federation for Pharmaceutical Sciences 137, 104982. Available at https://doi.org/10.1016/j.ejps.2019.104982.

Blue, R.S., Chancellor, J.C., Antonsen, E.L., Bayuse, T.M., Daniels, V.R., Wotring, V.E., 2019. Limitations in predicting radiation-induced pharmaceutical instability during long-duration spaceflight. Npj Microgravity 5 (1), 15. Available at https://doi.org/10.1038/s41526-019-0076-1.

Bola, S., Rashid, M., Hickey, S., 2017. Optimising the use of otowicks in otitis externa. The Journal of Laryngology and Otology 131 (9), 809–812. Available at https://doi.org/10.1017/S002221511700144X.

Brown, L.H., Krumperman, K., Fullagar, C.J., 2004. Out-of-hospital medication storage temperatures. Prehospital Emergency Care: Official Journal of the National Association of EMS Physicians and the National Association of State EMS Directors 8 (2), 200–206. Available at https://doi.org/10.1080/312703004362.

Carr, D., Liao, E., Kramer, W., Brennen, L., Wright, C., 2009. The effect of intranasal oxymetazoline on the safety and pharmacokinetics of intranasal ketamine hydrochloride. The Journal of Pain 10 (4), S40. Available at https://doi.org/10.1016/j.jpain.2009.01.169.

Class 2 Device Recall Paradigm. US Food and Drug Administration, March 3, 2025. https://www.accessdata.fda.gov/scripts/cdrh/cfdocs/cfres/res.cfm?id=212321 (Accessed 28.07.25).

Coliat, P., Erb, S., Diemer, H., et al., 2023. Influence of pneumatic transportation on the stability of monoclonal antibodies. Scientific Reports 13 (1), 21875. Available at https://doi.org/10.1038/s41598-023-49235-6.

Daniels, V.R., Bayuse, T.M., Mcguire, K.M., Antonsen, E.L., Putcha, L., 2017. Radiation impact on pharmaceutical stability: retrosspective data review. https://ntrs.nasa.gov/api/citations/20170009940/downloads/20170009940.pdf

Du, B., Daniels, V.R., Vaksman, Z., Boyd, J.L., Crady, C., Putcha, L., 2011. Evaluation of physical and chemical changes in pharmaceuticals flown on space missions. The AAPS Journal 13 (2), 299–308. Available at https://doi.org/10.1208/s12248-011-9270-0.

Elste, J.M., Ipema, H.J., Denton, C., et al., 2023. Light-sensitive injectable prescription drugs—2022. Hospital Pharmacy 58 (5), 448–475. Available at https://doi.org/10.1177/00185787221133804.

Fleischer, D.J., 2021. Pharmaceutical stability after exposure to vacuum. IAF/IAA Space Life Sciences Symposium (A1) Medical Care for Humans in Space (3).

Fleischman, M.L., Chung, J., Paul, E.P., Lewus, R.A., 2017. Shipping-induced aggregation in therapeutic antibodies: utilization of a scale-down model to assess degradation in monoclonal antibodies. Journal of Pharmaceutical Sciences 106 (4), 994–1000. Available at https://doi.org/10.1016/j.xphs.2016.11.021.

Friello, P., May 8, 2024, Pilot study of medications exposed to vacuum. In: Presented at: Aerospace Medical Association 94th Annual Scientific Meeting, Chicago, IL. https://ntrs.nasa.gov/citations/20240004440

Guo, J., Lewis, L.M., Billones, H., Torres, E., Kolhe, P., 2016. The effect of shipping stresses on vaccine re-dispersion time. Journal of Pharmaceutical Sciences 105 (6), 2009–2013. Available at https://doi.org/10.1016/j.xphs.2016.03.023.

Güngören, M.H., Romeijn, S., Dijkstra, J.A., Crul, M., 2024. Investigating the impact of drone transport on the stability of monoclonal antibodies for inter-hospital transportation. Journal of Pharmaceutical Sciences 113 (7), 1816–1822. Available at https://doi.org/10.1016/j.xphs.2024.04.002.

Hasanain, F., Guenther, K., Mullett, W.M., Craven, E., 2014. Gamma sterilization of pharmaceuticals–a review of the irradiation of excipients, active pharmaceutical ingredients, and final drug product formulations. PDA Journal of Pharmaceutical Science and Technology / PDA 68 (2), 113–137. Available at https://doi.org/10.5731/pdajpst.2014.00955.

Hawkins, S.F., Hankinson, J.B., Merritt, J.H., Medina, M.A., 1971. Effect of reduced barometric pressure on drug action and metabolism in mice. Biochemical Pharmacology 20 (9), 2221–2222. Available at https://doi.org/10.1016/0006-2952(71)90221-8.

Hedaya, M.A., 2024. Chapter 19 - Routes of drug administration. In: Dash, A.K., Singh, S. (Eds.), *Pharmaceutics, second ed.*. Academic Press, pp. 537–554. Available at https://doi.org/10.1016/B978-0-323-99796-6.00006-0.

Ioele, G., Tavano, L., Luca, M.D., Muzzalupo, R., Mancuso, A., Ragno, G., 2017. Light-sensitive drugs in topical formulations: stability indicating methods and photostabilization strategies. Future Medicinal Chemistry 9 (15), 1795–1808. Available at https://doi.org/10.4155/fmc-2017-0105.

Janga, K.Y., King, T., Ji, N., et al., 2018. Photostability issues in pharmaceutical dosage forms and photostabilization. AAPS PharmSciTech 19 (1), 48–59. Available at https://doi.org/10.1208/s12249-017-0869-z.

Jaworske, D.A., Myers, J.G., 2016. Pharmaceuticals exposed to the space environment: problems and prospects. NASA. https://ntrs.nasa.gov/api/citations/20160003684/downloads/20160003684.pdf (accessed 29.07.25)..

Kaessner, N., Nave, R., Roepcke, S., Facius, A., Lahu, G., 2012. Population pharmacokinetic metaanalysis of intranasal fentanyl spray as a means to enrich pharmacokinetic information for patients with cancer breakthrough pain. Int. Journal of Clinical Pharmacology and Therapeutics. 50 (9), 665–677. Available at https://doi.org/10.5414/CP201737.

Kast, J., Yu, Y., Seubert, C.N., Wotring, V.E., Derendorf, H., 2017. Drugs in space: pharmacokinetics and pharmacodynamics in astronauts. European Journal of Pharmaceutical Sciences: Official Journal of the European Federation for Pharmaceutical Sciences 109, S2–S8. Available at https://doi.org/10.1016/j.ejps.2017.05.025.

King, B.R., Goss, P.W., Paterson, M.A., Crock, P.A., Anderson, D.G., 2011. Changes in altitude cause unintended insulin delivery from insulin pumps. Diabetes Care 34 (9), 1932–1933. Available at https://doi.org/10.2337/dc11-0139.

Langbein, D., 1990. Fluid statics and dynamics in microgravity. Journal of Physics. Condensed Matter: An Institute of Physics Journal 2 (S), SA491–SA498. Available at https://doi.org/10.1088/0953-8984/2/S/078.

Lee, R., Kumar, R., Shah, J., Ong, J., Waisberg, E., Tavakkoli, A., 2025. Tear film and keratitis in space: fluid dynamics and nanomedicine strategies for ocular protection in microgravity. Pharmaceutics 17 (7), 847. Available at https://doi.org/10.3390/pharmaceutics17070847.

Link, M.M., 1965. Space Medicine in Project Mercury. Scientific and Technical Information Division, National Aeronautics and Space Administration. Accessed July 26, 2025. Available at https://www.jstor.org/stable/3101962?origin=crossref vol. 4003.

Linkuvienė, V., Ross, E.L., Crawford, L., et al., 2022. Effects of transportation of IV bags containing protein formulations via hospital pneumatic tube system: particle characterization by multiple methods. Journal of Pharmaceutical Sciences 111 (4), 1024–1039. Available at https://doi.org/10.1016/j.xphs.2022.01.016.

Maquille, A., Jiwan, J.L.H., Tilquin, B., 2008. Radiosterilization of drugs in aqueous solutions may be achieved by the use of radioprotective excipients. International Journal of Pharmaceutics 349 (1-2), 74–82. Available at https://doi.org/10.1016/j.ijpharm.2007.07.024.

McNulty, M.J., Berliner, A.J., Negulescu, N.G., McKee, L., Hart, O., Yates, K., et al., 2021b. Evaluating the cost of pharmaceutical purification for a long-duration space exploration medical foundry. bioRxiv. Available at https://doi.org/10.1101/2021.06.21.449300, https://www.biorxiv.org.

McNulty, M.J., Xiong, Y., Yates, K., Karuppanan, K., Hilzinger, J.M., Berliner, A.J., et al., 2021a. Molecular pharming to support human life on the moon, mars, and beyond. Critical Reviews in Biotechnology 41 (6), 849–864. Available at https://doi.org/10.1080/07388551.2021.1888070, http://www.tandfonline.com/loi/ibty20.

Mehta, P., Bhayani, D., 2017. Impact of space environment on stability of medicines: challenges and prospects. Journal of Pharmaceutical and Biomedical Analysis 136, 111–119. Available at https://doi.org/10.1016/j.jpba.2016.12.040.

Mehta, S.B., Cook, J., Liu, W., Brisbane, C., 2022. Risk mitigation of plunger-stopper displacement under low atmospheric pressure by establishing design space for filling-stoppering process of prefilled syringes: a design of experiment (DoE) approach. Journal of Pharmaceutical Sciences 111 (7), 2038–2048. Available at https://doi.org/10.1016/j.xphs.2022.01.028.

Memon, H., Ong, J., Waisberg, E., et al., 2024. Biophysics of ophthalmic medications during spaceflight: principles of ocular fluid dynamics and pharmacokinetics in microgravity. Life Sciences and Space Research 42, 53–61. Available at https://doi.org/10.1016/j.lssr.2024.04.007.

Menzen, T., 2024. Is lyophilization key for transportation and storage of biopharmaceutical drug products in space? Journal of Pharmaceutical Sciences 113 (7), 1695–1700. Available at https://doi.org/10.1016/j.xphs.2024.04.030.

NASA spaceflight human-system standard volume 2: human factors, habitability, and environmental health. May 1, 2025.

NASA Office of the Chief Health and Medical Officer (OCHMO), November 21, 2023. Pharmaceuticals and medications: a technical brief. national aeronautics and space administration. Pharmaceutical Care. Available at https://www.nasa.gov/sites/default/files/atoms/files/pharmaceuticals_and_medications_tech_brief_ochmo.pdf

Oakey, A., Waters, T., Zhu, W., et al., 2021. Quantifying the effects of vibration on medicines in transit caused by fixed-wing and multi-copter drones. Drones 5 (1), 22. Available at https://doi.org/10.3390/drones5010022.

Pachiyappan, J.K., Patel, M., Roychowdhury, P., et al., 2024. A review of the physiological effects of microgravity and innovative formulation for space travelers. Journal of Pharmacokinetics and Pharmacodynamics 51 (6), 605–620. Available at https://doi.org/10.1007/s10928-024-09938-3.

Perkins, S., Evans, A., King, A., 2020. Updated list of light-sensitive oral medications. Hospital Pharmacy 55 (6), 349–365. Available at https://doi.org/10.1177/0018578719844699.

Plante, I., Daniels, V., Young, M., Gaza, R., Wu, H., Reichard, J.F., 2025. The long-term stability of solid-state oral pharmaceuticals exposed to simulated intravehicular space radiation. Npj Microgravity 11 (1), 17. Available at https://doi.org/10.1038/s41526-025-00469-w.

Quality of biotechnological products: stability testing of biotechnological/biological products Q5C, November 30, 1995. https://database.ich.org/sites/default/files/Q5C%20Guideline.pdf

Samyuktha, A.P., Ramesh, Y., Chandra, Y.P., Penabaka, V., 2025. A review of lyophilization. GSC Biological and Pharmaceutical Sciences 30 (2), 007–021. Available at https://doi.org/10.30574/gscbps.2025.30.2.0035.

Van Santvliet, L., Ludwig, A., 2004. Determinants of eye drop size. Survey of Ophthalmology 49 (2), 197–213. Available at https://doi.org/10.1016/j.survophthal.2003.12.009.

Sarcan, E.T., Ozer, A.Y., 2020. Ionizing radiation and its effects on pharmaceuticals. Journal of Radioanalytical and Nuclear Chemistry 323 (1), 1–11. Available at https://doi.org/10.1007/s10967-019-06954-3.

Sardu, M.L., Poggesi, I., 2024. Pharmacokinetics of intranasal drugs, still a missed opportunity? Xenobiotica; The Fate of Foreign Compounds in Biological Systems 54 (7), 424–438. Available at https://doi.org/10.1080/00498254.2024.2349046, http://www.tandfonline.com/loi/ixen20.

Schimmerling, W., n.d. An assessment of how radiation incurred during a Mars mission could affect food and pharmaceuticals.

Shennara, K., June 4, 2025. Longitudinal analysis of pharmaceutical stability after exposure to 120,000 feet of altitude: does a simulated near-vacuum exposure change long-term medication stability? In: Presented at: 2025 AsMA-UHMS Annual Scientific Meeting, Atlanta, GA. https://ntrs.nasa.gov/citations/20250005450

Šklubalová, Z., Zatloukal, Z., 2005. Systematic factors affecting eye drop size. Pharm – International Journal of Pharmaceutical Sciences 60 (12), 917–921.

Stabenau, K., Pendleton, M., Mateus, J., Asadi, A., 2024. [460] Hard vacuum chamber testing of common medications in spaceflight. Aerospace Medicine and Human Performance 95 (8), 604–605.

Staniforth, J.N., Rees, J.E., 1982. Effect of vibration time, frequency and acceleration on drug content uniformity. The Journal of Pharmacy and Pharmacology 34 (11), 700–706. Available at https://doi.org/10.1111/j.2042-7158.1982.tb06203.x.

Theobald, K., Zhu, W., Waters, T., Cherrett, T., Oakey, A., Royall, P.G., 2023. Stability of medicines transported by cargo drones: investigating the effects of vibration from multi-stage flight. Drones 7 (11), 658. Available at https://doi.org/10.3390/drones7110658.

Tønnesen, H.H., 2001. Formulation and stability testing of photolabile drugs. International Journal of Pharmaceutics 225 (1-2), 1–14. Available at https://doi.org/10.1016/s0378-5173(01)00746-3.

United States Pharmacopeia. (n.d.). Stability considerations in dispensing practice. In General Chapter. United States Pharmacopeia. Available at https://doi.org/10.31003/USPNF_M99910_03_01.

United States Pharmacopeia, General Chapter, "Monitoring Devices-Time, Temperature, and Humidity". United States Pharmacopeia., Rockville, MD United States Pharmacopeia., Rockville, MD. 2023. Available at https://doi.org/10.31003/USPNF_M99837_01_01.

U.S. Pharmacopeial Convention. USP general chapter <659> packaging and storage requirements: revision bulletin. Published online May 2020. Available at https://www.uspnf.com/sites/default/files/usp_pdf/EN/USPNF/revisions/659_rb_notice.pdf

Veronica, N., Heng, P.W.S., Liew, C.V., 2022. Ensuring product stability – choosing the right excipients. Journal of Pharmaceutical Sciences 111 (8), 2158–2171. Available at https://doi.org/10.1016/j.xphs.2022.05.001.

Veronica, N., Heng, P.W.S., Liew, C.V., 2023. Relative humidity cycling: implications on the stability of moisture-sensitive drugs in solid pharmaceutical products. Molecular Pharmaceutics 20 (2), 1072–1085. Available at https://doi.org/10.1021/acs.molpharmaceut.2c00812.

Wang, W., 1999. Instability, stabilization, and formulation of liquid protein pharmaceuticals. International Journal of Pharmaceutics 185 (2), 129–188. Available at https://doi.org/10.1016/S0378-5173(99)00152-0.

Wang, W., Singh, S., Zeng, D.L., King, K., Nema, S., 2007. Antibody structure, instability, and formulation. Journal of Pharmaceutical Sciences 96 (1), 1–26. Available at https://doi.org/10.1002/jps.20727.

Waters, T.P., Sherman, L., Moxey, H., et al., 2024. Design, build and test of packaging for vibration control of medical goods delivered by drone. Journal of Physics: Conference Series 2647 (17), 172008. Available at https://doi.org/10.1088/1742-6596/2647/17/172008.

De Winter, S., Bronselaer, K., Vanbrabant, P., et al., 2016. Stability of drugs used in helicopter air medical emergency services: an exploratory study. Air Medical Journal 35 (4), 247–250. Available at https://doi.org/10.1016/j.amj.2016.02.001.

World Health Organization (WHO), 2018. Annex 10: stability testing of active pharmaceutical ingredients and finished pharmaceutical products. In: WHO Expert Committee on Specifications for Pharmaceutical Preparations: Fifty-Second Report. WHO Technical Report Series No. 1010. Geneva: World Health Organization.

Wotring, V.E., 2016. Chemical potency and degradation products of medications stored over 550 earth days at the international space station. The AAPS Journal 18 (1), 210–216. Available at https://doi.org/10.1208/s12248-015-9834-5.

Zannikos, P., Solanki, B., De Meulder, M., Badorrek, P., Hohlfeld, J.M., Singh, J., 2023. Pharmacokinetics of nasal esketamine in patients with allergic rhinitis with and without nasal decongestant pretreatment and in healthy subjects with and without nasal corticosteroid pretreatment. Adis, United States Clinical Pharmacokinetics 62 (9), 1315–1328. Available at https://doi.org/10.1007/s40262-023-01273-z, https://www.springer.com/journal/40262.

Chapter 19

# Sedation and general anesthesia

**Arthur Formanek, Nicolai Goettel, Taania Girgla, Yousef Ahmed, Daniel Berenson, Megan Hunt, and Kareem El-Tayeb**
*Department of Anesthesiology, Mass General Brigham, Harvard Medical School, Boston, MA, United States*

Perioperative medicine poses significant challenges during and after exposure to spaceflight. In addition to the difficulties of administration of anesthetic care within a space environment, the physiologic changes that persist after return to Earth pose challenges for spaceflight participants needing surgical or acute care. Despite the potential need for urgent procedural intervention immediately after return to Earth or while still in spaceflight, there is a paucity of primary literature on the effects of spaceflight physiology and perioperative medicine (Agnew et al., 2004). The relative lack of experience requires leaning on experience from animal models and human medical conditions that serve as physiologic analogs.

## 19.1 History of anesthetics in spaceflight and austere environments

The use of anesthesia peri-spaceflight has been a subject of ongoing study and concern since the early days of space exploration. Starting in the late 1940s, rhesus monkeys were among the first animals launched into suborbital flight, starting with the Albert monkeys in 1948 (Burgess and Dubbs, 2007b). These early missions faced significant challenges, as many animals died due to spacecraft malfunctions, hypoxia, or heat exposure, even in early mission phases (Burgess and Dubbs, 2007b). To reduce stress, many monkeys flown on these missions were anesthetized during flight, likely with barbiturates, as this was a standard veterinary practice at the time (Burgess and Dubbs, 2007b; Komorowski and Campbell, 2018).

Procedural sedation and postflight general anesthesia have had a variable record in primates. In 1959, the rhesus monkey Able, one of the first primates to survive spaceflight, died on return date +4 (R+4) after a dose of aerosolized trichloroethylene was administered for anesthesia prior to implanted electrode removal (Komorowski and Campbell, 2018). Trichloroethylene is known to increase the risk of cardiac dysrhythmias, but this death was unexpected (Waters et al., 2009). The autopsy did not reveal any findings, and the cause of mortality is simply reported as “anesthetic death” (II. Bio-Flight 2 Able, 1959).

The 1973–1996 Bion Project, a joint Soviet-Russian and American initiative, further highlighted the risks when an anesthetic-associated death of a rhesus monkey occurred after the final Bion 11 mission. Across 11 missions, 22 rhesus monkeys were sent into low Earth orbit for periods of 5–22 days (Burgess and Dubbs, 2007a). Upon return, most monkeys underwent general anesthesia with ketamine and isoflurane on R+2 or R+3 for biopsies, bone density measurements, and the retrieval of implanted monitors. The 11th mission deviated from this protocol, with two monkeys anesthetized on R+1 after a 15-day flight (Komorowski and Campbell, 2018). One of these monkeys, Multik, died shortly after the procedure. The postmortem revealed bronchopneumonia, and resuscitative drugs were ineffective, likely due to severe hypoxia (Komorowski and Campbell, 2018). The NASA Task Force concluded that the physiologic changes of spaceflight had reduced Mulitk’s ability to tolerate stress and anesthesia immediately after return to Earth presented an “unacceptable mortality risk” (Burgess and Dubbs, 2007a). Of note, the postevent analysis also revealed anesthetic risks, including “reward juice had been offered in the hour prior to anesthesia,” a “lack of preanesthetic laboratory testing and intraoperative monitoring,” and unnecessarily prolonged anesthetic due to poor logistical planning (Mackowski, 2022).

## 19.2 Relevant physiologic changes of spaceflight and implications

Microgravity-induced physiological changes have been described since the inception of human spaceflight, but the small sample size of spaceflight participants limits our ability to draw generalizable conclusions. Despite this, certain

**Procedural Medicine for Spaceflight. DOI:** https://doi.org/10.1016/B978-0-443-27552-4.00036-1

physiological changes that are consistently noted or are under active investigation may have significant implications on anesthetic and perioperative care. Therefore, both the onboard Crew Medical Officer (CMO) and the terrestrial medical support team must be aware of these considerations in case acute care or surgical intervention is needed during the peri-flight period.

## 19.3 Neurologic system

### 19.3.1 Neuromuscular changes

Muscle atrophy secondary to prolonged microgravity exposure is a well-established complication of spaceflight. Despite countermeasures, most spaceflight participants lose 10%–20% of muscle mass (LeBlanc et al., 2000). Without strict countermeasures, it is postulated that astronauts could suffer even greater losses with long-duration space missions (Musculo-Skeletal System: Bone and Muscle Loss, n.d.). The majority of muscle loss is seen in the muscle groups of the lower body and back due to relative inactivity and reduced weight-bearing that plateaus over several weeks (Barratt and Pool, 2008b).

This level of microgravity-induced muscle tissue wasting can be comparable to that seen in prolonged immobilization. Without sufficient gravitational load on muscles, nerve signals to the muscles can also decrease, ultimately mimicking denervation (Fitts et al., 2000). At the neuromuscular junction, this leads to destabilization of acetylcholine receptor clusters (AChRs) and changes from epsilon (mature, innervated) to gamma (immature, denervated) subunits (Leite et al., 2023). In animal models, the effect of gravitational unloading has been noted as disruptions in the clustering of AChRs, leading to less defined NMJs, extra-junctional AChRs, and immature AChRs (Royland and Weber, 1989). Though few studies have actually investigated the changes to the NMJ in spaceflight participants, they indicate a postflight downregulation in genes responsible for synapse and NMJ organization. In the MyoGravity project, an astronaut's satellite cells examined 30 hours after a 5-month ISS mission exhibited dramatic functional impairment attributed to gene downregulation (Di Filippo et al., 2024).

The implications of microgravity-induced muscular changes impact the choice of neuromuscular blockade, although, to date, neuromuscular blockade has not been administered in flight. Postflight anesthetic data from spaceflight participants is limited and challenging to obtain. Thus, it is unknown how spaceflight participants fare with depolarizing and nondepolarizing NMBs in the postflight period. However, based on known NMJ changes, it is prudent to avoid depolarizing NMB (i.e., succinylcholine) because of the risk of catastrophic hyperkalemia from upregulated AchRs in patients exposed to spaceflight for longer than 72 hours. After mobility and muscle mass have been restored, the NMJ may require 1–2 years to recover based on terrestrial data of patients after deconditioning (Martyn and Richtsfeld, 2006).

### 19.3.2 Cerebral fluid dynamics

Acute disruptions in cerebral blood flow in microgravity have been observed, though the long-term effects remain unclear (Du et al., 2021). Initial studies indicated increased cerebral blood flow secondary to cephalad fluid shifts (Bateman and Bateman, 2022), but recent studies have noted stagnant and even reversed flow in the internal jugular veins (Marshall-Goebel et al., 2021). Early studies using transcranial Doppler observed transient variability in middle cerebral artery flow velocity (Bondar et al., 1990; Klein et al., 2020). Some analog studies suggest that gravitational changes may alter regional cerebral perfusion pressure differently (Taylor et al., 2013). One study conducted on the Space Shuttle NeuroLab missions showed no significant change in cerebral autoregulation before, during, and after spaceflight (Iwasaki et al., 2007). Thus, the extent to which these changes in cerebral fluid dynamics affect cerebral autoregulation remains unclear at present (Blaber et al., 2011; Kermorgant et al., 2020). Therefore, clinicians should presume that cerebral autoregulation is altered and ensure tight hemodynamic control both in-flight and in recent spaceflight participants.

### 19.3.3 Intracranial pressure

Similar to cerebral autoregulation, intracranial pressure (ICP) during spaceflight continues to be actively studied(Tymko et al., 2017). The theoretical increase in ICP in microgravity is attributable to the approximate 2 L cephalad fluid shift, vascular adaptations, alterations in cerebrospinal fluid (CSF) dynamics, and increased $CO_2$ levels within the closed shuttle system (Michael and Marshall-Bowman, 2015). For context, the $CO_2$ levels on the ISS nominally range anywhere from 2500 to 3000 ppm, with a maximum allowable 24-hour average $CO_2$ concentration at 5250 ppm (Beard, 2020). In contrast, the concentration of carbon dioxide ($CO_2$) in outdoor air is 250–400 parts per million (ppm), and in occupied indoor spaces with good air exchange, the typical concentration is 400–1000 ppm (Carbon Dioxide Levels Chart, 2025).

Additionally, the lack of gravitational gradients disrupts normal CSF flow and resorption. Over 50% of spaceflight participants complain of headaches, which may be due to these physiological changes (Law et al., 2014).

Though ICP has not been directly measured in orbit, ICP changes have been directly studied in terrestrial analogs, such as head-down tilt studies and parabolic flights. These data suggest that ICP in microgravity is probably lower than ICP when supine on Earth but greater than ICP when upright on Earth (Lawley et al., 2017). Though this lends credence to the theoretical increase in average ICP in microgravity, it challenges the presumption of a unidirectional change in ICP and instead supports a more dynamic state, suggesting that ICP may stabilize at a clinically or sub-clinically elevated level as adaptation to microgravity occurs.

Some studies indicate that these ICP and CSF flow changes gradually normalize (Lawley et al., 2017), but the temporal pattern is not well characterized. Space medicine physicians should thus avoid medications and maneuvers that can contribute to increased ICP, drastic ICP changes, or increased IOP. Similar to idiopathic intracranial hypertension, neuraxial anesthesia in spaceflight participants is safe, as an uncal hernia is not a concern, but further increases in ICP should be avoided.

### 19.3.4 Sensorimotor dysfunction

Spaceflight participants often experience balance and postural difficulties for days to weeks postflight as they recalibrate to the effects of gravity (Buckey, 2023). During this time, it may be wise to avoid medications that could lead to vestibular disturbances, such as nitrous oxide, or lead to CNS depression, such as benzodiazepines.

### 19.3.5 Sensation

Some spaceflight participants have reported regional sensory loss both in flight and for months postflight. These have included distal peripheral neuropathy, upper-extremity numbness in a cape-like distribution, saddle anesthesia, and sciatica-related pain and numbness both (Clark et al., 2008). Imaging has revealed no pathology, and symptoms usually resolve within months.

### 19.3.6 Neurological implications of decompression sickness

Decompression sickness (DCS) can occur anytime there is a rapid transition to a lower-pressure environment, causing dissolved nitrogen to "bubble" out of the blood and get trapped in body spaces. In mild cases, this can present as joint pain (also known as "the bends"), but in more severe cases, the bubbles can present as an air embolus leading to respiratory compromise, Acute Coronary Syndome-like symptoms, or stroke. In the latter situation, the cerebral penumbra is the central focus of anesthetic care, with the goal of maintaining tight hemodynamic control, euglycemia, and normothermia. To the extent that conscious sedation is possible, it is recommended over general anesthesia (Liang et al., 2023). Modest delivery of hyperbaric $O_2$ is possible in orbit by increasing suit pressure combined with cabin pressure to a total of 1140 mmHg (1.5 bar; Padaki et al., 2019), but patients should be closely monitored for signs of oxygen toxicity (Ciarlone et al., 2019). While severe cases of DCS have not been reported in spaceflight due to careful and detailed EVA prebreathe protocols, DCS remains a very real risk.

## 19.4 Cardiovascular system

### 19.4.1 Fluid shifts

Microgravity results in significant fluid redistribution, which is perhaps one of the most well-known and immediate changes to physiology. During the first few hours in microgravity, ~1–2 L of body fluids shift from the lower extremities to the upper body and thorax (Moore and Thornton, 1987), which increases central blood volume and preload (Norsk, 2020).

Microgravity also removes the hydrostatic gradient in the venous vascular system, causing a significant shift of intravascular volume to the extravascular space (Clément, 2011). Plasma volume declines quickly and stabilizes at about 12% below normal in both short- and long-duration missions (Johnson, 1979). Overall, fluid redistribution occurs rapidly in the first day but reaches a new steady state within 3–12 days of flight (Leach et al., 1996). While changes in plasma volume may be adaptive in spaceflight, the effects on hemodynamics during the induction of anesthesia are unknown.

### 19.4.2 Cardiac remodeling

On Earth, gravity allows the heart to adopt an elongated shape. However, in space, the heart adopts a more spherical shape. The lack of gravity's downward force alters cardiac chamber dimensions, with the left ventricle becoming ~3% shorter and 0.5% wider (May et al., 2014). Since the heart does not need to pump against gravity, less force is required to circulate blood throughout the body, and the cardiac muscles and muscular layers of blood vessels are at risk of atrophy (Charles, 1999). Echocardiography has found LV mass to decrease by ~8%–10% following short-duration flight, but normalize after a few days, presumably due to adaptation of vascular volumes (Summers et al., 2005b).

Heart volume increases initially due to increased preload (i.e., blood volume flowing into the heart) but slowly decreases below its terrestrial size (Perhonen et al., 2001). This is presumably due to overall decreased fluid volume and reduced workload, given that blood does not need to be pumped against gravity. Collectively, this impairs cardiac function, but can also cause postflight orthostatic intolerance and decreased exercise tolerance.

### 19.4.3 Hemodynamic changes

Some data show minimal change in heart rate (HR) or arterial pressure (MAP) with exposure to microgravity, while other data suggest HR stabilizes at slightly higher preflight sitting baselines, and MAP stabilizes slightly lower than preflight sitting baselines. Specifically, a slight decrease in diastolic blood pressure (DBP) and a slight increase in systolic blood pressure (SBP) lead to an overall slightly reduced MAP, which is likely clinically insignificant (Fritsch-Yelle et al., 1996; Gundel et al., 2002; Hughson et al., 2012; Verheyden et al., 2009).

In contrast, alterations in preload and afterload during the first 24 hours in microgravity cause a 60%–70% increase in stroke volume, resulting in a 35% increase in cardiac output (CO; Prisk et al., 1995). CO subsequently falls briefly as circulating blood volume decreases, but after 2 weeks in microgravity, it increases again when HR rises, and the stroke volume remains elevated. Several studies show that CO levels remain elevated in microgravity (Prisk et al., 1995; Johnson et al., n.d.; Norsk et al., 2015). This increase in CO despite a relatively stable HR and BP suggests a decrease in systemic vascular resistance (SVR) (Norsk et al., 2006).

Curiously, central venous pressure (CVP) falls below preflight supine levels once in microgravity (Buckey et al., 1996). This was discovered in 1991, when CVP was measured invasively (via a 4 French catheter inserted into an arm vein and a fiberoptic central venous catheter) in astronauts preflight, during launch, and in microgravity (Buckey et al., 1996). This decrease in CVP is thought to occur due to elimination of hydrostatic gradients and tissue compressive forces, which reduces central and peripheral venous pressures (Buckey and Lan, 2022). Thus, microgravity alters the compliance of blood vessels such that the same or more blood volume is contained at a lower pressure. Paradoxically, despite a reduction in CVP, left ventricular end-diastolic pressure (LVEDP) is noted to be elevated (Macnamara et al., 2025), indicating that spaceflight-induced hemodynamic changes might not mimic the changes modeled in terrestrial analogs. One theory is that the spherical change in the heart allows for the same or more blood volume filling at a lower CVP, but this is difficult to directly measure and prove (Buckey et al., 1996).

The time for hemodynamics to return to preflight baselines is likely dependent on the time needed to reequilibrate intravascular blood volume. A summary of these changes can be seen in Table 19.1.

### 19.4.4 Autonomic dysregulation

Baroreceptor sensitivity (BRS) has been shown to decrease by approximately 30% in both short (4–5 day) and long-duration missions (Fritsch et al., 1992), which is thought to be due to reduced stimulation of the cardiovascular regulatory centers (Fritsch-Yelle et al., 1994). Thus, in microgravity, increased carotid transmural pressure is needed to elicit the same effect in HR. Interestingly, vagal parasympathetic BRS was shown to be more affected than sympathetic BRS (Cooke et al., 2000), which possibly explains the slight increase in HR noted in some astronauts.

This change in autonomic regulation is thought to contribute to the orthostatic hypotension faced by approximately 60% of spaceflight participants upon return to Earth (Barratt and Pool, 2008c). Other significant contributors to postflight orthostatic intolerance are a decreased intravascular volume, cardiac atrophy, and a relative vasoplegia in the immediate postflight period (Johnson et al., n.d.). The severity of orthostatic intolerance is related to the time spend in microgravity and can take days to even a few weeks to fully resolve (Lee et al., 2015; Meck et al., 2001).

### 19.4.5 Arrythmia

It is unclear whether microgravity has a dysarrhythmogenic effect, but some studies showed an inconsistent increase in QTc and a decrease in P-R interval in microgravity, but this does not manifest in short-duration flight (Smith et al., 1976;

**TABLE 19.1 Summary of hemodynamic changes in microgravity.**

| Hemodynamic change | |
|---|---|
| Heart rate (HR) | =or ↑ |
| Systolic blood pressure (SBP) | ↑ |
| Diastolic blood pressure (DBP) | ↓ |
| Mean arterial pressure (MAP) | =or ↓ |
| Intrathoracic pressure | ↓ |
| Systemic vascular resistance (SVR) | ↓ |
| Central venous pressure (CVP) | ↓ |
| Left ventricular end-diastolic pressure (LVEDP) | ↑ |
| Stroke volume (SV) | ↑ |
| Cardiac output (CO) | ↑ |

Smith, 1977). NASA has previously flown astronauts who received treatment for atrial fibrillation, atrial flutter, and/or supraventricular tachycardias without in-orbit issues (Johnston et al., 1975). A 56-year-old commercial spaceflight participant was able to safely travel to space despite a high burden of premature ventricular complexes (PVCs), which were managed with beta-blockade on the initial flight and by calcium-channel blockade in the subsequent flight (Jennings et al., 2010). Fortunately, Holter monitoring and 12-lead EKG is possible in-flight, with continuous telemetry often used during EVAs.

Incidental dysarrhythmias have been detected in spaceflight, including PVCs, premature atrial complexes, and nodal and ventricular bigeminy in initial Apollo missions (Johnston et al., 1975). One episode of 14–16 beat nonsustained ventricular tachycardia was noted in a cosmonaut in 1995 (Fritsch-Yelle et al., 1998) and another cosmonaut suffered refractory paroxysmal supraventricular tachycardia after an EVA which prompted earlier than planned return to Earth (Foale, 1999; Rossum et al., 1997).

### 19.4.6 Changes in blood volume and composition

Microgravity-induced loss of hydrostatic pressure causes a shift of ~10%–12% of plasma volume to the extravascular and intracellular compartments, which results in an increase in hematocrit (Johnson, 1979). This reduces erythropoietin production and decreases red blood cell mass (Alfrey et al., 1996). Other reported changes include an increase in antidiuretic hormone (Schrier et al., 1979) and a decrease in plasma albumin levels (Cirillo et al., 2003).

### 19.4.7 Vascular adaptations

Microgravity and space radiation may potentially accelerate atherosclerosis secondary to (1) increased oxidative stress, (2) induction of a chronic proinflammatory state, and (3) hypertriglyceridemia secondary to muscle atrophy-related altered plasma fat clearance (Krittanawong et al., 2023). Aortic and carotid stiffness/intimal thickening are also observed postflight (Hussain et al., 2024; Krämer et al., 2018).

An astronaut on a long-duration spaceflight mission was found to have an incidental left internal jugular vein thrombus, causing total occlusion. This thrombus did not resolve in flight despite enoxaparin and apixaban administration, but it resolved spontaneously a few days after return to Earth. This incident has led the spaceflight community to extensively investigate the vascular changes that could lead to such a serious complication (Aunõn et al., 2020).

All risk factors of Virchow's triad are present in microgravity (Kim et al., 2021). Firstly, the decrease in plasma volume and increase in hematocrit in microgravity lead to hypercoagulability of blood (Kunz et al., 2017; Martin et al., 1986). This change in blood composition may also lead to local hyper-viscosity and subsequent changes to hemorheological characteristics of blood flow (Martin et al., 1986; Lampe et al., 1992). Ultrasound of the jugular veins has revealed stagnant and even reverse flow in the left internal jugular vein (Marshall-Goebel et al., 2024). Though both left and right jugular veins are noted to be distended soon upon entry into microgravity (Marshall-Goebel et al., 2024; Lee et al., 2020a,b)—and throughout the entire duration of flight (Marshall-Goebel et al., 2024; Lee et al., 2020a,b; Martin et al., 2016)—there is a difference in flow bilaterally (Lan et al., 2022). Vessel injury is possible due to changes in wall shear forces secondary to stretch and distention (Kuzichkin et al., 2022). Studies also show that microgravity can cause endothelial dysfunction secondary to inflammatory responses (Kuzichkin et al., 2022; Navasiolava et al., 2020), leading to a procoagulant state and have demonstrated increases in the plasma levels of vWF, hs-CRP, fibrinogen, and circulating endothelial microparticles (Kuzichkin et al., 2022).

## 19.5 Pulmonary system

### 19.5.1 Pulmonary indices

Microgravity influences the lungs in several ways, driven by fluid redistribution, altered respiratory mechanics, and an upward shift of the diaphragm and abdominal viscera (Prisk, 2005). Lung volumes decrease as gravity does not provide a downward distension force (Ghani et al., 2023; Prisk, 2000). Some pulmonary parameters have been directly measured during short-duration Space Shuttle and ISS missions, summarized in Table 19.2.

- Tidal Volume (TV)—decreases approximately 15% compared to preflight standing and supine baselines and remains decreased while in-flight (Prisk, 2005; Ghani et al., 2023; Elliott et al., 1994; Guy et al., 1988).
- Vital Capacity (VC)—a 5% decrease is seen in the first 24 hours upon arrival to microgravity, but seems to return to preflight baseline by 72–96 hours in-flight (Prisk, 2005; Ghani et al., 2023; Prisk, 2000; Elliott et al., 1994; Guy et al., 1988).
- Residual Volume (RV)—decreases approximately 18% compared to preflight standing and supine baselines both. This is presumably due to more uniform alveolar emptying in the absence of the usual apical/basilar gravity-induced lung deformation (Prisk, 2005; Ghani et al., 2023; Prisk, 2000; Elliott et al., 1994; Guy et al., 1988).
- Respiratory Rate (RR)—initially increases in response to a decreased TV, but some reports show that it can eventually decrease by almost 30% compared to preflight baselines in longer duration missions (Prisk, 2000, 2005; Ghani et al., 2023; Elliott et al., 1994; Guy et al., 1988).
- Functional Reserve Capacity (FRC)—decreases approximately 15% compared to preflight standing baselines, but remains higher than preflight supine baselines. FRC remains stable in this intermediary range regardless of time in microgravity (Elliott et al., 1996).
- Functional Expiratory Volume in 1 second (FEV1)—decreases approximately 3%–5% in the initial few days but then eventually returns to normal and even improves compared to preflight baselines (Elliott et al., 1996).
- Diffusion Capacity (DLCO)—increases approximately 13%–28% presumably due to more uniform capillary filling and thereby increase in surface area that can support diffusion (Prisk et al., 1995).

**TABLE 19.2** Summary of pulmonary changes in microgravity.

| Pulmonary change | |
|---|---|
| Tidal volume (TV) | ↓ |
| Vital capacity (VC) | ↓ then= |
| Residual volume (RV) | ↓ |
| Respiratory rate (RR) | ↑ then ↓ |
| Functional residual capacity (FRC) | ↓ |
| Forced expiratory volume in 1s ($FEV_1$) | ↓ then= |
| Diffusing capacity for CO (DLCO) | ↑ |
| Ventilation-perfusion (V/Q) ratio | → 1 |
| Hypoxic ventilatory response | ↓ |
| Hypercapnic ventilatory response | = |

- Ventilation to Perfusion (V/Q) Ratio—the lack of a G-vector allows for more even distribution of blood flow and expansion of alveoli. This results in more uniformity in the V/Q ratio, a reduction in dead space, and thus relatively unchanged alveolar ventilation. This does not affect gas exchange significantly, though, because despite an improved V/Q ratio, regional variances in lung elasticity still persist (Verbandt et al., 2000).
- Control of Ventilation—there is a reduction in hypoxic but not hypercapnic ventilatory responses (Prisk et al., 2000).

The lungs maintain efficient gas exchange in microgravity, and the pulmonary system undergoes adaptive changes (Prisk et al., 1995; Prisk, 2014b).

### 19.5.2 Compliance, resistance, and respiratory mechanics

Microgravity-induced upward fluid shift results in increased thoracic blood volume and capillary pressure. Data shows that pulmonary capillary blood volumes remain high throughout orbit (Prisk, 2014a). Imaging and physiological measurements further confirm increased thoracic impedance and fluid accumulation (Venturoli et al., 1998; Wantier et al., 1998).

This fluid redistribution can lead to interstitial edema, though the literature is somewhat contradictory on this point (Prisk, 2014; Gharib, 1995; Glaister, 1970; Guy et al., 1992; Prisk, 2019; Prisk and West, 1998; Smith et al., 2024; Stepanek et al., 2023; West et al., 1997). Some reports indicate that interstitial edema is not significantly impacted but simulated microgravity animal studies (e.g., head-down tilt models) show histological evidence of fluid accumulation in airway tissues. These findings correlate with increased lung water content and thickened airway walls. Spirometry measurements in astronauts also reveal slight reductions in forced expiratory flow rates initially, suggesting edema-related airway narrowing. This fluid shift can decrease lung compliance and increase airway resistance, potentially impacting ventilation dynamics.

Overall, though, lung tissue shows show minimal functional impairment post spaceflight (Gharib, 1995; Glaister, 1970; Guy et al., 1992; Prisk, 2019; Prisk and West, 1998; Smith et al., 2024; Stepanek et al., 2023; West et al., 1997).

### 19.5.3 Upper airway changes

Microgravity-induced cephalad fluid shifts notoriously cause facial puffiness, jugular distension, and upper body edema. Astronauts frequently complain of nasal congestion and "facial fullness" that resolves once back on Earth (Khan et al., 2024). It had been theorized that this leads to upper airway edema, though direct measurements are still pending (Mittal et al., 2024). While direct in vivo measurements in astronauts are limited, indirect evidence and animal models strongly support the presence of airway edema in microgravity (Ballard, 2001; Ewald, 1994). However, a study evaluating sleep-disordered breathing in space revealed that there was a significant reduction in respiratory-related nighttime awakenings and an almost complete elimination of snoring in microgravity that returned to baseline postflight (Elliott et al., 2001). This indicates that though microgravity may cause upper airway edema, it may also reduce obstructive sleep apnea.

Additionally, microgravity affects multiple aspects of upper airway ciliary clearance, including ciliary beat frequency, mucus transport, and airway surface liquid properties (Baeyens, 2001; Ballard and Parker, 2007; Foster, 1998). Specifically, fluid redistribution leads to changes in mucous viscosity, which can hinder the movement of mucous along the ciliated epithelium. Also, in microgravity, gravitational clearance is absent, further increasing the risk of mucus stasis and respiratory complications in astronauts (West, 2001).

## 19.6 Gastrointestinal system

### 19.6.1 Motility

Electrogastrography has shown decreased bowel activity leading to ileus in the initial days of exposure to microgravity (Harm et al., 2002; Thornton et al., 1987; Thornton and Moore, 1986). Symptoms of reflux, dyspepsia, and constipation are frequently reported (Lane et al., 1999). It is theorized that decreased gastrointestinal motility is related to both space motion sickness (SMS) and microgravity effects on the autonomic nervous system (Thornton et al., 1987). Since these inciting factors persist upon reentry to gravity, reduced gastric emptying should be presumed in the initial days back on Earth. Additionally, some Russian studies demonstrated an increase in gastric acidity over time in long-duration missions (Smirnov and Ugolev, n.d.), which can make an aspiration event more catastrophic. As such, any intubation performed in microgravity should be performed as a rapid sequence with a presumed full stomach, regardless of the duration of NPO.

### 19.6.2 Absorption

Ultrasonography has shown slight pancreatic and hepatic enlargement in space (Tigranyan, 1985). These changes, combined with altered intestinal blood flow, have been shown to impair gut absorption and alter drug bioavailability in microgravity (Amidon et al., 1991; Smirnov and Ugolev, 1996; Yang et al., 2020). Thus, efficacy of oral medications can be unreliable, and intravenous (IV) or intramuscular (IM) drug administration should be preferred in the peri-flight period. Gut edema also disrupts the mucosal barrier, which can further compromise immune responses and contribute to poor wound healing (Sun et al., 2025).

### 19.6.3 Space motion sickness

SMS is perhaps one of the most well-described spaceflight-associated phenomena(Heer and Paloski, 2006). Most experts agree that it occurs due to changes in gravitational inputs to the central nervous system, although the exact etiology of SMS is not well defined (Crampton, 1990; Reschke et al., 1994). Some other postulated factors include changes in endolymph pressure secondary to flight-induced fluid shifts and changes to otolith structure and signaling (Simanonok and Charles, 1994).

SMS has been reported both on exposure to microgravity and reentry back to gravity after return. Symptoms can be transient and mild or prolonged and severe, leading to repeated emesis and dehydration. Onset ranges from immediate to 48 hours postexposure, both on exposure to microgravity and reentry back to gravity. Recovery of symptoms usually occurs by postflight day 2 while in orbit, but recovery from symptoms on reentry to gravity can take longer and usually correlates with time spent in orbit. Thus, SMS can significantly increase the risk of postoperative nausea/vomiting if peri-flight anesthesia is needed (Davis et al., 1988; Graybiel and Lackner, 1980).

Currently, preflight adaptation training and pharmacological intervention remain the standard of care for SMS (Bagian and Ward, 1994; Davis et al., 1993a; Jennings, 1998; Lackner and Graybiel, 1994). Currently, IM promethazine in doses of 25–50 mg has been shown to be most effective in treating SMS(Davis et al., 1993b). ISS kits have also been stocked with oral ondansetron, meclizine, and promethazine. Interestingly, promethazine's sedating side effect has not been largely reported in spaceflight (Bagian and Ward, 1994). Preflight oral prophylaxis and in-flight oral treatment with diphenhydramine or chlorpromazine have had varying success due to altered bioavailability in-flight (Boyd et al., 2009). Transdermal scopolamine with dextroamphetamine was previously used but caused rebound symptoms; it is reserved for emergent treatment. Postflight, IM promethazine is effective, but aggressive rehydration and multimodal oral agents are also recommended.

## 19.7 Practical considerations and concerns

### 19.7.1 Expected level of expertise of crew

Presently, a fully qualified physician is not required to be on board a space mission per federal and commercial regulatory policies (Medical Operations Team Activities, 2026). Depending upon the spacecraft orbital platform, crew members receive first-aid and paramedic training, and the CMO receives additional training in diagnostics and therapeutics. Some specific skills a CMO may be trained on include obtaining patient assessments and physical exams, understanding medical monitoring equipment, placing IV lines for blood samples and drug delivery (and intraosseous (IO) lines for emergencies), performing diagnostic ultrasounds, responding to medical emergencies, and providing basic life support and airway management. In addition, the CMO is expected to be familiar with all on-orbit medical kits, instruments, medications, and life-support systems. The ISS has been equipped with dental kits, eye wash kits, wound repair kits, basic surgery kits, IV drip infusions, automatic external defibrillators, and advanced airway equipment (which includes a small automated ventilator, an ambu bag, endotracheal tubes [ETTs], laryngeal mask airways [LMAs], oral and nasal airways, bougie, suction, and a direct laryngoscope; Barratt and Pool, 2008a; Medical Kit- Contents and Reference, 2015).

Nonphysician trained CMOs have performed well in low-Earth orbit (LEO) with the support of ground flight surgeons, but the ideal level of expertise for a CMO for longer duration missions is still debated. Exploration-class missions to the Moon, Mars, and beyond pose the unique challenges of (1) communication delay, (2) limited resupply, and (3) inability to evacuate (Antonsen et al., 2022). Longer duration missions also increase the total medical risk and increase the probability of a medical or surgical emergency (Antonsen et al., 2022). Experts in the field are actively using Probabilistic Risk Assessment Monte-Carlo simulation tools to characterize the medical risk profile and inform exploration medical capabilities and the scope of mission training (Kreykes et al., 2023). Currently, a list of over 100 medical and surgical conditions is under review, including penetrating injuries, wound infections, diverticulitis, pancreatitis, hernia, and more (Blue, 2019).

At present, CMOs in orbit undergo regular training to avoid attrition of knowledge or skills; however, the ability to operate independently of the ground flight surgeon is critical for exploration-class missions. Keeping this in mind, it would be ideal to have a fully trained physician who (1) has knowledge of all organ systems, (2) has the aptitude to manage medical emergencies, (3) is capable of providing acute and critical care, (4) is deft with ultrasound, and (5) has practical procedural and basic surgical skills.

### 19.7.2 Monitoring

All standards for basic anesthetic monitoring are possible in space and during EVAs, including electrocardiography, continuous telemetry, pulse oximetry, noninvasive blood pressure, end-tidal $CO_2$ measurements, and temperature assessments. All-in-one monitors, like the Tempus Pro, have been flown to space (Mission, 2024). Cables and wires of devices in microgravity can crowd the tight space in the capsule and interfere with medical care if they are in the way, underscoring the critical importance of design.

Additionally, CMOs and flight surgeons in the past have also monitored oxygen consumption to calculate metabolic rates, pulmonary function tests, urinalysis, and labs requiring blood draw. There are portable iSTAT (NASA) and Reflotrom (Russian) blood analyzers, as well as gas analyzers, in orbit (Nicogossian, 2016a). Advanced invasive monitoring provided by arterial lines, central lines, pulmonary artery (PA) lines, and transesophageal echocardiography has not been attempted to date, although these have been done in parabolic flights (Videbaek and Norsk, 1997; Buckey et al., 1995; Mukai et al., 1991; Schlegel et al., 1998). Of note, due to the relatively noisy environment on board, auscultation can prove to be challenging in space.

Diagnostic imaging is also used for monitoring in space. Ultrasound remains the main noninvasive diagnostic imaging tool in orbit used for monitoring, and various brands of portable ultrasound have been flown. Crew are trained on cardiac, pulmonary, vascular, abdominal, musculoskeletal, and ocular imaging as well as FAST exams. Just-in-time clinical training and remote expert guidance are often used to supervise image acquisition as needed. Recently, X-rays were obtained on parabolic and orbital flights (Lerner et al., 2023). Also, the first trans-nasal endoscopy was recently conducted on board a commercial spaceflight (Mittal et al., 2024).

### 19.7.3 Pharmacy

#### *19.7.3.1 Drug stability and packaging*

Reducing mass, volume, and waste are critical factors when determining which medications to pack for space, and currently, repackaging of pharmaceuticals is necessary to achieve this. However, the moment a medication is opened from its original packaging, the reported shelf life can no longer be relied upon. Additionally, it is well-known that environmental factors like humidity, oxygen, light, radiation, and exposure to temperature extremes can compromise shelf life, drug efficacy, and toxic byproduct development (Nicogossian, 2016b; Wotring, 2020). For example, the shelf life of an unopened propofol package is 24 months, but propofol is known to be sensitive to temperature extremes.

Until now, this has not been a significant concern for LEO missions—since resupply is possible—but reduced drug stability becomes a barrier for exploration-class missions that will likely not be able to resupply for years at a time. This becomes even more complicated given that repackaged medications are often opened, tested, and repackaged almost 1–2 years prior to launch. To address this, extensive drug storage and extended shelf life testing is underway (Blue, 2019).

### 19.7.4 Blood products

Administration of blood and other blood products would be ideal in cases of severe trauma in orbit; however, the ability to stock this is reasonably limited by the narrow shelf life of blood products. Other operational constraints, such as mass limitations and limited refrigeration, must also be considered. Hemoglobin-based oxygen carriers (artificial blood; Kirkpatrick et al., 2001) and rehydration of freeze-dried red blood cells are being explored (Elder et al., 2024). If proven to be as efficacious as pRBC administration on Earth, this could prove vital for exploration-class missions. A last option would be a floating blood bank analogous to the military model where one crewmember could donate blood to another if they are selected based on analogous blood type.

## 19.8 Airway management during human spaceflight

Airway management in human spaceflight presents unique challenges that are not encountered in terrestrial environments (Keller et al., 2000; Komorowski et al., 2016). During emergency airway management, the physiological changes to the airway and pulmonary system mean that even simple procedures like bag-valve-mask (BVM) ventilation or intubation can be significantly more difficult. For instance, reduced functional residual capacity (FRC) and upper airway edema may necessitate earlier and more aggressive airway intervention to maintain oxygenation and ventilation.

### 19.8.1 Airway management scenarios in spaceflight

The need for airway management in space can arise from several scenarios, including the following:

- Medical emergencies: anaphylaxis, pulmonary embolism, severe asthma exacerbation, or cardiac arrest.
- Trauma: accidental impacts during extravehicular activity (EVA), exercise, or in-flight emergencies.
- Decompression events: sudden changes in pressure can result in barotrauma or pulmonary injury.
- Sedation and anesthesia: will be required for certain procedures and may include airway protection and ventilation (Summers et al., 2005a).

## 19.9 Training and simulation for airway management

Training for airway management in space is a critical component of crew preparation, as airway management simulation improves knowledge and psychomotor skills (Kennedy, 2014). Astronauts undergo extensive medical training, including didactic instruction, hands-on practice, and simulation-based learning using high-fidelity mannequins and virtual reality in microgravity. Parabolic flights provide an opportunity to practice airway management techniques in a low-gravity environment, but current astronaut candidates do not routinely practice procedures in this analog (Warnecke et al., 2019).

## 19.10 Tools and techniques for airway management in microgravity

### 19.10.1 Bag-valve-mask ventilation

Bag-valve-mask (BVM) ventilation remains the first line of airway management during emergencies in space. However, its application in microgravity is challenging due to the lack of a stable surface to support the patient's head and the rescuer's inability to maintain a secure mask seal while floating. Crew members must use restraints, such as seat belts or straps, to stabilize the patient, and rescuers must anchor themselves to a fixed structure to apply sufficient force to ventilate effectively.

In response to these challenges, modified BVM devices with hands-free mask seals or mechanical ventilation assist devices have been developed and tested in analog environments (Bauman et al., 2010; Otten et al., 2014). Preflight crew training focuses on achieving a proper mask seal in microgravity conditions.

### 19.10.2 Supraglottic airway devices

Supraglottic airway devices, such as the Laryngeal Mask Airway (LMA), are particularly advantageous in space. LMAs are easier to insert compared to endotracheal tubes (ETT) and require minimal head and neck manipulation, which is crucial in microgravity. LMAs also provide effective ventilation without the need for precise placement, making them suitable for nonexpert crew members (Hinkelbein et al., 2021). Studies conducted in parabolic flight have demonstrated that supraglottic airway devices are reliable and effective under microgravity conditions (Warnecke et al., 2019).

### 19.10.3 Endotracheal intubation

Endotracheal intubation is the gold standard for securing the airway in critically ill or injured patients. However, in microgravity, lack of a downward G-vector impairs patient positioning, which is essential for direct laryngoscopy. To overcome this, crew members must stabilize the patient using restraints or a specially designed medical station that allows head elevation and alignment of the airway axes (Groemer et al., 2005; Rabitsch et al., 2006).

Video laryngoscopy provides a magnified view of the vocal cords and may improve success rates, especially for inexperienced operators, in some scenarios. Modern video laryngoscopes are lightweight, portable, and have been successfully tested in parabolic flight (Starck et al., 2020; Thierry et al., 2023).

### 19.10.4 Robot-assisted endotracheal intubation

This medical device integrates imaging, machine learning, and robotic manipulation to facilitate intubation in challenging scenarios, including anatomically difficult airways and remote or resource-limited environments such as spaceflight (Khan and Karmakar, 2023). Robotic platforms, such as the Kepler Intubation System and others, use high-definition video laryngoscopy combined with robotic arms to guide ETT placement (Biro et al., 2020; Cheng et al., 2018; Hemmerling et al., 2012; Wang et al., 2018). Robotic systems are also in development to assist less experienced providers, providing step-by-step guidance, real-time feedback, and precision control, reducing the risk of trauma to airway structures.

### 19.10.5 Cricothyrotomy and surgical airways

When conventional airway management fails, cricothyrotomy or tracheotomy may be required to secure the airway. Tools such as prepackaged cricothyrotomy kits with step-by-step components simplify the procedure for nonphysician crew members.

## 19.11 General anesthesia and sedation

Historically, most humans who undertook spaceflight would likely be classified with an American Society of Anesthesiologists (ASA) physical classification of 1 or 2, signifying normal health or mild systemic disease. Nevertheless, evidence has shown significant strains on physiologic homeostasis of otherwise healthy individuals during spaceflight. Furthermore, the rise of commercial spaceflight will inevitably increase the number of spaceflight participants with chronic medical conditions (Table 19.3).

Sedation and general anesthesia are a continuum; general anesthesia is simply defined as being unarousable to painful stimulus (Statement on Continuum of Depth of Sedation: Definition of General Anesthesia and Levels of Sedation/Analgesia, 2025). A common misconception is that a type of airway or respiratory support defines general anesthesia, but general anesthesia with a natural airway is frequently performed (endoscopy, etc.). Minimal sedation is defined as a normal response to verbal stimulation, moderate (conscious) sedation is a purposeful response to verbal or tactile

**TABLE 19.3** American Society of Anesthesiologists (ASA) physical status classifications.

| ASA class | Definition | Adult examples (not exhaustive) |
|---|---|---|
| ASA I | Healthy with no systemic disease and excellent functional capacity | 45-year-old cyclist, nonsmoker, taking no medications; 30-year-old with no chronic medical conditions but occasional social alcohol use |
| ASA II | Mild systemic illness without significant impact on daily activities or functional status | 68-year-old with well-controlled type 2 diabetes on metformin only; 42-year-old with well-controlled hypertension on a single medication; 35-year-old with mild intermittent asthma; 60-year-old with BMI 33 who walks daily and has no known end-organ disease |
| ASA III | Significant systemic disease that results in noticeable functional limitation but is not life-threatening in the short term | 63-year-old with COPD requiring daily inhalers; 70-year-old with stage 4 chronic kidney disease, not yet on dialysis; 49-year-old with stable angina who can climb one flight of stairs but becomes dyspneic with greater effort |
| ASA IV | Severe systemic illness that poses a persistent, ongoing threat to life, even at rest | 60-year-old with COPD requiring oxygen; 80 year old with unstable angina; 70-year-old with decompensated heart failure on pressors; 58-year-old with NSTEMI within the last month |
| ASA V | Critically ill, moribund patient who is unlikely to survive without the procedure | 67-year-old with urosepsis due to obstructing stone; 52-year-old with a ruptured thoracic aortic aneurysm; 30-year-old after penetrating trauma with ongoing hemorrhagic shock despite resuscitation |
| ASA VI | A patient who has been declared brain-dead and is undergoing organ procurement | 40-year-old declared brain-dead after a severe head trauma, scheduled for organ donation surgery |

The addition of "E" denotes emergency surgery.
**Source:** Based on the American Society of Anesthesiologists. ASA Physical Status Classification System. ASA; 2020. Accessed December 27, 2025. https://www.asahq.org/standards-and-guidelines/asa-physical-status-classification-system.

stimulation, and deep sedation is a purposeful response following repeated or painful stimulation (Statement on Continuum of Depth of Sedation: Definition of General Anesthesia and Levels of Sedation/Analgesia, 2025). Patients will transition readily between states of sedation and anesthesia regardless of "safety" of medication and mechanical support used based upon their own intrinsic physiologic state and response to pharmacology. Airway management is guided by the potential need for pulmonary support during sedation and general anesthesia.

## 19.12 Hemodynamic support

Anesthesiologists should be prepared for marked hemodynamic perturbations given the intravascular volume depletion noted in many astronauts. Furthermore, positive pressure ventilation may also decrease preload (Corp et al., 2021). Altered circulation, tissue perfusion, and fluid distribution in microgravity may create unpredictable responses to typical drugs employed in anesthetics. For instance, the central compartment may receive a higher blood supply than peripheral tissue, altering distribution of drugs (Dello Russo et al., 2022). Additionally, lower intravascular fluids may lead to decreased renal and liver function, attenuating drug clearance.

## 19.13 Hypnotics

### 19.13.1 Inhaled anesthetics

Volatile anesthetics, such as sevoflurane, isoflurane, and desflurane, are the mainstay of hypnotics utilized in terrestrial anesthesia. Physiologically, volatile agents carry the advantage of being quite stable. However, several drawbacks complicate their use in microgravity. First, with the exception of nitrous oxide and xenon, specialized vaporizers would be required to use them in space (Komorowski et al., 2013). Nitrous oxide has a relatively benign cardiovascular profile, but it is a strong oxidizer, and it cannot achieve general anesthesia alone. Xenon is currently impractical as an anesthetic because of its relative scarcity. Appropriate adjustment of inspired concentration of volatile anesthetics would be made based on the pressurization level of the spacecraft. Still, without end-tidal monitoring of the anesthetic, it may be difficult to properly titrate the agent. Furthermore, volatile agents typically increase cerebral blood flow, which may be disadvantageous given the possibility of increased ICP in many astronauts returning from spaceflight (Lawley et al., 2017; Slupe and Kirsch, 2018). Lastly, volatile anesthetics would be difficult to properly contain and prevent cabin contamination (Komorowski et al., 2013).

### 19.13.2 Intravenous anesthetic agents

IV anesthetics offer significant advantages in space compared to volatile agents. They are both easier to administer and less likely to cause environmental contamination. Additionally, most IV drugs can be easily titrated to achieve the desired effects. Many of the pharmacokinetic principles on Earth would still apply in space. For instance, propofol, a lipophilic hypnotic, would accumulate in fat stores and cause peripheral vasodilation (Larijani et al., 1989). Similarly, anesthesiologists should maintain vigilance towards cardio-depressant effects of any IV hypnotics (Komorowski et al., 2013). While agents such as ketamine or etomidate are frequently described as cardiac-stable, hemodynamic collapse may be achieved with any induction agent, and the crewmember providing anesthesia should never be lulled into a false sense of security (Komorowski et al., 2013). Ketamine could aid in maintaining spontaneous respiration, which may be advantageous when neuromuscular blockade is not necessitated by the surgery. Drawbacks to ketamine include the lack of reversibility, increased salivation, and dose-dependent increase or decrease in sympathetic tone (Komorowski et al., 2013). After use of etomidate, anesthesiologists should remain vigilant against adrenal suppression and aspiration given increased risk for postoperative nausea and vomiting (Forman and Warner, 2011). Selection of hypnotics would be governed by the availability of medications in the setting as well as the resources available to offset secondary effects, such as the need for mechanical ventilation.

### 19.13.3 Opioids

Opioids potentiate respiratory depression (Boom et al., 2012), and the effects of inducing hypercapnia in a spacecraft environment with baseline elevated $CO_2$ are unknown. Careful titration and monitoring are necessary to prevent drug accumulation, but opioids carry the advantage of being reversible. Short-acting opioids such as remifentanil may be particularly useful to decrease doses of hypnotic agents during anesthesia (Beers and Camporesi, 2004). Opioids also have the beneficial effect of reliably slowing respiratory rate, which is generally easier to monitor in comparison with tidal volume.

### 19.13.4 Paralytics and neuromuscular blockade

Neuromuscular blocking agents (NMBAs) are routinely used in general anesthesia to facilitate endotracheal intubation and improve surgical conditions. As with other drugs, the distribution and elimination of NMBAs and their reversal agents may be altered in space, necessitating careful monitoring with a nerve stimulator and dosing. With regard to the selection of paralytic drug, nondepolarizing agents would be preferable as succinylcholine, a depolarizing NBMA, could exacerbate an increase in intraocular and intracranial pressure. More concerning, upregulation of acetylcholine receptors from skeletal muscle deconditioning can result in marked hyperkalemia in response to pharmacologic depolarization. Nondepolarizing agents do not carry this risk and, if using rocuronium or vecuronium, may be quickly reversed with sugammadex.

### 19.13.5 Automated and remotely administered anesthesia

Providing general anesthesia in microgravity may prove challenging without an anesthesiologist as a crewmember. However, automated closed-loop systems for vital sign monitoring and anesthesia delivery can control hypnosis, analgesia, and muscle relaxation from induction to emergence in response to vital signs, analgoscore, processed EEG, and train-of-four monitoring (Hemmerling and Terrasini, 2012). These automated systems are currently limited by a lack of clinical intuition, an inability to adapt to nonstandard circumstances, and machine failures, among other issues. For example, Sedasys, an automated system for propofol sedation in gastroenterology procedures, was used until 2016 when it was removed from the market due to a lack of clinical efficacy and an inability to properly adapt to variable clinical circumstances (Goudra and Singh, 2017; Blaber et al., 2011). Ultimately, opportunities exist for further research and development of a system that both allows for telemedical delivery of general anesthesia and ensures safety.

### 19.13.6 Anticipating and solving problems under anesthesia in space

Concerted efforts to improve the safety of anesthetic care in the late twentieth century were inspired by safety protocols and crisis resource management (CRM) strategies pioneered in the aerospace industry (Burden, 2020; Kanjia et al., 2024). A subdiscipline of anesthesia CRM has created systematic approaches to decision-making (Gaba et al., 2015). Today, cognitive aids and simplified, standardized algorithms are available for a variety of problems experienced during anesthesia, from cardiac arrest to oxygen failure. Such algorithms can be a starting point for the development of similar tools for extraterrestrial anesthesia; however, the unique environment of space will mandate a number of changes.

## 19.14 Key differences

All disciplines of space medicine are impacted by elements of microgravity, mass restrictions on equipment, and distance from specialty care. Extraterrestrial anesthesia protocols, like all systems in the space environment, must balance inclusion of appropriate failsafes and backups without overemphasizing rare events that would require unrealistically extensive training and equipment. Prioritization is the guiding principle when adapting terrestrial protocols for spaceflight. Given this fact, it is advisable to focus on strengthening the primary, first alternative, and second alternative or failsafe plans, even to the neglect of accounting for all possible anesthetic complications with specific contingency plans.

First- and second-line drugs and dosages should be preselected and prepared for easy and rapid administration by either IV injection or infusion, if possible. Equipment, including backup equipment, must be stored so that it is readily accessible but also remains secured. The patient and care providers must be restrained so traction and pressure can be applied during laryngoscopy or line placement (Warnecke et al., 2019).

Selection of the medical system will prioritize supplies with a smaller footprint (e.g., a noninvasive blood pressure cuff instead of an arterial line) and devices that can do double duty (e.g., biochemical analyzers that can be combined with equipment for the mission's scientific objectives). Neuromuscular blockade (train-of-four) monitors could be sacrificed, and standardized sugammadex dosing nomograms used instead. A self-inflating bag can replace a dedicated mechanical ventilator, at the cost of occupying a crewmember's hands. On-demand 3D printing could be another source of unexpectedly needed equipment (Komorowski et al., 2018).

Since anticipating and solving problems in anesthesiology can occur on different timescales, communication delays will be differentially problematic. Even emergency surgeries require time to gather equipment, and during this interval, crew members can communicate with experts on Earth to craft the anesthetic plan, including airway management, vascular access, and the need for nerve blocks (Komorowski et al., 2018; Kamine et al., 2022). However, in true anesthesia crises such as hypoxia, hypotension, or cardiac arrest, the crew may need to act independently because the time delay in communicating with Earth before taking action could increase morbidity. Management of these hyperacute problems should be stressed during the training of potential astronaut anesthetists, which should include practice working collaboratively with AI tools (Komorowski et al., 2018; Cheung et al., 2023; Goh et al., 2024). Training must emphasize the most critical elements, and 20–50 repetitions of key skills may be a reasonable target for competence (Komorowski et al., 2013, 2018; Prekker et al., 2023).

## 19.15 Considerations for troubleshooting the airway in space

Rather than mastering the full spectrum of airway management, astronaut training and equipment should be streamlined. Emphasis should favor maintaining spontaneous ventilation with a natural airway, even under general anesthesia (e.g., ketamine; Komorowski et al., 2013; Lee et al., 2020a). Simulation training should emphasize the use of capnography and pulse oximetry for the rapid identification of inadequate oxygenation or ventilation with a natural airway. Use of supplemental oxygen should be considered in the context of the spacecraft's overall environmental control and life support systems. Astronaut anesthetists should be proficient in basic airway manipulation maneuvers like chin lift, jaw thrust, oropharyngeal airway placement, and bag-mask ventilation, and, if a natural airway fails, supraglottic devices provide a simple yet versatile next step (Komorowski et al., 2016). Training in supraglottic airways should highlight correct sizing (prelabeled for crew), confirmation with capnography, and management of leak or laryngospasm. Protocols may include succinylcholine for laryngospasm but must weigh the inability to manage malignant hyperthermia in space and concern for hyperkalemia with susceptible neuromuscular junctions after disuse during spaceflight. Continuous suction is required due to aspiration risk from gastric contents or microgravity-associated debris, and altered gastric motility may heighten this hazard.

Proceeding to endotracheal intubation with neuromuscular blockade will incur additional risk, as paralysis initiates a finite window before hypoxia develops, and thus the airway must be secured promptly. Video laryngoscopy improves first-pass success for inexperienced intubators (Prekker et al., 2023; Pourmand et al., 2023), and sugammadex permits rapid reversal of even profound neuromuscular blockade. Nevertheless, without backup personnel or advanced equipment, intubation attempts carry substantial risk. Techniques such as bronchoscopic or awake intubation, cricothyrotomy, and tracheostomy demand specialized training and are unrealistic for most spaceflight crew medical officers. If intubation fails, immediate reversal and return to spontaneous ventilation should be prioritized over repeated attempts, with consultation to Earth-based support guiding subsequent decisions.

Although the equipment and specific training in anesthesia crises will be highly limited, astronauts will have the advantage of extensive training in broad CRM. CRM concepts originally developed in aviation and later adopted in terrestrial anesthesiology emphasize team coordination, closed-loop communication, clear leadership, shared situational awareness, and efficient utilization of resources. By integrating these principles with thoughtful, streamlined anesthesia protocols, it should be possible to safely administer anesthesia.

## 19.16 Perioperative medicine after return to Earth

Changes from microgravity that could impact anesthetic management include autonomic dysfunction with cardiovascular deconditioning, baroreceptor reflex changes, lack of autonomic response to cardiovascular stress, decreased intravascular volume, SMS, orthostatic hypotension, and altered reaction to depolarizing agents due to muscle denervation (Agnew et al., 2004). Management should prioritize preload with fluids and alpha-agonists, but persistent hypotension may indicate autonomic dysfunction and require inotropes.

Autonomic instability may persist postoperatively, presenting as refractory hypotension or hypertension. These disturbances should be treated promptly with vasoactive medications, including alpha, beta, or V1 agonists and fluid boluses. The Trendelenburg position may be useful to augment cerebral perfusion in severe, refractory hypotension in partial gravity environments, but head-down positioning may worsen respiratory mechanics and should be used judiciously. Standard anesthetic-induced pulmonary complications include hypoventilation and increased intrapulmonary shunting from decreased FRC (Gropper, 2020), but recent exposure to microgravity will have altered pulmonary blood flow, creating more uniform but still heterogeneous V/Q matching (Prisk, 2014a,b).

A known case of venous thromboembolism in spaceflight (White and Wenthe, 2023) was likely a sequela of the known alterations in thrombin generation, increased fibrinogen, and endothelial injury that occur in microgravity and likely persist in the weeks after return to Earth. Abnormal hemostasis should be anticipated, and providers should acquire additional workup for coagulopathy, such as rotational thromboelastography or a disseminated intravascular coagulation (DIC) panel. Additionally, while internal bleeding could present differently in microgravity, the focused assessment with sonography for trauma (FAST exam) has been validated to assess for internal bleeding in porcine models on parabolic flights (Kirkpatrick et al., 2003).

Astronauts are at baseline at risk of SMS from vestibular dysfunction (Agnew et al., 2004), which will be additive to the risk of postanesthetic nausea and vomiting (PONV). Management may include standard PONV protocols, but clinicians should provide judicious, incremental doses of antiemetics, given the layering of mechanisms of action of many pharmaceutical agents and the increased sensitivity to side effects of spaceflight participants (Butterworth, 2022). Astronaut core temperatures increase up to 1°C after 2.5 months in space (Stahn et al., 2017), and given this increase in their thermal set point, they may be even more susceptible to the anesthetic-associated risk of hypothermia (Butterworth, 2022).

In sum, perioperative care of a patient returning from microgravity should consider the patient's history, physical examination, urgency of procedure, and duration of time in space. The surgeon and anesthesiologist should focus on autonomic regulation, pulmonary mechanics, hemostasis, pain management, PONV, and normothermia in the perioperative period.

## References

Agnew, J.W., Fibuch, E.E., Hubbard, J.D., 2004. Anesthesia during and after exposure to microgravity. Aviation Space and Environmental Medicine 75 (7), 571–580.

Alfrey, C.P., Udden, M.M., Leach-Huntoon, C., Driscoll, T., Pickett, M.H., 1996. Control of red blood cell mass in spaceflight. Journal of Applied Physiology 81 (1), 98–104. Available at https://doi.org/10.1152/jappl.1996.81.1.98, jap.physiology.org.

Amidon, G.L., DeBrincat, G.A., Najib, N., 1991. Effects of gravity on gastric emptying, intestinal transit, and drug absorption. The Journal of Clinical Pharmacology 31 (10), 968–973. Available at https://doi.org/10.1002/j.1552-4604.1991.tb03658.x.

Antonsen, E.L., Myers, J.G., Boley, L., Arellano, J., Kerstman, E., Kadwa, B., et al., 2022. Estimating medical risk in human spaceflight. NPJ Microgravity 8 (1). Available at https://doi.org/10.1038/s41526-022-00193-9, https://www.nature.com/npjmgrav/.

Aunõn, S.M., Pattarini, J.M., Moll, S., Sargsyan, A., 2020. Venous thrombosis during spaceflight. New England Journal of Medicine 382 (1), 89–90. Available at https://doi.org/10.1056/NEJMc1905875, http://www.nejm.org/medical-index.

Baeyens, W., 2001. Ciliary beat frequency in simulated microgravity. European Respiratory Journal 18 (5), 790–795.

Bagian, J.P., Ward, D.F., 1994. A retrospective study of promethazine and its failure to produce the expected incidence of sedation during space flight. The Journal of Clinical Pharmacology 34 (6), 649–651. Available at https://doi.org/10.1002/j.1552-4604.1994.tb02019.x.

Ballard, R.D., 2001. Fluid shifts and respiratory function in spaceflight. Respiratory Physiology 128 (1), 25–34.

Ballard, S.T., Parker, J.C., 2007. Long-term exposure to microgravity impairs mucociliary transport in the airway. American Journal of Respiratory Cell and Molecular Biology 36 (2), 192–199.

Barratt, M.R., Pool, S.L., 2008a. Principles of Clinical Medicine for Space Flight Principles of Clinical Medicine for Space Flight. Springer, New York, United States, pp. 1–596. Available at http://www.springerlink.com/openurl.asp?genre=book&isbn=978-0-387-98842-9.

Barratt, M.R., Pool, S.L., 2008b. Principles of Clinical Medicine for Space Flight. Springerlink, pp. 599.

Barratt, M.R., Pool, S.L., 2008c. Principles of Clinical Medicine for Space Flight. Springerlink, pp. 688–689.

Bateman, G.A., Bateman, A.R., 2022. A perspective on spaceflight associated neuro-ocular syndrome causation secondary to elevated venous sinus pressure. npj Microgravity 8 (1). Available at https://doi.org/10.1038/s41526-022-00188-6, https://www.nature.com/npjmgrav/.

Bauman, E.B., Joffe, A.M., Lenz, L., DeVries, S.A., Hetzel, S., Seider, S.P., 2010. An evaluation of bag-valve-mask ventilation using an ergonomically designed facemask among novice users: a simulation-based pilot study. Resuscitation 81 (9), 1161–1165. Available at https://doi.org/10.1016/j.resuscitation.2010.05.005.

Beard, B.L., 2020. Characterization of How CO2 Level May Impact Crew Performance Related to the HSIA Risk. NASA Ames Research Center.

Beers, R., Camporesi, E., 2004. Remifentanil update: clinical science and utility. CNS Drugs 18 (15), 1085–1104. Available at https://doi.org/10.2165/00023210-200418150-00004.

Biro, P., Hofmann, P., Gage, D., Boehler, Q., Chautems, C., Braun, J., et al., 2020. Automated tracheal intubation in an airway manikin using a robotic endoscope: a proof of concept study. Anaesthesia 75 (7), 881–886. Available at https://doi.org/10.1111/anae.14945.

Blaber, A.P., Goswami, N., Bondar, R.L., Kassam, M.S., 2011. Impairment of cerebral blood flow regulation in astronauts with orthostatic intolerance after flight. Stroke 42 (7), 1844–1850. Available at https://doi.org/10.1161/STROKEAHA.110.610576.

Blue, R., 2019. Development of an Accepted Medical Condition List for Exploration Medical Capability Scoping. Published online.

Bondar, R.L., Stein, F., Vaitkus, P.J., Johnston, K.W., Chadwick, L.C., Norris, J.W., 1990. Transcranial doppler studies of flow velocity in middle cerebral artery in weightlessness. The Journal of Clinical Pharmacology 30 (5), 390–395. Available at https://doi.org/10.1002/j.1552-4604.1990.tb03475.x.

Boom, M., Niesters, M., Sarton, E., Aarts, L., Smith, T.W., Dahan, A., 2012. Non-analgesic effects of opioids: opioid-induced respiratory depression. Current Pharmaceutical Design 18 (37), 5994–6004. Available at https://doi.org/10.2174/138161212803582469.

Boyd, J., Wang, Z., Putcha, L., 2009. Bioavailability of Promethazine during Spaceflight.

Buckey, J.C.S.P., 2023. Chapter 6: Balance: Neurovestibular Effects of Spaceflight and Their Operational Consequences. Oxford University Press.

Buckey, J.C., Lan, M., 2022. Why central venous pressure falls below supine levels in weightlessness. Journal of Applied Physiology (Bethesda, Md.: 1985). 133 (4), 970–972. Available at https://doi.org/10.1152/japplphysiol.00271.2022.

Buckey, J.C., Gaffney, F.A., Lane, L.D., Levine, B.D., Watenpaugh, D.E., Blomqvist, C.G., 1995. Circulatory filling pressures during transient microgravity induced by parabolic flight. NASA Technical Reports. Published online. Available at https://ntrs.nasa.gov/citations/19950008681.

Buckey, J.C., Gaffney, F.A., Lane, L.D., Levine, B.D., Watenpaugh, D.E., Wright, S.J., et al., 1996. Central venous pressure in space. Journal of Applied Physiology 81 (1), 19–25. Available at https://doi.org/10.1152/jappl.1996.81.1.19, jap.physiology.org.

Burden, A., 2020. The history of crises and crisis management in anesthesia: prevention, detection, and recovery. International Anesthesiology Clinics 58 (1), 2–6. Available at https://doi.org/10.1097/aia.0000000000000258.

Burgess, C., Dubbs, C., 2007a. Cosmos/bion: the age of the biosatellites. Animals in Space. Springer Praxis Books. Springer, New York, pp. 277–305. Available at http://doi.org/10.1007/978-0-387-49678-8_10.

Burgess, C., Dubbs, C., 2007b. Animals in Space. Springer Praxis Books. Springer, New York, pp. 25–59. Available at http://doi.org/10.1007/978-0-387-49678-8_2.

Butterworth, J.F., 2022. Morgan & Mikhail's Clinical Anesthesiology, seventh ed. McGraw-Hill LLC.

Carbon Dioxide Levels Chart, 2025. CO2 Meter. Available at www.co2meter.com/blogs/news/carbon-dioxide-indoor-levels-chart?srsltid=AfmBOop6qizQLdcp-1L5cIqfAucuWD-l_e0W-ICwI9D516Mis87LgoM2.

Charles, J.B. 1999. Cardiovascular deconditioning, in Extended Duration Orbiter Medical Project Final Report. Houston, TX: NASA.

Cheng, X., Jiang, G., Lee, K., Laker, Y.N., 2018. IntuBot: design and prototyping of a robotic intubation device. in: 2018 IEEE International Conference on Robotics and Automation (ICRA). IEEE, pp. 1482–1487. doi:10.1109/ICRA.2018.8460779.

Cheung, H.C., De Louche, C., Komorowski, M., 2023. Artificial intelligence applications in space medicine. Aerospace Medicine and Human Performance 94 (8), 610–622. Available at https://doi.org/10.3357/AMHP.6178.2023, http://ingentaconnect.com/content/asma/amhp;jsessionid=13tlejf2f0ojk.alice.

Ciarlone, G.E., Hinojo, C.M., Stavitzski, N.M., Dean, J.B., 2019. CNS function and dysfunction during exposure to hyperbaric oxygen in operational and clinical settings. Redox Biology 27. Available at https://doi.org/10.1016/j.redox.2019.101159, http://www.journals.elsevier.com/redox-biology.

Cirillo, M., De Santo, N.G., Heer, M., Norsk, P., Elmann-Larsen, B., Bellini, L., et al., 2003. Low urinary albumin excretion in astronauts during space missions. Nephron 93 (4), p102. Available at https://doi.org/10.1159/000070243.

Clark, J.B., Bloomberg, J.J., Bacal, K, 2008. Chapter 23: Neurologic concerns. Principles of Clinical Medicine for Space Flight. pp. 711–746.

Clément, G., 2011. Chapter 4: Cardiovascular system in space. Fundamentals of Space Medicine. Springer Science+Business, pp. 143–180.

Cooke, W.H., Ames, J.E., Crossman, A.A., Cox, J.F., Kuusela, T.A., Tahvanainen, K.U.O., et al., 2000. Nine months in space: effects on human autonomic cardiovascular regulation. Journal of Applied Physiology 89 (3), 1039–1045. Available at https://doi.org/10.1152/jappl.2000.89.3.1039.

Corp, A., Thomas, C., Adlam, M., 2021. The cardiovascular effects of positive pressure ventilation. BJA Education 21 (6), 202–209. Available at https://doi.org/10.1016/j.bjae.2021.01.002.

Crampton, G.H., 1990. Neurophysiology of Motion Sickness. CRC Press (Motion, Space Sickness GHC, eds.).

Davis, J.R., Vanderploeg, J.M., Santy, P.A., Jennings, R.T., Stewart, D.F., 1988. Space motion sickness during 24 flights of the space shuttle. Aviation Space and Environmental Medicine 59 (12), 1185–1189.

Davis, J.R., Jennings, R.T., Beck, B.G., 1993a. Comparison of treatment strategies for space motion sickness. Acta Astronautica 29 (8), 587–591. Available at https://doi.org/10.1016/0094-5765(93)90074-7.

Davis, J.R., J, R.T., B, B.G., B, J.P., 1993b. Treatment efficacy of intramuscular promethazine for space motion sickness. Aviation, Space, and Environmental Medicine Published online.

Dello Russo, C., Bandiera, T., Monici, M., Surdo, L., Yip, V.L.M., Wotring, V., et al., 2022. Physiological adaptations affecting drug pharmacokinetics in space: what do we really know? A critical review of the literature. British Journal of Pharmacology 179 (11), 2538–2557. Available at https://doi.org/10.1111/bph.15822.

Di Filippo, E.S., Chiappalupi, S., Falone, S., Dolo, V., Amicarelli, F., Marchianò, S., et al., 2024. The MyoGravity project to study real microgravity effects on human muscle precursor cells and tissue. npj Microgravity 10 (1). Available at https://doi.org/10.1038/s41526-024-00432-1, https://www.nature.com/npjmgrav/.

Du, J., Cui, J., Yang, J., Wang, P., Zhang, L., Luo, B., et al., 2021. Alterations in cerebral hemodynamics during microgravity: a literature review. Medical Science Monitor 27. Available at https://doi.org/10.12659/msm.928108.

Elder, C.A., Moore, J., Janis, B.R., Shacklette, S., Jones, C., Cantrell, R., et al., 2024. Rehydration outcomes for freeze-dried red blood cells in reduced gravity. Acta Astronautica 214, 64–71. Available at https://doi.org/10.1016/j.actaastro.2023.10.006, http://www.journals.elsevier.com/acta-astronautica.

Elliott, A.R., Prisk, G.K., Guy, H.J.B., West, J.B., 1994. Lung volumes during sustained microgravity on spacelab SLS-1. Journal of Applied Physiology 77 (4), 2005–2014. Available at https://doi.org/10.1152/jappl.1994.77.4.2005, jap.physiology.org.

Elliott, A.R., Prisk, G.K., Guy, H.J.B., Kosonen, J.M., West, J.B., 1996. Forced expirations and maximum expiratory flow-volume curves during sustained microgravity on SLS-1. Journal of Applied Physiology 81 (1), 33–43. Available at https://doi.org/10.1152/jappl.1996.81.1.33, jap.physiology.org.

Elliott, A.N.N.R., Shea, S.T.E.V.E.N.A., Dijk, D.E.R.K.-J.A.N., Wyatt, J.A.M.E.S.K., Riel, E.Y.M.A.R.D., Neri, D.A.V.I.D.F., et al., 2001. Microgravity reduces sleep-disordered breathing in humans. American Journal of Respiratory and Critical Care Medicine 164 (3), 478–485. Available at https://doi.org/10.1164/ajrccm.164.3.2010081.

Ewald, R., 1994. Effects of simulated microgravity on the respiratory system. Aviation, Space, and Environmental Medicine 65 (12), 1116–1120.

Fitts, R.H., Riley, D.R., Widrick, J.J., 2000. Physiology of a microgravity environment invited review: microgravity and skeletal muscle. Journal of Applied Physiology 89 (2), 823–839. Available at https://doi.org/10.1152/jappl.2000.89.2.823, https://www.physiology.org/journal/jappl.

Foale, C., 1999. Waystation to the Stars: The Story of Mir, Michael and Me. Headline Book Publishing.

Forman, S.A., Warner, D.S., 2011. Clinical and molecular pharmacology of etomidate. Anesthesiology 114 (3), 695–707. Available at https://doi.org/10.1097/aln.0b013e3181ff72b5.

Foster, J.S., 1998. Effects of simulated microgravity on the ciliary beat frequency of human respiratory epithelium. Aviation, Space, and Environmental Medicine 69 (8), 770–774.

Fritsch, J.M., Charles, J.B., Bennett, B.S., Jones, M.M., Eckberg, D.L., 1992. Short-duration spaceflight impairs human carotid baroreceptor-cardiac reflex responses. Journal of Applied Physiology 73 (2), 664–671. Available at https://doi.org/10.1152/jappl.1992.73.2.664.

Fritsch-Yelle, J.M., Charles, J.B., Jones, M.M., Beightol, L.A., Eckberg, D.L., 1994. Spaceflight alters autonomic regulation of arterial pressure in humans. Journal of Applied Physiology 77 (4), 1776–1783. Available at https://doi.org/10.1152/jappl.1994.77.4.1776, jap.physiology.org.

Fritsch-Yelle, J.M., Charles, J.B., Jones, M.M., Wood, M.L., 1996. Microgravity decreases heart rate and arterial pressure in humans. Journal of Applied Physiology 80 (3), 910–914. Available at https://doi.org/10.1152/jappl.1996.80.3.910.

Fritsch-Yelle, J.M., Leuenberger, U.A., D'Aunno, D.S., Rossum, A.C., Brown, T.E., Wood, M.L., et al., 1998. An episode of ventricular tachycardia during long-duration spaceflight. American Journal of Cardiology 81 (11), 1391–1392. Available at https://doi.org/10.1016/S0002-9149(98)00179-9, www.elsevier.com/locate/amjcard.

Gaba, D.M., Fish, K.J., Howard, S.K., Burden, A.R., Gaba, D.M. (Eds.), 2015. Crisis Management in Anesthesiology, second ed. Elsevier/Saunders.

Ghani, F., Cheung, I., Phillips, A., Tawhai, M., Ashworth, E., 2023. Lung volume, capacity and shape in microgravity: a systematic review and meta-analysis. Acta Astronautica 212, 424–435. Available at https://doi.org/10.1016/j.actaastro.2023.08.005.

Gharib, C., 1995. Effects of microgravity on respiratory system: an overview. European Journal of Applied Physiology and Occupational Physiology 70 (3), 183–190.

Glaister, D., 1970. The Effects of Gravity and Acceleration on the Lung. Technivison Services.

Goh, E., Gallo, R., Hom, J., Strong, E., Weng, Y., Kerman, H., et al., 2024. Large language model influence on diagnostic reasoning: a randomized clinical trial. JAMA Network Open 7 (10). Available at https://doi.org/10.1001/jamanetworkopen.2024.40969, https://jamanetwork.com/journals/jamanetworkopen.

Goudra, B., Singh, P.M., 2017. Failure of sedasys: destiny or poor design? Anesthesia & Analgesia 124 (2), 686–688. Available at https://doi.org/10.1213/ane.0000000000001643.

Graybiel, A., Lackner, J., 1980. Space motion sickness: skylab revisited. Aviation, Space, and Environmental Medicine 51, 814–822.

Groemer, G.E., Brimacombe, J., Haas, T., de Negueruela, C., Soucek, A., Thomsen, M., et al., 2005. The feasibility of laryngoscope-guided tracheal intubation in microgravity during parabolic flight: a comparison of two techniques. Anesthesia & Analgesia 101 (5), 1533–1535. Available at https://doi.org/10.1213/01.ane.0000181001.25777.53.

Gropper, M.A. (Ed.), 2020. Miller's Anesthesia, ninth ed. Elsevier.

Gundel, A., Drescher, J., Spatenko, Y.A., Polyakov, V.V., 2002. Changes in basal heart rate in spaceflights up to 438 days. Aviation Space and Environmental Medicine 73 (1), 17–21.

Guy, H.J., Prisk, G.K., West, J.B., 1988. Pulmonary function in microgravity. Spacelab 4 and Beyond. Acta Astronautica 17 (10), 1139–1143. Available at https://doi.org/10.1016/0094-5765(88)90198-1.

Guy, H.J., Prisk, G.K., West, J.B., 1992. Pulmonary function in microgravity. The Physiologist 35 (1).

Harm, D.L., Sandoz, G.R., Stern, R.M., 2002. Changes in gastric myoelectric activity during space flight. Digestive Diseases and Sciences 47 (8), 1737–1745. Available at https://doi.org/10.1023/A:1016480109272.

Heer, M., Paloski, W.H., 2006. Space motion sickness: incidence, etiology, and countermeasures. Autonomic Neuroscience: Basic and Clinical 129 (1-2), 77–79. Available at https://doi.org/10.1016/j.autneu.2006.07.014.

Hemmerling, T.M., Terrasini, N., 2012. Robotic anesthesia: not the realm of science fiction any more. Current Opinion in Anaesthesiology 25 (6), 736–742. Available at https://doi.org/10.1097/ACO.0b013e328359aa9f.

Hemmerling, T.M., Taddei, R., Wehbe, M., Zaouter, C., Cyr, S., Morse, J., 2012. First robotic tracheal intubations in humans using the Kepler intubation system. British Journal of Anaesthesia 108 (6), 1011–1016. Available at https://doi.org/10.1093/bja/aes034.

Hinkelbein, J., Ahlbäck, A., Antwerber, C., Dauth, L., DuCanto, J., Fleischhammer, E., et al., 2021. Using supraglottic airways by paramedics for airway management in analogue microgravity increases speed and success of ventilation. Scientific Reports 11 (1). Available at https://doi.org/10.1038/s41598-021-88008-x.

Hughson, R.L., Shoemaker, J.K., Blaber, A.P., Arbeille, P., Greaves, D.K., Pereira-Junior, P.P., et al., 2012. Cardiovascular regulation during long-duration spaceflights to the International Space Station. Journal of Applied Physiology 112 (5), 719–727. Available at https://doi.org/10.1152/japplphysiol.01196.2011, http://jap.physiology.org/content/112/5/719.full.pdf+html.

Hussain, I., Ullah, R., Simran, B.F.N.U., Kaur, P., Kumar, M., Raj, R., et al., 2024. Cardiovascular effects of long-duration space flight. Health Science Reports 7 (8). Available at https://doi.org/10.1002/hsr2.2305, onlinelibrary.wiley.com/journal/23988835.

II. Bio-Flight 2 Able, 1959. The Journal of Aviation Medicine 30 (12), 894–912.

Iwasaki, K.I., Levine, B.D., Zhang, R., Zuckerman, J.H., Pawelczyk, J.A., Diedrich, A., et al., 2007. Human cerebral autoregulation before, during and after spaceflight. Journal of Physiology 579 (3), 799–810. Available at https://doi.org/10.1113/jphysiol.2006.119636.

Jennings, R.T., 1998. Managing space motion sickness. Journal of Vestibular Research 8 (1), 67–70. Available at https://doi.org/10.3233/ves-1998-8110.

Jennings, R.T., Stepanek, J.P., Scott, L.R., Voronkov, Y.I., 2010. Frequent premature ventricular contractions in an orbital spaceflight participant. Aviation, Space, and Environmental Medicine 81 (6), 597–601. Available at https://doi.org/10.3357/ASEM.2742.2010.

Johnson, P.C., 1979. Fluid volumes changes induced by spaceflight. Acta Astronautica 6 (10), 1335–1341. Available at https://doi.org/10.1016/0094-5765(79)90125-5.

Johnson, R.L., Driscoll, T.B., LeBlanc, A.D., n.d. Blood volume changes. In: Johnston, R.S., Dietlein, L.F. (Eds.), Biomedical Results from Skylab. NASA SP-377.

Johnston, R.S., Dietlein, L.F., Berry, C.A., 1975. Biomedical results of APOLLO. NASA SP-368. NASA Special Publication, p. 368. Accessed February 10, 2025. https://ui.adsabs.harvard.edu/abs/1975NASSP.368.J.

Kamine, T.H., Siu, M., Kramer, K., Kelly, E., Alouidor, R., Fernandez, G., et al., 2022. Spatial volume necessary to perform open appendectomy in a spacecraft. Aerospace Medicine and Human Performance 93 (10), 760–763. Available at https://doi.org/10.3357/AMHP.6062.2022, http://ingentaconnect.com/content/asma/amhp;jsessionid=13tlejf2f0ojk.alice.

Kanjia, M.K., Kurth, C.D., Hyman, D., Williams, E., Varughese, A., 2024. Perspectives on anesthesia and perioperative patient safety: past, present, and future. Anesthesiology 141 (5), 835–848. Available at https://doi.org/10.1097/ALN.0000000000005164, http://journals.lww.com/anesthesiology/pages/default.aspx.

Keller, C., Brimacombe, J., Giampalmo, M., Kleinsasser, A., Loeckinger, A., Giampalmo, G., et al., 2000. Airway management during spaceflight: a comparison of four airway devices in simulated microgravity. Anesthesiology 92 (5), 1237–1241. Available at https://doi.org/10.1097/00000542-200005000-00010, http://journals.lww.com/anesthesiology/pages/default.aspx.

Kennedy, A.R., 2014. Biological effects of space radiation and development of effective countermeasures. Life Sciences in Space Research 1, 10–43. Available at https://doi.org/10.1016/j.lssr.2014.02.004.

Kermorgant, M., Nasr, N., Czosnyka, M., Arvanitis, D.N., Hélissen, O., Senard, J.-M., et al., 2020. Impacts of microgravity analogs to spaceflight on cerebral autoregulation. Frontiers in Physiology 11. Available at https://doi.org/10.3389/fphys.2020.00778.

Khan, M.J., Karmakar, A., 2023. Emerging robotic innovations and artificial intelligence in endotracheal intubation and airway management: current state of the art. Cureus 15 (7). Available at https://doi.org/10.7759/cureus.42625.

Khan, F.I., Somawardana, I., Dongre, R., Khan, N.S., Rashidi, K., Razmi, S., et al., 2024. Clearing new frontiers: sinonasal health and the future of spaceflight. Ear, Nose & Throat Journal. Available at https://doi.org/10.1177/01455613241274865.

Kim, D.S., Vaquer, S., Mazzolai, L., Roberts, L.N., Pavela, J., Watanabe, M., et al., 2021. The effect of microgravity on the human venous system and blood coagulation: a systematic review. Experimental Physiology 106 (5), 1149–1158. Available at https://doi.org/10.1113/EP089409, http://onlinelibrary.wiley.com/journal/10.1111/(ISSN)1469-445X.

Kirkpatrick, A.W., Dulchavsky, S.A., Boulanger, B.R., Campbell, M.R., Hamilton, D.R., Dawson, D.L., et al., 2001. Extraterrestrial resuscitation of hemorrhagic shock: fluids. Journal of Trauma - Injury, Infection and Critical Care 50 (1), 162–168. Available at https://doi.org/10.1097/00005373-200101000-00036, http://www.jtrauma.com.

Kirkpatrick, A.W., Hamilton, D.R., Nicolaou, S., Sargsyan, A.E., Campbell, M.R., Feiveson, A., et al., 2003. Focused assessment with sonography for trauma in weightlessness: a feasibility study. Journal of the American College of Surgeons 196 (6), 833–844. Available at https://doi.org/10.1016/s1072-7515(02)01906-3.

Klein, T., Sanders, M., Wollseiffen, P., Carnahan, H., Abeln, V., Askew, C.D., et al., 2020. Transient cerebral blood flow responses during microgravity. Life Sciences in Space Research 25, 66–71. Available at https://doi.org/10.1016/j.lssr.2020.03.003.

Komorowski, M., Campbell, M.R., 2018. The bion project and the history of post-spaceflight primate anesthesia. Aerospace Medicine and Human Performance 89 (4), 414–415. Available at https://doi.org/10.3357/AMHP.5060.2018, http://docserver.ingentaconnect.com/deliver/connect/asma/23756314/v89n4/s14.pdf?expires=1522904871&id=0000&titleid=72010555&checksum=06D8E033DF91A0B86209EB18911A6BB6.

Komorowski, M., Watkins, S.D., Lebuffe, G., Clark, J.B., 2013. Potential anesthesia protocols for space exploration missions. Aviation Space and Environmental Medicine 84 (3), 226–233. Available at https://doi.org/10.3357/ASEM.3427.2013, http://docserver.ingentaconnect.com/deliver/connect/asma/00956562/v84n3/s8.pdf.

Komorowski, M., Fleming, S., Hinkelbein, J., 2016. Anaesthesia in outer space: the ultimate ambulatory setting? Current Opinion in Anaesthesiology 29 (6), 649–654. Available at https://doi.org/10.1097/ACO.0000000000000390, http://journals.lww.com/co-anesthesiology/pages/default.aspx.

Komorowski, M., Fleming, S., Mawkin, M., Hinkelbein, J., 2018. Anaesthesia in austere environments: literature review and considerations for future space exploration missions. npj Microgravity 4 (1). Available at https://doi.org/10.1038/s41526-018-0039-y.

Krämer, B.K., Mang, J.F., Schubert, R., 2018. The effect of microgravity on central aortic blood pressure. American Journal of Hypertension 31 (11), 1180–1182. Available at https://doi.org/10.1093/ajh/hpy133, https://academic.oup.com/ajh/issue.

Kreykes, A.J., Suresh, R., Levin, D., Hilmers, D.C., 2023. Selecting medical conditions relevant to exploration spaceflight to create the IMPACT 1.0 medical condition list. Aerospace Medicine and Human Performance 94 (7), 550–557. Available at https://doi.org/10.3357/AMHP.6199.2023, http://ingentaconnect.com/content/asma/amhp;jsessionid=13tlejf2f0ojk.alice.

Krittanawong, C., Isath, A., Kaplin, S., Virk, H.U.H., Fogg, S., Wang, Z., et al., 2023. Cardiovascular disease in space: a systematic review. Progress in Cardiovascular Diseases 81, 33–41. Available at https://doi.org/10.1016/j.pcad.2023.07.009.

Kunz, H., Quiriarte, H., Simpson, R.J., Ploutz-Snyder, R., McMonigal, K., Sams, C., et al., 2017. Alterations in hematologic indices during long-duration spaceflight. BMC Hematology 17 (1). Available at https://doi.org/10.1186/s12878-017-0083-y, http://www.springer.com/medicine/internal/journal/12878.

Kuzichkin, D.S., Nichiporuk, I.A., Zhuravleva, O.A., Markin, A.A., Rykova, M.P., Zhuravleva, T.V., et al., 2022. Endothelial dysfunction markers and immune response indices in cosmonauts' blood after long-duration space flights. npj Microgravity 8 (1). Available at https://doi.org/10.1038/s41526-022-00237-0, https://www.nature.com/npjmgrav/.

Lackner, J.R., Graybiel, A., 1994. Use of promethazine to hasten adaptation to provocative motion. The Journal of Clinical Pharmacology 34 (6), 644–648. Available at https://doi.org/10.1002/j.1552-4604.1994.tb02018.x.

Lampe, L., Wienhold, K., Meyer, G., Baisch, F., Maass, H., Hollmann, W., et al., 1992. Effects of simulated microgravity (HDT) on blood fluidity. Journal of Applied Physiology 73 (4), 1366–1369. Available at https://doi.org/10.1152/jappl.1992.73.4.1366.

Lan, M., Van Akin, M., Anderson, A., Buckey, J.C., 2022. Microgravity-induced reduced jugular vein flow is more pronounced on the non-dominant side. Acta Astronautica 198, 746–755. Available at https://doi.org/10.1016/j.actaastro.2022.05.048, http://www.journals.elsevier.com/acta-astronautica.

Lane, H.W., Whitson, P.A., Putcha, L, 1999. Regulatory physiology: gastrointestinal function during extended duration space flight. In: Sawin, CF, Taylor, GR, Smith, WL (Eds.), Extended Duration Orbiter Medical Project Final Report, Vol. 2.

Larijani, G.E., Gratz, I., Afshar, M., Jacobi, A.G., 1989. Clinical pharmacology of propofol: an intravenous anesthetic agent. DICP 23 (10), 743–749. Available at https://doi.org/10.1177/106002808902301001.

Law, J., Van Baalen, M., Foy, M., Mason, S.S., Mendez, C., Wear, M.L., et al., 2014. Relationship between carbon dioxide levels and reported headaches on the International Space Station. Journal of Occupational & Environmental Medicine 56 (5), 477–483. Available at https://doi.org/10.1097/jom.0000000000000158.

Lawley, J.S., Petersen, L.G., Howden, E.J., Sarma, S., Cornwell, W.K., Zhang, R., et al., 2017. Effect of gravity and microgravity on intracranial pressure. The Journal of Physiology 595 (6), 2115–2127. Available at https://doi.org/10.1113/jp273557.

Leach, C.S., Alfrey, C.P., Suki, W.N., Leonard, J.I., Rambaut, P.C., Inners, L.D., et al., 1996. Regulation of body fluid compartments during short-term spaceflight. Journal of Applied Physiology 81 (1), 105–116. Available at https://doi.org/10.1152/jappl.1996.81.1.105, jap.physiology.org.

LeBlanc, A., Lin, C., Shackelford, L., Sinitsyn, V., Evans, H., Belichenko, O., et al., 2000. Muscle volume, MRI relaxation times (T2), and body composition after spaceflight. Journal of Applied Physiology 89 (6), 2158–2164. Available at https://doi.org/10.1152/jappl.2000.89.6.2158.

Lee, S.M.C., Feiveson, A.H., Stein, S., Stenger, M.B., Platts, S.H., 2015. Orthostatic intolerance after ISS and space shuttle missions. Aerospace Medicine and Human Performance 86 (12), 54–67. Available at https://doi.org/10.3357/AMHP.EC08.2015.

Lee, S., Onye, A., Latif, A., 2020a. Emergency anesthesia in resource-limited areas. Anesthesiology Clinics 38 (1), 213–230. Available at https://doi.org/10.1016/j.anclin.2019.10.011.

Lee, S.M.C., Martin, D.S., Miller, C.A., Scott, J.M., Laurie, S.S., Macias, B.R., et al., 2020b. Venous and arterial responses to partial gravity. Frontiers in Physiology 11. Available at https://doi.org/10.3389/fphys.2020.00863, http://www.frontiersin.org/Physiology/archive/.

Leite, A.P.S., Pinto, C.G., Tibúrcio, F.C., Muller, K.S., Padovani, C.R., Barraviera, B., et al., 2023. Acetylcholine receptors of the neuromuscular junctions present normal distribution after peripheral nerve injury and repair through nerve guidance associated with fibrin biopolymer. Injury 54 (2), 345–361. Available at https://doi.org/10.1016/j.injury.2022.11.047.

Lerner, D., Pohlen, M., Wang, A., Walter, J., Cairnie, M., Gifford, S., 2023. X-ray imaging in the simulated microgravity environment of parabolic flight. Aerospace Medicine and Human Performance 94 (10), 786–791. Available at https://doi.org/10.3357/amhp.6286.2023.

Liang, F., Wu, Y., Wang, X., Yan, L., Zhang, S., Jian, M., et al., 2023. General anesthesia vs conscious sedation for endovascular treatment in patients with posterior circulation acute ischemic stroke: an exploratory randomized clinical trial. JAMA Neurology 80 (1), 64–72. American Medical Association China. Available at https://doi.org/10.1001/jamaneurol.2022.3018, http://archneur.jamanetwork.com/issues.aspx.

Mackowski, M.P., 2022. Life in Space: NASA Life Sciences Research during the Late Twentieth Century. University of Florida Press. Available at http://doi.org/10.2307/j.ctv2j04sfm.

Macnamara, J.P., Hearon, C.M., Manferdelli, G., Shah, A.M., Tayon, K.G., Pandey, A., et al., 2025. Heart failure in zero gravity-external constraint and cardiac hemodynamics. JAMA Cardiology 10 (1), 98–100. Available at https://doi.org/10.1001/jamacardio.2024.4596, http://jamanetwork.com/journals/jamacardiology.

Marshall-Goebel, K., Macias, B.R., Laurie, S.S., Lee, S.M.C., Ebert, D.J., Kemp, D.T., et al., 2021. Mechanical countermeasures to headward fluid shifts. Journal of Applied Physiology 130 (6), 1766–1777. Available at https://doi.org/10.1152/japplphysiol.00863.2020, https://journals.physiology.org/doi/full/10.1152/japplphysiol.00863.2020.

Marshall-Goebel, K., Lee, S.M.C., Lytle, J.R., Martin, D.S., Miller, C.A., Young, M., et al., 2024. Jugular venous flow dynamics during acute weightlessness. Journal of Applied Physiology 136 (5), 1105–1112. Available at https://doi.org/10.1152/japplphysiol.00384.2023, https://journals.physiology.org/doi/epdf/10.1152/japplphysiol.00384.2023.

Martin, D.G., Convertino, Goldwater, Ferguson, Schoomaker, E.B., 1986. Plasma viscosity elevations with simulated weightlessness. Aviation, space, and environmental medicine 57 (5), 426–431.

Martin, D.S., Lee, S.M.C., Matz, T.P., Westby, C.M., Scott, J.M., Stenger, M.B., et al., 2016. Internal jugular pressure increases during parabolic flight. Physiological Reports 4 (24). Available at https://doi.org/10.14814/phy2.13068, http://physoc.onlinelibrary.wiley.com/hub/journal/10.1002/(ISSN)2051-817X/issues/.

Martyn, J.A.J., Richtsfeld, M., 2006. Succinylcholine-induced hyperkalemia in acquired pathologic states: etiologic factors and molecular mechanisms. Anesthesiology 104 (1), 158–169. Available at https://doi.org/10.1097/00000542-200601000-00022, http://journals.lww.com/anesthesiology/pages/default.aspx.

May, C., Borowski, A., Martin, D., Popovic, Z., Negishi, K., Hussan, J.R., et al., 2014. Affect of microgravity on cardiac shape: comparison of pre- and in-flight data to mathematical modeling. Journal of the American College of Cardiology 63 (12), A1096. Available at https://doi.org/10.1016/s0735-1097(14)61096-2.

Meck, J.V., Reyes, C.J., Perez, S.A., Goldberger, A.L., Ziegler, M.G., 2001. Marked exacerbation of orthostatic intolerance after long-vs.-short-duration spaceflight in veteran astronauts. Psychosomatic Medicine 63 (6), 865–873. Available at https://doi.org/10.1097/00006842-200111000-00003, http://www.psychosomaticmedicine.org/.

Medical Kit- Contents and Reference. NASA, 2015.

Medical Operations Team Activities, Retrieved 17 March 2026. Available at https://humans-in-space.jaxa.jp/en/.

Michael, A.P., Marshall-Bowman, K., 2015. Spaceflight-induced intracranial hypertension. Aerospace Medicine and Human Performance 86 (6), 557–562. Available at https://doi.org/10.3357/AMHP.4284.2015, http://docserver.ingentaconnect.com/deliver/connect/asma/23756314/v86n6/s10.pdf.

Mission, P.D. SpaceX. Published online September 10, 2024.

Mittal, B., Asadi, A., Rosenberg, M.J., 2024. Assessment of Acute Upper Airway Changes During Spaceflight Using Endoscopic Airway Examination and Ultrasound. NASA Human Research Program.

Moore, T.P., Thornton, W.E., 1987. Space shuttle inflight and postflight fluid shifts measured by leg volume changes. Aviation, Space, and Environmental Medicine 58 (9).

Mukai, C.N., Lathers, C.M., Charles, J.B., Bennett, B.S., Igarashi, M., Patel, S., 1991. Acute hemodynamic responses to weightlessness during parabolic flight. Journal of Clinical Pharmacology 31 (10).

Musculo-skeletal system: bone and muscle loss, n.d. European Space Agency. Available from: www.esa.int/Enabling_Support/Preparing_for_the_Future/Space_for_Earth/Space_for_health/Musculoskeletal_system_Bone_and_Muscle_loss#:~:text=Muscle%20loss%20of%2010%20to,landing%20(e.g.%20emergency%20egress).

Navasiolava, N., Yuan, M., Murphy, R., Robin, A., Coupé, M., Wang, L., et al., 2020. Vascular and microvascular dysfunction induced by microgravity and its analogs in humans: mechanisms and countermeasures. Frontiers in Physiology 11. Available at https://doi.org/10.3389/fphys.2020.00952.

Nicogossian, A.E., 2016a. Space Physiology and Medicine from Evidence to Practice. pp. 285.

Nicogossian, A.E., 2016b. Space Physiology and Medicine from Evidence to Practice. pp. 12.

Norsk, P., 2020. Adaptation of the cardiovascular system to weightlessness: surprises, paradoxes and implications for deep space missions. Acta Physiologica 228 (3). Available at https://doi.org/10.1111/apha.13434.

Norsk, P., Damgaard, M., Petersen, L., Gybel, M., Pump, B., Gabrielsen, A., et al., 2006. Vasorelaxation in Space. Hypertension 47 (1), 69–73. Available at https://doi.org/10.1161/01.hyp.0000194332.98674.57.

Norsk, P., Asmar, A., Damgaard, M., Christensen, N.J., 2015. Fluid shifts, vasodilatation and ambulatory blood pressure reduction during long duration spaceflight. Journal of Physiology 593 (3), 573–584. Available at https://doi.org/10.1113/jphysiol.2014.284869, http://onlinelibrary.wiley.com/journal/10.1111/(ISSN)1469-7793.

Otten, D., Liao, M.M., Wolken, R., Douglas, I.S., Mishra, R., Kao, A., et al., 2014. Comparison of bag-valve-mask hand-sealing techniques in a simulated model. Annals of Emergency Medicine 63 (1), 6. Available at https://doi.org/10.1016/j.annemergmed.2013.07.014.

Padaki, A., Police Reddy, A., Lehnhardt, K., 2019. The utility of hyperbaric oxygen therapy for human spaceflight – past, present, and future. Acta Astronautica 164, 192–196. Available at https://doi.org/10.1016/j.actaastro.2019.08.003.

Perhonen, M.A., Franco, F., Lane, L.D., Buckey, J.C., Blomqvist, C.G., Zerwekh, J.E., et al., 2001. Cardiac atrophy after bed rest and spaceflight. Journal of Applied Physiology 91 (2), 645–653. Available at https://doi.org/10.1152/jappl.2001.91.2.645, https://www.physiology.org/journal/jappl.

Pourmand, A., Terrebonne, E., Gerber, S., Shipley, J., Tran, Q.K., 2023. Efficacy of video laryngoscopy versus direct laryngoscopy in the prehospital setting: a systematic review and meta-analysis. Prehospital and Disaster Medicine 38 (1), 111–121. Available at https://doi.org/10.1017/S1049023X22002254, http://journals.cambridge.org/action/displayJournal?jid=PDM.

Prekker, M.E., Driver, B.E., Trent, S.A., Resnick-Ault, D., Seitz, K.P., Russell, D.W., et al., 2023. Video versus direct laryngoscopy for tracheal intubation of critically Ill adults. New England Journal of Medicine 389 (5), 418–429. Available at https://doi.org/10.1056/NEJMoa2301601, http://www.nejm.org/medical-index.

Prisk, G.K., 2000. Microgravity and the lung. Journal of Applied Physiology (Bethesda, Md: 1985) 89 (1), 385–396.

Prisk, G.K., 2005. The lung in space. Clinics in Chest Medicine 26 (3), 415–438. Available at https://doi.org/10.1016/j.ccm.2005.05.008.

Prisk, G.K., 2014a. Microgravity and the respiratory system. European Respiratory Journal 43 (5), 1459–1471. Available at https://doi.org/10.1183/09031936.00001414, http://erj.ersjournals.com/content/43/5/1459.full.pdf+html.

Prisk, G.K., 2014b. Pulmonary gas exchange in microgravity. Current Pharmaceutical Biotechnology 15 (1), 48–54.

Prisk, G.K., 2019. Pulmonary challenges of prolonged journeys to space: taking your lungs to the moon. Medical Journal of Australia 211 (6), 271–276. Available at https://doi.org/10.5694/mja2.50312, www.mja.com.au.

Prisk, G.K., West, J.B., 1998. Effects of spaceflight on the lung. Journal of Applied Physiology 85 (1), 81–86.

Prisk, G.K., Guy, H.J.B., Elliott, A.R., 1995. Pulmonary diffusing capacity, capillary blood volume, and cardiac output during sustained microgravity. Journal of Applied Physiology 79 (1), 129–136.

Prisk, G.K., Elliott, A.R., West, J.B., 2000. Sustained microgravity reduces the human ventilatory response to hypoxia but not to hypercapnia. Journal of Applied Physiology 88 (4), 1421–1430. Available at https://doi.org/10.1152/jappl.2000.88.4.1421, https://www.physiology.org/journal/jappl.

Rabitsch, W., Moser, D., Inzunza, M.R., Niedermayr, M., Köstler, W.J., Staudinger, T., et al., 2006. Airway management with endotracheal tube versus Combitube® during parabolic flights. Anesthesiology 105 (4), 696–702. Available at https://doi.org/10.1097/00000542-200610000-00014, http://journals.lww.com/anesthesiology/pages/default.aspx.

Reschke, M.F., Harm, D.L., Parker, D.E., 1994. Neurophysiologic aspects: space motion sickness. In: Nicogossian, A.E., Huntoon, C.L., Pool, S.L. (Eds.), Space Physiology and Medicine, 3rd edn. Lea & Febiger.

Rossum, A.C., Wood, M.L., Bishop, S.L., Deblock, H., Charles, J.B., 1997. Evaluation of cardiac rhythm disturbances during extravehicular activity. American Journal of Cardiology 79 (8), 1153–1155. Available at https://doi.org/10.1016/S0002-9149(97)00071-4.

Royland, J.E., Weber, L.J., 1989. The effect of suspension on nicotinic acetylcholine receptor number and affinity at the rat neuromuscular junction. Physiologist 32 (1).

Schlegel, T.T., Benavides, E.W., Barker, D.C., Brown, T.E., Harm, D.L., DeSilva, S.J., et al., 1998. Cardiovascular and Valsalva responses during parabolic flight. Journal of Applied Physiology 85 (5), 1957–1965. Available at https://doi.org/10.1152/jappl.1998.85.5.1957.

Schrier, R.W., Berl, T., Anderson, R.J., 1979. Osmotic and nonosmotic control of vasopressin release. American Journal of Physiology - Renal Fluid and Electrolyte Physiology. 5 (4), F321. Available at https://doi.org/10.1152/ajprenal.1979.236.4.f321.

Simanonok, K.E., Charles, J.B., 1994. Space sickness and fluid shifts: a hypothesis. The Journal of Clinical Pharmacology 34 (6), 652–663. Available at https://doi.org/10.1002/j.1552-4604.1994.tb02020.x.

Slupe, A.M., Kirsch, J.R., 2018. Effects of anesthesia on cerebral blood flow, metabolism, and neuroprotection. Journal of Cerebral Blood Flow and Metabolism 38 (12), 2192–2208. Available at https://doi.org/10.1177/0271678X18789273, http://journals.sagepub.com/loi/jcb.

Smirnov, K.V., Ugolev, A.M., n.d. Digestion and absorption. In: Leach Huntoon, C.S., Antipov, V.V., Grigoriev, A.I. (Eds.), Humans.

Smirnov K.V., Ugolev A.M., n.d. Digestion and absorption. In: Space Biology and Medicine. Book 2: Vol. 3, Humans in Spaceflight Ed. Carolyn S. Leach Huntoon. Vol 3. 2nd ed. AIAA; 1996:211-230.

Smith, R.F., 1977. Vector cardiographic changes during extended space flight (M093): observations at rest and during exercise. In: Johnston, R.S., Dietlein, L.F. (Eds.), Biomedical Results of Skylab, pp. 339–350.

Smith, R.F., Stanton, K., Stoop, D., Janusz, W., King, P.H., 1976. Quantitative electrocardiography during extended space flight: the second manned skylab mission. Aviation Space and Environmental Medicine 47 (4), 353–359.

Smith, M.B., Chen, H., Oliver, B.G.G., 2024. The lungs in space: a review of current knowledge and methodologies. Cells 13 (13), 1154. Available at https://doi.org/10.3390/cells13131154.

Stahn, A.C., Werner, A., Opatz, O., Maggioni, M.A., Steinach, M., Von Ahlefeld, V.W., et al., 2017. Increased core body temperature in astronauts during long-duration space missions. Scientific Reports 7 (1). Available at https://doi.org/10.1038/s41598-017-15560-w, www.nature.com/srep/index.html.

Starck, C., Thierry, S., Bernard, C.I., Morineau, T., Jaulin, F., Chapelain, P., et al., 2020. Tracheal intubation in microgravity: a simulation study comparing direct laryngoscopy and videolaryngoscopy† British Journal of Anaesthesia 125 (1), e47. Available at https://doi.org/10.1016/j.bja.2019.11.029, https://www.journals.elsevier.com/british-journal-of-anaesthesia.

Statement on continuum of depth of sedation: definition of general anesthesia and levels of sedation/analgesia. Available from: https://www.asahq.org/standards-and-practice-parameters/statement-on-continuum-of-depth-of-sedation-definition-of-general-anesthesia-and-levels-of-sedation-analgesia, (accessed 25.04.18).

Stepanek, J., Blue, R.S., Connolly, D., 2023. Pulmonary function in human spaceflight. Seminars in Respiratory and Critical Care Medicine 44 (5), 696–704. Available at https://doi.org/10.1055/s-0043-1770064.

Summers, R.L., Johnston, S.L., Marshburn, T.H., Williams, D.R., 2005a. Emergencies in space. Annals of Emergency Medicine 46 (2), 177–184. Available at https://doi.org/10.1016/j.annemergmed.2005.02.010, http://www.elsevier.com/inca/publications/store/6/2/3/2/7/8/index.htt.

Summers, R.L., Martin, D.S., Meck, J.V., Coleman, T.G., 2005b. Mechanism of spaceflight-induced changes in left ventricular mass. American Journal of Cardiology 95 (9), 1128–1130. Available at https://doi.org/10.1016/j.amjcard.2005.01.033, www.elsevier.com/locate/amjcard.

Sun, P., Gao, X., Wei, D., Ge, J., Deng, X., Chen, H., et al., 2025. The digestive system under microgravity environment: changes, mechanisms and the prospects of the future. Gastroenterology and Endoscopy 3 (2), 108–115. Available at https://doi.org/10.1016/j.gande.2024.12.002, https://www.sciencedirect.com/journal/gastroenterology-and-endoscopy.

Taylor, C.R., Hanna, M., Behnke, B.J., Stabley, J.N., McCullough, D.J., Davis, R.T., et al., 2013. Spaceflight-induced alterations in cerebral artery vasoconstrictor, mechanical, and structural properties: implications for elevated cerebral perfusion and intracranial pressure. FASEB Journal 27 (6), 2282–2292. Available at https://doi.org/10.1096/fj.12-222687, http://www.fasebj.org/content/27/6/2282.full.pdf+html.

Thierry, S., Jaulin, F., Starck, C., Ariès, P., Schmitz, J., Kerkhoff, S., et al., 2023. Evaluation of free-floating tracheal intubation in weightlessness via ice-pick position with a direct laryngoscopy and classic approach with indirect videolaryngoscopy. npj Microgravity 9 (1). Available at https://doi.org/10.1038/s41526-023-00314-y.

Thornton, W.E., Moore, T.P., 1986. Neurological studies: bowel sounds. NASA 235–238.

Thornton, W.E., Linder, B.J., Moore, T.P., Pool, S.L., 1987. Gastrointestinal motility in space motion sickness. Aviation, Space, and Environmental Medicine 58 (9).

Tigranyan, R.A., 1985. Metabolic aspects of problems in stress in spaceflight. Problemy Kosmicheskoi Biologii 52, 1–222.

Tymko, M.M., Boulet, L.M., Donnelly, J., 2017. Intracranial pressure in outer space: preparing for the mission to Mars. The Journal of Physiology 595 (14), 4587–4588. Available at https://doi.org/10.1113/jp274315.

Venturoli, D., Semino, P., Negrini, D., Miserocchi, G., 1998. Respiratory mechanics after 180 days space mission (EUROMIR'95). Acta Astronautica 42 (1-8), 185–204. Available at https://doi.org/10.1016/s0094-5765(98)00116-7.

Verbandt, Y., Wantier, M., Prisk, G.K., Paiva, M., 2000. Ventilation-perfusion matching in long-term microgravity. Journal of Applied Physiology 89 (6), 2407–2412. Available at https://doi.org/10.1152/jappl.2000.89.6.2407, https://www.physiology.org/journal/jappl.

Verheyden, B., Liu, J., Beckers, F., Aubert, A.E., 2009. Adaptation of heart rate and blood pressure to short and long duration space missions. Respiratory Physiology and Neurobiology 169, S13. Available at https://doi.org/10.1016/j.resp.2009.03.008, http://www.elsevier.com/inca/publications/store/6/2/2/7/2/7/.

Videbaek, R., Norsk, P., 1997. Atrial distension in humans during microgravity induced by parabolic flights. Journal of Applied Physiology 83 (6), 1862–1866. Available at https://doi.org/10.1152/jappl.1997.83.6.1862.

Wang, X., Tao, Y., Tao, X., Chen, J., Jin, Y., Shan, Z., et al., 2018. An original design of remote robot-assisted intubation system. Scientific Reports 8 (1). Available at https://doi.org/10.1038/s41598-018-31607-y.

Wantier, M., Estenne, M., Verbanck, S., Prisk, G.K., Paiva, M., 1998. Chest wall mechanics in sustained microgravity. Journal of Applied Physiology 84 (6), 2060–2065. Available at https://doi.org/10.1152/jappl.1998.84.6.2060, https://www.physiology.org/journal/jappl.

Warnecke, T., Tochtermann, F., Kerkhoff, S., Komorowski, M., Neuhaus, C., Hinkelbein, J., 2019. Airway management in microgravity: a systematic review. Acta Anaesthesiologica Scandinavica 63 (1), 2–7. Available at https://doi.org/10.1111/aas.13251.

Waters, E.M., Gerstner, H.B., Huff, J.E., 2009. Trichloroethylene. I. An overview. Journal of Toxicology and Environmental Health 2 (3), 671–707. Available at https://doi.org/10.1080/15287397709529469.

West, J.B., 2001. Pulmonary physiology and pathology in microgravity. Journal of Applied Physiology 91 (1), 353–369.

West, J.B., Elliott, A.R., Guy, H.J.B., Prisk, G.K., 1997. Pulmonary function in space. JAMA 277 (24), 1957–1961. Available at https://doi.org/10.1001/jama.277.24.1957, http://jama.jamanetwork.com/journal.aspx.

White, N.J., Wenthe, A., 2023. Managing hemostasis in space. Arteriosclerosis, Thrombosis, and Vascular Biology 43 (11), 2079–2087. Available at https://doi.org/10.1161/ATVBAHA.123.318783, https://www.ahajournals.org/journal/atvb.

Wotring, V., 2020. Spaceflight pharmacology. Principles of Clinical Medicine for Space Flight. Springer New York, France, pp. 815–840. Available at http://doi.org/10.1007/978-1-4939-9889-0.

Yang, J.-Q., Jiang, N., Li, Z.-P., Guo, S., Chen, Z.-Y., Li, B.-B., et al., 2020. The effects of microgravity on the digestive system and the new insights it brings to the life sciences. Life Sciences in Space Research 27, 74–82. Available at https://doi.org/10.1016/j.lssr.2020.07.009.

Chapter 20

# Local and regional anesthesia

Matthew A. Levin[1,2,3], Cassandra Zhi[4], Amran Asadi[5,6], and Ariana M. Nelson[7]

[1]*Department of Anesthesiology, Perioperative and Pain Medicine, Icahn School of Medicine at Mount Sinai, New York, NY, United States,* [2]*Department of AI and Human Health, Icahn School of Medicine at Mount Sinai, New York, NY, United States,* [3]*Department of Genetics and Genomic Sciences, Icahn School of Medicine at Mount Sinai, New York, NY, United States,* [4]*Department of Anesthesiology, Maimonides Medical Center, Brooklyn, NY, United States,* [5]*Space Exploration Technologies (SpaceX), Hawthorne, CA, United States,* [6]*Department of Anesthesiology and Perioperative Medicine, University of California Los Angels, Los Angeles, CA, United States,* [7]*Department of Anesthesiology & Perioperative Medicine, University of California Irvine, Irvine, CA, United States*

## 20.1 Anesthetic options in spaceflight

### 20.1.1 Indications for local or regional anesthesia in spaceflight

Common injuries encountered in the space environment include sprains or strains, which are frequently documented in the Lifetime Surveillance of Astronaut Health database (Lifetime Surveillance of Astronaut Health (LSAH), 2025). Occasional lacerations have also occurred, and repetitive nail trauma during extended EVA may induce subungual hematoma requiring trephination (hematoma evacuation via creation of an aperture for aspiration), or partial or full delamination (Bloom et al., 2020; Salter et al., 2006; Opperman et al., 2010). Skin rashes during spaceflight occur with relatively high frequency and may be due in part to abnormal skin cell turnover and immune dysregulation (Nguyen and Urquieta, 2023; Mehta et al., 2022). Soft tissue infections, such as abscesses, have a reasonable probability of occurring as bacteria commonly found on surfaces in the International Space Station (ISS) exhibit properties associated with increased virulence and antibiotic resistance, which increases risk of progression to abscess requiring treatment with incision and drainage.

The extent of most injuries documented to date has been relatively minor due to rigorous engineering and operational controls, as well as the relative speed withwhich definitive care in an emergency could be made available when operating in low earth orbit (LEO). An astronaut with a fracture (a low likelihood event) could theoretically be evacuated to Earth from LEO for surgery in as little as 4–24 hours. However, as the scope of space exploration expands to include both exploration and surface habitat construction on the moon and beyond, the risk of serious injury and the need to definitively treat in place can both be expected to increase. Surveys of astronaut bone mineral density (BMD) after exposure to microgravity demonstrate rates of decline of 0.5% to 1.5% on a *monthly* basis, which is comparable to the *annual* BMD loss in postmenopausal women, and most significantly affects load bearing regions of the skeleton (lumbar spine, femoral neck, etc.; Baran et al., 2022). This decline in BMD has been shown to translate to increased risk of skeletal fracture in long duration astronauts, with a hip fracture relative risk in males of approximately 7.2 times the general population at age 65 (Sibonga et al., 2024). Modeling efforts for future Mars design reference missions indicate that, with current countermeasures, 62% of astronauts on return would meet BMD criteria for osteoporosis at the femoral neck (Axpe et al., 2020). Risk of wrist fracture following a fall during a hypothetical lunar or martian surface EVA is estimated to be as high as 0.31%–1.28% for men and 0.83%–2.4% for women (Lewandowski et al., 2008). Martian surface EVAs are predicted to be at higher fracture risk compared to lunar EVA due to both higher gravitational forces, and exposure of astronauts to prolonged microgravity in transit. Some in the space medicine community opine that the lower gravitational pull of the lunar surface would actually increase risk of fracture compared to Earth as mid-course correction after a stumble is more difficult in an unfamiliar gravitational environment. Regardless of likelihood, if a fracture were to occur, delays of one or more weeks are predicted for lunar surface evacuation. Martian evacuation would also include delay for an appropriate transit window, which may exceed a year and therefore is unlikely to be more rapid than the initially planned return mission (Hilmers and Kreykes, 2023).

**Procedural Medicine for Spaceflight. DOI:** https://doi.org/10.1016/B978-0-443-27552-4.00022-1

#### 20.1.1.1 General versus neuraxial anesthesia

Although many of the predicted procedures for long duration exploration missions would be performed under general anesthesia (GA) or neuraxial anesthesia on Earth, these techniques are less feasible in flight. Utilization of GA in space is limited by equipment upmass, substantial operator skill required, extended recovery process, and potential for high-risk complications. Available commercial off-the-shelf equipment for delivery of inhaled anesthetics is currently not well suited to applications in the spaceflight environment. Anesthesia machines occupy substantial volume and would pose almost insurmountable integration challenges with vehicle avionics and life support systems. Most anesthetic vaporizers rely on the normal gravitational force for air liquid separation and in microgravity would be susceptible to dangerously high volatile output and/or obstructed outflow. Even if these challenges could be mitigated, the leakage of volatile anesthetics would contaminate the enclosed cabin space. Total intravenous anesthesia (TIVA) is a more viable option than volatile anesthetic for the maintenance of GA but also carries the risk of hemodynamic perturbations, albeit to a lesser degree. Crew may be more susceptible to cardiovascular instability from GA due to the expected physiologic changes induced by the weightless environment. There is also the concern for complications associated with airway management in a constrained and enclosed location, which can occur with either GA or TIVA, as the expected apnea from induction of anesthesia requires that crewmembers oxygenate and ventilate their anesthetized colleague (Drudi et al., 2012).

Neuraxial techniques involve the injection of local anesthetics to the spinal column in either the intrathecal (i.e. spinal) or epidural space. Spinal anesthesia terrestrially provides a dense block to the lower extremities, pelvis, and abdomen up to the approximate level of the nipple (T4 dermatome). The duration of surgical block is somewhat modifiable based on the local anesthetic selected but is typically in the range of 1–2.5 hours. More extended surgical procedures or prolonged analgesic coverage could be accomplished by placing a lumbar epidural or spinal catheter, which allows redosing at intervals. Although placement of an epidural catheter is more technically challenging, it does pose as an additional benefit to gradually increase the level of block to avoid a potentially catastrophic overdose, or avoid the hypotension associated with sympathectomy that occurs with the rapid-onset dosing of spinal anesthesia. In either case, spinal or epidural, a single procedure with a small quantity of medication can provide sufficient anesthesia for lower extremity fracture reductions, debridements, large soft tissue defect repairs, urologic and gynecological procedures, and some abdominal surgeries. However, microgravity-induced fluid shifts complicate neuraxial techniques, as at the onset of weightlessness all fluids contained in the human body shift superiorly. These cephalad fluid shifts are thought to increase intracranial pressure (ICP) which may change the expected distribution of medication in the cerebrospinal fluid (CSF) during neuraxial block. Hypothetically, although difficult to definitely study, this could result in unpredictable location and density of a neuraxial anesthetic block (Longo et al., 2019). The baricity of the local anesthetic relative to the CSF determines how medication spreads within the spinal canal (Hocking and Wildsmith, 2004; Hodkinson et al., 2017). It is hypothesized that the local anesthetic that is hypo or hyperbaric may behave unpredictably as solutions that are not isobaric require gravitational forces to spread in a reliable fashion. Isobaric local anesthetics may therefore be the best option for extraterrestrial neuraxial procedure due to the lack of reliance on gravity for proper spread (Uppal et al., 2017). An epidural anesthetic may be the most prudent of the neuraxial techniques as the operators will not be concerned about migration of medication within the CSF. However, changes to the spinal anatomy in weightlessness could result in a more challenging procedure for either spinal or epidural placement as spinal offloading will alter landmarks like the intervertebral discs and postural muscles (Nelson, Lacinski and Steller, 2024).

#### 20.1.1.2 Local and regional anesthesia

Anesthesia or analgesia for the extremities can also be achieved with local anesthetics via either infiltration, which requires minimal equipment and comparatively little training, or with regional anesthesia (RA) techniques for more severe trauma or more invasive procedures. RA is a collective set of techniques for targeted nerve blockade and offers profound anesthesia or analgesia without the risks associated with GA or neuraxial techniques (Kiberd et al., 2024). With the procedural knowledge of a few nerve blocks, upper and lower extremity surgery can be performed without the need for GA or neuraxial anesthesia. RA is often used terrestrially as the primary anesthetic for patients, most commonly in cases where rapid resumption of cognitive processing is desirable (i.e. outpatient surgery), for patients with substantial cardiac comorbidities or as an adjunctive postoperative pain control technique that is opioid sparing. Regional modalities include both "single-shot" and catheter-based techniques. With the "single-shot" approach, a one-time deposit of local anesthetic is concentrated in the vicinity of targeted nerves or within a fascial plane via hypodermic needle, with the duration of effect determined by the specific agent (or agents) injected. Catheter-based techniques, as the name implies, utilize introduction of a hollow flexible catheter through a hypodermic needle into the targeted area, where the needle is then removed, the catheter secured in place, and medication can be delivered as repeat bolus doses or via continuous infusion.

A nerve block can provide durable pain relief that is opioid sparing and can also be used as a temporizing bridge to treatment. Femoral nerve blockade with a catheter is often employed terrestrially in patients that are elderly before hip fracture repair to reduce pain while awaiting definitive surgery. Similarly, RA has found extensive application within military medicine. Soldiers wounded on the battlefield may endure protracted evacuation to a medical care facility, and placement in the field of a regional continuous nerve catheter provides pain control during evacuation or permits initial surgical management in a forward operating field setting.

For the future exploration astronaut, minimizing opioids may be crucial for mission integrity as opioid medications have a host of side effects including cognitive impairment, gastrointestinal side effects (nausea, constipation), respiratory depression, and development of tolerance over longer time periods. Many of these side effects would exacerbate baseline pathologic manifestations of microgravity such as space motion sickness associated with gravity transitions and the propensity towards constipation due to behavioral, dietary, and logistical factors. Similarly, a sedated crewmember may represent an impactful loss of overall crew capability both in terms of allocatable effort and available skillset. Furthermore, high vigilance, especially in small crews, is required due to the constant threat of spontaneous emergency aboard the spacecraft. For example, a sedated crewmember may not be able to rapidly don a spacesuit in the event of depressurization due to micrometeorite strike, increasing mortality risk.

The use of RA techniques is not without its limitations. Whereas GA can be utilized for any surgery, RA techniques are options only for specific corporal locations that have a peripheral nerve innervation which is a suitable target. This translates to mastery of a collection of related techniques, rather than a single procedure, to achieve approximately the same scope. Furthermore, the effectiveness of RA blocks is strongly influenced by clinician skill, which is an argument against their use in the space environment (Komorowski et al., 2013). However, proficiency in general anesthetic management requires years to attain and is difficult to simulate in spaceflight analogs, whereas a curated set of local/regional anesthesia techniques can not only be obtained more rapidly but also simulated in microgravity settings (Kiberd et al., 2024). Lastly, the skills required for mastery of RA are largely in the psychomotor and visuospatial reasoning domains, which are areas in which professional astronauts traditionally excel.

## 20.2 Spaceflight physiology impact on local/regional anesthesia

### 20.2.1 Neurological changes

The neurological changes associated with spaceflight, in particular long duration missions, has been a topic of considerable research interest. One notable pathophysiological mechanism under investigation is Spaceflight-Associated Neuro-Ocular Syndrome (or SANS). Approximately 70% of long duration astronauts have been observed to develop SANS-related pathology, which includes globe flattening, optic disc edema, folds within the retina and choroid, cotton wool spots, and retinal nerve fiber thickening (Brunstetter et al., 2022; Mader et al., 2011). Although the etiology is not yet known, and many potential mechanisms have been proposed, the primary hypothesis is that cephalad fluid shifts that occur following microgravity exposure lead to altered CSF dynamics, and consequently elevated ICP (Ong et al., 2023). However, this has never been conclusively demonstrated. Some similarities between the manifestations of SANS and idiopathic intracranial hypertension have been noted, which may be viewed as plausibly supportive of this hypothesis. No in-flight direct measurements of ICP have been made to date. Postflight measurements have been made in four astronauts following return with results ranging from borderline to elevated opening pressures (21–28 cm H20), although these were collected many days following return and only after a lumbar puncture was deemed clinically indicated (Ong et al., 2023; Lee et al., 2020). Although not a direct measurement of ICP, anatomical changes in brain and CSF volume/distribution have been shown with MRI, findings that lend additional credence to the assertion that ICP may be elevated (Roberts et al., 2017).

The implications of these spaceflight-induced changes on use of RA are theoretical only, as no investigation has been performed. In the case of ocular blocks used terrestrially for many ophthalmologic surgeries, peri- or retrobulbar injection of local anesthetic could lead to spikes in intraocular pressure or optic nerve compression, which may be particularly deleterious in the context of SANS-related retinal & optic nerve pathology. Whether this is truly the case in spaceflight has yet to be established and, as such, is still quite speculative.

### 20.2.2 Cardiovascular and respiratory changes

The cardiovascular system is among the bodily systems most altered by changes in gravitational forces and decrease in load bearing exercise. Astronauts tend to become hypovolemic, anemic, and have orthostatic hypotension on reintroduction to terrestrial 1G as a result of adaptation to a microgravity environment (Komorowski, Fleming and

Kirkpatrick, 2016). Initial acclimation to the microgravity environment takes place over 3–6 weeks, while the adaptation phase can take up to months (Graebe et al., 2004). With loss of the gravitational effects on extravascular fluid pooling towards dependent regions of the body, fluid is redistributed toward the upper body. Redistribution is thought to lead to transient sensed intravascular "hypervolemia" with a resultant diuresis (Komorowski, Fleming and Kirkpatrick, 2016). Stroke volume and cardiac output increase about 20%–50% because of this redistribution of fluid, returning to baseline after a period of adaptation. Risk of arrhythmia increases for reasons that are likely multifactorial. High-G acceleration loads during launch, reentry, and landing are known to be proarrhythmogenic, although abnormalities are typically transient and benign. Prolonged atrial stretch during microgravity exposure can lead to cardiac remodeling and is suggested as the pathophysiologic mechanism for earlier age of onset of atrial fibrillation in astronauts versus the terrestrial population (Khine et al., 2018). Other factors including catecholamine release associated with stress or prolonged exertion, dehydration, and electrolyte disturbances may also contribute. In 1987, an otherwise healthy astronaut with no cardiac history was evacuated after a 14-beat run of ventricular tachycardia (Komorowski, Fleming and Kirkpatrick, 2016). There is also a well-documented decrease in aerobic capacity of participants during spaceflight (Downs et al., 2015). With respect to RA, local anesthetics block voltage gated sodium channels and therefore have cardiotoxic effects, particularly in high doses. Of those in common use, bupivacaine exhibits the greatest cardiotoxicity and therefore ropivacaine is the preferred choice for RA.

While microgravity leads to changes in the distribution of pulmonary perfusion and ventilation, respiratory gas exchange is mostly preserved (Prisk, 2014). The environment itself, however, may lead to impacts that warrant special consideration. The ISS operates at a conventional Earth-like atmosphere, approximately 14.7 psi and 21% oxygen (Lavelle, 2022; Takada et al., 2023). Ambient carbon dioxide is generally at least five times higher than on Earth due to crew metabolic loads and incomplete removal during air revitalization and transiently will increase further with heightened crew exertional activity (Beard, 2020). The high frequency of space-related headaches are thought toresult at least in part from prolonged low-level hypercapnia (NASA-STD-3001 Technical Brief: Carbon Dioxide ($CO_2$) OCHMO-TB-004, 2023). Anecdotal reports of improvement during 100% oxygen breathing on mask, a preparatory procedure for EVA, support this association. Onset of hypercapnia-related symptoms has been noted to occur at lower levels than in the terrestrial setting. Further, despite adequacy of the bulk environmental parameters, formation of local "pockets" of oxygen-deficient, and $CO_2$-enriched air may occur around the head of an astronaut owing to lack of natural buoyant convective forces, particularly if located in areas of poor forced ventilation. For this reason, astronauts are trained to recognize their personal symptoms associated with hypoxia or hypercapnia, so that once recognized they may take corrective action.

Environmental hypercapnia may have specific implications for use of common local anesthetics. Carbon dioxide tension modulates vasomotor tone, with significant effects on cerebral perfusion. Cerebral blood flow changes 2%–4% per mmHg change in $CO_2$ over much of the physiologic range (Brian, 1998). Hypercapnia-induced increases in blood flow may potentiate both the uptake and delivery of local anesthetics to the CNS. It's not surprising then that hypercapnia is known to potentiate the effects of local anesthetic systemic toxicity (LAST; El-Boghdadly et al., 2018). In today's spacecraft environment, cerebral perfusion could be elevated at least 10%–20% above baseline on average. Acidosis, if present, may also lead to increased ionization and "trapping" of local anesthetic once having crossed the blood–brain barrier. These effects could plausibly make occurrence of LAST more likely at any given dose. "Safe" maximum dosages employed in common practice may need to be lowered to prevent this catastrophic consequence, at least until tolerance is demonstrated.

Although the ISS operates at an Earth-like atmosphere, spacecraft for future exploration class missions are likely to see a return to more "exotic" environments. Since the high nitrogen partial pressure of the ISS requires hours-long denitrogenation times to prevent decompression sickness with exposure to the low pressure EVA-suit, it is poorly suited for frequent surface EVAs. Apollo spacecraft overcame this obstacle by utilizing an approximately 5 psi 100% oxygen environment, such that denitrogenation occurred completely during transit to the lunar surface, and as such no additional denitrogenation time was needed (Sumonski and Tzccker , 1972). The Artemis campaign vehicles and habitats (e.g., Orion, HALO) used in the NASA mission to return to the lunar surface are intended to have environments enriched in oxygen concentration, reduced in nitrogen, and reduced in pressure relative to the ISS. Despite oxygen enrichment, this "exploration atmosphere" may still be relatively hypoxic due to low absolute pressure and capped oxygen concentration in an attempt to manage flammability risks. As a result, future astronauts are expected to live and work at the equivalent of a low-moderate altitude, yet another factor potentially affecting the CNS sensitivity to medication. Even if no increased sensitivity to LAST occurs, in these environments, hyperventilation and a relative respiratory alkalosis may present with symptoms of paresthesias, dyspnea, and palpitations, which can mimic LAST and confuse the clinical picture (Stepanek et al., 2023).

### 20.2.3 Metabolism, pharmacokinetics, and pharmacodynamics

The pharmacokinetics of local anesthetics are likely different in microgravity, but definitive characterization is elusive as many studies show conflicting results. The overall changes in blood volume and total body water impact the available volume of distribution and could cause higher than expected levels of local anesthetic in the blood for a given administered dosage (Kast et al., 2017). Also, liver function may be affected by altered perfusion, which could impact local anesthetic metabolism. Analog studies have shown no variation of clearance of IV lidocaine in the supine position (Kates et al., 1980), but an increase in clearance with –6° bedrest (Saivin et al., 1995), which has been used as a ground based experimental analog for the microgravitational offloading and fluid redistribution. However, head down tilt only accounts for a subset of the physiologic changes expected during spaceflight.

Data evaluating the pharmacodynamics of pharmaceuticals in space are also limited and to date none are specific to local anesthetics. Any information that is available is mainly extrapolated from known relationships between physiologic changes of the body in space and the effect of these changes in a normal gravity environment (Derendorf, 1994). Several antibiotic pharmacodynamic studies in simulated microgravity environments have shown differences in the extent and duration of the drug's actions (Kast et al., 2017), but more research is needed.

## 20.3 Regional anesthesia technique in spaceflight

### 20.3.1 Patient monitoring

Terrestrial standards set by the American Society of Anesthesiologists (ASA) for performance of any anesthetic, including RA, mandate monitoring of oxygenation, ventilation, circulation, and temperature (where changes in body temperature are expected; Standards for Basic Anesthetic Monitoring, 2020) Although not designed for austere settings, the ASA standard allows reasonable flexibility, and modern technology makes gold-standard monitoring feasible despite spacecraft mass and volume restrictions. Continuous monitoring of oxygenation is performed using small finger-tip pulse oximeters and ventilation through "continual observation of qualitative signs" in cases without sedation, or, when sedation is present, with in-line or sidestream capnography devices. Many small form factor electrocardiographic (ECG) devices also exist for the consumer market and can also assess heart rate. Blood pressure should be measured every 5 minutes and preferably automated to reduce crewmember workload. If not available, it may be feasible to use palpable radial pulse strength as a surrogate marker for clinical hypotension, as is taught in tactical combat casualty care field instruction, although this may not be completely reliable (Naylor et al., 2020). All-in-one emergency medical service monitors such as the Tempus Pro (Phillips), LifePak (Stryker), and X-series (Zoll), or the disaster-oriented WVSM (Athena GTX), would be more than sufficient to meet all monitoring requirements, but size may exceed upmass constraints and thus individual monitors may be chosen for each domain.

### 20.3.2 Ultrasound

Ultrasound can aid in diagnosis of several common spaceflight pathologies including SANS, deep venous thrombosis, cardiac dysfunction, altered volume status, barotrauma-induced pneumothorax, kidney and urinary issues, and musculoskeletal injuries (Drudi et al., 2012). It is also useful for procedural needle guidance, applicable to IV catheter insertions, and is the most common guidance technique for regional anesthetic blockade. The benefit of visual needle guidance in increasing success and minimizing vascular injury risk has largely led to the replacement of nerve-stimulation-based regional techniques with ultrasound for all but a few deep blocks where ultrasound visualization is poor.

The ISS currently hosts a "cart-style" ultrasound device, the HDI-5000 (Phillips ATL) that has been repackaged to meet space, power, and cooling requirements, as part of its Human Research Facility (Sargsyan et al., 2006). As one of only two clinical imaging modalities on station, the other being optical coherence tomography (for retinal imaging), it is an important tool both for research and clinical purposes but is large, complex to use, and is likely unsuitable for smaller exploration vehicles. Handheld ultrasounds, designed to be used for bedside/point-of-care ultrasonography (POCUS) in conjunction with a smart device, offer an appealing alternative. The only handheld POCUS device that has previously flown is the Butterfly IQ Ultrasound (Butterfly Network, Inc. Burlington, MA), first flown aboard CRS-22 to the ISS in June of 2021 and later aboard SpaceX's private Dragon missions (Dusang, 2021). It is interoperable with mobile platforms, has a highly integrated design, and involves a single probe. The original model had lower frame rates than other ultrasound machines, had no ECG gating in cardiac mode, and had limited adjustability of settings (Sargsyan et al., 2006). The Butterfly iQ3 (3rd generation) has twice the processing power compared to its predecessor and many new features such as biplane mode and NeedleViz mode, which allows for easier needle tip visualization during in-plane and out-of-plane techniques.

### 20.3.3 Medications

Local anesthetic selection is based upon the intended administration site and desired profile. Pharmacologic agents for use in spinal or epidural anesthesia must be preservative free, which unfortunately shortens their potential shelf life and usability for long duration exploration field care without planned resupply. Inactive chemical preservatives may provoke significant inflammatory reactions when administered intrathecally, and thus must be avoided for spinal use (Rauck, 2009). Otherwise, the choice of local anesthetic is typically determined by desired onset time, duration of action, and planned dose (Table 20.1; Taylor and McLeod, 2020). Note that elimination half-life correlates with, but is not the same as, the duration of effective surgical blockade, which will vary depending upon the concentration administered as well as the site of administration. The addition of a vasoconstrictor (i.e., epinephrine) to the local anesthetic serves to decrease systemic absorption, which prolongs the duration of anesthesia and reduces the risk of toxicity.

Most local anesthetics are suitable for "room temperature" storage, which should be preferred over any agent requiring refrigeration in an expeditionary setting. Lidocaine is a common anesthetic choice for short procedures because of its rapid (typically 2–5 minute) onset time. Short half-life anesthetics may also be useful in situations where pain relief is needed but prolonged blockade may mask important neurological examination findings, such as with attempted reductions of fractures or dislocations. Bupivacaine, on the other hand, is primarily useful when extended block durations are desired, but comes at the "cost" of relatively low safe maximum doses with higher potential for cardiac toxicity, and slower (20–30 m) onset times. A liposomal formulation of bupivacaine (Exparel © Pacira Biosciences) was introduced to help improve the safety profile, but per manufacturer specification should be refrigerated between 2°C and 8°C and is only suitable for room temperature storage for up to 30 days, limiting this formulation's use in spaceflight.

### 20.3.4 Effect of microgravity on block safety and effectiveness

The microgravity environment of space poses challenges to any procedure, and no in-vivo studies for peripheral nerve blocks have been performed extraterrestrially. However, peripheral nerve blocks have recently been demonstrated to be effective in a simulated underwater microgravity environment when performed correctly. Using a "meat model" of a muscle and nerve strapped to a free floating mannikin, with the investigators also free floating, there was no increased length of time taken to perform an ultrasound guided block in asimulated underwater environment versus standard dry land conditions ($p=0.751$; Kiberd et al., 2024). There was also no significant difference in the rate of accidental intraneural

**TABLE 20.1** Physiochemical parameters of common local anesthetics.

| Anesthetic | Class | Onset | Elimination half-life (min) | Max dose without vasoconstrictor (mg/kg) | Max dose with vasoconstrictor (mg/kg) |
|---|---|---|---|---|---|
| Chloroprocaine | Ester | Fast | 7 | 12 | 15 |
| Prilocaine | Amide | Slow | 100 | 8 | 8 |
| Lidocaine | Amide | Fast | 100 | 4.5 | 7 |
| Mepivacaine | Amide | Moderat-e | 115 | 4.5 | 7 |
| Ropivacaine | Amide | Moderat-e | 110 | 2.5 | 2.5 |
| Bupivacaine | Amide | Slow | 210 | 2.5 | 3 |

injections or block success rates (*P*>0.999). However, these results might not be applicable to a true space environment due to the difference in the operator skill as well as other unanticipated challenges.

Methods to ensure block efficacy are similar in space as on the ground. Proper visualization of the nerve and surrounding structures by ultrasound can provide standardization of the block and improve rates of success. Monitoring ofECG, pulse oximetry, and blood pressure can alert the crew tointravascular injection of local anesthetic, which canresult in tachycardia, hypertension, ringing in the ears, or numbness in the lips.

### 20.3.5 Block selection

The main blocks that would be required to fully cover the upper extremity are supraclavicular for the shoulder and upper arm, and the median, ulnar, and radial nerves at the level of the mid-forearm for more distal coverage (Yurgil et al., 2020). Supraclavicular nerve blockade provides coverage of the shoulder andspares more distal parts of the forearm but has an advantage of reduction of phrenic nerve blockade compared to interscalene block (Aliste et al., 2018). The lower extremities can be covered through neuraxial techniques but also through peripheral nerve blocks of the sciatic or the femoral nerve (Yurgil et al., 2020). A transversus abdominal plane (TAP) block is a relatively simple block to perform that would provide analgesia to the anterior musculature of the abdomen (Komorowski et al., 2016; see Table 20.2.

### 20.3.6 Continuous versus single-shot technique

Single-injection local anesthetic techniques are simpler to perform than continuous catheter placement. Although continuous catheters may provide better pain relief, they also have higher complication rates (Singhal and Taksande, 2024). Both techniques would require ultrasound, extension tubing, syringes, and medications, but the continuous nerve catheter requires the addition of a larger bore blunt needle that can accommodate the catheter being threaded to the site of interest, the perineural catheter itself, and occlusive dressings for the insertion site. Terrestrially, the continuous nerve catheter would ideally be attached to an infusion pump but in spaceflight this would likely be manually performed as intermittent boluses from a crewmember.

## 20.4 Safety considerations and potential complications

### 20.4.1 Spaceflight specific risks

Spaceflight medical systems have always prioritized minimizing upmass and volume (Norfleet, 2000) given the cost of bringing one kilogram (such as a liter of crystalloid) of material into space was approximately $22,000 USD in the year 2000. However, this was reduced to around $2,700 USD per kilogram by 2010 and has been further reduced due to the continually expanding spaceflight market (Jones, 2018). Also, the flight environment produces vibration, vacuum, as well as varied humidity and temperature that can impact the stability of pharmaceuticals. Studies have also shown that photolabile drugs such as ciprofloxacin and promethazine have a higher degree of degradation due to the higher radiation levels in flight. For example, the rate of promethazine degradation was 200 times greater in space than on Earth (Mehta and Bhayani, 2017). There have been no studies performed on the degradation and stability of local anesthetics in a microgravity environment.

### 20.4.2 Prevention and management of regional anesthesia complications

Microgravity requires that restraints are included in payload such as magnetic pads and Velcro to secure patient, operator, and operating equipment during RA (Drudi et al., 2012). This is also important for ACLS protocols in space, which may be needed to treat sequelae of local anesthetic toxicity (Hinkelbein et al., 2020). In microgravity, it is more difficult to consistently achieve adequate depth of chest compressions during cardiopulmonary resuscitation (CPR). On the Polaris Dawn mission, CPR was shown to significantly fatigue crewmembers, increase carbon dioxide production, and produce vector forces that impact the trajectory of the spacecraft (Menon et al., 2025).

LAST often occurs when a greater dose of local anesthetic is administered than intended, or after accidental injection of local anesthetic directly into a blood vessel. This risk is reduced with ultrasound guidance and by using the minimum amount of local anesthetic required, which is ideally well below the maximum allowable dose (Table 20.1). Intralipid 20% is the drug of choice for LAST and the standard dose is 1.5 mL/kg as an initial bolus, which can be repeated for 1–2 times for persistent asystole. An infusion at 0.25 mL/kg/min should then be started for 30–60 minutes and the rate increased to 0.50 mL/kg/min for refractory hypotension (Neal et al., 2018). In a standard 70 kg patient, 105–315 mL of intralipid may be required for initial treatment, with an additional 1050 mL per hour afterward as an infusion. With current launch prices,

**TABLE 20.2 Comparison of technical difficulty and relative safety of various regional techniques.**

| Technique | Target region | Technical difficulty | Relative safety | Key considerations |
|---|---|---|---|---|
| Spinal (subarachnoid) | Neuraxial | Moderate | Moderate | Hypotension, PDPH; avoid with coagulopathy |
| Epidural (lumbar/thoracic) | Neuraxial | Advanced | Moderate | Higher failure rate; risk of epidural hematoma |
| Interscalene (brachial plexus) | Upper limb | Advanced | Moderate | Phrenic nerve palsy common; pneumothorax rare |
| Supraclavicular (brachial plexus) | Upper limb | Advanced | Moderate | Pneumothorax risk; u/s reduces risk |
| Infraclavicular (brachial plexus) | Upper limb | Moderate | Moderate | Vascular proximity; catheter-friendly |
| Axillary (brachial plexus) | Upper limb | Easy | Safer | Lower pneumothorax risk; multiple nerves to target |
| PECS I/II | Truncal | Easy | Safer | Superficial plane blocks; minimal hemodynamic impact |
| Serratus anterior plane | Truncal | Easy | Safer | Analgesia for lateral thorax; superficial |
| Paravertebral (thoracic) | Truncal | Advanced | Higher risk | Pneumothorax, hypotension; neuraxial-adjacent |
| Erector spinae plane (ESP) | Truncal | Moderate | Safer | Superficial plane; broad spread, variable |
| Transversus abdominis plane (TAP) | Abdominal wall | Easy | Safer | Superficial plane; visceral sparing |
| Rectus sheath | Abdominal wall | Easy | Safer | Midline analgesia; superficial |
| Ilioinguinal/Iliohypogastric | Abdominal wall | Easy | Safer | Superficial; variable anatomy |
| Quadratus lumborum (QL) | Abdominal wall | Advanced | Moderate | Deeper plane; variable spread; hypotension rare |
| Femoral nerve | Lower limb | Easy | Safer | Quadriceps weakness; fall risk |
| Adductor canal (saphenous) | Lower limb | Easy | Safer | Motor-sparing; superficial |
| Fascia iliaca (supra-inguinal) | Lower limb | Easy | Safer | Good for hip analgesia; large volume |
| Sciatic (popliteal) | Lower limb | Moderate | Moderate | Foot/ankle surgery; vascular proximity |
| Ankle block (terminal branches) | Lower limb | Easy | Safer | Multiple injections; superficial |
| Superficial cervical plexus | Head/Neck | Easy | Safer | Superficial; avoid deep spread |
| Bier block (IVRA) | Hand | Easy | Moderate | Tourniquet dependent; LAST if cuff failure |

*ESP*, erector spinae plane block; *IVRA*, intravenous regional anesthesia (Bier block); *LAST*, local anesthetic systemic toxicity; *PDPH*, postdural puncture headache; *PECS*, pectoral nerve block; *QL*, quadratus lumborum block; *TAP*, transversus abdominis plane block; *u/s*, ultrasound.

sufficient intralipid for bolus plus a 3 hour infusion (~3500 ml) might cost $10,000 to launch. Considering its 24-month shelf life and the poorly characterized effect of the spaceflight environment on its stability, it may be prudent to more strictly limit the maximum dose of local anesthetic used rather than include intralipid in the medical system.

Aside from LAST, some individuals may have allergies to the preservatives in local anesthetics, which may also produce an anaphylactic response requiring ACLS. Ideally, all astronauts would be tested for local anesthetic allergies before flight, but allergen testing for astronauts flying to the ISS has historically been limited to only a murine allergen-specific IgE panel (Godard, 2024).

## 20.5 Operator preparedness

### 20.5.1 Operator training

When ultrasound is used for blocks, approximately 10 repetitions of a single block can help a clinician reach a 90% success rate (Luyet et al., 2010), but this is difficult to achieve without violating the principle of nonmaleficence on either volunteer subjects or vulnerable patients. However, high-resolution simulation would be ideal to educate operators on block performance before flight. The space environment can also disrupt circadian rhythms, whether from irregular work schedules, poor sleep, environmental factors, or even potentially the effect of microgravity on the circadian clock at the cellular level (Mallis and DeRoshia, 2005; Ranieri et al., 2015; Strauch et al., 2024; Pandi-Perumal and Gonfalone, 2016). Disruption of circadian rhythms can lead to significant performance decreases on otherwise routine tasks, and sleep deprivation has been shown to significantly affect performance on a simulated manual docking task (Piechowski et al., 2024). Competence might thus be tested when crew are fatigued during late evening hours in preflight crew training periods.

### 20.5.2 Augmented reality

Visualization of needle trajectory and tip can be improved with augmented reality, thereby reducing the risk of complications and improving block success (Luyet et al., 2010; Seybold et al., 2024). Typically, augmented reality consists of a computer generated image and a user view of the real world, often as an overlay on the ultrasound image (Johnson et al., 2017). Many specialties have already adopted augmented reality and incorporated it as a part of their training, (Marhofer and Eichenberger, 2023), and it may be especially impactful in procedural space medicine. The Autonomous Medical Officer Support ISS Technology Demonstration showed that the use of "in-flight just-in-time (JIT) instruction" can effectively teach nonphysician operators to perform kidney and bladder ultrasound without preflight training or real-time assistance (Ebert et al., 2022; Campbell et al., 1996).

## 20.6 Recipient preparedness

### 20.6.1 Patient cooperation

Prolonged space missions strain mental health and well-being (Arone et al., 2021) due to isolation and environs that are both unfamiliar and relatively dangerous. The visualization of a needle using ultrasound requires a patient to be motionless, thus patient cooperation is critically important (Campbell et al., 1996), but use of restraints can exacerbate baseline procedural anxiety. These circumstances may increase the necessity of anxiolytic techniques that ideally are noninvasive, require no additional equipment, and avoid the side effects of pharmacologic sedatives (Pagnini, 2024).

### 20.6.2 Relaxation techniques

Alternative techniques used as adjuvants for control of pain and anxiety in the perioperative period are meditation and audio-visual distraction (Kurdi and Gasti, 2018; Ayub et al., 2018; O'Connor and Pace, 2010). There is evidence that meditation is generally favorable in reducing anxiety but existing studies are heterogenous and of mixed quality (Rajjoub et al., 2024). Audiovisual technology results in significantly lower anxiety scores than placebo in children undergoing GA (Oktaviani et al., 2024) and immersive virtual reality (VR) via headsets can be similarly effective (Barry et al., 2022; Ledford et al., 2021; Faruki et al., 2019; Chiu et al., 2023; Faruki et al., 2022). VR is a promising modality for the reduction of both preoperative anxiety and intraoperative anesthetic requirements (Perucchini et al., 2025; Marhoon et al., 2024) but would obviously have a larger footprint than guided meditation.

## 20.7 Conclusion

Given the increasing duration of spaceflight missions, as well as the growing commercialization of the near Earth orbital environment, local anesthesia and RA are steadily gaining traction as critical capabilities for the future of human spaceflight. There is currently limited data on the uptake, distribution, metabolism, and excretion of local anesthetics in space, and these unresolved pharmacologic uncertainties, along with procedural considerations, are undeniably impacted by the spaceflight environment. Nevertheless, local and RA will likely be the preferred methods for providing anesthesia and analgesia in spaceflight given the risks of procedural sedation, neuraxial anesthesia, and GA. Therefore, the safety and efficacy of local anesthetic medications in low gravity environments, as well as training in their delivery, will be necessary for exploration class missions of the future. The complications and adverse effects of local anesthetic administration in microgravity require further characterization, and training protocols must be developed to ensure operators acquire the psychomotor skills necessary to perform these techniques in the austere space environment.

## References


Aliste, J., Bravo, D., Fernández, D., Layera, S., Finlayson, R.J., Tran, D.Q., 2018. A randomized comparison between interscalene and small-volume supraclavicular blocks for arthroscopic shoulder surgery. Regional Anesthesia and Pain Medicine 43 (6), 590–595. Available at https://doi.org/10.1097/AAP.0000000000000767, http://journals.lww.com/rapm/pages/currenttoc.aspx.

Arone, A., Ivaldi, T., Loganovsky, K., Palermo, S., Parra, E., Flamini, W., et al., 2021. The burden of space exploration on the mental health of astronauts: a narrative review. Clinical Neuropsychiatry 18 (5), 237–246. Available at https://doi.org/10.36131/cnfioritieditore20210502, https://www.clinicalneuropsychiatry.org/download/the-burden-of-space-exploration-on-the-mental-health-of-astronauts-a-narrative-review/?wpdmdl=2393&refresh=61d2de055396b1641209349.

Axpe, E., Chan, D., Abegaz, M.F., Schreurs, A.-S., Alwood, J.S., Globus, R.K., et al., 2020. A human mission to Mars: predicting the bone mineral density loss of astronauts. PLoS ONE 15 (1), e0226434. Available at https://doi.org/10.1371/journal.pone.0226434.

Ayub, K., Corrigan, R.A., Misra, J., Galitzine, S., 2018. Audio–visual distraction as an adjunct to standby anaesthesia in persons with paraplegia: a case series of five operations. Spinal Cord Series and Cases 4 (1). Available at https://doi.org/10.1038/s41394-017-0035-6.

Baran, R., Wehland, M., Schulz, H., Heer, M., Infanger, M., Grimm, D., 2022. Microgravity-related changes in bone density and treatment options: a systematic review. International Journal of Molecular Sciences 23 (15), 8650. Available at https://doi.org/10.3390/ijms23158650.

Barry, K.S., Nimma, S.R., Spaulding, A.C., Wilke, B.K., Torp, K.D., Ledford, C.K., 2022. Perioperative outcomes of immersive virtual reality as adjunct anesthesia in primary total hip and knee arthroplasty. Arthroplasty Today 18, 84–88. Available at https://doi.org/10.1016/j.artd.2022.09.015, http://www.journals.elsevier.com/arthroplasty-today.

Beard, B.L., 2020. Characterization of how CO2 level may impact crew performance related to the HSIA risk. NASA Ames Research Center. Available at https://hsi.arc.nasa.gov/publications/NASA_TM20205011433.pdf.

Bloom, A., Blanken, B., Schlakman, B., Arena, T., Mironov, Z., Vlahovic, T.C., 2020. A review of nail dystrophies for the practitioner. Advances in Skin and Wound Care 33 (1), 20–26. Available at https://doi.org/10.1097/01.ASW.0000613536.27194.3c, http://journals.lww.com/aswcjournal/pages/default.aspx.

Brian, J.E., 1998. Carbon dioxide and the cerebral circulation. Anesthesiology 88 (5), 1365–1386. Available at https://doi.org/10.1097/00000542-199805000-00029.

Brunstetter, T., Masso, S., Miller, E., et al. 2022. Spaceflight Associated Neuro-ocular Syndrome (SANS): clinical update. Available at https://ntrs.nasa.gov/citations/20220000516 (accessed 25.10.19).

Campbell, M.R., Billica, R.D., Jennings, R., Johnston, S., 1996. Laparoscopic surgery in weightlessness. Surgical Endoscopy 10 (2), 111–117. Available at https://doi.org/10.1007/BF00188354, link.springer.de/link/service/journals/00464/index.htm.

Chiu, P.L., Li, H., Yap, K.Y.L., Lam, K.M.C., Yip, P.L.R., Wong, C.L., 2023. Virtual reality-based intervention to reduce preoperative anxiety in adults undergoing elective surgery: a randomized clinical trial. JAMA Network Open 6 (10). Available at https://doi.org/10.1001/jamanetworkopen.2023.40588, https://jamanetwork.com/journals/jamanetworkopen.

Derendorf, H., 1994. Pharmacokinetic/pharmacodynamic consequences of space flight. The Journal of Clinical Pharmacology 34 (6), 684–691. Available at https://doi.org/10.1002/j.1552-4604.1994.tb02024.x.

Downs, M., Moore, A., Lee, S.M.C., Ploutz-Snyder, L., 2015. Evidence report: risk of reduced physical performance capabilities due to reduced aerobic capacity. National Aeronautics and Space Administration Lyndon B. Johnson Space Center. Available at https://humanresearchroadmap.nasa.gov/evidence/reports/Aerobic.pdf?rnd=0.283149571755505.

Drudi, L., Ball, C.G., Kirkpatrick, A.W., Saary, J., Marlene Grenon, S., 2012. Surgery in space: where are we at now? Acta Astronautica 79, 61–66. Available at https://doi.org/10.1016/j.actaastro.2012.04.014.

Dusang, K., 2021. TRISH demonstrates performance of portable handheld ultrasound in space. BCM. Available at https://www.bcm.edu/news/trish-demonstrates-performance-of-portable-handheld-ultrasound-in-space (accessed 25.10.19).

Ebert, D., Byrne, V., Walton, M., Sargsyan, A., 2022. Autonomous medical officer support (AMOS) ISS technology demonstration: enabling Earth-independnt procedure guidance. NASA. Available at https://ntrs.nasa.gov/api/citations/20220014327/downloads/Ebert_AMOS%20IWS%202023%20abstract_20220916.pdf.

El-Boghdadly, K., Pawa, A., Chin, K.J., 2018. Local anesthetic systemic toxicity: current perspectives. Local and Regional Anesthesia 11, 35–44. Available at https://doi.org/10.2147/LRA.S154512, https://www.dovepress.com/getfile.php?fileID=43573.

Faruki, A.A., Nguyen, T.B., Gasangwa, D.-V., Levy, N., Proeschel, S., Yu, J., et al., 2022. Virtual reality immersion compared to monitored anesthesia care for hand surgery: a randomized controlled trial. PLoS ONE 17 (9), e0272030. Available at https://doi.org/10.1371/journal.pone.0272030.

Faruki, A., Nguyen, T., Proeschel, S., Levy, N., Yu, J., Ip, V., et al., 2019. Virtual reality as an adjunct to anesthesia in the operating room. Trials 20 (1). Available at https://doi.org/10.1186/s13063-019-3922-2.

Godard, B., 2024. Allergy and hypersensitivity diseases in space: physiological changes, clinical aspects, and mechanisms with countermeasures. Journal of Allergy and Hypersensitivity Diseases 2, 100007. Available at https://doi.org/10.1016/j.jahd.2024.100007.

Graebe, A., Schuck, E.L., Lensing, P., Putcha, L., Derendorf, H., 2004. Physiological, pharmacokinetic, and pharmacodynamic changes in space. The Journal of Clinical Pharmacology 44 (8), 837–853. Available at https://doi.org/10.1177/0091270004267193.

Hilmers, D., Kreykes, A., 2023. Evolution of the IMPACT evidence library methods-what we have learned human research program exploration medical capability element. Available at https://ntrs.nasa.gov/api/citations/20230000496/downloads/Hilmers_Evolution%20of%20the%20IMPACT%20Evidence%20Library%20Methods%20%E2%80%93.pdf (accessed 25.10.16).

Hinkelbein, J., Kerkhoff, S., Adler, C., Ahlbäck, A., Braunecker, S., Burgard, D., et al., 2020. Cardiopulmonary resuscitation (CPR) during spaceflight - a guideline for CPR in microgravity from the German Society of Aerospace Medicine (DGLRM) and the European Society of Aerospace Medicine Space Medicine Group (ESAM-SMG). Scandinavian Journal of Trauma, Resuscitation and Emergency Medicine 28 (1). Available at https://doi.org/10.1186/s13049-020-00793-y.

Hocking, G., Wildsmith, J.A.W., 2004. Intrathecal drug spread. British Journal of Anaesthesia 93 (4), 568–578. Available at https://doi.org/10.1093/bja/aeh204.

Hodkinson, P.D., Anderton, R.A., Posselt, B.N., Fong, K.J., 2017. An overview of space medicine. British Journal of Anaesthesia 119 (suppl_1), i143. Available at https://doi.org/10.1093/bja/aex336.

Johnson, A.N., Peiffer, J.S., Halmann, N., Delaney, L., Owen, C.A., Hersh, J., 2017. Ultrasound-guided needle technique accuracy prospective comparison of passive magnetic tracking versus unassisted echogenic needle localization. Regional Anesthesia and Pain Medicine 42 (2), 223–232. Available at https://doi.org/10.1097/AAP.0000000000000549, http://journals.lww.com/rapm/pages/currenttoc.aspx.

Jones, H.W., 2018. The recent large reduction in space launch cost. In: 48th International Conference on Environmental Systems. Available at http://hdl.handle.net/2346/74082(accessed 25.10.19).

Kast, J., Yu, Y., Seubert, C.N., Wotring, V.E., Derendorf, H., 2017. Drugs in space: pharmacokinetics and pharmacodynamics in astronauts. European Journal of Pharmaceutical Sciences 109, S2. Available at https://doi.org/10.1016/j.ejps.2017.05.025, www.elsevier.com/locate/ejps.

Kates, R.E., Harapat, S.R., Keefe, D.L.D., Goldwater, D., Harrison, D.C., 1980. Influence of prolonged recumbency on drug disposition. Clinical Pharmacology and Therapeutics 28 (5), 624–628. Available at https://doi.org/10.1038/clpt.1980.213.

Khine, H.W., Steding-Ehrenborg, K., Hastings, J.L., Kowal, J., Daniels, J.D., Page, R.L., et al., 2018. Effects of prolonged spaceflight on atrial size, atrial electrophysiology, and risk of atrial fibrillation. Circulation. Arrhythmia and Electrophysiology. 11 (5), e005959. Available at https://doi.org/10.1161/CIRCEP.117.005959.

Kiberd, M.B., Brownbridge, R., Mackin, M., Werry, D., Bird, S., Barry, G., et al., 2024. Feasibility of ultrasound-guided nerve blocks in simulated microgravity: a proof-of-concept study for regional anaesthesia during deep space missions. British Journal of Anaesthesia 133 (6), 1276–1283. Available at https://doi.org/10.1016/j.bja.2024.07.034, https://www.sciencedirect.com/science/journal/00070912.

Komorowski, M., Fleming, S., Kirkpatrick, A.W., 2016. Fundamentals of anesthesiology for spaceflight. Journal of Cardiothoracic and Vascular Anesthesia 30 (3), 781–790. Available at https://doi.org/10.1053/j.jvca.2016.01.007, http://www.journals.elsevier.com/journal-of-cardiothoracic-and-vascular-anesthesia/.

Komorowski, M., Watkins, S.D., Lebuffe, G., Clark, J.B., 2013. Potential anesthesia protocols for space exploration missions. Aviation, Space, and Environmental Medicine 84 (3), 226–233. Available at https://doi.org/10.3357/ASEM.3427.2013Germany, http://docserver.ingentaconnect.com/deliver/connect/asma/00956562/v84n3/s8.pdf.

Kurdi, M.S., Gasti, V., 2018. Intraoperative meditation music as an adjunct to subarachnoid block for the improvement of postoperative outcomes following cesarean section: a randomized placebo-controlled comparative study. Anesthesia: Essays and Researches 12 (3), 618. Available at https://doi.org/10.4103/aer.aer_114_18.

Lavelle, H., 2022. Advanced oxygen generation. NASA. Available at https://www.nasa.gov/podcasts/houston-we-have-a-podcast/advanced-oxygen-generation/ (accessed 25.10.19).

Ledford, C.K., VanWagner, M.J., Sherman, C.E., Torp, K.D., 2021. Immersive virtual reality used as adjunct anesthesia for conversion total hip arthroplasty in a 100-year-old patient. Arthroplasty Today 10, 149–153. Available at https://doi.org/10.1016/j.artd.2021.07.001, http://www.journals.elsevier.com/arthroplasty-today.

Lee, A.G., Mader, T.H., Gibson, C.R., Tarver, W., Rabiei, P., Riascos, R.F., et al., 2020. Spaceflight associated neuro-ocular syndrome (SANS) and the neuro-ophthalmologic effects of microgravity: a review and an update. NPJ Microgravity 6 (1). Available at https://doi.org/10.1038/s41526-020-0097-9, https://www.nature.com/npjmgrav/.

Lewandowski, B.E., Myers, J.G., Sulkowski, C., Ruehl, K., Licata, A., 2008. Forecasting proximal femur and wrist fracture caused by a fall to the side during space exploration missions to the Moon and Mars. . Available at https://ntrs.nasa.gov/citations/20090004579.

Lifetime Surveillance of Astronaut Health (LSAH). NASA. Available at https://www.nasa.gov/directorates/esdmd/hhp/lifetime-surveillance-of-astronaut-health-lsah/ (accessed 25.10.19).

Longo, D.L., Stepanek, J., Blue, R.S., Parazynski, S., 2019. Space medicine in the era of civilian spaceflight. New England Journal of Medicine 380 (11), 1053–1060. Available at https://doi.org/10.1056/NEJMra1609012, http://www.nejm.org/medical-index.

Luyet, C., Schüpfer, G., Wipfli, M., Greif, R., Luginbühl, M., Eichenberger, U., 2010. Different learning curves for axillary brachial plexus block: ultrasound guidance versus nerve stimulation. Anesthesiology Research and Practice 2010, 1–7. Available at https://doi.org/10.1155/2010/309462.

Mader, T.H., Gibson, C.R., Pass, A.F., Kramer, L.A., Lee, A.G., Fogarty, J., et al., 2011. Optic disc edema, globe flattening, choroidal folds, and hyperopic shifts observed in astronauts after long-duration space flight. Ophthalmology 118 (10), 2058–2069. Available at https://doi.org/10.1016/j.ophtha.2011.06.021.

Mallis, M.M., DeRoshia, C.W., 2005. Circadian rhythms, sleep, and performance in space. Aviation, Space, and Environmental Medicine 76 (6 Suppl), B94–B107. Available at https://pubmed.ncbi.nlm.nih.gov/15943202/.

Marhofer, P., Eichenberger, U., 2023. Augmented reality in ultrasound-guided regional anaesthesia: useful tool or expensive toy? British Journal of Anaesthesia 131 (3), 442–445. Available at https://doi.org/10.1016/j.bja.2023.05.022.

Marhoon, F., Afaq, M., Abraham, D., John, S., Atulkar, A., Nawaz, F., et al., 2024. 497 use of virtual reality in adjunct to anesthesia in surgery and anesthesia requiring procedures: a systematic review and meta-analysis. Annals of Emergency Medicine 84 (4), S220. Available at https://doi.org/10.1016/j.annemergmed.2024.08.486.

Mehta, P., Bhayani, D., 2017. Impact of space environment on stability of medicines: challenges and prospects. Journal of Pharmaceutical and Biomedical Analysis 136, 111–119. Available at https://doi.org/10.1016/j.jpba.2016.12.040.

Mehta, S.K., Szpara, M.L., Rooney, B.V., Diak, D.M., Shipley, M.M., Renner, D.W., et al., 2022. Dermatitis during spaceflight associated with HSV-1 reactivation. Viruses 14 (4). Available at https://doi.org/10.3390/v14040789, https://www.mdpi.com/1999-4915/14/4/789/pdf.

Menon, A., Mampre, D., Isaacman, J., Poteet, S., Gillis, S., Mateus, J., et al., 2025. Cardiopulmonary resuscitation techniques on the dragon spacecraft: a polaris dawn mission study. Wilderness & Environmental Medicine 36 (1_suppl). Available at https://doi.org/10.1177/10806032251346746.

NASA-STD-3001 Technical Brief: Carbon Dioxide (CO2) OCHMO-TB-004. NASA Office of the Chief Health & Medical Officer (OCHMO). Available from: https://www.nasa.gov/wp-content/uploads/2023/12/ochmo-tb-004-carbon-dioxide.pdf, 2023 (accessed 25.10.19).

Naylor, J.F., Fisher, A.D., April, M.D., Schauer, S.G., 2020. An analysis of radial pulse strength to recorded blood pressure in the department of defense trauma registry. Military Medicine 185 (11-12), E1903. Available at https://doi.org/10.1093/milmed/usaa197, https://academic.oup.com/milmed/issue.

Neal, J.M., Woodward, C.M., Harrison, T.K., 2018. The American Society of regional anesthesia and pain medicine checklist for managing local anesthetic systemic toxicity: 2017 version. Regional Anesthesia and Pain Medicine 43 (2), 150–153. Available at https://doi.org/10.1097/AAP.0000000000000726, http://journals.lww.com/rapm/pages/currenttoc.aspx.

Nelson, A.M., Lacinski, R.A., Steller, J.G., 2024. Spaceflight-associated pain. Current Opinion in Anaesthesiology 37 (5), 581–587. Available at https://doi.org/10.1097/ACO.0000000000001401, http://journals.lww.com/co-anesthesiology/pages/default.aspx.

Nguyen, C.N., Urquieta, E., 2023. Contemporary review of dermatologic conditions in space flight and future implications for long-duration exploration missions. Life Sciences in Space Research 36, 147–156. Available at https://doi.org/10.1016/j.lssr.2022.10.004, http://www.journals.elsevier.com/life-sciences-in-space-research/.

Norfleet, W.T., 2000. Anesthetic concerns of spaceflight. Anesthesiology 92 (5), 1219–1222. Available at https://doi.org/10.1097/00000542-200005000-00006, http://journals.lww.com/anesthesiology/pages/default.aspx.

O'Connor, K., Pace, N., 2010. Audio-visual distraction during spinal anaesthesia for orthopaedic surgery. Journal of perioperative practice 20 (12), 428–429. Available at https://doi.org/10.1177/175045891002001201.

Oktaviani, J.R., Widjaja, V., Singgih, N.A., Pranowo Sampurno Secodiningrat, R.H., 2024. Audiovisual technology intervention for reducing preoperative anxiety in children undergoing general anesthesia: a systematic review and meta-analysis. Journal of Education and Health Promotion 13 (1). Available at https://doi.org/10.4103/jehp.jehp_1344_23, https://journals.lww.com/jehp/pages/results.aspx?txtKeywords=Audiovisual+Technology+Intervention+for+Reducing+Preoperative+Anxiety+in+Children+Undergoing+General+Anesthesia%3A+A+Systematic+Review+and+Meta+Analysis.

Ong, J., Mader, T.H., Gibson, C.R., Mason, S.S., Lee, A.G., 2023. Spaceflight associated neuro-ocular syndrome (SANS): an update on potential microgravity-based pathophysiology and mitigation development. Eye 37 (12), 2409–2415. Available at https://doi.org/10.1038/s41433-023-02522-y, https://www.nature.com/eye/.

Opperman, R.A., Waldie, J.M.A., Natapoff, A., Newman, D.J., Jones, J.A., 2010. Probability of spacesuit-induced fingernail trauma is associated with hand circumference. Aviation, Space, and Environmental Medicine 81 (10), 907–913. Available at https://doi.org/10.3357/ASEM.2810.2010.

Pagnini, F., 2024. Supporting the mind in space: psychological tools for long-duration missions. Interactive Journal of Medical Research 13, e66626. Available at https://doi.org/10.2196/66626.

Pandi-Perumal, S.R., Gonfalone, A.A., 2016. Sleep in space as a new medical frontier: the challenge of preserving normal sleep in the abnormal environment of space missions. FLASS, Canada Sleep Science 9 (1), 1–4. Available at https://doi.org/10.1016/j.slsci.2016.01.003. Available from: http://www.sciencedirect.com/science/article/pii/S1984006316000043/pdfft?md5=3551e151ec004f8ef3892f7d2e1010d8&pid=1-s2.0-S1984006316000043-main.pdf.

Perucchini, F., Baumeister, P., Fankhauser, C.D., 2025. Beyond anesthesia: a mini review of virtual reality as an adjunct in urological procedures. European Urology Focus 11 (3), 511–514. Available at https://doi.org/10.1016/j.euf.2025.01.007, https://www.sciencedirect.com/science/journal/24054569.

Piechowski, S., Kalkoffen, L.J., Benderoth, S., Wolf, O.T., Rittweger, J., Aeschbach, D., et al., 2024. Effects of total sleep deprivation on performance in a manual spacecraft docking task. NPJ Microgravity 10 (1). Available at https://doi.org/10.1038/s41526-024-00361-z, https://www.nature.com/npjmgrav/.

Prisk, G.K., 2014. Microgravity and the respiratory system. The European Respiratory Journal: Official Journal of the European Society for Clinical Respiratory Physiology 43 (5), 1459–1471. Available at https://doi.org/10.1183/09031936.00001414.

Rajjoub, R., Sammak, S.E., Rajjo, T., Rajjoub, N.S., Hasan, B., Saadi, S., et al., 2024. Meditation for perioperative pain and anxiety: a systematic review. Brain and Behavior 14 (7). Available at https://doi.org/10.1002/brb3.3640, http://onlinelibrary.wiley.com/journal/10.1002/.

Ranieri, D., Cucina, A., Bizzarri, M., Alimandi, M., Torrisi, M.R., 2015. Microgravity influences circadian clock oscillation in human keratinocytes. FEBS Open Bio 5, 717–723. Available at https://doi.org/10.1016/j.fob.2015.08.012, http://www.elsevier.com/wps/find/journaldescription.cws_home/726807/description#description.

Rauck, R.L., 2009. Compounding intrathecal drugs. Neuromodulation 1, 483–490. Available at https://doi.org/10.1016/B978-0-12-374248-3.00037-9, https://www.sciencedirect.com/book/9780123742483.

Roberts, D.R., Albrecht, M.H., Collins, H.R., Asemani, D., Chatterjee, A.R., Spampinato, M.V., et al., 2017. Effects of spaceflight on astronaut brain structure as indicated on MRI. New England Journal of Medicine 377 (18), 1746–1753. Available at https://doi.org/10.1056/NEJMoa1705129.

Saivin, S., Traon, A.P.-L., Cornac, A., Güell, A., Houin, G., 1995. Impact of a four-day head-down tilt (–6°) on lidocaine pharmacokinetics used as probe to evaluate hepatic blood flow. The Journal of Clinical Pharmacology 35 (7), 697–704. Available at https://doi.org/10.1002/j.1552-4604.1995.tb04110.x.

Salter, S.A., Ciocon, D.H., Gowrishankar, T.R., Kimball, A.B., 2006. Controlled nail trephination for subungual hematoma. American Journal of Emergency Medicine 24 (7), 875–877. Available at https://doi.org/10.1016/j.ajem.2006.03.029.

Sargsyan, A.E., Hamilton, D.R., Melton, S.L., Young, J., 2006. The international space station ultrasound imaging capability overview for prospective users. Wyle Laboratories, Life Sciences Group NASA Johnson Space Center, Houston. Available at https://citeseerx.ist.psu.edu/document?repid=rep1&type=pdf&doi=fcae14ee0cb74d030a63928dbe5e52ae4567d337.

Seybold, B., Gaier, N., Ofenloch, A., Boeckler, D., Kalenka, A., Fiedler-Kalenka, M.O., 2024. Learnability of ultrasound-guided locoregional anesthesia for carotid endarterectomy. Journal of Clinical Medicine 13 (24), 7557. Available at https://doi.org/10.3390/jcm13247557.

Sibonga, J.D., Spector, E.R., Yardley, G., Alwood, J.S., 2024. Evidence report: risk of bone fracture due to spaceflight-induced changes to bone. National Aeronautics and Space Administration Lyndon B. Johnson Space Center (JSC). Available at https://humanresearchroadmap.nasa.gov/Evidence/reports/Bone%20Fracture%20Evidence%20Report_07-18-2024.pdf.

Singhal, A., Taksande, K., 2024. Continuous catheter techniques versus single-injection nerve blocks: a comprehensive review of postoperative pain management strategies. Cureus 16 (9). Available at https://doi.org/10.7759/cureus.70040.

Standards for Basic Anesthetic Monitoring. Available at https://www.asahq.org/standards-and-practice-parameters/standards-for-basic-anesthetic-monitoring, 2020 (accessed 25.10.19).

Stepanek, J., Blue, R.S., Connolly, D., 2023. Pulmonary function in human spaceflight. Seminars in Respiratory and Critical Care Medicine 44 (5), 696–704. Available at https://doi.org/10.1055/s-0043-1770064.

Strauch, L., von der Wiesche, M., Noppe, A., Mulder, E., Rieger, I., Aeschbach, D., et al., 2024. Simulating microgravity with 60 days of 6 degree head-down tilt bed rest compromises sleep. NPJ Microgravity 10 (1). Available at https://doi.org/10.1038/s41526-024-00448-7.

Sumonski Jr, F.H., Tzccker, E.M., 1972. Apollo experience report - command and service module environmental control system. National Aeronautics and Space Administration. Available at https://ntrs.nasa.gov/api/citations/19720012252/downloads/19720012252.pdf.

Takada, K., Hornyak, D., Garr, J., Van Keuren, S., Faulkner, C., ElSherbini, A., 2023. Status of the advanced oxygen generation assembly. In: Meeting: 52nd International Conference on Environmental Systems, vol. 44. Available at https://ntrs.nasa.gov/citations/20230008013 (accessed 25.10.19).

Taylor, A., McLeod, G., 2020. Basic pharmacology of local anaesthetics. BJA Education 20 (2), 34–41. Available at https://doi.org/10.1016/j.bjae.2019.10.002.

Uppal, V., Retter, S., Shanthanna, H., Prabhakar, C., McKeen, D.M., 2017. Anesthesia and analgesia. Lippincott Williams and Wilkins Canada Hyperbaric Versus Isobaric Bupivacaine for Spinal Anesthesia: Systematic Review and Meta-Analysis for Adult Patients Undergoing Noncesarean Delivery Surgery, vol. 125. https://doi.org/10.1213/ANE.0000000000002254 15267598 5 1627 1637. Available at http://journals.lww.com/anesthesia-analgesia/toc/publishahead.

Yurgil, J.L., Hulsopple, C.D., Leggit, J., 2020. Nerve blocks: part II. Lower extremity. American Family Physician 101 (11), 669–679. Available at https://www.aafp.org/pubs/afp/issues/2020/0601/p669.html.

Yurgil, J.L., Hulsopple, C.D., Leggit, J.C., 2020. Nerve blocks: part I. Upper extremity. American Family Physician 101 (11), 654–664. Available at https://www.aafp.org/pubs/afp/issues/2020/0601/p654.pdf.

Chapter 21

# Pain diagnosis and management

Lonnie G. Petersen[1], Gwen E. Owens[2], and Ariana M. Nelson[3]
[1]*Department of Aeronautics and Astronautics, Massachusetts Institute of Technology, Cambridge, MA, United States,* [2]*Department of Anesthesiology, Critical Care and Pain Medicine, Harvard University, Cambridge, MA, United States,* [3]*Department of Anesthesiology & Perioperative Medicine, University of California Irvine, Irvine, CA, United States*

## 21.1 Introduction: what is pain, why do we need to manage it, and how is it different in space?

Pain is an unpleasant sensory and emotional experience that resembles or is associated with actual or perceived tissue damage (Raja et al., 2020). It is a complex phenomenon involving both physiological and psychological components, and if left unmanaged, it can have negative consequences on both. Managing pain during spaceflight and extraplanetary missions is a unique and critical challenge with many unanswered questions, such as the following:

- Given the changes in tissue composition, fluid filling, and regional pressures, will pain phenotypes present the same way in microgravity as in a terrestrial environment?
- Can long-term or acute exposure to space radiation cause damage to peripheral nerves that augment or decrease pain perception?
- Will pain susceptibility and perception change as deconditioning associated with long-term spaceflight manifests?
- Is the risk of chronic pain elevated?
- What analgesics are most appropriate for treating pain during missions where peak performance is always required, risk of emergency is omnipresent, and "operating heavy machinery" might be necessary at any given moment?

Adverse events that are manageable on Earth could lead to mission-critical situations in space; even relatively benign and common side effects of analgesics, such as constipation or diarrhea, could pose significant challenges in a confined, isolated, resource-poor environment (especially if confined in a spacesuit) compared to a fully staffed hospital ward with a well-stocked pharmacy. Additionally, the pharmacokinetics (PK) and pharmacodynamics (PD) of analgesics in microgravity are poorly understood, which inherently increases the risk of reduced efficacy or adverse events, such as unintentional overdose or constipation (Kast et al., 2017). Even with appropriate dosing, there are additional considerations for analgesic medications that cause sedation in a crew that must always remain capable of peak cognitive and physical performance.

While the origin and cause of pain can be elusive even in a familiar terrestrial setting, the differential diagnoses of acute and chronic pain are much broader in the setting of a space mission. Imagine a crewmember returning from an extravehicular activity (EVA) with a sore knee. Could this pain be due to mechanical bruising from the suit? Overloading of the joint? Decompression sickness? Or simple traumatic injury? Additionally, the diagnosis and treatment of pain will likely be performed by nonspecialized personnel with limited diagnostic tools at their disposal. Lastly, the biologic mechanisms and experience of pain, like many sensory experiences, are altered by the spaceflight environment, and thus its diagnosis and treatment must also advance with the evolving landscape of spaceflight missions (Patron et al., 2024; Sauer et al., 2023).

### 21.1.1 Sequelae of untreated pain

Treating and managing pain is crucial as it can significantly impact an individual's physical, emotional, and social well-being. Managing pain involves addressing the source whenever possible, as well as the experience of pain itself, often through medications, physical therapies, behavioral support, and/or interventional procedures. Effective pain management is critical to ensure human performance and quality of life, prevent transition to chronic pain, enhance recovery, and

**Procedural Medicine for Spaceflight. DOI:** https://doi.org/10.1016/B978-0-443-27552-4.00037-3

improve mental health and performance (Hsu et al., 2019). Pain can limit mobility and prevent daily activities and tasks; EVAs have been canceled or aborted due to reported pain (Ceniza-Bordallo et al., 2024). Moreover, pain can impair the ability to adhere to an exercise countermeasure program, causing decompensation to accelerate with detrimental and possibly irreversible impact to crew health. In one reported incident, a herniated disc in-flight prevented a crewmember from exercising (Scheuring et al., 2024). While the cervical symptoms and sequelae were mitigated, the overall deconditioning due to lack of exercise had a negative impact both in-flight and during postflight recovery. Effective pain management helps the body recover more quickly from injury. When pain is controlled, the crew can engage in rehabilitation more effectively, reducing recovery time and improving overall outcomes. Persistent pain activates stress responses and elevates levels of cortisol with potential impact on cardiovascular health, gastrointestinal issues, and weakened immune function (Cohen, Vase and Hooten, 2021).

If not mitigated, acute pain can transition into intractable chronic pain by altering pathways and neural transduction of the pain stimuli. Chronic pain, if left unmanaged, can lead to central sensitization (nervous system is more sensitive to pain stimuli), creating a vicious cycle. Acute and chronic pain can negatively impact not only physical function but also general well-being and increase the risk of anxiety and depression (De Heer et al., 2018; Aaron et al., 2025). This increases risk to team dynamics and poses a risk to both individual crew and mission objectives. Moreover, maintaining a circadian rhythm and high-quality sleep is often a challenge in space. Pain can disrupt sleep patterns, cause frequent awakenings, and reduce sleep quality. Poor sleep can augment perception of pain and discomfort, and this vicious cycle may lead to fatigue, irritability, and impaired cognitive function. Effective pain management can improve sleep quality, which in turn boosts mood, concentration, and overall performance.

### 21.1.2 Ethical considerations

The management of pain, in any form, is a cornerstone of ethical patient care, both on and off Earth. Clinicians must remain committed to minimizing pain and discomfort for our patients, wherever they may be. In situations where treatment is not an option and death is imminent, effective pain and anxiety management is also essential. With deep-space missions planned for the near future, palliative care in a remote, isolated environment with limited resources will be a reality and should be initiated with little hesitation and the utmost respect for crew life and dedication to the mission.

## 21.2 Overview of pain categories, mechanisms, and diagnosis

The treatment of pain depends on identifying the pain type and level, as these correspond to different underlying mechanisms and therefore require analgesics with distinct mechanisms of action. Often, more than one pain type is present, and pain can change in quality and severity over time. Monitoring of pain characteristics is, therefore, part of treatment and is used to guide management decisions.

### 21.2.1 Pain categories

#### *21.2.1.1 Nociceptive pain*

This is the most common type of pain, triggered by the activation of nociceptors (pain receptors) due to tissue damage. Nociceptors are specialized nerve endings found throughout the body that respond to harmful stimuli such as heat, cold, pressure, or chemical changes. It can be further divided into the following:

- Somatic pain: Often originating from skin, muscles, joints, or bones. It can be sharp or aching, but it is typically in a well-defined distribution.
- Visceral pain: Arising from internal organs like the heart, lungs, or intestines. It is often diffuse, deep, and difficult to localize.

#### *21.2.1.2 Neuropathic pain*

This type of pain is caused by damage or dysfunction in the nervous system itself, either in the peripheral nervous system or the central nervous system. Examples include diabetic neuropathy, radicular pain due to nerve root impingement (sciatica), or poststroke pain. Neuropathic pain is often described as burning, tingling, or shooting sensations, can be difficult to manage, and has a high risk of transitioning to chronic pain. Treatment will also initially include nonopioid therapies, although many of these will have concerning side effects such as sedation (Table 21.1).

**TABLE 21.1** Nonopioid analgesics for neuropathic pain.

| Drug class | Common examples | Main routes | Common considerations | Space considerations |
|---|---|---|---|---|
| **Anticonvulsants** | Gabapentin, pregabalin | **PO** | First line for neuropathic pain (postherpetic neuralgia, diabetic neuropathy). Sedation and dizziness possible. | Depressed cognitive function. |
| **Antidepressants (TCAs & SNRIs)** | Amitriptyline (TCA), nortriptyline, duloxetine (SNRI), venlafaxine | **PO** | Useful in neuropathic and chronic musculoskeletal pain. TCAs can also improve sleep onset and continuity. | Xerostomia may increase risk of caries, depressed cognitive function. |
| **Topical local anesthetics** | Lidocaine 5% patch | **TD/topical** | For localized neuropathic pain (postherpetic neuralgia). | No special considerations aside from shelf life and in-flight availability. |
| **Topical capsaicin** | Capsaicin cream, 8% patch | **TD/topical** | Depletes substance P; effective in peripheral neuropathic pain. Burning sensation common initially. | No special considerations. |
| **Muscle relaxants** | Cyclobenzaprine, tizanidine, baclofen | **PO** | Adjunct for chronic pain with muscle spasm. Sedation and hypotonia possible. | Depressed cognitive function. |
| **Alpha-2 adrenergic agonists** | Clonidine (patch or oral) | **PO, TD** | May reduce neuropathic pain intensity and aid sleep. | Depressed cognitive function, may exacerbate orthostatic intolerance post-flight. |
| **NMDA-receptor antagonists (low-dose, adjuvant)** | Ketamine (chronic pain infusion protocols) | **IV, IN** | Used under specialist supervision for refractory neuropathic pain. | Depressed cognitive function, psychiatric dissociation. |

*IN, intranasal; IV*, Intravenous; *NMDA*, N-methyl-D-aspartate; *PO*, per os (oral); *SNRI*, serotonin–norepinephrine reuptake inhibitor; *TCA*, tricyclic antidepressant; *TD*, transdermal.

#### 21.2.1.3 *Inflammatory pain*

Pain is a cardinal symptom of inflammation, and the increased activity and swelling of the tissue increases pain sensitivity. Causes include the following:

- Physical: mechanical or environmental damage to tissues resulting from trauma, burns/frostbite, ionizing radiation, or other probable risks of space or planetary missions.
- Infectious: pathogens (bacteria, viruses, fungi), all of which have been found in excess on the International Space Station (ISS) and must be expected in any and all future stations and habitats.
- Chemical: toxins, aerosols, which are of elevated concern in any confined environment with poor airflow but especially in weightless environments with no convection.

#### 21.2.1.4 *Nociplastic pain*

This type of pain is due to abnormal processing of pain signals in the nervous system and is often modulated by heightened stress levels (emotional, psychological, and behavioral factors impact this category of pain). The origin of nociplastic pain is usually not linked to clear physical injury or disease and can range from fibromyalgia to myofascial pain to nonspecific low back pain, which is common in spaceflight (Ceniza-Bordallo et al., 2024). Anxiety, depression, and real or perceived danger can exacerbate or even produce this type of pain. Given the heightened risk of nociplastic pain during spaceflight, crew-reported symptoms should trigger systematic evaluation for this phenotype and informed treatment planning, rather than premature reassurance or dismissal.

### 21.2.2 Pain mechanisms

#### 21.2.2.1 *Pain pathways*

Pain perception involves a complex pathway, and the signal conduction can be modulated or blocked at any level of conduction or perception:

- Transduction: Nociceptors detect harmful stimuli and convert them into electrical signals.
- Transmission: The signals travel along nerve fibers (A-delta fibers and C fibers) to the spinal cord.
- Perception: Signals are processed in the brain, where the pain is consciously perceived.
- Modulation: The body can alter pain perception through processes like endorphin release, which can reduce the sensation of pain.

#### 21.2.2.2 *Acute versus chronic pain*

If left untreated, pain can evolve from the acute to chronic state, a process that can be mitigated through modulation of both the transduction, the transmission, and the perception of pain (Oweidat, Kalagara and Sondekoppam, 2024).

- Acute pain is defined as short-term (typically <1 month) and is usually a direct result of an injury or illness.
- Subacute pain is defined as pain of 1–3 months duration.
- Chronic pain persists beyond the expected time for tissue healing (typically >3–6 months). Effective treatment is often complex and can require a combination of modalities, including pharmacologic, physical, and cognitive therapies.

### 21.2.3 Monitoring pain severity and effect of therapeutics

Pain is subjective, and the experience of pain varies from person to person, even given the same traumatic injury or stimulus. Accurate characterization of pain is critical to both diagnosing the cause and for monitoring the efficacy of intervention. Severity of pain is most commonly assessed clinically using a Numerical Rating Scale (see Fig. 21.1), but in research it is often assessed with the Visual Analog Scale (VAS). The VAS permits patients to make a mark along a line with anchors (0=no pain, 10=worst pain imaginable), which can then be measured, allowing for precise pain scoring on a continuous rather than an ordinal scale like the NRS. These unidimensional tools are simple, can be performed even by nonclinicians, and have been clinically validated in a variety of settings, which has resulted in wide international adoption (Karcioglu et al., 2018; Fig. 21.1). The spaceflight environment, therefore, will preferentially utilize these unidimensional pain measurement instruments to screen for pain and allow for standard and universal communication with ground control and across languages. However, multidimensional pain assessment tools are most appropriate for pain that persists beyond the acute phase. Multidimensional instruments can monitor the evolution of pain and its impact on mood and function, as well as provide a more nuanced evaluation of patient response to therapeutic interventions.

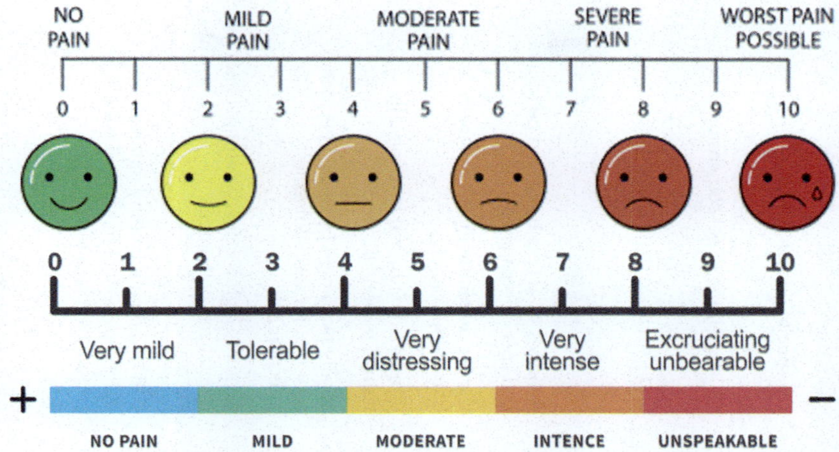


FIGURE 21.1 **Numeric rating scale of pain intensity.** Numeric rating scale (NRS) tool adopted from World Health Organization (WHO) guidelines and recommended for unidimensional assessment of pain intensity in spaceflight.
Modified from numeric rating scale tool in the World Health Organization (WHO) guidelines.

## 21.3 Analgesic use in spaceflight

Pharmaceuticals can be administered to the body through various pathways, and the choice of administration route depends on the desired effects, speed of action, drug properties, and patient/disease-specific factors (e.g., oral route is a poor choice for a nauseated patient; see Table 21.2. Moreover, the medication formulation options are impacted by the space environment, for example, the unsuitability of inhalational anesthetics in a weightless and confined habitat. Even if a closed-loop system was implemented to collect exhaled gases, the risk of contamination of the air in the confined cabin and thus impact on other crewmembers is significant. The Environmental Control and Life Support System (ECLSS) must be calibrated and equipped to clear away any spills to the cabin air. However, some drugs commonly delivered as sprays or aerosols on Earth can be converted to gel-suspension and still be delivered safely, for example, through the sublingual route (Sayed et al., 2013). Gel-suspension drugs generally have a much shorter shelf life than tablets because gel suspensions contain water, which promotes hydrolysis and are less compact, which facilitates oxidation. Therefore, gel suspensions commonly have a shelf life of 6–24 months, which is in contrast to most tablets that commonly have a shelf life of 2–5 years. However, the formulation of medications into dry gels such as aerogels is a promising area of pharmaceutical development for space medicine, as they will have a better stability and usability profile than traditional gel suspensions (Boruah and Sarma, 2025). Transdermal delivery is a commonly used administration route for analgesics and can be either in a gel or patch form applied to the skin. Patches are composed of a complex multilayer design including adhesives, a drug suspension or polymer matrix that ensures controlled slow release of the analgesic to the skin, and finally backing material, which isolates the patch and prevents oxidation. Most pain-relief drug patches, such as lidocaine or fentanyl, have a shelf life of 2–3 years, which, side effects notwithstanding, makes them sufficiently durable for spaceflight. With optimized design of the layers and cold storage, shelf life could be even further enhanced.

While each class has many specific drugs, given the limited capacity for pharmaceutical storage on board spacecraft, only a limited number of representative drugs across these classes should be selected for flight to optimize efficiency and efficacy. Few of these medications have been studied for dose-effect in microgravity, but the limited data available suggest that PKs and PDs are highly similar to terrestrial use.

### 21.3.1 First-line analgesics

The pharmaceuticals most commonly recommended as the starting point for pain management are nonopioid medications that might offer slightly less efficacy but markedly improved safety over opioids. These include oral medications like acetaminophen and non-steroidal antiinflammatory drugs (NSAIDs), which are useful for mild to moderate pain, especially due to musculoskeletal discomfort or inflammatory conditions. Topical agents like diclofenac gel, lidocaine patches, or capsaicin can be effective when pain is localized and limit systemic side effects. However, capsaicin and lidocaine will cause loss of sensation in the areas applied. For neuropathic pain specifically, medications such as gabapentin or tricyclic antidepressants like nortriptyline can be considered first-line therapy, though in the spaceflight

**TABLE 21.2** Analgesic administration routes.

| Route | Description | Example drugs | Common considerations | Space considerations |
|---|---|---|---|---|
| **Enteral (via gastrointestinal [GI] tract)** | | | | |
| Oral (PO) | Swallowed, absorbed through stomach or intestines. | Acetaminophen (paracetamol), amoxicillin | slower onset; subject to first-pass metabolism and slow gastric absorption if pain or nauseated. | No special considerations. |
| Sublingual | Placed under tongue; absorbed via oral mucosa. | Nitroglycerin, buprenorphine | Rapid onset; avoids first-pass metabolism. | No special considerations. |
| Buccal | Absorbed through cheek lining. | Buccal midazolam, nicotine lozenges | Good for patients unable to swallow. | Short shelf life of gel-suspension. |
| Rectal | Inserted into rectum; absorbed via rectal mucosa. | Diazepam, paracetamol suppositories | Useful if vomiting or unconscious; partial avoidance of first-pass effect. | Short shelf life of gel-suspension. |
| **Parenteral (bypass GI tract)** | | | | |
| Intravenous (IV) | Directly into bloodstream. | Morphine, antibiotics, IV fluids | Fastest onset; precise control; risk of infection. | Requires IV access by trained personnel and special equipment. |
| Intramuscular (IM) | Into muscle tissue. | Vaccines, ketorolac | Moderate onset; suitable for depot injections. | Can be administered by minimally trained personnel. |
| Subcutaneous (SC) | Into tissue beneath skin. | Insulin, heparin | Slower, sustained absorption; avoid in shock. | Can be administered by minimally trained personnel. |
| Intraarticular | Into a joint space. | Corticosteroids | Localized treatment; requires sterile technique. | Ideally performed by trained personnel, may require special needles. Infection risk. |
| Epidural or intrathecal (IT) | Epidural space is within the spinal column; IT is the cerebrospinal fluid (CSF). | Morphine, bupivacaine | Used for anesthesia and pain control; specialist use. | Requires highly trained personnel and specialized equipment. |
| Intraosseous (IO) | Into bone marrow. | Emergency fluids, drugs in pediatric emergencies | Used when IV access is difficult. | Can be administered by minimally trained personnel but is very painful in conscious patients. |
| **Inhalational** | | | | |
| Pulmonary (inhalation) | Through lungs as gas, aerosol, or powder. | Salbutamol, anesthetic gases | Rapid absorption due to large surface area. | Aerosols relatively contraindicated in spaceflight. |
| Nasal (intranasal) | Through nasal mucosa. | Naloxone, desmopressin | Useful for systemic or local effects. | Aerosols relatively contraindicated in spaceflight. |
| **Topical** | | | | |
| Dermal (cutaneous) | Applied to skin for local effect. | Antifungal creams, antibiotic ointments | Minimal systemic absorption. | Avoid ointment becoming free floating (see aerosol). |

| | | | | |
|---|---|---|---|---|
| Transdermal | Absorbed through skin for systemic effect. | Fentanyl patch, lidocaine patch, nicotine patch | Slow, continuous delivery; avoids GI tract. | Pharmacokinetics may be altered by fluid shifts in cutaneous tissues. |
| Ophthalmic (eye) | Applied to eyes. | Atropine drops, artificial tears | Local effect; avoid contamination. | Avoid medication becoming free floating. |
| Otic (ear) | Applied to ear canal. | Antibiotic ear drops | Local effect; ensure eardrum integrity. | Avoid medication becoming free floating. |
| Vaginal | Applied into vagina. | Antifungal pessaries, estrogen creams | Local or systemic effects possible. | No special considerations if viscous gel-suspension or tablet form. |
| **Other specialized** | | | | |
| Intraperitoneal | Into peritoneal cavity. | Dialysis fluids, some chemotherapies | Used for peritoneal dialysis or local drug delivery. | Requires trained personnel or remote guidance and specialized equipment. |
| Intracardiac | Directly into heart (rare). | Epinephrine (historical emergency use) | Now largely replaced by IV or IO routes. | Requires trained personnel or remote guidance and specialized equipment. |
| Implant/depot | Long-term slow release under skin or in muscle. | Contraceptive implants, depot antipsychotics | Provides sustained drug levels over weeks–months. | Can be administered by minimally trained personnel. |
| Intravitreal | Into vitreous humor of eye. | AntiVEGF drugs (e.g., ranibizumab) | Used for retinal diseases like macular degeneration. | Requires trained personnel or remote guidance. |

*CSF*, Cerebrospinal fluid; *GI*, gastrointestinal; *IM*, intramuscular; *IO*, intraosseous; *IT*, intrathecal; *IV*, intravenous; *PO*, per os (oral); *VEGF*, vascular endothelial growth factor.

environment their side effect of sedation precludes their use early in the algorithm (Schmidt et al., 2013). The overall goal when choosing first-line analgesics is to relieve symptoms while preserving crew function to ensure mission success (Table 21.3).

### 21.3.2 Opioids and ketamine

If optimization of first-line analgesics fails to adequately control pain, the addition of opioids and other controlled substances can be considered. Early use of opioids is indicated in traumatic injury and surgical management and may also be considered in cases of severe pain. Choice of opioids and related analgesics can be dictated by the patient's ability to tolerate PO, the rapidity of onset, duration of action, and expected duration of use. Conversion to long-acting opioids should be considered following an extended period of short-acting opioid use, usually 10–14 days. Additionally, early use of N-methyl-D-aspartate receptor-modifying agents like ketamine and methadone should be considered in patients with expected longer duration or higher dose opioid requirements to decrease overall opioid use and related side effects and reduce the likelihood of conversion to chronic pain. However, if opioids are considered, the patient's respiratory status should be closely monitored with pulse oximetry and assessment of respiratory rate (normal range is 12–20 breaths per minute; Apfelbaum et al., 2013).

In terrestrial use as well as for spaceflight, use of opioids even for short periods of time (days to weeks) is associated with a risk for opioid dependence and addiction. The development of addiction would put both the individual crew, mission objectives, and overall mission at risk. Moreover, the limited supply of these pharmaceuticals would quickly be depleted. Therefore, an emphasis on opioid-minimizing therapies is warranted, balanced with the objective and appreciation that inadequately controlled pain would also impair mission objectives and put crew at elevated risk. Of note, planned or potential use of even low-dose or short-term opioid analgesics requires the availability of naloxone for reversal in case of overdose, especially given the possible difficulties associated with resuscitation and airway management in space (Table 21.4).

General considerations when choosing the administration route (Fink et al., 2024):

- **Speed of onset:** Intravenous (IV) > intraosseous (IO) > inhalation > sublingual (SL) > intramuscular (IM) > subcutaneous (SC) > oral (PO) > transdermal
- **First-pass metabolism:** Bypassed by all nonoral routes (SL, IV, transdermal)
- **Duration:** Longest with the transdermal route, followed by extended-release formulations delivered by the oral route. For moderate to severe pain, patients will ideally be medicated to provide a stable baseline of analgesia and manage breakthrough pain via a more immediate-release formulation (e.g., IM injection).
- Final route selection will depend upon drug availability, drug properties, patient condition, desired speed, and effect site.

Special considerations for use in spaceflight: topical gels, patches, SL, IM, and PO medications can be self-administered and do not require specialized equipment or infusion pumps. IO, IM, and IV are fast and appropriate in acute situations; IV requires a higher level of training for administration than IO and IM.

## 21.4 Common etiologies of pain in spaceflight

### 21.4.1 Diagnosing pain phenotypes

Sources of pain that are expected in both short and longterm missions are principally nociceptive and neuropathic rather than nociplastic. Some of these pain generators will be more common in different mission segments. For example, most crew describe some degree of malaise and headache during the initial days of space flight, which is commonly accepted as a facet of space adaptation syndrome and is usually well managed with oral nonopioid analgesic (Van Oosterhout et al., 2024; Table 21.5). Opioid medications are generally reserved for the treatment of acute pain of intense severity (Table 21.6).

### 21.4.2 Low back pain

Back pain most often occurs during adaptation to the weightless environment. Low back pain occurs far more frequently during or after spaceflight than on Earth, with data suggesting it affects over half of astronauts (52%). Crewmembers classify most symptoms as mild (86%) or moderate (11%) in severity, but 3% report experiencing severe pain (Kerstman et al., 2012). This elevated incidence likely reflects a myriad of contributing factors. Spinal changes have been linked to gross anatomic changes such as atrophy of the supporting paraspinal muscles (Marfia et al., 2023), and compromised

**TABLE 21.3** Classes of nonopioid medications.

| Drug class | Common examples | Main routes (IM, IV, PO, SC, TD) | Common considerations | Space considerations |
|---|---|---|---|---|
| **NSAIDs** | Ibuprofen, ketorolac, diclofenac, naproxen | **PO, IM, IV, TD** (topical gels/patches) | First line for musculoskeletal pain, postoperative pain, inflammation. Avoid in renal impairment, gastrointestinal (GI) bleeding. | No special considerations. |
| **Acetaminophen (paracetamol)** | Acetaminophen/paracetamol | **PO, IV** | Mild–moderate pain or adjunct with NSAIDs. Safe in most patients; hepatotoxic in overdose. | No special considerations. |
| **COX-2 inhibitors** | Celecoxib, etoricoxib | **PO** | Similar efficacy to NSAIDs with lower GI toxicity. Useful for acute pain where NSAIDs are contraindicated. | No special considerations. |
| **Topical NSAIDs** | Diclofenac gel, ketoprofen gel | **TD/topical** | Effective for localized soft-tissue or joint pain; minimal systemic absorption. | No special considerations apart from risk of transfer of topical drug to cabin surface. |
| **Local anesthetics** | Lidocaine gel/patch, lidocaine injection, ropivacaine injection | **SC, IV, IM TD/ topical, perineural** | For localized acute pain or postprocedural discomfort. | No special considerations apart from risk of transfer of topical drug to cabin surface. |
| **NMDA antagonist (adjuvant)** | Low-dose ketamine | **IV, IM, SC** (rarely **PO**) | Short-term adjunct for severe acute pain, burns, or trauma; used in controlled settings. | Risk of depressed cognitive function. |

*COX-2*, Cyclooxygenase-2; *GI*, gastrointestinal; *IM*, intramuscular; *IV*, intravenous; *NMDA*, N-methyl-D-aspartate; *NSAIDs*, nonsteroidal antiinflammatory drugs; *PO*, per os (oral); *SC*, subcutaneous; *TD*, transdermal.

**TABLE 21.4** Overview of opioid classes.

| Drug class/potency | Common examples | Main routes (IM, IV, PO, SC, TD) | Common considerations | Space considerations |
|---|---|---|---|---|
| **Natural opiate** | Morphine | **IM, IV, SC, PO** | Gold standard for acute severe pain; titratable IV for postop or trauma. Concern for respiratory depression. | Moderate risk of respiratory depression and cognitive performance decrement. |
| **Semisynthetic opioids** | Hydromorphone, oxymorphone | **IV, IM, PO** | More potent than morphine; used when morphine not tolerated. | Respiratory depression (monitoring) and cognitive function (risk to crew, mission objectives and overall mission). |
| **Synthetic opioids (short-acting)** | Fentanyl, alfentanil, remifentanil | **IV, IM, TD, IN** | Rapid onset and short duration—used perioperatively, in anesthesia, or procedural sedation. | Respiratory depression (monitoring) and cognitive function (risk to crew, mission objectives and overall mission). |
| **Codeine (pro-drug)** | Codeine phosphate | **PO** | Combined with acetaminophen/NSAIDs for mild–moderate pain. Highly variable metabolism (CYP2D6). | Highly variable efficacy, mild risk of respiratory depression and cognitive performance decrement. |
| **Atypical opioids** | Tramadol, tapentadol | **PO, IV, IM** | Weak μ-agonist and norepinephrine or serotonin/norepinephrine reuptake inhibitors. Lower abuse risk; can cause seizures or serotonin syndrome. | Variable effect, mild risk of respiratory depression and cognitive performance decrement. |

*CYP2D6*, Cytochrome P450 2D6; *IM*, intramuscular; *IN*, intranasal; *IV*, intravenous; *NSAIDs*, nonsteroidal antiinflammatory drugs; *PO*, per os (oral); *SC*, subcutaneous; *TD*, transdermal.

**TABLE 21.5** Analgesic preferred for pain phenotype.

| Pain type | Preferred nonopioid analgesics | Typical routes |
|---|---|---|
| **Acute nociceptive (injury)** | NSAIDs, acetaminophen, COX-2 inhibitors, topical NSAIDs | PO, IV, IM, TD |
| **Localized musculoskeletal** | Topical NSAIDs, lidocaine gel/patch | TD |
| **Neuropathic (chronic)** | Gabapentin, pregabalin, amitriptyline, duloxetine, lidocaine patch, capsaicin | PO, TD |
| **Chronic musculoskeletal** | NSAIDs (as tolerated), muscle relaxants, antidepressants (SNRIs/TCAs) | PO, TD |

*COX-2*, Cyclooxygenase-2; *IM*, intramuscular; *IV*, intravenous; *NSAIDs*, nonsteroidal antiinflammatory drugs; *PO*, per os (oral); *SNRIs*, serotonin–norepinephrine reuptake inhibitors; *TCAs*, tricyclic antidepressants; *TD*, transdermal.

intervertebral disc integrity (Ramachandran et al., 2018). However, there is also likely a molecular component to higher rates of low back pain given the decreased water content within the nucleus pulposus and reduced osteoblast activity (Harrison et al., 2018; Treffel et al., 2020). These adaptations are dynamic and may vary depending on the chronology and duration of exposure to weightlessness, in addition to individual crewmember characteristics that may alter susceptibility to back pain at baseline (Penchev et al., 2021).

### 21.4.3 Pain due to trauma

Traumatic injury will be one of the more common pain etiologies in the spaceflight environment and can occur anywhere in the timeline from the acceleration of launch to the postural instability of landing. Astronauts are exposed to extreme biomechanical stresses, which may result in minor aches, scrapes, overexertion, and soreness after an EVA or after collisions with blunt objects in the habitat (Nelson et al., 2024). Although trauma associated pain is nociceptive rather than neuropathic or neuroplastic and will, in most cases, resolve spontaneously, it can cause crewmember impairment and risk to mission objectives and overall mission success. Aggressive management of acute pain has been shown to reduce the likelihood of transition to chronic pain (Cohen et al., 2021), and thus any spaceflight mission should be equipped with adequate supplies of analgesics to treat the expected number of minor traumatic injuries.

### 21.4.4 Neuropathic pain

Neuropathic pain has been most commonly reported by astronauts as paresthesia in the soles of the feet, which are often associated with changes in pain thresholds. This pattern has long been known to be associated with reduced sensitivity of slow-adapting receptors in the great toe and increased sensitivity of fast-adapting receptors in the heel. Astronauts with this receptor hypersensitivity demonstrate greater postural dysequilibrium than those without it, suggesting that these changes in sensation in the feet may contribute to the ambulatory difficulties astronauts experience after returning to the gravitational environment of Earth (Lowrey et al., 2014).

## 21.5 Treatment regimens

### 21.5.1 Regional anesthesia versus systemic pain management

Selection of medical therapeutics differs in the terrestrial and extraterrestrial environments due to the urgency of achieving mission objectives during spaceflight, and analgesics are among the most common classes of medications used in spaceflight (Wotring, 2015). However, analgesics commonly have side effects of sedation or even decreased reaction time, which often precludes the use of certain medications, such as strong opioids, in crewmembers needing to perform highly precise tasks. Therefore, the widely accepted WHO analgesic ladder (Fig. 21.2) should be adapted for the spaceflight environment to emphasize consideration of invasive procedures at an earlier point in the therapeutic algorithm (Fig. 21.3). Extraterrestrially, it may be appropriate to implement modalities like ultrasound or acupuncture even prior to initiation of weak opioids. Similarly, the assigned crew medical officer and supervising flight surgeons from ground control may consider performing even more invasive treatments like trigger point injections or nerve blocks prior to initiation of strong opioids.

It is always recommended to treat a disease or symptom as locally as possible. However, certain modalities are not yet validated in spaceflight. Epidural and spinal administration of local anesthetic, for example, can completely block pain transduction and are commonly used procedures for surgery or childbirth. However, successful and safe administration

**TABLE 21.6** Opioid analgesic options and considerations.

| Drug class/potency | Common examples | Main routes | Common considerations | Space considerations |
|---|---|---|---|---|
| **Strong opioids (long-acting formulations)** | Morphine SR, oxycodone CR, hydromorphone ER | **PO, SC, IV** | Mainstay for moderate to severe chronic pain. Use sustained release for baseline pain control with immediate release for breakthrough pain. | Depressed respiration (monitoring) and cognitive function (risk to crew, mission objectives, and overall mission) |
| **Transdermal opioids** | Fentanyl patch, buprenorphine patch | **TD** | Provide continuous pain control over 48–72 hours. Heat increases absorption. | Relatively mild, but risk to respiration and cognitive performance |
| **Methadone** | Methadone | **PO, IV** | Used for chronic pain or opioid rotation. Long half-life. Requires expert titration. Also for opioid dependence therapy. | Depressed respiration (monitoring), arrythmia (difficult to manage in space) and cognitive function (risk to crew, mission objectives, and overall mission) |
| **Tapentadol** | Tapentadol | **PO** | Dual mechanism: μ-opioid agonist + norepinephrine reuptake inhibition. Used for chronic musculoskeletal or neuropathic pain. | Depressed respiration (monitoring) and cognitive function (risk to crew, mission objectives, and overall mission) |
| **Buprenorphine (partial agonist)** | Buprenorphine (SL tablet, patch) | **SL, TD** | Ceiling effect for respiratory depression; useful in chronic pain and opioid dependence. | Depressed respiration (monitoring) and cognitive function (risk to crew, mission objectives, and overall mission) |
| **Combination opioid/ acetaminophen** | Oxycodone + acetaminophen (Percocet), hydrocodone + acetaminophen | **PO** | Moderate pain control. Watch for total daily acetaminophen dose. | Combination can reduce side effects. Relatively mild, but risk to respiration and cognitive performance |

*CR*, Controlled release; *ER*, extended release; *IV*, intravenous; *PO*, per os (oral); *SC*, subcutaneous; *SL*, sublingual; *SR*, sustained release; *TD*, transdermal.

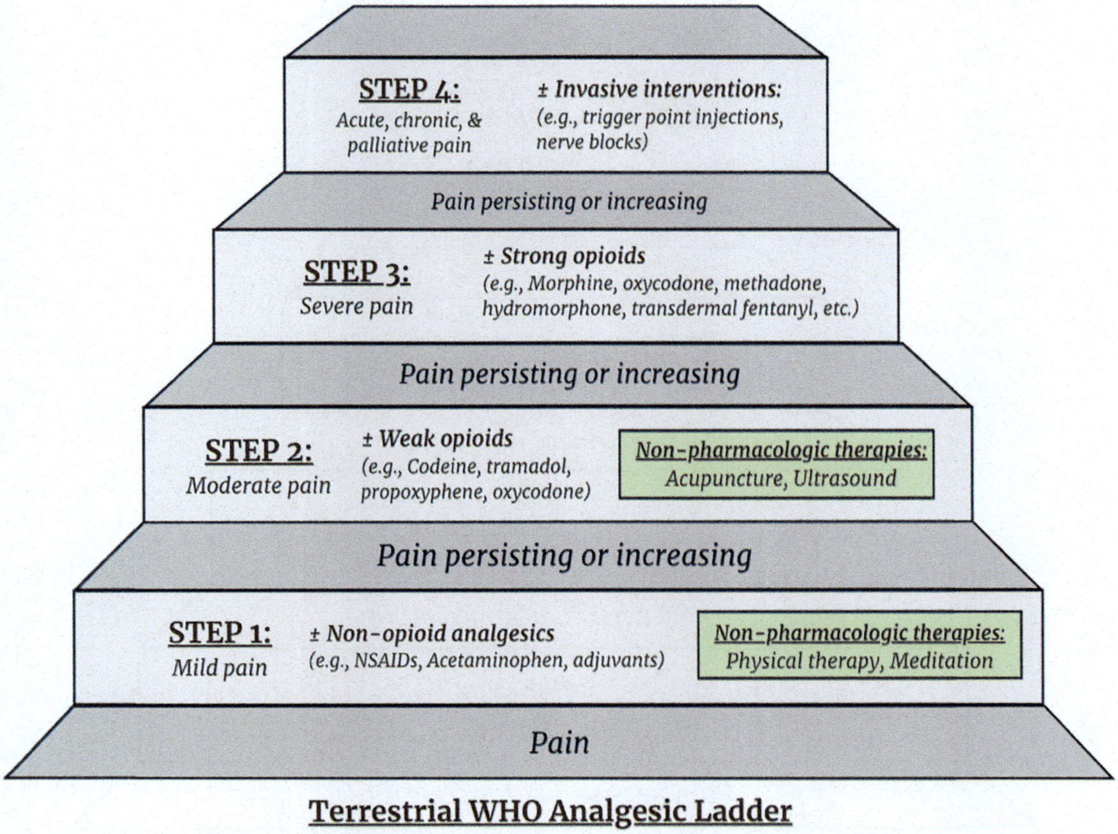


FIGURE 21.2 **Terrestrial analgesic ladder.** The World Health Organization (WHO) analgesic ladder illustrates the stepwise algorithm for treating painful conditions. Safe but relatively less potent options are considered initially with analgesics or therapeutics with increasing risk introduced if earlier steps in the algorithm are ineffective.
Modified from the World Health Organization Guidelines by Kimiyah Ghassemzadeh.

requires significant experience and has yet to be attempted in spaceflight. However, systemic analgesics' side effects can quickly escalate and pose a risk to crew health or overall mission. Therefore, nonpharmaceutical therapeutics and local anesthetics should be considered earlier in the treatment paradigm in space than in terrestrial medicine. On Earth, we often use nonpharmacologic options as supplementary or potentiating in treatment of intractable and/or chronic pain. Despite the lack of data on efficacy during spaceflight, we recommend that this treatment option be used earlier and to a larger extend during spaceflight because the adverse risk profile is very low.

Local or regional anesthesia is indicated when pain is limited to an area supplied by a targetable nerve or nerve bundle. The advantages include no depression of cognitive function and alertness; however, regional blocks can induce reduced function of the treated limb. Moreover, local and regional anesthesia does not cause depression of respiratory and cardiovascular function. Efficacy is very high when the block is performed correctly, but this leads to the most significant limitation: a high degree of technical skill and training is required to administer both peripheral and truncal blocks. Moreover, additional equipment and materials are needed, such as a large variety of needles, catheters, tubing, pumps, and ultrasound for guided placement. Additional risks of adverse events include infection at both the site of injection/catheter placement as well as systemic infection and risk of shock. For a complete overview on the application, indications, advantages, and limitations of local and regional anesthesia for pain in space, please see Chapter 21.

## 21.5.2 Systemic treatment paradigm

The ideal treatments used for pain in spaceflight will differ slightly from terrestrial management, but many of the pharmaceutical formulations and recommended doses are unchanged (Table 21.7).

**Mild pain, VAS 0–4:** First line of treatment should include nonnarcotic pharmaceuticals:

1. Nonopioid analgesic:
   - Topical NSAIDs

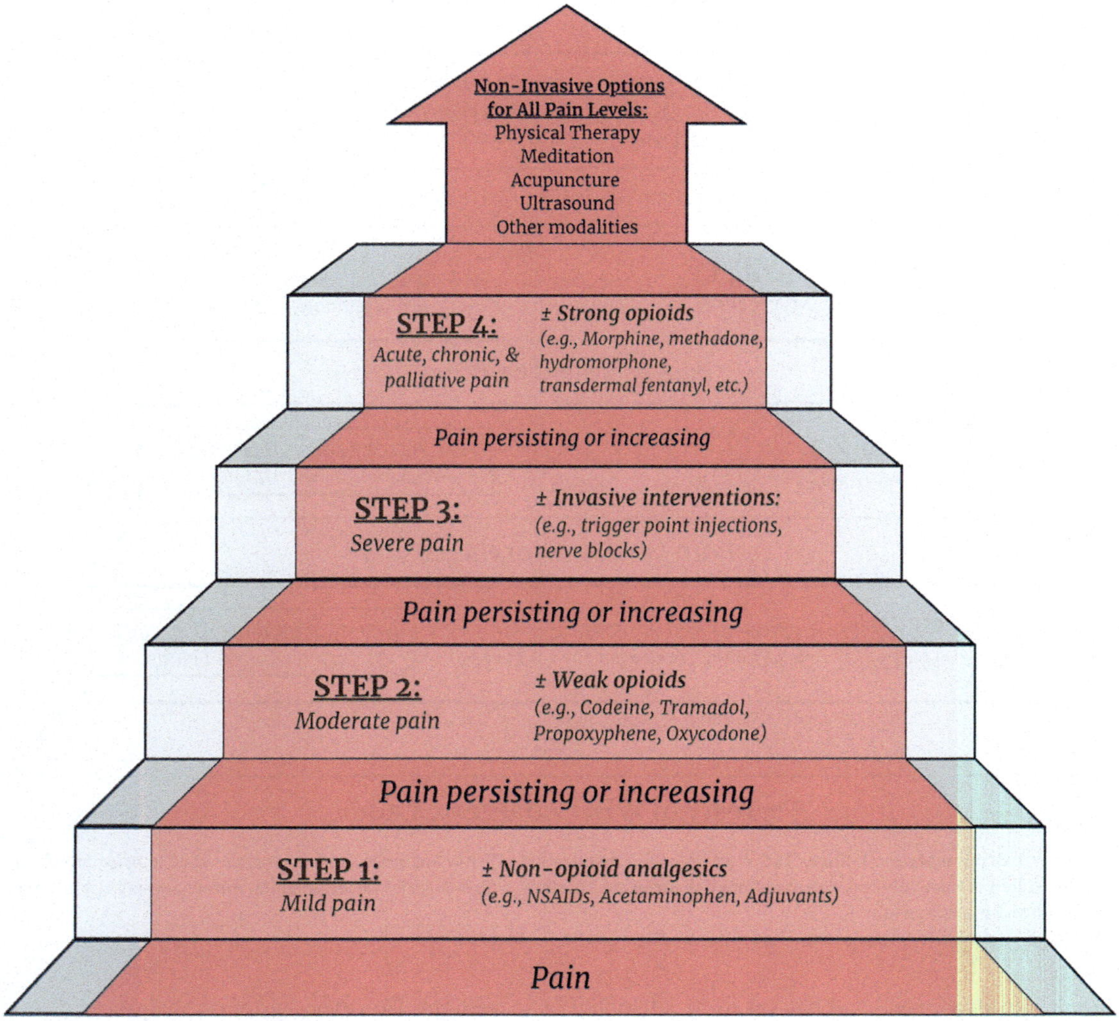


**FIGURE 21.3** **Extraterrestrial analgesic ladder.** An adapted version of the WHO analgesic ladder suggests use of noninvasive options for pain relief at all steps in the algorithm and recommends use of invasive therapeutic procedures prior to use of strong opioids.
Modified from the World Health Organization Guidelines by Kimiyah Ghassemzadeh.

**TABLE 21.7** Recommended treatment regimen based on type of pain.

| Pain type | Preferred opioids | Route |
|---|---|---|
| **Acute severe pain (trauma)** | Morphine, hydromorphone, fentanyl (IV/IM) | IV, IM, PCA |
| **Chronic pain** | Morphine SR, oxycodone CR, fentanyl patch | PO, TD, SC |
| **Breakthrough pain (on background opioid)** | Immediate-release morphine or fentanyl (SL/buccal) | PO, SL, buccal |
| **Neuropathic or mixed pain** | Methadone, tapentadol, tramadol | PO |
| **End-of-life/palliative care** | Morphine (various routes), fentanyl, hydromorphone | PO, SC, TD, rectal |

*CR*, Controlled release; *IM*, intramuscular; *IV*, intravenous; *PCA*, patient-controlled analgesia; *PO*, per os (oral); *SC*, subcutaneous; *SL*, sublingual; *SR*, sustained release; *TD*, transdermal.

- Oral acetaminophen
- Oral NSAIDs such as ibuprofen
- Both ibuprofen and acetaminophen have been used during spaceflight for decades with no reports of unusual adverse events or significantly altered PKs and dynamics. However, other NSAIDs might be preferable to ibuprofen given the terrestrial data that celecoxib has a superior renal and gastrointestinal safety profile and naproxen has a superior cardiovascular risk profile (Barcella et al., 2019; Qaseem et al., 2020; Tawfik et al., 2025)

2. Nonanalgesic pharmaceuticals should be added whenever appropriate to provide holistic treatment and mitigate adverse side effects. These include the following:
   - Antiemetics
   - Sleep aids, it is well documented that lack of or poor-quality sleep potentiates perception of pain (Irwin et al., 2023).
   - Laxatives/antacids and other ways of stabilizing gastrointestinal side effects of analgesics.
3. Nonpharmacological therapeutics and adjuvants such as the following:
   - Conservative measures: Ice/heat packs: something as simple as an icepack to alleviate swelling after minor injury and inflammation or a heat pack on sore muscles is recommended as initial treatment.
   - Ultrasound: specialized thermal, high-frequency ultrasound machines can generate heat and thereby increase blood flow in muscle and connective tissue to reduce soreness. It is frequently used in sports medicine with subjective benefits and a low risk of adverse events. Ultrasound is also an attractive therapeutic modality because it has been available in space for decades and is likely to be available on all future missions for diagnostic purposes.

**Moderate pain, VAS 4–6**: Second line of treatment does not replace the first line but instead is used in addition

1. Weak opioids and mixed agonist–antagonists (Table 21.5):

- Short-acting weak opioids such as the morphine pro-drug codeine (30–60 mg every 4 hours) or tramadol (50–100 mg every 4 hours) can be used in combination with NSAIDs. Adverse effects include constipation and nausea, which are easily manageable on Earth but can escalate and lead to more severe problems in space. Moreover, even weak opioids cause sedation and potentially confusion; their use during space missions for crew who are still in operation should therefore be considered carefully. Lastly, given the dramatically differential effects of codeine and tramadol on specific individuals due to differences in patient metabolism and the unique mechanisms of action of the pharmaceuticals, they should only be used in spaceflight if ground-tested in the specific individual prior to launch.
- The mixed opioid agonist/antagonist buprenorphine can be used in transdermal form as a patch for longer duration pain relief, as the side effects are generally milder than traditional strong opioids. Transdermal pain patches provide excellent baseline coverage, and breakthrough pain can be managed with additional short-acting tablets.

2. Adjuvant therapies:

- Corticosteroids are a modality with numerous side effects but also a myriad of applications in the setting of moderate to severe pain. Use of oral corticosteroids can lead to significant adverse effects, including bone demineralization, hyperglycemia, immune suppression, mood dysregulation, and insomnia (Cheisson et al., 2018). However, they do offer improvement in function in the setting of several pain generators, including radicular pain, joint pain, gout, or symptoms of infection, such as a sore throat (Rainer et al., 2016; Sadeghirad et al., 2017).
- Antihistamines, particularly first-generation antihistamines like diphenhydramine and chlorpheniramine, have been shown to provide direct analgesia and also reduce the need for other analgesics. Significant sedation does result, so this adjuvant should be used sparingly or topically (e.g., in a mouthwash for mucosal pain or applied to the skin for musculoskeletal pain).

**Severe pain, VAS 6–8:** Unlike the terrestrial setting, where the third line of treatment would commonly be strong opioids, caution is recommended for spaceflight use. Instead, regional anesthesia or trigger point injections could be substituted. These can be remotely guided or applied in combination with just-in-time training.

- Trigger point injections can be used to treat focal areas of taut muscle bands or myofascial pain by physically passing a needle in a fanning motion after insertion into the muscle tissue. This leads to improved pain control in areas of muscular inflammation within 24–48 hours and can typically be repeated every 2–6 weeks, although this interval could be shortened in spaceflight. Materials required include nonsterile gloves, skin cleansing agents (alcohol or chlorhexidine wipes), a 25–30 gauge needle (1–1.5 inches in length), and a 3–10 mL syringe. Although analgesia is attributed to the physical passes of the needle, the procedure is often better tolerated if supplemented by injection of a small amount of lidocaine or bupivacaine at each insertion site (0.5 mL of either 1% lidocaine or 0.25% bupivacaine). Steroid should not be included in the injectate (Benzon et al., 2025). Given their safety, these injections can be performed even by relatively inexperienced personnel, although care should be taken to avoid the highest morbidity adverse effect, pneumothorax, a pleural puncture that occurs if the needle is inserted too deeply in the thoracic region.

- Nerve blocks are injections of local anesthetic and/or steroids that provide analgesia by blocking nerve signal transmission and reducing inflammation around the nerve. In combination, this can be highly effective; however, correct application requires trained personnel to avoid vascular or nerve injury. See Chapter 20 for a more detailed description of these regional anesthesia procedures.
- Anxiolytics such as benzodiazepines are often used as adjuvant treatment both in the acute phase of pain to reduce stress response and to facilitate rest and sleep in longer-term pain management.
- Moderate to strong opioids categorically cause sedation. However, pain itself will reduce function and cognitive performance. Therefore, if pain is intractable and must be managed, then strong opioids should be used in the lowest possible dose. Sedative effects can be mitigated to some extent by central stimulants.

**Severe to excruciating pain, VAS 8–10**: The last line of treatment for severe pain is strong opioids. These are highly effective but possess significant side effects. It is critically important to balance the side effect profile to avoid respiratory depression or severe constipation (Table 21.8)

- Strong opioids such as oxycodone and fentanyl can be administered both as tablets and for more immediate effect through the IV or IM route, with the latter requiring little to no clinical skill to administer. Recommended regimen:
- Morphine: 5–10 mg every 4 hours with additional doses for breakthrough pain
- Oxycodone: 5–10 mg, which can be administered PO as tablets or solution, intranasal spray, rectally, SC, IM, or IV
- Fentanyl is a highly potent opioid with high efficacy even when administered transdermally through patches (typical starting dose of 37.5 mcg/hour). If it is delivered IM or IV, the patient will require close monitoring for respiratory depression for several hours after the first dose and with any increase in dosage.
- Naloxone (opioid antagonist): when managing severe or excruciating pain, respiratory depression can occur suddenly as a consequence of acute overdose or gradually as a consequence of the buildup of various pharmaceuticals with overlapping side effects. Naloxone should be administered if apnea occurs or the patient is unarousable after opioid administration. Objective clinical symptoms would include a respiratory rate of less than 8/min, oxygen saturation less than 90%, and/or if the patient is not responsive to mechanical stimulus (i.e., sternal rub). A dose or puff of naloxone should be administered intranasally (dose is 2–4 mg per 0.1 ml) and repeated in the opposite nostril if the initial dose is not completely effective. Alternative administration routes include IM (0.4 mg) or IV (0.1 mg), both of which can be repeated every few minutes until the patient stabilizes.

### 21.5.3 Complementary and integrative therapies

- Acupuncture: Evidence suggests benefits both for acute and chronic pain relief from this alternative medicine therapy. Because the application requires training and the treatment poses a risk of infection, caution is recommended and use only if sterility can be maintained. Microsystem-based acupuncture therapy can also be considered, which targets smaller areas of the body to modulate pain via neural pathways and at distant sites (i.e., auricular acupuncture for headache relief).
- Magnesium: This supplement has gained widespread use for chronic intractable or neuropathic pain. A dose of 200–400 mg orally once daily of elemental magnesium for any pain condition is the typical dose, although lower doses should be used

**TABLE 21.8** Risks of opioid medications.

| Risk/concern | Management/precaution |
|---|---|
| **Respiratory depression** | Monitor respiration frequency, saturation, sedation level, especially after initiation or dose changes. Have **naloxone** available. |
| **Sedation and confusion** | Titrate slowly; assess renal/hepatic function. |
| **Constipation** | Start **bowel regimen (stimulant + stool softener)** with all chronic opioids. |
| **Tolerance and dependence** | Rotate opioids or use adjuvants when escalation occurs. |
| **Diversion and misuse** | Use treatment agreements, PDMP checks, and lowest effective dose. |

The bold formatting in all tables indicates additional medications that should be available to mitigate potential risks or concerns.

initially to avoid the side effect of diarrhea. Its mechanism of action is believed to be via NMDA-receptor antagonism and modulation of pain conduction pathways over time (Shin, Na and Do, 2020).

- Transcutaneous Electrical Nerve Stimulation: This device is used terrestrially for pain relief of a variety of functional and chronic conditions. Adverse events are minimal, but it is unlikely to be considered in near-term missions given that clinical evidence of effectiveness is conflicting, the upmass of the equipment required, and the fact that it creates a radiofrequency field that can interfere with other equipment (Vance et al., 2014).
- Aromatherapy may bring alleviation in some cases of acute or chronic pain but is not recommended in the confined space of a space station or habitat due to impact on other crewmembers and the ECLSS.
- Music therapy as an adjuvant in pain management is frequently used and holds no risk for adverse events, though clinical evidence is lacking.
- Acupressure and manual massage therapy can provide significant relief of musculoskeletal pain, headache, and stress, but care must be taken to respect interpersonal boundaries between crewmembers if this therapeutic option is to be considered.

## 21.6 Conclusion

Effective pain diagnosis and management in spaceflight is best achieved by judiciously adapting familiar terrestrial principles to the austere environment of spaceflight. Diagnosis must consider high probability etiologies of pain, including increased risk of musculoskeletal strain or trauma relative to Earth, and classify the pain phenotype (nociceptive, neuropathic, or nociplastic) when formulating the treatment plan. First line therapies are similar to terrestrial regimens: acetaminophen, NSAIDs, and targeted topical agents are favored for their safety profiles and minimal neurocognitive effects. Interventional treatment options, including injection therapies, are introduced much earlier in the treatment algorithm, although safety must be prioritized given that all procedural interventions carry risk. Neuropathic pharmaceuticals like gabapentin or duloxetine will be used much more sparingly compared to terrestrial analgesic plans given their sedating side effects and protracted time to onset of effect. Other integrative approaches, including stretching protocols, acupressure or acupuncture, and behavioral countermeasures, should serve as foundational adjuvants. Ultimately, pain care in spaceflight must prioritize crew safety and performance over complete symptomatic elimination. In this context, clinicians should be less focused on stepping up the analgesic ladder and instead prioritize selecting the least impairing intervention that preserves crewmember quality of life while ensuring completion of mission objectives.

## References


Aaron, R.V., Ravyts, S.G., Carnahan, N.D., Bhattiprolu, K., Harte, N., McCaulley, C.C., et al., 2025. Prevalence of depression and anxiety among adults with chronic pain: a systematic review and meta-analysis. JAMA Network Open 8 (3). Available at https://doi.org/10.1001/jamanetworkopen.2025.0268, https://jamanetwork.com/journals/jamanetworkopen.

Apfelbaum, J.L., Silverstein, J.H., Chung, F.F., Connis, R.T., Fillmore, R.B., Hunt, S.E., et al., 2013. Practice guidelines for postanesthetic care: an updated report by the American Society of Anesthesiologists task force on postanesthetic care. Anesthesiology 118 (2), 291–307. Available at https://doi.org/10.1097/ALN.0b013e31827773e9.

Barcella, C.A., Lamberts, M., McGettigan, P., Fosbøl, E.L., Lindhardsen, J., Torp-Pedersen, C., et al., 2019. Differences in cardiovascular safety with non-steroidal anti-inflammatory drug therapy—a nationwide study in patients with osteoarthritis. Basic and Clinical Pharmacology and Toxicology 124 (5), 629–641. Available at https://doi.org/10.1111/bcpt.13182, http://onlinelibrary.wiley.com/journal/10.1111/(ISSN)1742-7843.

Benzon, H.T., Provenzano, D.A., Nagpal, A., Souza, D., Eckmann, M.S., Nelson, A.M., et al., 2025. Use and safety of corticosteroid injections in joints and musculoskeletal soft tissue: guidelines from the American Society of Regional Anesthesia and Pain Medicine, the American Academy of Pain Medicine, the American Society of Interventional Pain Physicians, and the International Pain and Spine Intervention Society. Regional Anesthesia & Pain Medicine. Available at https://doi.org/10.1136/rapm-2024-105656.

Boruah, M., Sarma, A., 2025. Dry gels: concept, current trends, and new avenues in drug delivery and biomedical application. Advanced Healthcare Materials 14 (29). Available at https://doi.org/10.1002/adhm.202500863, http://onlinelibrary.wiley.com/journal/10.1002/(ISSN)2192-2659.

Ceniza-Bordallo, G., Zimmermann, E., Vigouroux, M., Niburski, K., Fortin, M., Ouellet, J., et al., 2024. Low back pain during and after spaceflight: a systematic review with meta-analysis. Journal of Pain Research 17, 4103–4139. Available at https://doi.org/10.2147/JPR.S491060, https://www.dovepress.com/article/download/98034.

Cheisson, G., Jacqueminet, S., Cosson, E., Ichai, C., Leguerrier, A.M., Nicolescu-Catargi, B., et al., 2018. Perioperative management of adult diabetic patients. The role of the diabetologist. Anaesthesia Critical Care and Pain Medicine 37, S37. Available at https://doi.org/10.1016/j.accpm.2017.10.009, http://www.journals.elsevier.com/anaesthesia-critical-care-and-pain-medicine/.

Cohen, S.P., Vase, L., Hooten, W.M., 2021. Chronic pain: an update on burden, best practices, and new advances. The Lancet 397 (10289), 2082–2097. Available at https://doi.org/10.1016/S0140-6736(21)00393-7, http://www.journals.elsevier.com/the-lancet/.

Fink, P.B., Wheeler, A.R., Smith, W.R., Brant-Zawadzki, G., Lieberman, J.R., McIntosh, S.E., et al., 2024. Wilderness medical society clinical practice guidelines for the treatment of acute pain in austere environments: 024 update. 2. Wilderness and Environmental Medicine 35 (2), 198–218. Available at https://doi.org/10.1177/10806032241248422, https://journals.sagepub.com/home/wem.

Harrison, M.F., Garcia, K.M., Sargsyan, A.E., Ebert, D., Riascos-Castaneda, R.F., Dulchavsky, S.A., 2018. Preflight, in-flight, and postflight imaging of the cervical and lumbar spine in astronauts. Aerospace Medicine and Human Performance 89 (1), 32–40. Available at https://doi.org/10.3357/AMHP.4878.2018.

De Heer, E.W., Ten Have, M., Van Marwijk, H.W.J., Dekker, J., De Graaf, R., Beekman, A.T.F., et al., 2018. Pain as a risk factor for common mental disorders. Results from the Netherlands Mental Health Survey and Incidence Study-2: a longitudinal, population-based study. Pain 159 (4), 712–718. Available at https://doi.org/10.1097/j.pain.0000000000001133, https://journals.lww.com/pain/Pages/currenttoc.aspx.

Hsu, J.R., Mir, H., Wally, M.K., Seymour, R.B., Archer, K.R., Attum, B., et al., 2019. Clinical practice guidelines for pain management in acute musculoskeletal injury. Journal of Orthopaedic Trauma 33 (5) E158. Available at https://doi.org/10.1097/BOT.0000000000001430, http://journals.lww.com/jorthotrauma.

Irwin, M.R., Olmstead, R., Bjurstrom, M.F., Finan, P.H., Smith, M.T., 2023. Sleep disruption and activation of cellular inflammation mediate heightened pain sensitivity: a randomized clinical trial. Pain 164 (5), 1128–1137. Available at https://doi.org/10.1097/j.pain.0000000000002811, https://journals.lww.com/pain/Pages/currenttoc.aspx.

Karcioglu, O., Topacoglu, H., Dikme, O., Dikme, O., 2018. A systematic review of the pain scales in adults: which to use? American Journal of Emergency Medicine 36 (4), 707–714. Available at https://doi.org/10.1016/j.ajem.2018.01.008, http://www.journals.elsevier.com/american-journal-of-emergency-medicine/.

Kast, J., Yu, Y., Seubert, C.N., Wotring, V.E., Derendorf, H., 2017. Drugs in space: pharmacokinetics and pharmacodynamics in astronauts. European Journal of Pharmaceutical Sciences 109, S2. Available at https://doi.org/10.1016/j.ejps.2017.05.025, www.elsevier.com/locate/ejps.

Kerstman, E.L., Scheuring, R.A., Barnes, M.G., DeKorse, T.B., Saile, L.G., 2012. Space adaptation back pain: a retrospective study. Aviation, Space, and Environmental Medicine 83 (1), 2–7. Available at https://doi.org/10.3357/ASEM.2876.2012, http://docserver.ingentaconnect.com/deliver/connect/asma/00956562/v83n1/s3.pdf?expires=1330671432&id=67516688&titleid=8218&accname=Elsevier&checksum=ED17EA33A4B629846462B8A828BC0FC3.

Lowrey, C.R., Perry, S.D., Strzalkowski, N.D.J., Williams, D.R., Wood, S.J., Bent, L.R., 2014. Selective skin sensitivity changes and sensory reweighting following short-duration space flight. Journal of Applied Physiology 116 (6), 683–692. Available at https://doi.org/10.1152/japplphysiol.01200.2013, http://jap.physiology.org/content/116/6/683.full-text.pdf+html.

Marfia, G., Guarnaccia, L., Navone, S.E., Ampollini, A., Balsamo, M., Benelli, F., et al., 2023. Microgravity and the intervertebral disc: the impact of space conditions on the biomechanics of the spine. Frontiers in Physiology 14. Available at https://doi.org/10.3389/fphys.2023.1124991, http://www.frontiersin.org/Physiology/archive.

Nelson, A.M., Lacinski, R.A., Steller, J.G., 2024. Spaceflight-associated pain. Current Opinion in Anaesthesiology 37 (5), 581–587. Available at https://doi.org/10.1097/ACO.0000000000001401, http://journals.lww.com/co-anesthesiology/pages/default.aspx.

Van Oosterhout, W.P.J., Perenboom, M.J.L., Terwindt, G.M., Ferrari, M.D., Vein, A.A., 2024. Frequency and clinical features of space headache experienced by astronauts during long-haul space flights. Neurology 102 (7). Available at https://doi.org/10.1212/WNL.0000000000209224, http://www.neurology.org.

Oweidat, A., Kalagara, H., Sondekoppam, R.V., 2024. Current concepts and targets for preventing the transition of acute to chronic postsurgical pain. Current Opinion in Anaesthesiology 37 (5), 588–596. Available at https://doi.org/10.1097/ACO.0000000000001424, http://journals.lww.com/co-anesthesiology/pages/default.aspx.

Patron, M., Neset, M., Mielkozorova, M., Bisson, D.G., Vigouroux, M., Cata, J.P., et al., 2024. Markers of tissue deterioration and pain on earth and in space. Journal of Pain Research 17, 1683–1692. Available at https://doi.org/10.2147/JPR.S450180, https://www.dovepress.com/getfile.php?fileID=98888.

Penchev, R., Scheuring, R.A., Soto, A.T., Miletich, D.M., Kerstman, E., Cohen, S.P., 2021. Back pain in outer space. Anesthesiology 135 (3), 384–395. Available at https://doi.org/10.1097/ALN.0000000000003812, http://journals.lww.com/anesthesiology/pages/default.aspx.

Qaseem, A., McLean, R.M., O'Gurek, D., Batur, P., Lin, K., Kansagara, D.L., Clinical Guidelines Committee of the American College of Physicians; Commission on Health of the Public and Science of the American Academy of Family Physicians, Cooney, T.G., Forciea, M.A., Crandall , C.J., Fitterma, N. N., Hicks, L.A., Horwitch, C., Maroto, M., McLean, R.M., Mustafa, R.A., Tufte, J., Vijan, S., Williams Jr., J.W., 3 November, 2020. Nonpharmacologic and pharmacologic management of acute pain from non-low back, musculoskeletal injuries in adults: a clinical guideline from the American College of Physicians and American Academy of Family Physicians. Annals of Internal Medicine. 173 (9), 739-748. doi: 10.7326/M19-3602. Epub 2020 Aug 18. Erratum in: Ann Intern Med. April, 2023, 176(4), 584. Available at https://doi.org/10.7326/L23-0043. PMID:32805126.

Rainer, T.H., Cheng, C.H., Janssens, H.J.E.M., Man, C.Y., Tam, L.S., Choi, Y.F., et al., 2016. Oral prednisolone in the treatment of acute gout: a pragmatic, multicenter, double-blind, randomized trial. Annals of Internal Medicine 164 (7), 464–471. Available at https://doi.org/10.7326/M14-2070, http://annals.org/data/Journals/AIM/935159/AIME201604050-M142070.pdf.

Raja, S.N., Carr, D.B., Cohen, M., Finnerup, N.B., Flor, H., Gibson, S., et al., 2020. The revised international association for the study of pain definition of pain: concepts, challenges, and compromises. Pain 161 (9), 1976–1982. Available at https://doi.org/10.1097/j.pain.0000000000001939, https://journals.lww.com/pain/Pages/currenttoc.aspx.

Ramachandran, V., Wang, R., Ramachandran, S.S., Ahmed, A.S., Phan, K., Antonsen, E.L., 2018. Effects of spaceflight on cartilage: implications on spinal physiology. Journal of Spine Surgery 4 (2), 433–445. Available at https://doi.org/10.21037/jss.2018.04.07.

Sadeghirad, B., Siemieniuk, R.A.C., Brignardello-Petersen, R., Papola, D., Lytvyn, L., Vandvik, P.O., et al., 2017. Corticosteroids for treatment of sore throat: systematic review and meta-analysis of randomised trials. BMJ (Clinical Research Ed.). 358, j3887. Available at https://doi.org/10.1136/bmj.j3887.

Sauer, A.K., Vigouroux, M., Dougherty, P.M., Cata, J.P., Ingelmo, P.M., 2023. Pain experience and sensory changes in astronauts during and after short-lasting commercial spaceflight: a proof-of-concept study. Journal of Pain Research 16, 4253–4266. Available at https://doi.org/10.2147/JPR.S440630, https://www.dovepress.com/getfile.php?fileID=95109.

Sayed, S., Ibrahim, H.K., Mohamed, M.I., El-Milligi, M.F., 2013. Fast-dissolving sublingual films of terbutaline sulfate: formulation and in vitro/in vivo evaluation. Molecular Pharmaceutics 10 (8), 2942–2947. Available at https://doi.org/10.1021/mp4000713.

Scheuring, R.A., Effenhauser, R., Anderson, D.N., Hei, M.V., Holman, P., Helgeson, M., 2024. Cervical spine intervertebral disc herniation on board the international space station: diagnosis, treatment and operational mission impact. Evidence Report Document ID: 20240001305, National Aeronautics and Space Administration.

Schmidt, P.C., Ruchelli, G., Mackey, S.C., Carroll, I.R., 2013. Perioperative gabapentinoids choice of agent, dose, timing, and effects on chronic postsurgical pain. Anesthesiology 119 (5), 1215–1221. Available at https://doi.org/10.1097/ALN.0b013e3182a9a896.

Shin, H.J., Na, H.S., Do, S.H., 2020. Magnesium and pain. Nutrients 12 (8), 1–13. Available at https://doi.org/10.3390/nu12082184, https://www.mdpi.com/2072-6643/12/8/2184/pdf.

Tawfik, A.G., Gomez-Lumbreras, A., Del Fiol, G., Kawamoto, K., Trinkley, K.E., Reese, T., et al., 2025. Nonsteroidal anti-inflammatory drugs and risk of gastrointestinal bleeding: a systematic review and meta-analysis. Clinical Pharmacology and Therapeutics. Available at https://doi.org/10.1002/cpt.70054, http://onlinelibrary.wiley.com/journal/10.1002/(ISSN)1532-6535.

Treffel, L., Navasiolava, N., Mkhitaryan, K., Jouan, E., Zuj, K., Gauquelin-Koch, G., et al., 2020. DI-5-Cuffs: lumbar intervertebral disc proteoglycan and water content changes in humans after five days of dry immersion to simulate microgravity. International Journal of Molecular Sciences 21 (11). Available at https://doi.org/10.3390/ijms21113748.

Vance, C.G.T., Dailey, D.L., Rakel, B.A., Sluka, K.A., 2014. Using TENS for pain control: the state of the evidence. Pain Management 4 (3), 197–209. Available at https://doi.org/10.2217/pmt.14.13.

Wotring, V.E., 2015. Medication use by U.S. crewmembers on the International Space Station. The Journal 29 (11), 4417–4423. Available at https://doi.org/10.1096/fj.14-264838, http://www.fasebj.org/content/29/11/4417.full.pdf+html.

Section IV

# Post procedure care

Chapter 22

# Recovery and monitoring


Faye Chiou Tan[1], Sanghee Im[1,2], and Dana R. Levin[3,4,5]

[1]*Department of Physical Medicine and Rehabilitation, Baylor College of Medicine, Houston, TX, United States,* [2]*Department and Research Institute of Rehabilitation Medicine, Severance Hospital, Yonsei University College of Medicine, Seoul, South Korea,* [3]*Space Medicine Operations, Mission Operations Vast Inc., Long Beach, CA, United States,* [4]*Department of Emergency Medicine, University of California, Los Angeles, CA, United States,* [5]*Division of Space, Ecological, Arctic, and Resource-Limited Medicine, Massachusets General Hospital, Boston, MA, United States*


Prolonged exposure to microgravity has profound effects on the body systems that resist gravity. Bone density, muscle strength, and cardiovascular performance all decondition in microgravity, with the effects worsening the longer one remains in microgravity (Gabel et al., 2022; Ploutz-Snyder et al., 2018). NASA astronauts have successfully mitigated many of these effects of microgravity through exercise, leveraging the Advanced Resistive Exercise Device, which provides analogous forces to weightlifting up to 600 pounds, and the cycle ergometer with vibration isolation and stabilization system, which provides aerobic exercise similar to riding a bicycle up to 350 Watts (Petersen et al., 2016). These programs are time-consuming, taxing to spacecraft power and life support systems, require substantial mass and volume, carry a risk of injury, require discipline, and are custom-tailored to each astronaut with the specific prescriptions for resistance, aerobic loads, and exercises adjusted to meet individual needs. Studies have shown that 2 hours of mixed aerobic and resistance exercise each day, 6 days a week, is effective at countering many of the effects of microgravity (Sibonga et al., 2019; OCHMO-TB-030 NASA Office of the Chief Health & Medical Officer OCHMO Bone Loss Technical Brief, 2025). Lower-intensity workouts with less resistance and/or a greater focus on aerobics were not able to sufficiently offset the loss of muscle mass and bone density (Gaskill, 2024), though early studies with fewer sessions but higher intensity workouts have also shown promise (English et al., 2020). These findings are consistent with rehabilitation recommendations on Earth that emphasize that muscle hypertrophy and strengthening do not occur until muscle fatigue is achieved (Fish et al., 2003).

The importance of exercise in space underscores the complex challenge of managing a patient who has sustained an injury, illness, or medical intervention that limits their ability to exercise. This chapter explores this challenge through the physical medicine and rehabilitation physician's (physiatrist's) thought processes. This draws from recovery strategies used in intensive care settings on Earth where patients are often limited in their physical capabilities and subject to prolonged periods of bed rest, inducing deconditioning that is analogous (though not identical) to spaceflight.

## 22.1 Postintervention monitoring

### 22.1.1 Immediate postintervention monitoring

As is true on Earth, after a procedure in reduced or microgravity, monitoring the patient for changes in clinical status is key to detecting early decompensation. The best strategy for rehabilitation in this phase is ensuring stability and recovery from the acute effects of the procedure (Committee on Standards and Practice Parameters, 2019). As the patient recovers, paying specific attention to the presence of pain, nausea, vomiting, breathing status, hydration status, drainage from any wounds, and bleeding is essential to facilitate recovery (Apfelbaum et al., 2013).

It is reasonable to assume that adjustments to postprocedure outcomes and functional expectations may be necessary in microgravity. For example, baseline systolic blood pressures in space are lower than preflight or postflight. Fu et al. (2019) noted that preflight systolic blood pressures of 120 +/− 10 mmHg dropped to 106 +/− 9 mmHg in flight in a study of 12 astronauts. Another example is the apparent reorganization of the central nervous system in response to the unweighting of peripheral sensors in microgravity. Reduction in the magnitude of the Hoffman reflex (a quantitative version of the classic spinal stretch reflex), deemphasis of inputs from the otolith organs, and reductions in the baroreceptor reflex all suggest that the space environment may alter the effectiveness of some postprocedure exercises commonly used terrestrially.


**Procedural Medicine for Spaceflight. DOI:** https://doi.org/10.1016/B978-0-443-27552-4.00034-8

(Canadian space agency, no date; Stella et al., 2021). This reorganization is a normal adaptive response to microgravity and reverses upon returning to gravity.

This reorganization of sensory emphasis is not pathological and is the likely mechanism behind the reduction in space adaptation symptoms. However, the Hoffman reflex in particular may serve as a method to quantify the effectiveness of in space rehabilitation exercises. To measure the Hoffman reflex, a low-level electric shock is delivered over a peripheral nerve, which carries the stimulus to the spinal cord, where it synapses and activates the associated motor neuron. The magnitude of the resulting contraction can then be measured using electromyography. One of the first experiments conducted on the International Space Station found that the magnitude of this reflex rapidly decreased by about 35% in flight, and recovered by the 10th day after landing. This likely corresponds to the central nervous system shifting the emphasis placed on proprioceptive sensors. Since the test only takes a few minutes and changes occur relatively rapidly, measuring changes in the Hoffman reflex over time may help rehabilitation specialists monitor the effectiveness of an inflight rehabilitation regimen (Canadian space agency, no date; Stella et al., 2021).

### 22.1.2 Monitoring of vital signs and remote telemonitoring

Remote monitoring of vital signs, or telemedicine, has been utilized in spaceflight since the early Space Age, when heart rate measurements were transmitted in real time back to Houston during Project Mercury (Ettelson et al., 2006, Robinson et al., 1966). Now, more than 50 years later, continuous postprocedure vital sign monitoring is considered standard of care in the developed world. This frequent use of telemetric monitoring has been shown to detect "failure to rescue" situations in which a patient dies or suffers a preventable complication (Verrillo and Winters, 2018). Spaceflight presents unique challenges here since continuous monitoring often requires the use of consumable components and expertise on where to place the sensors and how to interpret the data. Less cumbersome telemetry systems are being developed to address these problems. Smart watches, nonconsumable wearables, and other wireless sensors have been explored, with three-lead electrocardiogram patches worn with little discomfort for up to seven consecutive days (Glazkova et al., 2020). Systems for monitoring vital signs solely through cameras are another promising technology that is rapidly being developed to an operational status (Canepa et al., 2025;Padaki et al., 2025).

### 22.1.3 Postanesthetic recovery

While most procedures that can be performed in space in the near term future will not require substantial anesthesia, some may benefit from sedation. The American Society of Anesthesiologists recommends continuous monitoring of "oxygenation, ventilation, circulation, level of consciousness, and temperature"(Committee on Standards and Practice Parameters, 2019). On Earth, postanesthetic recovery from the time of discontinuing anesthesia to return to baseline physiology is assessed using the Aldrete Scoring System (Ding, 2023). This system evaluates five areas from 0 to 2, with a maximum of 10 points:

- activity,
- respiration,
- circulation,
- consciousness,
- and oxygen saturation.

A fully recovered patient, with a score of 10, is able to move all extremities, breathe deeply and cough, has a blood pressure within +20 mmHg of preanesthetic level, is fully awake, and has an oxygen saturation >92% on room air. In the ambulatory setting, careful anesthetic drug selection with short half-lives and rapid recovery times can be used to minimize complications of pain, nausea, and vomiting (Komorowski et al., 2016). Ketamine is frequently used as the anesthetic of choice for this reason. It also is used for pain control (Sewbuckas, 2018). For this reason, as well as this particular drug's potential to moderate the effects of depression, the possible applications and administrative protocols for use of Ketamine in spaceflight are currently being considered by NASA (Kutz et al., 2025).

In addition to vital signs and consciousness, postprocedure complications often include nausea/vomiting 9.8%, upper airway complications (e.g., bronchospasm) 6.9%, hypotension 2.7%, dysrhythmias 1.4%, hypertension 1.1%, and cardiac events 0.6%. Less commonly, there are reports of visual disturbance and delirium (Glick et al., 2024).

Additionally, up to 70% of terrestrial patients experience urinary retention after procedures. Since spaceflight exposure alone is associated with a substantial incidence of urinary retention, space medical providers should anticipate this and be prepared to manage it (Stepaniak et al., 2007). This is most often temporary, but it may persist in older men with preexisting urinary retention and those with neurogenic bladders. Postprocedure urinary bladder ultrasound is likely to be useful and should be a routine part of postprocedure care in spaceflight. Post void residual volumes should be evaluated in addition to a patient's subjective symptoms. Medications such as Bethanicol (a cholenergic agonist used for post operative urinary retention) may be helpful with mild to moderate retention. However, retention over 400 ml will likely require catheterization (Glick et al., 2024). Given that space adaptation often causes urinary retention in itself, it may be desirable to delay any procedures until after the first few days of spaceflight if possible.

### 22.1.4 Enhanced recovery after surgery

Enabling a rapid recovery after surgery begins long before the flight itself. According to the Enhanced Recovery After Surgery protocol (ERAS) published in the Journal of the American Medical Association in 2017, it begins with general fitness and healthy nutritional support (prehabilitation). ERAS creates guidelines for various regions of the body and different types of surgeries and provides visual aids (flowcharts) outlining the recommended phases for rapid recovery (Ljungqvist et al., 2017; Fig. 22.1).

These phases begin with baseline nutritional support, medical optimization, and a complete medical history. In the preoperative phase, selective bowel preparation with preoperative carbohydrate drinks up to 2 hours before surgery is permitted in most instances in this paradigm. That is to say that the patient no longer needs to be "nothing by mouth" or NPO (nil per os) for as long as previously recommended. Prophylaxis against nausea and vomiting is also supported. During the intraoperative phase, ERAS recommends minimal use of drains and tubes, regional anesthesia, avoiding narcotics, maintaining temperature control, and using balanced fluids. Finally, early discontinuation of urinary catheters and intravenous (IV) fluids, multimodal nonopioid pain control, and early mobilization are evidence-based tenets of the ERAS protocol (Ljungqvist et al., 2017).

Pain control decreases insulin resistance and enhances early mobilization. This will be especially important in the context of space flyers. Judicious early mobilization to the greatest extent possible will help mitigate the deconditioning of rest, which is magnified by the microgravity environment. Care must be taken to avoid reinjury or rupture of the surgically repaired site, since the straps and harnesses used to simulate weight may apply unusual forces to the exercising astronaut. It is likely best to start with light mobility as soon as practicable after the procedure is complete, and gradually increase the loaded weight as tolerated back to the standard exercise regimen. If an astronaut cannot tolerate a more intensive 3-day/week regimen, it may be preferable to spread their exercise over a 6-day protocol initially and/or begin with minimal initial weight. However, return to normal "weight" bearing should be done as soon as possible to minimize deterioration and speed healing. Indeed, the advantage of microgravity is that the risk of falls is all but eliminated, and it is possible to titrate an individual's weight-bearing without the need for complex suspension harnesses.

Equally important is early oral intake of fluids and solids to encourage the return of bowel function after sedation or more invasive procedures. Protein and energy-rich nutritional supplements may be added to supplement the menu. As a result of ERAS protocols on Earth, there was faster recovery with a 30%–50% shorter hospital stay and 11% reduction in complications (Ljungqvist et al., 2017). Blood sugar monitoring in the postprocedural period will also be important given that tight glycemic control postoperatively decreases the rate of infection and mortality (Cheisson et al., 2018; Wang et al., 2018).

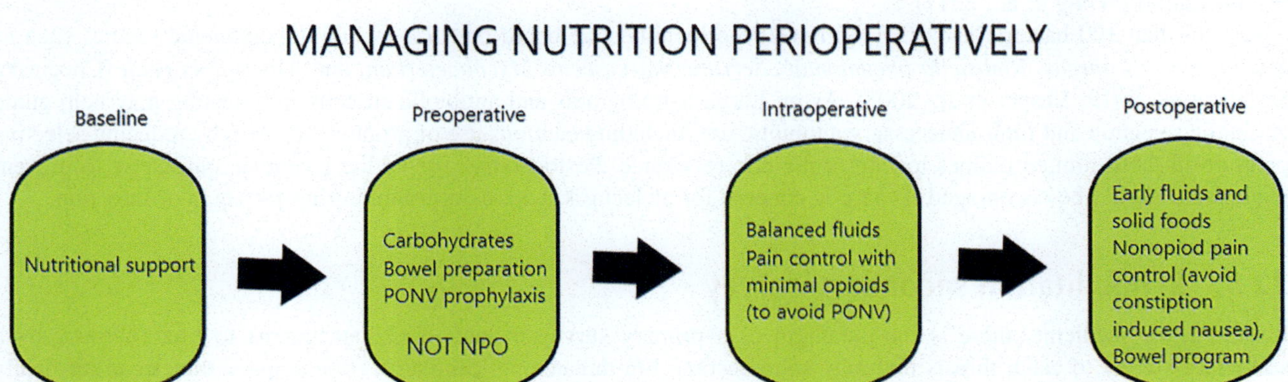


**FIGURE 22.1 Enhanced Recovery After Surgery (ERAS) guidelines.** Managing nutrition and bowel function perioperatively—ERAS guidelines. *PONV*, Post Operative Nausea and Vomitting.

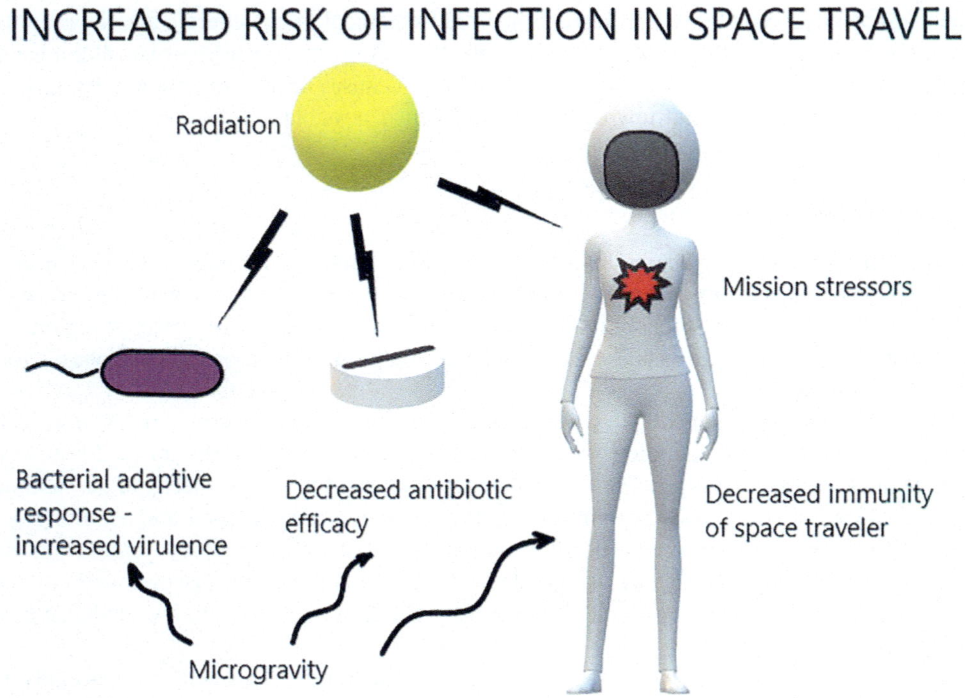


FIGURE 22.2 **Potential infection risks in space.** Radiation and microgravity cause increased bacterial virulence in lab studies.

## 22.2 Infection control and management

As on Earth, microorganisms are very much present within spacecraft. The spacecraft microbiome often resembles that of the crew that flies within it (Lang et al., 2017; Salido et al., 2025). While studies show a lack of biodiversity, changes have been observed in microbial behavior and human immune system gene expression. These changes in microbial behavior may represent a bacterial adaptive response to radiation and microgravity stressors (Bijlani et al., 2021; Raza et al., 2021). Some evidence suggests that these responses may increase antibiotic resistance and enhance biofilm formation, but the clinical significance of these changes is not clear (Cowen et al., 2024; Fig. 22.2; Mortazavi et al., 2014; Knox et al., 2016; Romsdahl et al., 2018; Bijlani et al., 2021).

Physical fitness, quarantine, vaccinations, and screening for latent disease, such as tuberculosis, are all valuable countermeasures against infection risk (Cowan et al., 2024). In-flight probiotics, nutrition, hand hygiene, and vitamins have been suggested, but the clinical significance is not clear (Cowan et al., 2024). NASA recommends surgical masking, cough etiquette, and hand sanitation for suspected respiratory illness (Weinstein and Mermel, 2013). Microbial control on ISS is further assisted by High Efficiency Particulate Air filtering and monitoring of food and water for microbial contamination (Wong et al., 2017).

Greater than 100 bacteria and 30 fungi have been identified on board the ISS, including pathogenic species such as *Staphylococcus aureus*, *Klebsiella pneumoniae*, *Serratia marcescens*, *Mycobacterium*, and *Aspergillus genii* (Checinska Sielaff et al., 2019; Vesper et al., 2008). As on Earth, antibiograms and antibiotic susceptibility testing may help guide antibiotic selection and limit unnecessary antibiotic use, including cautious use of prophylaxis, thereby reducing selective pressure in these limited biomes towards antibiotic resistance. Postoperative use of the incentive spirometer to prevent pneumonia should be encouraged if there is concern for atelectasis, with careful monitoring for signs of infection.

## 22.3 Rehabilitation secondary survey

In emergency medicine, there is the paradigm of a primary survey to halt immediate threats to life, followed by a secondary survey to catch threats that may soon become life-threatening. Similarly, once a space traveler is medically stabilized postprocedure, the provider must evaluate them for nonlife-threatening and occult medical conditions. Postoperative care and rehabilitation traumatology includes a thorough review of systems from head to toe. Essentially, this is a comprehensive systematic exam to detect slow bleeds, delayed presentation of injuries, signs of infection, discolorations, deformities, and other indicators of a developing issue. Often, the secondary survey must be repeated to monitor for change over time.

This is particularly important with patients who are sedated, recovering from sedation, experiencing altered mental status, or otherwise unable to examine themselves and report symptoms. In addition to issues of the body's systems, postprocedure patients may still have IV lines, drains, catheters, and other medical interventions still in place to facilitate healing and monitoring for the detection of potential issues.

Terrestrially, physicians often write orders and monitoring criteria for a nurse or other assistant to check. For near- and mid-term space travel, the CMO will likely have primary responsibility for this. However, as is the case with terrestrial medicine, no medical provider can function effectively 24/7, and it is important to offload any amenable tasks and monitoring to ground teams or other crewmembers to ensure the CMO is able to respond again when needed. In many cases, telemetered data from inputs and outputs can wait a few minutes or an hour for providers to respond to changes, allowing monitoring from the ground. Medication pumps may even be adjusted remotely. Most drains and some medications may fall into this category after stabilization is established. For those who cannot, other crewmembers can be taught the inspection routine in real time and instructed to notify the CMO, similar to a nurse or resident contacting the doctor on call terrestrially.

Another consideration for postprocedure recovery is that reduced or functionally absent gravity may alter how draining fluids present. Just as IV lines will require the use of pumps or compression devices to ensure delivery of appropriate doses and rates in nonterrestrial settings, negative pressure bulbs or 1-way valves will be needed for wound drains and catheters. Fluid will not collect at the bottom of these drains, or may take longer to settle in reduced-gravity settings. Observers must account for these changes so as not to over- or underestimate the fluid outputs.

In some cases, electronic sensors may be worth the cost in extra complexity for systems where highly accurate measurements are critical. However, engineering-grade precision is rarely necessary in medicine, where volume errors of a few milliliters are rarely significant. It is likely that most inspections will be amenable to modified procedures such as disconnecting a collection vessel and swinging it with the "down" end pointed out from the provider's hand, instituting a brief angular moment, and allowing fluids to collect at the "bottom" and permit an estimation of the output volume. Similarly, the use of a syringe or plunger system may enable estimation of volume by concentrating the solution into a smaller container with gauge markings, enabling a visual estimation of the volume within medically tolerable error, for example, "less than 10 ml" or "approximately 25 ml." Table 22.1 illustrates some representative differences between terrestrial and in-space postprocedure monitoring.

During the rehabilitation phase following the procedure, resumption of bowel and bladder function, skin protection, prevention of contracture, and pulmonary hygiene are crucial for preventing future complications. Managing bowel movements in space is more complex than on Earth, particularly since the toilet systems have not traditionally been designed for assistance, and there is presently no equivalent to a bedside commode. Additionally, laundry and cleaning facilities are limited in space. Soiled clothing or materials may not be cleanable and may need to be discarded and accounted for in the vehicle trash system design. These considerations are beyond the scope of this chapter but are important for providers to consider as they design procedures. One advantage of bowel movements is that the material is often self-adherent and likely to stay intact, though resumption of bowel movements after a procedure may lead to flatulence and the possibility of projectile fecal particles. Bowel incontinence and gas build-up can be reduced by limiting pain medications, using stool softeners, and scheduling timed voids. Wearing garments such as disposable diapers may help to limit projectile and incontinent feces from contaminating vehicle surfaces and limited clothing supplies.

Infection risk is best mitigated through frequent patient exams and reassessments. In general, any area where skin and/ or tissue function or integrity was disrupted is more prone to infection, particularly if catheters or other material are present. For these reasons, the most common areas of infection are the skin over the procedure site, the tissue where the procedure was performed, the lungs (pneumonia), and the urinary tract. Infection risk at the procedure site is best mitigated through the use of clean and sterile techniques, preserving blood flow to the area to the extent possible, and monitoring postprocedure for signs of cellulitis, increased drainage, fever, and pain. Pneumonia risk can be decreased through preventive measures, including preprocedure education and breathing exercises, use of incentive spirometers (if mass and volume allow), and/or observed and guided breathing exercises with the CMO acting as the "incentive" to encourage deep breaths and early mobilization (Boden et al., 2018; Pasquina et al., 2006). Urinary infections may be mitigated through avoiding the use of Foley catheters when possible, discontinuing indwelling Foley catheters as soon as possible, and checking PVR after a procedure to assess urinary retention. This is particularly important if medications that disrupt bladder emptying were used (e.g., anticholinergics or others that disrupt sympathetic/parasympathetic balance; Stepaniak et al., 2007; 36 OCHMO-MTB-003 Urinary Health. NASA, no date)

**TABLE 22.1** Example differences in postprocedure monitoring on Earth versus spaceflight.

| Earth | Space |
|---|---|
| Multiple trained clinical staff to reassess the patient, compare to physician set criteria, and carry out physician standing orders to adjust medications or call for help | 1–2 trained providers to reassess the patient and follow recovery.Offload amenable data and electronic pump measurement to ground teams or other crewmembers trained "as needed" in inspection procedures. |
| IV pole—IV bags to gravity | IV fluid pump |
| Drains/Foley catheter to gravity | Drains to suction bulb and/or one-way valves |
| | Modified volume estimation procedures (centrifugal "swinging" or use of syringe suction to estimate output volume) |
| Nonweight-bearing precautions | Restraint pressure precautions |
| Decubitus prevention-turn side back side | Restraint pressure precautions |
| Pneumatic compression stockings | N/A |
| CPR compressions—gravity assisted | CPR compressions—against firm surface |
| Blood pressure—normal 120 +/− 10 mmHg | Blood pressure—normal 106 +/− 9 mm Hg |
| DVT (lower body > upper body circulation) | DVT neck (lower body < upper body circulation) |
| Muscle stretch reflex—normal range H reflex | Muscle stretch reflex—H-reflex decreased |
| Manual muscle strength testing level 3 antigravity | Manual muscle strength testing mild resistance |
| Tilt test for orthostatic intolerance | Check for volume depletion (e.g., urine output, dehydration) |
| | Check for low hemoglobin (e.g., postoperative anemia, GI bleed) |
| Spasticity (antigravity muscles) worse with gravity | Spasticity improved with gravity elimination |
| Postvoid residuals—gravity assisted urination | Postvoid residuals—urine adheres to bladder wall |

*CPR*, cardiopulmonary resuscitation; *DVT*, deep vein thrombosis; *GI*, gastrointestinal; *H reflex*, Hoffmann reflex; *IV*, intravenous; *N/A*, not applicable.

Skin protection is another concern and can be assisted with the use of topical incontinence barrier creams (e.g., Camoseptine, Balmex, A+D, etc.) to prevent skin rash and breakdown in moisture and friction-prone areas. In reduced- and microgravity settings, gravity-dependent pressure sores are less of a concern but may still occur beneath tight restraining straps used during or after a procedure (Drudi et al., 2012). For this reason, the secondary survey after a procedure will require attention to the skin beneath any straps or harnesses.

Finally, as mentioned earlier, early mobilization and activity are critical. The first step may be to maintain their range of motion, particularly in manipulated tissues, through provider-mediated passive ranging with careful attention to avoid damaging recently sutured sites or other areas prone to damage. This should be quickly followed by active-assisted and independent range-of-motion and stretching exercises at the earliest point tolerated by the patient. These need not be maximal effort and should be guided by the patient's ability, pain levels, and the specific nature of the procedure. In general, providers should encourage patients to push themselves just a little further than they think they can, but providers must be careful to avoid movements and effort that may damage recently manipulated tissue, particularly in patients who are highly self-motivated, as may be the case with astronauts. The balance of encouragement vs permission to rest is highly individualized for each patient. Ground providers can upload specific instructional guidance and prescriptions tailored to the patient's needs and response to therapy, but the CMO must have some knowledge of what movements and level of effort are possible after a procedure to avoid injury to the patient while maximizing recovery and minimizing postprocedure and microgravity-induced deconditioning.

## 22.4 Venous thrombosis prevention and hemostasis

### 22.4.1 Increased risk of venous thrombosis in space

We do not yet have sufficient evidence to quantify the risk of deep vein thrombosis (DVT) in flight, but the two known cases, and the physiology associated with microgravity, suggest that the risk is increased over the terrestrial baseline of 1–2 cases per 1000 person-years. Considering that the total number of person-years spent on orbit as of this publication is on the order of 100 and that we have so far detected two DVTs, this increased risk is likely to be substantial and may be as high as 1 in 50 person-years (Bostick Laura et al., 2025).

Disorders of thrombosis and hemostasis have received heightened attention in recent years following the development of an internal jugular vein deep venous thrombosis in orbit, with subsequent examinations finding stagnant retrograde flow in a higher-than-expected proportion of astronauts on long-duration flights (Marshall-Goebel et al., 2019; Pavela et al., 2022). Changes in concentration of fibrinogen, platelet levels, and thrombin dysfunction may lead to hypercoagulability, while White and Wenthe noted that headward fluid shift, venous distention, and vascular congestion due to microgravity lead to sluggish, turbulent blood flow in the upper body circulation. There is also some suggestion of endothelial injury in flight and, certainly, there would be after procedural intervention (Kuzichkin et al., 2022). Thus, alterations in all three components of Virchow's triad (turbulence, vascular injury, hypercoagulability) may occur during spaceflight, likely increasing the baseline risk. The theoretical risks combined with the limited available population-based evidence for DVTs demonstrate that DVT risk is indeed likely to be higher in flight, a circumstance likely worsened by procedural interventions and postprocedure lack of mobility. Additionally, the findings of stagnant upper circulatory flow and the location of previously identified DVTs in flight suggest that spaceflight DVTs may present in a different circulatory region than the typical terrestrial lower extremity DVT (White and Wenthe, 2023; Harris et al., 2023; Pavela et al., 2022).

Providers must therefore be vigilant for this complication, prepared to recognize its symptoms, and have supplies on hand to intervene should it occur. The signs and symptoms most often cited include the following:

- Neck pain or tenderness (often unilateral, along the sternocleidomastoid),
- Neck swelling on the affected side,
- Erythema or warmth over the neck,
- Palpable cord along the jugular vein (uncommon but classic),
- Facial swelling or periorbital edema,
- Upper extremity swelling (especially arm or shoulder on the same side),
- Dilated superficial veins of the neck or chest,
- Headache,
- Facial fullness or pressure,
- Visual disturbances,
- Dizziness,
- Fever (especially if septic thrombophlebitis, e.g., Lemierre syndrome).

Symptoms may worsen with Valsalva or during exercise and may be masked by the cephalad fluid shift symptoms that typically occur during spaceflight. However, it is worth noting that the first DVT identified in spaceflight was discovered incidentally, and the astronaut reported no symptoms (Aunõn et al., 2020). The implication is that, similar to DVTs in terrestrial medicine, monitoring and detection depend on a high clinical suspicion and a low threshold for testing.

### 22.4.2 Complications of deep vein thrombosis (DVT)

As on Earth, the DVT itself typically is not dangerous, but it can lead to dangerous pathology that is far more difficult and resource-intensive to manage. Thus, preventing a DVT is ideal, but once it occurs, interventions must focus on preventing the associated downstream pathology. In this case, the major pathologies of concern include pulmonary embolism (PE), infection/septic thrombophlebitis (Lemierre's Syndrome), and compression of adjacent structures. Of these, PE carries the most significant and immediate life threat.

Terrestrially, the location of the DVT is associated with differences in the incidence of complications. Upper Extremity DVTs are thought to have a lower, but not absent, risk for clinically significant PEs. The incidence values are complicated by the rare occurrence and heterogeneity in study design and criteria for IJ. A recent review estimated the terrestrial occurrence of IJ DVT to be between 0.5% and 3.5% of all DVTs. This translates to a terrestrial incidence range of 1 IJ DVT in 28,600–400,000 person-years. This same review noted the incidence of clinically significant PEs in this IJ DVT population to be ~10%, which would place the terestrial event incidence as 1 PE from an IJ DVT every 286,000 to

4,000,000 person-years. This is compared to the incidence of lower extremity DVT (1–2 DVTs in 2000 person-years) with the incidence of significant PE from these being on the order of 1 in 2000 to 1 in 7000 (Drakos et al., 2020).

Clearly, these numbers are not applicable in spaceflight, and the interpretation of PE risk in the context of spaceflight carries far greater uncertainty due to limited evidence. As an exercise to put a ballpark number on potential spaceflight DVT and PE incidence we can assume the incidence of an IJ DVT is 1 in 50 person-years based on the 2025 NASA risk assessment (Bostick et al., 2025). If we also assume a 10% incidence for clinically significant PEs from IJ DVTs based on the Drakos study than the incidence of a clinically significant PE in space could be as high as 1 in 500 person-years.

Mortality rates for terrestrial PEs are as high as 30% without treatment which puts the potential mortality risk for untreated PE in space as high as 1 death in 1700 person-years. With anticoagulation the mortality rate drops to 5% for large PEs and negligable for the more common smaller cases representing a relative risk reduction of 83% to nearly 100% (Bria et al., 2023).

While, these rates are highly uncertain it does highlight that the risk is potentially substantial and can be mitigated with a low complexity, low mass, and low volume treatment. This likely justifies including screening and anticoagulation as part of any long duration medical system and support plan for long duration space missions in microgravity.

#### *22.4.2.1 Prevention of venous thrombosis*

While it is still early to identify definitive methods to prevent DVTs, terrestrial experience suggests that preventive strategies can be effective (Nicholson et al., 2020). Screening is a good place to start, since it serves both to gather additional data and catch operationally relevant occurrences. Before flight, candidates can be screened for risk factors that may further elevate their DVT risk, for example, prior DVT, cancer, thrombocytopenia, hormone replacement, congestive heart failure, pregnancy, and genetic risk factors. Mitigation strategies can be put in place for modifiable risk factors, and nonmodifiable risk can be considered by medical mission planners (Harris et. al, 2023). Age in particular is an important, nonmodifiable risk factor since 60% of venous thrombosis occurs in patients over 70 years of age (Nicholson et al., 2020). It may also be worth considering in-flight screening and surveillance protocols (Elias et al., 2024; Nicholson et al., 2020).

Another prevention technique commonly employed on Earth is mechanical and pharmacologic prophylaxis. This includes ambulation to enhance circulation and pneumatic or graduated compression stockings worn soon after the procedure (Nicholson et al., 2020). For example, with cephalad fluid shifting in microgravity and likely differences in location incidence for DVT, it is unclear if graduated compression stockings or pneumatic compression would reduce risk if utilized in the typical terrestrial manner. It is also unclear if pneumatic compression would be of greater benefit if applied to the arms rather than the legs if the upper body venous system is more prone to stasis in space. Since the risk appears to stem from changes in blood flow distribution, it may be worth investing in research on lower body negative pressure to achieve a more terrestrial distribution of blood and potentially reduce overall risk.

Pharmacologic prophylaxis with oral anticoagulants or low-molecular-weight heparin must demonstrate that the benefits outweigh the risks (Nicholson et al., 2020). These risks may vary by mission phase. For example, dynamic phases, such as launches and landings, may carry a higher risk than benefit, while the low dynamic phases in orbit or transit may change the risk calculation. It is worth investigating the effective risk reduction achievable with pharmacological interventions, weighed against the probability of DVT on a mission, the risks of the medications, and the mass/volume/storage requirements for anticoagulant medications. As of publication, there are no definitive guidelines for pharmacologic venous thrombosis prevention or prophylaxis in space. However, the European Space Agency (ESA) has formed a Topical Team to address coagulation, venous thrombosis, prophylaxis, and treatment for space travelers (Harris et al., 2023).

An additional measure for DVT prevention involves cardiovascular and resistance exercise. It remains to be seen if these are effective in reducing the risk of thromboembolic events in space. However, since venous stasis is a contributing factor and exercise helps limit stasis traditional exercise may have some effect (Bostick Laura et al., 2025). However, it may be worth ensuring that exercise include upper extremities in addition to lower extremities given the location of identified spaceflight DVTs.

### 22.4.3 Management of deep venous thrombosis and pulmonary embolism

In 2020, the American Society of Hematology provided primary guidelines for managing DVT and PE on Earth. The panel found that in uncomplicated cases of DVT and PE, treatment with oral anticoagulants outside of the hospital (i.e., at home) is acceptable (Ortel et al., 2020). For proximal DVT, such as one found in the IJ, anticoagulation alone is sufficient without thrombolytic therapy unless there is a possibility of loss of limb and/or extensive DVT. Thrombolytic therapy by catheter is recommended in those cases, which is not presently feasible in space and may prompt an early return on missions where this is feasible (Ortel et al., 2020).

Published experience on the management of a DVT in space is limited to one astronaut case study, which occurred 2 months into a mission on the ISS. This individual received enoxaparin 1.5 mg/kg/d, which was then changed to 1.0 mg/kg/d to stretch supplies. After resupply, they received apixaban 5 mg/bid for 42 days, then the dose was tapered at 3 months. Four days before reentry to Earth, anticoagulant therapy was halted (Aunõn et al., 2020; Harris et al., 2023). Consideration was also made for hemorrhage control related to anticoagulation, but it was ultimately not needed.

In 2025, NASA's Office of the Chief Health and Medical Officer convened a multidisciplinary Venous Thromboembolism (VTE) Working Group to review the status and progress of research and clinical activities intended to mitigate the risk of VTE during spaceflight. The group identified venous stasis and retrograde flow in the internal jugular veins—frequently observed in microgravity—as potential prethrombotic conditions unique to space. Based on in-flight ultrasound findings and terrestrial data, the current therapy of choice for VTE prevention and treatment on the International Space Station is the direct oral anticoagulant apixaban. The recommended treatment regimen consists of apixaban 10 mg twice daily for 7 days, followed by 5 mg twice daily, whereas a prophylactic dose of 2.5 mg twice daily is suggested for crewmembers with venous stasis and additional risk factors such as thrombophilia, a family history of unprovoked thrombosis, or the use of hormones that increase the risk of VTE. Anticoagulation should be discontinued at least five half-lives of the pharmaceutical compound or 48 hours prior to landing to reduce the risk of bleeding during the nominal high G and impact phases of landing or in case of trauma from off-nominal events (Bostick Laura et al., 2025).

The Working Group also recommended that the risk/benefit of hormonal therapies, such as oral contraception, should also be evaluated preflight to minimize risks. Progesterone-only contraceptives—such as a levonorgestrel intrauterine device or the Nexplanon implant carry a lower risk for venous thrombosis while still achieving menstrual cycle suppression, which may make them more appealing than estrogen-based therapies. Chapter 13 contains more information on contraceptive use in spaceflight.

### 22.4.4 Hemostasis post procedure

Hemostasis to stop bleeding is on the opposite end of the management of hypercoagulability. In spaceflight, any relative hypercoagulability and increased risk of thrombosis would, at first glance, seem to enhance hemostasis after a procedure. However, additional challenges unique to spaceflight can further complicate achieving hemostasis. These include hyperdistention of vessels in the head and neck due to fluid shift. The cross-sectional area of the jugular vein, for example, has been shown to increase 30%–225%. This increased venous pressure could make hemostasis more difficult and rebleeding postprocedure possible. In addition, early thrombocytopenia, anemia, and relative blood volume contraction of 10%–15% may make hemostasis and hemodynamic stability more challenging (White and Wenthe, 2023). Chapter 6 discusses hemostasis in spaceflight.

## 22.5 Wound healing and care

### 22.5.1 Phases and mechanism of wound healing

The three phases of wound healing are inflammation, proliferation, and remodeling (Bacci and Bani, 2022; Cialdai and Monici, 2013). The first phase, inflammation, is triggered by the coagulation cascade. Platelets release substances such as growth factors and fibronectin to assist in repair and regeneration (Locatelli et al., 2021). Neutrophils and macrophages cause stromal cells to differentiate into myofibroblasts that contract the wound. They also stimulate the second phase of healing, the proliferation phase. In this phase, epithelial and endothelial cells grow along with neoangiogenesis to reepithelialize the wound. In the third phase, remodeling, macrophages, and enzymes remove the clot and debris, allowing the formation of a healed scar (Bacci and Bani, 2022). Wound healing homeostasis is maintained by the epidermal proliferative unit (EPU). This consists of epidermal stem cells, transit-amplifying cells, and keratinocytes stored in stem cell niches near hair follicles (Bacci and Bani, 2022). While keratinocytes work at the level of the epidermis and produce keratin and collagen to give skin resilience, fibroblasts work in the dermis to make connective tissue. Fibroblasts have a multitude of functions in all three phases of wound healing; the details can be found in Cialdai and Monici (2013) and Cialdai et al. (2022).

### 22.5.2 Factors that inhibit wound healing

Many factors inhibit good wound recovery, such as conditions causing poor circulation (e.g., smoking, pressure on the sore, diabetes, etc.), persistence of inflammation (due to infection, burns, or other irritants), and overactivation of myofibroblasts (such as the process of keloid formation; Bacci and Bani, 2022) Additional factors include obesity, age, and hormone profile (Cialdai et al., 2022). There is also evidence to suggest that wound healing may be altered by exposure to the spaceflight environment. Changes in signaling within the EPU and in platelet count and function have both been observed in spaceflight research. Additionally, keratinocyte growth and migration are accelerated by microgravity in vitro, suggesting that wound healing may even be accelerated (Locatelli et al., 2021; Bacci and Bani, 2022; Radstake et al., 2023; Cialdai et al., 2022). However, these findings have not translated into clinically significant impacts. It is possible that these results are either artifacts of simulated microgravity or features of spaceflight adaptation that have, as yet, undetermined significance. What is clear to date is that minor injuries do not appear to have a significantly different clinical course from what is expected terrestrially (Scheuring et al., 2009).

### 22.5.3 Management and facilitation of wound healing

Wound healing requires an environment conducive to skin regeneration and scar formation. Strategies start with avoiding exposures known to inhibit wound healing (e.g., smoking cessation, control of diabetes, etc.). Additional advice involves keeping the wound clean, controlling active bleeding, preventing infection, and keeping wound edges moist to facilitate cell migration and proliferation. This latter step often uses petroleum-based gels to create a vapor barrier on the wound surface. Specific guidance will depend on the location, type, size, and cause of the wound, but in general, keeping it clean, monitoring for signs of infection, avoiding reinjury, and avoiding manipulation or repeat chemical exposures to allow the body to perform the healing process work well for otherwise healthy individuals. Looking to the future, the ESA has created a Topical Team that researches "Tissue Healing in Space" which includes examination of the healing of sutured wounds ("Suture in Space" project; Puhl et al., 2022)

## 22.6 Early mobilization and rehabilitation postprocedure

Rehabilitation efforts start with physical conditioning to build and maintain muscle mass and bone strength. These efforts start before the flight to prepare the crew and continue in-flight to maintain them. Setting a good starting point is one of the key elements in building resilience, as it helps limit the effects of deconditioning from unexpected events. It is equally important, though often less emphasized, to perform stretching exercises of muscles and joints to prevent contractures and maintain a full range of motion. Fortunately, we have not had to rehabilitate an astronaut after a substantial injury or procedure in space. However, this does not mean we must rely solely on analog data to project what best practice in the microgravity and reduced-gravity environment may look like.

### 22.6.1 Early mobilization

There are many lessons we can learn from terrestrial postoperative ICU protocols. For example, early mobilization is safe, reduces hospital length of stay, and decreases physical and psychological sequelae such as pain, nausea, and perceptions of helplessness (Eggmann et al., 2024; Lang et al., 2020; Sepúlveda et al., 2025). Exercise therapy has been shown to prevent postintensive care syndrome, a complex combination of physical, mental, and cognitive impairment (Liu et al., 2024). Because of studies like these, early postprocedure mobilization is now considered the standard of care (ERAS society guidelines, no date). However, with a highly motivated population, such as astronauts are likely to be, it is important to monitor them for overly aggressive self-rehabilitation that could lead to injury (Appelman et al., 2024).

Traditional contraindications to mobilization include the following:

- Unstable vital signs,
- Confusion/altered mental status,
- Agitation,
- Emotionally lability,
- Severe pain,
- Hemoglobin <7 g/dl (Hiser et al., 2025),
- Platelets <20,000 cells/mm$^2$.

The last two items on this list may not be realistic to assess quantitatively in space. Physical exam findings may serve as a partial substitute for such cases. Findings such as conjunctival pallor are effective indicators of severe anemia, with sensitivity and specificity for hemoglobin levels less than 8 mg/dL near 90% in some studies (Vyas et al., 2023). However, studies on platelet count exam findings show poor test characteristics (Cao et al., 2002; Furlan et al., 2016). The presence of tubes and drains should not deter mobilization. On Earth, a special walker is adapted to hold all the devices, but in microgravity, this may simply be a matter of taking care to avoid snagging any lines on spacecraft surfaces.

Terrestrially, functional mobilization includes performing tasks such as sitting and walking on the first day after surgery (Sepúlveda et al., 2025). In space, this should be adapted to mimic the activities the weight-bearing crew would be expected to do, such as exercise or managing the momentum of objects moved through the station from one location to another. Other solutions may mimic the strategies that have been employed to manage severe weakness in ICU setting when weight-bearing is not possible. These include neuromuscular electrical stimulation, passive and active range of motion exercises, or resistance training, mobility transfers, and proprioceptive neuromuscular facilitation techniques (Hiser et al., 2025; Sepúlveda et al., 2025). Cycle ergometers, flywheels, impedance devices, and in-space resistive exercise devices are likely to be well-suited for this purpose. Reduced weight-bearing status and gradual increases in loads may be an area of advantage for spaceflight since the general challenge is finding ways to simulate bearing weight in the first place. In reduced-gravity settings, lower body negative pressure and/or weighted suits may assist with this process.

One area of concern is a finding in ICU settings known as critical illness-related bone loss. Preventing bone loss in the ICU setting primarily includes nutritional support with calcium and vitamin D (Agu, 2024). Medications that block bone loss have also been used, along with early mobilization, weight-bearing, and resistance exercises, but evidence on efficacy remains scarce. While the causes are not well elucidated it is thought to be a combination of medication effects, malnutrition, reduced activity, and chronic inflammation, implying that the rate of bone loss may not be much faster than what already occurs in spaceflight and countermeasures such as exercise may be sufficient to mitigate the effects for the type of procedures likely to be needed in the next 1–2 decades (Cai et al., 2022).

## 22.6.2 Rehabilitation post procedure

On Earth, team members (nurses, physical therapists, physiatrist) work as "mobility champions" to identify and address barriers to functional mobility. This can be due to multiple causes, which all need to be systematically addressed. Thus, a head-to-toe complete review of systems is often employed on bedside rounds to identify potential limitations. A few common examples are cognition (sedation), postoperative pain, decreased exercise tolerance due to oxygen desaturation, and dizziness from postoperative anemia. Thus, having a physiatrist (physical medicine and rehabilitation physician) with knowledge of space medicine available for teleconsultation is likely to be helpful (Eggmann et al., 2024).

In general, the physiatrist's consultation on Earth begins with determining the patient's baseline. For space missions, the physiatrist could be part of the ground-based medical team and involved in training to assess the starting point of the patient. This is often done through assessing the metabolic equivalent of a task (MET) rates, where the metabolic loads of

**TABLE 22.2** Examples of activities and associated MET rates (Herrmann et al., 2024).

| MET rate | Activity description |
|---|---|
| 16.3 | Climbing steep hills (grade 10%–40%) at a brisk pace |
| 12 | Carrying >33 kg load up stairs |
| 9.3 | Climbing stairs quickly |
| 8 | Marching briskly, Monty Python Ministry of Silly Walks |
| 7 | Walking, 4.5 mph |
| 6 | Typical desk job |
| 5 | Showering |
| 4.5 | Stair climbing, slow pace |
| 4 | Walking, self-selected speed on a firm surface |
| 3.5 | Descending stairs |
| 2.5 | Bird watching, walking, and stopping |
| 2.1 | Walking, treadmill, less than 1.0 mph, 0% grade |
| 1.2 | Daily activities in space |
| 1 | Resting metabolic rate |

**Source:** Adapted from Herrmann, S.D., Willis, E.A., Ainsworth, B.E., Barreira, T.V., Hastert, M., Kracht, C.L., et al., 2024. 2024 Adult Compendium of Physical Activities: a third update of the energy costs of human activities. Journal of Sport and Health Science. 13 (1), 6–12.

each task are normalized to the patient's resting metabolic rate. A MET rate of 1 is defined as the amount of oxygen consumed when seated quietly and without stress. For most people, this is approximately 3.5 ml $O_2$/kg/min. This establishes the baseline for what a patient should be able to maintain immediately after a procedure. It is then a team effort to move the patient to the next MET level by increasing exercise tolerance and addressing medical limitations to function. Performing activities of daily living (dressing, face washing, etc.) is often around a MET rate of 2 terrestrially, but studies have demonstrated this is closer to a rate of 1.2 in microgravity. This likely reflects the reduced workload of not having to resist gravity (Ewert et al., 2022). Table 22.2 contains examples of activities and their associated MET rates.

Terrestrially, the activities of daily living are often considered part of the rehabilitation workout. However, in space, the reduced loads mean that additional exercise will likely be needed within 24 hours of the procedure. On Earth, once a patient reaches a MET rate of 3 (casual walking), they are transferred to inpatient rehabilitation to achieve a discharge MET of 5 (e.g., showering and walking up two flights of stairs). MET rates of 5–7 (e.g., "sitting desk job" on Earth). After achieving seven or more, a patient can be discharged from rehabilitation and resume normal activity without restrictions other than those needed to protect the procedure site. On Earth, this typically involves in-hospital training with weights no higher than 0.5 kg, followed by inpatient rehabilitation unit with 1.5–3 kg weights, arm and leg ergometers, and stairs designed to help the patient reach a MET of 5. The outpatient gym will have equipment similar to a regular commercial gym.

Adapting this to space will require knowledge of the specific exercise systems available on the vehicle. While many exercise systems may be capable of stressing an astronaut similarly to a terrestrial gym, the intended goals according to NASA's Baseline Values Assessment Document are closer to a MET rate of 7 (Ewert et al., 2022). Additionally, the postprocedure astronaut is unlikely to be as debilitated as a recovering ICU patient and will likely be able to progress more quickly. Immediately after the procedure, a patient may be encouraged to participate in daily activities roughly akin to a MET of 1.2. For the next 48 hours or so, they may be encouraged to perform exercises at about 25%–50% of their usual prescription as tolerated, approximating a MET rate of 2–3.5. Increasing activity from this point should be done in consultation with physicians familiar with the procedure and experts in rehabilitation to avoid overloading any sensitive areas and injuring the crewmember from overstrain. This is a key point to keep in mind with motivated populations. When tolerated, activity should be advanced in stages, likely adding 25% increases up to an exercise prescription corresponding to a MET rate of 7. At this point, they are likely to be receiving the necessary benefit to counter microgravity effects. A similar concept can serve as a baseline for reduced-gravity environments.

It is important to note that recovery of pain-free range of motion is required before strengthening exercises can begin, each individual recovers differently, and each procedure will require unique restrictions. While this description may serve as a basic outline, it will need to be carefully tailored to ensure the specific needs of the case are met. Passive and active range of motion of the limbs have been shown to improve mobility and reduce pain after terrestrial procedures and are likely to benefit patients in microgravity and reduced-gravity settings (Adler and Malone, 2012). Pre- and in-flight maintenance of full range of motion with stretching exercises can also help prevent strains and tears due to activities and shorten recovery time after an unexpected procedure (Chiou-Tan, 2022). An example specific to space travelers is Achilles tendon contracture. Decreased ankle movement leads to overuse of the plantar fascia and subsequent plantar fasciitis (Barratt et al., 2019). This is exacerbated by the microgravity neutral floating position with plantar flexion at the ankles. Achilles tendon and plantar fascia injury can be preempted with preflight and in-flight stretching exercises, which should be continued after any procedures (Adler and Malone, 2012).

## 22.7 Summary

There are unique challenges in postprocedural recovery and monitoring in microgravity and reduced-gravity environments. These result from reduced gravitational forces that are normally used to assist in postoperative care on Earth. In addition, there is an increased risk of venous thrombosis. Limited crew size, limited on-board expertise, and limited on-board equipment also present a challenge. On the other hand, microgravity reduces concerns about some postprocedural complications, such as nonweight-bearing status, decubitus ulcers, and reduced spasticity severity.

## References

36 OCHMO-MTB-003 Urinary Health, NASA. Available from: https://www.nasa.gov/ochmo-mtb-003-urinary-health-2/.

Adler, J., Malone, D., 2012. Early mobilization in the intensive care unit: a systematic review. Cardiopulmonary Physical Therapy Journal 23 (1), 5–13. Available at https://doi.org/10.1097/01823246-201223010-00002.

Agu, R., 2024. Managing bone loss in critical care: pathophysiology and treatment approaches. Journal of Anesthesiology and Pain Research 7 (6), 276.

Apfelbaum, J.L., Silverstein, J.H., Chung, F.F., Connis, R.T., Fillmore, R.B., Hunt, S.E., et al., 2013. Practice guidelines for postanesthetic care: an updated report by the American Society of Anesthesiologists task force on postanesthetic care. Anesthesiology 118 (2), 291–307. Available at https://doi.org/10.1097/ALN.0b013e31827773e9.

Appelman, B., Charlton, B.T., Goulding, R.P., Kerkhoff, T.J., Breedveld, E.A., Noort, W., et al., 2024. Muscle abnormalities worsen after post-exertional malaise in long COVID. Nature Communications 15 (1). Available at https://doi.org/10.1038/s41467-023-44432-3, https://www.nature.com/ncomms/.

Aunõn, S.M., Pattarini, J.M., Moll, S., Sargsyan, A., 2020. Venous thrombosis during spaceflight. New England Journal of Medicine 382 (1), 89–90. Available at https://doi.org/10.1056/NEJMc1905875.

Bacci, S., Bani, D., 2022. The epidermis in microgravity and unloading conditions and their effects on wound healing. Frontiers in Bioengineering and Biotechnology 10. Available at https://doi.org/10.3389/fbioe.2022.666434.

Barratt, M.R., Baker, E.S., Pool, S.L., 2019. Principles of Clinical Medicine for Space Flight. Springer, New York. Available at 10.1007/978-1-4939-9889-0.

Bijlani, S., Stephens, E., Singh, N.K., Venkateswaran, K., Wang, C.C.C., 2021. Advances in space microbiology. iScience 24 (5), 102395. Available at https://doi.org/10.1016/j.isci.2021.102395.

Boden, I., Skinner, E.H., Browning, L., Reeve, J., Anderson, L., Hill, C., et al., 2018. Preoperative physiotherapy for the prevention of respiratory complications after upper abdominal surgery: pragmatic, double blinded, multicentre randomised controlled trial. BMJ (Clinical Research Ed.) 360. Available at https://doi.org/10.1136/bmj.j5916.

Bostick Laura, L., Coffey Kristin, M., Kaouk Joanne L. Taoufik Sarah, D., Francisco David, R., 2025. Assessment of risk of venous thromboembolism during spaceflight. NASA/SP-20250001074/REV1. NASA National Aeronautics and Space Administration, Washington, DC.

Bria, K.E., Gage, B.F., Marques, S.F., Beasley, M., Kramer, K., Samant, M., et al., 2023. Survival outcomes in patients with intermediate and high-risk pulmonary embolism treated with low molecular weight or unfractionated heparin: time to start of anticoagulant therapy. Blood 142 (1).

Canadian Space Agency, H-Reflex flexes its muscles! https://www.asc-csa.gc.ca/eng/sciences/hreflex.asp (accessed 10.09.24).

Committee on Standards and Practice Parameters. 2019. American Society of Anesthesiologists, 2019. Standards for post anesthesia care. 2019. <https://www.asahq.org/standards-and-practice-parameters/standards-for-postanesthesia-care > (accessed 30.08.24).

Cai, Y., Kang, F., Wang, X., 2022. Critical illness and bone metabolism: where are we now and what is next? European Journal of Medical Research 27 (1). Available at https://doi.org/10.1186/s40001-022-00805-w.

Canepa, C.A., Levin, D.R., Padaki, A.S., 2025. Comparison of camera-acquired vital signs to conventional vital signs in a space-analog environment. Wilderness and Environmental Medicine 36 (1 Special Issue: Space Medicine & Health Systems). Available at https://doi.org/10.1177/10806032241291994.

Cao, X.T.P., Ngo, T.N., Wills, B., Kneen, R., Nguyen, T.T.H., Ta, T.T.M., et al., 2002. Evaluation of the World Health Organization standard tourniquet test and a modified tourniquet test in the diagnosis of dengue infection in Viet Nam. Tropical Medicine & International Health: TM & IH 7 (2), 125–132.

Checinska Sielaff, A., Urbaniak, C., Mohan, G.B.M., Stepanov, V.G., Tran, Q., Wood, J.M., et al., 2019. Characterization of the total and viable bacterial and fungal communities associated with the International Space Station surfaces. Microbiome 7 (1). Available at https://doi.org/10.1186/s40168-019-0666-x, http://www.microbiomejournal.com/.

Cheisson, G., Jacqueminet, S., Cosson, E., Ichai, C., Leguerrier, A.-M., Nicolescu-Catargi, B., et al., 2018. Perioperative management of adult diabetic patients. Postoperative period. Anaesthesia Critical Care & Pain Medicine 37, S27. Available at https://doi.org/10.1016/j.accpm.2018.02.023.

Chiou-Tan, F.Y., 2022. Musculoskeletal mimics of cervical radiculopathy. Muscle and Nerve 66 (1), 6–14. Available at https://doi.org/10.1002/mus.27553, http://onlinelibrary.wiley.com/journal/10.1002/(ISSN)1097-4598.

Cialdai, F., Monici, M., 2013. Wound healing: what happens in microgravity? Current Biotechnology 2 (3), 250–256. Available at https://doi.org/10.2174/22115501113026660025.

Cialdai, F., Risalliti, C., Monici, M., Cialdai, F., Monici, R.C.M., 2022. Role of fibrdoblasts in wound healing and tissue remodeling on Earth and in space. Frontier in Bioengineering and Biotechnology 10, 1–18 958381.

Cowen, D., Zhang, R., Komorowski, M., 2024. Infections in long-duration space missions. The Lancet Microbe 5 (9), 100875. Available at https://doi.org/10.1016/s2666-5247(24)00098-3.

Ding, D., 2023. Aldrete Scoring System StatPearls. FL StatPearls Publishing, Treasure Island, FL. Available at https://www.ncbi.nlm.nih.gov/books/NBK594237/.

Drakos, P., Ford, B.C., Labropoulos, N., 2020. A systematic review on internal jugular vein thrombosis and pulmonary embolism. Journal of Vascular Surgery. Venous and Lymphatic Disorders 8 (4), 662–666.

Drudi, L., Ball, C.G., Kirkpatrick, A.W., Saary, J., Grenon, S.M., 2012. Surgery in space: Where are we at now? Acta Astronautica 79, 61–66. Available at https://doi.org/10.1016/j.actaastro.2012.04.014.

ERAS Society Guidelines. Available from: https//erassociety.org/guidelines/ (accessed 30.08.24).

Eggmann, S., Timenetsky, K.T., Hodgson, C., 2024. Promoting optimal physical rehabilitation in ICU. Intensive Care Medicine 50 (5), 755–757. Available at https://doi.org/10.1007/s00134-024-07384-w, https://www.springer.com/journal/134.

Elias, A., Weber, T., Green, D.A., Harris, K.M., Laws, J.M., Greaves, D.K., et al., 2024. Systematic review of the use of ultrasound for venous assessment and venous thrombosis screening in spaceflight. NPJ Microgravity 10 (1). Available at https://doi.org/10.1038/s41526-024-00356-w, https://www.nature.com/npjmgrav/.

English, K.L., Downs, M., Goetchius, E., Buxton, R., Ryder, J.W., Ploutz-Snyder, R., et al., 2020. High intensity training during spaceflight: results from the NASA Sprint study. NPJ Microgravity 6 (1). Available at https://doi.org/10.1038/s41526-020-00111-x.

Ettelson, B., Reeves, J.A., Ettelson, B., Reeves, J.A., 2006. Space-Proven Medical Monitor: The Total Patient Care Package. Spacelabs Healthcare.

Ewert, M.K., Chen, T.T., Powell, C.D., 2022. TX life support baseline values and assumptions document. NASA STI Program Office NASA/TP-2015–218570/REV2. NASA Jonhson Space Center Houston.

Fish, D.E., Krabak, B.J., Johnson-Greene, D., DeLateur, B.J., 2003. Optimal resistance training: comparison of DeLorme with Oxford techniques. American Journal of Physical Medicine and Rehabilitation 82 (12), 903–909. Available at https://doi.org/10.1097/01.PHM.0000098505.57264.DB.

Fu, Q., Shibata, S., Hastings, J.L., Platts, S.H., Hamilton, D.M., Bungo, M.W., et al., 2019. Impact of prolonged spaceflight on orthostatic tolerance during ambulation and blood pressure profiles in astronauts. Circulation 140 (9), 729–738. Available at https://doi.org/10.1161/CIRCULATIONAHA.119.041050, http://circ.ahajournals.org.

Furlan, N.B., Tukasan, C., Estofolete, C.F., Nogueira, M.L., da Silva, N.S., 2016. Low sensitivity of the tourniquet test for differential diagnosis of dengue: an analysis of 28,000 trials in patients. BMC Infectious Diseases 16 (1).

Gabel, L., Liphardt, A.-M., Hulme, P.A., Heer, M., Zwart, S.R., Sibonga, J.D., et al., 2022. Incomplete recovery of bone strength and trabecular microarchitecture at the distal tibia 1 year after return from long duration spaceflight. Scientific Reports 12 (1). Available at https://doi.org/10.1038/s41598-022-13461-1.

Gaskill, M.L., 2024. Science in space NASA Melissa L. Gaskill, astronaut exercise. Available from: https://www.nasa.gov/missions/station/iss-research/astronaut-exercise/.

Glazkova, N., Podladchikova, T., Gerzer, R., Stepanova, D., 2020. Non-invasive wearable ECG-patch system for astronauts and patients on Earth. Acta Astronautica 166, 613–618. Available at https://doi.org/10.1016/j.actaastro.2019.01.036.

Glick, D.B., Hold, N.F., Glick, D.B., Hold, N.F., Mussmeier, N.A., 2024. Overview of post-anesthetic care for adult patients. Available from: http://www.uptodate.com/ (accessed 30.08.24).

Harris, K.M., Arya, R., Elias, A., Weber, T., Green, D.A., Greaves, D.K., et al., 2023. Pathophysiology, risk, diagnosis, and management of venous thrombosis in space: where are we now? NPJ Microgravity 9 (1). Available at https://doi.org/10.1038/s41526-023-00260-9, https://www.nature.com/npjmgrav/.

Herrmann, S.D., Willis, E.A., Ainsworth, B.E., Barreira, T.V., Hastert, M., Kracht, C.L., et al., 2024. 2024 adult compendium of physical activities: a third update of the energy costs of human activities. Journal of Sport and Health Science 13 (1), 6–12.

Hiser, S.L., Casey, K., Nydahl, P., Hodgson, C.L., Needham, D.M., 2025. Intensive care unit acquired weakness and physical rehabilitation in the ICU. British Medical Journal 388, e077292. Available at https://doi.org/10.1136/bmj-2023-077292.

Knox, B.P., Blachowicz, A., Palmer, J.M., Romsdahl, J., Huttenlocher, A., Wang, C.C.C., et al., 2016. Characterization of Aspergillus fumigatus isolates from air and surfaces of the International Space Station. mSphere 1 (5). Available at https://doi.org/10.1128/mSphere.00227-16, http://msphere.asm.org/content/msph/1/5/e00227-16.full.pdf.

Komorowski, M., Fleming, S., Hinkelbein, J., 2016. Anaesthesia in outer space: the ultimate ambulatory setting? Current Opinion in Anaesthesiology 29 (6), 649–654.

Kutz, C.J., Mistry, A.M., Dukes, C.H., 2025. Expanding ketamine application for treatment of acute suicidality in long-duration spaceflight. Aerospace Medicine and Human Performance 96 (6), 509–519. Available at https://doi.org/10.3357/amhp.6619.2025.

Kuzichkin, Nichiporuk, Zhuravleva, Markin, Rykova, Zhuravleva, et al., 2022. Endothelial dysfunction markers and immune response indices in cosmonauts' blood after long-duration space flights. NPJ Microgravity 8 (1). Available at https://doi.org/10.1038/s41526-022-00237-0.

Lang, J.M., Coil, D.A., Neches, R.Y., Brown, W.E., Cavalier, D., Severance, M., et al., 2017. A microbial survey of the International Space Station (ISS). PeerJ 5. Available at https://doi.org/10.7717/peerj.4029.

Lang, J.K., Paykel, M.S., Haines, K.J., Hodgson, C.L., 2020. Clinical practice guidelines for early mobilization in the icu: a systematic review. Critical Care Medicine 48 (11) E1121. Available at https://doi.org/10.1097/CCM.0000000000004574, http://journals.lww.com/ccmjournal/pages/default.aspx.

Liu, K., Tronstad, O., Flaws, D., Churchill, L., Jones, A.Y.M., Nakamura, K., et al., 2024. From bedside to recovery: exercise therapy for prevention of post-intensive care syndrome. Journal of Intensive Care 12 (1). Available at https://doi.org/10.1186/s40560-024-00724-4, https://jintensivecare.biomedcentral.com/.

Ljungqvist, O., Scott, M., Fearon, K.C., 2017. Enhanced recovery after surgery a review. JAMA Surgery 152 (3), 292–298. Available at https://doi.org/10.1001/jamasurg.2016.4952, http://jamanetwork.com/journals/jamasurgery/fullarticle/2595921.

Locatelli, L., Colciago, A., Castiglioni, S., Maier, J.A., 2021. Platelets in wound healing: what happens in space? Frontiers in Bioengineering and Biotechnology 9. Available at https://doi.org/10.3389/fbioe.2021.716184.

Marshall-Goebel, K., Laurie, S.S., Alferova, I.V., Arbeille, P., Auñón-Chancellor, S.M., Ebert, D.J., et al., 2019. Assessment of jugular venous blood flow stasis and thrombosis during spaceflight. JAMA Network Open 2 (11), e1915011. Available at https://doi.org/10.1001/jamanetworkopen.2019.15011.

Mortazavi, S.M.J., Motamedifar, M., Namdari, G., Taheri, M., Mortazavi, A.R., Shokrpour, N., 2014. Non-linear adaptive phenomena which decrease the risk of infection after pre-exposure to radiofrequency radiation. Dose-Response 12 (2), 233–245. Available at https://doi.org/10.2203/dose-response.12-055.Mortazavi, http://dose-response.metapress.com/media/eafy7jqqql1tw334gt7r/contributions/k/3/3/t/k33t316331422t29.pdf.

Nicholson, M., Chan, N., Bhagirath, V., Ginsberg, J., 2020. Prevention of venous thromboembolism in 2020 and beyond. Journal of Clinical Medicine 9 (8), 2467. Available at https://doi.org/10.3390/jcm9082467.

OCHMO-TB-030 NASA Office of the Chief Health & Medical Officer (OCHMO). Bone Loss Technical Brief. 2025. 2023 11, 29.

Ortel, T.L., Neumann, I., Ageno, W., Beyth, R., Clark, N.P., Cuker, A., et al., 2020. American society of hematology 2020 guidelines for management of venous thromboembolism: treatment of deep vein thrombosis and pulmonary embolism. Blood Advances 4 (19), 4693–4738. Available at https://doi.org/10.1182/bloodadvances.2020001830, https://ashpublications.org/bloodadvances/article/4/19/4693/463998/American-Society-of-Hematology-2020-guidelines-for.

Padaki, A.S., Zarzour, A.L., Keene, K.R., Canepa, C.A., Levin, D.R., Antonsen, E.L., 2025. Clinical validation of non-contact vital signs in an emergency department setting. Front Med Technol 7, 1728913. Available at https://doi.org/10.3389/fmedt.

Pasquina, P., Tramèr, M.R., Granier, J.-M., Walder, B., 2006. Respiratory physiotherapy to prevent pulmonary complications after abdominal surgery: a systematic review. Chest 130 (6), 1887–1899.

Pavela, J., Sargsyan, A., Bedi, D., Everson, A., Charvat, J., Mason, S., et al., 2022. Surveillance for jugular venous thrombosis in astronauts. Vascular Medicine 27 (4), 365–372. Available at https://doi.org/10.1177/1358863x221086619.

Petersen, N., Jaekel, P., Rosenberger, A., Weber, T., Scott, J., Castrucci, F., et al., 2016. Exercise in space: the European Space Agency approach to in-flight exercise countermeasures for long-duration missions on ISS. Extreme Physiology & Medicine 5. Available at https://doi.org/10.1186/s13728-016-0050-4.

Ploutz-Snyder, L.L., Downs, M., Goetchius, E., Crowell, B., English, K.L., Ploutz-Snyder, R., et al., 2018. Exercise training mitigates multisystem deconditioning during bed rest. Medicine and Science in Sports and Exercise 50 (9), 1920–1928. Available at https://doi.org/10.1249/MSS.0000000000001618.

Puhl, C., Caplin, N., Fogtman, A., Van Ombergen, A., 2022. Wound management and healing in space. Frontiers in Bioengineering and Biotechnology 10. Available at https://doi.org/10.3389/fbioe.2022.958515.

Radstake, W.E., Gautam, K., Miranda, S., Van Rompay, C., Vermeesen, R., Tabury, K., et al., 2023. Gravitational effects on fibroblasts' function in relation to wound healing. npj Microgravity 9. Available at https://doi.org/10.1038/s41526-023-00286-z.

Raza, S., Matuła, K., Karoń, S., Paczesny, J., 2021. Resistance and adaptation of bacteria to non-antibiotic antibacterial agents: physical stressors, nanoparticles, and bacteriophages. Antibiotics 10 (4), 435. Available at https://doi.org/10.3390/antibiotics10040435.

Robinson, D.H., Link, M.M., Link, M.M., 1966. Space medicine in project mercury. vol. 4003. Scientific and technical information division, national aeronautics and space administration, 1965. p. 144-145. Technology and Culture 7 (3), 446. Available at https://doi.org/10.2307/3101962.

Romsdahl, J., Blanchowicz, A., Chiang, A.J., Romsdahl, J., Blanchowicz, A., Chiang, A.J., et al., 2018. Characterization of Aspergillus miger isolated from the International Space Station. mSystems 3, e00112–e00118 mSystems. **3** (2018).

Salido, R.A., Zhao, H.N., McDonald, D., Mannochio-Russo, H., Zuffa, S., Oles, R.E., et al., 2025. The International Space Station has a unique and extreme microbial and chemical environment driven by use patterns. Cell 188 (7).

Scheuring, R.A., Mathers, C.H., Jones, J.A., Wear, M.L., 2009. Musculoskeletal injuries and minor trauma in space: incidence and injury mechanisms in U.S. astronauts. Aviation, Space, and Environmental Medicine 80 (2), 117–124.

Sepúlveda, P., Gallardo, A., Arriagada, R., González, E., Rocco, P.R.M., Battaglini, D., 2025. Protocolized strategies to encourage early mobilization of critical care patients: challenges and success. Critical Care Science 37. Available at https://doi.org/10.62675/2965-2774.20250128.

Sewbuckas, P., 2018. A guide to anesthesia during space flight. Conference reprint 15, 1–16.

Sibonga, J., Matsumoto, T., Jones, J., Shapiro, J., Lang, T., Shackelford, L., et al., 2019. Resistive exercise in astronauts on prolonged spaceflights provides partial protection against spaceflight-induced bone loss. Bone 128.

Stella, A.B., Ajčević, M., Furlanis, G., Manganotti, P., 2021. Neurophysiological adaptations to spaceflight and simulated microgravity. Clinical Neurophysiology 132 (2), 498–504. Available at https://doi.org/10.1016/j.clinph.2020.11.033.

Stepaniak, P.C., Ramchandani, S.R., Jones, J.A., 2007. Acute urinary retention among astronauts. Aviation, Space, and Environmental Medicine 78 (4 Suppl).

Verrillo, S.C., Winters, B.D., 2018. Review: continuous monitoring to detect failure to rescue in adult postoperative inpatients. Biomedical Instrumentation & Technology 52 (4), 281–287. Available at https://doi.org/10.2345/0899-8205-52.4.281.

Vesper, S.J., Wong, W., Kuo, C.M., Pierson, D.L., 2008. Mold species in dust from the International Space Station identified and quantified by mold-specific quantitative PCR. Research in Microbiology 159 (6), 432–435. Available at https://doi.org/10.1016/j.resmic.2008.06.001.

Vyas, N., Hendren, S., Tushar Sehgal, D.M., Monga, C., Ranjan, R., Chaturvedi, H., et al., 2023. The accuracy of physical examination to diagnose anemia among patients five years or older: a systematic review. Indian Journal of Hematology & Blood Transfusion: An Official Journal of Indian Society of Hematology and Blood Transfusion 39 (1), 90–101. Available at https://doi.org/10.1007/s12288-022-01543-z.

Wang, Yy, Hu, Sf, Ying, Hm, Chen, L., Li, Hl, Tian, F., et al., 2018. Postoperative tight glycemic control significantly reduces postoperative infection rates in patients undergoing surgery: a meta-analysis. BMC Endocrine Disorders 18 (1). Available at https://doi.org/10.1186/s12902-018-0268-9, http://www.biomedcentral.com/bmcendocrdisord/.

Weinstein, R., Mermel, L.A., 2013. Infection prevention and control during prolonged human space travel. Clinical Infectious Diseases 56 (1), 123–130. Available at https://doi.org/10.1093/cid/cis861, http://cid.oxfordjournals.org/content/by/year.

White, N.J., Wenthe, A., 2023. Managing hemostasis in space. Arteriosclerosis, Thrombosis, and Vascular Biology 43 (11), 2079–2087. Available at https://doi.org/10.1161/ATVBAHA.123.318783, https://www.ahajournals.org/journal/atvb.

Wong, W.C., Oubre, Mehta, S.K., 2017. Chapter 1: preventing infectious diseases in spacecraft and space habitats. Modeling the Transmission and Prevention of Infectious Disease, Advances in Environmental Microbiology 1–17.

Chapter 23

# Rehabilitation, modification of crew schedule, and fitness for duty

**Mark Rosenberg**
*Department of Neurology, Medical University of South Carolina, Charleston, SC, United States*

## 23.1 Introduction

Understanding the interconnected nature of physical performance, underlying health, and its surrounding environment is essential when considering the ability to rehabilitate crew in the spaceflight setting, particularly during or after periods of microgravity. In the event of an unexpected injury or medical complication, interventions ranging from pharmacological agents to procedures may be necessary. However, beyond initial stabilization, return to full capabilities is essential for mission success. This chapter focuses on rehabilitation in-flight to facilitate and expedite a timely recovery after a crewmember is treated or injured in space.

## 23.2 Preconditioning

Preconditioning is the process of enhancing an individual's functional capacity before illness or surgery. Patients undergoing multiple surgeries and those with potential susceptibilities to complications following medical interventions (i.e., chemotherapy) may benefit from this approach (Gillis et al., 2022). Several metrics have been shown to correlate positively with improved health outcomes, reduced hospital stays, and reduced overall burden of complications (Molenaar et al., 2023). In the context of spaceflight, preconditioning plays a crucial role in mitigating the deconditioning associated with microgravity and other exposures. High-intensity training, resistance training, and aerobic training have all demonstrated improvements in overall musculoskeletal mass burden, physical endurance, and delays in the loss of bone and muscle during spaceflight (O'Conor et al., 2020). Additionally, higher levels of cardiovascular fitness during spaceflight have been associated with reduced risk of latent and viral reactivation, suggesting a potential immunologic benefit (Agha et al., 2020). In addition to musculoskeletal benefits, there is some evidence of preconditioning benefits regarding neurologic and endocrine changes. Maintaining health and physical performance capacity is paramount for preventive measures in the event of unforeseen illness or injury.

## 23.3 General considerations

Beyond preconditioning, several factors should be considered as strategies for rehabilitation, including active countermeasures (O'Conor et al., 2020), nutritional planning (Chaloulakou et al., 2022), sleep optimization, and psychosocial support. Recognizing limitations, including available resources, distance from advanced healthcare support, concurrent mission responsibilities, and crew expertise, may impact health outcomes and mission success. There have been attempts to reduce microgravity-induced deconditioning on the International Space Station (ISS; English et al., 2019). Despite considerable efforts, including exhaustive exercise regimens, some physical deconditioning still occurs. Simultaneously, functional and neurophysiological changes have been observed in the neurovestibular, proprioceptive, and other tactile systems (Wuyts et al., 2025). Such changes are considered neuroplastic (Hupfeld et al., 2022); they are compensations, both structural and functional, in the brain that can be adaptive or maladaptive. Adaptive neuroplasticity enables astronauts to become accustomed to and compensate for the various sensory and motor demands imposed by microgravity

**Procedural Medicine for Spaceflight. DOI:** https://doi.org/10.1016/B978-0-443-27552-4.00027-0

(Mikheeva et al., 2021). Therefore, in addition to rehabilitating the illness/injury in question, the neurological adaptation to spaceflight must be accounted for and included in any rehabilitation plan.

Protocols used on the ground can be applied to facilitate recovery, including adequate nutritional supplementation, early mobilization, and substantial psychosocial support. In the remaining chapter, these approaches will be described in detail, though specific measures will depend upon the exact illness or injury in question. Furthermore, all recommended measures will also be applicable to environments of partial gravity, including Lunar and Martian environments (with respect to gravity). Telemedicine may be utilized to personalize and reconfigure exercise regimens to assist in ongoing recovery and optimization of function.

## 23.4 Musculoskeletal rehabilitation

Movement, particularly load-bearing exercise, increases the demand on the musculoskeletal system, which responds with rapid, sometimes observable changes. During periods of underutilization, the opposite can be observed, directly posing problems in the setting of long-duration spaceflight (LDSF). Chronic musculoskeletal disuse, as seen in spaceflight without sufficient countermeasures, is in some ways analogous to that observed in terrestrial populations that engage in low-frequency or low-intensity exercise. In aging Earth populations, for example, changes of up to 48% in quadriceps size can be seen in those who train who previously did not (Goodpaster et al., 2001). Likewise, environmental factors, including indoor location (Brito et al., 2024), available oxygen, ambient temperature (Girard et al., 2016), and even lighting (Crane et al., 2008), have been shown to be associated with power output and performance. These few examples highlight the importance of identifying how extrinsic and intrinsic triggers and stressors activate hormonal, neurologic, and biomechanical systems in altered gravity.

For this reason, among others, the musculoskeletal system has been closely studied since the Gemini and Spacelab era. Decades of data from the United States, Russian, European, and other space agencies have since demonstrated that humans operating in the space environment for extended periods will experience musculoskeletal deconditioning unless rehabilitative efforts are employed. Musculoskeletal atrophy fluctuates during the flight and can have effects extending beyond the musculoskeletal system, including the formation of urinary stones (Xiao et al., 2024) and mood changes (Abeln et al., 2015), among others.

In microgravity, normal bone resorption continues, while bone formation is reduced or absent. Bone quality is also compromised: microgravity impairs bone microarchitecture, mineralization, and maintenance, increasing fracture risk and potentially affecting bone strength even after return to gravity (Coulombe et al., 2020). A recent study in LDSF astronauts on 6-month missions found a threefold increase in the rate of bone resorption relative to formation (Walle et al., 2024). This imbalance leads to significant bone loss, particularly in the weight-bearing bones of the axial skeleton, with a pronounced loss of 1% and 1.5% per month in the hips and spine, respectively (Lang et al., 2004). Notably, the studies above were conducted in individuals who regularly engaged in cardiovascular activity (running/cycling) and weightlifting using the Advanced Resistive Exercise Device (ARED; Fig. 23.1, a vacuum tube and flywheel cable system that enables complex weightlifting exercises on the ISS. Therefore, it can be assumed that bone resorption rates may increase during in-flight postprocedural recovery periods when individuals are unable to exercise (Wang et al., 2023).

Beyond the day-to-day interventions to preserve the musculoskeletal system in space, consideration must be given to managing acute illness and injury. Terrestrially, rapid mobilization is essential following illness for improvements in morbidity and mortality and outcomes ranging from long-term musculoskeletal to neurologic disability (Winstein et al., 2016). Following medical or surgical intervention, illness, or injury, additional rehabilitation will be required.

### 23.4.1 Upper extremity injury

Upper extremity injuries may occur more frequently due to deconditioning and increased mechanical demands, particularly during extravehicular activities. Areas of concern include the shoulders and hands (Ward et al., 2025) due to increased stress from reduced range of motion (ROM; Fiedler et al., 2024). It is a common complaint seen in up to 32% of astronauts. However, note that this is in part due to the hard upper torso of space suits, which increases the risk of injury.

#### *23.4.1.1 Treatment*

Emphasize antiinflammatories, including heating pads (if available) and NSAIDS, with rest and activity modification. Once pain has declined to a manageable level, passive ROM utilization of the extremity should be initiated as soon as is tolerable (Liaghat et al., 2023). Then, introduce active-assisted ROM and finally active ROM. Use isometric and light

**FIGURE 23.1** Interim resistive exercise device (iRED)/advanced resistive exercise device (ARED). *Left:* Suni Williams on iRed. *Right:* Sandy Magnus on the ARED.
*Source:* Image Credit: NASA.

resistance exercises, then introduce eventual strength-based exercises, including the ARED. However, throughout rehabilitation, caution must be exercised to prevent the injury from being exacerbated by the reintroduction of activity.

## 23.4.2 Back injury

Back pain is extremely common in spaceflight, primarily due to the loss of axial load, which results in spine elongation (Penchev et al., 2021), swelling of the intervertebral disk space (Treffel et al., 2016), and straightening of the lower lumbar lordosis. However, beyond the above-mentioned findings of "space adaptation back pain," axial musculoskeletal underutilization leads to weakening of the paraspinal muscles, potentially resulting in spinal instability. If preexisting vertebral/spinal injuries occur, they may improve or worsen in spaceflight, depending on the degree of pathophysiologic involvement. One of the most common complaints is that around 70% of astronauts will report back pain, with studies indicating that 50% with preexisting disk herniation express more concerning symptoms (Bailey et al., 2022).

### *23.4.2.1 Treatment*

The approach to treatment and rehabilitation is generally similar to that of the upper extremity but with additional attention to any neurologic symptoms, including paresthesias, weakness, and significant changes in bowel/bladder function. Slight variations in urinary output are expected; however, caution is recommended after injury. If none are identified, then proceed with rehabilitation, including NSAIDs and heating pads. If the pain persists beyond 2 weeks, a holistic approach including integration of exercise regimens that do not require gravity (i.e., McKenzie method) and potential cognitive behavioral therapy for central pain modulation is recommended. If included in the medical kit, muscle relaxants may also be considered to reduce pain as an alternative to narcotics or benzodiazepines. However, the improvements from these medications are inconsistent, and they often cause cognitive deficits and disorientation. For these reasons, ground testing these medications is advisable. In all cases, maintaining as close to normal activity as possible, using gradually increasing ROM exercises, and progressively loading the axial skeleton back to baseline levels is the recommended approach.

## 23.4.3 Lower extremity injury

Lower extremity injuries are much less common than back or upper extremity injuries in spaceflight and account for approximately 5.5% of documented musculoskeletal injuries (Ward et al., 2025). Currently, countermeasures against lower extremity musculoskeletal atrophy include the T2 treadmill, cyclo-ergometer with vibration isolation and stabilization (CEVIS), and ARED. In flight, injury can occur during aerobic or load-bearing exercise; however, particular caution is required upon return to a gravitational environment where bone demineralization, proprioceptive changes, and altered body kinetics can increase gait abnormalities and subsequent injuries.

#### *23.4.3.1 Treatment*

The preconditioning stage will be especially important to prevent debilitating injury upon return to a gravity well. This is further complicated if gait is impaired (Comfort et al., 2021). Evolution of ROM from active-assisted to active ROM, with the use of isometric and light resistance exercises to eventual strength-based exercises, is recommended

### 23.4.4 Additional complications

Depending on the location and severity of the musculoskeletal injury, one should be attentive to potential involvement of the cardiovascular and nervous systems. If vascular supply is compromised, this may cause disproportionate pain, pallor, loss of pulses, swelling, and weakness (Mortensen et al., 2019). If identified, immediate evacuation should be considered. Conversely, if the patient demonstrates sensory changes (numbness, paresthesias), motor deficits (weakness, abnormal muscle activation, rigidity), changes in reflexes, neuropathic pain (electric-type), then there is potential for nerve involvement (Neal and Fields, 2010). If the spinal trauma is suspected, additional attention is required to assess autonomic changes (blood pressure variability, sweating reflex, bowel/bladder), as there is an elevated risk for decompensation (Ropper and Ropper, 2017).

## 23.5 Neurologic rehabilitation

The nervous system is essential for a proper response to internal and external stimuli, maintaining survivability in a varying environment. It is anatomically divided into the central nervous system (CNS) and peripheral nervous system (PNS), with a functional subset being the autonomic nervous system (ANS; Wehrwein et al., 2016). The CNS comprises the brain and spinal cord, with the PNS integrating all peripheral organs, muscles, and the sympathetic, parasympathetic, and enteric divisions. Like other parts of the body, the nervous system can undergo neurological deconditioning due to inactivity. Regarding the CNS, underutilization leads to impairments in neuroplasticity, the promotion of neurodegeneration, and the pruning of synaptic connections (Rosso et al., 2013). Depending on the location and severity, this can result in neurocognitive effects, reduced coordination, loss of motor control, and other effects. Peripherally, axonal degeneration and reductions in motor endplate connectivity result in muscle weakness and sensory changes. This is compounded by potential involvement of the ANS, through which the volatility of blood pressure and heart rate can worsen known dysautonomia following spaceflight.

Due to a strong concern for mission impact following neurological changes, studies on cognition, learning, locomotion, and vision have been explored in the context of animal models, dating back to the early days of Soviet spaceflight. As spaceflight evolved, investigations ranging from characterizing the behavior of the neurovestibular system in NeuroLab (Moore et al., 2003) to cognitive batteries on the ISS and beyond have been performed (Dev et al., 2024). More recently, five of NASA's delineated 22 Human Research Roadmap Risks pertain specifically to the nervous system (https://humanresearchroadmap.nasa.gov/risks), with a focus on performance reductions, sensorimotor changes, psychosocial challenges (which will be discussed later in this chapter), circadian desynchronization, and spaceflight-associated neuro-ocular syndrome. Therefore, maintaining neurocognitive performance and rehabilitation in the event of adverse events is not only critical to the health of the astronaut but also the success of the mission as a whole.

In microgravity, altered ambulation and neurosensory inputs lead to changes in several cognitive domains, including sensorimotor, neurovestibular, coordination, attention, and spatial orientation (Tays et al., 2021). The primary method of combating neurocognitive deconditioning is preconditioning, with astronauts undergoing intensive coordination exercises integrated into their robust physical activity regimen. Although this is not presently implemented as a countermeasure to the level that resistive exercise is for musculoskeletal deconditioning, the coordination required for running and cycling, as well as the proprioceptive components of strength training, likely provide some degree of neurocognitive conditioning in flight. Currently, NASA's Complement of Integrated Protocols for Human Exploration Research program is collecting sensorimotor data through cognitive batteries and coordination tests. Terrestrially, NASA and ESA are investigating proprioceptive training and electrical muscle stimulation (EMS) in a 6° head-down-tilt bedrest to assess whether protocolled EMS can aid in sensorimotor feedback; however, studies are still

**FIGURE 23.2** NASA astronaut Chris Cassidy wearing virtual reality for a study assessing effects on motion and orientation.
*Source:* Image Credit: NASA.

ongoing (https://ntrs.nasa.gov/citations/20240012134). Additionally, virtual reality is being investigated as a potential tool to assess oculomotor and neurovestibular function with plans to develop training for spaceflight (Fig. 23.2).

Lastly, beyond the standard high-neurocognitive load placed upon astronauts, crew are susceptible to several pathologies intrinsic to spaceflight in microgravity as well as complications of standard terrestrial pathologies. Rehabilitation is required postprocedurally and following spaceflight with special attention to the deconditioning effects on orthostatics, cardiovascular, and sleep. These range from benign to life-threatening. Therefore, similar to a significant musculoskeletal injury, rapid assessment and triage are required to determine whether the condition (and subsequent recovery) can be managed in space or requires a rapid return to Earth for escalation of care. Therefore, the following subsection will discuss prominent neurologic pathology and the actions required for their prevention and mitigation.

### 23.5.1 Neurovestibular dysfunction/space motion sickness

Experienced by approximately 60%–80% of astronauts, space motion sickness (SMS) is symptomatically similar to terrestrial motion sickness, including nausea, vomiting, malaise, fatigue, cognitive complaints, and dizziness. While not fully understood, it is thought to result from a combination of inappropriate proprioceptive feedback and dysfunction of the neurovestibular apparatus. Generally, time affords the best rehabilitation, with some evidence that repeated flying can reduce symptom severity.

#### *23.5.1.1 Treatment*

Symptomatic management with antiemetics is preferred, including scopolamine and promethazine. Comparative studies have demonstrated that promethazine is the preferred agent, given its rapid relief. First-generation antihistamines and serotonin antagonists (i.e., ondansetron) have no demonstrated consistent efficacy for SMS (Heer and Paloski, 2006).

### 23.5.2 Cardiovascular deconditioning

Management of cardiovascular deconditioning in orbit centers on structured exercise countermeasures, adjunctive nonpharmacological strategies, and careful assessment and monitoring, with protocols adapted from spaceflight and bedrest studies. Management of deconditioning in spaceflight requires structured exercise, nonpharmacological strategies, and close monitoring, with protocols adapted from spaceflight and bedrest analogues. In spaceflight, exercise is the cornerstone countermeasure, and most protocols will combine strength and aerobic protocols to maintain muscle and cardiovascular function.

#### *23.5.2.1 Treatment*

Robust countermeasures have not been established. Regardless of the duration of the mission, orthostatic intolerance develops following abnormalities in the baroreceptor reflex, coupled with fluid volume shifts. In orbit, this could be reduced by using lower-body negative-pressure suits and monitoring electrolyte levels (Hedge et al., 2022).

## 23.6 Conclusion

As humanity begins a new chapter in spaceflight with increased commercial involvement, in-flight rehabilitation protocols may vary depending on crew complement, crew health, and mission length. Nonetheless, prioritizing preconditioning wherever possible is crucial to preserving neurocognitive and musculoskeletal health. Having a supraphysiologic baseline could provide critical time to properly triage and treat illness or injury and would facilitate postprocedural rehabilitation. Rehabilitation should focus on both the crewmember's musculoskeletal and neurophysiological decrements, leveraging technologies such as physiologic monitors and exercise equipment where possible and in a timely manner, permitting symptom management.

## References


Abeln, V., MacDonald-Nethercott, E., Piacentini, M.F., Meeusen, R., Kleinert, J., Strueder, H.K., et al., 2015. Exercise in isolation–a countermeasure for electrocortical, mental and cognitive impairments. PLoS One 10 (5). Available at https://doi.org/10.1371/journal.pone.0126356.

Agha, N.H., Mehta, S.K., Rooney, B.V., Laughlin, M.S., Markofski, M.M., Pierson, D.L., et al., 2020. Exercise as a countermeasure for latent viral reactivation during long duration space flight. FASEB Journal Official Publication of the Federation of American Societies for Experimental Biology 34 (2), 2869–2881. Available at https://doi.org/10.1096/fj.201902327R.

Bailey, J.F., Nyayapati, P., Johnson, G.T.A., Dziesinski, L., Scheffler, A.W., Crawford, R., et al., 2022. Biomechanical changes in the lumbar spine following spaceflight and factors associated with postspaceflight disc herniation. The Spine Journal Official Journal of the North American Spine Society 22 (2), 197–206.

Brito, H., Lopes, H., Vaz de Carvalho, M., Carrilho, D., Carvalho, A., Araújo, D., 2024. The effects of nature-based vs. indoor settings on the adaptability, performance and affect of calisthenics exercisers. A registered report. Psychology of Sport and Exercise 73.

Chaloulakou, S., Poulia, K.A., Karayiannis, D., 2022. Physiological alterations in relation to space flight: the role of nutrition. Nutrients 14 (22) Available from. Available at https://doi.org/10.3390/nu14224896, http://www.mdpi.com/journal/nutrients/.

Comfort, P., McMahon, J.J., Jones, P.A., Cuthbert, M., Kendall, K., Lake, J.P., et al., 2021. Effects of spaceflight on musculoskeletal health: a systematic review and meta-analysis, considerations for interplanetary travel. Sports Medicine 51 (10), 2097–2114. Available at https://doi.org/10.1007/s40279-021-01496-9.

Coulombe, J.C., Senwar, B., Ferguson, V.L., 2020. Spaceflight-induced bone tissue changes that affect bone quality and increase fracture risk. Current Osteoporosis Reports 18 (1), 1–12. Available at https://doi.org/10.1007/s11914-019-00540-y.

Crane, D.K., Hensarling, R.W., Jung, A.P., Sands, C.D., Petrella, J.K., 2008. The effect of light color on muscular strength and power. Perceptual and Motor Skills 106 (3), 958–962.

Dev, S.I., Khader, A.M., Begerowski, S.R., Anderson, S.R., Clément, G., Bell, S.T., 2024. Cognitive performance in ISS astronauts on 6-month low earth orbit missions. Frontiers in Physiology 15. Available at https://doi.org/10.3389/fphys.2024.1451269.

English, K.L., Bloomberg, J.J., Mulavara, A.P., Ploutz-Snyder, L.L., 2019. Exercise countermeasures to neuromuscular deconditioning in spaceflight. Comprehensive Physiology 10 (1), 171–196. Available at https://doi.org/10.1002/cphy.c190005.

Fiedler, B., Jami, M., Rakauskas, T., Ahmed, A.S., 2024. Impact of spaceflight on upper extremity orthopedic health. Aerospace Medicine and Human Performance 95 (10), 777–783. Available at https://doi.org/10.3357/AMHP.6482.2024.

Gillis, C., Ljungqvist, O., Carli, F., 2022. Prehabilitation, enhanced recovery after surgery, or both? A narrative review. British Journal of Anaesthesia 128 (3), 434–448.

Girard, O., Brocherie, F., Morin, J.-B., Millet, G.P., 2016. Running mechanical alterations during repeated treadmill sprints in hot versus hypoxic environments. A pilot study. Journal of Sports Sciences 34 (12), 1190–1198. Available at https://doi.org/10.1080/02640414.2015.1101482.

Goodpaster, B.H., Carlson, C.L., Visser, M., Kelley, D.E., Scherzinger, A., Harris, T.B., et al., 2001. Attenuation of skeletal muscle and strength in the elderly: the Health ABC Study. Journal of Applied Physiology 90 (6), 2157–2165.

Hedge, E.T., Patterson, C.A., Mastrandrea, C.J., Sonjak, V., Hajj-Boutros, G., Faust, A., et al., 2022. Implementation of exercise countermeasures during spaceflight and microgravity analogue studies: developing countermeasure protocols for bedrest in older adults (BROA). Frontiers in Physiology 13. Available at https://doi.org/10.3389/fphys.2022.928313.

Heer, M., Paloski, W.H., 2006. Space motion sickness: incidence, etiology, and countermeasures. Autonomic Neuroscience Basic & Clinical 129 (1-2), 77–79.

Hupfeld, K.E., McGregor, H.R., Koppelmans, V., Beltran, N.E., Kofman, I.S., De Dios, Y.E., et al., 2022. Brain and behavioral evidence for reweighting of vestibular inputs with long-duration spaceflight. Cerebral Cortex 32 (4), 755–769. Available at https://doi.org/10.1093/cercor/bhab239.

Lang, T., LeBlanc, A., Evans, H., Lu, Y., Genant, H., Yu, A., 2004. Cortical and trabecular bone mineral loss from the spine and hip in long-duration spaceflight. Journal of Bone and Mineral Research the Official Journal of the American Society for Bone and Mineral Research 19 (6), 1006–1012.

Liaghat, B., Pedersen, J.R., Husted, R.S., Pedersen, L.L., Thorborg, K., Juhl, C.B., 2023. Diagnosis, prevention and treatment of common shoulder injuries in sport: grading the evidence - a statement paper commissioned by the Danish Society of Sports Physical Therapy (DSSF). British Journal of Sports Medicine 57 (7), 408–416. Available at https://doi.org/10.1136/bjsports-2022-105674.

Mikheeva, I., Mikhailova, G., Shtanchaev, R., Arkhipov, V., Pavlik, L., 2021. Influence of a 30-day spaceflight on the structure of motoneurons of the trochlear nerve nucleus in mice. Brain Research 1758.

Molenaar, C.J., van Rooijen, S.J., Fokkenrood, H.J., Roumen, R.M., Janssen, L., Slooter, G.D., 2023. Prehabilitation versus no prehabilitation to improve functional capacity, reduce postoperative complications and improve quality of life in colorectal cancer surgery. The Cochrane Database of Systematic Reviews 5 (5). Available at https://doi.org/10.1002/14651858.CD013259.pub3.

Moore, S.T., Clément, G., Dai, M., Raphan, T., Solomon, D., Cohen, B., 2003. Ocular and perceptual responses to linear acceleration in microgravity: alterations in otolith function on the COSMOS and Neurolab flights. Journal of Vestibular Research Equilibrium & Orientation 13 (4-6), 377–393.

Mortensen, S.J., Vora, M.M., Mohamadi, A., Wright, C.L., Hanna, P., Lechtig, A., et al., 2019. Diagnostic modalities for acute compartment syndrome of the extremities: a systematic review. JAMA Surgery 154 (7), 655–665. Available at https://doi.org/10.1001/jamasurg.2019.1050.

Neal, S., Fields, K.B., 2010. Peripheral nerve entrapment and injury in the upper extremity. American Family Physician 81 (2), 147–155.

O'Conor, D.K., Dalal, S., Ramachandran, V., Shivers, B., Shender, B.S., Jones, J.A., 2020. Crew-friendly countermeasures against musculoskeletal injuries in aviation and spaceflight. Frontiers in Physiology 11. Available at https://doi.org/10.3389/fphys.2020.00837.

Penchev, R., Scheuring, R.A., Soto, A.T., Miletich, D.M., Kerstman, E., Cohen, S.P., 2021. Back pain in outer space. Anesthesiology 135 (3), 384–395. Available at https://doi.org/10.1097/ALN.0000000000003812.

Ropper, A.E., Ropper, A.H., 2017. Acute spinal cord compression. The New England Journal of Medicine 376 (14), 1358–1369. Available at https://doi.org/10.1056/NEJMra1516539.

Rosso, A.L., Studenski, S.A., Chen, W.G., Aizenstein, H.J., Alexander, N.B., Bennett, D.A., et al., 2013. Aging, the central nervous system, and mobility. The Journals of Gerontology. Series A, Biological Sciences and Medical Sciences 68 (11), 1379–1386. Available at https://doi.org/10.1093/gerona/glt089.

Tays, G.D., Hupfeld, K.E., McGregor, H.R., Salazar, A.P., De Dios, Y.E., Beltran, N.E., et al., 2021. The effects of long duration spaceflight on sensorimotor control and cognition. Frontiers in Neural Circuits 15. Available at https://doi.org/10.3389/fncir.2021.723504.

Treffel, L., Mkhitaryan, K., Gellee, S., Gauquelin-Koch, G., Gharib, C., Blanc, S., et al., 2016. Intervertebral disc swelling demonstrated by 3D and water content magnetic resonance analyses after a 3-day dry immersion simulating microgravity. Frontiers in Physiology 7.

Walle, M., Gabel, L., Whittier, D.E., Liphardt, A.-M., Hulme, P.A., Heer, M., et al., 2024. Tracking of spaceflight-induced bone remodeling reveals a limited time frame for recovery of resorption sites in humans. Science Advances 10 (51). Available at https://doi.org/10.1126/sciadv.adq3632.

Wang, H., Wu, J., Wang, B., Qin, H., Fan, L., Wang, Y., et al., 2023. Progress in prevention and treatment of acute bone loss in orthopedics. International journal of endocrinology 2023. Available at https://doi.org/10.1155/2023/9373043.

Ward, G.H., Anderson, D.N., Scheuring, R.A., 2025. Peri-flight neuromusculoskeletal injuries in astronauts. Aerospace Medicine and Human Performance 96 (2), 155–167. Available at https://doi.org/10.3357/AMHP.6501.2025.

Wehrwein, E.A., Orer, H.S., Barman, S.M., 2016. Overview of the anatomy, physiology, and pharmacology of the autonomic nervous system. Comprehensive Physiology 6 (3), 1239–1278. Available at https://doi.org/10.1002/cphy.c150037.

Winstein, C.J., Stein, J., Arena, R., Bates, B., Cherney, L.R., Cramer, S.C., et al., 2016. Guidelines for adult stroke rehabilitation and recovery: a guideline for healthcare professionals from the american heart association/american stroke association. Stroke a Journal of Cerebral Circulation 47 (6). Available at https://doi.org/10.1161/STR.0000000000000098.

Wuyts, F.L., Deblieck, C., Vandevoorde, C., Durante, M., 2025. Brains in space: impact of microgravity and cosmic radiation on the CNS during space exploration. Nature Reviews Neuroscience 26 (6), 354–371. Available at https://doi.org/10.1038/s41583-025-00923-4.

Xiao, Y., Wang, J., Yin, S., Tang, Y., Cui, J., Yang, Z., et al., 2024. Association between physical activity and the prevalence of kidney stones in American adults: results from a multiyear national survey. Urology 189, 9–18.

Section V

# Other considerations

Chapter 24

# Diagnostic tools

**Christopher Zahner**[1] **and Michael Pohlen**[2]

[1]*Department of Pathology, University of Texas Medical Branch, Galveston, TX, United States,* [2]*Department of Radiology, University of California, San Diego, CA, United States*

## 24.1 Introduction

Human space exploration is rapidly transitioning from short-duration, low Earth orbit (LEO) missions to extended voyages that may last weeks, months, or even years, with destinations ranging from prolonged stays aboard the International Space Station (ISS) to lunar outposts and, eventually, Mars. As missions grow in duration and distance, maintaining crew health becomes increasingly complex. On Earth, physicians have access to extensive laboratory infrastructure and immediate expert consultation. In space, every element, from water and power to consumables and even gravity, is limited or entirely absent.

This chapter offers a comprehensive overview of diagnostic technologies reimagined for the spaceflight environment. We examine how conventional diagnostic systems must be adapted to function reliably under altered fluid dynamics, increased radiation exposure, limited resupply, and delayed communication. Where possible, we distinguish between diagnostic tools that have already been validated in orbit, such as point-of-care ultrasound, and those still under development, including portable molecular assays and low-field imaging systems.

Through an exploration of emerging innovations, including microfluidic assays, AI-driven automation, and synthetic biology, we outline the building blocks of safe, autonomous medical diagnostics for human spaceflight (McGregor et al., 2020; Blue et al., 2019).

## 24.2 Purpose

As space exploration moves farther from Earth, the need for reliable, self-sufficient diagnostic systems becomes increasingly critical. In the absence of traditional laboratory infrastructure and immediate expert consultation, crew members must rely on compact, resource-efficient technologies to monitor and maintain their health. However, most diagnostic systems used on Earth are not designed to operate under the extreme constraints of spaceflight, where gravity, consumables, and communication bandwidth are limited.

This chapter examines the primary challenges associated with translating Earth-based diagnostics into the space environment. We examine how engineers and scientists are redesigning traditional tools to operate under microgravity and radiation exposure, and how novel technologies, such as microfluidic platforms, synthetic biology for on-demand reagent production, and AI-powered automation, are emerging to fill these gaps. In addition to the technical solutions, we also consider the ethical, legal, and operational frameworks necessary to ensure diagnostic tools are implemented safely and equitably during long-duration missions.

The goal is to provide a cohesive overview of where space-based diagnostics stand today, what innovations are on the horizon, and what it will take to build systems capable of supporting autonomous healthcare beyond LEO.

## 24.3 How are diagnostic tools used in space medicine?

In practice, diagnostic tools serve a wide range of functions essential to maintaining crew health. Routine monitoring is a cornerstone of in-flight medical care, allowing for the regular assessment of hydration status, muscle metabolism, and

**Procedural Medicine for Spaceflight. DOI:** https://doi.org/10.1016/B978-0-443-27552-4.00020-8

immune function biomarkers. These frequent measurements help track physiological adaptations to microgravity and flag deviations that may signal emerging health concerns.

When specific symptoms arise, such as discomfort following extravehicular activity or subtle neurological changes, onboard systems enable targeted diagnostics and treatment. Point-of-care testing and portable imaging, such as ultrasound, allow crew members to evaluate potential issues like internal bleeding or organ dysfunction without the delay of ground-based consultation.

In emergency situations, rapid diagnostic capabilities are critical. Tools that can provide immediate blood counts, metabolic panels, or imaging data help guide urgent medical interventions and triage, ensuring that time-sensitive decisions are made effectively, even in the absence of real-time Earth-based support.

Across these use cases, the overarching challenge remains the same. Systems must be both highly automated and user-friendly. Given that most astronauts are not medical specialists, diagnostic platforms must deliver results that are easily interpreted under stressful conditions, ensuring actionable insights when they are needed most.

## 24.4 Overview of upcoming missions

Human spaceflight is entering a transformative period. NASA's Artemis program aims to return humans to the Moon and establish a sustained presence through the Lunar Gateway, while international partners and private companies are pursuing deep-space ventures, including missions to Mars. These missions may span durations from a few weeks to several years, with increasing crew autonomy and minimal access to ground-based medical support.

These extended timelines and expanded mission profiles introduce new demands for onboard medical care. While ISS astronauts have access to resupply and real-time consultation within hours, lunar missions may experience delays of several days, and Mars expeditions will face communication blackouts and round-trip delays measured in tens of minutes. Moreover, as commercial participation grows, astronaut crews may represent a wider range of ages and preexisting conditions. Diagnostic systems must be versatile, autonomous, and resilient, capable of handling both routine assessments and unexpected medical emergencies within the limited infrastructure of a spacecraft or habitat.

## 24.5 Unique challenges of spaceflight diagnostics

Diagnostic systems for spaceflight face a set of technical, physiological, and operational challenges that are fundamentally different from those encountered in terrestrial medicine. These must be addressed through reengineering, novel design, and strict resource optimization.

### 24.5.1 Effects of microgravity

Microgravity alters human physiology in ways that complicate diagnostic testing. Fluids shift toward the head, hormonal regulation is disrupted, and immune responses may be suppressed. These changes affect the interpretation of laboratory results, and often necessitate the establishment of new in-flight reference ranges for values such as white blood cell counts, inflammatory markers, and endocrine levels (Grimm et al., 2020; Crucian et al., 2018; Smith and Zwart, 2008). Since new reference values require extensive and robust data from many individuals, it is not practical to develop a set of revised clinical reference values for spaceflight at this time. Alternatively, and potentially more in line with the movement towards more personalized care, it may be possible to develop a personalized baseline for each astronaut and directly compare changes in flight to this baseline. Relatedly, physicians caring for astronauts may consider setting a threshold for how much change from the individual's baseline is meaningful and what relative change should trigger further investigation or intervention.

### 24.5.2 Sample handling and separation techniques

Sample handling is also affected. On Earth, many diagnostic processes rely on gravity to move reagents, sediment cells, or separate plasma from whole blood. In orbit, these gravitational effects are absent, requiring alternative methods. While centrifugation is a mechanical process that utilizes centrifugal force rather than gravity to separate fluid components, most terrestrial centrifuge systems are heavy, complex, and highly sensitive to imbalance, making them unsuitable for use in microgravity. Instead, space-adapted systems leverage technologies such as acoustic wave separation, capillary action, and microfluidic channels to manipulate fluids without relying on gravitational force (Moo-Young, 2011; Research Roadmap, 2020).

**TABLE 24.1** Handheld centrifuges.

| Device | Peak speed or RCF | Time to plasma or processing | Validation or comparison outcome |
|---|---|---|---|
| Paperfuge (Bond and Richards-Kortum, 2017) | ~125,000 rpm ≈30,000 g | Pure plasma <1.5 min; parasite isolation 15 min | Rapid, very high instantaneous g comparable to high-speed bench runs for microcapillary volumes. |
| CentREUSE (Lin et al., 2020) | ~10.5 RCF ±1.3 | 3–10 min depending on endpoint | Sedimentation for 1 mL suspension comparable to commercial centrifugation endpoints; no significant differences in compactness at defined times. |
| Torq Zero Delay Centrifuge (Hong et al., 2021) | Not stated as rpm; integrated system | Immediate point-of-collection processing | Clinical study (*n*=134) showed plasma from ZDisc agreed with reference PSTs for 14 analytes with mean % biases below clinical thresholds and low hemolysis. |
| TrueSpin (Iacovetti et al., 2023) | Device rotates on AA batteries | Separation in ~5 min | No significant difference versus reference in hemolysis or residual RBC concentration; improved short-term analyte stability after separation. |

*AA*, Double-A battery; *PSTs*, plasma separator tubes; *RBC*, red blood cells; *RCF*, relative centrifugal force; *rpm*, revolutions per minute.

In addition to these newer processing methods, lessons and technology from global health practitioners may provide low power, low mass alternatives. For example, hand-powered centrifuges may be able to achieve similar efficacy to benchtop centrifuges for many applications (Table 24.1).

### 24.5.3 Environmental and operational constraints

Traditional diagnostic devices are typically designed for use in hospitals that have stable power, a continuous water supply, and ample space. In space, every gram and watt must be justified. Instruments must be miniaturized, lightweight, and designed to operate with minimal power draw. They must also tolerate temperature fluctuations, launch-related vibration, and prolonged periods without maintenance (Human Research Roadmap, 2020).

Reagents present an additional challenge. Many liquid reagents degrade over time or are sensitive to heat, freezing, or radiation. To mitigate this, diagnostic systems increasingly rely on lyophilized reagents that remain stable during long-duration missions. There is also active exploration into in-situ production of essential reagents using synthetic biology approaches (Blue et al., 2019; Averesch et al., 2023).

In parallel, diagnostic hardware must be engineered to withstand cosmic radiation, which can disrupt electronic systems and compromise accuracy (Wotring, 2018; Table 24.2).

### 24.5.4 Communication constraints

Diagnostic autonomy becomes increasingly important as missions extend beyond Earth orbit. While astronauts on the ISS can consult with medical teams in real-time, missions to Mars will experience round-trip communication delays of up to 40 minutes. Signal blackouts may also occur due to planetary occlusion. Lunar missions, while generally within communication range, may still face intermittent delays or constraints that limit dynamic consultation, especially during surface operations or orbital occlusion.

As a result, diagnostic platforms must function independently. Systems must be intuitive for nonphysician crewmembers to operate and interpret. This includes step-by-step guidance for procedures such as blood draws, imaging, or multistep analyses, as well as real-time error checking. Embedded AI systems can provide decision support, suggest follow-up tests, and detect abnormalities across multiple data sources.

The inability to obtain rapid feedback from ground-based experts means diagnostic results must be immediately actionable. This shifts the design priority from high complexity instrumentation to reliability, clarity, and integration into medical workflows that astronauts can execute confidently under stress.

**TABLE 24.2** Diagnostic tool/assay and challenges.

| Diagnostic tool | Key challenges in space | Potential solutions | Citation |
|---|---|---|---|
| Clinical chemistry analyzers | High water use, gravity-dependent separation, high power draw, radiation-sensitive components | Microfluidics, low-power compact analyzers | Averesch et al. (2023) |
| Hematology analyzers | Microgravity alters fluid dynamics and cell sedimentation, complicating measurement accuracy. Frequent recalibration and reagent stability are also concerns. | Use of acoustic or electric focusing; sealed, precalibrated cartridges | Crucian et al. (2021) |
| PCR machines | Thermal instability, energy demands, radiation-sensitive enzymes | Closed cartridge systems, freeze-dried reagents | Parra et al. (2017) |
| Immunoassay systems | Complex wash steps, high reagent volumes, temperature-sensitive reagents | Capillary action microfluidics, lyophilized antibodies | |
| Microscopy for pathology | Gravity-dependent slide prep, bulky equipment, large datasets | Digital microscopes, AI-enhanced onboard interpretation | Horneck et al. (2010) |
| Advanced imaging (CT/MRI) | Mass, volume, power, magnetic, and radiation safety concerns | Cone beam CT, low-field MRI, modular or portable platforms | |

*AI*, Artificial intelligence; *CT*, computed tomography; *MRI*, magnetic resonance imaging; *PCR*, polymerase chain reaction.

## 24.5.5 Obstacles to implementing ground-based diagnostic tools in space

Diagnostic instruments developed for Earth are deeply embedded in a system of reliable gravity, abundant consumables, stable infrastructure, and continuous human oversight. Translating these tools into the spaceflight environment reveals a set of design assumptions that break down under the constraints of microgravity, limited power and volume, and crew autonomy.

Fluid handling is one of the most fundamental challenges. In terrestrial laboratories, gravity assists with sedimentation, fluid movement, and phase separation. In space, these processes behave unpredictably or not at all. Devices that rely on liquid reagents, wash cycles, or gravity-dependent separation must be replaced or reengineered. Microfluidic systems, passive capillary flow, and acoustic focusing are among the most promising alternatives for this purpose.

Power consumption and physical footprint are equally problematic. Clinical analyzers and imaging systems often consume significant electrical power and require substantial cooling or vibration isolation. In space, power must be rationed across all subsystems, and available volume is strictly limited. Equipment must be miniaturized, ruggedized, and able to function reliably through launch vibration, docking maneuvers, and long periods of inactivity.

Reagent stability presents another serious limitation. Many diagnostic assays require temperature-sensitive reagents that degrade over time or require cold storage. Spaceflight imposes variable temperatures, radiation exposure, and long-duration missions where resupply is not possible. Lyophilized reagents, freeze-dried enzymes, or in-situ biosynthesis of reagents may help overcome this constraint.

Calibration and quality control, often performed daily in hospital labs, are difficult or impossible to replicate in space. Automated onboard calibration routines and prevalidated cartridges may reduce error, but diagnostic systems must still be robust to mechanical drift, component degradation, and slight variations in user technique. Furthermore, many diagnostic systems assume the presence of trained operators with biomedical expertise. In space, diagnostic platforms must be intuitive and self-guiding, providing clear procedural instructions and support for interpreting results to nonspecialist crewmembers.

In contrast to lab-based diagnostics, imaging introduces separate limitations due to equipment mass, safety constraints, and the physics of image generation. While ultrasound is already deployed on the ISS with Earth-based operator guidance, higher-resolution imaging, such as CT and MRI, remains far from practical deployment due to mass, power, and magnetic safety limitations. However, there is steady progress. The Fram2 mission evaluated a portable X-ray device under microgravity analog conditions, providing early evidence that skeletal imaging may be feasible in future missions. Low-dose cone beam CT or low-field MRI platforms, if modular and shielded appropriately, may one day be integrated into longer-duration outposts or transit vehicles.

Waste disposal and consumable management are also essential considerations. Many Earth-based devices rely on disposable plastics, chemical waste reservoirs, and frequent maintenance cycles. These systems must be adapted or replaced with enclosed, low-waste designs that can function autonomously for months without external support.

The table below summarizes key obstacles and adaptive strategies for core diagnostic platforms under consideration for spaceflight deployment (Wilson et al., 2007; Castro-Wallace et al., 2017; Chiu et al., 2017; Jokerst, Emory and Henry, 2012; Boguraev et al., 2017; Heidt et al., 2020; Krakos, 2024).

### 24.5.6 Need for novel and innovative diagnostic strategies

To overcome the limitations of traditional diagnostic tools in space, multiple new strategies are being explored. These innovations aim to reduce resource demands, increase automation, and support clinical decision-making in the absence of continuous ground communication. Among the most promising are microfluidic lab-on-a-chip platforms, AI-supported diagnostic interpretation, synthetic biology for reagent generation, and noninvasive continuous monitoring technologies.

**Microfluidics** enables the miniaturization of entire diagnostic workflows onto a compact cartridge. These systems can handle blood, urine, or saliva with minimal sample volumes and no reliance on gravity. Fluids are guided through channels using capillary action, pressure gradients, or electric fields. Lab-on-a-chip devices have already demonstrated the ability to perform immunoassays, polymerase chain reaction (PCR), and even cell counts in microgravity analogs. Their compact size and reagent efficiency make them attractive for future missions.

**Artificial intelligence** is playing an increasingly important role in space-based diagnostics. AI systems can assist in image interpretation, anomaly detection, and data fusion across multiple physiological streams. For nonexpert users, AI can also guide diagnostic workflows, flag abnormal results, and recommend follow-up actions. Onboard decision-support systems reduce reliance on ground consultation and help standardize care, even in high-stress or emergency scenarios (Cheung, De Louche and Komorowski, 2023). However, the benefits of these systems must be balanced against constraints on processing power, memory, energy, heat rejection, and utility. The availability of applicable baseline data to train the system and how closely the "normal" of spaceflight tracks against analog datasets is also a challenge, given the small number and narrow demographics of space flyers to date. Chapter 28 covers this topic in more detail.

**Synthetic biology** represents a cutting-edge approach to onboard reagent management. Rather than launching and storing perishable chemicals, synthetic biology systems are the biological equivalent to additive manufacturing solutions like 3D printing. These can be used to produce enzymes, antibodies, or reagents on demand by leveraging the power of evolved or engineered biological systems. This includes freeze-dried genetic templates activated with water, or engineered organisms capable of biosynthesizing diagnostic inputs from basic feedstocks. While still experimental, this strategy could dramatically reduce payload mass and extend mission self-sufficiency (Averesch et al., 2023; Horneck et al., 2010).

**Noninvasive and continuous monitoring** is also an essential area of innovation. Devices that can monitor vital signs, biochemical markers, hydration status, or stress hormones without requiring blood draws can significantly enhance crew safety and autonomy. Wearable technologies, including biosensor patches, contactless optical monitors, and implantable sensors, are being developed for use in spaceflight. These tools reduce procedural burden and enable trend tracking over time, helping detect problems before symptoms arise (Chen and Wang, 2020; Heidt et al., 2020).

Each of these innovations contributes to a broader shift in space medicine: from reactive testing to proactive, integrated health monitoring. Future diagnostic platforms must be compact, rugged, intelligent, and capable of autonomous operation over extended periods with minimal resupply or external support.

## 24.6 Laboratory testing in space

Over the past two decades, several diagnostic platforms have been deployed aboard the ISS to assess feasibility, refine designs, and support crew health. These early efforts have provided foundational insights into how laboratory testing must be adapted for spaceflight.

Among the first significant tools was the Lab-on-a-Chip Application Development Portable Test System, which allowed astronauts to perform surface microbial testing using handheld cartridges. It was designed to rapidly detect bacterial endotoxins and fungal markers in the ISS environment, demonstrating that astronauts could successfully operate biochemical assays without Earth-based support (Wilson et al., 2007).

The WetLab-2 system, developed by NASA Ames, expanded molecular diagnostic capabilities by enabling real-time analysis of gene expression aboard the station. This system included a microgravity-optimized sample prep unit and a quantitative PCR module. Astronauts used it to monitor changes in gene expression related to immune function, muscle atrophy, and microbial behavior (Parra et al., 2017).

Another major milestone was the use of the MinION nanopore sequencer, which allowed direct DNA sequencing in microgravity for the first time. Originally designed for pathogen detection and environmental monitoring, the MinION has also been tested for its potential to support medical diagnostics through genotyping and microbial identification (Castro-Wallace et al., 2017).

In terms of clinical diagnostics, astronauts have performed urinalysis, blood glucose monitoring, hematocrit measurements, and white blood cell counts using compact, manually operated systems. These tests have helped monitor hydration status, immune system activity, and general physiological adaptation to microgravity (Blue et al., 2019). While comprehensive panels, such as full-chemistry or immunoassay panels, remain challenging, several platforms in development aim to address this gap.

Lessons from ISS operations have emphasized the importance of simplicity, automation, and minimal consumables. Sample preparation remains one of the most error-prone steps, and efforts to streamline fluid handling through microfluidics and sealed cartridge systems are ongoing. Astronauts have proven capable of performing complex tasks with proper training, but in long-duration or deep-space missions, greater diagnostic autonomy will be required.

These early in-flight experiences continue to inform the design of next-generation platforms that prioritize ruggedness, user-friendliness, and clinical utility within the constraints of space-based medical care.

## 24.6.1 Key challenges

### *24.6.1.1 Sample collection and preservation*

Drawing and isolating blood or other biological fluids in microgravity presents a persistent technical challenge. In the absence of gravity, fluids do not behave as they do on Earth, forming cohesive spheres influenced by surface tension, and may adhere to nearby surfaces or instruments. Air movement inside the spacecraft can further affect the position and stability of these droplets, potentially helping or hindering control. These physical dynamics pose significant risks of sample loss or environmental contamination. Specialized vacutainers and microsyringes have been engineered to seal samples immediately after collection, minimizing uncontrolled exposure. Once collected, preserving biological specimens adds another layer of complexity. Without conventional refrigeration, analytes can degrade rapidly. Potential solutions include chemical preservatives, lyophilization, or compact refrigeration units designed for spaceflight conditions, which must also consider power efficiency and vibration resistance.

### *24.6.1.2 Equipment footprint and workflow*

Terrestrial laboratories typically rely on multiple machines, each optimized for specific tests, which require large footprints and regular maintenance. In contrast, spaceflight requires small, multifunctional diagnostic platforms that can run multiple assays from a minimal sample volume. These systems must integrate preparation, testing, and waste management within a compact, self-contained device. They must also be robust to launch vibration, microgravity fluid behavior, and limited access for repairs. Given that most crew members lack formal laboratory training, devices must support intuitive workflows, provide visual cues, and require minimal user input while maintaining clinical accuracy.

### *24.6.1.3 Data interpretation and crew training*

Diagnostic systems in space must present results that are immediately understandable by nonexperts. Current systems use color-coded indicators, simple pass/fail criteria, or built-in reference ranges. Some platforms incorporate basic automated logic to flag abnormal results or confirm assay completion. More advanced artificial intelligence tools, such as trend analysis, anomaly detection, and cross-modality decision support, are still in development and not yet standard on missions (Cheung et al., 2023). These AI systems hold promise but will require extensive validation and crew training. Until then, user interfaces must prioritize clarity, redundancy, and embedded procedural guidance to reduce errors and increase crew confidence.

### *24.6.1.4 Time constraints*

Crew time is one of the most limited resources aboard any mission. Diagnostic procedures must be fast, streamlined, and predictable, ideally providing results within minutes. Complex multistep protocols, lengthy sample prep, or complicated interpretation workflows increase the likelihood of error and can delay medical decision-making. Diagnostic tools must therefore emphasize efficiency, automation, and low cognitive load, allowing testing to occur in parallel with other operational demands without compromising safety or mission flow.

### 24.6.2 Innovations driving laboratory testing

#### *24.6.2.1 Lab-on-a-chip integrations*

Lab-on-a-chip systems consolidate multiple diagnostic functions into a single microfluidic platform. These devices dramatically reduce the volume of blood and reagents required and eliminate the need for traditional fluid handling steps. They can be engineered to support a wide range of assays, including metabolic panels, electrolyte tests, and immunoassays, all from a single, sealed disposable cartridge (Chiu et al., 2017). The RAL Space "Lab-on-Chip" platform is one example currently under development for autonomous diagnostics in microgravity. It uses capillary-driven flow and onboard sensors to process and analyze samples without pipetting or centrifugation. This makes it well-suited to the low-resource environment of spaceflight, where simplicity, automation, and reagent stability are critical.

#### *24.6.2.2 Advanced optical methods*

Optical biosensors, spectrophotometry, and laser-based detection techniques are being developed as alternatives to the reagent-heavy methods of wet chemistry (Jokerst et al., 2012). These systems rely on absorbance, fluorescence, or interferometric signals to quantify biomarkers in real-time. By minimizing the need for chemical reagents and complex sample preparation, optical methods reduce system complexity, enabling faster, lower-risk testing. Their compact size and electronic readout make them well suited for integration into spacecraft diagnostics. Optical sensing is also more tolerant of vibration and orientation changes, improving reliability under launch and spaceflight conditions (Chen and Wang, 2020).

#### *24.6.2.3 Portable automated analyzers*

Next-generation analyzers build on proven point-of-care technologies, such as the Abbott i-STAT system, but significantly expand their capabilities. These compact platforms now support broader test menus, onboard data interpretation, and extended reagent shelf lives. To evaluate their suitability for spaceflight, many have been tested in terrestrial analog environments such as Antarctic research stations, high-altitude expeditions, and NASA's NEEMO undersea habitat. These settings simulate key mission constraints, such as isolation, limited medical access, and resupply challenges (Blue et al., 2019; Crucian et al., 2021). Separately, diagnostic platforms undergo hardware validation testing to simulate spaceflight stressors, such as launch vibration, temperature extremes, and power fluctuations, during NASA's payload certification process.

Compact analyzers are being developed to consolidate hematology, clinical chemistry, and limited molecular diagnostics into a single sealed cartridge, dramatically reducing crew time and consumables. While not yet flown in full clinical configuration, these rugged, user-friendly devices are well-suited for environments with minimal technical support and tight operational timelines.

#### *24.6.2.4 AI-powered automation*

Automation features are increasingly being incorporated into space-compatible diagnostic platforms. Most current implementations rely on basic rule-based algorithms, such as flagging abnormal values, verifying assay completion, or identifying operational errors like incomplete reagent mixing or expired cartridges. These quality control features help minimize user error and ensure reliability without requiring extensive training.

More sophisticated AI tools, such as trend analysis, cross-modality integration, and autonomous diagnostic suggestions, are still in development. These advanced systems are not yet certified for operational missions but are being actively prototyped for future deployment in long-duration, autonomous environments like Mars transit or surface outposts (Human Research Roadmap, 2020). Once validated, such AI integration could enable real-time clinical decision-making without reliance on Earth-based support, allowing astronauts to respond more effectively to evolving medical scenarios in flight.

#### *24.6.2.5 Tested diagnostic capabilities on the ISS*

Several diagnostic tests have already been conducted aboard the ISS. The WetLab-2 platform, developed by NASA Ames, enabled quantitative real-time PCR (qRT-PCR) to measure gene expression from biological samples in orbit (Parra et al., 2017). The MinION device from Oxford Nanopore Technologies has been used to sequence DNA and RNA in microgravity, successfully identifying microbial species from environmental swabs and synthetic samples (Castro-Wallace et al., 2017). These platforms demonstrated that in-flight molecular diagnostics are feasible without requiring the return of samples to Earth.

Additionally, blood analysis for hematocrit, electrolyte levels, and white blood cell counts has been performed using dry-chemistry analyzers operated directly by the crew with minimal ground-based assistance (Blue et al., 2019; Crucian et al., 2021). These early experiments have laid the foundation for more advanced, autonomous diagnostic systems in future missions.

### 24.6.3 Future directions

Looking ahead, future laboratory systems may evolve into "universal analyzers" that combine clinical chemistry, hematology, immunoassays, and molecular diagnostics into one compact module. Advances in microelectronics, microfluidics, and synthetic biology could enable these devices to process a single blood droplet in parallel channels, each dedicated to a different assay. Such integrated systems will be designed for rapid turnaround, minimal user input, and long-term reliability, being critical attributes for missions without immediate access to Earth-based laboratory facilities.

## 24.7 The future of space diagnostics

### 24.7.1 Anatomic pathology

Anatomic pathology remains essential for diagnosing conditions such as infections, tumors, and inflammatory processes. However, conventional histopathology workflows face major barriers in spaceflight environments.

#### *24.7.1.1 Biopsy and sample handling*

Performing a biopsy in microgravity requires specialized instruments that can stabilize and secure tissue immediately upon extraction, preventing sample loss or contamination. Once obtained, tissue must be preserved with fixatives that do not require refrigeration. While these tools are not yet space-certified, early prototypes are being evaluated in analog missions and simulated microgravity experiments (Ehteshami Bejnordi et al., 2017; Al-Janabi et al., 2012).

#### *24.7.1.2 Slide preparation and staining*

Traditional microtomes and staining trays depend on gravity for even fluid distribution. In space, automated or semiautomated slide preparation systems housed within sealed compartments could ensure uniform sectioning and staining. Although not currently deployed in orbit, such systems are a logical next step and may be adapted from terrestrial automation platforms widely used in digital histopathology (Al-Janabi et al., 2012).

#### *24.7.1.3 Digital microscopy and AI analysis*

Digital microscopy has become a standard component of modern pathology workflows on Earth, enabling high-resolution image capture, remote consultation, and AI-assisted interpretation (Ehteshami Bejnordi et al., 2017). These systems allow prepared slides to be scanned and viewed with diagnostic clarity, even in resource-limited environments. AI tools are already in clinical use for triaging cases, detecting patterns such as cancer or inflammation, and flagging diagnostic outliers (Bulten et al., 2020). In spaceflight settings, these capabilities offer enormous potential: high-resolution slide scanners could be used onboard to generate images for immediate review by crew or remote experts. In the future, embedded AI algorithms could assist in real-time decision-making when ground communication is delayed or unavailable. While full AI integration into space-based workflows has not yet occurred, the foundational technologies are mature and continue to rapidly advance.

#### *24.7.1.4 Future prospects*

Emerging imaging technologies, such as multiphoton and confocal microscopy, could enable three-dimensional tissue visualization with minimal sample preparation (Chen and Wang, 2020). These modalities, along with synthetic biology-derived fixatives and advanced onboard analytics, may eventually transform space-based pathology into a rapid, low-burden diagnostic tool. While not yet practical for near-term missions, they represent promising directions for future capability.

### 24.7.2 Imaging diagnostics

Imaging is pivotal in noninvasively assessing internal structures, identifying injuries, and monitoring chronic conditions. In the space environment, advanced imaging modalities must be compact, energy-efficient, and capable of functioning without extensive user training or gravitational assistance (Wong, 2020).

#### *24.7.2.1 Portable ultrasound systems*

Ultrasound is currently the mainstay of space imaging and is widely used aboard the ISS. It enables real-time assessment of abdominal organs, musculoskeletal injuries, and vascular flow. However, challenges in probe stabilization and image interpretation persist in microgravity. Enhanced ultrasound systems, supported by AI algorithms that provide real-time guidance on probe positioning and automated interpretation, are being developed to overcome these limitations and reduce the need for expert oversight (Sargsyan et al., 2005; Jones et al., 2009; Law and Macbeth, 2011).

#### *24.7.2.2 Fram2 portable X-ray demonstration*

A major milestone was achieved with the successful demonstration of a compact digital radiography unit aboard the ISS during the Fram2 mission (A Historic Image: The First X-ray in Space Is Here, 2025). This experiment confirmed the feasibility of acquiring diagnostic-quality X-ray images in microgravity using a lightweight, low-power system designed for austere environments. Future systems may combine portable radiographic capability with AI-based image interpretation and 3D reconstruction to support in-flight trauma assessment and disease monitoring (Wong, 2020).

#### *24.7.2.3 Low-field MRI prototypes*

While traditional MRI remains infeasible for spaceflight due to its size, power requirements, and the need for magnetic shielding, low-field MRI systems are being developed using smaller magnets and simplified cooling systems (Cooley et al., 2021). Although these devices offer lower resolution than conventional systems, they may still detect critical conditions such as intracranial hemorrhages, spinal injuries, or structural changes in soft tissue. If miniaturization efforts are successful, low-field MRI could become a valuable tool for future missions.

#### *24.7.2.4 Hybrid modalities and optical techniques*

Alternative imaging methods, such as optical coherence tomography and diffuse optical tomography, offer compact, low-power solutions that utilize nonionizing radiation. These techniques are especially suited for assessing superficial tissues and ocular structures, which are particularly susceptible to microgravity-induced changes (Mader et al., 2011; Lee et al., 2020). They may also support monitoring of skin, vascular, and neurological systems on long missions.

#### *24.7.2.5 Future innovations*

Continued progress in AI-guided image reconstruction, power-efficient hardware design, and further miniaturization of imaging components is expected to yield more capable onboard systems (Wong, 2020; Cooley et al., 2021). The long-term goal is to create an integrated, multimodality imaging suite that provides high-quality diagnostic insight with minimal user input. Such systems will support autonomous medical decision-making when Earth-based consultation is limited or delayed.

### 24.7.3 Implementation strategies for space-based diagnostics

Integrating diagnostic tools into space missions requires both technological innovation and a strategic approach to deployment, resource management, and crew training.

#### *24.7.3.1 Modular and scalable systems*

A modular design enables diagnostic capabilities to be tailored to the specific needs of each mission. Rather than relying on a single, monolithic device, a universal base station can support interchangeable modules, ranging from chemistry and hematology cartridges to imaging attachments, allowing for flexible test selection and simplified maintenance (Wong, 2020). This approach also conserves mass and power, as only the modules required for a given mission need to be included.

#### 24.7.3.2 *Integration of artificial intelligence*

AI is a cornerstone of next-generation space-based diagnostics. Advanced algorithms can analyze incoming data streams, flag anomalies, and provide step-by-step guidance for complex procedures (Cheung, De Louche and Komorowski, 2023). For instance, an AI platform might detect a subtle trend in a blood panel that suggests an early infection, prompting the user to perform a follow-up test. By integrating data from laboratory tests, imaging devices, and environmental sensors, these platforms provide a more comprehensive overview of crew health. However, AI is not intended to replace crew medical judgment; rather, it serves as a decision-support tool that enhances autonomous clinical management when immediate ground-based consultation is unavailable (Blue et al., 2019).

#### 24.7.3.3 *Resource optimization*

Every milliliter of fluid, watt of power, and gram of mass aboard a spacecraft must be accounted for, and diagnostic systems must reflect this reality. To minimize consumables, many modern diagnostic platforms utilize microfluidic and dry-chemistry techniques, which drastically reduce reagent volumes (Krakos, 2024). Devices should be engineered with energy efficiency in mind, incorporating low-power modes, power-on-demand functionality, and energy-saving standby settings. Waste management must also be considered early in the design process, with closed-loop systems that safely contain and recycle biohazardous waste, minimizing storage requirements and reducing the risk of contamination (Heidt et al., 2020).

#### 24.7.3.4 *Telemedicine and ground support*

Although real-time consultation becomes increasingly limited as missions extend deeper into space, structured telemedicine remains a key pillar of medical support. Diagnostic data can be transmitted in scheduled batches to Earth-based experts for asynchronous review and recommendations. A hybrid model, where onboard AI manages immediate needs while scheduled "consult windows" enable expert oversight, allows for continuity of care even in environments with long communication delays (Wong, 2020; McPhee et al., 2009).

#### 24.7.3.5 *Automation and user-friendly interfaces*

Automation reduces crew workload and mitigates the risk of human error. Systems with automated calibration, sample loading, and error checking routines help ensure test accuracy under stressful or time-constrained conditions. Interfaces must be intuitive and informative, providing clear, easily interpreted outputs such as color-coded alerts, trend visualizations, or simple numeric thresholds that support confident decision-making (Parra et al., 2017).

#### 24.7.3.6 *Long-term reliability and maintenance*

For missions lasting months or even years, diagnostic systems must be engineered with durability and in-flight maintainability in mind. Hardware should utilize radiation-hardened electronics, minimize moving parts, and incorporate ruggedized enclosures to withstand the stresses of launch and microgravity (Averesch et al., 2023). In the event of a malfunction, onboard 3D printing capabilities may enable the rapid production of replacement parts, thereby reducing system downtime and preserving mission continuity (Human Research Roadmap, 2020). Self-monitoring features, such as automated performance checks and calibration routines, enable the device to detect and resolve emerging issues before they impact patient care, thereby further reducing the burden on crew members.

### 24.7.4 Ethical and legal considerations

Implementing diagnostic systems in space raises essential ethical and legal questions that must be addressed in conjunction with technological advancements.

#### 24.7.4.1 *Informed consent and privacy*

Maintaining individual privacy while ensuring collective safety is challenging in a spacecraft's confined space. Crew members may undergo mandatory diagnostic testing for mission safety, which may potentially compromise their personal privacy. Protocols must be established to balance the need for constant monitoring with respect for individual rights, ensuring that sensitive health data are securely managed and confidential (Blue et al., 2019).

#### 24.7.4.2 Resource allocation

Finite resources characterize space missions. In the event of multiple medical emergencies, allocating limited diagnostic supplies, such as test cartridges or reagents, will require preestablished triage protocols based on clinical need. Ethical frameworks must guide these decisions to ensure fair distribution while prioritizing mission safety. The potential financial implications of misallocation must also be considered, as test wastage or overutilization can have significant cost ramifications.

#### 24.7.4.3 Psychological impacts

The psychological stress of receiving abnormal diagnostic results in an isolated, high-stakes environment cannot be underestimated. In space, where immediate access to specialist care is limited, uncertainty can exacerbate anxiety and impact overall crew morale. Integrating support with diagnostic systems is essential. It is critical that results are communicated clearly and that the crew has access to counseling and reassurance from onboard systems, as well as via scheduled telemedicine sessions with mental health professionals.

#### 24.7.4.4 Additional legal complexities

With multinational crews and private companies involved, the legal frameworks governing medical diagnostics in space are still evolving. Determining liability for diagnostic errors or device malfunctions in an environment with delayed communication and limited on-site expertise is a complex task. Establishing clear regulatory standards and international agreements is critical to ensure accountability and protect crew members and mission operators. Legal protocols must be developed to address issues such as device certification, medical malpractice, and data privacy in the unique context of spaceflight.

## 24.8 Conclusion

### 24.8.1 Summary of challenges and solutions

Developing diagnostic tools for near-term human spaceflight necessitates fundamentally reengineering terrestrial laboratory practices to operate effectively in microgravity, under severe resource constraints, and in a high-radiation environment. Traditional instruments, which rely on gravity, abundant water, and immediate expert support, must be replaced by innovative solutions, such as microfluidic lab-on-a-chip systems, AI-driven automated analyzers, and synthetic biology approaches for reagent production. These innovations overcome individual challenges and work synergistically to provide a comprehensive, autonomous diagnostic suite that ensures crew safety.

### 24.8.2 Importance for future missions

As space missions extend beyond LEO and progress toward lunar and Martian exploration, the reliability of onboard diagnostics will be a cornerstone of mission success. Rapid, accurate detection of infections, trauma, and chronic conditions can prevent catastrophic health events and ensure timely interventions, directly impacting individual patient outcomes and the overall feasibility of deep-space missions. Integrating advanced diagnostic systems, with robust automation and minimal resource demands, will be essential to support the increasingly diverse and demanding nature of future crew compositions.

### 24.8.3 Final thoughts

Space-based diagnostic innovation presents a multidisciplinary challenge that requires integrating engineering ingenuity, clinical expertise, ethical responsibility, and international regulatory cooperation. The technologies developed for space not only push the boundaries of what is medically possible in extreme environments but also have the potential to revolutionize healthcare on Earth, especially in remote or underserved regions where resources are similarly constrained. By harnessing these advancements, near-term human spaceflight can be made safer, more autonomous, and more resilient, paving the way for extended human presence in deep space.

## AI disclosure

During the preparation of this work, the author(s) used ChatGPT, Claude, and Open Evidence for reference review and assistance in locating relevant sources. After using this tool/service, the author(s) reviewed and edited the content as needed and take(s) full responsibility for the content of the publication.

## References

A historic image: the first X-ray in space is here, 2025. Available at https://kaimaging.com/wp-content/uploads/2025/04/2025-04-02-SpaceXray-first-image-PR.pdf.

Al-Janabi, S., Huisman, A., Diest, P.J.V., 2012. Digital pathology: current status and future perspectives. Histopathology 61 (1), 1–9. Available at https://doi.org/10.1111/j.1365-2559.2011.03814.x.

Averesch, N.J.H., Berliner, A.J., Nangle, S.N., Zezulka, S., Vengerova, G.L., Ho, D., et al., 2023. Microbial biomanufacturing for space-exploration—what to take and when to make. Nature Communications 14 (1). Available at https://doi.org/10.1038/s41467-023-37910-1, https://www.nature.com/ncomms/.

Blue, R.S., Bayuse, T.M., Daniels, V.R., Wotring, V.E., Suresh, R., Mulcahy, R.A., et al., 2019. Supplying a pharmacy for NASA exploration spaceflight: challenges and current understanding. Microgravity 5 (1). Available at https://doi.org/10.1038/s41526-019-0075-2, https://www.nature.com/npjmgrav/.

Boguraev, A.S., Christensen, H.C., Bonneau, A.R., Pezza, J.A., Nichols, N.M., Giraldez, A.J., et al., 2017. Successful amplification of DNA aboard the International Space Station. npj Microgravity 3 (1). Available at https://doi.org/10.1038/s41526-017-0033-9, https://www.nature.com/npjmgrav/.

Bond, M., Richards-Kortum, R., 2017. Diagnostics for global health: hand-spun centrifuge. Nature Biomedical Engineering 1 (1). Available at https://doi.org/10.1038/s41551-016-0017.

Bulten, W., Pinckaers, H., van Boven, H., Vink, R., de Bel, T., van Ginneken, B., et al., 2020. Automated deep-learning system for Gleason grading of prostate cancer using biopsies: a diagnostic study. The Lancet Oncology 21 (2), 233–241. Available at https://doi.org/10.1016/S1470-2045(19)30739-9, http://www.journals.elsevier.com/the-lancet-oncology/.

Castro-Wallace, S.L., Chiu, C.Y., John, K.K., Stahl, S.E., Rubins, K.H., McIntyre, A.B.R., et al., 2017. Nanopore DNA sequencing and genome assembly on the International Space Station. Scientific Reports 7 (1). Available at https://doi.org/10.1038/s41598-017-18364-0, www.nature.com/srep/index.html.

Chen, C., Wang, J., 2020. Optical biosensors: an exhaustive and comprehensive review. Analyst 145 (5), 1605–1628. Available at https://doi.org/10.1039/c9an01998g, http://pubs.rsc.org/en/journals/journal/an.

Cheung, H.C., De Louche, C., Komorowski, M., 2023. Artificial intelligence applications in space medicine. Aerospace Medicine and Human Performance 94 (8), 610–622. Available at https://doi.org/10.3357/AMHP.6178.2023, http://ingentaconnect.com/content/asma/amhp;jsessionid=13tlejf2f0ojk.alice.

Chiu, D.T., deMello, A.J., Di Carlo, D., Doyle, P.S., Hansen, C., Maceiczyk, R.M., et al., 2017. Small but perfectly formed? Successes, challenges, and opportunities for microfluidics in the chemical and biological sciences. Chem 2 (2), 201–223. Available at https://doi.org/10.1016/j.chempr.2017.01.009, http://www.cell.com/chem/home.

Cooley, C.Z., McDaniel, P.C., Stockmann, J.P., Srinivas, S.A., Cauley, S.F., Śliwiak, M., et al., 2021. A portable scanner for magnetic resonance imaging of the brain. Nature Biomedical Engineering 5 (3), 229–239. Available at https://doi.org/10.1038/s41551-020-00641-5, www.nature.com/natbiomedeng/.

Crucian, B., Valentine, R., Calaway, K., Miller, R., Rubins, K., Hopkins, M., et al., 2021. Spaceflight validation of technology for point-of-care monitoring of peripheral blood WBC and differential in astronauts during space missions. Life Sciences in Space Research 31, 29–33. Available at https://doi.org/10.1016/j.lssr.2021.07.003, http://www.journals.elsevier.com/life-sciences-in-space-research/.

Crucian, B.E., Choukèr, A., Simpson, R.J., Mehta, S., Marshall, G., Smith, S.M., et al., 2018. Immune system dysregulation during spaceflight: potential countermeasures for deep space exploration missions. Frontiers in Immunology 9. Available at https://doi.org/10.3389/fimmu.2018.01437, https://www.frontiersin.org/articles/10.3389/fimmu.2018.01437/full.

Ehteshami Bejnordi, B., Veta, M., Johannes van Diest, P., van Ginneken, B., Karssemeijer, N., Litjens, G., et al., 2017. Diagnostic assessment of deep learning algorithms for detection of lymph node metastases in women with breast cancer. JAMA the Journal of the American Medical Association 318 (22), 2199. Available at https://doi.org/10.1001/jama.2017.14585.

Grimm, D., Wehland, M., Corydon, T.J., Richter, P., Prasad, B., Bauer, J., et al., 2020. The effects of microgravity on differentiation and cell growth in stem cells and cancer stem cells. Stem Cells Translational Medicine 9 (8), 882–894. Available at https://doi.org/10.1002/sctm.20-0084, http://stemcellsjournals.onlinelibrary.wiley.com/hub/journal/10.1002/(ISSN)2157-6580.

Heidt, B., Siqueira, W.F., Eersels, K., Diliën, H., Van Grinsven, B., Fujiwara, R.T., et al., 2020. Point of care diagnostics in resource-limited settings: a review of the present and future of PoC in its most needed environment. Biosensors 10 (10). Available at https://doi.org/10.3390/BIOS10100133, https://www.mdpi.com/2079-6374/10/10/133.

Hong, K., Lacovetti, G., Rahimian, A., Hong, S., Epperson, J., Neal, C., et al., 2021. Clinical evaluation of the torq zero delay centrifuge system for decentralized blood collection and stabilization. Diagnostics (Basel) 11 (6).

Horneck, G., Klaus, D.M., Mancinelli, R.L., 2010. Space microbiology. Microbiology and Molecular Biology Reviews 74 (1), 121–156. Available at https://doi.org/10.1128/MMBR.00016-09Germany, http://mmbr.asm.org/cgi/reprint/74/1/121.

Iacovetti, G., Collier, B., Setzer, S., Peevler, M., Ragar, J., Hong, K., et al., 2023. Evaluation of a compact, portable centrifuge for separating microvolume blood samples at the point of collection. The Journal of Applied Laboratory Medicine 8 (6), 1042–1053. Available at https://doi.org/10.1093/jalm/jfad071.

Jokerst, J.C., Emory, J.M., Henry, C.S., 2012. Advances in microfluidics for environmental analysis. Analyst 137 (1), 24–34. Available at https://doi.org/10.1039/c1an15368d, http://pubs.rsc.org/en/journals/journal/an.

Jones, J.A., Sargsyan, A.E., Barr, Y.R., Melton, S., Hamilton, D.R., Dulchavsky, S.A., et al., 2009. Diagnostic ultrasound at mach 20: retroperitoneal and pelvic imaging in space. Ultrasound in Medicine and Biology 35 (7), 1059–1067. Available at https://doi.org/10.1016/j.ultrasmedbio.2009.01.002.

Krakos, A., 2024. Lab-on-chip technologies for space research — current trends and prospects. Microchimica Acta 191 (1). Available at https://doi.org/10.1007/s00604-023-06084-4.

Law, J., Macbeth, P.B., 2011. Ultrasound: from earth to space. McGill Journal of Medicine 13 (2), 59. Available at http://www.ncbi.nlm.nih.gov/pmc/articles/PMC3296555/.

Lee, A.G., Mader, T.H., Gibson, C.R., Tarver, W., Rabiei, P., Riascos, R.F., et al., 2020. Spaceflight associated neuro-ocular syndrome (SANS) and the neuro-ophthalmologic effects of microgravity: a review and an update. npj Microgravity 6 (1). Available at https://doi.org/10.1038/s41526-020-0097-9.

Lin, J.-C., Burnouf, T., Wu, Y.-W., Fan, Y.-J., 2020. Development of electrical powerless system for blood separation and coagulation time measurement. Health Technology 4. Available at https://ht.amegroups.org/article/view/6089/html.

Mader, T.H., Gibson, C.R., Pass, A.F., Kramer, L.A., Lee, A.G., Fogarty, J., et al., 2011. Optic disc edema, globe flattening, choroidal folds, and hyperopic shifts observed in astronauts after long-duration space flight. Ophthalmology 118 (10), 2058–2069. Available at https://doi.org/10.1016/j.ophtha.2011.06.021.

McGregor, H.R., Lee, J.K., Mulder, E.R., de Dios, Y.E., Beltran, N.E., Kofman, I.S., et al., 2020. Brain Connectivity and Behavioral Changes in a Spaceflight Analog Environment with Elevated CO2 bioRxiv. United States bioRxiv. Available at https://www.biorxiv.org.

McPhee, J.C., Charles, J.B.; N. Aeronautics and S. Administration, 2009. Human Health and Performance Risks of Space Exploration Missions: Evidence Reviewed by the NASA Human Research Program.

Moo-Young, M., 2011. Chapter 238 – Centrifugation Comprehensive Biotechnology 4, 489–500.

Parra, M., Jung, J., Boone, T.D., Tran, L., Blaber, E.A., Brown, M., et al., 2017. Microgravity validation of a novel system for RNA isolation and multiplex quantitative real time PCR analysis of gene expression on the International Space Station. PLoS One 12 (9). Available at https://doi.org/10.1371/journal.pone.0183480, http://journals.plos.org/plosone/article/file?id=10.1371/journal.pone.0183480&type=printable.

Roadmap, H.R., 2020.

Sargsyan, A.E., Hamilton, D.R., Jones, J.A., Melton, S., Whitson, P.A., Kirkpatrick, A.W., et al., 2005. FAST at MACH 20: clinical ultrasound aboard the International Space Station. Journal of Trauma - Injury, Infection and Critical Care 58 (1), 35–39. Available at https://doi.org/10.1097/01.TA.0000145083.47032.78.

Smith, S.M., Zwart, S.R., 2008. Chapter 3 nutritional biochemistry of spaceflight. Advances in Clinical Chemistry 46, 87–130. Available at https://doi.org/10.1016/S0065-2423(08)00403-4.

Wilson, J.W., Ott, C.M., zu Bentrup, K.H., Ramamurthy, R., Quick, L., Porwollik, S., et al., 2007. Space flight alters bacterial gene expression and virulence and reveals a role for global regulator Hfq. Proceedings of the National Academy of Sciences 104 (41), 16299–16304. Available at https://doi.org/10.1073/pnas.0707155104.

Wong, S., 2020. Diagnostics in space: will zero gravity add weight to new advances? Expert Review of Molecular Diagnostics 20 (1), 1–4. Available at https://doi.org/10.1080/14737159.2020.1699061, http://www.tandfonline.com/loi/iero20.

Wotring, V., 2018. Space Pharmacology: How Space Affects Pharmacology. Springer Science and Business Media LLC, pp. 1–13. Available at http://doi.org/10.1007/978-3-319-56637-5_68-1.

Chapter 25

# Preflight training and curriculum development for procedural skills

Chanel Fischetti[1], Emily Frisch[2], Laurel Kaye[1], and Alaina Brinley Rajagopal[3]

[1]*Department of Emergency Medicine, Massachusetts General Hospital, Boston, MA, United States,* [2]*Department of Obstetrics and Gynecology, Stanford Hospital, Palo Alto, CA, United States,* [3]*California Institute of Technology, Pasadena, CA, United States*

## 25.1 Background

Historically, the majority of human spaceflight missions have been undertaken in low Earth orbit (LEO), which has allowed for more immediate medical assistance, ground support, and evacuation capability (Melton et al., 2020; Duncan et al., 2008; Levin et al., 2023, 2024; Barratt and Pool, 2008). However, as human spaceflight missions extend beyond LEO, the challenges of providing timely medical assistance, evacuation, or resupply increase significantly due to communication delays and limited ground support. Furthermore, the composition of spaceflight crews is evolving to include more varied physiologies and training backgrounds (Levin et al., 2024).

To address this, space medicine curricula must be adaptable to the varying mission and crew profiles. This will likely require a combination of didactic instruction with hands-on, simulation-based methods tailored to mission parameters such as duration, crew composition, travel distance, existing complementary technologies, and onboard medical capability.

## 25.2 Historical curriculum design principles

In the past, crew skill sets and preflight medical training were developed using aggregated data on in-flight incidents and previously documented emergencies (Levin et al., 2023; Antonsen, 2017). Training emphasized management of predictable conditions based on crew health, team knowledge, skills (KSAs—Knowledge, Skills, and Abilities), and available medical resources in LEO.

Agencies such as the National Aeronautics and Space Administration (NASA) have since adopted formal risk management paradigms to refine this process. One example, the IMPACT (Informing Mission Planning via Analysis of Complex Tradespaces) Tool, developed by the Human Research Program (HRP), applies probabilistic risk assessment to determine which medical and operational skills provide the highest value for a given mission profile (Levin et al., 2024, 2022). This allows educators and planners to prioritize training content according to the likelihood and consequence of specific medical events.

In recent years, NASA has moved toward continuous, data-driven risk analysis through frameworks such as Risk-Informed Decision Making and Continuous Risk Management (CRM). These processes integrate real-time mission data and crew performance metrics to continually refine training needs, ensuring that medical competencies remain aligned with evolving mission parameters and risk profiles (CRM; Levin et al., 2025). These configurations were based on the KSAs and scope of practice established by NASA physician teams (Levin et al., 2025).

Operational tools such as the Integrated Medical Model and the Medical Extensible Database Probabilistic Risk Assessment Tool (MEDPRAT) now use predictive analytics to tailor medical training to each mission's anticipated health risks (Levin et al., 2025). Similarly, the Performance Risk Model helps identify where human performance limitations may require enhanced procedural or cognitive training. Together, these models inform the structure, frequency, and focus of simulation-based and scenario-driven instruction for crewmembers.

This evolution in risk modeling marks a broader shift in the purpose of preflight medical education from static emergency preparedness toward adaptive, mission-specific training. By quantifying medical and performance risks, these tools enable training teams to proactively address gaps in knowledge and skills, emphasizing flexibility, autonomous medical decision-making, and

**Procedural Medicine for Spaceflight. DOI:** https://doi.org/10.1016/B978-0-443-27552-4.00032-4

problem-solving. Such an approach is increasingly essential as missions extend beyond LEO and as commercial spaceflight introduces more variable crew backgrounds and capabilities in spaceflight (Blue et al., 2014).

## 25.3 Future planning for curriculum design principles

Future mission planning must take into account new complexities, such as varied mission duration (months versus years), crew composition, travel distance, onboard medical capability, and the anticipated gravitational profile.

### 25.3.1 Mission duration: long-duration versus short-duration spaceflight

On the International Space Station (ISS), most missions last 150–180 days (Blue et al., 2014). , As efforts are made towards a lunar base and more exploratory missions, including Mars exploration, training and preparation will need to adapt accordingly (Levin et al., 2024; Blue et al., 2014).

Government agencies, such as NASA, have gained significant expertise in optimal crew training for long-duration missions (over 3 months). However, with the exception of the Apollo program which lasted between 8 and 12 days, all of this experience has been in LEO (Schmidt et al., 2010; Musso, 2018). Missions in LEO benefit from nearly continuous, real time communication with Mission Control (Levin et al., 2023). Because of this, crewmembers do not necessarily need to retain all procedural skill sets in memory or in checklists, as they can draw on the significant experience and expertise of the flight surgeons, sonographers, and other experts on Earth (Martin et al., 2012; Elias et al., 2024). However, as government and commercial entities undertake longer-duration, exploration-class missions, there will be increasing delays and gaps in communication. Crewmembers may need to perform emergency procedures without assistance from the ground (Martin et al., 2012; Tokadlı and Dorneich, 2017).

The probability of medical events increases with larger crews and/or longer missions. Similarly, mission activities may introduce or change the probability and types of injuries and illnesses the crew may experience. For example, extravehicular activities may expose the crew to decompression sickness, and surface operations introduce risks of gravity induced injury such as falls. These factors significantly change the medical training needs for crew members. For example, lunar missions will require increased self sufficiency and different skill sets compared to LEO missions, while missions to Mars will need to function almost entirely autonomously (Martin et al., 2012; Landon et al., 2018)

Crew and CMOs will need additional skills . It is also likely that crew members will require more frequent refresher training. (Noe et al., 2011; Phelps et al., 2022; Rector et al., 2021).

In these instances, Just-In-Time-Training (JITT), augmented reality, and hands-on simulation sessions will be important (Blue et al., 2014). Similarly, anticipating and outlining capabilities and resources will be an important part of mission planning. Some of the more technical skill sets involved will require more advanced medical training than other skill sets (Levin et al., 2024).

### 25.3.2 Crew composition

For any mission scenario, preflight training and preparation must be reflective of the unique mission profile and needs, as well as the individualized baseline skill sets of the crewmembers. For exploration-class missions, a crewmember with advanced medical training (e.g., a physician) would be preferred given the diversity and level of proficiency of skills that might be needed. Training a nonmedical crewmember to obtain the same proficiency of surgical skills as a medical doctor is likely outside the scope of mission training due to the time and experience required to reach proficiency (Chung, 2005). However, there is wide variation among physician training, and not all specialties can be treated equally. For example, a nonsurgeon physician would need additional training to perform surgical procedures. However, it is likely that the base training received by all physicians who have graduated medical school will allow them to pick up cross trained skills faster and with more expertise.

The level of training should anticipate the expertise needed, risk approximated, equipment available, and probability of mission success contingent on predicted emergencies (Levin et al., 2023). Most risk models and analyses indicate that a level of training equivalent to medical school completion is required to substantially reduce mission risk in long-duration missions (Russell et al., 2023; McIntyre et al., 2022). Consensus from the 2012 Exploration Medical Capability (ExMC) Telemedicine Workshop indicates that, in fact, all crew "should be familiar with performing expected medical procedures and should know the location of supplies and equipment" (Blue et al., 2014).

Additional crewmembers with medical training (e.g., physician, physician assistant, nurse practitioner, nurse, paramedic, emergency medical technician) would also be valuable where an emergency or injury incapacitates the acting CMO crew member (Blue et al., 2014; McIntyre et al., 2022; Center LBJS. Human Research Program Integrated Research Plan, 2013;

Gick and Holyoak, 1980; Pena et al., 2010; Epstein, 2021). Substantiated studies have implied that crews with diverse, complementary skill sets, particularly in critical thinking, are better equipped to manage novel, high-stakes scenarios in isolated environments. This additional medically qualified crewmember would also provide valuable support for acute emergency scenarios involving multiple crewmembers, where procedural assistance is required, or if the CMO is incapacitated (Nowadly et al., 2019). Preflight training will need to be specific to a provider's skill level and background. For example, major surgery is outside the scope of practice of a nurse practitioner and would not be required, whereas a surgeon would need preflight and refresher training related to various surgical procedures that may be needed for a long-duration mission where a high level of autonomy is expected (i.e., exploration-class; Discussion Paper: Scope of Practice for Nurse Practitioners, 2022; Chung, 2005; Perez et al., 2013).

### 25.3.3 Microgravity or hypogravity adaptations

Additionally, strategies should consider the impact of altered gravitational environments, such as microgravity or lunar/Martian partial gravity, on procedural processes (Melton et al., 2001; Nishisaki et al., 2008). For instance, this might include practice of the procedure during parabolic flight gravitational profiles consistent with the intended mission.

Selection of procedures for which each crew member must train and then maintain proficiency will be related to a variety of factors including, but not limited to, (1) whether is it feasible within the mission gravitation profile (e.g., microgravity versus lunar gravity versus Martian gravity); (2) whether the provider has sufficient academic background, training, and experience to complete the procedure in altered-gravity; (3) whether sufficient resources will be available to complete the procedure; and (4) whether sufficient resources will be available to continue treatment after completion of the procedure. For example, it would not make sense to train any provider, regardless of academic background, to complete an emergency thoracotomy in the setting of trauma if resources such as a surgical suite, blood products, IV fluids, and vasopressors are not available to achieve patient stabilization after the procedure (DiGiacomo and Angus, 2017). Additionally, time optimization is also essential for astronaut preparation, given the extensive resources invested in preflight training.

Procedures that are straightforward on Earth may be more complex in microgravity or partial gravity. Preflight training must address these differences through gravity-specific modules. For example, stabilizing a patient and provider for CPR in microgravity might require astronauts to strap themselves and the patient to a surface, while also using modified chest compression techniques that rely on arm strength rather than body weight (Rehnberg et al., 2014). There is evidence to suggest hand movements may be slower in microgravity, and that proficiency may change as providers to adjust to the microgravity environment. Procedure times and methods might need to be adjusted to account for this (Jamšek et al., 2021). Even seasoned experts struggle with tasks such as intubation in simulated low-gravity conditions such as parabolic flight (Thierry et al., 2023; Warnecke et al., 2019).

To prepare for partial gravity environments such as the Moon (approximately one-sixth of the Earth's gravity) or Mars (one-third of the Earth's gravity), astronauts train in partial gravity simulators using counterweights, parabolic flights, or similar systems appropriate for the procedure being trained.s. For example, NASA uses the large water filled pool known as the Neutral Buoyancy Laboratory (NBL), to simulate microgravity for spacewalks(Swain et al., 2025). The facility allows astronauts to perform tasks in these settings, helping them understand how tool handling and body stabilization differ from Earth. Updates to the NBL include the possibility of integrating more advanced virtual reality modules and haptic feedback systems to improve procedural realism (Sinnott et al., 2019). Training protocols will need to be updated as these systems continue to evolve.

### 25.3.4 Onboard medical capabilities

Historically, NASA has implemented a structured training approach for crewmembers, ensuring that they demonstrate competency in various procedures before mission departure, with the greatest emphasis for the CMO. This process includes a preflight examination where the crewmember must perform procedures autonomously, utilizing only the resources that will be available during the mission (Blue et al., 2014). Preparatory training can incorporate various methods such as book learning, videos, auditory instruction, and simulations that present diverse scenarios and troubleshooting opportunities. Additionally, crewmembers may practice procedures on real subjects and showcase their skills through written, oral, and practical examinations.

The preparation for various procedures, and time dedicated to maintenance of competency, should also depend on the risk and likelihood that a particular procedure will be needed. For example, a procedure such as intravenous (IV) line insertion is fairly low-risk, simple to accomplish in altered-gravity, and requires minimal tools and training, even for nonmedical personnel. It is fairly likely that an IV line would be needed at some point for at least one crew member over the course of a 3-year mission (Nowadly et al., 2019). This is a procedure that would be worthwhile for all crewmembers to maintain some level of proficiency for a long-duration mission (Komorowski et al., 2016). It is also worth noting that for procedures that are likely to be needed, multiple crewmembers besides the CMO should maintain some level of proficiency, particularly if they are procedures that the CMO would not be able to perform on themselves while well, or especially if incapacitated.

### 25.3.5 Travel distance: retention and decay of skills

Upon achieving the necessary capabilities, it is essential for crewmembers to maintain their skills throughout the mission, especially for rare procedures that may require immediate recall, such as CPR (cardiopulmonary resuscitation; Spooner et al., 2007; Jay et al., 2003). Most research suggests that retention is enhanced when training is followed by frequent delayed testing (Foale et al., 2005; Mulavara et al., 2010). Specifically, motor skill retention was higher postflight with a structured recovery plan with regular sessions, rather than a sporadic, less frequent approach (Blue et al., 2014; Bloomberg et al., 2015). Retention was notably higher when training was delivered just before the task execution (Pieters and Zaal, 2019). Additional strategies include gamification, which may improve knowledge acquisition and retention, engagement, and enjoyment, all of which could prove valuable for long-duration missions (Blue et al., 2014; Brinley Rajagopal et al., 2020; Rajagopal et al., 2020; Brinley Rajagopal et al., 2020; Ahmed et al., 2015; McGaghie et al., 2011).

For future mission planning and to optimize skill retention, several strategies tailored to crewmembers' skill sets, preferences, and mission profiles can be implemented. While crewmembers would ideally be selected based on relevant skill sets, allowing them to concentrate on proficiency and performance rather than basic skills (Chung, 2005), this may not always be possible as spaceflight becomes more available and democratized. Therefore, preflight training will be crucial to ensure all crewmembers can perform necessary tasks as expected. Alternatively, as vehicles become larger it may be possible to have dedicated crew members who undergo required training and allow passengers to be more specialized or become passive participants similar to airlines. Effective strategies for maintaining procedural skills during extended missions may include the following:

- Selection based on background (academic, degrees, skills, experience),
- Comprehensive preflight training,
- Detailed procedure documents and checklists (Kerner et al., 2017),
- Access to viewable and practical refresher modules (Blue et al., 2014),
- In-flight Objective Structured Clinical Examinations (OSCEs; Patrício et al., 2013),
- Rapid review sessions based on procedure requirements and mission profiles (Yang et al., 2014),
- Organ system-based learning systems (Custers, 2010),
- Techniques for enhancing memory retention (Custers, 2010).

Memory retention has been studied for decades, particularly in areas such as the sciences and medicine, where large volumes of information must be learned in a short period of time and subsequently maintained for rapid recall and application. Evidence suggests a loss of one-fourth to one-third of knowledge in the first year, with retention of approximately half of knowledge by the second year (Custers, 2010). Learning retention is maximized through the systematic use of distributed/spaced practice with increasing intervals between sessions, variable practice scenarios, and frequent testing (Bahrick, 2000; Bahrick and Hall, 1991; Naveh-Benjamin, 1990; Landauer and Ainslie, 1975; Pashler et al., 2007). For maximal retention of medical knowledge, some have suggested a frequency of review of 1 hour per week, or 5–10 days of concentrated review two times per year (Custers, 2010; DuBois et al., 1969). Additionally, a tailored and focused program would help eliminate unnecessary information and concentrate on only on the most mission-essential preparation.

Onboard refresher and JITT training modules will likely be key components for skill maintenance during long-duration missions beyond LEO. These are likely to use various electronic modalities, including training documents, test modules (e.g., OSCEs), and video demonstrations. Crew on the ISS have access to various training programs that are recorded and easily accessible for crewmembers to use as refreshers in-flight (Blue et al., 2014). A larger, more comprehensive database will be needed to enable greater autonomy on longer, exploration-class missions. While various models and phantoms can be valuable training tools on Earth, their volume and mass will likely preclude their use on spacecraft beyond LEO. These modalities could be replaced using advanced and novel technologies using computer vision, haptics, and virtual reality (Izard et al., 2018; Brent Woodland et al., 2024). However, if replaced, it is important that haptic and tactile feedback is an integral part of the training (Mergen et al., 2024; Rangarajan et al., 2020).

The major challenge of completing these procedures rapidly and effectively in a spaceflight environment will depend on the fidelity of preflight training, the additional challenges presented by the spaceflight environment (microgravity, atmospheric/habitability challenges, etc.), and the ability for crewmembers to accurately and rapidly recall their training. The strategies presented above will assist with the completion of these tasks; however, in-situ aids may also improve the fidelity of procedures. These aids could include checklists, auditory assistance (e.g., auditory cues for next steps), and video assistance (e.g., demonstration of the procedure) that may be prerecorded by the crewmembers themselves or by experts in the procedure. For time-sensitive scenarios such as a traumatic incident, rapid recall will likely be the most effective strategy, emphasizing the need for review of critical procedures on an ongoing basis.

The first part of mitigating skill decay is understanding the time frame during which skills are lost. Predictive models can estimate how specific skills, such as suturing or CPR, might decay over time and recommend refreshers at optimal intervals (Wright et al., 2012). Onboard training tools, such as compact haptic trainers, can allow crewmembers to practice and maintain these skills in-flight. For instance, astronauts on Mars missions might use a portable haptic device to refresh training on chest tube insertion, a rarely needed but life-saving skill, at regular intervals. These refreshers can ensure readiness for emergencies. It is further critical to maintain up-to-date databases that can supply crewmembers with JITT training. Providing immediate, procedure-specific, training modules when a medical situation arises ensures that crew can refresh their knowledge and skills at critical moments. This approach has been recognized as beneficial in maintaining proficiency during long-duration missions (Phelps et al., 2022; Phelps et al., 2021).

### 25.3.6 Government versus commercial

With the democratization of space travel also comes increased diversity in mission profiles, risk tolerances, and crew training/preexisting skill sets. Most of the known and publicly available astronaut training comes from NASA and other public or governmental agencies; however, commercial entities are developing their own unique training protocols based on vehicle specifics, mission needs/durations, and crew expertise (Table 25.1). For example, space hotels are anticipated to

**TABLE 25.1** Comparison of risk tolerance, selection criteria, and mission length between government and commercial space sectors (Bowman, 2024; Krittanawong et al., 2023; Smith, Kelley and Basner, 2020).

| Category | Government/agency-led space missions (e.g., NASA, ESA, Roscosmos) | Commercial/private space missions |
|---|---|---|
| Astronaut selection criteria | • Stringent medical, psychological, and physical fitness standards.<br>• Typically career astronauts with advanced STEM or piloting backgrounds.<br>• Selection by national agencies based on mission role and specialization (Bowman, 2024; Defense Space Strategy Summary, 2020). | • Broader medical and demographic inclusion (e.g., civilians, researchers, tourists).<br>• Companies define their own criteria; may emphasize customer experience or mission sponsorship.<br>• No current unified global standard for commercial medical certification (Schweinsberg and Fennell, 2024). |
| Training duration and intensity | • 1–2 years of pre-mission training including EVA, robotics, language, and emergency procedures.<br>• Ongoing recurrent training for spaceflight readiness and mission-specific tasks (Astronaut Selection and Training (National Aeronautics and Space Administration (NASA)), no date). | • Significantly shorter, often weeks to a few months of mission-specific familiarization.<br>• Focus on safety, microgravity adaptation, and emergency basics; less emphasis on mission operations or science (Ax-4 Crew Launch-Ready After Completing SpaceX Training. Axiom Space, no date). |
| Mission length | • Long duration (3–12 months typical for ISS, future lunar missions 30–180 days, Mars ≥500 days; Bowman, 2024; Krittanawong et al., 2023). | • Short-duration (suborbital: minutes; orbital: typically ≤30 days).<br>• Future commercial space station stays projected at 1–30 days (Krittanawong et al., 2023). |
| Mission objectives | • Primarily scientific discovery, engineering validation, technology demonstration, and planetary exploration (Bowman, 2024; Schweinsberg and Fennell, 2024; Defense Space Strategy Summary, 2020). | • Primarily commercial profit, tourism, brand, or private research.<br>• Often part of a business ecosystem (e.g., Axiom Station, space hotels; Schweinsberg and Fennell, 2024; Falcon User's Guide-Version 8, 2025). |
| Medical standards and oversight | • NASA/ESA medical certification (Class II/III), ongoing biomedical monitoring, strict exclusion criteria for chronic disease, vision, and cardiovascular health. | • Private providers may adopt simplified or modified standards aligned with FAA's minimal safety oversight for commercial participants; still evolving (Falcon User's Guide-Version 8, 2025; Human Space Flight. Federal Aviation Administration, no date). |

launch as early as the late 2020s (Bowman, 2024; Krittanawong et al., 2023). The goals, preparation, and astronauts selected for these varied missions must be taken into consideration (Schweinsberg and Fennell, 2024).

## 25.4 Training implementation

Preflight training for space medicine must continue to balance didactic (classroom) instruction with hands-on methods such as simulations. The most basic approaches to training preflight can be described by known methodologies such as didactic lectures, online educational modules, and practice workshops. There are benefits, but also drawbacks, when relying on each of these, as described below.

The fundamental training methodologies include the following:

1. **Didactic (classroom) lectures:** These can be in-person or virtual, relying heavily on auditory learning styles. Recorded lectures facilitate course review, repetition, and asynchronous learning.
2. **Online/electronic learning modules:** These provide flexibility for where training occurs, allowing for asynchronous learning and course repetition.
3. **Workshops:** Focused on procedural and hands-on skills, these workshops are essential for meeting the procedural scope of practice requirements for medical interventions (Levin et al., 2022). Each session can be customized to address specific skill needs.
4. **Simulation:** A simulation-based curriculum can be integrated for on-site training and testing. Simulations provide opportunities for active engagement, dynamic interaction, and immediate feedback for both trainers and instructors. Simulations may include new technologies like virtual reality.

Currently, NASA standards require a physician crewmember on any lunar or planetary mission outside LEO lasting longer than 210 days (Blue et al., 2014). Fundamental knowledge is best conveyed in classrooms through didactics and discussion, equipping students to participate in practice simulations. This can later evolve into debriefs and post hoc analysis of performance to help students identify their weaknesses and improve in follow-on practice sessions. Important concepts, such as taking or relaying a medical history, a thoughtful assessment of the patient (basics like airway, breathing, and circulation), and being able to recognize life threats and perform basic high-value interventions like airway maneuvers and Heimlich maneuvers form important parts of emergency preparation and information prior to workshop or simulation training.

There are blended approaches that combine didactics, online lectures, and workshops, optimizing timing flexibility and catering to individual attention and learning styles. As crewmembers will each have unique preexisting skill sets and medical knowledge, pretests for training can help gauge existing learner skill sets, thereby freeing time for other topics and competencies needed for mission success or safety (Grantcharov and Reznick, 2008). The preflight curriculum should be personalized for individual astronauts and the entire crew's collective abilities. Checklists and practical OSCEs are existing options to test and ensure that crewmembers are competent in the required medical training needs for in-flight emergencies (O'Tuathaigh et al., 2019).

### 25.4.1 The utilization of simulation

High-fidelity simulation affords a realistic experience for preflight training and is necessary for refinement, retention, and consolidation of knowledge. Simulators and medical scenarios can provide prompted case studies by which real-time assessment, evaluation, and adjustments can be made preflight (McHenry et al., 2020; Krihak, 2023). Simulations are commonly used in terrestrial medical training and have demonstrated success in skill retention, even with skills as basic as CPR (McHenry et al., 2020). Real, translatable skills can be honed with case repetition and hands-on practice (McGaghie et al., 2011). Table 25.2 is a summary and product of NASA's ExMC and HRP with the stated goal of addressing the risk of the "Inability to Adequately Recognize or Treat an Ill or Injured Crewmember" (Exploration Medical Capability Medical Scenarios (National Aeronautics and Space Administration), 2021). These are scenarios for which preflight simulation could be helpful for the predicted needed hands-on skills for preflight practice.

Medical decision-making and training protocols are designed to support the evaluation, diagnosis, treatment, and management of both anticipated and unexpected medical conditions. Scenarios are used to explore and test medical capabilities, focusing on operational concerns and Earth-independent treatments.

Simulations will play an essential role in honing situational awareness, delegating team skill sets, and assessing core medical KSAs in emergent situations. Terrestrial medicine requires healthcare providers to continually maintain required life support skills, such as basic life support training and advanced life support training, which are core competencies and strategic methods by which training can be monitored, maintained, and progressed. It will also be critical to host

**TABLE 25.2** Medical scenarios released by Exploration Medical Capability and Human Research Program for in-flight medical training (Exploration Medical Capability Medical Scenarios (National Aeronautics and Space Administration), 2021).

| Scenario | Description | Expected training |
|---|---|---|
| Exercise scenario | Routine exercise on a treadmill with health monitoring. | Use of exercise equipment, interpretation of exercise variables, maintaining proper form, and health monitoring. |
| Dental examination | Routine dental exam, including history review, imaging, and treatment planning. | Conducting dental evaluations, interpreting radiologic findings, using dental tools effectively. |
| Headache management | Self-management of headache with over-the-counter medications. | Symptom tracking, safe use of medications, medical system operation. |
| Urinary tract infection (UTI) | Diagnosing and treating UTI symptoms, including dysuria and suprapubic tenderness. | Patient history taking, physical exams, lab result interpretation, ultrasound use. |
| Dental abscess | Diagnosis and treatment of infection such as a periapical abscess. | Ultrasound imaging, use of dental equipment, antibiotic administration, and differential diagnosis skills. |
| Arm injury | Treatment of trauma resulting from equipment malfunction, e.g., laceration or fracture. | Emergency response protocols, wound care, bleeding control, and physical stabilization techniques. |
| Decompression sickness (DCS) | Management of acute symptoms related to pressure changes. | Recognizing symptoms, oxygen therapy, and emergency stabilization procedures. |
| Acute radiation syndrome (ARS) | Treatment of symptoms resulting from radiation exposure. | Radiation monitoring, administration of countermeasures, and symptom management. |
| Kidney stone management | Handling cases of kidney stones leading to urosepsis or septic shock. | Diagnostic imaging, infection control, and medication administration. |
| Death of crew family member | Addressing behavioral health impacts after traumatic personal news. | Behavioral health support, long-term psychological management, and operational focus retention techniques. |

**TABLE 25.3** Example skill applications.

| Skill | Short-duration implementation | Long-duration implementation |
|---|---|---|
| Focused assessment with sonography (FAST ultrasound) | Crew conducts ultrasound under live telemedicine supervision; ground provides interpretation. | Crew independently performs and interprets ultrasound to detect internal bleeding; uploads data asynchronously for review. |
| Advanced airway management | Uses supraglottic device with teleguidance from flight surgeon. | Performs intubation, ventilator setup, and sedation management autonomously, referencing onboard protocols. |

simulations in which the designated CMO is possibly incapacitated, a reality that should be anticipated for longer-duration missions, to ensure overall mission safety Emergency Medical Procedures Manual for the International Space Station (ISS) [partial] (2016). Part of this will be to standardize proficiency in medical skills and tailor it to assigned missions. Table 25.3 models proficiency levels (1–5) for short-duration and long-duration missions, following NASA's human-systems KSA framework (1=minimal proficiency and 5=fully autonomous expert-level competency; Levin et al., 2023). Table 25.4 provides nuanced examples of the differences in skill competency needs for short- and long-duration missions.

Effective emergency triage, diagnosis, and management during space missions depend heavily on making accurate and timely diagnoses in-flight. Point-of-care ultrasound (POCUS) is one of the few portable diagnostic imaging and

**TABLE 25.4** NASA medical proficiency matrix for short- versus long-duration missions.

| KSA category | Short-duration missions (e.g., ISS, ≤6 mo) | Long-duration missions (e.g., Lunar/Mars, >6 mo) |
|---|---|---|
| Preventive medicine & health monitoring | Level 3: Routine health surveillance, environmental monitoring, infection prevention; relies on ground medical oversight. | Level 5: Fully autonomous preventive medicine; crew performs health surveillance, interprets biomarkers, implements countermeasures with minimal ground input. |
| Diagnostic capability | Level 3: Can perform basic history, exam, ultrasound, ECG with teleguidance; limited independent differential diagnosis. | Level 5: Independent diagnostic reasoning using limited resources and delayed comms; interprets imaging and labs autonomously. |
| Acute stabilization & emergency response | Level 3–4: Competent in CPR, bleeding control, AED use, splinting; guided by ground for advanced procedures. | Level 5: Performs advanced life support, airway management, chest tube insertion, shock management independently without evacuation. |
| Procedural skills & equipment proficiency | Level 3: Operates standard onboard medical kits, ultrasound, automated analyzers with supervision. | Level 5: Performs and maintains complex procedures/equipment, including repairs, calibration, and contingency improvisation. |
| Pharmaceutical & supply management | Level 2–3: Uses prelabeled medications; relies on ground for substitutions and expirations. | Level 5: Full pharmacologic stewardship: dosage calculation, substitution protocols, stability management for aged stock. |
| Telemedicine/decision support use | Level 4: Proficient in real-time telemedical collaboration and digital documentation. | Level 5: Uses asynchronous/delayed telemedicine; autonomous decision-making augmented by onboard AI or protocols. |
| Skill maintenance & training retention | Level 3: Periodic JIT refreshers and short-cycle task rehearsal. | Level 5: Uses self-directed simulation, virtual reality/augmented reality modules, and autonomous assessment tools for long-term skill retention. |
| Behavioral health & team medical care | Level 3: Participates in psychological monitoring; relies on ground mental-health support. | Level 5: Provides autonomous peer counseling, stress mitigation, and crisis management without real-time ground help. |
| System & resource management | Level 2–3: Executes predefined checklists; reports anomalies to ground. | Level 5: Integrates medical logistics, resource allocation, and risk-based decision-making into mission planning autonomously. |
| Cross-training & redundancy (nonphysician competence) | Level 2: Basic first-aid and triage under physician or ground guidance. | Level 4–5: Multirole proficiency; nonphysicians perform essential diagnostics and procedures if CMO (Crew Medical Officer) incapacitated. |

therapeutic modalities currently available in-flight for space missions (Kuypers, 2013; Sargsyan et al., 2006). Hands-on training in POCUS is practical and essential, as the technology is already available for most missions. It possesses both research and clinical uses, enabling both diagnosis and therapeutic interventions. For example, a skill set such as peripheral IV needle insertion can also be enhanced with ultrasound guidance. While most injectable medications do have other parenteral options, such as subcutaneous or intramuscular injection, titratable medications warrant more immediate intravenous access. For other needle guidance procedures, such as central lines or peripheral nerve blocks, ultrasound guidance is essential (Rando et al., 2014). However, this skill set often requires practice and repetition (Wang et al., 2025). In the near future, the addition of artificial intelligence (AI) tools may also supplement this skill set by improving the identification of landmarks and access points (Schellman et al., 2025; Viderman et al., 2022).

While advances in AI have been integrated into some sonographic (ultrasound) machine systems, image acquisition and interpretation are largely the responsibility of the sonographer, and terrestrial training for an AI model may not be entirely applicable or as accurate in the microgravity or partial gravity environment (operator; Boman et al., 2024; Tenajas et al., 2023). This highlights the importance of hands-on skills training and subsequent practice for retention. JITT reference cards, ground support, and AI are adjuncts, but in cases of communications latency, emergency, and other resource limitations, a basic working knowledge of diagnostic and procedural skills involving a portable ultrasound machine will need to be established by preflight training (Krihak et al., 2019). Other diagnostic and interventional tools must be considered similarly.

For more advanced procedures and their attendant skill sets, such as emergency surgery, there is a notable difference in how skilled and trained physicians perform these compared to those without the same formal training (Levin et al., 2022). Even among board-certified physicians from different residency and training backgrounds, it is not expected that an emergency medicine physician will complete a surgical task to the same standard of expertise or measured performance as a trained surgeon (Levin et al., 2022; Kuypers, 2013). As such, it may not be expected that any single crew member will possess the entirety of skills necessary for any in-flight emergency (Levin et al., 2022; Kuypers, 2013). Teleguidance assistance and evacuation will help supplement care during long-duration missions as distance and communication delays permit (Levin et al., 2022).

## 25.5 Future directions in preflight training

The evolution of space exploration demands continual advancements in preflight training methodologies. Emerging technologies, mission-specific challenges, and the evolution of space medicine will shape how astronauts prepare for long-duration missions beyond LEO. These advances are transforming traditional training approaches, making them more adaptive, efficient, and tailored to the unique demands of spaceflight and to the needs of each individual (Tokuç and Varol, 2023). Some such elements that will likely play an even larger role in the future of preflight training include robotics and AI, virtual reality, higher-fidelity simulations, real-time feedback, and increased personalization. While each of these is already at play to some extent in medical training, their use will likely expand well past their current benchmarks.

On Earth, simulators, virtual reality, and advanced haptic technologies would serve as valuable tools for education and performance enhancement (Bloomberg et al., 2015; Rangarajan, et al., 2020). These technologies hold significant potential for use in long-duration space missions, in particular for when required skills exceed the available KSAs (Bingham et al., 2024; Lelevé et al., 2020). Although mass and space constraints are critical considerations, advancements in portability make haptic systems a viable addition to in-flight missions, particularly for emergency scenarios and skill reinforcement (Bingham et al., 2024).

### 25.5.1 Robotics and AI

While existing training paradigms rely primarily on CMO and crew KSAs, rapid technological advancements may enable greater automation to reduce task loads and cognitive strain. Increased automation may also reduce training requirements, as robotics or software with advanced algorithms can eliminate the need for in-depth crew training for many tasks. As an example, pharmacy inventory, medication dose calculations, substitution protocols, stability management, etc., could all be managed electronically, reducing the burden for crew training and skill maintenance. Similarly, these tools could provide auditory checklists and procedural guidance during procedures or emergencies in order to reduce cognitive load and decrease the incidence of recall-associated errors.

Robotics and autonomous diagnostic tools will likely play an increasingly important role in enhancing astronauts' autonomy and safety, especially during extended missions. The integration of robotic systems capable of performing medical procedures is a significant focus (Casilla-Lennon et al., 2020). These robots are engineered to execute tasks such as biopsies or minimally invasive surgeries with high accuracy, especially in cases where the operator might not have completed a full surgical residency training. The development of surgical robots designed for microgravity environments aims to enable remote or autonomous surgical interventions when immediate medical support is unavailable. One such example, the Miniaturized In Vivo Robotic Assistant (MIRA) developed by Virtual Incision was designed to be controlled remotely by surgeons on Earth while it operates in microgravity (Pantalone, 2023). In early 2024, MIRA successfully completed a simulated surgical procedure aboard the ISS, marking the first instance of a robot performing surgery in a microgravity environment.

The coming years will likely see increased experimentation with robotic medical devices, followed by implementation, and training should reflect this. Depending on the extent of robotic autonomy, crewmembers would at a minimum need to be trained in the setup and troubleshooting of the robotic device and, at most, would require a fully trained, competent surgeon upskilled for operating in microgravity. Challenges associated with near-term implementation of some of these robotic systems include mass and volume constraints, maintenance and repair issues, and the differing system needs regarding autonomy for LEO versus exploration-class missions. Most of these technologies are at various stages of development and will likely require further refinement before true in-flight implementation.

Robotics has also begun to incorporate AI, and use cases are likely to drastically increase in the years to come (Yip et al., 2023). Devices are now able to incorporate computer vision, medical image analysis, navigation, precision manipulation, and autonomous planning driven by machine learning (Yip et al., 2023). In the context of space, robotics and AI have come together recently through the development of autonomous clinical support systems for space missions (Levin et al., 2023; Yule et al., 2023). These systems are designed to integrate physiological data monitoring with AI-driven analysis to provide real-time patient-specific advice for managing medical emergencies.

A proposed Precision Space Health (PSH) system would autonomously monitor, aggregate, and assess biomedical data, predict personalized health outcomes, and provide timely, actionable insights for individual crewmembers and their medical officers (Scott et al., 2023). The PSH system envisages integrating continuous biomonitoring technologies with AI-driven predictive analytics to detect early signs of health issues, enabling preventive measures before conditions become critical. This approach emphasizes personalized medicine tailored to each astronaut's unique physiological responses to the space environment. Trainees would need to know how to interface with and debug potential failures of this system.

### 25.5.2 AI-driven simulations and adaptive learning systems

While the importance of simulation has been discussed, future training will likely see the increased implementation of AI in designing and adapting simulation scenarios to meet crew and mission needs. For example, AI can create training simulations that dynamically respond to an astronaut's performance similar to how medical students train in simulations run by an experienced physician. Just as a human instructor can adjust the simulation complexity in real-time to create situations that are targeted to a students specific knowledge deficits and provide directed feedback, a well designed AI driven simulation may be able to adjust training modules when the trainee is millions of miles from the nearest medical school (Komasawa and Yokohira, 2023).

AI can further be incorporated into decision-making scenarios themselves (Lacinski et al., 2024). The Medical Extensible Dynamic Probabilistic Risk Assessment Tool (MEDPRAT) is a computational engine designed to work with a medical evidence database to model medical risks for various mission profiles. (Robinson et al., 2018). The tool was built to update the capabilities of NASA's Integrated Medical Model for a wider range of mission profiles. Designed with extensibility in mind, the tool is applicable across a broad array of exploration missions and leverages both spaceflight and terrestrial medical datasets. While it was intended to inform mission risk assessment, mission planning, and medical system design, its architecture is intentionally flexible and could potentially be combined with other models to aid in decision support and triaging available resources while on mission. (Robinson et al., 2018).

Despite ongoing advances in AI and decision-support technology, NASA's current JITT programs for CMOs do not yet incorporate Clinical Diagnostic Support Systems (CDSS). This omission reflects several well-established constraints, including power and computational demands, challenges with heat rejection, and the added mass and volume such systems would require. Current medical training and procedural strategies have proven sufficient for existing mission needs, reducing the benefit of introducing complex AI-based CDSS at this stage. Moving forward, the development of AI-powered CDSS could consider mission phases, medical risks, and logistical constraints. This includes designing systems capable of supporting Earth-independent medical operations for diagnostics, treatment, and risk mitigation in remote space environments. This can also be further adapted in cases where medical needs are balanced against greater crew or mission safety risks. For instance, during a spacecraft decompression event, a trainee might need to decide whether to treat a crew member with an open fracture or secure oxygen supplies, with AI evaluating the decision-making process against mission safety priorities.

Another benefit of AI-powered learning systems is the ability to analyze individual performance data to identify skill gaps, monitor improvement rates, and predict when skills might decay (Kluge and Frank, 2014). For example, a surgeon training for space missions in simulated microgravity conditions might show full/unchanged proficiency in laparoscopic techniques but slower suturing rates. Learning analytics could recommend additional training, schedule targeted refreshers, and predict the point at which the astronaut's skills may degrade to a critical level without practice.

It is important to note that AI is not always accurate and/or precise, and has significant costs in engineering time, power, processing, heat load, memory, and training (Evans and Snead, 2024). It is not currently relied upon as an independent tool but can be helpful in conjunction with well-rounded training and equipment. It has been shown to be helpful for predicting outcomes in numerous clinical scenarios, including diagnosis, triage, and radiographic interpretation (Thierry et al., 2023; Evans and Snead, 2024). However, there may be uncertainty in the results. AI has demonstrated biases that are not always addressed in training sets, which often lack spaceflight-associated pathology and diagnostic/imaging presentations in microgravity (Chen et al., 2023). It is likely that AI models trained on terrestrial data may exhibit flaws or failures in altered-gravity environments. Additionally, given that the astronaut population is relatively small and does not reflect the health of the general population and that individuals take on specific risks and exposures during missions and training, there are bound to be gaps in machine learning approaches that rely on large datasets. This increases the risk of unpredictable errors, and training should emphasize anticipating and rapidly responding to them.

As with everything in spaceflight (and most other scenarios), it is important to understand what you plan to use the tool for and why a specific tool may be needed. It is critical to ask what purpose the AI model serves, what the costs are in terms of the limited available resources it requires, and to make sure that the model is truly the best option available. In some cases an AI model may be an excellent choice but it is only one option and often it is neither the best performer nor the most efficient use of limited resources. In most present-day educational and clinical use cases, rules-based models, human operators, static cue cards, and other options perform as well or better than AI, use significantly fewer resources, and are more efficient. It is worth considering AI, especially as we gain a better understanding of what the various forms of AI models are good for and as they improve, but the choice should be an informed one that avoids the temptation to reach for the most exciting new technology and carefully considers all available options.

### 25.5.3 Haptics and tactile feedback

High-fidelity haptic simulators have been incorporated into training fine motor skills critical to medical procedures in space (Abdi et al., 2020). These simulators replicate the tactile resistance and feedback of real-life scenarios. For example, a haptic device for lumbar puncture simulates the differential resistance encountered by the needle at each depth of entry, offering realistic feedback that improves technique. Empirical evidence suggests that incorporating haptic feedback into teleoperated systems can improve task execution, reduce error rates, and shorten completion times (Abdi et al., 2020). By providing a sense of touch, haptic interfaces enable operators to perform delicate manipulations with greater confidence and accuracy, particularly during intricate medical interventions.

Wearable haptic devices are another evolving innovation that provides real-time feedback during training. Unlike specific simulators, such as the lumbar puncture example above, the adaptability of a wearable haptic device, such as a glove, can create opportunities for well-rounded training scenarios. Combined with virtual and augmented reality, these devices can also simulate the unique challenges of microgravity, such as tools floating away, or the difficulty of physically stabilizing a patient.

### 25.5.4 Biofeedback systems

Biofeedback technologies have growing potential as tools for astronaut training and curriculum enhancement. Rather than relying on continuous physiologic monitoring in-flight, which can be uncomfortable and is not routinely employed, these systems can be applied selectively during training to help astronauts recognize and regulate stress responses in real time. By linking physiologic feedback (e.g., pulse, blood pressure, heart rate variability, respiration rate, or muscle tension) to performance under simulated high-pressure scenarios, trainees can develop strategies for composure and cognitive control that translate to improved technical proficiency and decision-making in mission settings (Bio-monitor Shirt, no date; Jimenez et al., 2024). Integrating such biofeedback modules into training curricula could therefore strengthen stress resilience, enhance retention of procedural skills, and promote adaptive performance during both normal and emergency operations.

### 25.5.5 Curriculum updates

To maintain efficacy, training curricula must constantly evolve alongside advancements in space medicine, feedback from past missions, and the introduction of new technologies. Postmission analysis provides invaluable insights for refining training protocols. For instance, data from ISS missions and microgravity trainers have highlighted difficulties in performing ultrasound-guided procedures due to equipment drift in microgravity (Dulchavsky et al., 2012; Kiberd et al., 2024). Near-miss incidents, or delayed diagnoses, due to insufficient training in interpreting ultrasound images can guide the design of future scenarios. These findings have led to the development of stabilization aids and revised training for POCUS (Dulchavsky et al., 2012).

## 25.6 Conclusion

As human spaceflight extends beyond LEO into exploration-class missions, training programs must evolve to meet increasingly complex operational and medical demands. Future curricula should emphasize adaptable, scenario-based learning that integrates simulation, JITT, and emerging technologies such as AI-driven decision support and biofeedback tools as adjuncts. Training should be mission-specific, reflecting the medical risks and resource constraints unique to each flight profile, while maintaining core competencies for emergency response and autonomous care. Ensuring mission success and crew health will require continuous skill maintenance, evidence-based curriculum design, and proactive incorporation of innovative training modalities as they mature (Emergency Medical Procedures Manual for the International Space Station (ISS) [partial], 2016).

## AI disclosure

During the preparation of this work, the author(s) used ChatGPT in order to help formulate the tables. After using this tool/service, the author(s) reviewed and edited the content as needed and take(s) full responsibility for the content of the publication.

## References


Abdi, E., Kulic, D., Croft, E., 2020. Haptics in teleoperated medical interventions: force measurement, haptic interfaces and their influence on user's performance. IEEE Transactions on Biomedical Engineering 67 (12), 3438–3451. Available at https://doi.org/10.1109/tbme.2020.2987603.

Ahmed, M., Sherwani, Y., Al-Jibury, O., Najim, M., Rabee, R., Ashraf, M., 2015. Gamification in medical education. Medical Education Online 20 (1), 29536. Available at https://doi.org/10.3402/meo.v20.29536.

Antonsen, E., 2017. Risk of adverse health outcomes and decrements in performance due to in-flight medical conditions. 20170004604.

Astronaut Selection and Training (National Aeronautics and Space Administration (NASA)), n.d.

Axiom Space, n.d. Ax-4 crew launch-ready after completing spaceX training. Available at https://www.axiomspace.com/mission-blog/ax4-spacex-training

Bahrick, H.P., Hall, L.K., 1991. Preventive and corrective maintenance of access to knowledge. Applied Cognitive Psychology 5 (1), 1–18. Available at https://doi.org/10.1002/acp.2350050102.

Bahrick, H.P., 2000. Long-Term Maintenance of Knowledge. Oxford University Press (OUP), pp. 347–362. Available at http://doi.org/10.1093/oso/9780195122657.003.0022.

Barratt, M.R., Pool, S.L., 2008. Principles of Clinical Medicine for Space Flight Principles of Clinical Medicine for Space Flight. Springer, New York, United States, pp. 1–596. Available at http://www.springerlink.com/openurl.asp?genre=book&isbn=978-0-387-98842-9.

Bingham, L.K., Garcia, A.D., Kincaid, J.A., Weno, B.M., Foreman, C.D., Bell, B.N., et al., 2024. Immersive technologies for human-in-the-loop lunar surface simulations. In: IEEE Aerospace Conference Proceedings 9798350304626 IEEE Computer Society United States. http://doi.org/10.1109/AERO58975.2024.10521209.

Bio-monitor Shirt, n.d. 1408. at https://nlsp.nasa.gov/view/lsdapub/lsda_hardware/IDP-LSDA_HARDWARE-0000000000000612.

Bloomberg, J.J., Peters, B.T., Cohen, H.S., Mulavara, A.P., 2015. Enhancing astronaut performance using sensorimotor adaptability training. Frontiers in Systems Neuroscience 9 (September). Available at https://doi.org/10.3389/fnsys.2015.00129, http://journal.frontiersin.org/article/10.3389/fnsys.2015.00129/pdf.

Blue, R.S., Bridge, L.M., Chough, N.G., Cushman, J., Khpal, M., Watkins, S., 2014. Identification of medical training methods for exploration missions.

Boman, R., Penkala, S., Chan, R.H.M., Joshua, F., Cheung, R.T.H., 2024. Integration of artificial intelligence with medical diagnostic sonography. Sonography 11 (3), 184–189. Available at https://doi.org/10.1002/sono.12441.

Bowman, A., 2024. Commercial Space Frequently Asked Questions. National Aeronautics and Space Administration (NASA). Available at https://www.nasa.gov/humans-in-space/leo-economy-frequently-asked-questions/?utm_source=chatgpt.com.

Brent Woodland, M., Ong, J., Zaman, N., Hirzallah, M., Waisberg, E., Masalkhi, M., et al., 2024. Applications of extended reality in spaceflight for human health and performance. Acta Astronautica 214, 748–756. Available at https://doi.org/10.1016/j.actaastro.2023.11.025.

Brinley Rajagopal, A., Sudario, G., Boysen Osborn, M., Weiland, D., 2020. Family game show-style didactic for teaching nervous system disorders during emergency medicine training. Journal of Education and Teaching in Emergency Medicine 5 (2). Available at https://doi.org/10.5070/m552047746.

Brinley Rajagopal, A., Jasperse, N., Boysen Osborn, M., 2020. Simulated mass casualty incident triage exercise for training medical personnel. Journal of Education and Teaching in Emergency Medicine 5 (4). Available at https://doi.org/10.5070/m554050366.

Casilla-Lennon, M.M., Hittelman, A.B., Jose, J.M., 2020. New robotic systems. Minimally Invasive and Robotic-Assisted Surgery in Pediatric Urology. Springer International Publishing, United States, pp. 405–417. Available at https://link.springer.com/book/10.1007/978-3-030-57219-8.

Center LBJS, 2013. Human Research Program Integrated Research Plan.

Chen, R.J., Wang, J.J., Williamson, D.F.K., Chen, T.Y., Lipkova, J., Lu, M.Y., et al., 2023. Algorithmic fairness in artificial intelligence for medicine and healthcare. Nature Biomedical Engineering 7 (6), 719–742. Available at https://doi.org/10.1038/s41551-023-01056-8, https://www.nature.com/natbiomedeng/.

Chung, R.S., 2005. How much time do surgical residents need to learn operative surgery? The American Journal of Surgery 190 (3), 351–353. Available at https://doi.org/10.1016/j.amjsurg.2005.06.035.

Custers, E.J.F.M., 2010. Long-term retention of basic science knowledge: a review study. Advances in Health Sciences Education 15 (1), 109–128. Available at https://doi.org/10.1007/s10459-008-9101-y.

Defense Space Strategy Summary, June 2020.

DiGiacomo, J.C., Angus, L.D.G., 2017. Thoracotomy in the emergency department for resuscitation of the mortally injured. Chinese Journal of Traumatology 20 (3), 141–146. Available at https://doi.org/10.1016/j.cjtee.2017.03.001.

Discussion Paper: Scope of Practice for Nurse Practitioners, 2022. Available at https://storage.aanp.org/www/documents/advocacy/position-papers/Scope-of-Practice.pdf?_gl=1*1aura0q*_gcl_au*MTM5NDExMjI0Ny4xNzM1NjU1MzE4.

DuBois, A.B., Nemir, P., Schumacher, C.F., Hubbard, J.P., 1969. Graduate medical education in basic sciences. Academic Medicine 44 (11), 1035–1043. Available at https://doi.org/10.1097/00001888-196911000-00006.

Dulchavsky, S.A., Garcia, K., Hamilton, D.R., Melton, S., Sargsyan, A.E., 2012. Advanced ultrasound for the space program and on earth. International Space Station Benefits for Humanity 27–31.

Duncan, J.M., Bogomolov, V.V., Castrucci, F., Koike, Y., Comtois, J.M., Sargsyan, A.E., 2008. Organization and management of the International Space Station (ISS) multilateral medical operations. Acta Astronautica 63 (7-10), 1137–1147. Available at https://doi.org/10.1016/j.actaastro.2007.12.001.

Elias, A., Weber, T., Green, D.A., Harris, K.M., Laws, J.M., Greaves, D.K., et al., 2024. Systematic review of the use of ultrasound for venous assessment and venous thrombosis screening in spaceflight. Microgravity 10 (1). Available at https://doi.org/10.1038/s41526-024-00356-w, https://www.nature.com/npjmgrav/.

Epstein, D., 2021. Range: Why Generalists Triumph in a Specialized World. Penguin.

Evans, H., Snead, D., 2024. Why do errors arise in artificial intelligence diagnostic tools in histopathology and how can we minimize them? Histopathology 84 (2), 279–287. Available at https://doi.org/10.1111/his.15071.

Exploration Medical Capability Medical Scenarios (National Aeronautics and Space Administration), October 2021.

Emergency Medical Procedures Manual for the International Space Station (ISS) [partial], 2016 (National Aeronautics and Space Administration), March 14, 2016.

Falcon User's Guide- Version 8, March 2025. Available at https://www.spacex.com/assets/media/falcon-users-guide-2025-05-09.pdf?utm_source=chatgpt.com.

Foale, C.M., Kaleri, A.Y., Sargsyan, A.E., Hamilton, D.R., Melton, S., Martin, D., et al., 2005. Diagnostic instrumentation aboard ISS: just-in-time training for non-physician crewmembers. Aviation, Space, and Environmental Medicine 76 (6), 594–598.

Gick, M.L., Holyoak, K.J., 1980. Analogical problem solving. Cognitive Psychology 12 (3), 306–355. Available at https://doi.org/10.1016/0010-0285(80)90013-4.

Grantcharov, T.P., Reznick, R.K., 2008. Teaching procedural skills. BMJ 336 (7653), 1129–1131. Available at https://doi.org/10.1136/bmj.39517.686956.47.

Human Space Flight, no date. Federal Aviation Administration. Available at https://www.faa.gov/space/human_spaceflight.

Izard, S.G., Juanes, J.A., García Peñalvo, F.J., Estella, J.M.G., Ledesma, M.J.S., Ruisoto, P., 2018. Virtual reality as an educational and training tool for medicine. Journal of Medical Systems 42 (3). Available at https://doi.org/10.1007/s10916-018-0900-2, https://link.springer.com/journal/10916.

Jamšek, M., Kunavar, T., Blohm, G., Nozaki, D., Papaxanthis, C., White, O., et al., 2021. Effects of simulated microgravity and hypergravity conditions on arm movements in normogravity. Frontiers in Neural Circuits 15. Available at https://doi.org/10.3389/fncir.2021.750176.

Jay, G.D., Lee, P., Goldsmith, H., Battat, J., Maurer, J., Suner, S., 2003. CPR effectiveness in microgravity: comparison of three positions and a mechanical device. Aviation, Space, and Environmental Medicine 74 (11), 1183–1189.

Jimenez, R., Yurk, D., Dell, S., Rutledge, A.C., Fu, M.K., Dempsey, W.P., et al., 2024. Resonance sonomanometry for noninvasive, continuous monitoring of blood pressure. PNAS Nexus 3 (7). Available at https://doi.org/10.1093/pnasnexus/pgae252.

Kerner, T., Schmidbauer, W., Tietz, M., Marung, H., Genzwuerker, H.V., 2017. Use of checklists improves the quality and safety of prehospital emergency care. European Journal of Emergency Medicine 24 (2), 114–119. Available at https://doi.org/10.1097/MEJ.0000000000000315, http://journals.lww.com/euro-emergencymed/pages/default.aspx.

Kiberd, M.B., Brownbridge, R., Mackin, M., Werry, D., Bird, S., Barry, G., et al., 2024. Feasibility of ultrasound-guided nerve blocks in simulated microgravity: a proof-of-concept study for regional anaesthesia during deep space missions. British Journal of Anaesthesia 133 (6), 1276–1283. Available at https://doi.org/10.1016/j.bja.2024.07.034, https://www.sciencedirect.com/science/journal/00070912.

Kluge, A., Frank, B., 2014. Counteracting skill decay: four refresher interventions and their effect on skill and knowledge retention in a simulated process control task. Ergonomics 57 (2), 175–190. Available at https://doi.org/10.1080/00140139.2013.869357.

Komasawa, N., Yokohira, M., 2023. Simulation-based education in the artificial intelligence era. Cureus 15 (6). Available at https://doi.org/10.7759/cureus.40940.

Komorowski, M., Fleming, S., Kirkpatrick, A.W., 2016. Fundamentals of anesthesiology for spaceflight. Journal of Cardiothoracic and Vascular Anesthesia 30 (3), 781–790. Available at https://doi.org/10.1053/j.jvca.2016.01.007, http://www.journals.elsevier.com/journal-of-cardiothoracic-and-vascular-anesthesia/.

Krihak, M., 2023. Exploration Medical Capability Clinical Decision Support Use Cases for CDSS Test Bed.

Krihak, M., Middour, C., Reyes, D., Nusbaum, D., Antonsen, E., 2019. Communication Bandwidth Considerations for Exploration Medical Care During Space Missions.

Krittanawong, C., Singh, N.K., Scheuring, R.A., Urquieta, E., Bershad, E.M., Macaulay, T.R., et al., 2023. Human health during space travel: state-of-the-art review. Cells 12 (1). Available at https://doi.org/10.3390/cells12010040, https://www.mdpi.com/journal/cells.

Kuypers, M.I., 2013. Emergency and wilderness medicine training for physician astronauts on exploration class missions. Wilderness and Environmental Medicine 24 (4), 445–449. Available at https://doi.org/10.1016/j.wem.2013.03.024, http://www.sciencedirect.com/science/journal/10806032.

Lacinski, R.A., Steller, J.G., Anderson, A., Nelson, A.M., 2024. Harnessing Artificial Intelligence for Medical Diagnosis and Treatment During Space Exploration Missions.

Landauer, T.K., Ainslie, K.I., 1975. Exams and use as preservatives of course-acquired knowledge. Journal of Educational Research 69 (3), 99–104. Available at https://doi.org/10.1080/00220671.1975.10884842.

Landon, L.B., Slack, K.J., Barrett, J.D., 2018. Teamwork and collaboration in long-duration space missions: going to extremes. American Psychologist 73 (4), 563–575. Available at https://doi.org/10.1037/amp0000260, http://www.apa.org/pubs/journals/amp/.

Lelevé, A., McDaniel, T., Rossa, C., 2020. Haptic training simulation. Frontiers in Virtual Reality 1. Available at https://doi.org/10.3389/frvir.2020.00003.

Levin, D., Nelson, N.G., McIntyre, L., Anderson, A., Steller, J.G., Hilmers, D.C., 2023. A Pilot Project for Quantifying the Effect of Medical Provider Knowledge, Skills, and Abilities on Outcomes for Spaceflight Using a Probabilistic Risk Assessment Tool. 20230011811.

Levin, D.R., Siu, M., Kramer, K., Kelly, E., Alouidor, R., Fernandez, G., et al., 2022. Time cost of provider skill: a pilot study of medical officer occupied time by knowledge, skill, and ability level. Aerospace Medicine and Human Performance 93 (11), 816–821. Available at https://doi.org/10.3357/AMHP.6093.2022, http://ingentaconnect.com/content/asma/amhp;jsessionid=13tlejf2f0ojk.alice.

Levin, D.R., Steller, J., Anderson, A., Lemery, J., Easter, B., Hilmers, D.C., et al., 2023. Enabling human space exploration missions through progressively Earth Independent Medical Operations (EIMO). IEEE Open Journal of Engineering in Medicine and Biology 4, 162–167. Available at https://doi.org/10.1109/OJEMB.2023.3255513, https://ieeexplore.ieee.org/xpl/RecentIssue.jsp?punumber=8782705&_ga=2.23991482.1496039865.1593798416-2073569860.1593798416.

Levin, D.R., Nelson, A.M., Zahner, C., Stratton, E.R., Anderson, A., Steller, J., 2024. A method to determine capabilities and resources for spacecraft medical systems. Aerospace Medicine and Human Performance 95 (7), 403–408. Available at https://doi.org/10.3357/AMHP.6365.2024, http://ingentaconnect.com/content/asma/amhp;jsessionid=13tlejf2f0ojk.alice.

Levin, D.R., McIntyre, L., Steller, J.G., Nelson, A., Zahner, C., Anderson, A., et al., 2025. Predicting exploration crew medical officer training needs: applying evidence-based predictive analytics to space medicine training. Wilderness & Environmental Medicine 36 (1_suppl). Available at https://doi.org/10.1177/10806032241292535.

Martin, A., Sullivan, P., Beaudry, C., Kuyumjian, R., Comtois, J.M., 2012. Space medicine innovation and telehealth concept implementation for medical care during exploration-class missions. Acta Astronautica 81 (1), 30–33. Available at https://doi.org/10.1016/j.actaastro.2012.06.021, http://www.journals.elsevier.com/acta-astronautica.

McGaghie, W.C., Issenberg, S.B., Cohen, E.R., Barsuk, J.H., Wayne, D.B., 2011. Does simulation-based medical education with deliberate practice yield better results than traditional clinical education? A meta-analytic comparative review of the evidence. Academic Medicine 86 (6), 706–711. Available at https://doi.org/10.1097/ACM.0b013e318217e119, http://www.lww.com/product/?1040-2446.

McHenry, N., Hunt, T., Young, W., Gardner, A., Bontz, B., Bhagavatula, U., et al., 2020. Evaluation of Pre-flight and on Orbit Training Methods Utilizing Virtual Reality. AIAA Scitech 2020 Forum 1 American Institute of Aeronautics and Astronautics Inc, AIAA United States 10.2514/6.2020-0168.

McIntyre, L., Myers, J.G., Leinweber, L., et al., 2022. A Model Based Approach to Estimating Human Spaceflight Medical Risk.

Melton, S., Simmons, S.C., Smith, B.A., Hamilton, D.R., 2020. Telemedicine. Principles of Clinical Medicine for Space Flight. Springer, New York, United States, pp. 253–271. Available at http://doi.org/10.1007/978-1-4939-9889-0.

Melton, S., Beck, G., Hamilton, D., Chun, R., Sargsyan, A., Kirkpatrick, A., 2001. How to test a medical technology for space: trauma sonography in microgravity. McGill Journal of Medicine 6 (1). Available at https://doi.org/10.26443/mjm.v6i1.554.

Mergen, M., Graf, N., Meyerheim, M., 2024. Reviewing the current state of virtual reality integration in medical education - a scoping review. BMC Medical Education 24 (1). Available at https://doi.org/10.1186/s12909-024-05777-5.

Mulavara, A.P., Feiveson, A.H., Fiedler, J., Cohen, H., Peters, B.T., Miller, C., et al., 2010. Locomotor function after long-duration space flight: effects and motor learning during recovery. Experimental Brain Research 202 (3), 649–659. Available at https://doi.org/10.1007/s00221-010-2171-0.

Musso, G., 2018. Habitability Issues in Long Duration Space Missions. Advances in Human Aspects of Transportation.

Naveh-Benjamin, M., 1990. The acquisition and retention of knowledge: exploring mutual benefits to memory research and the educational setting. Applied Cognitive Psychology 4 (4), 295–320. Available at https://doi.org/10.1002/acp.2350040406.

Nishisaki, A., Scrattish, L., Boulet, J., et al., 2008. Effect of recent refresher training on in situ simulated pediatric tracheal intubation psychomotor skill performance. Advances in Patient Safety: New Directions and Alternative Approaches Vol 3 Performance and Tools.

Noe, R.A., Dachner, A., Saxton, B., Keeton, K.E., EASI, N., 2011. Team training for long-duration missions in isolated and confined environments: a literature review, an operational assessment, and recommendations for practice and research. Research Networkk 44.

Nowadly, C.D., Trapp, B.D., Robinson, S.K., Richards, J.R., 2019. Resuscitation and evacuation from low earth orbit: a systematic review. Prehospital and Disaster Medicine 34 (5), 521–531. Available at https://doi.org/10.1017/S1049023X19004734, http://journals.cambridge.org/action/displayJournal?jid=PDM.

O'Tuathaigh, C.M.P., Idris, A.N., Duggan, E., Costa, P., Costa, M.J., Arrieta, A., 2019. Medical students' empathy and attitudes towards professionalism: relationship with personality, specialty preference and medical programme. PLoS One 14 (5), e0215675. Available at https://doi.org/10.1371/journal.pone.0215675.

Pantalone, D., 2023. Surgery in the next space missions. Life 13 (7), 1477. Available at https://doi.org/10.3390/life13071477.

Pashler, H., Bain, P.M., Bottge, B.A., et al., 2007. Organizing instruction and study to improve student learning. IES Practice Guide. NCER 2007-2004. National Center for Education Research.

Patrício, M.F., Julião, M., Fareleira, F., Carneiro, A.V., 2013. Is the OSCE a feasible tool to assess competencies in undergraduate medical education? Medical Teacher 35 (6), 503–514. Available at https://doi.org/10.3109/0142159X.2013.774330.

Pena, G.P., de Souza Andrade-Filho, J., 2010. Analogies in medicine: valuable for learning, reasoning, remembering and naming. Advances in Health Sciences Education 15 (4), 609–619. Available at https://doi.org/10.1007/s10459-008-9126-2.

Perez, R.S., Skinner, A., Weyhrauch, P., Niehaus, J., Lathan, C., Schwaitzberg, S.D., et al., 2013. Prevention of surgical skill decay. Military Medicine 178 (10), 76–86. Available at https://doi.org/10.7205/milmed-d-13-00216.

Phelps, S.E., Martin, K., Costes, S.V., et al., 2021. TOPICAL: Development of Medical Capabilities and Technologies for Health Monitoring, Diagnostics, and Treatment during Human Exploration Spaceflight.

Phelps, S.E., Easter, B., Lehnhardt, K., 2022. Developing the foundations of an exploration class medical system: bridging the capability gap between LEO and Mars. AIAA Science and Technology Forum and Exposition, AIAA SciTech Forum 2022. Available at https://doi.org/10.2514/6.2022-0580.

Pieters, M.A., Zaal, P.M.T., 2019. Training for long-duration space missions: a literature review into skill retention and generalizability. IFAC-PapersOnLine 52 (19), 247–252. Available at https://doi.org/10.1016/j.ifacol.2019.12.099, http://www.journals.elsevier.com/ifac-papersonline/.

Rajagopal, A.B., Slader, M.J., Boysen-Osborn, M., 2020. Build your own eye: a method for teaching ocular anatomy and pathophysiology. Journal of Education and Teaching in Emergency Medicine 5 (3). Available at https://doi.org/10.5070/m553049156.

Rando, K., Castelli, J., Pratt, et al., 2014. Ultrasound-guided internal jugular vein catheterization: a randomized controlled trial. Heart, Lung and Vessels 6 (1).

Rangarajan, K., Davis, H., Pucher, P.H., 2020. Systematic review of virtual haptics in surgical simulation: a valid educational tool? Journal of Surgical Education 77 (2), 337–347. Available at https://doi.org/10.1016/j.jsurg.2019.09.006, http://www.elsevier.com.

Rector, T., Cripe, C., Casler, J., 2021. Training and Learning for Long Duration Spaceflight. Springer Science and Business Media LLC, pp. 564–571. Available at http://doi.org/10.1007/978-3-030-90179-0_73 1499.

Rehnberg, L., Ashcroft, A., Baers, J.H., Campos, F., Cardoso, R.B., Velho, R., et al., 2014. Three methods of manual external chest compressions during microgravity simulation. Aviation, Space, and Environmental Medicine 85 (7), 687–693. Available at https://doi.org/10.3357/ASEM.3854.2014, http://docserver.ingentaconnect.com/deliver/connect/asma/00956562/v85n7/s2.pdf.

Robinson, J., Myers, J., Goodenow, D., et al., 2018. Introduction to Medical Extensible Dynamic Probabilistic Risk Assessment Tool (MEDPRAT).

Russell, B.K., Burian, B.K., Hilmers, D.C., Beard, B.L., Martin, K., Pletcher, D.L., et al., 2023. The value of a spaceflight clinical decision support system for earth-independent medical operations. NPJ Microgravity 9 (1). Available at https://doi.org/10.1038/s41526-023-00284-1, https://www.nature.com/npjmgrav/.

Sargsyan, A.E., Hamilton, D.R., Melton, S.L., Young, J., 2006. The International Space Station ultrasound imaging capability overview for prospective users. NASA/TP. 2006-213731.

Schellman, R., Hampton, M., Rhodes, M., Chapman, M., Thames, J., Nguyen, S., et al., 2025. Bridging an education gap: ultrasound-guided peripheral IVs. Journal of PeriAnesthesia Nursing 40 (1), 18–21. Available at https://doi.org/10.1016/j.jopan.2024.03.018.

Schmidt, G.R., Landis, G.A., Oleson, S.R., Borowski, S.K., Krasowski, M.J., 2010. HERRO: ascience-oriented strategy for crewed missions beyond LEO. 48th AIAA Aerospace Sciences Meeting Including the New Horizons Forum and Aerospace Exposition. American Institute of Aeronautics and Astronautics Inc, United States 10.2514/6.2010-629.

Schweinsberg, S., Fennell, D., 2024. Space tourism: a historical and existential perspective. Sustainability 16 (1), 79. Available at https://doi.org/10.3390/su16010079.

Scott, R.T., Sanders, L.M., Antonsen, E.L., Hastings, J.J.A., Park, S.M., Mackintosh, G., et al., 2023. Biomonitoring and precision health in deep space supported by artificial intelligence. Nature Machine Intelligence 5 (3), 196–207. Available at https://doi.org/10.1038/s42256-023-00617-5, https://www.nature.com/natmachintell/.

Sinnott, C.B., Liu, J., Matera, C., Halow, S., Jones, A.E., Moroz, M., et al., 2019. Underwater virtual reality system for neutral buoyancy training: development and evaluation. In: Proceedings of the ACM Symposium on Virtual Reality Software and Technology, VRST. 9781450370011 Association for Computing Machinery United States. https://doi.org/10.1145/3359996.3364272.

Smith, M.G., Kelley, M., Basner, M., 2020. A brief history of spaceflight from 1961 to 2020: an analysis of missions and astronaut demographics. Acta Astronautica 175, 290–299. Available at https://doi.org/10.1016/j.actaastro.2020.06.004.

Spooner, B.B., Fallaha, J.F., Kocierz, L., Smith, C.M., Smith, S.C.L., Perkins, G.D., 2007. An evaluation of objective feedback in basic life support (BLS) training. Resuscitation 73 (3), 417–424. Available at https://doi.org/10.1016/j.resuscitation.2006.10.017.

Swain, P., Swain, A., Santos, F., Hughes, L., Lindsay, K., Lomax, I., et al., 2025. Hypogravity simulation using the variable gravity suspension system: a technical report. Experimental Physiology. Available at https://doi.org/10.1113/ep092172.

Tenajas, R., Miraut, D., Illana, C.I., Alonso-Gonzalez, R., Arias-Valcayo, F., Herraiz, J.L., 2023. Recent advances in artificial intelligence-assisted ultrasound scanning. Applied Sciences 13 (6), 3693. Available at https://doi.org/10.3390/app13063693.

Thierry, S., Jaulin, F., Starck, C., Ariès, P., Schmitz, J., Kerkhoff, S., et al., 2023. Evaluation of free-floating tracheal intubation in weightlessness via ice-pick position with a direct laryngoscopy and classic approach with indirect videolaryngoscopy. NPJ Microgravity 9 (1). Available at https://doi.org/10.1038/s41526-023-00314-y.

Tokadlı, G., Dorneich, M.C., 2017. Development of a functionality matrix for a cognitive assistant on long distance space missions. Proceedings of the Human Factors and Ergonomics Society Annual Meeting 61 (1), 247–251. Available at https://doi.org/10.1177/1541931213601545.

Tokuç, B., Varol, G., 2023. Medical education in the era of advancing technology. Balkan Medical Journal 40 (6), 395–399. Available at https://doi.org/10.4274/balkanmedj.galenos.2023.2023-7-79.

Viderman, D., Dossov, M., Seitenov, S., Lee, M.H., 2022. Artificial intelligence in ultrasound-guided regional anesthesia: a scoping review. Frontiers in Medicine 9. Available at https://doi.org/10.3389/fmed.2022.994805, http://journal.frontiersin.org/journal/medicine.

Wang, L.W., Liu, C.H., Li, W.Y., Chiang, W.C., Chiu, Y.L., Ma, M.H.M., et al., 2025. Is ultrasound training sustainable? A systematic review of competency retention in healthcare trainees. Medical Education. Available at https://doi.org/10.1111/medu.15751, http://onlinelibrary.wiley.com/journal/10.1111/(ISSN)1365-2923.

Warnecke, T., Tochtermann, F., Kerkhoff, S., Komorowski, M., Neuhaus, C., Hinkelbein, J., 2019. Airway management in microgravity: a systematic review. Acta Anaesthesiologica Scandinavica 63 (1), 2–7. Available at https://doi.org/10.1111/aas.13251.

Wright, C.F., Wei, Y., Higgins, J.P., Sagoo, G.S., 2012. Non-invasive prenatal diagnostic test accuracy for fetal sex using cell-free DNA a review and meta-analysis. BMC Research Notes 5. Available at https://doi.org/10.1186/1756-0500-5-476.

Yang, A., Goel, H., Bryan, M., Robertson, R., Lim, J., Islam, S., et al., 2014. The picmonic® learning system: enhancing memory retention of medical sciences, using an audiovisual mnemonic web-based learning platform. Advances in Medical Education and Practice 5, 125–132. Available at https://doi.org/10.2147/AMEP.S61875, https://www.dovepress.com/getfile.php?fileID=19939.

Yip, M., Salcudean, S., Goldberg, K., Althoefer, K., Menciassi, A., Opfermann, J.D., et al., 2023. Artificial intelligence meets medical robotics. Science 381 (6654), 141–146. Available at https://doi.org/10.1126/science.adj3312.

Yule, S., Robertson, J.M., Mormann, B., Smink, D.S., Lipsitz, S., Abahuje, E., et al., 2023. Crew autonomy during simulated medical event management on long duration space exploration missions. Human Factors 65 (6), 1221–1234. Available at https://doi.org/10.1177/00187208211067575, https://journals.sagepub.com/home/HFS.

Chapter 26

# Procedural Guidance*

Andrew W. Kirkpatrick[1,2,3], Jessica L. McKee[4], Smith L. Johnston, III[5,6], and Douglas R. Hamilton[7,8]

[1]*TeleMentored Ultrasound Supported Medical Interventions (TMUSMI) Research Group, Calgary, AB, Canada,* [2]*Professor of Critical Care Medicine, Departments of Critical Care Medicine and Surgery Foothills Medical Centre, Calgary, AB, Canada,* [3]*Surgery Departments of Critical Care Medicine and Surgery Foothills Medical Centre, Calgary, AB, Canada,* [4]*Clinical Director and Project Manager Consultant Zoll Medical Corporation TeleMentored Ultrasound Supported Medical Interventions (TMUSMI) Research Group Edmonton, AB, Canada,* [5]*Retired NASA JSC Medical Officer FMR CMO AXIOM Space, Inc, Houston, TX, United States,* [6]*Professor of Aerospace Medicine at the School of Medicine Central Florida University Orlando, FL Director of Health Innovation Solamed Solutions, LLC Woodstock, GA, United States,* [7]*Clinical Associate Professor Medicine Adjunct Professor Electrical Engineering, University of Calgary, Calgary, AB, Canada,* [8]*General Internal Medicine, University of Calgary Foothills Medical Center - Health Sciences Center Calgary, AB, Canada*

## 26.1 Procedural medicine

### 26.1.1 Background

A procedure is a series of steps for doing something, and a medical procedure can be defined as a course of action intended to achieve a result in the delivery of healthcare. Medical procedures can be subdivided into subcategories, such as diagnostic or therapeutic, invasive and noninvasive, major and minor, or basic versus advanced life support levels. The critical basis of procedural medicine for spaceflight is that a task needs to be accomplished and often is done to enhance or restore the probability of mission success. Knowledge or stored information is useless unless it can be acted on to achieve the relevant task (Kirkpatrick et al., 1997; Campbell, 1999). Life-saving procedures or interventions (LSIs) include advanced airway management, hemorrhage control, and vascular and body cavity access with resuscitation techniques (Gerhardt et al., 2011). These are often not well-defined (Cameron and Andrews, 2022) and are closely linked to the patient's clinical needs and the logistical constraints of the environment. In the extreme austere domain of space medicine, any procedure that ultimately saves a life or prevents a mission abort can be considered life or mission-saving.

## 26.2 The challenges of procedural medicine in space

As is well detailed throughout this book, although space itself is hostile to living species, with extremes of cold, radiation, and lack of oxygen, the lack of gravity is a minor impediment to conducting procedures as long as it has been planned for. To date, many procedural studies have been conducted in microgravity, in which weightlessness is provided by balancing centripetal versus centrifugal forces in low Earth orbit (LEO) during 'freefall' conditions, or in parabolic flight (Kirkpatrick et al., 2009a, 2009b). Prior to such studies being conducted, it was predicted from early Russian work and low-fidelity analog models, such as neutral water buoyancy, that there would be inherent difficulties with human proprioception, organ behavior, and fluid management (Satava, 1988; Yaroshenko et al., 1967) in microgravity. Fortunately, subsequent research and experience have provided reassurance that if logistics are planned for, and the operators, subjects, and necessary supplies are restrained, complex procedures can be completed in weightlessness with high success rates, as long as experienced operators perform the procedures (Campbell et al., 2005; Kirkpatrick et al., 2003; Campbell et al., 1993a; Campbell, 2002; Kirkpatrick, 2019). A remaining challenge is what to do when experienced operators are NOT available. The planners for Space Station Freedom (a precursor concept design for the International Space Station from the 1980s) faced this challenge, as they envisioned having surgical capabilities compatible with a regional trauma hospital (Houtchens and Tonnesen, 1987; Houtchens, 1983). In terms of anticipated medical events during human space exploration, injury and toxic exposure have been ranked at the highest levels of probable incidence

* Co-author: Levine.

Procedural Medicine for Spaceflight. DOI: https://doi.org/10.1016/B978-0-443-27552-4.00013-0

multiplied by impact for risk to mission and health (Billica et al., 1996). The NASA Human Research Roadmap lists "red" unmitigated risks for lunar surface and Mars missions, including injury during extra-vehicular activity and bone fractures (Human Research Roadmap, 2025). Likewise, acute surgical conditions may arise in even the healthiest, well-screened individuals (Ball et al., 2012; Lazow et al., 2024; Pantalone et al., 2021). Then the International Space Station (ISS) designs were finalized 10 years later hese plans were dramatically downscaled. This was part of an engineering trade analysis that included, among other considerations, an evience based risk analysis showing a vanishingly small risk of surgical pathology requiring on orbit intervention. The use of modern medical probabalistic risk analysis tools such as NASA's Integrated Medical Model, and our experience on orbit to date confirm these early assessments Walton 2020.

Today, crews on board the ISS rely on medical support from ground based experts in space medicine with two on board Crew Medical Officers (CMO). The CMOs are not required to be a physician (though they often are), and receive between 12 and 60 hours of medical training for each mission largely to familiarize them with the ISS medical system resources and checklists, and teach them how to communicate and work with ground based flight surgeons(Campbell, 2002; Roslee, 2021). NASA includes 2 CMOs in case one of them is the crewmember requiring medical care.

While a broadly trained surgical or critical care specialist might be the best choice to deliver complex interventions the majority of illnesses and injuries do not require complex procedural interventions. Particularly in healthy and pre-screening individuals such as government sponsored and near term future commercial astronauts. The success of the CMO model on ISS means there is no guarantee that a surgeon, or even a physician, will be selected for an exploration-class mission (Kirkpatrick et al., 2009c). The choice to include a physician will likely be informed by an evidence based analysis of the mission risks traded against the mission objectives using a process akin to the probabalistic risk analysis informed expert reviews employed by space agancies today (Walton 2020).

While the ISS was planed and built surgical capabilities, it *was* planned with state-of-the-art ultrasound capabilities. Since Space Shuttle era CMOs did not routinely receive comprehensive ultrasound training, this set the stage for creating the remote telementored ultrasound (RTMUS) paradigm, which enabled the entire remote telementored (RTM) procedural guidance that has become the backbone of space medicine and a valuable "spin-off" for terrestrial medicine.

## 26.3 Remote telementored ultrasound

Before the turn of the 21st century, it was unknown whether clinical ultrasound was feasible in space for many reasons, including for the frequent use case on Earth of imaging in gravity-dependent intraabdominal locations after traumatic injury (Scalea et al., 1999). However, sonographic studies with anesthetized swine during parabolic flights confirmed that ultrasound in space was indeed effective and, for some findings, might even be more sensitive than on Earth (Kirkpatrick et al., 2003; Melton et al., 2001; Kirkpatrick et al., 2007; Hamilton et al., 2004).

This is primarily due to the properties of surface tension, which becomes the primary physical force determining fluid behavior in the absence of gravity and enhances the fluid/solid and fluid/gas boundaries that ultrasound is well suited to detect (Kirkpatrick et al., 2009b, 2003; Campbell et al., 1993b). However, the logistical challenge of not having an ultrasonographer onboard the ISS remained a substantial problem. Drs. Ashot Sargsyan, Scott Dulchavsky, and Douglas Hamilton reasoned that, as ultrasound is traditionally provided on Earth as a two-step procedure, with images first being obtained by a technician and later assessed by a radiologist (Kirkpatrick et al., 2003), this paradigm could be radically, geographically, and operationally extended using informatics (Sargsyan et al., 2005). Thus was born the RTMUS paradigm (Hamilton et al., in press) as an extreme solution to many of the problems facing medical imaging in space medicine. The RTMUS program received a boost as a small silver lining to the tragic Space Shuttle Columbia Disaster in 2003. This catastrophic transport loss to and from the ISS meant that most of the previously planned research was canceled, leaving the evolving RTMUS program to receive the most on-orbit hours of any research before or since. This research program confirmed that point-of-care ultrasound can be helpful in a myriad of situations when necessary, and that motivated but untrained caregivers can be very accurate when guided by remote experts. The RTMUS paradigm examined the potential of out-of-the-box ultrasound to examine numerous novel indications onboard the ISS (Sargsyan et al., 2005; Fincke et al., 2005; Chiao et al., 2005).

The RTMUS paradigm developed for space medicine also introduced the concept of the patient performing their own exams, a paradigm called Remote Telementored Self-Performed Ultrasound (RTMSPUS; Sargsyan et al., 2005). Its terrestrial utility became greatly appreciated during the recent COVID pandemic. While astronauts routinely performed ultrasound on their co-astronauts while being mentored, there were also instances of astronauts being mentored from Earth to ultrasound themselves, thus initiating the RTMSPUS concept (Sargsyan et al., 2005). This can now be recognized as a dramatic example of personalized medicine, where patients can be empowered to participate and contribute to their health maintenance and well-being (Kirkpatrick et al., 2022a, 2022b). These are all concepts that are less developed for terrestrial care but have immense potential for the future (Kirkpatrick and McKee, 2022).

## 26.4 Remote telementored ultrasound as a terrestrial space medicine spin-off

As the ability for RTMUS to support the health of a small number of astronauts became apparent, it was intuitive to attempt to translate this paradigm from the extreme to mainstream (Kirkpatrick, 2019). An example of such work was when in 2006, the Canadian Space Agency and Telesat Canada, sponosored a fixed teleultrasound link between the Emergency Department in the Rocky Mountains in Banff, AB, Canada, and an Emergency Department Tele-ultrasound Console at the Foothills Hospital in Calgary, AB, Canada (Hamilton et al., in press; Dyer et al., 2008; Al-Kadi et al., 2009). This essentially replicated the ISS paradigm, where a less experienced or even inexperienced care provider was mentored in real-time by a remote ultrasound expert in Calgary. This demonstrated that remote telementoring could greatly benefit patients in a real-world emergency clinical setting, such as making the first remotely mentored diagnosis of a pneumothorax, identifying prehospital massive intraabdominal hemorrhage, and guiding a direct-to-OR resuscitation (Dyer et al., 2008). However, the paradigm required a surgeon to be physically present to respond to a console, causing delays. This situation was logistically unsustainable in the real clinical practice, and ultimately became a barrier to the adoption of real-time RTMUS into mainstream practice.

## 26.5 Hand-held technologies and remote telementored ultrasound

A solution may be as simple as refocusing on hand-held technologies that most people carry with them (McBeth et al., 2013). The analogy for space medicine is to simply, but thoroughly, utilize the equipment already available onboard any spacecraft without increasing the logistical complexity. Technological advances in the 21st century have dramatically freed users from the delay of having to respond in person to a console, using hand-held smartphones or tablets to connect responders and caregivers (McBeth et al., 2010). Since 2010, investigators have described the utility of these advances when combined with naïve ultrasound users to complete various mentored ultrasound tasks. The most valuable lesson learned, regardless of the type of naïve participant, is that using simple language without medical jargon was invaluable to success (McBeth et al., 2013, 2011; Kirkpatrick et al., 2016). Using this paradigm, it has been noted that remote mentoring can occur in orbiting aircraft (Sargsyan et al., 2005), across countries and oceans (Biegler et al., 2013), and can involve multidisciplinary mentoring teams (Kirkpatrick et al., 2013). In 2011, it was envisioned that pregnant woman with limited access to specialist care could be screened remotely by obstetricians for prenatal risks, such as breech pregnancies, with relatively modest investments in infrastructure (McBeth et al., 2011). A contemporary systematic review of the literature over a decade later found that among 31 studies on this paradigm, most demonstrate a benefit for pregnant women.(Kariman et al., 2024).

## 26.6 Telementored ultrasound-supported medical interventions

Ultrasound is one type of medical information stream that may be communicated between on-site caregivers/patients and remote experts. Thus, with the efforts made to understand and practically refine RTMUS, it has become apparent that the future should embrace all digital medical technologies to provide a comprehensive assessment of patient health and physiology. In challenging environments, this will require both evaluation and, if necessary, remotely guided interventions to provide the best health outcomes. While ultrasound can make powerful, immediate point-of-care diagnoses in catastrophic situations (Gillman and Kirkpatrick, 2012), without being married to life-saving interventions, lives cannot be saved. Thus, to pursue this goal, the TeleMentored Ultrasound-Supported Medical Interventions (TMUSMI) Research Group at the University of Calgary (Kirkpatrick, 2019), as well as other investigators, have endeavored to explore the feasibility and logistics of remotely telementoring (RTM) less experienced care providers to perform guided medical interventions in simulated environments ranging from diagnostic to minimally invasive to open surgical techniques (Kirkpatrick et al., 2015a, 2015b; Netzer et al., 2019; Kirkpatrick et al., 2018). These procedures have included examples of tube thoracostomy, wound clamp application, tourniquet placement, and overall supervision of trauma resuscitations (Kirkpatrick et al., 2018; Chen et al., 2022; Kirkpatrick et al., 2022c, 2022d). In all these simulated environments, it was found that with good communication between experienced mentors and naive mentees, on-site caregivers could consistently perform beyond expectations in complex procedures.

## 26.7 Competing or complementary paradigms? Telerobotic versus telementored versus video modeled versus action recognition autonomous procedures for space

In February 2024, a miniaturized surgical robot onboard the ISS was telerobotically controlled from Nebraska, United States, to manipulate rubber bands on the ISS as a demonstration of telerobotic capabilities and the promise this holds

(Nicioli and Fisher, 2024). This experiment built upon simulations conducted in the underwater National Oceanographic and Atmospheric Administration "Aquarius" underwater habitat, with telerobotic surgery being used to remotely suture wound lacerations, among other emergency procedures (Doarn et al., 2009). It has been theorized by many that an autonomous surgical robot capable of performing procedures would be required to provide surgical capabilities during distant space exploration (Pantalone et al., 2021; Haidegger et al., 2011). However, that capability does not yet exist. Likewise, RTM and RTMUS, while the current backbone of space medicine onboard the ISS, will not be able to support exploration-class missions due to communications latency. A major challenge for telerobotic and telementored approaches to delivering procedural guidance during long-duration spaceflight is the fundamental physical limitation of the speed of light, where exploration beyond LEO will be associated with communication delays of up to approximately 45 minutes for bidirectional communication with a Mars mission (Kirkpatrick et al., 1997; Pantalone et al., 2021; Cheung et al., 2023).

This reality suggests that procedural telementoring for Mars exploration and beyond will have to utilize Earth-independent telementoring techniques as the only potential viable paradigm Levin et al., 2023 As a result, other novel approaches will be required to allow autonomous procedural telementoring in deep-space. Video modeling (VM) is a form of behavioral modeling that involves the demonstration of desired behaviors, outcomes, and attitudes through active visual representations (Kirkpatrick et al., 2009a).

Although VM is used every day by the non medical public, remarkably little has been reported in terms of exploring and quantifying the capabilities of VM in both space and terrestrial medical care.

It is, however, an active area of research.. For example, the initial Trauma TeleHelper for Operational Medical Procedure Support and Offline Network is intended to develop:

1. A voice-controlled mobile application with video references for procedural guidance,
2. Audio narration of each video using procedure mentoring scripts,
3. And an artificial intelligence (AI)-guided intervention system using AR overlay and voice command to create immersive video modeling (Birch et al., 2023).

While this technology is not yet deployable, it presents potential for autonomous and semi-autonimous guidance during deep-space exploration.

The authors have also conducted initial randomized trials exploring the relative attributes of RTM versus VM for LSIs such as tourniquet application, wound packing, tube thoracostomy, and wound clamping (Kirkpatrick et al., 2022c, 2022d). These early studies concluded that both techniques have positive attributes. RTM has excellent flexibility and allows troubleshooting by a mentor when a mentee potentially misunderstands or is about to commit an error. However, VM was often quicker and had the primary positive attribute of being autonomous. A follow-up concept advanced by the authors was rich immersive VM, in which AI could be combined with VM to create the ideal training video of the task that must be solved. In such a way, rather than having an infinite library of procedures and medical scenarios, AI could generate an individualized training video using known patient parameters. The unsolved challenges involve inputting the physiologic data to prime the machine-learning algorithms and following the progress through an LSI's diagnostic/procedural steps. However, Dr. Wachs of Purdue University addresses this challenge by linking action recognition software to life-saving procedures. The software is many times more accurate than random chance at correctly detecting and predicting the next steps of a simulated LSI, and with continued data input, it will further greatly improve in performance over time (Wachs et al., 2023; Liu and Liang, 2025). Thus, a future autonomous procedural guide for space medicine would ideally consist of guiding the responder in how to place diagnostic/physiologic monitors in order to allow the AI/machine-learning engine to assess a patient, diagnose life-threatening conditions, and compute physiologic thresholds. If intervention is required, it would then select an appropriate preexisting training video for the responder that outlines all the steps, and especially pitfalls, to avoid so as to complete the LSI. However, there is much still to be learned. Cheung et al. (2023) cautioned that space-induced physiological changes, limited size and breadth of space biomedical datasets, and changes in disease characteristics due to weightlessness may render models invalid when transferred from Earth to space.

## 26.8 Procedural guidance in spaceflight

The scope of procedures that may be envisioned is almost unlimited. Although medical support in spaceflight is inherently limited (Stepanek et al., 2019), the goal is to provide care as close to terrestrial standards as possible. Therefore, there are multiple indications for mentoring of medical procedures in space, including providing comprehensive health maintenance, performing early assessments of minor conditions, and responding to catastrophic medical conditions. This may include medical support during emergency egress as well as palliative care if treatment appears futile or if insufficient management resources are available. In 2025, nobody should die alone no matter how isolated or remote they are (Kirkpatrick and McKee, 2022).

The Health Maintenance Facility (HMF) onboard the ISS provides in-flight preventive, diagnostic, and therapeutic care for the majority of medical conditions expected to be encountered in flight. In a hypothetical situation where the CMO onboard the ISS was not familiar with a topic related to their own health or the health of any other crewmember, it would be logical to seek guidance from the terrestrial flight surgeons in Mission Control. Information can be exchanged using onboard communications, which include two-way voice and video communications. If an assessment is required, it can be guided from the ground using RTM. In this way, the scope of potential procedures is limited only by the equipment carried on the relevant spacecraft and the imagination of those using that equipment, even if the use is "off-label." The HMF is designed to manage the day to day medical needs of ISS astronauts and a limited set of on orbit emergecies. Anything beyond this scope will be evacuated beck to Earth, as was demonstrated with the medical evacuation of Astronaut Mike Fincke on January 15th, 2026 (Mogensen Roslee, 2021). However, even limited equipment can be repurposed in creative ways. An example previously mentioned was ultrasound, envisioned initially as a cardiovascular research tool for the ISS, but which, due to its applicability to almost all organ systems, has been referred to as the "prime diagnostic tool" for "planetary medical capability" (Roslee, 2021).

## 26.9 Supplies, equipment, and procedural operators

The logistical supplies in terms of medical equipment and medical consumables that will be utilized for procedural guidance of medical techniques will need to be determined before flight. Given the limitations inherent in spaceflight, there will always be a trade-off regarding the most likely scenarios to be treated and practical mass and volume constraints. Potentially, just-in-time 3D printing of medical supplies and the 3D biofabrication of tissue are examples of the technologies that might enable greater flexibility in providing in-flight supplies. These topics are more fully covered in other chapters (Van Ombergen et al., 2023; Majumder and Ghosh, 2023). As discussed, the scope of procedural telementoring includes any procedures known to terrestrial medicine that might be necessitated by conditions that could occur in the setting of spaceflight that can be managed with the equipment available. Therefore, the scope is vast, and there will be a need for robust real-time two-way voice, and typically one-way video communication. Almost any procedure can be telementored if needed, but the application of human communication principles is critical. Some suggested high-level principles of procedural telementoring are presented (Table 26.1), not as being already comprehensively validated in spaceflight, but as an initial set of principles garnered from two decades of experience.

## 26.10 Pitfalls of remote human procedural guidance: the humans

Effective RTM requires good communication as well as mutual trust and understanding between the mentor and the mentee. The stresses of spaceflight, isolation, and fatigue can challenge any astronaut. Conflicts between astronauts and

**TABLE 26.1 Principles of procedural guidance for spaceflight.**

| |
|---|
| Restraint and reliable physical relationship of the following:<br>– Patients or subjects<br>– Operators<br>– Supplies |
| Organized equipment to follow the procedural conduct of the procedure. |
| Trash and consumed supplies management. |
| Civility and the introduction of the mentee and mentor are appropriate to the situation's urgency. |
| Clarified roles of mentee and mentor. |
| A clarified understanding of prior training and expected competencies of all participating. |
| Clarified acceptance of questions and further explanations when required. |
| Simple language with clear breaks and delineation of steps within a more complex task. |
| Maximum use of any prerehearsal, standardized nomenclature, and phraseology, if possible. |
| Immediate "safe-words" for halting the procedure if unsafe or erroneous steps are observed. |

ground controllers have not been rare in space exploration (Palinkas, 2001; Nasr, 2020), with Apollo 7 nearly ending in a crew mutiny (Niler, 2018). NASA's Human Research Roadmap documents concerns regarding inadequate cooperation, coordination, communication, and psychosocial adaptation within a team as unmitigated risks for both lunar and Martian missions (Human Research Roadmap, 2025). Remote telemedical guidance and oversight from Earth is accepted as the default management strategy for an off-nominal medical event in LEO (Martin et al., 2012). While we are unaware of documented space conflicts arising astronauts being guided by remote experts, terrestrial examples are likely quite illustrative (Table 26.2). It has also been assumed that decision support technologies will aid less experienced caregivers and reduce their perceived stress, but these fundamental questions are poorly studied (Kirkpatrick et al., 2023).

Investigators have noted that when mentors, who may be technically brilliant, are not good communicators, the mentored outcomes are poor (McBeth et al., 2011). When mentors do not understand the operational restrictions of the environment in which mentees must operate, results are also poor (Chen et al., 2022). Where mentors do not understand the capabilities and are felt to slow down, hinder, or overcontrol mentees, the results are poor (Netzer et al., 2019; Kirkpatrick et al., 2020). If mentors are perceived as "them," rather than "us" as a team, communication and resultant clinical outcomes will be limited. It should be emphasized that this domain has been little researched and requires much study to objectify and formalize standards and requirements, both in space and on Earth. TMUSMI has conducted initial studies using virtual mentoring networks made feasible by standardized communication protocols that allow multidisciplinary mentoring with technical experts, translators, and prehospital nonphysicians all providing input (Kirkpatrick et al., 2018). This borrows from the traditional "mission control" paradigm, where any other expertise could be added modularly as needed (such as a psychologist) but only further research will clarify whether this is helpful. The most contemporary high-level summary of the North American work in mature EMS systems confirms that mentoring does not seem beneficial for those who are confident in their skills, but it is invaluable for those who lack experience, recent training in those skills, or both. Formal scoping reviews of the best, but still limited, data reveal that standardization coupled with short, concise, validated scripts might improve efficacy, safety, and engagement in telementored procedures (Hampton et al., 2020). This review further concluded that improvements in this domain would require multidisciplinary input, practice, and deliberate research efforts to address barriers in optimum care (Hampton et al., 2020).

## 26.11 The discipline of the remote medical mentor as a specialist

The authors are of the opinion that the concept of remote mentoring in order to provide procedural guidance, first developed for ultrasound but now broadened for any prehospital medical interaction, may eventually become one of the greatest gifts to humanity from space exploration (Kirkpatrick, 2019; Kirkpatrick et al., 2023, 2015c). If this paradigm is dramatically broadened to support global healthcare, many more experts will be required, thus legitimizing a new specialty in medicine, that of the "Remote Medical Mentor." Currently, the specialty of "remote medicine" does not exist. However, in the future, this should become a respected and necessary specialty that cuts across many other disciplines and becomes critical to human health. To this end, all relevant characteristics, including the technology, techniques, lexicon, administration, remuneration, medicolegal issues, and issues of scope, need to be refined. This current textbook represents an important step in initiating this process.

**TABLE 26.2** Human factors limitations of mentor/mentee interaction with procedural telementoring.

| |
|---|
| Poor communication |
| Poor understanding of the operational environment |
| Poor knowledge of the mentee's level of capabilities |
| Over-controlling |
| Excessive details |
| Slow mentoring |
| Lack of empathy with mentees |
| Lack of "team" approach |

## Competing interest

All authors declare that they have no competing interests.

## Authors' contribution

A.W.K. and D.H. conceived and drafted the manuscript; all authors critically revised the manuscript, contributed crucial scientific knowledge, and gave final approval.

## Acknowledgments

None.

## Conflict of interest

All authors declare no conflict of interest except for A.W.K., who has consulted for the Zoll, Acelity (3M/KCI), and Innovative Trauma Care Corporations. A.W.K. is the principal investigator of the COOL trial: Closed or Open after Laparotomy for Source Control Laparotomy in Severe Complicated Intra-Abdominal Sepsis (https://clinicaltrials.gov/ct2/show/NCT03163095).

## References

Al-Kadi, A., Dyer, D., Ball, C.G., McBeth, P.B., Hall, R., Lan, S., et al., 2009. User's perceptions of remote trauma telesonography. Journal of Telemedicine and Telecare 15 (5), 251–254.

Ball, C.G., Kirkpatrick, A.W., Williams, D.R., Jones, J.A., Polk, J.D., Vanderploeg, J.M., et al., 2012. Prophylactic surgery prior to extended-duration space flight: is the benefit worth the risk? Canadian Journal of Surgery. Journal Canadien de Chirurgie 55 (2), 125–131.

Biegler, N., McBeth, P.B., Tiruta, C., Hamilton, D.R., Xiao, Z., Crawford, I., et al., 2013. The feasibility of nurse practitioner-performed, telementored lung telesonography with remote physician guidance - 'a remote virtual mentor. Critical ultrasound journal 5 (1), 5.

Billica, R.D., Simmons, S.C., Mathes, K.L., McKinley, B.A., Chuang, C.C., Wear, M.L., et al., 1996. Perception of the medical risk of spaceflight. Aviation, Space, and Environmental Medicine 67, 467–473.

Birch, E., Couperus, K., Gorbatkin, C., Kirkpatrick, A.W., Wachs, J., Candelore, R., et al., 2023. Trauma THOMPSON: clinical decision support for the frontline medic. Military Medicine 188 (Suppl 6), 208–214.

Cameron, P., Andrews, T., 2022. Life saving interventions in resuscitation of the injured patient. Injury 53 (7), 2387–2388.

Campbell, M.R., 1999. Surgical care in space. Aviation, Space, and Environmental Medicine 70, 181–184.

Campbell, M.R., 2002. A review of surgical care in space. Journal of the American College of Surgeons 194, 802–812.

Campbell, M.R., Billica, R.D., Johnson, S.L., 1993a. Animal surgery in microgravity. Aviation, Space, and Environmental Medicine 64, 58–62.

Campbell, M.R., Billica, R.D., Johnston, S.L., 1993b. Surgical bleeding in microgravity. Surgery, Gynecology & Obstetrics 177, 121–125.

Campbell, M.R., Williams, D.R., Buckey, J.C., Kirkpatrick, A.W., 2005. Animal surgery during spaceflight on the Neurolab shuttle mission. Aviation, Space, and Environmental Medicine 76, 589–593.

Chen, J., Dobron, A., Esterson, A., Fuchs, L., Glassberg, E., Hoppenstein, D., et al., 2022. A randomized, controlled, blinded evaluation of augmenting point-of-care ultrasound and remote telementored ultrasound in inexperienced operators. The Israel Medical Association Journal: IMAJ 24 (9), 596–601.

Cheung, H.C., De Louche, C., Komorowski, M., 2023. Artificial intelligence applications in space medicine. Aerosp Med Hum Perform 94 (8), 610–622.

Chiao, L., Sharipov, S., Sargsyan, A.E., Melton, S., Hamilton, D.R., McFarlin, K., et al., 2005. Ocular examination for trauma; clinical ultrasound aboard the International Space Station. The Journal of Trauma 58 (5), 885–889.

Doarn, C.R., Anvari, M., Low, T., Broderick, T.J., 2009. Evaluation of teleoperated surgical robots in an enclosed undersea environment. Telemedicine Journal and e-Health: The Official Journal of the American Telemedicine Association 15 (4), 325–335.

Dyer, D., Cusden, J., Turner, C., Boyd, J., Hall, R., Lautner, D., et al., 2008. The clinical and technical evaluation of a remote telementored telesonography system during the acute resuscitation and transfer of the injured patient. The Journal of Trauma 65 (6), 1209–1216.

Fincke, E.M., Padalka, G., Lee, D., van Holsbeeck, M., Sargsyan, A.E., Hamilton, D.R., et al., 2005. Evaluation of shoulder integrity in space: first report of musculoskeletal US on the International Space Station. Radiology 234 (2), 319–322.

Gerhardt, R.T., Berry, J.A., Blackbourne, L.H., 2011. Analysis of life-saving interventions performed by out-of-hospital combat medical personnel. The Journal of Trauma 71 (1 Suppl), S109–S113.

Gillman, L.M., Kirkpatrick, A.W., 2012. Portable bedside ultrasound: the visual stethoscope of the 21st century. Scandinavian Journal of Trauma, Resuscitation and Emergency Medicine 20, 18.

Haidegger, T., Sandor, J., Benyo, Z., 2011. Surgery in space: the future of robotic telesurgery. Surgical Endoscopy 25 (3), 681–690.

Hamilton, D.R., Kohli, S.S., McBeth, P.B., Moore, R., W., J., Hamilton, K.J. et al., in press. Low Earth orbit communication satellites: a positively disruptive technology that could change the delivery of health care in rural and northern Canada. Journal of Medical Internet Research.

Hamilton, D.R., Sargsyan, A.E., Kirkpatrick, A.W., Nicolaou, S., Campbell, M., Dawson, D.L., et al., 2004. Sonographic detection of pneumothorax and hemothorax in microgravity. Aviation, Space, and Environmental Medicine 75 (3), 272–277.

Hampton, L., Brindley, P., Kirkpatrick, A., McKee, J., Regehr, J., Martin, D., et al., 2020. Strategies to improve communication in telementoring in acute care coordination: a scoping review. Canadian Journal of Surgery. Journal Canadien de Chirurgie 63 (6), E569–E577.

Houtchens, B., 1983. System for the Management of Trauma and Emergency Surgery in Space: Final Report. Houston, Texas.

Houtchens, B.A., Tonnesen, A.S., 1987. Minor Surgery and Anesthesia Capabilities for Space Station Health Maintenance Facility (HMF): Final Report. NASA Johnson Space Center, Houston, Texas 1987.

Human Research Roadmap: A Risk Reduction Strategy for Human Space Exploration: National Aeronautics and Space Administration; 2025. Available from: https://humanresearchroadmap.nasa.gov/.

Kariman, S.S., van den Heuvel, J.F.M., Adriaanse, B.M.E., Oepkes, D., Bekker, M.N., 2024. The potential of tele-ultrasound, handheld and self-operated ultrasound in pregnancy care: a systematic review. Prenatal Diagnosis.

Kirkpatrick, A.W., 2019. Point-of-care resuscitation research: from extreme to mainstream: Trauma Association of Canada Fraser Gurd Lecture 2019. Journal of Trauma and Acute Care Surgery 87 (3), 571–581.

Kirkpatrick, A.W., McKee, J.L., 2022. Empowering catastrophic far-forward self-care: nobody should die alone without trying. Journal of Military and Veterans' Health.

Kirkpatrick, A.W., Campbell, M.R., Novinkov, O.L., Goncharov, I.B., Kovachevich, I.V., 1997. Blunt trauma and operative care in microgravity: a review of microgravity physiology and surgical investigations with implications for critical care and operative treatment in space. Journal of the American College of Surgeons 184, 441–453.

Kirkpatrick, A.W., Hamilton, D.R., Nicolaou, S., Sargsyan, A.E., Campbell, M.R., Feiveson, A., et al., 2003. Focused assessment with sonography for trauma in weightlessness: a feasibility study. Journal of the American College of Surgeons 196 (6), 833–844.

Kirkpatrick, A.W., Jones, J.A., Sargsyan, A., Hamilton, D.R., Melton, S., Beck, G., et al., 2007. Trauma sonography for use in microgravity. Aviation, Space, and Environmental Medicine 78 (4 Suppl), A38–A42.

Kirkpatrick, A.W., Keaney, M., Hemmelgarn, B., Zhang, J., Ball, C.G., Groleau, M., et al., 2009a. Intra-abdominal pressure effects on porcine thoracic compliance in weightlessness: implications for physiologic tolerance of laparoscopic surgery in space. Critical Care Medicine.

Kirkpatrick, A.W., Keaney, M., Kmet, L., Ball, C.G., Campbell, M.R., Kindratsky, C., et al., 2009b. Intraperitoneal gas insufflation will be required for laparoscopic visualization in space: a comparison of laparoscopic techniques in weightlessness. Journal of the American College of Surgeons 209 (2), 233–241.

Kirkpatrick, A.W., Ball, C.G., Campbell, M., Williams, D.R., Parazynski, S.E., Mattox, K.L., et al., 2009c. Severe traumatic injury during long duration spaceflight: light years beyond ATLS. Journal of Trauma Management and Outcomes 3, 4.

Kirkpatrick, A.W., Blaivas, M., Sargsyan, A.E., McBeth, P.B., Patel, C., Xiao, Z., et al., 2013. Enabling the mission through trans-atlantic remote mentored musculoskeletal ultrasound: case report of a portable hand-carried tele-ultrasound system for medical relief missions. Telemedicine Journal and e-Health: The Official Journal of the American Telemedicine Association 19 (7), 530–534.

Kirkpatrick, A.W., McKee, J.L., McKee, I., Panebianco, N.L., Ball, C.G., 2015a. Remote telementored ultrasound-directed compression to potentially accelerate hemostasis in exsanguinating junctional vascular injuries. Journal of Special Operations Medicine 15 (4), 71–74.

Kirkpatrick, A.W., Tien, H., LaPorta, A.T., Lavell, K., Keillor, J., Wright Beatty, H.E., et al., 2015b. The marriage of surgical simulation and telementoring for damage-control surgical training of operational first responders: a pilot study. The Journal of Trauma and Acute Care Surgery 79 (5), 741–747.

Kirkpatrick, A.W., Hamilton, D., Beckett, A., LaPorta, A., Brien, S., Glassberg, E., et al., 2015c. The need for a robust 24/7 subspecialty "clearing house" response for telementored trauma care. Canadian Journal of Surgery. Journal Canadien de Chirurgie 58 (3), S85–S87.

Kirkpatrick, A.W., McKee, I., McKee, J.L., Ma, I., McBeth, P.B., Roberts, D.J., et al., 2016. Remote just-in-time telementored trauma ultrasound: a double-factorial randomized controlled trial examining fluid detection and remote knobology control through an ultrasound graphic user interface display. American Journal of Surgery 211 (5), 894–902 e1.

Kirkpatrick, A.W., McKee, J.L., Netzer, I., McBeth, P.B., D'Amours, S., Kock, V., et al., 2018. Transoceanic telementoring of tube thoracostomy insertion: a randomized controlled trial of telementored versus unmentored insertion of tube thoracostomy by military medical technicians. Telemedicine Journal and e-Health: The Official Journal of the American Telemedicine Association.

Kirkpatrick, A.W., McKee, J.L., Netzer, I., McKee, I.A., McBeth, P., Wachs, J.P., et al., 2020. A Randomized trial of mentored vs nonmentored military medics compared in the application of a wound clamp without prior training: when to shut up and just watch!. Military Medicine 185 (Suppl 1), 67–72.

Kirkpatrick, A.W., McKee, J.L., Ball, C.G., Ma, I.W.Y., Melniker, L.A., 2022a. Empowering the willing: the feasibility of tele-mentored self-performed pleural ultrasound assessment for the surveillance of lung health. Ultrasound J 14 (1), 2.

Kirkpatrick, A.W., McKee, J.L., Couperus, K., Colombo, C.J., 2022b. Patient self-performed point-of-care ultrasound: using communication technologies to empower patient self-care. Diagnostics (Basel) 12 (11).

Kirkpatrick, A.W., McKee, J.L., Tomlinson, C., Donley, N., Ball, C.G., Wachs, J., 2022c. A randomized controlled pilot trial of video-modelling versus telementoring for improved hemorrhage control wound packing. American Journal of Surgery.

Kirkpatrick, A.W., Tomlinson, C., Donley, N., McKee, J.L., Ball, C.G., Wachs, J.P., 2022d. To watch before or listen while doing? A randomized pilot of video-modelling versus telementored tube thoracostomy. Prehospital and Disaster Medicine: The Official Journal of the National Association of EMS Physicians and the World Association for Emergency and Disaster Medicine in Association With the Acute Care Foundation 37 (1), 71–77.

Kirkpatrick, A.W., McKee, J.L., Barrett, R., Couperus, K., Wachs, J., 2023. Considering human cognitive architecture in stressful medical prehospital interventions might benefit care providers. Canadian Journal of Surgery. Journal Canadien de Chirurgie 66 (6), E522–E534.

Lazow, S.P., Siu, M., Brown, L., Kamine, T.H., 2024. Percutaneous cholecystostomy for acute cholecystitis during spaceflight. Aerospace Medicine and Human Performance 95 (10), 771–776.

Levin, DR, Steller, J, nderson A, A, Lemery, J, Easter, B, Hilmers, DC, Lehnhardt, KR, 2023. Enabling Human Space Exploration Missions Through Progressively Earth Independent Medical Operations (EIMO). IEEE Open J Eng Med Biol 4, 162–167. Available at 10.1109/OJEMB.2023.3255513.

Liu, X., Liang, G. (Eds.), 2025. Action Recognition and Action Anticipation Tasks in the Trauma THOMPSON Challenge Technical Report. Springer Nature Switzerland, Cham.

Majumder, N., Ghosh, S., 2023. 3D biofabrication and space: a 'far-fetched dream' or a 'forthcoming reality'? Biotechnology Advances 69, 108273.

Martin, A., Sullivan, P., Beaudry, C., Kuyumjian, R., Comtois, J.R., 2012. Space medicine innovation and telehealth concept implementation for medical care during exploration-class missions. Acta Astronautica 81, 30–33.

McBeth, P.B., Crawford, I., Blaivas, M., Hamilton, T., Musselwhite, K., Panebianco, N., et al., 2011. Simple, almost anywhere, with almost anyone: remote low-cost telementored resuscitative lung ultrasound. The Journal of Trauma 71 (6), 1528–1535.

McBeth, P., Crawford, I., Tiruta, C., Xiao, Z., Zhu, G.Q., Shuster, M., et al., 2013. Help is in your pocket: the potential accuracy of smartphone- and laptop-based remotely guided resuscitative telesonography. Telemedicine Journal and e-Health: The Official Journal of the American Telemedicine Association 19 (12), 924–930.

McBeth, P.B., Hamilton, T., Kirkpatrick, A.W., 2010. Cost-effective remote iPhone-teathered telementored trauma telesonography. The Journal of Trauma 69 (6), 1597–1599.

Melton, S., Beck, G., Hamilton, D., Chun, R., Sargsyan, A., Kirkpatrick, A.W., 2001. How to test a medical technology for space: trauma sonography in microgravity. McGill Journal of Medicine 6, 66–73.

Mogensen, Jackie Flynn, Cameron, Claire, NASA identifies which astronaut triggered the unprecedented medical evacuation of the ISS. Scientific American. Feb. 2026 Available at https://www.scientificamerican.com/article/nasa-astronaut-mike-fincke-identified-as-suffering-a-medical-event-on-iss/. Accessed Mar. 21, 2026.

Nasr, M., 2020. Evolution of the flight crew and mission control relationship. AIAA Scitech 2020 Forum.

Netzer, I., Kirkpatrick, A.W., Nissan, M., McKee, J.L., McBeth, P., Dobron, A., et al., 2019. Rubrum coelis: the contribution of real-time telementoring in acute trauma scenarios-a randomized controlled trial. Telemedicine Journal and e-Health: The Official Journal of the American Telemedicine Association.

Nicioli, T. and Fisher, K., 2024. Surgery in space: Tiny remotely operated robot completes first simulated procedure at the space station. CNN. Available from: https://www.cnn.com/2024/02/13/world/mira-robot-first-surgery-in-space-scn/index.htm

Niler, E., 2018. The Apollo Mission That Nearly Ended With a Mutiny in Space: A&E Television Networks; [updated June 24 2019 Jan 17 2025]. Available from: https://www.history.com/news/apollo-7-near-mutiny-ground-control-astronauts.

Van Ombergen, A., Chalupa-Gantner, F., Chansoria, P., Colosimo, B.M., Costantini, M., Domingos, M., et al., 2023. 3D bioprinting in microgravity: opportunities, challenges, and possible applications in space. Advanced Healthcare Materials 12 (23) e2300443.

Palinkas, L.A., 2001. Psychosocial issues in long-term space flight: overview. Gravitational and Space Biology Bulletin: Publication of the American Society for Gravitational and Space Biology 14 (2), 25–33.

Pantalone, D., Faini, G.S., Cialdai, F., Sereni, E., Bacci, S., Bani, D., et al., 2021. Robot-assisted surgery in space: pros and cons. A review from the surgeon's point of view. NPJ Microgravity 7 (1), 56.

Roslee, M.F., 2021. Review of medical capabilities requirements for manned missions on Lunar and Martian surfaces base activities. REACH 23–24 (100042).

Sargsyan, A.E., Hamilton, D.R., Jones, J.A., Melton, S., Whitson, P.A., Kirkpatrick, A.W., et al., 2005. FAST at MACH 20: clinical ultrasound aboard the International Space Station. The Journal of Trauma 58 (1), 35–39.

Satava, R.M., 1988. Surgery in space. Phase I: basic surgical principles in a simulated space environment. Surgery 103, 633–637.

Scalea, T.M., Rodriguez, A., Chiu, W.C., Brenneman, F.D., Fallon, W.F., Kato, K., et al., 1999. Focused assessment with sonography for trauma (FAST): results from an international consensus conference. The Journal of Trauma 46, 466–472.

Stepanek, J., Blue, R.S., Parazynski, S., 2019. Space medicine in the era of civilian spaceflight. The New England Journal of Medicine 380 (11), 1053–1060.

Wachs, J., Jiang, N., Zhuo, Y., Couperus, K., Colombo, C. and Birch, E. et al., 2023. The Trauma THOMPSON challenge proposal. Proceedings of the International Conference on Medical Image Computing and Computer-Assisted Intervention (MICCAI 2023), Vancouver.

Walton, ME, Kerstman, EL, 2020. Quantification of Medical Risk on the International Space Station Using the Integrated Medical Model. Aerosp Med Hum Perform 91 (4), 332–342. Available at 10.3357/AMHP.5432.2020.

Yaroshenko, G.L., Terentyev, B.G., Mokrov, M.N., 1967. Characteristics of surgical intervention in conditions of weightlessness. Voenno-Meditsinskii Zhurnal 10, 69–70.

Section VI

# The future

Chapter 27

# Major invasive surgery

**Stefanie P. Lazow**
*Department of Surgery, Division of Acute Care Surgery, University of Vermont Medical Center, Burlington, VT, United States*

## 27.1 Introduction

As humanity embarks on exploration-class missions in the future, the likelihood of a mission-critical pathology requiring major invasive surgery will increase in tandem with mission length and at-risk person-years in space. For low Earth orbit or cislunar space, stabilizing measures followed by medical evacuation to Earth for definitive care remain an option. However, for some lunar missions and missions further out, such as to Mars, evacuation to Earth will not be possible in medically meaningful timeframes (Anderson et al., 2024; Levin et al., 2023; Rubin, 2019). As the number of person-years in space increases, a certain risk threshold for mission-critical surgical diagnoses may be crossed that supports the development of capacity for major invasive surgery. This chapter will review the anticipated milestones in space exploration that will support the need for major surgery capacity in space, as well as highlight the goals and challenges of its development.

## 27.2 Historical examples of remote surgical capacity

There is precedent in the history of austere medicine for surgery performed in remote locations with limited resources and nonsurgeon personnel where evacuation is similarly prohibitive. On the USS Sea Dragon submarine in 1942, a nonsurgeon Pharmacist's Mate (who notably did have prior surgical training and experience) performed a successful open appendectomy in the wardroom with tablespoons serving as retractors (Bornmann and Herman, 1996). The submarine was in a combat zone and unable to safely evacuate. In 1961, in a remote location in Antarctica without feasible evacuation options, Russian surgeon Leonid Ivanovich Rogozov performed an appendectomy on himself under local anesthesia with a mirror and two assistants holding retractors (Rogozov, 1964). This case was the initial foundation for ongoing restrictions by Australia that mandate that all physicians on their Antarctic crews undergo prophylactic appendectomy prior to arrival (Davis et al., 2016). While a wide range of surgeries have been performed in Antarctica between 1904 and 2022 including appendectomies, laparotomies, and multiple orthopedic and dental procedures (Tissot et al., 2023), these surgeries were performed infrequently, and the development of improved evacuation capability is currently being favored over surgical capacity in Antarctica. Furthermore, military personnel in combat zones routinely provide surgical care in austere environments with limited personnel, though this is often with the goal of damage-control stabilization prior to evacuation (Baker et al., 2021).

## 27.3 Milestones in space travel supporting the need for surgical capacity

In the next few decades, it is anticipated that major milestones will be reached in human space exploration with the goal of returning humans to the lunar surface by the late 2020s. Subsequent missions are tasked with building an orbiting lunar space station to serve as a springboard to Mars and eventual off-world settlements (NASA's Plan for Sustained Lunar Exploration and Development, 2020; Artemis, 2024). These missions will both increase the at-risk person-years for emergent surgical pathology arising due to longer mission duration as well as make evacuation prohibitive due to increased mission distance from Earth. With increased distance from Earth, the evacuation duration will be longer, and astronauts may not be able to reach definitive care within the necessary timeline for treatment. During Artemis missions on the Moon, evacuation to Earth may take as long as 14 days (Anderson et al., 2024), which will be prohibitive for certain

**Procedural Medicine for Spaceflight. DOI:** https://doi.org/10.1016/B978-0-443-27552-4.00017-8

pathologies. Martian missions are anticipated to last 2–3 years, such that evacuation to Earth will not be logistically possible within a medically meaningful timeframe (2009).

Several risk-reduction strategies have therefore been suggested to reduce the risk of the development of surgical pathologies. These include preflight screening, such as performing ultrasounds to assess for gallstone burden to assess the risk of in-flight cholecystitis (Lazow et al., 2024), as well as optimization of nonoperative management strategies, such as antibiotic therapy and minimally invasive nonsurgical procedures. Another initially proposed strategy for risk-reduction was prophylactic surgery prior to spaceflight for the most common general surgery pathologies, such as appendicitis and cholecystitis. However, a risk/benefit probabilistic assessment study using NASA's Integrated Medical Model (IMM) did not support routine prophylactic appendectomy or cholecystectomy, and as such this is not currently recommended (Reyes et al., 2021; Ball et al., 2012). This topic warrants further investigation, as a recent review demonstrated that nonoperative temporizing drainage (i.e., percutaneous cholecystostomy) for the management of cholecystitis has high rates of long-term failure (Lazow et al., 2024), and clinical trials have shown high rates of recurrent appendicitis despite antibiotic treatment (Flum et al., 2020). Longer-term failure of these nonoperative management strategies will become more likely as mission distance and duration increase, which supports a reconsideration of prophylactic surgery for certain mission profiles.

As of this writing, humanity has accumulated around 200 person-years in orbit (2025) without any evacuations required for surgical pathology so far. Therefore, if an event were to occur tomorrow, the risk of an evacuation to Earth for surgical pathology might be estimated at no more than 0.007 events per person-year. However, the commercial human spaceflight sector is developing rapidly with anticipated increased at-risk person-years. Even with risk mitigation strategies, a certain risk threshold may be reached in the mid-term future that supports the development of surgical capacity. It behooves us to begin considering what such training might look like now to ensure we are prepared for when it is needed.

## 27.4 Risk of surgical pathology in spaceflight

At minimum, the capacity for major invasive surgery in space should include both emergency general surgery and trauma hemorrhage control operations to address conditions that are either high probability and/or have a high morbidity and mortality risk. While the risk of specific surgical pathologies in space cannot be quantified, given that no surgical events have occurred so far in the history of spaceflight, risk is extrapolated from terrestrial data below.

## 27.5 Emergency general surgery

The most common terrestrial emergency surgeries are appendectomy for appendicitis, cholecystectomy for cholecystitis, adhesiolysis for bowel obstruction, bowel resection (small or large bowel) for bowel ischemia or perforation, and omental patch for peptic ulcer perforation (Pucher et al., 2021; Wohlgemut et al., 2020). Based upon data from the Australian Antarctic program from 1947 to 1979, the risk of appendicitis is reported at 0.015 events per person-year (1 case every 67 person-years) with 0.003 events per person-year requiring evacuation for definitive treatment or 1 evacuation every 333 person-years (Campbell et al., 2004). Using NASA's IMM, the risk of a member of an astronaut analog population developing appendicitis is reported at 0.0009 events per person-year (Reyes et al., 2021). Per the IMM, the risk of developing cholecystitis is 0.0013 events per person-year for women (1 case every 769 person-years) and 0.0008 for men (1 case every 1250 person-years; Reyes et al., 2021). Initial development of surgical capacity should focus on these higher probability conditions. Population-based studies have reported the risk of small bowel obstruction at 0.006 events per person-year (Scott et al., 2012), perforated diverticulitis at 0.00003 events per person-year (Humes et al., 2009), and perforated peptic ulcer disease at 0.00007 events per person-year (Thorsen et al., 2013). While these are low likelihood pathologies, consequences can be categorized as severe or catastrophic. These pathologies are highly morbid and have significantly increased mortality rates if definitive surgical management is delayed (Sell et al., 2021). Furthermore, it is possible that both the likelihood and progression of emergency surgery pathology may be influenced by unique features of long-duration spaceflight, such as microgravity, radiation exposure, and anticipated changes in gastrointestinal microbiome and immune dysregulation (Krittanawong et al., 2023).

## 27.6 Trauma

Hemorrhage secondary to trauma is theorized to be one of the most likely causes of potentially preventable mortality in spaceflight (Pantalone et al., 2022). Furthermore, the risk of traumatic injury, such as crush or penetrating injury, will increase with longer-duration spaceflight and more frequent extravehicular activity. Objects in microgravity still possess mass and can generate significant force when accelerated (McCuaig and Houtchens, 1992; Pantalone et al., 2022), leading to increased likelihood of blunt injury, though this will be limited relative to a higher gravity setting. While some of this risk can be mitigated with engineering design and mission planning, there will still be some risk of injury.

If terrestrial trauma data is extrapolated, the anticipated risk of requiring surgery for traumatic injury in spaceflight is overall low. For chest trauma, the need for operative intervention is rare following both blunt trauma (<10%) and penetrating trauma to the chest (15%–30%; Edgecombe et al., 2023), though less invasive procedures such as tube thoracostomy placement may be required. For abdominal trauma, operative intervention is traditionally mandated in penetrating injury to the abdomen, especially in the absence of advanced cross-sectional imaging, but it is only performed in 10%–20% of blunt trauma with solid organ injury (Kanlerd et al., 2022). However, low terrestrial rates of surgical intervention for blunt abdominal solid organ injury are achieved because of effective nonoperative management strategies, such as balanced transfusion, which will be limited in spaceflight, and interventional radiology for embolization, which would require development to be a viable option in spaceflight. Therefore, developing capacity for laparotomy for hemorrhage control may still be justified to prevent mortality despite the overall low probability of typical operative traumatic pathology.

## 27.7 Development of major surgical capacity

The requirements for developing major invasive surgical capacity in space can be categorized into multiple domains, including team personnel, perioperative care and anesthesia, operating room specifications, surgical equipment, and fluid management. Goals and challenges for the development of open and minimally invasive surgical capacity will be highlighted below.

## 27.8 Team personnel

Planned lunar missions will have a crew of four people while proposed Martian missions will have a crew of six people (2009; Rubin, 2019). Currently on the International Space Station, two members of the crew serve as the designated chief medical officer (CMO), and planned lunar missions also plan for two CMOs (Rubin, 2019). The CMO currently receives mandatory preflight medical training but is not required to be a physician (Pantalone et al., 2022). In longer-duration spaceflight, the crewmember serving as the CMO is more likely to be a trained physician though unlikely to be a surgeon (Pantalone et al., 2022). In order to alleviate the need for a skilled surgeon as part of the crew, telementoring and telerobotics have been proposed (Doarn et al., 2009; Kamine et al., 2022; Otto et al., 2010). Even in low earth orbit, high bandwidth communication is not a guarantee depending upon satellite redundancy, weather stability, and solar activity. Furthermore, telementoring is unlikely to be feasible as mission distance from Earth increases due to increased communication delay. For example, there is a round-trip communication delay between Earth and Mars ranging from 10 to 40 minutes (Perseverance Rover, 2020). This communication delay is expected to lead to significant impairment in surgical performance (Kamine et al., 2022; Korte et al., 2014). Given these communication constraints, crew autonomy will necessarily increase as the distance from Earth grows, mandating a higher intrinsic crew skill level and capacity for earth-independent medical operations, including surgery (Levin et al., 2023).

In the terrestrial setting, the basic surgical team unit is comprised of a four to five-person team including a surgeon, a surgical assistant and/or scrub technician, an anesthesiologist, and a circulating nurse. The equivalent surgical team unit in spaceflight could be reduced to three members (plus the fourth crewmember serving as patient) with some redundancy to allow other mission-critical tasks to be completed in parallel. The proposed roles would be as follows: (1) sterile CMO as surgeon, (2) sterile crewmember as surgical assistant and scrub technician, and (3) nonsterile backup CMO as anesthesiologist and circulating nurse with ability to perform other mission-critical tasks during surgery if needed.

The CMO will serve as the surgeon and ideally will have the broad skill set of a trauma/general surgeon, with additional training in the surgical management of high-probability pathologies in other fields, such as orthopedics, ENT, and neurosurgery. Since simulation studies have suggested that there is significant cost to crew time when novices are asked to perform complex medical tasks, an expert medical officer would ideally be placed in this role if mission planning otherwise allows (Levin et al., 2022). If this is not possible, the acting surgeon should be a nonsurgeon physician with special training in basic emergency surgery. The eventual development of advanced autonomous robotic surgical technology could allow for offloading of surgical skill. However, the CMO surgeon would ideally still have a surgical skill set to allow for damage-control open surgery if needed. Since the CMO role as surgeon will be critical for surgical capacity, the CMO cannot also serve as backup for other mission-critical tasks (such as piloting) in case the pilot were to be the injured patient.

A second crewmember will also be sterile and serve in the surgical assistant and technician roles (i.e., retracting and handling sterile instruments). The crewmember serving as a surgical assistant does not need to be a fully trained surgeon or physician, but they should also have some basic surgical training.

The backup CMO will be nonsterile and function as both anesthesiologist and circulating nurse, with the tasks of perioperative monitoring and equipment troubleshooting. Some of the perioperative monitoring cognitive task burden could potentially be offloaded to an autonomous digital-based clinical system if developed (Khan and Shah, 2025). From an anesthesia perspective, this role typically involves intravenous (IV) access, provision of anesthetic agents, airway management including use of airway adjuncts (oral or nasopharyngeal airway, laryngeal mask airway, as well as intubation if needed), intraoperative monitoring, and resuscitation. In order to reduce the required number of medically skilled crewmembers, the preoperative advanced procedural care (i.e., access and airway management) could be performed by the CMO surgeon prior to the operation (Komorowski et al., 2021). Finally, since this third crewmember is not sterile, they would also be free to perform other mission-critical tasks during surgery such as ship maintenance and communication with earth if the mission has only a total four-person crew.

## 27.9 Perioperative care and anesthesia

The perioperative care domain involves the provision of anesthesia, airway management, perioperative cardiopulmonary monitoring, and resuscitation. These topics are covered in detail elsewhere in this textbook for more minor procedures.

For anesthesia, regional approaches will be favored over general anesthesia when possible, due to their lower risk profile and smaller skill set requirements. Regional blocks can also be performed with the use of ultrasound in microgravity (Kiberd et al., 2024). Smaller procedures could also be performed under sedation without an advanced airway. If general anesthesia is needed, especially for emergent thoracoabdominal surgery, then there must be the capacity to place an advanced airway and provide ventilation either with extended bag-valve mask, which would be labor-intensive, or with a ventilator. Small, portable, automatic, and human-powered ventilators are currently in development at NASA for use in this setting (Human Medicine and Biotechnology, no date). Hyperoxia without a secured airway will lead to increased fire risk, especially if there is the use of electrocautery. This risk can be mitigated with the use of a supraglottic airway or cuffed endotracheal tube, though this increases the required crew skill level, and minimally invasive surgery to separate the surgical field from the oxygen source and contain ignition sources.

Cardiopulmonary monitoring, both intraoperatively and postoperatively, includes cardiac telemetry and blood pressure monitoring with an automatic cuff or an arterial line. A defibrillator should be available in the event of cardiac arrest, though this will increase fire risk in the oxygen-rich environment and may cause radiofrequency interference with navigation and control systems (Beebeejaun and Roby, 2019). To ensure adequate oxygenation and ventilation if the patient requires ventilatory support, oxygen saturation and end-tidal $CO_2$ monitoring will be necessary, though end-tidal $CO_2$ will likely need to be recalibrated in the hypercarbic spaceship environment. In the postoperative setting, cardiopulmonary monitoring will need to be continued until the patient's physiology normalizes. The crewmember in the nursing role during the surgery could continue to provide nursing care and monitoring in the postoperative setting, or such responsibility could be transferred to the CMO once available, depending upon other task needs. Alternatively, artificial intelligence-based monitoring systems may again allow this process to be automated, reducing task load (Khan and Shah, 2025).

IV access will be required for the provision of IV medication and fluid resuscitation, though studies have shown that this requires IV tubing with a pump in microgravity (Nowak et al., 2019). Ultrasound may be useful to assist with difficult IV access. Given that traumatic hemorrhagic shock is a relatively high-risk threat to crew life, it will be critical that blood transfusion and resuscitation are possible. Due to restrictions on blood product shelf life and challenges with resupply, a floating blood bank akin to the military walking blood bank has been suggested as a solution (Nowak et al., 2019). A floating blood bank would involve either selecting two crewmembers with O-negative blood types to serve as universal donors or selecting crewmembers with matched blood type pairs to facilitate within-crew transfusion.

A multitude of medications required for safe major surgery must be stored for the duration of the mission, given the challenges of resupply. These medications include, but are not limited to, anesthetic agents (e.g., neuromuscular blockers, sedatives, analgesics, and local anesthesia), physiological modulators (e.g., antihypertensives, vasopressors, heart rate modulators), antibiotics, and antiemetics. Volatile anesthetic agents should be avoided in an enclosed environment (Komorowski et al., 2021).

## 27.10 Operating room specifications

While terrestrial requirements dictate that the volume of an operating room with a presumed 8-foot ceiling must be at least 90.6 $m^3$ in the United States (Langlands, 2018), a prior simulation-based study reported that the minimum spatial volume needed to perform an open appendectomy is much smaller at 2.9–4.3 $m^3$ (Kamine et al., 2022). Ideally, the actual volume designated for the operating room in the spaceflight or habitat environment would be between these two values, allowing adequate space for both anesthesia and surgery. In near-future space vehicles such as Haven-1 with 43 $m^3$ of habitable volume (Haven-1 Features, 2025), Sierra Space's first-stage Life Habitat with 285 $m^3$ of habitable volume (Space, 2025), and Starship with ~1000 $m^3$ of pressurized volume, volume will be constrained with multiple competing demands. Surgical theater volume would ideally be multipurposed for other tasks when not in use for surgery.

Other terrestrial Joint Commission operating room requirements that would still be relevant for surgery in space include the need for overhead lights to allow for adequate visualization of the operative field, an operating room temperature control between 72°F–78°F with maximum relative humidity of 60%, and a ventilation system with a minimal total air exchange of four times per hour (Interpreting Joint Commission Standards: FAQs, 2024).

A sterile operative field should be created. There is robust data from premodern settings and modern austere environments that postoperative infection rates exceed 50%–60% after major surgery in nonsterile fields (Smith et al., 2012; Teodorescu et al., 2021). These infection rates are significantly higher than the expected rate of wound infection following surgery with sterile technique in contaminated or dirty fields, which are reported at ~15%–40% (Herman et al., 2023). Due to the risk burden of postoperative infection and need for antibiotic therapy, sterility during major surgery should ideally be maintained in the spaceflight environment. A sterilization system will be needed for repeated cleaning of surgical instruments; however this will require power and create waste heat that will need to be managed. The ability to sterilize and recycle instruments would reduce the storage space required for redundant equipment. 3D printing of instruments is another option that is further detailed later in this chapter. Additionally, a surgical scrub will need to be devised to sterilize the surgeon's hands, likely involving a waterless, gel-based product to minimize fluid containment issues in microgravity or a sterile, antiseptic wipe such as those containing benzalkonium chloride. Similarly, a surgical prep wipe will be required to create a sterile operative field.

## 27.11 Surgeon and patient positioning

Critically, microgravity poses unique challenges for surgeons, patients, and equipment mobility during surgery. The patient will need to be positioned with restraints to maintain the surgical exposure. Similarly, providers will need to be restrained to maintain their relative position to the surgical field. Moreover, a temporary storage system, such as the space equivalent of a Mayo stand or tray, will need to be devised for all surgical equipments to allow for the alternating use of instruments with intermittent immobilization when not in use during the operation. Mechanical or adhesive restraint systems will need to be devised that do not include magnets, due to the risk of interference with critical spacecraft systems, or Velcro, due to the risk of serving as a fomite. An alternative approach to mitigate these mobility issues would be the creation of artificial gravity with the use of centrifugal force via a rotating surgery pod, though this introduces other challenges with Coriolis forces and gravity gradients that may cause motion sickness and reduce surgeon task efficiency (Clément et al., 2015).

## 27.12 Surgical equipment

For emergent presentations with hemodynamic instability, open surgery techniques will be favored due to physiologic intolerance of minimally invasive techniques using insufflation and potential need for staged damage-control operations. For open invasive surgery, the following equipment will be needed at a minimum: gauze and/or towels, retractors, a scalpel with blades, a dissecting instrument, forceps, suture with needles, and a needle driver. Surgical staplers would be useful, though not strictly required, for bowel surgery and skin closure. Most of these instruments required for open surgery could potentially be 3D printed if the spacecraft has 3D printing capacity to limit storage utilization (Wong and Pfahnl, 2016), though this would take time for prepping if printing were performed on request per surgery. An alternative option would be to have 1–2 sets of printed basic surgical instruments available at all times that could be recycled by melting down the instruments after use. For open surgery, overhead lights or a headlamp worn by the surgeon will be required.

For urgent, nonemergent presentations with hemodynamic stability, minimally invasive techniques such as laparoscopic or robotic approaches would be preferred (Pantalone et al., 2021), though these techniques require more advanced surgical skill and equipment. Studies in parabolic flight have demonstrated that minimally invasive abdominal surgery is possible in microgravity and does require use of insufflation for abdominal domain (Kirkpatrick et al., 2009). Anticipated benefits of minimally invasive techniques include improved fluid containment within the enclosed body cavity, better lighting within the cavity and reduced need for overhead lights, as well as smaller incisions with reduced postoperative pain and faster recovery times. Minimally invasive surgery also offers an opportunity to offload crew task burden and skill requirements with robotic technology. While surgical robots will not be able to be safely controlled by a surgeon on Earth due to prohibitive communication delays as mission distances from Earth increase, the eventual development of autonomous robotic surgical assistance could further offload the skill requirements of the crew surgeon (Pantalone et al., 2021). The disadvantages of minimally invasive techniques involve increased space and equipment requirements including longer instruments, a camera and screen for viewing the surgical field, and insufflation equipment (Kirkpatrick et al., 2009).

## 27.13 Fluid containment

In the microgravity environment, fluid containment strategies are required to prevent blood, enteric contents, and irrigation from escaping and contaminating the spacecraft during open surgery. This risk is reduced with the use of minimally invasive techniques within an enclosed body cavity. For open surgery, one possible solution is the Aqueous Immersion Surgical System, an enclosed dome system placed over the relevant body cavity with a fluid medium and ports for instrument placement, similar to current laparoscopic systems (Hayden et al., 2013). Fluid containment may also be facilitated by precise suction instrumentation with or without the creation of an enclosed surgical space. A team at the University of Louisville has designed a closed-loop surgical fluid management system, which is a multipurpose handheld instrument for suctioning, irrigation, and illumination in microgravity (Barrow et al., 2020). Future versions of this device may also incorporate laser cautery for hemostasis and tissue dissection, given concerns that conventional monopolar and bipolar electrocautery may cause radiofrequency interference with avionics (Barrow, et al., 2020). However, the overall role for electrocautery in space remains unclear given the inherent fire risk. More investigation will be required to optimize fluid containment and cautery systems.

## 27.14 Conclusion

In summary, anticipated near-future advances in human spaceflight are expected to lead to increased mission distance from Earth and at-risk person-years for mission-critical surgical pathology. With these changes, evacuation to Earth for definitive surgical care will be logistically prohibitive, and the benefits of the development of surgical capacities in space may outweigh the costs for certain mission profiles. In the resource-constrained space environment, the development of surgical capacity will entail optimization of team personnel, perioperative monitoring strategies, operating theater parameters, and surgical equipment. Given the complexity of performing major surgery in both microgravity and partial gravity environments, extensive study and development will be necessary to continue to adapt terrestrial techniques.

## References

Anderson, A., Parmar, P., Levin, D., Hilmers, D., 2024. Quantifying the risk of medical evacuation in spaceflight. Acta Astronautica 224, 458–464. Available at https://doi.org/10.1016/j.actaastro.2024.08.026.

Artemis, 2024. Available at https://www.nasa.gov/humans-in-space/artemis/#:~:text=With%20NASA%27s%20Artemis%20campaign%2C%20we%20are%20exploring,of%20the%20lunar%20surface%20than%20ever%20before.

Astronaut/Cosmonaut Statistics. WorldSpaceFlight. Updated March 2026. Available at https://www.worldspaceflight.com/bios/stats.php.

Baker, J.B., Northern, D.M., Frament, C., Baker, D.A., Remick, K., Seery, J., et al., 2021. Austere resuscitative and surgical care in support of forward military operations—joint trauma system position paper. Military Medicine 186 (1-2), 12–17. Available at https://doi.org/10.1093/milmed/usaa358.

Ball, C.G., Kirkpatrick, A.W., Williams, D.R., Jones, J.A., Polk, J.D., Vanderploeg, J.M., et al., 2012. Prophylactic surgery prior to extended-duration space flight: is the benefit worth the risk? Canadian Journal of Surgery 55 (2), 125–131. Available at https://doi.org/10.1503/cjs.024610.

Barrow, B.E., Pantalos, G.M., Roussel, T.J., 2020. Design and evaluation of a multifunctional surgical device for space-based applications. Acta Astronautica 175, 118–127. Available at https://doi.org/10.1016/j.actaastro.2020.05.037.

Beebeejaun, A., Roby, H., 2019. Dangers of defibrillation in flight. Journal of the Australasian Society of Aersopace Medicine 12 (1), 1–4.

Bornmann, R.C., Herman, J.K., Operating Under Pressure. 1996.

Campbell, M.R., Johnston, S.L., Marshburn, T., Kane, J., Lugg, D., 2004. Nonoperative treatment of suspected appendicitis in remote medical care environments: implications for future spaceflight medical care. Journal of the American College of Surgeons 198 (5), 822–830. Available at https://doi.org/10.1016/j.jamcollsurg.2004.01.009.

Clément, G.R., Bukley, A.P., Paloski, W.H., 2015. Artificial gravity as a countermeasure for mitigating physiological deconditioning during long-duration space missions. Frontiers in Systems Neuroscience 9 (June). Available at https://doi.org/10.3389/fnsys.2015.00092.

Davis, C.R., Trevatt, A.E.J., Dixi, A., Datta, V., 2016. Systematic review of clinical outcomes after prophylactic surgery. Annals of the Royal College of Surgeons of England 98 (6), 353–357. Available at https://doi.org/10.1308/rcsann.2016.0089.

Doarn, C.R., Anvari, M., Low, T., Broderick, T.J., 2009. Evaluation of teleoperated surgical robots in an enclosed undersea environment. Telemedicine and e-Health 15 (4), 325–335. Available at https://doi.org/10.1089/tmj.2008.0123.

Edgecombe, L., Sigmon, D., Galuska, M., Angus, L., Thoracic Trauma, 2023.

Flum, D.R., Davidson, G.H., Monsell, S.E., Shapiro, N.I., Odom, S.R., Sanchez, S.E., et al., 2020. , A randomized trial comparing antibiotics with appendectomy for appendicitis. Journal of Medicine 383 (20), 1907–1919. Available at https://doi.org/10.1056/NEJMoa2014320.

Haven-1 Features, 2025. Available from: https://www.vastspace.com/haven-1.

Hayden, J.A., Pantalos, G.M., Burgess, J.E., Antaki, J.F., 2013. A hermetically sealed, fluid-filled surgical enclosure for microgravity. Aviation, Space, and Environmental Medicine 84 (12), 1298–1303. Available at https://doi.org/10.3357/ASEM.3751.2013.

Herman TF, Popowicz P, Bordoni B. Wound Classification. StatPearls, 2023.

Human Medicine and Biotechnology, no date. Available from: https://technology.nasa.gov/patent/MSC-TOPS-86.

Humes, D.J., Solaymani-Dodaran, M., Fleming, K.M., Simpson, J., Spiller, R.C., West, J., 2009. A population-based study of perforated diverticular disease incidence and associated mortality. Gastroenterology 136 (4), 1198–1205. Available at https://doi.org/10.1053/j.gastro.2008.12.054.

Interpreting Joint Commission Standards: FAQs, 2024. Available from: http://jointcommission.org/standards/standard-faqs.

Kamine, T.H., Siu, M., Kramer, K., Kelly, E., Alouidor, R., Fernandez, G., et al., 2022. Spatial volume necessary to perform open appendectomy in a spacecraft. Aerospace Medicine and Human Performance 93 (10), 760–763. Available at https://doi.org/10.3357/AMHP.6062.2022, http://ingentaconnect.com/content/asma/amhp;jsessionid=13tlejf2f0ojk.alice.

Kamine, T.H., Smith, B.W., Fernandez, G.L., 2022. Impact of time delay on simulated operative video telementoring: a pilot study. Aerospace Medicine and Human Performance 93 (2), 123–127. Available at https://doi.org/10.3357/AMHP.5972.2022.

Kanlerd, A., Auksornchart, K., Boonyasatid, P., 2022. Non-operative management for abdominal solidorgan injuries: a literature review. Chinese Journal of Traumatology 25 (5), 249–256. Available at https://doi.org/10.1016/j.cjtee.2021.09.006.

Khan, M.M., Shah, N., 2025. AI-driven wearable sensors for postoperative monitoring in surgical patients: a systematic review. Computers in Biology and Medicine 196 (Pt B), 110783. Available at https://doi.org/10.1016/j.compbiomed.2025.110783.

Kiberd, M.B., Brownbridge, R., Mackin, M., Werry, D., Bird, S., Barry, G., et al., 2024. Feasibility of ultrasound-guided nerve blocks in simulated microgravity: a proof-of-concept study for regional anaesthesia during deep space missions. British Journal of Anaesthesia 133 (6), 1276–1283. Available at https://doi.org/10.1016/j.bja.2024.07.034.

Kirkpatrick, A.W., Keaney, M., Kmet, L., Ball, C.G., Campbell, M.R., Kindratsky, C., et al., 2009. Intraperitoneal gas insufflation will be required for laparoscopic visualization in space: a comparison of laparoscopic techniques in weightlessness. Journal of the American College of Surgeons 209 (2), 233–241. Available at https://doi.org/10.1016/j.jamcollsurg.2009.03.026.

Komorowski, M., Thierry, S., Stark, C., Sykes, M., Hinkelbein, J., 2021. On the challenges of anesthesia and surgery during interplanetary spaceflight. Anesthesiology 135 (1), 155–163. Available at https://doi.org/10.1097/aln.0000000000003789.

Korte, C., Sudhakaran Nair, S., Nistor, V., Low, T.P., Doarn, C.R., Schaffner, G., 2014. Determining the threshold of time-delay for teleoperation accuracy and efficiency in relation to telesurgery. Telemedicine and e-Health 20 (12), 1078–1086. Available at https://doi.org/10.1089/tmj.2013.0367.

Krittanawong, C., Singh, N.K., Scheuring, R.A., Urquieta, E., Bershad, E.M., Macaulay, T.R., et al., 2023. Human health during space travel: state-of-the-art review. Cells 12 (1). Available at https://doi.org/10.3390/cells12010040.

Langlands, B., 2018. The 2018 FGI Guidelines: Proposed Revisions to Procedure, Operating, and Imaging Room Requirements. Available at https://www.fgiguidelines.org/wp-content/uploads/2017/08/SLS17_FGI_ExamProcedureOperatingImaging_170721.pdf.

Lazow, S.P., Siu, M., Brown, L., Kamine, T.H., 2024. Percutaneous cholecystostomy for acute cholecystitis during spaceflight. Aerospace Medicine and Human Performance 95 (10), 771–776. Available at https://doi.org/10.3357/AMHP.6352.2024.

Levin, D.R., Siu, M., Kramer, K., Kelly, E., Alouidor, R., Fernandez, G., et al., 2022. Time cost of provider skill: a pilot study of medical officer occupied time by knowledge, skill, and ability level. Aerospace Medicine and Human Performance 93 (11), 816–821. Available at https://doi.org/10.3357/AMHP.6093.2022.

Levin, D.R., Steller, J., Anderson, A., Lemery, J., Easter, B., Hilmers, D.C., et al., 2023. Enabling human space exploration missions through progressively Earth Independent Medical Operations (EIMO). IEEE Open Journal of Engineering in Medicine and Biology 4, 162–167. Available at https://doi.org/10.1109/OJEMB.2023.3255513, https://ieeexplore.ieee.org/xpl/RecentIssue.jsp?punumber=8782705&_ga=2.23991482.1496039865.1593798416-2073569860.1593798416.

Mars Architecture Steering Group. Human Exploration of Mars Design Reference Architecture 5.0. 2009 [cited 2024 12/08/2024]; Available at https://www.nasa.gov/wp-content/uploads/2015/09/373665main_nasa-sp-2009-566.pdf.

McCuaig, K.E., Houtchens, B.A., 1992. Management of trauma and emergency surgery in space. Journal of Trauma - Injury, Infection and Critical Care 33 (4), 610–626. Available at https://doi.org/10.1097/00005373-199210000-00018.

NASA's Plan for Sustained Lunar Exploration and Development, 2020. Available from: https://www.google.com/url?sa=t&source=web&rct=j&opi=89978449&url=https://www.nasa.gov/wp-content/uploads/2020/08/a_sustained_lunar_presence_nspc_report4220final.pdf&ved=2ahUKEwjw-9yuqbWQAxX5kyYFHU2RHcEQFnoECAwQAQ&usg=AOvVaw0AlC_oQuaRLL0-GOLOPKqE.

Nowak, E.S., Reyes, D.P., Bryant, B.J., Cap, A.P., Kerstman, E.L., Antonsen, E.L., 2019. Blood transfusion for deep space exploration. Transfusion 59 (10), 3077–3083. Available at https://doi.org/10.1111/trf.15493.

Otto, C., Comtois, J.M., Sargsyan, A., Dulchavsky, A., Rubinfeld, I., Dulchavsky, S., 2010. The Martian chronicles: remotely guided diagnosis and treatment in the arctic circle. Surgical Endoscopy 24 (9), 2170–2177. Available at https://doi.org/10.1007/s00464-010-0917-1.

Pantalone, D., Faini, G.S., Cialdai, F., Sereni, E., Bacci, S., Bani, D., et al., 2021. Robot-assisted surgery in space: pros and cons. A review from the surgeon's point of view. npj Microgravity 7 (1). Available at https://doi.org/10.1038/s41526-021-00183-3.

Pantalone, D., Chiara, O., Henry, S., Cimbanassi, S., Gupta, S., Scalea, T., 2022. Facing trauma and surgical emergency in space: hemorrhagic shock. Frontiers in Bioengineering and Biotechnology 10. Available at https://doi.org/10.3389/fbioe.2022.780553.

Pucher, P.H., MacKenzie, H., Tucker, V., Mercer, S.J., 2021. A national propensity score-matched analysis of emergency laparoscopic versus open abdominal surgery. British Journal of Surgery 108 (8), 934–940. Available at https://doi.org/10.1093/bjs/znab048.

Reyes, D.P., Carroll, D.J., Walton, M.E., Antonsen, E.L., Kerstman, E.L., 2021. Probabilistic risk assessment of prophylactic surgery before extended-duration spaceflight. Surgical Innovation 28 (5), 573–581. Available at https://doi.org/10.1177/1553350620979809.

Rogozov, L.I., 1964. Self Operation Soviet Antarctic Expedition Inf Bull. pp. 223–224.

Rover., P., 2020., Available from: https://science.nasa.gov/mission/mars-2020-perseverance/.

Rubin, D., 2019. Recommendations for a Medical System Concept of Operations for Gateway Missions. Available from: https://ntrs.nasa.gov/api/citations/20200001341/downloads/20200001341.pdf.

Scott, F.I., Osterman, M.T., Mahmoud, N.N., Lewis, J.D., 2012. Secular trends in small-bowel obstruction and adhesiolysis in the United States: 19882007. American Journal of Surgery 204 (3), 315–320. Available at https://doi.org/10.1016/j.amjsurg.2011.10.023.

Sell, N.M., Stafford, C.E., Goldstone, R.N., Kunitake, H., Francone, T.D., Cauley, C.E., et al., 2021. Delay to intervention for complicated diverticulitis is associated with higher inpatient mortality. Journal of Gastrointestinal Surgery 25 (11), 2920–2927. Available at https://doi.org/10.1007/s11605-021-04972-9.

Smith, P.W., Watkins, K., Hewlett, A., 2012. Infection control through the ages. American Journal of Infection Control 40 (1), 35–42. Available at https://doi.org/10.1016/j.ajic.2011.02.019.

Sierra Space, 2025. The Future of Commercial Space Stations. Available at: https://www.sierraspace.com/commercial-space-stations/life-space-habitat/.

Teodorescu, D.L., Okajima, S., Moten, A., Teodorescu, M.H.M., El Hechi, M., Gutierrez-Arango, S., et al., 2021. A paradigm shift in critical care infrastructure in complex settings: evaluating an ultraportable operating room to improve field surgical safety. Military Medicine 186 (Suppl 1), 295–299. Available at https://doi.org/10.1093/milmed/usaa311.

Thorsen, K., Søreide, J.A., Kvaløy, J.T., Glomsaker, T., Søreide, K., 2013. Epidemiology of perforated peptic ulcer: age- and genderadjusted analysis of incidence and mortality. World Journal of Gastroenterology 19 (3), 347–354. Available at https://doi.org/10.3748/wjg.v19.i3.347.

Tissot, C., Lecordier, M., Hitier, M., 2023. Surgical epidemiology of Antarctic stations from 1904 to 2022: a scoping review. International Journal of Circumpolar Health 82 (1). Available at https://doi.org/10.1080/22423982.2023.2235736.

Wohlgemut, J.M., Ramsay, G., Jansen, J.O., 2020. The changing face of emergency general surgery: a 20-year analysis of secular trends in demographics, diagnoses, operations, and outcomes. Annals of Surgery 271 (3), 581–589. Available at https://doi.org/10.1097/SLA.0000000000003066.

Wong, J.Y., Pfahnl, A.C., 2016. 3D printed surgical instruments evaluated by a simulated crew of a mars mission. Aerospace Medicine and Human Performance 87 (9), 806–810. Available at https://doi.org/10.3357/AMHP.4281.2016.

Chapter 28

# Clinical decision support systems for human spaceflight

**Benjamin Easter[1] and Curtis Lowery III[2]**
[1]*Department of Emergency Medicine, University of Colorado School of Medicine, Aurora, CO, United States,* [2]*Emergency Department, Minerva Medical Group, University of Texas Health Houston, Houston, TX, United States*

## 28.1 Introduction

A clinical decision support system (CDSS) is any tool designed to aid a medical officer during the decision process and/or provide procedural guidance. These can be as simple as a laminated reference card or as complex as a stacked series of artificial intelligence models capable of independent agency. The type of CDSS will change depending on provider experience, acuity of the patient, and the capabilities of the spacecraft. Whatever the underlying design, the best CDSS is always the one that a medical officer is willing and able to use and that improves patient outcomes. In terrestrial settings, CDSS tools are effective in reducing diagnostic errors, improving treatment plans, reducing task loads, and optimizing patient outcomes (Elhaddad & Hamam, 2024). Often these systems operate in the background, performing seemingly simple but often overlooked tasks such as drug interaction checking, medication dose/timing alerts, rapid interpretation and/or visualization of data, documentation and proofreading, inventory tracking, and/or suggesting reference or training materials (Russell et al., 2023; Beard et al., 2021; Jia et al., 2016; Poon, et al.,2010; Sutton, 2020; Peris-Lopez et al., 2011; Holland et al., 2020).

NASA's current paradigm for spaceflight medical operations involves a team of ground-based operators supporting an in-flight crew medical officer (CMO). Historically, the core team has involved the following:

- A crew surgeon,
- A deputy crew surgeon,
- A biomedical engineer,
- An astronaut strength, conditioning, and rehabilitation specialist,
- The office of the chief medical officer,
- The on-call flight surgeon,
- And a team of specialists available for emergency consultations should the need arise.

This level of near real-time support enables the CMO to receive between 12 and 40 hours of training while preserving a high level of onboard medical capability (Fig. 28.1).

However, this level of support is only possible for near-Earth missions with continuous, high channel capacity communications. Limitation of ground station or relay satellite availability, spacecraft mass/power/volume, and distance from Earth all result in communication delays that limit real-time care to only the expertise available onboard the vehicle itself (Krittanawong et al., 2022). Table 28.1 provides an example of how the mission changes the need for support.

This potential on board skills gap is one area where CDSS may be valuable in space, enabling CMOs to augment their own knowledge and skills and operate at a higher level than might be traditionally expected based on their training (Russell et al., 2023; Levin et al., 2023; Levin et al., 2025). Similarly, medical care often requires a team approach. The team NASA uses is one version of this, as is the more familiar terrestrial doctor–nurse–technician–specialist team. Without real-time communication, and with small crews, the complex task loads associated with medical care must be handled by 1–2 individuals. It is well established that high stress, high task load environments lead to mistakes in even

**Procedural Medicine for Spaceflight. DOI:** https://doi.org/10.1016/B978-0-443-27552-4.00024-5

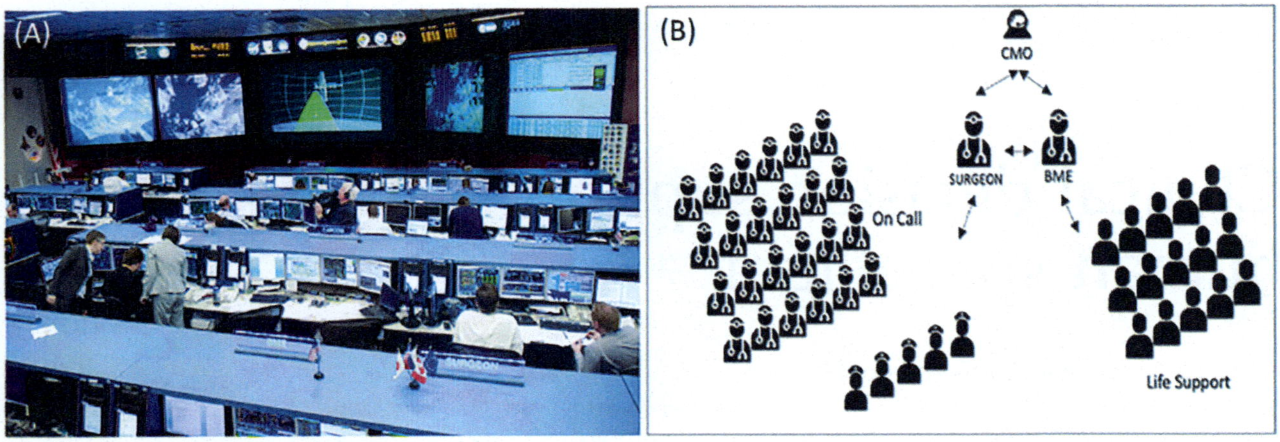


FIGURE 28.1 NASA mission control and medical support for the International Space Station.
Adapted from Russell, B.K., Burian, B.K., Hilmers, D.C., Beard, B.L., Martin, K., Pletcher, D.L., et al., 2023. The value of a spaceflight clinical decision support system for earth-independent medical operations. Nature Research, Npj Microgravity. 9 (1), https://doi.org/10.1038/s41526-023-00284-1. Available from: https://www.nature.com/npjmgrav/.

**TABLE 28.1 Comparison of medically relevant differences between representative design reference missions (Rubin et al., 2019; Krihak et al., 2019; Levin et al., 2023).**

| Challenge | Low-Earth orbit | Cis-lunar mission | Mars mission |
|---|---|---|---|
| Distance from definitive care | 400–600 kilometers | 360–405 thousand kilometers | 55–400 million kilometers |
| Two-way communication lag | Minimal | Up to a few seconds | Up to 44 minutes |
| Loss of signal | Rare, <1 hour | Rare but possible on the far side of the moon, <1 hour | Likely. Up to 2 weeks during opposition of the planets |
| Potential for resupply | Weekly or monthly | A few times/year | Up to 3 years between missions |
| Mission duration | 6–12 months | 1–6 months | 3.5 years |
| Habitable volume | ~1000 $m^3$ | ~9–200 $m^3$ | 134–1100 $m^3$ |
| Medical system allocation | ~2 $m^3$ (ISS) | ~0.25 $m^3$ (based on ISS, 0.2% of vehicle habitable volume) | ~0.25–2.2 $m^3$ (based on ISS, 0.2% of vehicle habitable volume for NASA Gateway and SpaceX Starship) |

highly trained individuals (In-Flight Breakup During Test Flight Scaled Composites SpaceShipTwo, N339SS Near Koehn Dry Lake, California, 2015; Levin et al., 2023). Since a CMO may not be a medical expert or at least not an expert in all fields of medicine, this is a substantial risk to onboard medical capability on long-duration missions. Finally, CDSS may be used to preserve previously trained skills. Over long missions, infrequent practice may degrade CMO skill. Since degraded medical skills may have high consequences, CDSS can mitigate these risks through interactive skills practice, just-in-time training, and real-time "over the shoulder" procedural guidance (Elhaddad & Hamam 2024; Bajwa et al., 2021; Blue et al., 2019; Reichard et al., 2023). For these reasons and capabilities, CDSS may be a critical component for enabling medical care on exploration class missions with small crews. This chapter provides a brief overview of how CDSS can be tailored to different space mission types (known as design reference missions or DRMs) and adapted to support astronauts with different levels of medical expertise (Blue et al., 2014; Canadian Space Agency 2024; Levin et al., 2022; Levin et al., 2023; Levin et al., 2025).

## 28.2 Foundational concepts of clinical decision support systems in spaceflight

To conceptualize what a spaceflight CDSS would look like, it is necessary to understand how and why it may be used. A space CDSS has three main functions: reduce mission risk, reduce crew task loading, and enhance crew capability with regard to managing medical events. The main factors driving the need for and capabilities of a CDSS in space are as follows:

- Level of onboard expertise,
- Time to return to definitive care,
- Availability of real-time communications and/or length of communication delay,
- Duration of mission,
- Number of crewmembers,
- Mission activity risk (e.g., does it include space suited operations, or will the crew engage in activities likely to cause injury).

As noted above, short missions close to Earth carry little risk and enable rapid return to more definitive care if needed. These missions will need little more than cue card-based emergency protocols to enable crew to prepare the ship and the patient for an emergency return. Lunar missions may experience a few seconds of time delay and require a few days to a few weeks to return an injured crewmember to Earth. A crew will have to manage the patient for this duration but will still have rapid two-way communication with Earth. For these missions, cue cards may still be sufficient for many activities, but a system capable of responding to the mission context and taking on some of the tasks associated with care such as documentation, monitoring and alerts, cueing communications, and self-organizing the presented information/care recommendations would significantly aid a CMO. Lunar missions will also serve as a proving ground for CDSS technology, given the dramatically increased complexity associated with missions to Mars.

Mars missions will need to operate largely autonomously from a medical perspective. One-way communication delays of up to 22 minutes will make procedural and specialist guidance from Earth impractical without high levels of onboard expertise (Hamilton et al., 2008). Even flying a physician may not be sufficient, given that the skills a CMO is likely to need are beyond the scope of any single terrestrial specialty (Levin et al., 2025). These missions may benefit from a fully integrated "intelligent" CDSS capable of predictive analytics, real-time health monitoring, diagnosis support, and procedural guidance. While this type of system does not yet exist, components that can support it, and the pathway for what it might look like, have already been developed.

A CDSS must adapt to the environment, vehicle, mission plan, and expertise of those using it, as well as any data processing and transmission limitations (Beard et al., 2021; Krihak et al., 2020). CDSS systems may also serve different purposes in different phases of the mission. The precise tasks will be governed by mission planning needs, but there are several areas where a CDSS system may be beneficial. Table 28.2 lists several common domains of CDSS assistive capability and the mission phases in which these specific functions may benefit practitioners.

While it is tempting to assume each of these capabilities requires computationally intensive artificial intelligence models, simple cue cards, spreadsheets, or rules based programming may be sufficient for many use cases, particularly true in cases where computational resources, thermal management, or power may be limited. As a general rule of thumb, the simplest solution is typically the best. Medical system designers should consider what is the least complex, most reliable, yet effective solution to meet mission goals and seek to optimize that solution rather than depend on innovation. In the context of CDSS, the mission goals should always be to reduce medical risk, reduce crew task load, and generally make the crew's life easier. Even the most capable CDSS is worthless if the crew is unwilling or unable to use it when it is needed.

When considering static CDSS models, one must present information in a clear, concise, and easy-to-follow manner. This often requires consulting with designers to identify the clearest size, color, and position of text and images. It is also important to consider the ability of people to follow instructions in a language different from their primary one or in degraded visibility/high-task loads. Static CDSS systems must also be integrated with inventory management and vehicle storage locations, so that items can be located quickly and without difficulty. When designing such systems, it is critical to conduct usability tests to identify how instructions may be misinterpreted and to train crew with the materials they will use to build familiarity and retrieval cues.

### 28.2.1 Artificial intelligence-based clinical decision support systems

With the caveats described above, advances in artificial intelligence (AI) systems in recent years have made them powerful tools with great potential for a context aware, dynamically responsive CDSS. AI is a catchall term that refers to a variety

**TABLE 28.2** Clinical decision support system (CDSS) domains of capability and the mission phases in which they may apply.

| | Vehicle design | Crew selection | Training and health promotion | In-flight routine care | In-flight acute care | Landing and crew recovery | Postflight rehabilitation |
|---|---|---|---|---|---|---|---|
| Probabilistic risk analysis | x | x | | x | x | | x |
| Inventory assessment and adjustment by DRM | x | | | x | x | | |
| Health monitoring and anomaly detection (sleep, nutrition, medication use, vital signs, laboratory data, image analysis, etc.) | | | x | x | x | x | x |
| Acuity assessment and triage | | | | x | x | x | |
| Differential diagnosis and condition coding | | x | x | x | x | x | x |
| Emergency medical response and stabilization | | | | | x | x | |
| Clinical guidelines and protocol guidance | | | x | x | x | | x |
| Just-in-time training | | | | x | x | x | x |
| Skills maintenance | | | x | x | x | | |
| Pharmacy interaction, contraindication, and dose timing checking | | | x | x | x | x | x |
| Inventory management and supplies tracking | | | | x | x | | |
| Medication, follow-up, and activity reminders | | | x | x | x | | x |
| Procedure guidance | | | x | x | x | | x |
| Clinical documentation | | x | x | x | x | x | x |
| Clinical communications with ground teams | | | | x | x | | |
| Flagging research eligibility | | x | x | x | x | x | x |
| Flagging anomalous or concerning trends in data | | x | x | x | x | x | x |

**TABLE 28.3 Constraints of the spaceflight environment relevant to artificial intelligence models.**

| Constraint | Example |
|---|---|
| **Environmental** | Cosmic radiation (e.g., bit flips)Limited thermal rejection of waste heatAltered gravity may affect procedures designed for terrestrial medicinePharmaceutical and nutrient degradation due to radiation, humidity, $CO_2$, and timeDistance-induced communication lag |
| **Operational** | Limited on board maintenance expertiseLimited replacements for damaged system componentsLimited crew timeLimited ability to update the systemCrew skill erosion from lack of practice |
| **Vehicle design** | Limited mass allocationLimited volume allocationLimited electrical power allocationLimited processor time and powerSmall habitable volume for privacy, infection isolation, escape from contaminationCDSS needs traded with other ship systems |
| **Data availability** | Limited bandwidthLimited data channel capacityConnectivity interruption or absenceGround system latency-induced communication lagLimited data for machine learning training setsCompetition with other vehicle telemetry needs |
| **Model limitations** | Limited in-space datasets (terrestrial presentation of an illness may not reflect the illness's presentation in space) Incomplete data availabilityHallucination risk (specific to large language models, LLMs) |

**Source:** From Levin, D., Russell, B. and Campbell, S., 2024. Clinical decision support systems in spaceflight operations, in Shelhamer, M.J. and Antonsen, E. (eds.) Systems medicine for human spaceflight. Singapore: World Scientific Publishing Company, pp. 255–290.

of computational models each capable of responding in real time to input data and making autonomous decisions. As with everything in the spaceflight world, the benefits of these systems must be carefully traded against the limitations and required resources. Different models will be better for different tasks; for example, convolutional neural networks are excellent for identifying images, large language models are fantastic at interpreting and communicating in natural language, and expert systems are great for consistency and explainability. However, all AI models are subject to a few common constraints: they require power, cooling, processor time, and access to data and must have functional interfaces with both crew and vehicle. Table 28.3 presents some limitations of the spaceflight environment that are relevant to AI models.

None of these limitations is insurmountable, but the time, effort, system complexity, and reliability must be considered when deciding to rely on a CDSS to reduce mission risk as part of a medical system. Since the consequences of an inadequate medical system can be life threatening, medical devices are often considered mission or safety critical, and the reliability of these systems must be high with sufficient built-in redundancy to ensure mission success.

A few of these models, and their pluses and minuses, will be discussed here. Pretrained transformers, such as large language models, can interpret natural language questions and translate medical decision trees or procedural guidance into training-level-appropriate guidance. These models can also be associated with voice interpretation to allow CMOs to have a conversation with the CDSS, and rule-based CDSS utilize preprogrammed logic and clinical algorithms to guide astronauts through diagnostics and treatment protocols, often embedded within prewritten procedures and direct contact with ground-based medical specialists through private medical conferences to standardize care and reduce error in high-risk environments. Knowledge-based systems incorporate structured medical data, such as symptom checklists or pharmacologic interactions, to support clinical reasoning. Machine learning-based CDSS can analyze biosensor inputs and historical medical data to predict adverse events or adapt recommendations in real time. Workflow-integrated CDSS, which combine procedural prompts with task tracking and documentation, help maintain situational awareness and guide astronauts step by step through complex procedures, which are essential when operating with limited medical expertise and no real-time ground support.

#### *28.2.1.1 Types of artificial intelligence models*

It is important to remember that AI is not a monolith, and at its core, AI is a tool. Just as the world's best screwdriver would make a terrible suture needle, it is important to understand where each AI model performs well and where it may fail. Table 28.4 summarizes some of the more common computational models used in AI systems, along with their strengths and limitations.

It is important to recognize that no single extant model is sufficient to perform all the functions of a CDSS. High reliability systems will likely incorporate multiple models working in concert. Such a system may combine a Bayesian

**TABLE 28.4** Major artificial intelligence model types, potential spaceflight applicability, strengths, and key limitations.

| Category | Model examples | Capabilities | Key strengths | Key weaknesses | Potential applicability in spaceflight |
|---|---|---|---|---|---|
| **Rule-based systems** | Expert systems<br>Guideline engines | Flow chart like deterministic logic based on predefined rules (e.g., If this, then that). Can be set up with Boolean logic gates to enable complex behavior. | Transparent<br>Explainable<br>Low computing power | Dependent on predefined scenarios<br>Limited adaptability | Procedural medicine<br>Checklists,<br>Contraindication, allergy, drug interaction checks |
| **Classical supervised machine learning** | Logistic regression<br>Random forest<br>Gradient boosting | Predicts risk from structured data | Strong performance on tabular data<br>Relatively transparent<br>Efficient | Potentially inaccurate with changes from training datasets<br>Requires calibration | Triage<br>Patient deterioration prediction based on trends,<br>Treatment prioritization for multiple patients |
| **Bayesian models** | Bayesian networks<br>Probabilistic inference | Explicit modeling of uncertainty | Functions with incomplete data<br>Integrates expert knowledge with probabilistic inference | Requires substantial expert input<br>Limited scalability for data with many dependent variables | Diagnostic reasoning under uncertainty<br>Prediction with intermittent sensor data or partial information |
| **Convolutional neural networks (CNNs)** | Computer vision | Image and spatial signal interpretation | Image interpretation and recognition | Requires massive amounts of data to train<br>Potentially inaccurate with changes from training datasets<br>Requires calibration | Medical imaging interpretation,<br>Exam assistance,<br>Facial recognition and tracking |
| **Recurrent neural networks (RNNs/LSTMs)** | | Temporal pattern recognition | Captures physiological trends and trajectories | Lacks transparency<br>Training instability | Continuous monitoring of vitals,<br>Anomaly detection and prediction of data taken over time |
| **Transformer models** | Bidirectional encoder representations from transformers (BERT) | Integrates multiple sources and types of data over time to infer a predicted outcome | Natural Language<br>Multimodal fusion<br>Tracking trends in multiple data types across time | Computational and memory intensive Opaque reasoning | Multimodal crew health monitoring to recognize and interpret state changes, e.g., nausea after launch,<br>Complex diagnostic reasoning |
| **Large language models (LLMs)** | Modern chatbot assistants | Natural language understanding and generation | Conversational communication<br>Data summarization and explanation | Nondeterministic outputs<br>May confidently provide incorrect answers (hallucinations) | Procedural guidance<br>Documentation Interface layer, when used with constraints to limit hallucination |

model layered with a deterministic expert system to enable rule-based processing with incomplete data and a tightly constrained large language model (LLM) to communicate with the crewmember and/or guide procedural interventions.

Such a system might be able to interview a patient using a tightly constrained LLM with each answer guiding the next question or test request. The collected data would be fed to a Bayesian model that fills in as many datapoints as possible in an expert system and identifies the most likely outcome. Additional transformer style models may track trends in vital signs and recognize deviations from normal in the context of recent events and a convolutional neural network may interpret ultrasound data collected by the CMO. These data can then all be combined through the Bayesian network to generate a most likely diagnosis, and a second, tightly constrained LLM can interpret the data and communicate it to the patient. A third LLM/Bayesian expert system combination may then guide the patient or CMO through the steps of a procedure to manage the diagnosed illness.

The optimal model combination may also need to change depending on who is using the system, since an expert physician may not need the same level of guidance and may wish to see more data and more complex data visualizations than a geologist cross-trained as a CMO. Another factor is the acuity and context of use. Just as a preventive medicine doctor, a general surgeon, and an emergency physician all use different mental models to approach a patient, the same is likely true of a CDSS. The AI model stacks used to monitor long-term health, diagnose a minor issue, manage a major emergency, or guide a medical procedure are likely to be different. LLMs may be able to compensate for some of this, but a CDSS will need to be able to recognize what function is being performed and ideally switch to the optimal mode.

Such systems remain notional, and there are many details and complexities to work out. However, the concepts presented above illustrate the complexity required for a functional CDSS and the engineering, testing, validation, and processing power that may be required to construct one that functions at the level needed to support health care in space.

## 28.2.2 Types of decision support

CDSS in spaceflight can provide three main types of decision support: alerting systems, diagnostic support, and therapeutic support. In terrestrial hospitals, alerting systems are able to monitor vital signs, medication interactions and dosage, ventilators, and numerous other medical devices. If any of these devices or orders are noted to be outside the prescribed parameters, the CDSS is able to alert the medical team of a potential problem (Park et al., 2022). Alerting systems in spaceflight could function similarly by continuously monitoring an astronaut's vital signs and environmental factors through wearables and noninvasive environmental sensors to detect abnormalities that may indicate a developing medical issue. For instance, continuous heart rate and blood pressure monitoring can identify early signs of cardiovascular strain, ensuring timely medical assessment. These alerting mechanisms are crucial in spaceflight, where early detection and intervention can prevent conditions from escalating in an environment with limited medical resources. However, it is also important to consider the potential for false-positive notifications, which may contribute to alarm fatigue if the signal-to-noise ratio becomes suboptimal. This may drive astronauts to turn the system off or simply ignore it.

### *28.2.2.1 Diagnostic support*

***Diagnosis*** involves processing data from various sources, such as laboratory tests, imaging modalities, and expert medical guidance and integrating this information into a pattern (Sutton et al., 2020). This pattern can then be translated into actionable information for the onboard medical expert to accurately deliver treatment for a specific medical condition. Given the constraints of space travel, CDSS should be tailored specifically for the mission and may need to be equipped with onboard capabilities anywhere from a deck of flowcharts to blood tests, ultrasound scans, and a stacked series of pretrained AI models to identify illnesses accurately and quickly. For example, an ultrasound assessment of a suspected deep vein thrombosis (DVT) could be interpreted by the CDSS, in conjunction with the CMO, which would then provide the CMO guidance on potential treatment options. CDSS can integrate multiple physiological signals to identify systemic issues. A case study from NASA's Exploration Medical Capability (ExMC) CDSS Concept of Operations 5.4 describes an advanced functionality in which the system detected two seemingly unrelated events, a persistent headache and decreased exercise tolerance, and correlated them to a potential increased $CO_2$ concentration, ultimately identifying a system malfunction and guiding appropriate $CO_2$ scrubber maintenance (Krihak et al., 2021). For future spaceflight missions, additional data categories may be incorporated to provide a more comprehensive assessment of crew health and performance. These may include cabin atmosphere variability, exercise capacity and compliance, changes in medication usage, real-time monitoring of voice patterns and tone to assess interpersonal dynamics, and tracking of food intake and overall nutritional status. Integrating these parameters into support systems can enhance situational awareness and enable more informed medical decision-making in isolated and resource-limited environments.

### 28.2.2.2 *Therapeutic support*

Therapeutic intervention/guidance is another critical function of CDSS, offering evidence-based recommendations for medications, procedural interventions, and rehabilitation strategies. A CDSS can provide step-by-step guidance for medical procedures, such as the administration of intravenous fluids, wound management, or even ultrasound-guided procedures to augment limited onboard expertise (Baloescu et al., 2025). If an astronaut develops an infection, the system could theoretically analyze available antibiotics and their resistance profiles to recommend the most effective treatment. This is similar to how EHR software integrates local antibiograms at individual hospitals to target the ideal antibiotic treatments for specific infections (Katzman et al., 2019). Therapeutic support extends beyond acute interventions and is able to assist in chronic condition management as well, ensuring that astronauts maintain optimal health over extended missions by guiding exercise regimens, mental health evaluations, nutritional adjustments, and medication adherence.

An example of this is the ExMC scenario 5.5, which was developed to evaluate the risk of bone fractures and recommend mitigation strategies during long-term spaceflight. Spaceflight-induced bone loss is an adaptive response to reduced g-force loading on the skeleton, which results in bone resorption and overall decreased bone density over time. The resultant bone density loss results in a higher risk of potential fracture. During this scenario, the CDSS was able to collect data during daily exercise (vital signs, calories expended, resistive loads, etc.), data on bone mass, mineral density, geometry, and bone microarchitecture as well as dietary intake and mineral excretion (urine samples). Additionally, the CDSS analyzed the crew's range of motion and coordination during daily tasks and exercises through video interpretation. In this scenario, the CDSS collected data throughout the mission duration to evaluate the level of spaceflight-induced bone loss, predict the risk of fracture, and provide mitigation strategies prior to arrival on Mars (Krihak et al., 2021). Recommendations included dietary modifications, increased exercise load, and increased fluid intake. Daily feedback and alterations are tailored to each individual crewmember to optimize performance and minimize risk during the mission. The CDSS could also analyze the existing medical supplies versus the remaining mission duration and recommend when to use/not use various consumables.

In order for the CDSS to be effective in spaceflight, it must integrate seamlessly with onboard spacecraft systems and sensors to enable comprehensive health monitoring and management. Integration with environmental control and life support systems ensures that CDSS can track oxygen levels, carbon dioxide buildup, cabin pressure, and radiation exposure, all of which can directly affect astronaut health. The CDSS should operate in the background to minimize human task loads, directly interfacing with medical and spacecraft diagnostic sensors and incorporating results into its decision algorithms. A CDSS linked to crew activity tracking may allow it to provide personalized schedule adjustments to help mitigate fatigue, optimize physical conditioning, and aid in task planning. As missions extend beyond low-Earth orbit (LEO), CDSS tools may help crewmembers manage complex health care in the setting of limited and/or delayed support from Earth.

## 28.3 Clinical decision support systems from low-Earth orbit to lunar and deep space/Mars

As of the drafting of this chapter in 2025, we are entering what has been dubbed the new "golden age" of space travel (Solovyev, 2018; Devika, 2024). Future human space missions will span a spectrum of environments, from LEO to the Moon, and eventually to Mars and deep space. LEO missions, such as those conducted by NASA aboard the International Space Station (ISS) and planned commercial ventures like SpaceX's Starship, VAST's Haven program, and Blue Origin's Orbital Reef, benefit from proximity to Earth, allowing for real-time telemedicine and rapid return in emergencies (SpaceX and SpaceX, n.d.; Haven-1, n.d.; Haven-2 VAST, n.d.; Blue Origin 2023). Lunar missions, including NASA's Artemis program and planned commercial lunar surface habitats, will face a communication delay of ~5–14 seconds (round trip) and require more robust onboard medical capabilities to manage delayed medical consultations and prolonged stays (Parisi et al., 2023; NASA, 2024). As missions extend to Mars and deep space, with distances ranging from 55–400 million km (35–249 million miles) and communication delays of up to 22 minutes one-way, the need for autonomous medical care increases dramatically. Future Mars missions will necessitate advanced onboard diagnostics, surgical capabilities, and AI-driven decision support to manage medical emergencies independently (Sanders & Scott, 2024; Babocs et al., 2024).

### 28.3.1 Overview of low-Earth orbit mission profile

LEO (orbit of less than 2000 km) presents a unique testing ground for space-based clinical decision support, with the ISS serving as the primary example (Guzman, 2024). While the crew onboard the ISS are physically isolated from Earth-based medical support, they are able to communicate nearly instantaneously with mission control. This allows for a lower level of crew medical training since hundreds of medical experts are readily available to assist with any concern. Future commercial orbital platforms (e.g., Vast's Haven 1 and 2) will expand this landscape, introducing new medical considerations based on mission duration, crew medical history, crew training levels, and support infrastructure. While physiological adaptations such as spaceflight-associated neuroocular syndrome and bone loss are significant in long-duration missions, shorter but more frequent stays on commercial platforms with less experienced and less healthy crew with more limited medical training may result in more acute conditions and challenges in managing them, such as differentiating space adaptation syndrome from an intracranial hemorrhage. The presence of robust ground support allows for medical guidance in near real time, significantly shaping clinical decision-making. For each of the design reference missions (LEO, Lunar, and Mars), three factors will have significant influence on the autonomy of the crews:

Communication delay: In LEO, communication delays are minimal, typically limited to a few seconds due to the close proximity to Earth. This enables real-time procedural guidance from ground-based medical teams, ensuring a high level of remote support for astronaut health care (Parisi et al., 2023).

Resupply time: With platforms such as the ISS, there is frequent resupply of spare parts and other resources from visiting vehicles to maintain the vehicle and crew health (Parisi et al., 2023).

Return time to Earth: Evacuation from the ISS to Earth can be achieved within hours via the Crew Dragon, or Soyuz spacecraft. This short return window often makes medical evacuation the preferred response to serious conditions rather than prolonged onboard management.

## 28.3.2 Clinical decision support strategies for low-Earth orbit

In LEO, medical decision support strategies leverage real-time communication with Earth-based medical teams, allowing for telemedicine-driven interventions. NASA astronauts aboard the ISS receive continuous medical oversight from flight surgeons at Mission Control, who can guide them through procedures, interpret medical data, and provide real-time recommendations. This strategy minimizes the need for advanced onboard diagnostics and treatment capabilities, as emergency evacuations can be executed within hours if necessary. Commercial spaceflight companies like SpaceX, Blue Origin, and Axiom Space are developing orbital missions for private astronauts and tourists, presenting new challenges for medical decision support. Unlike NASA-trained crews, commercial space travelers may have varied medical problems and may lack medical expertise, necessitating more intuitive and user-friendly medical technologies. AI-driven diagnostic tools, voice-activated medical guidance, and wearable health monitors are being explored to enhance in-flight medical support for nonprofessional astronauts (NASA & Office of Inspector General 2024; Getteau, 2025). Commercial space operators may adopt hybrid medical support models, where onboard human or AI-assisted diagnostics work in tandem with ground-based telemedicine to ensure rapid and effective medical decision-making. As commercial spaceflight grows, LEO missions will require increasingly automated and accessible medical solutions to support a broader range of crew compositions with varying levels of medical training.

## 28.3.3 Case studies in low-Earth orbit medical emergencies

A real-life example of a medical event in LEO includes a NASA crewmember aboard the ISS who developed an internal jugular vein thrombosis (deep vein thrombosis or DVT) identified through ultrasound (Auñón-Chancellor et al., 2020). Decision support involved remote consultation with terrestrial specialists, initiation of anticoagulation therapy in orbit under close monitoring, daily crew-guided imaging assessments with Earth-based review to adjust treatment accordingly, resupply of anticoagulants to the ISS, and successful completion of the mission without requiring evacuation. A second, more time-critical, scenario could be where a crewmember is repositioning a massive object and gets crushed between the hull and the object, resulting in blunt chest trauma. In this case, the CMO would then immediately contact ground control for real-time support. Vital sign deviations would trigger onboard CDSS alerts, flagging possible respiratory compromise. However, instead of autonomous action, the system would primarily serve as a triage and workflow assistant: potentially guiding ultrasound probe placement (if the CMO is inexperienced), documenting findings, and facilitating communication with Earth-based experts. A diagnosis of pneumothorax would be confirmed in partnership with ground-based flight surgeons, who would then supervise interventions, through real-time video supervision, such as needle decompression or oxygen administration remotely. Depending on the severity, the patient might be monitored conservatively or stabilized for evacuation to Earth within hours (Yadao et al., 2022). The proximity to definitive care and the constant expert input significantly reduce the burden of autonomous decision-making for the crew, making the CDSS an efficiency and safety enhancer rather than a frontline diagnostic tool.

Future commercial space ventures will include diverse international crews with varying medical training and primary languages (e.g., the SpaceX Fram2 or Axiom crew missions; NASA, n.d.). A potential challenge could arise with an international crew experiencing language barriers during a medical emergency. Consider a passenger on an orbital tourism mission presenting with severe abdominal pain. Communication barriers may exist due to a non-English-speaking participant struggling to describe pain symptoms in addition to the sociocultural norms around symptoms, presentation, and expectations of care. Decision-making challenges may arise as ground medical teams interpret translated symptoms and guide untrained crewmembers in evaluation and management. Risk stratification would be required to determine if the condition is a benign gastrointestinal problem or a serious emergency necessitating immediate deorbit. Concepts akin to Star Trek's "universal translator" or Douglas Adams' "Babel Fish" are becoming practical tools in the hospital setting, allowing medical teams to understand and respond to emergencies regardless of language barriers (Zappatore and Ruggieri, 2023). Integrating such systems into medical procedures would improve efficiency and patient outcomes in commercial spaceflight settings.

### 28.3.4 Overview of lunar missions

The next step beyond LEO is the Moon, where NASA's Artemis program, along with commercial lunar exploration efforts, aims to establish a sustainable human presence (NASA, 2024a). Lunar missions operate at a distance of ~380,000 km from Earth, with a communication delay of ~5–14 seconds round trip (due to communication platform latencies; NASA, 2024a). This relatively short delay still allows for some near real-time interaction with Earth-based medical teams but complicates the immediate response to emergencies and the ability to engage in step-by-step procedural guidance. Additionally, returning from the Moon takes days rather than hours, meaning that medical contingencies must be addressed onboard the spacecraft or lunar base, similar to Antarctic over-winter expeditions (Antarctic Stations (NSF) – NASA, n.d.). For sustained lunar missions, medical autonomy becomes critical. Crews will require enhanced diagnostic tools, such as portable ultrasound and AI-assisted decision-making software, to guide them in medical procedures. Pharmaceuticals must be carefully managed to ensure long-term viability, and contingency plans for medical emergencies must include stabilizing critical patients until they can return to Earth (Medical System Foundation Overview for Long-Duration Lunar Orbit and Lunar Surface Operations (LDLOLS) Missions., 2023). Evacuation from the Moon is a complex process requiring launch from the lunar surface, docking with an orbiting spacecraft, and transit back to Earth. This process can take several days to weeks, making immediate evacuation for medical emergencies impractical. As a result, medical care must focus on definitive treatment strategies rather than rapid stabilization for evacuation. Additionally, lunar gravity (1/6th of Earth's gravity) presents unknowns regarding long-term physiological effects, particularly for organ function and wound healing. Commercial entities like SpaceX and Blue Origin, alongside NASA, will need to develop medical systems capable of supporting astronauts for weeks to months in a lunar environment.

A short-term lunar settlement presents a unique operational setting for clinical decision support, distinct from LEO due to the Moon's gravity, radiation exposure, and communication delays. Unlike the ISS, where real-time medical guidance is possible, lunar medical care must incorporate greater autonomy and advanced decision-support tools (Terai & Nitta, 1991). The extended duration of lunar missions further necessitates both preventative care as well as increased emergent care capabilities.

### 28.3.5 Clinical decision support strategies for lunar exploration and settlement

For lunar exploration and settlement, clinical decision support strategies must bridge the gap between real-time telemedicine and autonomous medical care due to the increased communication delay and multi-day return time to Earth. Crewmembers will require greater medical training, and decision support systems must integrate AI-driven diagnostics, augmented reality (AR) medical guidance, and portable imaging tools, such as ultrasound and portable X-ray machines, to aid in medical assessments (Lerner et al., 2023). Commercial ventures, including SpaceX's Starship lunar missions, Blue Origin's Blue Moon lander, and planned private lunar habitats, will need similar medical infrastructure, particularly for long-duration stays. Given the limited real-time support from Earth, lunar medical care must emphasize crew autonomy. CMOs require more extensive medical training than their LEO counterparts, and onboard AI-driven CDSS must be capable of guiding complex procedures. AR overlays and interactive holographic guidance can assist nonmedical personnel in performing medical procedures. AI-assisted clinical decision-making tools will play a critical role in guiding astronauts through complex medical scenarios, while machine learning algorithms can help predict and mitigate health risks. Wearable biosensors and remote patient monitoring will also be key to tracking crew health trends over extended missions. Continuous monitoring will allow for tracking of astronauts' radiation dosage, thermoregulation, $CO_2$ environment, and physical health, enabling timely feedback and supporting early detection of potential health issues

(Getteau, 2025). Given the Moon's reduced gravity and unknown effects on long-term physiology, lunar medical systems must be adaptable and capable of addressing both anticipated and unforeseen medical conditions. Lunar medicine will transition from the stabilization-focused model of LEO to one emphasizing self-sufficiency. Protocols must be developed for managing chronic conditions, trauma, and radiation-induced illnesses without immediate Earth return. Telemedicine will still play a role, but clinical decision-making will be increasingly dependent on onboard expertise and CDSS automation. The combination of AI, telemedicine, and enhanced onboard medical training will be essential for ensuring crew safety as humans establish a sustained presence on the Moon (Medical System Foundation Overview for Long-Duration Lunar Orbit and Lunar Surface Operations (LDLOLS) Missions, 2023).

### 28.3.6 Case studies in lunar medical emergencies

Using the same real-life ISS example presented previously, consider an astronaut with a clot in the internal jugular vein. How might diagnosis and treatment be different on a Lunar expedition? The diagnosis of the medical condition would likely be the same with onboard imaging (portable ultrasound) used to confirm the clot's presence and resolution over time. These images would need to be largely acquired autonomously; real-time ultrasound feedback would be impossible with the anticipated communications delay, and step-by-step guidance would be delayed (Asachi et al., 2023). Without substantial preflight training or AI augmentation, these images are likely to be suboptimal for diagnosis of subtle findings. Unlike in LEO, where real-time consultations were available, lunar mission medical officers must rely on slightly delayed messaging for Earth-based specialist input while managing the condition autonomously. For the DVT, the diagnosis and treatment would likely proceed in a similar fashion to the case in LEO. Given the weight and spatial constraints of lunar medical kits, anticoagulation therapy must be carefully managed to balance clot resolution with bleeding risk in reduced gravity. Resupply, like in LEO, will be prescheduled and additional medications can be sent with the next cargo resupply. Without rapid return to Earth, longitudinal monitoring of clot progression and potential complications (e.g., embolism risk or proximal propagation of the thrombus) must be conducted onboard, integrating predictive modeling to guide therapy adjustments.

Alternatively, the more time-critical scenario of blunt chest trauma resulting in rib fractures and traumatic pneumothorax will likely necessitate a more robust system. A CDSS integrated with physiologic sensors may be able to detect early signs of deterioration by monitoring vital signs autonomously. This would enable the CMO to attend to other tasks and/or set up for a procedure without having to divide their attention.

### 28.3.7 Overview of deep space/Mars missions

Deep space and Mars missions present a drastically different medical environment from LEO, requiring greater autonomy in clinical decision-making. Unlike LEO, where real-time Earth-based support is possible, deep space missions must contend with significant communication delays from 5 to 20 minutes, each way. This precludes real-time medical consultation and necessitates increased onboard diagnostic and procedural autonomy. Additionally, early Mars missions will likely need to launch with all necessary medical supplies for the entire duration. This constraint necessitates careful inventory management, modular and multipurpose medical equipment, innovations such as 3D printing of medical tools, and in-situ pharmaceutical production. Using current chemical rocket technology, return from Mars is possible when the two planets are in appropriate alignment, approximately every 22 months. Accordingly, return to earth is not possible on demand. Medical conditions that require emergent medical care must be managed entirely onboard, shifting the paradigm from stabilization and evacuation (LEO and Lunar) to long-term treatment and recovery in space. Future Artemis lunar missions, gateway operations, and eventual crewed Mars expeditions will necessitate robust autonomous medical systems and enhanced crew medical training to manage health concerns independently. Missions beyond Luna expose astronauts to prolonged microgravity, increased ionizing and galactic cosmic radiation exposure, as well as the psychological effects of extreme isolation. The mission duration, weeks to months for lunar expeditions and years for Mars, compounds these risks. Unlike short-duration commercial orbital flights, deep-space travel requires crewmembers to function as both patients and caregivers, managing complex medical conditions with limited external support.

### 28.3.8 Deep space/Mars clinical decision support systems

Deep space and Mars expedition CDSS must be designed to function autonomously, given communication delays, resource constraints, and the need for on-site medical decision-making. Astronauts must receive substantial medical training preflight for these long-duration missions. However, even the best physician loses competency without practice and active learning (Cecilio-Fernandes et al., 2017). The support system could also be designed to provide continuing space repetition training throughout the flight with AR/VR medical emergency simulations and continuing education lessons to keep the crew's clinical acumen sharp and ready for the most likely or consequential conditions (NASA, n.d.; Perez et al., 2013). AI-enhanced spaced repetition teaching tools are already being used to educate emergency medicine residents (Tarzia and Patel, 2024). These systems should integrate AI-driven diagnostic tools, predictive analytics, and real-time physiological monitoring to assist crewmembers in managing medical events. Advanced telemedicine capabilities, AR for procedural guidance, and offline-accessible medical databases will enhance crew autonomy in delivering care. Simulation-based training and adaptive learning modules will ensure crew proficiency in utilizing CDSS effectively. A tiered decision-making framework, incorporating immediate CDSS-informed crew-led interventions and delayed consultation with Earth-based medical experts, will optimize medical responses, ensuring timely and evidence-based care in deep space missions.

### 28.3.9 Cases in deep-space medical emergencies

Again examining the past ISS example, if an astronaut developed an internal jugular deep venous thrombosis, how would diagnosis and treatment occur during a Mars expedition? On this mission, there would likely be at least one highly trained medical crewmember, ideally a physician, who would be able to effectively operate an ultrasound device and be assisted by AI support as needed. Image acquisition and interpretation would occur on the Martian surface with further consultation with an Earth-based physician delayed by over 1 hour (message transit time plus Earth-based data review and consultations). In this relatively nonemergent scenario, treatment initiation could be delayed until recommendations from Earth are received. Additionally, an alternative dosing regimen might need to be considered due to the lack of immediate resupply on Mars mission profiles.

Consider the more emergent prior theoretical example of a direct blow to the chest, resulting in a traumatic pneumothorax. Diagnosing and managing a pneumothorax on a Mars mission is extremely time-critical and requires substantial medical autonomy as well as immediate medical decision-making to prevent the death of the crew. Given the communication delay with Earth, CDSS would assist in early recognition by integrating real-time physiological monitoring, including respiratory rate, oxygen saturation, and end-tidal $CO_2$ levels. If recent trauma is noted and deviations from baseline vitals suggest a pneumothorax, the system would alert the CMO and recommend ultrasound and/or X-ray imaging (Sze-To et al., 2021). AI-assisted analysis of the ultrasound exam would help confirm the diagnosis by detecting the absence of lung sliding or identifying a "lung point" sign (Husain et al., 2012). Once diagnosed, the CDSS would provide step-by-step procedural guidance for management based on severity. For a stable pneumothorax, it could recommend supplemental oxygen to enhance nitrogen washout and establish continuous monitoring thresholds to detect resolution or worsening (Chadha & Cohn, 1983). If tension pneumothorax is diagnosed, the support system would initiate an emergency alert and guide the crew through needle decompression, offering procedural instructions via AR or an interactive interface (Jalota & Sayad, 2025). It would help identify anatomical landmarks and ensure proper needle placement, followed by recommendations for thoracostomy and chest tube insertion. The crew and CDSS would continue monitoring the patient by analyzing trends in vital signs and imaging data to detect complications or recurrence of the pneumothorax. The system could also use video/audio monitoring to evaluate for patient pain level and recommend when repeat pain medication dosing is appropriate (El-Tallawy et al., 2024). The CMO along with the CDSS and ground control would assess fitness for return to duties, the risk of barotrauma upon reentering a pressurized suit, and whether extended rest or reintegration into mission tasks is advisable.

## 28.4 Future directions

Predictive analytics will revolutionize space medicine by continuously analyzing biomedical data (through wearables, waste sampling, video, and audio) and identifying health issues before symptoms emerge. For example, AI tools could more rapidly detect cardiac arrhythmias, radiation-induced cellular damage, or musculoskeletal degeneration, allowing astronauts to take preventive measures before a condition worsens (Medical System Foundation Overview for Long-Duration Lunar Orbit and Lunar Surface Operations (LDLOLS) Missions., 2023). This capability is particularly crucial in environments where immediate medical evacuation is impossible, such as a Mars mission or lunar base. Adaptive training using AI-driven simulations will allow astronauts to receive just-in-time medical education tailored to their real-time health status and mission conditions. Rather than relying on static pre-mission training, AI could adjust training protocols dynamically, ensuring that astronauts maintain proficiency in medical procedures they are most likely to need. AR and virtual reality (VR) integration has the potential to further enhance medical autonomy by providing real-time overlays to guide nonmedical crewmembers through complex medical procedures, including surgery. AR could display step-by-step procedural guidance, while VR simulations could allow astronauts to rehearse life-saving interventions in a fully immersive, zero-risk environment.

### 28.4.1 Advancements in artificial intelligence-driven telemedicine and health monitoring

Several ongoing research areas are shaping the next generation of autonomous space healthcare. AI-driven medical simulators and holographic medical tutors are being developed to enhance medical training, allowing crewmembers to practice medical scenarios using advanced interactive models (Blue et al., 2014; Moro et al., 2020). In telemedicine, NASA and ESA are exploring AI-enhanced remote robotic surgery, which could allow Earth-based surgeons to operate on astronauts using teleoperated robotic systems, even with communication delays (Pantalone et al., 2021). Future AI-driven health monitoring systems will integrate biosensors and wearables to provide continuous real-time analytics on astronauts' vital signs, hydration levels, muscle atrophy, and even psychological well-being. These systems could automatically trigger alerts and recommend interventions, such as hydration adjustments, vitamin supplementation, or resistance-based countermeasures for bone density loss. One of the most exciting advancements in space medicine is the development of closed-loop medical intervention systems, where AI not only detects medical anomalies but also initiates appropriate treatments. For example, an AI-controlled insulin pump could regulate blood sugar levels in diabetic astronauts based on their eating habits and individual response to insulin, and automated drug-delivery systems could administer medication based on real-time physiological data as well as historic personalized response data (Cambuli and Baroni, 2023). These autonomous interventions will be crucial for handling medical emergencies with less immediate human oversight, particularly during deep-space missions where communication with Earth is delayed or disrupted.

### 28.4.2 In-situ medical manufacturing: 3D printing pharmaceuticals, medical devices, and organs

As space missions extend in duration and distance, resupplying medical supplies from Earth becomes impractical. The future of space health care will likely involve in-situ pharmaceutical production using 3D printing technology to manufacture medications on demand (Seoane-Viaño et al., 2022). This approach could eliminate the need for large stockpiles of perishable drugs, reducing payload weight and ensuring astronauts always have access to fresh, customized medications tailored to their specific health needs. Onboard pharmaceutical synthesis could also adapt drug formulations in real time, compensating for changes in astronaut metabolism due to prolonged microgravity exposure. 3D printing and bioreactors will allow astronauts to fabricate surgical tools, prosthetics, and even bioengineered tissue patches for wound healing (George et al., 2017; Weng et al., 2021; Waldburger et al., 2021). This capability would enable crews to produce medical instruments on demand, reducing dependency on Earth-based supply chains. In the longer term, bioprinting and organ fabrication may become viable solutions for treating organ damage or severe injuries in space. Stem cell-based organ printing could enable astronauts to generate replacement tissues or entire organs, revolutionizing long-term space medicine and making permanent space habitation more feasible (Rezapour Sarabi et al., 2023; NASA, n.d.).

### 28.4.3 Multiplanetary healthcare: preparing for lunar and Martian colonies

As humanity expands beyond Earth, developing multiplanetary healthcare systems will be invaluable for maintaining astronaut health in isolated environments. Future Moon and Mars colonies will require self-sufficient healthcare infrastructures that minimize reliance on Earth. Key challenges include establishing a stable medical supply chain, determining the feasibility of robotic-assisted surgery in low gravity and mitigating radiation exposure risks to prevent long-term health complications. The ultimate vision for space medicine includes AI-powered Martian hospitals, robotic assistants, and fully autonomous CDSS, capable of diagnosing and treating some medical conditions. Offloading some of this manual labor and cognitive load will allow for smaller human crews to perform more of the tasks essential for human exploration. Data gathered during these initial expedition-class missions will be used to train future AI models, thereby enhancing care delivery during subsequent missions. Such advancements will not only enable sustainable human life beyond Earth but also contribute to revolutionary breakthroughs in terrestrial medicine, benefiting health care on Earth as well.

## AI disclosure

During the preparation of this work, the author(s) used OpenAI in order to correct grammar, formatting, and spelling. After using this tool/service, the author(s) reviewed and edited the content as needed and take(s) full responsibility for the content of the publication.

## References


Antarctic Stations (NSF) - NASA, n.d. NASA. https://www.nasa.gov/antarctic-stations-nsf/.

Asachi, P., Ghanem, G., Burton, J., Aintablian, H., Chiem, A., 2023. Utility of ultrasound in managing acute medical conditions in space: a scoping review. The Ultrasound Journal 15 (1). Available at https://doi.org/10.1186/s13089-023-00349-y.

Auñón-Chancellor, S.M., Pattarini, J.M., Moll, S., Sargsyan, A., 2020. Venous thrombosis during spaceflight. New England Journal of Medicine 382 (1), 89–90. Available at https://doi.org/10.1056/nejmc1905875.

Babocs, D., Preda, A., Christiansen, R., 2024. Reaching mars: medical risks and potential surgical conditions in the Martian environment and during long-duration spaceflight. Journal of Space Safety Engineering. Available at https://doi.org/10.1016/j.jsse.2024.09.002.

Bajwa, J., Munir, U., Nori, A., Williams, B., 2021. Artificial intelligence in healthcare: transforming the practice of medicine. Future Healthcare Journal 8 (2), e188–e194. Available at https://doi.org/10.7861/fhj.2021-0095.

Baloescu, C., Bailitz, J., Cheema, B., et al., 2025. Artificial intelligence–guided lung ultrasound by nonexperts. JAMA Cardiology. Available at https://doi.org/10.1001/jamacardio.2024.4991 Published online January 15.

Beard, B., Russell, B., Krihak, M.L., Burian, B., Noel, J., Beaugrand, D., et al., 2021. Supporting crew medical decisions on deep space missions: a real-time performance monitoring capability. NASA Technical Report Server (NTRS), IAAS Leiden, NE.

Blue Origin, 2023. LEO Destinations I Blue Origin December 6. Available at https://www.blueorigin.com/destinations.

Blue, R.S., Bayuse, T.M., Daniels, V.R., Wotring, V.E., Suresh, R., Mulcahy, R.A., et al., 2019. Supplying a pharmacy for NASA exploration spaceflight: challenges and current understanding. NPJ Microgravity 5. Available at https://doi.org/10.1038/s41526-019-0075-2.

Blue, R.S., Bridge, L.M., Chough, N.G., Cushman, J., Khpal, M., Watkins, S., et al., 2014. Identification of Medical Training Methods for Exploration Missions. Available at https://ntrs.nasa.gov/api/citations/20130013532/downloads/20130013532_Final.pdf.

Cambuli, V.M., Baroni, M.G., 2023. Intelligent insulin vs. artificial intelligence for type 1 diabetes: will the real winner please stand up? International Journal of Molecular Sciences 24 (17), 13139. Available at https://doi.org/10.3390/ijms241713139 PMID:37685946;PMCID:PMC10488097.

Canadian Space Agency, 2024. Remote Healthcare in Space and on Earth: A Shared Reality that Calls for Innovative Solutions. February 5. Available at https://www.asc-csa.gc.ca/eng/health/remote-healthcare-in-space-and-on-earth.asp.

Cecilio-Fernandes, D., Cnossen, F., Jaarsma, D.A., Tio, R.A., 2017. Avoiding surgical skill decay: a systematic review on the spacing of training sessions. Journal of Surgical Education 75 (2), 471–480. Available at https://doi.org/10.1016/j.jsurg.2017.08.002.

Chadha, T.S., Cohn, M.A., 1983. Noninvasive treatment of pneumothorax with oxygen inhalation. Respiration 44 (2), 147–152. Available at https://doi.org/10.1159/000194541 PMID:6836190.

Devika, R.W., 2024. We're in the golden age of space exploration. March 8Theweek.. March 8https://theweek.com/science/golden-age-of-space-exploration-is-now.

El-Tallawy, S.N., Pergolizzi, J.V., Vasiliu-Feltes, I., Ahmed, R.S., LeQuang, J.K., El-Tallawy, H.N., et al., 2024. Incorporation of "artificial intelligence" for objective pain assessment: a comprehensive review. Pain and Therapy 13 (3), 293–317. Available at https://doi.org/10.1007/s40122-024-00584-8 PMID:38430433;PMCID:PMC11111436.

Elhaddad, M., Hamam, S., 2024. AI-driven clinical decision support systems: an ongoing pursuit of potential. Cureus. Available at https://doi.org/10.7759/cureus.57728.

George, M., Aroom, K.R., Hawes, H.G., Gill, B.S., Love, J., 2017. 3D printed surgical instruments: the design and fabrication process. World Journal of Surgery 41 (1), 314–319. Available at https://doi.org/10.1007/s00268-016-3814-5 PMID:27822724;PMCID:PMC6287965.

Getteau, C.M., 2025. Wearable tech for space station research - NASA. January 6NASA. January 6https://www.nasa.gov/humans-in-space/wearable-tech-for-space-station-research/.

Guzman, A., 2024. Commercial space Frequently asked questions - NASA. April 10NASA. April 10https://www.nasa.gov/humans-in-space/leo-economy-frequently-asked-questions/.

Hamilton, D., Smart, K., Melton, S., Polk, J.D., Johnson-Throop, K., 2008. Autonomous medical care for exploration class space missions. Journal of Trauma and Acute Care Surgery 64 (4), S354–S363. Available at https://doi.org/10.1097/ta.0b013e31816c005d.

Haven-1—VAST. https://www.vastspace.com/haven-1, n.d.

Haven-2—VAST. https://www.vastspace.com/haven-2, n.d.

Holland, W.C., et al., 2020. Interrupted time series of user-centered clinical decision support implementation for emergency department–initiated buprenorphine for opioid use disorder. Academic Emergency Medicine 27, 753–763.

Husain, L.F., Hagopian, L., Wayman, D., Baker, W.E., Carmody, K.A., 2012. Sonographic diagnosis of pneumothorax. Journal of Emergencies, Trauma, and Shock 5 (1), 76–81. Available at https://doi.org/10.4103/0974-2700.93116 PMID:22416161;PMCID:PMC3299161.

In-Flight Breakup During Test Flight Scaled Composites SpaceShipTwo, N339SS Near Koehn Dry Lake, California, 2015.

Jalota, S.R., Sayad, E., 2025. Tension pneumothorax. StatPearls [Internet]. StatPearls Publishing, Treasure Island (FL). Jan-. Available from: https://www.ncbi.nlm.nih.gov/books/NBK559090/.

Jia, P., Zhang, L., Chen, J., Zhao, P., Zhang, M., 2016. The effects of clinical decision support systems on medication safety: an overview. PLoS ONE 11, 1–17.

Katzman, M., Kim, J., Lesher, M.D., Hale, C.M., McSherry, G.D., Loser, M.F., et al., 2019. Customizing an electronic medical record to automate the workflow and tracking of an antimicrobial stewardship program. Open Forum Infectious Diseases 6 (8). Available at https://doi.org/10.1093/ofid/ofz352.

Krihak, M., Middour, C., Reyes, D., Nusbaum, D., 2019. Communication bandwidth considerations for exploration medical care during space missions. NASA Johnson Space Center (JSC), Houston, TX. NASA Technical Report Server (NTRS).

Krihak, M., Russell, B., Shetye, S., Shaw, T.L., 2021. Exploration medical capability clinical decision support system concept of operations. March 1NASA Technical Reports Server (NTRS). March 1https://ntrs.nasa.gov/citations/20220017856.

Krihak, M., Schmitt, B., Noel, J., Winther, S., Toscano, W., Bardina, J., et al., 2020. Medical data architecture platform and recommended requirements for a medical data system for exploration missions. NASA Headquarters Human Research Program Investigators' Workshop, Galveston, TX. NASA Technical Report Server (NTRS).

Krittanawong, C., Singh, N.K., Scheuring, R.A., Urquieta, E., Bershad, E.M., Macaulay, T.R., et al., 2022. Human health during space travel: state-of-the-art review. Cells 12 (1), 40. Available at https://doi.org/10.3390/cells12010040.

Lerner, D., Pohlen, M., Wang, A., Walter, J., Cairnie, M., Gifford, S., 2023. X-Ray imaging in the simulated microgravity environment of parabolic flight. Aerospace Medicine and Human Performance 94 (10), 786–791. Available at https://doi.org/10.3357/amhp.6286.2023.

Levin, D.R., Jon, S., Arian, A., Jay, L., Benjamin, E., David, C.H., et al., 2023. Enabling human space exploration missions through progressively earth independent medical operations (EIMO). IEEE open journal of engineering in medicine and biology 4, 162–167. Available at https://doi.org/10.1109/OJEMB.2023.3255513.

Levin, D.R., McIntyre, L., Steller, J.G., Nelson, A., Zahner, C., Anderson, A., et al., 2025. Predicting exploration crew medical officer training needs: applying evidence-based predictive analytics to space medicine training. Wilderness & Environmental Medicine 36 (1_suppl). Available at https://doi.org/10.1177/10806032241292535.

Levin, D.R., Siu, M., Kramer, K., Kelly, E., Alouidor, R., Fernandez, G., et al., 2022. Time cost of provider skill: a pilot study of medical officer occupied time by knowledge, skill, and ability level. Aerospace Medicine and Human Performance 93 (11), 816–821. Available at https://doi.org/10.3357/AMHP.6093.2022.

Medical System Foundation Overview for Long-Duration Lunar Orbit and Lunar Surface Operations (LDLOLS) Missions, 2023. In Exploration Medical Capability Element, Human Research Program. Available at https://ntrs.nasa.gov/api/citations/20230001041/downloads/Medical%20System%20Foundation%20Overview%20for%20Long-Duration%20Lunar%20Orbit_Draft%20(1).pdf.

Moro, C., Phelps, C., Jones, D., Stromberga, Z., 2020. Using holograms to enhance learning in health sciences and medicine. Medical Science Educator 30 (4), 1351–1352. Available at https://doi.org/10.1007/s40670-020-01051-7 PMID:34457800;PMCID:PMC8368738.

NASA & Office of Inspector General, 2024. NASA's Management of Risks to Sustaining ISS Operations through 2030. . Available at https://oig.nasa.gov/wp-content/uploads/2024/09/ig-24-020.pdf.

NASA selects Axiom space for third private Astronaut Station Mission - NASA, n.d. NASA. https://www.nasa.gov/news-release/nasa-selects-axiom-space-for-third-private-astronaut-station-mission/.

NASA, 2024. 2024 Moon to Mars Architecture Concept Review 1. In NASA. Available at https://www.nasa.gov/wp-content/uploads/2024/12/acr24-artemis-enables-priority-science.pdf?emrc=38eace.

NASA, n.d. Training continuum for Long-Duration, Deep Space Transport (DST) Missions. https://ntrs.nasa.gov/api/citations/20180001210/downloads/20180001210.pdf

NASA, n.d. Space Station Research Facility. https://www.nasa.gov/mission/station/research-explorer/facility/?#id=7599.

NASA, 2024. Artemis - NASA December 17. Available at https://www.nasa.gov/humans-in-space/artemis/.

Pantalone, D., Faini, G.S., Cialdai, F., Sereni, E., Bacci, S., Bani, D., et al., 2021. Robot-assisted surgery in space: pros and cons. A review from the surgeon's point of view. NPJ Microgravity 7 (1), 56. Available at https://doi.org/10.1038/s41526-021-00183-3 PMID:34934056;PMCID PMC8692617.

Parisi, M., Panontin, T., Wu, S., McTigue, K., Vera, A., 2023. Effects of communication delay on human spaceflight missions. July 20NASA Technical Reports Server (NTRS). July 20https://ntrs.nasa.gov/citations/20230010972.

Park, H., Chae, M.K., Jeong, W., Yu, J., Jung, W., Chang, H., et al., 2022. Appropriateness of alerts and physicians' responses with a medication-related clinical decision support system: retrospective observational study. JMIR Medical Informatics 10 (10), e40511. Available at https://doi.org/10.2196/40511.

Perez, R.S., Skinner, A., Weyhrauch, P., Niehaus, J., Lathan, C., Schwaitzberg, S.D., et al., 2013. Prevention of surgical skill decay. Military Medicine 178 (10S), 76–86. Available at https://doi.org/10.7205/milmed-d-13-00216.

Peris-Lopez, P., Orfila, A., Mitrokotsa, A., van der Lubbe, J.C.A., 2011. A comprehensive RFID solution to enhance inpatient medication safety. International Journal of Medical Informatics 80, 13–24.

Poon, E.G., et al., 2010. Effect of bar-code technology on the safety of medication administration. Obstetrical & Gynecological Survey 65, 629–630.

Reichard, J.F., Phelps, S.E., Lehnhardt, K.R., Young, M., Easter, B.D., 2023. The effect of long-term spaceflight on drug potency and the risk of medication failure. NPJ Microgravity 9 (1). Available at https://doi.org/10.1038/s41526-023-00271-6.

Rezapour Sarabi, M., Yetisen, A.K., Tasoglu, S., 2023. Bioprinting in microgravity. ACS Biomaterials Science & Engineering 9 (6), 3074–3083. Available at https://doi.org/10.1021/acsbiomaterials.3c00195 PMID:37155968;PMCID:PMC10265578.

Rubin, D.M., Kerry, L., Kris, F.N., 2019. Recommendation for a medical system concept of operations for Gateway missions. NASA Headquarters Human Research Program Investigators' Workshop, Galveston, TX. National Aeronautics and Space Administration (NASA) Johnson Space Center. NASA Technical Report Server (NTRS).

Russell, B.K., Burian, B.K., Hilmers, D.C., Beard, B.L., Martin, K., Pletcher, D.L., et al., 2023. The value of a spaceflight clinical decision support system for earth-independent medical operations. Nature Research, New Zealand npj Microgravity 9 (1). Available at https://doi.org/10.1038/s41526-023-00284-1, https://www.nature.com/npjmgrav/.

Sanders, L.M., Scott, R.T., 2024. Artificial intelligence medical support for long-duration space missions. February 2NASA Technical Reports Server (NTRS. February 2https://ntrs.nasa.gov/citations/20240000754.

Seoane-Viaño, I., Ong, J.J., Basit, A.W., Goyanes, A., 2022. To infinity and beyond: strategies for fabricating medicines in outer space. International Journal of Pharmaceutics X 4, 100121. Available at https://doi.org/10.1016/j.ijpx.2022.100121.

Solovyev, L., 2018. The Golden Age of Space Exploration: Unraveling the Mysteries of the Cosmos (Op-Ed) August 16. Available at Space.com, https://www.space.com/41497-golden-age-space-exploration-expert-voices.html.

SpaceX, n.d. https://www.spacex.com/vehicles/starship/.

Sutton, R.T., 2020. An overview of clinical decision support systems: benefits, risks, and strategies for success. NPJ Digital Medicine. Available at https://ntrs.nasa.gov/citations/20200001702.

Sutton, R.T., Pincock, D., Baumgart, D.C., Sadowski, D.C., Fedorak, R.N., Kroeker, K.I., 2020. An overview of clinical decision support systems: benefits, risks, and strategies for Success. Npj Digital Medicine 3 (1). Available at https://doi.org/10.1038/s41746-020-0221-y.

Sze-To, A., Riasatian, A., Tizhoosh, H.R., 2021. Searching for pneumothorax in x-ray images using autoencoded deep features. Scientific Reports 11 (1). Available at https://doi.org/10.1038/s41598-021-89194-4.

Tarzia, B., Patel, S., 2024. 180 AI-enhanced spaced repetition in emergency medicine education. Annals of Emergency Medicine 84 (4), S83–S84. Available at https://doi.org/10.1016/j.annemergmed.2024.08.185.

Terai, M., Nitta, K., 1991. The basic health care system for the lunar base crew. Acta Astronautica 25 (4), 239–243. Available at https://doi.org/10.1016/0094-5765(91)90077-i.

Waldburger, L., Schaller, R., Furthmüller, C., Schrepfer, L., Schaefer, D.J., Kaempfen, A., 2021. 3D-printed hand splints versus thermoplastic splints: a randomized controlled pilot feasibility trial. International Journal of Bioprinting 8 (1), 474. Available at https://doi.org/10.18063/ijb.v8i1.474 PMID:35187278;PMCID:PMC8852196.

Weng, T., Zhang, W., Xia, Y., Wu, P., Yang, M., Jin, R., et al., 2021. 3D bioprinting for skin tissue engineering: current status and perspectives. Journal of Tissue Engineering 12, 20417314211028574. Available at https://doi.org/10.1177/20417314211028574 PMID:34345398;PMCID:PMC8283073.

Yadao, S., Lamture, Y., Huse, S., 2022. Uses of antibiotics alone in case of uncomplicated appendicitis. Cureus. Available at https://doi.org/10.7759/cureus.28488.

Zappatore, M., Ruggieri, G., 2023. Adopting machine translation in the healthcare sector: a methodological multi-criteria review. Computer Speech & Language 84, 101582. Available at https://doi.org/10.1016/j.csl.2023.101582.

Chapter 29

# Evacuation and space emergency medical services

Meagan Rosenberg[1] and Tovy Haber Kamine[2]

[1]*General Surgery, University of Massachusetts Chan Medical School—Baystate, Springfield, MA, United States,* [2]*Trauma, Acute Care Surgery, and Surgical Critical Care, University of Massachusetts Chan Medical School—Baystate, Springfield, MA, United States*

## 29.1 Introduction

Since the successful launch of Yuri Gagarin into orbit in 1961, there have been over 1200 crewed flights with more than 700 astronauts who have spent time in low Earth orbit (LEO; Moskowitz and Wolf, 2025). Throughout the history of human spaceflight, there have been 21 fatalities, 4 evacuations, and numerous minor medical incidents (Summers et al., 2005; Significant Incidents & Close Calls in Human Spaceflight, 2019). While the fatalities, to date, have been secondary to catastrophic engineering failures resulting in nonsurvivable injuries, three of the four cosmonaut evacuations, including those from Mir EO-2, Salyut 5, and Salyut 7, were motivated by medical concerns (Summers et al., 2005; Significant Incidents & Close Calls in Human Spaceflight, 2019). There have also been one confirmed medical evacuation from the International Space Station (ISS) with Crew 11 one deep vein thrombosis where evacuation was likely considered in 2020, and an unreleased medical emergency on the Crew 8 landing that was likely related to events in space (Garcia, 2024NASA Communications Feb. 25th, 2026 Aunon-Chancellor et al., 2020 )The remainder of publically available medical complications experienced while in mission have been largely due to minor medical illness, minor traumatic injuries, or environmental exposure such as contaminants in the air supply or barometric pressure changes (Summers et al., 2005).

This is a remarkable safety record, with an evacuation rate of just over 1 in 300 missions (Moskowitz and Wolf, 2025; Summers et al., 2005; Significant Incidents & Close Calls in Human Spaceflight, 2019). As the population spending time in LEO increases, however, so does the risk of requiring a medical evacuation. The number of space launches has more than doubled in the last 5 years with a record number of 250 crewed and uncrewed space launches planned in 2024. The number of crewed spaceflights from May 2023 to May 2024 is nearly 60% of the total launches since 2020. This has largely been driven by a decrease in the cost of launching a spacecraft, which economists anticipate will only become more financially feasible with some projections of costs as low as $10/kg by the 2030s (Zhang and Wang, 2022; NASA Facts: Astronaut Selection and Training, 2017).

The demographics of spaceflight participants are also changing. Since Dennis Tito funded a seat for himself in 2001 aboard Soyuz TM-32 on a supply mission to the ISS, there has been continued commercial investment in space with crewed activities ranging from tourism to manufacturing to science (Zhang and Wang, 2022). While Tito and others aspiring to be commercial spaceflight participants have thus far undergone extensive health screening and training prior to flight, it is unclear how these screening criteria and training requirements may change over the next 10–20 years as space travel becomes increasingly accessible. A continuation of the current trajectory of space travel growth will result in a dramatic increase in the overall person-years spent in space during that same 10–20 year span. Even if current medical standards for spaceflight are maintained, the increase in person-years will increase the probability of a medical emergency occurring. If medical standards are relaxed to accommodate a wider range of personnel, even a small increase in the degree of medical risk multiplied over a large number of person-years will equate to a substantially higher number of potential medical emergencies and evacuations.

Companies like Vast, Sierra Space, Blue Origin, and Axiom are planning independent space stations over the next decade with crews ranging from 4–8 people (NASA Facts: Astronaut Selection and Training, 2017; Clark, 2021; Axiom Station, no date). SpaceX is building free-flying vehicles that support or will support missions for days to weeks at a time.

**Procedural Medicine for Spaceflight. DOI:** https://doi.org/10.1016/B978-0-443-27552-4.00023-3

NASA, China, Russia, and other international players are planning to return humans to the lunar surface. These broad mission profiles present a challenge to medical response that will likely require more than one model for rescue. It is critical to prepare for this new reality by identifying medical evacuation systems capable of responding to these varied platforms and a method for assessing which mode of evacuation is best suited to a given scenario.

## 29.2 Current emergency medical care framework

### 29.2.1 Astronaut medical screening criteria

The rigorous selection criteria for astronauts serve as a means of mitigating risk for in mission medical emergencies. Candidates undergo a 2-year-long training and evaluation period. At baseline, they must have vision correctable with lenses to 20/20, a resting blood pressure of no greater than 140/90, and a high degree of physical fitness including the ability to swim 3 pool lengths—75 meters total—in their flight suits and tread water in those suits for 10 minutes (NASA Facts: Astronaut Selection and Training, 2017; NASA Astronaut Medical Standards, Selection and Annual Recertification, 2021). Beyond initial screening, astronauts undergo yearly medical examinations, including blood chemistries, evaluations of musculoskeletal and dermatologic health, several ophthalmological tests, and cardiopulmonary screening to include blood pressure and electrocardiogram (Lewis, 2023b). Routine cancer screening is also performed, including colonoscopy or proctosigmoidoscopy, prostate-specific antigen in males, and pap smears and mammograms in females (MR089S Annual Medical Examination, 2004).

### 29.2.2 Medical training

While the majority of NASA astronauts have some background inscientific study, only a minority are medical professionals. A crew medical officer (CMO) designation was established to provide training to a crewmember to take lead on the provision of medical care in mission. To serve as CMO, the crewmember must undergo 40–70 hours of medical training conducted within 18 months prior to departure for their mission (McCarroll et al., 2017). Previously, for missions that were planned for 210 days or greater, NASA required that the CMO be a trained physician (McCarroll et al., 2017; Williams, 2014). In the NASA Technical Standards 3001 revision B in 2022, they transitioned from requiring physician-level experience on the basis of distance from Earth and instead now base that requirement on the Probabilistic Risk Assessment of the mission (41 NASA Technical Standard: NASA STD-3001, 2022). Currently, there is no protocol nor requirement for maintaining technical skills, such as intravenous and intraosseous line placement, during long-duration missions (Stewart and Drudi, 2011). In addition to the CMO, flight surgeons serve as a ground-based adjunct for health care during a mission (Lewis, 2023a). There is an extensive medical kit stocked on the ISS. Included items range from basic over the counter medications such as antihistamines, analgesics, eye drops, and stool softeners, as well as medical supplies including wound care equipment, advanced airway management supplies, and temporary tooth fillings (3.001 Medical Kit – Contents and Reference, 2015).

### 29.2.3 Experience with in-flight medical incidences

During the course of human space travel, there have been 21 fatalities related to five events, all of which involved catastrophic vehicle failures, most during launch or landing (Summers et al., 2005). Based on NASA data of medical events for astronauts while not in space using the Integrated Medical Model, the estimated rate of medical events requiring evacuation would be 0.017 per person-year (Summers et al., 2005). The actual evacuation rate as of 2016 for medical-related issues was 0.022 evacuation events per person-year if the Salyut-Soyuz 21 evacuation with medical concern due to a mechanical malfunction is included (Anderson et al., 2024). If the Soyuz 21 event is excluded, the rate drops further to 0.015 events/person-year (Anderson et al., 2024). The same rate of 2.2% is true for the International Space Station, with approximately 135 person-years of habitation time and 3 publicly known high-risk medical conditions that were evacuated or where evacuation was likely considered. However, it is worth noting that the risk of 1 evacuation every 45 person-years is based on limited data and individuals with a high degree of personal health, i.e. professional astronauts. This value is useful as a rough figure for planning but is likely to change as we gain more experience in space.

The most common in-flight medical incidents are ones that can be managed conservatively or with available medications. These include complications from environmental changes, such as space motion sickness and insomnia, musculoskeletal complaints such as back pain, minor abrasions, and musculoskeletal trauma, and others including sinus congestion, constipation, headaches, upper respiratory infection, and corneal irritation (Hodkinson et al., 2017). More uncommon pathologies have included renal and urinary diseases (renal and ureteral calculi, acute urinary retention, and urinary tract infections), cardiac dysrhythmias, infections (including gastroenteritis, prostatitis, and otitis media), dermatitis, decompression sickness, near-drowning events, and foreign body aspiration (Fritsch-Yelle et al., 1998; Jones et al., 2005; Scheuring et al., 2009).

Based on astronaut data and analog populations, NASA has developed multiple medical condition lists which include those conditions of greatest concern and/or likelihood for missions to LEO and exploration class missions. These conditions are numerous and include entities such as diverticulitis, appendicitis, herpes reactivation, dental infection, and sepsis as well as traumatic or environmental hazards including radiation sickness, severe decompression illness, ocular trauma, hemorrhage, burns, and traumatic limb injury (Hodkinson et al., 2017; Barratt and Pool, 2008; Watkins et al., 2011; Greensite, 1969; Pressley and Thangavelu, 2017; Canepa and Levin, 2023; Weor, 1966; Shaw, 2023; Atkinson, 2009; Man et al., 2022; Lang et al., 2017; Fischer et al., 1967; Chaloulakou et al., 2022).

## 29.3 Proposed evacuation models

### 29.3.1 Orbital lifeboat

An orbital lifeboat model proposes the use of a multipurpose spacecraft docked to a space station available for evacuation. This is the current model in use for the ISS and all previous space stations, which would use the Soyuz, Dragon, or other docked vehicles for emergency evacuation. Logistical complexity is reduced compared to a ground-launched rescue model, and it has a quicker response time with fewer steps to definitive care. The immediate availability of the lifeboat model is counterbalanced by its lack of tailored medical design. This model provides no additional medical personnel, nor does it provide medical supplies beyond the baseline supplies already available on the main vehicle. This contrasts with the ability of a ground-launched vehicle to tailor the medical supply stock to the emergency at hand. It should again be acknowledged that reentry is physically straining, and the transition from zero-gravity back to gravity has physiological impacts which could be deleterious for a tenuous patient.

Specialized orbital lifeboats called Crew Return Vehicles or Assured Crew Return Vehicles have been proposed for evacuation from the ISS, aside from Soyuz, including the HL-20 Personnel Launch System. None, however, has been implemented to date. Currently, evacuation via an orbital lifeboat (i.e., Soyuz or Dragon) would mean the evacuation of the entire crew and abandonment of the orbital platform so as not to leave the crew without safe haven. Theoretically, a secondary orbital lifeboat, such as the previously proposed Crew Return Vehicle, could be docked to the space station in addition to the primary launch vehicle, allowing for evacuation of only those personnel in need of definitive medical care. Either approach would be a viable means of evacuation for support crew in the event of a catastrophic failure of their station.

### 29.3.2 Orbital life raft

This model proposes a small purpose-built spacecraft that does not carry additional medical supplies. The primary difference between the lifeboat and life raft model is that the orbital life raft is a single-use vehicle that serves no secondary purpose beyond evacuation. As with the lifeboat model, this would allow for rapid mobilization and evacuation of ill or injured personnel. The Man Out of Space Easiest (MOOSE) model is an example of an orbital life raft for single-person evacuation which was proposed but never operationalized (Greensite, 1969). Small vehicles, however, are not ideal for patients who have limited mobility or positioning requirements such as those associated with a spinal or long bone injury.

Most of the proposed orbital life raft models such as MOOSE are for single-person evacuation. In a medical scenario, this poses a higher level of risk than average. In many cases, this may be feasible following stabilization on the space station. However, in the case of under-triaging the severity of the medical event or decompensation during reentry, it could lead to an incapacitated patient in a lone vehicle without medical support while awaiting retrieval.

### 29.3.3 Orbital ambulance

The orbital ambulance is similar to the orbital lifeboat as it is a prepositioned orbital vehicle; however, this proposed model would stock medical supplies to augment those available on the shuttle or station where the index emergency occurred. It can be compared to a medevac helicopter based near the patient with the ability to provide medical care during evacuation to definitive care. This model would allow for more rapid evacuation than either of the ground-launch options but with less complex ability to provide care than a ground-launched ambulance. There are technical considerations for docking with the orbital ambulance, but they would be less complex than coordinating a ground-launch and return. As with all other models, there is concern about the perturbations of landing and the deleterious effect they could have on a tenuous patient. As with all other uncrewed evacuation models, the orbital ambulance again leaves the crew in the position of sending an ill or injured crewmember back alone or with another crewmember, thus leaving the mission crew down two members.

Halberg and Robinson at the University of California, Davis, have proposed the redesign of the Boeing X-37B spaceplane as a means of medical evacuation from the ISS (Etan Halberg et al., 2016). Their proposed internal reconfiguration would allow for a crew of three—the pilot and two additional crew members—as well as predetermined medical equipment for patient stabilization prior to descent. Additional work evaluating the feasibility of a space ambulance have curated a list of recommendations for future crewed vehicles that would optimize it for use also as an evacuation method including crew capacity of three or more, ability to reenter without space suits, accommodation for one supine passenger, stowage for critical care equipment for predescent stabilization, low-g force reentry, and additional considerations for ergonomics and communications (Nowadly et al., 2019).

Pressley and Thangavelu have proposed a model called Space Orbiting Lifeboat And medical Care during Evacuation (SOLACE) as a combined emergency room and ambulance (Pressley and Thangavelu, 2017). While conceived with the needs of a deep-space mission in mind, many of the proposed components would function in a LEO evacuation situation as well. They lean on a "plug-and-play" modular approach for adaptation of a vehicle serving another purpose until medical need warrants transition of the space. The project explains in detail an ascent vehicle from a lunar or Martian surface and a design of the larger SOLACE orbiting body, which are not relevant to a LEO evacuation situation. Their proposal for a reentry vehicle, however, is relevant to LEO evacuation. They propose the use of a vehicle similar to NASA's Crew Return Vehicle with lifting body reentry to minimize G force strain on the occupants and which is designed for 6–7 crew members. There is additional emphasis on an ideal vehicle having autonomous control capabilities in the case of a completely incapacitated crew. An orbital ambulance paradigm such as SOLACE has the benefit of an immediately available site for medical management and stabilization of a critical patient. However, this model raises the question of sacrificing vehicle footprint for a permanent medical care area versus modular components used to convert an existing space into a medical care area. Additionally, it still requires turnover of stocked medications and supplies as they reach their expiration dates.

A recent abstract describes some of the potential skillsets that may be needed by spaceflight medical technicians (SMT), a crucial adjunct for evacuation preparations (Canepa and Levin, 2023). Use of an organized SMT structure within the orbital ambulance paradigm would mean a more highly trained medical care team with a high degree of familiarity with the nuances of space health care, available resources, and physiological changes that accompany a weightless environment. An orbiting SMT can address expiring medications and supplies by returning expired items with them when they rotated off duty in orbit. The next stationed SMT would arrive with the necessary replacement supplies. While this paradigm would ensure a highly trained individual for stabilization of a critical patient, it could pose significant financial cost for maintaining an in-orbit, on-call SMT force.

### 29.3.4 Ground launch lifeboat

A potential ground-based solution for medical evacuation is the launch of an uncrewed, small vehicle designed to rendezvous with an on-orbit vehicle or space station, load the patient, and return to Earth. The United States Coast Guard Monitor Boat implements an analogous rescue structure that launches to aid distressed ships. This paradigm has never been operationalized but has conceptual heritage in the "Rescue Gemini" vehicle proposed in 1966 by McDonnell Douglas for rescuing astronauts from orbit (Weor, 1966). An uncrewed rescue vehicle avoids the need to launch additional crew and decreases the weight and thus cost of a launch. While evacuees are likely to be patients who have been sufficiently resuscitated and deemed stable enough to tolerate reentry, diagnostic uncertainty could result in a patient decompensating without personnel available to provide care while awaiting ground crew retrieval.

A ground-launched system allows a single vehicle or small group of vehicles to respond to a variety of orbital platforms to provide evacuation coverage to multiple stations. However, this model requires a second spacecraft launch with all the associated complexities, costs, and time requirements. While this does decrease the overall needed number of medically trained crew capable of a space launch, it will require an on-call group of mission control personnel to coordinate the uncrewed launch as well as a crew on standby for capsule recovery. As an example, the current record for rapid insertion of any spacecraft into a defined orbit, the uncrewed Victus Nox mission, was just over 27 hours from announcement of need to launch for a prepared but unfueled spacecraft (Shaw, 2023).

### 29.3.5 Ground launch ambulance

A ground-launched ambulance requires similar logistical complexity to the ground-launched lifeboat. However, it carries the advantage of launching with a crew of trained medical personnel capable of providing stabilizing care on orbit and/or during at least part of the undocking, entry, descent, and landing process as well as after landing until recovery is accomplished. It would function similarly to a critical care transport vehicle and crew. This method was operationalized, although never enacted, through the STS-400 mission, during which the Space Shuttle Endeavor was ready for launch on short notice to aid the Space Shuttle Atlantis if a problem occurred during the STS-125 mission to repair the Hubble Space Telescope (Atkinson, 2009). Both a pro and con to this model is the necessary development of a network of medical professionals uniquely experienced in addressing space-related medical emergencies and facile in providing care in a zero-gravity situation. Having a pool of experienced and zero-gravity-qualified medical professionals would be beneficial to the well-being of potential patients. The development of such a network, however, would be time-intensive and expensive in terms of training. Additionally, the ambulance could carry with it additional medical supplies not routinely stocked on a mission vehicle. Although these additional supplies are likely to be a predetermined package due to weight and balance logistics, it could theoretically be possible to tailor the ambulance's load toward the medical emergency. For example, there may be one preplanned and approved supply load for an infection-related emergency and a different supply load preplanned and approved for a trauma-related emergency. Storing supplies so that they are ready for expedient departure, however, does pose logistical challenges.

Disadvantages, however, include the high volume of on-call personnel needed for this paradigm. In addition to personnel necessary to manage the logistics of launching the ambulance, a ground-launched ambulance necessitates a pool of on-call, space-ready, medically trained personnel. These individuals would require both spaceflight and paramedic-level medical care training. Both the degree of training and the depth of the on-call pool would present a significant financial investment. Additionally, atmosphere entry and exit are high-risk periods. Additional medical crew increases the risk that they too may experience harm.

### 29.3.6 Analysis of cis-lunar orbital urgent care

While not an evacuation method, an orbital urgent care model proposes the development of an orbiting station designed specifically for medical care. This model would provide the highest level of timely patient care and stabilization in an emergency. It is a model that does not work in isolation but would partner with evacuation methods such as the orbital ambulance to provide patient transportation. It bears resemblance to the U.S. Navy hospital ships which travel to provide care in areas with poor medical access. Using a station specifically for medical care means that there can be significantly more space dedicated to diagnostic and treatment capacities. It would be possible to stabilize critical patients who would otherwise not survive with more limited medical capabilities and who would not be able to tolerate the stress of reentry. The development and launch of an orbital urgent care, however, is the most resource-intensive of the proposed models. In addition to the up-front cost of building and outfitting the station, it would require adequate around-the-clock staffing and frequent supply drop-offs to maintain a stock of medications and blood products that have a short shelf life.

Providing on-call staffing for an orbital urgent care would require significant training as well as financial investment. The staffing for an orbital urgent care would be broader than for any of the other crewed paradigms given the expectation of delivering a higher level of medical care. Staffing would likely include at least one medical doctor trained in emergency and critical care and nursing staff. If operative procedures were anticipated, personnel with anesthesiology and surgical technician training would be required. Other ancillary staff could include radiology technicians and pharmacists. However, a number of roles may be fulfilled remotely to provide specialized care from a distance while minimizing crew in orbit.

## 29.4 Cis-lunar emergency medical service considerations

### 29.4.1 Environmental impacts on physiology and care implementation

The effects of the space environment have a marked impact on nearly all the body's systems. While there are still many unknowns, the knowledge that we do have is of vital importance in caring for critically ill or injured astronauts.

There are several musculoskeletal considerations with particular relevance for potential trauma victims. A micro- or zero-gravity environment results in imbalanced bone resorption at a rate of 1%–2% bone mineral density loss per month with variation dependent on the bone in question (Man et al., 2022). Bone density loss is greatly decreased but not fully mitigated by resistive exercises (Lang et al., 2017). This bone loss may correlate with a higher risk for fractures in a traumatic event. It is important to consider effective means of fracture reduction and immobilization in the event of a long bone fracture, and one may need an elevated suspicion for a spinal injury in the traumatic setting. Transport of a patient under spinal precautions is a consideration that would need to be incorporated into the design of a purpose-made evacuation vehicle.

The cardiovascular system experiences body fluid redistribution in microgravity with a loss of plasma volume by 10% (Fischer et al., 1967; Chaloulakou et al., 2022). Red blood cell mass subsequently decreases and, when coupled with decreased erythropoietin activity and increased red blood cell lysis, results in a phenomenon known as space anemia (Trudel et al., 2020; Lansiaux et al., 2024). The clinical impact of these laboratory findings remains unknown and requires further investigation regarding relevance, particularly in the traumatic setting where blood loss is a concern.

In addition to the physiological changes experienced by the patient, the microgravity environment changes the way in which certain types of care can be administered. One such example is cardiopulmonary resuscitation. On Earth, weight and force generated by the human compressing the chest allows for resuscitation, but that cannot be generated in a microgravity environment without restraints or stabilizers. Several methods have been proposed and are discussed at greater length elsewhere in this textbook. The handstand method proposes a backboard restraint of the patient with the rescuer in a handstand position over the patient in a cabin of appropriate size. With feet against the ceiling, appropriate pressure can be generated to allow for chest compression and recoil (Rehnberg et al., 2014). The Evetts-Russomano method involves strapping the compressor to the same secured backboard as the patient. Finally, the reverse bearhug involves wrapping the arms around the patient from behind with interlocked hands over the sternum to provide compressions, bearing some similarity to upward abdominal thrusts for asphyxiation (Rehnberg et al., 2014; Smyth, 2017). It is important to note, however, that the current utility of CPR in space is extremely limited as many of the adjuncts that contribute to CPR success and postreturn management of spontaneous circulation care are not available. The indications for CPR in LEO are likely limited only to conditions where a cardiac arrest would respond to external defibrillation. In most scenarios, it would take greater time to secure the patient for CPR than it would to retrieve the defibrillator and administer an early shock.

### 29.4.2 Training considerations

With the implementation of any crewed medical rescue and evacuation system, there arises the need for a trained, space-ready emergency medical service (EMS) team. Training would require both spaceflight and medical training. For a ground-based EMS system, a call pool would need to be maintained of trained personnel ready to launch on short notice. For spaceflight training, it may be possible to abbreviate the standard astronaut training program as they will not have prolonged duties aboard a ship or station. The exception, however, would be in the case of staffing aboard an orbital urgent care where personnel would need training to appropriately maintain the station and adapt to the lifestyle while stationed there. Launch and reentry are high-risk periods, and therefore the EMS team would need to be extremely comfortable in that setting and managing technical issues that could arise. This degree of training as well as the necessary call pool depth would represent a significant financial investment.

The necessary medical training for cis-lunar EMS personnel poses a unique problem. There exist multiple well-studied and widely implemented terrestrial trauma and cardiac support training courses. No such equivalent yet exists for space-based medical providers. While there may be some crossover in the terrestrial-based training courses, not all of it will be practical. As space travel volume increases and we anticipate the resultant medical emergencies, it is imperative that we create a training program that utilizes the relevant portions of terrestrial training courses with the practical knowledge thus far garnered on medical care provided in a zero-gravity setting.

Space-based medical providers will additionally need extensive knowledge about the physiological changes experienced by those spending time in a zero-gravity environment, as previously discussed. Similarly, they will require training to adapt usual medical interventions to a zero-gravity environment, ranging from running an intravenous infusion to providing chest compressions during cardiopulmonary resuscitation. An additional aspect of training that is vital for EMS personnel will be ingenuity in a resource-constrained environment. One of the benefits of having a ground-launch EMS would be the ability to stock targeted supplies within the weight, balance, and safety constraints of launch. The increasing utilization of additive manufacturing or 3D printing provides another potential tool for equipping crew with certain medical supplies, although this is an area with limited research thus far (Gaskill, 2025; Wong and Pfahnl, 2014).

Skill maintenance is another consideration. The current incidence of medical emergencies in LEO and cis-lunar space is low, which means that space-trained EMS personnel are unlikely to be frequently required, at least for the immediate future. As such, there is concern for skill attrition. A structured continuing education curriculum should be considered to keep members of the team mission-ready.

### 29.4.3 Role for ongoing research

Any initial iteration of a cis-lunar emergency medical service system is unlikely to be the same version that persists as the frontier of space travel expands with time. By beginning efforts early to develop a versatile and adaptive response system, there arises the opportunity for foundational research to meet the need as soon as it arises. Further tradespace analysis of the presented options will allow for characterization and quantification of their strengths and weaknesses, informing the direction of that future research.

## References


3.001 Medical Kit – Contents and Reference, 2015. Available at https://www.nasa.gov/wp-content/uploads/2015/03/medical_kit_checklist_-_full_release.pdf.

41 NASA Technical Standard, 2022. NASA STD-3001. 1. Available at https://www.nasa.gov/wp-content/uploads/2020/10/2022-01-05_nasa-std-3001_vol.1_rev._b_final_draft_with_signature_010522.pdf.

Anderson, A., Parmar, P., Levin, D., Hilmers, D., 2024. Quantifying the risk of medical evacuation in spaceflight. Acta Astronautica 224, 458–464. Available at https://doi.org/10.1016/j.actaastro.2024.08.026.

Atkinson, N., 2009. The STS-400 Shuttle Rescue Mission Scenario. Universe Today.

Aunon-Chancellor, Serena M, Pattarini, James M, Moll, Stephan, Sargsyan, Ashot, 2020. Venous Thrombosis during Spaceflight. New England Journal of Medicine 382 (1), 89–90. Available at 10.1056/NEJMc1905875.

Axiom Station, n.d. Available at https://www.axiomspace.com/axiom-station.

Barratt, M.R., Pool, S.L., 2008. Principles of Clinical Medicine for Space Flight Principles of Clinical Medicine for Space Flight. Springer New York, United States, pp. 1–596. Available at http://www.springerlink.com/openurl.asp?genre=book&isbn=978-0-387-98842-9.

Canepa, C.A., Levin, D., 2023. Required skillsets and training needs for orbital emergency medical technicians [Abstract]. 2023 Annual Aerospace Medical Association Meeting.

Chaloulakou, S., Poulia, K.A., Karayiannis, D., 2022. Physiological alterations in relation to space flight: the role of nutrition. Nutrients 14 (22), 4896. Available at https://doi.org/10.3390/nu14224896.

Clark, S., 2021. Blue Origin and Sierra Space team up for commercial space station. Spaceflight Now. Available at https://spaceflightnow.com/2021/10/25/blue-origin-and-sierra-space-team-up-for-commercial-space-station.

Etan Halberg, E., Robinson, S.K., Onishi, R., Blaesser, N.J., 2016. An ISS space ambulance based on X-37B technology. AIAA Space and Astronautics Forum and Exposition, SPACE 2016. American Institute of Aeronautics and Astronautics Inc, AIAA United States.

Fischer, C.L., Johnson, P.C., Berry, C.A., 1967. Red blood cell mass and plasma volume changes in manned space flight. JAMA: The Journal of the American Medical Association 200 (7), 579–583. Available at https://doi.org/10.1001/jama.1967.03120200057007.

Fritsch-Yelle, J.M., Leuenberger, U.A., D'Aunno, D.S., Rossum, A.C., Brown, T.E., Wood, M.L., et al., 1998. An episode of ventricular tachycardia during long-duration spaceflight. American Journal of Cardiology 81 (11), 1391–1392. Available at https://doi.org/10.1016/S0002-9149(98)00179-9, www.elsevier.com/locate/amjcard.

Garcia, Mark A.Oct. 25th, 2024NASA Provides Update on Agency's SpaceX Crew-8 Health. Available at https://www.nasa.gov/blogs/spacestation/2024/10/25/nasa-provides-update-on-agencys-spacex-crew-8-health/. Accessed Mar. 7th, 2026.

Gaskill, M.L., 2025. 3D printing: saving weight and space at Llaunch. Available at https://www.nasa.gov/missions/station/iss-research/3d-printing-saving-weight-and-space-at-launch/.

Greensite, A.L., 1969. Analysis and design of space vehicle flight control systems. 16.

Hodkinson, P.D., Anderton, R.A., Posselt, B.N., Fong, K.J., 2017. An overview of space medicine. British Journal of Anaesthesia 119 (suppl_1), i143. Available at https://doi.org/10.1093/bja/aex336.

Jones, J.A., Jennings, R., Pietryzk, R., Ciftcioglu, N., Stepaniak, P., 2005. Genitourinary issues during spaceflight: a review. International Journal of Impotence Research 17 (S1), S64. Available at https://doi.org/10.1038/sj.ijir.3901431.

Lang, T., Van Loon, J.J.W.A., Bloomfield, S., Vico, L., Chopard, A., Rittweger, J., et al., 2017. Towards human exploration of space: the THESEUS review series on muscle and bone research priorities. NPJ Microgravity 3 (1). Available at https://doi.org/10.1038/s41526-017-0013-0, https://www.nature.com/npjmgrav/.

Lansiaux, E., Jain, N., Yatin Chodnekar, S., Siddiq, A., Ibrahim, M., Yèche, M., et al., 2024. Understanding the complexities of space anaemia in extended space missions: revelations from microgravitational odyssey. Frontiers in Physiology 15. Available at https://doi.org/10.3389/fphys.2024.1321468.

Lewis, R.E., 2023a. Flight Surgeons. National Aeronautical and Space Administration. Available at https://www.nasa.gov/directorates/esdmd/hhp/flight-surgeons/.

Lewis, R.E., 2023b. Medical Examination Requirements (MER) for former astronauts. Available at https://www.nasa.gov/general/medical-examination-requirements-mer-for-former-astronauts/.

Man, J., Graham, T., Squires-Donelly, G., Laslett, A.L., 2022. The effects of microgravity on bone structure and function. npj Microgravity 8 (1). Available at https://doi.org/10.1038/s41526-022-00194-8, https://www.nature.com/npjmgrav/.

McCarroll, M.L., Ahmed, R.A., Schwartz, A., Gothard, M.D., Atkinson, S.S., Hughes, P., et al., 2017. Medical judgement analogue studies with applications to spaceflight crew medical officer. BMJ Simulation and Technology Enhanced Learning 3 (4), 163–168. Available at https://doi.org/10.1136/bmjstel-2017-000210.

Moskowitz, Wolf, Z., 2025. Everyone Who Has Ever Been to Space, Charted. Scientific American. Available at https://www.scientificamerican.com/article/everyone-who-has-ever-been-to-space-charted/.

MR089S Annual Medical Examination, 2004. Available at https://nlsp.nasa.gov/explore/lsah/MR089S.pdf.

NASA Astronaut Medical Standards, Selection and Annual Recertification, 2021. Available at https://www.nasa.gov/wp-content/uploads/2023/04/ochmo-std-100.1a.pdf.

NASA Communications, Feb. 25th, 2026NASA's SpaceX Crew-11 Astronaut Update. Available at https://www.nasa.gov/blogs/commercialcrew/2026/02/25/nasas-spacex-crew-11-astronaut-update/. Accessed Mar. 25th, 2026.

NASA Facts: Astronaut Selection and Training, 2017. Available at https://www.nasa.gov/wp-content/uploads/2017/05/606877main_fs-2011-11-057-jsc-astro_trng.pdf.

NBC News, no date. Available at https://www.nbcnews.com/science/space/nasa-astronaut-hospitalized-return-space-station-rcna177386.

Nowadly, C.D., Trapp, B.D., Robinson, S.K., Richards, J.R., 2019. Resuscitation and evacuation from low earth orbit: a systematic review. Prehospital and Disaster Medicine 34 (5), 521–531. Available at https://doi.org/10.1017/S1049023X19004734, http://journals.cambridge.org/action/displayJournal?jid=PDM.

Pressley, W.G., Thangavelu, M., 2017. Addressing the space-based medical facility capability gap with project SOLACE: space orbiting lifeboat and medical care during evacuation. AIAA SPACE and Astronautics Forum and Exposition, SPACE 2017. American Institute of Aeronautics and Astronautics Inc, AIAA United States 0. 9781624104831.

Rehnberg, L., Ashcroft, A., Baers, J.H., Campos, F., Cardoso, R.B., Velho, R., et al., 2014. Three methods of manual external chest compressions during microgravity simulation. Aviation, Space, and Environmental Medicine 85 (7), 687–693. Available at https://doi.org/10.3357/ASEM.3854.2014.

Scheuring, R.A., Mathers, C.H., Jones, J.A., Wear, M.L., 2009. Musculoskeletal injuries and minor trauma in space: incidence and injury mechanisms in U.S. astronauts. Aviation, Space, and Environmental Medicine 80 (2), 117–124. Available at https://doi.org/10.3357/ASEM.2270.2009.

Shaw, K., 2023. Space systems command successfully launches VICTUS NOX for U.S. space force: sets new tactically responsive space record. Available at https://www.vandenberg.spaceforce.mil/News/Article-Display/Article/3526687/space-systems-command-successfully-launches-victus-nox-for-us-space-force-sets/.

Significant Incidents & Close Calls in Human Spaceflight, 2019. Available at https://sma.nasa.gov/SignificantIncidents/.

Smyth, C., 2017. Handstand to restart heart helps astronauts solve pressing problem. The Times. Available at https://www.thetimes.com/uk/science/article/handstand-to-restart-heart-helps-astronauts-solve-pressing-problem-sr3n9qzsv.

Stewart, G.E., Drudi, L., 2011. Medical education for exploration class missions NASA aerospace medicine elective at the kennedy space centre. McGill Journal of Medicine 13 (2), 55–58. Available at http://www.medicine.mcgill.ca/mjm/cross13022.pdf Canada.

Summers, R.L., Johnston, S.L., Marshburn, T.H., Williams, D.R., 2005. Emergencies in space. Annals of Emergency Medicine 46 (2), 177–184. Available at https://doi.org/10.1016/j.annemergmed.2005.02.010, http://www.elsevier.com/inca/publications/store/6/2/3/2/7/8/index.htt.

Trudel, G., Shafer, J., Laneuville, O., Ramsay, T., 2020. Characterizing the effect of exposure to microgravity on anemia: more space is worse. American Journal of Hematology 95 (3), 267–273. Available at https://doi.org/10.1002/ajh.25699.

Watkins, Barr, Kerstman, E., 2011. The space medicine exploration medical condition list. 18th IAA Humans in Space Symposium.

Weor, F., 1966. Rescue teams for manned testing of Gemini spacecraft. AIAA. Available at https://doi.org/10.2514/6.1966-2235.

Williams, R.S., 2014. Nasa space flight human-system standard volume 1, revision a: crew health. Available at http://standards.globalspec.com/std/9858810/nasa-std-3001-vol-1.

Wong, J.Y., Pfahnl, A.C., 2014. 3D printing of surgical instruments for long-duration space missions. Aviation, Space, and Environmental Medicine 85 (7), 758–763. Available at https://doi.org/10.3357/ASEM.3898.2014, http://docserver.ingentaconnect.com/deliver/connect/asma/00956562/v85n7/s13.pdf.

Zhang, Y., Wang, L., 2022. Progress in space tourism studies: a systematic literature review. Tourism Recreation Research 47 (4), 372–383. Available at https://doi.org/10.1080/02508281.2020.1857522.

Venous Thrombosis during Spaceflight., N Engl J Med, 382, 89-90. Available at https://doi.org/10.1056/NEJMc1905875.

Chapter 30

# The future of human habitation in space

Ryan Zahalka[1] and Dana R. Levin[1,2,3]

[1]*Division of Space, Ecological, Arctic, and Resource-Limited Medicine, Massachusets General Hospital, Boston, MA, United States,* [2]*Space Medicine Operations, Mission Operations Vast Inc., Long Beach, CA, United States,* [3]*Department of Emergency Medicine, University of California, Los Angeles, CA, United States*

## 30.1 Introduction

In the first 60 years of spaceflight, humans learned how to survive and work in space. We landed on the moon, conducted thousands of science experiments, developed technologies that have improved our ability to predict weather, navigate, communicate, perform search and rescue, and expand our knowledge of the cosmos. We built large stations that support a continuous human presence in orbit, and learned to set aside competition for cooperation with international partners, leveraging government, private, and academic partnerships to achieve what was once impossible. However, this is just scratching the surface of what we can achieve.

The next two decades of human spaceflight are poised to realize some of the most profound and ambitious dreams of humanity. After 30 years of service as a testbed, training ground, and international laboratory, the lessons from International Space Station will be transitioned to a series of successor platforms. These stations and vehicles may be able to support a small "village" in Earth orbit, building on the lessons of the International Space Station and incorporating entirely new capabilities. Hardware and mission plans are underway to establish permanent lunar stations, build an economy in orbit, and set us on a path to explore planets in our solar system previously viewed only by robots and through telescopes. These long promised dreams are now transitioning from concepts to computer aided design (CAD) files, hardware, and operational plans. NASA and its international partners have stood up programs like Artemis, Gateway, and the Commercial Low Earth Orbit Destinations Program. China has developed a robust space program focused on planetary exploration. Already there are two active stations in Earth orbit, the ISS and Tiangong, with plans for as many as four commercial stations and two additional government stations to launch by 2030.

While it is unlikely all of these stations will be realized, it is equally unlikely all of them will fail, and most are intended to grow over their operational lifespans. The ISS was built to teach us how to work in space but the next decades will teach us how to live.

## 30.2 Intentional design

From a medical perspective, it is important to anticipate the capabilities and resources needed to support the next generation of spaceflight. The World Health Organization defines health as "a state of complete physical, mental, and social well-being and not merely the absence of disease or infirmity." However, terrestrial medicine has traditionally been reactive, spreading, and developing as people inhabit new areas with little preplanning. This has led to incredible but uneven advances, with resources and training more often geared towards a pathology-based model rather than promoting health.

For example, we focus our attention and understanding of health care on hospitals and emergency rooms providing care for the sick rather than developing the systems, services, and cultural norms that support the well. This is certainly not a universal truth in all cultures, and all developed societies have robust water, food, and waste management programs. However, the perception of these services as health care tends to be eclipsed by pharmaceuticals and procedures to restore health.

Both schools of thought for space health, the space medicine fellowships and the aerospace medicine residencies, emphasize promoting health through preparation, system design, training, and preventive care. To this end, the Space

**Procedural Medicine for Spaceflight. DOI:** https://doi.org/10.1016/B978-0-443-27552-4.00030-0

Medicine Fellowship programs list eight core competency domains with only two including elements of pathology-based reactive care (defined in Table 30.1). The American Board of Preventive Medicine (ABPM) is less explicit in their competency outlines but clearly emphasizes prevention by housing aerospace medicine under the larger umbrella of preventive medicine and including the pathology model of "clinical medicine" in only 2 of the 21 milestones used to guide training (ACGME, 2020; American Board of Preventive Medicine, 2025).

The emphasis on anticipatory design to reduce risk complements pathology-based care highlights that professionals in space medicine focus ensuring the system is capable of supporting crew well before the crew ever sets foot in the spacecraft. This same philosophy should be applied to the future of healthcare in space. We must begin outlining the needs of tomorrow now to ensure space-based health care systems are up to the challenges and increased scope before this scope is needed.

## 30.3 From mission planning to deployment to public health

Space exploration to date has largely been mission focused. Missions have clearly defined goals for which crewmembers are trained. The crew departs from the terrestrial base, spends a defined time in space completing their objectives, and returns. This paradigm made a lot of sense in the early days of the space program when missions were a few days to a few weeks, but "missions" to the ISS more closely resemble a deployment model. In this context, a semipermanent asset is strategically positioned for an extended time often to serve as a local station to facilitate missions within its operational area. In this view, the ISS is a semipermanent forward operating base, and the experiments and extravehicular activities conducted from it may be seen as individual "missions." The next generation of stations in orbit is being designed more for long-term habitation, akin to a frontier town forming around a fort. Lunar operations appear to be following a similar trajectory with the Apollo era missions transitioning to Artemis and Gateway era outposts and eventual colonies. This is an imperfect analogy, but it illustrates the transition presently occurring. A similar scenario may be seen in the context of the US National Science Foundation stations in Antarctica. While the initial outposts of the early 20th century were little more than shelters from the cold and supply depots, the present-day McMurdo Station more closely resembles a small town, supporting a population up to 1500 people during the peak season. It is not difficult to see the parallels with space travel. Indeed, Antarctica may serve as a model from which to begin considering the future of Space Medicine (United States Antarctic Program, 2025).

## 30.4 Learning from history: Antarctica as a case study

Exploration in Antarctica began as excursions from ships. Crews sailing to the continent, conducting missions, and returning to their ports of call. Eventually, supply depots and semipermanent structures were built to facilitate longer stays and deeper exploration of the continent, geology, and biome. Later, these were followed by more permanent structures such as Discovery Hut on Ross Island, meant to serve as bases to which visiting crews could return year after year. Finally, permanent structures were built, modified, and operated for continuous year-round occupation by comparatively large populations. Today, stations like McMurdo run like a small town, serving as a home on the ice from which to send smaller expeditions to seasonal areas of interest, such as the dry valleys or other field camps. Areas of more regular visits, like the South Pole or Palmer Peninsula, have their own small stations, somewhat akin to the Discovery Hut of the early 20th century.

Medical care on the continent was similar. Initially, expeditions took a surgeon or physician with them. These medical officers would often serve multiple roles and participate in the mission objectives as much or more often than their medical duties. However, as the expeditions became longer and the size of the expedition parties grew, health care was needed more frequently, and the expeditions could support more specialized services. Today, McMurdo functions with a small hospital that has a staff of 4–10 people including physicians, physician assistants, nurses, a pharmacist, and paramedics. These providers trade off for deployments of 3–9 months at a time and render care for the population of McMurdo as well as serving as more "definitive care" for the field camps and other stations.

The hospital is not a large facility, containing space for a 2-bed trauma bay, several exam rooms, and 4–6 inpatient beds. The inpatient facility is designed and supplied to support four critical care patients for up to 3 days. It also has a dental suite, a clinical lab, a pharmacy, a hyperbaric chamber, and a physical therapy room. Most of the time, it functions as an outpatient clinic with rare "admissions" usually for observation, isolation, or to facilitate an evacuation. These latter functions occur only a few times a year, if at all, but are frequent enough to justify the capabilities.

More interestingly, there is no present-day surgical capability beyond minor procedures and emergency stabilization. McMurdo did have surgical capabilities until at least the 1970s, while it was under US Department of Defense

**TABLE 30.1** Core competency domains in space medicine.

| Domain | Description |
|---|---|
| Systems Medicine | Systems Medicine is concerned with the interactions of the vehicle, on board equipment, and human occupants. It covers how the spacecraft affects humans as well as how humans affect the spacecraft. Since it is most prominent during the design, development, and mission planning phases, it is often agnostic to the individual crewmembers. However, operational systems medicine concerns do play a role when managing off-nominal situations and unusual mission plans. Some of the major subtopics in this domain include medical system/vehicle design, engineering requirements writing and review, device/system development testing, interface design, occupant protection, and industrial design/space architecture. |
| Medical Risk Assessment | This domain addresses measuring and understanding the medical risks to individuals, the crew as a whole, the vehicle, and the mission. It provides the tools for shared decision making and informed consent through balancing probability of condition occurrence, consequence of each outcome, and the necessity of each mission objective with organizational and individual risk tolerance. It requires an understanding of medical evidence interpretation, probability and statistics, long- and short-term risk, and the various mitigation strategies including candidate screening and certification, engineering controls, operational interventions, and administrative controls. |
| Ground Segment Clinical Operations | This refers to clinical care for flight crew that occurs in the preflight and postflight phases. In both phases, there is a need to focus on health promotion and maintenance, behavioral and emotional well-being, and acute care. While a Spaceflight Physician's primary clinical specialty forms the basis for their clinical capabilities, Space Medicine training fills in knowledge gaps to facilitate an understanding of how medical conditions relating to the trainee's primary specialty may be affected by preflight and postflight operations. This is intended to enable Space Medicine specialists to identify areas of concern, promote and maintain crew health and well-being, and manage health threats and acute conditions. This domain also includes an understanding of how a trainee's primary specialty contributes to event medicine, disaster preparedness, logistics and supply chains, emergency response, launch operations, and rescue and recovery operations, and rehabilitation including clearance for operating vehicles, and family/societal reintegration concerns. It also ensures Space Medicine practitioners are competent in telemedicine, radio communications (including delayed communications), and team dynamics. |
| Spaceflight Clinical Operations | This domain focuses on how medical and psychiatric comorbidities, the spacecraft environment, spaceflight adaptation, operational considerations, and individual characteristics interact to alter diagnostics, treatment plans, and recovery in the spaceflight environment. While the majority of in-flight issues relate to spaceflight and spacecraft environmental adaptation, general and specialist medical and behavioral health conditions do occur and as mission durations increase, mission profiles become more complex, and spaceflight participants diversify as accessibility to space increases, such conditions will occur with increased frequency and severity just as they do on Earth. It is intended to augment and fill in gaps in a trainee's primary specialty knowledge to facilitate the management of injury, illness, and crew member adaptation to the spaceflight environment. This domain also covers the importance of ethical data collection and use, integration with researchers and other operations teams, telemedical and radio communications, advance planning, long range logistics and supply chains, real-time risk analysis, as well as long- and short-term spaceflight health considerations. |
| Salutogenesis | Salutogenesis focuses on identifying mechanisms to increase an individual's quality of life. This is contrasted with the traditional pathology model of medicine which seeks to identify and remove disease. These are synergistic approaches that can and should be practiced together; however, much of Space Medicine either deals with healthy people or those with stable chronic illnesses entering an environment that threatens their health and stability. For this reason, the focus on promoting quality is of greater relative importance than many other specialties. This domain focuses on training physicians to identify what matters most to crew and ground team members, what may be missing from their lives on mission, and how vehicle design, operational considerations, crew dynamics, crew ground interactions, and ground team dynamics can be leveraged to maintain or even improve health and wellness on a mission. |

(*Continued*)

**TABLE 30.1** (Continued)

| Domain | Description |
|---|---|
| Clinical Support Services | Maintaining and responding to crew health and performance needs in space requires multifaceted resourcing managed by a diverse team that extends beyond physicians. While Spaceflight Physicians are not expected to be experts in each field, they must recognize when and how such disciplines may be needed and have sufficient familiarity to communicate, coordinate, direct, and advocate for the appropriate clinical support services. These include, but are not limited to, pharmacology, exercise physiology, physical/occupational therapy, psychology, and science and research. |
| Training and Education | The capacity to translate and communicate complex health information, risk assessments, and management plans in ways that both medical and nonmedical personnel understand is key to the successful practice of Space Medicine. This domain ensures that trainees have the necessary tools in scientific and medical communication to design formal and informal curricula, assess successful knowledge uptake, address questions, and develop individualized and group learning plans for a diversity of domains. This includes crew medical officers, crewmembers and patients, family members, engineering and design teams, ground teams, medical professionals, researchers, the public, and any other mission relevant personnel. |
| Administration | This domain helps Spaceflight Physicians better understand how the space medicine role fits within the larger ecosystem of spaceflight and health care, in general. It provides trainees with the tools to develop their practice, organize teams, coordinate logistics, navigate organization dynamics, and perform advocacy and outreach within their organization as well as with regulatory bodies and standard-setting organizations. It also covers the regulatory environment, relevant legal requirements, and where to find best practice standards and guidelines for both medical and nonmedical aspects of Space Medicine. It is also concerned with helping Spaceflight Physicians to develop such standards and advocate for change to facilitate better systems of care. |

**Source:** Reproduced with permission from the Committee, Space Medicine Practice Certification and Fellowship Accreditation (2025) ©.

administration as part of Operation Deep Freeze (U.S. Navy, 1967). Once it transitioned to civilian control, this capability was removed in favor of stabilizing surgical cases and transporting them back to Christchurch, New Zealand (Robins, 2024; Levin, 2025). This was likely done because resource/training costs of maintaining high-level surgical capability on station were not worth the benefit due to the small number of surgical cases, particularly since most surgical cases do not require immediate intervention after emergency stabilization (Tissot et al, 2023). According to the University of Texas Medical Branch Center for Polar Medicine, which currently holds the contract for providing polar medical care, the physician on station is required to be a:

*"Medical Doctor Degree, with a minimum of 3 years of practice experience, hold a current license to practice medicine in the US or a recognized US territory and must hold current Drug Enforcement Agency controlled substance registration certificate. Certification with American Heart Association in Advanced Cardiac Life Support (ACLS) required. Certification in Advanced Trauma Life Support (ATLS) preferred and Basic Cardiac Life Support (BCLS) required every two years. Must be a US citizen and hold a valid US passport. Competency in basic procedures needed in trauma situations (i.e., splinting, suturing, IV insertion, airway management)."*

However, UTMB also desires that these physicians hold:

*"Board-Certification in Emergency Medicine or Family Medicine with experience working in remote environments with limited medical support"* (University of Texas Medical Branch Center for Polar Medical Operations, 2025).

This model is likely to be similar to what will happen for space health care over the next 2 decades. While most medical care relies on an exchange of information that can be delivered telemedically as long as robust communication channels are in place, procedural medicine will still require a high degree of hands on skill. The intuitive sense of this statement is supported by probabilistic risk assessment from NASA's Integrated Medical Model, which estimates as many as 50% of the conditions most likely to drive evacuation from ISS may require emergency medical procedures. An early version of NASA's model for assessing risk on lunar missions is similar, with more than 70% of evacuation driving conditions potentially requiring procedural intervention. This is not intended to be alarmist about the number of evacuations. Both platforms estimate the rate for *considering* an evacuation and predict evacuation will be considered on less than 1/3 of one percent of missions, with the rate for conditions that potentially require a procedure even lower. However, they do illustrate that many conditions contributing to this risk do so because they require a relatively minor procedure, such as splinting, laceration repair, or a temporary airway intervention (Anderson et al., 2024; Walton and Kerstman, 2020). This suggests that as the duration of missions and the number of people increase, it may become more cost-effective to provide care on site than to abort a mission back to Earth.

## 30.5 Preventive planning

With an economically justifiable "why" (e.g., reducing the need for evacuation as in space person-years increase), we can turn our attention to what preparations might need to be made. As with the rest of this chapter, this section is largely speculative. It is included since it may highlight some of the gaps between what is known today and what might be needed in the future. In this way, it may be helpful as a guide for research or operational development.

### 30.5.1 Training

The current training program for in space practitioners is approximately 40 hours of training designed for the crew medical officer (CMO) to work in tandem with a telemedical provider on the ground (Russell et al., 2023). This is sufficient for a CMO to perform a basic exam and carry out basic interventions using on board supplies. A recent study using an early version of NASA's IMPACT tool investigated predicted CMO training needs for future mission profiles. The study used a probabilistic risk assessment model to estimate the relative risk increase for each trained capability if it were not available. Capabilities that increased risk 5% when not available were included (Levin et al., 2025). The study evaluated five mission profiles, which are listed in Table 30.2.

According to this study, exam skills, management decisions for basic conditions, and ground communications/ consultation capabilities were common to all profiles. Most profiles also required basic procedural skills such as eye irrigation, wound dressing, use of suction, and venipuncture/intravenous (IV) access. The two profiles with the most person-years, the ISS-like profile 1 and the early Mars mission-like profile 3, are both more procedure-intensive. Table 30.3 lists all capabilities identified in this study as procedures.

The capabilities in the three-person-year ISS profile are consistent with the type of procedural training presently required for an ISS mission, which lends some credence to the results. The 15-person-year Mars mission-like profile includes additional capabilities a CMO must learn. It is reasonable to assume that the longer the mission, the more a CMO must be able to manage. It is also worth noting that the capabilities in this study were limited to those required to treat the 120 conditions included in the IMPACT model (Kreykes et al., 2023). Using Antarctica as a guide, CMOs over the next 20 years may need to be part of a small team and, while required procedures are not likely to rise to the level of invasive

**TABLE 30.2 Mission profiles from predicting exploration crew medical officer training needs: applying evidence-based predictive analytics to space medicine training (Levin et al., 2025).**

| Profile | 1 | 2 | 3 | 4 | 5 |
|---|---|---|---|---|---|
| Analogous to | ISS | Artemis 3 | Early Mars mission | Short orbital mission in a large vehicle | Lunar flyby in a large vehicle |
| *Crew* | 6 | 4 | 6 | 60 | 20 |
| *Males/ females* | 3/3 | 2/2 | 3/3 | 30/30 | 10/10 |
| *Duration* | 6 months | 30 days total, 15 on surface | 30 months total, 15 months on surface | 3 days | 14 days |
| *EVA surface* | N/A | 2 crew, 10 each | 4 crew, 30 each | N/A | N/A |
| *EVA space* | 2 crew, 2 each | N/A | N/A | N/A | 16 crew, 1 each |
| *Person-years* | 3 | 0.3 | 15 | 0.5 | 0.8 |

**Source:** From Levin, D.R., McIntyre, L., Steller, J.G., Nelson, A., Zahner, C., Anderson, A., et al., 2025. Predicting Exploration crew medical officer training needs: applying evidence-based predictive analytics to space medicine training. Wilderness & Environmental Medicine. 36 (1_suppl), https://doi.org/10.1177/10806032241292535.

**TABLE 30.3** Predicted crew medical officer (CMO) training needs for various mission profiles (Levin et al., 2025). *Research procedures not considered essential by the model. ^ Activities treated as procedures in the model since they may require hand on intervention, however, they are not often thought of as medical procedures in the terrestrial sense.

| Profile | 1 | 2 | 3 | 4 | 5 |
|---|---|---|---|---|---|
| Analogous to | ISS | Artemis 3 | Early Mars mission | Short orbital mission in a large vehicle | Lunar flyby in a Large vehicle |
| *Diagnostic* | Adjust ECLSS^ | | Adjust ECLSS^ | | |
| | | | Venipuncture/intravenous access | | |
| | | | Electrocardiogram | | |
| | | | | | |
| *Intervention* | Cranio electrotherapy stimulation* | Cranio electrotherapy stimulation* | Cranio electrotherapy stimulation* | Cranio electrotherapy stimulation* | Cranio electrotherapy stimulation* |
| | Eye irrigation | Eye irrigation | Eye irrigation | Eye irrigation | Eye irrigation |
| | Wound dressing | Wound dressing | Wound dressing | Wound dressing | Wound dressing |
| | Medical suction^ | Medical suction^ | Medical suction^ | Medical suction^ | Medical suction^ |
| | Pressure point massage | Pressure point massage | Pressure point massage | Pressure point massage | Pressure point massage |
| | Venipuncture/intravenous access | Venipuncture/intravenous access | Venipuncture/intravenous access | Venipuncture/intravenous access | Venipuncture/intravenous access |
| | Noninvasive oxygen therapy | | Noninvasive oxygen therapy | | |
| | Temporary dental filling | | Temporary dental filling | | |
| | Dental suction^ | | Dental suction^ | | |
| | Infection control^ | | Infection control^ | | |
| | Antiblister adhesive application^ | | Cast wrist/hand | | |
| | Eye foreign body removal | | Oral sutures | | |
| | | | Dental extraction | | |
| | | | Hyperbaric oxygen therapy^ | | |
| | | | Wound irrigation | | |
| | | | Epistaxis management | | |
| | | | Nasal packing | | |
| | | | Oral tranexamic acid | | |
| | | | Reduce fracture | | |
| | | | Procedural/other sedation | | |
| | | | Sling/swath shoulder | | |
| | | | Splint wrist/hand | | |
| | | | Hemorrhage control | | |
| | | | Multilayer wound closure | | |
| | | | Joint/tendon injection | | |

**Source:** From Levin, D.R., McIntyre, L., Steller, J.G., Nelson, A., Zahner, C., Anderson, A., et al., 2025. Predicting exploration crew medical officer training needs: applying evidence-based predictive analytics to space medicine training. Wilderness & Environmental Medicine. 36 (1_suppl), https://doi.org/10.1177/10806032241292535.

surgery, it is likely they will need to perform a broader range of minor procedures to treat and to stabilize patients in order to minimize the need for early evacuations.

## 30.5.2 Resilient systems

A resilient medical system has the ability to absorb, adapt, and recover from unexpected circumstances, such as equipment failures, operational disruptions, environmental changes, and/or novel medical events, while maintaining an acceptable level of crew health protection. In practice, this means a system must have three primary characteristics:

- Redundancy,
- Restorability,
- Adaptability.

These characteristics may be complementary or mutually exclusive, and the optimal strategy for each may vary depending on the capability or resource in question. For instance, an urban hospital in a large metropolitan area may have a robust supply chain to purchase additional IV fluids to meet increased demands, while a spacecraft may rely on a system to generate IV fluids from potable water

For a truly space-faring population, these characteristics must be carefully considered and balanced for each medical system capability to ensure that the population has consistent and reliable access to the medical capabilities most essential to their health.

### *30.5.2.1 Redundancy*

Redundancy is the inclusion of additional resources beyond the minimum required to ensure reliability, safety, or continuity of function in the presence of failure. It is a critical characteristic of resilience in medical systems and ensures that alternative resources, personnel, and/or equipment are available in case of equipment failure, resource depletion, skill erosion, or loss of human operators. At the most basic level, medical systems contain the five elements:

- Human operators,
- Consumable resources,
- Nonconsumable resources,
- Training and knowledge support,
- Records and data systems.

Each element treats redundancy in a different way, but in general, redundancy comes in the form of a fully formed backup, such as an additional CMO or additional stash of supplies. The level of redundancy for equipment and personnel is often determined by a balance of predicted population needs, anticipated failure rates, and cost. Cost in this context may be any limited resource (e.g., mass, volume, risk, money, etc.). The challenge in this case is to include sufficient redundancy to prevent loss of capability but not oversupply which may waste resources and drive up costs. In some cases, carrying too many backups may inadvertently increase risk by expending limited storage space or financial resources on supplies at the expense of equipment necessary to reduce a more significant risk. This is often a complex and dynamic calculation that considers many factors beyond the scope of this chapter.

Redundancy may be most effective for devices that are complex and difficult to replace (e.g., a human crew medical officer or an ultrasound probe) or items that are low mass and easily lost or broken, such as a scalpel blade. On larger spacecraft with cheaper launch costs, as will be the case over the coming decades, storage will be less of an issue, and carrying spares or additional personnel will be an important component of a resilient medical system. However, the ability to carry spares will need to be traded against the ability to restore functions/supplies to determine the optimal mix for each vehicle or mission.

### *30.5.2.2 Restorability*

Complementary to redundancy, restorability is the capacity to recover function following a failure or exhaustion of the available supplies. Depending on the function, this may be achieved through repair, reconfiguration, or replacement with the latter potentially occurring through resupply or in-place manufacture. Each of these techniques has advantages and disadvantages. For example, repair and reconfiguration require the availability of appropriate tools, and system components may also need to be engineered or modified to permit serviceability. In addition, the crew must be trained to effect the necessary repairs. This may not be realistic for some systems, particularly when the crew is small and the components require a high level of expertise, since these can carry substantial costs in time or funding. Items that are relatively easy to understand but rely on an assembly of components may be more amenable to repair, such as a medical suction device.

#### 30.5.2.2.1 Resupply

Resupply is another potential alternative for restorability. Items that are consumable or that expire, such as pharmaceuticals or gauze pads, are an excellent target for resupply. However, supply chains to space are often costly and complex which makes them subject to unexpected disruptions, delays, and shortages. In some cases, such as on a Mars mission, resupply for damaged equipment or expired medication may not be possible in a reasonable timeframe.

#### 30.5.2.2.2 Regeneration

The corollary to resupply is regeneration or in situ manufacturing. This is a powerful but often complex concept that we are just beginning to explore. Again, consumable items that have a limited shelf life, such as pharmaceuticals, would potentially be amenable to this. However, manufacturing is complex. In addition to the methods and mechanisms to produce the active pharmaceutical ingredient (API), manufacturers must ensure a high degree of purity in the produced compound, ensure the absence of dangerous byproducts, accurately measure/weigh the produced API, and potentially ensure the compound is sterile. This is to say nothing of the methods for creating a pill or solution, nor for the packaging and storage of the compound once produced. Terrestrially, this requires substantial infrastructure for both biological and chemical productions, such as large amounts of base reactants and a robust regulatory system. Analogous systems and components must be present for manufacturing compounds in space as well. However, recent advances have miniaturized much of this process making it theoretically possible to produce pharmaceutical and nutritional compounds on board a spacecraft. One of the most promising potential methods would be to genetically engineer bacteria or yeast to produce the desired compound and build a miniaturized and automated purification process. The technology for this exists today and indeed some pharmaceuticals (including insulin and many antibiotics) are produced in a similar fashion albeit with much larger infrastructure.

For a spacecraft, frozen samples of engineered, nonpathogenic microooganisms could be stored on board. When onboard stores of the drug or nutrient are nearing depletion, the crew could defrost a sample, feed it, harvest the produced compound, purify it, test it, and package it in a suitable container and refreeze a portion of the microbes for later. This process is not unlike producing a sourdough loaf and storing the starter in a fridge. This would significantly reduce the need to carry supplies on board. However, it will require a high degree of redundancy to ensure that a "crop" is not lost, and it is presently resource intensive and requires some crew expertise in managing microbial colonies.

Another present-day example is the Water Recovery System (WRS) in use on ISS today which uses solar energy to regenerate potable water from the waste, exhaled air, and the excreted matter of ISS astronauts. Similarly, NASA has developed the IVGen system for producing on demand 0.9% saline IV fluid, by combining 1000 grams of potable water and 9 grams of sodium chloride. This system was demonstrated on ISS in 2010 (McQuillen et al., 2011). A miniaturized version has been designed and proposed for exploration class missions (Mars Campaign Office, 2020). While sterile water can be produced through the WRS, the IVGen system does require a supply of sodium chloride since there is currently no technology for reprecipitating NaCl from human excretions. That said, NaCl does not expire, and 1 kg would be sufficient to generate 111 liters of IV fluid. Consumption needs for managing conditions on a station would likely be on the order of a few liters per year so this supply would last a long time.

The other commonly cited additive manufacturing capability is 3D printing which can use a variety of readily available compounds to produce tools and supplies based on digital files. The advantage of these systems is that tools and supplies could be custom designed and printed as needed. A digital file designed and tested on Earth could be transmitted to the station if the tool is not present in the existing database

Finally, personnel skills may be considered a nonconsumable resource that has an expiration date. A lack of practice can atrophy the medical skills of even the most experienced physician, and in space, medical personnel will need to maintain their skills over multiple years with relatively infrequent patient contacts. In the vein of regenerating onboard resources, regular simulations and tests, possibly assisted by dynamic AI models and virtual reality, may be a valuable countermeasure.

### 30.5.3 Adaptability

The final major component of a resilient medical system is adaptability. Adaptability is the ability of a system to adjust its structure, processes, or strategies in response to internal or external change to maintain effectiveness. No design can anticipate every possible contingency and no prevention can prevent 100% of adverse events. A resilient system must be able to adapt and pivot to handle these unexpected events. This may mean including a database of additional training or staffing providers with sufficient creativity to adjust their processes to new circumstances. While this is a difficult thing to quantify, developing lists of alternatives for supplies and training providers in resource poor environments may help breed such creativity. Presently, humans are necessary for this kind of adaptability, but it is not difficult to imagine a near-term future where some of this may be handled by complex artificial intelligence models capable of answering questions based on databases that include alternative resources.

For example, a well-designed, probabilistic risk assessment tool tied to an inventory management system may be able to assess remaining supplies and predict potential future needs. By comparing future need with remaining supplies, this could form the basis for a tool that could participate in rationing and triaging by suggesting alternative treatment strategies intended to preserve available supplies of single use items and optimize the risk reduction based on remaining equipment. It might also be able to estimate the risk/benefit associated with treating a crew member with a condition requiring high resource use. Such a system could attempt to quantify the probability of successful treatment against the increased risk to other crewmembers by consuming supplies and recommend guidelines for changes in care up to and including estimating the point of medical futility.

## 30.6 Long-term preventive care in space habitation

The history of medicine in spaceflight is fundamentally a narrative of preventive care. For the past 70 years, the vast majority of humans in space have undergone meticulous screening for illnesses and vulnerabilities and have been trained and selected based on positive indicators of optimal health and performance. Additionally, over the past 30 years, medicine in spaceflight has expanded its focus to include the long-term monitoring of astronaut health following spaceflight exposure, along with its possible chronic effects on health and performance.

However, when exposure to spaceflight becomes indefinite, and members of the space habitation community are not as strictly screened, preflight prevention must become augmented by in-flight screening and prevention. Just as we get annual checkups on Earth, astronauts living in space will likely benefit from regular follow-up visits to assess known health threats and the effectiveness of employed countermeasures. Routine terrestrial preventive measures, such as childhood and elderly vaccination schedules, cancer screening, and metabolic disease surveillance and management are likely to become necessary routine parts of space medicine as well, driving the need to develop procedure suites for techniques like colonoscopies, cervical cancer screenings, more detailed body imaging, blood tests, and a host of other common medical procedures.

Preventive care for space habitation has the potential to establish new screening protocols that are not currently recommended due to altered risk profiles and limited medical capabilities. For instance, hypertension in space may present differently than terrestrially, since the absence of gravity reduces strain and fluid volumes. New guidelines may be needed, and existing guidelines will need to be confirmed.

## 30.7 Reproduction and pediatrics

If humans are ever to truly live in space, we must confront the challenges and develop the capabilities to manage reproductive health and eventually pediatric care. While such a concept may seem distant, the profound moral and ethical issues surrounding human reproduction in spaceflight necessitate early and ongoing consideration by the medical community.

While the ethical issues surrounding human gestation and pediatric development in partial gravity or microgravity are incredibly complex and not the focus of this chapter, it is crucial to begin addressing how a space colony medical system might be designed to manage human gestation, birth, and pediatric care.

The chapters in this book address limited obstetric emergencies, primarily related to pregnancy prevention and pathologies of early, unintended pregnancy. However, for human reproduction to continue in a space colony, there will be a need for advanced obstetric capabilities such as fetal monitoring, cesarean section, postnatal hemorrhage management, and, in the most extreme cases, rescue hysterectomy. There will also be a need for advanced fertility support and technology, as well as for adapting existing techniques and procedures to the spaceflight environment.

Similarly, all the procedures described in this book and elsewhere in space medicine literature will need to be further adapted to treating children. This will require adjusted equipment, pharmaceutical management, and sedation techniques to address the unique needs of a pediatric population. Standards for child development will be needed, along with a better understanding of the risks posed during conception, pregnancy, birth, neonatal development, and child development.

Countermeasures will need to be developed to address these effects, and spacecraft systems will need methods to accommodate changes in crew complement and adapt to the changing physiology of a growing child. Aside from potentially detrimental effects on gestation, pediatric milestones, and skeletal development, the altered development of native-born space colonists in non-terrestrial environments could completely isolate the population from returning to Earth or other higher-gravity habitations. For example, a colonist born and raised with 1/6 Earth gravity on the moon may not be able to maintain posture or normal respiratory mechanics on Earth and may be uncomfortable or experience debilitating orthostatic symptoms even on Mars.

## 30.8 Conclusion

The transition from mission-based space exploration to indefinite space habitation, and ultimately, colonization, represents a profound shift in the demands placed upon a space medical system. The procedural interventions described in the preceding sections of this textbook are the essential foundation for addressing acute medical and surgical needs in space. However, a comprehensive medical system for long-term habitation requires a paradigm change that moves beyond immediate procedural capability to embrace principles of operational resilience and long-term health maintenance.

Developing a resilient medical system for indefinite space habitation hinges on the careful balance of redundancy, resupply, and regeneration. Redundancy in complex equipment and in medically trained personnel will be critical to mitigate the risk of failure or incapacitation in an environment where ground support is significantly delayed. Resupply, already a complex logistical challenge, must be meticulously planned to account for supply chain lags and unexpected medical events. Most crucially, regeneration of medical resources—from simple IV fluids to complex sterile consumables and even the maintenance of clinical skills through artificial intelligence, augmented reality, and virtual reality assisted simulations—will determine the level of autonomy a habitat achieves.

Furthermore, the long-term presence of humans in a hostile environment necessitates a complete reevaluation of preventive care. Altered risk profiles due to chronic radiation exposure and gut microbiome changes demand new screening protocols and tailoring existing ones to the unique constraints of spaceflight. Finally, the leap from space habitation to a genuine space colony requires addressing the complex medical and ethical challenges of human reproduction and pediatric care in partial or microgravity which are prerequisites for a self-sustaining off-world population.

The future of human health in space is not just about having the right tools for an emergency; it is about establishing a semiautonomous, resilient, and adaptive ecosystem of care that ensures the long-term well-being of a diverse population. The procedures described throughout this textbook are a step on this path, but they must be combined with other elements of clinical medicine and with systems medicine to truly encompass what is needed to build successful space medicine systems (Shelhamer and Antonsen, 2024; Barratt et al., 2020). This knowledge must be integrated with principles of operational resilience, and existing gaps must be filled to prepare space medicine professionals and to continue laying the foundation for securing humanity's future among the stars.

## References

ACGME, 2020. Aerospace medicine milestones. Available from: https://www.acgme.org/globalassets/pdfs/milestones/aerospacemedicinemilestones.pdf.

American Board of Preventive Medicine, 2025. 27 Aerospace medicine content outline. Available from: https://www.theabpm.org/become-certified/exam-content/aerospace-medicine-content-outline/.

Anderson, A., Parmar, P., Levin, D., Hilmers, D., 2024. Quantifying the risk of medical evacuation in spaceflight. Acta Astronautica 458–464.

Barratt, M.R., Baker, E.S., Pool, S.L., 2020. In: Barratt, M.R., Baker, E.S., Pool, S.L. (Eds.), Principles of Clinical Medicine for Space Flight (second ed.). Springer Nature.

Kreykes, AJ, Suresh, R, Levin, D, Hilmers, DC, 2023. Selecting Medical Conditions Relevant to Exploration Spaceflight to Create the IMPACT 1.0 Medical Condition List. Aerosp Med Hum Perform 94 (7), 550–557. Available at https://doi.org/10.3357/AMHP.6199.2023.

Levin, D., 2025. Long Beach Personal Conversations with Dr. 2025.

Levin, D.R., McIntyre, L., Steller, J.G., Nelson, A., Zahner, C., Anderson, A., et al., 2025. Predicting exploration crew medical officer training needs: applying evidence-based predictive analytics to space medicine training. Wilderness & Environmental Medicine 36 (1_suppl). Available at https://doi.org/10.1177/10806032241292535.

Mars Campaign Office, 2020. National Aeronautics and Space Administration IntraVenous Fluid GENeration for Exploration Missions Mini.

McQuillen, J.B., McKay, T.L., Griffin, D.W., Brown, D.F., Zoldak, J.T., 2011. National Aeronautics and Space Administration Final Report for Intravenous Fluid Generation (IVGEN) Spaceflight Experiment.

Navy, U.S., 1967. Port Hueneme, CA U.S. Navy Operation Deep Freeze '67 (McMurdo Station cruisebook).

Robins, C., 2024. Medicine on the Ice: Two University of Iowa Medicine Alumni Met by Chance while Serving Patients at McMurdo Station. Antarctica.

Russell, B.K., Burian, B.K., Hilmers, D.C., Beard, B.L., Martin, K., Pletcher, D.L., et al., 2023. The value of a spaceflight clinical decision support system for earth-independent medical operations. NPJ Microgravity 9 (1). Available at https://doi.org/10.1038/s41526-023-00284-1.

Shelhamer, M., Antonsen, E., 2024. Systems Medicine For Human Spaceflight. World Scientific Publishing, London.

Space Medicine Practice Certification and Fellowship Accreditation Committee Space, 2025. Space medicine practice certification and fellowship accreditation: space med core competencies version 1.0. Los Angeles, CA.

Tissot, C., Lecordier, M., Hitier, M., 2023. Surgical epidemiology of Antarctic stations from 1904 to 2022: a scoping review. International Journal of Circumpolar Medicine 82 (1).

United States Antarctic Program, 2025. 27 United States Antarctic program protal. Available from: https://www.usap.gov/usapgov/

University of Texas Medical Branch Center for Polar Medical Operations, 2025. 27 UTMB UTMB health center for polar medical operations. Available from: https://liveutmb.sharepoint.com/sites/collaboration/webfiles/Shared%20Documents/Forms/AllItems.aspx.

Walton, M., Kerstman, E., 2020. Quantification of medical risk on the international space station using the integrated medical model. Aerospace Medicine and Human Performance 332–342.

# Index

*Note*: Page numbers followed by "*f*" and "*t*" refer to figures and tables, respectively.

## B



## C

## J



## K



## L

## M

## Q



## R

## S

## T



## U

Printed and bound by CPI Group (UK) Ltd, Croydon, CR0 4YY
14/09/2026
02216412-0002